D1391247

NANOSCALE MATERIALS IN CHEMISTRY

NANOSCALE MATERIALS IN CHEMISTRY

Second Edition

Edited by

Kenneth J. Klabunde and Ryan M. Richards

A JOHN WILEY & SONS, INC., PUBLICATION

Published by John Wiley & Sons, Inc., Hoboken, New Jersey
Published simultaneously in Canada

For general information on our other products and services or for technical support, please contact our Customer Care Department within the United States at (800) 762-2974, outside the United States at (317) 572-3993 or fax (317) 572-4002.

Wiley also publishes its books in a variety of electronic formats. Some content that appears in print may not be available in electronic formats. For more information about Wiley products, visit our web site at www.wiley.com.

Library of Congress Cataloging-in-Publication Data:

Nanoscale materials in chemistry.—2nd ed. / [edited by] Kenneth J. Klabunde and Ryan M. Richards.
p. cm.
Includes index.
ISBN 978-0-470-22270-6 (cloth)
1. Nanostructured materials. I. Klabunde, Kenneth J. II. Richards, Ryan.
TA418.9.N35N345 2009
660—dc22

2008053437

Printed in the United States of America

10 9 8 7 6 5 4 3 2 1

To Sarah, Sydney, Maya, Erik,
and Tyler

CONTENTS

CONTRIBUTORS xi

PART I INTRODUCTION TO NANOMATERIALS 1

1 **Introduction to Nanoscale Materials in Chemistry, Edition II** 3
Ryan M. Richards

2 **Unique Bonding in Nanoparticles and Powders** 15
Keith P. McKenna

3 **Particles as Molecules** 37
C. M. Sorensen

PART II NEW SYNTHETIC METHODS FOR NANOMATERIALS 71

4 **Microwave Preparation of Metal Fluorides and their Biological Application** 73
David S. Jacob, Jonathan Lellouche, Ehud Banin, and Aharon Gedanken

5 **Transition Metal Nitrides and Carbides** 111
Piotr Krawiec and Stefan Kaskel

6 **Kinetics of Colloidal Chemical Synthesis of Monodisperse Spherical Nanocrystals** 127
Soon Gu Kwon and Taeghwan Hyeon

7 **Nanorods** 155
P. Jeevanandam

PART III NANOSTRUCTURED SOLIDS: MICRO- AND MESOPOROUS MATERIALS AND POLYMER NANOCOMPOSITES 207

8 **Aerogels: Disordered, Porous Nanostructures** 209
Stephanie L. Brock

9 **Ordered Microporous and Mesoporous Materials** 243
Freddy Kleitz

10 **Applications of Microporous and Mesoporous Materials** 331
Anirban Ghosh, Edgar Jordan, and Daniel F. Shantz

PART IV ORGANIZED TWO- AND THREE-DIMENSIONAL NANOCRYSTALS 367

11 **Inorganic–Organic Composites** 369
Warren T. Ford

12 **DNA-Modified Nanoparticles: Gold and Silver** 405
Abigail K. R. Lytton-Jean and Jae-Seung Lee

PART V NANOTUBES, RIBBONS, AND SHEETS 441

13 **Carbon Nanotubes and Related Structures** 443
Daniel E. Resasco

PART VI NANOCATALYSTS, SORBENTS, AND ENERGY APPLICATIONS 493

14 **Reaction of Nanoparticles with Gases** 495
Ken-Ichi Aika

15 **Nanomaterials in Energy Storage Systems** 519
Winny Dong and Bruce Dunn

PART VII UNIQUE PHYSICAL PROPERTIES OF NANOMATERIALS 537

16 Optical and Electronic Properties of Metal and Semiconductor Nanostructures 539
Mausam Kalita, Matthew T. Basel, Katharine Janik, and Stefan H. Bossmann

PART VIII PHOTOCHEMISTRY OF NANOMATERIALS 579

17 Photocatalytic Purification of Water and Air over Nanoparticulate TiO_2 581
Igor N. Martyanov and Kenneth J. Klabunde

18 Photofunctional Zeolites and Mesoporous Materials Incorporating Single-Site Heterogeneous Catalysts 605
Masakazu Anpo, Masaya Matsuoka, and Masato Takeuchi

19 Photocatalytic Remediation 629
Shalini Rodrigues

PART IX BIOLOGICAL AND ENVIRONMENTAL ASPECTS OF NANOMATERIALS 647

20 Nanomaterials for Environmental Remediation 649
Angela Iseli, HaiDoo Kwen, and Shyamala Rajagopalan

21 Nanoscience and Nanotechnology: Environmental and Health Impacts 681
Sherrie Elzey, Russell G. Larsen, Courtney Howe, and Vicki H. Grassian

22 Toxicity of Inhaled Nanomaterials 729
John A. Pickrell, L. E. Erickson, K. Dhakal, and Kenneth J. Klabunde

INDEX 771

CONTRIBUTORS

Ken-Ichi Aika, The Open University of Japan, Setagaya-ku, Tokyo, Japan

Masakazu Anpo, Department of Applied Chemistry, Graduate School of Engineering, Osaka Prefecture University, Osaka 599-8531, Japan

Ehud Banin, The Institute for Advanced Materials and Nanotechnology, Bar-Ilan University, Ramat-Gan 52900, Israel

Matthew T. Basel, Kansas State University, Department of Chemistry and Terry C. Johnson Center for Basic Cancer Research, Manhattan, Kansas

Stefan H. Bossmann, Kansas State University, Department of Chemistry and Terry C. Johnson Center for Basic Cancer Research, Manhattan, Kansas

Stephanie L. Brock, Department of Chemistry, Wayne State University, Detroit, Michigan

K. Dhakal, Comparative Toxicology Laboratories, Department of Diagnostic Medicine/Pathobiology, Kansas State University, Manhattan, Kansas

Winny Dong, Chemical and Materials Engineering, California State Polytechnic University, Pamona, California

Bruce Dunn, Department of Materials Science and Engineering, University of California, Los Angeles, Los Angeles, California

Sherrie Elzey, Department of Chemical and Biochemical Engineering, University of Iowa, Iowa City, Iowa

L.E. Erickson, Department of Chemical Engineering, Kansas State University, Manhattan, Kansas

Warren T. Ford, Department of Chemistry, Oklahoma State University, Stillwater, Oklahoma

Aharon Gedanken, Department of Chemistry, Kanbar Laboratory for Nanomaterials, Nanotechnology Research Center, Bar-Ilan University, Ramat-Gan 52900, Israel

Anirban Ghosh, Artie McFerrin Department of Chemical Engineering, Texas A&M University, College Station, Texas

Vicki H. Grassian, Departments of Chemical and Biochemical Engineering and Chemistry and the Nanoscience and Nanotechnology Institute at the University of Iowa, Iowa City, Iowa

Courtney Howe, Department of Nanoscience and Nanotechnology Institute at the University of Iowa, Iowa City, Iowa

Taeghwan Hyeon, National Creative Research Initiative Center for Oxide Nanocrystalline Materials, and School of Chemical and Biological Engineering, Seoul National University, Seoul 151-744, Korea

Angela Iseli, NanoScale Corporation, Manhattan, Kansas

David S. Jacob, Department of Chemistry, Kanbar Laboratory for Nanomaterials, Nanotechnology Research Center, Bar-Ilan University, Ramat-Gan 52900, Israel

Katharine Janik, Kansas State University, Department of Chemistry and Terry C. Johnson Center for Basic Cancer Research, Manhattan, Kansas

P. Jeevanandam, Department of Chemistry, Indian Institute of Technology Roorkee, Roorkee-247667, India

Edgar Jordan, Artie McFerrin Department of Chemical Engineering, Texas A&M University, College Station, Texas

Mausam Kalita, Kansas State University, Department of Chemistry and Terry C. Johnson Center for Basic Cancer Research, Manhattan, Kansas

Stefan Kaskel, Institute of Inorganic Chemistry, Technical University of Dresden, Dresden, Germany

Kenneth J. Klabunde, Department of Chemistry, Kansas State University, Manhattan, Kansas

Freddy Kleitz, Canada Research Chair on Functional Nanostructured Materials, Department of Chemistry, Laval University, Québec, Canada

Piot Krawiec, Instituto de Tecnología Química, UPV-CSIC, Valencia, Spain

HaiDoo Kwen, NanoScale Corporation, Manhattan, Kansas

Soon Gu Kwon, National Creative Research Initiative Center for Oxide Nanocrystalline Materials, and School of Chemical and Biological Engineering, Seoul National University, Seoul 151-744, Korea

Jonathan Lellouche, Department of Chemistry, Kanbar Laboratory for Nanomaterials, Nanotechnology Research Center, Bar-Ilan University, Ramat-Gan 52900, Israel

Russell G. Larsen, Departments of Chemistry and the Nanoscience and Nanotechnology Institute at the University of Iowa, Iowa City, Iowa

Jae-Seung Lee, Massachusetts Institute of Technology, Cambridge, Massachusetts

Abigail K. R. Lytton-Jean, Massachusetts Institute of Technology, Cambridge, Massachusetts

Igor N. Martyanov, Department of Chemistry, University of Ottawa, Ottawa, Ontario, Canada

Masaya Matsuoka, Department of Applied Chemistry, Graduate School of Engineering, Osaka Prefecture University, Osaka, 599-8531, Japan

Keith P. McKenna, London Centre for Nanotechnology and University College London, London, UK

John A. Pickrell, Comparative Toxicology Laboratories, Department of Diagnostic Medicine/Pathobiology, Kansas State University, Manhattan, Kansas

Shyamala Rajagopalan, NanoScale Corporation, Manhattan, Kansas

Daniel E. Resasco, School of Chemical, Biological, and Materials Engineering, University of Oklahoma, Norman, Oklahoma

Ryan M. Richards, Department of Chemistry and Geochemistry, Colorado School of Mines, Golden, Colorado

Shalini Rodrigues, Macungie, Pennsylvania

Daniel F. Shantz, Artie McFerrin Department of Chemical Engineering, Texas A&M University, College Station, Texas

C. M. Sorensen, Department of Physics, Kansas State University, Manhattan, Kansas

Masato Takeuchi, Department of Applied Chemistry, Graduate School of Engineering, Osaka Prefecture University, Osaka, 599-8531, Japan

PART I

INTRODUCTION TO NANOMATERIALS

1

INTRODUCTION TO NANOSCALE MATERIALS IN CHEMISTRY, EDITION II

RYAN M. RICHARDS

1.1 Introduction, 4
1.2 Systems with Delocalized Electrons, 4
1.3 Systems with Localized Electrons, 6
1.4 Instrumentation Introduction, 8
1.5 Conclusion, 9
Problems, 9
Answers, 10

Nanoscale Materials in Chemistry covers a broad area of science and engineering at the core of future technological development. In particular, the challenges of energy and sustainability are certain to be interrelated with breakthroughs in this area. Among current buzz words (i.e. green, bio-, eco-), "nano" has been used (and abused) to describe an amazingly broad spectrum of systems that has led to frustration for many scientists. The National Nanotechnology Initiative has defined nanotechnology as "working at the atomic, molecular and supramolecular levels, in the length scale of approximately 1–100 nm range, in order to understand and create materials, devices and systems with fundamentally new properties and functions because of their small structure" (www.nano.gov). Naturally, this broadly defined area of science and engineering has a significant "chemistry" component. This book aims to explore the chemistry, both traditional and emerging, that is associated with nanoscale materials.

Nanoscale Materials in Chemistry, Second Edition. Edited by K. J. Klabunde and R. M. Richards

This book is intended to function as both a teaching text for upper-level undergraduate or graduate courses and a reference text, and both fundamental and applied aspects of this field are covered in the chapters. It is intended that each chapter be able to stand on its own to allow instructors to select those topics most appropriate for their course. Additionally, each chapter contains several problems designed by the authors to challenge students and enhance their comprehension of the material.

In this short introduction, we introduce the field of nanoscience in a very general sense and provide background that may be useful to readers not familiar with this area. More in-depth discussions of each topic are provided in the individual chapters, but we have found that an initial superficial introduction to the most common phenomena and instrumentation, followed by the problems provided at the end of the chapter, helps students to understand the broader picture and begin to explore the literature.

1.1 INTRODUCTION

Nanoscience is the natural progression of science exploring the nature of matter between atoms and molecules (defined by quantum mechanics) and condensed matter (defined by solid state chemistry/physics). Thus, one of the central questions in nanoscience is "at what point in diminishing the size of a material does it begin to act more like an atom or molecule?" or, conversely, "how many atoms (in a cluster) does it take to begin observing bulk-like (solid state) behavior?"

With regard to nanoscale materials, there are three general classifications that can be used (at least for inorganics): (1) materials with delocalized electrons (metals or conductors), (2) materials with localized electrons (insulators) and (3) materials with new structures (usually atomically defined) and properties (or new forms of matter) due to their nanostructure (C60 or carbon nanotubes). Semiconductors fall somewhere in between classifications 1 and 2 depending on their band gap. Although these classes of materials will be discussed in detail in the chapters of this book, a quick review of general materials properties and the effects of reducing size is provided here to give readers and students an opportunity to begin thinking about nanoscience in terms of atoms/molecules (i.e. chemistry).

1.2 SYSTEMS WITH DELOCALIZED ELECTRONS

One of the principal concepts influencing the chemistry and physical properties of nanoscale materials with delocalized electrons is the quantum confinement effect. A metal can be thought of as a regular lattice of charged metal ions in a sea of quasi-delocalized electrons. The most important property of metals is their ability to transport electrons. Electrons can become mobile only if the energy band they are associated with is not fully occupied. If molecular orbital theory is used to generate

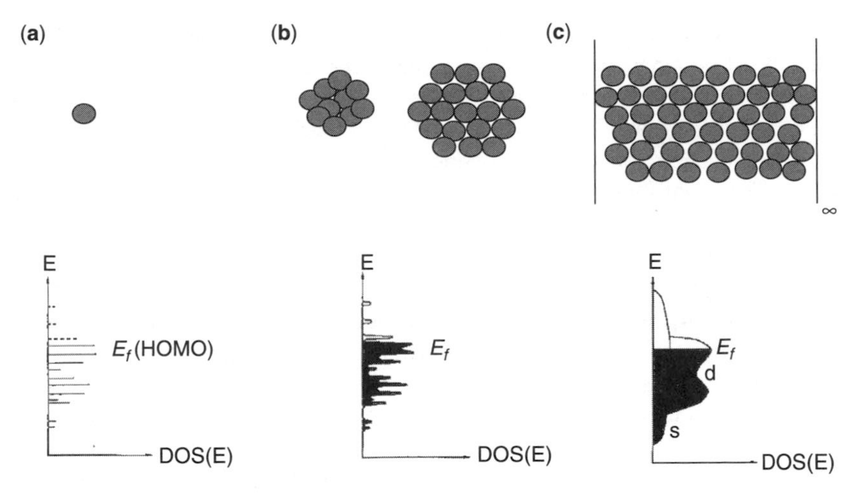

Figure 1.1 Development of the band structure of a metal: (a) molecular state, (b) nanocluster, and (c) bulk with s and d bands. (From Schmid, G. *Nanoscale Materials in Chemistry*, ed. K. J. Klabunde. New York: Wiley, 2001.)

the band structures, bulk metal possesses an indefinitely extended molecular orbital. The relationship between the molecular orbital of a finite molecular system and the indefinite situation in a bulk metal is that the highest occupied molecular orbital (HOMO) becomes the Fermi energy E_f of the free electron (Fig. 1.1). The Fermi energy depends only on the density of the electrons. If we assume that all levels up to the E_f are occupied with a total of N electrons, it can be estimated that the average level spacing is $\delta \sim E_f/N$ and therefore is inversely proportional to the volume $V = L^3$ ($L =$ side length of particle) or $\delta \propto E_f(\lambda_f/L)^3$ where $\lambda_f =$ wavelength of an electron with E_f. The wave character of the electron is assumed here, including that the allowed values for the wavelength λ are quantized (i.e. for an electron in a box of side L, only discreet values for the energy are allowed). The properties generally associated with bulk metals require a minimum number of electronic levels or a band.

The electrons in a three-dimensional metal spread as waves of various wavelengths usually called the DeBroglie wavelength.

$$\lambda = h/mv$$

where $\lambda =$ electronic wavelength, $h =$ Planck's constant, $m =$ mass of electron, and $v =$ speed of electron.

Delocalization of electrons in the conductivity band of a metal is possible as long as the dimension of the metal particle is a multiple of the DeBroglie wavelength. Thus, the smallest metal particles must have a dimension on the order of λ. Smaller particles have electrons localized between atomic nuclei and behave more like molecules. The transition between these two situations is gradual. Thus, for metals or systems with delocalized electrons, upon decreasing size we ultimately reach a size where the band structure disappears and discreet energy levels occur and we have to apply quantum mechanics; this is commonly referred to as the phenomenon of quantum confinement. The quantization effect represents one of the most exciting areas of modern science and has already found numerous applications in fields ranging from electronics to biomedicine.

Quantization refers to the restriction of quasi-freely mobile electrons in a piece of bulk metal and can be accomplished not only by reduction of the volume of a bulk material but also by reducing the dimensionality. A quantum well refers to the situation in which one dimension of the bulk material has been reduced to restrict the free travel of electrons to only two dimensions. Restricting an additional dimension then only allows the electrons to travel freely in one dimension and is called a quantum wire, while restricting all three dimensions results in a quantum dot.

1.3 SYSTEMS WITH LOCALIZED ELECTRONS

The effects of reducing size are very different for materials with localized electrons where defects are the most significant contributor to their properties. Naturally, due to the localization of electrons, the surface contains defects due to edges, corners, "f" centers, and other surface imperfections (Fig. 1.2). Defects can arise from a variety

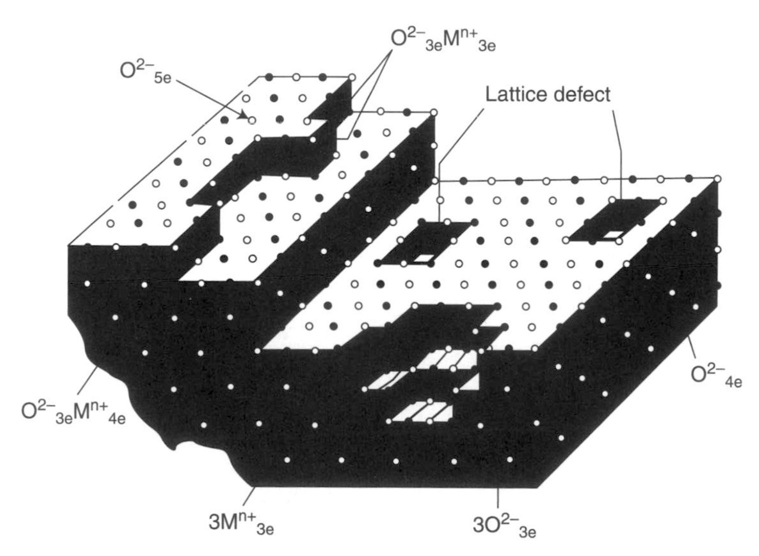

Figure 1.2 A representation of the various defects present on metal oxides. (From Dyrek, K. and Che, M. *Chem. Rev.* 1997, 305–331. With permission.)

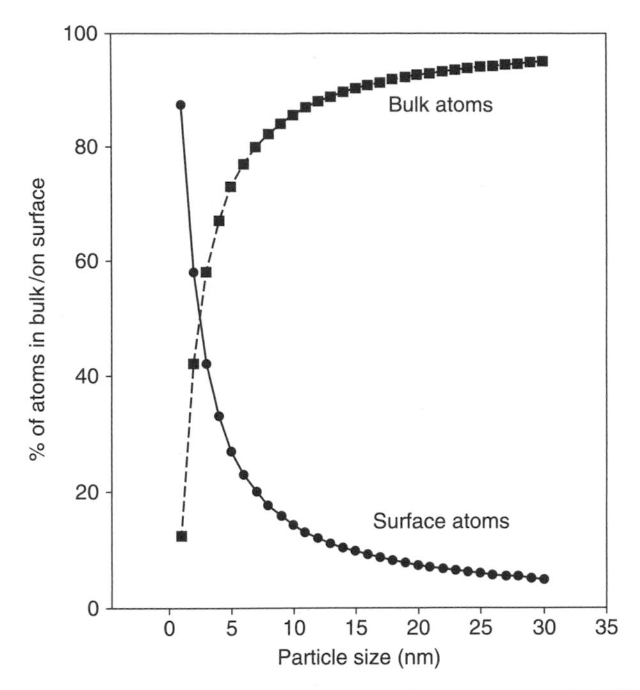

Figure 1.3 Calculated surface to bulk atomic ratio (for Fe). (From K. J. Klabunde et al. *J. Phys. Chem.* 1996, 100: 12142–12153. With permission.)

of causes: they may be thermally generated, or may arise in the course of fabrication of the solid, incorporated either unintentionally or deliberately. Defects are important because they are much more abundant at surfaces than in bulk, and in nanoscale materials they become predominant due to the large surface area (Fig. 1.3). Because of the number of atoms at the surface and the limited number of atoms within the lattice, the chemistry and bonding of nanoparticles is greatly affected by the defect sites present.

The defects that occur in the solids with localized electrons are grouped into the following classes: point, linear, planar, and volumetric defects. Point defects are a result of the absence of one of the constituent atoms (or ions) on the lattice sites, or their presence in interstitial positions. The most common defects are coordinatively unsaturated sites, for example, materials with a rock salt structure prefer to be bound to six neighbors, so those atoms on the surface are five-coordinate, on edges four-coordinate, and corners three-coordinate. Thus, using MgO as an example, the coordinatively unsaturated sites possess Lewis base character. Further, the crystallographic facet on the surface (as given by Miller indices) can also dramatically influence the properties of the system. For example, the (100) facet of MgO consists of alternating Mg cations and O anions and is thermodynamically favored, however, the (111) facet consisting of alternating layers of cations and anions has a polar surface and therefore different physical and chemical properties (Fig. 1.4).

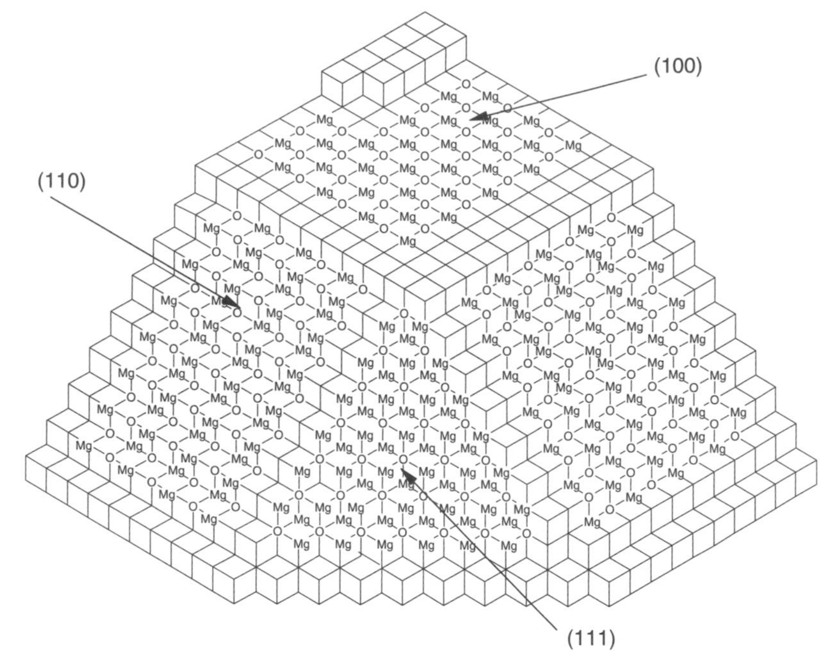

Figure 1.4 Schematic depiction of the (100), (110), and (111) facets of MgO.

1.4 INSTRUMENTATION INTRODUCTION

Developments in instrumentation, in particular microscopy, have allowed scientists to observe materials and phenomena with angstrom-level resolution, leading to a much deeper understanding of nanostructured materials. The two types of electron microscopy, scanning electron microscopy (SEM) and transmission electron microscopy (TEM), utilize an electron beam rather than light to resolve images (Fig. 1.5). In general, a TEM can be envisioned as a process similar to a film projector in which a beam passes through a sample and projects an image onto a screen. Conversely, an SEM is more comparable to shining a flashlight around a room and gaining a sense of the topography. With this technique resolution limits are generally on the nanometer length scale but the advantage lies in the topographical information gained. A great deal of caution should be taken when assessing the data provided by these techniques because they are operator biased in that they only show a small portion of the overall sample. While electron microscopy images provide valuable information regarding size, shape, composition, etc., they should be corroborated by a "bulk" technique such as powder x-ray diffraction (XRD) to demonstrate that the information in the microscopy image is representative of the whole sample.

In addition to electron microscopy techniques, developments in scanning probe microscopies have also allowed visualization and even the ability to manipulate matter at a new level. This class of microscopies acquires data by using a physical

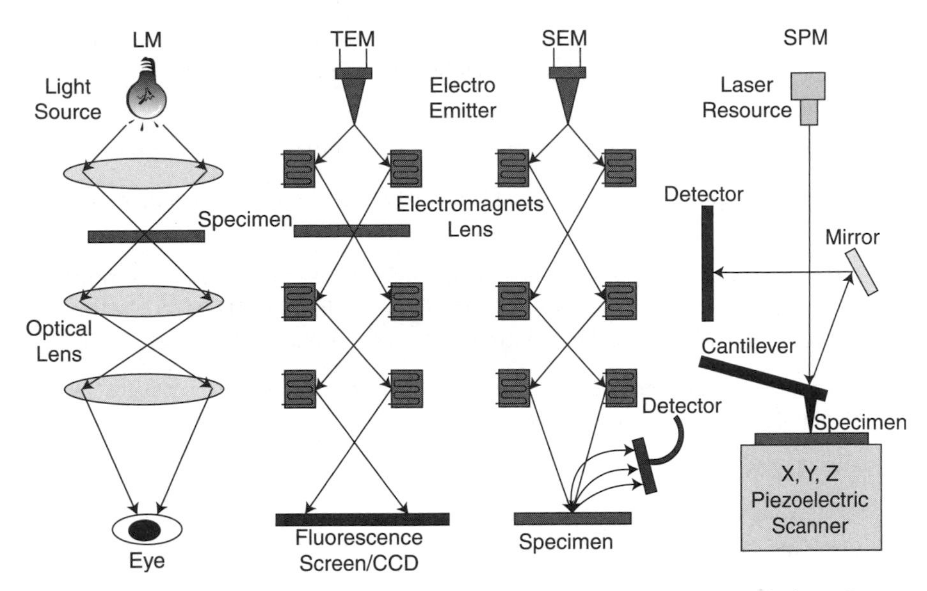

Figure 1.5 Schematic depiction of traditional light microscopy, transmission electron microscopy, scanning electron microscopy, and scanning probe microscopy.

probe to scan the surface (Fig. 1.5). Generally, the probe is moved mechanically across the surface in a raster scan, providing line by line data of the probe location and the interaction with the surface. There are now countless types of scanning probe microscopies, with the two most common being atomic force microscopy (AFM) and scanning tunneling microscopy (STM).

1.5 CONCLUSION

Hopefully, this short introduction instigates some class discussion about nanoscale materials in chemistry and facilitates an entry into the topics covered in this book and the literature. The editors have attempted to gather chapters covering a breadth of topics both fundamental and applied to provide the reader with an understanding of this important area of science and engineering. The contributors have been selected for their expertise in the subject area and have all done an excellent job of sharing their knowledge and making their topic accessible to a broad readership.

PROBLEMS

1. Draw figures illustrating the relationship of the density of states versus energy for (a) bulk, (b) quantum well, (c) quantum wire, and (d) quantum dots.
2. For a 2.1 nm cube of MgO (100) calculate the number of five-, four-, and three-coordinate sites.

3. Use the magic numbers equation $\Sigma(10n^2 + 2)$ to show the number of atoms in clusters with one, two, and three shells and calculate the number of surface atoms in each.

4. List five types of scanning probe microscopies and give a brief description of each.

5. Find the 12 principles of Green Chemistry and discuss how nanotechnology might have an impact on these areas.

6. Investigate the following analytical techniques and provide a brief description of each, including the information provided and limitations: TEM, SEM, powder XRD, XPS, EXAFS, XANES, nitrogen physisorption.

7. Explain what happens to the melting point and specific heat of metals as the size changes from the bulk to the nanoscale.

8. Provide MO diagrams for Li_2, Li_{20}, and bulk Li.

9. Describe what is meant by "bottom up" and "top down" preparations of nanoscale materials.

10. Describe the two paradigms of colloidal stabilization, steric and electrostatic, and provide an example of each.

ANSWERS

1. From K. J. Klabunde, editor. *Nanoscale Materials in Chemistry*, 1st edn. New York: Wiley Interscience, 2001, p. 22.

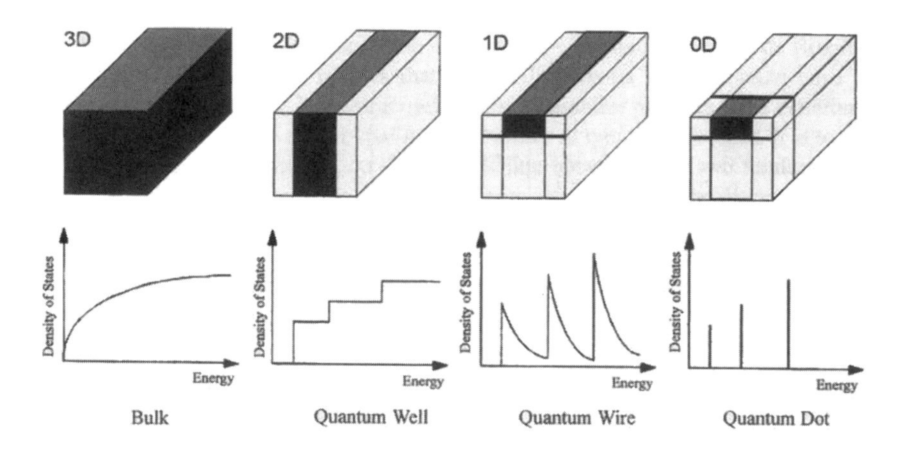

2. For MgO 100, d spacing is ~2.1 Å, thus a 2.1 nm cube is ~$10 \times 10 \times 10$ units or 1000 MgO that break down as follows:
5 coordinate = faces = $8 \times 8 \times 6$ (# faces)

4 coordinate = edges = 8×12 (# edges)

3 coordinate = corners = 8

leaving a core of $8 \times 8 \times 8$

Note: some students have an easier time starting with corners, then edges and faces. Also, this is a simplified exercise, in reality the highly unsaturated sites are often hydroxylated and the overall charge of the molecule is balanced.

3. 1 shell = 13 atoms (12 on surface)

2 shells = 55 atoms (42 on surface)

3 shells = 147 atoms (92 on surface)

4. Any of the following surface probe microscopies are possible:

AFM, atomic force microscopy

- Contact AFM
- Non-contact AFM
- Dynamic contact AFM
- Tapping AFM

BEEM, ballistic electron emission microscopy

EFM, electrostatic force microscope

ESTM, electrochemical scanning tunneling microscope

FMM, force modulation microscopy

KPFM, kelvin probe force microscopy

MFM, magnetic force microscopy

MRFM, magnetic resonance force microscopy

NSOM, near-field scanning optical microscopy (or SNOM, scanning near-field optical microscopy)

PFM, piezo force microscopy

PSTM, photon scanning tunneling microscopy

PTMS, photothermal microspectroscopy/microscopy

SAP, scanning atom probe

SECM, scanning electrochemical microscopy

SCM, scanning capacitance microscopy

SGM, scanning gate microscopy

SICM, scanning ion-conductance microscopy

SPSM, spin polarized scanning tunneling microscopy

SThM, scanning thermal microscopy

STM, scanning tunneling microscopy

SVM, scanning voltage microscopy

SHPM, scanning Hall probe microscopy

Of these techniques AFM and STM are the most commonly used followed by MFM and SNOM/NSOM.

5. The 12 principles of Green Chemistry:

This is a rapidly developing area of science and instructors may find it helpful to use recent literature reports to illustrate these points.

(1) *Prevention*: It is better to prevent waste than to treat or clean up waste after it has been created. (Catalysis by nanoscale particles.)

(2) *Atom Economy*: Synthetic methods should be designed to maximize the incorporation of all materials used in the process into the final product. (Optimizing the number of surface atoms and their activity comes from making nanoscale materials.)

(3) *Less Hazardous Chemical Syntheses*: Wherever practicable, synthetic methods should be designed to use and generate substances that possess little or no toxicity to human health and the environment. (There have been several reports of nanoscale catalysts that have allowed processes to become more green, using water as solvent or no solvents, eliminate by products, etc.)

(4) *Designing Safer Chemicals*: Chemical products should be designed to effect their desired function while minimizing their toxicity. (Nanotoxicity and nanoparticle lifecycle in the environment needs to be closely studied.)

(5) *Safer Solvents and Auxiliaries*: The use of auxiliary substances when used.

(6) *Design for Energy Efficiency*: Energy requirements of chemical processes should be recognized for their environmental and economic impacts and should be minimized. If possible, synthetic methods should be conducted at ambient temperature and pressure. (Nanocatalysis.)

(7) *Use of Renewable Feedstocks*: A raw material or feedstock should be renewable rather than depleting whenever technically and economically practicable.

(8) *Reduce Derivatives*: Unnecessary derivatization (use of blocking groups, protection/deprotection, temporary modification of physical/chemical processes) should be minimized or avoided if possible, because such steps require additional reagents and can generate waste.

(9) *Catalysis*: Catalytic reagents (as selective as possible) are superior to stoichiometric reagents. (Many nanoscale catalysts exhibit improved selectivity and/or activity as compared to bulk systems.)

(10) *Design for Degradation*: Chemical products should be designed so that at the end of their function they break down into innocuous degradation products and do not persist in the environment.

(11) *Real-time Analysis for Pollution Prevention*: Analytical methodologies need to be further developed to allow for real-time, in-process monitoring and control prior to the formation of hazardous substances.

(12) *Inherently Safer Chemistry for Accident Prevention*: Substances and the form of a substance used in a chemical process should be chosen to minimize the potential for chemical accidents, including releases, explosions, and fires.

From P. T. Anastas, J. C. Warner. *Green Chemistry: Theory and Practice*. New York: Oxford University Press, 1998, p. 30. By permission of Oxford University Press.

6. TEM: Under vacuum conditions focuses an electron beam through a sample dispersed on a grid. The resulting image can provide Å level resolution including lattice fringes. Most useful for determining particle size, size distribution and shape. Can also provide a great deal of additional information from diffraction techniques.

SEM: Provides more topography information than TEM but resolution is generally limited to nm scale.

Powder XRD: Provides information regarding unit cell, long range order, bonding and lattice. Can be helpful to determine the phase of a material (for example anatase vs rutile for TiO_2) and Scherrer eqn can be applied to estimate particle size.

XPS, X-ray photoelectron spectroscopy: Can be used to study first few layers of surface and provide the energies of the orbitals. This information can then be used to determine oxidation state of the element.

EXAFS and XANES are generally performed at a synchotron facility and provide information about neighboring atoms of the atom under investigation.

Nitrogen physisorption: Used to determine surface area, pore size, pore volume, pore size distribution and further textural properties through analysis of adsorption/desorption isotherms.

7. Students are directed to Chapter 8 in the first edition for comprehensive discussion. (See K. J. Klabunde, editor. *Nanoscale Materials in Chemistry*, 1st edn. New York: Wiley Interscience, 2001, pp. 263–277). In general, for free nanoparticles the melting point is always lower than the bulk value. Specific heats are generally enhanced as compared to the bulk.

8. From K. J. Klabunde, editor. *Nanoscale Materials in Chemistry*, 1st edn. New York: Wiley Interscience, 2001, p. 16.

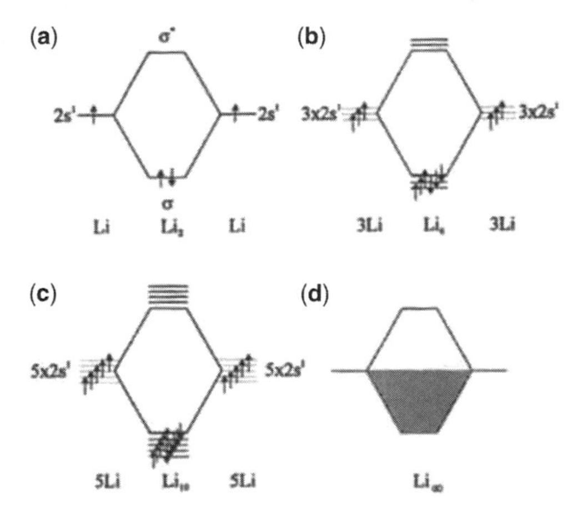

9. "Top down" approaches to nano refer to those in which larger systems are broken down until they reach they nanoscale while "bottom up" involves building nanoscale materials by putting together atoms or molecules.

10. The two general modes of colloidal stabilization are electrostatic (left) and steric (right). In the electrostatic mode there is a bilayer of anions (often halides) and a second layer of cations (for example tetra alkylammonium). In the steric stabilization there is a single bulky molecule attached to the surface (usually a P, N, or S donor, alkyl thiols are common).

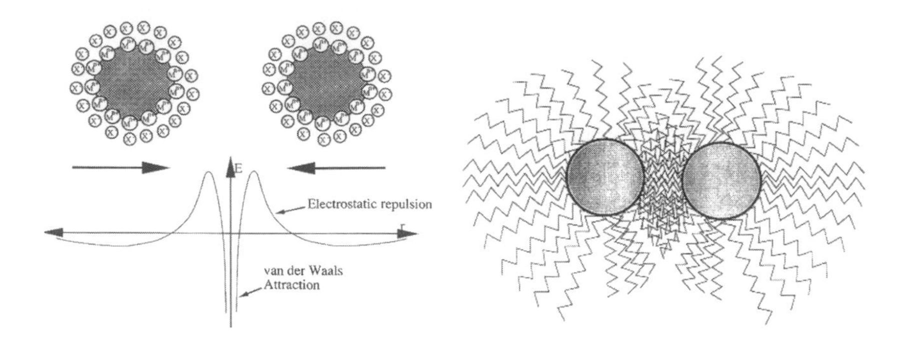

2

UNIQUE BONDING IN NANOPARTICLES AND POWDERS

Keith P. McKenna

2.1 Introduction, 15
2.2 Background, 17
 2.2.1 Size and Structure of Nanoparticles, 17
 2.2.2 Novel Properties of Nanoparticles, 20
2.3 Case Study 1: Magnesium Oxide Nanoparticles, 21
 2.3.1 Introduction, 21
 2.3.2 Properties of MgO Nanoparticles, 22
 2.3.3 Summary, 26
2.4 Case Study 2: Gold Nanoparticles, 26
 2.4.1 Introduction, 26
 2.4.2 Properties of Au Nanoparticles, 27
 2.4.3 Summary, 31
2.5 Concluding Remarks and Future Outlook, 31
Suggested Further Reading, 32
References, 33
Problems, 34
Answers, 35

2.1 INTRODUCTION

Nanoparticles (NPs) form a new class of materials possessing unique properties that are characteristic of neither the molecular nor the bulk solid-state limits. They have become the focus of considerable fundamental and applied research leading to

Nanoscale Materials in Chemistry, Second Edition. Edited by K. J. Klabunde and R. M. Richards

important technological applications in areas such as heterogeneous catalysis, optical communications, gas sensing, nanoelectronics, and medicine. NPs come in a wide range of sizes and shapes, with varied electronic, optical, and chemical properties. However, throughout this diversity a universal concept is applicable: the properties of NPs are intimately connected to their nanoscale size and atomic-scale structure. Understanding these properties requires careful consideration of the nature of bonding both between the constituent atoms of NPs and between atoms and molecules in their environment. In these respects theoretical models have played a central role and have provided interpretations for many experimental observations.

It is useful at the outset to explain some nomenclature. The terms nanoparticles, nanocrystallites, and clusters are often encountered in the literature and are frequently used interchangeably. A particle of matter is normally referred to as an NP if its extension in all three dimensions is less than 100 nm. To put this size into perspective it is about one thousandth of the width of a human hair. A nanocrystallite is generally understood to possess crystalline order in addition to nanoscale size, although not necessarily the crystal structure characteristic of the corresponding bulk material. Finally, clusters are particles containing a very small number of atoms such that it is no longer possible to clearly distinguish "bulk" atoms from those at the surface. There is no universally understood definition but a general rule is a few hundred atoms or smaller.

There is considerable variety in the types of NP systems that have been fabricated and studied. Aside from differences in their size and shape one important variable is their composition. Almost every element in the periodic table, together with various alloys and compounds, can form NPs. They can be metallic, semiconducting, or insulating and typically their properties are very different to those of the corresponding bulk material. For example, small metallic NPs behave like insulators as there is a sizable gap near the Fermi energy (see Question 1). Another source of variety is the environment of the NP. While isolated NPs can be produced, for example, by condensing a vapor in an inert gas, they are more commonly supported on substrates, collected into powders, or embedded inside another material. For example, supported NPs can be fabricated simply by collecting already formed NPs from a solution gas or by directly depositing atoms or molecules onto a surface (e.g. by evaporation or molecular beam epitaxy).

An important idea that underpins much of nanotechnology is that by controlling composition, size, and structure at the nanoscale one can engineer almost any desired properties. This is particularly true for NPs as demonstrated by their many varied technological applications. They also have considerable fundamental interest as one can study the emergence of bulk properties (i.e. electronic, chemical, structural, thermal, mechanical, and optical) as the number of atoms in the particle increases from one to many thousands. The aim of this chapter is to provide a broad introduction to NP systems, with particular attention paid to structure and its connection to various properties (so-called structure-property relationships). This chapter is divided into three main parts. First in Section 2.2, some general issues concerning the atomic structure of NPs and its dependence on various factors, such as particle size and the presence of a support, are discussed. Associated properties of NP systems, such as chemical and electronic, are also described, together with some of their important technological applications. Although Section 2.2 is intentionally fairly general, important issues are explored in more detail by considering two particular NP systems as case studies. These systems are chosen to represent prototypes for ceramic (MgO, Section 2.3) and metallic (Au, Section 2.4) materials, which are very important for numerous applications.

2.2 BACKGROUND

2.2.1 Size and Structure of Nanoparticles

The size and atomic-scale structure of NPs depend on their history; that is, how they are fabricated and the temperature and environments to which they are subsequently exposed. Numerous techniques can be used to produce NPs, for example, evaporation or deposition onto surfaces, wet chemistry synthesis, and gas-phase aggregation. They are also ubiquitous in nature, for example, as soot particles in the atmosphere and dust in interstellar space, and they are even produced by certain types of bacteria. One often distinguishes between two types of NP structures: those of low potential energy, which are close to thermodynamic equilibrium, and those of higher potential energy, which are formed by kinetically limited processes. It is often possible to transform the latter into the former by suitable thermal annealing; however, kinetically controlled structures can be preferable for applications if they exhibit structural features with desirable properties. The most common approach to theoretically modeling NPs is to determine the atomic configuration that has the lowest potential energy. Even though in reality NPs are characterized by a statistical distribution of structures, such models are often useful as they can be indicative of more general trends.

The distinguishing feature of NPs, irrespective of their detailed structure, is that they possess a large surface area relative to their volume and a large fraction of atoms that are under-coordinated compared to the bulk. These surface atoms are responsible for many of the unique properties of NPs and some insights into their structure can be gained by considering macroscopic surfaces. The orientation of a crystalline surface is conveniently defined in terms of Miller indices of the form ijk (see Reference 1 for example). In many cases the atomic structure of a given surface is as would be expected if one cut the bulk crystal in two and separated the resulting halves. In such cases there is usually only a small relaxation of atoms near the surface from their bulk positions due to the perturbation associated with missing bonds. On the other hand, some surfaces reconstruct more significantly [e.g. Si(111) or Au(111)] to adopt atomic structures with translational periodicity different to the bulk. To assess the thermodynamic stability of different surface structures one can compare surface formation energies, γ_{ijk}, defined as the cost in energy associated with forming a unit area of a surface (ijk).

Perhaps the simplest way one can imagine constructing an NP is to cut it from the bulk crystalline lattice by forming a polyhedron out of various surface planes. This is the philosophy behind the Wulff construction (2, 3) which gives the particle shape that minimizes the potential energy given various surface energies, γ_{ijk}. This geometrical procedure is illustrated in two dimensions in Figure 2.1a. The approach can also be applied to NPs that are supported upon a substrate. In this situation there is often a preferred NP orientation such that atoms maximize their adhesion to the substrate. The adhesion energy, γ_{ad} (per unit area), can be defined and used in an analogous way to surface energies. This is known as a Wulff–Kaichew construction and is illustrated in Figure 2.1b (4). A further application of this method is to situations

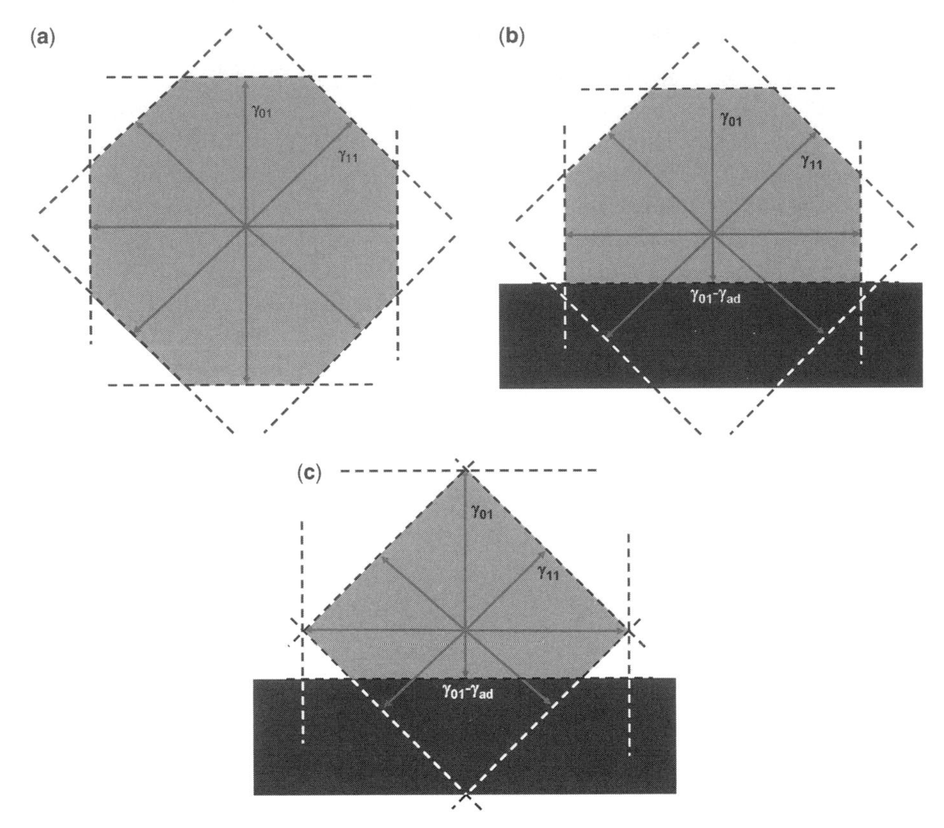

Figure 2.1 (a) An illustration of the Wulff construction in two dimensions. Vectors are drawn in directions [ij] with their length proportional to the surface energy γ_{ij}. Additional lines (or planes in 3D) are constructed to lie at the end of each vector and are oriented perpendicular to them. The polygon (polyhedron) formed by the intersection of the lines (planes) is the particle shape that minimizes the potential energy. (b) For NPs supported on a substrate the adhesion energy, γ_{ad}, reduces the extension of the vector directed towards the surface (Wulff–Kaichew construction). The strength of the adhesion energy relative to the surface energy determines the wetting of the NP on the surface. (c) In an atmosphere or liquid molecules may interact with the NP and modify its surface energies. As a consequence the shape that minimizes the potential energy can change.

where molecules (in the ambient atmosphere) can interact with facets of the particle modifying the energy of surfaces. In this case one can replace the surface energies with surface free energies which are a function of molecular concentrations and temperature. As a consequence the equilibrium shape of the NP will depend on its environment (e.g. see Fig. 2.1c). These effects have been studied experimentally using, for example, transmission electron microscopy (TEM) to observe the shape change of Pt NPs caused by H_2S (5).

The NP shape predicted by the Wulff construction does not depend on its size because it is assumed that its facets are sufficiently large that interfaces between them, that is, edges and vertices, make a negligible contribution to the total energy. It is also assumed that the NP possesses the same crystalline structure as the bulk material. However, for particles smaller than about 10 nm the number of surface atoms can represent a significant fraction of the total number of atoms and in this case the NP may lower its energy by adopting structures very different to the bulk. The thermodynamically favored structure results from a delicate balance between volume and surface energy contributions and determining this structure is a complex problem. NPs can, in principle, be imaged on an atomic level using experimental techniques, such as x-ray diffraction, TEM, and scanning tunneling microscopy (STM). However, in practice this is very difficult and a large part of the information we have about the atomic structure of NPs comes from theory.

Numerous computational methods have been developed to find atomic configurations that minimize the potential energy. Techniques such as simulated annealing, Monte Carlo and genetic algorithms (6) are designed to explore the atomic configuration space sufficiently such that the global minimum of the potential energy surface can be found. Empirical interatomic potentials parameterized on the basis of experimentally determined properties are often used for these calculations as the computational cost of quantum mechanical calculations is in many cases prohibitive. Examples include the embedded atom model for metals, the Lennard–Jones potential for noble gases, and the shell model for metal oxide systems. Detailed investigations into the dependence of structure on the number of atoms in the NP have been made for a wide range of systems (see Reference 7 for a review).

Typically one can identify ranges in size over which various structural types prevail, with the Wulff construction being valid only for very large particles. To compare the stability of NPs as a function of size a useful quantity is the binding energy divided by the number of atoms it contains. It is often found that NPs containing certain "magic" numbers of atoms are particularly stable. These magic numbers usually form a sequence corresponding to truncated morphologies with few low coordinated atoms. Experimentally, magic number sequences can be observed as peaks in the mass spectra of NPs (8). For small clusters, containing of the order of tens of atoms or less, structures result from a complex interplay between geometric and electronic effects. The configurations that have the lowest energy are often quite unusual, involving linear, planar, or even tubular arrangements of atoms. These structures are often easily perturbed by molecules that may adsorb on them or by nearby surfaces.

To summarize, large NPs with low surface to volume ratio have bulk-like crystalline structures and are terminated by low energy surfaces (the Wulff construction).

Smaller NPs can adopt a range of different geometrical structures due to a competition between bulk and surface energies, while the structure of very small clusters can be dominated by electronic effects.

2.2.2 Novel Properties of Nanoparticles

At finite temperature NPs can behave quite differently to macroscopic systems. For example, melting temperature is normally an intensive property but for NPs it decreases with decreasing particle size (e.g. for metals, see Reference 9 and Questions 2 and 3). Melting starts at the surface and then propagates into the interior and the dependence on particle size can be understood as a result of reduced atomic coordination. At finite temperature the structure of NPs can change dynamically as atoms diffuse between facets and are exchanged between the NP and the support. As a result of such processes a collection of NPs may undergo Ostwald ripening (where larger NPs grow at the expense of smaller ones) or sintering (where mobile clusters coalesce). Therefore, atomic diffusion can be an efficient mechanism for particles to move and to change their size and shape (10), which on the macroscale are attributes normally associated with liquids rather than solids.

The interaction of molecules with NPs is of interest for a wide range of applications but has been studied most intensively for heterogeneous catalysis. Examples of systems employed for catalysis include transition metals, noble metals, and metal oxides NPs for reactions such as oxidation, hydrocarbon reforming, and hydrogenation. While the precise nature of activity in many systems is still debated, it has long been recognized that atoms with reduced coordination, such as those found at surfaces, have modified chemical properties (e.g. molecular adsorption energies) and can be active sites for various reactions. Therefore, NPs are attractive catalysts, in part, simply because of their high surface area. They contain atoms with a range of different coordination numbers, for example, corresponding to atoms in closely packed facets, edges, and vertices. Moreover, the concentration of these chemically active sites can, in principle, be controlled by engineering the size and shape of NPs. A further advantage of the high surface area to mass ratio of NPs is that less active material is needed to produce a working catalyst (see Question 4). This is commercially very important because many materials that are used are very expensive; for example, Pt is used in automotive catalysts.

While the increased surface area of NPs is a relatively simple consequence of reducing their size, NPs also exhibit nontrivial size effects that can be important for chemical reactivity. For example, the electronic charge associated with atoms or ions can be different in small NPs and electronic structure can be modified due to confinement of electronic states. Another important factor in determining chemical activity is the interaction between NPs and their support, which can involve a number of interrelated effects.

1. *Geometrical*: The structure of NPs is influenced by the support and they may be strained.

2. *Electronic*: As a result of NP-support bonding or electrostatic effects electronic structure can be modified and in some cases there can be charge transfer either from or to the support.
3. *Chemical*: For example, the NP may be partially oxidized or reduced on interaction with the support.

Besides their unique chemistry NPs exhibit many other properties that find a myriad of technological applications. For example, TiO_2 NPs are used in coatings for self-cleaning glass due to their photocatalytic properties and also in invisible sun creams. The luminescence spectra of semiconductor NPs are very sensitive to particle size, a fact related to quantum confinement of electronic states, and have high quantum yields making then suitable for light-emitting devices. Metallic NPs, such as Au and Ag, also have size- and shape-dependent optical spectra connected to the excitation of plasmon resonances and have applications in medicine and photonics (see Section 2.4.2). The magnetic properties of NPs are also important for high density data storage applications. Magnetic moments can be enhanced on low coordinated atoms at the surface and materials that are normally nonmagnetic, for example, Au, can become magnetic when in the form of small clusters. These are just a few examples that indicate the wide-ranging properties and applications of NPs but there are many more that have not been mentioned. In the next two sections the morphology and associated properties of MgO and Au NPs will be discussed in more detail.

2.3 CASE STUDY 1: MAGNESIUM OXIDE NANOPARTICLES

2.3.1 Introduction

Magnesium oxide (MgO) is an insulating ceramic material that finds numerous applications in a range of areas such as electronics, chemistry, and telecommunications. It is also commonly employed as a substrate for surface science studies of supported nanostructures because its bulk surfaces are well ordered and relatively inert. However, in NP form MgO exhibits quite different optical and chemical properties, which are attractive for applications in heterogeneous catalysis and photonics. NPs of MgO can be produced very easily simply by burning magnesium in air. Examining the structure of NPs made in this way, for example, using TEM, one finds nanocrystallites that are very cubic in shape and usually about 50 nm in diameter. Many of the unusual optical and chemical properties of MgO NPs are connected to the increased concentration of low coordinated ions on their surface. Therefore, it is desirable for both experimental studies and applications to increase the concentration of these sites by producing smaller NPs and numerous techniques can be used. However, before considering NPs we briefly review the properties of MgO in the bulk.

MgO is a highly ionic material that crystallizes in the rock salt structure with lattice constant 4.21 Å at atmospheric pressure. In the crystalline environment electronegative O atoms accept two electrons from the electropositive Mg atoms. Although

O^{2-} is not stable in the gas phase, it is stabilized in the crystal by the electrostatic potential produced by the surrounding ions (Madelung potential). The cohesive energy of MgO, dominated by this electrostatic contribution, is very large (around 40 eV per MgO unit), explaining its high melting temperature (~3000 K). The electronic structure of MgO consists of a valence band comprised of O 2p states and a conduction band comprised of Mg 2s states. The band gap is 7.8 eV, making it a very good electrical insulator and the bottom of the conduction band is about 1 eV above the vacuum level (i.e. it has a negative electron affinity), which also means it is a good electron emitter.

The most stable surface of MgO is the nonpolar (100) surface with a calculated formation energy of 1.25 Jm^{-2}. The structure is almost exactly as one would expect if the bulk crystal was cut in two and most of the structural relaxation near the surface is confined within two atomic planes. There is a small contraction of the spacing between two outermost planes together with a small surface rumpling. The rumpling involves a slight outward relaxation of anions and an inward relaxation of cations (e.g. see Reference 11). Other surfaces such a (110) or (111) can be produced, by cleaving, for example. However, these surfaces have a tendency to microfacet exposing energetically favorable (100) surfaces separated by steps and edges (12, 13). Polar surfaces can be stabilized by adsorbed species; for example, OH^- can form in the presence of water vapor. For very thin films the transformation of the rock salt structure into a hexagonal graphite-like structure can be favored as a result of a balance between bulk and surface energies (14).

2.3.2 Properties of MgO Nanoparticles

2.3.2.1 Morphology In general, the morphology of MgO NPs depends on the method and conditions used for their fabrication. As mentioned above burning Mg in air yields almost perfectly cubic particles truncated by (001) surfaces, which is the thermodynamically favored structure as predicted from the Wulff construction. Smaller MgO nanocrystallites, with 1 to 20 nm diameter, can be produced using chemical vapor deposition (CVD), with a high degree of size selectivity, or by chemical decomposition of magnesium hydroxide (15) and sol-gel techniques. Exotic nanoflower structures have also been fabricated in kinetically driven growth conditions. Although the overall morphology of MgO NPs can be determined using experimental methods, such as TEM, it can be very difficult to determine atomic-scale structure. Therefore, theoretical models can be very useful to identify structural features and their contribution to the properties of NPs.

The highly ionic nature of MgO means that quite accurate empirical potentials can be constructed. The polarizable shell model potential is the most widely used for MgO and also for a wide range of other ionic materials. It is instructive to discuss the main elements of this potential in order to understand the nature of interactions between the ions. The dominating contribution to the interaction is electrostatic and in the simplest approximation can be represented by associating a point charge (usually the formal charge) with each ion. In addition there is a short-range repulsive term due to the overlap of electron density between the ions (Born–Mayer) and a weakly attractive dispersion term. Together these three terms give the following interionic potential as a function of ion separation, r,

$$V_{ab}(r) = \frac{Z_a Z_b e^2}{r} + A_{ab} \exp\left(-\frac{r}{\rho_{ab}}\right) - \frac{C_{ab}}{r^6},$$

where a and b label the ionic species (Mg^{2+} and O^{2-}), Z is their formal charge, and A, ρ, and C are constants chosen to reproduce properties of bulk MgO (e.g. see Reference 16). It is also necessary to take into account the high polarizability of the anions which partially screen the repulsive interactions between cations. This is achieved by representing the O^{2-} ion by two charges (a core and a shell) connected by a spring so that in an electric field a dipole is induced. This model for MgO has been shown to be accurate for many properties of the bulk and surfaces by comparison to experiment and quantum mechanical calculations. It has also been used to investigate finite temperature effects such as diffusion and melting using molecular dynamics (17).

Empirical potentials have been used to find the structure of small MgO clusters using genetic algorithms and have shown that hexagonal ring and tube structures can be preferred over cubes for very small sizes, for example, see Figure 2.2a (18). This is in agreement with experimental mass spectra results, which exhibit magic number peaks spaced by three MgO units for clusters containing less than 30 MgO units (19, 20). For larger NPs the rock salt structure with cubic morphology is favored. Figure 2.2b shows the predicted structure of a $(MgO)_{500}$ NP, which has a diameter of about 2 nm. Surface rumpling is apparent on the facets in addition to their slight overall curvature. The curvature is a result of the structural relaxation associated

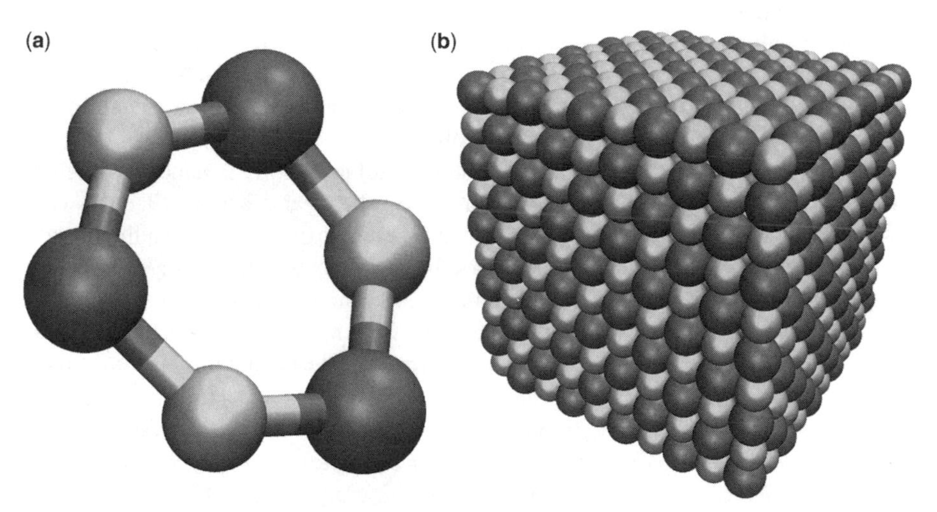

Figure 2.2 (a) The hexagonal ring-like structure of the $(MgO)_3$ cluster and (b) the cubic morphology of a larger 2 nm MgO NP.

with ions of reduced coordination at edges and corners. The corner atoms are affected the most strongly, relaxing inwards and decreasing the distance to their neighbors by 0.15 Å (7%). In addition the electrostatic potential at corner ions is different by 3 V compared to bulk ions and the O ions are strongly polarized.

So far we have considered MgO NPs in isolation; however, when they are formed, for example, in a CVD chamber (21), they are usually collected onto a support. As they land on the growing pile they may rotate and translate with respect to other nanocrystallites before finding a configuration corresponding to a local minimum in the potential energy. A powder produced in this way may have high densities of different types of topological feature, including interfaces between the (100) terraces of two different nanocrystallites, which are rotated with respect to each other. Transmission electron microscopy studies of MgO smoke particles have revealed a preference for alignment of nanocrystallites with commensurate (100) surfaces. However, a small number of nanocrystallites were found to be misaligned at angles corresponding to high site coincidence (100) twist grain boundaries (22). In addition, one may also find very metastable interfaces between nanocrystallites, which exist as a consequence of geometrical constraints imposed by surrounding nanocrystallites (23). If the powder is subjected to high temperature annealing after collection, a degree of sintering may occur and most of the metastable configurations should transform into lower energy configurations.

2.3.2.2 Electronic, Chemical, and Spectroscopic Properties While bulk MgO surfaces are relatively chemically inert, MgO nanopowders have been shown to be active for numerous reactions including oxidation, dehydrogenation, alkylation, and isomerization (24). This chemical activity is often associated with low coordinated ions on the surface, such as steps and corners, and their ability to trap charge. As such features are found with high concentrations in ultra-high surface area powders they have potential catalytic applications. In addition, MgO nanopowders exhibit optical excitation and luminescence bands not found on corresponding bulk MgO surfaces and may have application for energy-efficient lighting materials. These effects are also connected to low coordinated ions at the surface; therefore, spectroscopy can provide a site-specific experimental probe.

The trapping of electrons and holes at the surface of MgO nanopowders can produce a wide range of species that are very chemically active. Electrons and holes can be produced thermally, mechanically, electrically, and by irradiation with particles or photons. Irradiation produces both holes, which are trapped at low coordinated anions, and electrons, which can either trap at low coordinated surface cations or be ionized into the surroundings. The presence of O^-, that is, a hole localized on a surface oxygen ion, has been confirmed experimentally by electron paramagnetic resonance (EPR) (25). Electron trapping at Mg ions forming Mg^+ has also been observed (26). In addition to charge trapping at intrinsic surface features, adsorbed molecules may also trap charge. For example, in an oxygen atmosphere O_2 can trap electrons (e.g. under UV irradiation) forming an O_2^- molecule (27). Protons (formed by dissociation of H_2, for example) adsorb at oxygen ions forming OH^- and can also act as electron traps.

Theoretical calculations have been invaluable in determining the nature of various charge-trapping centers on the surface of MgO NPs. As charged defects can induce long-range polarization of the surrounding ions it is computationally prohibitive to model these systems at a fully quantum mechanical level. An alternative, known generally as QM/MM (quantum mechanics/molecular mechanics), is to treat a small region of the system quantum mechanically (QM region) while the rest is modeled at an empirical potential level (MM region) (e.g. see References 23, 28 and references within). The accuracy of this approach relies on a careful treatment of the interface between QM and MM regions. A further complication is that standard DFT approaches are well known to underestimate the tendency of charge to localize. To correct this deficiency various approaches have been proposed. One commonly used method is hybrid-DFT which mixes a proportion (often 25%) of exchange calculated at the Hartree-Fock level into the DFT functional. Due to the highly ionic nature of MgO QM/MM methods are very accurate and have been verified by applications to bulk and surface color centers, for example, where many experimental studies have been done.

Applications of these methods to MgO nanopowders have confirmed that a wide range of structural features containing three-coordinated oxygen ions, such as corners, step corners, and kinks, are able to trap holes. This trapping is associated with splitting of localized surface states from the bulk MgO valence band caused by structural relaxation and changes in the Madelung potential. Similarly three-coordinated magnesium ions can trap electrons, whereas the MgO(001) surface has a negative electron affinity (Fig. 2.3). The g-tensors of these paramagnetic species have also been calculated allowing direct comparison to experimental EPR spectra. Many of these features have been summarized in a recent paper (23) where interfaces between NPs in the powder were also considered. Here, it was found that interfaces can

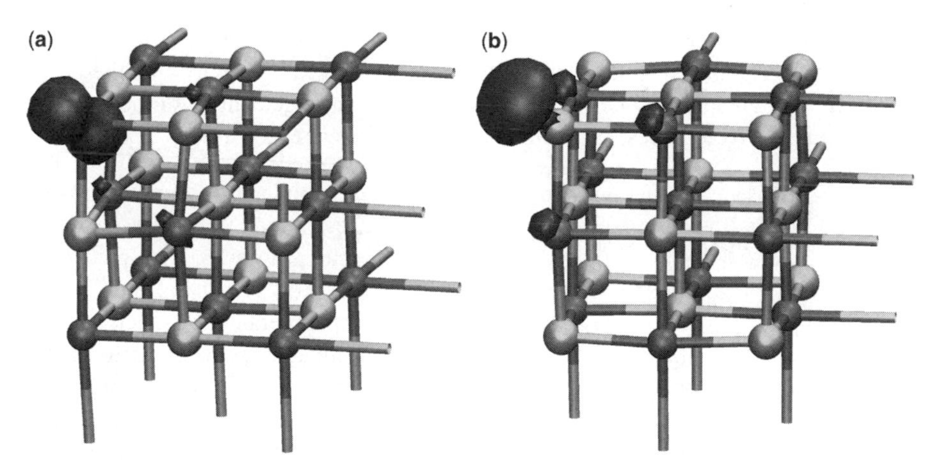

Figure 2.3 Spin-density distributions for charge trapped at low coordinated ions on the surface of MgO NPs. (a) A hole localized at an O corner and (b) an electron trapped at a Mg corner.

also provide places where holes can be trapped but in a delocalized form; therefore, they are EPR invisible.

The fact that local electronic structure is modified at surface features of MgO NPs also has consequences for their optical properties. Bulk MgO is transparent to photons with energies less than 7.7 eV (~160 nm), although defects (known as color centers) do give some absorption in the visible part of the spectrum. However, in nanopowders additional bands are observed at 6.6 eV, 5.3 to 5.7 eV, and 4.6 eV, which have been attributed to terrace, edge, and corner excitations, respectively (29) and this is confirmed by theoretical calculations. Following excitation, electrons and holes can recombine, leading to photoluminescence. Photon emission is observed experimentally at 3.3 to 3.4 eV. This is interpreted as charge recombination at three-coordinated ions with a large Stokes shift (>1 eV) due to structural relaxation. A more extreme example of structural modification under excitation is the photodesorption of low coordinated ions which has been observed experimentally and explained theoretically (30).

2.3.3 Summary

As the structural and electronic properties of MgO have been well studied in the bulk, at surfaces, and for numerous defects and impurities therein, it represents a useful model system on which to explore the properties of metal oxide NPs. Experimental and theoretical studies have shown that low coordinated ions, which have increased concentrations in NPs, have different electronic and optical properties compared to the bulk and can act as charge traps. The trapping of electrons and holes in insulating oxide materials, which can be both very useful and harmful, is an important issue of wide interest. Commonly used experimental probes such as x-ray- and UV-photoelectron spectroscopy and transmission electron microscopy (TEM), for example, may cause charge to be trapped in the material (often powder) directly, or as a consequence of secondary processes. Although the Coulomb interaction favors the recombination of electrons and holes, they may become separated and trapped in configurations that are thermally stable over long time scales. Associated with trapped charge at the NP surfaces is an increase in chemical reactivity, and the possibility to tailor chemical properties by controlling the arrangement of surface ions on the NP has received interest for numerous applications in catalysis.

2.4 CASE STUDY 2: GOLD NANOPARTICLES

2.4.1 Introduction

Gold is the most well known of all precious metals, forming the basis of many currencies and also used for decorative jewellery. Gold has historically been viewed as a very chemically inert material; chlorine is one of only a few substances that react with it. This property, coupled with its high electrical conductivity, makes it an ideal material for corrosion resistant wires and contacts in electronic device applications. Therefore, it came as some surprise when it was found that Au in NP form was an active heterogeneous catalyst for numerous reactions, such as the oxidation of CO, the partial oxidation of hydrocarbons, and the hydrogenation of carbon oxides (e.g. see Reference 31 and references within). This has since been the focus of a great deal of research. While the exact origin of activity is still debated it has become clear that it results from a complex interplay between both geometric and electronic factors. Au NPs also exhibit quite different optical properties from the bulk. The plasmon resonance that dominates their absorption spectrum is very sensitive to both size and shape (32) and is useful for applications such as plasmonic waveguides, surface enhanced spectroscopies, and biological markers.

As in the previous section, before considering Au NPs we briefly review the properties of the bulk material. Gold adopts the fcc crystal structure with lattice constant 4.08 Å and remains solid up to temperatures of about 1000°C. Its electronic structure consists of a wide free electron-like *sp* band that is hybridized with a narrower *d* band. The Fermi level is positioned just above the filled *d* band explaining its high electronic mobility. Gold surfaces have been studied in detail both experimentally and theoretically. The (111) and (100) surfaces are the most thermodynamically favorable, which are both observed to be reconstructed. These reconstructions involve surface atoms adjusting into more closely packed arrangements and the corresponding unit cells can be quite large.

2.4.2 Properties of Au Nanoparticles

2.4.2.1 Morphology Au NPs can be fabricated using a wide range of techniques that can result in an equally wide range of morphologies. Ionized mass selected clusters can be produced by plasma sputtering in an inert gas and may subsequently be deposited onto various substrates (33). Alternatively, gold can be evaporated directly onto a substrate allowing NPs to nucleate and grow at defects or steps, for example. Wet synthesis methods, which are the preferred industrial route due to their cost and scalability, can also be used to form NPs on oxide supports (31).

To experimentally determine the atomic structure of small gold clusters is very difficult. Indirect information in the form of mass spectra indicates that certain cluster sizes are particular stable (e.g. see Reference 7). Such magic number clusters show up as peaks in the mass spectra and theoretical calculations have been very useful to interpret these experiments. For very small Au clusters ($N < 20$) the lowest energy structures have been determined at a quantum mechanical (DFT) level. A tendency towards planar atomic arrangements has been found which has been connected to relativistic effects in the electronic structure of Au. Small Au clusters have also been studied on various substrates where they tend to interact strongly with defects such as vacancies. For example, on oxide surfaces Au clusters can exchange electrons with surface color centers, making them partially charged and highly reactive.

For larger Au NPs many theoretical calculations have been made using empirical interatomic potentials. A number of different models have been developed to represent the many-body character of bonding in metals, for example, Finnis–Sinclair, Gupta, and glue models. Here, we discuss the embedded atom method (EAM), which has many similarities with the models mentioned above but can be considered as more general (34). In the EAM the total energy takes the following form:

$$E = \sum_i E_i = \sum_i F(\rho_i) + \frac{1}{2}\sum_{j \neq i} \phi(r_{ij}),$$

where F is a function that gives the energy associated with embedding atom i in the electron density associated with all surrounding atoms (ρ_i), and ϕ is a short range repulsive potential that depends on the separation between atoms, r_{ij}. This expression for the total energy is inspired by the ideas of density functional theory, namely that the total energy of a system is a unique functional of the electronic density. The next step is made by assuming that the electron density at atom i can be written as a linear sum of contributions from other atoms,

$$\rho_i = \sum_{j \neq i} f(r_{ij}),$$

where f describes the electronic density around each atom. In general F, f, and ϕ can be regarded as parameters that can be varied to fit various physical properties. However, quantum mechanical models of electronic structure can give insights into appropriate functional forms.

Using these many body potential models atomic structures of unsupported Au clusters that have the lowest potential energy have been determined using various global optimization methods. These studies indicate that although three-dimensional morphologies are generally favored the bulk-like fcc structure is not always adopted. For example, in the range 10 to 500 atoms a number of structures with fivefold symmetry, for example, icosahedra (Fig. 2.4) and decahedra, have been found in addition

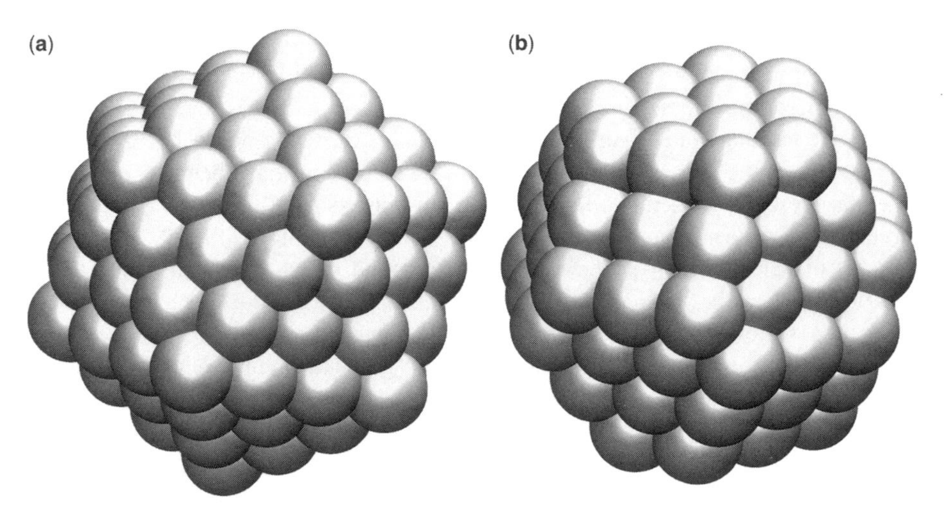

Figure 2.4 The atomic structure of Au NPs with different structural motifs: (a) icosahedron (Au_{147}) and (b) truncated octahedron (Au_{116}).

to some with very low symmetry. For larger NPs it becomes computationally impossible to employ global minimization algorithms. Instead one can compare the energies of different structures as a function of size (7). For Au, icosahedra are favored for fewer than 100 atoms, decahedra dominate until about 500 atoms, at which point fcc Wulff polyhedra take over. The ordering of these motifs is explained by a strain contribution to the total energy for non-fcc structures which grows linearly with particle size and is much larger for icosahedra than decahedra. At the NP surface, under-coordinated atoms relax inwards, decreasing their average bond lengths. This is predicted by empirical and first principles theoretical models and has recently been observed experimentally using coherent x-ray diffraction (35).

For Au NPs supported on the MgO(100) surface TEM has shown that even very small clusters are well faceted and fcc ordered (36). This is explained by the strong adhesion and small lattice mismatch ($\sim$3%) between Au and MgO, which favors epitaxial Wulf–Kaischew-like morphologies. For larger lattice mismatch (e.g. Pd/MgO) NPs may be significantly strained and contain dislocations. To model these effects theoretically requires careful treatment of the metal-substrate interaction (37). In contrast for weakly interacting supports, such as graphite, strained decahedral and icosahedral clusters have been observed by TEM (33).

2.4.2.2 Au NPs for Heterogeneous Catalysis While metallic NPs in general are of interest for numerous catalytic reactions, Au has received particular attention since it was found to be active for oxidation of CO to CO_2. This activity, which is significant only for NPs smaller than about 8 nm, is exciting because it persists to very low temperature (down to 200 K). In contrast, traditional Pt- or Pd-based catalysts for CO oxidation that are used in car exhausts only work at elevated temperatures; therefore, most CO pollution occurs during the initial few minutes after the engine is switched on. Au NPs have also been demonstrated to be active for a range of other reactions, including epoxidation of propylene, reduction of nitrogen oxides, and the selective oxidation of hydrocarbons, which is important in the manufacture of pharmaceutical compounds, for example.

Evidence suggests that more than one effect may contribute to the chemical activity of Au NPs. For example, it has been demonstrated experimentally that activity for CO oxidation increases with decreasing particle size (e.g. see Reference 38 for a summary of some data), suggesting a correlation with the number of exposed surface atoms. However, on a supported NP one can distinguish several different types of surface atom that are potential candidates for active sites. Near the interface between the NP and support there are Au atoms coordinated by both atoms of the support material and other Au atoms. The sites of this type accessible for molecular adsorption and reaction lie on the perimeter of the NP. One reason for believing these perimeter atoms may be important is the observation that the choice of support materials (often metal oxides such as TiO_2 are used) can strongly affect catalytic performance. To explain the origin of this activity a number of different ideas have been proposed. For example, Au can become polarized and partially charged, either positive or negative, by the oxide surface and by interaction with defects such as oxygen vacancies. Supporting this idea, quantum mechanical calculations have shown that small anionic

Au clusters are able to oxidize CO with low barriers to reaction (39). Another suggestion is that perimeter atoms provide places where molecules can adsorb in favorable configurations. For example, one molecule can be adsorbed on the oxide and the other on a nearby Au atom. Quantum mechanical calculations for Au supported on the MgO(100) surface have demonstrated that such effects can also lead to low barriers to reaction (40).

Besides atoms in direct contact with the support Au NPs possess other atoms with low coordination at edges and vertices which have also been proposed as active sites (38). On the close packed Au(111) surface adsorption energies for CO and O_2 are low because the *d* band states that participate in bonding are too low in energy to interact significantly with the frontier orbitals of the molecules. However, as the Au atom coordination is reduced the local electronic structure is modified such that the *d* band is narrowed and moved to higher energy. Therefore, at low coordinated Au atoms adsorption energies are higher and barriers to reaction are lower.

While it is clear that Au NP structure strongly influences the adsorption and reaction of molecules, the reverse is also true: molecules can influence NP structure. For example, it has recently been shown that supported Pd NPs cyclically exposed to CO and NO undergo reversible changes in morphology driven by molecular adsorption (41). For large NPs changes in morphology can be rationalized in terms of modified surface energies as discussed in Section 2.2.1, but for small clusters atomic-scale models are needed. A theoretical study of CO adsorption on an unsupported Au_{79} cluster suggests that morphology transformations should also be expected for Au (42). As CO favors adsorption on low coordinated Au atoms the NP is predicted to adopt compact structures that expose higher numbers of low coordinated atoms, as illustrated in Figure 2.5. These effects may play an important role in catalysis; however, in realistic reactor conditions the situation can be quite complicated as molecules and their by-products may block certain sites and kinetically hinder diffusion processes.

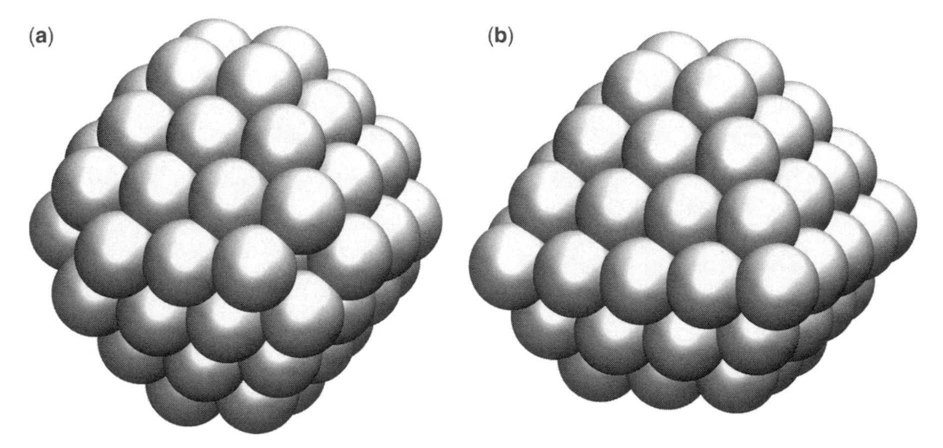

Figure 2.5 Transformation of the atomic structure of the Au_{79} NP in the presence of CO molecules. (a) The equilibrium truncated octahedral structure in vacuum and (b) the equilibrium structure for an average CO coverage of four molecules (42).

2.4.2.3 Optical Properties The interaction of electromagnetic radiation with Au NPs can stimulate collective excitations of conduction electrons known as plasmons. At certain frequencies a resonance condition can be fulfilled resulting in intense scattering in the visible part of the spectrum. The frequency and width of the resonance depend on both the size and shape of the NPs and the dielectric medium in which they are embedded (43). Although these effects are inherently quantum in nature they can be modeled using classical electromagnetism, for example, Mie theory, and are of interest for a wide range of applications. For example, ordered arrays or chains of NPs can be used as nanoscale photonic waveguides allowing one to beat the diffraction limit of light (e.g. see Reference 44 and references within). They can be employed as biomolecular sensors due to the sensitivity of the plasmon resonance peak position on the dielectric media. They also have potential application as biological markers. For example, wet chemistry methods can be used to produce Au NPs that are passivated by a layer of molecules, for example, thiols. By binding relevant functional groups to these NPs they can be attached to specific molecules or proteins which can then be easily identified using optical microscopes.

As Au NPs find more and more applications there has been increasing concern about their possibly harmful interaction with biological organisms (nanotoxicity). Recent studies indicate that prolonged exposure of cells to high concentrations of NPs does cause cell damage. These effects appear to be related to the penetration of NPs into cells and the generation of active oxygen species (45).

2.4.3 Summary

The perception of gold as a highly inert material has been transformed by the discovery that nanosized gold particles are chemically active for numerous reactions. They also possess unique electronic and optical properties which can be tuned for specific applications by controlling their size and atomic structure. Surface science studies and theoretical models that have helped to understand these complex structure-property relationships suggest that confinement of electronic states, under-coordination of atoms, charging, and NP-support interactions all play important roles. These effects are not only relevant to Au but to other metallic NP systems as well, such as transition metals. However, Au is a particularly interesting example because its bulk surfaces are inert whereas bulk transition metal surfaces are active catalysts. Many of the applications of Au NPs, for example, in biology, gas sensing, and catalysis, involve interactions with molecules that can influence NP structure. The dynamical interplay between molecules and NPs is, in general, both complex and nonequilibrium in nature. Understanding these processes is an important problem that underpins many applications in nanotechnology and which presents a challenge for both theory and experiment.

2.5 CONCLUDING REMARKS AND FUTURE OUTLOOK

One of the reasons NPs are attractive for a wide range of applications is the potential to engineer their electronic, chemical, and optical properties through their structure. Varying the composition of the NP provides an additional degree of freedom and is receiving increasing interest for a wide range of systems. The structure and composition of NPs are closely interrelated as one can influence the other. For example, atoms of one type may prefer to segregate to particular locations on the NP surface modifying its structure. Multicomponent NP systems often exhibit properties that are characteristic of neither component in isolation and can be tailored for various applications. For example, doped semiconductor and metal oxide NPs have interesting optical properties and bimetallic NPs (e.g. AuPt, AuPd) are highly active catalysts.

The work on NPs to date has shown us that theoretical modeling and experiment must work closely together to unravel the precise nature of structure-property relationships. Recent advances in techniques such as TEM (33), STM, and atomic force microscopy mean that atomic-scale resolution of NP structure is now becoming possible. This is complemented by increasingly powerful computers that allow for quantum mechanical calculations on larger and more complex NP systems (e.g. including the substrate or interfaces between NPs). One of the key remaining challenges (for experiment and theory) concerns the atomic-scale dynamics of NPs. This is important for understanding NP growth, an important step towards controlling structure, and for understanding the role of temperature in real applications. A related issue concerns the interaction of NPs with their environment, for example, molecules in a surrounding liquid or gas, which can influence their structure and properties. To address this problem *in situ* experimental studies are becoming increasingly common and can be complemented by similar developments in theoretical modeling.

SUGGESTED FURTHER READING

K. J. Klabunde. editor. *Nanoscale Materials in Chemistry*. New York: Wiley-Interscience, 2001.

D. G. Pettifor. *Bonding and Structure of Molecules and Solids*. Oxford: Oxford University Press, 1995.

J. Venables. *Introduction to Surface and Thin Film Processes*. Cambridge: Cambridge University Press, 2000.

G. A. Ozin, A. C. Arsenault. *Nanochemistry: A Chemical Approach to Nanomaterials*. Royal Society of Chemistry, 2005.

R. L. Johnston. *Dalton Trans*. **2003**, 4193.

C. Noguera. *Physics and Chemistry at Oxide Surfaces*. Cambridge: Cambridge University Press, 1996.

K. P. McKenna, P. V. Sushko, A. L. Shluger. *J. Am. Chem. Soc*. **2007**, *129*: 8600.

F. Beletto, R. Ferrando. *Rev. Mod. Phys*. **2005**, *77*: 371.

D. Astruc, editor. *Nanoparticles and Catalysis*. New York: Wiley-VCH, 2007.

L. M. Molina, B. Hammer. *Phys. Rev. B* **2004**, *69*: 155424.

S. Utamapanya, K. J. Klabunde, J. R. Schulp. *Chem. Materials* **1991**, *3*: 175.

S. I. Stoeva, B. L. V. Prasad, U. Sitharaman, P. Stoimenov, V. Zaikovski, C. M. Sorensen, K. J. Klabunde. *J. Phys. Chem. B. (invited paper for A. Henglein Special Issue)* **2003**, *107*: 441.

REFERENCES

1. M. P. Marder. *Condensed Matter Physics*. New York: Wiley, 2000.
2. G. Wulff, Z. Kristallogr. 1901, *34*: 449.
3. C. Herring. *Phys. Rev.* **1951**, *82*: 87.
4. W. L. Winterbottom. *Acta Metall.* **1967**, *15*: 303.
5. P. J. F. Harris. *Nature* **1986**, *323*: 792.
6. R. L. Johnston. *Dalton Trans.* **2003**, 4193.
7. F. Beletto, R. Ferrando. *Rev. Mod. Phys.* **2005**, *77*: 371.
8. K. Clemenger. *Phys. Rev. B* **1985**, *32*: 1359.
9. P. Buffat, J. Borel. *Phys. Rev. A* **1976**, *13*: 2287.
10. N. Combe, P. Jensen, A. Pimpinelli. *Phys. Rev. Lett.* **2000**, *85*: 110.
11. V. E. Henrich, P. A. Cox. *The Surface Science of Metal Oxides*. Cambridge: Cambridge University Press, 1996.
12. G. W. Watson, E. T. Kelsey, N. H. de Leeuw, D. J. Harris, S. C. Parker. *J. Chem. Soc., Faraday Trans.* **1996**, *92*: 433.
13. A. Wander, I. J. Bush, N. M. Harrison. *Phys. Rev. B* **2003**, *68*: 233405.
14. J. Goniakowski, C. Noguera, L. Giordano. *Phys. Rev. Lett.* **2004**, *93*: 215702.
15. E. Giamello, M. C. Paganini, D. M. Murphy, A. M. Ferrari, G. J. Pacchioni. *Phys. Chem. B* **1997**, *101*: 971.
16. G. V. Lewis, C. R. A. Catlow. *J. Phys. C: Solid State Phys.* **1985**, *18*: 1149.
17. R. Ferneyhough, D. Fincham, G. D. Price, M. J. Gillan. *Modelling Simul. Mater. Sci. Eng.* **1994**, *2*: 1101.
18. C. Roberts, R. L. Johnston. *Phys. Chem. Chem. Phys.* **2001**, *3*: 5024.
19. W. A. Saunders. *Phys. Rev. B* **1998**, *37*: 6583.
20. P. J. Ziemann, A. W. Castleman Jr. *J. Chem. Phys.* **1991**, *94*: 718.
21. S. Stankic, M. Müller, O. Diwald, M. Sterrer, E. Knözinger, J. Bernardi. *Angew. Chem. Int. Ed.* **2005**, *44*: 4917.
22. P. Chaudhari, J. W. Matthews. *J. Appl. Phys.* **1971**, *42*: 3063.
23. K. P. McKenna, P. V. Sushko, A. L. Shluger. *J. Am. Chem. Soc.* **2007**, *129*: 8600.
24. H. Hattori. *Chem. Rev.* **1995**, *95*: 537.
25. O. Diwald, M. Sterrer, E. Knozinger, P. V. Sushko, A. L. Shluger. *J. Chem. Phys.* **2002**, *116*: 1707.
26. M. Chiesa, M. C. Paganini, E. Giamello, C. Di Valentin, G. Pacchioni. *Angew. Chem. Int. Ed.* **2003**, *42*: 1759.
27. M. Che, A. J. Tench. *Adv. Catal.* **1982**, *31*: 77.
28. P. V. Sushko, J. L. Gavartin, A. L. Shluger. *J. Phys. Chem. B* **2002**, *106*: 2269.
29. S. Stankic, J. Bernardi, O. Diwald, E. Knozinger. *J. Phys. Chem. B* **2006**, *110*: 13866.
30. W. P. Hess, A. G. Joly, K. M. Beck, M. Henyk, P. V. Sushko, P. E. Trevisanutto, A. L. Shluger. *J. Phys. Chem. B* **2005**, *109*, 19563.
31. M. Haruta. *Catalysis Today* **1997**, *36*: 153.
32. S. Link, M. A. El-Sayed. *J. Phys. Chem. B* **1999**, *103*: 4212.

33. Z. Y. Li, N. P. Young, M. Di Vece, S. Palomba, R. E. Palmer, A. L. Bleloch, B. C. Curley, R. L. Johnston, J. Jiang, J. Yuan. *Nature* **2008**, *451*: 46.
34. M. S. Daw, M. I. Baskes. *Phys. Rev. B* **1984**, *29*: 6443.
35. W. J. Huang, R. Sun, J. Tao, L. D. Menard, R. G. Nuzzo, J. M. Zuo. *Nature Materials* **2008**, *7*: 308.
36. L. D. Marks. *Rep. Prog. Phys.* **1994**, *57*: 603649.
37. J. Goniakowski, C. Mottet, C. Noguera. *Phys. Stat. Sol. (B)* **2006**, *243*: 2513.
38. N. Lopez, T. V. W. Janssens, B. S. Clausen, Y. Xu, M. Mavrikakis, T. Bligaard, J. K. Nørskov. *J. Catalysis* **2004**, *223*: 232.
39. B. Yoon, U. Landman, A. Wörz, J.-M. Antonietti, S. Abbet, K. Judai, U. Heiz. *Science* **2005**, *307*: 403.
40. L. M. Molina, B. Hammer. *Phys. Rev. B* **2004**, *69*: 155424.
41. M. A. Newton, C. Belver-Coldeira, A. Martínez-Arias, M. Fernández-García. *Nature Materials* **2007**, *6*: 528.
42. K. P. McKenna, A. L. Shluger. *J. Phys. Chem. C (Letter)* **2007**, *111*: 18848.
43. S. Lal, S. Link, N. J. Halas. *Nature Photonics* **2007**, *1*: 641.
44. J. R. Krenn. *Nature Materials* **2003**, *2*: 210.
45. N. Lewinski, V. Colvin, R. Drezek. *Small* **2008**, *4*: 26.
46. R. Kubo. *J. Phys. Soc. Jpn.* **1962**, *17*: 975.
47. W. H. Qi, M. P. Wang, M. Zhou, W. Y. Hu. *J. Phys. D* **2005**, *38*: 1429.

PROBLEMS

1. Due to the finite number of valence electrons in a metallic NP its electronic states are discrete and separated in energy (e.g. see Reference 46) and as a consequence they become electrically insulating below a critical temperature. Estimate the metal-insulator transition temperature for a 2 nm and a 10 nm lithium NP (hint: compare the gap near the Fermi energy to the available thermal energy). Some properties of Li you may find helpful are $E_F = 4.74$ eV, $\rho = 535$ kg m^{-3}, and $M = 6.941$ g mol^{-1}.

2. Derive an expression for the average binding energy per atom of a metallic NP and the dependence on its diameter (e.g. see Reference 47).

3. Using the equation derived above estimate the melting temperatures of a 1 nm and 10 nm diameter Au NP. Note the cohesive energy of Au is 3.9 eV.

4. The specific surface area of an NP is defined as the total surface area per unit mass, usually measured in m^2 g^{-1}. Calculate the specific surface area for a cubic MgO NP with 3 nm diameter. Note the density of MgO is 3.58 g cm^{-3}.

5. Calculate the relative proportion of 3C (corner), 4C (edge), 5C (terrace), and 6C (bulk) ions in MgO NPs assuming perfect cubes of length L and lattice constant a.

6. Comment on how the answers to Problems 4 and 5 may be modified for NPs in a powder.

7. In Reference 21 the optical absorption spectrum of an MgO powder is measured and a band observed at 4.6 eV is attributed to three-coordinated anions. Considering the electronic structure and morphology of the NPs what is the origin of the higher frequency band at 5.4 eV?

8. (a) Determine the nearest neighbor coordination number for atoms in the (100) and (111) surfaces of an fcc crystal (assuming no reconstruction).

   (b) Calculate the surface area per atom for each of these surfaces (with nearest neighbor separation, d).

9. If we assume the binding energy per atom in the problem above is linearly proportional to its coordination number, what is the ratio of (111) and (100) surface energies?

10. Low coordinated atoms on metallic NPs have been proposed to be catalytically active sites. How many atoms with a coordination of six or lower are there on icosahedral, octahedral, and truncated octahedral Au NPs.

ANSWERS

1. First we need to estimate the number of atoms in the NP. This can be done by assuming it is approximately spherical in shape and has a similar density to bulk Li. This gives $N = (N_a/M)(4/3\pi r^3 \rho)$, where N_a is Avogadro's number. From Reference 46 the average spacing between energy levels at the Fermi energy is: $\Delta = 4E_F/3N$. Note that Li is a simple metal with one valence electron per atom. Finally we need to equate the thermal energy, $k_B T$ to the spacing between energy levels to obtain: $T_{met\text{-}ins} = \Delta/k_B = (E_F M)/(N_a r^3 \pi \rho k_B)$. Putting in the values we find that $T_{met\text{-}ins}(2\text{ nm}) = 47$ K and $T_{met\text{-}ins}(10\text{ nm}) = 0.4$ K.

2. There are various ways to do this but here is one example. Assume a spherical particle of diameter, D, with nearest neighbor distance between atoms, d. The total number of atoms in the particle is approximately: $N = (D/d)^3$, although this number is actually smaller due to the finite packing fraction. The number of atoms on the surface of the NP can be estimated as: $N_s = 4(D/d)^2$. The total binding energy of the NP is $E = NE_c - N_s\Delta$, where E_c is the bulk cohesive energy and Δ is the average energy cost associated with each surface atom. To obtain a simpler expression it is useful to assume that Δ can be expressed as a fraction of the bulk cohesive energy, that is, $\Delta = fE_c$ (e.g. where $f < 1$ and is related to the number of broken bonds). Rearranging we find $E/N = E_c[1 - C(d/D)]$, where C is a constant related to the geometry of the NP and is approximately equal to unity for three-dimensional NPs.

3. We need to equate thermal energy to the binding energy per atom. The empirical relationship that allows one to estimate melting temperatures is $T_m = (0.032/k_B) * E_b$ (e.g. see Reference 47). For Au, using $d = 2.88$ Å we find that $T_m(1\text{ nm}) = 1031$ K and $T_m(10\text{ nm}) = 1407$ K.

4. The surface area of a cube with diameter D is $A = 6D^2$. The mass associated with this cube is $m = D^3\rho$. Therefore, the specific surface area, $A/m = 6/(D\rho)$. Using $\rho(\mathrm{MgO}) = 3.58\ \mathrm{g\ cm^{-3}}$ and $D = 3$ nm then $A/m = 559\ \mathrm{m^2 g^{-1}}$.

5. There are eight 3C ions per cube, $[(L/a) - 1] * 24$ 4C ions per cube, $[2(L/a - 1)]^2 * 6$ 5C per cube, $2(L/a - 1)^3$ 6C per cube. The total number of ions per cube is $(L/a)^3$ and their relative proportions can be easily evaluated. For example, an MgO cube with 1 nm diameter has about 7% 3C atoms, while a 5 nm cube has less than 0.1%.

6. Interfaces between nanocubes in a powder will reduce the overall specific surface area and the number of low coordinated ions.

7. The band at 5.4 eV has higher intensity than that at 4.6 eV. If the band at 5.4 eV is due to excitation of four-coordinated ions (edges) then the difference in intensity can be understood in terms of the relative populations of three- and four-coordinated ions (see Problem 5).

8. (**a**) $Z = 8$ for atoms in (100) surfaces and $Z = 9$ for atoms in (111) surfaces.

(**b**) The area per atom in a (100) surface is $A_{100} = d^2$. The area associated with each atom in a (111) surface is $A_{111} = [\sqrt{3/2}] * d^2$.

9. The surface energy for the (100) surface is $\gamma_{100} = C(12 - 8)/A_{100}$ and for the (111) surface is $\gamma_{111} = C(12 - 9)/A_{111}$, where C is a constant. Therefore, $\gamma_{111}/\gamma_{100} = (3/4) * [2/\sqrt{3}] = 0.87$.

10. Icosahedral NPs have 12 6C atoms, octahedral NPs have six 4C atoms, and truncated octahedral NPs have 24 6C atoms.

3

PARTICLES AS MOLECULES

C. M. SORENSEN

3.1 Introduction, 37
3.2 Nanoparticle Molecules, 38
3.3 Stoichiometric Cluster Compounds, 40
  3.3.1 Semiconductor Cores, 40
  3.3.2 Metallic Cores, 40
3.4 Nanoparticles, 42
  3.4.1 Types of Nanoparticles, 43
  3.4.2 The Capping Ligand, 44
3.5 Interparticle Interactions, 46
  3.5.1 Colloidal Solutions, 47
  3.5.2 Solubility of Nanoparticle Molecules, 49
  3.5.3 Nanoparticle Interactions, 50
3.6 Superlattices, 55
  3.6.1 Superlattice Melting, 59
3.7 Conclusions, 60
References, 60
Problems, 63
Answers, 64

3.1 INTRODUCTION

Recent advances in synthetic chemistry have given rise to a wide variety of nanoparticles (NPs) with a high degree of both chemical and physical uniformity (1–4). These nanoparticles are surface ligated with a variety of organic compounds and these

Nanoscale Materials in Chemistry, Second Edition. Edited by K. J. Klabunde and R. M. Richards

ligands cause colloids of the nanoparticles to be stable against irreversible aggregation. Often these colloids act as solutions, with the nanoparticles displaying temperature- and solvent-dependent solubility. Also in many cases the precipitating solid is a two- or three-dimensional superlattice of the nanoparticles.

This chapter promotes the perspective that this new class of nearly monodisperse nanoparticles, which have been called quantum dots, nanoclusters, monolayer protected nanoparticles, etc., can be viewed and described as large molecules. I will give justification for this perspective in terms of stoichiometry, properties, their solutions, and their assemblies. I will not describe their electronic and optical properties, which are very fascinating and with which further support for their classification as molecules can be found.

3.2 NANOPARTICLE MOLECULES

I start by asking: what is a molecule? I answer that a molecule is a discrete, complete entity, the smallest piece of a compound. A compound is a combination of elements whose atoms are bound together with the same definite proportions, that is, the same stoichiometry. This is different than a mixture, where the proportions are indefinite and the atomic binding is weak. A molecule if further divided would have radically different properties. Moreover, molecules tend to keep their identities when the compound is melted or vaporized or dissolved. In the solid the molecules often sit as discrete entities either at lattice points of a crystal or at random in the amorphous solid.

As an example, consider carbon tetrachloride, CCl_4. Carbon tetrachloride is a compound because we always find it with the same stoichiometeric ratio C/Cl of 1/4. The atoms are covalently bonded together so that the carbon tetrachloride molecule is a definite entity. We can boil CCl_4 and the molecules will fill space keeping their integrity as CCl_4 units, that is, molecules. If we freeze CCl_4, we will find the molecules arranged as entities in a lattice. If we dissolve CCl_4 in acetone once again we will find that the molecules keep their identities. Yet if we look wider, we find this perfect example of molecular definition and identity is often broken by some of our favorite molecules. Certainly water is a compound because of its definite stoichiometry, but the water molecule looses its integrity in the many forms of solid ice in which the hydrogens are shared via hydrogen bonds so that the identity of the H_2O molecule is confused. These ices can be thought of as huge entities (huge molecules?) bound together via a combination of covalent and hydrogen bound linkages. Similarly salt, NaCl, is a compound and its molecule is a single NaCl unit; yet dissolve salt in water and it dissociates, and the salt crystal is one huge macro-continuum of an ionic lattice. More relevant to nanoparticles, consider polymer molecules. These have the requisite stoichiometry yet can be huge, with essentially any molecular weight the synthesizer chooses.

Here we contend that nearly monodisperse nanoparticles can be fit under the rather broad definition of molecule. From the discussion above we see that all molecules have stoichiometry. The near monodispersity of the nanoparticles of the last two decades (indeed, the reason they have become so interesting) is the analogue to compound stoichiometry. For example, a 5.0-nm diameter gold particle ligated with dodecane thiol can be written

$$Au_{3870}(C_{12}SH)_{365} \tag{3.1}$$

where the standard deviation on the numbers is about 10%. This is close to the uniformity in mass of most compounds given the variety of stable isotopes most elements enjoy. We think of molecules as having an exact stoichiometry but then we remember that this is hard to achieve for polymeric molecules where a spread in polydispersity of 5% is considered very narrow. NPs are entities which if divided would not only be smaller but in some manner would be different than before division because of the well known size-dependent properties of these materials. True, the change would not be as drastic a change as one would get dividing H_2O. Consider, however, that division of a polymer molecule changes the properties but, like the nanoparticle, in a gradual manner. Moreover, the relatively large size of the NP of Equation (3.1) should not exclude it from the molecule club since we readily recall that polymer molecules can have large molecular weights, well beyond 10^6, and hence large physical dimension. The nanoparticle in Equation (3.1) has a molecular weight of approximately 800,000. The NP of Equation (3.1) can be dissolved in toluene, and when it is, it keeps its identity (unlike any ionic salt molecule when dissolved in water). Careful drying of such a solution will yield two- and three-dimensional superlattices analogous to the molecular lattices of "normal" molecules like CCl_4. Other properties of such particle molecules continue the analogy. Thus, the electronic properties evolve from the bonding-antibonding orbitals of the simple diatomic molecule to the energy bands of the bulk solid with the HOMO-LUMO gap becoming the band gap of the solid. Many polymers and biomolecules such as proteins have solvent- and temperature-induced conformational changes. In analogy the ligands of NPs such as in Equation (3.1) have been shown to have similar changes.

The analogies of nanoparticles to molecules can be classified into two categories:

1. Truly stoichiometric cluster compounds. These have between several tens to a thousand atoms, hence sizes in the range of one to a few nanometers.
2. Nanoparticles with narrow size distributions. Here the stoichiometry is approximate. The size ranges from a nanometer to typically several nanometers although examples exist to tens of nanometer scales. The number of atoms involved ranges from approximately 100 to 10,000.

One might say that chemistry rules the first category and physics the second. We say this because cluster compounds are just that, true compounds, with composition and size determined by the chemical reactivities and propensities of the constituent atoms. On the other hand, if the nanoparticles of the second category are to be viewed as molecular analogues, the first requirement is near stoichiometry, which is achieved by the narrow distributions of sizes. This in turn is achieved physically by one of three methods:

1. Nucleation kinetics that forces a sudden burst of particle creation from which a narrow distribution can ensue.

2. Digestive ripening, a heat treatment during which the particle size distribution narrows for poorly understood yet physical reasons.
3. Size selective precipitation/chromatography.

In category 2 above we have a new class of molecules; we can call them "nanoparticle molecules" and their compounds "stoichiometric particle compounds (SPC)." Yet, notwithstanding the arguments above, there are differences with the canonical picture of molecules and *vive la difference*! Unlike the molecules of the past, nanoparticle molecules are not tightly bound by rules of valency. Thus, their stoichiometry can range continuously over broad ranges of composition. Moreover, this freedom allows nanoparticle molecules to range broadly over size as well. And it is here that Nature has been kind because nanoparticles display size-dependent properties. Thus, we find ourselves with a new molecular category, a category with stoichiometries, sizes, and properties that we can choose and tailor to our needs, and a category with which we can create a whole new universe of materials. With these nanoparticle molecules, we have a new kind of chemical matter.

3.3 STOICHIOMETRIC CLUSTER COMPOUNDS

3.3.1 Semiconductor Cores

Stoichiometric cluster compounds involving II-VI semiconductor materials have a long history. For example, $[Cd_{10}(SCH_2CH_2OH)_{16}]^{+4}$ (5), $[Cd_4S_{10}(SPh)_{16}]^{4-}$, $[Cd_8S(CH_2CHCH_3CH_3)]^{2+}$, $Cd_{17}S_4(PhS)_{28}]^{2-}$ (6, 7), $Cd_{10}S_4(SPh)^{12}$ (8), $Cd_{32}S_{14}(SPh)_{36} \cdot DMF_4$ (9), and $Cd_{17}S_4(SCH_2CH_2OH)_{26}$ (10) have been synthesized. These cluster compounds have approximately 1.0- to 1.5-nm cores that are essentially chunks of the corresponding bulk material, for example, CdS with wurtzite, zinc blende, or cubic sphalerite structures. These cores are covalently bonded to their ligand shells which via the sulfur are organic extensions of those cores. As such they are complete entities with a definite stoichiometry. They are soluble in organic solvents and display emission spectra blue shifted from the bulk. Other examples include (11, 12) stoichiometric copper-selenium compounds with cores of Cu_2Se_7, $Cu_{50}Se_{20}$, $Cu_{73}Se_{35}$, and $Cu_{140}Se_{70}$ and similarly $Ni_{34}Se_{22}$ with stoichiometric butyl and alkylphosphine ligands and $Cu_{146}Se_{73}(PPh_3)_{30}$. The colors of these compounds range from red to brown to black with increasing core size. These systems are also soluble in organics. A great many other examples exist, see Reference 13.

3.3.2 Metallic Cores

Another important class of cluster compounds is those with metallic cores like gold, platinum, and palladium, see the review by Schmid (14). The stability and hence the stoichiometry of these materials relies to first order on a magic number of atoms in the core. These magic numbers are based on close packing of hard spheres around a central sphere in successive layers. These structures can be created for

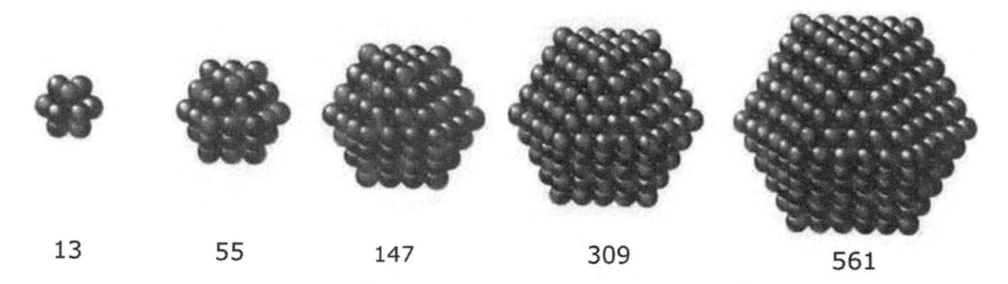

Figure 3.1 Hexagonal close-packed full-shell "magic number" clusters of spherical atoms. The series starts as a single atom and then 12 others can be placed around it all touching the central atom, thus giving 12-fold coordination. This is the first shell. The number of atoms in the nth shell is $10n^2 + 2$. The number of atoms in the total cluster is $(5/3)n(n + 1)(2n + 1) + 2n + 1$.

finite-sized clusters by starting with a single particle and then surrounding it with 12 other same-sized particles, each touching the first, now central, particle to create a cluster with 13 total spheres. Given this 13-sphere cluster, another layer of particles can be added such that the additional spheres all fit in threefold coordination sites of the underlayer. This second layer would have 42 spheres to yield a 55-particle cluster. This process can be continued as illustrated in Figure 3.1. For small clusters this arrangement yields energetically favorable clusters especially when the shells are full and hence the numbers 13, 55, 147, 309... are "magic" numbers.

Perhaps the best examples are the cluster compounds synthesized by Schmid $Au_{55}(PPh_3)_{12}C_{16}$ (15, 16), which have a metal core diameter of 1.4 nm and a total overall diameter of 2.3 nm. These clusters are soluble in organic liquids of low polarity. Since the triphenylphosphines are weakly ligating, they may be replaced. Thus, the clusters may be made water soluble by exchanging the triphenylphosphine ligand with $Ph_2PC_6H_4SO_3Na$.

Clusters with larger magic numbers are increasingly hard to make; however, clusters of $Pt_{309}Phen_{36}O_{30}$, $Pd_{561}Phen_{36}O_{200}$, $Pd_{1415}Phen_{54}O_{1000}$, and $Pd_{2057}Phen_{78}O_{1600}$ with a variety of aromatic ligands and anions have been reported. Now recognize that the stoichiometry becomes inexact with a typical spread of $\pm 5\%$ in the numbers of atoms in the core, ligands around the core, and anions associated with the ligands. The numbers reported for the core favor the magic numbers out of a reasonable sense that they might represent the most stable clusters. This polydispersity in number may be joined by polydispersity in structure within the core. For example, Moiseev's group (17, 18) reports evidence for at least three kinds of metal cores in $Pd_{561}phen_{60}Oac_{180}$, nearly perfect packed fcc metal, icosohedral multiple twinned structures and roughly amorphous. Given all this, it has been argued that these cluster compounds cannot be considered as stoichiometrically well-defined molecules. Moreover, the ligands are no longer an integral part of the particle as they were for the semiconductor particles described above and hence are more labile and their number less certain. We see a general trend that as the clusters become larger, the stoichiometry becomes less exact and the ligands become more independent of the core.

Recently Jadzinski et al. (19) reported x-ray crystallography analysis of a thiolated gold molecular cluster with an exact stoichiometry. The cluster was made via borohydride reduction of gold chloride in the presence of *p*-thiobenzene acid, which became the ligand. The molecule contained 102 gold atoms and 44 ligands. It was shown that 79 of the gold atoms formed a truncated-decahedron inner core. Beyond that there were layers of gold atoms singly coordinated and then doubly coordinated with the ligand. The ligands not only interacted with the gold but with each other via both parallel and perpendicular ring stacking and the sulfur interacting with the ring. Most of the ligands assembled into chains extending from one pole of the roughly spherical molecule to the other. Finally the existence of this particular conformation for the molecule was ascribed to an electronic shell closing in which the 109 gold atoms donated 109 electrons, 44 of which engaged in bonding one of the 44 ligands. This leaves 58 electrons free to pair up and occupy 29 delocalized orbitals to yield a stable zero angular momentum situation. This cluster compound is a small version of gold nanoparticles that we will discuss below.

3.4 NANOPARTICLES

The division between cluster compounds and nanoparticles is rather tenuous. The cluster compounds described above have either definitive covalent linkages or special magic numbers that define their narrow size distribution, hence their exact or nearly exact stoichiometry. With increased size, however, these qualities appear to become less important and the particles begin to act more like conventional particles that could be any size. Then the narrow size distribution must rely on something else. This "something else" has been controlled nucleation and subsequent growth, fractionation of polydisperse as prepared systems or digestive ripening to narrow the distribution.

Another important distinction between cluster compounds and nanoparticles is, as perhaps first pointed out by Andres et al. (20), the flexibility of the latter in regard to ligand exchange. Since cluster compounds have a definite stoichiometry between their cores and their ligand shells, these are not separable entities. On the other hand, nanoparticles have cores and ligand shells that are typically quite independent. Thus, the ligand shell can be modified or replaced, through place exchange reactions (21, 22), with appropriate chemistry and this fact lends great flexibility to control their properties.

Nanoparticles can be created by simple chemical reactions that yield insoluble products in the liquid phase, the ancient chemical procedure of precipitation. Precipitation occurs because the nanoparticles so produced are unstable against either coarsening or aggregation and subsequently readily fall to the bottom of the vessel. The key to the current nanoparticle revolution was thus arresting these growths and that was accomplished by capping the particles with surface active agents. These capping agents can stick to the surface via either covalent bonds or less energetic mutual attractions such as ligation. The cluster compounds described above are examples of capping covalently, for example, $Cd_{10}S_4$ capped with 16 thiolphenyls. As the particle regime is entered with bigger sizes, one finds the capping agents are more labile and surface mobile, hence implying that they are not covalently connected but rather ligated to the surface. I will use the term "ligands" to describe these surface agents.

Brus and coworkers (23) were the first to show that nanoparticle growth could be controlled and aggregation arrested by capping. They studied CdSe capped with phenylselenide. The particles, "molecular particles," were redispersable in organic solvents and the solubility could be controlled via the capping ligand. Subsequently Brust et al. (24) invented a phase transfer method for producing gold nanoparticles via borohydride reduction in the presence of capping ligands, typically alkyl thiols. This method has been used extensively. Capping and controlled growth were also achieved early in the development of this field by Murray, Kagan, and Bawendi (25), who used nucleation and subsequent controlled growth in the presence of ligands to produce narrowly dispersed CdSe nanoparticles. Most recently digestive ripening, discovered by our group (26–29), has proven very useful for creating ligated nanoparticle molecules of narrow size distribution and bulk quantities.

Early in the development of this field, Whetten's group made quite small nanoparticles that they called "gold-cluster molecules" (30–33). These nanoparticle molecules, which lie at the boundary between cluster compounds and nanoparticles, were made via borohydride reduction of gold chloride in the presence of organic thiols. Depending on ligand and temperature a mix of sizes was made that could be separated into specific fractions that had molecular weights of, for example, 8, 14, 22, 30, 40, 65, and 180 kDa, corresponding to diameters of 1.1 to 3.1 nm, and 40 to 900 gold atoms. They were ligated by alkane thiols and thus soluble in organic solvents, or in one case of mole weight 10.4 kDa, glutathione, which were soluble in water (33). The fact that these fractions appeared from the rapid borohydride reduction of the gold salt implies that the sizes of these fractions have special symmetries, lower energies, or both to make them result in greater abundance than other sizes.

There is a very great variety of nanoparticle molecules known and described in the literature, and it is certainly well beyond the scope of this chapter to review them. Good reviews do exist; see References 1 to 4. Here I classify these into four major groups.

3.4.1 Types of Nanoparticles

1. *Gold.* Gold nanoparticles are the most common, largely because gold is relatively inert, yet it is the most electrophillic of metallic elements and this makes ligation propitious. Figure 3.2 shows a cartoon of what one might call a canonical gold nanoparticle molecule and Table 3.1 gives some useful statistics. There is a great variety in sizes and ligands. Sizes can range from approximately 1 to 20 nm, although 4 to 7 nm, is most common. The most common ligands are the alkyl thiols; the electrophilic gold binds the lone pairs of the sulfur readily. Alkane chain lengths from C6 to C16 are typical and aromatics have been used as well. Other ligands such as amines and phosphines have been used. Narrow sizes dispersions have been obtained through fractionation or digestive ripening (26–29).

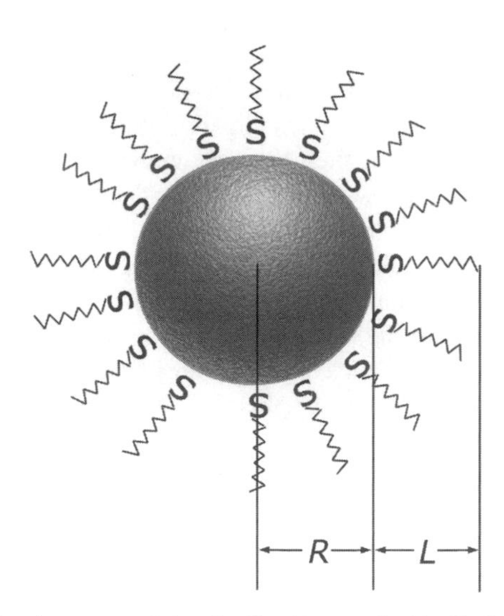

Figure 3.2 Schematic of a nanoparticle of radius R capped with alkyl thiol ligands of length L.

TABLE 3.1 Gold Nanoparticle Statistics

Gold atomic volume	16.9×10^{-24} cc = 16.9×10^{-3} nm^3
Number of gold atoms in a spherical gold nanoparticle	$31\,d^3$ (nm)
Number of thiols on the surface of a spherical gold nanoparticle	$14.7\,d^2$ (nm)

2. *Other Metals*. Silver, similar to gold, copper, palladium, platinum, iron, cobalt, nickel, and various bimetallics have been synthesized. As for gold, the most common sizes range in diameters of 3 to 7 nm. Ligands include the alkyl thiols for the more "noble" of the metals, but these ligands are too reactive for the transition metals iron, cobalt, and nickel where long chain fatty acids can be used.
3. *Semiconductors*. Perhaps most common are the cadmium chalcogenides, CdS, CdSe, CdTe ligated with trioctylphosphine oxide, hexadecylamine, etc.
4. *Metal Oxides*. Iron, cobalt, nickel oxides, and some ferrites ligated with long chain fatty acids.

3.4.2 The Capping Ligand

The surface capping ligands keep the nanoparticles from irreversibly aggregating and largely determine their solution and superlattice properties. With regard to gold and thiols, the properties of self-assembled monolayers, SAMs, of organic thiols adsorbed on bulk gold surfaces have been studied for some time, and this has been used to understand the ligation of thiols at the curved surfaces of the nanoparticles. For example, the footprint of a SAM alkylthiol on gold is 0.214 nm^2 and this value is used for the nanoparticle as well. Leff et al. (34) have shown that the size of gold nanoparticles prepared via the Brust–Schiffrin method depends on the ligand to gold ratio and gave surface energy arguments to explain this. A considerable amount of work from our laboratory has shown how the ligand is very active in digestive ripening, a process whereby the nanoparticle size distribution can be narrowed significantly by digesting solutions under reflux in the presence of the ligand (26–29). We found size dependence on the nonmetal of the ligand and a slight dependence on the chain length. Comparing alkyl thiol ligands to alkyl ammonium ligands showed a remarkable reversible change between spherical particles and flat plates (35).

The lability of the ligands allows for their controlled exchange. So-called "place exchange" reactions have been described (22, 36). With this method, one can change the ligand end group (for a thiol, opposite the sulfur) functionality. For example, one can change the solubility from organic to aqueous by place exchanging ligands terminated with methylene groups with alcohol or carboxylic acid groups. In one application, so-called mixed monolayer protected clusters involved gold capped with a mixture of octylthiol and 11-thioundecanoic acid (37). These could be made to aggregate at low pH due to hydrogen bonding between COOH groups and disperse at high pH due to the negatively charged COO^- terminal groups. Another use is to put chemically active groups that can bind with complementary groups on other nanoparticles.

Control of the ligand capping can also allow formation of amphiphilic nanoparticle molecules by capping with both hydrophilic and hydrophobic ligands. Such mixed coatings have been reported in the literature (38–40) and phase equilibria of such nanoparticle molecules have been studied with simulations to yield a potentially rich phase diagram (41). Mixed coatings portend the prospect to have anisotropic interactions between nanoparticles.

Experimental and simulation studies have shown that the ligand chains can undergo an order to disorder transition with increasing temperature. The transition temperature increases with increasing chain length. The ordered state is one of chain bundling, often in a zigzag all-trans conformation and often extending radially outward from nanoparticle faces. These assemblies are called 3d SAMs to distinguish them from their 2d analogues. The transition can have a latent heat (42) but has also been seen to evolve continuously with temperature (21).

Structuring of the ligands has been demonstrated by coating NPs with two different ligands that in the bulk would phase separate (43). Octanethiol (OT) and mercaptopropanoic (MPA) acid were used as capping agents for 3.2-nm gold particles. These ligands phase separated on the surface of the NPs into parallel bands of "latitude" as small as 0.5 nm in width; see Figure 3.3. Changes in core size, ligand lengths, and molar ratios modified the patterns. This nano-phase separation implies that the ligands are mobile on the surface of the nanoparticle.

The nanoparticle molecules can be made chemically reactive by placing appropriate functional groups at opposite ends of the capping ligands. For example,

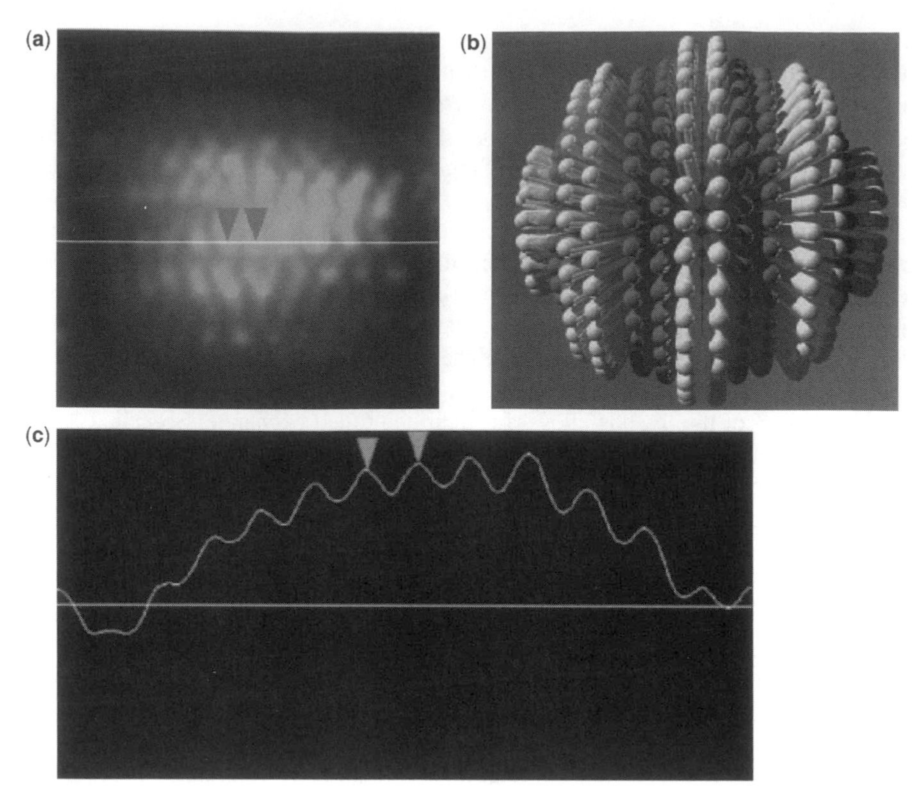

Figure 3.3 Nanoparticle molecules with phase-separated, ordered (rippled) domains on their ligand shell. (a) STM image of OT/MPA (2 : 1 molar ratio) gold nanoparticle with rippled ligand surface. (b) Schematic drawing of rippled nanoparticle. (c) Surface plot of the ligand shell contour. From A. M. Jackson et al. *Nature Materials* **2004**, *3*, 330. With permission.

alpha-omega dithiols can link to two metal nanoparticles (20, 44). Then nanoparticle molecules can bind together into roughly spherical clumps, as reported in Reference 44. With such dithiol linking, Sidhaye et al. (45) were able to reversibly expand and contract the spacing between linked nanoparticles with an optically induced cis/trans conformation change in the linking molecule. Another tack is to put matching functional groups on the nonligating ends of the ligands, for example, hydrogen bond donors and acceptors (46). Gold nanoparticles have been ligated with mercaptoalkyl oligonucleotides that can detect, via binding, complementary nucleotides bound to other gold nanoparticles (47).

3.5 INTERPARTICLE INTERACTIONS

Nanoparticles viewed as molecules have a great wealth of possible interparticle interactions that will influence their behavior. To date NP systems have been studied almost exclusively as solutions and the precipitation of the NP from the solution into two- and three-dimensional superlattices. The melting and vaporization aspects of solid or liquid systems have seen far less study likely because the thermal stability of most NP molecules is not good and decomposition could result before the phase transition. There are some melting and liquid studies, however. I start with a general description of colloidal solutions and then I describe in some detail the current understanding of NP interactions.

3.5.1 Colloidal Solutions

It is well known that colloidal suspensions can share many features with simple molecular systems such as gas, liquid, and solid crystalline and amorphous glass phases. This is particularly true when the colloid is nearly monodisperse for then the interparticle interactions, which are usually size dependent, are nearly all the same and hence the phase boundaries, which depend on the interactions, are distinct. Indeed, as the size distribution of a colloid narrows, one could claim that the colloidal suspension transforms into a solution, just as the different particles, by becoming alike or even identical, are transforming to molecules. Unlike simple molecular systems, which by their definition have no variety and are not dissolved in a medium, particle colloids and solutions can vary the interactions via changing size, surface groups, solvent, etc., and thereby change the phase diagram.

Figure 3.4, which is borrowed (somewhat modified) from a recent review (48), shows the possible phase diagrams that can occur for a colloidal system. On the left

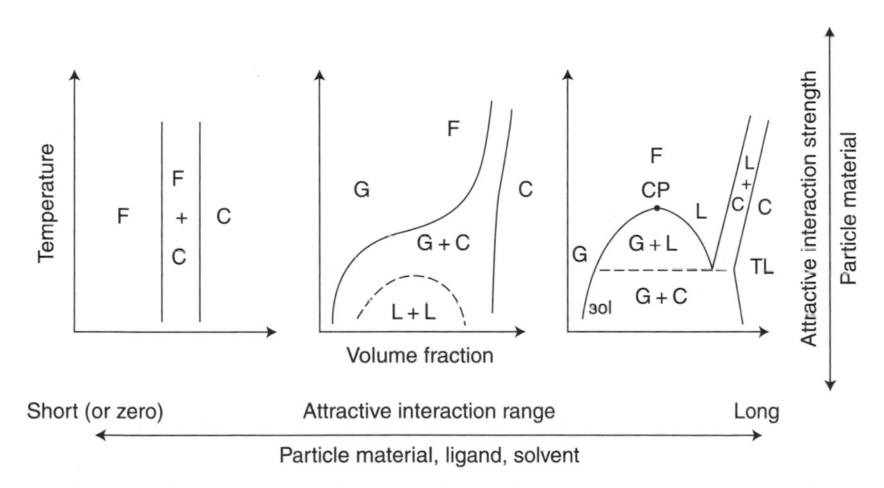

Figure 3.4 (Left) Phase diagram for purely hard sphere interaction potential, which shows only fluid (F) and crystalline (C) phases. (Middle) A short-range (relative to "particle" size) interaction is added to the hard sphere potential. Then a dilute gas (G) phase can appear, as well as a metastable liquid-liquid coexistence (L + L). (Right) The attractive interaction is long range and a complete phase diagram occurs, with gas, liquid, and crystalline solid phases. The triple line (TL) temperature increases with increasing attractive interaction strength.

is the phase diagram for a purely hard sphere system with no attractive interactions, which shows only fluid (F) and crystalline (C) phases. The transition from liquid to crystal is driven by entropy. For volume fractions greater than 0.545 there is more translational freedom, hence greater entropy, for hard spheres in a close packed solid than the amorphous glass. The fluid to crystalline equilibrium for a solution is the saturated dissolved solute, which could be single nanoparticles of the SPC, in equilibrium over the precipitated solid, which for SPC could be a superlattice. Achieving a solid to liquid transition at high volume fractions can be difficult, however, because dense systems can experience kinetic slowing and arrest.

Addition of short-range attractive interactions brings on a metastable liquid-liquid (L) coexistence comprised of coexisting high and low concentration colloidal solutions. One can imagine that as the attractive interaction is turned on the metastable coexistence curve rises up from low temperature and pushes the fluid to the fluid-crystal region to the left in Figure 3.4. This is metastable relative to the fluid-crystalline equilibrium, the fluid phase of which develops a low concentration gaseous phase (G) at low temperature. As the attractive interaction increases in range relative to the size of dissolved colloidal entity, the liquid-liquid coexistence rises to higher temperature, into an equilibrium regime, and a triple line (TL) appears. Now the phase diagram looks like that of a simple atomic system with three-phase equilibria, the phase diagram on the right.

The relative range of the attractive interaction between the colloidal particles is a key parameter that affects the phase diagram. It is likely that one can control this key parameter in solutions of nanoparticles by adjusting combinations of the particle size and ligand shell depth (i.e. ligand length, see Fig. 3.2) and hence range throughout the possibilities of Figure 3.4. Another key parameter is the strength of the attractive interaction (i.e. the depth of the interparticle potential well) which controls the effective temperature, hence the position of the triple line. For particles the strength of the interaction depends on both the particle and ligand shell Hamaker constants relative to the solvent. The triple line goes to higher temperature the greater the interaction strength. We have included these properties as global parameters along the margins of Figure 3.4.

The phenomena displayed in Figure 3.4 are for spherically symmetric potentials, and once this symmetry is relaxed, the complexity, hence opportunities, in the phase diagram expand significantly. One might say that Figure 3.4 is the argon atom limit.

Comparison to gas-liquid-solid systems is very useful, but it must be remembered that a colloid is much more complex. Recall that in a real gas the molecules move in straight lines between collisions; that is, they move ballistically. In a solution the particles move diffusively. The pressure of a real gas is replaced by the osmotic pressure of the particles in the solution. Given this, it is not surprising that since the real gas pressure has a virial expansion so does the osmotic gas of the solution as expressed by the van't Hoff equation:

$$P = kTc(1 + B_2c) \tag{3.2}$$

where P is the osmotic pressure, k is Boltzmann's constant, c the concentration, and B_2 the second virial coefficient. B_2 is equal to 4 for a hard sphere potential and in general is positive for repulsive and negative for attractive potentials.

As an example, it is useful to recall protein solutions. Protein molecules can be quite large; for example, a lysozyme molecule is an ellipsoid with dimensions $3 \times 4 \times 5$ nm, similar to nanoparticles (they are in fact nanoparticles). Protein aqueous solutions have seen significant study in the recent past because it is desirable to form protein crystals from solution for structural analysis. Such a system contains not only the protein molecule and the water solvent but usually dissolved ions which dissociate and a variable pH. Moreover, the protein molecule may have a variety of surface states that affect its interaction with other protein molecules as well as the water. The lesson here is that often the system is too complex and an effective interparticle interaction potential must be prescribed. Such a procedure often works because the number concentration of the large protein molecules or nanoparticles is far less than that of the other constituents.

3.5.2 Solubility of Nanoparticle Molecules

There appear to be no quantitative studies of nanoparticle molecule solubility, although essentially every researcher knows that the ligand end groups greatly affect the type of solvent that will keep the colloid stable. The old rule of "like dissolves like" applies. Thus, alkane-coated nanoparticles will not be soluble in polar solvents, for example, water, but will be in another alkane. Conversely, if the alkane ligands are terminated with carboxylic acid groups, the nanoparticles will be soluble in water but not in the alkane.

It is interesting that here we encounter the double identity of nanoparticle molecules. Are they suspended particles in a colloid or dissolved molecules in a solution? We contend that if they satisfy the condition of near stoichiometry discussed above to classify them as molecules, then their stable suspensions are more than that, they are solutions as well. We remark here that gravity can now get in the way, for even a monodisperse, hence stoichiometric, system of particles will settle out if the thermal energy, kT, is not large enough to keep the monomers suspended. This happens for particles on the order of 50 nm.

We presented qualitative observations of solubility of 5-nm gold nanoparticles ligated with dodecanethiol dissolved in toluene (49). A rough temperature versus concentration phase diagram showed the expected greater solubility with increasing temperature. In other work (28) we showed that 5-nm gold ligated with alkylthiols ranging over C8, C10, C12, C16 became more soluble with increasing chain length at room temperature; the C8 thiol being essentially insoluble. Calculation of the van der Waals attractive potential for two gold cores separated by one ligand length yielded values of 5 kT, 2.2 kT, 2 kT, and 0.6 kT, for the different chain lengths, respectively, at room temperature.

Nucleation and growth of 3d superlattices from nanoparticle solution has seen recent study (50–52). Superlattice cluster size versus time was studied after the system was destabilized via either place exchanging to a less soluble ligand, creating

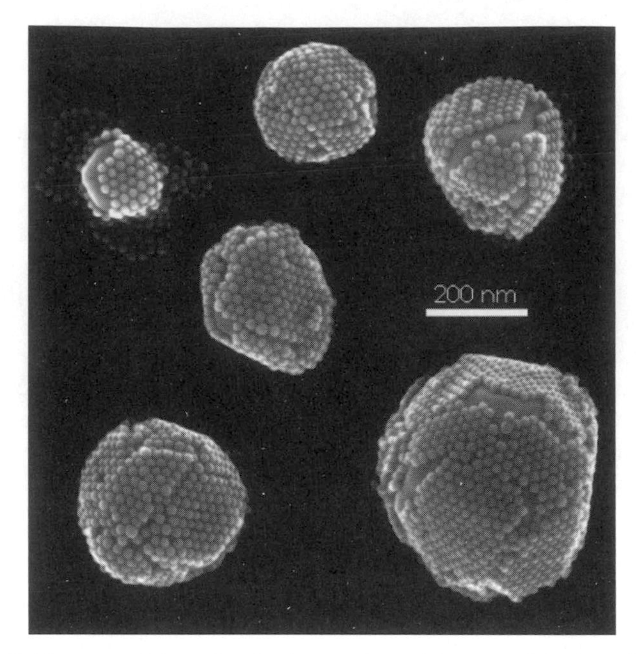

Figure 3.5 Gold nanoparticle molecule clusters formed from precipitation of the system of monomers.

the nanoparticles in a poor solvent, or quenching temperature from a one-phase to two-phase regime. The classic diffusion limited theory of LaMer and Dinegar (53) was not successful in describing the growth kinetics. Roughly spherical superlattices did nucleate from the solution, as shown in Figure 3.5 (50). Work currently in progress in our laboratory (52) has shown that the size of the nanoparticle molecule clusters decreases with depth of quench, as shown in Figure 3.6. This is similar to precipitation of molecular and ionic solids.

3.5.3 Nanoparticle Interactions

The discussion above can be simply summarized to say that solution phase behavior is much more complex than pure component solid-liquid-gas phase behavior because the interparticle interactions have a much greater variety. The effective interparticle interaction in solution can have:

1. *Excluded Volume Effects.* These could be effective hard sphere potentials representing the finite size of the particles. Such potentials are important for the crystallization transition; the left side of Figure 3.4 is an example. For ligated nanoparticles, the finite size is likely better represented by a soft sphere potential, the softness due to the steric interactions of the ligands (see below).
2. *London–van der Waals Attractive Forces.* These forces depend on the dielectric constants of the particle materials and the surrounding solvent via the

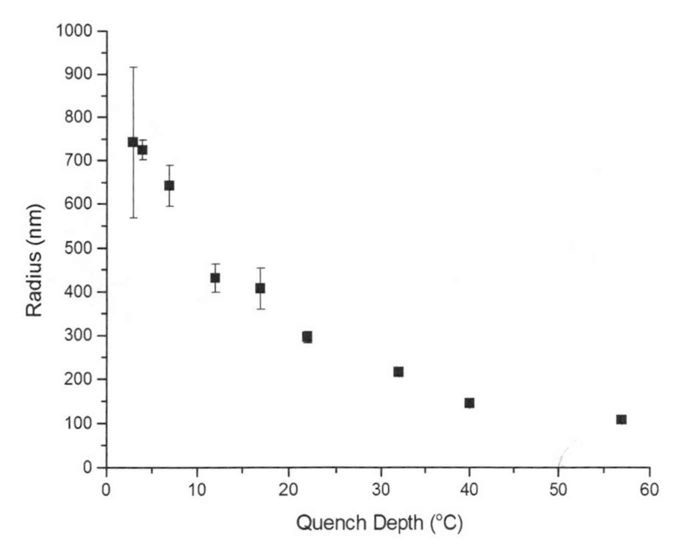

Figure 3.6 Radius of gold nanoparticle molecule clusters formed by temperature quenching stable solutions of the molecules in a mixture of 2-butanone and *tert*-butyltoluene to various depths below the saturated solution temperature.

Hamaker constant. The Hamaker constant for a metal is about 2 eV and about two orders of magnitude larger than for nonmetals. For two spheres of radius $R = d/2$ the van der Waals potential is (54)

$$V_{rdw}(r) = -\frac{A}{12}\left[\frac{1}{x^2-1} + \frac{1}{x^2} + 2\ln\left(1 - \frac{1}{x^2}\right)\right] \tag{3.3}$$

where A is the Hamaker constant of the particle and $x = r/d$ where r is the center-to-center distance. This has limits of

$$V_{rdw}(r) \sim s^{-1} \quad \text{for} \quad r \gtrsim d \tag{3.4}$$

$$\sim r^{-6} \quad \text{for} \quad r \gg d \tag{3.5}$$

In Equation (3.4), s is the separation between the two particles surfaces, see Figure 3.7.

3. *Solvent-Mediated Forces.* These break into two classes:
   a. *Electrostatic.* Dissolved ions form counter ion layers around charged particles to yield a screened Coulomb potential. This combined with the van der Waals potential yields the classic DLVO potential (55).
   b. *Ligand–Solvent Interactions.* Simply said the rule of "like dissolves like" qualitatively describes these interactions. To estimate the free energy of mixing of the ligands in the presence of the solvent when ligand layers

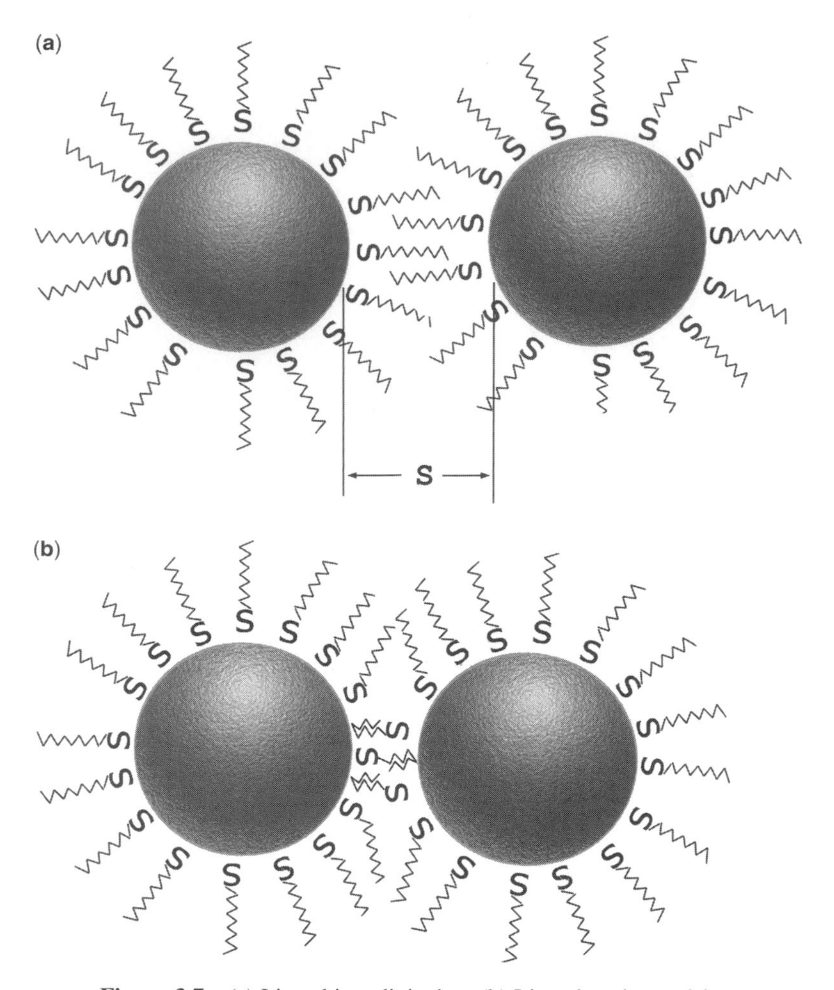

Figure 3.7 (a) Ligand interdigitation. (b) Ligand steric repulsion.

from two different nanoparticles start overlapping, one needs to consider two different regimes (56). In the first regime, the ligand chains undergo interpenetration, and in the second regime the chains undergo interpenetration and compression. These two regimes are shown schematically in Figure 3.7a and 3.7b, respectively. These two regimes can be distinguished as:

Regime 1: $1 + \ell < x < 1 + 2\ell$ (Interpenetrations only)

Regime 2: $x < 1 + \ell$ (Interpenetration and Compression)

where ℓ is the scaled contour length of the ligand chains, that is, $\ell = L/d$ Free energy of mixing in both regime 1 and regime 2 are known in the literature in terms of the Flory χ-parameter between the solvent and the

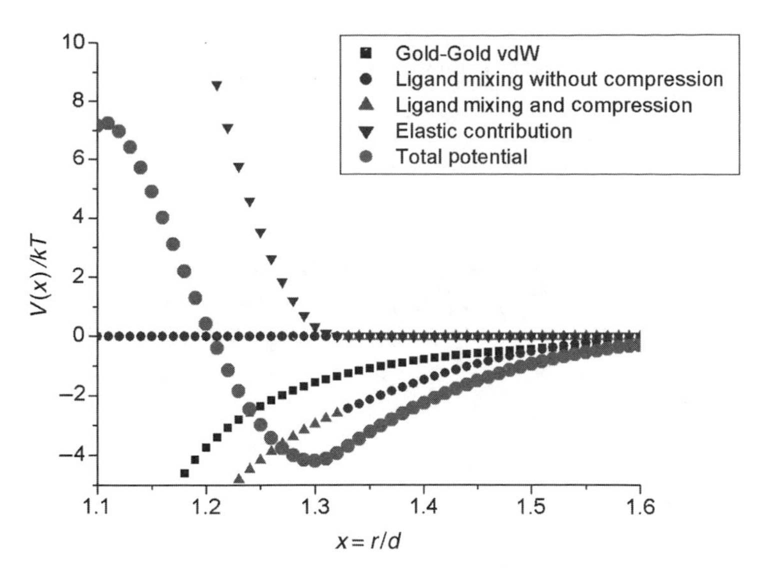

Figure 3.8 Various components of the interaction potential between two 5-nm gold nanoparticle molecules with dodecanethiol ligands.

tethered chains. In terms of our scaled variables, one can write this in regime 1 as

$$\frac{V_2}{kT} = \frac{\pi d^3}{2v_m}\phi_{av}^2\left(\frac{1}{2} - \chi\right)[x - (1 + 2\ell)]^2; \quad 1 + \ell < x < 1 + 2\ell \qquad (3.6)$$

where v_m is the volume of a solvent molecule and ϕ_{av} is the average volume fraction of the ligand segments in the tethered layer.

$$\frac{V_3}{kT} = \frac{\pi d^3}{v_m}\phi_{av}^2\left(\frac{1}{2} - \chi\right)\left[3\ln\left(\frac{L}{x-1}\right) + 2\left(\frac{x-1}{\ell}\right) - \frac{3}{2}\right]; \quad x < 1 + L \qquad (3.7)$$

Note that when $x - 1 + \ell$; $V_2 = V_3$ as expected.

4. *Ligand–Ligand Interactions.* An elastic contribution to the potential due to loss of conformational entropy as the ligands begin to overlap, a steric repulsion, is also known in the literature (57, 58). In terms of the scaled variables it can be written as

$$\frac{V_4}{kT} = \frac{1}{2}\frac{\pi v d^2}{\ell}[x - (1 + \ell)]^2; \quad x < 1 + \ell \qquad (3.8)$$

where v is the number of ligands per unit area of the nanoparticles.

5. *Dipolar Forces.* These can be either electric or magnetic and in each case significantly larger for nanoparticles than for atoms or normal-sized molecules.

Figure 3.8 shows the interaction potential between two 5-nm gold nanoparticle molecules with dodecanethiol ligands, the sum of Equations (3.3), (3.6), (3.7), and (3.8).

I now describe how these various interactions affect nanoparticle solutions and the phases that can be obtained from these solutions.

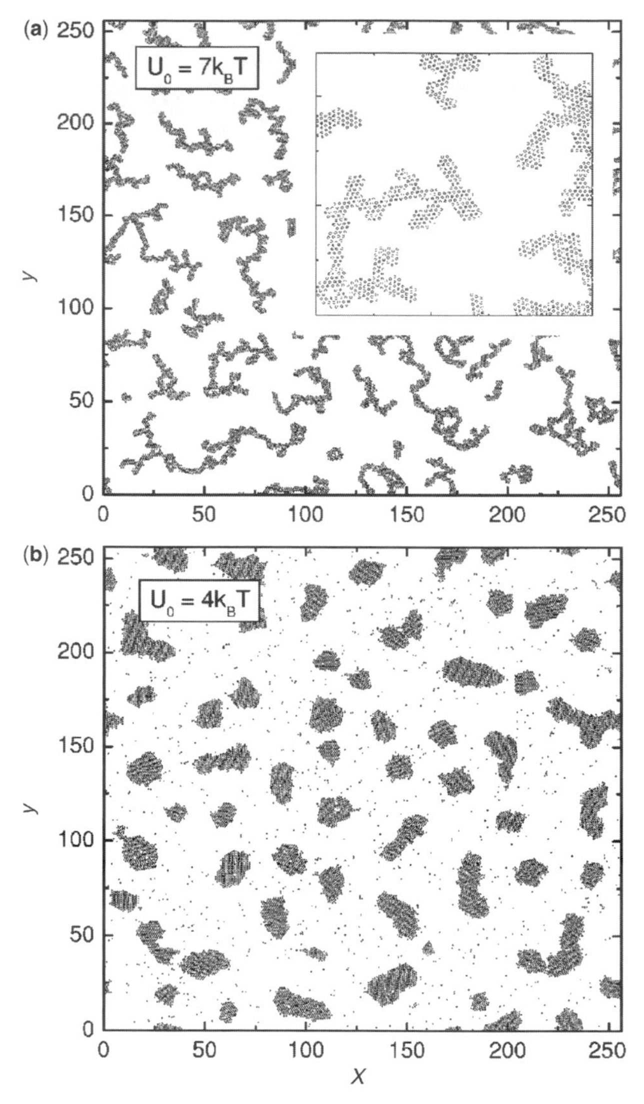

Figure 3.9 Results from 2d random aggregation simulations in which the depth of the potential of interaction was either 7 kT (a) or 4 kT (b). At 4 kT the aggregates are roughly circular crystallites in equilibrium with a monomer phase, that is, a dissolved phase. At 7 kT the aggregates are fractal, with a fractal dimension of 1.4, the DLCA value, over large length scales, but they retain a dense crystal packing over small length scales. These are called fat fractals.

The hard sphere interaction causes a liquid to solid transition at a solution volume fraction of 0.545, as described above with reference to Figure 3.4. The resulting solid has an fcc lattice. This is slightly, about 0.001 kT, more stable than its close packed counterpart at 0.74, the hcp lattice. Binary hard sphere systems can form five different lattices of different stoichiometry depending on the size ratio.

Addition of the van der Waals interaction can cause formation of condensed phases at much lower volume fractions than 0.545. If strong relative to the thermal energy kT, the phases are ramified aggregates, usually fractal, and these can lead to gels. Variation of the relative strength of the interaction can yield more compact structures like "fat fractals" and lattices, illustrated in Figure 3.9 (59). Ohara et al. (60) showed that since the van der Waals interaction is size dependent, bigger nanoparticles will nucleate to superlattices first when the solution is destabilized and size segregation can occur. Korgel and Fitzmaurice (61) showed that van der Waals interaction with a substrate when the solution is dried can compete with the nanoparticle-nanoparticle interaction. Then depending on the relative strength of the two interactions, the morphology of the resulting dried layers can be controlled.

Inclusion of screened Coulomb electrostatic interaction along with the van der Waals and hard sphere interactions can engender the entire description of Figure 3.4. This has been realized in many ways in protein solutions, but not yet for nanoparticle solutions.

3.6 SUPERLATTICES

Perhaps one of the most molecular things nanoparticle molecules do is form two-and three-dimensional crystals in which the nanoparticles sit at lattice sites. Such crystals of nanoparticles, which are often crystals themselves, can form from solution and are called "superlattices." The formation of a superlattice is usually called "self-assembly," which seems to give the particles some degree of free will. However, if we view the nanoparticles as molecules, we realize that self-assembly is simply crystallization, a process common for atoms and molecules.

The primary keys to superlattice formation are first a narrow size distribution; usually a standard deviation of 10% but much better is 5%. A narrow distribution allows the lattice to keep its long range order and also means a narrow distribution of interaction potentials so the material can actually act analogously to a molecular system. Second, it is necessary for the particles to have a significant attractive interaction relative to the thermal energy kT. This can be controlled by the Hamaker constant of the core material, which is about two orders of magnitude greater for metals than nonmetals, and the ratio $\ell = L/d$ of the ligand length L to the core diameter d. This determines the relative closest approach distance. For example, Prasad et al. (28) varied ligands from C8 to C16 thiols for approximately 5.5-nm gold particles and saw the smaller the ligand the less soluble the nanoparticles were in toluene and the more likely 3d superlattices formed. Thomas, Kulkarni, and Rao (62) found large 2d arrays of Pd NPs formed when d/L was between 1.5 and 3.8, whereas outside of this range the structures were disordered.

Superlattices are now very common, and it is not my intent to provide a comprehensive review of all the work. From a historical perspective I note that Bentzon et al. (63) first observed superlattices of iron oxide nanoparticles. It was description of CdSe superlattices by Murray, Kagan, and Bawendi (25), however, that effectively established this new paradigm of molecular solids. The superlattices of Murray et al., were formed of CdSe nanoparticles with sizes ranging from 1.5 to 10 nm. Any given lattice used a size in this range narrowed to $\pm 4\%$. The CdSe particles were ligated with trioctylphosphine oxide or selenide. Faceted superlattice crystals up to 50 μm in size could be grown by gentle evaporation of the solvents. Control of the spacing between nanoparticles was achieved by exchanging the octyl ligands for butyl or hexadecyl ligands (64). Spectrographic evidence was given for interparticle coupling.

Many examples now exist in the literature for both 2d and 3d nanoparticle superlattices, including gold (65–67), silver (68), palladium and platinum (69, 70), and magnetic materials like iron, cobalt, FePt, $CoFe_2O_4$ (71, 72); see also References 1–4. Examples are given in Figures 3.10 and 3.11. The lattice spacing is typically much less than the combined length of the ligands to imply significant interdigitation, as sketched in Figure 3.7(a). Systematic variation of the chain length has shown that the gap between particles increases by 0.12 nm per carbon atom on the ligand chain (73). This agrees well with the increase of the linear length of an

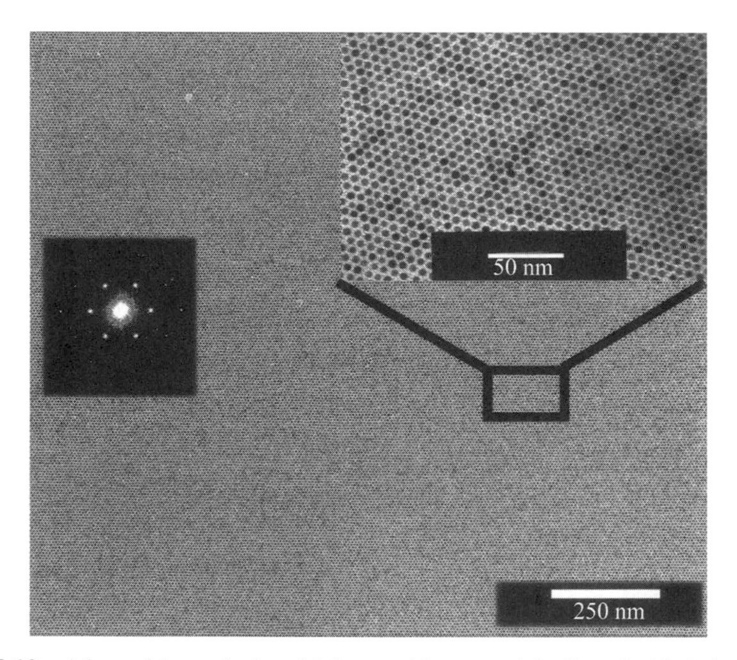

Figure 3.10 A large, 2d superlattice of 5.5-nm gold nanoparticles ligated with dodecanethiol on a silicon nitride surface. Note hexatic, close-packed structure (like pennies on a table top) and the spacing between the nanoparticles, which is filled with the alkane chains of the ligands, which keeps the gold particles from touching, which would lead to irreversible aggregation.

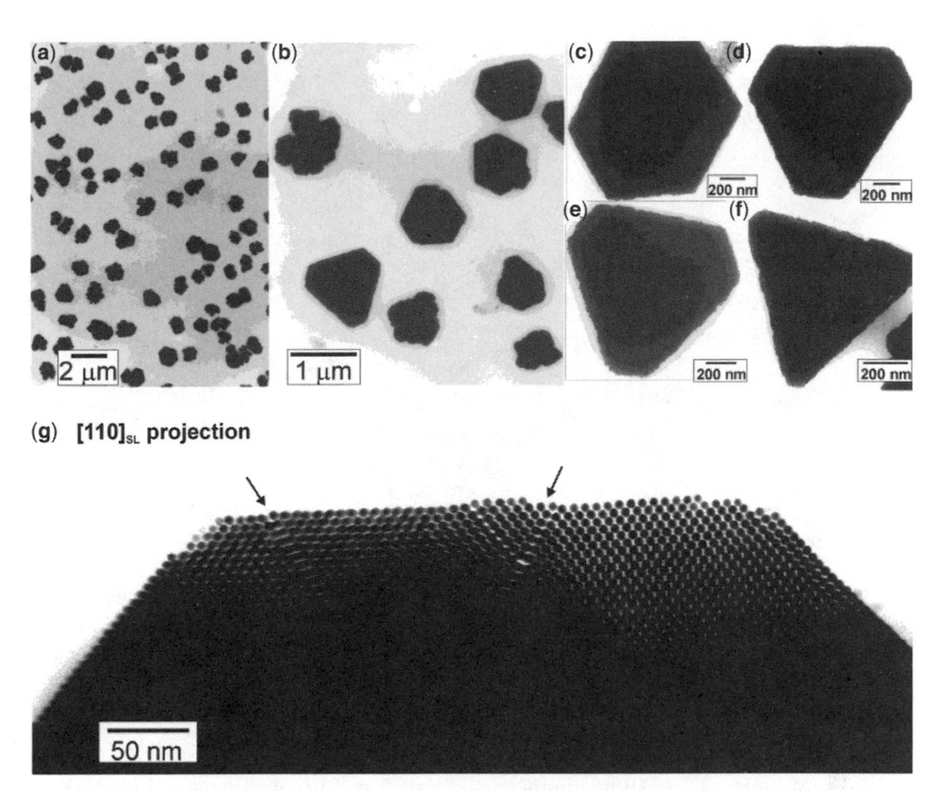

Figure 3.11 TEM micrographs of nanoparticle superlattices of Au nanoparticles prepared by the inverse micelle method: (a) and (b) low-magnification images; (c)–(f) regularly shaped nanoparticle superlattices; (g) magnified image of a superlattice edge (note the perfect arrangement of the Au nanoparticles).

alkane chain of 0.126 nm per carbon atom given the bond length of 0.154 nm and the bond angle of 109.47 degrees and the assumption of complete interdigitation. Multilayers have been shown to form by addition of particles to either the threefold sites on the hexatic lattice, as expected, or a more unusual addition at twofold sites, as shown in Figure 3.12 (29, 74). Lin and coworkers (75, 76) demonstrated a kinetically driven 2d superlattice formation at the interface between an evaporating solvent and the air above. Essentially the liquid evaporated faster than the nanoparticles could move via diffusion out of the way of the falling interface and once the interface hit them, they were stuck by surface forces. Lin et al. (77) have also created 2d free-standing superlattices with structural integrity that show elastic properties.

Most 3d nanoparticle superlattices have a close-packed twofold coordination. Nature forms crystal lattices of lower coordinations in a great variety of ionic crystals. Recently Kalsin et al. (78) assembled an ionic lattice of oppositely charged nanoparticles. The nanoparticles were gold ligated with mercaptoundecanoic acid and silver ligated with *N,N,N*-trimethyl(11-mercaptoundecyl) ammonium chloride salt. The nanoparticles were essentially the same size, each about 5 nm in diameter.

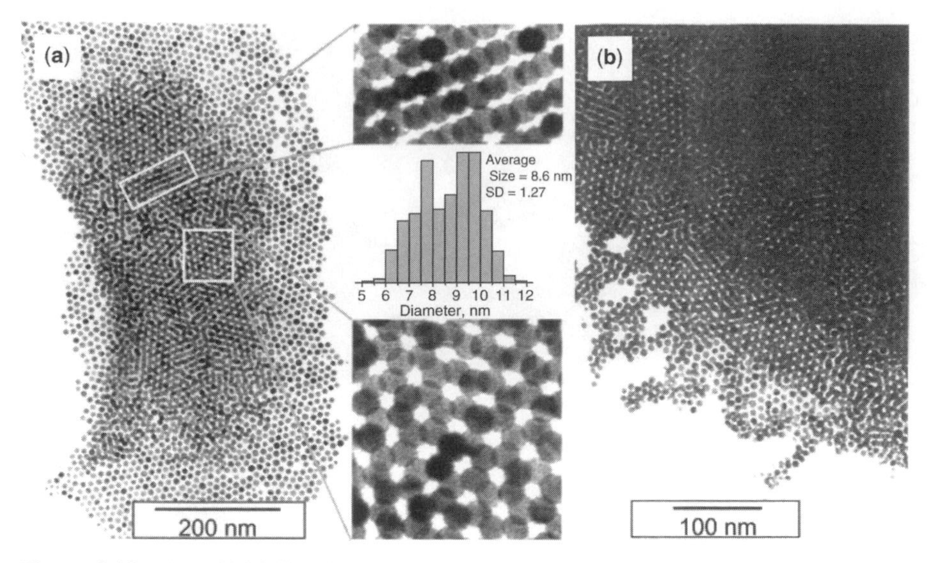

Figure 3.12 Au colloid digestively ripened with $C_{12}H_{25}NH_2$. (a) and (b) are pictures taken from different samples. The different types of ordering observed are highlighted. 2d superlattices show second layers adding at either the threefold sites or the twofold sites of the lattice below.

A diamond-like (four nearest neighbors) sphalerite lattice was achieved despite the presence of only spherically symmetric dispersion and electrostatic interactions. This was explained as due to a screened Coulomb potential with screening length greater than the four oppositely charged nearest neighbors but less than the 12 same charged second nearest neighbors. This delicate combination of different relative scales of the interactions made the sphalerite lattice more energetically favorable than the usual close-packed lattices.

Systems of two different particle types and sizes have led to a wealth of possible superlattices with stoichiometry dependent on the particle size ratio (79–84). Murray and coworkers (81–84) showed that in addition to size ratio a key experimental component is to charge tune the nanoparticles with either oleic acid or trioctylphosphine to yield superlattice stoichiometries of AB, AB_2, AB_3, AB_4, AB_5, AB_6, and AB_{13}, a greater diversity than found in nature for micron-sized particles (see Fig. 3.13). This charge tuning is successful because at the nanoscale (not the microscale) all the interactions above can contribute with comparable weight. These, combined with substrate interactions and the inherent nonequilibrium nature of the evaporative process to create the superlattices, yields the diversity.

Molecular dynamics simulations have been used to study the structure and dynamics of superlattices (31, 85, 86). Ligand alkyl chain bundling is found as a function of temperature. For example, for a gold nanoparticle with 1289 atoms (diameter ca. 3.5 nm) ligated with decane thiol there was chain bundling below 300 K, and partial bundling up to 375 K, where an effective chain melting occurs at roughly the bulk melting point of the alkane. These changes were reversible. Infrared spectrographic

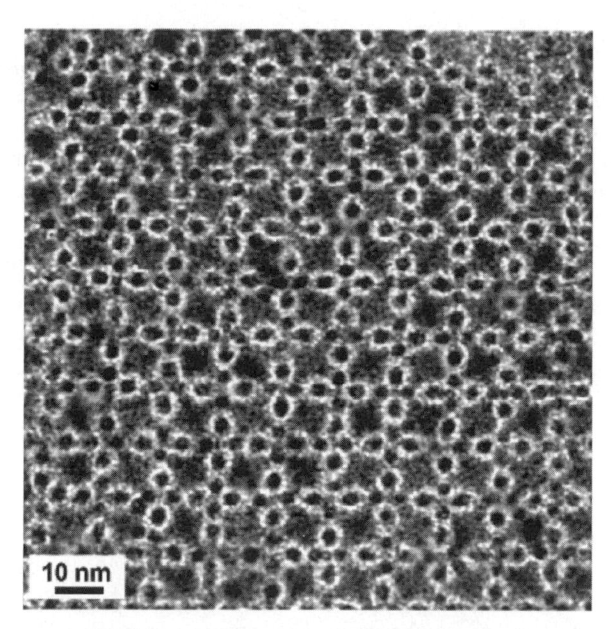

Figure 3.13 Example of a binary superlattice with AB13 stoichiometry composed of 5.8-nm PbSe and 3.0-nm Pd nanoparticles.

studies of alkane thiols bound to gold NPs for chain lengths from 3 to 24 have shown liquid-like disorder of the shorter chains and crystalline packing for longer chains (87). The molecular dynamics studies also showed structural change with temperature from bcc orthorhombic to bcc. Relative ligand length also affected the superlattice structure such that the structure was fcc for $L/R < 0.6$, bcc for $0.60 < L/R < 0.66$ and bct for $L/R > 0.66$. The van der Waals forces between the ligands were important. Very similar results were seen experimentally for superlattices of approximately 4-nm silver nanoparticle capped with octyl and dodecylthiol (88), but theoretical analysis implicated core-core attractive forces were important as well as ligand-ligand forces.

3.6.1 Superlattice Melting

There appear only a few discussions of superlattice melting in the literature (89–91). Such melting is expected to involve both the melting of the lattice and the possible order to disorder transition of the chains of the attached ligand molecules. Pradeep et al. (89, 90) studied 3d superlattices of 4.0-nm silver nanoparticles ligated with either octyl or octyldecylthiol. These superlattices showed reversible melting with x-ray diffraction and DSC measurements at about 400 K with the C_8SH slightly lower in temperature. For the $C_{18}SH$ superlattice the DSC showed transitions at 340 and 399 K, which were ascribed to ligand melting and superlattice melting, respectively. Respective enthalpies were on the order of 130 and 10 J/g. Melting and subsequent recrystallization seemed to change the system and reproducibility was hard to gain. Some of this was ascribed to changes in ligand conformations via interdigitation. A liquid system of nanoparticles, 2.0-nm Pt ligated with a very large ligand, *N,N*-dioctyl-*N*-(3-mercaptoptopyl)-*N*-methylammonium sulfonate, was shown to freeze at −20°C and remelt at 30°C with DSC scans (92). It is reasonable to speculate that this lower temperature melting was a result of the small NP size compared to the large ligand and the concomitant small van der Waals force between NPs.

3.7 CONCLUSIONS

This chapter has discussed a number of aspects of nanometer-sized particles that lend weight to the argument that these clusters have attributes similar to the canonical molecules of the past and hence can be viewed simply as large molecules. The key attribute of a molecule is its stoichiometry, which the smallest of these nanoparticles possess and which the largest possess within some small uncertainty.

Fundamental aspects of this new class of nanoparticle molecules are their controllable size and their core-shell structure. With controlled size, we can control their properties. The core-shell structure gives us extra latitude in property control because these two aspects can be, for the most part, varied separately. Thus, nanoparticle molecules offer the possibility to new materials.

REFERENCES

1. C. P. Collier, T. Vossmeyer, J. R. Heath. *Annu. Rev. Phys. Chem.* **1998**, *49*, 371.
2. J. D. Aiken, R. G. Finke. *J. Mole. Catal. A: Chem.* **1999**, *145*, 1.
3. M.-C. Daniel, D. Astruc. *Chem. Rev.* **2004**, *104*, 293.
4. J. P. Wilcoxon, B. L. Abrams. *Chem. Soc. Rev.* **2006**, *35*, 1162.
5. P. J. Strickler. *J. Chem. Soc. Chem. Commun.* **1969**, *1969*, 655.
6. G. S. H. Lee, D. C. Craig, I. Ma, M. L. Scudder, T. D. Bailey, I. G. Dance. *J. Am. Chem. Soc.* **1988**, *110*, 4863.
7. I. G. Dance, A. Choy, M. L. Scudder. *J. Am. Chem. Soc.* **1984**, *106*, 6285.
8. E. Farneth, N. Herron. *Chem. Mater.* **1992**, *4*, 916.
9. N. Herron, J. C. Calabrese, W. E. Farneth, Y. Wang. *Science* **1993**, *259*, 1426.
10. T. Vossmeyer, G. Reck, L. Katsikas, E. T. K. Haupt, B. Schulz, H. Weller. *Science* **1995**, *267*, 1476.
11. N. Zhu, D. Fenske. *J. Chem. Soc., Dalton Trans.* **1999**, 1067.
12. H. Krautshied, D. Fenske, G. Baum, M. Semmelmann. *Angew. Chem. Int. Ed. Engl.* **1993**, *32*, 1303.
13. J. S. Bradley, G. Schmid. In *Nanoparticles*, Ed. G. Schmid. Weinheim: Wiley, 2004.
14. G. Schmid. *Chem. Rev.* **1992**, *92*, 1709.
15. G. Schmid. *Angew. Chem.* **1978**, *90*, 417.
16. G. Schmid. *Chem. Ber.* **1981**, *114*, 3634.

17. V. V. Volkov, G. Van Tendeloo, G. A. Tsirkov, N. V. Cherkashina, M. N. Vargaftik, I. I. Moiseev, V. M. Novotortsev, A. V. Kvit, A. L. Chuvilin. *J. Cryst. Growth* **1996**, *163*, 377.
18. V. Oleshko, V. Volkov, W. Jacob, M. Vargftik, I. Moiseev, G. Van Tendeloo. *Z. Phys. D.* **1995**, *34*, 283.
19. P. D. Jadzinski, G. Calero, C. J. Ackerson, D. A. Bushnell, R. D. Kornberg. *Science* **2007**, *318*, 430.
20. R. P. Andres, J. D. Bielefeld, J. I. Hendeerson, D. B. Janes, V. R. Kolagunta, C. P. Kubiak, W. J. Mahoney, R. G. Osifchin. *Science* **1996**, *273*, 1690.
21. R. Mukhopadhyay, S. Mitra, M. Johnson, V. R. R. Kumar, T. Pradeep. *Phys. Rev. B* **2007**, *75*, 75414.
22. M. J. Hostetler, A. C. Templeton, R. W. Murray. *Langmuir* **1999**, *15*, 3782.
23. L. Steigerwald, A. P. Alivisatos, J. M. Gibson, T. D. Harris, R. Kortan, A. J. Muller, A. M. Thayer, D. C. Douglas, L. E. Brus. *J. Am. Chem. Soc.* **1988**, *110*, 3046.
24. M. Brust, M. Walker, D. Bethel, D. J. Schiffrin, R. Whyman. *Chem. Commun.* **1994**, *801.*
25. C. B. Murray, C. R. Kagan, M. G. Bawendi. *Science* **1995**, *270*, 1335.
26. X. M. Lin, C. M. Sorensen, K. J. Klabunde. *J. Nanopart. Res.* **2000**, *2*, 157.
27. S. Stoeva, C. M. Sorensen, K. J. Klabunde, I. Dragieva. *J. Am. Chem. Soc.* **2002**, *124*, 2305–2311.
28. B. L. V. Prasad, S. I. Stoeva, C. M. Sorensen, K. J. Klabunde. *Langmuir* **2002**, *18*, 7515.
29. B. L. V. Prasad, S. I. Stoeva, C. M. Sorensen, K. J. Klabunde. *Chem. Mater.* **2003**, *15*, 935.
30. R. L. Whetten, J. T. Khoury, M. M. Alvarez, S. Murthy, I. Vezmar, Z. L. Wang, P. W. Stephens, C. L. Cleveland, W. D. Luedtke, U. Landman. *Adv. Mater.* **1996**, *8*, 428.
31. S. A. Harfenist, Z. L. Wang, M. M. Alvarez, I. Vezmar, R. L. Whetten. *J. Phys. Chem.* **1996**, *100*, 13904.
32. R. L. Whetten, M. N. Shafigullin, J. T. Khoury, T. G. Schaaff, I. Vezmar, M. M. Alvarez, A. Wilkinson. *Acc. Chem. Res.* **1999**, *32*, 397.
33. T. G. Schaaff, G. Knight, M. N. Shafigullin, R. F. Borkman, R. L. Whetten. *J. Phys. Chem. B* **1998**, *102*, 10643.
34. D. V. Leff, P. C. Ohara, J. R. Heath, W. M. Gelbart. *J. Phys. Chem.* **1995**, *99*, 7036.
35. S. Stoeva, V. Zaikovski, B. Prasad, P. S. Stoimenenov, C. M. Sorensen, J. K. Klabunde. *Langmuir* **2005**, *21*, 10280.
36. M. Montalti, L. Prodi, N. Zacharoni, R. Baxter, G. Teobaldi, F. Zerbetto. *Langmuir* **2003**, *19*, 5172.
37. J. Simard, C. Briggs, A. K. Boal, V. M. Rotello. *Chem. Commun.* **2000**, *2000*, 1943.
38. S. Westenhoff, N. A. Kotov. *J. Am. Chem. Soc.* **2002**, *124*, 2448.
39. E. R. Zubarev, J. Xu, A. Sayyad, J. D. Gibson. *J. Am. Chem. Soc.* **2006**, *128*, 4958.
40. T. Song, S. Dai, K. C. Tam, S. Y. Lee, S. H. Goh. *Langmuir* **2003**, *19*, 4798.
41. C. R. Iocovella, M. A. Horsch, Z. Zhang, S. C. Glotzer. *Langmuir* **2005**, *21*, 9488.
42. T. P. Ang, T. S. A. Wee, W. S. Chin. *J. Phys. Chem. B* **2004**, *108*, 11001.
43. A. M. Jackson, J. W. Meyerson, F. Stellacci. *Nature Materials* **2004**, *3*, 330.
44. I. Hussain, Z. Wang, A. I. Cooper, M. Brust. *Langmuir* **2006**, *22*, 2938.

45. D. S. Sidhaye, S. Kashyap, M. Sastry, S. Hotha, B. L. V. Prasad. *Langmuir* **2005**, *21*, 7979.
46. R. Shenhar, V. M. Rotello. *Acc. Chem. Res.* **2003**, *36*, 549.
47. R. Elganian, J. J. Storhoff, R. C. Mucuc, R. L. Letsinger, C. A. Mirkin. *Science* **1997**, *277*, 1078.
48. V. J. Anderson, H. N. W. Lekkerkerker. *Nature* **2002**, *416*, 811.
49. X. M. Lin, G. M. Wang, C. M. Sorensen, K. J. Klabunde. *J. Phys. Chem.* **1999**, *103*, 5488.
50. O. C. Compton, F. E. Osterloh. *J. Am. Chem. Soc.* **2007**, *129*, 7793.
51. B. Abecassis, F. Testard, O. Spalla. *Phys. Rev. Lett.* **2008**, *100*, 115504.
52. H. Yan, C. M. Sorensen. Unpublished data.
53. V. K. LaMer, R. H. Dinegar. *J. Am. Chem. Soc.* **1950**, *72*, 4847.
54. H. C. Hamaker. *Physica* **1937**, *4*, 1058.
55. R. J. Hunter. *Introduction to Modern Colloid Science*. Oxford: Oxford University Press, 2003.
56. S. R. Raghavan, J. Hou, G. L. Baker, S. A. Khan. *Langmuir* **2000**, *16*, 1066.
57. R. Evans, J. B. Smitham, D. H. Napper. *Colloids Polymer Sci.* **1977**, *255*, 161.
58. A. Ulman. *J. Mater. Ed.* **1989**, *11*, 205.
59. A. Chakrabarti, D. Fry, C. M. Sorensen. *Phys. Rev. E* **2004**, *69*, 031408.
60. P. C. Ohara, D. V. Leff, J. R. Heath, W. M. Gelbart. *Phys. Rev. Lett.* **1995**, *75*, 3466.
61. B. A. Korgel, D. Fitzmaurice. *Phys. Rev. Lett.* **1998**, *80*, 3531.
62. P. J. Thomas, G. U. Kulkarni, C. N. R. Rao. *J. Phys. Chem.* **2000**, *104*, 8138.
63. M. D. Bentzon, J. van Wonterghem, S. Morup, A. Tholen, C. J. W. Koch. *Philos. Mag. B* **1989**, *60*, 169.
64. C. B. Murray, D. J. Norris, M. G. Bawendi. *J. Am. Chem. Soc.* **1993**, *115*, 8706.
65. X. M. Lin, H. M. Jaeger, C. M. Sorensen, K. J. Klabunde. *J. Phys. Chem.* **2001**, *105*, 3353.
66. J. Fink, C. J. Kiely, D. Bethell, D. J. Schiffrin. *Chem. Mater.* **1998**, *10*, 922.
67. S. Stoeva, B. Prasad, S. Uma, P. Stoimenov, V. Zaikovski, C. M. Sorensen, K. J. Klabunde. *J. Phys. Chem. B* **2003**, *107*, 7441.
68. A. B. Smetana, K. J. Klabunde, C. M. Sorensen. *J. Colloid Int. Sci.* **2005**, *284*, 521.
69. J. E. Martin, J. P. Wilcoxon, J. Odinek, P. Provencio. *J. Phys. Chem. B* **2002**, *106*, 971.
70. Z. Yang, K. J. Klabunde, C. M. Sorensen. *J. Phys. Chem. C* **2007**, *111*, 18143.
71. S. Yamamuro, D. F. Farrell, S. A. Majetich. *Phys. Rev. B* **2002**, *65*, 224431.
72. T. Hyeon. *Chem. Comm.* **2003**, *927*.
73. J. E. Martin, J. P. Wilcoxon, J. Odinek, P. Provencio. *J. Phys. Chem. B* **2000**, *104*, 9475.
74. D. Zanchet, M. S. Moreneo, D. Ugarte. *Phys. Rev. Lett.* **1999**, *82*, 5277.
75. S. Narayanan, J. Wang, X. M. Lin. *Phys. Rev. Lett.* **2004**, *93*, 135503.
76. T. P. Bigioni, X. M. Lin, T. T. Nguyen, E. I. Corwin, T. A. Witten, H. M. Jaeger. *Nature Mater.* **2006**, *5*, 265.
77. K. E. Mueggenburg, X. M. Lin, R. H. Goldsmith, H. M. Jaeger. *Nature Mater.* **2007**, *6*, 656.
78. A. M. Kalsin, M. Fialkowski, M. Paszewski, S. K. Smoukov, K. J. M. Bishop, B. A. Grzybowski. *Science* **2006**, *312*, 420.
79. C. J. Kiely, J. Fink, M. Brust, D. Bethell, D. J. Schiffrin. *Nature* **1998**, *396*, 444.

80. C. J. Kiely, J. Fink, J. G. Zheng, M. Brust, D. Bethell, D. J. Schiffrin. *Adv. Mater.* **2000**, *12*, 640.
81. F. X. Redl, K.-S. Cho, C. B. Murray, S. O'Brien. *Nature* **2003**, *423*, 968.
82. E. V. Shevchenko, D. V. Talapin, N. A. Kotov, S. O'Brien, C. B. Murray. *Nature* **2006**, *439*, 55.
83. E. V. Shevchenko, D. V. Talapin, C. B. Murray, S. O'Brien. *J. Am. Chem. Soc.* **2006**, *128*, 3620.
84. B. D. Rabideau, R. T. Bonnecaze. *Langmuir* **2005**, *21*, 10856.
85. W. D. Luedtke, U. Landman. *J. Phys. Chem.* **1996**, *100*, 13323.
86. U. Landman, W. D. Luedtke. *Faraday Trans.* **2004**, *125*, 1.
87. M. J. Hostetler, J. J. Stokes, R. W. Murray. *Langmuir* **1996**, *12*, 3604.
88. B. A. Korgel, D. Fitzmaurice. *Phys. Rev. B* **1999**, *59*, 14191.
89. N. Sandhyarani, T. Pradeep, J. Chakrabarti, M. Yousuf, H. K. Sahu. *Phys. Rev. B* **2000**, *62*, R739.
90. N. Sandhyarani, M. P. Antony, G. P. Selvam, T. Pradeep. *J. Chem. Phys.* **2000**, *113*, 9794.
91. N. K. Chaki, K. P. Vijayamohanan. *J. Phys. Chem. B* **2005**, *109*, 2552.
92. S. C. Warren, M. J. Banholzer, L. S. Slaughter, E. P. Giannelis, F. J. DiSalvo, U. B. Wiesner. *J. Am. Chem. Soc.* **2006**, *128*, 12074.
93. A. Ulman. *An Introduction to Ultrathin Organic Films*. San Diego, CA: Academic Press, 1991.

PROBLEMS

1. Derive the results in Table 3.1. The footprint of the thiol on gold is 0.124 nm^2 (93). Calculate the number of silver atoms in a silver nanoparticle.

2. Use Table 3.1 (the results of Problem 1) to prove Equation (3.1).

3. Find the longest length of a Cn alkane chain given the carbon-carbon single bond distance is 0.154 nm and the bond angle is the tetrahedral angle, 109.47 degrees.

4. Consider a 5.0-nm diameter gold nanoparticle molecule ligated with dodecylthiol. If the thiols stretch out radially from the nanoparticle, how much surface area do they have per thiol at the edge of this ligand shell? The data in Equation (3.1) will be useful. Compare this to the fact that each thiol covers 0.124 nm^2 on the surface of the gold nanoparticle.

5. Prove Equations (3.4) and (3.5) given Equation (3.3).

6. Calculate the depth of the van der Waals interaction potential between two 5.0-nm gold nanoparticles with surface to surface separations equal to the lengths of octyl, dodecyl, and hexadecyl alkane chains and compare these to the thermal energy at room temperature. The Hamaker constant for gold is 1.95 eV.

7. Calculate the depth of the van der Waals interaction potential between two 6.0-nm gold nanoparticles with surface to surface separations equal to the length of a dodecyl alkane chain and compare this to the result from Problem 6 for two 5.0-nm particles with the same C12 ligand.

8. It is an interesting question to ask why gold particles with a density of 19.3 g/cc do not fall out of solution. There are two ways to answer this question and these two ways must be considered for a complete understanding. As an example, consider gold particles with no ligand shell in water.
   (a) Use the Stokes law of hydrodynamic drag to calculate the terminal velocity of fall for 10-μm and 100-nm diameter gold spheres in water. The viscosity of water is 0.01 poise (g/cm · s).
   (b) Calculate the height h at which the gravitational potential energy of gold spheres of diameter 10 and 100 nm is equal to the thermal energy at room temperature.
   (c) With the results above, answer the question of how or why gold nanoparticles do or do not fall out of solution.

ANSWERS

1. *Gold.*

$$\text{Atomic mass} = 196.97\ \text{g}$$

$$\text{Density} = 19.32\ \text{g/cc}$$

$$\text{Molar volume} = \frac{196.97}{19.32} = 10.2\ \text{cc}$$

$$\text{Atom volume} = 10.2\ \text{cc}/6.022 \times 10^{23} = 1.69 \times 10^{-23}\ \text{cc}$$

Number of atoms in a sphere of diameter d

$$N = \left(\frac{\pi d^3}{6}\right) \Big/ 1.69 \times 10^{-23} = 30.93 \times 10^{21}\, d^3\ \text{(cm)} = 30.93\, d^3\ \text{(nm)}$$

Surface Ligands. Alkyl thiols have a 0.214 nm^2 footprint (93). The number of thiols on a sphere of diameter d is

$$N = \pi d^2/0.214\ \text{nm}^2 = 14.68\, d^2\ \text{(nm)}$$

Silver.

$$\text{Atomic mass} = 107.9\,\text{g}$$
$$\text{Density} = 10.50\,\text{g/cc}$$
$$\text{Molar volume} = \frac{107.9}{10.50} = 10.276\,\text{cc}$$
$$\text{Atomic volume} = 10.276\,\text{cc}/6.022 \times 10^{23}$$
$$= 1.706 \times 10^{-23}\,\text{cc}$$

Number of atoms in a sphere of diameter d

$$N = \left(\frac{\pi d^3}{6}\right) \Big/ 1.706 \times 10^{-23}$$
$$= 30.68 \times 10^{21}\, d^3(\text{cm})$$
$$= 30.68\, d^3(\text{nm})$$

2.

$$N\,(\text{Au atoms}) = 31\, d^3(\text{nm})$$
$$= 31 \times 5^3 = 3875$$
$$N\,(\text{thiols}) = 14.7\, d^2(\text{nm})$$
$$= 14.7 \times 5^2 = 368$$

3.

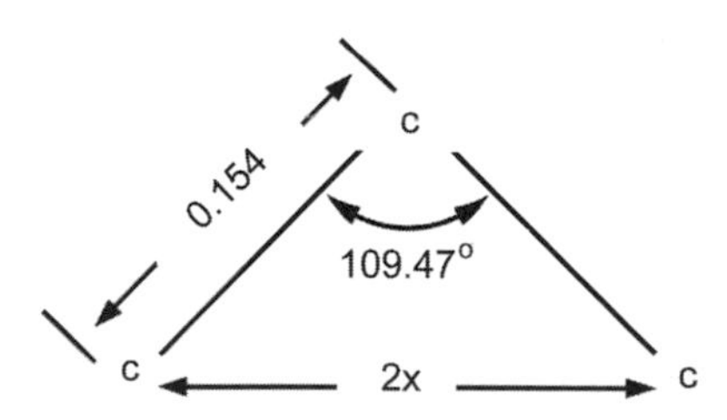

Law of cosines

$$c^2 = a^2 + b^2 - 2ab\cos\theta$$
$$a = b = 0.154$$
$$c^2 = 2a^2(1 - \cos\theta)$$
$$(2x)^2 = 2(0.154)^2(1 - \cos 109.47)$$
$$4x^2 = 2(0.154)^2(1 - (-1/3))\,\text{nm}$$
$$x^2 = \frac{2}{3}(0.154)^2$$
$$x = \sqrt{\frac{2}{3}}(0.154) = 0.126\,\text{nm}$$

Length of $Cn = (n-1)(0.126)$ nm.

4. From Equation (3.1) (or Problem 2) we have for a 5-nm gold particle 365 C12 ligands. The ligand length is $11(0.126) = 1.39$ nm (see Problem 3). Thus, the total diameter of the nanoparticle molecules is

$$d = 5 + 2(1.39) = 7.78\,\text{nm}.$$

Surface area

$$A = \pi d^2 = \pi \times 8^2 = 190\,\text{nm}^2$$

Area per C12 ligand

$$\begin{aligned} A(\text{C12}) &= 190/365 \\ &= 0.52\,\text{nm}^2 \end{aligned}$$

This is more than twice the area at the thiol head group ($0.214\,\text{nm}^2$) to imply the methylene ends have motional freedom.

5. Equation (3.3) is

$$V(r) = -\frac{A}{12}\left[\frac{1}{x^2-1} + \frac{1}{x^2} + 2\ell n\left(1 - \frac{1}{x^2}\right)\right]$$

where $x = r/d$.

(**a**) When $r \gtrsim d$, let $r = d + s$ then $x = 1 + s/d$. Let $x = 1 + \delta$, $\delta = s/d \ll 1$. Then

$$\begin{aligned} x^2 &= (1+\delta)^2 \simeq 1 + 2\delta \\ x^{-2} &\simeq 1 - 2\delta \end{aligned}$$

Substitute into $V(r)$

$$\begin{aligned} V(r) &= -\frac{A}{12}\left[\frac{1}{1+2\delta-1} + 1 - 2\delta + 2\ell n(1-(1-2\delta))\right] \\ &= -\frac{A}{12}\left[\frac{1}{2\delta} + 1 - 2\delta + 2\ell n 2\delta\right] \end{aligned}$$

recall $\delta \ll 1$ so $1/2\delta$ dominates

$$V(r) \simeq -\frac{A}{24\delta} = -\frac{Ad}{24s}$$

(**b**) When $x \gg 1$, expand on $x^{-2} \ll 1$

$$V(r) = -\frac{A}{12}\left[\frac{1}{x^2(1-x^{-2})} + \frac{1}{x^2} + 2\ell n\left(1 - x^{-2}\right)\right]$$

Use $\left(1-x^{-2}\right)^{-1} \simeq 1+x^{-2}+x^{-4}$ (Binomial expansion)

$$\ell n(1-x^{-2}) \simeq -x^{-2}-\frac{1}{2}\left(-x^{-2}\right)^{2} \quad \text{(Taylor expansion)}$$

Then

$$\begin{aligned} V(r) &\simeq -\frac{A}{12}\left[\frac{1}{x^2}\left(1+x^{-2}+x^{-4}\right)+\frac{1}{x^2}+2\left(-x^{-2}-\frac{1}{2}\left(-x^{-2}\right)^{2}\right)\right] \\ &= -\frac{A}{12}\left[\frac{1}{x^2}+\frac{1}{x^4}+\frac{1}{x^6}+\frac{1}{x^2}-\frac{2}{x^2}-\frac{1}{x^4}\right] \\ &= -\frac{A}{12}\frac{1}{x^6}=\frac{Ad^6}{12}\frac{1}{r^6} \end{aligned}$$

6. Center to center separation $r=d+s$. Then $x=1+s/d$. For this situation we take s to be the length of the ligand.
Cn ligand length is $(n-1)\times 0.126$ nm, see page 27 or Problem 2. Thus:

$$\begin{aligned} s &= 7\times 0.126 = 0.88\,\text{nm}, \;\text{thus}\; x=1.18 \text{ for C8} \\ &= 11\times 0.126 = 1.39\,\text{nm}, \;\text{thus}\; x=1.28 \text{ for C12} \\ &= 15\times 0.126 = 1.89\,\text{nm}, \;\text{thus}\; x=1.37 \text{ for C16.} \end{aligned}$$

The Hamaker constant for gold is 1.95 eV. Use Equation (3.3) for these x values. Thermal energy at room $T=298$ K is kT $=1.38\times 10^{-23}$ joules/K $\times$ 298 K/ 1.6×10^{-19} joules/eV $=0.0257$ eV.
Then:

$$\begin{aligned} V(1.18) &= -0.12\,\text{eV} = -4.14\,\text{kT} \\ V(1.28) &= -0.047\,\text{eV} = -1.84\,\text{kT} \\ V(1.37) &= -0.025\,\text{eV} = -0.96\,\text{kT}. \end{aligned}$$

We see relatively strong interaction for C8 and weak for C16.

7. $r=d+s, \quad x=1+s/d.$

$d=6\,\text{nm}, \quad s=1.39\,\text{nm}\ (C12)$

so $x=1.25$. Use Equation (3.3) to find

$V(1.23)=-0.073\,\text{eV}=-2.83\,\text{kT}$

This is about 50% larger than for $d=5$ nm gold particles.

8. **(a)** Stokes drag force F

$$F=6\pi\eta a v$$

where η is the medium viscosity, a the particle radius, and v the particle velocity. Falling under gravity one has

$$mg = 6\pi\eta av$$

where g is the acceleration of gravity and m is the particle mass.

$$m = \frac{4\pi}{3}\rho a^3$$

where ρ is the particle mass density. These equations combine to yield a terminal velocity of fall given by

$$v_T = \frac{2}{9}(\rho - 1)\frac{ga^2}{\eta}$$

The subtraction of the density of water, 1 g/cc, from ρ accounts for buoyancy. For $a = 10$ nm $= 10^{-6}$ cm

$$\begin{aligned} v_T &= \frac{2}{9}(19.3 - 1)\frac{980 \cdot (10^{-6})^2}{0.01} \\ &= 3.95 \times 10^{-7}\text{cm}/s \simeq 4 \times 10^{-7}\text{cm}/s \\ &\simeq 0.3\ \text{mm/day}. \end{aligned}$$

For a $= 100$ nm v_T increases by 10^2 because $v_T \sim a^2$.

$$\begin{aligned} v_T(100\text{nm}) &= 4 \times 10^{-5}\ \text{cm/sec} \\ &= 3\ \text{cm/day} \end{aligned}$$

(b) Gravitational potential energy $= mgh$
Thermal energy $= kT$

$$\begin{aligned} mgh &= kT \\ h &= \frac{kT}{mg} \\ &= \frac{3kT}{4\pi\rho a^3\,g} \\ &= \frac{3(1.38 \times 10^{-16})(298)}{4\pi \cdot 19\cdot 3 \cdot 980}a^{-3} \end{aligned}$$

If $a = 10$ nm, $h(10\ \text{nm}) \cong 0.5$ cm
If $a = 100$ nm, $h(100\ \text{nm}) \simeq 5 \times 10^{-4}$ cm

(c) The terminal velocity of fall sets the time scales for the nanoparticle solution to come to equilibrium. Given test tube length scales of approximately 1 cm, these time scales are approximately 1000 hours and 10 hours for $a = 10$ nm and 100 nm, respectively. The equilibrium state to which the kinetics leads is determined by Boltzmann statistics

$$e^{-mgh/kT}$$

There is a fight between gravitational energy pulling the particles down and the thermal energy, kT, keeping them suspended. For $a = 10$ nm kT will not allow setting below approximately 0.5 cm and, as seen above, this takes a long time to get to. Thus, 10 nm particles don't settle. For $a =$ 100 nm the settling $1/e$ point can be small $h \simeq 5$ μm and obtained in less than a day. Thus 100-nm particles settle out.

PART II

NEW SYNTHETIC METHODS FOR NANOMATERIALS

4

MICROWAVE PREPARATION OF METAL FLUORIDES AND THEIR BIOLOGICAL APPLICATION

David S. Jacob, Jonathan Lellouche, Ehud Banin, and Aharon Gedanken

4.1 Introduction, 74
4.2 Microwaves, 75
4.3 Ionic Liquid, 76
4.3.1 Fundamental Aspects of Ionic Liquids, 76
4.3.2 Types of Ionic Liquids, 77
4.3.3 Properties of Ionic Liquids, 77
4.3.4 Challenge in the Application of Ionic Liquids, 77
4.3.5 Applications of Ionic Liquids, 78
4.4 Bacterial Biofilms, 80
4.5 Experimental Setup, 82
4.6 Experimental Procedure and Results, 82
4.6.1 Anisotropic Structures, 83
4.6.2 Core-Shell Metal Fluoride, 93
4.7 MgF_2 Nanoparticles as Novel Antibiofilm Agents, 95
4.7.1 MgF_2-Coated Glass Surfaces Inhibit Bacterial Colonization, 96
4.7.2 MgF_2 Coating Provides Long-Lasting Antibiofilm Activity, 98
4.8 Conclusions, 100
References, 101
Problems, 107
Answers, 107

Nanoscale Materials in Chemistry, Second Edition. Edited by K. J. Klabunde and R. M. Richards

4.1 INTRODUCTION

New material fabrication strategies are of fundamental importance in the advancement of science and technology (1). In the area of nanotechnology, one-dimensional nanometer-sized materials, such as nanowires, nanotubes, nanorods, and other nanoparticles with different morphologies, have attracted considerable attention because of their intrinsic size-dependent properties and resulting applications (2–4). Controlling the size, shape, monodispersity, yield of the desired size, as well as the shape and the various useful properties of nanocrystals with different chemistry has become a challenge for material chemists (5). However, nanoparticles of inorganic fluorides have received less attention compared to other classes of compounds, such as metals, oxides, and semiconductors. It is also worth mentioning that nanofluoride materials were not included in the nanomaterials handbook. Researchers have found that nanofluoride materials have the following interesting physical properties:

1. Most metal fluorides have transmission in a wide spectral range, from 200 to 6000 nm.
2. They have high thermal conductivity, better mechanical properties, and high moisture resistance.
3. Their single crystals are used in active and passive elements for tunable laser systems.
4. Nanoceramics made of fluorides are used as scintillators, etc.
5. Metal fluorides have been used as probes for cytological, microbiological, and medical studies.
6. Nanoparticles of fluorides with a high surface are used as catalysts and sorbents.

Thus, the main inspiration for researchers is to develop a simple general synthetic method that can yield nanometal fluorides of a desired form with wide application. In this chapter we focus our attention on the preparation of various types of nanofluorides with different morphologies and their biological application in the field of biofilms, specifically the use of MgF_2 nanoparticles as novel antibiofilm agents.

The many different methods reported on the preparation of metal fluorides can be divided into two categories:

1. Physical methods, which include (a) a vapor phase condensation technique under ultra-high vacuum conditions (6), (b) mechanical milling (7), (c) laser dissipation (8), and (d) molecular-beam epitaxy (9).
2. Chemical methods, which include (a) pyrolysis of desired fluorinated precursor materials (10), (b) sol-gel synthesis (11), and (c) nonaqueous sol-gel synthesis (8).

The fluorinating reagents that are reported used are HF (but handling this is difficult; 12), HF-pyridine (13, 14), Et_3N-HF (15, 16), and polyhydrogen fluoride (17, 18). In this chapter we combine nonaqueous, sol-gel fluorine chemistry and microware radiation as the energy heating source for the synthesis of different-shaped nanometric metal fluorides. The fluorinating agent used in our synthesis is an ionic liquid (IL). There is only one report where IL is used as a fluorinating agent, and it is for the synthesis of organofluorine compounds (19). The method described in this chapter is an innovative, nonaqueous sol-gel reaction in which the first step is the dissolution of the metal salts in the IL. The nanosized metal fluoride products for the different solutions of the metal salts in the IL are obtained from a microwave radiation reaction, which leads to products having different morphological structures. We found that in our reactions the IL acts as a template, stabilizing agent, capping agent, and structure-directing agent. The method of microwave radiation heating, the properties of the solvent (IL), and the experimental details, are outlined briefly in this chapter. The preparation and formation of different types of metal fluorides are also explained. The products obtained are investigated using XRD (x-ray diffraction patterns), and the morphology is studied with HRSEM (a high resolution scanning electron microscope), and HRTEM (a high resolution transmission electron microscope). The other characterization methods of these metal fluorides are beyond the scope of this chapter.

In this chapter we also address the core-shell structure of metal fluorides for the case of FeF_2, where the FeF_2 is the core and carbon is the shell. According to Fedorov et al. (10), water will react very slowly with fluorides, yielding HF. The carbon shell is important because it will prevent this dissolution. Finally the use of metal fluorides to produce sterile abiotic surfaces and block bacterial biofilm formation is discussed.

4.2 MICROWAVES

Microwaves are electromagnetic in nature, consisting of two components, a magnetic and an electric field (20). Microwave radiation is positioned between infrared radiation and radio frequencies with wavelength ranges of 1 mm to 1 m, which correspond to frequencies of 300 GHz to 300 MHz, respectively. The extensive application of microwaves in the field of telecommunications means that only specially assigned frequencies are allowed to be allocated for industrial, scientific, or medical purposes (20). Industrial microwave ovens are operated at either 2.45 GHz or 915 MHz, while most common domestic microwave ovens operate at 2.45 GHz.

Microwave heating is a uniform heating effect, although localized superheating does occur. Heating by microwave is based on the absorption and transference of electromagnetic energy into heat by different liquids and solids (20–25). In general, during the interaction of microwaves with materials, three different types of behavior of a material can be observed, depending on the type of material used, (1) electrical conductors (metals), (2) insulators (quartz, porcelain), and (3) dielectric materials (water, organic and ionic liquids). When a strongly conducting material is exposed to microwave radiation, microwaves are reflected from its surface and the material is not effectively heated by the microwaves. In response to the electric field of microwave radiation, electrons move freely on the surface of the material, and the flow of electrons can heat the material through an ohmic (resistive) heating mechanism. In the case of an insulator, microwaves can penetrate the material without any absorption, loss, or heat generation, whereas in the case of dielectric materials, the reorientation of either permanent or induced dipoles during microwave radiation, which is electromagnetic in nature, can give rise to the absorption of microwave energy and generate heat due to the so-called dielectric heating mechanism. Because of its dependence on the frequency, the dipole may move in time to the field, either lag behind it, or remain unaffected. When the dipole lags behind the field, the interaction between the dipole and the field leads to an energy loss by heating, which is a dielectric heating mechanism. The electric field component is responsible for this dielectric heating mechanism since it can cause molecular motion either by migration of the ionic species (conduction mechanism) or the rotation of dipolar species (dipolar polarization mechanism). In a microwave field, the electric field component oscillates very quickly at 4.9×10^9 times per second at 2.45 GHz, and the strong agitation, provided by the cyclic reorientation of molecules, can result in intense internal heating.

In this chapter we will discuss in detail the use of IL as a solvent for material synthesis under microwave radiation. An overview of the IL and its typical applications are outlined in the following section. ILs have important properties such as high fluidity, low melting temperature, an extended temperature range in the liquid state, nonflammability, high ionic conductivity, the ability to dissolve a variety of materials, and most importantly, nonmeasurable vapor pressure (26–44). Moreover, the key advantage in using ILs as a solvent in microwave reactions is the presence of large positive organic molecules with high polarizability, which helps to absorb microwave radiation, thus leading to very high heating rates and temperatures (45–47).

In most of the synthesized materials of different compounds in ionic liquids, we obtained one-dimensional structures. The ionic-conductive nature and polarizability of the ionic liquids helps in the movement and polarization of ions under the rapidly changing electric field of the microwaves. This results in high heating and in the transient, anisotropic microdomains for the reaction system, which assists the anisotropic growth of the nanostructures.

4.3 IONIC LIQUID

The term *ionic liquid* is commonly used for the molten salts whose melting point is below 100°C (48). In particular, the salts that are liquid at room temperature are called room temperature ionic liquids. The earliest known ionic liquid (published in 1914; 49, 50) was ethyl ammonium nitrate $EtNH_3^+NO_3^-$, which has a melting point of 12°C. It was these initially developed ionic liquids (molten salts) that were used as electrolytes to study the electrochemical behavior of other compounds. Recently, ionic liquids with interesting properties have been synthesized and used as solvents, and studied in different areas of chemistry (26–44).

4.3.1 Fundamental Aspects of Ionic Liquids

Ionic liquids are salts that consist of cations and anions (51–56). Most commonly, the cations are bulky organic, symmetric, and asymmetric molecules (imidazolium, pyridinium, pyrolidinium), and the inorganic anion molecules are (PF_6^-, BF_4^-, and $[(CF_3SO_2)_2N]^{-1}$; 57] By changing the cation and the anion, the physical properties of the ionic liquids, such as melting, viscosity, density, and hydrophobicity, are modified. To match the reaction conditions according to the required properties, these ionic liquids can be designed, and hence they are called *designer solvents* (58, 59).

4.3.2 Types of Ionic Liquids

Different types of cation-anion association yield special types of ionic liquids having different physical properties.

1. Cations: The types of cations used in synthesizing ionic liquids are tetraalkylammonium, trialkylsulfonium, tetra-alkylphosphonium, 1-3-dialkylimidazolium, *N*-alkylpyridinium, *N-N*-dialkylpyrrolidinium, *N*-alkylthiazolium, *N-N*-dialkyltriazolium, *N-N*-dialkyloxazolium and *N-N*-dialkylpyrazolium. The most commonly used cations are 1-3-dialkylimidazolium and *N*-alkylpyridinium, due to their important physicochemical properties (60).
2. Anions: These are classified into two groups: mononuclear anions and polynuclear anions. The mononuclear anions generally lead to neutral, stoichiometric ionic liquids. Examples of the anions are BF_4^-, PF_6^-, $N(FSO_2)_2^-$, $N(CF_3SO_2)_2^-$, $C(CF_3SO_2)_3^-$, $CH_3SO_3^-$, and $CF_3SO_3^-$. Commonly used anions are BF_4^-, PF_6^-, and $N(CF_3SO_2)_2^-$. Polynuclear anions are air and water sensitive, for example, $Al_2Cl_7^-$, $Al_3Cl_{10}^-$, $Au_2C_{17}^-$, $Fe_2Cl_7^-$, and $Sb_2F_{11}^-$ (60).

4.3.3 Properties of Ionic Liquids

The most important properties of ionic liquids are the salts that are liquid at room temperature. They possess good thermal stability, and do not decompose over a large temperature range. They have no vapor pressure, and have a high ionic conductivity and a large electrochemical window (61). The nature of the cations and the anions can affect physical properties such as the melting point, viscosity, density, and hydrophobicity of the ionic liquids (57). The physical properties, for example, the viscosity of the ionic liquids, can be changed, depending on alkyl chains such as long chain, highly branched, or compact alkyl groups (57, 62, 63). The type of anion can also affect the viscosity of the ionic liquids. For example, for a 1-3-dialkylimidazolium cation with a different anion, the viscosity trend increases from $N(CF_3SO_3)_2^- < BF_4^- < PF_6^- < Cl^-$ (57, 60). The solubility properties are also affected by different cations and anions. Generally the cation (1-alkyl-3-methylimidazolium) with a long alkyl chain length from butyl to octyl increases the hydrophobicity and the viscosity, whereas the density and the surface tension decrease (57, 60).

4.3.4 Challenge in the Application of Ionic Liquids

Changes in the physical properties of ionic liquids with different cations and anions have attracted the attention of many industries and academicians for their application in different fields of science, mainly in chemistry. Since there are different complex and simple ionic liquids, selecting a particular type of ionic liquid with the required properties necessitates a specific system. The major challenges that can significantly affect the physical properties of the ionic liquids are trace impurities (64, 65). These impurities, which are found mainly in ionic liquids, are water, solvents, and the salts that remain after the synthesis process, as the purification of ionic liquids is a tedious process compared to that of other organic materials. The major problem lies in the fact that they are nonvolatile, they cannot be distilled, and they are not even sufficiently solid to recrystallize (64, 65). Even determining the trace amount of impurities and the type of impurities requires several types of analyses. Hence, care should be taken to carry out reactions in ionic liquids. The level of the impurities should be noted, and the side reactions due to the presence of impurities should be monitored.

4.3.4.1 Structural Stability The structural stability of different ionic liquids is explained in terms of interactions between cations and anions, as well as their symmetry. These interactions, and the force between the cations and the anions, lead to changes in the physical properties of the ionic liquids (57). Lengthening the alkyl chain of the cation decreases the symmetry of the cation and increases the van der Waals forces, as it requires more energy for molecular motion (57). Hence, shorter alkyl chains are preferable because of their low viscous properties. The functionalization of the side chain also affects the interaction between the cations and anions, which ultimately changes the physical properties. The hydrogen bonding and the dipole-dipole between the anion and the cation also affect the intermolecular forces and the stability properties (57).

4.3.4.2 Thermal Stability According to a literature search, it is known that ionic liquids are thermally stable (300°C to 400°C). The imidazolium salts are reported to be thermally stable up to 300°C. Above 300°C, there is a cleavage of the C-N bond of the imidazolium nitrogen and the carbon of the alkyl chain (57, 66–68). It was found that different anions also affect the thermal properties of the ionic liquids (66–68). Ionic liquids with less nucleophilic or coordinating anions show a higher thermal stability (66–69).

4.3.5 Applications of Ionic Liquids

4.3.5.1 Capping of Nanomaterials Ionic liquids are used in different reactions under different conditions for the synthesis of different nanomaterials. For the preparation of one-dimensional structures, ionic liquids are found to be stabilizing agents that prevent the aggregation of the synthesized product (70). The important property that makes ionic liquids an appropriate solvent for the synthesis of nanosized structure materials is the low interface tension, which results in high nucleation rates. This also leads to a very weak Ostwald ripening, resulting in very small nanoparticles (38, 70). The low interface energies of ionic liquids allow for the good stabilization or solvalization of the molecular species (70). Ionic liquids with hydrophobic and directional polarizability enable them to orient themselves parallel or perpendicular to the dissolved species (nanoparticles), thus forming a capping that prevents the aggregation of the species (70). There are also examples of thiol-functionalized ionic liquids where these ionic liquids are used as a capping agent to obtain thiol-functionalized metal nanoparticles (71). Ionic liquids, which can form well-extended hydrogen bond systems in a liquid state, are highly structured. This property enables the spontaneous, well-defined, and extended ordering of the nanoscale structure to give different morphology species (70).

4.3.5.2 Templates Due to their designer solvent nature, ionic liquids can be synthesized with different cations and anions for a particular application (51–59). The physical properties of ionic liquids change by varying the alkyl chain length of the cation (57). It is known that ionic salts with short alkyl chain lengths are liquid at room temperature and are formed into glass upon cooling, whereas the longer alkyl chains are low melting solids that show an enantiotropic mesomorphism with an extensive thermotropic mesophage range (72). Because of the long alkyl chain of ionic liquids, they tend to form a particular type of aggregation of the ionic liquid molecules. In the case of imidazolium salts, the long alkyl chain orients itself parallel to the imidazole plane in the crystalline state, where the chains interdigitate each other to induce a mesophase formation with a bilayer lamellar arrangement (two-dimensional) of a repeating layer distance in super-microsize (73). Thus, the two-dimensional ionic crystal arrangement of the ionic liquid acts as a well-defined template for the super-micropores. The long alkyl chain cation with hydrophilic anions forms a viscous gel-phase nanostructure, termed *ionogel*, on the addition of an appropriate concentration of water (74). The ionogel materials constrain the water channels, which assists their potential utility as templates for the formation of nanostructured particles (74). The arrangement of long alkyl chain ionic liquid molecules forms a framework that acts as a supporting agent for the synthesis of different morphologies of inorganic materials. It was found in the literature that surfactant-templated mesoporous materials were always found to collapse after removal of the templates, whereas in the case of ionic liquids, the pore structure of the inorganic materials is maintained with a high order, even after removal of the ionic liquid (73).

4.3.5.3 Structural Directing Agent To obtain anisotropic structures of nanoparticles, two practical approaches are preferred, hard- and soft-template methods (74). The former is an example of a rigid organizing matrix for *in situ* synthesis, which includes the fabrication of ordered templates. In this case, the metal of interest is deposited in the pores of any nanoporous template (75–79). In the soft-template method, well-arranged molecules, such as micelles, vesicles, liquid crystals, and microemulsions, are used to direct the growth and control the size of the nanoparticles (80–85). There are literature reports on the control of the shape and size of nanoparticles using ionic liquids. It was found that ionic liquids as a soft template exhibit a wide variety of structural motifs, which help in developing different morphological nanoparticles. It was further found that different concentrations of ionic liquids give different morphological structures of nanoparticles (74). Ionic liquids also act as a surfactant or a capping agent and bind to a particular plane of the nanocrystal due to unique physical properties that hinder the growth of the nanocrystal along the binding site. However, the unbound side of the nanocrystal shows preferential growth, and hence we get different anisotropic structures (74, 86).

4.4 BACTERIAL BIOFILMS

The increased resistance of bacteria to antibiotic therapy is a growing concern for doctors and medical officials worldwide. In the last two decades bacteria have developed resistance to almost all the commercially available antibiotics and the number of new antibiotics expected to enter the market is limited. One of the modes by which bacteria exert this resistance is their ability to develop biofilms. Biofilms are bacterial communities encased in a hydrated polymeric matrix. Biofilm development is known to follow a series of complex but discrete and well-regulated steps (Fig. 4.1): (1) microbial attachment to the surface, (2) growth and aggregation of cells into microcolonies, (3) maturation, and (4) dissemination of progeny cells for new colony formation (87, 88).

An important characteristic of microbial biofilms is their innate resistance to immune system and antibiotic killing (89, 90). This has made microbial biofilms a common and difficult-to-treat cause of medical infections (87, 91, 92). It has recently been estimated that over 60% of the bacterial infections currently treated in hospitals are caused by bacterial biofilms (91). Several chronic infections (e.g. respiratory infections caused by *Pseudomonas aeruginosa* in the cystic fibrosis lung, Staphylococcal lesions in endocarditis, and bacterial prostatitis, primarily caused by *Escherichia coli*) have been shown to be mediated by biofilms (93). More notably, biofilms (particularly of *Staphylococcus aureus*, *P. aeruginosa*, and *E. coli*) are also a major cause of infections associated with medical implants (94, 95). The number of implant-associated infections approaches 1 million per year in the United States alone, and their direct medical costs exceed $3 billion annually (96). Thus, there is an urgent need to find novel approaches to eradicate biofilms.

The inherent resistance of biofilms to killing and their pervasive involvement in implant-related infections has prompted the search for surfaces/coatings that inhibit bacterial colonization. Kingshott and colleagues have studied the effects of different

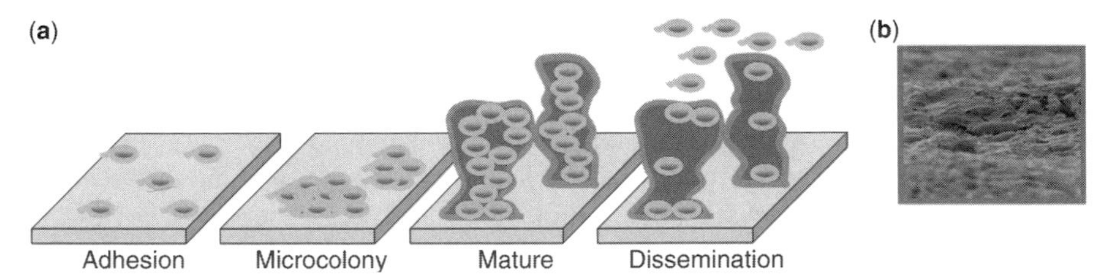

Figure 4.1 Presentation of bacterial biofilm development on abiotic surfaces. (a) Adhesion initially involves reversible association with the surface. As this proceeds bacteria undergo irreversible attachment with the substrate through cell surface adhesions. In later stages bacteria will start secreting a protective extracellular matrix and form microcolonies that develop into mature biofilms. These structures protect the bacteria from host defenses and systemically administered antibiotics. (b) An electron micrograph of a biofilm-infected catheter.

polyethylene glycol (PEG) attachment strategies on the adhesion of Gram-negative bacteria (*Pseudomonas* sp.). They demonstrated that the coated surface reduced the level of adhesion by two to four orders of magnitude in the first 5 h (97). Busscher, Kaper, and Norde (98) have covalently attached poly(ethylene oxide) brushes to glass surfaces and examined *S. epidermidis* adhesion. The coated surface dramatically reduced bacterial adhesion in the first 4 h. An alternative approach to reduce bacterial adhesion is based on coatings that have been designed to release a flux of antibacterial agents (active surfaces). These include antibiotics, antibodies, silver and nitric oxide (reviewed in Reference 99). These surfaces show promising results in reducing surface colonization and preventing biofilm formation.

A new approach comes from advance in nanotechnology that offer an opportunity for the discovery of novel compounds with antibiofilm activity. One such example is nanosilver crystals, which show potent antimicrobial activity and have been used to coat catheters (100). Furthermore the use of nanofabrication surface techniques can change surface properties such as charge, wetability, and topography, in addition to the ability to add functional activities. The precise control of surface composition and chemical functionality, along with novel methods such as microwave synthesis, discussed in this chapter, promote our ability to attach nanoparticles to a wide range of surfaces and can offer important new tools to combat bacterial biofilm persistence.

Figure 4.2 Experimental setup for microwave reactions.

4.5 EXPERIMENTAL SETUP

A modified domestic microwave oven with a refluxing system (Kenwood Microwave, 2.45 GHz, 900 W) is used to carry out the reactions. The experimental setup is shown in Figure 4.2. In all the experiments the microwave is operated at a specific cycling mode: on 21 seconds, off 9 seconds, with the total power always at 900 W. The cycling mode was chosen in order to avoid the bumping of the solvent. The reactions were carried out in a closed hood with air-exhaust ventilation.

4.6 EXPERIMENTAL PROCEDURE AND RESULTS

Preparation of metal fluorides under microwave irradiation: A general procedure for the synthesis of metal fluoride is as follows, $BMIBF_4$ (1-butyl-3-methylimidazolium tetrafluoroborate), the ionic liquid (IL) solvent (precursor for fluoride ions), and the different metal salts at a weight ratio of 10 : 1, are mixed in a round-bottom flask fixed with a water condenser. All the reaction mixtures were heated in the above-mentioned

$$\text{a) } Fe(III)NO_3.9H_2O \xrightarrow[\text{(BMIBF}_4\text{) IL, time(2min)}]{\text{MW}} Fe_2O_3$$

$$\text{b) } Fe(III)NO_3.9H_2O \xrightarrow[\text{(BMIBF}_4\text{) IL+ time(10min)}]{\text{MW}} FeF_2$$

$$\text{c) } Co(II)NO_2.3H_2O \xrightarrow[\text{(BMIBF}_4\text{) IL, time(10min)}]{\text{MW}} CoF_2$$

$$\text{d) } Zn(II)NO_3.6H_2O \xrightarrow[\text{(BMIBF}_4\text{) IL, time(10min)}]{\text{MW}} ZnF_2$$

$$\text{e) } La(III)NO_3.6H_2O \xrightarrow[\text{(BMIBF}_4\text{) IL, time (5min)}]{\text{MW}} LaF_3$$

$$\text{f) } Y(III)NO_3.nH_2O \xrightarrow[\text{(BMIBF}_4\text{) IL, time(5min)}]{\text{MW}} YF_3$$

$$\text{g) } Sr(II)NO_3 \xrightarrow[\text{(BMIBF}_4\text{) IL, time(5min)}]{\text{MW}} SrF_2$$

$$\text{h) } Mg(II)CH_3COO.4H_2O \xrightarrow[\text{(BMIBF}_4\text{) IL, time(2min)}]{\text{MW}} MgF_2$$

Scheme 4.1 Formation of metal fluorides and metal oxide in IL under microwave (MW) heating.

domestic microwave oven at different time intervals from 2 to 10 min reaction time. The origin of the fluoride ions in the final product is from the anion of the ionic liquid, which is also a solvent for our reaction systems. At the end of the reaction time the products obtained were washed several times with acetone and ethanol to remove the IL and other organic impurities, and then centrifuged for 10 min at 9000 rpm. The washed products were dried under vacuum. The different metal fluorides that were synthesized by this method are FeF_2, ZnF_2, CoF_2, SrF_2, MgF_2, LaF_3, and YF_3.

The precursors used in the synthesis of the nanosized metal fluorides are listed in Scheme 4.1.

4.6.1 Anisotropic Structures

4.6.1.1 Nanobar FeF_2 The microwave reaction of iron nitrate ($Fe(III)NO_3.9H_2O$) in IL yields nanostructure iron fluorides. The reaction that occurs is the thermal decomposition of the metal nitrate and the IL, which leads to the formation of metal fluoride. The formation of the metal fluoride occurs in two steps under microwave heating: (1) the formation of the metal oxide and (2) formation of FeF_2, with respect to irradiation time. Over short reaction time, metal oxide is the product, whereas over longer time, metal fluoride is the principal product. It is found that a morphological change occurs also during the course of the reaction. Nanosized metal oxide particles are obtained first, and then further modified into nanobar-shaped FeF_2 when the reaction irradiation time is increased. The formation of the metal oxide and the metal fluoride and their morphologies are presented in Scheme 4.2.

Ten percent of water was added to the reaction mixture and to the nine hydrated water molecules that the metal nitrate contains, to examine whether the addition of water can influence the structure of the FeF_2 nanobars. We found a different product, that is, nanoparticles of FeF_2, in the range of 100 to 200 nm. The formation of nanoparticles of Fe_2O_3 and the FeF_2 (with the addition of water) can be explained on the basis of the capping properties of the ionic liquid. With the addition of water the surface tension of the IL decreases (101), and the IL acts as a surfactant, resulting in the formation of nanoparticles of Fe_2O_3. However, with the formation of the bars of FeF_2, the IL acts as a structure-directing agent to give one-dimensional nanostructures.

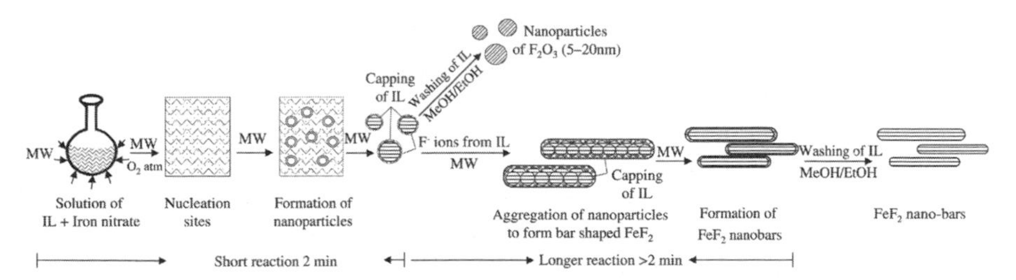

Scheme 4.2 The reaction products and their morphologies obtained under microwave radiation with respect to time for the iron nitrate system.

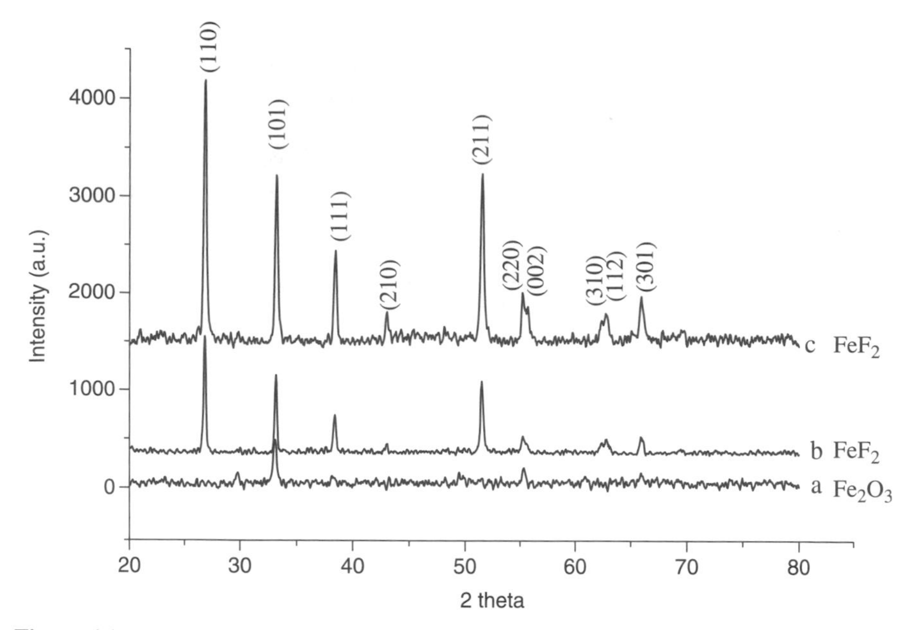

Figure 4.3 (a) XRD pattern of Fe_2O_3 nanoparticles; (b) XRD pattern of FeF_2 (bar-shaped nanoparticles); (c) XRD pattern of FeF_2 nanoparticles.

Scheme 4.2 illustrates how the nanoparticles are capped with IL, and then agglomeration of nanoparticles in one-dimension takes place with the help of IL to get bar-shaped nanoparticles. The XRD patterns of the products are presented in Figure 4.3. Figure 4.3a is the XRD pattern of Fe_2O_3 nanoparticles, which match the PDF-39-0238. Figures 4.3b and c present the XRD patterns of FeF_2, bars and nanoparticles, respectively. These patterns match well with the PDF-45-1062 (tetragonal FeF_2). The XRD pattern of the nanoparticle of FeF_2 (Fig. 4.3c) exhibits high intense diffraction lines, as well as a broad FWHM (full width at half maxima), as compared to that of the nanobars of FeF_2 (Fig. 4.3b). The morphology of the product is presented in Figure 4.4.

HRTEM images of nanoparticles of Fe_2O_3 are shown in Figure 4.4a. The single nanoparticles are in the range of 5 to 20 nm from the HRTEM images. The observed spacing of the fringes of nanoparticle Fe_2O_3 is 2.7 Å, fitting well with the d value of the (222) plane. The nanobars of FeF_2 morphology are seen in the HRSEM image in Figure 4.4b, and the HRSEM image of the nanoparticle of FeF_2 (reaction with the addition of water) is presented in Figure 4.4c.

4.6.1.2 Nanoneedle YF_3 The microwave and IL methods developed by our group have been found to be very efficient and simple, compared to other methods employed in the past for the fabrication of rare earth fluorides. To obtain yttrium fluoride nanoneedles we first had to prepare yttrium nitrate, which is obtained by refluxing yttrium oxide in a nitric acid. In the next step, the yttrium nitrate is dissolved in the IL, and then heated in a domestic microwave oven. We examined the product during the course of

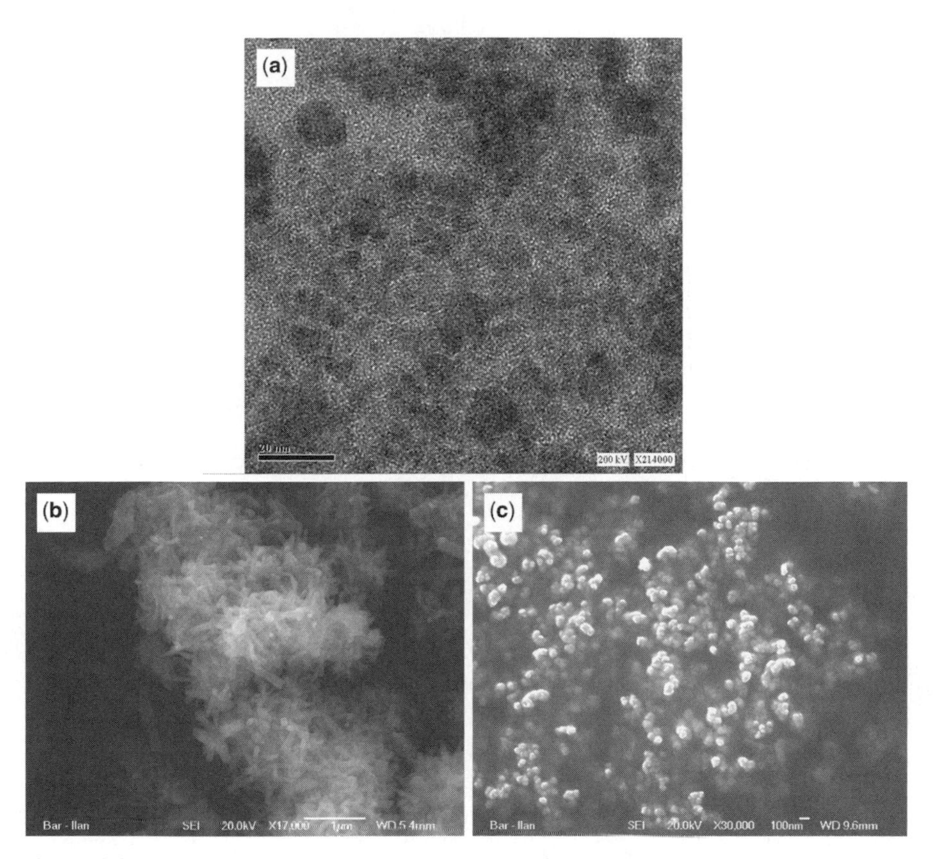

Figure 4.4 (a) HRTEM images of nanoparticles of Fe_2O_3; (b) HRSEM image of FeF_2 nanobars; (c) HRSEM image of the nanoparticle of FeF_2.

the reaction. The XRD patterns are illustrated in Figures 4.5a and b. We found that at shorter reaction times (2 min) the XRD depicts the mixture of yttrium oxide (PDF-76-151) and yttrium oxyfluoride (PDF-71-2100), which is shown in Figure 4.5a, whereas we also see an emerging peak of yttrium fluorides. These small diffraction peaks match well with the XRD data file (PDF-1-70-1935, orthorhombic YF_3), as shown in Figure 4.5b. After 5 min of reaction, the XRD pattern completely changes to that of the yttrium fluoride pattern, and the impurity peaks disappear. Thus, we can learn that even after a short reaction time, yttrium fluoride is already obtained. From the XRD results it is clear that this is already the second system in which the formation of the final product, the metal fluoride, is preceded by an oxidation stage, after which fluorination takes over, yielding the desired product. The formation of the YF_3 nanoparticle is presented in Scheme 4.3. Here also we try to show how the nanoparticles are aggregated with the help of IL to form the needle structure of YF_3. Since the solubility products of the metal oxide and of the metal fluoride in the IL are not known, any attempt to explain why the metal fluoride is not obtained directly will be speculative. The morphology of the final product is presented in Figures 4.6a and b, and the

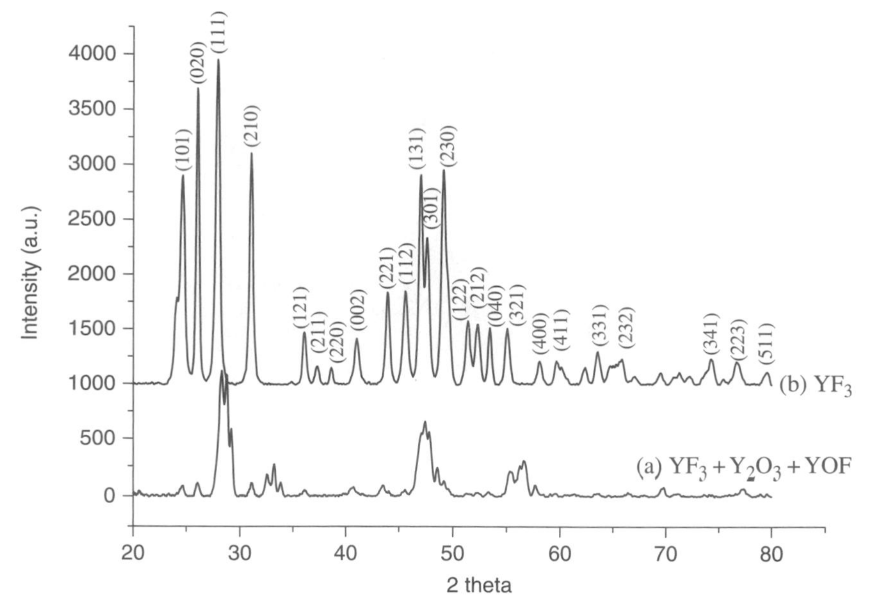

Figure 4.5 (a) XRD pattern for a mixture of yttrium oxide (PDF-76-151) and yttrium oxy-fluoride (PDF-71-2100); (b) XRD pattern of orthorhombic YF_3 (PDF-1-70-1935).

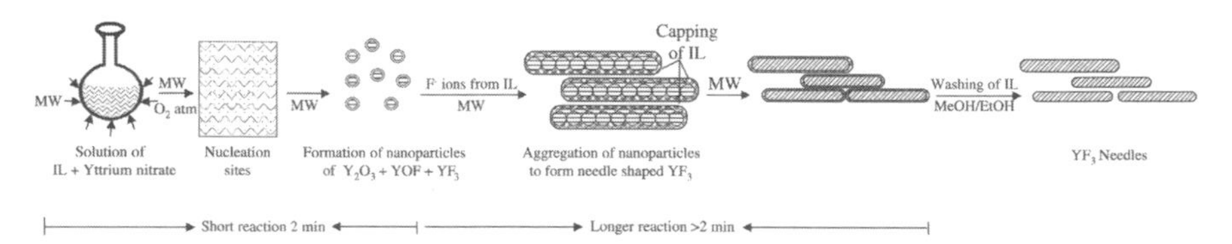

Scheme 4.3 The reaction products and their morphologies obtained under microwave radiation with respect to time for the yttrium nitrate system.

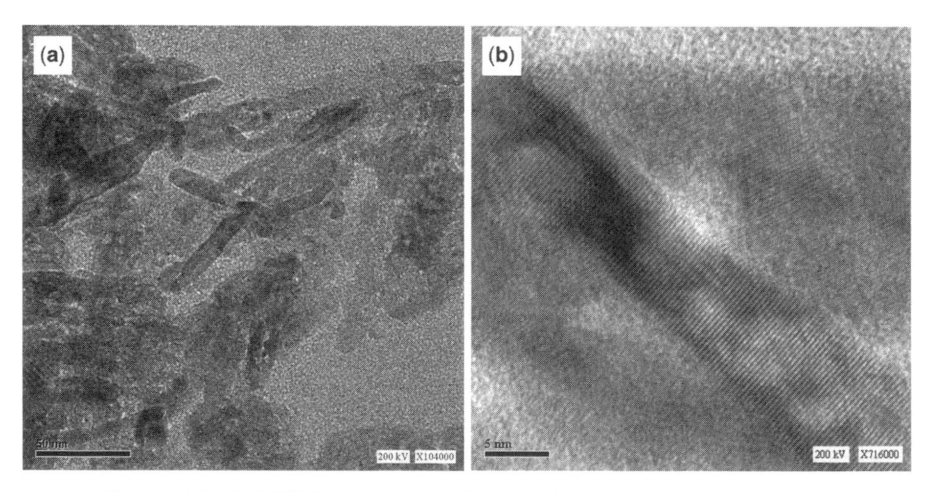

Figure 4.6 HRTEM images of needle-shaped structure YF_3 nanoparticles.

mixture of products obtained in a shorter time is not shown. The HRTEM images show the needle-shaped structure of the YF_3. The length of the nanoneedles is about 100 to 150 nm and the width is about 5 to 10 nm, as calculated from the HRTEM images. The specific growth along the c direction is explained as a result of the ionic liquid binding to the particular planes of the crystal and inhibiting their growth, whereas at the unbinding site, the crystal grows preferentially to give a one-dimensional structure.

4.6.1.3 Nano-Aggregated CoF_2 Needles In this reaction, cobalt nitrate is a precursor for the preparation of cobalt fluoride. The cobalt nitrate is dissolved in the IL solvent and then heated in a microwave oven. The reaction is monitored at different time intervals by XRD. The XRD patterns are shown in Figures 4.7a and b. In this system, we found that the products of shorter (5 min) reaction time and longer reaction time (10 min) reveal the same XRD patterns. The pattern of these two samples matches

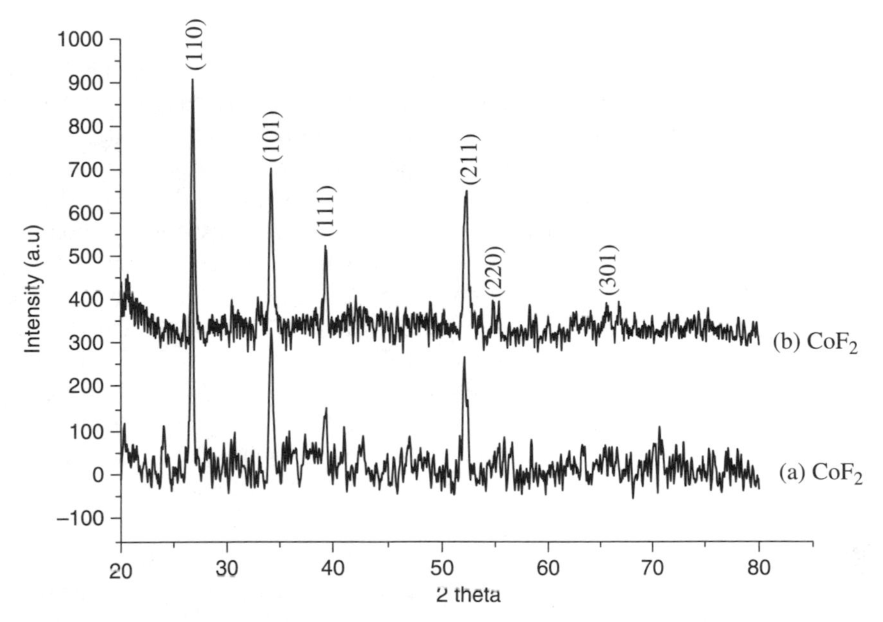

Figure 4.7 (a) Shorter (5 min) reaction product; (b) longer (10 min) reaction product of CoF_2 (PDF-1-71-653).

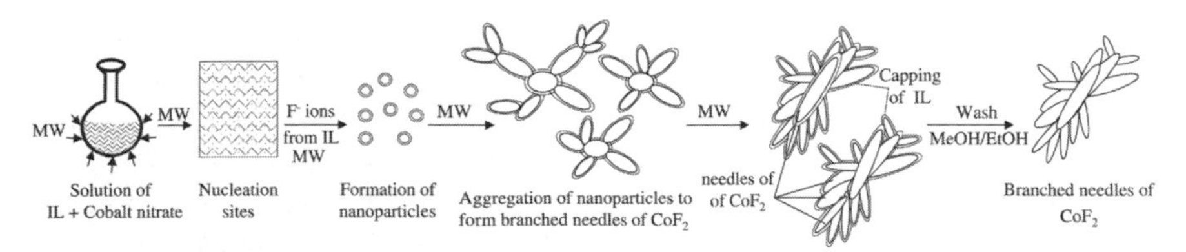

Scheme 4.4 The reaction product and its morphology obtained under microwave radiation for cobalt nitrate system.

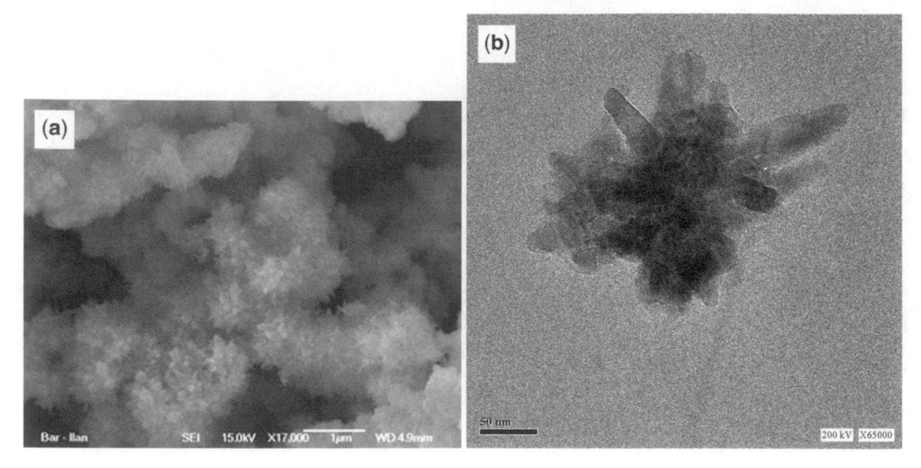

Figure 4.8 (a) HRSEM and (b) HRTEM images of the CoF_2 nanoparticles.

well with the PDF data (PDF-1-71-653) of CoF_2. In this reaction, the rate of formation of fluoride is faster, compared to the other systems where the reaction proceeds through the oxidation and then fluorination of the desired metallic ion. The formation of branched needles of CoF_2 is presented in Scheme 4.4, which also confirms that the time factor influences the structure of the product, to get longer one-dimensional structures. However, we can conclude from the XRD patterns that at shorter reaction time the product obtained is less crystalline, compared to those products obtained at longer heating periods. The intensity of the diffraction peaks obtained at longer heating times is stronger than those obtained at short reaction times. The HRSEM and HRTEM images of the CoF_2 nanoparticles are presented in Figure 4.8a and b. The HRSEM and HRTEM images show the aggregated needle structure of the CoF_2. The d spacing (3.32 Å) obtained from the HRTEM image of the CoF_2 is analogous to the d value of the (110) plane. In this reaction, the formation of aggregated nanoparticles precedes the one-dimensional preferential growth, which occurs on the nanoparticles to form aggregated needles of CoF_2.

4.6.1.4 Nano-Anisotropic ZnF_2 This reaction is very similar to the reaction of the cobalt system, as the rate of formation of the fluoride final product is very fast. Hence, we could not confirm from the XRD data whether the reaction first creates the metallic oxide, or directly forms the metallic fluoride. This is similar to what we have observed in the reaction leading to the synthesis of cobalt fluoride. Thus, a solution of zinc nitrate in IL yields the same product for short and long reaction times. Trace amounts of an unknown product can be observed in the XRD pattern of the short reaction, whereas this impurity is not detected in the longer heating reaction. The XRD pattern matches well with the PDF data (PDF-1-89-5014) of ZnF_2. The XRD pattern of the ZnF_2 is shown in Figure 4.9. The anisotropic nanostructures obtained in this reaction are shown by the HRSEM images (Fig. 4.10). We can see from the HRSEM images different morphologies such as nanoparticles, bar-, tripod-, and

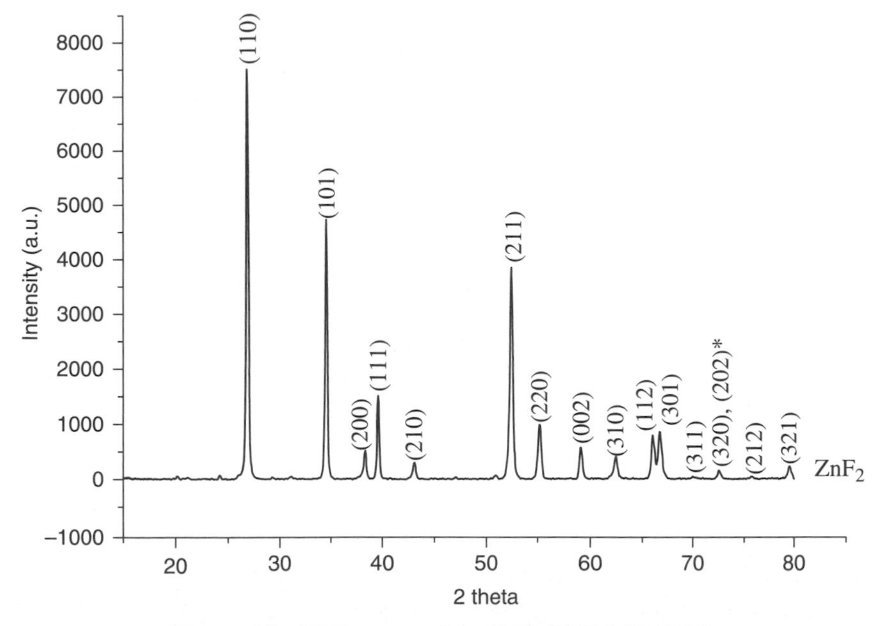

Figure 4.9 XRD pattern of the ZnF_2 (PDF-1-89-5014).

Figure 4.10 HRSEM images of the anisotropic ZnF_2 nanostructure.

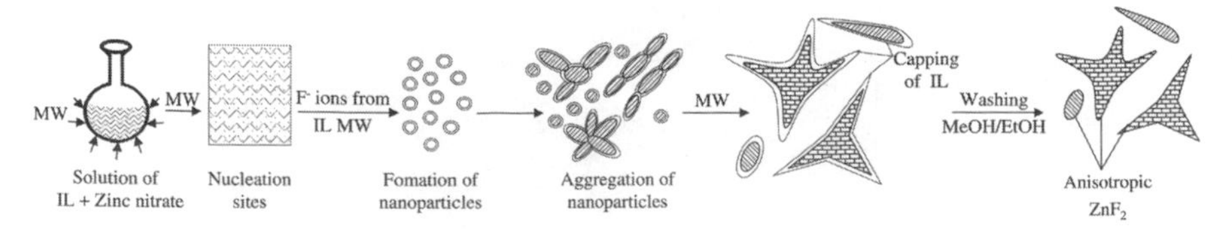

Scheme 4.5 The reaction product and its morphology obtained under microwave radiation for the zinc nitrate system.

star-shaped ZnF_2 nanostructures, which are obtained in the same reaction. In this reaction, the IL acts as a structure-directing agent to get various arrangements of the structure of the nanosized ZnF_2. The formations of different structures of ZnF2 are presented in Scheme 4.5.

4.6.1.5 Nano-Oval LaF_3 The microwave synthesis of LaF_3 from the decomposition of lanthanum nitrate in the IL is a very fast reaction. Within 5 min the reaction is completed and we obtain the desired final product, LaF_3. In this reaction, the product analyzed after 2 min is identical to the product obtained after 5 min of microwave reaction, as evident from the XRD results. The only difference is the crystallinity of the final product, which increases in the longer reaction time. The XRD pattern of LaF_3 matches well with the XRD data of PDF-32-483, shown in Figure 4.11. The HRTEM images of the final product LaF_3 are presented in Figure 4.12. It can also be confirmed from the HRTEM images that the rate of formation of LaF_3 is faster, resulting in an oval-shaped structure. We attribute this specific shape to the lack of

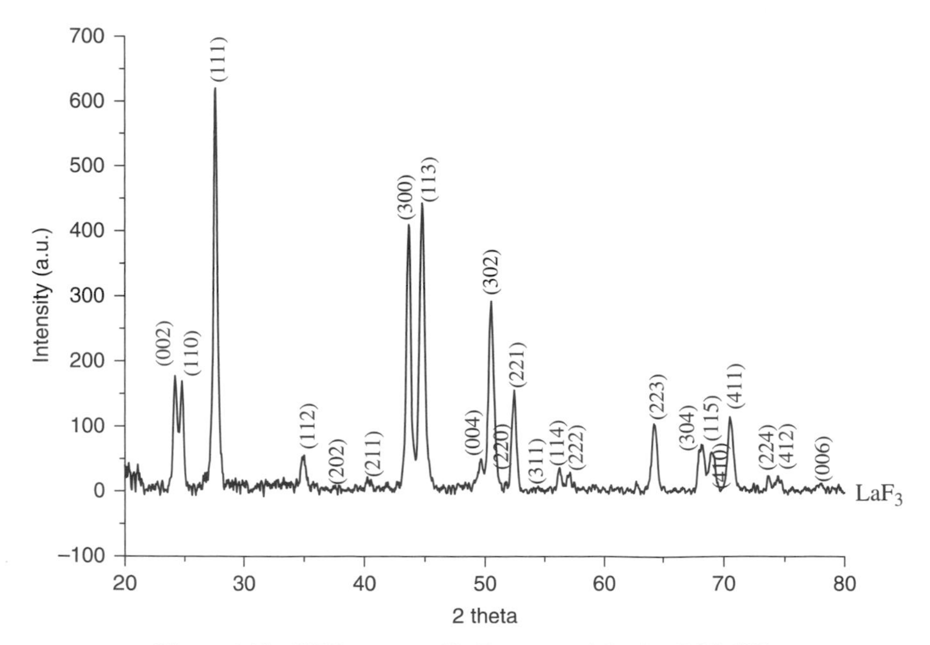

Figure 4.11 XRD pattern of LaF_3 nanoparticles (PDF-32-483).

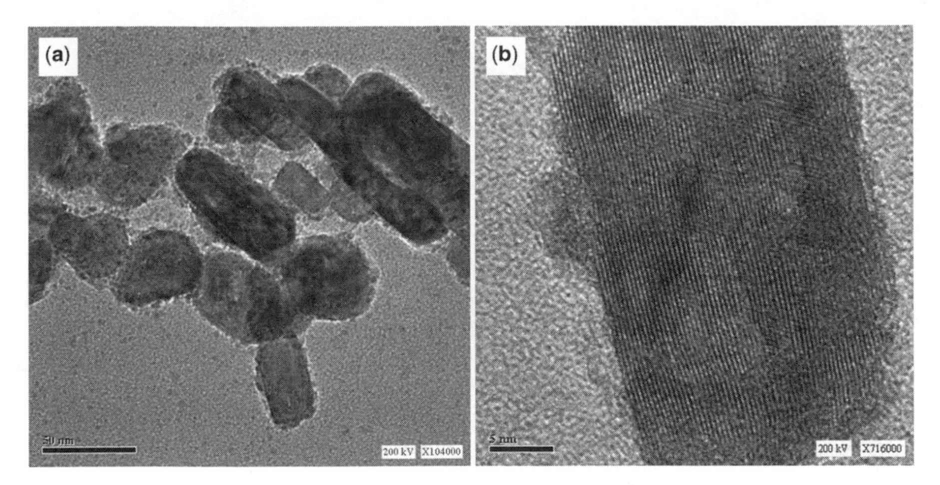

Figure 4.12 HRTEM images of LaF_3 nanoparticles.

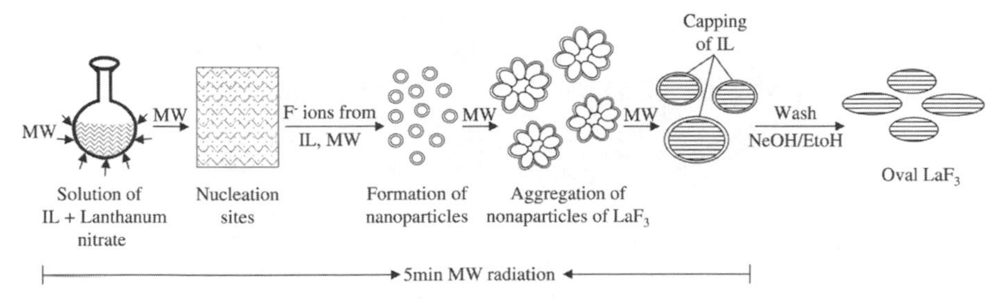

Scheme 4.6 The reaction product and its morphology obtained under microwave radiation for the lanthanum nitrate system.

time for the solvent to rearrange around the nucleation seed and direct it toward one-dimensional growth, resulting in oval-shaped particles. These particles are formed from small nanoparticles that are fused during microwave heating. This is shown in Scheme 4.6.

4.6.1.6 Nanoparticles of MgF_2 In this reaction magnesium acetate is used instead of the nitrate salt. Magnesium acetate is dissolved in the IL solvent and is heated under microwave radiation. The reaction yields insoluble magnesium fluoride. In parallel, the cation decomposes to form carbon. By optimizing the reaction time, the concentration of the precursors, and the volume of the IL (2 min, 1 g magnesium acetate, and 10 g IL, respectively) we have obtained spherical magnesium fluoride nanoparticles with an average size of 240 to 340 nm (Fig. 4.13). XRD measurements were employed to establish the chemical identity of the observed nanoparticles and in all the reactions we achieved a crystalline phase of MgF_2. The particle size calculated from the Scherrer equation ($d = k\lambda/\beta \cos\theta$) is $\sim$350 nm. This is comparable with the size of the particles measured by scanning electron (SEM) (Fig. 4.14). The formation of nanoparticles of MgF_2 is shown in Scheme 4.7.

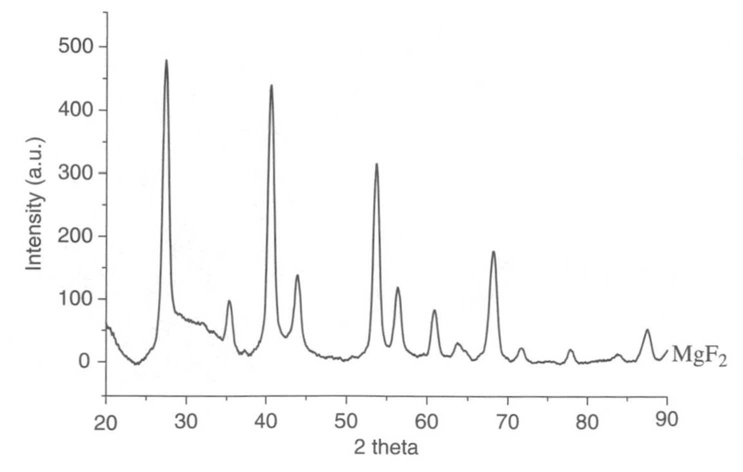

Figure 4.13 XRD pattern of MgF_2 nanoparticles (PDF-06-290, tetragonal).

Figure 4.14 SEM image of MgF_2 nanoparticles.

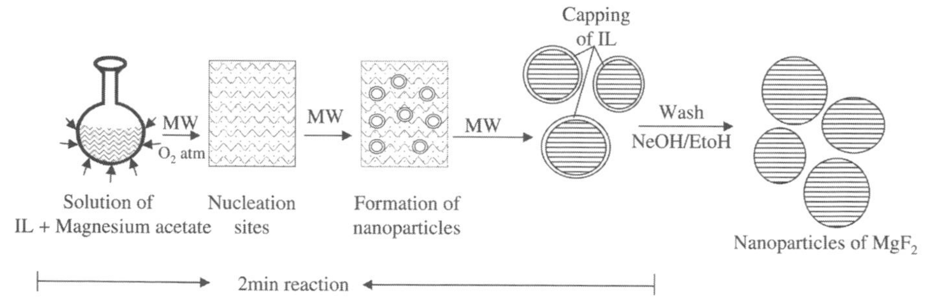

Scheme 4.7 The reaction product and its morphology obtained under microwave radiation for the magnesium acetate system.

4.6.2 Core-Shell Metal Fluoride

Most of the metal fluorides react slowly with water to produce HF (10). This process occurs at high temperatures as well as at room temperature, and is termed pyrohydrolysis. Pyrohydrolysis starts on the surface and continues in the bulk. It is also found that fluorides are characterized by strong water vapor adsorption on their surface, which leads to hydrolysis. Lattice defects and fractal structure also accelerate the process of pyrohydrolysis. To solve the above problem of hydrolysis, one of the methods employed in nanotechnology is to coat the nanoparticles with a protective layer, forming a core-shell structure. There are various types of polymeric coatings on the nanoparticles to protect them from oxidation and to form stable nanoparticles. In this chapter we describe the *in situ* formation of carbon coating on the surface of the FeF_2 during the course of microwave reaction.

Carbon-coated nanomaterials, especially metals, are of great interest due to their stable nature toward oxidation and degradation, and they show potential applications (102–108), whereas bare metal nanoparticles can be easily oxidized under ambient conditions. It is well known that a carbon shell coated on a nanomaterial provides excellent protection against air oxidation (4). There are various known methods for obtaining carbon coating on metal nanoparticles, the most popular being the arc-discharge technique (109). In general, the various methods for obtaining carbon-coated metal nanomaterials are very chaotic, and closed systems with very harsh conditions, such as high pressure, are required to attain the desired products (110, 111). In this chapter we explain a simple and quite efficient method to form a core-shell structure of FeF_2. For other metal fluorides, we have occasionally observed a carbon coating. However, it was not observed for all particles, and we cannot recommend it as a general technique for all metal fluorides.

In the example of FeF_2 the microwave reaction in an IL proceeds through the ionization of $Fe(NO_3)_3$ to give Fe^{3+} ions. The Fe^{3+} ions hydrolyze first to form the Fe_2O_3 nanoparticles as the reaction was not under inert conditions, and moreover, the solution contains the hydration water. As the reaction continues, the boiling temperature of IL is reached due to its ionic nature, as well as it being a good receptor of microwave radiation. The IL decomposed on the surface of the nanoparticles, as well as in the reaction cell. The decomposition of the cationic part of IL leads to the formation of carbon, and from the anionic decomposition we get fluoride ions. The residual carbon reduces the Fe^{3+} ions to Fe^{2+} ions at a high temperature, and the Fe^{2+} ions further react with the fluoride ions to form FeF_2, which is insoluble in the IL. The

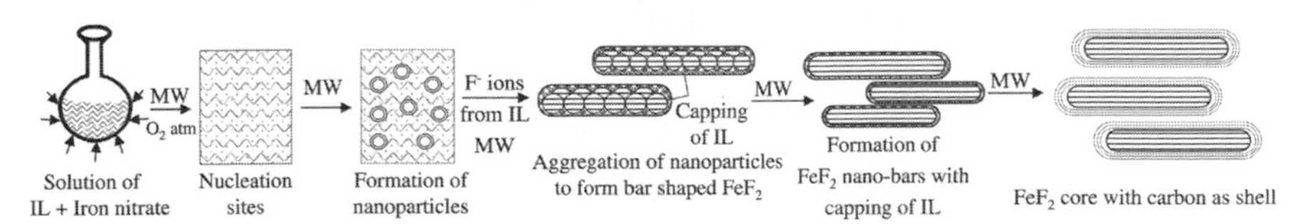

Scheme 4.8 The core-shell structure formation: step one, is the capping of the FeF_2 nanoparticles with IL and then step two, decomposition of IL on the nanoparticles to get a shell of carbon during the course of reaction under microwave radiation.

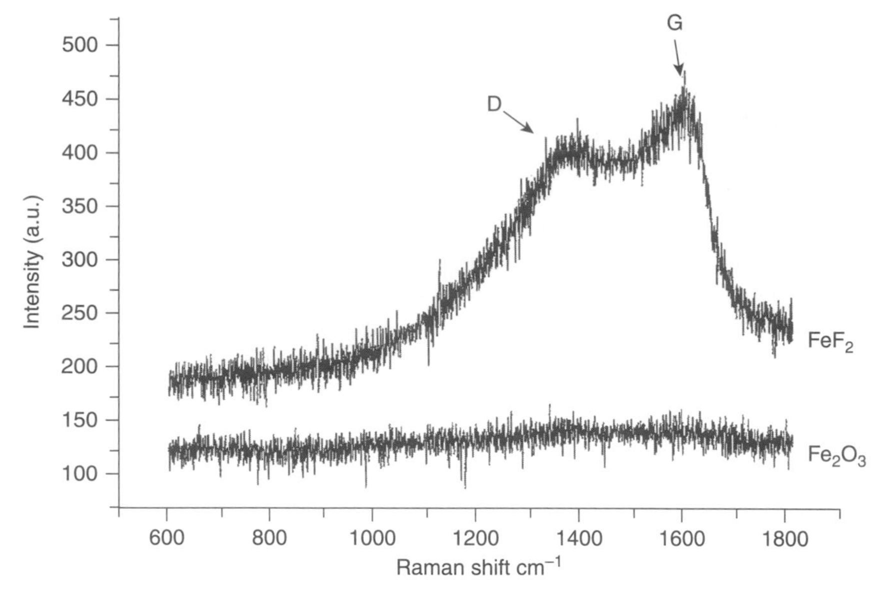

Figure 4.15 Raman spectra illustrate the carbon band for FeF_2 nanoparticles and its absence in Fe_2O_3 nanoparticles samples.

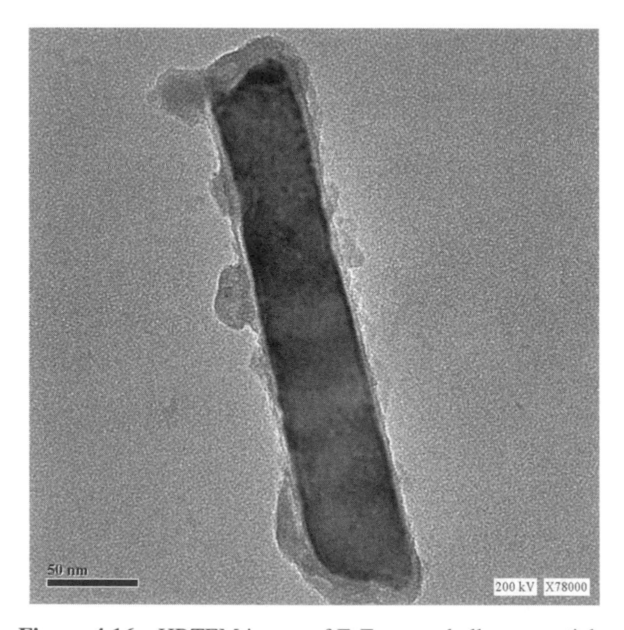

Figure 4.16 HRTEM image of FeF_2 core-shell nanoparticle.

IL is introduced in excess quantities in the reaction and acts as a structure-directing agent. It attaches to the nanoparticles on a particular plane, which inhibits the growth along this direction, whereas the unwrapped plane grows to form the bar-shaped structures of the FeF_2. When the microwave heating is continued, the IL bonded on the surface of the full-grown nanoparticles decomposes and a layer of amorphous/graphitic carbon is formed. The formation of carbon-coated FeF_2 is presented in Scheme 4.8. The formation of the carbon is confirmed using Raman analysis, showing the two characteristic bands of carbon at $1354\ cm^{-1}$ (D-band, disordered carbon) and at $1589\ cm^{-1}$ (G-band, graphitic carbon). The layered structure on the nanobar is depicted in the HRTEM image. Raman spectra of both bare Fe_2O_3 and the carbon coated FeF_2 nanobar core-shell structure are shown in Figure 4.15. The HRTEM image of carbon-coated FeF_2 is presented in Figure 4.16.

In all the microwave reactions described here, IL is used as a solvent, and depending on the thermodynamic reaction, the fast and slow formation of metal fluorides of different morphologies is obtained.

4.7 MgF_2 NANOPARTICLES AS NOVEL ANTIBIOFILM AGENTS

Fluorides are well known for their antimicrobial activity (112). This activity is mediated via three major mechanisms: (1) the formation of metal fluoride complexes, especially with aluminum and beryllium cations, which intimately interact with F-ATPase and nitrogenase enzyme and inhibit their activity; (2) formation of HF, which disturbs the proton movement through the cell membrane and finally (3) F^- or HF directly bind and inhibit specific enzymes in the cell (112). For example, enolase (an important enzyme in glycolysis) is inhibited by a complex of F^- and Mg^{2+} at micromolar concentrations in low pH (112). To date the antimicrobial and antibiofilm activity of magnesium fluoride nanoparticles have not been characterized. We utilized our simple and fast method to synthesize magnesium fluoride nanocrystalline particles (MgF_2 NPs) and examined their ability to inhibit bacterial biofilm development. Since the antimicrobial assays were carried out in bacteriological media we further characterized the stability of the nanoparticles in water, tryptic soy broth (TSB), and TSB + glucose (TSB-Glu) growth media. The MgF_2 NPs were very stable ($Ksp = 5.16 \times 10^{-11}$) in all aqueous solutions. We first characterized the antimicrobial activity of the MgF_2 nanoparticles using a 96-well growth assay. The assay was done with two common biofilm-forming pathogens, *S. aureus* (Gram positive) and *E. coli* (Gram negative). Different concentrations of MgF_2 NPs were added to 10^5 bacteria/ml solution in either TSB growth media (for *E. coli*) or TSB-Glu (for *S. aureus*). The antimicrobial activity (i.e. inhibition of bacterial growth) was determined by measuring the absorbance (OD_{595}) after 18 h of incubation at 37°C. The tested concentrations of MgF_2 nanoparticles were calculated based on $Ksp_{(MgF2)} = 5.16 \times 10^{-11}$ in order to achieve working concentrations of F^- that range from nanomolar to micromolar (at a pH $\sim$ 7). MgF_2 only slightly inhibited the growth of *E. coli* (from 0.65 to 0.4 OD_{595}) at the highest tested concentration (12 mg/ml), while the growth of

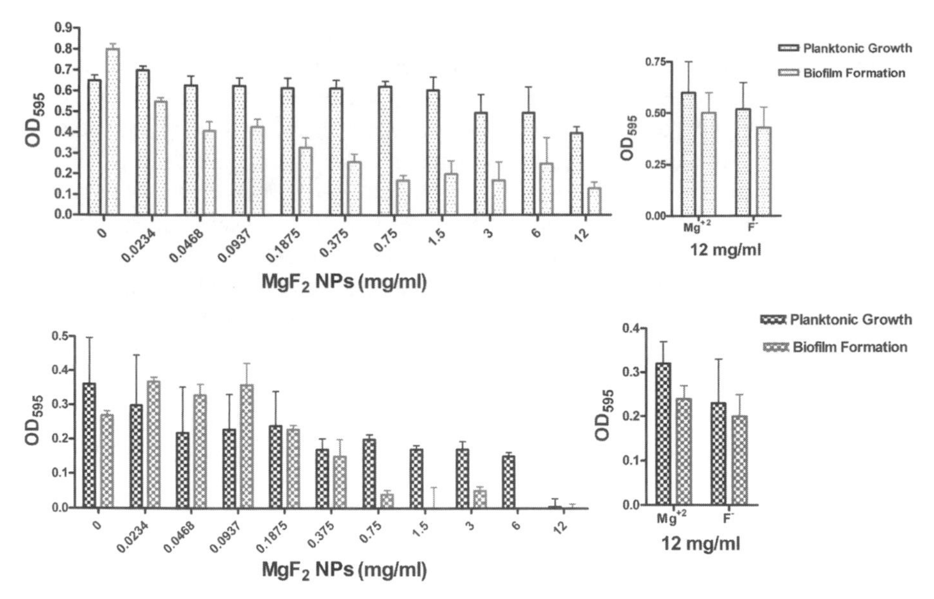

Figure 4.17 Antibacterial and antibiofilm activity of MgF_2 NPs in solution. (Top) *E. coli* and (Bottom) *S. aureus*. Concentrations of nanoparticles in the bacterial broth medium range from 12 mg/ml (which contains approx. 1.7 μg/ml of F^-) to 23.4 μg/ml (which contains approx. 3.4 ng/ml of F^-). Mg^{2+}- and F^--treated controls are also presented at the highest tested concentration.

S. aureus was completely inhibited at a concentration of 6 mg/ml (Fig. 4.17). These results show that *E. coli* is fairly resistant to the antimicrobial activity of the particles compared to the *S. aureus* strain.

We also examined the ability of free MgF_2 nanoparticles to block biofilm formation. We utilized a 96-well static biofilm assay and crystal violet staining (113) to determine the amount of biofilm biomass of *E. coli* and *S. aureus* cultures grown with and without MgF_2 nanoparticles. *E. coli* biofilm formation decreased by almost 90%, *S. aureus* biofilm development was completely inhibited even at concentrations that did not completely inhibit growth. These results prove a high biofilm specific activity (i.e. a modest anti-planktonic growth activity vs. a high antibiofilm activity) of the MgF_2 particles against *E. coli* and *S. aureus* strains (Fig. 4.17).

4.7.1 MgF_2-Coated Glass Surfaces Inhibit Bacterial Colonization

Based on the results obtained with free MgF_2 nanoparticles we further examined the ability of MgF_2-coated surfaces to restrict bacterial colonization and biofilm formation. Glass surfaces were coated by adding coupons cut from standard microscope glass slide directly into the chemical reaction. Stable coating was formed by the high microwave energy irradiation. To remove all organic components we washed the coated glasses; first with methanol to dissolve IL residues and then with double distilled water (DDW) to remove the methanol solvent. The coated glass coupons were then incubated for 18 h in bacterial suspension of *E. coli* or *S. aureus*

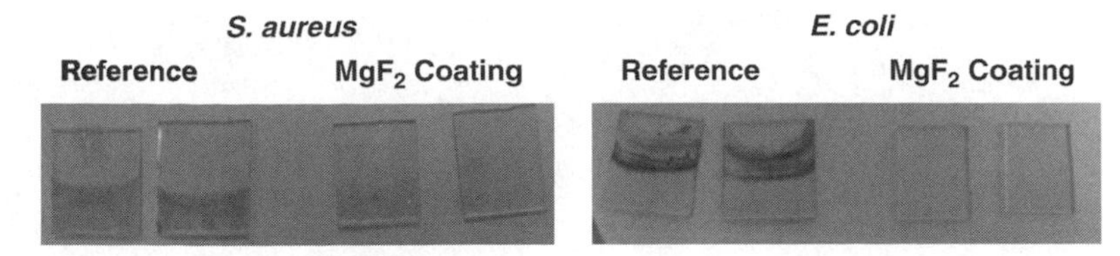

Figure 4.18 Photos of MgF_2 NPs coated and noncoated glass coupons after overnight incubation with *S. aureus* and *E. coli*. Biofilm biomass was observed by staining with crystal violet. High biomass (i.e. blue staining) is observed mainly in the air–liquid interface of the untreated glass surface. This is not seen in the MgF_2-coated glass coupons. (See color insert.)

(i.e. 1×10^5 cells/ml). After incubation for 18 h, the coated slides were removed from the bacterial suspension and washed several times with DDW to remove unattached cells. The biofilm biomass was quantified using crystal violet staining to evaluate the biofilm biomass. This procedure revealed that uncoated glass supported massive biofilm formation while the coated glass restricted bacterial colonization (Fig. 4.18). When the biofilms were stained with rhodamine to allow fluorescent microscope visualization, the air-liquid interface of the untreated glass revealed a large number of biofilm colonies while the coated surface had no observed biofilm (i.e. no fluorescence was detected). Light imaging showed the coated surfaces contained particles (i.e. MgF_2 aggregates) and no bacterial cells (Fig. 4.19).

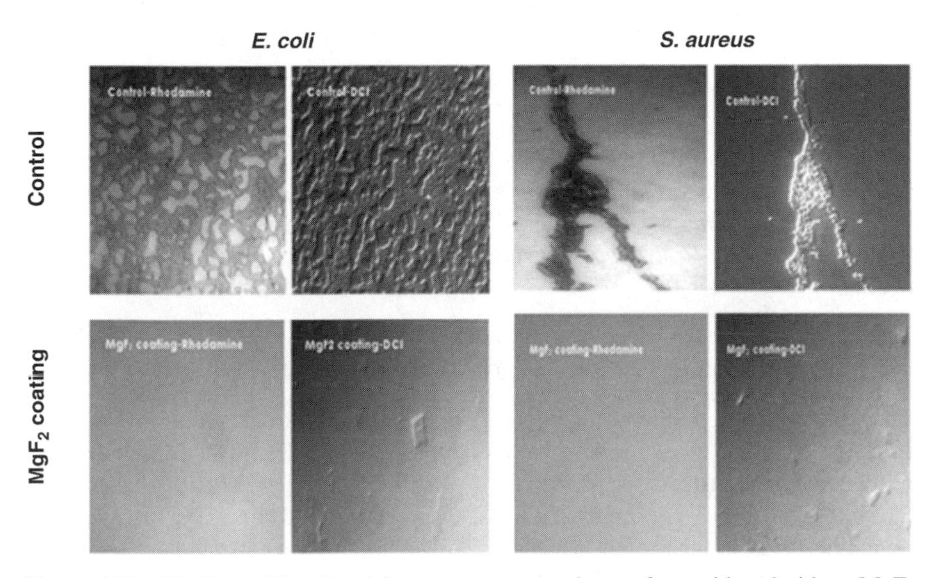

Figure 4.19 Biofilms of *E. coli* and *S. aureus* grown on glass surfaces with and without MgF_2 coatings. Biofilms were grown for 18 h, the cells were stained with rhodamine (100 μM) to allow fluorescent microscope observation, and images were acquired by Zeiss AxioImager Z1 microscope with Apotom apparatus. Biofilm formation was evident on the untreated glass surface in both *E. coli* and *S. aureus*. MgF_2 coating blocked bacterial colonization. Light microscope imaging revealed residues of the synthesis of the nanoparticles and MgF_2 aggregates on the surface can be observed.

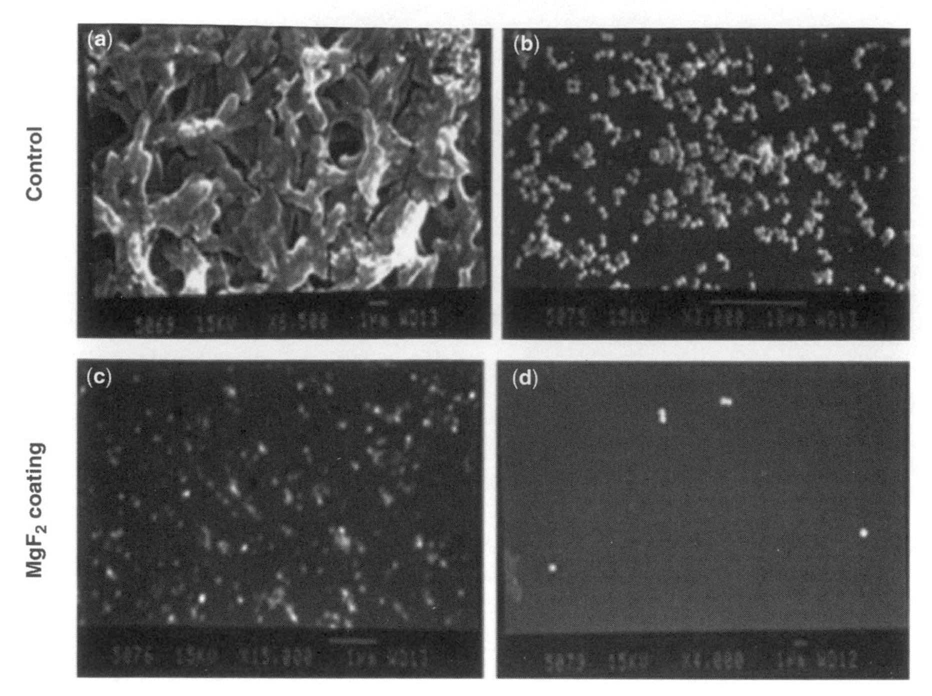

Figure 4.20 SEM images of MgF_2-coated glass after incubation with *E. coli* and *S. aureus*. Biofilm formation on uncoated glass surfaces after 18 h incubation with (a) *E. coli* and (b) *S. aureus*.

We also utilized scanning electron microscopy to evaluate the morphology of *E. coli* and *S. aureus* biofilms and the effect the coated surfaces have on biofilm formation. As presented in Figure 4.20, dense bacterial colonization can be seen in the untreated glass controls while no biofilm formation is observed ion coated surfaces. These results suggest that MgF_2 nanoparticles are extremely effective antibiofilm agents.

4.7.2 MgF_2 Coating Provides Long-Lasting Antibiofilm Activity

The antibacterial and antibiofilm properties of the coated surfaces were also tested over time. MgF_2-coated glass pieces were inoculated daily with a freshly grown bacteria suspension for up to 3 days. At the end of each inoculation/incubation cycle (24 h), the TSB/TSB-Glu was inspected for bacterial growth and the glass surfaces were transferred to a fresh bacterial suspension. Several slides were taken out of the experiment and stained with crystal violet to evaluate biofilm biomass that had formed thus far. This cycle was repeated for 3 days. The protocol allowed us to unambiguously test the ability of the MgF_2 coating to generate sufficient antibiofilm activity over time without getting depleted. As can be seen in Figures 4.21 and 4.22, the coated surfaces are able to restrict *S. aureus* biofilm formation throughout the entire 3 days (even when

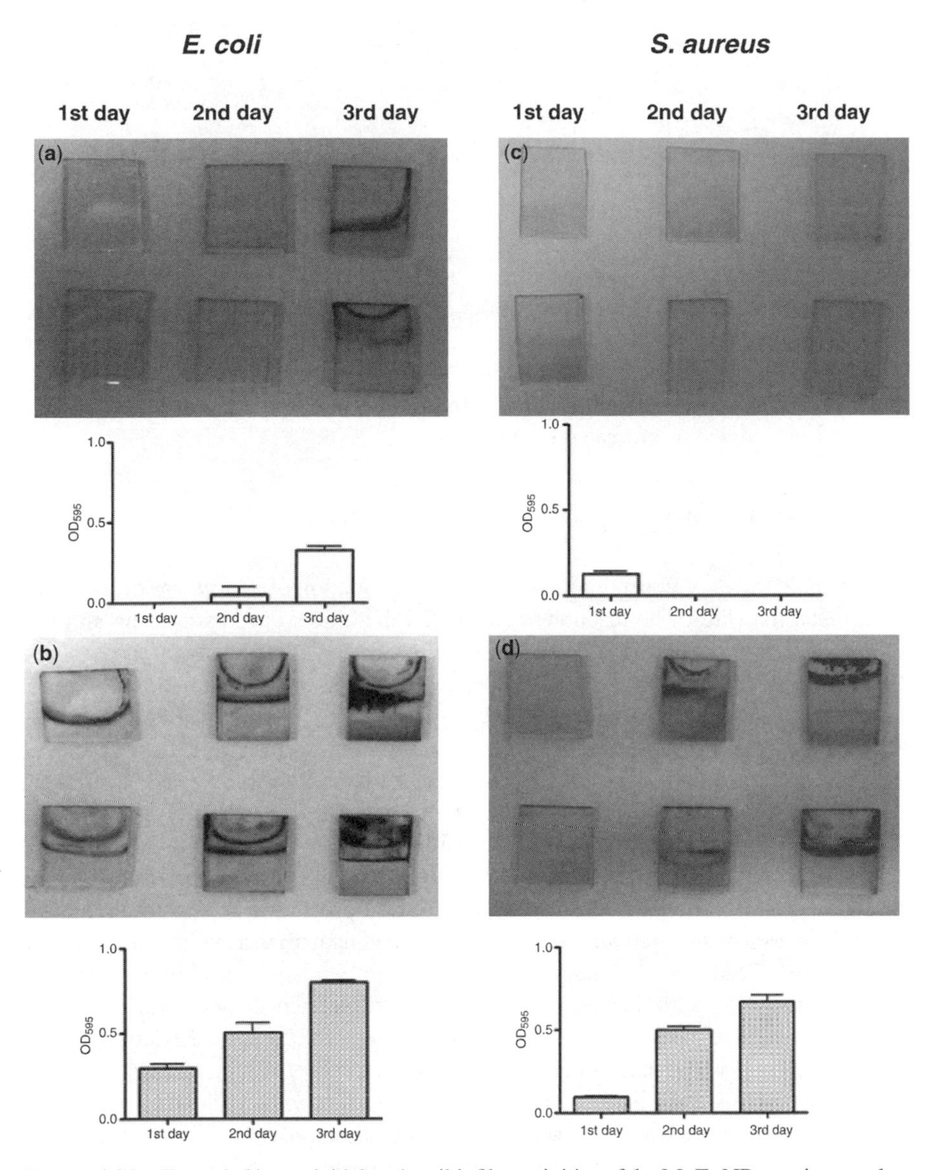

Figure 4.21 Extended bactericidal and antibiofilm activities of the MgF_2 NPs coating on glass toward *E. coli* and *S. aureus* strains. Crystal violet biomass staining of the biofilms formed after 24 h, 48 h, and 72 h on *E. coli* and *S. aureus* strains on MgF_2-coated surface (a) and (c) compared with control biofilm formation (b) and (d). (See color insert.)

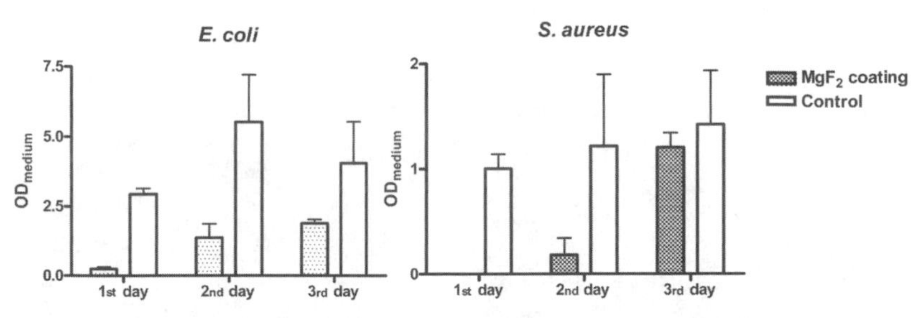

Figure 4.22 The slow release of MgF_2 NPs inn the broth media and the high killing effect on the planktonic population.

growth is not inhibited). *E. coli* biofilm formation is inhibited until day 2 and reduced by $\sim$75% compared to untreated control by day 3.

4.8 CONCLUSIONS

This chapter presents an original reaction for the formation of a large variety of nanosized metal fluorides. The reaction is conducted in a microwave oven using an ionic liquid as a solvent and as the source of the fluoride ion. Different morphologies of the products were presented. In addition, we have described a novel antibacterial and antibiofilm activity of nanosized MgF_2 particles. A relatively simple synthesis protocol was used to fabricate the material and to coat standard glass surfaces. The kill rates of the nanosized materials were better compared to that of magnesium and fluoride ions and prove the importance of the nanosized form of these materials. The MgF_2 nanoparticles coating was capable of killing both Gram-negative and Gram-positive bacteria in solution and attached to an abiotic surface. Additionally, the MgF_2 coating provided long-lasting antibiofilm protection for up to three days. A major question that remains open is the mechanism of action of the particles. Christian and Warburg (114) found that in addition F^-, Mg^{2+}, and phosphate were also important for bacterial inhibition and they formulated a model for predicting the antimicrobial activity based on the concentrations of these three key components. The model suggested that fluoride binds directly in the F^-/HF form to enzymes and regulatory proteins and disrupts their activity. Interestingly the nanosized MgF_2 particles exhibit enhanced antibacterial and antibiofilm activity. We speculate that the spherical nanosized form of the particles increase bacteria-nanoparticle interactions. This results in a high local concentration of MgF_2 NPs, which is toxic to the cells. Our preliminary results with SEM imaging of bacterial cells exposed to the MgF_2 suggest the particles may be causing membrane damage to the cells (data not shown). Future work will be required to further elucidate the mechanism of action and evaluate the ability of the MgF_2 particles to block bacterial colonization of other biofilm-forming bacteria. In addition variations in the coated polymer, size reduction of the particles, and the use of different metal such as AlF_3 and GaF_3 should also be examined. We believe this study sets the stage for future applications where nanoparticles can be utilized to coat surfaces and block bacterial infection.

REFERENCES

1. Li Y, Peng Q, Wang X, Zhuang J. **2005**. A general strategy for nanocrystal synthesis. *Nature 437*:121–124.
2. Alivisatos AP, Dittmer JJ, Huynh WU. **2002**. Hybrid nanorod-polymer solar cells. *Science 295*:2425–2427.
3. Cui Y, Duan X, Huang Y, Lieber CM, Wang J. **2001**. Indium phosphide nanowires as building blocks for nanoscale electronic and optoelectronic devices. *Nature 409*:66–69.
4. Iijima S. **1991**. Helical microtubules of graphitic carbon. *Nature 354*:56–58.
5. Alivisatos AP, Manna L, Scher EC. **2003**. Shape control and applications of nanocrystals. *Philos Trans R Soc A 361*:241–255.
6. Manoharan PT, Ramasamy S, Thangadurai P. **2004**. Pb-207 MAS NMR and conductivity identified anomalous phase transition in nanostructured PbF_2. *Eur Phys J B 37*:425–432.
7. Sakka S. 2005. *Handbook of Sol-Gel Science and Technology*. Boston: Kluwer.
8. Grob U, Kemnitz E, Rudiger S. **2007**. Non-aqueous sol-gel synthesis of nano-structured metal fluorides. *J Fluorine Chem 128*:353–368.
9. Berry AD, Ling LJ, Purday AP. **1992**. Sodium fluoride thin films by chemical vapor deposition. *Thin Solid Films 209*:9–16.
10. Fedorov PP, Kuznetsov SV, Osiko VV, Tkatchenko EA. **2006**. Inorganic nanofluorides and nanocomposites based on them. *Uspekhi Khim. 75*:1193–1211.
11. Grob U, Kemnitz E, Krishna Murthy J, Rudiger S, Winfield JM. **2006**. Sol-gel-fluorination synthesis of amorphous magnesium fluoride. *J Solid State Chem 179*: 739–746.
12. Becerra R, Bowers A, Denot E, Ibáñez LC. **1960**. Steroids. CXXXIX. New fluorination procedures. Part I. The addition of Br-F and I-F to cyclohexene and a variety of unsaturated steroids. *J Am Chem Soc 82*:4001–4007.
13. Kerebe I, Nojima M, Olah GA, Olah JA, Vankar YD, Welch JT. **1979**. Synthetic methods and reactions. 63. Pyridinium poly(hydrogen fluoride) (30% pyridine–70% hydrogen fluoride): A convenient reagent for organic fluorination reactions. *J Org Chem 44: 3872–3881.*
14. Chi DY, Katzenellenbogen JA, Kilbourn MR, Welch MJ. **1987**. A rapid and efficient method for the fluoroalkylation of amines and amides. Development of a method suitable for incorporation of the short-lived positron emitting radionuclide fluorine-18. *J Org Chem 52:658–664.*
15. Alvernhe G, Haufe G, Lawrent A. **1987**. Triethylamine tris-hydrofluoride [(C2H5)3 N.3HF]: a highly versatile source of fluoride-ion for the halofluorination of alkenes. *Synthesis 6*:562–564.
16. Sekiya A, Shibakami M, Tamura M. **1995**. Potassium fluoride-poly(hydrogen fluoride) salts as fluorinating agents for haofluorination of alkenes. *Synthesis 5*:515–517.
17. Li XY, Olah GA, Prakash GKS, Wang Q. **1993**. Poly-4-vinylpyridinium poly(hydrogen fluoride): A solid hydrogen fluoride equivalent reagent. *Synthesis 7*:693–699.
18. Bucsi I, Marco AI, Olah GA, Prakash GKS, Rasul G, Török B. **2002**. Stable dialkyl ether/poly(hydrogen fluoride) complexes: dimethyl ether/poly(hydrogen fluoride), a new, convenient, and effective fluorinating agent. *J Am Chem Soc 124*:7728–7736.

19. Hagiwara R, Ito Y, Matsubara S, Matsumoto K, Oshima K, Yoshino H. **1993**. Fluorination with ionic liquid EMIMF(HF)(2.3) as mild HF source. *J Fluorine Chem 127*:29–35.
20. Pozar DM. 1993. *Microwave Engineering*. Wokingham, UK: Addison-Wesley.
21. Diaz-Ortiz A, Hoz A, Moreno A. **2005**. Microwaves in organic synthesis. Thermal and non-thermal microwave effects. *Chem Soc Rev. 34*:164–178.
22. Gaillard P, Stuerga D. **1996**. Microwave a thermal effects in chemistry: A myth's autopsy. Part I: Historical background and fundamentals of wave-matter interaction. *J Microwave Power Electromag Ener 31*:87–99.
23. Gaillard P, Stuerga D. **1996**. Microwave athermal effects in Chemistry: A myth's autopsy Part II: Orienting effects and thermodynamic consequences of electric field. *J Micro P Electromag Ener 31*:101–113.
24. Kidwai M. **2001**. Dry media reactions. *Pure Appl Chem. 73:147–151.*
25. Strauss C, Trainor R. **1995**. Developments in microwave-assisted organic chemistry. *Aust J Chem 48*:1665–1692.
26. Baudequin C, Baudoux J, Cahard D, Gaumont A, Levillain J, Plaquevent J. **2003**. Ionic liquids and chirality: Opportunities and challenges. *Tetrahedron: Asymmetry 14*: 3081–3093.
27. Beckmann EJ, Blanchard LA, Brennecke JF, Hancu D. **1999**. Green processing using ionic liquids and CO_2. *Nature 399*:28–29.
28. Bon SAF, Carmichael AJ, Haddleton DM, Seddon KR. **2000**. Copper(I) mediated living radical polymerisation in an ionic liquid. *Chem Commun 14*:1237–1238.
29. Borissenko N, El Abedin SZ, Endres F. **2004**. Electrodeposition of nanoscale silicon in a room temperature ionic liquid. *Electrochem Commun 6*:510–514.
30. Broker GA, Farina LM, Holbrey JD, Reichert WM, Rogers RD, Swatloski RP, Visser AE. **2002**. On the solubilization of water with ethanol in hydrophobic hexafluorophosphate ionic liquids. *Green Chem 4*:81–87.
31. Bukowski M, Endres F, Hempelmann H, Natter H. **2003**. Electrodeposition of nanocrystalline metals and alloys from ionic liquids. *Angew Chem Int Ed 42*:3428–3430.
32. Chan TH, Law MC, Wong KY. **2004**. Organometallic reactions in ionic liquids. Alkylation of aldehydes with diethylzinc. *Green Chem 6*:241–244.
33. Deev A, Klamt A, Marsh K, Tran E, Wu A. **2002**. Room temperature ionic liquids as replacements for conventional solvents: A review. *Korean J Chem Eng 19*:357–362.
34. Du JM, Gao HX, Han BX, Jiang T, Liu ZM, Wang Y, Zhang JL. **2004**. Aqueous/ionic liquid interfacial polymerization for preparing polyaniline nanoparticles. *Polymer 45*:3017–3019.
35. Dupont J, Fichtner PEP, Fonseca GS, Teixeira S, Umpierre AP. **2002**. Transition-metal nanoparticles in imidazolium ionic liquids: Recycable catalysts for biphasic hydrogenation reactions. *J Am Chem Soc 124*:4228–4229.
36. Dupont J, Suarez P, Souza R. **2002**. Ionic liquid (molten salt) phase organometallic catalysis. *Chem Rev 102*:3667–3691.
37. Hardacre C, Holbreg DJ, Katdare SP, Seddon KR. **2002**. Alternating copolymerisation of styrene and carbon monoxide in ionic liquids. *Green Chem 4*:143–146.
38. Holbrey JD, Rogers RD, Spear SK, Swatloski RP. **2002**. Dissolution of cellose with ionic liquids. *J Am Chem Soc 124*:4974–4975.

39. Keim W, Wasserscheid P. **2000**. Ionic liquids: New "solutions" for transition metal catalysis. *Angew Chem Int Ed 39*:3772–3789.
40. Kon Y, Min E, Wu M, Zhao D. **2002**. Ionic liquids: Applications in catalysis. *Catal Today 74*:157–189.
41. Maderia Lau R, Rantwijk F, Seddon KR, Sheldon RA, Sorgedrager MJ. **2002**. Biocatalysis in ionic liquids. *Green Chem 4*:147–151.
42. Welton T. **1999**. Room-temperature ionic liquids. Solvents for synthesis and catalysis. *Chem Rev. 99*:2071–2083.
43. Wilkes JS. **2004**. Properties of ionic liquid solvents for catalysis. *J Mol Catal A Chem 214*:11–17.
44. Xu L, Chen W, Xiao J. **2000**. Heck reaction in ionic liquids and the in situ identification of n-hetrocyclic carbene complexes of palladium. *Organometallics 19*:1123–1127.
45. Adam D. **2003**. Microwave chemistry: Out of the kitchen. *Nature 421*:571–572.
46. Kim D, Lee JK, Lee S, Song CE. **2002**. Microwave-assisted Kabachnik–Fields reaction in ionic liquid. *Bull Korean Chem Soc 23*:667–668.
47. Oliver Kappe C. **2004**. Controlled microwave heating in modern organic synthesis. *Angew Chem Int Ed 43*:6250–6284.
48. Chum HL, Koch VR, Miller LL, Osteryoung RA. **1975**. Electrochemical scrutiny of organometallic iron complexes and hexamethylbenzene in a room-temperature molten-salt. *J Am Chem Soc 97*:3264–3265.
49. Walden P. **1914**. Molecular weights and electrical conductivity of several fused salts. *Bull Acad Imper Sci* (St. Petersburg) 405–422.
50. Sugden S, Wilkins H. **1929**. CLVXIL the preacher and chemical constitution. Part VII: Fused metals and salts. *J Chem Soc 29*:1291–1298.
51. Broker GA, Huddleston JG, Reichert WM, Roger RD, Visser AE, Willauer HD. **2001**. Characterization and comparison of hydrophilic and hydrophobic room temperature ionic liquids incorporating the imidazolium cation. *Green Chem 3*:156–164.
52. Chan BKM, Chang NH, Grimmett RM. **1977**. The synthesis and thermolysis of imidazole. quaternary salts. *Aust J Chem 30*:2005–2013.
53. Dyson PJ, Grossel MC, Srinivasan N, Vine T, White AJP, Williams DJ, Welton T, Zigras TJ. **1997**. Organometallic synthesis in ambient temperature chloroaluminate(III) ionic liquids. Ligand exchange reactions of ferrocene. *J Chem Soc Dalton Trans 19*: 3465–3469.
54. Hurley FH, Weir TPJ. **1951**. Electrodeposition of metals from fused quaternary ammonium salts. *J Electrochem Soc 98*:203–206.
55. Hussey CL, Levisky JA, Wilkes JS, Wilson RA. **1982**. Dialkylimidazolium chloroaluminate melts: A new class of room-temperature ionic liquids for electrochemistry, spectroscopy and synthesis. *Inorg Chem 21*:1263–1264.
56. Wilkes JS, Zaworotko MJJ. **1992**. Air and water stable 1-ethyl-3-methylimidazolium based ionic liquids. *J Chem Soc Chem Commun 13*:965–967.
57. Handy ST. **2005**. Room temperature ionic liquids: Different classes and physical properties. *Current Org Chem 9*:959–988.
58. Freemantle M. **1998**. Designer solvents: Ionic liquids may boost clean technology development. *Chem Eng News 76*:32–37.

59. Hagiwara R, Ito Y. **2000**. Room temperature ionic liquids of alkylimidazolium cations and fluoroanions. *Fluorine Chem 105*:221–227.
60. Magna L, Olivier-Bourbigou H. **2002**. Ionic liquids: Perspectives for organic and catalytic reactions. *J Mol Catal A Chem 182*:419–437.
61. Chauhan S, Chauhan SMS, Jain N, Kumar A. **2005**. Chemical and biochemical transformations in ionic liquids. *Tetrahedron 61*:1015–1060.
62. Bartsch RA, Dzyuba S. **2001**. New room-temperature ionic liquids with C-2-symmetrical imidazolium cations. *Chem Commun 16*:1466–1467.
63. Compton S, Ensor D, Ray L, Swartling D. **2000**. Preliminary investigation into modification of ionic liquids to improve extraction prarameters. *Bull Biochem Biotechnol 13*:1.
64. Holbrey JD, Seddon KR, Wareing R. **2001**. A simple colorimetric method for the quality control of 1-alkyl-3-methylimidazolium ionic liquid precursors. *Green Chem 3*:33–36.
65. Seddon KR, Stark A, Torres M. **2000**. Influence of chloride, water, and organic solvents on the physical properties of ionic liquids. *Pure Appl Chem 72*:2275–2287.
66. Dam J, Hakvoort G, Jansen J, Reedijk J. **1975**. Thermochemistry of nickel(II) imidazole complexes. *J Inorg Nucl Chem 37*:713–718.
67. De Long HC, Fox DM, Gilman JW, Trulove PC. **2005**. TGA decomposition kinetics of 1-butyl-2,3-dimethylimidazolium tetrafluoroborate and the thermal effects of contaminants. *J Chem Thermodynamics 37*:900–905.
68. Grant DM, Kuhlmann K. **1964**. Spin-spin coupling in the tetrafluoroborate ion. *J Phys Chem 68*:3208–3213.
69. Brukental I, Felner I, Gedanken A, Gottlieb HE, Jacob DS, Lavi R, Makhluf S, Nowik I, Persky R, Solovyov LA. **2005**. Sonochemical synthesis and characterization of $Ni(C_4H_6N_2)_6(PF_6)_2$, $Fe(C_4H_6N_2)_6(BF_4)_2$, and $Ni(C_4H_6N_2)6(BF_4)2$ in 1-butyl-3-methylimidazole with hexafluorophosphate and tetrafluoroborate. *Eur J Inorg Chem 13*:2669–2677.
70. Antonietti M, Kuang D, Smarsly B, Zhou Y. **2004**. Ionic liquids for the convenient synthesis of functional nanoparticles and other inorganic nanostructures. *Angew Chem Int Ed 43*:4988–4992.
71. Demberelnyamba D, Kim KS, Lee H. **2004**. Size-selective synthesis of gold and platinum nanoparticles using novel thiol-functionalized ionic liquids. *Langmuir 20*:556–560.
72. Holbrey JD, Seddon KR. **1999**. The phase behaviour of 1-alkyl-3-methylimidazolium tetrafluoroborates; ionic liquids and ionic liquid crystals. *J Chem Soc Dalton Trans 13*:2133–2139.
73. Antonietti M, Zhou Y. **2003**. A series of highly ordered, super-microporous, lamellar silicas prepared by nanocasting with Ionic liquids. *Chem Mater 16*:544–550.
74. Dietz DL, Firestone MA, Miller DJ, Seifert S, Trasobares S, Zaluzec NJ. **2005**. Ionogel-templated synthesis and organization of anisotropic gold nanoparticles. *Small 7*:754–760.
75. Bachtold A, Birk H, Fokkink LGJ, Henny M, Huber R, Kruger M, Schmid C, Schonenberger C, Staufer U, van der Zande BMI. **1997**. Template synthesis of nanowires in porous polycarbonate membranes: Electrochemistry and morphology. *J Phys Chem B 101*:5497–5499.
76. Bohmer MR, Fokkink LGJ, Schonenberger C, van der Zande BMI. **2000**. Colloidal dispersions of gold rods: Synthesis and optical properties. *Langmuir 16*:451–458.

77. Crowley TA, Erts D, Holmes JD, Lyons DM, Morris MA, Olin H, Ziegler KJ. **2003**. Synthesis of metal and metal oxide nanowire and nanotube arrays within a mesoporous silica template. *Chem Mater 15*:3518–3522.
78. Jirage K, Kang M, Martin CR, Nishizawa M. **2001**. Investigations of the transport properties of gold nanotubule membranes. *J Phys Chem B 105*:1925–1934.
79. Martin CR. **1996**. Membrane-based synthesis of nanomaterials. *Chem Mater 8*: 1739–1746.
80. Alfredsson V, Andersson M, Kjellin P, Palmqvist AEC. **2002**. Macroscopic alignment of silver nanoparticles in reverse hexagonal liquid crystalline templates. *Nano Lett 2*:1403–1407.
81. Bender CM, Gao JX, Murphy CJ. **2003**. Dependence of the gold nanorod aspect ratio on the nature of the directing surfactant in aqueous solution. *Langmuir 19*:9065–9070.
82. Derre A, Faure C, Neri W. **2003**. Spontaneous formation of silver nanoparticles in multilamellar vesicles. *J Phys Chem B 107*:4738–4746.
83. Esumi K, Ghosh SK, Kundu S, Mandal M, Pal T. **2002**. UV photoactivation for size and shape controlled synthesis and coalescence of gold nanoparticles in micelles. *Langmuir 18*:7792–7797.
84. Firestone MA, Seifert S, Tiede DM. **2000**. Magnetic field-induced ordering of a polymer-grafted biomembrane-mimetic hydrogel. *J Phys Chem B 104*:2433–2438.
85. Huang LM, Mitra AP, Wang HT, Wang ZB, Yan YH, Zhao D. **2002**. Cuprite nanowires by electrodeposition from lyotropic reverse hexagonal liquid crystalline phase. *Chem Mater 14*:876–880.
86. Hu X, Qi R, Wang W, Zhu Y. **2004**. Microwave-assisted synthesis of single-crystalline tellurium nanorods and nanowires in ionic liquids. *Angew Chem Int Ed 43*:1410–1414.
87. Costerton JW, Hall-Stoodley L, Stoodley P. **2004**. Bacterial biofilms: From the natural environment to infectious diseases. *Nature Rev Microbiol 2*:95–108.
88. Kaplan HB, Kolter R, O'Toole G. **2000**. Biofilm formation as microbial development. *Annu Rev Microbiol 54*:49–79.
89. Davies D. **2003**. Understanding biofilm resistance to antibacterial agents. *Nature Rev Drug Discov 2*:114–122.
90. Costerton JW, Fux CA, Stewart PS, Stoodley P. **2005**. Survival strategies of infectious biofilms. *Trends Microbiol 13*:34–40.
91. Costerton JW, Greenberg EP, Stewart PS. **1999**. Bacterial biofilms: A common cause of persistent infections. *Science 284*:1318–1322.
92. Costerton JW, Fux CA, Hall-Stoodley L, Stoodley P. **2003**. Bacterial biofilms: A diagnostic and therapeutic challenge. *Expert Rev Anti Infect Ther 1*:667–683.
93. Parsek MR, Singh PK. **2003**. Bacterial biofilms: An emerging link to disease pathogenesis. *Annu Rev Microbiol 57*:677–701.
94. Arciola CR, Costerton JW, Montanaro L. **2005**. Biofilm in implant infections: Its production and regulation. *Int J Artif Organs 28*:1062–1068.
95. Costerton JW, Donlan RM. **2002**. Biofilms: Survival mechanisms of clinically relevant microorganisms. *Clin Microbiol Rev 15*:167–193.
96. Darouiche RO, Engl N. **2004**. Treatment of infections associated with surgical implants. *J Med 350*:1422–1429.

97. Bagge-Ravn D, Gadegaard N, Gram L, Kingshott P, Wei J. **2003**. Covalent attachment of poly(ethylene glycol) to surfaces, critical for reducing bacterial adhesion. *Langmuir* *19*:6912–6921.
98. Busscher HJ, Kaper HJ, Norde W. **2003**. Characterization of poly(ethylene oxide) brushes on glass surfaces and adhesion of *Staphylococcus epidermidis*. *J Biomater Sci Polym Ed* *14*:313–324.
99. Hetrick EM, Schoenfisch MH. **2006**. Reducing implant-related infections: Active release strategies. *Chem Soc Rev 35*:780–789.
100. Ballesteros A, Gill I. **1998**. Encapsulation of biologicals within silicate, siloxane, and hybrid sol-gel polymers: An efficient and generic approach. *J Am Chem Soc* *120*:8587–8598.
101. Guggenbichler JP, Samuel U. **2004**. Prevention of catheter-related infections: The potential of a new nano-silver impregnated catheter. *Int J Antimicrob Agents 23*:S75–S78.
102. Liu W, Wang H, Yu M, Zhang Y, Zhao T. **2006**. The physical properties of aqueous solutions of the ionic liquid [BMIM][BF_4]. *J Solution Chem 35*:1337–1346.
103. Caruso F, Davis SA, Donath E, Möhwald H, Sukhorukov GB. **1998**. Novel hollow polymer shells by colloid-templated assembly of polyelectrolytes. *Angew Chem Int Ed* *37*:2202–2205.
104. Chan B, Lorents DC, Malhotra R, Ruoff RS, Subramoney S. **1993**. Single-crystal metals encapsulated in carbon nanoparticles. *Science 259*:346–348.
105. Choi CJ, Kim BK, Kim JC, Wang ZH, Zhang ZD. **2003**. Characterization and magnetic properties of carbon-coated cobalt nanocapsules synthesized by the chemical vapor-condensation process. *Carbon 41*:1751–1758.
106. Chowdari BVR, Dong ZL, Shaju KM, Sharma N, Subba Rao GV, White TJ. **2004**. Carbon-coated nanophase $CaMoO_4$ as anode material for Li ion batteries. *Chem Mater* *16*:504–512.
107. Jung YS, Lee KT, Oh SM. **2003**. Synthesis of tin-encapsulated spherical hollow carbon for anode material in lithium secondary batteries. *J Am Chem Soc 125*:5652–5653.
108. Kirkpatrick EM, Majetich SA, McHenry ME. **1995**. Synthesis, structure, properties and magnetic applications of carbon-coated nanocrystals produced by a carbon arc. *Mater Sci Eng A 204*:19–24.
109. Fostiropoulos K, Huffman DR, Krätschmer W, Lamb LD. **1990**. Solid C-60 a new form of carbon. *Nature 347*:354–358.
110. Gedanken A, Pol SV, Pol VG. **2004**. Reactions under autogenic pressure at elevated temperature (RAPET) of various alkoxides: Formation of metals/metal oxides-carbon core-shell structures. *Eur J Chem A 10*:4467–4473.
111. Hao GM, Ling J, Liu Y, Zhang XG. **2003**. Preparation of carbon-coated Co and Ni nanocrystallites by a modified AC arc discharge method. *Mater Sci Eng B 100*:186–190.
112. Clock SA, Marquis RE, Mota-Meira M. **2003**. Fluoride and organic weak acids as modulators of microbial physiology. *FEMS Microbiol Rev 26*:493–510.
113. Kolter R, O'Toole GA. **1998**. Initiation of biofilm formation in *Pseudomonas fluorescens* WCS365 proceeds via multiple, convergent signaling pathways: A genetic analysis. *Mol Microbiol 28*:449–461.
114. Christian W, Warburg O. **1942**. Isolierung und Kristallisation des Gärungsferments Enolase. *Biochem Z 310*:384–421.

PROBLEMS

1. What are ionic liquids? How can an ionic liquid act as a surfactant? Give some examples of ionic liquids as surfactants.
2. Calculate the change in surface area when spherical particles of 500 μm size are ground to cubic 50 nm sizes? What will be the effect on the rate of reaction if we use 50 nm particles instead of 500 μm?
3. Plot the graph from the data given below. Match the pattern of the graph with the PDF data using a search match program. Calculate the particle size from the XRD plot of the given data. Give the (hkl) plane for the peaks in the XRD plot and calculate the Miller index of the (hkl) planes from the given data.
4. Why is Raman spectroscopy the best tool to detect different types of carbon?

ANSWERS

1. Ionic liquids are salts that are liquid at room temperatures. The ionic liquid consists of organic cation and inorganic anion. The polar cation contains an hydrophobic tail and hydrophilic (PF_6^-) or hydrophilic (BF_4^-) anions, ionic liquid acts as surfactant.
   Example: 1-Dodecyl-3-methylimidazolium bromide, 1-dodecyl-3-methylimidazolium hexafluorophosphate, 1-hexadecyl-3-methyl-imidazolium chloride.
2. The change in surface areas of one particle is 0.785 μm^2. The surface area is calculated by the famous formula $4\pi R^2$, where R is the radius of the particle. The rate of a reaction is affected by the physical size of the particles. Decreasing the particle size from 500 μm to 50 nm will increase the number of particles for a given weight. With small particles the rate of reaction will increase because the surface area of the particles has been increased.
3. *Hints*: (1) Use Excel for the plot; (2) match the peaks using search match program; (3) use Scherrer equation to calculate the size of the particles, $\lambda = 0.154$ nm; (4) to find Miller index, first find the intercepts, then reciprocal of the intercepts and convert it into the lowest integer.
4. Different types of carbon have various bonding state which shows strong correlation between shape and width of the vibrational signals and the structure of the amorphous arrangement and hence because of the resonance phenomena Raman spectroscopy is the best tool to detect different types of carbon. The assignment of the G and D bands of the Raman spectrum of carbon is not in debate. On the other hand the interpretation of solid state NMR is still in doubt.

Data for XRD Plot

2θ	Intensity	2θ	Intensity	2θ	Intensity	2θ	Intensity	2θ	Intensity	2θ	Intensity
25	8.3952	28.75	78.72586	39.95	112.845	43.7	126.77567	48.55	4.84244	52.3	17.23977
25.05	11.20139	28.8	77.73205	40	139.78452	43.75	132.58185	48.6	5.31529	52.35	17.71262
25.1	11.74091	28.85	76.93824	40.05	168.99071	43.8	136.52137	48.65	6.58814	52.4	20.25214
25.15	12.01376	28.9	76.01109	40.1	201.26356	43.85	138.26089	48.7	5.12766	52.45	21.59166
25.2	12.01995	28.95	74.88394	40.15	232.33642	43.9	138.93375	48.75	6.06718	52.5	22.06451
25.25	13.6928	29	76.02346	40.2	269.9426	43.95	137.93993	48.8	5.87337	52.55	23.73737
25.3	14.36565	29.05	76.29632	40.25	306.41546	44	133.07945	48.85	5.61289	52.6	24.47689
25.35	15.03851	29.1	76.96917	40.3	339.02164	44.05	128.15231	48.9	6.48574	52.65	25.48307
25.4	16.04469	29.15	76.57536	40.35	373.02783	44.1	123.75849	48.95	5.42526	52.7	27.62259
25.45	18.45088	29.2	76.51488	40.4	398.36735	44.15	116.76468	49	4.43145	52.75	32.36211
25.5	19.05707	29.25	74.98773	40.45	419.70687	44.2	112.37087	49.05	4.9043	52.8	34.50163
25.55	19.99659	29.3	75.06058	40.5	433.71306	44.25	103.57705	49.1	5.71049	52.85	41.30782
25.6	19.86944	29.35	75.06677	40.55	441.25258	44.3	91.91657	49.15	5.91668	52.9	49.51401
25.65	21.34229	29.4	75.07296	40.6	442.5921	44.35	80.38943	49.2	7.32286	52.95	58.78686
25.7	20.14848	29.45	72.47914	40.65	438.66495	44.4	70.92895	49.25	6.26238	53	69.79305
25.75	22.02133	29.5	70.28533	40.7	429.47114	44.45	62.8018	49.3	6.0019	53.05	83.2659
25.8	20.96085	29.55	71.15818	40.75	410.54399	44.5	55.54132	49.35	4.94142	53.1	101.07209
25.85	21.70037	29.6	70.83104	40.8	385.08351	45.65	17.15028	49.4	3.41428	53.15	119.67827
25.9	23.03989	29.65	71.30389	40.85	355.22303	45.7	16.22313	49.45	3.08713	53.2	144.48446
25.95	26.37941	29.7	70.37674	40.9	323.62922	45.75	17.09598	49.5	2.09332	53.25	170.49065
26	26.5856	29.75	69.58293	40.95	287.0354	45.8	15.96884	49.55	1.4995	53.3	196.9635
26.05	29.25845	29.8	68.45578	41	252.10826	45.85	14.70836	49.6	1.97236	53.35	222.03635
26.1	31.39797	29.85	67.32864	41.05	218.78111	45.9	15.78121	49.65	0.97854	53.4	245.97587
26.15	33.47083	29.9	66.20149	41.1	185.12063	45.95	15.92073	49.7	1.11806	53.45	269.24873
26.2	36.07701	29.95	66.34101	41.15	156.59348	46	15.39358	49.75	3.25758	53.5	286.38825
26.25	39.6832	30	67.21386	41.2	132.39967	46.05	14.59977	49.8	2.86377	53.55	301.72777
26.3	45.15605	37.5	15.74186	41.25	113.67252	46.1	14.87262	49.85	2.53662	53.6	309.66729

26.35	52.76224	37.55	14.61471	41.3	99.27871	46.15	13.87881	49.9	2.94281	53.65	316.60681
26.4	58.50176	37.6	13.75423	41.35	87.61823	46.2	14.085	49.95	2.41566	53.7	316.81299
26.45	67.04128	37.65	14.29375	41.4	79.29108	46.25	12.49118	50	3.35518	53.75	313.88585
26.5	73.11413	37.7	12.69994	41.45	71.36393	46.3	11.2307	50.05	2.62803	53.8	306.02537
26.55	80.78699	37.75	10.37279	41.5	64.70345	46.35	10.23689	50.1	2.76755	53.85	293.16489
26.6	91.12651	37.8	10.24564	41.55	60.30964	46.4	9.57641	50.15	2.70707	53.9	277.43774
26.65	103.13269	37.85	10.58516	41.6	58.11583	46.45	10.84926	50.2	2.37993	53.95	254.71059
26.7	117.40555	37.9	11.52468	41.65	57.05535	46.5	10.32212	50.25	4.38611	54	231.11678
26.75	136.4784	37.95	11.6642	41.7	55.1282	46.55	10.9283	50.3	4.3923	54.05	205.4563
26.8	158.55125	38	13.33706	41.75	52.46772	46.6	11.93449	50.35	5.33182	54.1	182.06249
26.85	186.95744	38.05	11.54324	41.8	50.14057	46.65	11.67401	50.4	6.13801	54.15	157.93534
26.9	219.29696	38.1	10.54943	41.85	47.81343	46.7	10.34686	50.45	6.34419	54.2	133.40819
26.95	255.43648	38.15	12.75562	41.9	47.48628	46.75	9.41972	50.5	5.48371	54.25	112.54771
27	292.84267	38.2	13.62847	41.95	45.29247	46.8	8.29257	50.55	6.0899	54.3	96.42057
27.05	334.11552	38.25	16.23466	42	44.43199	46.85	7.89876	50.6	8.09609	54.35	82.82675
27.1	367.58837	38.3	16.70751	42.05	43.50484	46.9	7.90494	50.65	9.03561	54.4	69.76627
27.15	400.52789	38.35	16.78036	42.1	43.44436	46.95	7.71113	50.7	9.10846	54.45	61.10579
27.2	427.06741	38.4	16.51988	42.15	42.11721	47	7.71732	50.75	9.04798	54.5	53.84531
27.25	451.2736	38.45	19.2594	42.2	41.25673	47.05	8.1235	50.8	8.05417	54.55	47.78483
27.3	469.81312	38.5	20.93226	42.25	40.92959	47.1	8.26302	50.85	7.52702	54.6	44.85768
27.35	477.61931	38.55	20.87178	42.3	39.80244	47.15	7.40254	50.9	7.46654	54.65	40.26387
27.4	480.75883	38.6	20.74463	42.35	40.60863	47.2	7.94206	50.95	8.00606	54.7	38.07006
27.45	475.89835	38.65	20.21748	42.4	39.14815	47.25	7.81492	51	7.21225	54.75	35.00958
27.5	466.3712	38.7	20.49034	42.45	40.28767	47.3	6.88777	51.05	7.81843	54.8	32.81576
27.55	450.44405	38.75	20.49652	42.5	39.42719	47.35	5.82729	51.1	6.69129	54.85	29.75528
27.6	430.58357	38.8	22.03604	42.55	39.96671	47.4	5.63348	51.15	7.16414	54.9	28.02814
27.65	403.38976	38.85	23.97556	42.6	41.70623	47.45	5.63966	51.2	7.83699	54.95	27.10099
27.7	370.99594	38.9	25.04842	42.65	41.04575	47.5	5.17918	51.25	7.77651	55	26.70718
27.75	335.40213	38.95	25.92127	42.7	43.05193	47.55	7.65204	51.3	7.71603	55.05	25.6467

(*Continued*)

2θ	Intensity	2θ	Intensity	2θ	Intensity	2θ	Intensity	2θ	Intensity	2θ	Intensity
27.8	296.40832	39	24.79412	42.75	44.92479	47.6	6.72489	51.35	6.85555	55.1	26.05288
27.85	260.08117	39.05	26.93364	42.8	45.79764	47.65	6.66441	51.4	6.19507	55.15	25.3924
27.9	227.35402	39.1	31.27316	42.85	45.40383	47.7	6.73726	51.45	6.40126	55.2	25.93192
27.95	196.96021	39.15	33.41268	42.9	47.61001	47.75	8.01012	51.5	6.80745	55.25	25.20478
28	170.69973	39.2	32.28554	42.95	47.6162	47.8	7.74964	51.55	7.54697	55.3	25.1443
28.05	148.63925	39.25	31.42506	43	49.08905	47.85	8.35582	51.6	8.21982	55.35	24.41715
28.1	133.24544	39.3	34.29791	43.05	51.29524	47.9	8.36201	51.65	9.29267	55.4	25.75667
28.15	119.65162	39.35	35.57076	43.1	51.36809	47.95	8.23486	51.7	9.96553	55.45	26.62952
28.2	112.19114	39.4	40.31028	43.15	53.90761	48	7.10772	51.75	10.17171	55.5	28.50238
28.25	103.864	39.45	41.51647	43.2	57.44713	48.05	7.58057	51.8	10.5779	55.55	30.77523
28.3	99.53685	39.5	43.38932	43.25	63.18665	48.1	8.12009	51.85	11.65075	55.6	32.78142
28.35	92.54304	39.55	47.26218	43.3	69.12617	48.15	7.79294	51.9	12.32361	55.65	36.5876
28.4	88.74922	39.6	50.06836	43.35	72.79903	48.2	8.8658	51.95	12.66313	55.7	41.59379
28.45	86.48874	39.65	53.40788	43.4	79.87188	48.25	8.73865	52	13.73598	55.75	47.06664
28.5	84.22826	39.7	57.48074	43.45	84.87807	48.3	7.0115	52.05	14.4755	55.8	51.87283
28.55	81.03445	39.75	63.55359	43.5	90.55092	48.35	6.35102	52.1	14.41502	55.85	55.61235
28.6	80.64064	39.8	69.95978	43.55	102.55711	48.4	5.49054	52.15	14.48787	55.9	61.35187
28.65	77.64682	39.85	79.2993	43.6	112.29663	48.45	4.1634	52.2	14.76073	55.95	68.95806
28.7	77.71968	39.9	93.10548	43.65	118.90281	48.5	4.50292	52.25	14.76691	56	78.36424

*Intensity (a.u.)

5

TRANSITION METAL NITRIDES AND CARBIDES

PIOTR KRAWIEC AND STEFAN KASKEL

5.1 Introduction, 111
  5.1.1 Crystal Structure and Composition, 112
  5.1.2 Electrical and Magnetic Properties, 114
5.2 Applications, 114
  5.2.1 Bulk Materials Applications, 114
  5.2.2 Nanosized Transition Metal Nitrides and Carbides: Advantages and Specific Applications, 116
5.3 Synthesis, 120
  5.3.1 Bulk Materials and Powders (Industrial Techniques), 120
  5.3.2 Nanomaterials: State of the Art and New Concepts, 121
5.4 Summary, 123
References, 124
Problems, 124
Answers, 125

5.1 INTRODUCTION

Due to their excellent mechanical properties and high-temperature durability, metal nitrides and carbides have many industrial and technological applications (1). While more details about applications will be given in the next chapter, here the chemical and physical properties of these interesting compounds are addressed.

Nanoscale Materials in Chemistry, Second Edition. Edited by K. J. Klabunde and R. M. Richards

In addition to transition metal compounds, the most commonly used nitrides and carbides (e.g. SiC, Si_3N_4, and AlN) are also discussed and compared.

5.1.1 Crystal Structure and Composition

Carbon and nitrogen form compounds with all transition metals, except for the late elements of the second and third row. Early groups of the transition metal compounds represent simple structures, similar to those of the pure transition metal. Other structures are more complicated and were systematized first by the Hägg, who states that the geometrical rules of atomic arrangement depend on the radius ratio:

$$r = r_x/r_M$$

where r_x is the atomic radius of a nonmetal and r_M is the atomic radius of a metal.

For r lower than 0.59 simple structures, such as face-centered cubic (fcc), hexagonal close-packed (hcp), and simple hexagonal (hex), are predicted. Nonmetal ions are placed in the interstitial sites of octahedral or cubic prismatic geometry. For r larger than 0.59 the metallic arrangement becomes distorted and more complex structures are formed (Fig. 5.1).

However, more details on the exact crystallographic structure determination are given by the analysis of their electronic properties. Engel–Brewer theory assumes that for metals the changes in structure depend on the number of valence *sp* electrons

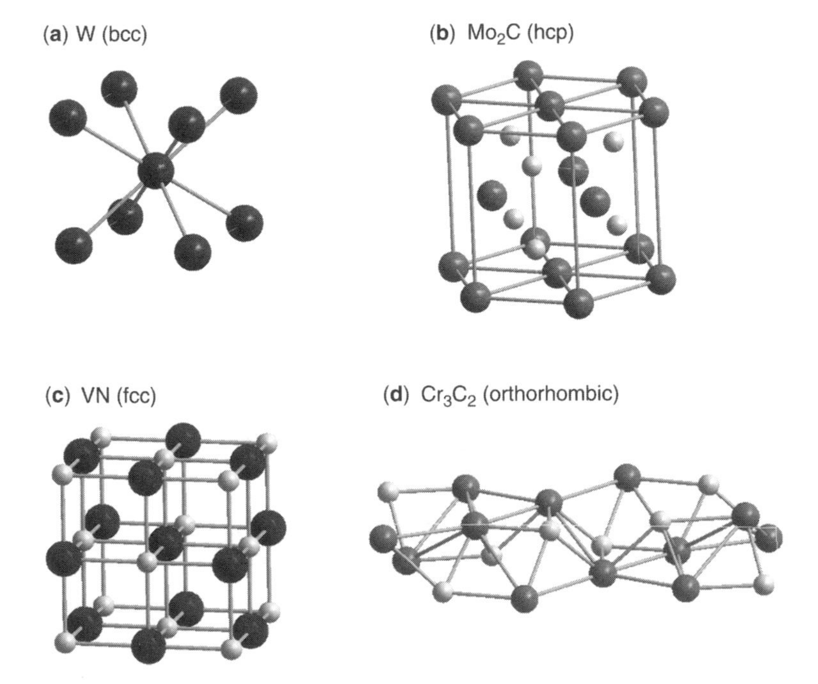

Figure 5.1 Examples of different crystal structures in metals, metal nitrides, and carbides.

TABLE 5.1 Crystal Structures of Transition Metals, and Structure Progression in Nitrides and Carbides

e/a	Structure	Metal	Nitride or Carbide
Increasing ↑	Body-centered cubic (bcc)	Mo, W	Mo, W
	Hexagonal close-packed (hcp)	Ru, Os	Mo_2C, W_2C
	Face-centered cubic (fcc)	Pd, Pt	γ-Mo_2N, β-W_2N

per atom (e/a). A similar trend can be observed for nitrides and carbides. The presence of nitrogen or carbon atoms in the metallic framework increases the concentration of *sp* electrons causing the structural transformation from bcc to hcp and fcc (Table 5.1).

Early transition metal compounds (groups 3 to 6) have MX or M_2X stoichiometry (where X = C, N), while, for the late groups stoichiometry shifts towards M_3X or M_4X. The increased metal content in carbides or nitrides with the increasing group number means also lower affinity towards carbon or nitrogen and lower stability. The high stability of the group 4 to 6 compounds is reflected in their high melting points (Table 5.2) and many of them are superior to the ceramic materials like SiC or Si_3N_4. Beside binary nitrides (MX_n) ternary ($M_IM_{II}X_n$) and quarternary ($M_IM_{II}M_{III}X_n$) nitrides and carbides are also known in the form of solid solutions where one transition metal can be randomly substituted in the lattice by the other (e.g. V_xMoN_y). Also, nitrogen or carbon can be substituted by oxygen and called, respectively, oxynitrides and oxycarbides. Finally, many carbide and nitride compounds are non-stoichiometric, showing defects in the nitrogen or carbon network.

TABLE 5.2 Physical and Chemical Properties of Common Transition Metal Nitrides and Carbides

		Melting Point (K)			Hardness (kg mm^{-2})		
Group	Element	Metal	Nitride	Carbide	Metal	Nitride	Carbide
4	Ti	1933	3220 (TiN)	2903 (TiC)	55	2100	3200
5	V	2190	2619 (VN)	3103 (VC)	55	1500	2600
	Cr	2130	2013 (Cr_2N)	2168 (Cr_3C_2)	230	1100	1300
6	Mo	2883	2223 (Mo_2N)	2793 (Mo_2C)	250	1700	1500
	W	3680	873 (WN)	3049 (WC)	360	–	2400
8	Fe	1810	943 (Fe_4N)	1923 (Fe_3C)	66	–	840
Others		4073 (diamond)	2173 (Si_3N_4)	2573 (SiC)	7600 (diamond)	1700 (Si_3N_4)	2580 (SiC)

Nitrides and carbides are also considered among the hardest materials. Table 5.2 shows data for the measured Vickers microhardness for these compounds. Such measurements are performed using a diamond indenter with square geometry. The indenter is forced towards the surface of the material and the diagonal of the micro-indentation is measured. In all cases, carbides and nitrides are significantly harder than the pure metals and are also comparable or superior to that of ceramic materials.

5.1.2 Electrical and Magnetic Properties

In contrast to the physical properties, transition metal carbides and nitrides possess electric and magnetic properties that are often similar to metals. For example, electrical resistivities of Ti or W are 39 and 5.39 $\mu\Omega$ cm at room temperature, while their respective carbides have only slightly higher resistivities of 68 and 22 $\mu\Omega$ cm. For comparison the electrical resistivity of the hard SiC ceramics is significantly higher (1000 $\mu\Omega$ cm).

Carbides and nitrides of Fe, Co, and Ni are ferromagnetic as their parent metals. The other compounds which are paramagnetic, show typically lower susceptibilities compared to the pure metals.

5.2 APPLICATIONS

5.2.1 Bulk Materials Applications

5.2.1.1 Structural Materials Due to their specific properties (high-temperature strength and hardness) transition metal nitrides and carbides have found application in harsh abrasive conditions. Most common is WC, which is used for the fabrication of "cemented carbide," where WC is bonded in the Co matrix. In such composites, WC defines the hardness, while Co is a plastic matrix that allows relatively easy shaping and processing of such materials. Cemented carbides are commonly used in cutting tools like saws or drillers (Fig. 5.2).

For high-temperature applications, carbides are used as pure-material sintered parts or in a Co/Mo/W/carbide sintered composite. They outperform standard alloys and superalloys in rocket nozzles and jet engine parts where erosion resistance at

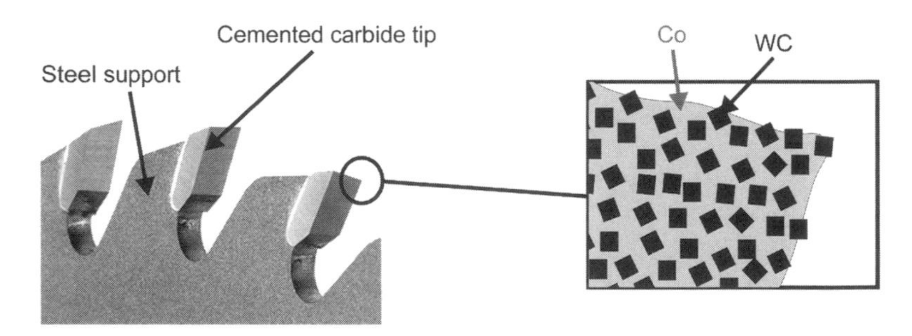

Figure 5.2 Cemented carbide used in cutting tools.

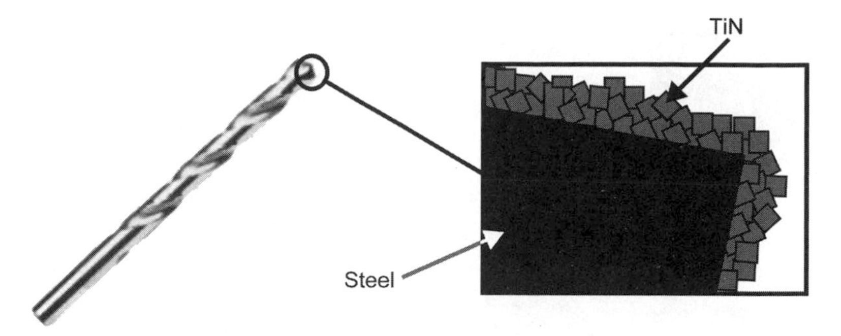

Figure 5.3 TiN-coated drill.

temperatures up to 2500°C is crucial. In particular VC and TiC retain high strengths up to 1800°C and therefore can be used as high-temperature structural materials.

5.2.1.2 Coatings on Cutting Tools Many transition metal nitrides and carbides are applied commercially as hard coatings. For example, TiN is commonly used in coating of steel drills to increase the lifetime and hardness. One can easily recognize TiN coatings because they have a shiny golden color, which also clearly reflects their metallic character (Fig. 5.3). In another case such coatings may be applied to press stamps and significantly prolong their lifetime.

5.2.1.3 Catalysis In 1972 Boudart (2) for the first time demonstrated that transition metal carbides and nitrides can be used as catalysts in reactions that are specific to the expensive transition metals like platinum or palladium (prices of platinum and WC differ by orders of magnitude). Since then many potential applications in catalysis have been reported for isomerization, hydrotreating (HDN, hydrodenitrogenation; HDS, hydrodesulfurization) and dehydrogenation reactions. For example, HDN and HDS processes are very important in the petrochemical industry for removal of sulfur and nitrogen compounds from gasoline and diesel. Nitrides and carbides were also found to be more resistant to poisoning than industrially used Co-Mo sulfide catalysts. However, the first commercial application came in 1997, when MoO_xN_y was used as a catalyst for the hydrazine decomposition in satellite thrusters and outperformed a standard catalyst based on iridium (3).

$$N_2H_4 \longrightarrow N_2 + 2H_2$$

The similar catalytic properties of transition metal carbides and nitrides to those of noble metals can be attributed to their similar electronic properties and structure. The valence electron count of WC is similar to that of Pt (1).

5.2.1.4 Carbide-Derived Carbons (CDC) An interesting property of the transition metal carbides is that the metal can be selectively leeched from the carbide network by chlorine to form a nanosolid porous carbide-derived carbon structure (CDC)

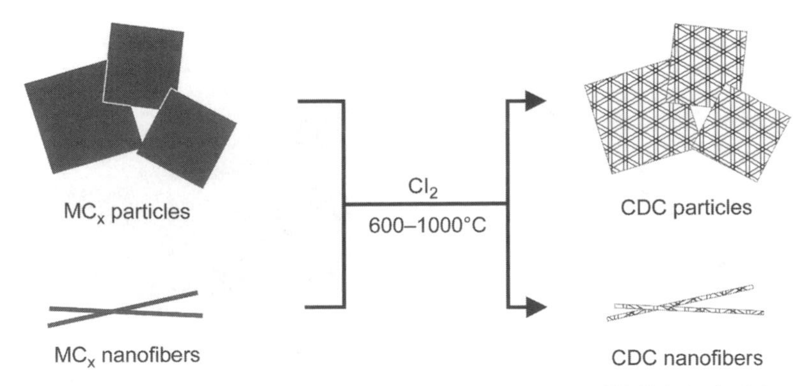

Figure 5.4 Different morphologies of carbide-derived carbons (CDC) inherited from their metal carbide precursors.

with high specific surface areas and pore diameters between 0.3 and 6 nm. Such materials can be used in the separation processes and immobilization of large molecules (enzymes), and their properties can be tuned by using different metals and different metal removal conditions.

$$TiC_{(solid)} + 2Cl_{2(gas)} \xrightarrow{600-1000^{\circ}C} C_{(solid-porous)} + TiCl_{4(gas)}$$

One can also define the shape of CDC material on the nanoscale by changing the morphology of the carbide precursor (MC_x) (Fig. 5.4).

Another possibility of carbon formation is the hydrothermal treatment of carbides (100 MPa, 300°C to 400°C in H_2O). However, in this case a thin, uniform carbon film is produced on the surface of the carbide. Recently, it was also shown that SiC can be transformed into nanocrystalline diamond during silicon etching with Cl_2 (by proper adjustment of the reaction parameters). It has not yet been tested if it also can be applied on the transition metal carbides.

Carbide-derived carbons are considered interesting materials for electrodes and hydrogen storage.

5.2.2 Nanosized Transition Metal Nitrides and Carbides: Advantages and Specific Applications

5.2.2.1 Nanopowders Reduction of the particle size has a strong effect on the properties of materials. In the powder form, the specific surface area increases dramatically with decreasing particle size. Typically, the relation between the particle size and specific surface area can be predicted based on simple geometric considerations assuming cubic or spherical particle shape (Fig. 5.5).

The increased specific surface area is very important in heterogeneous catalysis since reactions take place at the gas-solid or liquid-solid interface. For example, the dehydrogenation rate of butane over VN significantly improves with the increasing

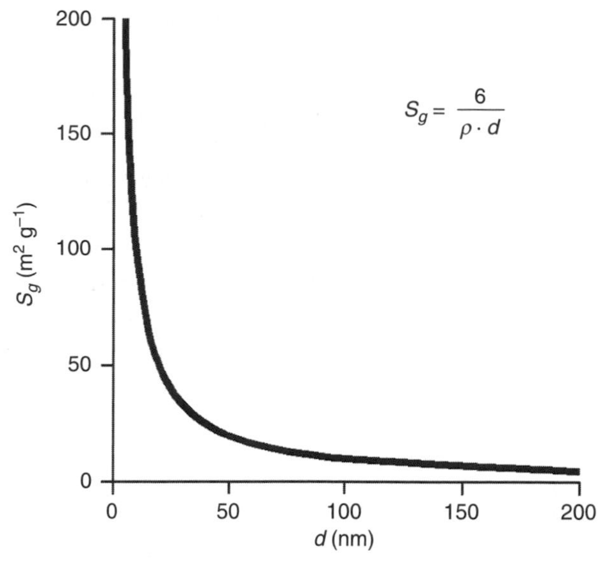

Figure 5.5 Specific surface area S_g (m^2g^{-1}) as a function of VN particle diameter d (nm) assuming density $\rho = 6.1$ g cm^{-3}.

specific surface area (Table 5.3; Reference 4).

$$\xrightarrow[450°C]{VN}$$

Due to the large specific surface area, nanopowders are also more reactive than micrometer-sized materials and some of the nano-sized nitrides and carbides can be completely oxidized if directly exposed to air. For example, Mo_2N with specific surface area of 120 m^2g^{-1} can be rapidly oxidized upon exposure to air (powder starts to glow red), while low specific surface area material (<1 m^2g^{-1}) will show no reaction to the same treatment. In order to prevent rapid oxidation, nanoparticulate materials can be slowly passivated in inert gas (N_2, He) containing low amounts of oxygen (0.5% to 1%). In such conditions only a thin layer of oxide is formed and nitrogen cools down the system in order to slow down the reaction. It is known that

TABLE 5.3 Butane Dehydrogenation Rate over VN Catalyst with Different Specific Surface Areas (4)

S_g (m^2g^{-1})	d_S (nm)	R (nmol g^{-1} s)
14	70	89
29	34	166
59	18	256

S_g, specific surface area; d_S, nanoparticle diameter calculated according to the equation presented in Figure 5.5; R, gravimetric reaction rate.

nanopowders of metals are explosive and should be handled with care. The same concerns the nanopowders of the other transition metal carbides and nitrides.

Recently, nanocrystalline VN has been discovered as an excellent material for supercapacitors (5). Compared to standard capacitors (where charge is stored due to the pure electrostatic attraction) in supercapacitors energy is stored at the electrolyte/solid interface in the form of the electrical double layer and due to the reversible redox reaction on the nanoparticle surface (Fig. 5.6):

$$VN_xO_y + OH^- \longleftrightarrow VN_xO_y||OH^- + VN_xO_y-OH$$

where $VN_xO_y||OH^-$ represents the electrical double layer formed by hydroxyl ions adsorbed on nonspecific sites, wherein a large increase in the specific capacitance arises primarily due to the successive oxidation by hydroxyl species (OH^- from electrolyte). While the charge capacities in standard units are typically at the level of microfarad (10^{-6}), modern supercapacitors with similar external dimensions can reach capacities of kilofarad (10^3). They are of extreme interest, since they can be used to effectively store electric energy for mobile applications. Supercapacitors clearly outperform lithium-based batteries on the charging speed and cycle lifetime and are planned to be used as electrical energy storage units for city buses in Shanghai. In this respect, VN can be an economic alternative for the expensive supercapacitors that are base on Ru nanoparticles.

5.2.2.2 Bulk Nanomaterials One of the most effective ways to densify nanopowders into bulk nanomaterials is SPS (spark plasma sintering). The powder is compressed and a high density current is passed through it in order to heat it up to the

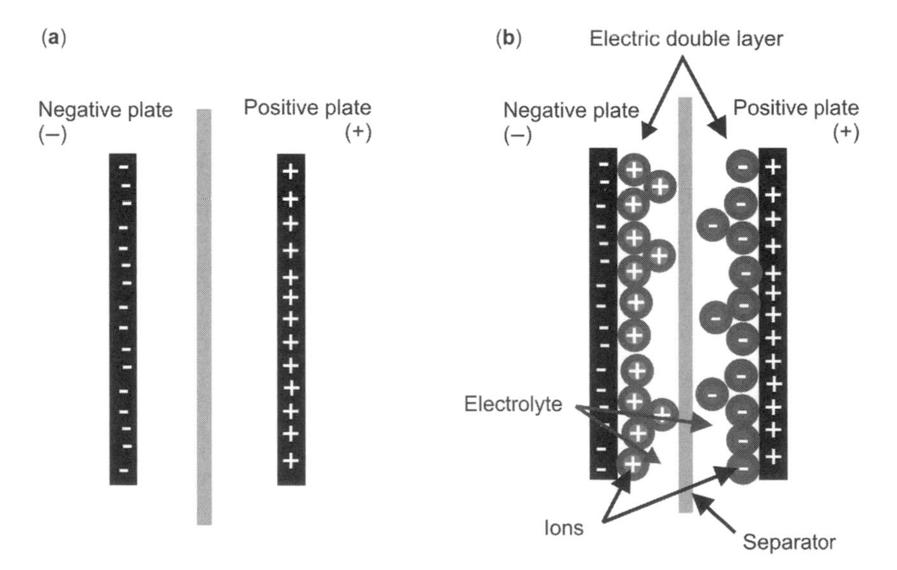

Figure 5.6 Schematic drawings of standard capacitor (a) and supercapacitor (b).

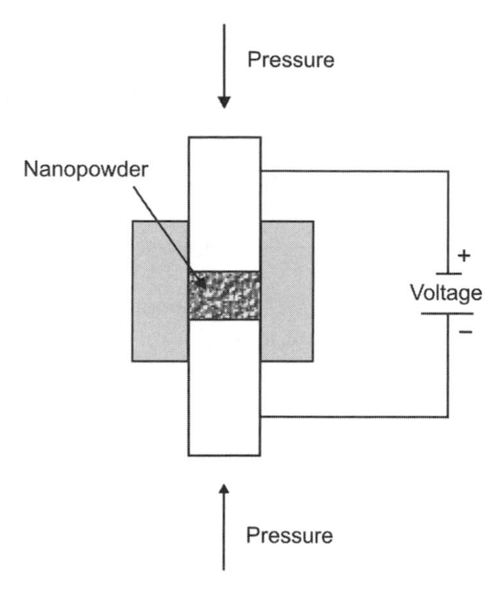

Figure 5.7 Spark plasma sintering (SPS) of nanopowders.

desired sintering temperature (Fig. 5.7). In this respect advantage can be taken of the metallic character of transition metal nitrides and carbides and they can be heated easily by passing current through them. Moreover, very fast heating ramps can be achieved using this method, which is extremely important for proper densification. Nanopowders can be sintered at lower temperatures since they are energy-rich systems (high specific surface area) and generate high intrinsic stress:

$$\sigma = -\frac{2\gamma}{r}$$

where σ is the sintering stress, γ is the surface tension, and r is the pore radius (smaller pores are present between nanoparticles). The estimated sintering stress is of the order of 500 MPa, compared to that of $\sim$5 MPa for conventional micrometer-sized powder compact. The externally applied pressure of 100 MPa used in most of the densification processes fails further to reasonably enhance the sintering kinetics.

The bulk sintered nanomaterials are known to have different mechanical properties with respect to materials with micrometer-sized grains (6). In principle, most of the reports suggest higher hardness with decreasing grain size, but also enhanced plasticity and toughness can be observed. Interestingly, the effect of superplasticity can appear for some of the bulk nanomaterials and they can be deformed beyond the point where they would normally break. It is of great importance how the nanopowders are prepared before sintering. The correct processing is crucial to achieve proper mechanical properties and a proper preparation is necessary. No large pores (surrounded by many particles) should be present in the compacted powder since they are extremely difficult to densify and if they remain in the material they can significantly lower its mechanical properties.

5.3 SYNTHESIS

5.3.1 Bulk Materials and Powders (Industrial Techniques)

Transition metal nitrides can be prepared easily by nitridation of metals, oxides, and other precursors containing the transition metal (e.g. sodium or ammonium salts):

$$M + xNH_3 \longleftrightarrow MN_x + 3x/2H_2$$

Instead of ammonia also a mixture of hydrogen and nitrogen or other precursors containing nitrogen (e.g. hydrazine) can be used.

A similar approach can be used for the synthesis of carbide materials. If instead of nitrogen, a carbon source is used, the transition metal carbides can be obtained. Typically the source of carbon can be hydrocarbons (methane, ethane, propane) or solid carbon. High surface area carbons are preferred, since they increase the reaction speed substantially.

$$M + xC \longleftrightarrow MC_x$$

Both nitridation and carburization are typically carried out at high temperatures (e.g. VC is prepared at ~1000°C using methane and VN at ~800°C using ammonia).

Due to the highly exothermic nature of such reactions, some of them can be self-propagating processes (Fig. 5.8). The powder mixture containing metal and carbon is locally preheated only to initiate the reaction and then the external heating is switched off. The temperature rises due to the heat generated internally at the reaction

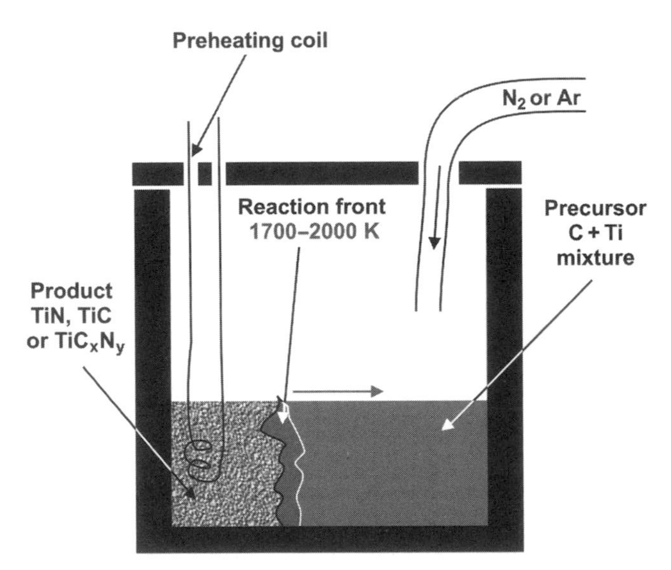

Figure 5.8 Self-propagating synthesis of titanium nitride, carbide, or carbonitride via carbothermal reduction.

front where metal carbide or nitride forms. This economic method is especially known in industry for the preparation of SiC materials, but transition metal carbides and nitrides are also prepared in this way. Carbides are typically prepared in an inert atmosphere, while in the case of nitrides the reaction proceeds in a nitrogen atmosphere. However, due to the high temperatures generated during the self-propagating synthesis, the materials obtained have particle sizes in the range of micrometers. The typical problem faced in the preparation of carbides is residues of carbon. This can be the effect of either carbon precursor (solid carbon) not completely reacted or free carbon deposition from the hydrocarbons at high temperature.

5.3.2 Nanomaterials: State of the Art and New Concepts

State-of-the-art procedures used for the preparation of nanometer-sized carbides and nitrides are presented in this chapter, with an introduction to the most recent research in this area. This section is divided into three parts based on the physical state of substrates used for the synthesis.

5.3.2.1 Gas-Phase Reactions Gas-phase reactions are typically used for the preparation of thin films or nano-sized transition metal carbides and nitrides. High volatility of metal chloride precursors is useful and nitrides can be obtained by reaction with ammonia (M is the corresponding transition metal):

$$MCl_x + yNH_3 \longrightarrow MN_y + zHCl$$

Not only chlorides, but also organometallic precursors (typically having nitrogen atoms connected directly to metal) can be used. This process is called the metal-organic chemical vapor deposition (MOCVD) process. For example, amido complexes of the general formula $M(NMe_2)_x$, $M(NEt_2)_x$ are used as metal precursor, with the main advantage being very low corrosiveness. This is especially important in coating of electronic materials or metals where chlorine atmosphere causes severe corrosion.

Alternatively, physical vapor deposition processes can be used to achieve thin coatings of nitides. Typically metal atoms are sputtered from the metal electrode in the presence of gas plasma and, depending on the atmosphere, different products can be obtained. This method is often used for the preparation of thin layers of TiN, where Ti electrode and nitrogen gas are used.

Transition metal carbides can be prepared by similar methods, and typically hydrocarbons are used as carbon source:

$$MCl_x + y/zC_zH_h \longleftrightarrow MC_y + nH_2$$

This process may lead to the deposition of free carbon, but it can also be controlled by the proper adjustment of the reaction parameters. Atmosphere (reducing or inert), gas flow, and temperature can be used to control the particle size of deposited materials. Different methods can also be used for heating of the system (cold- and hot-wall reactors).

5.3.2.2 Gas–Solid Reactions Metal oxides are popular precursors for the preparation of high surface area nitrides and carbides. It has been found that the specific surface area of the product strongly depends on the ammonia or methane flow, and the method is often referred to as a temperature programmed reduction process (TPR):

$$V_2O_5 + 10/3NH_3 \longrightarrow 2VN + 5H_2O + 2/3N_2$$

The faster the flow of gas the higher the specific surface area that can be achieved. This is often attributed to the speed of water removal in high gas flow, which can shift the reaction equilibrium. Nitridation of V_2O_5 proceeds via several steps, where first the oxide is reduced to VO and then topotactic substitution (of oxygen with nitrogen or carbon) takes place. In contrast to this, the nitridation of MoO_3 to form Mo_2N does not follow a topotactic substitution mechanism (1).

Another way to enhance the specific surface area is prestructuring of the precursor. In this case prestructured porous oxides with high specific surface areas can be used, as well as chemically modified oxides templated with organic molecules (e.g. long chain amines). Such organic molecules are removed in the course of the thermal treatment in ammonia (7, 8). For example, in the case of prestructured precursors, specific surface areas up to 190 m^2g^{-1} have been reported for the VN product, while in the case of standard oxide precursor surface areas below 90 m^2g^{-1} were reported for the same material (9).

More complex molecular precursors can also be used for the preparation of high surface area materials. By complexing $TiCl_4$ with different organic ligands (THF, CH_3CN, Et_2O, Et_3N) it was shown that a large degree of control over the nitride morphology can be obtained. After nitridation with ammonia (where ligand is subsequently removed at high temperatures), titanium nitrides with specific surface areas from 37 to 230 m^2g^{-1} can be obtained (10).

5.3.2.3 Solid–Solid Reactions, Template Synthesis In industrial production of carbides, carbon is typically used as an inexpensive precursor. However, by such methods bulk materials are obtained with large particles (micro- or millimeter scale). If instead of bulk carbon one uses nanostructured materials with a finely dispersed transition metal precursor, nano-sized transition metal carbides can be achieved. Such a strategy was used by Yu et al. (11), who incorporated titanium precursor into the walls of nanostructured carbon; after heating in inert atmosphere TiC nanocrystals could be obtained. The nanoporous template acts as a carbon source and at the same time it sterically hinders growth of the carbide particles to the confined space (Fig. 5.9).

For the preparation of nanoparticulate transition metal nitrides, as a solid nitrogen precursor (and structure-directing substrate at the same time) nanostructured C_3N_4 can be used. Porous matrix is then infiltrated with the solution of metal chloride and pyrolyzed at high temperature (Fig. 5.10). During the heating, metal chloride reacts with C_3N_4 to form nitrides and subsequently the matrix decomposes to gaseous products. The fine powder can be obtained and easily redispersed in solvents (12).

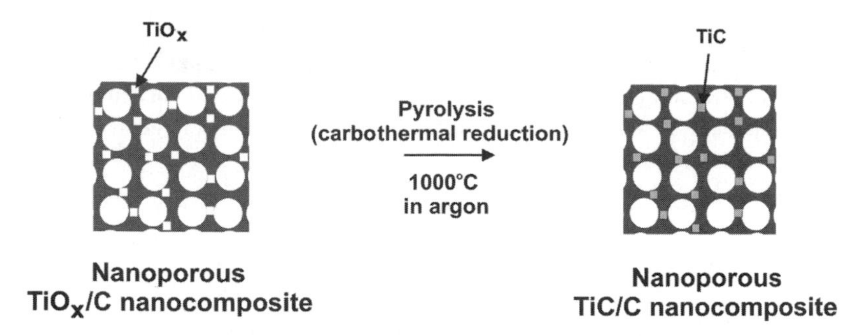

Figure 5.9 Praparation of TiC nanoparticles within the pore walls of nanoporous carbon via a carbothermal reduction process.

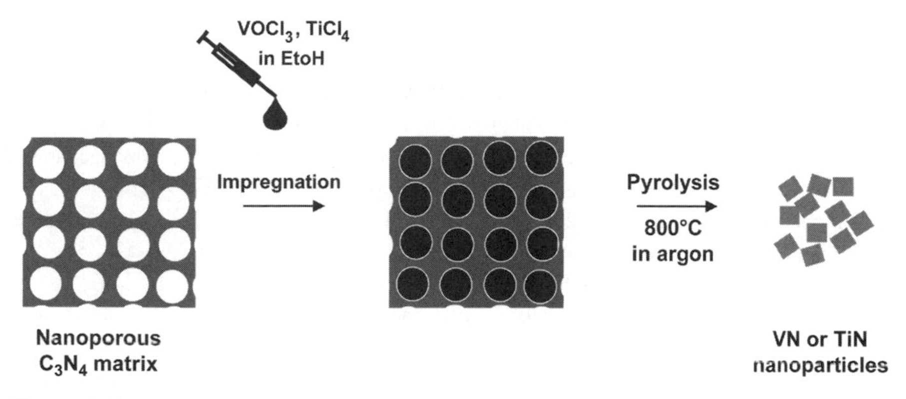

Figure 5.10 Preparation of transition metal nitride nanoparticles using nanoporous C_3N_4 matrix.

5.4 SUMMARY

Transition metal carbides and nitrides find broad interest in chemistry and technology. In the form of nanopowders they can be used in electronics and for catalysis. As catalysts they have similar properties to those of the expensive noble metals. They are extremely hard and therefore are often used in cutting tools or in harsh abrasive conditions. However, nanoparticulate nitrides and carbides are more reactive due to heir higher specific surface area. Some of them oxidize rapidly in air and are typically passivated with a thin layer of oxide before exposure to ambient conditions.

Different methods have been discussed for the preparation of transition metal nitride and carbide nanoparticles, including state-of-the-art procedures and more recently reported methods.

In the case of bulk, sintered nanoparticulate materials, typically mechanical properties can be increased (toughness, hardness, strength) in the dense product. However, in order to achieve this effect nanopowders have to be properly prepared; after sintering no pores should be present. Nanopowders are often sintered at lower temperatures and recent reports suggest SPS (spark plasma sintering) as a very promising technique. In addition to this, superplastic behavior was reported for nanoparticulate sintered solids, and such materials can be easily processed by mechanical deformation at relatively low temperatures.

REFERENCES

1. Oyama, S. T. 1996. *The Chemistry of Transition Metal Carbides and Nitrides*. London: Blackie Academic & Professional.
2. Levy, R. B., Boudart, M. **1973**. Platinum-like behaviour of tungsten carbide in surface catalysis. *Science 181*: 547–549.
3. Rodrigues, J. A. J., Cruz, G. M., Bugli, G., Boudart, M., DjegaMariadassou, G. **1997**. Nitride and carbide of molybdenum and tungsten as substitutes of iridium for the catalysts used for space communication. *Catal. Lett. 45*: 1–3.
4. Kwon, H., Choi, S., Thompson, L. T. **1999**. Vanadium nitride catalysts: Synthesis and evaluation for n-butane dehydrogenation. *J. Catal. 184*: 236–246.
5. Choi, D., Blomgren, G. E., Kumta, P. N. **2006**. Fast and reversible surface redox reaction in nanocrystalline vanadium nitride supercapacitors. *Adv. Mater. 18*: 1178–1182.
6. Mukhopadhyay, A., Basu, B. **2007**. Consolidation microstructure property relationships in bulk nanoceramics and ceramic nanocomposites: A review. *Int. Mater. Rev. 52*: 257–288.
7. Krawiec, P., De Cola, P. L., Gläser, R., Weitkamp, J., Weidenthaler, C., Kaskel, S. **2006**. Oxide foams for the preparation of high surface area vanadium nitride catalysts. *Adv. Mater. 18*: 505–508.
8. Oyama, S. T., Kapoor, R., Oyama, H. T., Hofmann, D. J., Matijevic, E. **1993**. Topotactic synthesis of vanadium nitride solid foams. *J. Mater. Res. 8*: 1450–1454.
9. Kapoor, R., Oyama, S. T. **1992**. Synthesis of high surface area vanadium nitride. *J. Solid State Chem. 99*: 303–312.
10. Kaskel, S., Schlichte, K., Chaplais, G., Khanna, M. **2003**. Synthesis and characterisation of titanium nitride based nanoparticles. *J. Mater. Chem. 13*: 1496–1499.
11. Yu, T., Deng, Y., Wang, L., Liu, R., Zhang, L., Tu, B., Zhao, D. Y. **2007**. Ordered mesoporous nanocrystalline titanium-carbide/carbon composites from in situ carbothermal reduction. *Adv. Mater. 19*: 2301–2306.
12. Fischer, A., Antonietti, M., Thomas, A. **2007**. Growth confined by the nitrogen source: Synthesis of pure metal nitride nanoparticles in mesoporous graphitic carbon nitride. *Adv. Mater. 19*: 264–267.

PROBLEMS

1. Assuming the body-centered structure (bcc) of metal (M), which of the following compounds MX or MX_2 (where X = C, N) is more likely to progress its structure to hexagonal or face-centered cubic?

2. Which physical property makes transition metal carbides and nitrides a good component for cutting tools? In this respect why is cobalt used as a matrix material for their sintering?

3. Why are nanoparticles of transition metal carbides and nitrides considered interesting substitutes for platinum in heterogeneous catalysis?

4. Which solid precursors of carbon and nitrogen can be used for the preparation of nanoparticles of transition metal carbides and nitrides?

5. Which precursor would be more suitable for the chemical vapor deposition of VN on Fe surface: $VOCl_3$ or $V(NMe_2)_2$?

6. Why can nanopowders be sintered at lower temperatures?

ANSWERS

1. Due to the increasing *sp* electron density (larger number of N or C atoms per metal atom), MX_2 structures are more likely to progress to hexagonal or face-centered ones.

2. Transition metal carbides and nitrides are very hard materials, stable at high temperatures, and therefore can be used in harsh abrasive conditions. However, they are also very brittle and difficult to process. The plastic cobalt matrix increases the toughness and prevents the cutting tools from breaking (cracking).

3. They are much cheaper and have been proved to catalyze reactions specific for Pt, Pd, and Ir.

4. For preparation of carbides nanoporous carbon can be used, while nitrides can be made with nanoporous carbonitride precursor.

5. Due to the highly corrosive nature of chloride towards Fe, an organometallic precursor would be preferred.

6. Smaller pores between nanoparticles cause higher internal sintering stress (by orders of magnitude) and therefore allow densification at lower temperatures.

6

KINETICS OF COLLOIDAL CHEMICAL SYNTHESIS OF MONODISPERSE SPHERICAL NANOCRYSTALS

Soon Gu Kwon and Taeghwan Hyeon

6.1 Introduction, 128
6.2 Basic Concepts of Size Distribution Control, 129
  6.2.1 Burst Nucleation, 129
  6.2.2 Size Focusing, 132
6.3 Theoretical Description of the Crystallization Process, 134
  6.3.1 Derivation of the Nucleation Rate Equation, 134
  6.3.2 Derivation of the Growth Rate Equation, 136
  6.3.3 Features of the Growth Process and Ostwald Ripening, 140
6.4 Kinetics of the Hot Injection Method, 142
  6.4.1 The Synthetic Procedure, 142
  6.4.2 Size Distribution Control, 143
6.5 Kinetics of the Heat-Up Method, 145
  6.5.1 The Synthetic Procedure, 145
  6.5.2 Size Distribution Control, 147
6.6 Summary, 150
References, 151
Problems, 151
Answers, 152

Nanoscale Materials in Chemistry, Second Edition. Edited by K. J. Klabunde and R. M. Richards

6.1 INTRODUCTION

Studies on the synthesis of monodisperse particles date back to the 1940s. In these early days, the research was focused on the synthesis of micrometer-sized particles and the characterization of their size-dependent properties, such as light scattering, hydrodynamic behavior, and catalytic activity (1–3). For the last 20 years, tremendous progress has been made in the synthesis of monodisperse nanocrystals. On the nanometer scale, many material properties, which are usually regarded as being intrinsic for bulk material, become extremely sensitive to its dimensions. In other words, size control became a new way of tuning the properties of nanostructured materials. Consequently, the synthesis of uniformly sized nanocrystals is critical, because their size uniformity is directly correlated with the homogeneity of their properties. Usually, nanocrystals with a relative standard deviation (σ_r) of the size distribution of less than 5% are said to be monodisperse. For example, monodisperse iron nanocrystals with a mean diameter of 4 nm should contain between 2400 and 3300 iron atoms. Consequently, the synthesis of monodisperse nanocrystals is very much challenging. Recently, several different colloidal chemical methods have been developed for the synthesis of monodisperse nanocrystals. Among them, crystallization in organic media has been most popularly used, and monodisperse nanocrystals of various materials, including II-VI semiconductors, transition metals, metal alloys, and metal oxides have been synthesized (Fig. 6.1; Reference 4). Studies on the crystallization mechanism have also been intensively pursued.

In this chapter, we introduce the methodology of size distribution control in nanocrystal synthesis, focusing on homogeneous crystallization in organic solution. The

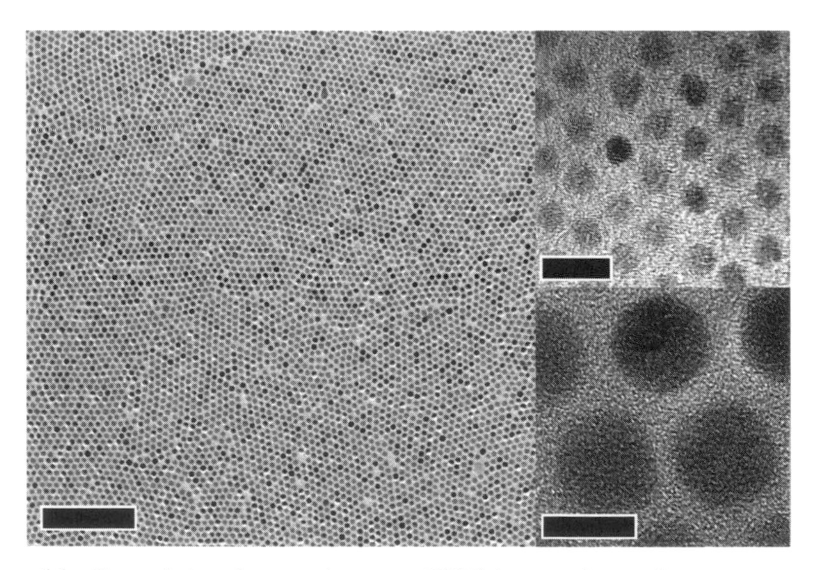

Figure 6.1 Transmission electron microscopy (TEM) images of monodisperse nanocrystals. They are iron oxide nanocrystals synthesized by the heat-up method introduced in Section 6.5. The scale bar in the left picture is 2 μm and the ones on the right are 10 nm.

sections are organized as follows: We present a description of the size distribution control process in Section 6.2. In doing so, we tried to give a conceptual picture rather than a rigid analytic model. The essential concepts in the size distribution control process are explained with as little mathematics as possible. In Section 6.3, we proceed to give a deeper understanding of the crystallization process to support the discussions in the previous section. We investigate the theoretical model for nucleation and growth and derive equations describing these processes. Inevitably, this requires some mathematical formulations. For those readers who want to save time and avoid the effort required to follow the series of equations, it is recommended that they skip Sections 6.3.1 and 6.3.2 and go directly to Section 6.3.3. Lastly, we discuss two representative synthetic methods, hot injection and heat-up, in Sections 6.4 and 6.5, respectively. Furthermore, the theoretical concepts introduced in Sections 6.2 and 6.3 are applied to explain the size distribution control mechanism in these two synthetic methods.

6.2 BASIC CONCEPTS OF SIZE DISTRIBUTION CONTROL

6.2.1 Burst Nucleation

For the growth of a crystal in solution, there must be a substance that acts as a seed onto which crystallization can occur (5). The nuclei can be introduced externally or generated in the solution. If the crystallization proceeds with preexisting seeds, this process is called heterogeneous nucleation. This terminology indicates that the reaction system is in the heterogeneous phase in the beginning. On the other hand, in homogeneous nucleation, the system consists of a single phase in the beginning and the nucleus formation takes place in the course of the crystallization process.

To see how the nucleation process influences the size distribution of the crystal particles, consider a reaction system in which homogeneous nucleation occurs, as shown in Figure 6.2. In the system depicted in the upper part of the figure, the nucleation occurs randomly all the time during the crystallization process, and all the particles have different growth histories. As shown in the figure, this results in the uncontrolled growth of the particles and a broad size distribution. This example clearly shows that effective size distribution control cannot be achieved for the growing particles under conditions of random nucleus generation. Therefore, a controlled nucleation process is a necessary prerequisite for the size distribution control. At this point, one could imagine an ideal case in which all of the particles nucleate at once and grow under the same conditions. In such a case, all of the particles would have the same growth history and the size distribution would be monodisperse (see the lower part of Fig. 6.2).

A situation very similar to this idea can be realized by the seed mediated growth method. This method utilizes preformed uniform seed particles as nuclei. Heterogeneous nucleation using these seed particles imitates the single nucleation event. The growth of the seeds in the well-agitated solution yielded particles conserving the initial uniformity. Although this method seems to be attractive at first glance, it has one problem; it presumes the existence of uniform particles for the preparation of

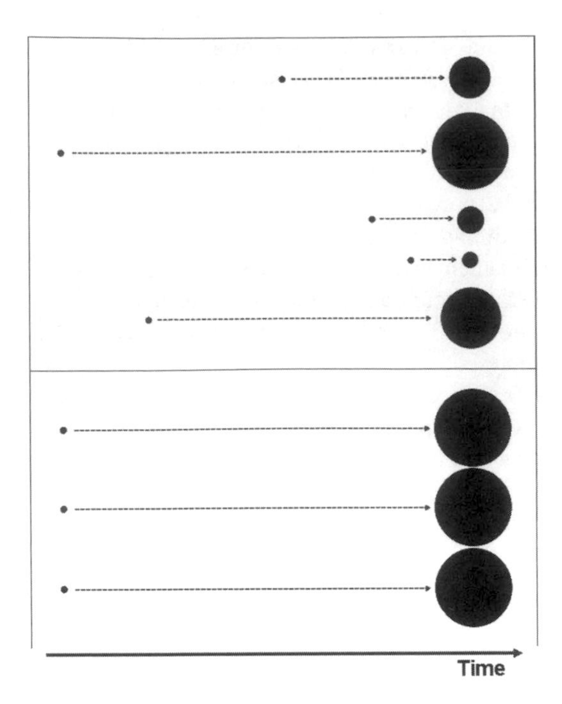

Figure 6.2 Crystallization reaction systems in which homogeneous nucleation occurs randomly (upper) and at once (lower). The dots (•) to the left of the arrows indicate the formation of nuclei. Each arrow and black circle shows the growth of the corresponding particle and its size.

other uniform particles. Consequently, the synthesis of uniform seed particles still remains the main problem.

One of the solutions to the problem of achieving controlled nucleation is to exploit the kinetics of homogeneous nucleation. The homogeneous nucleation reaction has a very high energy barrier compared to that of heterogeneous nucleation, and an extremely high supersaturation level is necessary to commence the homogeneous nucleation process in the solution. The crystallization process starts with nucleation, which lowers the supersaturation level. Consequently, the nucleation process terminates by itself. Due to this self-regulating nature, the homogeneous nucleation process takes place only for a short time, during which the supersaturation level stays very high. This short duration of the homogeneous nucleation process, or "burst nucleation," is near to the ideal single nucleation event (2, 3). However, there is an important difference between them, which makes the situation complex. Comparing the schematic of burst nucleation (Fig. 6.3) with that of the single nucleation event (Fig. 6.2, the lower part), it can be seen that burst nucleation itself does not guarantee monodispersity. In fact, however short the duration is, a broad size distribution at the end of the nucleation period is inevitable. The explanation for this is as follows: The high supersaturation level promotes not only homogeneous nucleation but also crystal growth. Consequently, in the course of the nucleation process, the nuclei formerly formed grow very fast, while those just formed retain their initial size. As in the case of random

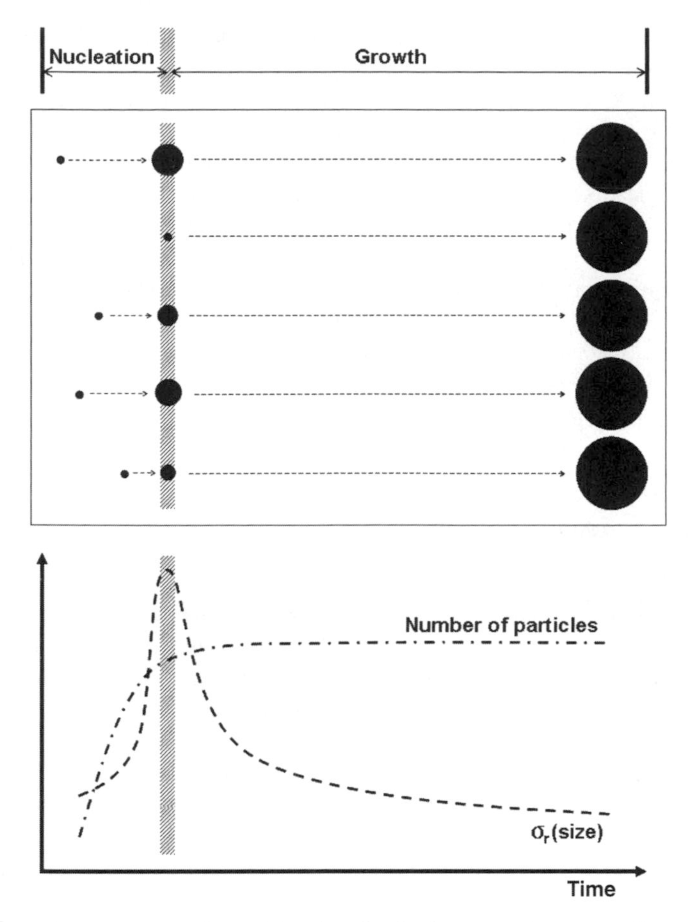

Figure 6.3 Schematic illustration of the size distribution control process. The vertical thick line (shaded) indicates the point in time at which the nucleation process is terminated, dividing the nucleation and growth periods. The arrows and the black circles share the same symbolism in Figure 6.2. In the lower part of the figure, the time evolution of the number of particles and the relative standard deviation of the size distribution, σ_r(size), are shown.

nucleation, the size distribution broadens when the nucleation and growth reaction proceed simultaneously (note that not only nucleation but also the growth reaction occurs in the "nucleation" period in Fig. 6.3).

Given that burst nucleation does not guarantee the formation of monodisperse particles, why is it so important? The answer has already been given above. Burst nucleation satisfies the necessary condition for the size distribution control. If the duration of the nucleation process is much shorter than that of the growth process, the crystal particles share almost the same growth history throughout their crystallization. The characteristics of burst nucleation are depicted in the lower part of Figure 6.3. In the nucleation period, the number of particles in the solution rapidly increases. This increase is accompanied by the broadening of the size distribution, and the plot of σ_r(size) reaches its maximum at the end of the nucleation period. Then, the process goes into the growth period in which the number of particles remains stable and the size distribution is narrowed. These changes indicate that the size distribution control mechanism works in the growth process, which is discussed in the next section.

Before closing this section, it is worth mentioning another role of burst nucleation, although it is not directly correlated with monodispersity. The word "burst" implies that burst nucleation is not only short but also very violent. In fact, violent "burst" nucleation is critical to obtain nanocrystals rather than microparticles. Let's consider the synthesis of monodisperse spherical iron particles from the crystallization of a solution containing 1.0×10^{-3} mol of iron precursor whose molar volume is $\sim 7.0 \times 10^{-6}\ \mathrm{m^3\ mol^{-1}}$. When the number of particles in the solution is 10^2, 10^6, 10^{12}, 10^{16}, and 10^{18}, the particle diameter would be 510 μm, 23 μm, 240 nm, 11 nm, and 2.4 nm, respectively. Consequently, it is very obvious that rapid and violent nucleation leading to the formation of large numbers of nuclei is critical for the synthesis of nanocrystals.

6.2.2 Size Focusing

If one looks carefully at the schematic in the upper part of Figure 6.3, one can see that some strange things happen during the growth period. That is, the small particles grow faster, whereas the large particles grow slower, reaching the same size in the end. Although this process might look unusual at first glance, it is the essence of the size distribution control mechanism during the growth period, and actually works in the nanocrystal synthesis of various materials.

First, let's look at how a size-dependent growth process can achieve the desired monodispersity. Theoretical studies showed that, under certain conditions, the curve for the growth rate vs. size resembles a hyperbola (see Section 6.3.2), as shown in Figure 6.4. In this figure, the growth rate is defined as the increment of the radius per unit time. The graph for the growth rate as a function of the radius has a negative slope. That is, the larger particle grows slower, as they do in Figure 6.3. Using the plots in Figure 6.4, we can see how particles of different sizes, namely, particles 1 and 2, grow under this condition. The initial diameter of particle 1 is r_1^0 and its growth rate is v_1^0. Then, during the first unit time step, particle 1 grows from r_1^0 to $r_1^0 + v_1^0$, which is designated as r_1^1. Because v_1^0 is the edge length of the first square on the left in the figure, the right bottom corner of this square locates at r_1^1. Similarly, $r_1^1 + v_1^1$ is equal to r_1^2 and its position can be found by drawing the second square on the right side of the first square. Consequently, the radii of particles 1 and 2 after four units of time are obtained by the graphical method, namely, drawing squares. The result is very interesting. In the lower part of Figure 6.4, it is clearly seen that the difference in the particle radii is remarkably reduced as the particles grow. Moreover, following the same argument in this example, it can be easily shown that any negative-slope graph other than a hyperbola also yields a similar result (see Problem 1). Generally, the growth rate function of the negative slope induces the narrowing of the size distribution, which is referred to as *size focusing* (6).

In the following discussion, we explain why the growth rate has such a negative size dependency. Let's consider a highly supersaturated solution, which guarantees

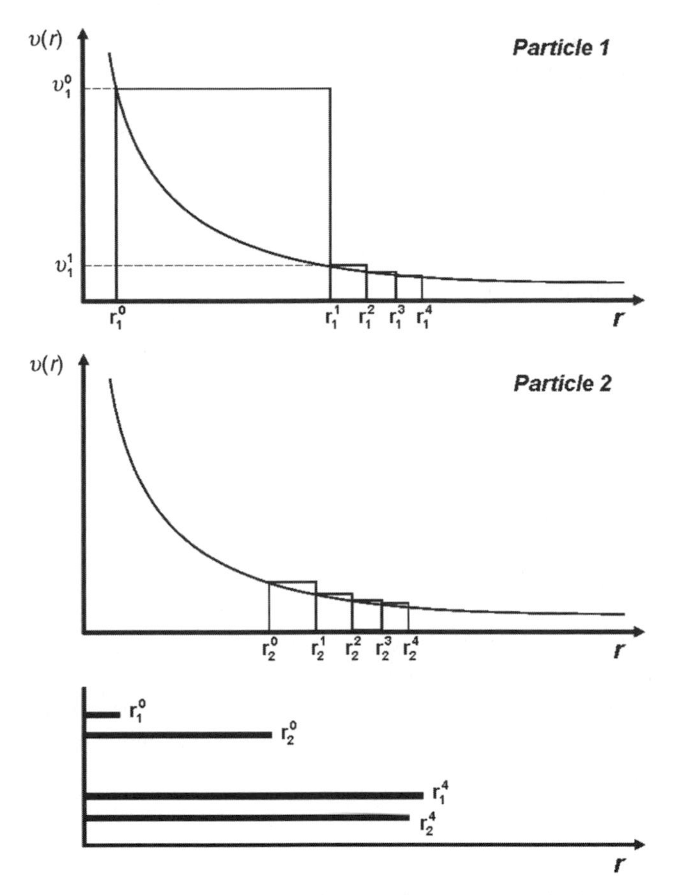

Figure 6.4 Plots for the growth rate vs. particle radius (upper and middle). $v(r)$ and r indicate the growth rate and the particle radius, respectively. The graphs for $v(r)$ in both plots are identical. r_1^n and r_2^n are the radii of particle 1 and particle 2 at nth growth step. The explanation for the squares is given in the text. For the comparison, radii of particle 1 and particle 2 at the initial ($n = 0$) and final ($n = 4$) steps are shown together as a bar chart (lower).

the so-called *diffusion-controlled growth* process (2, 3). In this growth mode, the rate of increase of the particle volume is equal to the rate of solute diffusion from the solution onto the particle surface. The diffusion rate is written as dM/dt where M is the amount of solute. From the principle of diffusion mass transport, we know that $dM/dt \propto r$ where r is the particle radius. On the other hand, the amount of solute M required to increase the radius from r to $r + dr$ is proportional to the particle surface area, so $M \propto r^2$. In this condition, the growth rate, dr/dt, of the larger particle is slower than that of the smaller one, because of the shortage of solute supply. Here is an example. Let's consider two particles with radii of r and $5r$. To grow those particles to $r + dr$ and $5r + dr$, it requires M and $25M$ moles of solute, respectively. However, the mass transport process dictates that, while the smaller particle takes M mol of solute from the solution, the larger particle gets only $5M$ mol. As a result, the growth of the larger particle is retarded by the rate of solute diffusion, from which the name diffusion-controlled growth is derived.

According to theoretical investigations, size focusing takes effect only when the crystal growth occurs in diffusion-controlled mode (7, 8). However, this mode is an extreme case of the crystal growth mechanism. Normally, crystallization is a competition between two opposite reactions, precipitation, which is phase transfer from solute to solid, and dissolution, which is phase transfer from solid to solute. On the other hand, in this growth mode, dissolution occurs to a negligible extent and all of the solutes that diffuse onto the crystal surface immediately precipitate. Therefore, to sustain the diffusion-controlled growth mode, a very strong driving force for precipitation is needed. However, as the growth of the crystal particles proceeds, the driving force is exhausted and the growth mode is no longer in the diffusion-controlled regime. Actually, in many synthetic reactions, size focusing continues for no longer than a few minutes. When the crystallization driving force is lowered, some of the crystal particles dissolve, while others keep growing, which leads to the broadening of the size distribution. This process is called *ripening*. Ripening is a slow process compared to size focusing. Therefore, the size distribution remains narrowed for a while even after size focusing ends. Usually, the synthetic procedure of nanocrystals is terminated at this point of time.

It should be noted that size focusing requires the absence of additional nucleation as the prerequisite condition, as mentioned previously. In fact, the growth rate function has a negative slope in the nucleation period. However, the continuous generation of new nuclei disturbs the focusing effect. This means that the sooner the nucleation period ends, the more beneficial it is for size focusing. This is often referred to as separation of nucleation and growth (2, 3).

6.3 THEORETICAL DESCRIPTION OF THE CRYSTALLIZATION PROCESS

6.3.1 Derivation of the Nucleation Rate Equation

In this section, we introduce a new terminology, the monomer, which is the minimum building unit of a crystal and can both be solvated in solution and precipitate to form crystal. This monomer concept is more convenient and simpler than the conventional term, solute for the theoretical treatment of the nucleation process, because often the solute in solution and the constituent of crystal are not identical. In the remainder of this chapter, the monomer is referred to as M in the equations.

The nucleation reaction can be regarded as the phase transition of the monomer from solution to crystal. Then, the reaction rate of nucleation can be written in the Arrhenius form:

$$\frac{dN}{dt} = A \exp\left[-\frac{\Delta G_N}{kT}\right] \tag{6.1}$$

where N, A, k, and T are the number of nuclei, the pre-exponential factor, the Boltzmann constant, and temperature, respectively. ΔG_N is the free energy of nucleation. The derivation of ΔG_N is based on thermodynamic considerations. When a nucleus of radius r forms from the homogeneous solution, the change in the free energy ΔG is

$$\Delta G = 4\pi r^2 \gamma + \frac{4}{3}\pi r^3 \Delta G_V. \tag{6.2}$$

where γ is the surface free energy per unit area and ΔG_V is the free energy per unit volume of crystal. ΔG_V is expressed as the difference between the free energy of the monomer in crystal and solution:

$$\begin{aligned}\Delta G_V &= \frac{RT(\ln C_o - \ln C)}{V_m} \\ &= -\frac{RT \ln S}{V_m}\end{aligned} \tag{6.3}$$

where V_m is the molar volume of the monomer in crystal. C is the monomer concentration in the solution, and C_o is the equilibrium monomer concentration in bulk crystal. S is supersaturation, which is defined as the ratio C/C_o. S represents the driving force for both the nucleation and growth reactions.

In the homogeneous solution, nucleation is accompanied by the formation of an interface between solution and crystal at the cost of an increase in the free energy. On the other hand, the monomer in crystal has a smaller free energy than that in solution if the solution is supersaturated ($S > 1$). Therefore, there are two opposite tendencies in the nucleation reaction. The first is the increase in the free energy caused by the formation of the interface, which is reflected in the first term on the right-hand side of Equation (6.2). This term is always positive. The other tendency is the decrease in the free energy caused by the formation of the crystal, which is shown in the second term, which is negative when $S > 1$.

In Figure 6.5, the graph for Equation (6.2) is shown. Because the contributions from the surface and the volume of the nucleus are second- and third-order curves, respectively, their summation has a maximum point at $r = r_c$. The physical meaning of this graph is as follows: In the region where $r < r_c$, the only direction in which the free energy is decreased is that corresponding to the reduction of r. Consequently, any nucleus smaller than r_c dissolves away spontaneously. However, if a nucleus is larger than r_c, its growth is thermodynamically favored. As a result, r_c is the minimum radius of a nucleus that can exist in the solution. The value of r_c can be found by using $d\Delta G/dr = 0$ at $r = r_c$.

$$r_c = -\frac{2\gamma}{\Delta G_V} \tag{6.4}$$

Inserting this relation into Equation (6.2) and using Equation (6.3), we obtain the expression for ΔG_N.

$$\begin{aligned}\Delta G_N &= \frac{16\pi\gamma^3}{3(\Delta G_V)^2} \\ &= \frac{16\pi\gamma^3 V_m^2}{3(RT \ln S)^2}\end{aligned} \tag{6.5}$$

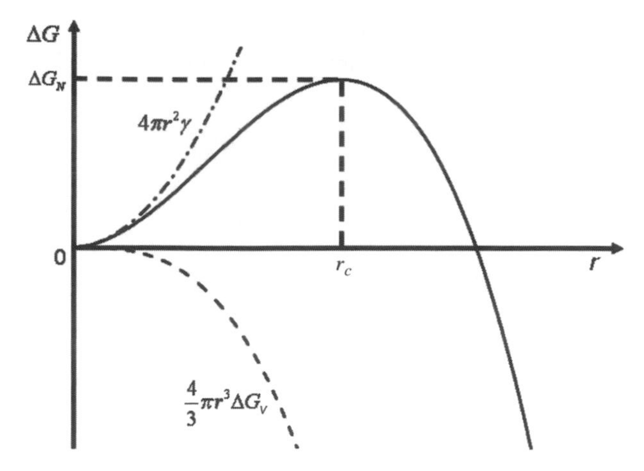

Figure 6.5 The plot for the crystallization free energy vs. particle radius (solid line). The contributions from the surface (dash-dot) and the volume (dash) are plotted separately. r_c and ΔG_N are the critical radius and the corresponding free energy, respectively.

Finally, by inserting Equation (6.5) into Equation (6.1), the nucleation rate equation is obtained (5).

$$\frac{dN}{dt} = A \exp\left[-\frac{16\pi\gamma^3 V_m^2}{3k^3T^3N_A^2(\ln S)^2}\right] \tag{6.6}$$

where N_A is Avogadro's number.

6.3.2 Derivation of the Growth Rate Equation

The growth of the crystal particle in the solution occurs via two processes. The first is the transport of monomers from the bulk solution onto the crystal surface and the second is the reaction of monomers on the surface. In this section, we derive separate equations describing these mechanisms, and then combine them to obtain the general growth rate equation.

The description of the first process begins with Fick's law of diffusion:

$$J = -D\frac{dC}{dt} \tag{6.7}$$

where J and D are the monomer flux and the diffusion constant, respectively. The concentration gradient near the surface of a nanocrystal particle is shown in Figure 6.6, along with a schematic of the diffusion layer structure. Let x be the radial distance from the center of the particle. The monomer concentration is C_s at the surface of the crystal ($x = r$). At a certain point far enough from the particle ($x = r + \delta$), the concentration reaches C_b, the bulk concentration of the solution. Now, consider a spherical

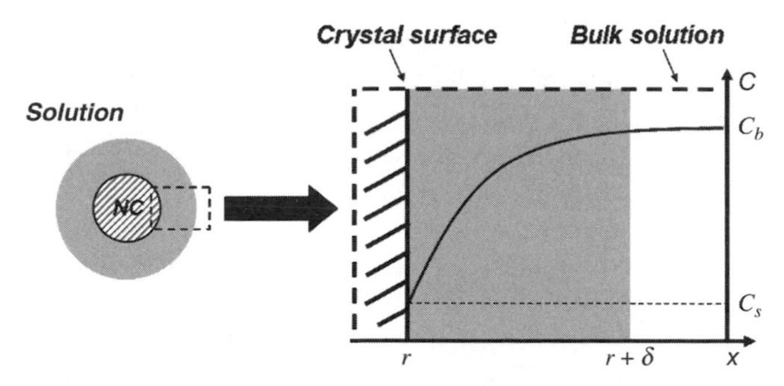

Figure 6.6 Left: a schematic illustration of diffusion layer structure near the surface of a nanocrystal (NC). The shaded area indicates the diffusion layer. Right: Plot for the monomer concentration as a function of distance x.

surface of radius x whose center is identical to that of the nanocrystal. From Equation (6.7), the diffusion rate of the monomers, dM/dt, through this surface is

$$\begin{aligned}\frac{dM}{dt} &= J \cdot \mathrm{A} \\ &= -4\pi x^2 D \frac{dC}{dx}\end{aligned} \tag{6.8}$$

In the steady state, dM/dt is constant at any x. After dividing both sides with x^2, the equation can be integrated from r to $r + \delta$ and from C_s to C_b for left and right-hand sides, respectively. Then, letting $r \ll \delta$, we have

$$\frac{dM}{dt} = 4\pi Dr(C_s - C_b) \tag{6.9}$$

The increasing rate of the particle volume is equal to the monomer supply rate. In equation form, this can be expressed as:

$$\frac{dM}{dt} = -\frac{4\pi r^2}{V_m}\frac{dr}{dt} \tag{6.10}$$

Equating (6.9) and (6.10), we get

$$\frac{dr}{dt} = \frac{V_m D}{r}(C_b - C_s) \tag{6.11}$$

This is a growth rate equation obtained by considering only the mass transport process (9). Although simplified, this equation is a good approximation when C_s varies little with r as is the case on the micrometer scale. The hyperbola shape of the graph in Figure 6.3 is derived from this equation (Problem 2). However, on the nanometer scale, C_s is very sensitive to r and Equation (6.11) is far from accurate.

To modify Equation (6.11), we need an expression of C_s as a function of r. This can be obtained by considering the second process of the crystal growth. On the crystal surface, the precipitation and dissolution reactions occur simultaneously. Let M^S and M^C be the monomer in the solution and in the crystal, respectively. Then, the equation for those two reactions is

$$nM^S \underset{k_d}{\overset{k_p}{\rightleftharpoons}} M_n^C \tag{6.12}$$

where k_p and k_d are the reaction rate constants for the precipitation and dissolution, respectively. Their reaction rates are expressed as follows:

$$\frac{dM^C}{dt} = 4\pi r^2 k_p C_s \tag{6.13}$$

$$\frac{dM^S}{dt} = 4\pi r^2 k_d \tag{6.14}$$

Note that precipitation is assumed to be a first-order reaction of C_s, while dissolution is independent of it. To obtain the expression for C_s, we utilize the relation between dM/dt, dM^C/dt, and dM^S/dt.

$$\frac{dM}{dt} = \frac{dM^S}{dt} - \frac{dM^C}{dt} \tag{6.15}$$

By inserting Equations (6.9), (6.13), and (6.14) into this equation and rearranging, we get an equation for C_s.

$$C_s = \frac{DC_b + k_d r}{D + k_p r} \tag{6.16}$$

However, because k_p and k_d in this equation are also functions of r, we have to know these functions to complete the derivation of the growth rate equation.

The relation between the chemical potential of a particle and its radius is known as the Gibbs–Thomson relation. Let μ_C° and $\mu_C(r)$ be the chemical potentials of bulk crystal and a particle with radius r, respectively. Then, their difference $\Delta\mu$ is

$$\Delta\mu = \frac{2\gamma V_m}{r} \tag{6.17}$$

That is, the smaller particle has a higher chemical potential (Problem 3). The effect of the change in the chemical potential of the particle on the reaction kinetics of precipitation/dissolution is depicted in Figure 6.7. In this figure, the smallest particle has the highest chemical potential and the precipitation reaction faces a higher energy barrier than the dissolution reaction (left). As the particle size increases, the chemical

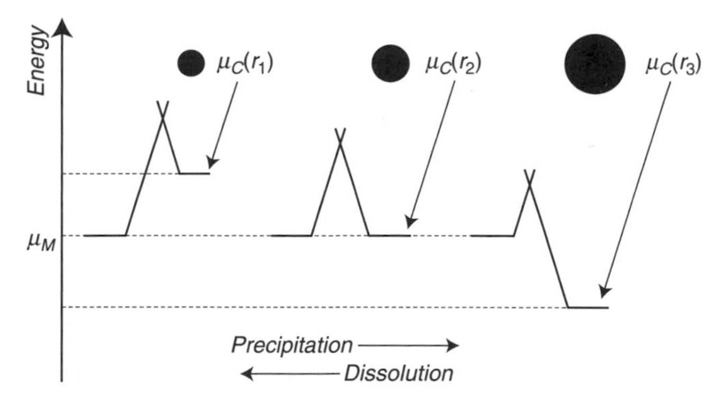

Figure 6.7 Schematics of energy vs. reaction coordinate for particles of three different sizes. The direction from left to right corresponds to the precipitation reaction and the reverse to dissolution. μ_M and $\mu_C(r)$ are the chemical potential of monomers in the solution and particle with radius r, respectively. Black circles indicate the sizes of corresponding particles.

potential of the particle is lowered and the situation is reversed, that is, the barrier for dissolution is higher than the barrier for precipitation (right). Consequently, the increase of r results in a higher k_p and lower k_d.

The example above can be expressed by equations as follows: In general, the reaction constant is proportional to the term $\exp(-\Delta E/RT)$, where ΔE is the molar activation energy. Then, the equations for k_p and k_d reflecting the change of the activation energy according to the chemical potential of the particle are

$$k_p = k_p^\circ \exp\left[-\alpha \frac{\Delta\mu}{RT}\right] \tag{6.18}$$

$$k_d = k_d^\circ \exp\left[(1-\alpha)\frac{\Delta\mu}{RT}\right] \tag{6.19}$$

In these equations, α is the transfer coefficient and the degree sign ($^\circ$) indicates that the parameter is for bulk crystal. Inserting Equation (6.17) into them yields

$$k_p = k_p^\circ \exp\left[-\alpha \frac{2\gamma V_m}{rRT}\right] \tag{6.20}$$

$$k_d = k_d^\circ \exp\left[(1-\alpha)\frac{2\gamma V_m}{rRT}\right] \tag{6.21}$$

Lastly, putting together Equations (6.11), (6.16), (6.20), and (6.21), we have the general growth rate equation (7, 8):

$$\frac{dr^*}{d\tau} = \frac{S - \exp(1/r^*)}{r^* + K\exp(\alpha/r^*)} \tag{6.22}$$

where the three normalized dimensionless variables are defined as follows:

$$r^* = \frac{RT}{2\gamma V_m} r \tag{6.23}$$

$$\tau = \frac{R^2 T^2 D k_d^\circ}{4\gamma^2 V_m k_p^\circ} t \tag{6.24}$$

$$K = \frac{RT}{2\gamma V_m} \frac{D}{k_p^\circ} \tag{6.25}$$

6.3.3 Features of the Growth Process and Ostwald Ripening

The features of the growth rate function are very important for understanding the size distribution control mechanism. So it is worthwhile to deal with them in a separate section. In Figure 6.8, the graphs of Equation (6.22) are shown. Generally, there is a maximum in the graph, and it converges to zero when $r^* \to \infty$ and diverges to negative infinity when $r^* \to 0$. The negative value of the growth rate means that the particle is dissolving. At this point, please recall that the mass transport process allows smaller particles to grow faster than larger ones. One the other hand, according to the Gibbs–Thomson effect, smaller particles tend to dissolve rather than grow, because they are more thermodynamically unstable than larger particles. The growth rate graphs in Figure 6.8 describe how these two opposite effects compete with each other. Let $r^* = r^*_{\max}$ be the position of the maximum in the growth rate. When $r^* < r^*_{\max}$, the Gibbs–Thomson effect is overwhelming and the particles are too unstable to grow. In the region where $r^* > r^*_{\max}$, the two tendencies take effect simultaneously and the balance between them is largely affected by two parameters, K and S.

K is a ratio between the diffusion rate and the precipitation reaction rate (7, 8). When $K \ll 1$, the diffusion rate is far slower than the reaction rate and the overall growth kinetics is controlled by the diffusion process. As mentioned previously, this growth mode is called diffusion-controlled growth. On the other hand, in *reaction-controlled growth*, $K \gg 1$ and the kinetics is controlled by the reaction rate.

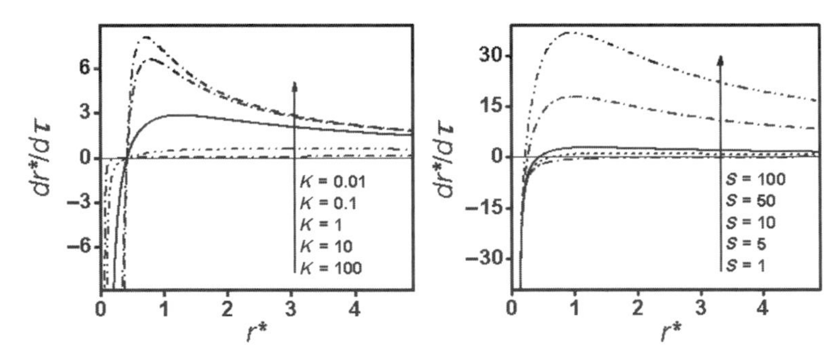

Figure 6.8 Graphs of the growth rate function in Equation (6.22) for various values of K (left) and S (right). $S = 10$ in the left plot and $K = 1$ in the right. The values of K and S are listed in the order of the corresponding graphs in the arrow direction.

On the left of Figure 6.8, one can see that the graph in the region where $r^* > r^*_{max}$ acquires a more hyperbola-like shape as K approaches to zero, which is reminiscent of Figure 6.4. The increase of S enhances the driving force for the precipitation reaction, making the growth kinetics more dependent on the diffusion rate than on the reaction rate. Therefore, increasing S has a similar effect to decreasing K, as shown in the right-hand graph in Figure 6.8. To summarize, the optimal condition for size focusing is satisfied when the precipitation reaction is much faster than the diffusion rate ($K \ll 1$) and/or when the level of supersaturation is very high ($S \gg 1$).

In the course of the growth process, the main parameter affecting the time evolution of the particle size distribution is the level of supersaturation, S. In the synthesis of nanocrystals, the supersaturation level is high at the early stage of the growth process. The shape of the growth rate function at that time is somewhat like that on the left of Figure 6.9. As shown in Figure 6.8, when S is large, r^*_{max} is small and the negative slope of the graph is steep in the region where $r^* > r^*_{max}$. As a result, the whole of the size distribution curve is situated in the region where the growth rate curve has a steep negative slope, namely, the *focusing region*. In this condition, size focusing takes effect for all particles in the system, as described in Figure 6.4. However, the consumption of the monomers by the particles lowers the supersaturation level. Consequently, as the growth process proceeds, the shape of the growth rate function changes to the one on the right of Figure 6.9, causing some of the particles to be outside of the focusing region. When the supersaturation level is low, the broadening of the size distribution occurs mainly via the *Ostwald ripening* process. In this process, a considerable number of the smaller particles dissolve, and the larger particles grow by receiving monomers from the dissolving particles (10). However, in some cases in which the particle size distribution is very broad initially, Ostwald ripening could contribute to improving the size uniformity by eliminating the smaller particles.

In the following two sections, we will discuss the kinetics of two representative synthetic procedures for monodisperse spherical nanocrystals, which are the hot injection and heat-up methods. Although there are other synthetic methods for monodisperse nanocrystals, their formation kinetics have not been clearly elucidated yet (11–13).

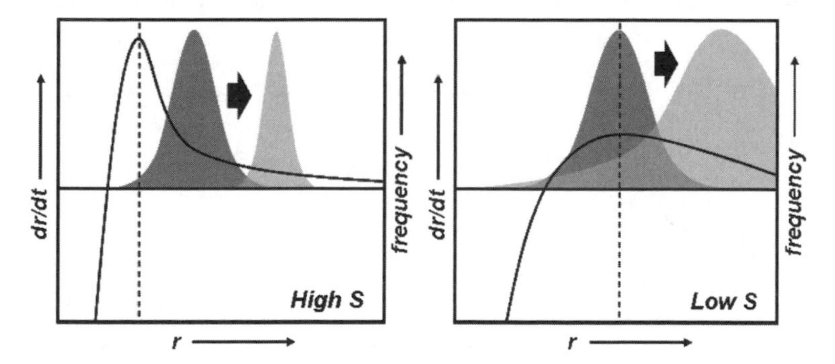

Figure 6.9 Schematics of plots for the growth rate (solid line) and the particle size distribution (shaded area). The arrows mean the time evolution of the size distribution and the vertical dashed lines indicate the positions of the maximum growth rate point.

6.4 KINETICS OF THE HOT INJECTION METHOD

The hot injection method is one of the earliest synthetic methods for monodisperse nanocrystals and has been widely used for the synthesis of uniform nanocrystals of various materials (4, 14–16). In this section, we will discuss the kinetics of the synthesis of monodisperse CdSe nanocrystals via the hot injection method.

6.4.1 The Synthetic Procedure

In 1993, the Bawendi group published the legendary paper in which the use of the hot injection method for the synthesis of uniform nanocrystals of cadmium chalcogenides was introduced (14). The size of the CdSe nanocrystals could be tuned from 1.2 to 12 nm by varying the experimental conditions. The CdSe nanocrystals synthesized exhibited the strong quantum confinement effect. This pioneering work has been extended to the synthesis of nanocrystals of various materials, including semiconductors, metals, and metal oxides (14). Later various modified hot injection methods have been developed to synthesize high quality monodisperse CdSe nanocrystals (16).

CdSe is a compound II-VI semiconductor composed of Cd^{2+} and Se^{2-} ions. In the original synthetic procedure described by the Bawendi group, dimethyl cadmium [$(Me)_2Cd$] and tri-*n*-octylphosphine selenide (TOPSe) were used as the precursors for Cd and Se, respectively. Later, other cadmium compounds such as cadmium acetate [$Cd(Ac)_2$] and CdO were used alternatively, because $(Me)_2Cd$ is extremely toxic and pyrophoric. Although the exact reaction pathway has not been clearly elucidated, it is thought that atoms of Cd and Se are released via the thermal decomposition of the precursors.

Surfactants are crucial in the colloidal chemical synthesis of nanocrystals. Popularly used surfactants include tri-*n*-octylphosphine (TOP), tri-*n*-octylphosphine oxide (TOPO; 6, 14), hexadecylamine (17), fatty acids, and phosphonic acids (18–20). In general, the polar head groups of the surfactants bind to the surface of the nanocrystals, as illustrated in Figure 6.10 (21). The surfactants play crucial roles in the nanocrystal synthesis (14–16). First, the surfactant capping prevents the agglomeration of the nanocrystals, endowing them with a good colloidal stability. Nanocrystals have a

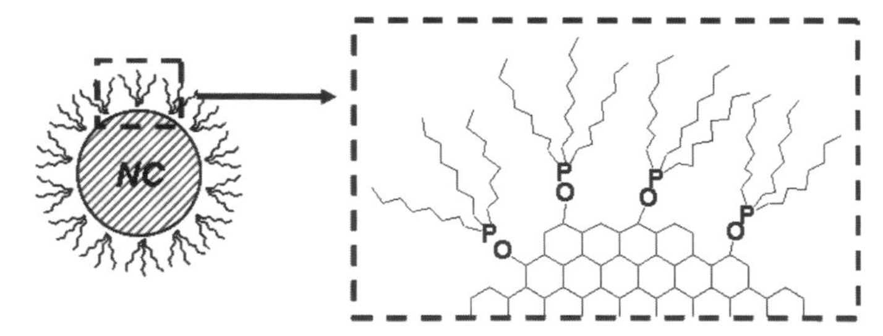

Figure 6.10 A nanocrystal capped with TOPO molecules. The surface is magnified in the right-hand box. The honeycomb-like structure in the box represents the wurtzite structure of the CdSe crystal.

strong tendency to aggregate to relieve their high surface free energy derived from the Gibbs–Thomson effect. The long alkyl chains in the tail groups of the surfactants prevent the agglomeration of the nanocrystals via so-called steric stabilization. These hydrophobic alkyl chains of the surfactant make the nanocrystals hydrophobic, thus allowing them to be well dispersed in the organic solvent. Second, the surfactants control the nucleation and growth rates during the nanocrystal formation. For example, as the bulkiness of the surfactant tail decreases from TOP/TOPO to tri-*n*-butyl, -ethyl and -methylphosphine/-phosphine oxide, the crystal growth temperature can be lowered from 280°C to 230°C, 100°C, and 50°C (14). The dense surfactant layer can block the approach of the monomers in the solution to the nanocrystal surface, thus retarding the crystallization reaction. Lastly, the surfactant layer protects the nanocrystal surface against oxidation. Moreover, the coordination of the surfactant head groups onto the surface defects and dangling bonds can passivate the charge trap sites and improve the optical properties of the CdSe nanocrystals.

The synthetic reaction for CdSe nanocrystals is, in any case, basically the precipitation reaction of Cd and Se. The reaction temperature is usually higher than 300°C. Although the precipitation reaction can occur at much lower temperatures, the synthesis is generally performed at such a high temperature because both the nucleation and crystallization rates need to be extremely fast to obtain monodisperse nanocrystals, as discussed in Section 6.2. In addition, the high temperature leads to the *in situ* annealing of the nanocrystals, improving their crystallinity (21). Coordinating solvents such as a mixture of TOPO and TOP are usually used as the precipitation medium (6, 14, 17). Alternately, instead of a coordinating solvent, long-chain hydrocarbon solvents containing strongly binding surfactants, for example, oleic acid dissolved in 1-octadecene, are used (20). To initiate the reaction, a stock solution containing the precursor is injected rapidly into the hot surfactant solution. The injection leads directly to the precipitation of CdSe nanocrystals. Between the injection and the precipitation, there is little or no induction time. This immediacy is characteristic of the hot injection method and is essential for the formation of monodisperse nanocrystals, as is discussed below.

6.4.2 Size Distribution Control

In the hot injection process, the high reaction temperature, the reactive precursors, and the rapid injection cooperatively render the reaction system extremely highly supersaturated at the start. Soon after the injection, the reaction kinetics afterwards runs on a steep downhill slope of the chemical potential, along with a rapid decrease in the supersaturation level. Then, how does the nanocrystal size distribution evolve as the reaction proceeds?

To characterize the size distribution control process in nanocrystal synthesis, we need to trace the temporal change of the following variables: the number concentration of the nanocrystals, N, the mean value of their size distribution, $<r>$ and its relative standard deviation, $\sigma_r(r)$. The experimental results are shown in Figure 6.11. The left-hand plot shows that the nucleation rate is extremely high at the start of the reaction. After the burst increase, the number concentration of nanocrystals soon reached the

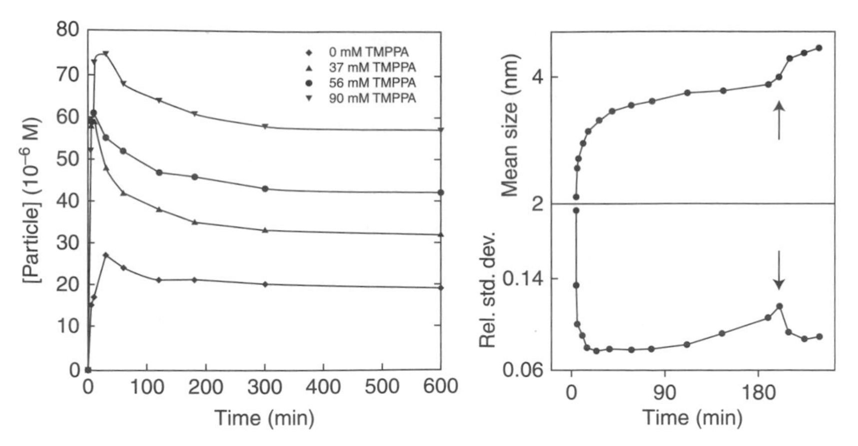

Figure 6.11 The temporal change of the number concentration of CdSe nanocrystals for various surfactant concentrations (left). The surfactant was bis-(2,2,4-trimethylpentyl) phosphinic acid (TMPPA). The time evolution of the mean size and the relative standard deviation are shown together (right). The arrows indicate the additional precursor injection time. In all plots, the injection time is set as zero ($t = 0$). Reprinted with permission from reference 22 (left) and reference 6 (right). Copyright 2005 and 1998, American Chemical Society.

maximum point, which can be regarded as the end of the nucleation period. The decrease in the number of nanocrystals after the maximum point is due to the dissolution of the smaller nanocrystals via the Ostwald ripening process, which indicates the lowering of the supersaturation level in the solution. In the right-hand plot, we can see that the high growth rate is correlated with the narrowing of the size distribution in the early stage of the growth process. After this rapid narrowing, the size distribution slowly broadens simultaneously with the increase in the mean size, which is another characteristic of Ostwald ripening. The crucial role of the high supersaturation level in the formation of uniformly sized nanocrystals is confirmed by the additional precursor injection. As shown in the figure, this injection reversed the Ostwald ripening process for a while by increasing the supersaturation level. Consequently, the experimental data prove that burst nucleation and size focusing actually take effect in the hot injection process.

A numerical simulation using the theoretical model introduced in Section 6.3 can successfully reproduce the hot injection process (the left of Fig. 6.12). Using the simulation result, we can observe the main events of the size distribution control process in detail (the right of Fig. 6.12; Reference 4). The reaction process is divided into two periods, the nucleation and growth periods, which are similar to those in Figure 6.3. In the nucleation period, the nucleation reaction proceeds while retaining a high $\sigma_r(r)$. Because the activation energy for the nucleation reaction is much higher than that for the precipitation reaction, the nucleation stops at a supersaturation level that is still high enough for the precipitation reaction to occur. Consequently, the system evolves spontaneously from the nucleation period to the growth period. In the growth period, there is no nucleation and the supersaturation level is high, which

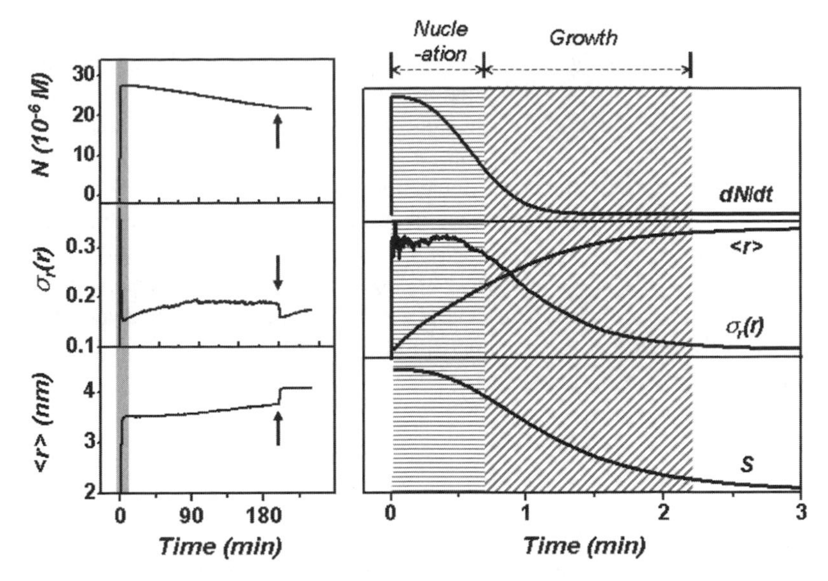

Figure 6.12 The computer simulation result reproducing the hot injection process in Figure 6.11 (left). The arrows indicate the additional precursor injection time. The vertical grey stripe indicates the first few minutes after the injection. The temporal changes of various parameters in that period are shown together (right). dN/dt is the nucleation rate. Shaded areas with the horizontal and slanted stripes are the nucleation and growth period, respectively.

are exactly the conditions required for size focusing. As a result, the focusing effect takes place and the value of $\sigma_r(r)$ decreases with increasing mean size of the nanocrystals. To summarize, in the hot injection process, the high initial supersaturation level induces both burst nucleation and subsequent size focusing with the consumption of the monomers.

6.5 KINETICS OF THE HEAT-UP METHOD

Heat-up is a simple but effective method of synthesizing highly uniform nanocrystals, which yields a degree of size uniformity comparable to that of the best result from the hot injection method. This method is adopted mainly for the synthesis of metal oxide nanocrystals. In this section, we describe the synthetic procedure of iron oxide nanocrystals via the heat-up method as a representative example (23–25).

6.5.1 The Synthetic Procedure

Both the heat-up and hot injection methods utilize precipitation in organic solutions for the formation of nanocrystals. These two synthetic methods share some common features. For example, the same surfactants are generally used that allow for colloidal stability and reaction rate control. Long-chain hydrocarbons such as

1-octadecene are frequently used as solvents. However, there is a clear difference between these two synthetic procedures. In the heat-up method, there is no operation that abruptly induces high supersaturation. The size distribution control mechanism of this method relies mainly on the reaction kinetics of the precursors and the heat procedure.

The precursors used for the iron oxide nanocrystal synthesis via the heat-up method are various iron carboxylate complexes, including the most widely used iron-oleate complex. Generally, when heated, metal carboxylate complexes thermally decompose at temperatures near 300°C or higher to produce metal oxide nanocrystals along with some byproducts, such as CO, CO_2, H_2, water, ketones, esters, and various hydrocarbons. It is thought that the decomposition reaction proceeds via the formation of thermal free radicals from metal carboxylate (26, 27):

$$\mathrm{M-OOCR} \longrightarrow \mathrm{M^{\bullet} + RCOO^{\bullet}} \tag{6.26}$$

$$\mathrm{M-OOCR} \longrightarrow \mathrm{MO^{\bullet} + RC^{\bullet}O} \tag{6.27}$$

In the synthesis of the nanocrystals, a homogeneous iron-oleate solution prepared at room temperature is heated to 320°C, which is the thermal decomposition temperature of iron-oleate complex, and held at that temperature (24). As shown in Figure 6.13, the curve for the reaction extent of the thermal decomposition of iron-oleate complex in the solution shows a sigmoidal shape, which is typical of autocatalytic reactions. Interestingly, there is a time lag between the onset of the reaction extent curve and the initiation of the nanocrystal formation. As shown in the right of the figure, when the solution temperature just reaches 320°C, there is a trace amount of nanocrystals in the solution while about half of the iron-oleate complex has already been decomposed. This implies that iron oxide crystal is not a direct product of the thermal decomposition of iron-oleate complex. Rather, when iron-oleate complex is thermally

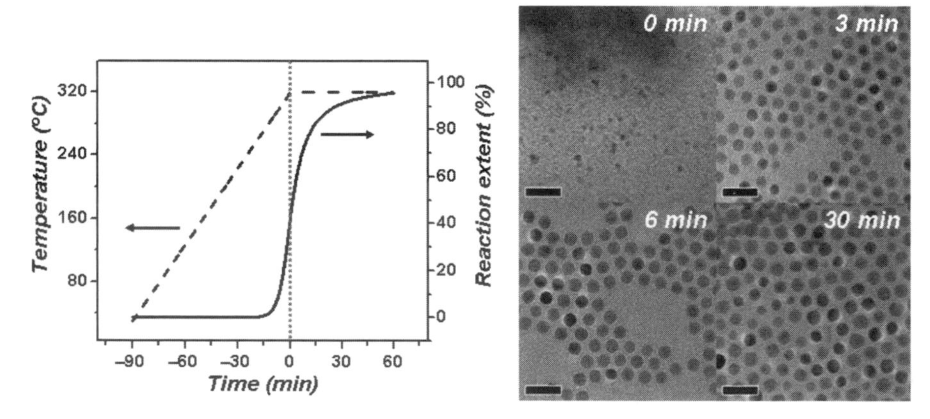

Figure 6.13 The temporal change of the solution temperature and the reaction extent of the thermal decomposition (left). The time when solution temperature just reached 320°C is set as zero ($t = 0$) and indicated by a vertical dotted line. The TEM images of the nanocrystals in the solution at different times of aging at 320°C are shown (right). All scale bars are 20 nm.

decomposed in the solution, it seems to be converted into more labile species, supposedly, polyiron oxo clusters, that can precipitate as iron oxide. In other words, the intermediate species produced by the decomposition of iron-oleate complex seems to act as the monomer for the iron oxide precipitation reaction (28).

As shown in Figure 6.13, the nanocrystals initially generated grow rapidly and become uniform in size within a few minutes. Further aging at 320°C induces a slow ripening process. During the heat process, the reduction of Fe^{3+} ions to Fe^{2+} ions occurs by *in situ* generated H_2. As a result, the iron oxide nanocrystals that are produced are generally composed of γ-Fe_2O_3 and Fe_3O_4. When the reaction temperature is higher than 380°C, extensive reduction occurs to produce nearly pure Fe nanocrystals (24, 29). The size of the nanocrystals can be controlled by varying the reaction conditions such as the temperature, aging time, surfactant, precursor concentration, and the precursor to surfactant ratio. In general, it is very hard to obtain monodisperse iron oxide nanocrystals smaller than 5 nm using the heat-up method, whereas the hot injection method can produce nanocrystals as small as 1 to 2 nm.

The most attractive characteristics of the heat-up method are its simplicity and reliability. The synthetic procedure can be easily scaled up to yield nanocrystals in quantities as high as several tens of grams (24). In the following section, we will discuss how this very simple heat-up method can produce monodisperse nanocrystals.

6.5.2 Size Distribution Control

The time evolutions of both the number concentration and the size distribution of the iron oxide nanocrystals are shown in Figure 6.14. Interestingly, although the procedures used to initiate the precipitation reaction of the heat-up and hot injection processes are very much different, the precipitation reactions themselves proceed in a similar way once they are started. Comparing Figure 6.11 and Figure 6.14, it can be seen that the nucleation and growth process of the iron oxide nanocrystals via the heat-up method is very similar to that observed in the hot injection process in that there is a sudden increase in the number concentration of the nanocrystals and a rapid narrowing of the size distribution accompanied by a high growth rate. In other words, burst nucleation and size focusing also take place in the heat-up process, as in the hot injection process. Considering this similarity in the nanocrystal formation kinetics in these two seemingly different synthetic processes, a high supersaturation level is critical for the nucleation and growth process to occur in the heat-up process. In the hot injection process, the synthetic reaction is initiated intentionally in the high supersaturation condition. On the other hand, in the heat-up process for the synthesis of iron oxide nanocrystals, the actual monomer is not iron-oleate complex but the intermediate species produced by the thermal decomposition of iron-oleate complex, as mentioned above. As a result, in the heat-up process, the solution is not supersaturated at the start and the high supersaturation is induced in a manner different from the hot injection process.

The induction of the high supersaturation and the initiation of burst nucleation are explained by the theory of homogeneous nucleation. As discussed in Section 6.3.1, the interfacial free energy acts as an energy barrier for the nucleation reaction. Because the

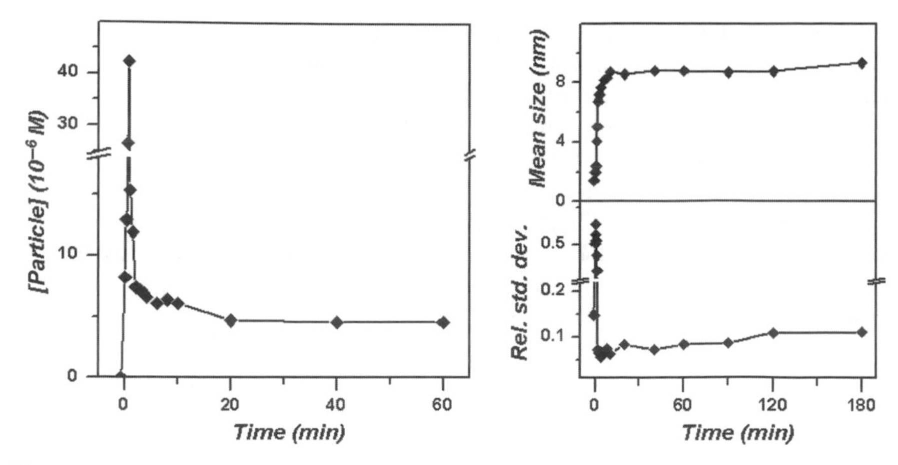

Figure 6.14 The temporal change of the number concentration of iron oxide nanocrystals (left). The time evolution of their mean size and the relative standard deviation are shown together (right). The time is set as zero when solution temperature just reached 320°C, just the same as in Figure 6.13.

surface to volume ratio of a nucleus is very high, the energy barrier is high enough to prevent the nucleation reaction even at considerably high supersaturation levels. As a result, as the temperature increases, the monomers (the intermediate species) generated from the thermal decomposition of iron-oleate complex accumulate in the solution until the supersaturation is high enough to overcome the energy barrier. The nucleation process is initiated suddenly during the heat procedure as if a switch is turned on. However, as in the case of the hot injection process, the monomer consumption of the nuclei lowers the supersaturation level and the nucleation is terminated soon.

We performed a numerical simulation to investigate the size distribution control mechanism in the heat-up process (28). As shown in thc left-hand plots of Figure 6.15, the simulation results matched very well with the experimental results shown in Figure 6.14. In the right-hand plots of Figure 6.15, the formation process of the nanocrystals can be divided into three periods. In the first period, only the accumulation of the monomers in the solution takes place and the nucleation process is suppressed by the energy barrier for the nucleation reaction. In the second period, the nucleation process is initiated suddenly. The combination of the rapid nucleation and growth of the nanocrystals leads to the increase of the relative standard deviation of the size distribution of the nanoparticles in this period. In the third period, the nucleation process is terminated due to the decrease in the supersaturation level and only the growth process proceeds. In this period, the conditions required for size focusing, namely no additional nucleation and high supersaturation, are both satisfied and fast narrowing of the size distribution occurs.

Notably, the characteristics of the nucleation and growth periods in the heat-up process are the same as those in the hot injection process (Figs. 6.12 and 6.15). In both cases, burst nucleation and size focusing are kinetically driven by the high supersaturation level at the start of the nucleation period. With respect to the size distribution

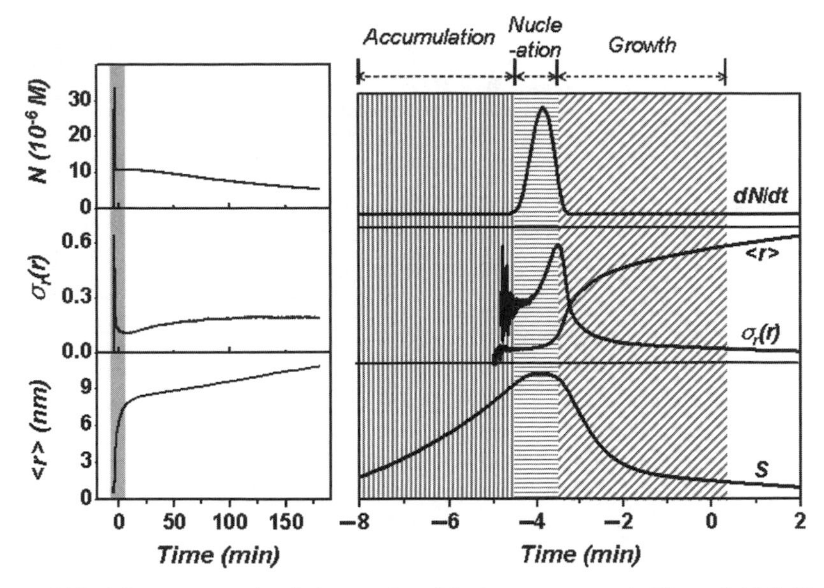

Figure 6.15 The computer simulation results of the heat-up process in Figure 6.14 (left). The temporal changes of various parameters near $t = 0$ are shown in the right plots. The symbols and notations are the same as those in Figure 6.12 except for the shaded area with the vertical stripes, which indicates the accumulation period.

control, the essential difference between the two methods is the way in which the high supersaturation condition is achieved before the initiation of the precipitation process. In the hot injection method, it is accomplished by an external operation, namely, the rapid injection of the reactive precursors. In the heat-up method, the accumulation of the intermediate species caused by the thermal decomposition of iron-oleate complex induces the high supersaturation level.

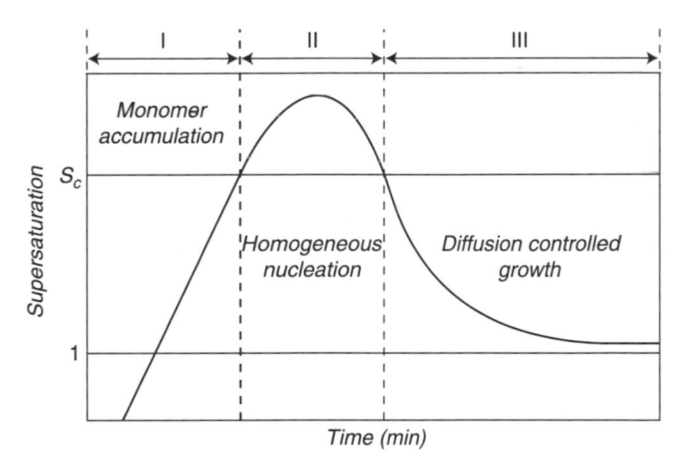

Figure 6.16 The LaMer diagram. S_c is the critical supersaturation, the minimum supersaturation level for the homogeneous nucleation.

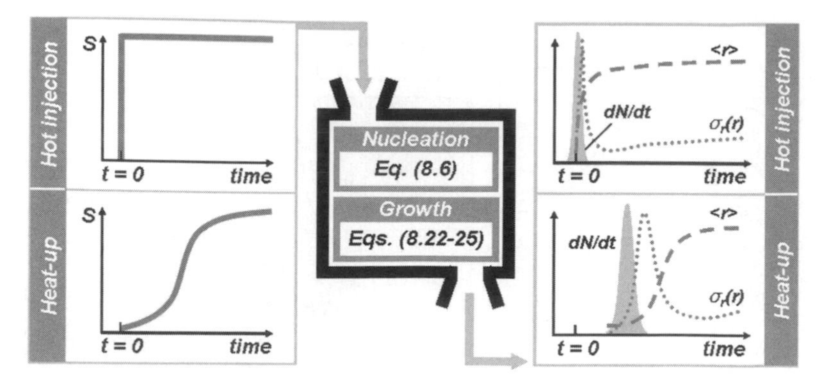

Figure 6.17 Schematic of the size distribution control mechanism of the hot injection and heat-up methods. In the left boxes, the monomer supply modes are shown as the plots of supersaturation vs. time. In the right boxes, the resulting time evolutions of the nucleation rate, the mean size, and the relative standard deviation are shown. The injection time and the start of the heat procedure are set as $t = 0$ in the hot injection and the heat-up processes, respectively.

It is very interesting that both the heat-up and hot injection methods can be well fitted to the LaMer model, a classical theory for the formation of uniform particles (2, 3). The LaMer diagram shows schematically how the rapid nucleation and the separation of nucleation and growth are achieved under the condition of continuous monomer supply (Fig. 6.16). In the diagram, the whole particle formation process is divided into three stages; the prenucleation stage (stage I), the nucleation stage (stage II), and the growth stage (stage III). These three stages coincide with the accumulation of the monomers in the first period, the burst nucleation in the second period, and the size focusing in the third period, respectively, in the current heat-up process, as depicted in Figure 6.15. Schematically, the hot injection process can be regarded as a special case of the LaMer model in which stage I is omitted. Consequently, the heat-up and hot injection processes share stages II and III in the LaMer diagram.

6.6 SUMMARY

In this chapter, we described the basic concepts of the control of the size distribution and the theory of the nucleation and growth reactions of two representative synthetic methods for monodisperse nanocrystals, the hot injection and heat-up methods. Mechanistic studies reveal that the size distribution control processes of these two methods are virtually the same, even though the synthetic procedures are quite different. The size distribution control mechanisms of these two synthetic methods are summarized schematically in Figure 6.17. Given that the nucleation and growth models are governed by Equations (6.6) and (6.22) to (6.25), the two different monomer supply modes induce very similar size distribution control mechanisms, which comprise burst nucleation followed by a size focusing growth process.

REFERENCES

1. V. K. LaMer, R. H. Dinegar. *J. Am. Chem. Soc.* **1950**, *72*, 4847.
2. T. Sugimoto. *Adv. Colloid Interfac. Sci.* **1987**, *28*, 65.
3. T. Sugimoto. *Monodispersed Particles*. Amsterdam: Elsevier Science, **2001**.
4. J. Park, J. Joo, S. G. Kwon, Y. Jang, T. Hyeon. *Angew. Chem. Int. Ed.* **2007**, *46*, 4630 and references therein.
5. J. W. Mullin. *Crystallization* (4th edn). Oxford: Oxford University Press, **2001**.
6. X. Peng, J. Wickham, A. P. Alivisatos. *J. Am. Chem. Soc.* **1998**, *120*, 5343.
7. D. V. Talapin, A. L. Rogach, M. Haase, H. Weller. *J. Phys. Chem. B* **2001**, *105*, 12278.
8. D. V. Talapin, A. L. Rogach, E. V. Shevchenko, A. Kornowski, M. Haase, H. Weller. *J. Am. Chem. Soc.* **2002**, *124*, 5782.
9. H. Reiss. *J. Chem. Phys.* **1951**, *19*, 482.
10. A. S. Kabalnov, E. D. Shchukin. *Adv. Colloid Interfac. Sci.* **1992**, *38*, 69.
11. S. Stoeva, K. J. Klabunde, C. M. Sorensen, I. Dragieva. *J. Am. Chem. Soc.* **2002**, *124*, 2305.
12. B. L. V. Prasad, S. I. Stoeva, C. M. Sorensen, K. J. Klabunde. *Langmuir* **2002**, *18*, 7515.
13. B. L. V. Prasad, S. I. Stoeva, C. M. Sorensen, K. J. Klabunde. *Chem. Mater.* **2003**, *15*, 935.
14. C. B. Murray, D. J. Norris, M. G. Bawendi. *J. Am. Chem. Soc.* **1993**, *115*, 8706.
15. C. B. Murray, C. R. Kagan, M. G. Bawendi. *Annu. Rev. Mater. Sci.* **2000**, *30*, 545.
16. C. de Mello Donega, P. Liljeroth, D. Vanmaekelbergh. *Small* **2005**, *1*, 1152.
17. D. V. Talapin, A. L. Rogach, A. Kornowski, M. Haase, H. Weller. *Nano Lett.* **2001**, *1*, 207.
18. L. Qu, Z. A. Peng, X. Peng. *Nano Lett.* **2001**, *1*, 333.
19. Z. A. Peng, X. Peng. *J. Am. Chem. Soc.* **2001**, *123*, 183.
20. W. W. Yu, X. Peng. *Angew. Chem. Int. Ed.* **2002**, *41*, 2368.
21. J. E. B. Katari, V. L. Colvin, A. P. Alivisatos. *J. Phys. Chem.* **1994**, *98*, 4109.
22. J. V. Embden, P. Mulvaney. *Langmuir* **2005**, *21*, 10226.
23. N. R. Jana, Y. Chen, X. Peng. *Chem. Mater.* **2004**, *16*, 3931.
24. J. Park, K. An, Y. Hwang, J.-G. Park, H.-J. Noh, J.-Y. Kim, J.-H. Park, N.-M. Hwang, T. Hyeon. *Nat. Mater.* **2004**, *3*, 891.
25. W. W. Yu, J. C. Falkner, C. T. Yavuz, V. L. Colvin. *Chem. Commun.* **2004**, 2306.
26. A. S. Rozenberg, V. R. Stepanov. *Russ. Chem. Bull. Int. Ed.* **1996**, *45*, 1336.
27. F. Kenfack, H. Langbein. *Thermochim. Acta.* **2005**, *426*, 61.
28. S. G. Kwon, Y. Piao, J. Park, S. Angappane, Y. Jo, N.-M. Hwang, J.-G. Park, T. Hyeon. *J. Am. Chem. Soc.* **2007**, *129*, 12571.
29. D. Kim, J. Park, K. An, N.-K. Yang, J.-G. Park, T. Hyeon. *J. Am. Chem. Soc.* **2007**, *129*, 5812.

PROBLEMS

1. Using the graphical method depicted in Figure 6.4, verify that the steeper the graph, the faster the size focusing process occurs.

2. Assuming C_b and C_s are constants, from Equation (6.11) derive that;

$$\frac{d(\sigma_r^2)}{dt} = 2V_m D(C_b - C_s)\left[1 - \bar{r} \cdot \overline{\left(\frac{1}{r}\right)}\right]$$

where

$$\bar{r} = \frac{1}{N}\sum_{i=1}^{N} r_i$$

and

$$\overline{\left(\frac{1}{r}\right)} = \frac{1}{N}\sum_{i=1}^{N} \frac{1}{r_i}$$

N is the number of particles in the system. According to the arithmetic-geometric-harmonic mean inequalities, $\overline{(1/r)}$ is always greater than $1/\bar{r}$ and, thus, the right-hand side of the above equation is negative. This shows that the size distribution is always narrowed when the growth rate is proportional to $1/r$ and C_b and C_s are constants. (*Hint*: use the relation $\sigma_r^2 = \overline{r^2} - (\bar{r})^2$.)

3. Using the relations $dG/dn = \Delta\mu$ and $dG = \gamma dA$, derive Equation (6.17). (*Hint*: for a spherical particle, $dA = 8\pi r dr$.)

4. What are the roles of the surfactants in the colloidal chemical synthesis of nanocrystals?

5. What are the two most important requirements (or conditions) for the formation of monodisperse nanocrystals?

6. Referring to the LaMer plot in Figure 6.16, explain what happens in each of the three stages.

ANSWERS

2. From the definition of σ_r, we have

$$\frac{d(\sigma_r^2)}{dt} = \frac{d}{dt}\left[\frac{1}{N}\sum_{i=1}^{N} r_i^2 - \left(\frac{1}{N}\sum_{j=1}^{N} r_j\right)^2\right]$$

$$= \frac{1}{N}\sum_{i=1}^{N} 2r_i\frac{dr_i}{dt} - \frac{2}{N^2}\sum_{j=1}^{N} r_j \sum_{k=1}^{N}\frac{dr_k}{dt}$$

Inserting Equation (6.11), we can get

$$\frac{d(\sigma_r^2)}{dt} = \frac{V_m D}{N}(C_b - C_s)\sum_{i=1}^{N} 2r_i \cdot \frac{1}{r_i} - \frac{2V_m D}{N^2}(C_b - C_s)\sum_{j=1}^{N} r_j \sum_{k=1}^{N} \frac{1}{r_k}$$

Rearranging leads to

$$\frac{d(\sigma_r^2)}{dt} = 2V_m D(C_b - C_s)\left[1 - \frac{1}{N}\sum_{j=1}^{N} r_j \cdot \frac{1}{N}\sum_{k=1}^{N} \frac{1}{r_k}\right]$$

By the definition of $\bar{r}$ and $\overline{(1/r)}$, this equation can be rewritten as the form given in the problem.

3. From the definition, we know that

$$dG = \Delta\mu\, dn = \gamma\, dA$$

For a spherical particle, the following relation is valid.

$$dn = \frac{4\pi r^2}{V_m} dr$$

Combining the equations presented so far, we have

$$\Delta\mu\, dn = \Delta\mu \frac{4\pi r^2}{V_m} dr$$

and

$$\Delta\mu\, dn = 8\pi r \gamma\, dr$$

By equating those two equations, we get the result.

4. (1) To prevent the aggregation of the nanocrystals; (2) control of the size and shape of the nanocrystals; (3) to provide their solubility in a wide range of solvents; (4) exchanged with another coating of organic molecules having different functional groups or polarity.

5. (1) Burst nucleation: inhibition of additional nucleation during growth; complete separation of nucleation and growth.
(2) Diffusion-controlled growth: size focusing works.

6. The three stages coincide with the accumulation of the monomers in the first period, the burst nucleation in the second period, and the size focusing growth in the third period, respectively.

7

NANORODS

P. Jeevanandam

7.1 Introduction, 156
 7.1.1 Interesting Physicochemical Properties and Applications of Nanorods, 156
 7.1.2 Different Classes of Nanorods, 157
7.2 New Synthetic Methods, 158
 7.2.1 Seed-Mediated Synthesis, 158
 7.2.2 Template-Based Methods, 159
 7.2.3 Synthesis Using Micelles, 165
 7.2.4 Electrochemical Methods, 165
 7.2.5 Solvothermal and Hydrothermal Syntheses, 166
 7.2.6 Synthesis through Decomposition of Precursors, 167
 7.2.7 Photochemical Synthesis, 168
 7.2.8 Catalytic Growth Method, 170
 7.2.9 Sol-Gel Synthesis, 172
 7.2.10 Conversion of Nanosheets and Nanotubes to Nanorods, 173
 7.2.11 Green Chemical Synthesis, 173
 7.2.12 Biomimetic Synthesis, 174
 7.2.13 Synthesis in Restricted Dimensions, 175
 7.2.14 Miscellaneous Methods, 176
 7.2.15 Nanorods on Substrates, 178
 7.2.16 Superlattices of Nanorods, 183
 7.2.17 Core-Shell Nanorods, 184
 7.2.18 Functionalized Nanorods, 187
 7.2.19 Nanorod Composites, 188
7.3 Conclusion and Future Outlook, 190
References, 190
Problems, 195
Answers, 196

Nanoscale Materials in Chemistry, Second Edition. Edited by K. J. Klabunde and R. M. Richards

7.1 INTRODUCTION

Nanoscale materials belong to a unique family of compounds that form a bridge between molecules and condensed matter (1). Recently, there has been tremendous interest in the area of nanoscale materials, as these materials are increasingly finding applications in the area of medical diagnostics, health care, environmental remediation, high density data storage, etc. The most interesting thing about nanoscale materials is that their physicochemical properties change with size and shape; for example, band gap, melting point, magnetic properties, and specific heat can change with size. A large number of nanoscale materials have been synthesized and their properties studied (2). In this chapter, new synthetic routes to nanorods are discussed; only nanorods that are inorganic in nature will be discussed; for nanorods of organic materials and polymers the reader can find suitable material elsewhere.

Nanorods are one-dimensional nanostructures that belong to the general class of nanoscale materials. They provide an opportunity to investigate the fundamental understanding on the effect of size and shape on the magnetic, electronic, optical, and chemical properties of materials. For example, one can study how electrons, phonons, or photons are transported if they are confined to move in one direction. Nanorods offer a chance to attach multiple functionalities along the length of the rods making them versatile for applications. Nanorods are believed to be the building blocks of the next generation of electronic and molecular devices. They are expected to be useful in the area of optoelectronics and nanophotonics. For example, semiconductor nanorods or nanowires (length, 1 to 50 μm) show interesting optical properties such as wave guiding. Synthesis of nanorods in an easy and reproducible manner with good size distribution, crystallinity, and high purity is very important. Let us begin our discussion by defining what is a nanorod. A nanorod is a particle with nanoscale dimension in which the length of the particle can vary from 10 nm up to a few micrometers but the width is of the order of nanometers (10 to 100 nm). Aspect ratio is an important parameter for nanorods and it is defined as the ratio of length to width. In general, nanorods possess aspect ratio less than 10 while nanowires possess aspect ratio greater than 10, although this definition is not strictly followed in many cases.

7.1.1 Interesting Physicochemical Properties and Applications of Nanorods

Nanorods of materials possess interesting properties compared to other shapes. They possess enhanced photochemical and photophysical properties. From the optical, electrical, and magnetic properties point of view, nanorods are different from other shapes of the same material. For example, a gold nanorod possesses higher extinction for visible and near-infrared radiation compared to a sphere. Gold nanorods show enhancement of fluorescence intensity by a factor of 10^6 compared to that of the metal. α-Fe_2O_3 nanorods show a magnetic transition from a canted antiferromagnetic state to antiferromagnetic state at 166 K while the corresponding nanotubes show a three-dimensional magnetic ordering at $T > 300$ K. Nanorods usually possess enhanced thermal stability compared to nanotubes of the same material.

Nanorods find applications in sensing, catalysis, bioimaging [traditional diagnostic methods such as antigen detection, polymerase chain reaction (PCR), virus isolation, etc., are time consuming], solar cells, optoelectronic devices, electromagnetic shielding, microwave applications, photothermal therapy, thermoelectric devices, low temperature soldering applications, photothermal therapeutic applications, gas storage, composite materials, plasmonic sensors, drug delivery, gas ionization device applications, field emission applications, photovoltaic applications, light emitting diodes, biological detection, etc. Application of nanorods in superconductivity has been realized (3). For example, MgO nanorods have been used as pinning centers in high temperature superconductors to produce materials with high critical density. Nanocomposites with nanorods incorporated provide good mechanical strength. Nanorods possess anisotropic magnetic properties and they can be exploited for the new generation of data recording based on spintronics (4). Nanorods possess aerodynamic advantages compared to other shapes; that is, if one can deliver nanorods with high aspect ratio inside a chamber, they can stay longer in space compared to spheres of the same material. This particular property is currently being exploited for smoke reduction (5).

7.1.2 Different Classes of Nanorods

The following are the major types of nanorods.

7.1.2.1 Metals Among the nanorods of metals, silver and gold have been explored the most. They are very stable, easy to synthesize, and provide better reproducibility. There is much literature available on the synthesis of metal nanorods (6). Other metal nanorods include selenium, copper, and nickel.

7.1.2.2 Alloys Alloys are an important class of nanorods, because of their applications in direct methanol fuel cells, electrooxidation of methanol, etc. The list of nanorods of alloys that have been synthesized include bimetallic nanorods such as Ag-Cu, Ag-Ni, Fe-Ni, Fe-Pt, Co-Pt, Fe-B, Zn-Ni, Pt-Ru, Pt-Ni, Si-Ge, Se-Te, Ni-Co, and Cu-Co, as well as ternary alloys such as Pt-Ru-Ni.

7.1.2.3 Metal Oxides Among the nanoscale materials, nanoscale metal oxides are an interesting class of compounds (7). Metal oxides possess interesting properties that are structure related, such as magnetism, ferroelectricity, superconductivity, giant magnetoresistance, etc. Metal oxide nanorods hold considerable promise in the area of photoelectrochemical, photocatalytic, superconductivity, sensors, and photovoltaic applications. The list of synthesized oxide nanorods include MgO, ZnO, CuO, CdO, Ga_2O_3, In_2O_3, SnO_2, RuO_2, V_2O_5, Sb_2O_5, Co_3O_4, MoO_3, MoO_2, MnO_2, IrO_2, $CdWO_4$, $BaCrO_4$, $BaWO_4$, $BaTiO_3$, $SrTiO_3$, and $PbZr_{0.52}Ti_{0.48}O_3$.

7.1.2.4 Metal Sulfides and Selenides Nanorods of metal sulfides are interesting from the point of view of photoelectrochemical conversion, photoelectrochemical cells, etc. Nanorods of many metal sulfides have been synthesized, including CdS,

ZnS, CuS, In_2S_3, Bi_2S_3, MoS_2, $CdSeS$, SnS, FeS_2, PbS, Ag_2S, Sb_2S_3, La_2O_2S (La = Eu, Gd), Fe_7S_8, Cu_2S, Tl_2S, NiS, Mo_2S_3, $CoSbS_2$, Cu_3BiS_3, α-MnS, $AgBiS_2$, β-La_2S_3, Cu_3SnS_4, WS_2, Ag_3CuS_2, $CdIn_2S_4$, $PbSnS_3$, $CuInS_2$, $AgInS_2$, and $CuFeS_2$. The list of selenides include CdSe, ZnSe, PbSe, $CuInSe_2$, and α-MnSe.

7.1.2.5 ***Metal Nitrides and Carbides*** Nanorods of nitrides are useful in the area of magnetism, abrasives, etc. The list includes Ta_3N_5, TiN, InN, GaN, AlN, Si_3N_4, α-Si_3N_4, BN, MoN, and SiC_xN_y. The nanorods of metal carbides include TiC, NbC, and Fe_3C.

7.2 NEW SYNTHETIC METHODS

The synthesis of nanorods can be classified into two major types: physical methods and chemical methods. The physical methods involve a "top-down" approach (8) and include sputtering, pulsed laser deposition, laser ablation, thermal evaporation, high energy ball milling, arc method, nanolithography, and ion-beam implantation. These methods require expensive apparatus. The chemical methods involve a "bottom-up" approach. In this chapter, we will be concerned only with the chemical methods.

Chemical methods provide better control to synthesize nanorods with required aspect ratio and uniform size distribution. The methods are simple, versatile, and economically viable. There are many chemical methods that can be used to synthesize nanorods, the most important of which are discussed below. It is impossible to cover all the topics related to synthesis of nanorods in depth due to lack of space and only the salient features of the synthetic methods with specific examples will be discussed in this chapter.

7.2.1 Seed-Mediated Synthesis

Seed-mediated synthesis was pioneered by Murphy et al. (9). Nanorods of metals such as silver and gold have been synthesized using this method. The synthesis is carried out at room temperature and in air. The method involves making the seeds of the metal first, followed by the addition of a growth solution. The seed solution is prepared by adding to a metal salt solution (e.g. $HAuCl_4$), a stabilizing agent such as sodium citrate along with a strong reducing agent such as $NaBH_4$. The growth solution contains the metal salt, a surfactant [e.g. cetyl trimethyl ammonium bromide (CTAB)] which can be a structure-directing agent, along with a weak reducing agent (e.g. ascorbic acid). The presence of the surfactant in excess of the critical micellar concentration is important; without the surfactant, only spheres have been produced. The general schematic of the seed-mediated synthetic procedure is given in Scheme 7.1.

The aspect ratio of the rods can be controlled from about 2 to 25 by varying the concentration of the reagents and the seeds. Smaller seeds lead to the formation of lengthier metal nanorods. Another important parameter is the presence of other metal ions during the growth of the nanorods. For example, addition of small amounts

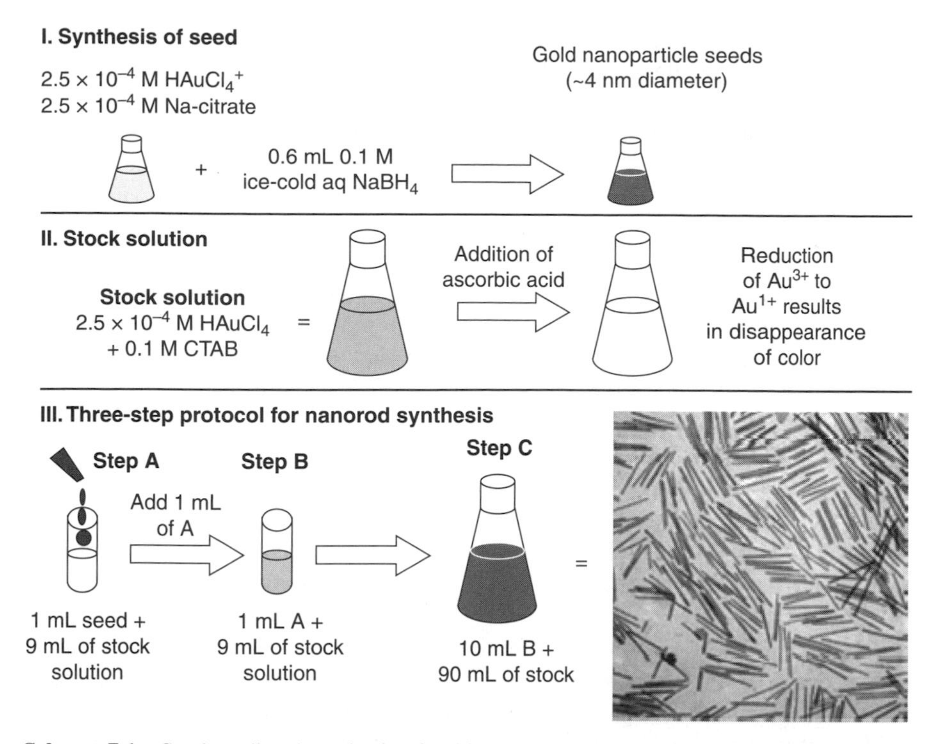

Scheme 7.1 Seed-mediated synthesis of gold nanorods. The seeds are prepared in step I. A stock solution is prepared in step II and the protocol for nanorod synthesis is shown in step III. (Reprinted with permission from C. J. Murphy et al. *J. Phys. Chem. B* **2005**, 109, 13857. Copyright (2005) American Chemical Society.)

of silver ion during the growth of gold nanorods leads to more yield. A proposed mechanism for the growth of gold nanorods in the presence of higher concentration of CTAB is that AgBr is formed on the surface of the growing rods, inhibiting the growth along those directions in which AgBr is adsorbed. The effect of the counterion of the surfactant has been studied, too. While CTAB produced nanorods, CTAC (cetyl trimethyl ammonium chloride) and CTAI (cetyl trimethyl ammonium iodide) produced spheres and mixtures of various shapes, respectively.

7.2.2 Template-Based Methods

One of the challenges in the synthesis of nanorods is to produce nanorods of uniform size and distribution in an easy and reproducible manner. If a template is provided for the synthesis of nanorods, the dimension/geometry of the template dictates the dimension of the nanorods. Templates have been commonly used to produce isolated oriented nanorods. The most commonly used templates are anodic aluminum oxide (AAO) or polycarbonate membrane (ion-track etched). Other templates include three-dimensional microporous materials, such as zeolites, mica, glass, block-copolymer, and even carbon nanotubes. A template should contain uniform sized pores, be chemically inert, and easily removable. Some of the important template-based synthetic methods are described below.

7.2.2.1 ***Sol-Gel Template Process and Sol-Gel Electrophoresis*** Using sol-gel chemistry, synthesis of nanostructured materials inside the templates such as anodic aluminum oxide has been possible. The sol-gel template method can provide highly crystalline materials. Using a stable sol during the preparation of nanorods is important. For example, to prepare TiO_2 nanorods by the sol-gel template process (10), a sol stable for at least three days is prepared first. A solution is prepared by dissolving titanium isopropoxide in ethanol. A second solution prepared by mixing ethanol in water with acetyl acetone is added to the first solution to obtain a sol. The anodic aluminum oxide membrane is then dropped into this sol followed by room temperature drying and calcination in air at 400°C for 24 h. The anodic aluminum oxide membrane can be easily dissolved using NaOH solution. Nanorods of TiO_2 with dimensions 200 to 250 nm are obtained when the ethanol content in the sol is increased. Synthesizing ZnO nanorods or nanofibers is easy, too (10). Zinc acetate solution in ethanol is prepared first by boiling and then $LiOH \cdot H_2O$ is added followed by ultrasonication to get a sol. Now, the template is dipped in and dried at room temperature followed by calcination in air.

Synthesis of WO_3 nanorods can also be carried out inside the pores of the template by first preparing a solution by dissolving WCl_6 in oxygen-free ethanol (10). A second solution prepared by mixing 2,4-pentanedione with water is added to the first solution. A blue sol is obtained. Now the template can be dipped inside the sol, followed by drying in air, and then calcination to get WO_3 nanorods. Other fibrils of MnO_2, Co_3O_4, and SiO_2 have been prepared. Immersion time of the template inside the sols is very important. The template has to be dipped inside the sol for longer than five seconds for rods to form. With short immersion time, only nanotubes are formed. Even minor changes in the temperature of the dipping process can make a change in the morphology of the rods. For example, in the case of TiO_2 sol, dipping at 20°C leads to TiO_2 nanorods, while at 15°C thin-walled tubules are formed. The proposed mechanism (10) indicates that the positively charged sol particles adsorb onto the negatively charged pore walls. This leads to enhanced concentration of adsorbed sol particles, leading to faster gelation compared to that in the solution. The nanorods can also exist as bundles inside the pores of the templates. The surface sites, which are Lewis acids, can attract to the oxide sites (or hydroxide sites on the other rods).

Electrophoretic deposition is used for depositing thin films from colloidal dispersion (11). Nanosized particles in a sol can be stabilized by steric or electrostatic means, and they develop a charge on the surface. When an electric field is applied, these charged particles move in response to the field and this motion is called electrophoresis; for example, positively charged particles will deposit at the cathode. Nanorod arrays have been synthesized by a combination of sol preparation and electrophoretic deposition. The conditions employed for the growth of nanorod arrays by electrophoretic deposition are summarized in Table 7.1 (11). First, the sol is brought in contact with the template. Then a potential is applied (typically

TABLE 7.1 Conditions Employed for the Electrophoretic Deposition of Nanorods

Sol	Precursor(s)	Solvents and Other Chemicals	Approximate pH
TiO_2	Titanium (IV) isopropoxide	Glacial acetic acid, water	2
SiO_2	Tetraethyl orthosilicate	Ethanol, water, hydrochloric acid	2
Nb_2O_5	Niobium chloride	Ethylene glycol, ethanol, citric acid, water	1
V_2O_5	Vanadium pentoxide	Hydrogen peroxide, water, hydrochloric acid	2.7
$Pb(Zr,Ti)O_3$	Lead (II) acetate, titanium isopropoxide, zirconium *n*-propoxide	Glacial acetic acid, ethylene glycol	4
$BaTiO_3$	Titanium isopropoxide, barium acetate	Glacial acetic acid, ethylene glycol	4
$SrNb_2O_6$	Strontium nitrate, niobium chloride	Ethylene glycol, ethanol, citric acid, water	1
Indium tin oxide	Indium chloride, tin (IV) chloride	Ethylene glycol, ethanol, citric acid, water	1

Source: Reprinted with permission from G. Cao, *J. Phys. Chem. B* **2004**, 108, 19921. Copyright (2004) American Chemical Society.

about 5 V) and the deposition is carried out. The excess sol is wiped off followed by drying of the template at ~100°C. Finally, the membrane is fired to remove the template, yielding dense nanorods. Examples of nanorod arrays synthesized using sol-electrophoretic deposition include SiO_2, Nb_2O_5, V_2O_5, $BaTiO_3$, and $Sr_2Nb_2O_7$.

7.2.2.2 Nanotubes as Templates Carbon nanotubes (CNTs) are the most prominent template for confining and growing nanorods. When carbon nanotubes are used as the template, there is a significant increase in the surface area of the material to be grown inside compared to a flat surface. The incorporation of metals inside carbon nanotubes has been achieved by introducing the metal precursors along with the source of carbon. Extreme conditions such as high temperature or arc evaporation are often required. Sometimes metals cannot wet the interior of carbon nanotubes if they possess higher surface tension. Depositing the metals inside the inner surface of carbon nanotubes through chemical vapor deposition is an alternative method. Specific examples of synthesis of nanorods using carbon nanotubes as templates are discussed below.

Single crystalline β-Ag_2Se nanorods have been synthesized using carbon nanotubes as templates (12). First, Ag/C nanocables are synthesized from $AgNO_3$, K_2CO_3, and NH_2SO_3H under hydrothermal conditions. The Ag/C nanocables are first dispersed in water. Then $N_2H_4 \cdot H_2O$ and Se powder are added; hydrazine acts as the reducing and coordinating agent. The contents are subjected to hydrothermal treatment and the product is washed with a dilute solution of KCN to obtain nanorods of β-Ag_2Se inside the carbon nanotubes (Fig. 7.1). Crystalline ZnO nanorods with diameters in the range 20 to 40 nm and length 250 to 1000 nm have been synthesized with carbon nanotubes as templates (13). For example, acid-treated multiwalled carbon nanotubes (MWCNTs) (acid treatment leads to opening of the tubes) are stirred with a saturated solution of $Zn(NO_3)_2{\cdot}6H_2O$. The contents are filtered, washed with water, dried, and calcined at 500°C under an inert atmosphere. Heating the calcined samples in air at 750°C leads to burning of carbon, producing nanorods of zinc oxide. The conversion of carbon nanotubes by reacting with a volatile oxide species leads to nanorod formation (14). For example, GaP nanorods can be obtained by reacting Ga_2O with carbon nanotubes in a phosphorus vapor atmosphere. Appropriate amounts of Ga_2O, CNT, and P in an evacuated quartz ampoule, if heated at 1000°C leads to the formation of GaP nanorods, and most of them are single crystals. The reaction is represented as $Ga_2O + C$ (nanotubes) + 2P (g) → 2GaP (nanorods) + CO (g). Platinum metal-filled carbon nanotubes have been prepared by impregnation of carbon nanotubes with $H_2PtCl_6 \cdot 6H_2O$ followed by heat treatment at 500°C under H_2 or by using 0.1 M $NaBH_4$ (15); no platinum metal is observed on the outside wall of the carbon nanotube.

Nanorods with diameters 10 to 200 nm and lengths up to a few microns have been synthesized from metal oxides such as V_2O_5, WO_3, MoO_3, Sb_2O_5, MoO_2, RuO_2, and IrO_2 using carbon nanotubes as templates (16). Acid-treated carbon nanotubes on treatment with oxide precursors (e.g. alkoxide, HVO_3, H_2WO_4, H_2MoO_4, $SbCl_5$) are dried at 100°C, followed by calcination at 450°C. Finally, the calcined samples are heated at 700°C in air to remove carbon. Rutile and anatase nanorods have been

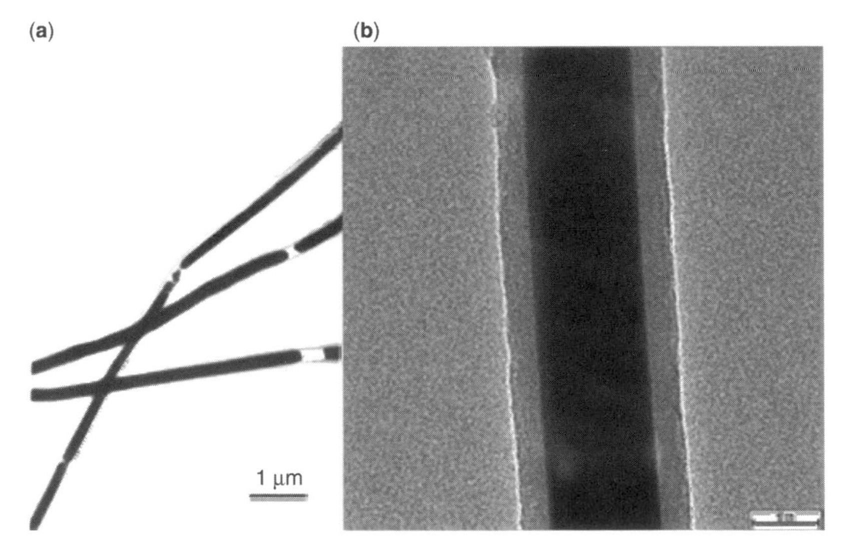

Figure 7.1 TEM image of β-Ag_2Se nanorods inside carbon nanotubes (a), and its magnified image (b). (Reprinted with permission from D. Ma et al. *Inorg. Chem.* **2006**, 45, 4845. Copyright (2006) American Chemical Society.)

synthesized starting from titanium, iodine, and CNTs according to the reaction $TiI_4(g) + C(s) \rightarrow TiC(s) + 2I_2(g)$; heat treatment at 525°C or 800°C leads to anatase and rutile nanorods, respectively (17). Transition metal carbide nanorods can be synthesized by the reaction of CNTs with volatile halides (e.g. titanium and niobium iodides; Reference 18). In the beginning, TiC coating is formed uniformly inside the template followed by inward growth of TiC to produce the nanorod. TiC, NbC, Fe_3C, SiC, BC_x (2 to 30 nm diameter, lengths up to 20 μm) can be synthesized by the above approach.

Formation of strings of nanorods of Au on multiwalled carbon nanotubes (MWCNTs) have been observed using a layer-by-layer assembly approach (LBL; Reference 19). First, the carbon nanotubes are wrapped with negatively charged polyelectrolyte such as polystyrene sulfonate. This is followed by adsorption of a positively charged polyelectrolyte [poly(diallyl dimethyl) ammonium chloride]. Gold nanorods synthesized by seed-mediated synthesis with the help of CTAB followed by ligand exchange with poly(vinyl) pyrrolidone (PVP) have been used. The Au nanorods assemble on both sides using CNTs as the template, as shown in Figure 7.2. Carbon nanotubes are expensive and hence alternative templates will be useful.

7.2.2.3 Liquid Crystals as Templates Liquid crystals can serve as templates for the synthesis of nanorods. Lyotropic liquid crystals have been used the most (20). The template is not affected by the introduction of starting reagents. After the decomposition of the template the nanorods can be recovered. Several examples are described below.

Using lamellar liquid crystals of $C_{12}E_4$ [tetraethylene glycol monodecyl ether (Brij30®)], zinc sulfide nanorods (diameter 60 nm and width ~80 to 380 nm) have been synthesized (21). The reactant concentration, surfactant: water molar ratio in the liquid crystal assembly affects the size of the nanorods. It is possible to do *in situ* templating of nanorods in liquid crystals (22). Metal sulfides such as PbS can be templated *in situ* in the reverse hexagonal phase liquid crystals. Using a polycation modified SDS/decanol system, a multilamellar liquid crystal template can be produced and used for growing nanorods (23). A traditional template can be coupled with a liquid crystal template. For example, anodic aluminum oxide and a hexagonal phase of lyotropic liquid crystals containing the metal ions of interest may be

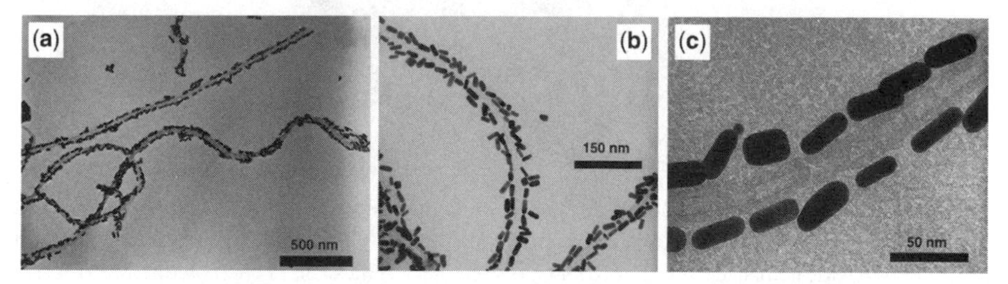

Figure 7.2 Gold nanorods assembled using carbon nanotube as a template. (a) to (c) show increasing magnifications. (Reproduced with permission from M. A. Correa-Duarte et al. *Angew. Chem. Int. Ed.* **2005**, 44, 4375. Copyright Wiley-VCH Verlag GmbH & Co.)

employed. By electrochemical reduction of metal ions in the liquid crystals from aqueous solution into the pores of anodic aluminum oxide, alloy nanorods with high ordering can be obtained. Zn-Ni alloy nanorods have been produced in this way (24; Fig. 7.3); a double template leads to a reduction of the size of the nanorods.

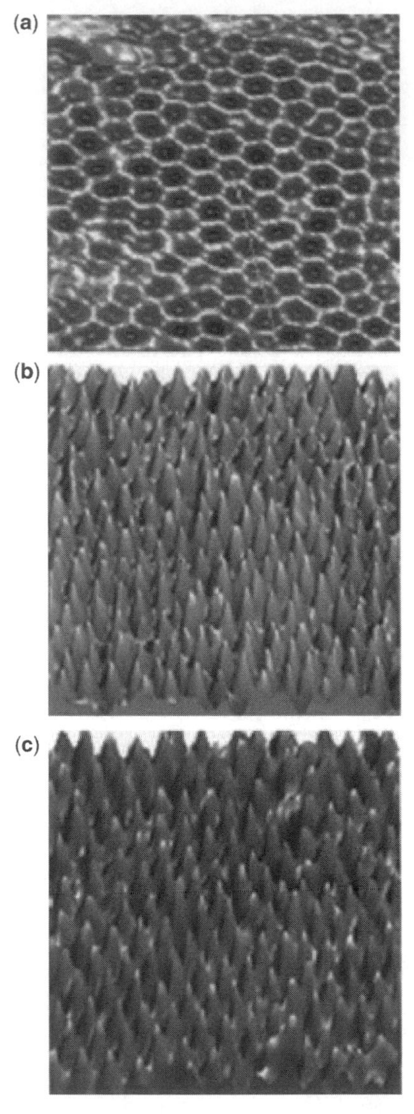

Figure 7.3 AFM images of anodic aluminum oxide membrane (a), and Zn–Ni alloy nanorods (b, c) obtained using double templates (liquid crystal and AAO). (b) and (c) were obtained with different deposition charges, 0.6 C and 0.9 C, respectively. The scale of each picture is 1.6×1.6 μm^2. (Reprinted from A. Foyet et al. *J. Electroanal. Chem.* **2007**, 604, 137. Copyright (2007) Elsevier. With permission.)

Although the template-based synthetic methods offer many advantages, there are some disadvantages. In all cases, the templates have to be removed and so these methods may not be suitable for making large quantities of nanorods. Also, a track-ion etched polycarbonate membrane may possess intersecting pores that will affect the homogeneity of the rods produced.

7.2.3 Synthesis Using Micelles

When we have small water pools in a continuous oil phase an emulsion is produced that can be stabilized by the introduction of surfactants such as cetyl trimethyl ammonium bromide. The microemulsions thus produced can serve as nanoreactors for the synthesis of nanorods. The ratio of water to surfactant affects the size of the nanoreactor and this in turn can influence the size and shape of nanorods. Nanorods of oxides, metals, and semiconductors have been synthesized; examples include CeO_2, $BaTiO_3$, $BaCrO_4$, Ag, and CdSe (25).

High concentrations of surfactant are necessary during the synthesis of nanorods by the reverse micelles method. For example, MnOOH and Mn_3O_4 nanorods have been synthesized only above 0.2 M concentration of the surfactant (26). The length of the nanorods increases with increase in the surfactant concentration while the diameter remains essentially constant. The pH and the ratio of the reactants can control the agglomeration of the nanorods. For example, calcium phosphate nanorods have been synthesized using reverse micelles of calcium bis(2-ethyl hexyl) phosphate in water in cyclohexane, NH_4HPO_4, and a triblock copolymer (27). The $[Ca]:[PO_4]$ ratio and pH control whether nanorods are agglomerated or not. When the ratio is 1.1 and $pH = 8.2$, bundles are produced (2 nm width and >300 μm length). On the other hand, when $pH = 9$ and the $[Ca]:[PO_4]$ ratio is 1.66, discrete nanofilaments are produced (100 to 500 nm length and 10 to 15 nm diameter). Nanorods of solid solutions have also been prepared by the reverse micelle method. For example, tungsten doped MoS_2 nanorods ($Mo_{0.95}W_{0.05}S_2$) have been synthesized from a trisulfide precursor on pyrolysis.

It is possible to do sol-gel synthesis in a reverse micelle. Iron oxide nanorods have been synthesized using this simple method (28). The aspect ratio of the nanorods can be controlled by varying the water to surfactant/ligand (e.g. oleic acid) ratio during the gelation process. The phase of the nanorods can be controlled by varying the atmosphere, temperature, etc. Using ultrasound can be helpful during synthesis of nanorods by the reverse micelle method. For example, Ag nanorods were synthesized in sodium bis(2-ethyl hexyl) sulfosuccinate/isooctane reverse micelles by using a mild ultrasound irradiation. In the presence of ultrasound, the spherical micelles transform to ellipsoidal; the sonication time may be used to tune the size of the nanorods.

7.2.4 Electrochemical Methods

Electrochemical methods have been used mainly to deposit metals or semiconductors into the templates. One does not need expensive instrumentation, and the synthesis can be carried out under ordinary temperatures and pressures. A general scheme for the

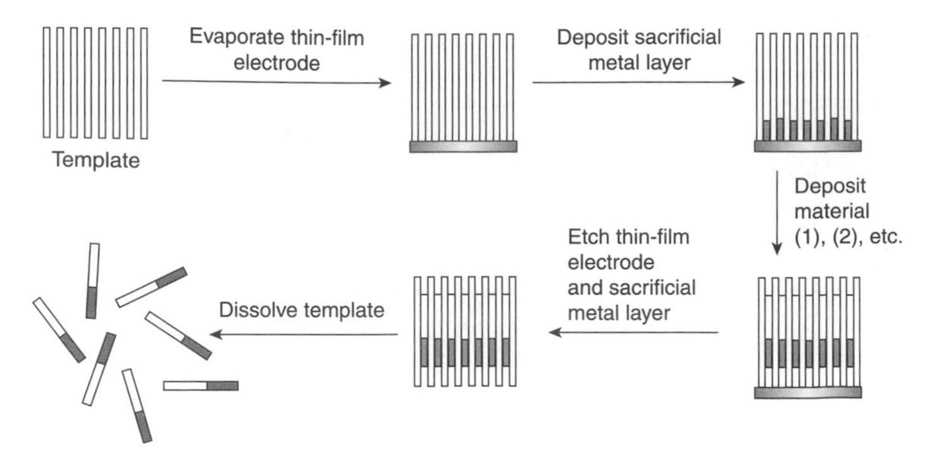

Scheme 7.2 Steps involved in an electrochemical method to synthesize nanorods. (Reproduced with permission from C. R. Mirkin et al. *Angew. Chem. Int. Ed.* **2006**, 45, 2672. Copyright Wiley-VCH Verlag GmbH & Co.)

synthesis of nanorods by electrochemical methods is given in Scheme 7.2. First a thin film of metal is deposited on the template, which will serve as the working electrode for the deposition. This is followed by the deposition of the sacrificial metal. Then the deposition of the materials of interest is carried out electrochemically. Finally, the templates are removed by chemical treatment to get nanorods.

Martin and coworkers (29) pioneered the electrochemical deposition of nanorods of metals such as Ag, Au, Co, Cu, Ni, Pt, Pd, and Zn using hard templates such as anodic aluminum oxide (AAO). The metal ions in the solutions are reduced by applying a negative potential and the morphology of the rods is controlled by two parameters, the pore size of the template, and the amount of the charge passed, since the length of the nanorods will be decided by this. It is possible to deposit multiple elements (e.g. grow multisegmented rods) within the pores of the template. One can adopt pulsed electrochemical deposition with a bath containing multiple ions with well-separated redox potentials. It is also possible to deposit semiconductors into the pores of the template; for example, ZnO nanorods or nanowires can be prepared by applying a cathodic current to an aqueous solution containing zinc nitrate. The major drawbacks of electrochemical methods are (1) modification of the template electrochemically, for example, plating to make it a working electrode is inconvenient, and (2) to make nanorods, the metal ions in the solution should be easily reducible; if we cannot reduce a metal electrochemically, this approach cannot be used.

7.2.5 Solvothermal and Hydrothermal Syntheses

The solvothermal and hydrothermal syntheses involve heating the reactants in water or a solvent at high temperatures and pressures (30). The role of solvent (or water) is that of a pressure-transmitting medium and the solubility of the reactants is pressure and temperature dependent. A sealable Teflon-lined container, called a bomb, is used to keep the solvent and the reactants inside. After sealing, the container is kept at high temperatures inside an oven (temperatures vary from 100°C to 500°C). The pressure of the container depends on the level of filling of the solvent (or water). Solvothermal (hydrothermal) conditions provide unique supercritical conditions that can lead to unique or unexpected morphologies of products. The method is simple, economical, robust, and most of the time the conversion efficiency is close to 100%. Various experimental parameters such as concentration of reagents, pH, and introduction of additives can be varied to tune the morphologies of the products. The effect of various experimental parameters on the reaction equilibria seems to be the key. Solvothermal and hydrothermal syntheses have been used to synthesize a variety of nanorods which are summarized in Table 7.2.

During solvothermal synthesis, sometimes layered precursors are used as the starting materials and template molecules such as amines are used. The layered precursors with template amine molecules in the interlamellar space on solvothermal treatment lead to the transformation of a two-dimensional structure into a one-dimensional structure. Simple conditions such as using different acidic solvents (e.g. H_2SO_4, HCl, salicylic acid) can lead to nanorods of different aspect ratio. The major drawback of these methods is that the mechanism of synthesis is sometimes not clearly established and reproducibility may be an issue.

7.2.6 Synthesis through Decomposition of Precursors

This method involves decomposition of precursor(s) in a coordinating or noncoordinating solvent or ligand. Two examples are discussed below.

TABLE 7.2 Nanorods Synthesized by Hydrothermal or Solvothermal Methods

Type	Compounds
Metals	Se, Te, Co
Metal hydroxides	$La(OH)_3$, β-FeOOH, $Mg(OH)_2$, MnOOH, GaOOH, $Dy(OH)_3$, $Ni(OH)_2$, $Eu(OH)_3$
Metal oxides	ZnO, $Zn_{1-x}Cd_xO$, Fe_3O_4, TiO_2, Co_3O_4, In_2O_3, Cu_2O, $PbWO_4$, $PbCrO_4$, SnO_2, MnO_2, Mn_2O_3, Mn_3O_4, CeO_2, $W_{18}O_{49}$, γ-LiV_2O_5, LiV_3O_8, $CdWO_4$, $CoWO_4$, U_3O_8, γ-Al_2O_3, CuO, $LaBO_3$, $HgWO_4$, VO_2, $LiMnO_2$, La_2O_3, $ZnWO_4$, Fe_3O_4, α-Fe_2O_3, $SrSnO_3$, $Pb(Zr,Ti)O_3$, WO_3, $CuSb_2O_6$, $NiFe_2O_4$, Eu_2O_3, $LiAlO_3$, Pr_6O_{11}, $PbCrO_4$, $BaFe_{12}O_{19}$, $LaVO_4$, $BaTa_2O_6$
Metal sulfides	CdS, PbS, $Mn_xZn_{1-x}S$, Bi_2S_3, Ag_2S, Sb_2S_3, CuS, Tl_2S, β-La_2S_3, Cu_3SnS_4, $CuInS_2$, HgS, $AgInS_2$, Sb_2S_3, Ag_2S
Metal selenides	CdS_xSe_{1-x}, CdSe, $CuInSe_2$, α-MnSe, PbSe, Tl_2Se, $NiSe_2$, ZnSe
Metal tellurides	ZnTe, Bi_2Te_3, CdTe, $NiTe_2$, $CoTe_2$
Miscellaneous	$BaCO_3$, $SmPO_4$, Sn_4P_3, Zn doped SnO_2, Zn doped CdS, Nd doped TiO_2, Mn doped ZnS, Nd doped TiO_2, Mn,Cr,Co doped ZnO

Nanorods of CdS and CdSe can be synthesized using decomposition of precursors in the presence of a coordinating solvent (31). To synthesize CdS nanorods, a cadmium complex of thiosemicarbazide ($NH_2CSNHNH_2$) dissolved in tri-*n*-octyl phosphine (TOP) is rapidly injected into tri-octyl phosphine oxide (TOPO) kept at 300°C. TOPO acts as the coordinating solvent. The solution is cooled to 70°C and then methanol is added to get CdS nanorods, which can be redispersed in toluene. Cadium tetramethyl thiourea complex mixed with thiosemicarbazide can also be used as the precursor. To obtain CdSe nanorods, cadmium selenosemicarbazide, generated *in situ* by reacting cadmium acetate and selenosemicarbazide dissolved in TOP, can be rapidly injected into TOPO kept at 300°C followed by cooling and addition of methanol. ZnO nanorods have been produced by the decomposition of zinc acetate in the presence of a noncoordinating solvent such as diphenyl ether or octadecene and oleyl amine (32); aminolysis occurs when oleylamine is injected into the hot solution containing zinc acetate at about 120°C to yield ZnO nanorods.

7.2.7 Photochemical Synthesis

Photochemical synthesis can be considered as a one-pot synthesis. It is mild, efficient, and an environmentally friendly method. Let us take some examples where photochemical synthesis has been successfully used. Kim et al. (33) were able to synthesize gold nanorods with control over aspect ratio using photochemistry in the presence of Ag^+ ions. An aqueous solution consisting of CTAB and tetradodecyl ammonium bromide has been used as the growth solution along with $HAuCl_4 \cdot 3H_2O$ as the source of gold. A small amount of acetone and cyclohexane is used to loosen the micellar structure. Different amounts of 0.01 M $AgNO_3$ aqueous solution is added to the above solution and it is irradiated with UV light (~254 nm). The resulting precipitate after dispersion in water yields Au nanorods (Fig. 7.4). The formation of gold nanorods is indicated by the appearance of a longitudinal band in the UV-Vis spectrum around 600 to 800 nm. Depending on the amount of Ag^+ ions (15.8 to 32 μl), Au nanorods with aspect ratio 2.8 to 4.8 can be synthesized. Silver ions play a crucial role in the formation of nanorods and a sample prepared without silver nitrate consists of only spherical particles. Increase in Ag^+ content in the solution leads to a decrease in the diameter of the rods. If irradiation is continued for a longer time, shorter rods are produced at a given concentration of Ag^+. A combination of crystal aggregation and stabilization of a particular crystal face has been suggested to be the mechanism. Silver ions are reduced to silver nanoclusters when irradiated and they are oxidized back into Ag^+ in the presence of $HAuCl^-$. There is competition between photoreduction by UV irradiation to silver and its oxidation to Ag^+ in the presence of $HAuCl_4^-$; Au^{3+} gets reduced in the process. This leads to fresh surfaces of Au nanocrystals followed by growth along a particular direction.

A photochemical reaction of ketone to synthesize gold nanorods in a micellar solution of CTAB consisting of $HAuCl_4$, $AgNO_3$, acetone, and ascorbic acid is possible (34). The presence of ketone is crucial for the formation of the rods in this case. The UV irradiation leads to the formation of ketyl radicals and the radicals reduce the Au^+ ions to Au^o. The Au^o act as the nuclei for the anisotropic growth of Au nanocrystals in

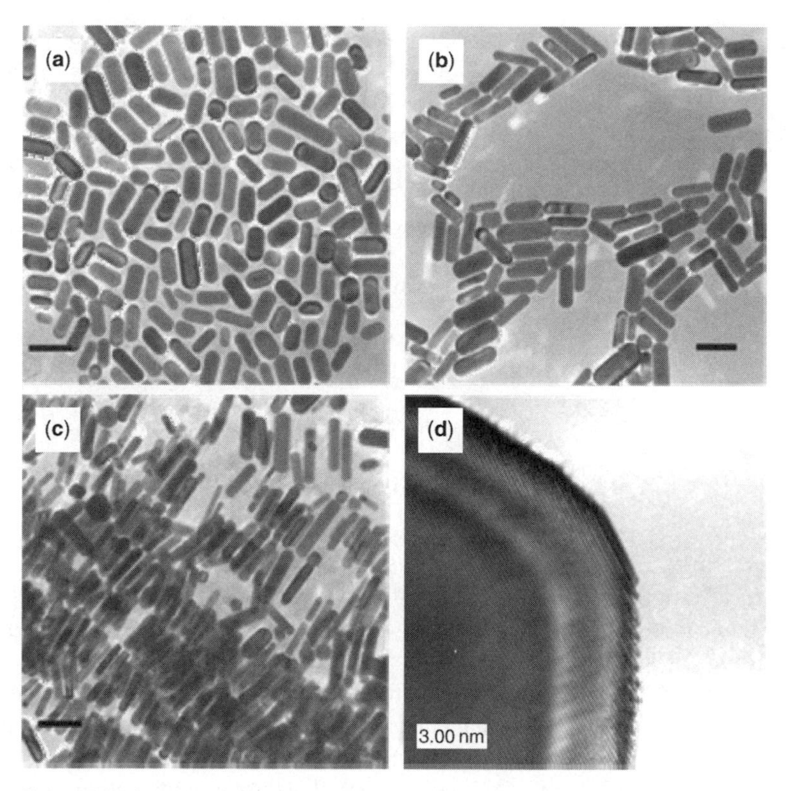

Figure 7.4 TEM images of Au nanorods prepared by photochemical synthesis with varying amounts of $AgNO_3$: (a) 15.8 μL, (b) 23.7 μL, (c) 31.5 μL. (d) A rod at higher magnification. (Reprinted with permission from F. Kim et al. *J. Am. Chem. Soc.* **2002,** 124, 14316. Copyright (2002) American Chemical Society.)

the presence of ascorbic acid and the growth solution, consisting of $AuBr_2^-$, AgBr clusters, and CTAB. The proposed mechanism is given below (34).

$$Au^{3+} \xrightarrow{\text{Ascorbic acid}} Au^{+}$$

$$(CH_3)_2CO \xrightarrow{h\nu} (CH_3)_2CO^{*}$$

$$(CH_3)_2CO^{*} + RH \longrightarrow (CH_3)_2^{\bullet}COH + {}^{\bullet}R$$

$$Au^{+} + (CH_3)_2^{\bullet}COH \longrightarrow Au^{o} + (CH_3)_2CO + H^{+}$$

$$nAu^{o} \longrightarrow Au_n(\text{Nanorods})$$

CTAB is denoted by RH and $R^{\bullet}$ is the radical generated from CTAB by hydrogen abstraction by $(CH_3)_2^{\bullet}COH$ which reduces Au^+ to Au^o.

Apart from nanorods of metals, photochemical synthesis of nanorods of other materials such as Bi_2Se_3 has been achieved using anodic aluminum oxide as the template (35). The starting materials are $Bi(NO_3)_3$, sodium selenosulfate, EDTA, and ascorbic acid. Under UV irradiation, sodium selenosulfate decomposes slowly and releases Se^{2-} ions, homogeneously. The template is soaked in the solution and irradiated with UV light for 3 h at room temperature leading to the formation of Bi_2Se_3 nanorods. The mechanism for the formation of the nanorods is as follows (35).

$$\text{Bi-EDTA} \longrightarrow Bi^{3+} + EDTA^{3-}$$

$$H_2O \xrightarrow{h\nu} H^{\bullet} + OH^{\bullet}$$

$$2H^{\bullet} + SeSO_3^{2-} \xrightarrow[\text{Ascorbic acid}]{h\nu} Se^{2-} + 2H^{+} + SO_3^{2-}$$

$$2Bi^{3+} + 3Se^{2-} \longrightarrow Bi_2Se_3$$

$$nBi_2Se_3 \longrightarrow (Bi_2Se_3)_n$$

Small particles of Bi_2Se_3 adsorb on the bottom and surface of the template which grow further into nanorods; pH, complexing agents, reducing agents, and irradiation time affect the growth of nanoparticles.

The energy and intensity of the UV light used during the growth of gold nanorods has been studied (36). Under similar experimental conditions (aqueous solution consisting of CTAB, tetraoctyl ammonium bromide, $HAuCl_4 \cdot 3H_2O$, $AgNO_3$, acetone, cyclohexane is used) 300 nm UV light produces longer nanorods with narrower size distribution as compared to irradiation with 254 nm light. Also, high intensity accelerates the growth of Au nanorods. Longer irradiation time reduces the concentration of rods, with a concordant increase in the number of spherical particles. The advantage of the photochemical method compared to the electrochemical method is that spherical particles are not present. If only photoreduction method is employed, it takes a long time to grow nanorods in some cases (e.g. 30 h). If chemical and photochemical reactions are combined, the synthesis time can be reduced considerably, as elucidated by the above examples.

7.2.8 Catalytic Growth Method

The growth of nanorods during synthesis is facilitated by using catalysts such as transition metals (e.g. Ag, Ni, Cu, and Ti). The nanorods grow by a well-accepted vapor-liquid-solid (VLS) growth mechanism (37). The metal vapor from the high temperature region inside a tubular furnace is carried to the low temperature region by the flow of an inert gas where it deposits on the substrate. The metal droplets on the substrate act as the nucleation centers for the growth of nanorods. In the VLS mechanism, first nucleation and growth of alloy droplets occur and the nanorods grow due to supersaturation. Since the catalyst droplets serve as the nuclei for further growth, the dimension of the rods depends on the size of the catalyst droplets. Some specific examples are given below.

ZnO nanorods on various substrates such as Si, SiO_2, and sapphire have been prepared using Au as the catalyst (38). A thin layer of catalyst (~5 nm thick) is deposited on the substrate, forming islands on the substrate. A mixture of ZnO and graphite is then vaporized and it condenses on the particles, forming ZnO nanorods. Using different catalysts can lead to remarkably different results (39). For example, Au catalyst during ZnO nanorods synthesis leads to a homogeneous distribution of rods, while using Pt leads to the formation of nanoribbons in addition to nanorods.

Self-catalyzed growth of nanorods is also possible. For example, MgO nanorods with spherical metal particles at the tips have been produced (Fig. 7.5) by the reaction of commercial magnesium ribbons with oxygen under a flow of argon/oxygen mixture at 900°C (40). Magnesium metal is not only the reactant but also a catalyst and the reaction involved is 2Mg (vapor) + O_2 (g) → 2MgO (s).

Figure 7.5 MgO nanorods produced by a self-catalytic process in which Mg metal acts as a catalyst. (Reprinted from M. Zhao et al. *Materials Letters* **2006**, 60, 2017. Copyright (2005) Elsevier. With permission.)

7.2.9 Sol-Gel Synthesis

In sol-gel synthesis, a sol is prepared first from the suitable precursors. The sol becomes a gel on aging. The gel after drying and calcination leads to nanorods. The sol-gel process has been used mainly for the synthesis of metal oxide nanorods, with a few exceptions. The advantages of the sol-gel process compared to other methods are (1) better composition control, (2) good homogeneity, (3) low processing temperature, and (4) adaptability for easier fabrication. Typical examples of nanorods synthesized by the sol-gel process along with the chemicals used are summarized in Table 7.3. Let us illustrate sol-gel synthesis of nanorods by taking two examples. Nanorods of $K_2Ti_4O_9$ and $K_2Ti_6O_{13}$ have been synthesized by mixing CH_3OK and $Ti(OC_2H_5)_4$ in ethanol (41). The molar ratio of CH_3OK to $Ti(OC_2H_5)_4$ is varied. A required volume of HCl is added to control the hydrolysis and condensation reactions. The sol after keeping for about 4 to 5 days is dried to get a xerogel. The xerogel on calcination leads to $K_2Ti_4O_9$ nanorods if the ratio of CH_3OK to $Ti(OC_2H_5)_4$ is 1:1. If the CH_3OK to $Ti(OC_2H_5)_4$ ratio is 1:2, the final product is $K_2Ti_6O_{13}$ nanorods.

The second example is the synthesis of γ-alumina nanorods from boehmite nanofibers using a modified sol-gel process (42). First, a solution of aluminum isopropoxide in anhydrous ethanol is prepared. To this, ethanol with 4% water is added leading to a viscous liquid after 15 h. The viscous liquid heated at 600°C leads to γ-alumina nanorods (diameter <10 nm; length 50 to 200 nm). Since in the synthesis, less water is taken, only partial hydrolysis takes place. The removal of one water molecule from two AlO(OH) octahedra leads to the formation of Al_2O_3 nanorods, $(CH_3CH_2CH_2O)_3Al + 2H_2O \rightarrow AlO(OH) + 3CH_3CH_2CH_2OH$.

Sol-gel synthesis can also be carried out without the hydrolysis step. The nonhydrolytic sol gel process pioneered by Hyeon et al. (43) has been used to synthesize nanorods of metal oxides. A metal alkoxide reacts with a metal halide at high temperature in the presence of a coordinating solvent leading to nanoparticles of the

TABLE 7.3 Typical Examples of Nanorods Synthesized by the Sol-Gel Process

Nanorod	Chemicals Used
$CoSb_3$	Cobalt (II) chloride, antimony (III) chloride, ethanol, citric acid
$(K_{0.5}Bi_{0.5})Ba_{0.6}TiO_3$	Bismuth nitrate, potassium nitrate, barium acetate, titanium tetrabutoxide
$Ba_5Nb_4O_{15}$	Barium nitrate, niobium (V) ethoxide, ethylene glycol, citric acid, ethanol
Cd doped ZnO	Zinc acetate dihydrate, 2-methoxy ethanol, monoethanolamine, cadmium acetate
In_2O_3	Indium chloride, ammonia, polyethylene octyl phenyl ether (OP-10)
Fe_2O_3	Iron (III) chloride, oleic acid, benzyl ether, propylene oxide
V doped ZnO	Zinc (II) acetyl acetonate, vanadyl (IV) acetyl acetonate, ethylene glycol, methyl alcohol
$Ce_{1-x}Gd_xO_{2-\delta}$	Gadolinium nitrate, cetyl trimethyl ammonium bromide, aqueous ammonia, cerium (III) chloride, sulfuric acid

corresponding metal oxide. The size, shape, and phase depend on the temperature of the reaction and the halide used. The reaction between metal alkoxide and metal halide leads to an M-O-M bridge along with the elimination of alkyl halide, R-X.

$$mMX_n + nM(OR)_n \longrightarrow M_mM_nO_{nm} + nmRX$$

MX_n is the metal halide and $M(OR)_n$ is an alkoxide. The rate of alkyl halide elimination plays a crucial role in the nucleation and growth of nanocrystals. Examples of nanorods synthesized by this method include HfO_2 (aspect ratio, 2.3) prepared by the reaction of $HfCl_4$ and $Hf(OiPr)_4$ at 400°C, in trioctyl phosphine oxide under argon, and nanorods of $Hf_{0.66}Zr_{0.34}O_2$ with aspect ratio 3.6, prepared by the reaction of $Hf(O^iPr)_4$, $HfCl_4$, and $ZrCl_4$ in the ratio of 2 : 1 : 1.

7.2.10 Conversion of Nanosheets and Nanotubes to Nanorods

Conversions of nanosheets into nanorods have been possible for metal oxides such as ZnO (44). The self-assembled nanosheets are first prepared by a solution phase hydrothermal synthesis. The reactant solution contains zinc nitrate, hexamethylene tetramine and hydrazine, which helps in the formation of precursor nanosheets in an orderly fashion. At shorter time scales ($\sim$4 h), nanosheets are formed and after 8 h, hollow microspheres are formed. The nanosheets are calcined at 400°C to get ZnO nanorods, as depicted in Figure 7.6. This is an example where one nanostructure transforms to a completely different morphology.

Conversion of nanotubes to nanorods has also been demonstrated (45). For example, $NaHTi_3O_7$ nanotubes have been converted to $Na_2Ti_6O_{13}$ nanorods by calcination. The $NaHTi_3O_7$ nanotubes are prepared from TiO_2 and NaOH under hydrothermal conditions. The release of structural water molecules from the interlayer space is thought to be the mechanism for the conversion of nanotubes to nanorods (Fig. 7.7).

7.2.11 Green Chemical Synthesis

In recent times, emphasis is being placed on environmentally friendlier synthetic routes to make new materials. These methods are called green chemistry routes, which do not involve toxic starting reagents and by-products. Examples of green chemical synthesis are as follows.

α-MnO_2 nanorods have been photochemically deposited on functionalized polystyrene beads by immobilization of permanganate ions under alkaline conditions (46). Synthesis of nanorods by hydrothermal conditions is also sometimes considered green chemical synthesis provided the chemicals used are nontoxic and environmentally benign. For example, nanorods of α-MoO_3 can be synthesized by decomposition followed by condensation of peroxomolybdic acid (47). Using reagents such as hydrogen peroxide is promising in the area of green chemical synthesis. For example, zinc oxide nanorods can be produced by the reaction $Zn + H_2O_2 \rightarrow ZnO + H_2O$ (48). The advantages of such reactions are that no by-product is formed and possibly no impurities will be present in the final product.

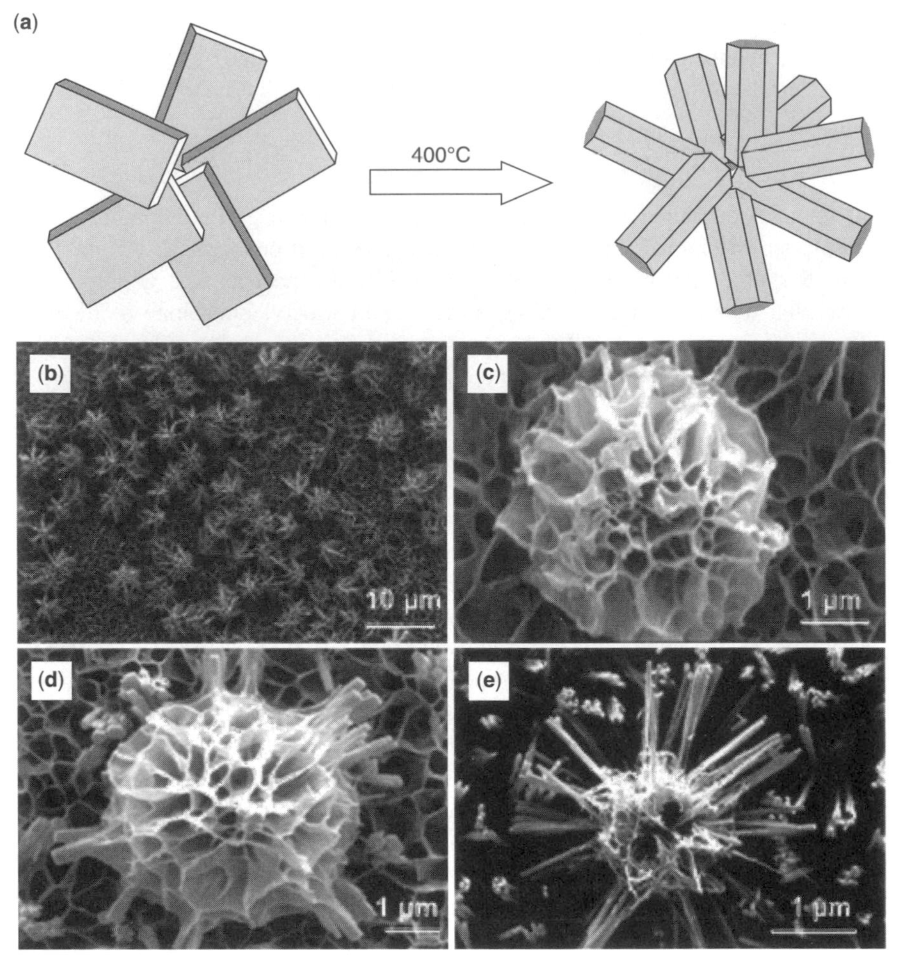

Figure 7.6 Synthesis of ZnO nanorods by the transformation of nanosheets into nanorods. (Reprinted with permission from S. Konar et al. *J. Phys. Chem. B* **2006**, 110, 4054. Copyright (2006) American Chemical Society.)

7.2.12 Biomimetic Synthesis

Nature follows unique means of making new materials. To achieve perfection in orientation and morphology, natural biomaterials (e.g. sea shells) use organic molecules. This approach has been used successfully in the preparation of inorganic materials, particularly nanorods (49). By using a biomimetic approach, one can synthesize nanorods by chemical modification of surfaces with analogous organic molecules nature employs. One can get novel structures like spiral columns, multisection rods, etc. Knowledge on the morphological control of biomaterials can be extended to synthesize novel nanostructures. For example, by using citrate ions, helical nanorods of zinc oxide and columns have been produced (reference 50; Fig. 7.8); one can control the orientation, defect structure, and other textural properties. Hydroxyapatite is a

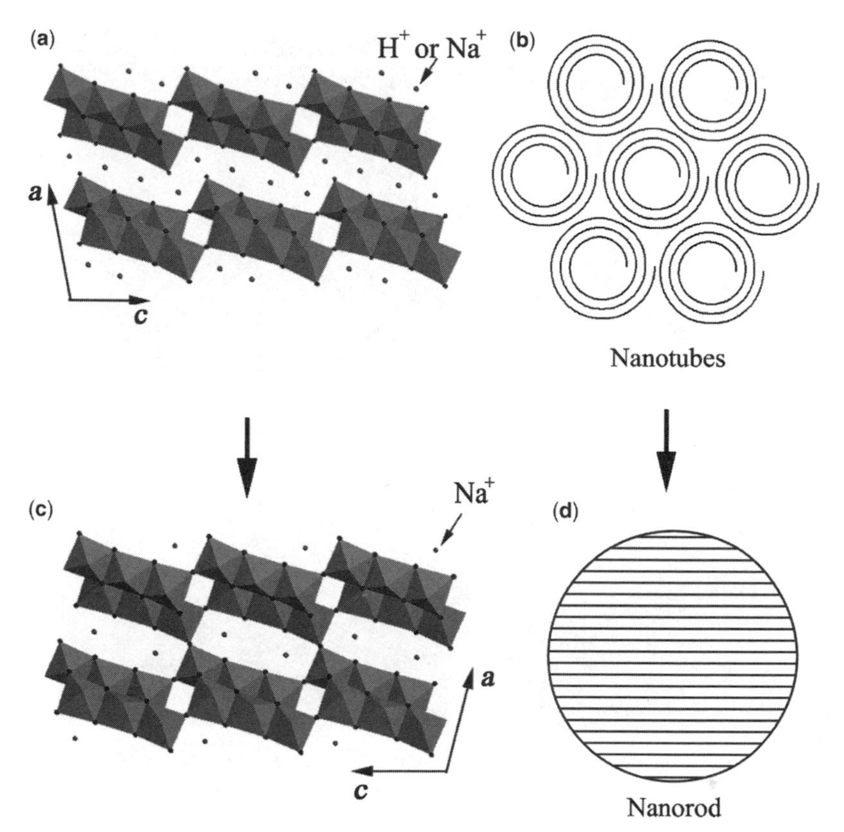

Figure 7.7 Transformation of $NaHTi_3O_7$ nanotubes into $Na_2Ti_6O_{13}$ nanorods: (a) structure of $NaHTi_3O_7$; (b) cross-section of nanotubes; (c) structure of $Na_2Ti_6O_{13}$; and (d) cross-section of nanorods. The octahedrons in the figure represent TiO_6. (Reprinted from K. R. Zhu et al. *Solid State Communications* **2007**, 144, 450. Copyright (2007) Elsevier. With permission.)

biomineral that can act as a scaffold for the synthesis of nanorods. Reduction of silver ions can occur on the surface of hydroxyapatite, which leads to the formation of silver nanorods that are bound on the surface of the hydroxyapatite (51). The reduction of metal ions occurs by transfer of electrons from hydroxyl groups to the metal ion. Ag_2CrO_4 nanorods have been synthesized using an emulsion-liquid membrane method (52); the membrane phase contains a surfactant (Span-80), a carrier (N7301), and kerosene as the solvent. The aqueous phase consists of $AgNO_3$ and $K_2Cr_2O_7$ solutions. During anaerobic growth, bacteria such as *Bacillus selenitireducens* use tellurium oxyanions as respiratory electron acceptors leading to the formation of tellurium metal in the form of nanorods (53).

7.2.13 Synthesis in Restricted Dimensions

It is possible to synthesize nanorods inside a restricted space such as the pores in a mesoporous material. Mesoporous materials are different from conventional templates

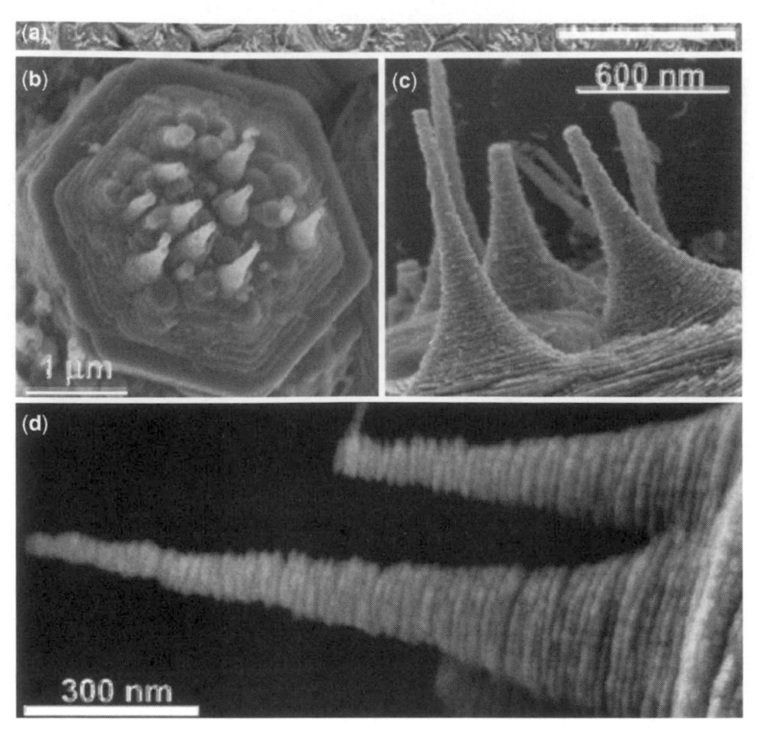

Figure 7.8 SEM images of helical ZnO nanorods oriented on ZnO crystals. The tilted arrangement of the helical nanorods on ZnO can be clearly noticed at higher magnifications (c and d). (Reprinted with permission from Z. R. Tian et al. *J. Am. Chem. Soc.* **2002**, 124, 12954. Copyright (2002) American Chemical Society.)

such as anodic aluminum oxide in that their surface chemistry can be suitably tailored. Let us take an example to illustrate synthesis in the restricted dimension. GaN nanorods have been synthesized within the pores of SBA-15, a mesoporous aluminosilicate with well-defined pores. The procedure is as follows (54).

First, the silanol groups (Si-OH) inside the mesoporus silica are converted into Si-OCH_3 groups by reacting with methyl trimethoxysilane so that the pores are converted from hydrophilic to hydrophobic nature. Then, $GaCl_3$ dissolved in anhydrous toluene is impregnated into the pores of SBA-15 powder under nitrogen atmosphere. After impregnation, the SBA-15 powder containing $GaCl_3$ is heated first under nitrogen and then under NH_3 atmosphere at 900°C to produce GaN nanorods. It is possible to obtain free-standing GaN nanorods by removing the silica framework by dissolving with an aqueous HF solution. The GaN nanorods inside the SBA-15 and the free-standing GaN nanorods after removing the matrix are shown in Figure 7.9.

7.2.14 Miscellaneous Methods

There are many miscellaneous methods to prepare nanorods. A few of them are discussed briefly below.

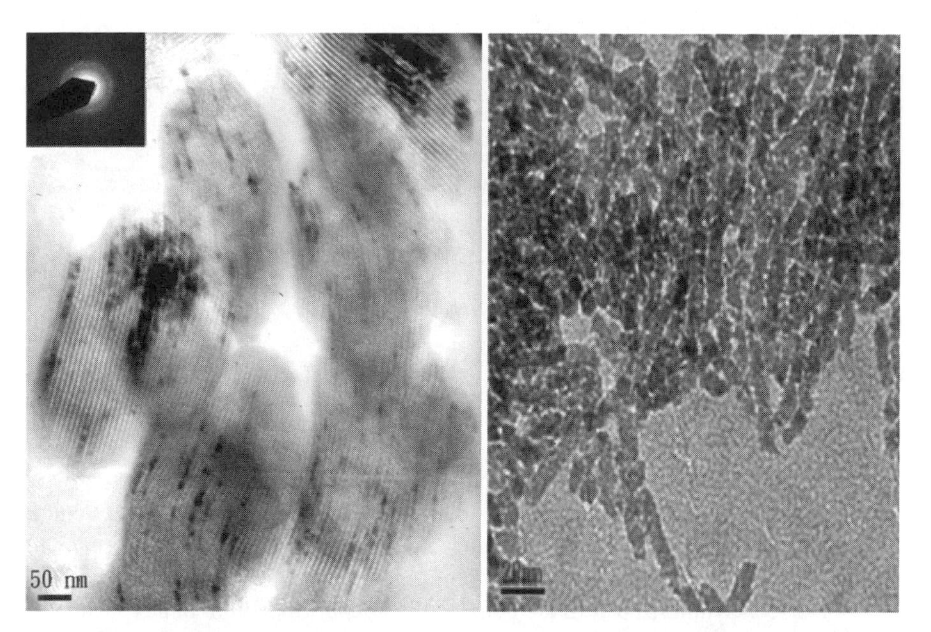

Figure 7.9 TEM image of GaN nanorods incorporated in SBA-15. The dark streaks indicate the nanorods inside the mesoporous material. The TEM image of the free-standing GaN nanorods obtained after the removal of the matrix is shown on the right side. (Reprinted with permission from C. T. Yang et al. *J. Phys. Chem. B.* **2005**, 109, 17842. Copyright (2005) American Chemical Society.)

- *Precursor method*: A lamellar precursor with needle-like morphology can be converted into required metal hydroxide or metal oxide with the retention of its morphology (55). This method does not require a template and the reaction involves milder conditions. Simple starting materials such as magnesium chloride hexahydrate and magnesium oxide can be used to synthesize a precursor (magnesium oxychloride) which can be converted into magnesium hydroxide nanorods under mild NaOH treatment conditions. The hydroxide nanorods can be converted into MgO nanorods by calcination in air (Fig. 7.10). The source of MgO influences the synthesis of nanorods; nanocrystalline MgO leads to faster synthesis with high aspect ratio for the products compared to macrocrystalline MgO.
- *Microwave synthesis*: It has been possible to synthesize a few nanorods of materials using microwave methods (56). This method is faster compared to other methods and the only condition is that one of the reactants should absorb microwave radiation. If the frequency of the microwave radiation is such that the electric field changes its sign at a speed of the order of the relaxation time of the dipoles, energy is absorbed, which leads to rapid homogeneous heating of the reactants. Nanorods synthesized by microwave methods include ZnS, CdSe, Bi_2S_3, Sb_2S_3, Ag, Fe, and rare earth oxides such as Pr_2O_3, Nd_2O_3, Sm_2O_3, Eu_2O_3, Gd_2O_3, Tb_2O_3, and Dy_2O_3.

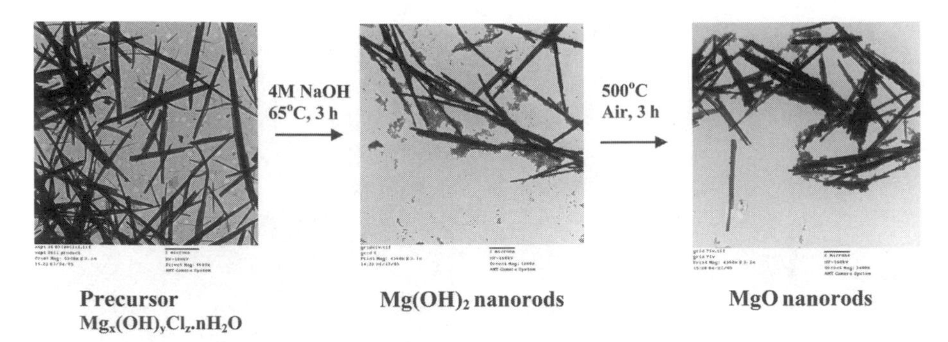

Figure 7.10 A precursor approach for the synthesis of MgO nanorods. The precursor on NaOH treatment leads to $Mg(OH)_2$ nanorods, which on calcination yield MgO nanorods.

- *Sonochemical synthesis*: This topic has been discussed in detail elsewhere (57) and only a short note is given here. Acoustic cavitation in a liquid leads to unique conditions such as high temperature and pressure during the collapse of the bubbles, apart from producing impinging micro-jets. A variety of nanorods have been synthesized with the help of ultrasound irradiation. In a typical sonochemical synthesis of nanorods, the precursor(s) of a particular metal(s) in a high boiling solvent such as decalin along with a surfactant such as cetyltrimethylammonium *p*-toluene sulfonate (CTAPTS) or a macromolecule such as cyclodextrin are used. Nanorods that have been produced by sonochemical methods include copper and magnetite.

7.2.15 Nanorods on Substrates

From the spatial orientation and arrangement of nanorods on different substrates, new properties and applications may arise, for example, in the area of nanophotonics. Aligned nanorod arrays of metals such as silver serve as substrates for surface enhanced IR absorption spectroscopy (58). Well-aligned arrays of various nanostructured oxides on substrates are highly desirable for device applications, such as nanopiezoelectric generators, dye-sensitized solar cells, and nanosensor arrays. Usually, expensive techniques such as sputtering and lithography are used. Chemical methods offer economically viable, alternative routes. A brief discussion on different methods is given below.

7.2.15.1 By Self-Assembly For implementation of nanostructures into nanoelectronic or microelectronic devices, self-assembly is important. Self-assembly of nanostructures can lead to interesting properties due to collective interactions. For example, a side-by-side assembly of nanorods leads to blue shift of the longitudinal plasmon band and the inter-nanorod distance affects the strength of plasmon coupling (59). Some specific examples of self-assembly are discussed below.

Layer-by-layer self-assembly can be used to grow nanorods on a substrate (60). Au nanorods in a multilayered structure can be deposited on an ITO substrate.

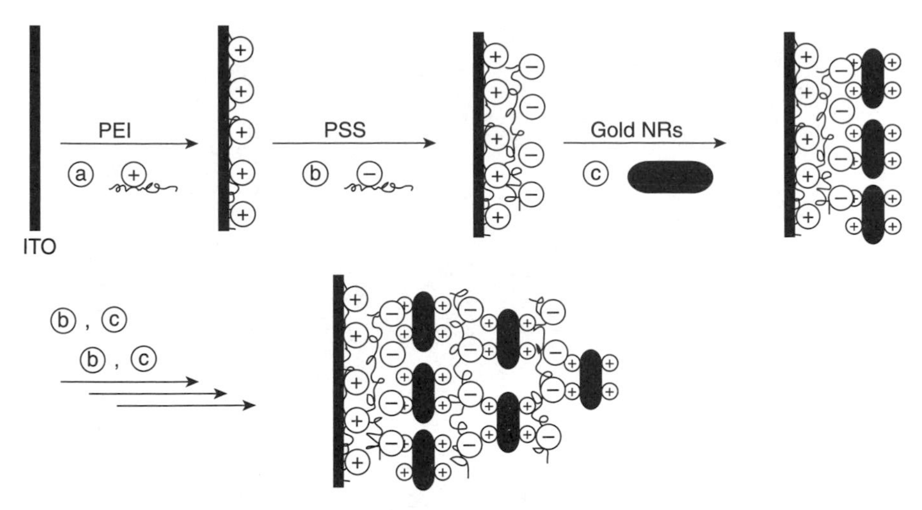

Scheme 7.3 Formation of multilayered gold nanorods on an ITO substrate using a layer-by-layer assembly approach. (Reprinted with permission from X. Hu et al. *J. Phys. Chem. B* **2005**, *109*, 19385. Copyright (2005) American Chemical Society.)

Polyethylene imine acts as an adhesive layer and polystyrene sulfonate (an anionic polyelectrolyte) has been used according to Scheme 7.3. First a solution containing Au nanorods is prepared using seed-mediated growth. Then, the clean ITO slides are immersed in polyethyleneimine solution followed by polystyrene sulfonate solution. After rinsing and drying, the slides are immersed in the Au nanorod solution. The Au rods self-assemble and deposit on substrate, as shown in Figure 7.11.

FePt nanorods of high quality have been produced by decomposition of $Fe(CO)_5$ and reduction of $Pt(acac)_2$ in a confined cylindrical mesophase consisting of surfactants such as oleic acid and oleyl amine at high concentrations and slow heating (61). Self-assembly of FePt nanorods can be carried out on a silicon wafer by evaporating a hexane dispersion. Hydrothermal synthesis has also been used for self-assembly by precoating the substrate with aluminum metal (62). Under hydrothermal conditions, the aluminum metal transforms to a hydrotalcite-like phase which provides a lattice matched substrate for the growth of oxide nanorods such as ZnO (Fig. 7.12). The substrates can be diverse, flat, or curved, such as silicon, polystyrene beads, carbon nanotube array, etc.

By making the Au nanorods from hydrophilic to hydrophobic, the nanorods can be made to self-assemble on a substrate by simple evaporation (63). Au nanorods prepared using hexadecyl trimethylammonium bromide are hydrophilic. But they can be made hydrophobic by treating them with mercaptopropyltrimethoxysilane and octadecyltrimethoxy silane. Depending on the concentration of the nanorods in the solution, they assemble either parallel to the substrate (at low concentration) or perpendicular to the substrate (at higher concentration). Nanorod arrays can also be prepared by electrochemical self-assembly on substrates (64). For example, Mn, Co doped ZnO nanorod arrays on Cu substrates have been prepared by electrochemical self-assembly

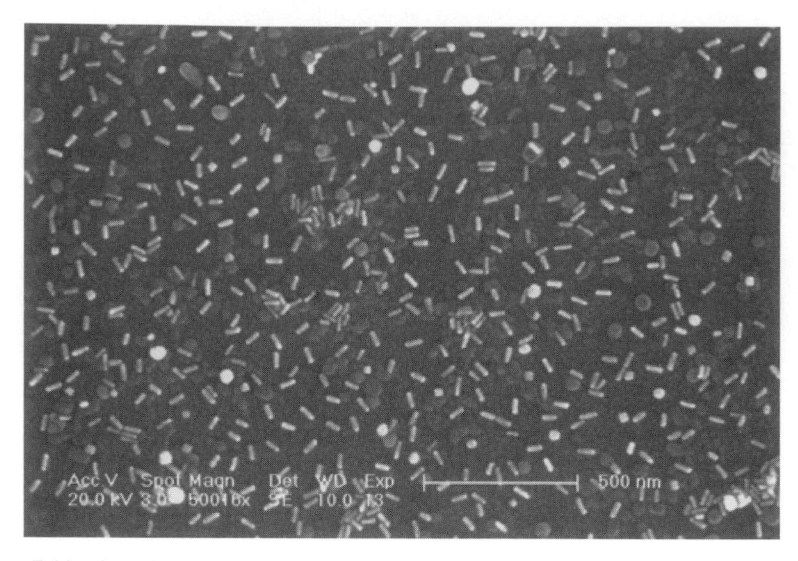

Figure 7.11 SEM image of a monolayer of gold nanorods assembled on an ITO substrate derivatized with polyethyleneimine and polystyrene sulfonate. (Reprinted with permission from X. Hu et al. *J. Phys. Chem. B* **2005,** 109, 19385. Copyright (2005) American Chemical Society.)

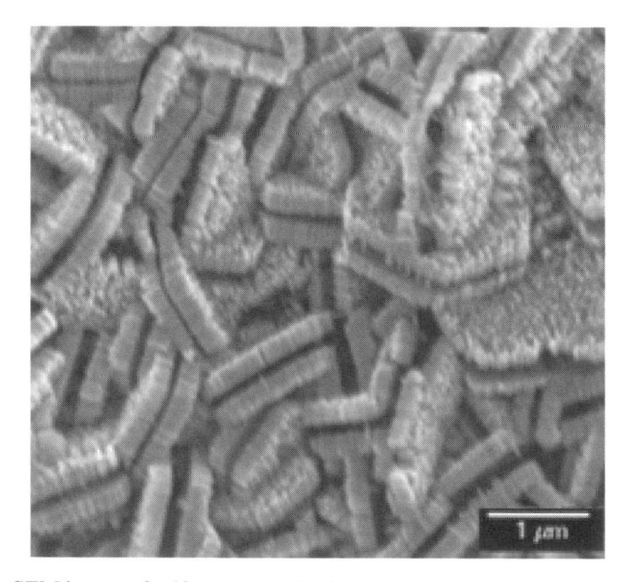

Figure 7.12 SEM image of self-assembled ZnO nanorods on a hydrotalcite template in a hexagonal array. (Reprinted with permission from Y. W. Koh et al. *J. Phys. Chem. B* **2004**, 108, 11491. Copyright (2004) American Chemical Society.)

in a solution of $ZnCl_2$, $MnCl_2$, $CoCl_2$, KCl, and tartaric acid at temperatures close to 90°C. Electrochemical deposition is a simple, quick method and the orientation and density of the nanorods can be controlled by adjusting the electrochemical parameters such as deposition potential, current density, and the concentration of reagents. Sometimes a quick self-assembly of nanorods is required. For this purpose, a rapid self-assembly of silver nanorods on substrates has been reported (65). The substrate such as quartz is modified with a universal polymer modifier, poly(4-vinyl pyridine). Au nanorods coated with SiO_2, prepared by a modified Stöber's process, are allowed to interact with the substrate. The pyridyl groups of the polymer interact with the hydroxyl groups on the Au-SiO_2 structures leading to the self-assembly of nanorods.

7.2.15.2 Chemical Bath Deposition Instead of growing nanorods by electrochemical deposition, it is possible to synthesize nanorods by using a simple chemical bath. The method is simple, economical, and oriented nanorods on a specific substrate can be synthesized. Metal oxide nanorods were easily synthesized by this method, for example, ZnO and In_2O_3. ZnO nanorods have been synthesized under aqueous conditions with nanorods growing along a particular direction (e.g. 0001; Reference 66). By varying the concentration of reagents (e.g. OH^- ions), the aspect ratio of the rods can be controlled. While growing nanorods on a substrate, the seed-mediated approach can be used (67). For example, on a substrate such as PET, a seed layer of ZnO prepared using the sol-gel process can be coated followed by the chemical bath deposition of ZnO. This way one can grow highly oriented metal oxide nanorods. Another example of the seed-mediated growth approach is the deposition of silver nanorods on a glass substrate, as shown in Scheme 7.4 (68). First, glass substrates are surface modified with 3-(aminopropyl) triethoxysilane. The slides are dipped in a solution containing silver seeds. After drying in a stream of N_2 the slides are dipped in a solution containing CTAB followed by the addition of the growth solution (it contains $AgNO_3$, ascorbic acid, and NaOH). The nanorods grow within 10 min and the silver nanorods-coated glass substrates are ready for use.

Chemical bath deposition can also be used to produce core-shell nanorod film electrodes, for example, TiO_2@CdS (69). It is possible to couple chemical bath deposition and the microwave radiation procedure. The advantage of this process is that there is no need for the heat treatment of the samples after the synthesis, since the microwave does the job. ZnO nanorods with an orientation perpendicular to the substrate can be synthesized in 12 min by this method, starting from zinc nitrate (or zinc acetate) and urea (or hexamethylene tetramine; Reference 70).

7.2.15.3 Chemical Vapor Deposition In chemical vapor deposition, the precursor molecules containing the elements of interest are decomposed (the molecular precursor is vaporized into a flow reactor) in the gas phase to produce the product. To make a compound that contains more than one element, one can use a precursor that contains the elements in the correct ratio. We can grow thin films of nanorods of metals, semiconductors, and insulators using chemical vapor deposition. There are many parameters that can affect the growth of the films; temperature, pressure, chemistry of the precursor molecule, etc. Two examples are given below.

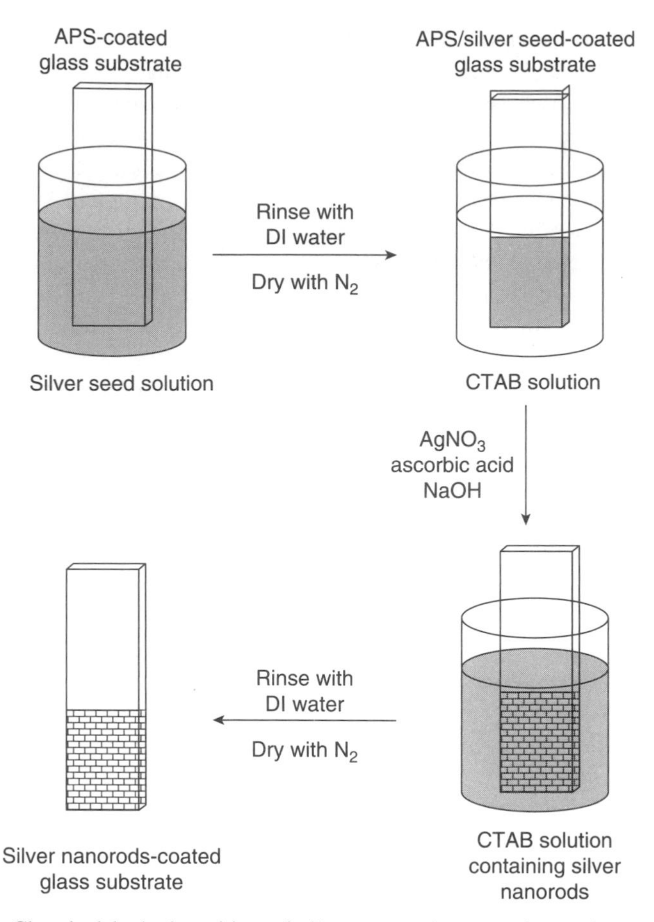

Scheme 7.4 Chemical bath deposition of silver nanorods on a glass substrate by a seed-mediated growth approach. (Reprinted with permission from K. Aslan et al. *J. Phys. Chem. B* **2005**, *109*, 3157. Copyright (2005) American Chemical Society.)

ZnMgO nanorods have been grown on Si(111) substrate using diethyl zinc and $[Mg(H_2O)_6](NO_3)_2$ (71). First, the silicon substrate is sprayed with an aqueous solution of $[Mg(H_2O)_6](NO_3)_2$ and a layer of this material is grown under vacuum. $[Mg(H_2O)_6](NO_3)_2$ acts as a source of magnesium as well as the oxidant. Diethyl zinc is introduced using nitrogen as the carrier gas and the substrate temperature is heated to about 450°C. ZnMgO nanorods are deposited on the substrate. Another example is the deposition of Bi_2O_3 nanorods on gold-coated Si(100) substrate using argon as the carrier gas at 450°C (72). Problems associated with chemical vapor deposition involve the toxicity, volatility, and pyrophoric nature of the precursors and it is sometimes difficult to control the mixing of reactants inside the reaction chamber with the desired ratio.

7.2.15.4 Galvanic Exchange Reactions Galvanic exchange reactions can lead to core-shell nanorods (see Section 7.2.17) on substrates. For example, a palladium shell has been produced on silver nanorods (73). First, silver nanorods are synthesized directly on surfaces (silicon or glass) using the seed-mediated approach. The surface-grown nanorods are then treated with palladium salt solution (or platinum salt solution). This leads to a galvanic exchange reaction producing a shell of metal on the surface of the nanorod, that is, a palladium shell on a silver nanorod. The nanorods keep their morphologies intact after the galvanic exchange reactions. The reactions involved are as follows (73).

$$2Ag^{o}\ (s) + PtCl_4^{2-}\ (aq) \longleftrightarrow Pt^{o}\ (s) + 2AgCl\ (s\ or\ aq) + 2Cl^{-};\ E^{o} = 0.508\ V$$

$$2Ag^{o}\ (s) + PdCl_4^{2-}\ (aq) \longleftrightarrow Pd^{o}\ (s) + 2AgCl\ (s\ or\ aq) + 2Cl^{-};\ E^{o} = 0.401\ V$$

These alloy-type nanorods may find applications in the areas of catalysis and sensing.

7.2.16 Superlattices of Nanorods

A superlattice means a collection of uniformly sized nanocrystals arranged in a periodic manner. When nanorods are produced with uniform size and shape distribution, they can self-assemble in solution or on a substrate leading to superlattice formation. Superlattices of nanorods possess interesting properties. For example, the plasmon resonance in gold nanorods can be tuned by assembling the nanorods into a three-dimensional superlattice (74). Also, nanorod superlattices show interesting properties such as strong nonlinearity in the current-voltage curve and current oscillations. Superlattices based on magnetic nanorods are interesting from the point of view of high density magnetic recording.

One can produce superlattices of nanorods by a variety of means. The most important is the decomposition of a precursor in the presence of surfactants. Under this category, superlattices of Co nanorods and CoO nanorods superlattice have been produced (75). For example, on decomposing a cobalt-based coordination compound in the presence of ligands (e.g. long chain amine/acid) cobalt nanorods assembled as a superlattice on a substrate have been produced. Pencil-shaped CoO nanorods produced by the thermal decomposition of cobalt-oleate self-assemble into superlattice structures (76), shown in Figure 7.13. An organometallic compound may be used as a precursor to produce a superlattice of nanorods of a metal. For example, $Co(\eta^3\text{-}C_8H_{13})(\eta^4\text{-}C_2H_{12})]$ in the presence of stearic acid and hexadecyl amine produces cobalt nanorods that self-assemble into three-dimensional superlattices (77).

Sometimes, nanorods can self-assemble into superlattices that behave like liquid crystals. Examples include the smectic-type superlattices of Cu_2S nanorods formed on thermal decomposition of copper thiolate at low temperature (140°C to 200°C; Reference 78). Under appropriate conditions, nematic-type superlattices of CdSe nanorods are produced from a concentrated solution of the nanorods in a low boiling solvent such as cyclohexane (79). When a solution of nanorods is destabilized by adding a nonsolvent, liquid crystal-like assemblies (nematic, smectic) of a

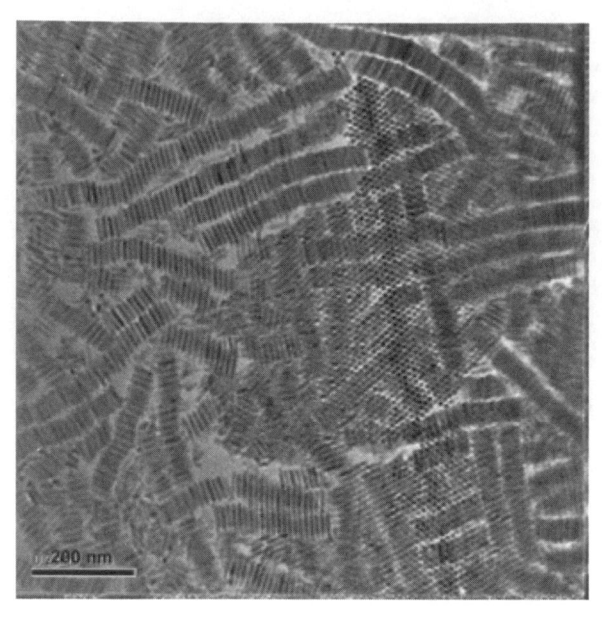

Figure 7.13 TEM image of self-assembled CoO nanorods, prepared by decomposition of cobalt oleate (size: 9 nm × 46 nm) observed after the slow evaporation of a hexane solution of nanorods. (Reprinted with permission from K. An et al. *J. Am. Chem. Soc.* **2006,** 128, 9753. Copyright (2006) American Chemical Society.)

semiconductor such as CdSe are produced (80). Controlled evaporation can lead to interesting formation of nanorod superlattices. For example, the controlled evaporation of solution containing nanorods trapped between a smooth substrate and a piece of highly oriented pyrolytic graphite (HOPG) can lead to crystallization of nanorod superlattices (81); superlattices with dimension 2 μm × 2 μm can be produced. Evaporation coupled with application of an electric field can help in aligning the nanorods perpendicular to the substrate (82); CdS nanorod superlattices can be produced by this method. Inducing a strain can lead to the spontaneous formation of nanorod superlattices. For example, a colloidal route has been devised to make CdS-Ag_2S nanorod superlattices (83). The partial ion exchange of Cd^{2+} ions by Ag^+ causes strain in the lattice leading to spontaneous formation of organized structures, that is, superlattices (Fig. 7.14).

7.2.17 Core-Shell Nanorods

Core-shell nanorods represent composite nanorods in which the core and shell dimensions can be independently varied. The physicochemical properties (e.g. band gap, near-infrared absorption) can be tuned by varying the dimension of the core and the shell. The major synthetic routes for the preparation of the core-shell nanorods are:

1. The solvothermal/hydrothermal method
2. Decomposition of precursors in the presence of surfactants

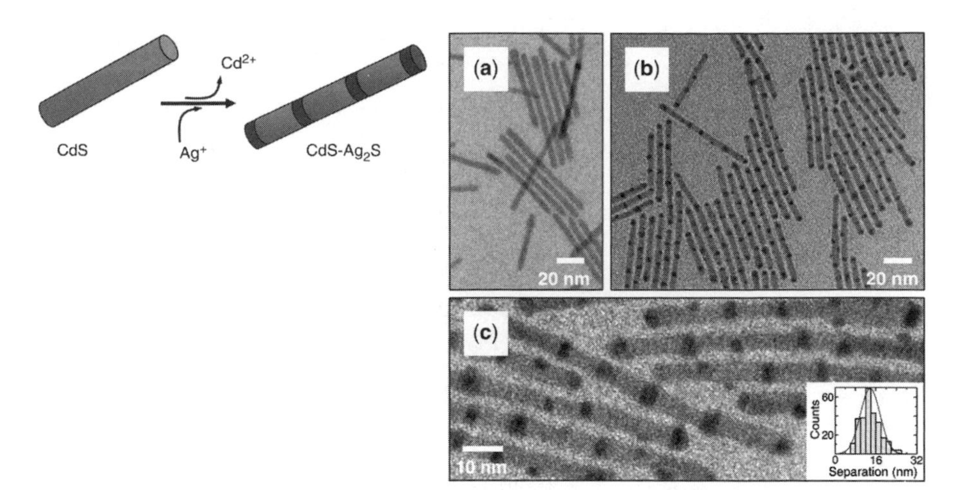

Figure 7.14 TEM images of superlattices formed via partial cation exchange of Cd^{2+} ions by Ag^+ ions. (a) CdS nanorods, (b) and (c) formation of superlattices. The spacing between Ag_2S segments is shown as a histogram. (From R. D. Robinson et al. *Science* **2007**, 317, 355. Reprinted with permission from AAAS.)

3. Seed-mediated synthesis of the core followed by chemical deposition (e.g. using a mild reducing agent)
4. Chemical bath deposition
5. Reaction under autogenic pressure at elevated temperatures
6. Phase-controlled hydrolysis method
7. Epitaxial growth
8. Wet chemical synthesis
9. Bio-templating
10. Galvanic exchange
11. Sonochemical synthesis

The core-shell nanorods that have been prepared using these methods are summarized in Table 7.4. A few specific cases are discussed below.

$Au_{core}Ag_{shell}$ nanorods have been prepared starting from Au nanorods (84). The Au nanorods prepared electrochemically are coated with silver by the reduction of $AgCl_4^-$ with hydroxyl amine. The thickness of the shell can be varied by adjusting the concentrations of $AgCl_4^-$ and NH_2OH. The TEM images show the shell clearly (Fig. 7.15). In some cases, first the nanorods are synthesized by the solvothermal method. The nanorods are functionalized with a suitable ligand and then a shell is grown on the surface. For example, CdS nanorods prepared by solvothermal synthesis are surface modified with citric acid followed by a treatment with zinc nitrate and sodium sulfide solutions to coat ZnS on CdS nanorods (85).

For the preparation of CdSe/ZnS core-shell nanorods (86), first CdSe nanorods are prepared by decomposing a cadmium precursor, $Cd(CH_3)_2$, and selenium metal in the

TABLE 7.4 Examples of Core-Shell Nanorods Prepared by Different Methods

Core-Shell Nanorod	Preparation Method(s)
Au/Ag	Chemical deposition
Ag/Pt,Pd	Galvanic exchange
Au/Ag,Pd	Using mild reducing agent
Au/SiO_2	Seed-mediated growth and TEOS hydrolysis
$Ni/V_2O_5.nH_2O$	Template-based electrochemical deposition
CdS/CdO	Metal organic chemical vapor deposition (MOCVD)
Fe_2O_3/ZnO	Solution-phase controlled hydrolysis
$ZnO/Zn_{0.8}Mg_{0.2}O$	Metal organic vapor-phase epitaxy
CdS/ZnS	Solvothermal synthesis
ZnO/ZnS	Hydrothermal, MOCVD, coprecipitation
CdSe/CdS	Seeded growth
TiO_2/CdS	Chemical bath deposition
CdSe/ZnS	Surfactant assisted synthesis
CdSe/ZnSe	Surfactant assisted synthesis
Nb_2O_5/C	Reaction under autogenic pressure
$W_{18}O_{49}/C$	Biotemplated synthesis
$BaCrO_4/PbCrO_4$	Template method
$CePO_4:Tb/LaPO_4$	Ultrasound irradiation

presence of a mixture of hexylphosphonic acid, tetradecyl phosphonic acid, and tri-*n*-octyl phosphine oxide at about 350°C under inert atmosphere. The CdSe nanorods along with hexadecyl amine is heated to high temperature and the shell precursor solution [the solution consists of $Zn(Et)_2$, trioctylphosphine and hexamethyl disilathiane] is added drop-wise to grow the ZnS shell. It is possible to grow shells of oxides (e.g. SiO_2) on the metal nanorods (87). Tetraethyl orthosilicate is added to a

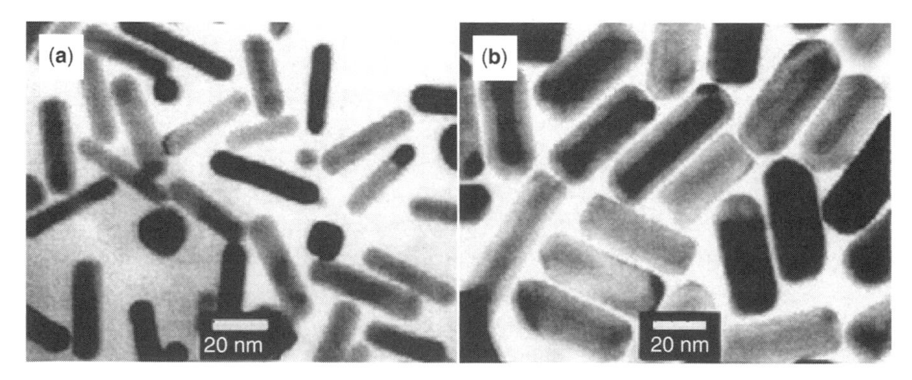

Figure 7.15 TEM images of Au-Ag core-shell nanorods prepared by the reduction of $AgCl_4^{3-}$ on the surface of Au nanorods. (a) $Au_{core}Ag_{shell(thin)}$ nanorods, (b) $Au_{core}Ag_{shell(thick)}$ nanorods. (Reprinted with permission from C. S. Ah et al. *J. Phys. Chem. B* **2001**, 105, 7871. Copyright (2001) American Chemical Society.)

solution containing gold nanorods and the pH is adjusted to close to 10 using ammonia. A shell of SiO_2 grows on the surface of the nanorods.

7.2.18 Functionalized Nanorods

Nanorods possess high surface area and there is a lot of scope for functionalization of the surface. First of all, why do we need to functionalize the nanorods? Functionalized nanorods find applications in such diverse areas as glucose and protein sensing, imaging, gene delivery, photovoltaics, photoelectrochemical, etc. Using functionalized nanorods leads to improved biological activity compared to commonly used methods. Functionalization can lead to increased solubility, which might help in studying the physical and chemical behavior of individual nanorods. Functionalization can also help in linking the nanorods with each other through rational design and one can also attach the nanorods on a suitably modified substrate. Functionalized nanorods can serve as morphological seeds, for example, they can direct the growth of a single crystalline polymer oriented along the nanorod direction (template-directed growth; Reference 88). Functionalization can help in assembling the nanorods inside a template (89); Most functionalization has been done with polymers. It is possible to have a high grafting density of polymer molecules on the surface (e.g. about 3000 chains per gold nanorod, with dimension 10 nm $\times$ 45 nm). The polymer functionalized nanorods render the nanorods soluble in nonaqueous solvents and also favor the self-assembly of nanorods on hydrophobic substrates. Many examples of functionalization can be given but a brief account is given below.

Gold nanorods have been functionalized with a copolymer (90). The copolymer contains sulfonate and maleic acid groups which have been conjugated with an azide molecule. Using the functionalized gold nanorods one can do chemistry; for example, a copper catalyzed dipolar cycloaddition reaction has been carried out by attaching the functionalized nanorods to tripsin. Functionalization with a polymer can lead to biocompatibility. For example, the biomedical application of CTAB-coated gold nanorods is limited since CTAB is absorbed nonspecifically by the cells (91). However, replacing the CTAB with a polymer such as oligoethylene glycol with folate termination makes the nanorods biocompatible, that is, they accumulate specifically on the tumor cells.

Functionalization of nanorods with polyelectrolytes has been carried out by layer-by-layer deposition (92). First, CTAB-coated nanorods are prepared. Since these nanorods are positively charged, they can adsorb cationic and anionic polyelectrolytes. Functionalization of nanorods with dyes is possible; a fluorescent dye, 4-chloro-7-nitrobenzofurazan has been functionalized on the surface of TiO_2 nanorods (93). Functionalization with a photoactive molecule such as ruthenium(II) tris(bipyridine) is also possible (94). A thiol derivative of the bipyridyl complex ($Ru(bpy)_3^{2+}$-C_5-SH) in dodecane thiol is used for the functionalization of gold nanorods. Functionalization of block magnetic nanorods is very useful (95), for example, in the separation of proteins. Consider a triblock nanorod consisting of only two metals, Ni and Au. If the Au blocks are functionalized with a thiol (e.g. 11-amino-1 undecane thiol) followed by covalent attachment of nitrostreptavidin, then one can use them for the separation of proteins. Bioconjugation can also be considered as functionalization; for example, bioconjugation of antibodies with Au nanorods.

7.2.19 Nanorod Composites

Nanocomposites are materials in which nanoparticles (in this case, nanorods) are dispersed in a continuous matrix. The matrix may be a polymer, nanorods, or other nanoparticles. Nanorod composites find applications in diverse areas such as efficient charge storage, removal of contaminants (e.g. surfactant) from water, emissivity control devices, and metallodielectrics, and so on. A number of methods such as electroless deposition, the sol-gel method, the hydrothermal method, solution casting, carbothermal reduction, the template-based method, the sonochemical method, and electrospinning can be used to prepare composite nanorods. Nanorod composites are different from core-shell nanorods. In core-shell nanorods, the coating is uniform, whereas in the nanorod composite (consisting of a nanorod and a nanoparticle on a surface), fine nanoparticles are dispersed on the surface of the nanorods. Some specific examples of the preparation of nanocomposites consisting of nanorods are described below.

On ZnO nanorods, silver nanoparticles have been deposited using electroless deposition using ammoniacal $AgNO_3$, $SnCl_2$, and CF_3COOH (96). Stannous chloride reduces Ag^+ to Ag^o and Sn^{2+} gets oxidized to Sn^{4+}. A SEM image of the nanocomposite is shown in Figure 7.16. The silver nanoparticles are uniformly dispersed

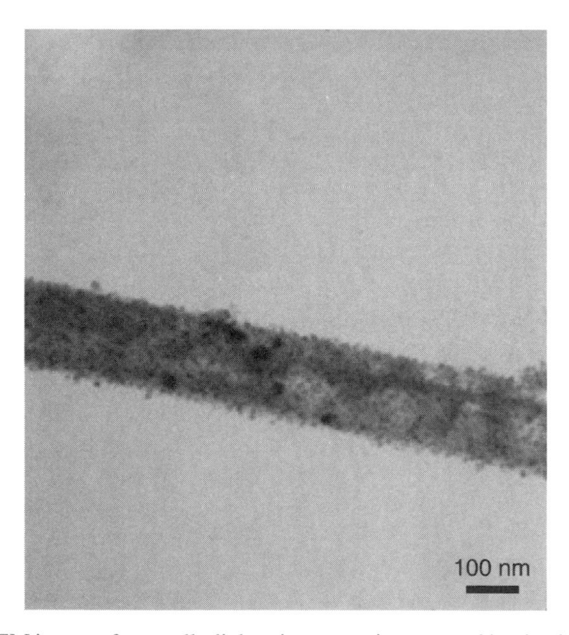

Figure 7.16 TEM image of a metallodielectric composite prepared by the deposition of silver nanoparticles, by electroless deposition, on ZnO nanorods. (Reprinted from X. Ye et al. *Materials Letters* **2008**, 62, 666. Copyright (2007) Elsevier. With permission.)

on the surface of the ZnO nanorod. SiO_2/TiO_2 nanorod composite membranes have been prepared using the sol-gel method (97). $Ti(OC_4H_9)_4$ and $Si(OC_2H_5)_4$ are used as precursors and ethanol as a solvent. Aqueous HCl is added slowly and the hydrolysis of the alkoxides is over in about an hour. A clear sol obtained after sonication is used for filling an alumina template membrane. After a certain immersion time, the template membrane is removed from the sol, dried in air and then calcined to get SiO_2/TiO_2 composite nanorods. Semiconductive Bi_2S_3 nanorods may be loaded in a polymer such as polyvinylidene fluoride (a ferroelectric polymer) using a solution cast method (98). The nanorods can be sonicated in a solution containing the polymer and a suitable solvent (e.g. *N,N*-dimethyl formamide) above ambient temperature. The solution can be cast on a glass slide to make a composite material.

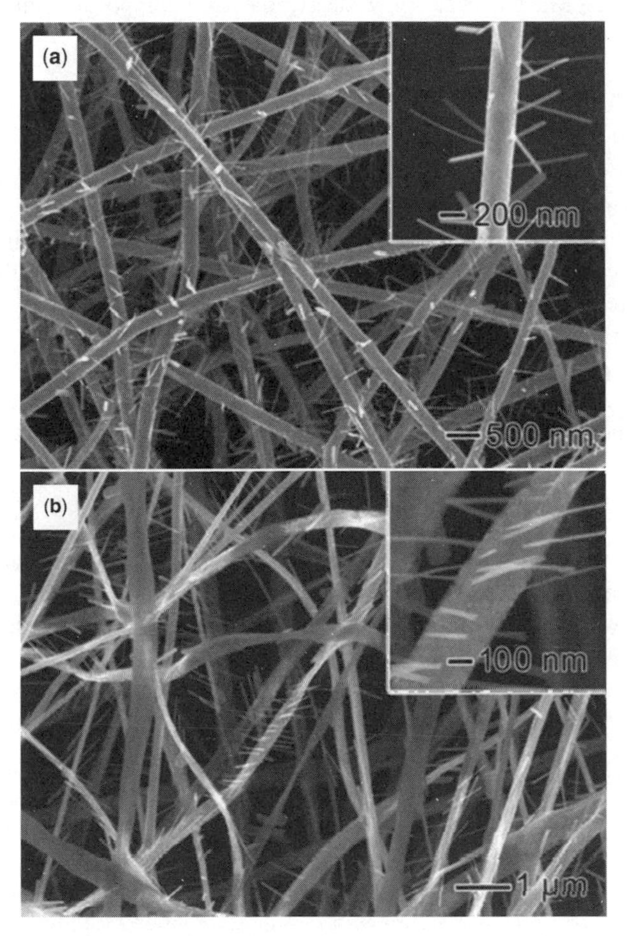

Figure 7.17 SEM images of nanocomposites with hierarchial nanostructures prepared by electrospinning followed by calcination: (a) V_2O_5-Ta_2O_5 nanorods on TiO_2 nanofibers, and (b) V_2O_5 nanorods on SiO_2 nanofibers. (Reprinted with permission from R. Ostermann et al. *Nano Lett.* **2006**, 6, 1297. Copyright (2006) American Chemical Society.)

Nanorod composites can also be prepared using metal nanoparticles dispersed on nanorods of metal oxides (99). For example, Au nanoparticles are deposited on ZnO nanorods to make Au/ZnO nanorod composites on silicon substrate by photochemical reduction of $HAuCl_4$/ethanol solution using 365 nm irradiation. Single crystal V_2O_5 nanorods can be grown on TiO_2 fibers (rutile) by calcination of nanofibers containing V_2O_5, TiO_2 (both in amorphous forms) and a polymer, polyvinyl pyrrolidone (100). First, uniform nanofibers of V_2O_5/TiO_2 are prepared by electrospinning of a sol solution consisting of 2-propanol, titanium isopropoxide, vanadium oxytriisopropoxide, and acetic acid. After keeping the fibers in air, they are calcined at temperatures between 375°C and 575°C. In a similar way, V_2O_5-Ta_2O_5-TiO_2 and V_2O_5-SiO_2 nanofibers can be prepared by electrospinning 2-propanol solutions of $VO(OiPr)_3$, $Ti(OiPr)_4$, and $TaO(OiPr)_3$ or $Si(OEt)_4$ and $VO(OiPr)_3$, respectively, followed by calcination. The SEM images show clearly the nanocomposite nature of the rods (Fig. 7.17).

7.3 CONCLUSION AND FUTURE OUTLOOK

Synthesis of nanorods has developed into a mature science. A variety of methods is available for the synthesis of nanorods, most of which have been well optimized to yield reproducible results. One has to choose between the available synthetic methods considering the chemical nature of the nanorods, purity required, and availability of chemicals, among other factors. Using simple chemistry, remarkable results can be achieved. Since nanochemistry and related areas of research have become very much interdisciplinary in nature, many nonchemists would like to synthesize nanorods for further studies. This has become a reality now since simple and easy methods to prepare high quality nanorods are available in the literature; mere basic chemistry knowledge is required.

The future of nanotechnology depends on inventing synthetic methods for producing new nanoscale materials as well as developing improved methods of existing syntheses. Chemists have a big role to play in this endeavor. Synthesis of nanorods is a part of the ongoing technology-driven research and it will certainly lead to exciting discoveries in the area of nanotechnology.

REFERENCES

1. K. J. Klabunde (Ed.), *Nanoscale Materials in Chemistry*, Wiley, New York, **2001**.
2. (a) G. Cao (Ed.), *Nanostructures and Nanomaterials: Synthesis, Properties and Applications*, Imperial College Press, London, **2004**; (b) G. A. Ozin, A. C. Arsenault, *Nanochemistry: A Chemical Approach to Nanomaterials*, Royal Society of Chemistry, Cambridge, **2005**; (c) L. M. Liz-Marzan, P. V. Kamat (Eds.), *Nanoscale Materials*, Kluwer Academic, Boston, **2003**; (d) C. N. R. Rao, A. Muller, A. K. Cheetham (Eds.), *Chemistry of Nanomaterials*, Wiley-VCH, Weinheim, **2004**; (e) S. J. Hurst, E. K. Payne, L. Qin, C. A. Mirkin, *Angew. Chem. Int. Ed.* **2006**, *45*, 2672; (f) G. Cao, D. Liu, *Adv. Colloid. Interf. Sci.* **2008**, 136, *45*; (g) J. Chen, B. J. Wiley, Y. Xia,

Langmuir **2007**, *23*, 4120; (h) G. R. Patzke, F. Krumeich, R. Nesper *Angew. Chem. Int. Ed.* **2002**, *41*, 2446.

3. P. Yang, C. M. Lieber, *Science* **1996**, *273*, 1836.
4. A. Murayama, M. Sakuma, *Appl. Phys. Lett.* **2006**, *88*, 122504/1.
5. R. Mulukutla, P. S. Malchesky, R. Maghirang, J. S. Klabunde, K. J. Klabunde, O. Koper, *PCT Int. Appl.* **2006**.
6. (a) M. Grzelczak, J. Perez-Juste, P. Mulvaney, L. M. Liz-Marzan, *Chem. Soc. Rev.* **2008**, *37*, 1783; (b) C. J. Murphy, A. M. Gole, S. E. Hunyadi, C. J. Orendorff, *Inorg. Chem.* **2006**, *45*, 7544; (c) J. Perez-Juste, I. Pastoriza-Santos, L. M. Liz-Marzan, P. Mulvaney, *Coord. Chem. Rev.* **2005**, *249*, 1870; (d) N. R. Jana, *Small*, **2005**, *1*, 875.
7. K. J. Klabunde, J. V. Stark, O. Koper, C. Mohs, D. G. Park, S. Decker, Y. Jiang, I. Lagadic, D. Zhang, *J. Phys. Chem.* **1996**, *100*, 12142.
8. G.-C. Yi, C. Wang, W. Park, *Semicond. Sci. Technol.* **2005**, *20*, S22.
9. C. J. Murphy, T. K. Sau, A. M. Gole, C. J. Orendorff, J. Gao, L. Gou, S.E. Hunyadi, T. Li. *J. Phys. Chem. B* **2005**, *109*, 13857.
10. (a) B. B. Lakshmi, P. K. Dorhout, C. R. Martin, *Chem. Mater.* **1997**, *9*, 857; (b) B. B. Lakshmi, C. J. Patrissi, C. R. Martin, *Chem. Mater.* **1997**, *9*, 2544.
11. G. Cao, *J. Phys. Chem. B* **2004**, *108*, 19921.
12. D. Ma, M. Zhang, G. Xi, J. Zhang, Y. Qian, *Inorg. Chem.* **2006**, *45*, 4845.
13. W. Hua-Qiang, X. W. Wei, M. W. Shao, J. S. Gu, *J. Cryst. Growth* **2004**, *265*, 184.
14. C. Tang, S. Fan, M. L. Chapelle, H. Dang, P. Li, *Adv. Mater.* **2000**, *12*, 1346.
15. T. Kyotani, L. Tsai, A. Tomita, *Chem. Commun.* **1997**, 701.
16. B. C. Satishkumar, A. Govindaraj, M. Nath, C. N. R. Rao, *J. Mater. Chem.* **2000**, *10*, 2115.
17. H. Dai, E. W. Wong, Y. Z. Lu, S. Fan, C. M. Lieber, *Nature* **1995**, *375*, 769.
18. E. W. Wong, B. W. Maynor, L. D. Burns, C. M. Lieber, *Chem. Mater.* **1996**, *8*, 2041.
19. M. A. Correa-Duarte, J. Perez-Juste, A. Sanchez-Iglesias, M. Giersig, L. M. Liz-Marzan, *Angew. Chem. Int. Ed.* **2005**, *44*, 4375.
20. Y. Zhu, X. Guo, Y. Shen, M. Mo, X. Guo, W. Ding, Y. Chen, *Nanotechnology* **2007**, *18*, 195601/1.
21. D. Zhang, L. Qi, H. Cheng, J. Ma. *J. Colloid Interf. Sci.* **2002**, *246*, 413.
22. N. M. Huang, R. Shahidan, P. S. Khiew, L. Peter, C. S. Kan, *Colloids Surf. A* **2004**, *247*, 55.
23. Q. Tong, S. Kosmella, J. Koetz, *Prog. Colloid Polym. Sci.* **2006**, *133*, 152.
24. A. Foyet, A. Hauser, W. Schaefer, *J. Electroanal. Chem.* **2007**, *604*, 137.
25. L. Mai, W. Guo, Y. Gu, W. Jin, Y. Dai, W. Chen, *Abstracts of 59th Southeast Regional ACS Meeting,* Greenville, SC, 2007.
26. X. Dong, X. Zhang, B. Liu, H. Wang, Y. Li, Y. Huang, Z. Du, *J. Nanosci. Nanotechnol.* **2006**, *6*, 818.
27. S. Sadasivan, D. Khushalani, S. Mann, *Chem. Mater.* **2005**, *17*, 2765.
28. K. Woo, H. J. Lee, J. P. Ahn, Y. S. Park, *Adv. Mater.* **2003**, *15*, 1761.
29. (a) J. C. Hulteen, C. J. Patrissi, D. L. Miner, E. R. Crosthwait, E. B. Oberhauser, C. R. Martin, *J. Phys. Chem. B* **1997**, *101*, 7727; (b) C. J. Brumlik, V. P. Menon, C. R. Martin, J. Mater. Res. **1994**, 9, 1174.

30. (a) Y. Qian, D. Yu, Z. Liu, *Trans. Mater. Res. Soc. Jpn.* **2004**, *29*, 2233; (b) W.-T. Yao, S. H. Yu, Int. J. Nanotechnol. **2007**, *4*, 129.
31. P. S. Nair, G. D. Scholes, *J. Mater. Chem.* **2006**, *16*, 467.
32. Z. Zhang, S. Liu, S. Chow, M. Y. Han, *Langmuir* **2006**, *22*, 6335.
33. F. Kim, J. H. Song, P. Yang, *J. Am. Chem. Soc.* **2002**, *124*, 14316.
34. K. Nishioka, Y. Niidome, S. Yamada, *Langmuir* **2007**, *23*, 10353.
35. S. Xu, W. Zhao, J. M. Hong, J.J. Zhu, H. Y. Chen, *Mater. Lett.* **2004**, *59*, 319.
36. O. R. Miranda, T. S. Ahmadi, *J. Phys. Chem. B* **2005**, *109*, 15724.
37. P. Yang, C. M. Lieber, *Appl. Phys. Lett.* **1997**, *70*, 3158.
38. J. Grabowska, K. K. Nanda, E. McGlynn, J. P. Mosnier, M. O. Henry, *Surf. Coat. Technol.* **2005**, *200*, 1093.
39. C. Andreazza-Vignolle, P. Andreazza, D. Zhao, *Superlattices and Microstructures* **2006**, *39*, 340.
40. M. Zhao, X. L. Chen, W. J. Wang, Y. J. Ma, Y. P. Xu, H. Z. Zhao, *Mater. Lett.* **2006**, *60*, 2017.
41. S. O. Kang, H. S. Jang, Y. Kim, K. B. Kim, M. J. Jung, *Mater. Lett.* **2007**, *61*, 473.
42. S. C. Kuiry, E. Megen, S. D. Patil, S. A. Deshpande, S. Seal, *J. Phys. Chem. B* **2005**, *109*, 3868.
43. B. Koo, J. Park, Y. Kim, S. H. Choi, Y. E. Sung, T. Hyeon, *J. Phys. Chem. B.* **2006**, *110*, 24318.
44. S. Konar, Z. R. Tian, *J. Phys. Chem. B* **2006**, *110*, 4054.
45. K. R. Zhu, Y. Yuan, M. S. Zhang, J. M. Hong, Y. Deng, Z. Yin, *Solid State Communications* **2007**, *144*, 450.
46. S. Jana, S. Praharaj, S. Panigrahi, S. Basu, S. Pande, C. H. Chang, T. Pal, *Org. Lett.* **2007**, *9*, 2191.
47. L. Fang, Y. Shu, A. Wang, T. Zhang, *J. Phys. Chem. C* **2007**, *111*, 2401.
48. Y. Zhao, Y. U. Kwon, *Chem. Lett.* **2004**, *33*, 1578.
49. (a) L. Liu, Q. Wu, Y. Ding, H. Liu, J. Qi, Q. Liu, *Aust. J. Chem.* **2004**, *57*, 219; (b) S. A. Morin, F. F. Amos, S. Jin, *PMSE* **2007**, *96*, 227 (Preprint).
50. Z. R. Tian, J. A. Voigt, J. Liu, B. McKenzie, M. J. McDermott, *J. Am. Chem. Soc.* **2002**, *124*, 12954.
51. S. K. Arumugam, T. P. Sastry, B. Sreedhar, A. B. Mandal, *J. Biomed. Res.* **2007**, *80A*, 391.
52. L. Liu, Q. Wu, Y. Ding, H. Liu, J. Qi, Q. Liu, *Aust. J. Chem.* **2004**, *57*, 219.
53. S. M. Baesman, T. D. Bullen, J. Dewald, D. Zhang, S. Curran, F. S. Islam, T. J. Beveridge, R. S. Oremland, *Appl. Environ. Microbiol.* **2007**, *73*, 2135.
54. C. T. Yang, M. H. Huang. *J. Phys. Chem. B* **2005**, *109*, 17842.
55. P. Jeevanandam, R. S. Mulukutla, Z. Yang, H. Kwen, K. J. Klabunde, *Chem. Mater.* **2007**, *19*, 5395.
56. (a) S. Bhattacharyya, A. Gedanken, *J. Phys. Chem. C* **2008**, *112*, 659; (b) J.-S. Liu, J.-M. Cao, Z. Q. Li, G. B. Ji, M. B. Zheng, *Mater. Lett.* **2007**, *61*, 4409.
57. Y. Mastai, A. Gedanken, *Chem. Nanomaterials* **2004**, *1*, 113.
58. C. L. Leverette, S. A. Jacobs, S. Shanmukh, S. B. Chaney, R. A. Dluky, Y. P. Zhao, *Appl. Spectros.* **2006**, *60*, 906.

59. P. K. Jain, S. Eustis, M. A. El-Sayed, *J. Phys. Chem. B* **2006**, *110*, 18243.
60. X. Hu, W. Cheng, T. Wang, Y. Wang, E. Wang, S. Dong, *J. Phys. Chem. B* **2005**, *109*, 19385.
61. M. Chen, T. Pica, Y. B. Jiang, P. Li, K. Yano, J. P. Liu, A. K. Datye, H. Fan, *J. Am. Chem. Soc.* **2007**, *129*, 6348.
62. Y. W. Koh, M. Lin, C. K. Tan, Y. L. Foo, K. P. Loh, *J. Phys. Chem. B* **2004**, *108*, 11419.
63. K. Mitamura, T. Imae, N. Saito, O. Takai. *J. Phys. Chem. B* **2007**, *111*, 8891.
64. G. R. Li, D. L. Qu, W. X. Zhao, Y. X. Tong, *Electrochem. Commun.* **2007**, *9*, 1661.
65. C. Wang, Z. Ma, T. Wang, Z. Su, *Adv. Funct. Mater.* **2006**, *16*, 1673.
66. B. Cao, W. Cai, *J. Phys. Chem. C* **2008**, *112*, 680.
67. S. H. Yi, S. K. Choi, J. M. Jang, J. A. Kim, W. G. Jung, *J. Colloid Interf. Sci.* **2007**, *313*, 705.
68. K. Aslan, Z. Leonenke, J. R. Lakowicz, C. D. Geddes, *J. Phys. Chem. B* **2005**, *109*, 3157.
69. H. Jia, H. Xu, Y. Hu, Y. Tang, L. Zhang, *Electrochem. Commun.* **2007**, *9*, 354.
70. A. M. Peiro, C. Domingo, J. Peral, X. Domenech, E. Vigil, M. A. Hernandez-Fenollosa, M. Mollar, B. Mari, J. A. Ayllon, *Thin Solid Films* **2005**, *483*, 79.
71. J. R. Wang, Z. Z. Ye, J. Y. Huang, Q. B. Ma, X. Q. Gu, H. P. He, L. P. Zhu, J. G. Lu, *Mater. Lett.* **2008**, *62*, 1263.
72. H. W. Kim, J. W. Lee, S. H. Shim, *Sensors Actuators B* **2007**, *126*, 306.
73. G. W. Slawinski, F. P. Zamborini, *Langmuir* **2007**, *23*, 10357.
74. T. J. Norman, C. D. Grant, J. Z. Zhang, *Nanoparticle Assemblies and Superstructures* **2006**, 193.
75. F. Wetz, K. Soulantica, M. Respaud, A. Falqui, B. Chaudret, *Mater Sci. Eng. C* **2007**, *27*, 1162.
76. K. An, N. Lee, J. Park, S. C. Kim, Y. Hwang, J. G. Park, J. Y. Kim, J. H. Park, M. J. Han, J. Yu, T. Hyeon, *J. Am. Chem. Soc.* **2006**, *128*, 9753.
77. F. Dumestre, B. Chaudret, C. Amiens, M. Respaud, P. Fejes, P. Renaud, P. Zurcher, *Angew. Chem. Int. Ed.* **2003**, *42*, 5213.
78. T. H. Larsen, M. Sigman, A. Ghezelbash, C. R. Doty, B. A. Korgel, *J. Am. Chem. Soc.* **2003**, *125*, 5638.
79. L. S. Li, A. P. Alivisatos, *Adv. Mater.* **2003**, *15*, 408.
80. D. V. Talapin, E. V. Shevchenko, C. B. Murray, A. Kornowski, S. Foerster, H. Weller, *J. Am. Chem. Soc.* **2004**, *126*, 12984.
81. S. Ahmed, K. M. Ryan, *Nano Lett.* **2007**, *7*, 2480.
82. K. M. Ryan, A. Mastroianni, K. A. Stancil, H. Liu, A. P. Alivisatos, *Nano Lett.* **2006**, *6*, 1479.
83. R. D. Robinson, B. Sadtler, D. O. Demchenko, C. K. Erdonmez, L. W. Wang, A. P. Alivisatos, *Science* **2007**, *317*, 355.
84. C. S. Ah, S. D. Hong, D. J. Jang, *J. Phys. Chem. B* **2001**, *105*, 7871.
85. A. Datta, S. K. Panda, S. Chaudhuri, *J. Phys. Chem. C* **2007**, *111*, 17260.
86. T. Mokari, U. Banin, *Chem. Mater.* **2003**, *15*, 3957.
87. J. J. Zhang, Y. G. Liu, L. P. Jiang, J. J. Zhu, *Electrochem. Commun.* **2008**, *10*, 355.
88. C. S. T. Laicer, T. Q. Chastek, T. P. Lodge, T. A. Taton, *Macromolecules* **2005**, *38*, 9749.

89. Q. Zhang, S. Gupta, T. Emrick, T. P. Russell, *J. Am. Chem. Soc.* **2006**, *128*, 3898.
90. A. Gole, C. J. Murphy, *Langmuir*, **2008**, *24*, 266.
91. T. B. Huff, L. Tong, Y. Zhao, M. N. Hansen, J. X. Cheng, A. Wei, *Nanomedicine* **2007**, *2*, 125.
92. A. Gole, C. J. Murphy, *Chem. Mater.* **2005**, *17*, 1325.
93. M. N. Tahir, P. Theato, P. Oberle, G. Melnyk, S. Faiss, U. Kolb, A. Janshoff, M. Stepputat, W. Tremel, *Langmuir* **2006**, *22*, 5209.
94. M. Jebb, P. K. Sudeep, P. Pramod, K. G. Thomas, P. V. Kamat, *J. Phys. Chem. B* **2007**, *111*, 6839.
95. B. K. Oh, S. Park, J. E. Millstone, S. W. Lee, K. B. Lee, C. A. Mirkin, *J. Am. Chem. Soc.* **2006**, *128*, 11825.
96. X. Ye, Y. Zhou, J. Chen, Y. Sun, Z. Wang, *Mater. Lett.* **2008**, *62*, 666.
97. H. Zhang, X. Quan, S. Chen, H. Zhao, Y. Zhao, *Appl. Surf. Sci.* **2006**, *252*, 8598.
98. Y. Shen, C.-W. Nan, M. Li, *Chem. Phys. Lett.* **2004**, *396*, 420.
99. J. J. Wu, C. H. Tseng, *Appl. Catal. B* **2006**, *66*, 51.
100. R. Ostermann, D. Li, Y. Yin, J. T. McCann, Y. Xia. *Nano Lett.* **2006**, *6*, 1297.
101. N. R. Jana, L. Gearheart, C. J. Murphy, *Adv. Mater.* **2001**, *13*, 2001.
102. T. K. Sau, C. J. Murphy, *Langmuir* **2004**, *20*, 6414.
103. B. Nikoobakht, M. A. El-Sayed, *Chem. Mater.* **2003**, *15*, 1957.
104. S. Ghosh, M. Ghosh, C. N. R. Rao, *J. Clust. Sci.* **2007**, *18*, 97.
105. F. Sediri, N. Gharbi, *J. Phys. Chem. Solids* **2007**, *68*, 1821.
106. X. W. Zhu, Y. Q. Li, Y. Lu, L. C. Liu, Y. B. Xia, *Mater. Chem. Phys.* **2007**, *102*, 75.
107. U. N. Maiti, P. K. Ghosh, S. F. Ahmed, M. K. Mitra, K. K. Chattopadhyay, *J. Sol-Gel Sci. Technol.* **2007**, *41*, 87.
108. S. Maensiri, C. Masingboon, V. Promarak, S. Seraphin, *Opt. Mater.* **2007**, *29*, 1700.
109. J. S. Lee, S. Kim, *J. Am. Ceram. Soc.* **2007**, *90*, 661.
110. Y. W. Chen, Y. C. Liu, S. X. Lu, C. S. Xu, C. L. Shao, C. Wang, J. Y. Zhang, Y. M. Lu, D. Z. Shen, X. W. Fan, *J. Chem. Phys.* **2005**, *123*, 134701.
111. S. M. Chang, R. A. Doong, *J. Phys. Chem. B.* **2006**, *110*, 20808.
112. L. Yu, X. Zhang, *Mater. Chem. Phys.* **2004**, *87*, 168.
113. D. Golberg, F. F. Xu, Y. Bando, *Appl. Phys. A* **2003**, *76*, 479.
114. F. Liu, J. Y. Lee, W. Zhou, *J. Electrochem. Soc.* **2006**, *153*, A2133.
115. F. Liu, J. Y. Lee, W. Zhou, *Adv. Funct. Mater.* **2005**, *15*, 1459.
116. D. Seo, C. Yoo, J. Jung, H. Song, *J. Am. Chem. Soc.* **2008**, *130*, 2940.
117. F. Liu, J. Y. Lee, W. Zhou, *J. Phys. Chem. B* **2004**, *108*, 17959.
118. S. C. Chang, M. H. Huang, *J. Phys. Chem. C* **2008**, *112*, 2304.
119. H. Zhang, D. Yang, X. Ma, Y. Ji, S. Z. Li, D. Que, *Mater. Chem. Phys.* **2005**, *93*, 65.
120. K. Yu, J. Zhao, X. Zhao, X. Ding, Y. Zhu, Z. Wang, *Mater. Lett.* **2005**, *59*, 2676.
121. K. Manzoor, V. Aditya, S. R. Vadera, N. Kumar, T. R. N. Kutty, *J. Phys. Chem. Solids* **2005**, *66*, 1164.
122. C. Pacholski, A. Kornowski, H. Weller, *Angew. Chem. Int. Ed.* **2002**, *41*, 1188.
123. Y. Hu, T. Mei, J. Guo, T. White, *Inorg. Chem.* **2007**, *46*, 11031.

124. B. P. Khanal, E. R. Zubarev, *Angew. Chem. Int. Ed.* **2007**, *46*, 2195.
125. S. O. Obare, N. R. Jana, C. J. Murphy, *Nano Lett.* **2001**, *1*, 601.
126. J. Zhang, Z. Liu, B. Han, T. Jiang, W. Wu, J. Chen, Z. Li, D. Liu, *J. Phys. Chem. B* **2004**, *108*, 2200.
127. M. E. Pearce, J. B. Melanko, A. K. Salem, *Pharm. Res.* **2007**, *24*, 2335.
128. S. J. Lee, V. Anandan, G. Zhang, *Biosensors Bioelectronics*, **2008**, *23*, 1117.
129. B. Wildt, P. Mali, P. C. Searson, *Langmuir* **2006**, *22*, 10528.
130. R. Gonzalez-McQuire, J. Y. C. Ching, E. Vignaud, A. Lebugle, S. Mann, *J. Mater. Chem.* **2004**, *14*, 2277.
131. K. K. Caswell, J. N. Wilson, U. H. F. Bunz, C. J. Murphy, *J. Am. Chem. Soc.* **2003**, *125*, 13914.
132. A. Thomas, B. Premlal, M. Eswaramoorthy, *Mater. Res. Bull.* **2006**, *41*, 1008.
133. L. Zhu, Q. Li, J. Li, X. Liu, J. Meng, X. Cao, *J. Nanopart. Res.* **2007**, *9*, 261.
134. X. Jia, W. He, X. Zhang, H. Zhao, Z. Li, Y. Feng, *Nanotechnology* **2007**, *18*, 075602.
135. A. Thomas, M. Schierhorn, Y. Wu, G. Stucky, *J. Mater. Chem.* **2007**, *17*, 4558.

PROBLEMS

1. What is the role of Ag^+ ions during the growth of Au nanorods in a seed-mediated growth procedure? Explain.

2. Apart from carbon nanotubes as template, can we use any other nanotubes as templates for growing nanorods? Explain, with examples.

3. Explain with an example how multisegmented metallic nanorods can be synthesized by electrochemical deposition.

4. Why are amines used in the solvothermal synthesis of nanorods? Take two examples and explain the role of the amines.

5. Can we introduce dopant(s) during the synthesis of nanorods by the sol-gel method? If so, give two examples and the procedure to make the doped nanorods.

6. Give an example where metal oxide nanorods can be grown inside a mesoporous silicate.

7. Can we have self-assembly of nanoparticles to nanorods? Explain, with an example.

8. How can we make a core-shell nanorod with a metal core and a polymer shell? Explain, with an example.

9. Is it possible to functionalize nanorods with biological molecules? Take one example and explain how functionalization can be done.

10. Briefly explain with an example how to synthesize a mesoporous nanorod.

ANSWERS

1. In the seed-mediated growth procedure for the synthesis of Au nanorods, the Au seeds are prepared first by reducing an aqueous solution of $HAuCl_4$ containing a surfactant (e.g. CTAB) with $NaBH_4$. The seed solution is added to the growth solution, which contains the surfactant, $HAuCl_4$, and small amounts of $AgNO_3$, followed by the addition of ascorbic acid. The aspect ratio of the rods can be controlled by varying the seed to metal salt ratio only in the presence of silver ions (101). The presence of Ag^+ is important for the formation of short nanorods (aspect ratio ~5; References 102 and 103). Short Au nanorods are important from the point of view of sensing and optical applications since they possess both transverse and longitudinal absorptions in the visible region of the electromagnetic spectrum. In the absence of Ag^+, larger nanorods (aspect ratio ~20) are produced, with lower yield (see Fig. P-1). Also, the yield of the nanorods increases drastically in the presence of Ag^+ ions. For example, the presence of about 5% silver ions during the synthesis, increases the yield to about 100%; in the absence of Ag^+ ions, only about 20% to 40% yield is obtained.

   On addition of Ag^+ ions during the seed-mediated growth, AgBr is formed on the surface of CTAB-coated gold nanocrystals. The source of Br^- is the surfactant and the silver ions are not reduced to Ag^o by ascorbic acid; only Au^{3+} ions are reduced. On adsorption of AgBr, there is a decrease in the head group charge of the surfactant (CTAB) leading to a decrease in the repulsion between the head groups of the surfactant molecules. This, in turn, causes the formation of the soft surfactant templates, with small aspect ratio (~5). Also, the adsorption of AgBr on the surface of Au nanocrystals slows or restricts the subsequent

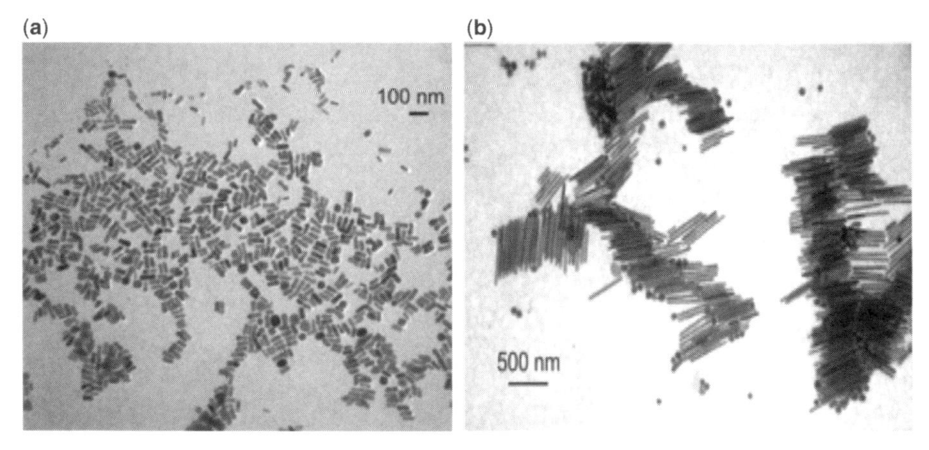

Figure P-1 TEM images of (a) Au nanorods synthesized in the presence of silver nitrate (6×10^{-5} M), and (b) in the absence of silver nitrate. The other synthetic conditions are $[Au]_{seed} = 5 \times 10^{-7}$ M, $[CTAB] = 9.5 \times 10^{-2}$ M, $[HAuCl_4] = 4 \times 10^{-4}$ M, and [ascorbic acid] $= 6.4 \times 10^{-4}$ M. (Reprinted with permission from T. K. Sau et al. *Langmuir* **2004**, 20, 6414. Copyright (2004) American Chemical Society.)

growth of Au nanorods. This explains the formation of short nanorods in the presence of Ag^+ ions. Adsorption of AgBr on the surface of Au particles stabilizes the rods and spheroids that are formed during the synthesis; in the absence of silver ions the spheroids destabilize to form nanospheres. This explains the formation of nanorods in high yield. Using Ag^+ ions also produces Au nanorods with different crystallography compared to those prepared in the absence of Ag^+ ions. According to HRTEM studies on Au nanorods, in the presence of Ag^+ ions, the $\{111\}$ faces are on the long sides of the rods while in the absence of Ag^+ ions, Au nanorods with $\{100\}$ facets along the long side of the rods are produced.

2. Although carbon nanotubes (single walled and multiwalled) are the most common nanotubes to be used as templates for the growth of nanorods, there are a few other nanotubes that can be used as templates. Two examples are discussed below.

VO_x/titanate composite nanorods have been synthesized using titanate nanotubes, $K_xH_{2-x}Ti_3O_7$, as templates employing hydrothermal conditions (112). The titanate nanotubes are prepared as follows. About 0.5 g of anatase TiO_2 and 32 mL of 10 M KOH are subjected to an autoclave treatment at 200°C for 20 h. After cooling, the product is filtered under vacuum, treated with 1 M HCl, washed with water and then dried at 60°C. $V_2O_5.nH_2O$ sols are prepared by dissolving 1 g of V_2O_5 in 50 mL of 30% H_2O_2 with a control of temperature using an ice bath. A bright orange solution forms after 20 min which changes to a red-brown gel after 24 h. The VO_x/titanate composite nanorods are prepared as follows. About 0.1 g of titanate nanotubes is added to $V_2O_5.nH_2O$ sol (titanate nanotubes : $V_2O_5.nH_2O$ weight ratio = 1 : 10) and stirred for ~24 h to get a brown mixture. The mixture is transferred to an autoclave, and kept at 200°C for 48 h. A black green powder comprising VO_x/titanate composite nanorods (see Fig. P-2) is obtained after drying the product at 60°C.

BN nanotubes have been used as the templates to grow Fe–Ni alloy (Invar) and Co nanorods (113). The advantage of using BN nanotubes over carbon nanotubes as the templates is that the former is an electrical insulator that serves as a shield for metallic nanorods, while the latter is semiconducting or metallic. Also, BN nanotubes possess enhanced thermal stability compared to that of carbon nanotubes. The procedure to grow nanorods of Fe–Ni alloy and cobalt using BN as the template is as follows. First carbon nanotubes on Fe–Ni alloy or cobalt substrates are prepared by chemical vapor deposition. The idea of using carbon nanotubes in the beginning of the growth process is that it is easy to wet the interior of carbon nanotubes with a metal compared to boron nitride nanotubes. Nanoparticles of Fe–Ni or Co are encapsulated at the tip ends of the carbon nanotubes. The carbon nanotubes with the encapsulated metal nanoparticles (Fe–Ni or Co) are kept in a graphite crucible and heated to temperatures slightly above the melting point of the metals (1723 to 1973 K) followed by heating in a N_2 atmosphere for 30 min with B_2O_3 kept in the lower zone of the crucible. The carbon to boron nitride conversion takes place under these conditions and Fe–Ni alloy nanorods (diameter 30 to 300 nm; length approximately several microns) and Co nanorods (diameter 20 to 70 nm; length ~7 μm) encapsulated BN

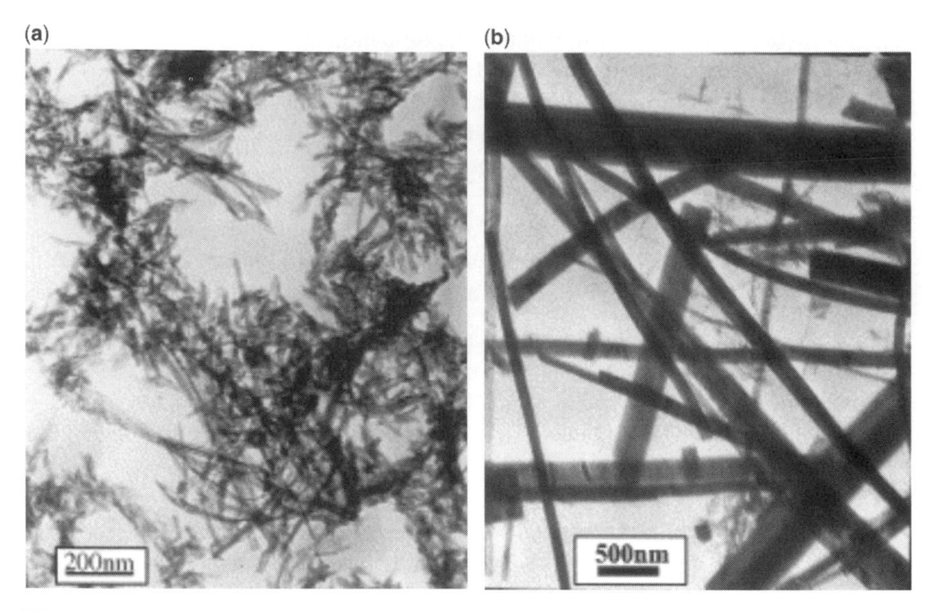

Figure P-2 TEM images of (a) titanate ($K_xH_{2-x}Ti_3O_7$) nanotubes used as the template, and (b) VO_x/titanate composite nanorods prepared by hydrothermal treatment of the template. (Reprinted from L. Yu et al. *Materials Chemistry and Physics*, **2004**, 87, 168. Copyright (2004) Elsevier. With permission.)

nanotubes are produced. Ordered structures within the metallic nanorods have been observed on using BN nanotubes as the templates and the lattice mismatch between BN and the metal is also very small (e.g. in the case of Co, the mismatch is ~0.3%).

3. By sequential electrodeposition of different metals into a template such as anodic aluminum oxide (AAO), multisegmented metallic nanorods can be synthesized. Examples of multisegmented metallic nanorods synthesized by this means include Pt-RuNi-Pt-RuNi-Pt (114), Pt-Ru, Pt-Ru-Pt and their analogs (115), Ag-Au-Ag (116), Ni-Pt, Ni-Pt-Ni and their analogs (117). The synthesis of multisegmented Pt-Ru nanorods by template electrochemical deposition is briefly described below.

   The AAO membranes are first deposited on one side with copper metal by electrochemical deposition with 1 M $CuSO_4$; copper acts as the working electrode. The platinum metal is electroplated for a predetermined time, and the membrane is washed with deionized water and then deposition with Ru metal is carried out for a predetermined time. Depending on how many metallic segments are required, the sequence can be repeated. Platinum is electroplated at 50°C using an aqueous solution of 0.01 M H_2PtCl_6 and 0.2 M H_2SO_4 while ruthenium is electroplated, also at 50°C, using $Ru(NO)Cl_3$ (concentration = 3.5 g/L) and NH_2SO_3H (concentration = 10 g/L). The length of the segments can be controlled by passing appropriate current for a predetermined time; Ru requires more charge per unit length compared to platinum. The position of the segment can also be controlled in a precise manner. The copper layer is removed by dipping the metal-coated AAO membranes in a HCl solution of $CuCl_2$ (0.05 M). Finally, the template (AAO) is removed by treatment with 0.5 M NaOH. The free nanorods are centrifuged, washed with water and then with ethanol to remove the impurities. Very good quality multisegmented metallic nanorods with precise control of length and position can be produced by this method, as can be seen in the FESEM image shown in Figure P-3.

4. The role of amines during the solvothermal synthesis of nanorods is at least threefold: (1) they act as reducing agents, (2) they act as structure-directing agents and they can control the size and morphology of the nanoparticles, and (3) they can make the nanorods stable in air and redispersible in nonaqueous solvents. Let us take two examples to illustrate the role of amines during solvothermal synthesis of nanorods.

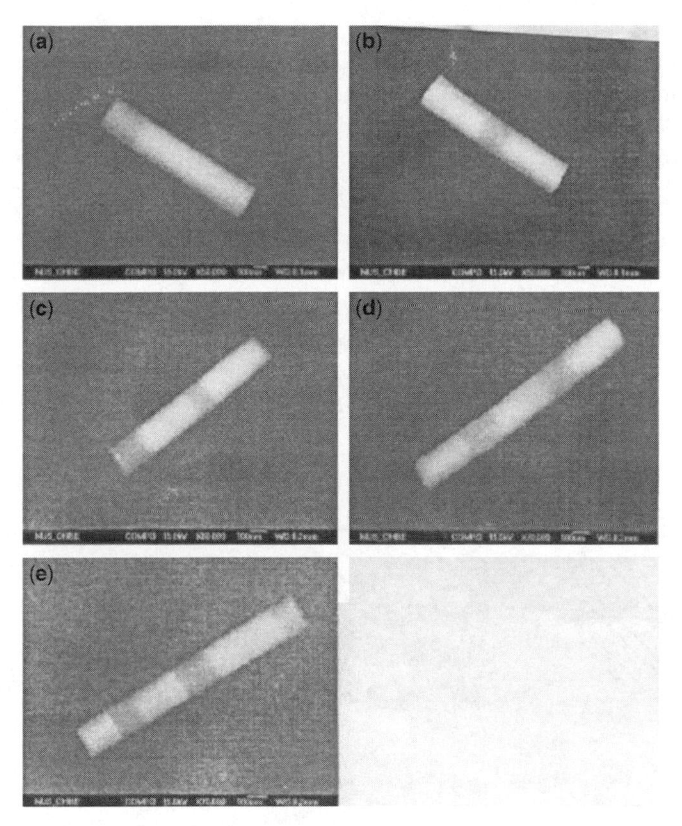

Figure P-3 FESEM images of multisegmented metallic nanorods: (a) Pt-Ru, (b) Pt-Ru-Pt, (c) Pt-Ru-Pt-Ru, (d) Pt-Ru-Pt-Ru-Pt, and (e) Pt-Ru-Pt-Ru-Pt-Ru. The dark segment is Ru and the bright one is Pt. (Reproduced with permission from F. Liu et al. *Adv. Funct. Mater.* **2005**, 15, 1459. Copyright Wiley-VCH Verlag GmbH & Co.)

Nickel nanorods (diameter 12 to 15 nm; length, 50 to 100 nm) have been synthesized by a solvothermal decomposition of nickel acetate in the presence of *n*-octylamine (nickel acetate to *n*-octylamine molar ratio is 1 : 300) at 250°C (104). The formation of Ni nanorods is favored by the presence of *n*-octyl amine; it reduces, under solvothermal conditions, the Ni^{2+} ions to Ni^{o} and also acts as a shape-controlling agent to produce metallic nickel nanorods. In the absence of linear alkyl amines, only NiO nanoparticles are produced. Using a similar approach, in the presence of *n*-octylamine, nanorods of ruthenium and rhodium metals have been produced starting from corresponding acetyl acetonate precursors, $Ru(acac)_3$ and $Rh(acac)_3$. The metallic nanorods are stable in air because of the amine coating and can be redispersed in hydrocarbon solvents.

The second example involves the use of an amine that acts as a templating agent as well as a reducing agent. Solvothermal treatment of V_2O_5 and benzylamine at 180°C leads to the formation of $VO_2(B)$ nanorods (diameter 20 to 100 nm; length 2.5 μm; Reference 105). The amine reduces V^{5+} to V^{4+}, under solvothermal conditions. The proposed mechanism for the formation of nanorods of $VO_2(B)$ is as follows. First, a precursor with a lamellar structure in which benzylamine molecules occupy the interlamellar space is formed as shown in Figure P-4. The structure is stabilized by the presence of benzylamine molecules. When the benzylamine molecules are displaced by the solvothermal treatment, the lamellar structure collapses leading to the formation of nanorods, as shown in the figure.

5. It is possible to introduce dopant(s) during the sol-gel synthesis of nanorods. Examples of nanorods synthesized by this method include Li/Mg doped ZnO (106), Cd doped ZnO (107), V doped ZnO (108), Gd doped CeO_2 (109), In doped ZnO (110), and Zr doped TiO_2 (111). The basic idea is to introduce the appropriate precursor(s) of the dopant(s) during the sol-gel process. The synthetic procedure for two of the examples, Li/Mg doped ZnO nanorods on glass substrates (106), and vanadium doped MgO nanorods are described below.

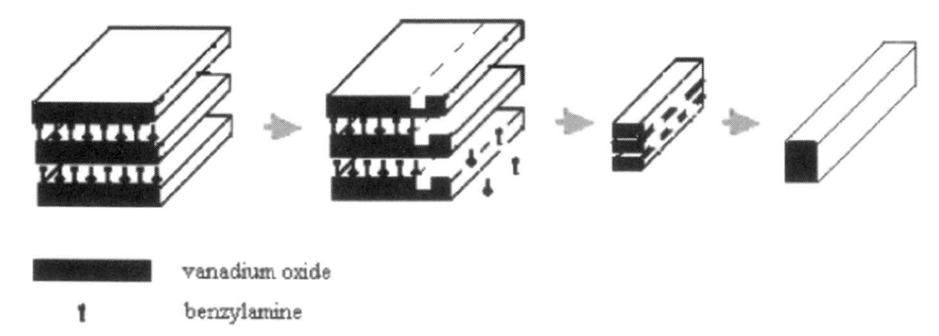

Figure P-4 Conversion of a lamellar precursor to $VO_2(B)$ nanorods by solvothermal synthesis in the presence of benzylamine. (Reprinted from F. Sediri et al. *J. Phys. Chem. Solids* **2007**, 68, 1821. Copyright (2007) Elsevier. With permission.)

The starting materials ($Zn(CH_3COO)_2{\cdot}2H_2O$, $LiCl{\cdot}H_2O$, and $MgCl_2{\cdot}6H_2O$) are dissolved in ethylene glycol monomethyl ether followed by the addition of monoethanolamine. Monoethanolamine acts as a stabilizer. The molar ratio of zinc acetate to monoethanolamine is kept at unity. The solution is heated to 60°C and stirred at this temperature for 2 h, resulting in a homogeneous clear solution. This solution is kept for about 24 h to get a sol and it is used for spin coating on glass substrates. The spin-coated substrates are heated at 310°C to remove the organic residues. The spin-coating of the slides with the sol and the heating steps are repeated multiple times. The slides are finally calcined at ~610°C to obtain Li/Mg doped ZnO nanorods on glass substrates. Another example of sol-gel synthesis, vanadium doped ZnO nanorods, $Zn_{0.95}V_{0.05}O$ and $Zn_{0.9}V_{0.1}O$ (108), is discussed below.

About 0.9 mmol of zinc (III) acetylacetonate hydrate and appropriate quantities of vanadyl (IV) acetylacetonate are dissolved in a 250 mL 1 : 4 volume mixture of ethylene glycol and methyl alcohol. An ultrasonic bath is used to hasten the dissolution. The contents are stirred for 2 h to get a homogeneous solution. To this, a 5 wt% aqueous solution of polyvinyl alcohol (mol. wt. = 72,000) is added until a viscous gel is produced. The gel is first dried at 273 K for a day followed by calcination at 600°C for 1 h to obtain vanadium doped ZnO nanorods.

6. This is illustrated by considering highly crystalline In_2O_3 nanorods incorporated in SBA-15, a mesoporous silicate (118). First, a solution of $In(NO_3)_3.xH_2O$ is prepared by dissolving 0.5 g of the salt in 3 mL of methanol. This solution is added drop-wise to 0.1 g of SBA-15 in a round bottom flask. The contents are stirred at room temperature for about 24 h, filtered, washed with methanol, followed by drying at 65°C. The SBA-15 impregnated with In^{3+} ions is placed in a ceramic boat inside a tubular furnace, purged with nitrogen, and then heated at 700°C for 4 h. A pale yellow powder is obtained after cooling the contents of the boat to room temperature. Free-standing In_2O_3 nanorods can be obtained by removing the SBA-15 template with 2.0 M NaOH. The nanorods incorporated in SBA-15 and the free-standing In_2O_3 nanorods are shown in the TEM images shown in Figure P-5. The In_2O_3 nanorods possess uniform diameter restricted by the dimension of the pores of SBA-15 (~6 nm). After removing the template, short nanorods with length less than 50 nm are produced.

7. There are many examples of the self-assembly of nanoparticles to nanorods: (1) self-assembly of CdS nanoparticles to CdS nanorods by oriented attachment (119), (2) self-assembly of TiO_2 nanoparticles to TiO_2 nanorods (120), (3) self-assembly of doped ZnS nanoparticles to single crystalline ZnS nanorods (121), (4) self-assembly of ZnO nanoparticles to ZnO nanorods (122, 123) and (5) self-assembly of CeO_2 nanoparticles to CeO_2 nanorods. The self-assembly of ZnO nanoparticles (nanodots) to nanorods (122) is discussed below.

Quasi-spherical ZnO nanoparticles are prepared from zinc acetate dihydrate in a basic alcoholic solution. About 0.01 mol of $Zn(CH_3COO)_2{\cdot}2H_2O$ is dissolved in ~125 mL methanol and vigorously stirred at 60°C. To this, 65 mL of 0.03 M KOH solution in methanol is added drop wise and the contents are stirred at 60°C

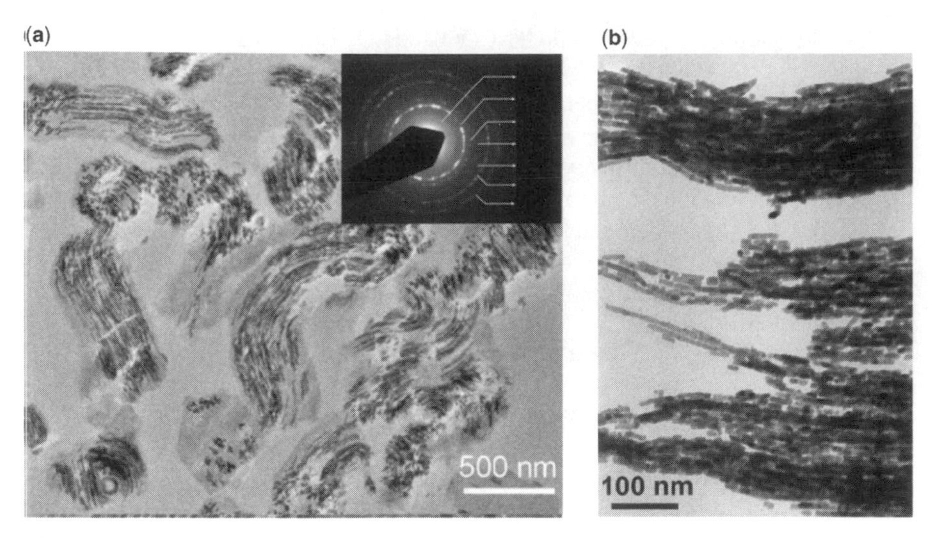

Figure P-5 TEM images of (a) In_2O_3 nanorods incorporated in SBA-15, and (b) free-standing In_2O_3 nanorods obtained after removal of the SBA-15 frame work. (Reprinted with permission from S. C. Chang et al. *J. Phys. Chem. C.* **2008**, 112, 2304. Copyright (2008) American Chemical Society.)

for 2 h. Quasi-spherical ZnO particles (size ~3 nm) are obtained (Fig. P-6a). The shape of the ZnO particles is highly sensitive to the concentration of the precursors. After the preparation of the sol containing the quasi-spherical particles, the concentration is increased by solvent evaporation (the concentration increases by about 10-fold) and finally the contents are refluxed for about a day. Formation of ZnO nanorods is observed (Fig. P-6b) after the heating (refluxing) step. It is proposed that the nanorods are formed by oriented attachment of the quasi-spherical ZnO nanoparticles. HRTEM studies reveal that the spherical ZnO nanoparticles attach to each other, epitaxially, along the (002) axis (i.e. *c*-axis).

8. A polymer shell on a metal nanorod imparts better stability and better solubility in nonaqueous solvents. A few examples of metal-core polymer-shell nanorods can be found in the literature. They include Au_{core}–$Polystyrene_{shell}$ (124, 125), and Ag_{core}–$Polystyrene_{shell}$ (126). One of the examples (124), in which Au nanorods are encapsulated with polystyrene is discussed below.

Gold nanorods are first prepared by a seed-mediated growth method using CTAB as the structure-directing agent. The CTAB groups on the surface of the Au nanorods are then ligand exchanged with 4-mercapto phenol groups. This is carried out by a drop-wise addition of 4-mercaptophenol in tetrahydrofuran into the aqueous solution containing CTAB-coated Au nanorods. The Au nanorods functionalized with thiophenol groups, $Au(SC_6H_4OH)_n$, are dispersed in CH_2Cl_2 and allowed to react with carboxy biphenyl terminated polystyrene with the addition of 4-(*N*,*N*-dimethylamino) pyridinium-4-toluene sulfonate (DPTS) and 1,3-diisopropyl carbodiimide (DIPC). The carboxy biphenyl

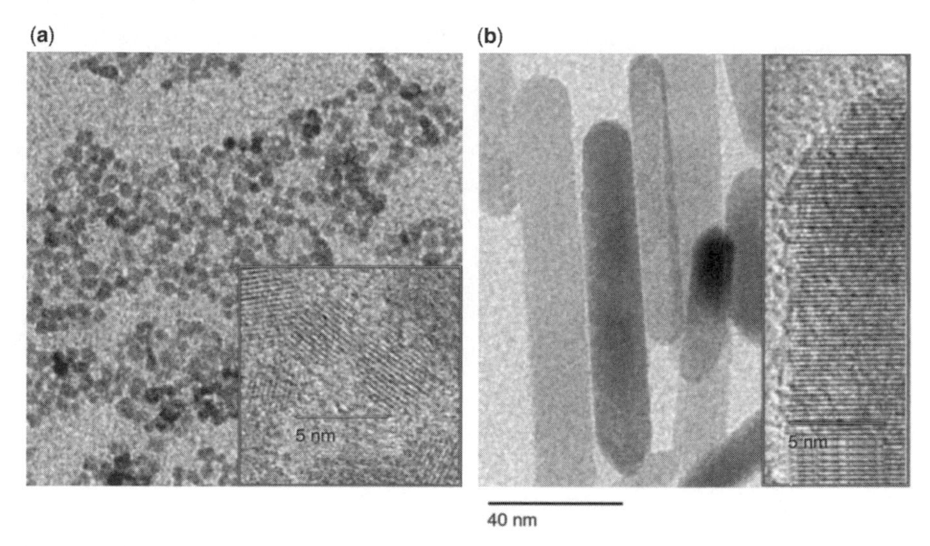

Figure P-6 TEM images of (a) quasi-spherical ZnO nanoparticles, and (b) ZnO nanorods obtained by the self-assembly of the quasi-spherical ZnO nanoparticles. (Reproduced with permission from C. Pacholski et al. *Angew. Chem. Int. Ed.* **2002,** 41, 1188. Copyright Wiley-VCH Verlag GmbH & Co.)

terminated polystyrene covalently attaches on the surface of the Au nanorods to yield Au_{core}–$Polystyrene_{shell}$ nanorods (Fig. P-7a). An interesting phenomenon observed for the Au_{core}–$Polystyrene_{shell}$ nanorods is that they spontaneously assemble into rings on a carbon-coated TEM grid prepared from a solution of nanorods dispersed in CH_2Cl_2 (Fig. P-7b).

9. It is very much possible to functionalize nanorods with biomolecules. Nanorods functionalized with biomolecules find applications in areas such as DNA detection, targeted gene delivery, etc. (127). Many examples can be given, such as (1) functionalization of avidin and biotin on Au nanorods (128), (2) functionalization of a multisegmental Au-Ni-Au nanorod with a thiolated KE2 antibody (129), (3) amino acid functionalized hydroxyapatite nanorods (130), (4) biotin-streptavidin functionalized Au nanorods (131), etc. One of the examples in which a multisegmented metallic nanorod is functionalized with an antibody (protein) is discussed in detail. Au-Ni-Au nanorods have been selectively functionalized on the Au segment with a protein such as KE2 human HLA primary antibody (129). The thiolated KE2 antibody is selectively bound to the Au segments and the functionalization is illustrated in Scheme P-1.

First, the KE2 antibody is thiolated using *N*-succinimidyl-*S*-acetyl thioacetate in dimethyl sulfoxide. The succinimide group is coupled to the primary amines on the protein (antibody) which leads to the formation of covalent amide bonds with the loss of hydroxy succinimide. The sulfhydryl group is thus protected. The deprotection, through deacetylation, is carried out using EDTA and hydroxyl

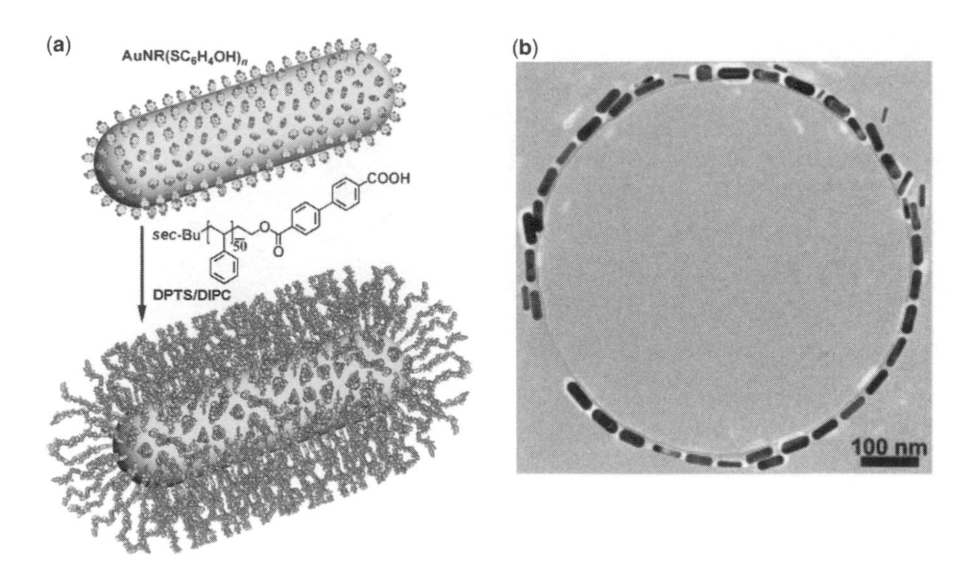

Figure P-7 (a) Preparation of an $Au_{core}-Polystyrene_{shell}$ nanorod, and (b) TEM image indicating the spontaneous assembly of the core-shell nanorods into a ring. (Reproduced with permission from B. P. Khanal et al. *Angew. Chem. Int. Ed.* **2007,** 46, 2195. Copyright Wiley-VCH Verlag GmbH & Co.)

amine and the thiolated protein is ready to be functionalized on the Au segment of Au-Ni-Au nanorods.

The Au-Ni-Au nanorods or nanowires are prepared by an electrochemical template synthesis. They are suspended in (3-aminopropyl) triethoxy silane in ethanol, sonicated and washed with ethanol followed with $NaHCO_3$ solution. The siloxane group has a high affinity for the Ni segment and its attachment to the Au segments is nonspecific. Methoxypoly(ethylene glycol) succinate *N*-hydroxy succinimide ester and $NaHCO_3$ are added to the nanorod (nanowire) suspension, sonicated, followed by washing with $NaHCO_3$. The PEG moiety is attached to the amino group of the siloxane through *N*-hydroxy succinimide coupling reaction with the amine. PEG attachment provides a hydrophilic environment on the Ni segment that minimizes the protein (antibody) adsorption on the Ni segment. The surface-modified nanorods are then suspended in the thiolated protein solution followed by incubation at 4°C. The thiol functional groups on the protein displace any siloxane groups that were attached nonspecifically on the Au segment. The siloxane linkage on the Ni segments is not displaced by the thiolated protein and thus only selective functionalization on the Au segments takes place. The protein functionalized nanorods/nanowires are washed with phosphate-buffered saline.

10. A mesoporous nanorod is a nanorod with pores in the mesopore size region (2 nm to 50 nm) along the length of the nanorod. Mesoporous nanorods have been synthesized using polycarbonate membrane as the template (132), using ultrasound

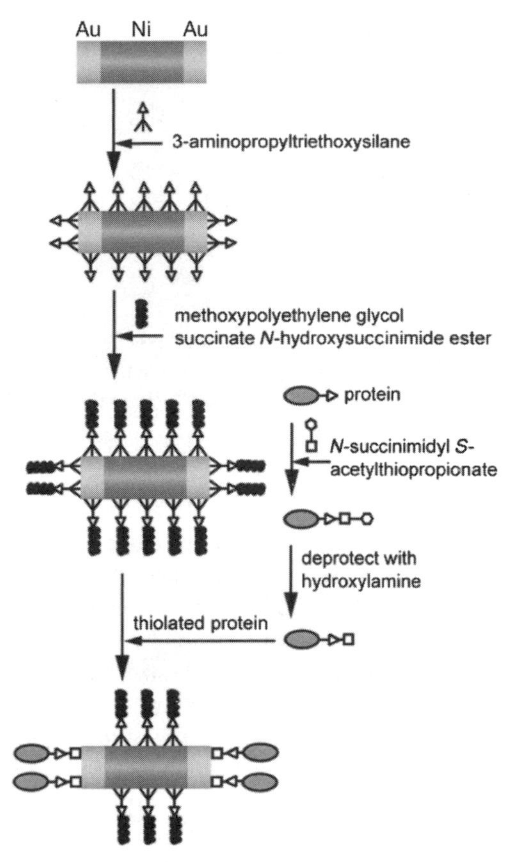

Scheme P-1 Selective functionalization of the Au segment in a multisegmented Au-Ni-Au nanorod (nanowire) with an antibody. (Reprinted with permission from B. Wildt et al. *Langmuir* **2006**, 22, 10528. Copyright (2006) American Chemical Society.)

irradiation (133), and using microwave and a block copolymer (134). The synthesis of mesoporous silica nanorods (135) is taken as an example and briefly described below.

A two-dimensional cylindrical confinement of spherical micelles formed by a copolymer inside a porous anodic aluminum oxide template leads to the formation of mesoporous SiO_2 nanorods. About 0.1 g of a block copolymer, polyethylene-co-butylene-block-polyethylene oxide (PHB-PEO), is dissolved in about 1 mL of ethanol with mild heating. The copolymer forms spherical micelles under this condition. To this about 0.5 g of tetraethyl orthosilicate (TEOS) and 0.25 g of HCl is added and the solution is stirred for 2 h. The porous anodic aluminum oxide templates are loaded now with a drop of the above precursor solution. After filling the template, the excess precursor solution containing the polymer and silica is wiped out. The template is aged at 25°C for about 24 h followed by calcination at ~550°C. Finally, the template is removed with 5% H_3PO_4 at

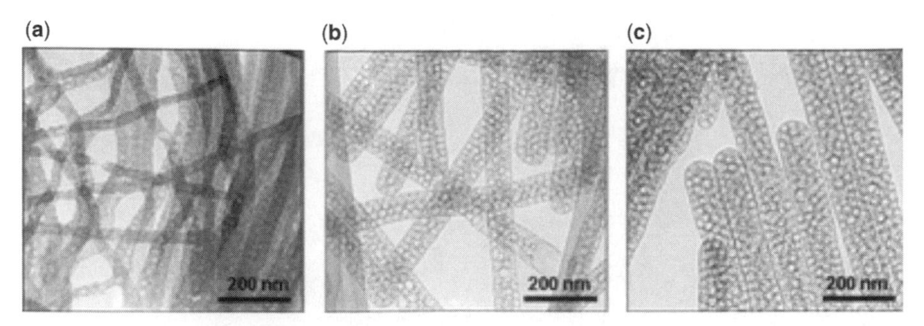

Figure P-8 TEM images of mesoporous silica nanorods synthesized by the 2D confinement of spherical micelles using different sizes of the porous aluminum oxide template: (a) 25 nm, (b) 60 nm, and (c) 90 nm. (From A. Thomas et al. *J. Mater. Chem.* **2007**, 17, 4558. Reproduced by permission of the Royal Society of Chemistry.)

65°C for 12 h to obtain free mesoporous silica nanorods (Fig. P-8). Even when there is confinement, the spherical micelles assemble in a close packed way inside the nanosized channels of the template; the pore size of the micelles increases from ~13 nm to ~20 nm on confinement. This is attributed to the soft nature of the template and also the favorable van der Waals forces. The concentration of the copolymer affects the micelle size. For a concentration of 20 wt% copolymer, the pores are uniform throughout the nanorods. However, at higher or lower concentration of the copolymer, the pores are not uniformly distributed. The interaction between the matrix of the template and the copolymer has been attributed as the reasons for this trend.

PART III

NANOSTRUCTURED SOLIDS: MICRO- AND MESOPOROUS MATERIALS AND POLYMER NANOCOMPOSITES

8

AEROGELS: DISORDERED, POROUS NANOSTRUCTURES

STEPHANIE L. BROCK

8.1 Introduction, 210
8.2 Porous Materials and Aerogels, 211
8.3 General Synthetic Methods for Aerogel Preparation, 212
  8.3.1 Sol-Gel Reactions and Aerogel Formation, 212
  8.3.2 "Dressing Up" Your Aerogel: Co-doping versus Functionalization After Gelation or Drying, 215
8.4 Silica Aerogels, 216
  8.4.1 Synthesis, 216
  8.4.2 Properties and Applications, 218
  8.4.3 Silica Composites: A Sampling, 220
8.5 Beyond Silica: Other Oxide Aerogels, 222
  8.5.1 Electronically Insulating Oxides, 223
  8.5.2 Electronically Conducting or Magnetic Oxides, 225
8.6 New Directions: Non-Oxides, 226
  8.6.1 Carbon Aerogels: Synthesis and Properties, 227
  8.6.2 Metal Chalcogenide Aerogels: Synthesis and Properties, 229
8.7 Concluding Remarks and Future Outlook, 233
Further Reading, 234
References, 234
Problems, 236
Answers, 237

Nanoscale Materials in Chemistry, Second Edition. Edited by K. J. Klabunde and R. M. Richards

8.1 INTRODUCTION

Aerogels are disordered nanoparticle networks with interconnected pore structures that span the regime from microporous (<2 nm pores) to mesoporous (2 to 50 nm) to macroporous (>50 nm). The intimate relationship of the pore and matter network that is characteristic to the aerogel structure can be seen in Figure 8.1. The name "aerogel," first coined by Kistler in 1931 (1), is believed to refer to the fact that aerogels are prepared by replacing the solvent in a wet gel with air while maintaining the gel structure—hence "aero" gel. This may not sound very challenging, but in fact, drying a gel often results in compaction and loss of porosity, resulting in a "xerogel" (from the Greek *xeros* meaning dry). Aerogels represent a unique form of matter. With porosities up to 99.8%, aerogels are the lowest density solids around, and can be prepared with densities as low as $0.004\,\mathrm{g \cdot cm^{-1}}$—less dense than any liquid, and only a few times more dense than a typical gas at STP. This leads to some unusual properties as well as actual and potential applications. In particular, the high surface area and the accessibility provided by the presence of relatively large and interconnected pores makes these materials excellent platforms for catalysis and sensing, while the architecture contributes to unprecedented low thermal conductivity and a high degree of elasticity (see Fig. 8.2). Moreover, the intriguing appearance of aerogels, which has led to descriptors such as "frozen smoke," and the Rayleigh scattering, which imbues transparent systems with different coloration depending on the source of the light and color of the background, has captured the imagination of scientists and artists alike (2). In this chapter, we will provide an overview of this unusual class of materials in terms of their chemical classification, properties, and

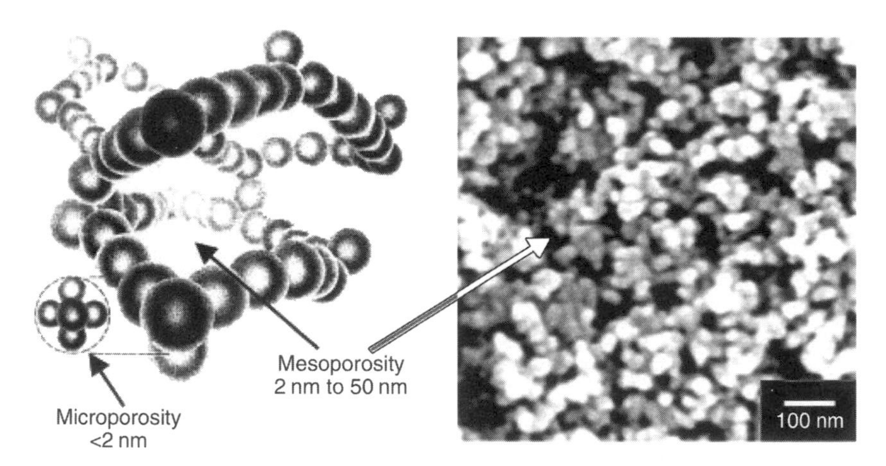

Figure 8.1 An architectural model (left) and transmission electron micrograph (right) of an aerogel showing the nanostructured network and pore structure. (Left panel reproduced with permission from D. R. Rolison and B. Dunn, *J. Mater. Chem.* **2001**, *11*, 963. Copyright 2001 Royal Society of Chemistry. Right panel reproduced with permission from M. L. Anderson et al., *Adv. Eng. Mater.* **2000**, 2, 481. Copyright 2000 Wiley-VCH.)

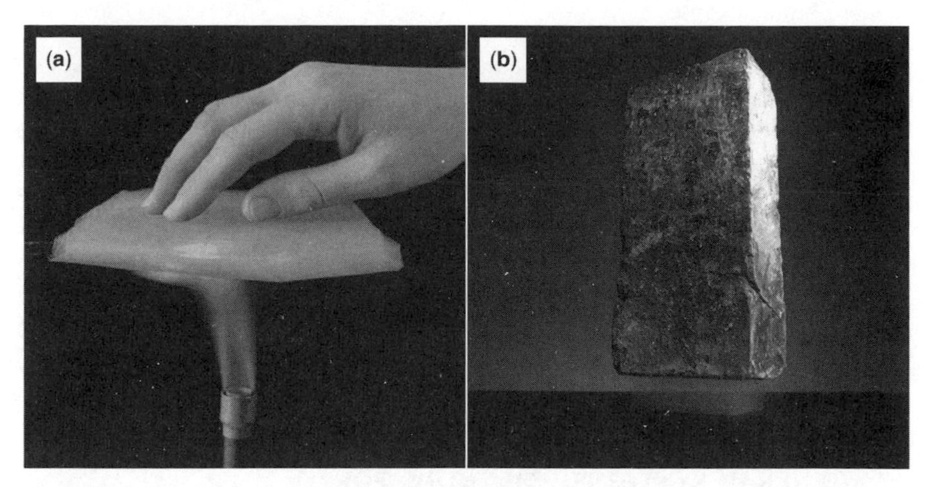

Figure 8.2 (a) Aerogels exhibit very low thermal transport, enabling a thin monolithic layer to protect flesh from a flame (for a short while, anyway). Photo courtesy of Lawrence Berkeley Lab. (b) Aerogels are very strong (if brittle) and can be prepared with a high degree of transparency. The brick (2.5 kg) supported by an aerogel (2 g) almost appears to be floating. Photo courtesy of NASA/JPL.

applications. As full coverage of the extensive body of work done in the aerogel field is beyond the scope of this chapter, selected examples will be provided to illustrate key points. Let us begin by putting these materials into context by comparing them to other members of the family of porous materials to which they belong.

8.2 POROUS MATERIALS AND AEROGELS

Aerogels set the benchmark for porosity in any solid material and have even made it into the *Guinness Book of World Records* as the lightest solid material on earth. In general, porous materials are used in a wide range of applications and can be differentiated by their characteristic pore size and pore structure, the degree of order in the framework, and the chemical nature of the framework. Materials with ordered pore structures include zeolites or molecular sieves. These are crystalline materials, typically aluminosilicates, with well-defined micropores. Mesoporous materials such as the MCM (Mobil Crystalline Material) class of oxides also exhibit ordered pore structures, but the size of the pores is larger, in the lower mesoporous range, and the framework itself is usually amorphous. The pore structures accessible in MCM materials depend on the surfactant template and the corresponding phase diagram for micellar aggregation of that surfactant, but these generally fall into the hexagonal (1-D pores) or cubic (wormhole gyroid) structure. Like zeolites, these are largely aluminosilicates and the pore size distribution is narrow.

In contrast, aerogels do not have an ordered pore structure or a narrow distribution of pores. While many see beauty in order, the disordered structure of aerogels may have significant advantages in terms of function (3, 4). Thus, because the pore size extends to the larger end of the mesoporous regime, and into the macroporous regime, molecular diffusion is facilitated. Furthermore, these pores are interconnected, enabling multiple points of entrance and egress. Thus, if pathways become blocked due to deposition of a solid phase or collapse of the pore wall, the molecular transport is not significantly affected, as is the case for the nonconnected pores of MCM materials. Functionally, the aerogel architecture, which embraces a football field-sized surface area within a cubic inch and an interconnected rapid transit system of tunnels, is ideal for applications that involve transport of matter or charge on the nano-scale. For this reason, aerogel materials are being investigated for applications ranging from thermal insulation to battery materials. However, before discussing properties and applications of aerogels in more detail, an overview of how these materials are made is in order.

8.3 GENERAL SYNTHETIC METHODS FOR AEROGEL PREPARATION

The general processes for making and modifying aerogels have some basic similarities regardless of the specific chemistry of the system. These are described in Sections 8.3.1 and 8.3.2, respectively. Details pertinent to the particular kinds of aerogels prepared will be presented in later sections.

8.3.1 Sol-Gel Reactions and Aerogel Formation

The simplest definition of an aerogel is *a dry gel that retains to a large extent the structure of the wet gel from which it is derived.* Wet gels can be made from a variety of processes, commonly referred to as sol-gel methods (5). The general feature of these processes is that they involve condensation of molecular-scale units to form a colloidal suspension, a nanoparticle "sol," in which the particles then condense together to form a network that spans the solvent. This is the swollen polymer, or "gel." A specific example of this process for the prototypical case of silica prepared from tetraethoxysilane (TEOS) is illustrated in Scheme 8.1. A characteristic of a gel is the ability to take the shape of the container, or mold, in which it is prepared. Depending on the firmness of the gel, this can lead to free-standing macroscale monoliths. Since the solvent is an integral part of the gel (actually supporting the network) retaining the pore characteristics upon solvent removal can be challenging. Typically, the gel structure is allowed to ripen (age) in order to strengthen interparticle interactions prior to drying. This is usually done for several hours to several days. During the aging process, the gel continues to transform. In some cases, syneresis (the expulsion of solvent from the gel) occurs, resulting in a discernible compaction of the gel body.

Drying by a simple process of evaporation on the benchtop, or vacuum removal of solvent, results in significant pore collapse. The loss of porosity results from [1] capillary forces that arise from interactions between the solvent and pore walls and induce collapse upon solvent evaporation, and [2] cross-linking of chemically active groups

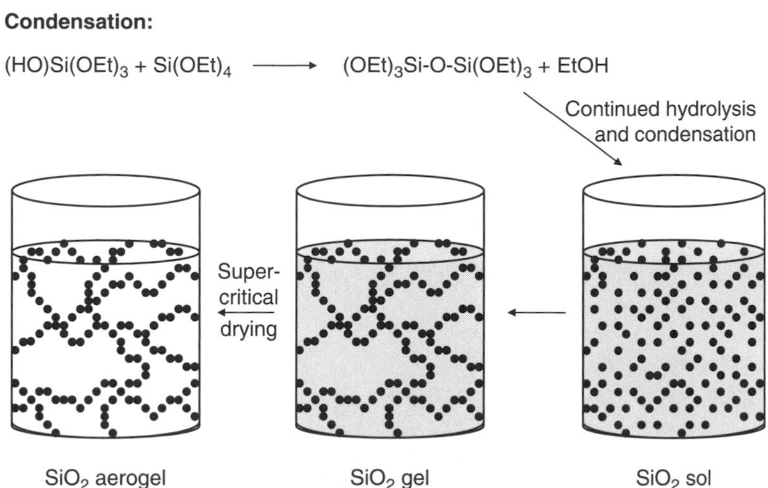

Scheme 8.1 Sol-gel reactions for silica aerogel formation.

across the pore void, thereby making the collapse irreversible. One way to get around this problem is to eliminate the liquid-gas interface in the system, and this can be achieved by removing the solvents when they are in their supercritical state. The supercritical point on a phase diagram is that place where the liquid and gas no longer exist as separate phases (see Fig. 8.3), thus solvent can be removed without inducing a phase change. The arrows and numbers in Figure 8.3 represent a possible pathway for achieving this. First, the temperature is ramped up past the supercritical point (T_c) in a sealed vessel, which also drives up the pressure past the critical pressure, P_c, inducing the supercritical state. Second, the pressure is released while maintaining $T > T_c$ and thus transforming the supercritical fluid directly into the vapor phase and releasing it from the chamber. Since the pressures required are upward of 6 MPa for common solvents, an autoclave is required. Whereas alcohols (the natural by-products of silica condensation from silicon alkoxides) require relatively high temperatures to achieve the supercritical state (EtOH: $T_c = 243°C$), CO_2 becomes supercritical near room temperature ($T_c = 31.1°C$). The "cold" supercritical drying method (CO_2 method) also has the advantage of yielding materials with higher surface areas and porosity, since at high temperature ("hot" supercritical drying) coarsening of the matrix (aging) is facilitated. Autoclaves for hot and cold supercritical drying are commercially available, so anyone can avail themselves of this method.

Conventional supercritical drying usually requires a significant investment of time due to the aging process (predrying) as well as the washing steps that may be necessary to remove by-products or, if cold methods are desired, to replace the native solvent with CO_2. In order to speed up the process, several groups have developed rapid

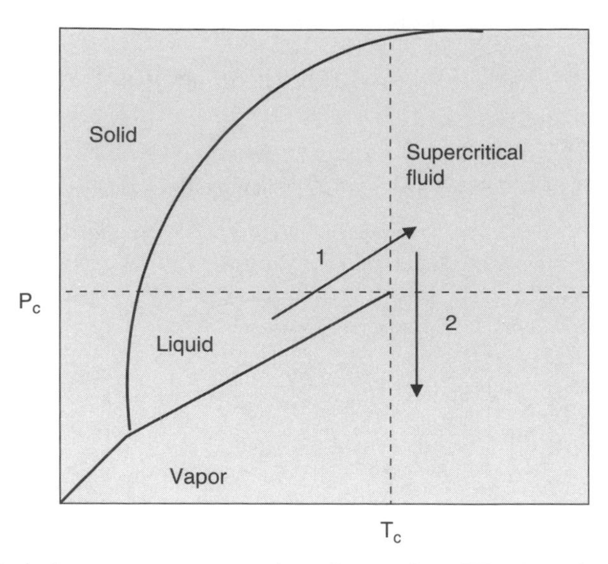

Figure 8.3 Typical pressure-temperature phase diagram. P_c and T_c refer to the critical pressure and temperature, respectively. The arrows and numbers (1, 2) illustrate a pathway for supercritical drying by (1) transforming the liquid to a supercritical fluid by increasing temperature in a closed vessel ($T > T_c$), thereby also increasing pressure ($P > P_c$); and (2) release of pressure while maintaining $T > T_c$, thereby transforming the supercritical fluide to the gas phase for removal.

supercritical extraction (RSCE) techniques. A team of scientists at Lawrence Livermore National Laboratory (LLNL) developed a method similar to injection molding in which the precursor is placed in a strong mold and heated rapidly to induce gelation and speed up aging (6). The heating, and the associated pressure created by heating a closed system, enables the critical point to be surpassed, at which point the pressure is released. More recently Gauthier et al. (7) have developed an RSCE method employing a hydraulic hot press. The chemical precursors are placed in a mold between two pieces of kapton film, then placed in the hot press where the temperature and pressure is ramped to supercritical conditions, and then the pressure is dropped rapidly so the supercritical fluid can escape. Both of these methods have the advantage of being rapid (precursors to aerogels in just a few hours instead of a few days). These methods also have the advantage of enabling precise formation of monoliths with desired shapes and sizes; however, they do require a more specialized setup.

An alternate approach that involves no special apparatus is to neutralize the pore walls chemically by coating them with a terminating group. In the case of silica, this is usually an organosilane. Originally developed by Brinker and coworkers (8, 9), these materials show a very intriguing property when dried under ambient pressure or under vacuum. First, they shrink due to the capillary forces, but because cross-linking across the pore is obviated, once the solvent is gone they spring back into shape. Of course, sometimes the initial forces are such that large molded pieces crack. For this reason, this approach is not suitable where large monolithic aerogels are needed. This method has been adopted by Cabot in their commercial process to make aerogels (Nanogel®).

8.3.2 "Dressing Up" Your Aerogel: Co-doping versus Functionalization After Gelation or Drying

Aerogel formation enables modification of the network at a number of different instances, thereby allowing the intrinsic properties to be adjusted according to the desired properties and applications. To functionalize, or "dress up," a plain aerogel, secondary species can be co-doped at the outset or introduced to the surface at the wet gel or aerogel stages. Usually, the former method has a profound impact on the gel network in terms of porosity, surface area, and strength. This is either because the co-dopant modifies the intrinsic condensation steps that lead to gelation (by acting as catalyst) or because it participates actively in cross-linking the gel. Nevertheless, this represents a common way to incorporate molecules or ions into the gel. For example, fluorescent oxygen sensors have been generated by incorporation of appropriate fluorophores into silica, and electrocatalysts by incorporating metal ions into organic gels, both at the sol stage. To some extent, the native morphology of the gel can be retained by adding the co-dopant at the very last minute, that is, just prior to gelation. This is particularly important when the dopant is similar in size to the nano-scale components of gel network. Rolison and coworkers refer to this method as

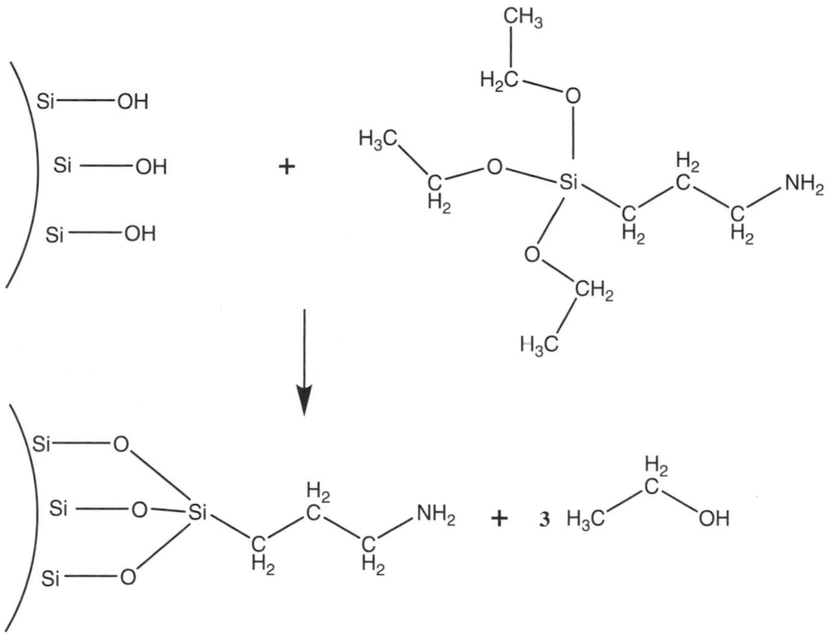

Scheme 8.2 Chemical modification of a silica surface via condensation of aminopropyl (triethyoxysilane) with surface silanols.

nanogluing and they have used it to incorporate a variety of secondary phases into silica aerogels, including metal nanoparticles such as Pt and Au, oxide nanoparticles like TiO_2, vulcanized carbon, and polymer nanoparticles (10).

Postmodification can be achieved at either the wet gel stage or after drying (aerogel stage). Postgelation modification typically takes advantage of the chemical characteristics of the solid surface to covalently link a secondary phase. For example, the silanol (Si-OH) functionalities on the surface of silica gels can be treated with modified alkoxysilanes that have at least one nonhydrolyzable group, for example, $RSi(OR')_3$, to generate organically modified gels (11). This is shown in Scheme 8.2 for formation of amine-terminated silica gels. Thus, it is possible to change the nature of the surface considerably, depending on the –R functionality. This enables further chemistry (tethering of sensing molecules, etc.) to be achieved in subsequent steps. Covalently linked functionalities are retained throughout the drying cycle, producing chemically modified aerogels.

Modification after drying typically involves deposition of chemical components from the gas phase by a process of chemical vapor infiltration. Thus, at temperatures $>500°C$, treatment of aerogels with acetylene leads to homogeneously deposited carbon, treatment with silanes leads to deposition of elemental Si, and treatment with ferrocene or iron carbonyls leads to iron and iron oxides (12).

8.4 SILICA AEROGELS

The most extensively studied chemical system for aerogel formation is the silica system. The sensational images associated with aerogels—free-standing transparent monoliths insulating a hand from a flame or holding up a brick (Fig. 8.2), are almost all of silica materials. The reason for this lies in the well-developed and facile sol-gel chemistry of silica that lends itself to formation of robust monoliths, as well as the intriguing physical properties and associated applications of such materials (11, 13).

8.4.1 Synthesis

The general process of silica gel formation is shown in Scheme 8.1. The facility of gel formation in the silica system is related to the kinetics and thermodynamics of the hydrolysis and condensation steps, with the morphology of the gel network dictated principally by the relative rates of the two steps. The reaction kinetics can be effectively tuned by altering the mechanism of hydrolysis, achieved through choice of catalyst (acid or base). When the catalyst is an acid, condensation becomes the rate-determining step. This leads to a polymer-type aerogel with little chain-branching. In this case, gel formation proceeds by a mechanism of reaction-limited cluster aggregation (RLCA), in which small molecular-scale clusters form from rapid hydrolysis and then are linked together by terminal silanols to form the polymer. When a base is employed as the catalyst, it is the hydrolysis that is rate determining. This leads to the formation of particles on the nanometer length scale. These grow, and then are linked together, by the reaction of additional monomers, yielding a colloidal gel through the mechanism of reaction-limited monomer cluster growth. The gel point is reached when a solvent-spanning continuous network is formed; however, the processes of hydrolysis and condensation (and their reverse processes) continue as long as the gel is maintained within the mother liquor. Thus, after the gel point, the residual monomer continues to react with the network just as surface Si-OH and Si-OR groups along the pore walls condense together.

These processes occur in tandem with the process of Ostwald ripening, in which hydrolysis and condensation reactions operate in equilibrium, leading to dissolution of silica from high energy regions and redeposition at low energy regions. As the latter tend to be curved surfaces, redeposition occurs preferentially at particle-particle interfaces (necks), the most fragile part of the structure, and in small pores. Because this leads to increased gel strength, as-prepared gels are usually allowed to sit undisturbed for some period of time (hours to weeks) to age. However, this strengthening phenomenon is often played off against the desire for high surface area and porosity, since aging leads to compaction of the gel network and a decrease in the number of micropores. Strengthening of the network can also be achieved by removing the mother liquor and replacing it with a fresh solution of a silicon alkoxide in alcohol followed by aging. The increased concentration of alkoxide facilitates deposition on the silica framework, thereby making a more robust architecture. These steps are particularly important when monoliths are desired.

Because alcohol is the by-product from sol-gel processing of silicon alkoxides, the hot supercritical drying method or the RSCE method can be applied directly to the wet gel. As indicated above, the high temperatures associated with supercritical fluid formation in alcohols enhances the rates of aging and ripening, and thus results in significant changes in the gel network in the process of drying, although this is somewhat minimized in RSCE due to the rapidity of the method. In contrast, cold CO_2 drying is less destructive, but requires additional washing and solvent exchange steps in order to replace the alcohol solvent (which is not miscible with liquid CO_2) with an intermediate solvent (e.g. acetone) and eventually CO_2. If ambient pressure drying is desired, chemical neutralization of the network surface during gelation, or immediately after gelation, is needed. This is generally achieved by postgelation treatment with a modified organosilane, such as trimethylchlorosilane (8, 9). The Cl functionality is hydrolyzed by reaction with surface Si-OH species, leading to a densely methylated surface that springs back into shape after drying. Alkyl group functionalization also confers hydrophobicity on the aerogel (native aerogels are hydrophilic).

Despite the fact that gelation can be conducted under either acidic or basic conditions, silica aerogels are most often formed from base-catalyzed reactions. This is because the polymeric framework of the acid-catalyzed gel contains a larger portion of smaller pores than the base-catalyzed colloidal analog. Removal of solvent from these smaller pores without gel compaction is difficult, even under supercritical conditions. One successful approach to access ultra-low-density aerogels with polymeric frameworks is to use a two-step acid/base gelation process such as that used by Tillotson and Hrubesh (14). The first step is to perform acid catalysis of tetramethoxysilane (TMOS) with hydrochloric acid to produce a sol from which the alcohol by-product is distilled off. In the second step, base catalysis of the sol with ammonium hydroxide in nonprotic solvents such as acetonitrile or acetone is used to produce a gel. Supercritical extraction of the solvent and residual alcohol at 300°C yields highly transparent aerogels with densities as low as $0.004\ g \cdot cm^{-3}$. This is an order of magnitude lower than a base-catalyzed aerogel produced under similar conditions. The fact that the polymer-type microstructure, typically associated with acid-catalyzed aerogels, dominates, suggests the conditions under which sol formation occurs also dictate the gel formation processes. The addition of the base-catalysis step appears to result in a strengthening of the gel that is essential for successful transformation to the aerogel.

It is important to note that the as-prepared aerogels are not pure silica networks, but retain a significant quantity of unhydrolyzed alkoxides, particularly when dried from the mother liquor (hot drying). In the case where organosilane functionalities are used to obviate the need for supercritical drying, the material also has a particularly high organic content, rendering the aerogels hydrophobic. Depending on the properties and applications desired, this may or may not be a good thing. If needed, silica aerogels can be heated in air at temperatures upwards of 300°C to remove residual organics.

8.4.2 Properties and Applications

Aerogels have unusual mechanical properties, being very elastic, although also very brittle. For this reason, along with excellent transparency and low thermal conductivity (more on these properties below) silica aerogels were the material of choice for NASA's Stardust mission to collect interstellar particles from the tail of the comet Wild 2 (see Fig. 8.4; Reference 15). In this case, aerogels were prepared with a density gradient in order to effectively slow down the particles without generating significant heat (to avoid melting them), eventually trapping these hypervelocity particles. The transparency of the aerogel allowed the particles and their tracks to be visualized when the capsule was successfully returned to earth on January 15, 2006. A team of scientists is now involved in characterizing this "space dust" and adsorbed organics encountered during the *Discovery* mission for clues pertaining to the development of the solar system, and even life here on earth.

Because silica aerogels can be prepared with refractive indices in the critical region near 1 and a high degree of transparency, they have found application for detecting Çerenkov radiation (16). The degree of transparency of aerogels is sensitively dependent on the size of scattering centers (i.e. silica particles and pores) that make up the network. Because these are on the nanoscale, Rayleigh scattering is common. However, this can be minimized by decreasing the particle and/or pore size and preparing materials with a high degree of uniformity. Thus, aerogels with transparencies on the order of 70% over the visible region of the spectrum can be obtained (17). The refractive index of aerogels is also tunable and depends on the density so that indices ranging from 1.007 to 1.24 can be accessed. This range of refractive index is difficult to achieve with other materials, including compressed gases and liquids, and is far lower than typical solids (a typical glass has a refractive index of 1.5). Thus, aerogels essentially serve as radiators for photons generated when elementary particles are traveling

Figure 8.4 The "tennis racket" module of NASA's Mission Stardust with aerogel tiles for capturing hypervelocity particles from the tail of comet Wild 2. Courtesy of NASA/JPL.

at greater than the speed of light (Çerenkov radiation). The ability to collect this light and differentiate its energy spectrum is critical for discerning between the various types of subatomic particles. Thus, large quantities of aerogel tiles have been prepared and exploited for high energy physics experiments and cosmology. The low refractive index also translates to a low dielectric constant, which has led to interest in the microelectronics industry for exploiting these materials as interconnect dielectrics.

A little closer to home and everyday life, aerogels are being developed commercially for applications including building insulation, clothing insulation, and the insulation of oil pipelines. Highly porous aerogels exhibit thermal conductivities considerably lower than conventional insulating materials, including polyurethanes and even air (11). This is because the pore structure results in restricted diffusion of air through the structure, thereby limiting the capacity of that air to deliver thermal energy. For mesopores, gas transport usually occurs via Knudsen diffusion, in which the mean free path of the gas molecules is long relative to the pore size, resulting in frequent wall collisions. Surface diffusion, in which the molecules are physisorbed on the pore walls and transported by site-to-site hopping, is also common in mesopores, whereas the smaller micropores require activated transport. In addition to reducing the thermal transport by the gaseous component, the minimal solid content also reduces thermal transport in the solid state. Altogether, net thermal conductivities (solid and gas phase transport) of as low as $12\ \mathrm{mW \cdot m^{-1} \cdot K^{-1}}$, less than half that of air or a polyurethane foam, can be achieved. Thus, critical applications, such as preventing oil from turning to sludge as it passes through pipelines in cold climates, can be addressed by aerogel insulation. It is often advantageous for these materials to be hydrophobic, thus preventing moisture from gaining access and potentially corroding metal components being insulated. Hydrophobation also acts to protect the silica network itself, which can break down over time in a humid atmosphere due to reactivity of the surface silanols. When the aerogels are sufficiently translucent, aerogels can be exploited for daylighting applications as Cabot has done with their Nanogel® material. Incorporation into skylights results in diffuse light and improvements in insulating ability by 75% to 350%. Aspen Aerogels markets fiber-reinforced aerogels for incorporation into buildings that can increase the thermal performance of walls up to 40%.

8.4.3 Silica Composites: A Sampling

The extent to which silica aerogels have been modified is impressive. Here we will give a few selected examples of silica composite aerogels that have not been alluded to previously and describe how the incorporation of different components radically changes the properties (and hence applications) of silica aerogels.

8.4.3.1 Mixed Metal Oxide Aerogels and Composites Incorporation of a secondary metal oxide phase into silica using co-condensation strategies can be challenging because the kinetics of hydrolysis and condensation of the secondary phase may be quite different than that of silica, resulting in heterogeneity. Clapsaddle et al. (18) have gotten around this by using metal salts, which form aqua acids in solution, as the secondary phase precursor, and controlling the pH (and hence the condensation rate) by using epoxides. As shown in Scheme 8.3 for a hypothetical metal salt MA_3 ($M =$ trivalent metal; $A = -1$ anion), epoxides can act as scavengers for the activated proton of the water ligands attached to the metal, thus activating the metal for condensation. Because pH control is critical and the acidity of the aqua complex varies depending on the ion, epoxides with different rates of reaction can be used to compensate. Thus, silicon-metal mixed oxide gels were successfully formed by reaction of

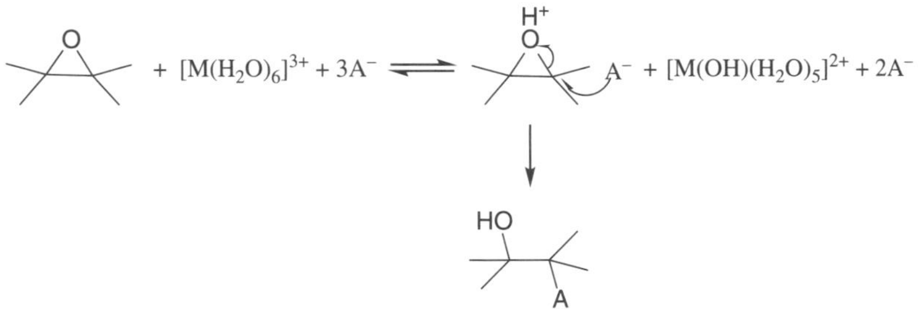

O + $[M(H_2O)_6]^{3+} + 3A^-$ ⇌ H^+ O A^- + $[M(OH)(H_2O)_5]^{2+} + 2A^-$

↓

HO

A

Scheme 8.3 Proton scavenging by epoxide produces hydroxylated metals (M) from aqua acids, enabling condensation reactions.

tetramethoxysilane (TMOS) with salts of early transition metals, most of the lanthanoids, as well as Al, Ga, In, and Sn. Most impressively, these were formed with silica as the minor component, making up anywhere from 15% to 45% of the composite. Gels were transformed to aerogels by cold supercritical drying. Monoliths could be obtained and surface areas ranged from 90 to nearly 800 m^2/g, depending on the identity of the secondary metal and the epoxide employed.

As an example of a composite formed by postmodification of aerogels, RuO_2/silica aerogels were prepared by cryogenic decomposition of RuO_4 onto a preformed silica aerogel. Unlike previous attempts to deposit RuO_2, the resulting materials were conductive, and this observation, together with electron microscopy images, suggested that nanoparticles of conducting RuO_2 have formed a cohesive network on the insulating SiO_2 surface. This "wired" silica aerogel can function as an electrocatalyst, oxidizing chloride ions to Cl_2 (19).

8.4.3.2 Hybrid Organic–Inorganic Aerogels Just as silica can be married to a second inorganic phase to make a bicontinuous network, as illustrated by the RuO_2/silica materials in the previous section, it is also possible to couple silica to an organic polymer. This can be achieved in several ways. For example, Novak, Auerback, and Verrier (20) explored silica gelation in the presence of preformed polymers such as poly(2-vinylpyridine), as well as conducting polymerization reactions of monomers (*N*,*N*-dimethylacrylamide) simultaneously with the hydrolysis/condensation of the silica. Cold supercritical drying produced composites with interpenetrating networks of organic and inorganic polymer. Notably, the poly(2-vinylpyridine)-containing materials exhibited both improved transparency and reduced brittleness.

The ability to improve the mechanical properties of aerogels by incorporating polymers has been expanded upon by Leventis (21), who adopted a strategy in which the polymer is conformally coated over the surface of the aerogel with the specific goal of strengthening the joints (necks) while retaining much of the porosity. In one case, amine-terminated gels (such as those represented in Scheme 8.2) were treated with epoxies, which were then thermally cross-linked (Scheme 8.4), followed by supercritical drying. Although the density of the aerogels was increased two to three times over a nonfunctionalized aerogel, it was possible to increase the modulus and stress at breakpoint by two orders of magnitude (22). While porosity is necessarily reduced by this approach, the surface areas remain reasonably high and the average pore diameter is on the order of 10 to 20 nm. Thus, the materials remain porous, having all the desirable features of an aerogel, but are considerably stronger.

8.4.3.3 Incorporation of Active Enzymes into Aerogels As shown in the previous two sections, incorporation of inorganic and organic materials into silica aerogels is well documented. A question that has only recently been addressed is whether living biological materials can be incorporated into aerogels and retain their function upon doing so. While there is a long history of enzyme incorporation into wet silica gels; keeping the enzyme functional during the processing needed to form aerogels has proved a challenge. Rolison and coworkers (23) recently resolved this issue

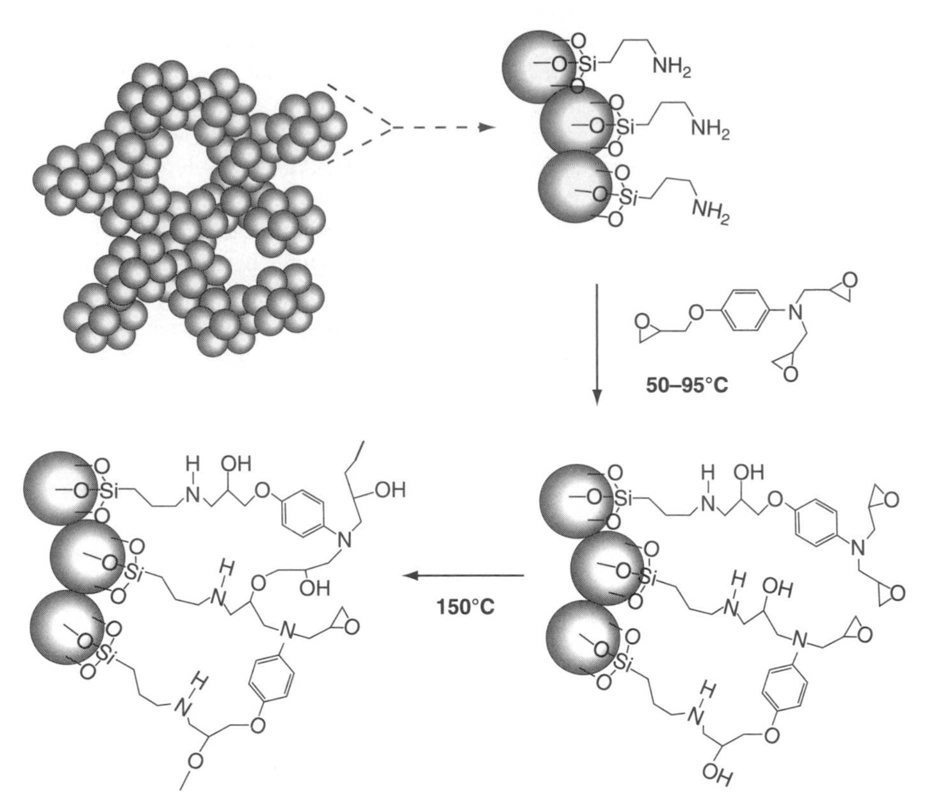

Scheme 8.4 Reaction of tri-epoxy with surface amines (generated as shown in Scheme 8.2) on silica gels and subsequent thermal cross-linking. (Reproduced with permission from M. A. B. Meador et al., *Chem. Mater.* **2005**, *17*, 1085. Copyright 2005 American Chemical Society.)

through [1] formation of a protein superstructure by mixing Au nanoparticles with cytochrome c and [2] nanogluing the superstructure into silica followed by cold supercritical drying. It appears that the enzyme deposits onto the nanoparticles in the first step, forming a thick shell that, while damaged on the surface during the supercritical processing step, nevertheless retains an internal environment that enables the structure, and hence activity, to be maintained. The resulting materials actually function as an optical NO sensor, due to the shift in the wavelength of the Soret band of active cytochrome c when NO binds. This work opens up a whole new field of aerogel endeavors—bio-aerogels.

8.5 BEYOND SILICA: OTHER OXIDE AEROGELS

It is logical to assume that the chemistry applied to silica should be generally applicable to other metals, and this is often true. This enables a greater range of properties and potential applications to be realized. An overview of oxide materials and composites with selected examples will be provided. For the sake of categorization, the oxides will be classified as electronically insulating (Section 8.5.1) or conducting (Section 8.5.2).

8.5.1 Electronically Insulating Oxides

SiO_2 is inherently electronically insulating, and a number of other oxide aerogels share this trait (e.g. Al_2O_3), or fall into the category of wide-bandgap semiconductors (e.g. TiO_2). Such phases often exhibit Lewis acidic and/or basic characteristics that make them useful as catalysts or catalyst supports. As previously indicated, the high surface area of the aerogel, combined with mesoporosity, makes these systems advantageous for applications that require facile transport of molecular species to and from active surfaces, such as catalysis. The insulating or wide-bandgap semiconductor oxide aerogels that have been prepared make up an extensive list, including Al_2O_3, GeO_2, SnO_2, TiO_2, ZrO_2, and Nb_2O_5 (11). We will focus here on the specific examples of Al_2O_3 and TiO_2 and their composites, discussing these materials in terms of selected potential applications.

8.5.1.1 Thermally Insulating Al_2O_3 Aerogels and Catalytic Composites The application of sol-gel hydrolysis and condensation reactions to Al alkoxides has proven to be more complex than for silica, due to the propensity of Al to form complex ions (clusters) with oxide and hydroxide. For this reason, the production of high quality aerogel monoliths of Al_2O_3, although intensively investigated, has been a challenge. However, Poco and coworkers have succeeded, using aluminum sec(butoxide) as a precursor in a two-step process involving initial prehydrolysis in water and alcohol followed by introduction of acid to induce gelation (24). The LLNL RSCE method was used to achieve aerogel monoliths, and these both proved to be stronger and to exhibit lower thermal transport at elevated temperature than silica aerogels of comparable density. The LLNL group has also created alumina aerogel monoliths by the treatment of Al salts with epoxides followed by supercritical CO_2 drying (see Scheme 8.3). The morphology can be tuned, depending on the anion (A^-) of the salt, from colloidal to weblike, with the latter exhibiting improved mechanical properties over the colloidal aerogels (25).

For many catalytic applications, monolithicity is not a critical factor, but strong interactions between the catalyst and support, as well as a strong resistance to thermal sintering, is critical. The range of materials incorporated with alumina in the composite, as well as the range of catalytic processes investigated, is extensive (26). Generally, alumina-containing aerogel composites are prepared from co-gelled species in which metal ions or alkoxides are incorporated with Al alkoxides followed by supercritical drying and (usually) postthermal treatment. Composites that include active transition, rare earth, or alkaline earth metal oxides have all been prepared. Noble metals often are reduced, producing finely divided metal on the high surface area alumina support. Much of the catalytic data, which includes reactions such as hydrocarbon combustion, selective catalytic reduction of NO and dehydration of methanol, is empirical. However, strong indications persist that the enhanced catalytic activity noted in aerogels is not just a function of the surface area, but can also be attributed to the intrinsic aerogel architecture.

8.5.1.2 Composite TiO_2 Aerogels for Catalysis and Photocatalysis Although it is a wide-bandgap semiconductor, titania is an excellent photocatalyst and can be sensitized using dyes to absorb in the visible spectrum, forming the basis of the Grätzel solar cell. It also serves as a reducible support, enabling its use in oxidation catalysis, particularly in the oxidation of CO by supported Au nanoparticles. Recently, Pietron and coworkers (27, 28) have explored the use of titania aerogels to achieve improved activities over traditional and nontraditional (nanoparticulate) titania supports.

Titania aerogels can be prepared from suitable alkoxides, in a process similar to that employed for silica, followed by supercritical drying. In order to produce solar cell electrodes, the titania aerogels were calcined, ground, and incorporated into thin films along with suitable dyes. A preliminary evaluation suggests that photocurrents similar to those obtained from state-of-the-art cells can be obtained with the aerogel thin films, holding out the promise of significant improvements in efficiency upon optimization (27). This again underscores the fact that the aerogel structure offers unique reactivity that cannot just be chalked up to surface area or monolithicity.

To make catalysts for CO oxidation, the incorporation of Au nanoparticles into titania aerogels was achieved by the introduction of preformed ligand-capped Au nanoparticles of 2 to 3 nm in diameter into the titania sol (3, 28). The use of precapped particles enables the size of the Au nanoparticles to be controlled throughout the gelation and supercritical drying process. This is critical because reports in the literature suggest that 5 nm Au particles are not active for CO oxidation (they are too large; Reference 29). Unfortunately, the need to heat the composite Au/TiO_2 aerogel to remove the capping ligands does result in Au particle growth, yielding 6 nm particles (from 2 to 3 nm particle precursors) in the composites (Fig. 8.5a). Surprisingly, these

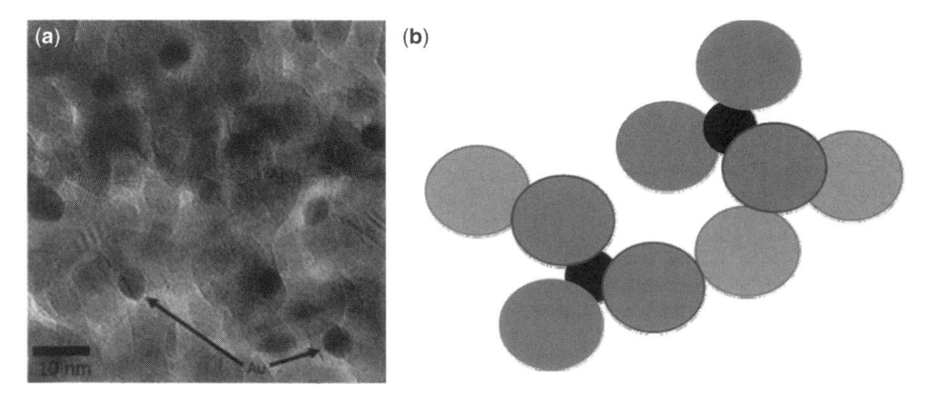

Figure 8.5 (a) Transmission electron micrograph of calcined Au/TiO_2 aerogels showing the Au particles (dark) within a matrix of TiO_2 particles (light). (b) A cartoon showing the way that co-gelation between similarly sized particles contributes to enhanced contacts between Au and TiO_2 nanoparticles, thereby augmenting catalytic activity. (Reproduced with permission from J. J. Pietron et al., *Nano Lett.* **2002**, *2*, 545. Copyright 2002 American Chemical Society.)

remain highly catalytically active—as active as the smaller particles in the conventionally prepared catalysts. Since the activation of O_2 is presumed to occur at the Au-TiO_2 interface, the authors suggested that by incorporating the Au into the network of comparably sized titania particles, they were able to maximize interfaces, thereby obviating the need for very small particles (Fig. 8.5b). In practice, the sol-gel step allows intimate mixing of nanoscale components resulting in strong interactions between the support and active catalyst that is not achieved by incipient wetness approaches, where the nanoscale catalyst ends up perched on a much larger support particle.

8.5.2 Electronically Conducting or Magnetic Oxides

In contrast to the main group or early (Group 4) transition metal oxides, those oxides of the middle transition metals are often electronically conducting and have a range of accessible oxidation states, which leads to unique electrochemical and magnetic properties and hence, a range of possible applications. Examples representing these two properties and associated applications will be presented.

8.5.2.1 Battery Applications: V_2O_5 Aerogels Battery applications require the ability to rapidly intercalate and de-intercalate ions into the host electrode with concomitant charge transfer. It has long been recognized that nanoscale grains and high surface areas are important for rapid transport of charge and matter (ions). Thus, aerogels are a logical system for investigation (30). A variety of systems have been examined, including MnO_2, MoO_3, $RuO_2{-}TiO_2$, and V_2O_5. As an illustrative example, a brief overview of vanadia aerogels will be provided.

Vanadia aerogels can be prepared by simple hydrolysis of $V(O)(O^iPr)_3$ in water. Unlike the analogous reaction with silica, no catalyst is needed, and the gelation occurs in less than a minute. Aging and supercritical extraction leads to aerogels with surface areas up to 450 m^2/g and exhibiting a ribbon morphology (Fig. 8.6). Electrodes based on vanadia aerogels exhibit excellent reversible insertion of ions with high capacities and high equilibrium potentials. Notably, over five Li ions can be incorporated per V_2O_5 unit, significantly greater than the value of two Li ions per V_2O_5 unit found in dense architectures, despite the fact that the oxidation state change per V is only one ($V^{5+} \rightarrow V^{4+}$). The increased ion capacity is purported to be due to high vacancy counts associated with the amorphous, high surface area aerogel structure. Thus, the aerogel architecture may enable exploitation of radically different designs for battery electrodes (30).

8.5.2.2 Magnetic Applications: $NiFe_2O_4$ (Nickel Ferrite) Aerogels Iron-containing oxides are among the most common magnetic materials we encounter. There has been considerable interest in exploring the nanoscale properties of magnetic materials because of a strong interest in miniaturization of magnetic devices. However, when particles are created with sizes <100 nm, their coercivity (ability to remain magnetized in the presence of an opposing field) changes. Understanding these changes, and how these materials will behave when they are integrated into connected structures, is critical. Since aerogels represent 3-D nanostuctures, they are an ideal system in which to study these phenomena.

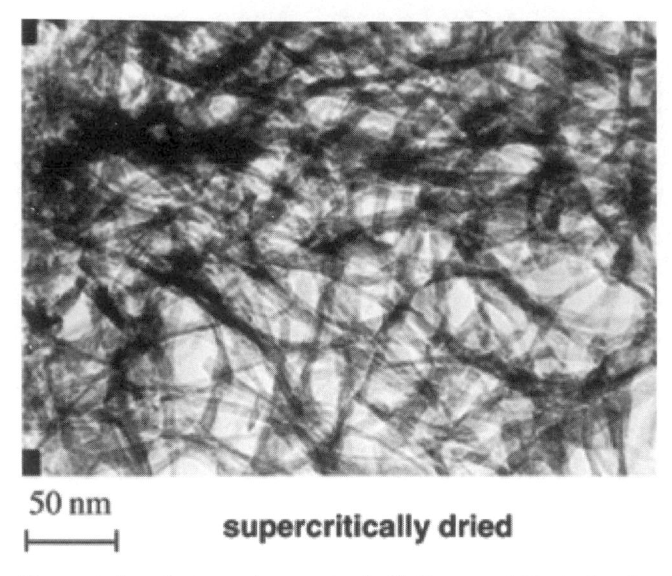

Figure 8.6 Transmission electron micrograph of a V_2O_5 aerogel illustrating the characteristic ribbon morphology. (Reproduced with permission from D. R. Rolison and B. Dunn, *J. Mater. Chem.* **2001**, *11*, 963. Copyright 2001 The Royal Society of Chemistry.)

Over the years, a variety of iron-oxide aerogels have been prepared from the reaction of a suitable base with iron salts; however, recent reports of iron oxides (binary and ternary) have exploited the epoxide approach originally developed by Gash, Satcher, and Simpson (31; see Section 8.4.3.1 and Scheme 8.3). Very recently, Pettigrew et al. (32) have used this approach for the preparation of $NiFe_2O_4$ aerogels composed of highly uniform nanoparticle building blocks. Critical to the success is a two-stage temperature processing, first in air, then in argon. Air heating yields uniformly shaped nanoparticles, whereas heating in the inert atmosphere leads to high crystallinity. By changing the temperature under which the inert heating is done, the size of the components can be adjusted from 4.5 to 8.8 nm, and the coercivity of the particles increases dramatically with increasing size. Despite the need for a high temperature treatment, surface areas of approximately 80 to 180 m^2/g can be achieved and the average pore diameter remains near the middle of the mesoporous regime (15 to 26 nm). In addition to traditional magnetic devices, these materials also have the potential to be active in magnetic separation of gases (i.e. separation of O_2 from air), since the network is highly porous (therefore gas permeable) and highly magnetic (33).

8.6 NEW DIRECTIONS: NON-OXIDES

Having a detailed understanding of the parameter-space accessible with oxides, the question arises as to whether the chemistry is, in fact, limited to oxides. The traditional sol-gel method is based on hydrolysis and condensation for formation of oxide bonds, but could this be adapted to other elements? If gels can be made, will aerogels be possible? Kistler, the originator of aerogels, says yes: "the ability to form an aerogel is a general property of gels" (34). Indeed, he experimented with a wide variety of natural and artificial gels, including egg albumen, stannic oxide, and nitrocellulose and pronounced "we see no reason why this list may not be extended indefinitely" (1). So, the question is, what are the limitations of sol-gel reactions? In this section we describe how sol-gel methods can be applied to two different non-oxide systems for formation of aerogels, carbon and metal chalcogenides (chalcogen = S, Se, Te), and their properties and applications.

8.6.1 Carbon Aerogels: Synthesis and Properties

Organic gels are quite well known and encountered in our daily lives (JELL-O® anyone?). Because a simple definition of a gel is a swollen polymer, this term can be attributed to a wide range of synthetic and natural materials. Of course, these gels are not composed solely of carbon, but have (at minimum) hydrogen as well as other heteroatoms (oxygen features prominently here as well). How then to prepare a pure carbon aerogel? The trick is to pyrolyze organic aerogels, thereby transforming the organic network into a conductive carbon one. Organic aerogels are commonly formed by addition-condensation reactions of resorcinol (R) and formaldehyde (F), as depicted in Scheme 8.5 (35). Addition of formaldehyde to resorcinol (typically catalyzed by base) results in the formation of mono-, di-, or tri-hydroxymethylated derivatives ($-CH_2OH$). The hydroxymethylated groups can then condense together to form ether linkages (CH_2-O-CH_2) or with an unsubstituted site on the phenyl ring to form methylene linkages (CH_2), liberating water in both cases. Thus, like the metal oxide gels prepared by sol-gel chemistry, RF gels are also predicated on condensation reactions involving oxygen. The RF hydrogel results from a high degree of cross-linking to form primary particles that then aggregate into the gel network. The size of the primary particle aggregates depends on the initial concentration of monomers, such that colloidal gels arise at high RF concentrations (the primary particles are thicker than the necks that connect them) and polymeric gels arise at low concentrations (the primary particles and interconnected necks are comparable in size). Critical point drying yields an organic aerogel and pyrolysis of the organic aerogel in an inert atmosphere at temperatures between 500°C and 2500°C yields carbon aerogels. These can be prepared as mesoporous monoliths with surface areas ranging from 400 to 1200 m^2/g.

The interest in carbon aerogels stems in large part from their high electronic conductivity ($>25\ S \cdot cm^{-1}$), which differentiates them from the majority of oxide aerogels, where the networks are inherently insulating, and carbon powders or cloths, which suffer from the absence of a 3-D macroscopic bonded network. This intrinsic conductivity coupled with a tunable pore structure and high surface area has made these materials attractive as electrodes, particularly those in supercapacitors where charge is stored at the interface (36). Additionally, the adsorptive properties of carbon aerogels (both physisorption and chemisorption) has led to

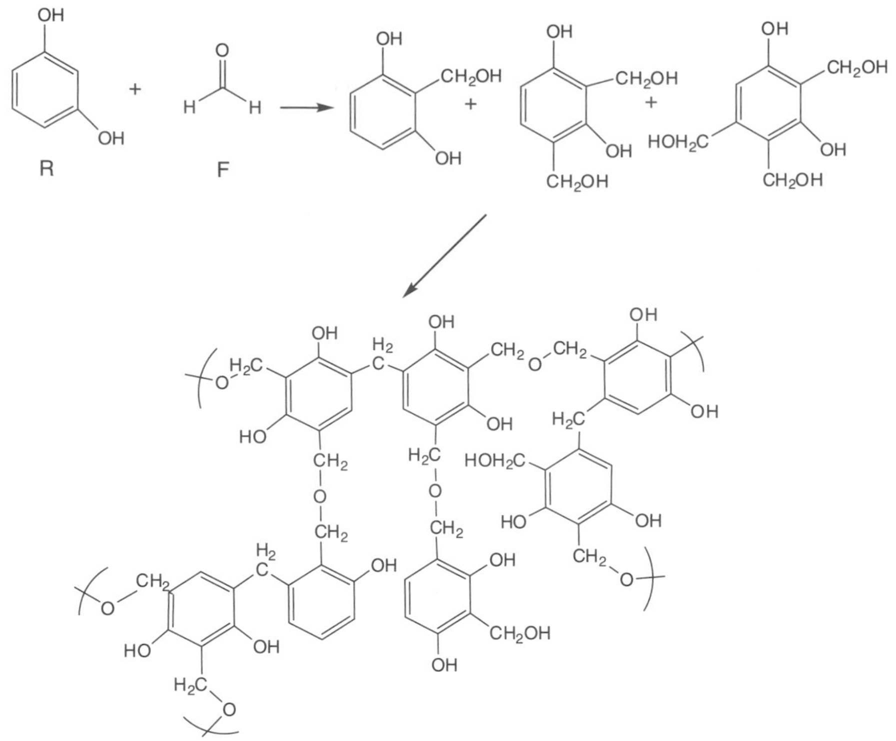

Scheme 8.5 Formation of an organic R-F gel by condensation of resorcinol (R) and formaldehyde (F).

their study for air purification (removal of VOCs; Reference 37) and hydrogen storage (38).

Like oxide aerogels, carbon aerogels can be prepared with a variety of dopants to modify the properties. Of particular interest is the incorporation of metals to tune the electrochemical properties, impart catalytic activity, or both (electrocatalysis; Reference 39). For example, Pt catalysts supported on carbon aerogels are promising electrocatalysts for proton exchange membrane fuel cells (40). Practically, metal doping is achieved by either [1] direct addition of an appropriate metal salt into the RF mixture followed by gelation, aerogel formation, and pyrolysis (41); [2] addition of a metal salt to an RF gel prepared from a resorcinol derivative with a metal binding moiety, followed by aerogel formation and pyrolysis (42); or [3] deposition of metal ions onto a conventionally-prepared RF aerogel, followed by pyrolysis (43), or directly onto a carbon aerogel (43, 44). As is the case for oxide-based materials incorporation of a secondary phase (the metal) has a profound effect on the morphology, surface area, and pore characteristics, and this is most dramatic when the metal is incorporated during gelation because the metal can catalyze the polymerization process. A wide variety of metals has been incorporated using these approaches (39), including noble metals (Pt, Pd, Ag, Au) and first row transition metals (Cr, Fe, Co, Ni, Cu).

The final state of the metal depends on its oxophilicity, method of incorporation, and processing conditions, but is typically finely divided metal or metal oxide.

8.6.2 Metal Chalcogenide Aerogels: Synthesis and Properties

In an effort to further extend the physical properties that can be incorporated into aerogels, a logical extension is to move down the group 16 elements from oxygen to sulfur, selenium and/or tellurium (chalcogens). Metal chalcogenides form the large class of II–VI and IV–VI semiconductors with direct bandgaps that extend from the UV, through the visible, and into the IR. Among oxide aerogels, there are very few with bandgaps in the visible region, tending to fall into categories of small-bandgap semiconductors or metals (V_2O_5, RuO_2), or large-bandgap semiconductors or insulators (TiO_2, SiO_2). Thus, chalcogenide aerogels have the potential to function in applications involving absorbance or emission from the solar spectrum, such as photocatalysis and photovoltaics (45, 46). The formation of metal chalcogenide aerogels is fairly new, with the first system reported in 2004 (47). To date, there are three demonstrated approaches to these materials: [1] replace water in the hydrolysis reactions with H_2S (thiolysis); [2] preforming metal chalcogenide nanoparticles and condensing them together into a gel matrix; [3] linking metal chalcogenide ions or clusters via secondary metals (Pt^{2+}). These approaches are discussed in more detail below, along with the properties of the resultant metal chalcogenide aerogels.

8.6.2.1 Thiolysis Reactions When H_2O is replaced by H_2S in a sol-gel reaction, the result is a *thiolysis* reaction, rather than *hydrolysis*. Thiolysis-based sol-gel reactions can be described by a general series of equations, as shown in Scheme 8.6 for the case of the thiolysis of zinc *tert*-butoxide (O^tBu). While this approach has been applied to a wide range of metals, including Ti, Nb, and W, formation of a gel is not a common outcome (45). Instead, precipitates, either crystalline or amorphous, typically form. To date, there is only one demonstrated example of an aerogel formed from thiolysed gels: GeS_2 (48). The GeS_2 gel is formed from the reaction of germanium ethoxide in toluene with gaseous H_2S. Cold supercritical drying leads to amorphous, sulfur-rich material, $GeS_{2.4}$, with an impressive surface area of 755 m^2/g.

Thiolysis:

$$Zn(OBu^t)_2 + xH_2S \longrightarrow Zn(OBu^t)_{2-x}(SH)_x + xHOBu^t$$

Condensation:

$$2\ Zn(OBu^t)_{2-x}(SH)_x \longrightarrow (OBu^t)_{2-x}(HS)_{x-1}Zn\text{-}S\text{-}Zn(HS)_{x-1}(OBu^t)_{2-x} + H_2S$$

Overall:

$$Zn(OBu^t)_2 + H_2S \longrightarrow ZnS + 2\ HOBu^t$$

Scheme 8.6 Equations for thiolysis and condensation of a metal alkoxide (zinc *tert*-butoxide) to yield a metal sulfide gel (zinc sulfide).

Given the sol-gel chemistry demonstrated by this approach, this would appear to be a fertile area for the development of new chalcogenide aerogels. Nevertheless, there are some distinct drawbacks. In addition to problems balancing reaction kinetics and thermodynamics to drive gel formation (in lieu of precipitation), oxygen contamination is a common problem. Even when gels form, the networks that have been prepared to date are very weak and do not lead to monoliths (45).

8.6.2.2 Nanoparticle Condensation An alternative strategy for preparing metal chalcogenide gels is to separate the particle formation and condensation steps. That is, prepare discrete nanoparticles of the material of interest and then condense them together into a gel network. This methodology was originally developed by Gacoin and coworkers for formation of CdS gels (49). The mechanistic basis for gel formation was the oxidative removal of surface thiolate functionalities from the surface of the particles, leading to reactive sites on the particles that could then condense into a gel network. This is illustrated in Scheme 8.7. In 2004, this method was applied by Mohanan and Brock (47) to the formation of the first CdS aerogel by cold supercritical drying of the wet gel, and has since been extended to aerogels of PbS, ZnS, CdSe, CdS (45, 50), and bicomponent systems formed from CdSe/ZnS core/shell particles (46, 51). These materials exhibit typical surface areas of 100 to 250 m^2/g and the characteristic broad pore size distribution expected from aerogels.

In addition to viewing these materials as chalcogenide analogs of traditional aerogels, it is also valid to think of them as an assembly of nanoparticle building blocks. Given the strong interest in the optical properties of semiconducting nanoparticles (Chapter 16), gelation can be viewed as a method for assembling discrete nanoparticles into integrated solid state structures. A key question is then whether

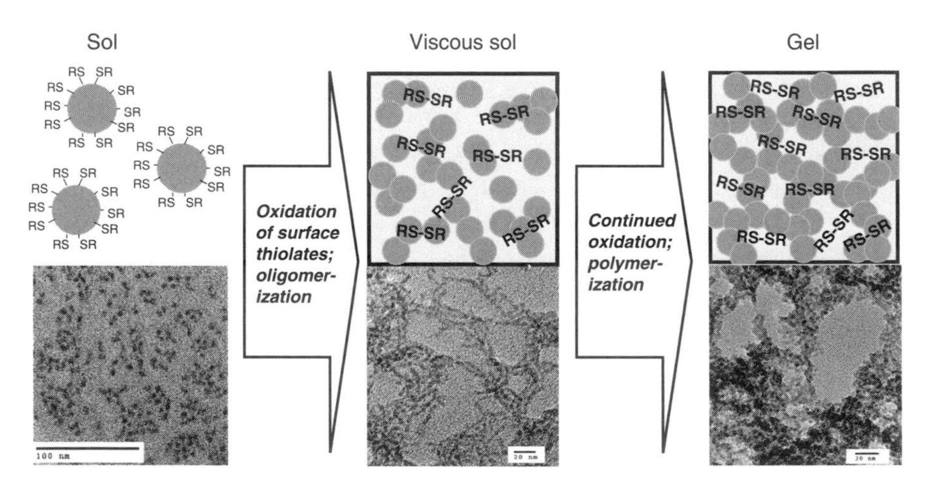

Scheme 8.7 General mechanism for gelation of metal chalcogenide nanoparticles capped with thiolate functionalities (RS-) by oxidative removal of capping groups as disulfide (RS-SR) and subsequent nanoparticle condensation. (Reproduced with permission from I. U. Arachchige and S. L. Brock, *Acc. Chem. Res.* **2007**, *40*, 801. Copyright 2007 American Chemical Society.)

the size-dependent bandgap and intense luminescence properties of the individual quantum dots would be maintained within the connected bulk. Surprisingly, it is found that aerogels retain a similar degree of quantum confinement to the starting particles, which is attributed to the low dimensionality of the tortuous solid network (in fact, aerogels are fractal solids with noninteger dimensionalities). Consequently, the extent of quantum confinement is reduced when dense structures (xerogels obtained by benchtop drying) are obtained, manifest in a red-shift of the absorbance band (a decrease in the bandgap energy; Fig. 8.7). If, however, the true zero-dimensional nature of the quantum dot can be maintained in the solid (achieved by wrapping the optically active CdSe core with a transparent ZnS shell prior to gelation), there is no density dependence at all. This latter strategy has the benefit of producing materials with outstanding luminosity, due to the passivation of defect sites on the aerogel surface.

A distinct advantage of this method is the ability to control the morphology of the aerogel by controlling the shape of the precursor particles from which the gels form.

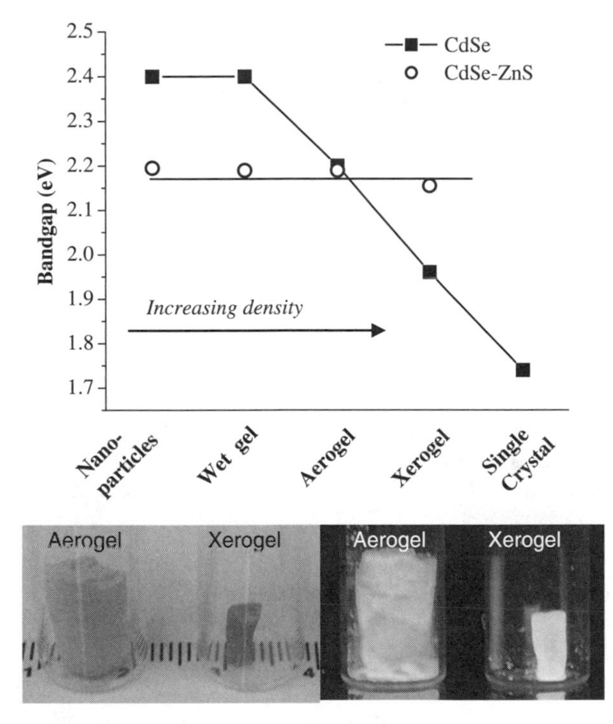

Figure 8.7 Top: Effect of density on the optical bandgap for aerogels prepared from naked CdSe nanoparticles and CdSe-ZnS core-shell nanoparticles. Bottom: CdSe-ZnS core-shell aerogel and xerogel monoliths under normal (left) and UV (right) illumination. The native aerogel and xerogel are orange under normal light and fluorescent green under UV stimulation. (Graph reproduced with permission from I. U. Arachchige and S. L. Brock, *Acc. Chem. Res.* **2007**, *40*, 801. Pictures reproduced with permission from I. U. Arachchige and S. L. Brock, *J. Am. Chem. Soc.* **2007**, *129*, 1840. Copyright 2007 American Chemical Society.) (See color insert.)

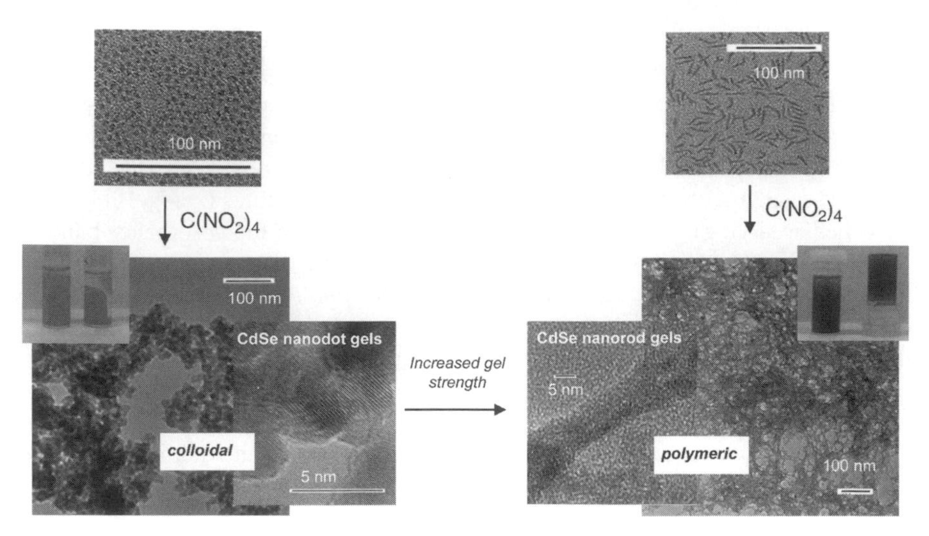

Figure 8.8 Effect of nanoparticle precursor shape on morphology (colloidal vs. polymeric) and strength of the gel network generated by oxidative treatment of thiolate-capped CdSe nanoparticles with tetranitromethane. Note that the nanodot gels undergo syneresis and contract during aging, whereas the nanorod gels are firm, take the shape of their container, and can be inverted. (Reproduced with permission from H. Yu et al., *J. Am. Chem. Soc.* **2008**, *130*, 5054. Copyright 2008 American Chemical Society.)

Thus, spherical CdSe particles lead to colloidal aerogels analogous to those formed with silica under base-catalyzed conditions, whereas rod or branched CdSe particles lead to polymeric aerogels analogous to acid-catalyzed silica (Fig. 8.8; Reference 52). The polymeric network is more robust than the colloidal network, leading to little or no gel shrinkage during aging (Fig. 8.8) and manifesting in a significantly greater viscosity enhancement for polydimethylsiloxane (PDMS)-polymer aerogel composites than for the corresponding colloidal aerogel composites, relative to native PDMS. That is, the polymeric networks appear to be inherently stronger. The polymeric aerogels also exhibit twice the surface area of the colloidal aerogels (239 vs 109 m^2/g).

8.6.2.3 Cluster Addition In 2007, Kanatzidis and coworkers demonstrated an alternative method of making chalcogenide gels and aerogels by linking molecular metal chalcogenide ions or clusters together with Pt ions, as shown in Scheme 8.8 (53). The displacement of terminal alkylammonium ions with bridging Pt^{2+} ions leads to polymerization and formation of an amorphous gel network. Cold supercritical drying in turn produces aerogels with surface areas ranging from 100 to 330 m^2/g. This approach has been demonstrated for sulfide and selenide clusters of Ge and Sn ($[M_4Q_{10}]^{4-}$, $[M_2Q_6]^{4-}$ and $[MQ_4]^{4-}$, M = Ge, Sn; Q = S, Se). Depending on the chemical composition, materials with small to moderate bandgaps can be obtained (0.2 to 2.0 eV) as bulk monoliths. One intriguing property demonstrated for these chalcogenide aerogels is their ability to selectively remove heavy ions, such as mercury,

$$(R_4N)_4[Ge_4S_{10}] + 2\ K_2PtCl_4$$
$$\downarrow$$
$$Pt_4[Ge_4S_{10}] + 4\ KCl + 4\ R_4NCl$$

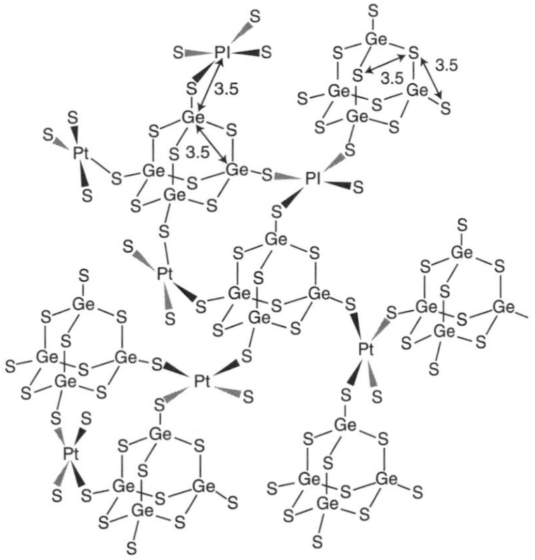

Scheme 8.8 Formation of chalcogenide gels via condensation of a molecular cluster (Ge_4S_{10}) with Pt. Gel structure image from S. Bag et al., *Science* **2007**, *317*, 490. (Reproduced with permission from AAAS.)

from solution. Thus, in a solution of 1 : 1 concentration Hg^{2+} and Zn^{2+}, nearly all of the Hg^{2+} is adsorbed relative to just 40% of the Zn^{2+}. This highlights the soft Lewis basic behavior of the chalcogenide, which leads to preferential adsorption of soft (polarizing) cations, like Hg^{2+}, over moderate to hard ions, like Zn^{2+}. This demonstrates another good reason to explore non-oxide systems, since adsorbents based on oxide (a hard Lewis base) are not selective for heavy metals.

8.7 CONCLUDING REMARKS AND FUTURE OUTLOOK

The unique pore and matter architecture of aerogels leads to unique properties, and hence, applications, that are only now being realized. Traditional silica aerogels are now finding applications in both niche (Çerenkov radiation detection, space dust collection) and large-scale (insulation, daylighting) applications. Nontraditional aerogels, in which the chemical foundation of the aerogel is either a modified silica, another oxide or composite, or a non-oxide (carbon or metal chalcogenide) are relatively recent players. However, the breadth of properties realized within these architectures is likely to lead to future applications that embrace a host of technologies. Perhaps most important in the current climate is the likely application of these materials in alternative energy and environmental remediation. Thus, conductive aerogels show unprecedented charge storage capacities suitable for battery applications; semiconducting aerogels are promising photoconductors and may represent a revolutionary technology for photovoltaic devices; catalytic architectures show promise for abatement of noxious gases and chalcogenide aerogels can remove toxic heavy ions from water streams.

The future of aerogels lies in [1] improvement of the properties of current aerogel systems for specific applications and [2] the development of aerogels based on new chemistry. In the case of [1], key issues include the development of inexpensive aerogel processing techniques and improvement of mechanical properties in silica-based systems. With respect to non-oxides, we need to develop a series of strategies, analogous to what we now have on hand for silica, for functionalization. The creation of aerogel structures based on nontraditional architectures (point 2) involves developing new chemistries (or, often, rediscovering old ones) to realize Kistler's prediction that just about any gel can be converted to an aerogel. While the applications we can envision from aerogels look bright, the exploratory investigation of new chemical systems may lead to applications that we cannot imagine, providing a motivation for chemists, physicists, and materials scientists to continue to study this intriguing class of materials.

FURTHER READING

Anderson, M. L.; Stroud, R. M.; Morris, C. A.; Merzbacher, C. I.; Rolison, D. R., "Tailoring Advanced Nanoscale Materials Through Synthesis of Composite Aerogel Architectures," *Adv. Engin. Mater.* **2000**, *2*, 481–488.

Arachchige, I. U.; Brock, S. L., "Sol-Gel Methods for the Assembly of Metal Chalcogenide Quantum Dots," *Acc. Chem. Res.* **2007**, *40*, 801–809.

Brinker, C. J.; Scherer, G. W., *Sol-Gel Science*. Academic Press, San Diego, California, 1990.

Brock, S. L.; Arachchige, I. U.; Kalebaila, K. K., "Metal Chalcogenide Gels, Xerogels and Aerogels," *Comments Inorg. Chem.* **2006**, *27* (5–6), 1–24.

Fricke, J., "Aerogels," *Sci. Amer.* **1988**, *258*, 92–97.

Hüsing, N.; Schubert, U., "Aerogels—Airy Materials: Chemistry, Structure, and Properties," *Angew. Chem. Int. Ed. Engl.* **1998**, *37*, 22–45.

Pierre, A. C.; Pajonk, G. M., "Chemistry of Aerogels and Their Applications," *Chem. Rev.* **2002**, *102*, 4243–4265.

Rolison, D. R., "Catalytic Nanoarchitectures: The Importance of Nothing and the Unimportance of Periodicity," *Science* **2003**, *299*, 1698–1701.

Rolison, D. R.; Dunn, B., "Electrically Conductive Oxide Aerogels: New Materials in Electrochemistry," *J. Mater. Chem.* **2001**, *11*, 963–980.

REFERENCES

1. Kistler, S. S., *Nature* **1931**, *127*, 741.
2. Michalou(di)s, I., *J. Non-Cryst. Solids* **2004**, *350*, 61.

3. Rolison, D. R., *Science* **2003**, *299*, 1698.
4. Long, J. W.; Rolison, D. R., *Acc. Chem. Res.* **2007**, *40*, 854.
5. Brinker, C. J.; Scherer, G. W., *Sol-Gel Science*. Academic Press, San Diego, California, 1990.
6. Poco, J. F.; Coronado, R. P.; Pekala, R. W.; Hrubesh, L. W., *Mater. Res. Soc. Symp. Proc.* **1996**, *431*, 297.
7. Gauthier, B. M.; Bakrania, S. D.; Anderson, A. M.; Caroll, M. K., *J. Non-Cryst. Solids* **2004**, *350*, 238.
8. Smith, D. M.; Deshpande, R.; Brinker, C. J., *Mater. Res. Soc. Symp. Proc.* **1992**, *271*, 567.
9. Prakash, S. S.; Brinker, C. J.; Hurd, A. J.; Rao, S. M., *Nature* **1995**, *374*, 439.
10. Morris, C. A.; Anderson, M. L.; Stroud, R. M.; Merzbacher, C. I.; Rolison, D. R., *Science* **1999**, *284*, 622.
11. Hüsing, N.; Schubert, U., *Angew. Chem. Int. Ed. Engl.* **1998**, *37*, 22.
12. Hunt, A. J.; Ayers, M. R.; Cao, W., *J. Non-Cryst. Solids* **1995**, *185*, 227.
13. Pierre, A. C.; Pajonk, G. M., *Chem. Rev.* **2002**, *102*, 4243.
14. Tillotson, T. M.; Hrubesh, L. W., *J. Non-Cryst. Solids* **1992**, *145*, 44.
15. Tsou, P., *J. Non-Cryst. Solids* **1995**, *186*, 415.
16. Cantin, M.; Casse, M.; Koch, L.; Jouan, R.; Mestran, P.; Roussel, D.; Bonnin, F.; Moutel, J.; Teichner, S. J., *Nucl. Instrum. Methods* **1974**, *118*, 177.
17. Rigacci, A.; Ehrburger-Dolle, F.; Geissler, E.; Chevalier, B.; Sallée, H.; Achard, P.; Barbieri, O.; Berthon, S.; Bley, F.; Livet, F.; Pajonk, G. M.; Pinto, N.; Rochas, C., *J. Non-Cryst. Solids* **2001**, *285*, 187.
18. Clapsaddle, B. J.; Sprehn, D. W.; Gash, A. E.; Satcher, J. H. J.; Simpson, R. L., *J. Non-Cryst. Solids* **2004**, *350*, 173.
19. Ryan, J. V.; Berry, A. D.; Anderson, M. L.; Long, J. W.; Stroud, R. M.; Cepak, V. M.; Browning, V. M.; Rolison, D. R.; Merzbacher, C. I., *Nature* **2000**, *406*, 169.
20. Novak, B. M.; Auerback, D.; Verrier, C., *Chem. Mater.* **1994**, *6*, 282.
21. Leventis, N., *Acc. Chem. Res.* **2007**, *40*, 874.
22. Meador, M. A. B.; Fabrizio, E. F.; Ilhan, F.; Dass, A.; Zhang, G.; Vassilaras, P.; Johnston, J. C.; Leventis, N., *Chem. Mater.* **2005**, *17*, 1085.
23. Wallace, J. M.; Rice, J. K.; Pietron, J. J.; Stroud, R. M.; Long, J. W.; Rolison, D. R., *Nano Lett.* **2003**, *3*, 1463.
24. Poco, J. F.; Satcher, J. H., Jr.; Hrubesh, L. W., *J. Non-Cryst. Solids* **2001**, *285*, 57.
25. Baumann, T. F.; Gash, A. E.; Chinn, S. C.; Sawvel, A. M.; Maxwell, R. S.; Satcher, J. H. J., *Chem. Mater.* **2005**, *17*, 395.
26. Pajonk, G. M., *Catal. Today* **1997**, *35*, 319.
27. Pietron, J. J.; Stux, A. M.; Compton, R. S.; Rolison, D. R., *Solar Ener. Mater. Solar Cells* **2007**, *91*, 1066.
28. Pietron, J. J.; Stroud, R. M.; Rolison, D. R., *Nano Lett.* **2002**, *2*, 545.
29. Bond, G. C.; Thompson, D. T., *Catal. Rev.* **1999**, *41*, 319.
30. Rolison, D. R.; Dunn, B., *J. Mater. Chem.* **2001**, *11*, 963.
31. Gash, A. E.; Satcher, J. H., Jr.; Simpson, R. L., *Chem. Mater.* **2003**, *15*, 3268.

32. Pettigrew, K. A.; Long, J. W.; Carpenter, E. E.; Baker, C. C.; Lytle, J. C.; Chervin, C. N.; Logan, M. S.; Stroud, R. M.; Rolison, D. R., *ACS Nano* **2008**, *2*, 784.
33. Long, J. W.; Logan, M. S.; Carpenter, E. E.; Rolison, D. R., *J. Non-Cryst. Solids* **2004**, *350*, 182.
34. Kistler, S. S., *J. Phys. Chem.* **1932**, *36*, 52.
35. Al-Muhtaseb, S. A.; Ritter, J. A., *Adv. Mater.* **2003**, *15*, 101.
36. Pekala, R. W.; Mayer, S. T.; Poco, J. F.; Kaschmitter, J. L., *Mater. Res. Soc. Symp. Proc.* **1994**, *349*, 79.
37. Fairén-Jiménez, D.; Carrasco-Marín, F.; Moreno-Castilla, C., *Langmuir* **2007**, *23*, 10095.
38. Kabbour, H.; Baumann, T. F.; Satcher, J. H. J.; Saulnier, A.; Ahn, C. C., *Chem. Mater.* **2006**, *18*, 6085.
39. Moreno-Castilla, C.; Maldonado-Hódar, F. J., *Carbon* **2005**, *43*, 455.
40. Smirnova, A.; Dong, X.; Hara, H.; Vasiliev, A.; Sammes, N., *Int. J. Hydrogen Energy* **2005**, *30*, 149.
41. Maldonado-Hódar, F. J.; Ferro-García, M. A.; Rivera-Utrilla, J.; Moreno-Castilla, C., *Carbon* **1999**, *37*, 1199.
42. Baumann, T. F.; Fox, G. A.; Satcher, J. H. J.; Yoshizawa, N.; Fu, R.; Dresselhaus, M. S., *Langmuir* **2002**, *18*, 7073.
43. Saquing, C. D.; Cheng, T.-T.; Aindow, M.; Erkey, C., *J. Phys. Chem. B* **2004**, *108*, 7716.
44. Miller, J. M.; Dunn, B., *Langmuir* **1999**, *15*, 799.
45. Brock, S. L.; Arachchige, I. U.; Kalebaila, K. K., *Comments Inorg. Chem.* **2006**, *27* (5–6), 1.
46. Arachchige, I. U.; Brock, S. L., *Acc. Chem. Res.* **2007**, *40*, 801.
47. Mohanan, J. L.; Brock, S. L., *J. Non-Cryst. Solids* **2004**, *350*, 1.
48. Kalebaila, K. K.; Georgiev, D. G.; Brock, S. L., *J. Non-Cryst. Solids* **2006**, *352*, 232.
49. Gacoin, T.; Malier, L.; Boilot, J.-P., *Chem. Mater.* **1997**, *9*, 1502.
50. Mohanan, J. L.; Arachchige, I. U.; Brock, S. L., *Science* **2005**, *307*, 397.
51. Arachchige, I. U.; Brock, S. L., *J. Am. Chem. Soc.* **2007**, *129*, 1840.
52. Yu, H.; Bellair, R.; Kannan, R. M.; Brock, S. L., *J. Am. Chem. Soc.* **2008**, *130*, 5054.
53. Bag, S.; Trikalitis, P. N.; Chupas, P. J.; Armatas, G. S.; Kanatzidis, M. G., *Science* **2007**, *317*, 490.

PROBLEMS

1. Write suitable mechanisms for (**a**) acid-catalyzed and (**b**) base-catalyzed hydrolysis of tetraethoxysilane (TEOS). How do these catalysts affect the morphology of the resultant aerogels?

2. Why are aerogels more commonly prepared from base catalysis rather than acid catalysis? Describe how the acid catalysis method can be amended so that a highly porous polymeric silica aerogel can be achieved.

3. Aerogel formation can be achieved by supercritical drying of the wet gel. What is the difference between "hot" and "cold" supercritical drying? Into which of these approaches do the RSCE methods fall? Indicate the general pros and cons of these methods.

4. Prepare a scheme depicting how you could chemically modify a silica surface with trimethylsilyl groups. Explain how this modification can enable aerogel formation using ambient-pressure drying. How do the properties of this aerogel compare to "native" silica aerogels? How could you characterize this material to prove that the modification had taken place?

5. Monolithic aerogels are often quite fragile. Describe two methods that can be employed to prepare stronger colloidal silica aerogels. Indicate how these methods might otherwise impact the native properties of the aerogel.

6. Aerogels have overall thermal conductivities that are significantly lower than that of air ($\kappa_g^\circ = 26\,\text{mW}\cdot\text{m}^{-1}\cdot\text{K}^{-1}$), and are excellent thermal insulators for this reason. For a silica aerogel of density $\rho = 250\,\text{kg}\cdot\text{m}^{-3}$, the thermal conductivity from gaseous and solid components is on the order of $15\,\text{mW}\cdot\text{m}^{-1}\cdot\text{K}^{-1}$. Given the thermal conductivity of the solid component of the aerogel (κ_s) can be approximated as $0.064\rho - 6.0$, what is the gas phase conductivity in the aerogel (κ_g)? Explain how it is that the gas phase conductivity of air in the pores of the aerogel is so much smaller than native air.

7. Recent efforts have focused on developing electronically conducting aerogels. Provide two methods by which this can be achieved. What is the motivation for preparing these kinds of networks and what kinds of applications are envisioned?

8. Write chemical equations pertaining to the stepwise redox intercalation of 2 moles of Li into V_2O_5. How does the matter and pore architecture of the aerogel contribute to this process and enable uptake of even more Li?

9. Magnetic aerogels may prove useful for separating O_2 from air. Considering the magnetic properties of O_2 relative to those of the other principal components of air (N_2, Ar), describe how this may be achieved. Why should aerogels be more suitable to this application than bulk (non-nanostructured) materials?

10. Metal chalcogenide aerogels prepared by condensation of metal chalcogenide nanoparticles (quantum dots) exhibit unique optical properties. What are these properties and how does the aerogel dimensionality determine the extent of quantum confinement? You may wish to consult Chapter 16 on semiconducting nanoparticles in answering this question.

ANSWERS

1. (a)

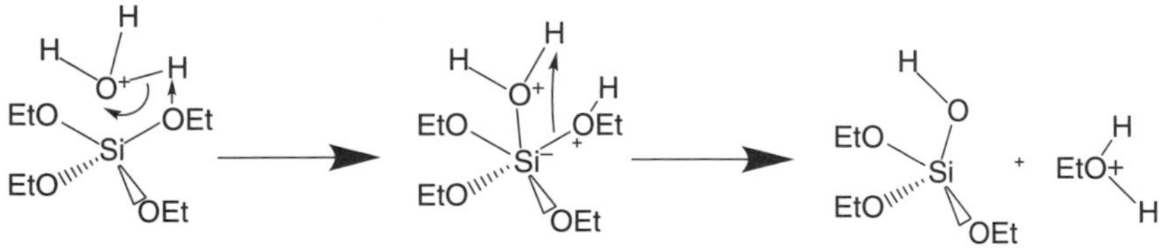

(b)

$$\mathrm{OH^-} + \mathrm{Si(OEt)_4} \longrightarrow \mathrm{(EtO)_3SiOH} + {}^{-}\mathrm{OEt}$$

Acid catalysis typically leads to polymeric gels by reaction limited cluster aggregation due to formation of small clusters under rapid hydrolysis followed by slow condensation of these clusters into the gel network.

Base catalysis typically leads to colloidal gels by reaction-limited monomer cluster growth in which slow hydrolysis restricts the concentration of active species for condensation. These tend to react with nucleated clusters to make large (nm) sized particles and then these particles link together to form the gel.

2. Relative to base-catalyzed gels, those catalyzed with acid have a larger number of small pores, and it is difficult to dry them without significant pore collapse. However, a two-step process in which the precursors are initially hydrolyzed with acid, followed by gelation under basic conditions, does lead to robust polymeric aerogels. In fact, this process yields the lowest density aerogels on record.

3. "Hot" supercritical drying usually involves a solvent such as an alcohol that has a relatively high supercritical temperature (T_c) that must be attained for supercritical drying (several hundred degrees centigrade). In contrast, cold supercritical drying employs CO_2, which has a T_c of just 31°C. The RSCE methods are hot methods and involve gelation within a mold and supercritical extraction of the alcohol by-product of gelation.

In general, the solvent to be removed supercritically in hot methods is the by-product of gelation, typically an alcohol if an alkoxide is employed. Thus, unlike CO_2 drying, there is no need for solvent exchange steps to replace the by-product solvent with CO_2. On the other hand, the high temperatures employed in hot methods facilitate aging, often resulting in more coarse (less porous) aerogels than what is achieved by cold supercritical drying.

The key advantage of the RSCE methods is that they are very fast. Although a hot method, significant coarsening can be avoided because the gelation and aging take place rapidly. However, the RSCE methods do require building the systems, whereas conventional hot and cold supercritical dryers can be purchased commercially.

4.

$$\mathrm{Si{-}OH} + \mathrm{Cl{-}Si(CH_3)_3} \longrightarrow \mathrm{Si{-}O{-}Si(CH_3)_3} + \mathrm{HCl}$$

The above scheme shows a possible way of modifying the silica with a tri-methyl-silyl group (the use of an alkoxy functionality instead of Cl would also work). The process of converting Si-OH groups to Si-$(CH_3)_3$ groups renders the surface non-reactive (condensation is not possible). Thus, during supercritical drying, there can be no cross-linking across the pore wall to form Si-O-Si when the pores are drawn inward by capillary forces. The result is that the gel contracts, and then springs back once the solvent has been completely removed. The methyl groups also render the surface hydrophobic, in contrast to the native Si-OH functionalized material. A variety of techniques can be used to confirm successful methylation, including IR (Si-C stretches and bends and $-CH_3$ modes should be detected in the product, but not the initial gel) and ^{1}H and ^{13}C solid state NMR.

5. One strategy for strengthening the gel body is to age the gel, preferentially in a fresh solution of an alkoxysilane. Ripening will lead to dissolution of silica from low-curvature bodies and deposition at the highly curved interfaces, thereby strengthening the weak points (the joints) of the network. The addition of a fresh solution of alkoxysilane provides a high concentration of precursors for deposition. Again, deposition is expected to be more rapid at the interfaces. Although stronger, the resultant aerogels are also likely to have fewer small pores (these also have high curvature and will fill in rapidly) and higher densities/lower surface areas due to the thicker solid matrix.

Another possibility is to "brace" the gel network by cross-linking it with an organic polymer. This can be achieved by simulataneous gelation and polymerization, or gelation around a preformed polymer, to form an interpenetrating network. Alternatively, the gel can be surface functionalized with monomers that are subsequently cross-linked. This latter strategy has the advantage that the polymer is chemically tethered to the inorganic network, providing support. These hybrid gels likewise suffer from a reduced surface area due to an increased density, thanks to the additional component. One might imagine that the surface functionalization would have less impact on porosity than the simultaneous gelation/polymerization (or gelation around a polymer) method, since the former leads to more of a conformal coverage of the inorganic matrix, rather than filling in the pores of the matrix.

6. For an aerogel of density $250\,\mathrm{kg\cdot m^{-3}}$, $\kappa_s = (0.064 \times 250) - 6.0 = 10\,\mathrm{mW\cdot m^{-1}\cdot K^{-1}}$. Therefore, $\kappa_g = 15\,\mathrm{mW\cdot m^{-1}\cdot K^{-1}} - 10\,\mathrm{mW\cdot m^{-1}\cdot K^{-1}} = 5\,\mathrm{mW\cdot m^{-1}\cdot K^{-1}}$.

The reason the conductivity of air is so much lower in the aerogel than when it is not encompassed in a solid network is that the diffusion of air is reduced in the former case, provided the pores are sufficiently small (micro-mesoporous). This restricts the ability of the pore air molecules to effectively transmit the heat.

7. One method to render silica aerogels conductive is to conformally coat them with a secondary conductive phase. This was demonstrated by Rolison and co-workers by solution-phase treatment of silica aerogels with RuO_4 at low temperatures, which subsequently decomposes to plate RuO_2 on the aerogel surface (19).

Electron microscopy suggests there is a continuous network of RuO_2 nanoparticles deposited on the surface of the aerogel, and this is responsible for the conductivity.

An alternate method to make conductive aerogels is to employ a network that is inherently conductive, like carbon. Carbon aerogels can be prepared by pyrolysis of resorcinol–formaldehyde organic aerogels. The resultant aerogels have conductivities exceeding $25 \cdot S \cdot cm^{-1}$.

The motivation for preparing conductive aerogels stems from the desire to be able to transport electrons through the nanostructure (solid network). When coupled to the high surface area and interconnected pore structure, this architecture enables ions or molecules to be transmitted to the surface where they can be reduced or oxidized by the conductive network. For this reason, such materials are being explored for charge storage and electrocatalytic applications.

8.

$$Li + (V^{5+})_2O_5 \longrightarrow LiV^{4+}V^{5+}O_5$$

$$Li + LiV^{4+}V^{5+}O_5 \longrightarrow Li_2(V^{4+})_2O_5$$

The matter and pore architecture enables rapid transport of Li ions and associated electrons to the active surface of the V_2O_5 enabling complete reduction over a short time period. The first two Li equivalents are formally intercalated with charge transfer, but it is possible for the lattice to adsorb the charge from over three additional Li ions. Presumably, these are associated with vacancies in the amorphous solid network.

9. O_2 is a paramagnetic molecule with two unpaired spins, whereas Ar and N_2 are both diamagnetic. Thus, one can anticipate that O_2 will be attracted to the magnetic field of the aerogel, whereas N_2 and Ar will be repelled. This should result in much slower diffusion of O_2 through the magnetic matrix than N_2 or Ar, thereby facilitating its separation.

Critical to separations is being able to expose the molecules to be separated to a high surface area sorbent, thereby maximizing their exposure time and therefore enabling differences in retention to be magnified. For this reason, a high surface area porous magnetic framework can be expected to be much more effective for reasonable column lengths than dense magnets.

10. Metal chalcogenide aerogels exhibit a blue shift in their absorbance onset relative to bulk materials. These properties reflect the quantum confinement effects of the nanoparticle building blocks (quantum dots), in which the bandgap is size dependent for semiconductors with a particle size smaller than the natural Bohr radius of the exciton (electron–hole pair). The extent of quantum confinement depends on the dimensionality of the material; hence, quantum dots are "0" dimensional (confined in three dimensions) and exhibit the strongest size-bandgap relationships, whereas quantum wires are confined in just two dimensions and quantum wells (thin films) in one. Quantum wires and wells thus do not have as wide an apparent bandgap for the critical limiting dimension as do quantum dots. Properties are maintained in the aerogel, despite being connected in 3-D, because the aerogel network is inherently low-dimensional, due to the tortuous solid network (fractal solid with a noninteger dimensionality). More dense structures (i.e. xerogels) show a lower degree of quantum confinement, with a bandgap decreasing towards the bulk value, suggesting that dimensionality is related to density. In contrast, if the quantum dots are wrapped with a shell of a wider bandgap material (i.e. formation of CdSe-ZnS core-shell structures), there is no density dependence on the bandgap because the CdSe quantum dots remain completely confined.

9

ORDERED MICROPOROUS AND MESOPOROUS MATERIALS

FREDDY KLEITZ

9.1 Introduction, 244

9.2 Porous Solids, 247
- 9.2.1 Concepts and Definitions, 247
- 9.2.2 Characterization of Porous Solids, 249

9.3 Synthesis Tools for the Formation of Nanostructured Porous Materials, 257
- 9.3.1 Template-Assisted Synthesis, 257
- 9.3.2 Inorganic Polymerization and Consolidation Processes, 260
- 9.3.3 Hydrothermal Syntheses, 262
- 9.3.4 Self-Assembled Structure-Directing Species, 262
- 9.3.5 Cooperative Self-Assembly and Hybrid Interfaces, 264
- 9.3.6 Coassembly of Nanocrystalline Precursors, 266
- 9.3.7 Solid Replication at the Nanoscale (*Hard Templating*), 267
- 9.3.8 Introduction of Guest Species, 269

9.4 Ordered Microporous Materials, 269
- 9.4.1 Zeolites and Zeotypes, 269
- 9.4.2 Metal Organic Frameworks (*Porous Coordination Solids*), 274

9.5 Ordered Mesoporous Materials, 283
- 9.5.1 Ordered Mesoporous Molecular Sieves (M41S Materials), 283
- 9.5.2 Synthesis Pathways for Ordered Mesoporous Materials, 284
- 9.5.3 Ordered Mesoporous Silica, 288
- 9.5.4 Functionalization of Mesoporous Materials, 296
- 9.5.5 Nonsiliceous Mesostructured and Mesoporous Materials, 302
- 9.5.6 Morphology, 309

9.6 Conclusion, 311

Further Readings, 312

References, 313

Problems, 323

Answers, 325

Nanoscale Materials in Chemistry, Second Edition. Edited by K. J. Klabunde and R. M. Richards

9.1 INTRODUCTION

The physical macroscopic notion of porosity or pores in a solid and the phenomenon of absorption of a fluid in a porous object are both quite familiar to all of us. Further, wide varieties of natural or synthetic solids, compounds, species, and materials are known to be of porous nature, for example, minerals, wood, cellulose fibers, seashells,

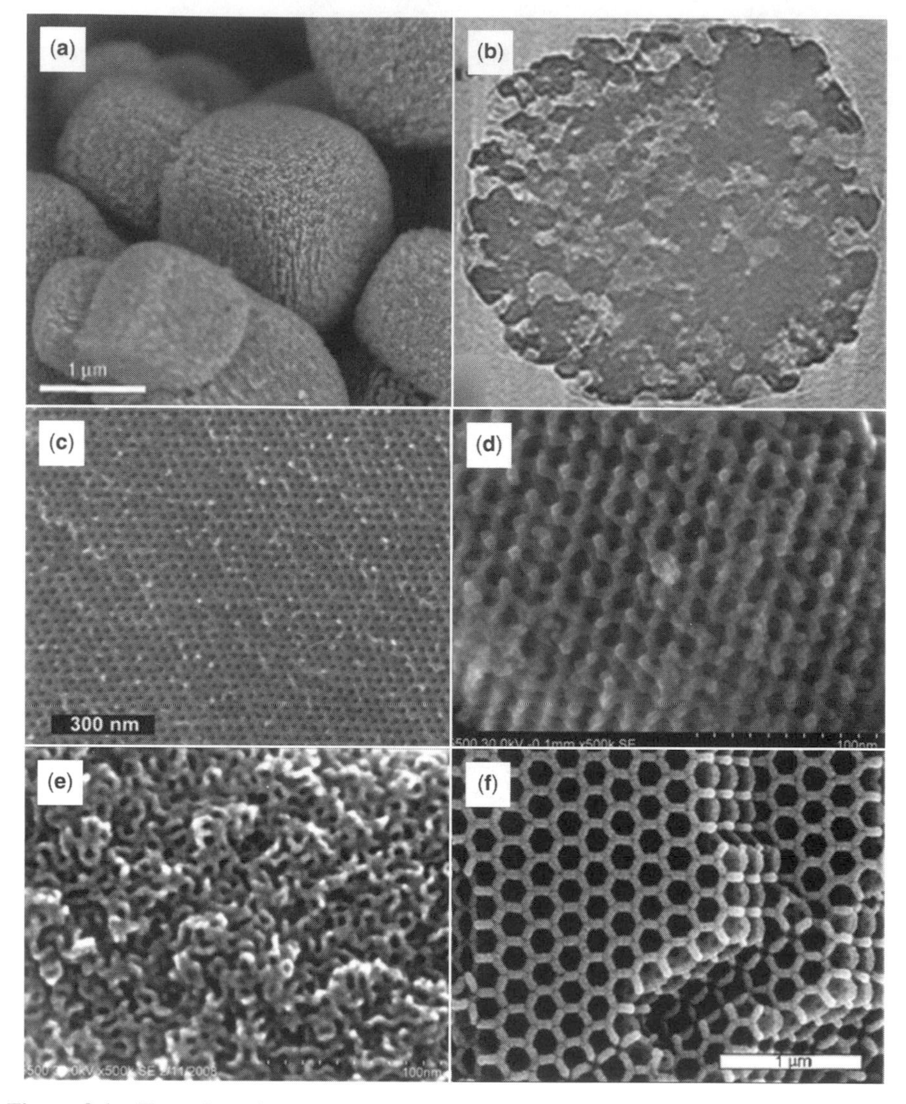

Figure 9.1 Examples of nanoporous materials illustrated by electron microscopy images: (a) SEM image of amphiphilic organosilane-directed mesoporous zeolite crystals (LTA). (Reproduced with permission from M. Choi et al., *Nature Mater*. **2006**, *5*, 718. Copyright 2006 Nature Publishing Group.) (b) ZSM-5 zeolite synthesized with carbon black as a porogen

bones, porous glass, gels, carbons, and inverse opals. The role of porous matter is obvious in nature where soil and sediments, for instance, exhibit complex porosity and often high functionality, whereby the porosity determines the water permeability. In constructions also, the porosity of building materials influences exchange of humidity and thermal insulation. Nevertheless, a rational and scientific approach in the synthesis of porous solids only began to appear in the first half of the twentieth century with the preparation of high surface area materials, such as carbons or silica gels, and the simultaneous development of practical methods to investigate the properties of porous solids, for example, gas adsorption and porosimetry techniques. Nowadays, porous materials became indispensable tools which find diverse applications in separation, catalysis, storage, sensing, or energy conversion. Thus, synthetic porous materials have been applied extensively in the chemical, oil and gas, food, and pharmaceutical industries. Advanced porous ceramics and composite materials are now applied in medical engineering, for example, as bone tissues. However, the size and the volume of the pores in a given material have a dominant influence on the properties of the solid, such as adsorption, diffusion, storage capacity, exclusion, density, mechanical stability, confinement and exclusion effects, and reactivity. The materials classified as nanoporous are very diverse in terms of composition, structure, and porosity. To simplify, a solid may be identified as *nanoporous* if it exhibits a pore system with pore width within the range of 1 to 100 nm. This interval of size is somewhat arbitrary and chosen as it is also a typical range associated with nanoscale materials showing special properties related to their size. Indeed, organized pores of nanoscale dimensions represent a type of *special* nanostructure and as a consequence they confer unique features on these materials (1–3). Many kinds of synthetic porous materials, such as ordered microporous and mesoporous materials, controlled pore glasses, gels, pillared clays, anodic alumina, porous polymers, carbon nanotubes, and so forth, have been comprehensively described in the literature (4–6). In general, one important characteristic is that nanoporous materials are high surface area materials. The only way to generate such materials with the desired high surface area is through a structuring of the solid at the nanometer level. In many cases, this is achieved with synthesis procedures that rely on structuring with the aid of porogen

Figure 9.1 (*Continued*) (2-D section of TEM tomographic reconstruction image). (Reproduced with permission from A. H. Janssen et al., *Microporous Mesoporous Mater.* **2003**, *65*, 59. Copyright 2003 Elsevier.) (c) SEM image of a nanoporous polystyrene-poly(*N*,*N*-dimethylacrylamide) monolithic polymer material. (Reproduced with permission from D. A. Olson et al., *Chem Mater.* **2008**, *20*, 869. Copyright 2008 American Chemical Society.) (d) High-resolution SEM image (HR-SEM) of a mesoporous $Co_3O_4/CoFe_2O_4$ composite prepared by nanocasting. (Reproduced with permission from H. Tüysüz et al., *J. Am. Chem. Soc.* **2008**, *130*, 11510. Copyright 2008 American Chemical Society.) (e) HR-SEM of block copolymer-templated ordered mesoporous silica material with 3D cubic structure (KIT-6). (Reproduced with permission from H. Tüysüz et al., *J. Am. Chem. Soc.* **2008**, *130*, 11510. Copyright 2008 American Chemical Society.) (f) SEM image of macroporous TiO_2 photonic crystals exhibiting a skeleton structure. Polystyrene (PS) opal was used as the template. (Reprinted with permission from F. Marlow and W. T. Dong, *ChemPhys. Chem.* **2003**, *5*, 549. Copyright 2003, Wiley-VCH Verlag.)

additives, which are most frequently either single organic molecules or supramolecular aggregates. In addition, distinct procedures whereby solvents remain trapped within the porous structures are also methods widely implemented, for example, the sol-gel process and solvothermal treatments. Furthermore, although some of the specific properties of the materials will depend on the degree of crystalline order, high surface area materials may be achieved with either crystalline or amorphous structures. Examples of various nanoporous solids are presented in Figure 9.1. Solids such as photonic crystals, porous polymers, and sol-gel materials obtained by phase separation methods often contain larger pores with dimensions in the range of hundreds of nanometers up to several micrometers.

Nanostructured porous materials are now emerging as key elements for the development of future science and technologies, including miniaturized electronic, magnetic or optical devices, environmentally friendly catalysts, materials for pollutant removal, and biocompatible implants and transport systems. Within this context, the nanopore size range offers extremely vast potential to construct elaborated functional systems with tailor-made properties in view of specific applications. As a few pertinent examples of viable applications, one may highlight the use of nanoporous materials as adaptable separation or decontamination media, highly selective sorbents, permselective membranes, systems for energy storage or energy conversion, recyclable catalysts for industrial processes, low *k*-dielectrics, sensors, biomaterials for drug delivery or medical imaging applications, and so forth. Furthermore, such a pore dimension inducing confinement effects allows the restriction of the growth of crystals, the inclusion of quantum objects, the shift of the phase behavior of fluids, the creation of compatible composite interfaces, or the production of environment-sensitive devices. Nanoreactors for performing size- and shape-selective chemical conversions are also being developed on the basis of nanoporous solids. In addition, cooperative and/or complementary chemical processes may eventually be conducted in the confined space of nanopores, acting as spatially functionalized cavities in close analogy with enzymatic active sites.

The content of this chapter focuses on three types of ordered nanoporous solids that are of major scientific and technical interest, namely, zeolites and zeotypes, porous coordination solids, also designed as metal-organic frameworks (MOFs), and ordered mesoporous materials. In many cases, so-called templating pathways are used to synthesize these materials, allowing for a high degree of control over their structural and textural properties. These categories of solids are generally viewed as high performance materials for applications in catalysis, separation, or storage. As specified by Schüth and Schmidt (7), their regular pore system with high surface area gives prospects for the conception of materials with additional functionalities, and thus can be used to introduce guests (e.g. molecules or particles) that are stabilized by the solid framework and spatially organized. In addition, the inorganic or organic-inorganic frameworks of these hosts can exhibit several appealing properties.

Before developing on the different materials in more details, general considerations on porous solids and the main methods to characterize nanoporous systems will be introduced. Applications of microporous and mesoporous materials are addressed in Chapter 10. Other important nanomaterials such as sol-gel materials, porous polymers, carbon nanotubes, organic nanotubes, and photonic crystals are treated elsewhere (2, 5, 8–12).

9.2 POROUS SOLIDS

9.2.1 Concepts and Definitions

As a simple convention, a porous solid can be defined as a solid exhibiting cavities or channels that are deeper than they are wide. It follows that the diversity in porous materials is extreme (4). Solids are known with pore dimensions ranging from small narrow angstrom-sized pores, being voids with molecular dimension, to pore diameters of several tenths of nanometers up to large submicrometer and micrometer pores or even greater, the diversity in composition being almost unlimited (oxides, silicates, clays, metals, ceramics, carbons, composites, polymers, organics, biological, etc.).

Initially, porous materials were defined in terms of their adsorption properties and thus distinguished by the pore size range. Pore size usually refers to *pore width*, that is, the diameter or distance between opposite walls in a solid. According to the IUPAC definition (13), porous solids are then divided into three classes: microporous (<2 nm), mesoporous (2 to 50 nm) and macroporous (>50 nm) materials (Fig. 9.2).

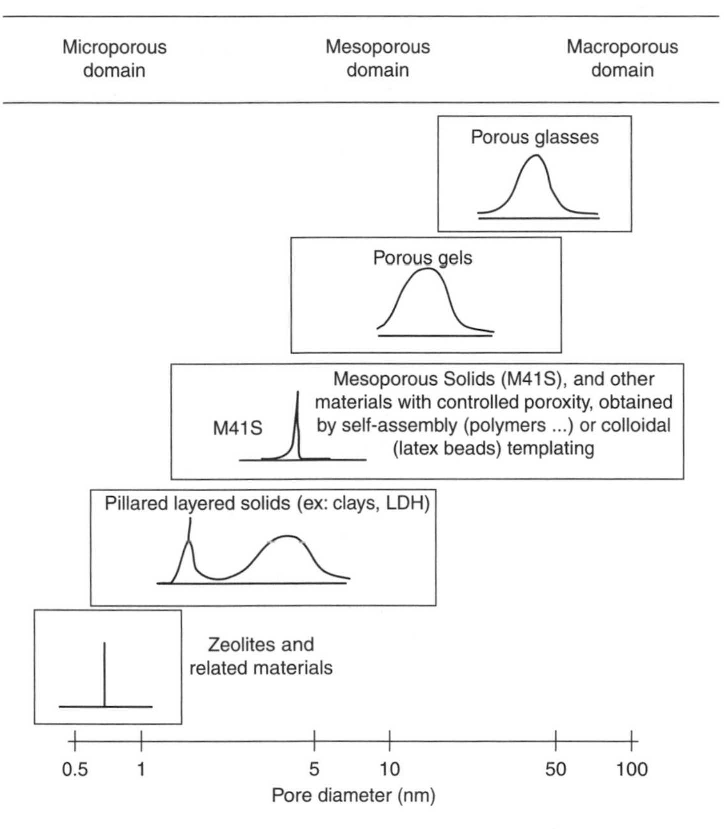

Figure 9.2 Pore size domains for microporous, mesoporous, and macroporous solids, illustrated with examples showing the respective pore size distribution. [Adapted from Behrens (24). Reproduced with permission from G. J. A. A. Soler-Illia et al., *Chem. Rev.* **2002**, *102*, 4093. Copyright 2002 American Chemical Society.]

Additionally, the term *nanoporous*, referring to pores in the nanometer size range (1 to 100 nm), is increasingly being used. Examples of microporous solids are zeolites, which are crystalline aluminosilicates (14–16). Related compounds, called zeotypes, usually possess similar structures with different framework compositions, for example, aluminophosphates and other metal phosphates. Because of their crystalline network, zeolitic materials exhibit an extremely narrow and uniform pore size distribution, justifying their use for size-specific applications in adsorption, molecular sieving, and shape-selective catalysis (17–21). Consequently, zeolites are used to a huge extent in the oil and chemical industries, detergents, and purification and drying systems. In contrast, sol-gel-derived porous oxides are noncrystalline solids that offer advantages in terms of processing, that is, tailoring on a macroscopic size scale to create membranes, monoliths, thin films, aerosols, or fibers (4, 8, 22). These materials have disordered pore systems and exhibit, therefore, a rather broad pore size distribution. However, they present a distinctive advantage with the rich silane chemistry developed for the functionalization of the surfaces. These amorphous porous gels and glasses are most often mesoporous or macroporous solids and are commonly used in separation processes, for example, as stationary phases in chromatography and as catalyst supports.

In addition to pore diameter, also pore shape, pore connectivity, and surface properties are key attributes of a porous material. Pore shape, for example, can be determining in some situations, such as when dealing with shape-selective molecular sieving. In the early studies about porous solids and porosity, ideal regular pore shapes were most often assumed, but later on more realistic pore shapes have also been introduced, such as slits, ink-bottle pores, interstices between globular particles, disordered networks, and fractal geometries. For the following, it is useful to introduce more definitions. Porosity, particularly, is a clearly defined value corresponding to the ratio of the pore volume to the overall volume of particle or granule. Surface area, also, refers to the area of total surface as determined by a stated method. Surface area is usually given as weight-specific surface area, that is, the area of total surface per mass of a given solid (usually expressed in m^2/g). Note that the area of a rough or macropore surface corresponds to the external surface, and the area of micro- or mesoporous walls represents the internal area. Consequently, porosity is differentiated from roughness (for which features are wider than deep). Finally, the notion of tortuosity is also useful and it is attributed to the ratio of path available for diffusion to distance across a porous bed (23).

Methods for creating porosity are very diverse. Most often, species acting as a porogen are introduced in the reaction mixture, around which a solid structure organizes. Porogens can be organic additives, solvents, single molecules or ions, organogelators, supramolecular assemblies, micelles, polymers, etc. In addition, the porogens can consist of solid matter such as colloidal particles, porous inorganic or organic materials, nanoparticles, latex beads, or even biomineral structures that will be used in synthesis procedures based on so-called hard templating or nanomolding strategies. Ideally, a synthetic porous material should have a narrow pore size distribution centered at the required dimension, this feature being critical for size-specific applications, and a readily tunable pore size and pore connectivity allowing flexibility for host-guest interactions. Furthermore, the porous material should exhibit sufficient stability, with high surface area and large pore volume (24, 25). Besides, the materials should possess the suitable functionalities to operate in a desirable manner in the application chosen. Finally, the possibility of using and regenerating the materials in safe, cost-effective, and environmentally friendly conditions should be considered crucial for the development of processes exploiting porous media.

9.2.2 Characterization of Porous Solids

Comprehensive characterization of porous materials with respect to pore size, surface area, porosity, and pore size distribution is normally required to optimize the applications. In general, porous materials are studied by a variety of techniques, including gas adsorption measurements, mercury porosimetry (liquid intrusion-extrusion), adsorption from solution, x-ray diffraction (XRD), small angle x-ray scattering (SAXS), small angle neutron scattering (SANS), electron microscopy (scanning and transmission), solid state nuclear magnetic resonance (NMR) spectroscopy, and thermogravimetric analysis, to list a few. However, each method has a certain length scale of applicability for pore size analysis (26). Among these methods, gas adsorption is probably the most popular because it allows evaluation of a wide range of pore sizes (0.35 nm up to 100 nm), including the full range of micropores, mesopores, and small macropores (23, 27). Gas adsorption measurements are thus very commonly employed to determine specific surface area, pore volume, and pore size distributions of solids. This technique is convenient to use, nondestructive, and is not excessively cost intensive.

Other techniques, such as XRD or SAXS and high resolution TEM are central for the determination of structures, lattice spacings, and unit cell dimensions, and for the elucidation of symmetry, short- and long-range order, and phase. Furthermore, some independent information about pore size and pore wall thickness may also be extracted from TEM or XRD if high-resolution data are available.

A comprehensive review of the different techniques employed to characterize porous materials and guests in nanopores is out of the scope of this chapter. Therefore, in the following, it was the choice of the author to focus on some of the most frequently implemented techniques that are relevant for the characterization of ordered microporous and mesoporous materials considering porosity, structure, and frameworks.

9.2.2.1 Porosity and Gas Physisorption The term *adsorption* described originally the condensation of gas on a free surface as opposed to its penetration into the bulk of a solid, that is, *absorption*. This distinction between the two processes has tended to disappear, and today the uptake of a gas by a porous material is commonly referred to as adsorption (or simply sorption), regardless which actual physical mechanism takes place (25).

Physisorption of probe molecules is used to determine surface area and characterize the pore size distribution of solid catalysts and materials. Basically, the process of adsorption that takes place on a solid surface involves adsorption processes between a solid (adsorbent) and a gas (adsorptive). Note that the word *adsorbate* is distinct from *adsorptive* and corresponds more specifically to the matter in the adsorbed state. Adsorption of a gas, such as nitrogen, argon, or CO_2, by a porous solid is expressed quantitatively by an adsorption isotherm, representing the amount of gas adsorbed at a fixed temperature as a function of pressure. As mentioned previously, porous materials are most frequently characterized in terms of pore diameter derived from the gas adsorption data, and IUPAC conventions have been proposed for classifying pore sizes and gas adsorption isotherms with associated hysteresis loops (13). The first step in the interpretation of a physisorption isotherm is the inspection of the shape of the isotherm. The six types of isotherms, shown in Figure 9.3, are characteristic of adsorbents that are either microporous (type I), nonporous or macroporous (types II and III), or mesoporous (types IV and V). Type VI, being less common, indicates a stepwise adsorption of layers on a highly uniform surface (28). This classification is a useful starting point for the identification of the adsorption and pore filling

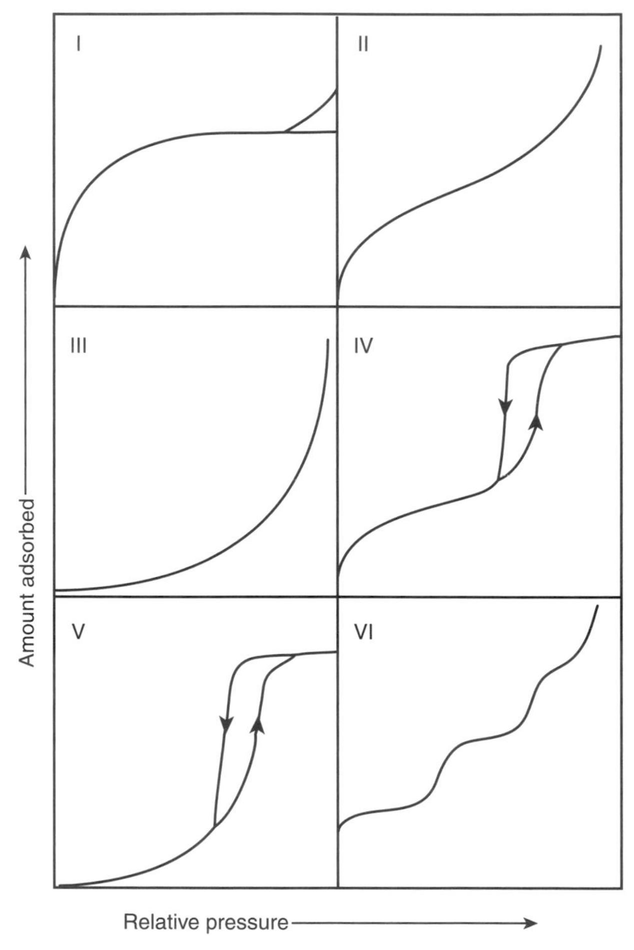

Figure 9.3 IUPAC classification of the physisorption isotherms. (Reprinted with permission from T. J. Barton et al., *Chem. Mater*. **1999**, *11*, 2633. Copyright 1999 American Chemical Society.)

mechanisms and, accordingly, choosing which computational procedures would be more appropriate to give useful quantitative information. The shape of the isotherm gives direct information about the adsorbent-adsorbate interactions (weak or strong interactions), monolayer-multilayer adsorption, filling and emptying of the pores, pore structure (size and shape), and layer by layer adsorption. In particular, the adsorption process in mesopores (type IV) is dominated by multilayer adsorption and capillary condensation, whereas filling of micropores is controlled by stronger interactions between the adsorbate molecules and the pore walls. Another characteristic of the type IV isotherms is the occurrence of H1 or H2 hysteresis loops. The H1 hysteresis loop is indicative of a narrow distribution of rather uniform mesopores with cylindrical-like pore geometry. The more frequent H2 hysteresis loop is typically attributed to percolation effects in complex pore networks, pore blocking effects, or cavitation effects in ink-bottle shaped pores (29, 30).

The concept of adsorption related to an exposed surface was first developed by Irving Langmuir (31). He suggested that adsorption corresponds to a dynamic equilibrium between a gas and a solid surface resulting in a surface layer that is only one molecule thick. Brunnauer, Emmett, and Teller (32) extended this theory and introduced the concept of multi-molecular layer adsorption, if the first adsorbed layer acts as substrate for further adsorption. From their model (the BET model), it is possible to derive the quantity of gas, n_m, necessary for monolayer coverage of the surface (called the BET monolayer capacity), and from this quantity n_m, the specific surface area, S_{BET}, is normally calculated (28). It is important to remember that the range of validity for the BET model is usually within a relative pressure range $P/P_0 = 0.05$ to 0.30 (the range of linearity of the BET plot used to extract n_m; 23, 28, 33). It follows that the values of the BET surface area must be considered with care, since linearity of the BET plot is not always observed. Also, in the case of micropores, multilayer adsorption does not occur, which makes the BET model not valid for microporous materials. As a result, the analysis of the adsorption isotherm through the BET model should only be strictly acceptable for nonporous, mesoporous, and macroporous materials. Nevertheless, BET area values are often reported in the literature for microporous solids (e.g. zeolites or carbon molecular sieves). It should be kept in mind that these figures do not represent the true internal area of these microporous solids, and they should therefore be more regarded as "apparent BET" or equivalent surface areas, which may only serve as fingerprint characteristics. Furthermore, because adsorption in micropores occurs at very low pressures by a pore-filling mechanism, a proper analysis of this region of the isotherm demands very precise measurements of pressures (below 10^{-6} atm) (23, 25, 27, 33).

Pore systems of solids may vary substantially both in size and shape. Therefore, it is somewhat difficult to determine the pore width and, more precisely, the pore size distribution of a solid. Most methods for obtaining pore size distributions make the assumption that the pores are nonintersecting cylinders or slit-like pores, while often porous solids actually contain networks of interconnected pores. To determine pore size distributions, several methods are available, based on thermodynamics (34), geometrical considerations (35–37), or statistical thermodynamic approaches (34, 38, 39). For cylindrical pores, one of the most commonly applied methods is the one described in 1951 by Barrett, Joyner, and Halenda (the BJH model; Reference 40), adapted from the Kelvin equation and taking into account the thickness of a multilayer of adsorbed nitrogen on the pore walls. However, compared to other more recent methods that rely on a localized description and molecular levels of interactions, such as density functional theory, the thermodynamically based BJH model, and other procedures based on a modified Kelvin equation, seem to underestimate substantially the pore diameter (by 20% to 30% for pores smaller than 10 nm; 39, 41). The validity of macroscopic models based on the Kelvin equation is questionable because it does not allow for the effect of surface curvature and adsorption forces on the pore condensation. In addition, other relevant aspects related to pore fluid confinement and hysteresis behavior are overlooked (27, 42). In addition, the BJH model is clearly not applicable for microporous materials, where multilayer adsorption does not take place. The alternative and more accurate way to determine pore size distributions is therefore to apply density functional theory methods (DFT; 38, 39, 43).

Nonlocal density functional theory (NLDFT) describes the configuration of adsorbed molecules in pores on a molecular level and thus provides detailed information about the local fluid structure near curved solid walls as compared to the bulk fluid. As the fluid in the pore is not of constant density, because it is subjected to adsorption forces near the pore walls, the fluid is considered to be inhomogeneous. The DFT method is used to create equilibrium density profiles of the adsorbate confined in a pore at given chemical potential and temperature, for all locations inside the pores. This local density of the pore fluid is obtained by minimizing the free energy potential (grand thermodynamic potential), which includes attractive and repulsive contributions of the fluid-fluid and fluid-wall interactions (23, 27, 38, 39, 43). So, in this approach, the adsorption and desorption isotherms in pores are calculated based on the intermolecular potentials of adsorbate-adsorbate and adsorbate-adsorbent interactions, the latter one (fluid-solid) being dependent on the pore model. The DFT methods were developed for pore size analysis taking into account the particular characteristics of the hysteresis, that is, the pore shape. Nonintersecting pores of different size are assumed to be of the same regular shape (cylinders, slits, or spheres). Correspondingly, pore size distributions are calculated for a given pore geometry, using a series of theoretical isotherms (kernels) for pores of the respective geometry with different diameters. In principle, The NLDFT method may be applied over the complete range of nanopore sizes when suitable kernels are available.

A comparative plot between pore sizes determined by the BJH model and NLDFT is shown in Figure 9.4. It can be seen that the deviation is most noticeable for small pore sizes. However, it should be remembered that both models make the assumption that pores are rigid and of well-defined shape, but in reality the pores could be quite irregular or composed of complex interconnected voids of various sizes. Also the choice of the isotherm branch used in the calculations can affect the extracted values of pore size.

9.2.2.2 Structures The porous solids described in this chapter are periodically ordered materials. Nevertheless, a distinction can be made between the ordering of the atoms, as is observed in the case of crystalline zeolites, for example, and long-range mesoscopic order associated with arrangements of pores, as observed for mesoporous materials such as MCM-41 or SBA-15 silicas (see Section 9.5). In both cases,

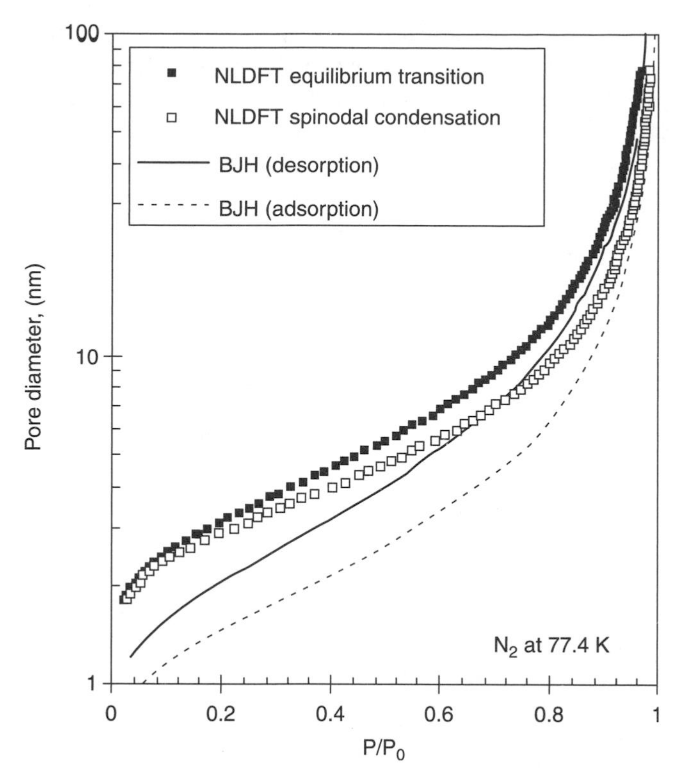

Figure 9.4 Comparison of pore diameters obtained from capillary hysteresis of nitrogen in cylindrical pores at 77.4 K. Equilibrium desorption and spinodal condensation pressures predicted by the NLDFT method in comparison with the results of the BJH method. (Reprinted with permission from A. V. Neimark and P. I. Ravikovitch., *Microporous Mesoporous Mater.* **2001**, *44–45*, 697. Copyright 2001 Elsevier.)

XRD and TEM investigations at different length scales can deliver information on crystal structure, symmetry, periodic lattice spacings, phase purity, arrangement of pores, and other types of nanoscopic organization. These two techniques are thus essential for proper identification and characterization of ordered microporous and mesoporous materials. In addition, scanning electron microscopy (SEM) is most often used to examine the morphology, size, and size distribution of nanoporous particles or crystals.

9.2.2.2.1 Powder X-Ray Diffraction (XRD) When a beam of x-rays of a given wavelength strikes a crystalline solid or a material possessing periodic long-range order constructive interference occurs when Bragg conditions are fulfilled. Structural periodicity on the nanometer scale gives rise to intense reflections arising from constructive interference (reflection is commonly used, although it is a diffraction phenomenon; Reference 44). It follows that sets of characteristic reflections of intensity I for (hkl) planes will be observed in a diffractogram when the glancing angle satisfies Bragg's law. Since the position of the reflections is defined by the symmetry and unit cell parameter of the solid under investigation, information about size and symmetry of the lattice can be obtained from the diffractogram. In addition, the intensity of the reflections is dependent on the scattering strength and position of each atom that composes the periodic solid. The respective scattering strengths are measured by scattering factors, which are related to the electron density and the size of the atoms. The overall intensity of a wave diffracted by a plane (*hkl*) is then given by the structure factor, F_{hkl}. The main use of powder XRD is the rapid and reliable identification of substances. In general, comparison of the position and intensity of the reflections observed with databanks enables a qualitative identification. In addition, quantitative analysis can be performed with the use of an internal standard.

The long-range ordering of periodic mesoporous materials can be evidenced using low angle powder XRD, whereby Bragg reflections can be detected at low values of the 2theta angle (less than 10°). From the positions of these reflections, the spacing between lattice planes can be determined and information about the size and symmetry of the mesoscopic lattice is obtained. In analogy with liquid crystal phases, mesostructured materials exhibit hexagonal, lamellar, or cubic phases. Hexagonal MCM-41 silica, for example, usually shows four or five reflections depending on the sample preparation. Here, the only reflections observed are due to the ordered array of aligned pore walls and can be indexed to a specific space group. Usually, no reflections are observed at wide angles. Figure 9.5 shows a typical low angle diffractogram obtained for silica exhibiting a 3D cubic pore mesostructure.

Often, nanoparticle guests are introduced inside the nanopores of a host material or the framework of the porous material, for example, transition metal oxides or alumina, may be constructed with nanocrystals. It is then possible to evaluate the size of these

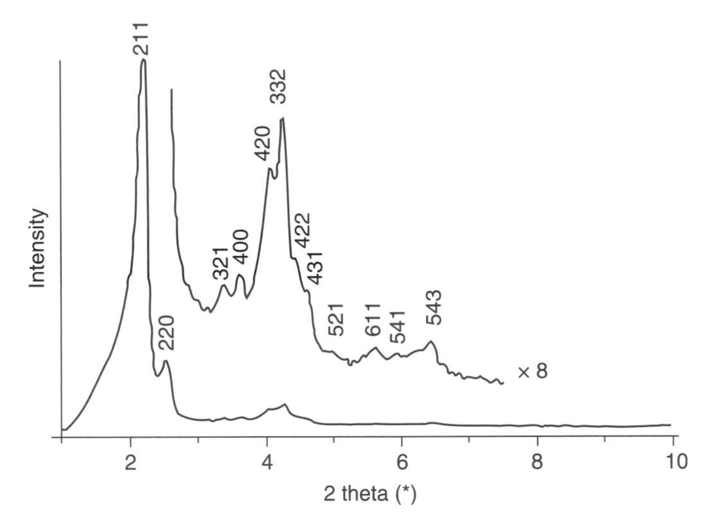

Figure 9.5 Example of typical powder x-ray diffractogram measured for an ordered mesoporous silica. The diffraction pattern shows reflections at low 2 theta angles indexed to the cubic *Ia3d* space group (3D cubic MCM-48 phase). The diffractogram is recorded at low angles with a conventional laboratory instrument.

nanoparticles using the Scherrer equation, which relates the size of the crystallites with the integral width of the reflections (45). Furthermore, the presence and position of the guest species located within the pores or on the pore surface of a porous material may be probed by XRD, since variations in the scattering intensity contrast are observed upon loading with guests, eventually affecting the intensity distribution of the XRD diffraction peaks.

9.2.2.2.2 Electron Microscopy Electron microscopes are instruments that use a focused beam of electrons to examine objects on a very fine scale. This examination gives information on topography (surface features and texture), morphology (shape and size of particles), composition, and crystallographic data (structural arrangements). In order for the electrons to hit the sample, both the microscope column and sample chamber must be kept under vacuum.

Transmission electron microscopy (TEM) and electron diffraction (ED): Transmission electron microscopy provides images of very small structures, and allows resolution down to the angstrom scale depending on the equipment. A transmission electron microscope works much like a slide projector. A beam of electrons, instead of light for the projector, is shone through the sample (specimen) and parts of it are transmitted and projected as a sharp image on a fluorescent screen. Scattered or diffracted electrons from the samples will form the image contrast. The wavelength of the electron beam applied limits the resolution of the microscope. The high energy electron beam provided in high voltage microscope systems allows reaching the domain of the structural resolution, where the resolution distance is in the range of the unit cell size. At present, high resolution electron microscopy (HREM) enables visualization at a quasi-atomic length scale (1 to 2 Å). Electron diffraction and transmission electron microscopy are combined to study crystalline or periodic materials. The results gained from both techniques are complementary, since the formation of an image and diffraction are intimately related. The image of a specimen obtained on a screen or a photographic film by TEM is formed by scattered or diffracted electrons. By varying the focus mode of the projection system of the electron microscope, one can obtain either an image or a diffraction pattern on the screen or film. In imaging mode, one generates either a bright field image or a dark field image depending on the position of the microscope objective aperture (*contrast aperture*). In the dark field image, only beams corresponding to a selective reflection *hkl* contribute to the image formation. The dark field imaging allows working in *diffraction mode* with the formation of an electron diffraction pattern. The resulting pattern contains information about the phases present (lattice spacing) and sample orientation. The electron diffraction pattern can be described as a plane section of the diffracted lattice in the reciprocal space, and Bragg's fringes can be indexed.

TEM analyses are often conducted to study the crystalline organization of microporous zeolites and to visualize the pore structure of ordered mesoporous materials (46–48). Moreover, high resolution TEM allows the examination of the nature and location of guests such as nanoparticles, clusters, coatings, etc. (46, 47, 49, 50).

Scanning electron microscopy (SEM): Scanning electron microscopy is widely employed to observe the surface of bulk samples and is also used to determine the size and shape of particles. A beam of electron is scanned across the sample, ionization occurs near the surface; backscattered and secondary electrons produced can exit the sample and be examined. The image is then formed sequentially on a screen. The resolution of the SEM reaches 10 to 20 Å for high performance equipments. Due to shading and the composition-dependent yield of secondary electrons, topographical and material contrast is obtained. Information about crystal size and size distribution of crystalline microporous samples, such as zeolites and metal-organic frameworks, is routinely collected by SEM. In addition, recent progress with SEM analysis proved also that information about pore structure can be extracted in the case of mesoporous materials when high resolution field emission SEM (FESEM) analyses are performed. Newer instruments based on FESEM use much lower voltages, which allows a higher resolution in the images (51).

9.2.2.3 Framework and Surfaces Since compositions and structures are very diverse, surface and framework properties are also extremely varied. In terms of compositions, coordination, and chemical environments, several methods are particularly informative for the characterization of nanoporous solids, such as nuclear magnetic resonance methods (NMR), UV-visible spectroscopy, Fourier-transform infrared spectroscopy (FTIR), Raman spectroscopy, x-ray absorption spectroscopies, x-ray photoelectron emission spectroscopy (XPS), and electron paramagnetic resonance (EPR) (4, 6). Among them, solid state NMR techniques are largely employed and will be briefly described in the following.

Basically, nuclear magnetic resonance (NMR) is the study of the properties of molecules containing magnetic nuclei by applying a magnetic field and observing the frequency of a resonant electromagnetic field. In the case of solids, NMR is complementary to XRD and affords information on the local structural and chemical environment of atoms. Solid-state NMR is a powerful method for the characterization of crystalline zeolitic structures, the amorphous framework of mesoporous materials, and also facilitates the study of the organization of micellar aggregates or organic molecules confined in an inorganic matrix. However, compared to NMR spectra recorded in solution, a lower resolution is generally observed with broad linewidths for solids originating from the interactions of the spins of the nuclei with the surrounding lattice. These interactions are either direct dipole-dipole magnetic interactions between neighboring nuclear spins, anisotropy of the chemical shift arising from orientation of the static molecule, or quadrupolar interactions when $I > 1/2$. A technique reducing the linewidth in spectra is the magic-angle-spinning method (MAS). The dipole-dipole interactions and the chemical shift anisotropy show a $1-3\cos^2\theta$ dependence. The *magic angle* is the angle at which $1-3\cos^2\theta = 0$, and corresponds to $\theta = 54.74°$. The sample is spun at high speed at the magic angle relative to the applied magnetic field. Rapid motion at this angle averages all dipole-dipole interactions and chemical shift anisotropy to zero (52). Solid-state techniques, such as ^{29}Si MAS NMR and ^{13}C CP-NMR (CP for cross polarization) allow the acquisition of NMR spectra of solids with high resolution and sensitivity.

Solid-state NMR spectroscopy is a useful method for the investigation of chemically modified surfaces of solids. It provides information about the atomic environment of elements that have a permanent nonzero nuclear magnetic moment within

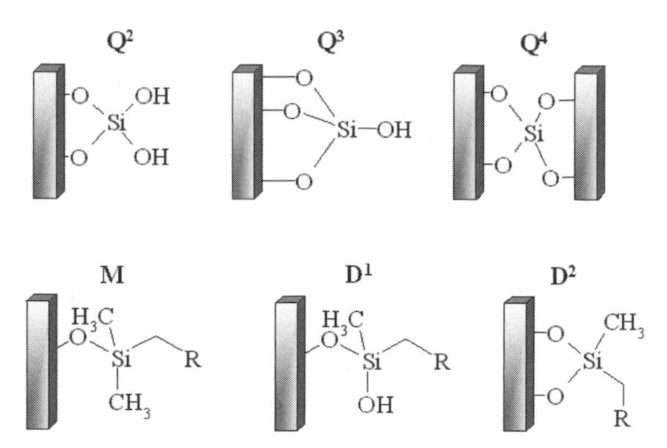

Figure 9.6 Schematic drawing representing different silicon atom species visible in ^{29}Si NMR according to their coordination environment.

the pore walls and at the surface. More specifically, ^{13}C CP NMR provides information on the chemical nature of immobilized organic species. For example, it finds application in the confirmation of whether the desired modification of a surface has occurred. Silicon nuclear magnetic resonance spectroscopy (^{29}Si MAS NMR) is a technique that can provide information about the condensation degree of a silica network and the local coordination of silicon atoms in the pore walls.

Here, the diverse chemistry of silicon is classified according to the number of silanol and siloxane functionalities (Fig. 9.6). Silicon atoms are also referred to as Q^n groups ($n = 1$ to 4), where n defines the number of neighboring silicon atoms. A Q^4 group corresponds to a tetraethylorthosilicate where all four oxygen atoms are bound to another silicon atom, whereas the silicon atom of a Q^2 group is oxygen-linked only to two other silicons. For example, atoms involved only in siloxane bridges (Q^4), single silanol (Q^3), and geminal silanol groups (Q^2) can be distinguished by this method. When the silicon atom is also bound to carbon, they are classified as M^n, D^n, or T^n groups depending on the number of Si–C bonds. From the characteristic peaks in a ^{29}Si NMR spectrum, the ratios between the different Q^n species can be calculated. The degree of condensation of a silica matrix is then often expressed by the ratio Q^4/Q^3, that is the ratio of $Si(OSi)_4$ to $Si(OSi)_3OR$ ($R = H$, CH_3, CH_2CH_3, etc.). This technique is especially helpful to assess the amount of silanol groups (SiOH) present in the materials. NMR methods, and other spectroscopic techniques in general, are used extensively for the characterization of organic functional groups grafted on pore surface or other organic species that may be occluded in cavities.

9.3 SYNTHESIS TOOLS FOR THE FORMATION OF NANOSTRUCTURED POROUS MATERIALS

9.3.1 Template-Assisted Synthesis

One approach to generate materials with tailored uniform pores in the submicrometer size range is based on the use of specific templates or *imprints* following biological models. One of the first successful templating procedures was achieved in 1949 by the use of bioorganic molecules in order to create artificial antibodies (53). The principles of templating have, since then, been adopted for the synthesis of many organic and inorganic materials (54–56). The word *template* is frequently used as a general expression for any species that is used to purposely manipulate the structural or morphological features of a product. In most cases, *templating* comprises the use of a synthesis solution and one or more chemical species, molecule, or assembly of molecules, acting as porogen and directing the formation of the templated phase in a suitable way. The solutions from which the templated solid is formed usually contain precursors that permit some solidification processes to occur, for example, precipitation from solution, sol-gel synthesis, redox processes leading to metal deposition, hydrothermal synthesis, or polymerization of organic monomers. During the solidification process, morphological construction should occur by direct imprinting of the shape and texture of the template. Accordingly, a template can be described as a central structure around which a network will form. Normally, the cavity created after the removal of the template is expected to retain morphological and stereochemical features of the central structure (55, 57, 58).

In fact, when templates are used in the synthesis of porous solids, two different modes in which the template can act may be distinguished: (1) a soft template which is a soluble precursor included into a growing solid particle, either as isolated entity or assemblies of entities (e.g. supramolecular aggregates such as micelles or liquid crystal mesophases), or (2) a solid structure belonging thus to the hard matter domain and providing a scaffold with voids. In the latter case, if the rigid scaffold is removed after filling of its voids, either a high surface area material consisting of isolated small particles or a porous solid network is generated, depending on the three-dimensional connectivity of this template. This mode, often called hard templating or nanocasting will be discussed in detail separately in Section 9.3.7.

Depending on the cases, the dimensions of the templated pores can range from molecular scale, where molecules are used as templates, to the macroscopic scale, where, for example, latex beads are precursors of the pore system. Obviously, the mechanisms by which a pore system develops will be very diverse. However, a favorable interaction between the templating species and the solid to be templated is mandatory, since otherwise the templating species would simply not be integrated and phase separation would occur. The smaller the size scale, the more imperative are such favorable interactions, because the interface between the solid and the template is increased with increasing dispersion (55).

The concept of template-assisted syntheses in relation to organic structure-directing agents became popular in zeolite science in the 1980s, although zeolite beta, which was obtained with the help of an organic additive, had been reported in 1968 (59). Zeolites and zeotype materials are commonly synthesized in the presence of a single small organic template or structure-directing agent, typically quaternary alkyl ammonium ions (60, 61). Amines, linear or cyclic ether, and coordination compounds have also been often used as organic additives. During the synthesis of these materials, the templating species is supposed to remain chemically and thermally stable in the reaction medium, although syntheses exist whereby the template species is actually generated *in situ* (e.g. controlled decomposition of organic precursors). The organic species are often occluded in the microporous voids of the synthesized material, contributing to the stability of the mineral backbone. These stabilizing interactions between guest and framework can be of Coulombic, H-bonding, or van der Waals types (62, 63). The porosity is subsequently created by the removal of the template, usually via thermal treatment or solvent extraction. The term template suggests a perfect correspondence between the shape of the molecule used as a template and the shape of the cavity remaining after template removal. When such a perfect transcription is not proven, it seems preferable, however, to use the term structure-directing agent (SDA) or "organic additive" (55). Even though examples of true templating have been demonstrated, templating processes with exact geometrical correspondence remain rare for zeolites, since the internal cavities usually do not conform rigorously to the molecule shape (Fig. 9.7). In most cases, individual organic molecules do not act as genuine templates but more typically direct structures by participating in the ordering of the reagents, or even simply act as space filling agents occluded in the porous product. Moreover, additional factors such as the influence of the temperature, pH, concentration, stirring rate, and the composition of the reaction vessels can be of importance.

Differently, if mesoporous materials are to be created by a templating pathway, it is not possible to rely on small molecules as templates, since they would not lead to the formation of pores in the mesopore size range. On the other hand, supramolecular assemblies (64) are the result of the association of a large number of small molecular building units into a specific phase exhibiting a well-defined organization, for example, micelles, mesophases, liquid crystalline phases, and polymolecular architectures in layers (65, 66). Consequently, in contrast to zeolite synthesis where single molecules act as templates, the generation of ordered mesoporous materials is made possible via templating pathways using self-assembled supramolecular aggregates of surfactant molecules. In that case, there will be a geometrical correlation between the amphiphilic molecules array size and shape and the final pore size and geometry in the generated mesophase. Different types of organic self-assembling templates are available for creating such nanoporous structures; these can be mainly separated into two principal classes: (1) molecular-based organized systems, including surfactant and organogelators, and (2) polymeric systems such as block-copolymers and dendrimers. Ideally, after removal of the core species from the surrounding matrix,

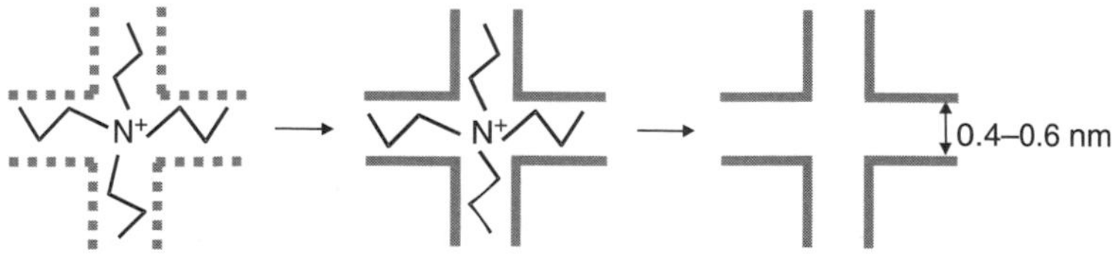

Figure 9.7 Simplified representation of the formation of microporous molecular sieves using a small single molecule organic structure-directing agent. (Adapted and reprinted with permission from T. J. Barton et al., *Chem. Mater.* **1999**, *11*, 2633. Copyright 1999 American Chemical Society.)

the shape of the voids that remain should here reflect the shape of the template. Usually, efficient template removal and faithful imprinting will depend largely on the nature of the interactions between the template and the embedding matrix, and the ability of the matrix to adapt to the template. The intimate template-matrix association required for supramolecular templating of inorganic mesophases is generally facilitated by the flexibility of amorphous inorganic networks with low structural constraints (e.g. small inorganic oligomers), and by large radius of curvature of the organic template. In general, the amount of organic template trapped in a mesoporous matrix can be determined rather easily by thermogravimetric methods. Furthermore, thermogravimetry can be coupled with differential thermal analysis, or differential scanning calorimetry, combined with mass spectrometry or gas chromatography, for example, and associated with separated solid state NMR, FTIR, and Raman spectroscopic analyses to achieve a complete characterization of the occluded species and host-guest interactions (67, 68).

Other texturing agents can be used to create organized porous materials. Colloidal suspension of spherical particles, such as colloidal polystyrene spheres (latex beads) for instance, can yield macroporous structures when applied as an inverse opal template. Even biological structures such as viruses, bacterias, or cells were proposed to obtained structured or textured inorganic frameworks with complex shape, either by using them directly as templates to form a mineral phase, or as hosts in which the growth of inorganic species is restricted (54).

9.3.2 Inorganic Polymerization and Consolidation Processes

The methods employed to prepare nanostructured porous oxide molecular sieves are often similar to the ones commonly used for sol-gel-derived oxides (8). The sol-gel process can be carried out at room temperature and there is a wide range of industrial applications for the materials produced. Optical coatings, fiber glass, and chromatographic column are just a few examples of typical sol-gel-derived materials. Starting from stable sols (colloidal suspensions), inorganic fibers, thin films, powders, and bulk materials (monoliths) exhibiting homogeneous pore structures can easily be prepared. Since silica (SiO_2) is the most simple and most frequently encountered sol-gel system, the sol-gel process will be described here only for silicate species. Compared to transition metals, silicon has a higher electronegativity and has no possibility to exhibit multiple coordination states, resulting in a lower reactivity and higher flexibility of frameworks. Sol-gel processes related to other compositions are treated in References 8 and 69.

For a nonmolecular silicon source, silica is obtained as a gel formed from a nonhomogeneous solution and subsequently treated hydrothermally. Silica can be prepared, for instance, by acidification of a basic aqueous solution, and when the reaction conditions are properly adjusted, a porous silica gel is obtained. Most frequently, two types of chemical reactions are involved: silicate neutralization producing silicic acids, followed by condensation polymerization of the silicic acid species. In the case of molecular silicon sources, solvent and catalyst are usually first combined to form a homogeneous solution to which a silicon alkoxide is then added. In both

$$\equiv Si-OR + H_2O \xrightarrow{\text{hydrolysis}} \equiv Si-OH + ROH \qquad \text{(Eq. 1)}$$

$$\equiv Si-OR + HO-Si\equiv \xrightarrow{\text{alcohol condensation (alcoxolation)}} \equiv Si-O-Si\equiv + ROH \qquad \text{(Eq. 2)}$$

$$\equiv Si-OH + HO-Si\equiv \xrightarrow{\text{water condensation (oxolation)}} \equiv Si-O-Si\equiv + H_2O \qquad \text{(Eq. 3)}$$

Scheme 9.1

cases, the first step for inorganic polymerization is the formation of silanol groups. This occurs either by a neutralization reaction or by hydrolysis of the alkoxysilane. Hydrolysis is generally catalyzed by a mineral acid (HCl) or a base (NaOH, NH_3). Depending on the synthesis conditions, hydrolysis may be only partial or go to completion. In both cases, the hydrolyzed species formed will undergo condensation processes producing siloxane bonds (Si-O-Si) and alcohol or water. This type of condensation reactions leads to formation of oligomeric species, which form chains, rings, or branched structures and continues building larger polymeric silicates (8). Note, here, that the alcohol produced by condensation can participate in reverse reactions on silanols (i.e. esterification) and siloxanes (i.e. alcoholysis or hydrolysis). Scheme 9.1 summarizes the principal reactions taking place in the process.

The overall reaction proceeds as a polycondensation forming soluble higher molecular weight polysilicates. This resulting colloidal dispersion is the *sol*. The polysilicates eventually link together to form a 3-D network which spans the container and is usually filled with solvent molecules, called the *gel*, or precipitate as precipitated silica. Sol-gel polymerization is complex and usually proceeds in a number of overlapping steps: polymerization to form primary particles, growth of the particles, and ultimately, aggregation (8, 22). Each of these steps depends on pH (i.e. the nature and the concentration of the catalysts), temperature, concentration, and cosolvent effects. In basic solutions, particles grow in size with decreasing number, whereas in acidic solution or in the presence of flocculating salts, particles aggregate into three-dimensional networks and eventually form gels.

The reactions that led to gelation do not stop at the gelation point, since there are still oligomers in solutions that are free to diffuse and react. In fact, gradual changes in the structure and properties of the gel continue to take place. This so-called aging of the gel reflects strengthening, stiffening, and shrinking of the gel network. Also, processes of reprecipitation may coarsen the pore structure and phase separation may occur. Changes during aging are categorized as: (1) polymerization, that is, ongoing condensation; (2) coarsening, that is, dissolution and reprecipitation (Ostwald ripening) in order to reduce the net curvature of the solid phase and interfacial area (small particles dissolve and small pores are filled); and (3) phase transformation (8, 22). Aging may be carried out under ambient conditions or at elevated temperatures. In the case of porous silica, the presence of silanol groups in pore walls is usually a result of incomplete condensation during synthesis. The aging of a powder material at a fixed temperature (usually between almost room temperature and 140°C) in its mother liquor is a common procedure that is applied to consolidate inorganic frameworks or induce changes in pore dimension or structure.

9.3.3 Hydrothermal Syntheses

Nanoporous materials can be synthesized by precipitation or gel transformation of inorganic phases conducted during a prolonged period in sealed vessels (autoclaves) heated at elevated temperatures (usually less than 200°C). Under such conditions, pressure is autogenerated. In general, oxide molecular sieves, for example, zeolites and zeotypes, are prepared by hydrothermal synthesis methods, which involve both chemical and physical transformations within an amorphous oxide gel, most often in the presence of template species or SDAs. The gel eventually crystallizes to form a material in which the template species and/or solvent molecules are trapped within the channels and cages of an oxide host framework. The global free energy change, in the case of zeolite synthesis particularly, is usually quite small, so that the product outcome is most frequently kinetically controlled. The porous material is finally obtained on removal of the guest molecules from the oxide framework.

This synthesis procedure is usually called hydrothermal synthesis when aqueous reaction mixtures are used; however, ordered nanoporous materials such as porous coordination solids or metal-organic frameworks are often obtained under similar conditions, but using polar organic solvents instead of water. There, the term *solvothermal synthesis* seems to be more appropriate. For metal-organic frameworks, the precursors are typically combined as a dilute solution in polar solvents such as water, alcohols, acetone, or acetonitrile and heated in sealed vessels such as Teflon-lined stainless steel bombs or glass tubes, generating the autogeneous pressure. In addition to the synthesis of microporous molecular sieves, the synthesis of ordered mesoporous silica materials and some transition metal oxides (e.g. ZrO_2) is occasionally carried out via hydrothermal treatment of reaction mixtures to induce the precipitation of the desired surfactant-containing mesophase.

9.3.4 Self-Assembled Structure-Directing Species

In the case of ordered mesoporous oxides, the templating relies on supramolecular arrays: micellar systems formed by surfactants or block copolymers. Surfactants consist of a hydrophilic part, for example, ionic, nonionic, zwitterionic or polymeric groups, often called the head, and a hydrophobic part, the tail, for example, alkyl or polymeric chains. This amphiphilic character enables surfactant molecules to associate in supramolecular micellar arrays. Single amphiphile molecules tend to associate into aggregates in aqueous solution due to hydrophobic effects. Above a given critical concentration of amphiphiles, called the critical micelle concentration (CMC), formation of an assembly, such as a spherical micelle, is favored. These micellar nanometric aggregates may be structured with different shapes (spherical or cylindrical micelles, layered structures, etc.; Fig. 9.8; Reference 70). The formation of micelles,

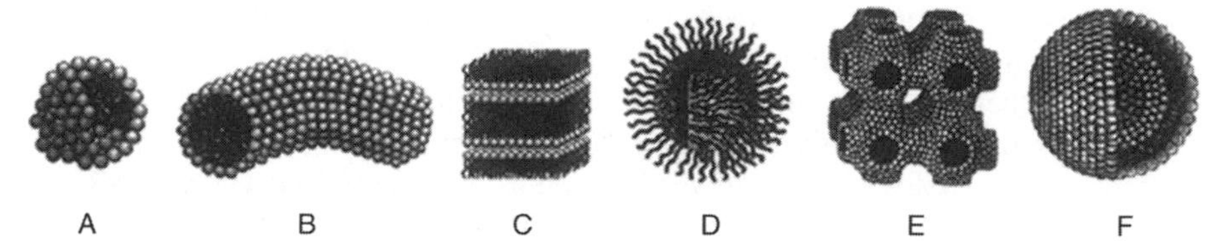

Figure 9.8 Micellar structures (A = spherical micelle, B = cylindrical micelle, C = bilayer lamellae, D = reverse micelle, E = bicontinuous cubic phase, F = vesicular-liposomes). (Reproduced with permission from D. F. Evans and H. Wennerström, eds., *The Colloidal Domain: Where Physics, Chemistry, Biology and Technology Meet*. Wiley-VCH, Weinheim, 1984. Copyright 1984 Wiley.)

the shape of the micelles, and their aggregation into liquid crystals, all depend on the surfactant concentration. At very low concentration, the surfactant is present as free molecules dissolved in solution and adsorbed at interfaces. At slightly higher concentration, that is, the CMC, the individual molecules form small spherical aggregates. At higher concentrations, spherical micelles eventually coalesce to form elongated cylindrical rod-like micelles. These transition concentrations are usually affected by temperature. With further increasing concentrations, liquid crystalline phases eventually form (cubic, hexagonal, or lamellar as the concentration increases). The details of the sequence might vary, depending on the surfactant, surfactant chain length, and surfactant counterion binding strength, but in general the sequence is valid for most of the systems.

The first syntheses of ordered mesoporous materials were traditionally carried out using low molecular weight ionic surfactants, such as cationic alkytrimethylammonium halides (C_nTA^+, with $n = 8$ to 20), but also anionic alkylsulfonates ($C_nSO_3^{2-}$, with $n = 12$ to 18) or long chain alkylphosphates-based surfactants were employed. Although synthesis conditions do influence the structure of the resulting porous solid, the type of surfactant is certainly a dominating factor in directing the formation of a specific structure. In other words, the architecture of the final materials will rely directly on the nature of the surfactant molecules, that is, the morphology of the micellar aggregates and the types of interactions at the inorganic-organic interface. Predictions about the inorganic-surfactant phase behavior can be made based on models developed for dilute surfactant systems. More precisely, the influence a certain surfactant exerts in the synthesis of a mesoporous material can in many cases be rationalized by the packing parameter concept originally developed by Israelachvili. The packing parameter concept (71) is based on a model that relates the geometry of the individual surfactant molecule to the shape of the supramolecular aggregate structures most likely to form (72). This parameter, called g, depends on the nature and molecular geometry of the surfactant, for example, the number of the carbons in the hydrophobic chain, degree of chain saturation, and size and charge of the polar head group. Spherical aggregates are formed preferentially by surfactants bearing large polar head groups. If, on the other hand, the head groups can pack tightly, rod-like or lamellar packing will be favored. In principle, the larger the value of the packing parameter g, the lower is the curvature in the aggregate (73). Changes in micellar curvature may be achieved by altering the surfactant chain length, introducing electrolytes or by adding polar and nonpolar organic additives.

On the other hand, organogelators are low-weight organic molecules that are able to form thermoreversible physical gels in a variety of solvents and at very low concentrations (<1% w/w). The solvent molecules are immobilized, and strongly isotropic structures are formed mainly in the shape of fibers, but also ribbons, platelets, or cylinders. Several examples demonstrate the use of organogels to direct the formation of porous silica and titania fibers (74–76). However, these molecular templates are rarely employed for the preparation of ordered micro- and mesoporous materials.

Concerning polymeric self-assembled templates, two families are most relevant for the design of nanoporous solids. The most important are amphiphilic block copolymers, which belong to an important family of polymeric surfactants, widely used in detergent and emulsifying technologies. Amphiphilic block copolymers are able to self-assemble similarly to their molecular surfactant counterparts in various morphologies (spherical or cylindrical micelles, lamellar structures, hexagonal structures, gyroids, micellar cubic mesophases, etc.; Reference 77). The organized systems formed by these amphiphilic block copolymers are excellent templates for structuring inorganic networks. In general, micellization behavior is driven by hydrophilic/hydrophobic effects, block size, and chain conformation. Furthermore, the size, shape, and curvature of the aggregates formed in solution by these types of amphiphiles is usually determined by the degree of polymerization of each block, the volume fraction of each block, and the degree of incompatibility between the blocks (78). Diblock or triblock copolymers are the most often used, in which hydrophilic blocks (polyethylene oxide, PEO, or polyacrylic acid, PAA) are associated with hydrophobic blocks (polypropylene oxide, PPO or polystyrene, PS, for example). The aggregation behavior of such di- or triblock copolymers is known to not only depend on the block copolymer concentration, but importantly also strongly on temperature (79–81). The self-assembly characteristics of these specific block copolymers can be tuned by adjusting solvent compositions, synthesis temperature, molecular weight, or polymer architecture. They can be used under acidic conditions to produce via the H-bonding-based pathways an assortment of mesoporous solids with large pores, large wall thickness, and various framework structures and compositions (see Section 9.5).

The other groups of polymer templates that may be used are dendrimers, which are multifunctional polymers of very high structural definition and characteristic solubility and viscosity. Although by far less exploited than amphiphilic block copolymers, their high structural definition, coupled with flexibility in size and functions and thermal stability, could make dendrimers very interesting templates. This was verified with the use of dendrimers as building blocks for assembling nanostructured materials (82) and as a template to prepare mesoporous silica and mesoporous titanium oxide (83, 84).

9.3.5 Cooperative Self-Assembly and Hybrid Interfaces

Ordered mesoporous materials typically result from a process of inorganic polymerization during which supramolecular templating is simultaneously proceeding. Most often, however, a so-called *cooperative self-assembly* takes place between the templating species and the mineral network precursors, with synchronized self-assembly and inorganic network formation, yielding highly organized mesoscopic architectures. Thus, the surfactant-inorganic hybrid mesophase forms cooperatively from the species present in solution, which are not in a liquid crystalline state prior to mixing of the precursors. The key feature in the synthesis of mesostructured materials is to achieve a well-defined segregation of the organic and inorganic domains at the nanometer scale. Here, the nature of the hybrid interface plays a fundamental role. The most relevant thermodynamic factors affecting the formation of a hybrid interface have been discussed by Huo et al. (85, 86) in their description of the charge density matching model. The free energy of the mesostructure formation (ΔG_{ms}) is composed of four main terms, which represent, respectively, the contribution of the inorganic-organic interface (ΔG_{inter}), the inorganic framework (ΔG_{inorg}), the self-assembly of the organic molecules (ΔG_{org}), and the contribution of the solution (ΔG_{sol}). In the cooperative assembly route, template concentration may be well below those necessary for obtaining liquid crystalline assemblies or even micelles. Thus, the creation of a compatible hybrid interface between the inorganic walls and the organic templates (ΔG_{inter}) is essential for the generation of a well-ordered hybrid structure with appropriate curvature. From the kinetics point of view, the formation of the organized hybrid mesostructure is viewed as resulting from the balance between organic-inorganic phase separation, organization of the SDA, and inorganic polymerization. Hence, two aspects are fundamental for fine-tuning mesophase formation: the reactivity of the inorganic precursors (e.g. rate of polymerization, pH, and isoelectric point) and the interactions involved to generate the hybrid interface.

A generalized cooperative mechanism of formation was proposed based on the specific electrostatic interactions between an inorganic precursor I and a surfactant head group S. The hybrid inorganic-surfactant mesophases obtained are strongly dependent on the interactions between the surfactants and the inorganic precursors (i.e. the hybrid interface).

In the case of ionic surfactants, the formation of the mesostructured material is mainly governed by electrostatic interactions. In the simplest case, the charges of the surfactant (S) and the inorganic species (I) are opposite under the synthesis conditions. Along with the S^+I^- interaction, cooperative interaction between inorganic and organic species can also be achieved by using the reverse charge matching interactions S^-I^+. With these two direct synthesis routes identified, S^+I^- and S^-I^+, two other synthesis paths, considered to be indirect, also yield hybrid mesophases from the self-assembly of inorganic and surfactant species. Synthesis routes involving interactions between surfactants and inorganic ions with similar charges are possible through the mediation of ions with the opposite charge ($S^+X^-I^+$ or $S^-M^+I^-$). The $S^+X^-I^+$ path takes place under acidic conditions, in the presence of halide anions ($X^- = Cl^-$, Br^-) and the $S^-M^+I^-$ route is characteristic of basic media, in the presence of alkali metal ions ($M^+ = Na^+$, K^+). Besides the syntheses based on ionic interactions, the assembly approach has been extended to pathways using neutral (S^0) (87) or nonionic surfactants (N^0) (88). In the approaches denoted (S^0I^0) and (N^0I^0), hydrogen bonding is considered to be the main driving force for the formation of the mesophase (Fig. 9.9).

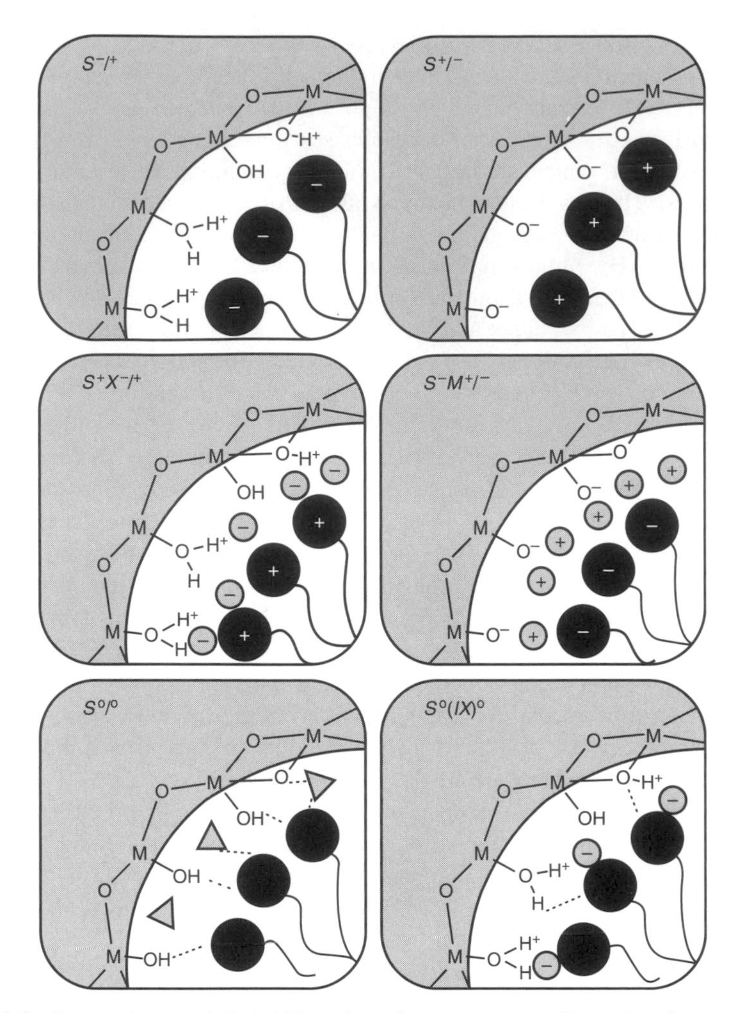

Figure 9.9 Inorganic-organic hybrid interfaces for mesostructure formation. *S* corresponds to the surfactant species and *I* to the inorganic framework. M^+ and X^- are corresponding counterions. Solvent molecules are represented as triangles. See text for details. (Reproduced with permission from G. J. A. A. Soler-Illia et al., *Chem. Rev.* **2002**, *102*, 4093. Copyright 2002 American Chemical Society.)

9.3.6 Coassembly of Nanocrystalline Precursors

One other method for the assembly of open frameworks is based on the use of preformed nanobuilding blocks (NBB). In this approach, a nanometric inorganic component is formed first by inorganic polymerization or precipitation reactions. These resulting NBB can subsequently be assembled and linked by organic connectors taking advantage of functional groups hanging on the particle surface and adequate volume fraction of the different building units. Here, the ability to control the dynamics of the precipitation of the nanometric building blocks is the key feature when syntheses are carried out under these conditions. For non-silica systems especially, the control of the hydrolysis and condensation usually involves obtaining dense metal-oxo-clusters as precursor species. Interestingly, the formation of nanoparticles can take place not only in solutions but also inside micelles, vesicles, or emulsion systems, eventually leading to materials with relatively complex shapes. This pathway is exploited successfully to generate nonsiliceous hybrid porous networks; a comprehensive account on the subject can be found in reviews by Sanchez and coworkers (89, 90).

9.3.7 Solid Replication at the Nanoscale (*Hard Templating*)

As an alternative to surfactants, the template may also be a (solid) material with structural pores in which another solid is created, serving thus as a rigid scaffold. A sort of nanoreplication is thus realized using solid molds. In this method, a porous solid material is used as a rigid matrix, that is, its pores are filled with one or more precursor species that will react inside the pores to form the desired material. The matrix is subsequently removed to yield the product as its negative replica. In analogy to macroscopic techniques, this process can be termed *nanocasting*, that is, casting on the nanometric length, or *nanomolding*, the porous matrix being the mold and the replicated product being the (nano)cast. In this nanocasting approach, the template is used as a true mold, with characteristic dimensions in the nanometer range.

Using this strategy, disordered porous carbons were obtained in the mid-1980s using silica particles as templates to produce a polymer, which was then carbonized (91). Since then, hard templating approaches have afforded a large variety of periodically ordered materials, such as carbons, noble metals, metal oxides, mesoporous zeolites, polymers, chalcogenides, or nitrides. Ordered mesoporous silica and aluminosilicate materials are now serving extensively as porous structure matrices (55, 92, 93). Ordered mesoporous carbon materials are also increasingly being utilized as hard templates, although their synthesis often involves a first hard templating step starting from porous silica (see Section 9.5.5.3). In addition to these mesoporous hard templates, plenty of other structures can be exploited for nanocasting, such as zeolites, carbon molecular sieves, porous polymers, silica spheres, silica or carbon porous monoliths, colloidal crystals, colloidal core-shell particles, and latex beads (10, 55, 92, 94–97).

Rigid templates may be very diverse, but a relatively general procedure for nanoreplication can be given. In the first step of a structure replication process, the pores of the matrix are filled with the precursor for the desired product, for example, sucrose or other polymerizable organic compounds for carbon, or metal salts such as nitrates for metal oxides. This is usually realized by impregnation with a precursor solution, either by a "wet impregnation" or by the "incipient wetness" technique. The interaction of the precursor species with the pore walls is crucial, and it may involve hydrogen bonding, coordination of metal ions (e.g. by silanol groups), Coulomb interaction, and van der Waals forces. Other approaches to efficiently infiltrate the precursor species into the pores of the matrix include vapor-phase infiltration and infiltration in the liquid/molten state in absence of solvent. After the infiltration of the precursors and, if needed, the subsequent removal of the solvent, the formation of the wanted product is typically accomplished at elevated temperatures.

Depending on the pore network connectivity of the templating solid, different materials are formed after removal of the scaffold, as illustrated schematically in Figure 9.10. A solid that consists of one fully continuous solid phase and one fully continuous pore system (i.e. a bicontinuous solid) will usually lead to the formation of a porous solid that is a negative of the original template. Conversely, a solid that does not have a continuous pore system, but rather many different disconnected pores, will lead preferentially to the formation of a finely divided high surface area solid (small particles), in which possible pores are rather textural, that is, there are voids between individual particles (55). Porous activated carbon, for example, can be employed as a template to generate varieties of high surface area oxides. In this

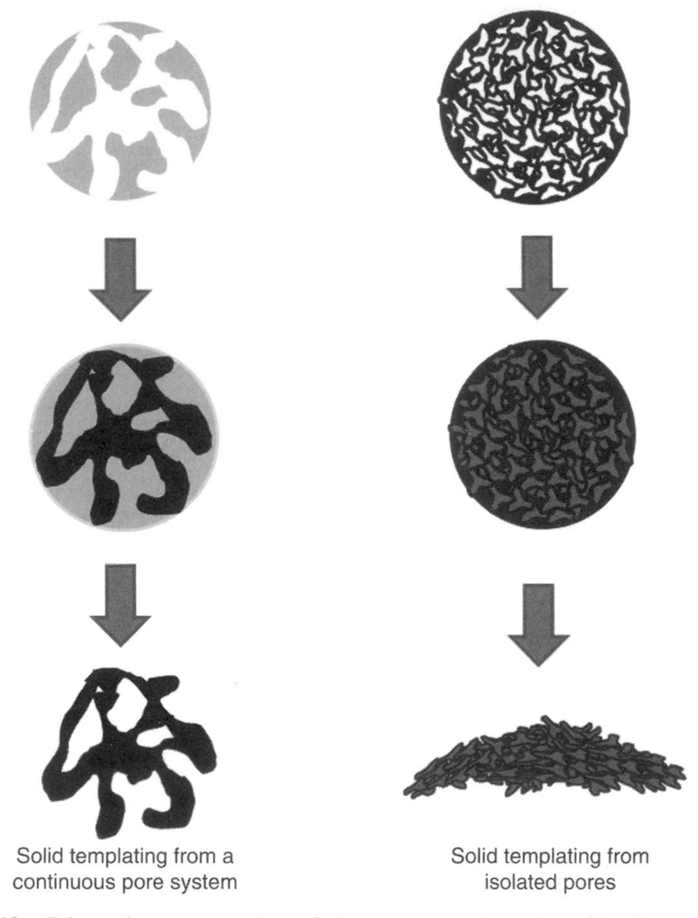

Figure 9.10 Schematic representation of the two general modes of hard templating, as described by Schüth. *Left*: In the case of a rigid template exhibiting a continuous pore system and with high loading of precursors, a continuous porous solid is then obtained with voids corresponding to the original template. *Right*: If the pores of the template are not 3D interconnected and/or filling of the pore system is not sufficiently high, then small particles are obtained with high surface areas and textural porosity. (Reproduced with permission from F. Schüth, *Angew. Chem. Int. Ed.* **2003**, *42*, 3604. Copyright 2003 Wiley-VCH Verlag.)

case, the synthesis process proceeds by impregnating an activated carbon with concentrated aqueous salt solutions or neat liquid alkoxides, drying, and subsequent combustion of the carbon. Several oxides with high surface areas can be synthesized using this approach, which is also suitable for preparing complex ternary oxides such as spinels and perovskites (98).

The nanocasting method has several interesting advantages when compared to direct soft-templating strategies. First, many ordered porous materials simply cannot be synthesized by direct soft templating. For example, only a limited range of mesoporous metal oxides and metals has been prepared successfully by utilization of amphiphilic structure-directing agents. Second, the formation of a product is possible at much higher temperatures. In some cases, high temperature treatments are necessary to increase the degree of crystallinity, form a specific crystalline phase, graphitize, or increase the stability of the materials. High temperatures can be reached during the nanocasting procedure, thus enabling the synthesis of crystalline products without losing the mesostructure ordering; the rigid matrix skeleton preventing the collapse of the pore system (93).

9.3.8 Introduction of Guest Species

Briefly, multitudes of pathways are available to introduce guest species inside the nanopores of microporous or mesoporous solids. Most frequently, functionalities are added to the materials by the incorporation of active sites inside the inorganic walls or by deposition of species on the inner surface of the pores. Guests such as clusters, nanoparticles, organic groups, oxide layers, polymers, and biomolecules, may be included *in situ* during synthesis, or subsequently, by postsynthesis procedures such as adsorption, ion exchange, impregnation, or assembly within the pore system. These pathways for introducing guests may be coupled with some reactions of species with the framework, for example, on surface silanol groups. Electron microscopy, gas adsorption, powder XRD, and solid state NMR are methods that can provide a relatively straightforward assessment of the effect of the inclusion of guests or functionalities on the porosity and other properties of the parent nanoporous host. In addition, procedures to modify nanoporous materials in a sequential and spatially controlled way are emerging. Several strategies have been proposed to permit spatial control of the functionalization and selective positioning of active sites (99–102). In this regard, XRD and electron microscopy are some of the methods adapted for the analysis of spatial modifications. XRD patterns, for example, are sensitive to the positioning of guest species in the pore system and high resolution TEM provides a direct visualization of the guests (103, 104).

9.4 ORDERED MICROPOROUS MATERIALS

9.4.1 Zeolites and Zeotypes

9.4.1.1 Description and Frameworks An important number of natural solids are characterized by an inorganic framework that contains cavities, cages, or channels in which water or inorganic cations are occluded. Among these solids, zeolites (from the

Greek, *zein*, to boil, and *lithos*, stone) represent a large family of microporous tectosilicates having pore sizes smaller than 1.4 nm. Zeolitic materials are constructed of negatively charged aluminosilicate host frameworks that are sufficiently porous to accommodate a variety of different countercations (as charge compensating ions) and, in many cases, guest molecules that can be reversibly adsorbed and desorbed. Zeolites have considerable impact as catalysts and adsorbents in the chemical and petroleum industries. Although mostly aimed at industrial catalytic processes, zeolites are also present in everyday life, for instance in phosphate-free cleaning products, filters, or isolating materials. These applications are possible because these solids are very efficient as selective ion-exchange agents and sorbents, owing to their well-defined porosity and high mobility of water and cations (17, 18).

The details of the structures and compositions of the different zeolites have been extensively discussed in the literature (see Further Reading and References for detailed surveys). Thus, only a brief description will be provided in the following. Zeolites exhibit three-dimensional framework structures with uniformly sized pores of molecular dimensions, typically ranging from 0.3 to 1 nm in diameter, and pore volumes ranging from 0.1 to 0.35 cc/g. The formal composition of aluminosilicate zeolites is considered to be $M_{2/n}O$, Al_2O_3, $xSiO_2$, yH_2O, where M is a mobile cation of valence n and $x \geq 2$. Zeolites are usually composed of TO_4 tetrahedra (with T = Si, Al, Ga, etc.) and each of the oxygen atoms is shared between adjacent tetrahedra. In the case of purely siliceous zeolites, each oxygen atom at the corner is shared by two SiO_4 tetrahedra and the overall charge is balanced in the absence of defects. When silicon is locally replaced by heteroelements, such as aluminum, charge compensating ions, such as K^+, Na^+, or Ca^{2+} and protons, H^+, are necessary. The size of the pores and dimensionality of the channel system are normally determined by the arrangement of the TO_4 tetrahedra. Precisely, the size of rings that are formed by connecting various numbers of TO_4 units is determining the pore size. An 8-member ring corresponds to a ring comprised of 8 TO_4 units and is considered to be a small pore opening (ideally, 0.41 nm in diameter), a 10-ring is then a medium one (0.55 nm), and a 12-ring is a large one (0.74 nm). Depending on the arrangement of these various rings, different structures and pore openings, such as cages, channels, chains, and sheets, are formed.

Nowadays, the term zeolite includes all microporous solids based on silica and exhibiting crystalline walls, as well as materials where a fraction of Si atoms has been substituted by another element, T, such as a trivalent (T = Al, Fe, B, Ga, ...) or a tetravalent (T = Ti, Ge, ...) metal. Crystalline microporous phosphates are known as zeotypes or as "related microporous solids" (14, 54). At present, there are 179 confirmed zeolite framework types. For the structure types, three-letter codes are used, which were adopted from the name of the first material reported with a specific structure. As an example, FAU is given for the structure of faujasite and its synthetic equivalents X and Y, and MFI for the structure of ZSM-5 or silicalite-1 (105). Figure 9.11 shows prominent examples of zeolite frameworks, for example, FAU, LTA, and MFI types (pentasil).

Besides the vast exploitation of zeolites for catalysis and separation, the zeolite cages structure can host and confine different guest species in an environment and

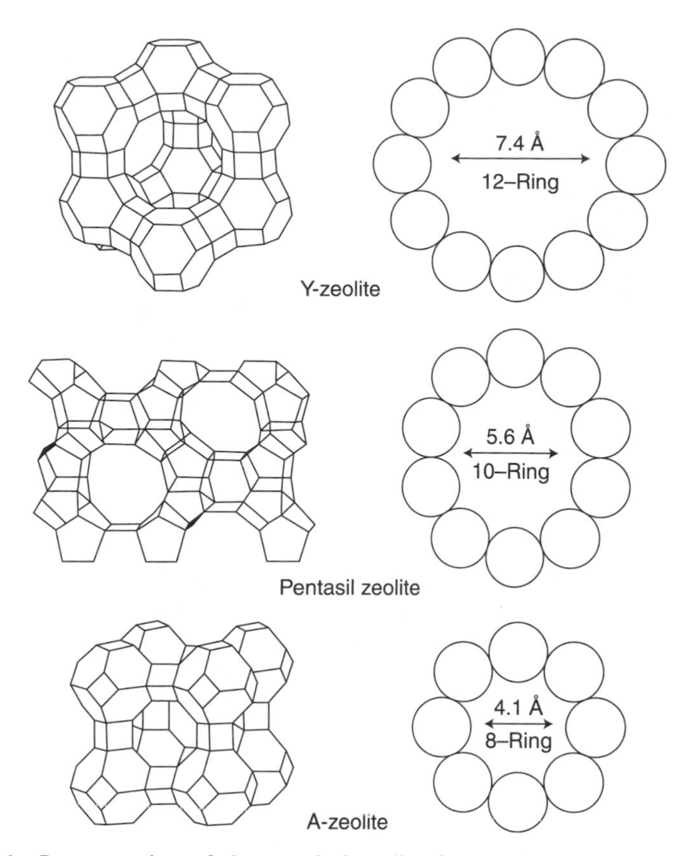

Figure 9.11 Representation of three typical zeolite frameworks: FAU (zeolite Y), MFI (ZSM-5 and silicalite-1), and LTA (zeolite A), shown with their respective 12-, 10-, and 8-member rings.

state that can result in modified chemical and physical properties of the guest. There are consequently many examples of zeolites and zeotypes that have been loaded with various active species, such as dyes, chromophors, or quantum dots, leading to host-guest materials with interesting optical and electronic properties (106). Also related to this strategy is the concept of catalyst design by inclusion of transition metal coordination complexes within the framework pores of zeolites, that is, the so-called "ship-in-bottle" approach (107 and references therein). This is nicely illustrated by the example of a cobalt phthalocyanine complex which was housed within the framework of zeolite Y. Owing to spatial restrictions, the occluded guest molecule appears relatively distorted, thereby inducing an increase of its reactivity and hydrophobicity. In addition, there is a noticeable enhancement in the magnetic moment of the complex upon encapsulation compared to its value in the free state. In another example, the dimer of copper acetate hosted within zeolite Y is shown to be in a compressed state with the Cu–Cu distance being significantly reduced from its normal value. These two examples are illustrated in Figure 9.12.

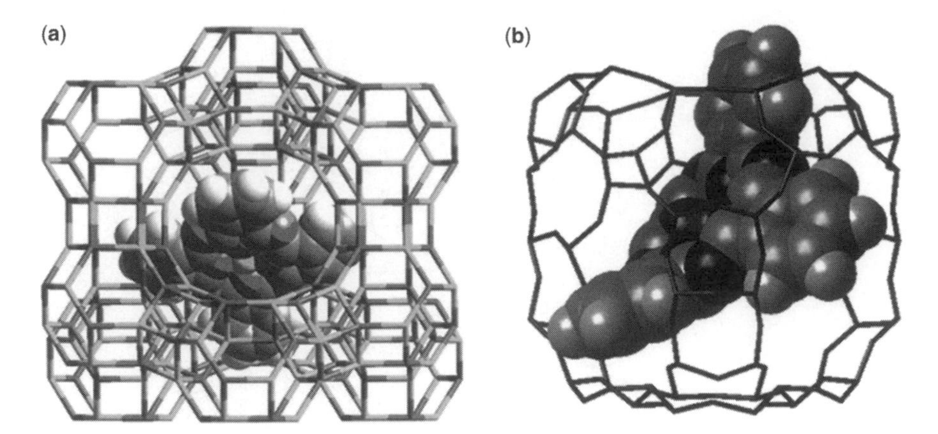

Figure 9.12 Representation of the "ship-in-bottle" inclusion of a metal complex within the large cage of zeolite Y. (a) Picture representing a cobalt phthalocyanine confined in zeolite Y. Because of spatial restrictions, the guest molecule is distorted, resulting in an increase of its hydrophobicity and reactivity. (Reproduced with permission from J. M. Thomas et al., *Angew. Chem. Int. Ed.* **2005**, *44*, 6456. Copyright 2005 Wiley-VCH Verlag.) (b) Molecular graphics representation of the cobalt phthalocyanine complex placed within the supercage of zeolite Y and having the minimum energy geometry in saddle-type distortion (shown is the orientation with minimum interaction energy with the walls of the supercage). (Reproduced with permission from S. Ray and S. Vasudevan, *Inorg. Chem.* **2003**, *42*, 1711. Copyright 2003 American Chemical Society.)

9.4.1.2 Synthesis of Zeolites and Structural Design The first synthesis procedures involved aluminosilicate gels as precursors and strong alkaline conditions, using various mineral bases. A great variety of novel structures have then been obtained by introducing organic templating molecules (generally N-based molecular compounds) into the reaction mixtures. Nowadays, crystalline molecular sieves are generally obtained by hydrothermal crystallization of a heterogeneous gel, which consists of a liquid and a solid phase. Typically, the reaction medium contains a source of the species building the framework (Si, Al, P, etc.), mineralizing agents (hydroxide or fluoride ions), mineral cations or soluble organic species (cations or neutral molecules), and solvent (usually water). The possible sources of silica, are numerous and include colloidal silica, fumed silica, precipitated silica, and silicon alkoxides. Typical alumina sources comprise sodium aluminate, aluminum nitrate, aluminum hydroxide, boehmite (AlOOH), and alumina. Zeolite syntheses are sensitive to many parameters, namely, reagent type, compositions, sequence of addition, mixing, and crystallization temperature and time. By varying the chemical composition of the gels, the reaction conditions, and the templating agent, it is thus possible to achieve a great range of molecular sieve compositions.

A typical hydrothermal zeolite synthesis can be summarized as follows: Amorphous precursors containing silica and alumina are mixed together with a cation source, usually in alkaline media. This aqueous reaction mixture is then heated (at $T < 200°C$), most often in a sealed autoclave. For some time at the fixed synthesis temperature, the reactants remain in an amorphous state; this is called the "induction period." After that, crystalline zeolite products begin to be detected, and gradually the amorphous material is replaced by zeolite crystals. Finally, the zeolite crystals are recovered by filtration, washed, and dryed. During the hydrothermal reaction in the presence of the mineralizing agent (most commonly an alkali metal hydroxide), the crystalline zeolite product containing Si-O-Al linkages is formed. The anions such as hydroxide or fluoride aid in the dissolution of the reactive silica species in the gel and their transfer to the growing crystals. The overall zeolite synthesis process occurs with numerous complex chemical reactions and organic-inorganic interactions taking place. Comprehensive accounts of zeolite nucleation and crystal growth mechanisms are found in the specialized literature (62, 63, 108, 109).

As discussed earlier in this chapter, zeolites and related microporous crystalline solids are synthesized in the presence of organic additives, usually single organic molecules, acting as templating or structure-directing species for the network formation. These species ultimately reside within the intracrystalline voids. However, there actually are several different ways by which an additive may really operate in a zeolite synthesis (54, 55), and the additive may be

- Acting as a space-filling agent occupying the voids between the framework constituents. In this case, the agent will contribute to the energetic stabilization of a more open structure as opposed to the creation of a denser structure.
- Preorganizing solution species to favor the nucleation of a specific framework structure.
- Acting as a true template, with the framework being formed around the organic molecules which determine shape and size of the voids in the structure.

Aiming for better insights into the different possibilities, Gies and Marler (110) have carried out an extensive study in which many different organic additives were used in the synthesis of all-silica molecular sieves. In such a case, charge interactions are absent and van der Waals contact between the organic molecules and the zeolitic material is considered to be the governing factor. Under these conditions, it was shown that the shape and size of the molecule used in the synthesis does indeed control the shape and the size of the cavity or channels generated in the framework. It was demonstrated that guest molecules with similar shapes and volumes, regardless of their differences in chemical properties, lead to the same types of silica framework structures, that is, globular guest molecules lead to frameworks with cages or cavities, linear chain guest molecules, such as straight chain alkyl amines, lead to one-dimensional channel-type pore systems, and branched chain guest molecules produce intersecting channel pore systems. Moreover, a linear correlation was established between the size of the guest and the size of the cage.

An example of true templating in zeolite synthesis is the preparation of the MFI-type materials mentioned above, for example, ZSM-5 (with Si/Al ratio varying between 10 and 500) or silicalite-1 (the pure siliceous form), in the presence of tetrapropylammonium ions. The MFI structure is characterized by the presence of two interconnected channels, the opening of which is delimited by 10-ring units. In this case, the structure of the as-made material shows that the tetrapropylammonium cation is located exactly in the intersections between the straight and sinusoidal channels of the framework. The four propyl groups extend into the four channels and are linked to the nitrogen atom at each intersection (55, 111). Also, when distinct organic molecules are included in an otherwise identical synthesis mixture, zeolites with totally different crystal structures may be formed. For example, when *N,N,N*-trimethyl 1-adamantammonium hydroxide is used, a zeolite that was named SSZ-24 is obtained, while ZSM-5 is produced by using tetrapropylammonium hydroxide. Clearly, here, the geometry of the structure-directing agent has a direct impact on the geometry of the zeolite synthesized. Another interesting case is SSZ-26, which is a zeolite with intersecting 10- and 12-ring pores, and synthesized by *a priori* design using a specific propellane-based structure-director. It was established experimentally and by theoretical calculations that the geometry of the pore sections of SSZ-26 fits perfectly with that of the organic SDA molecules and one of these molecules is present at each intersection (112). Nevertheless, one should also keep in mind that, in addition to templating or structure direction, organic species may also affect the pH value and complexation equilibria in the synthesis reaction mixture. Furthermore, it should be pointed out that the inclusion of quantities of cations, such as Al or Zn, to zeolite synthesis mixtures can also play a role in directing a particular structure, most likely by inducing changes in framework charges and bond angles and lengths, and therefore produce significantly different zeolite structures although identical structure-directing agents could be used (113).

Modifying pore size and pore shape of these microporous materials is thought to be promising to allow for a control of the transport of species into and out of the porous medium. Several strategies for zeolite design based, for example, on tailoring special templates, controlling ring sizes (e.g. by addition of Ge in the synthesis gel), and adjusting conditions for crystal growth are currently being developed, and new zeolitic materials with optimized framework structures and crystal dimensions are obtained (114). The latest developments in the field of zeolite science also include the designed synthesis of nanosized zeolite particles, shape and morphology control of zeolite crystals, and modern templating strategies for producing zeolites with mesoporous transport pore systems (97, 115, 116).

9.4.2 Metal Organic Frameworks (*Porous Coordination Solids*)

The publication in the 1990s of the synthesis and properties of the materials designated as MOF-2 and MOF-5 by Yaghi and coworkers opened up the door for major development in the science of nanoporous solids (117), although a synthesis of what would be categorized as a metal-organic framework was previously reported in 1965 (118). Metal-organic frameworks (MOFs) are highly porous nanostructured materials, in which metal ions act as coordination centers and are linked together with a range of polyatomic organic bridging ligands. In essence, MOFs are obtained

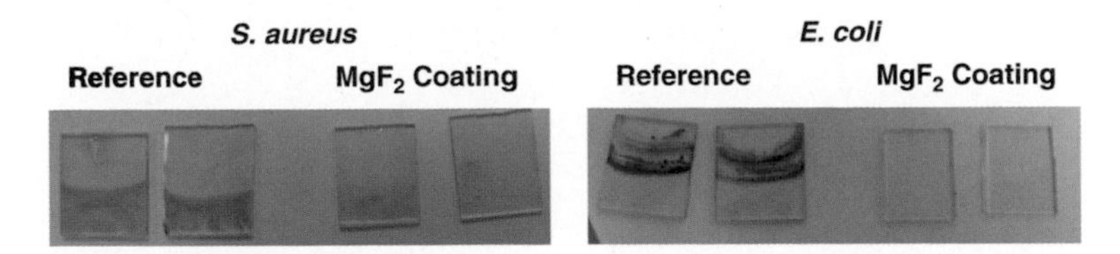

Figure 4.18 Photos of MgF_2 NPs coated and noncoated glass coupons after overnight incubation with *S. aureus* and *E. coli*. Biofilm biomass was observed by staining with crystal violet. High biomass (i.e. blue staining) is observed mainly in the air–liquid interface of the untreated glass surface. This is not seen in the MgF_2-coated glass coupons.

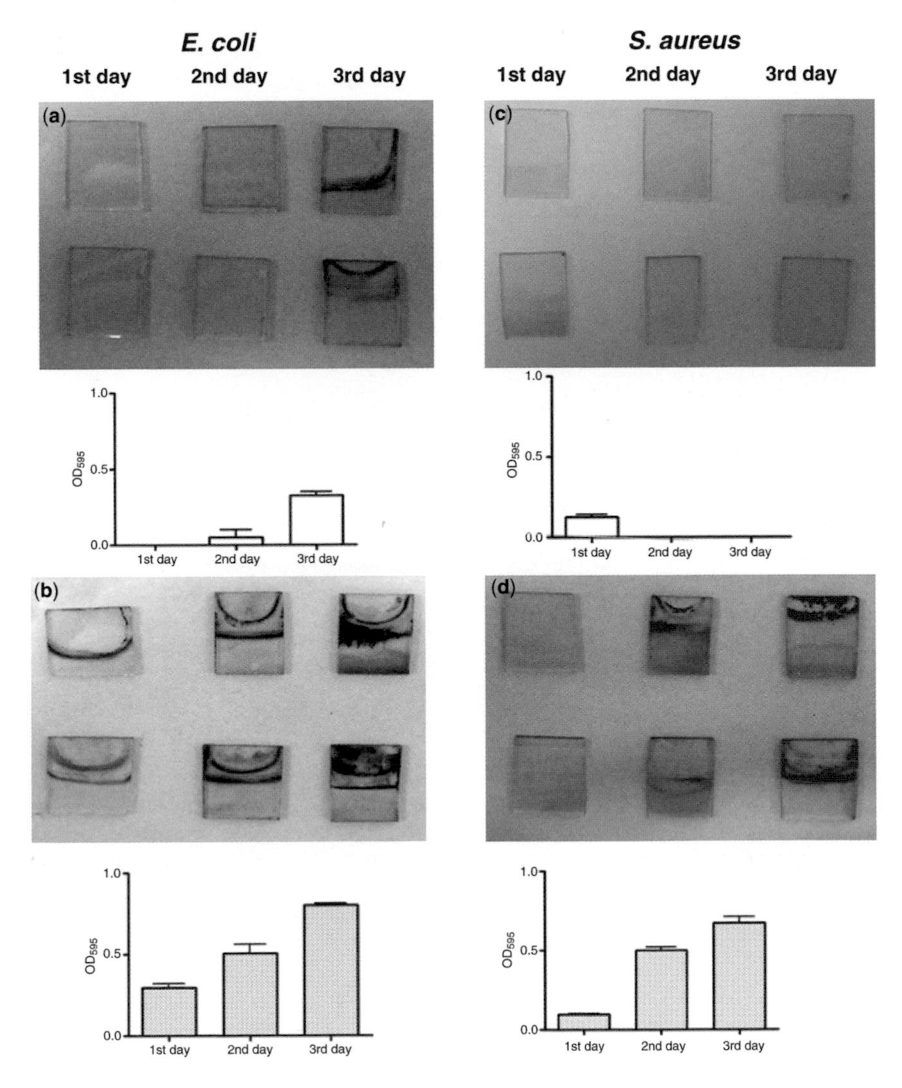

Figure 4.21 Extended bactericidal and antibiofilm activities of the MgF_2 NPs coating on glass toward *E. coli* and *S. aureus* strains. Crystal violet biomass staining of the biofilms formed after 24 h, 48 h, and 72 h on *E. coli* and *S. aureus* strains on MgF_2-coated surface (a) and (c) compared with control biofilm formation (b) and (d).

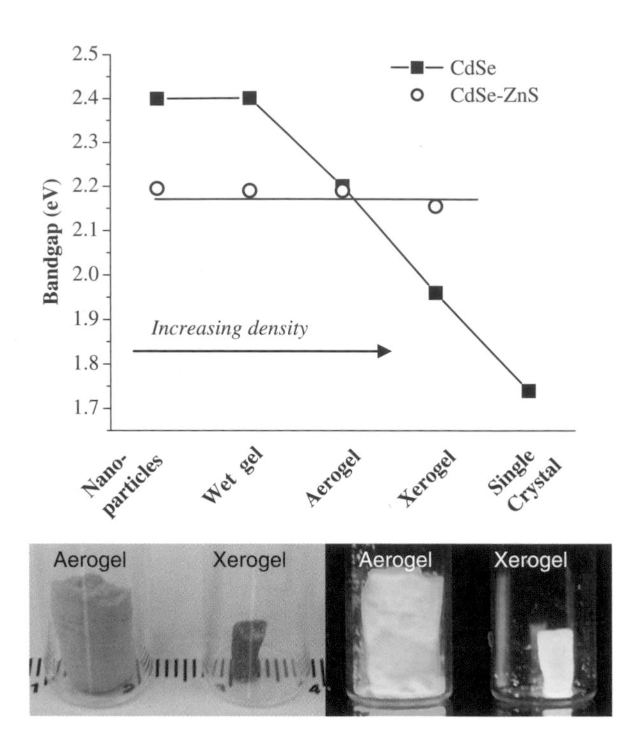

Figure 8.7 Top: Effect of density on the optical bandgap for aerogels prepared from naked CdSe nanoparticles and CdSe-ZnS core-shell nanoparticles. Bottom: CdSe-ZnS core-shell aerogel and xerogel monoliths under normal (left) and UV (right) illumination. The native aerogel and xerogel are orange under normal light and fluorescent green under UV stimulation.

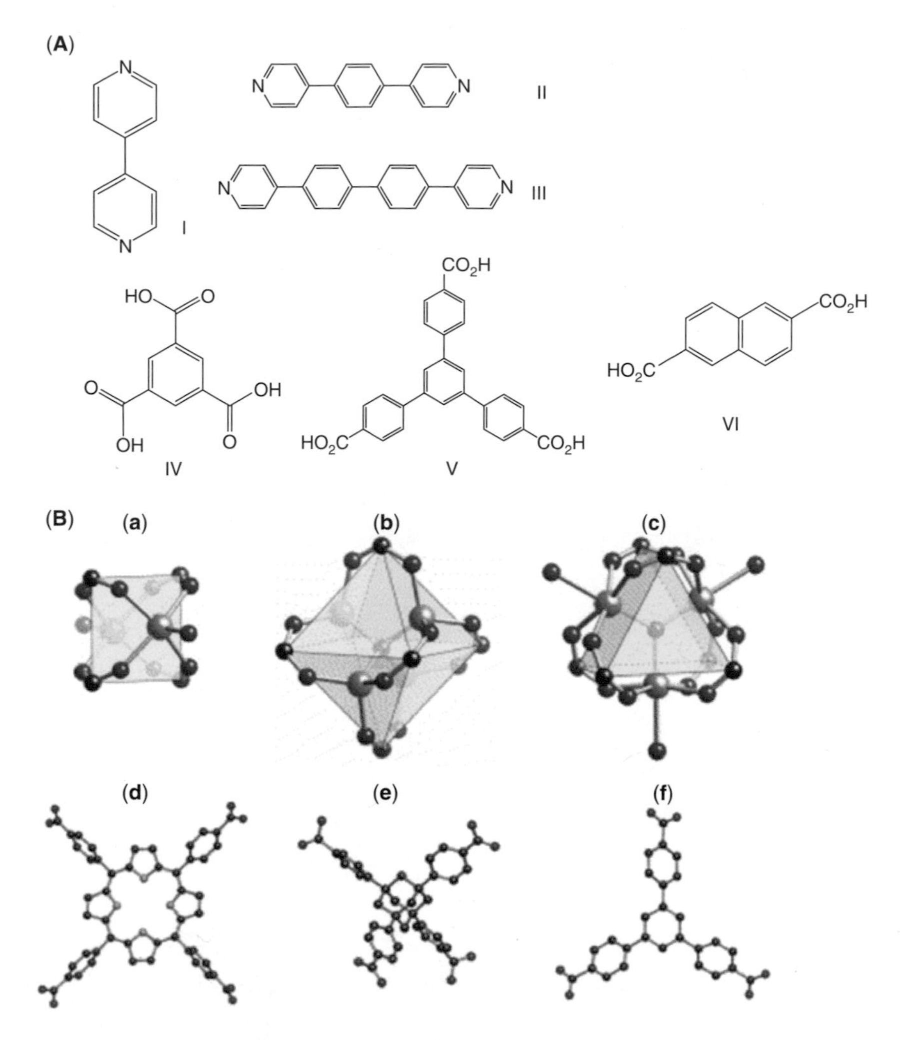

Figure 9.13 Examples of typical ligands and secondary building units (SBUs) involved in the formation of metal-organic frameworks (MOFs). (A) Classes of ligand used for framework construction: 4,4′-bipyridyl ligand (I) and analogs (II) and (III), 1,3,5-benzenetricarboxylate ligand (IV) with the extended analog (V), and 2,6-naphthalenedicarboxylic acid (VI). (B) *Top*: Some inorganic SBUs most commonly found in metal carboxylate MOFs: (a) square paddlewheel, with two terminal ligand sites, (b) octahedral basic zinc acetate cluster, and (c) the trigonal prismatic oxo-centered trimer, with three terminal ligand sites. The SBUs are reticulated into MOFs by linking the carboxylate carbons with organic units, but can also be linked by exchange of the terminal ligands. *Bottom*: Some examples of organic SBUs comprise the conjugate bases of (d) square tetrakis(4-carboxyphenyl)porphyrin, (e) tetrahedral adamantane-1,3,5,7-tetracarboxylic acid, and (f) trigonal 1,3,5-tris(4-carboxyphenyl)benzene. The metals are represented as blue spheres, carbon as black spheres, oxygen as red spheres, nitrogen as green spheres.

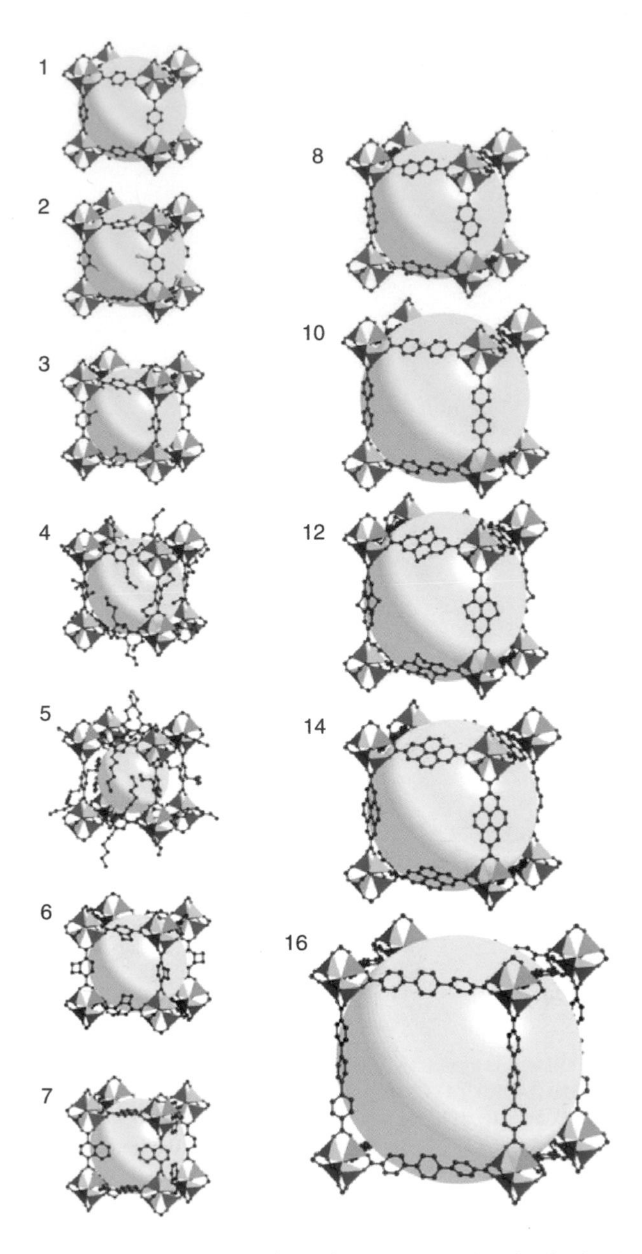

Figure 9.14 Single-crystal x-ray structures of IRMOF-*n* ($n =$ through 7, 8, 10, 12, 14, and 16), labeled respectively. Color scheme is as follows: Zn (blue polyhedra), O (red spheres), C (black spheres), Br (green spheres in 2), amino groups (blue spheres in 3). The large yellow spheres represent the largest van der Waals spheres that would fit in the cavities without touching the frameworks.

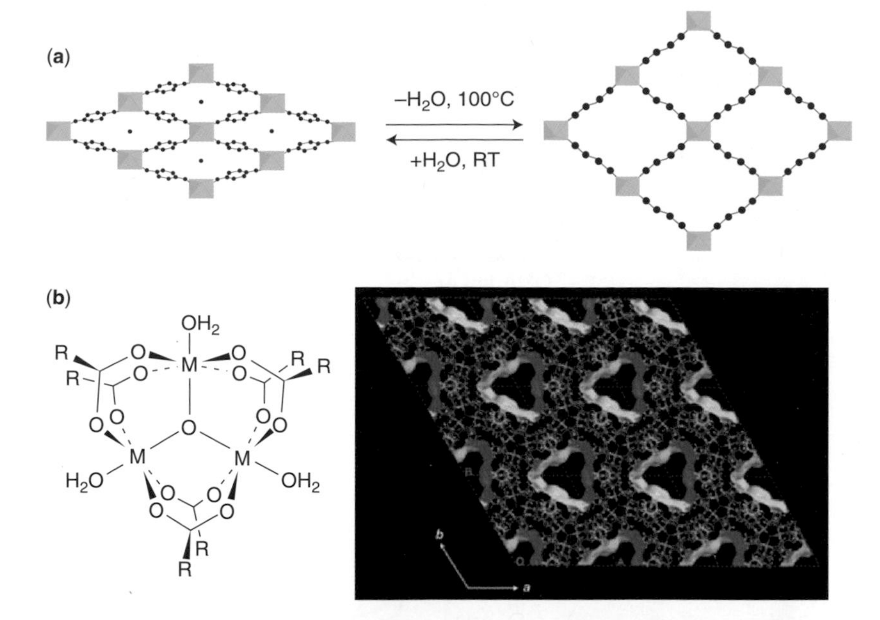

Figure 9.16 (a) Schematic illustration of flexibility in the MIL-53-type metal-organic framework. Shown is the breathing process observed upon reversible hydration and dehydration of MIL-53 (Al, Cr) with change in temperature. (b) Homochiral metal-organic porous material. *Left*: The oxo-bridged trinuclear metal carboxylates (M = divalent or trivalent transition metal ions, O_2CR = organic carboxylate anions). *Right*: The chiral pore system of the open-framework material *D*-POST-1. The view is down the c-axis showing the large chiral channels. The accessible surface in the channels is shown in blue.

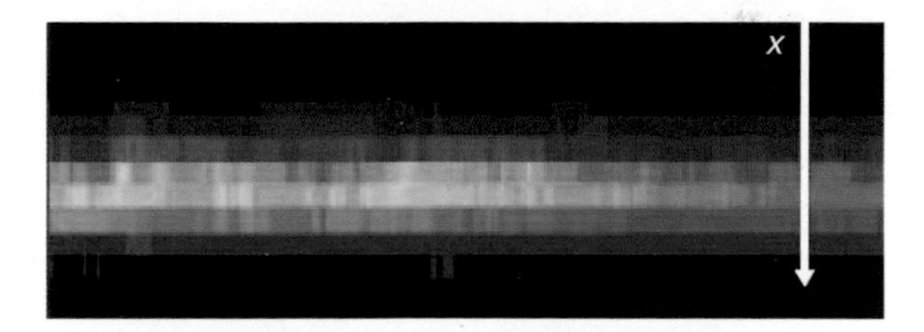

Figure 11.19 Rainbow layer-by-layer film of ten layers each of green, yellow, orange, and red CdTe nanoparticles alternating with polyelectrolytes.

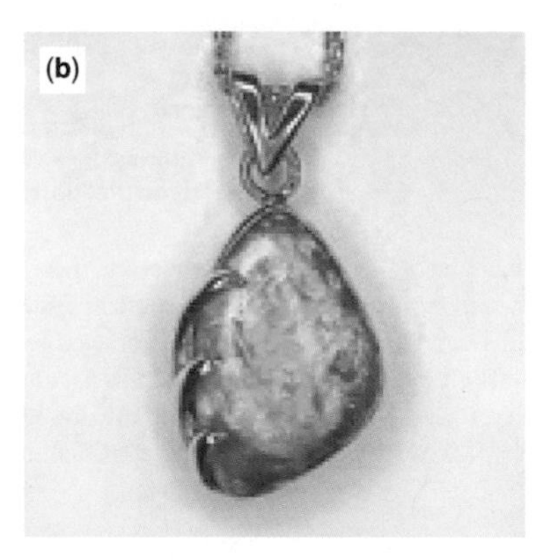

Figure 11.22 A natural opal (http://www.costellos.com.au/opals/types.html).

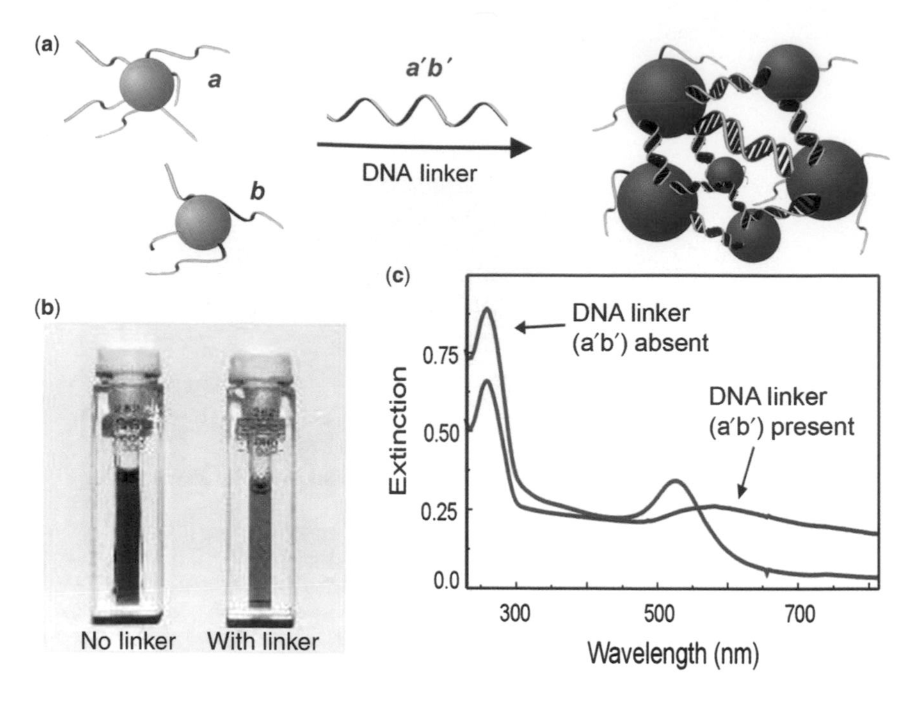

Figure 12.1 (a) DNA-AuNPs assembled using a complementary DNA linker. (b) 13 nm DNA-AuNPs appear red in color without linker DNA and turn blue when linker DNA induces nanoparticle assembly. (c) Extinction spectrum of dispersed and assembled DNA-AuNPs.

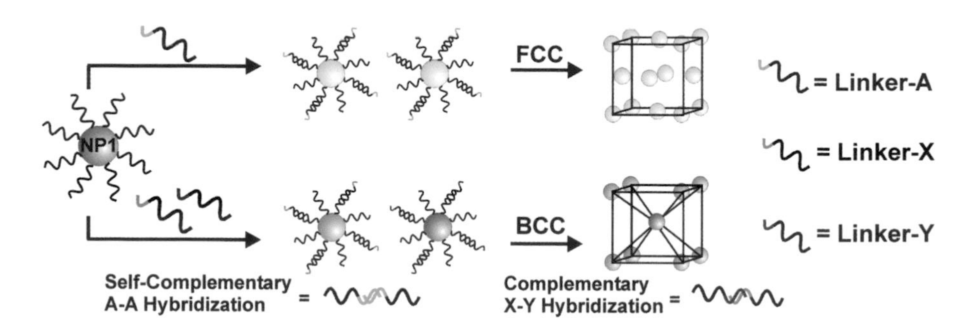

Figure 12.3 The green overhang of linker A is self-complementary; therefore, DNA-AuNPs hybridized to linker A are self-complementary and behave as a single component system. Maximizing the number of particle-particle interactions results in a close-packed structure depicted by an FCC unit cell. The red and blue overhang of linker X and linker Y are complementary and DNA-AuNPs hybridized to linkers X and Y are complementary and behave as a binary system. Maximizing the number of particle-particle interactions results in a non-close-packed structure depicted by a BCC unit cell.

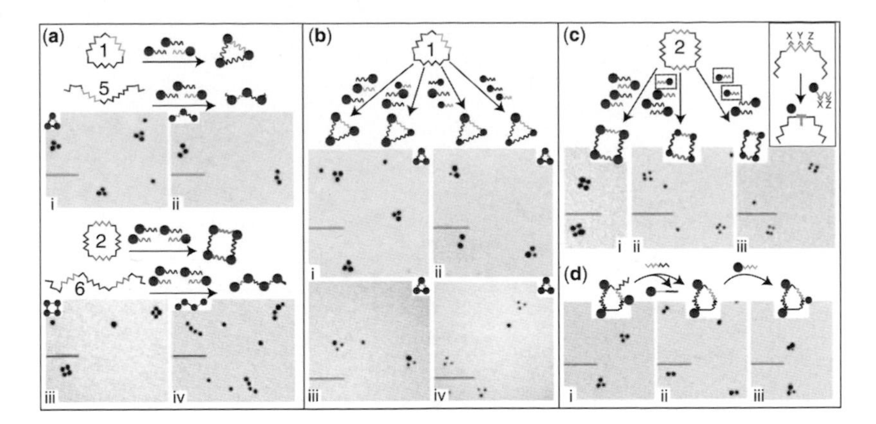

Figure 12.6 (a) **1** and **2** organize Au particles into triangles and squares; **5** and **6** result in open linear assemblies of three and four particles. (b) **1** generates triangles of (i) three large (15 nm, red), (ii) two large/one small (5 nm, purple), (iii) one large/two small, and (iv) three small particles. (c) **2** assembles four Au particles into (i) squares (15 nm particles), (ii) trapezoids, and (iii) rectangles (5 nm). Inset: use of a loop shortens the template's arm. (d) Write/erase function with **1** by (i) writing three Au particles (15 nm) into triangles, (ii) removal of a specific particle using an eraser strand, and (iii) rewriting with a 5 nm particle. Bar is 50 nm.

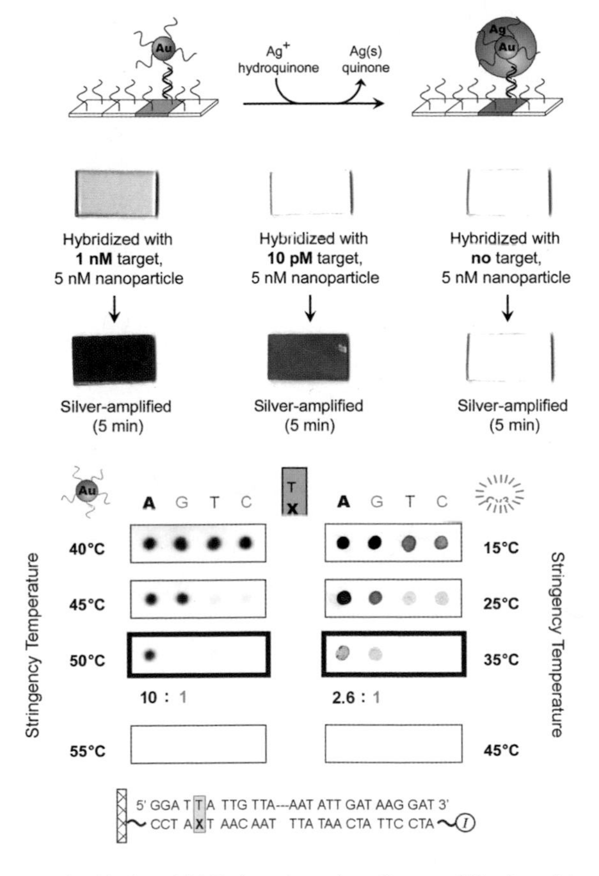

Figure 12.9 Scanometric chip-based DNA detection using silver amplification of the detection signal is illustrated on the left. A direct SNP detection/discrimination comparison of DNA-AuNP probe versus fluorophore-labeled probes is shown on the right. The DNA-AuNP system shows a selective advantage of approximately 4 : 1.

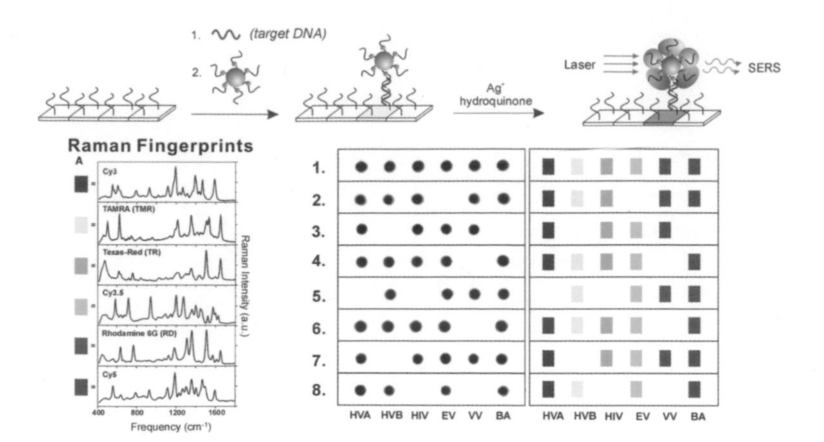

Figure 12.10 Raman-encoded multiplexed detection of six DNA targets using DNA-AuNP, where each Raman spectrum, or color, corresponds to a specific target of interest. Note that after silver staining, all the spots turn black, which makes it impossible to see the difference of the targets. By using SERS, however, the spots can be scanned with a single wavelength excitation laser and observed with dye (and thus, target)-specific Raman spectra.

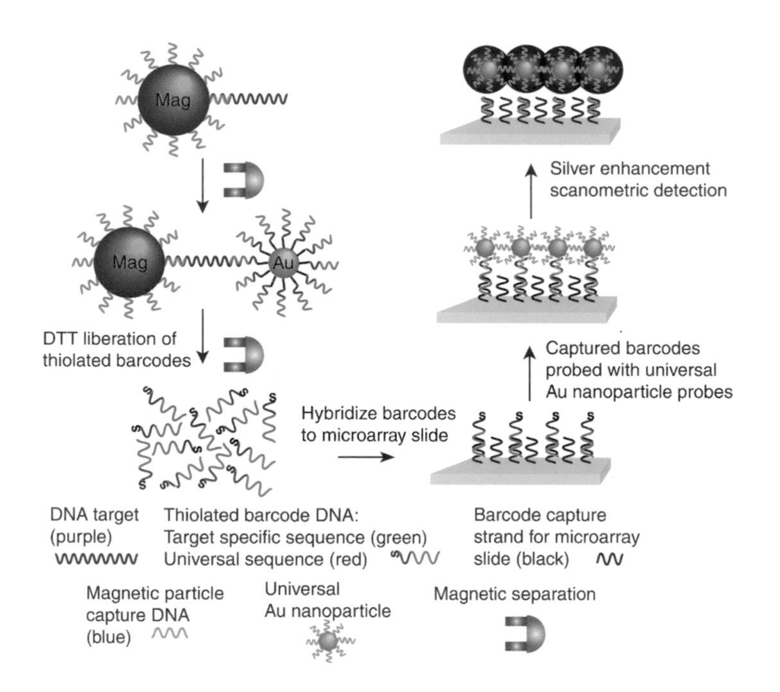

Figure 12.12 Biobarcode assay for DNA detection. Magnetic microparticles (Mag) hybridize to specific targets. AuNPs functionalized with target-specific barcode strands are then added. After magnetic separation, the barcodes are released from the surface of the AuNPs by DTT. Subsequently the barcodes are detected using the scanometric method.

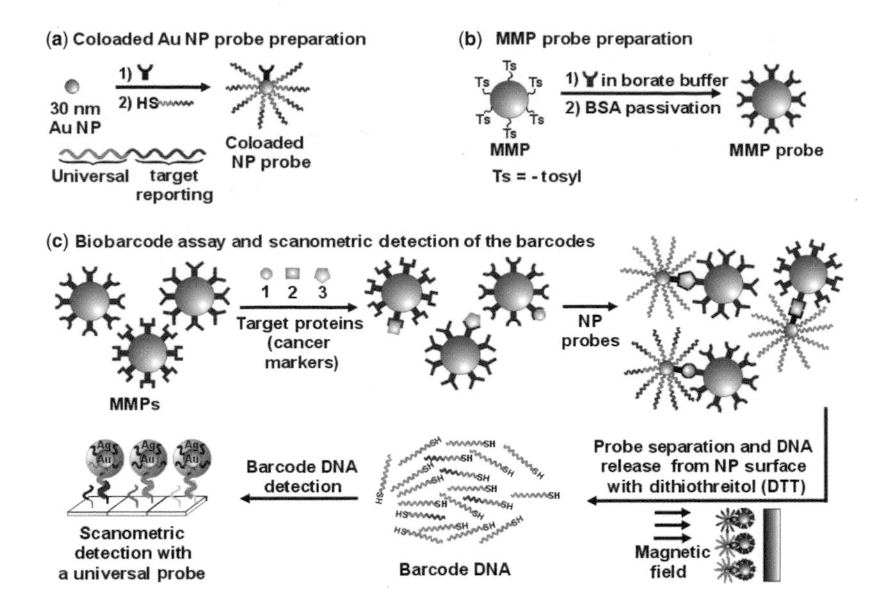

Figure 12.14 Schematic representation of the biobarcode assay for multiplexed protein detection. The general concept for multiplexed DNA detection is similar, with the appropriate recognition elements.

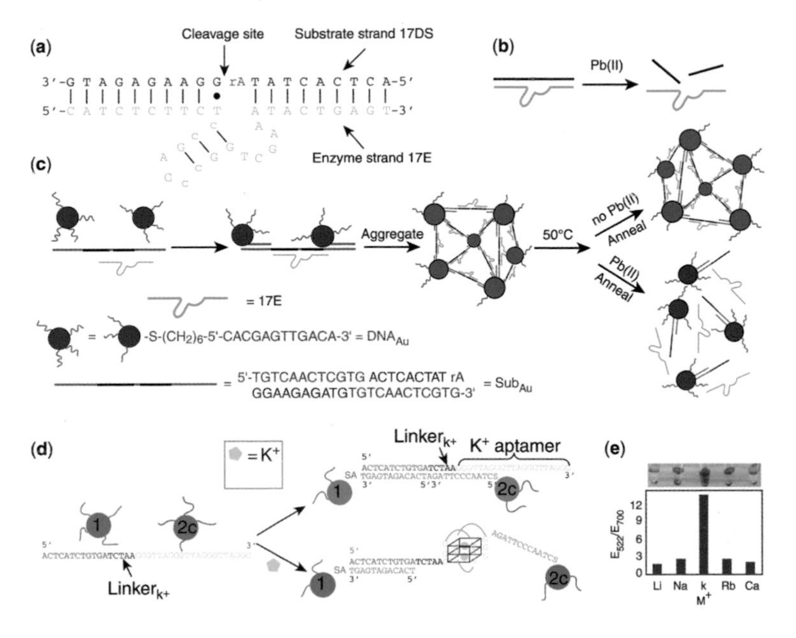

Figure 12.16 (a) Secondary structure of the DNAzyme system that consists of an enzyme strand and a substrate strand. Except for a ribonucleoside adenosine at the cleavage site (rA), all other nucleosides are deoxyribonucleosides. (b) Cleavage of substrate stand by the enzyme strand in the presence of Pb(II). (c) Schematics of DNAzyme-directed assembly of gold nanoparticles and their application as biosensors for metal ions such as Pb(II). In this system, the DNAzyme strand has been extended on both the 3′ and 5′ ends for 12 bases, which are complementary to the 12-mer DNA attached to the 13 nm Au NPs. (d) Materials responsive to K^+. Particles 1 and 2c are assembled by $linker_{K+}$ in the absence of K^+ to form aggregates. In the presence of K^+, the assembly is inhibited and the color remains red. (e) Colorimetric result in the presence of 250 mM of different monovalent metal ions.

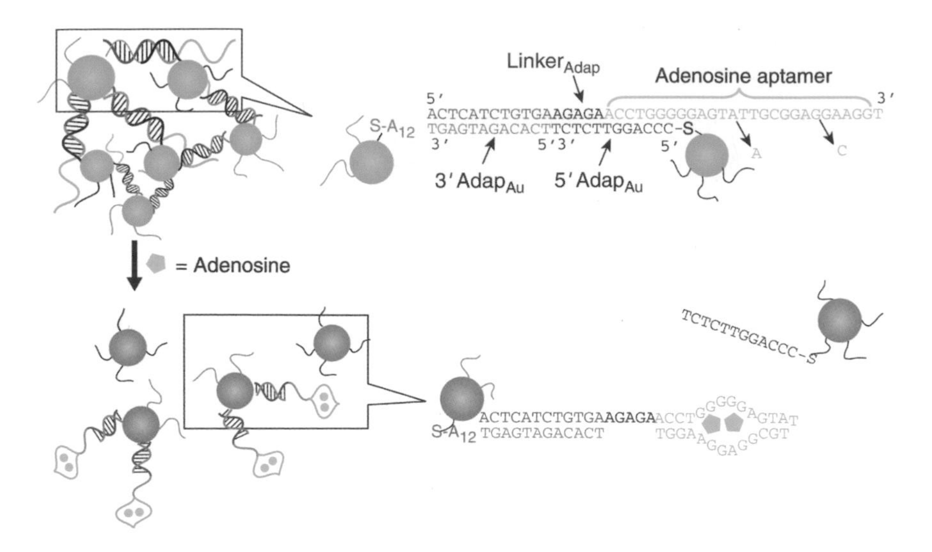

Figure 12.17 Schematic representation of colorimetric detection of adenosine. The DNA sequences are shown on the right side of the figure. The A_{12} in $3'Adap_{Au}$ denotes a 12-mer polyadenine chain. In a control experiment, a mutated linker with the two mutations shown by the two short black arrows was used. The detection of cocaine is performed in a similar manner. The drawing is not to scale.

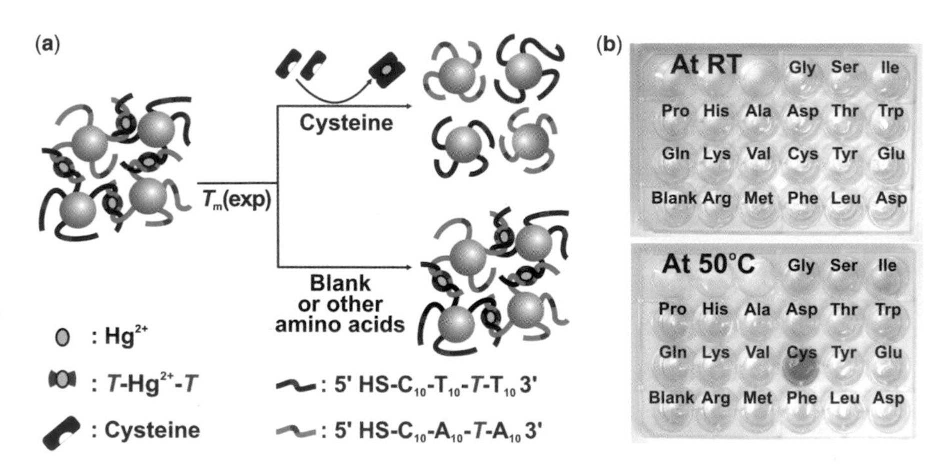

Figure 12.18 (a) Colorimetric detection of cysteine using DNA-AuNPs in a competition assay format. (b) The colorimetric response of the DNA-AuNP/Hg^{2+} complex aggregates in the presence of the various amino acids (each at 1 μM) at RT and 50°C.

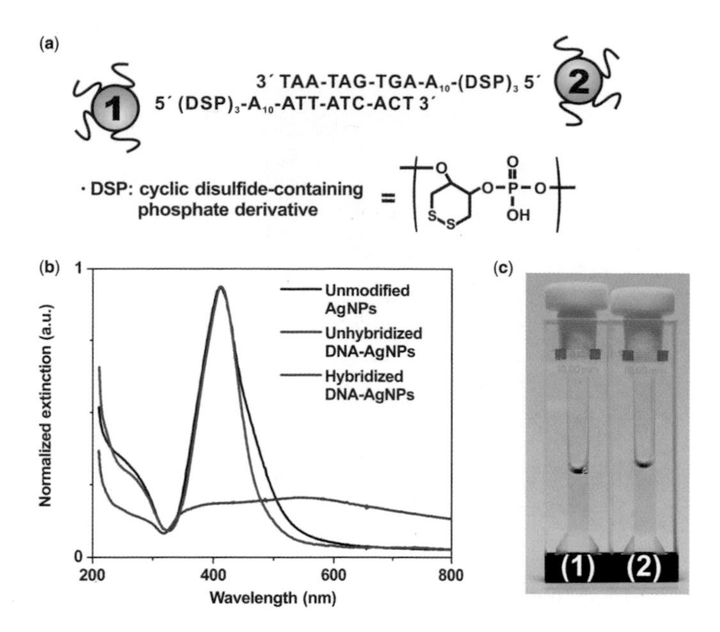

Figure 12.19 (a) Schematic illustration of the hybridization of two complementary DNA-AgNPs. (b) UV-vis spectra of unmodified AgNPs (black line), unhybridized DNA-AgNPs (blue line), and hybridized DNA-AgNPs (red line). Note that the wavelength at which the maximum of the extinction of AgNPs is obtained remains the same after DNA functionalization. After hybridization, however, the band of DNA-AgNPs broadens and red-shifts significantly from 410 nm to 560 nm. (c) Colorimetric change responsible for the assembly process of DNA-AgNPs. The intense yellow color of the unhybridized AgNPs (1) turns to pale red (2) as the particle aggregation proceeds. Heating of (2), however, results in the return of the solution color to yellow (1).

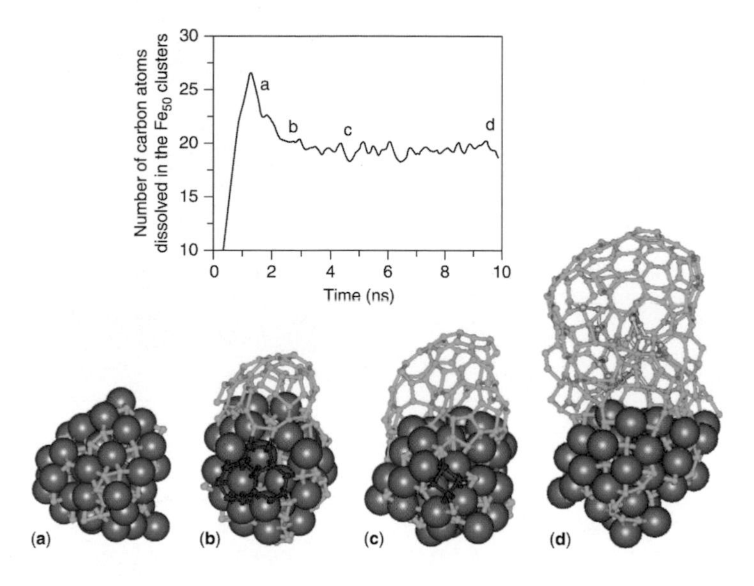

Figure 13.6 (Bottom) Simulated stages during the nucleation and growth of an SWNT on the surface of an iron cluster. The Fe atoms are shown as blue spheres while the C atoms are shown in the ball-and-stick representation mode. The red C atoms form a small graphitic island that later dissolves into the cluster while the grey atoms inside the nanotube are defects. (Top) Number of C atoms dissolved in the iron cluster during SWNT nucleation and growth. The a to d points in the plot refer to the a to d stages shown in the model.

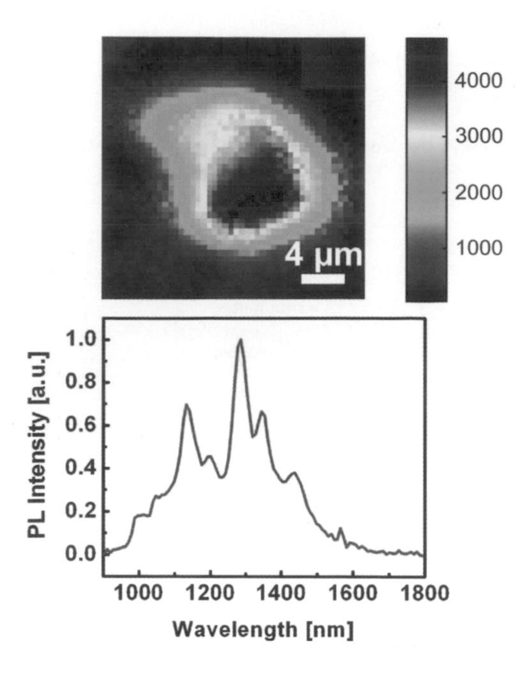

Figure 13.19 (Top) High-magnification NIR fluorescence image of a single Raji cell treated with SWNT-Rituxan conjugate showing NIR fluorescence over the cell. (Bottom) NIR emission spectrum recorded on a SWNT-Rituxan-treated Raji cell.

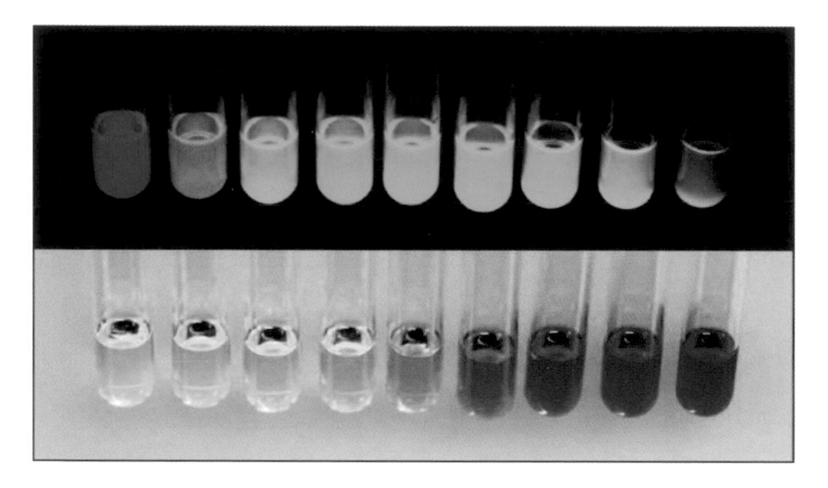

Figure 16.15 Colloidal suspension of CdSe quantum dots of increasing size from left (approximately 1.8 nm) to right (approximately 4.0 nm). Bottom: Samples viewed in ambient light vary in color from green-yellow to orange-red. Top: The same samples viewed under long-wave ultraviolet illumination vary in their color of emission from blue to yellow. (Reproduced with permission from E. M. Boatman et al., 2005. *J. Chem. Ed.* 82: 1697–1699. Copyright 2005 American Chemical Society.)

by design using metal-organic building blocks linked with organic spacers. Such a synthesis principle has been extensively exploited to generate countless framework structures, with remarkable diversity of networks (119–125). Today several hundred different MOFs have been identified. Practically, these porous coordination solids are prepared by combining solutions containing suitable organic ligands and metal ions, whereby both the organic and inorganic parts are designed synthetically in order to exploit the directional nature of metal-ligand interactions. Specific framework topologies are thus built resulting from self-assembly of the organic and inorganic units. The nitrogen sorption isotherms of MOF demonstrate a type I behavior implying that these solids are generally microporous materials. Although high surface areas are known for activated carbons and zeolites, the absence of any dead volume in MOF primarily affords the highest measured porosities and specific surface areas. Furthermore, MOF materials are very versatile, and some exist which exhibit selective sorption of small molecules (117a), or inclusion of large molecules or nanostructures such as the carbon fullerene C_{60} (126). MOFs are now prepared on pilot plant batch scale and the demonstration of useful sorption properties for applications in gas storage and separation has been made.

9.4.2.1 Multicomponent Frameworks and Structures MOF materials are robust solids with high thermal and mechanical stabilities. The conception of such MOFs has generated a great number of structures, which mostly have two components, namely, (transition) metal oxide units and organic units that are linked together to form extended porous frameworks. Most often, the metal oxide units are either zero-dimensional or one-dimensional, corresponding to discrete or rod-like geometric units named secondary building units (SBUs). These building units are simple geometric figures representing the inorganic clusters or coordination spheres that are linked together by the (typically linear) organic components to assemble the product framework. It should be kept in mind that these building units are, however, not generally introduced directly, although many of them have been observed in molecular compounds, but these are rather formed *in situ* under specific synthetic conditions. Conversely, the branched organic links with several coordinating functionalities could constitute preformed units. A large majority of the ligands involved in MOF chemistry are bipyridyl- and carboxylate-based. Since the organic unit can also be viewed as a secondary building unit, MOF structures are thus essentially composed of inorganic and organic SBUs. Some examples of typical organic ligands and SBUs which are commonly encountered in the metal-carboxylate MOFs are shown in Figure 9.13. This general concept of SBUs has generated an impressive cataloging of more than 2000 MOF structures. One major outcome of using the concept of SBUs is the ability to produce MOF with permanent porosity.

The conceptual approach by which an MOF is designated and assembled is termed reticular synthesis and is based on identification of how building blocks come together to form a net, or reticulate. The success of an SBU in the design of open frameworks relies both on its rigidity and directionality of bonding, which must be maintained during the assembly process. By using different metal-organic subunits, for example, different metal carboxylate clusters, various motifs can be created, leading to a large

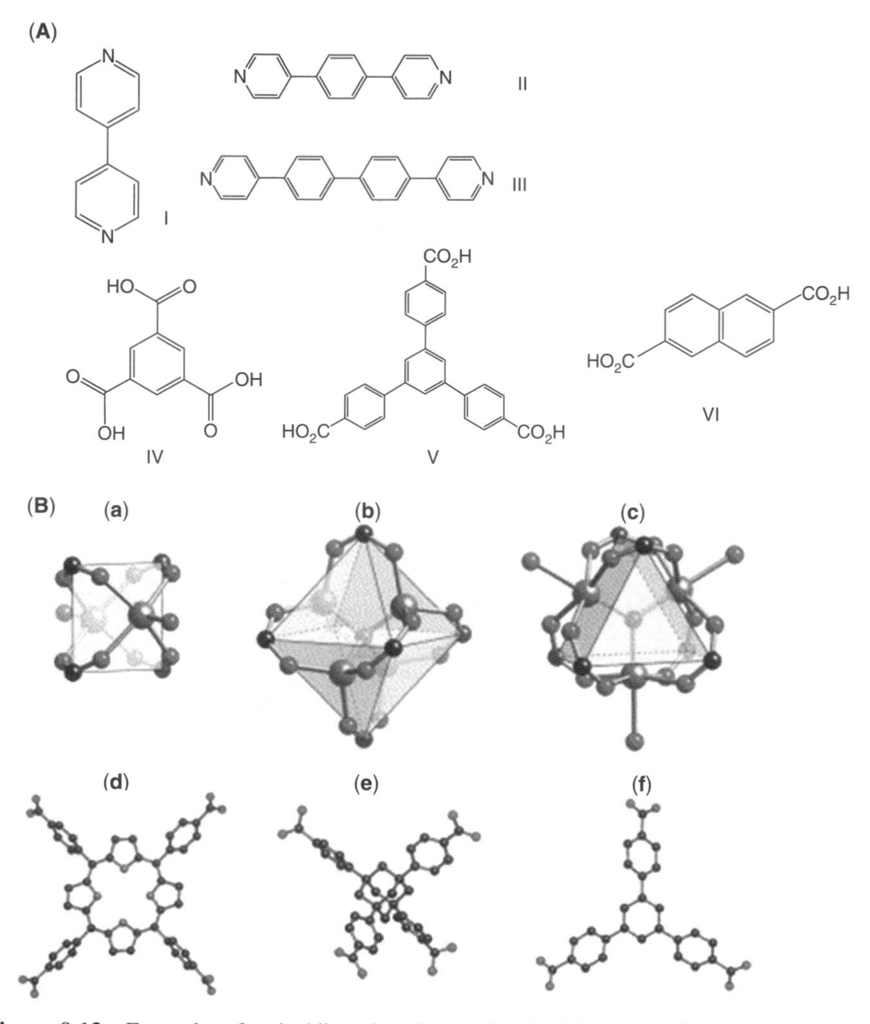

Figure 9.13 Examples of typical ligands and secondary building units (SBUs) involved in the formation of metal-organic frameworks (MOFs). (A) Classes of ligand used for framework construction: 4,4′-bipyridyl ligand (I) and analogs (II) and (III), 1,3,5-benzenetricarboxylate ligand (IV) with the extended analog (V), and 2,6-naphthalenedicarboxylic acid (VI). (Reproduced with permission from M. J. Rosseinsky, *Microporous Mesoporous Mater*. **2004**, *73*, 15. Copyright 2004 Elsevier.) (B) *Top*: Some inorganic SBUs most commonly found in metal carboxylate MOFs: (a) square paddlewheel, with two terminal ligand sites, (b) octahedral basic zinc acetate cluster, and (c) the trigonal prismatic oxo-centered trimer, with three terminal ligand sites. The SBUs are reticulated into MOFs by linking the carboxylate carbons with organic units, but can also be linked by exchange of the terminal ligands. *Bottom*: Some examples of organic SBUs comprise the conjugate bases of (d) square tetrakis(4-carboxyphenyl)porphyrin, (e) tetrahedral adamantane-1,3,5,7-tetracarboxylic acid, and (f) trigonal 1,3,5-tris(4-carboxyphenyl)benzene. The metals are represented as blue spheres, carbon as black spheres, oxygen as red spheres, nitrogen as green spheres. (Reproduced with permission from J. L. C. Rowsell and O. M. Yaghi, *Microporous Mesoporous Mater*. **2004**, *73*, 3. Copyright 2004 Elsevier.) (See color insert.)

number of materials with additional variability introduced by using different organic spacer molecules. Transition metals, as well as some main group metals, Zn, Cr, Cu, Ni, Al, Ti, V, Fe, Cd, Rh, etc., are the most exploited candidates as metal centers. In addition, porous coordination solids based on lanthanide frameworks, for example, Ln and Sm, are also becoming accessible. Since the content in metal component easily reaches values between 20 and 40 wt%, it is desirable to inspect the local metal cluster arrangements and environments, using methods such as x-ray photon spectroscopy (XPS), extended x-ray absorption fine structure (EXAFS), and x-ray absorption near edge structure (XANES). On the other hand, the size and the chemical environment of the void spaces are mostly defined by the length and functionalities of the organic units. The solids classified as MOFs normally exhibit the following characteristics: framework robustness resulting from strong bonding, linking units that are accessible for modification by organic reactions, and a geometrically defined framework structure. The latter property implies that these solids should be highly crystalline, and powder XRD is routinely utilized to characterize the crystallinity and phase purity of these materials. Permanent porosity is best evidenced by techniques such as gas adsorption measurements. Here, it is worth mentioning that the metal-organic frameworks that are sufficiently stable to display permanent porosity usually show pore volumes and window diameters exceeding substantially those of zeolitic materials (117). The MOF pore systems can be described as channel-like (1D space), layered (2D space), or intersecting channels (3D space) such as in the case of the isoreticular analogs of the MOF-5 material that will be discussed in the following.

A series of representative MOF compounds are described by the formula $Zn_4O(L)_3$ where L is a rigid linear dicarboxylate. These materials display the same cubic topology as the prototypical material labeled MOF-5 which is the framework generated when octahedral $Zn_4O(CO_2)_6$ clusters are linked along orthogonal axes by phenyl rings (117b). Figure 9.14 shows a representation of Zn_4-$O(BDC)_3$ framework of MOF-5 (BDC = 1,4-benzenedicarboxylate) consisting of the linked $Zn_4O(CO)_6$ clusters. MOF-5 exhibits a structure where the metal-carboxylate unit is an octahedral SBU linked by benzene units to produce a primitive cubic net, which gives rise to a highly porous structure with 3D intersecting channels. The synthesis of this structure is straightforward from a solution of Zn^{2+} and benzene-1,4-dicarboxylic acid under conditions that generate *in situ* the tetrazinc cluster. The virtual representation of a sphere of 1.85 nm in diameter shows the open space of the cavity, which is filled with solvent molecules in the as-synthesized form of the material. The solvent molecules can subsequently be removed and the framework of the material is retained. MOF-5 is stable and can be treated at 300°C for 24 h without being damaged and microporosity is confirmed by nitrogen adsorption measurements.

Once the reaction conditions for a given inorganic SBU have been established, it is then possible to insert into the synthesis preparation the organic link desired. This procedure is followed to yield isoreticular analogs of MOF-5, labeled from 1 to 16 (Fig. 9.14). The series of 16 isoreticular metal-organic frameworks (IRMOFs), that is, having the same underlying net topology, are produced in crystalline form, differing in the polarity, reactivity, and pendant groups on the aromatic link (127).

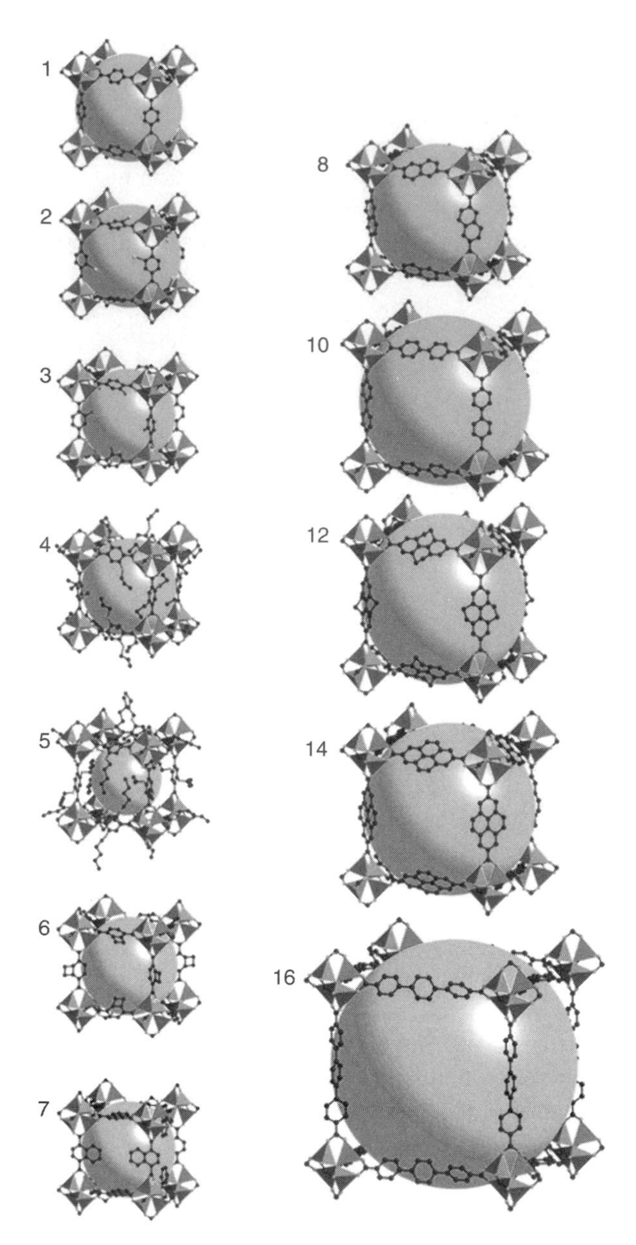

Figure 9.14 Single-crystal x-ray structures of IRMOF-*n* ($n =$ through 7, 8, 10, 12, 14, and 16), labeled respectively. Color scheme is as follows: Zn (blue polyhedra), O (red spheres), C (black spheres), Br (green spheres in 2), amino groups (blue spheres in 3). The large yellow spheres represent the largest van der Waals spheres that would fit in the cavities without touching the frameworks. (Reproduced with permission from M. Eddaoudi et al., *Science* **2002**, *295*, 469. Copyright 2002 American Association for the Advancement of Science.) (See color insert.)

The expansion of the network is achieved by alternatively substituting benzene-1,4-dicarboxylate by naphthalene-2,6-dicarboxylate, biphenyl-4,4′-dicarboxylate, pyrene-2,7-dicarboxylate, or terphenyl-4,4″-dicarboxylate in the synthesis, giving rise to a series of structures with calculated pore sizes up to 2.88 nm (i.e. mesoporous), and fractional free volumes up to 91%. Some of the MOFs thus obtained show the lowest crystal density of any material reported to date. Intrinsic characteristics of the metal-carboxylate bonds are responsible for the robustness of these frameworks. First, the bonding energy is relatively large due to enhanced electrostatic attraction, and second, the size of the carboxylate group facilitates bridging and chelation of metal cations to yield rigid, geometrically defined clusters (122, 123).

A large majority of studies have been conducted on MOFs such as the IRMOFs series consisting of discrete SBUs. Nevertheless, other periodic arrays of large channels can be formed as well when linear links are used to connect infinite inorganic rods. Quite often, an auxiliary bridging moiety, such as hydroxide, is present in the backbone of these "infinite SBUs" (119, 123, 128).

9.4.2.2 Synthesis of Metal-Organic Frameworks The use of the molecular ligands for building porous frameworks offers advantages in order to tailor both the porosity and the reactivity of the materials. Notably, the extensive coordination chemistry of metal ions allows for numerous combinations of metal and ligand building blocks that can be manipulated at the molecular level in order to direct the assembly of a given target material. Furthermore, the molecular building units show rather good solubility and the assembly reactions often proceed at reasonable temperatures. This latter characteristic is important since the synthesis is typically performed by precipitation of the product from a solution of the precursors. MOF synthesis is usually performed over reasonable time scale (hours to days), with the possibility of solvent recycling, and facile quality control monitoring by powder XRD or electron microscopy. In general, the synthesis of MOFs is relatively simple based on soluble salt precursors of the inorganic metal compound, for example, metal nitrates, sulfates, and acetates. The organic components, which are commonly mono-, di-, tri-, or tetra-carboxylic acids, are supplied in an organic polar miscible solvent, often amines (typically triethylamine) or amides (typically dimethylformamide). After combination of these inorganic and organic ingredients under stirring, the metal-organic structures form by self-assembly at temperatures in the interval between room temperature synthesis and up to 200°C under solvothermal conditions, and within a few hours. As a result, the framework assembly occurs in a single step. After this synthesis step, careful filtration and drying follow, because of the high content of occluded solvent (50 to 150 wt%). The solvent contained in the channels of a neutral MOF after synthesis plays in some way the role of the template in porous oxides, and must thus be removed to allow access to the pores. To do so, thermal activation is performed recovering the guest-free MOF materials. The temperatures of decomposition observed for carboxylate-based MOF frameworks, for instance, are in the range of 300°C to 500°C, leaving some interval of stability above the temperature required for guest removal.

Usually, either room temperature solvent diffusion methods or higher-temperature ($T < 200°C$, autogeneous pressure) solvothermal conditions are used to condense the organic and inorganic components. Possible limitations due to decreased solubility of the building blocks are quite often circumvented by using the solvothermal techniques, and, besides, mixtures of solvents are often chosen to tune the solution polarity and kinetics of solvent-ligand exchange. Nevertheless, optimal synthesis conditions must be found that are mild enough to maintain the functionality and conformation of the organic linker, yet reactive enough to ascertain the metal-organic bonds (129).

9.4.2.3 Properties Metal-organic frameworks present promising properties, including nanoporosity with virtually no inaccessible bulk volume in the solid, fully exposed metal sites, enormous weight-specific surface areas, and high mobility of guest species in their ordered nanopores. These properties may be, moreover, combined with framework flexibility and strength. By incorporating functions into the organic building block, such as reactive groups, redox centers, or chirality, a targeted property can be achieved periodically throughout the bulk material. A specific property can also be generated as a cooperative effect, such as the magnetic coupling of paramagnetic metal centers, or the alignment of asymmetric linkers, as is achieved in the case of MOFs based on Zn^{2+} or Cd^{2+} connected with bifunctional ligands resulting in nonlinear optical behavior (130). Moreover, in the case of MOFs containing paramagnetic transition metals or luminescent lanthanides, the optical, electronic, or magnetic properties of the framework may be altered by guest interactions, which would make sensing applications feasible, for instance.

Sorption. One of the most promising perspectives of applications of MOFs is the area of gas storage and separation (involving methane, ethylene, hydrogen, CO_2). To establish the realization of permanent porosity after guest removal, the measurement of gas isotherms is indispensable. The isotherm shape is typically of type I with little or no hysteresis, demonstrating that microporous structures exist under reversible physisorption of small molecules. The sorption measurement of nitrogen or argon probes allowed surface areas of these materials to be quantified and considerable weight-specific surface area values are regularly measured. For example, in the case of MOF-177 (126); values of $4500\,m^2/g$ are reported and for MIL-101 (131) immense areas up to $5900\,m^2/g$ are given. However, these surface areas calculated according to Langmuir or BET equations should only be regarded as equivalent or apparent surface areas, since these models are not necessarily applicable. MOFs are microporous solids in which the process of pore filling does not proceed through the formation of a monolayer (the volume of which is simply converted in surface area in the above models). Total filling of the huge micropore volume could result in substantially overestimated areas. In addition, the adsorption sites in these materials could be viewed as being quite diverse. Some characteristic type I nitrogen adsorption isotherms obtained for different MOFs are presented in Figure 9.15.

Flexibility and guest-responsive systems. The access to the porosity of rigid microporous materials such as zeolites is normally restricted by the dimensions of the windows giving access to the porosity, rather than the size of the cavities themselves. It has become clear, however, that the permanent porosity of MOFs can behave quite differently to that of the rigid zeolites. Unlike zeolites, carbons, or oxides, a number of coordination compounds have frameworks that possess a noticeable degree of

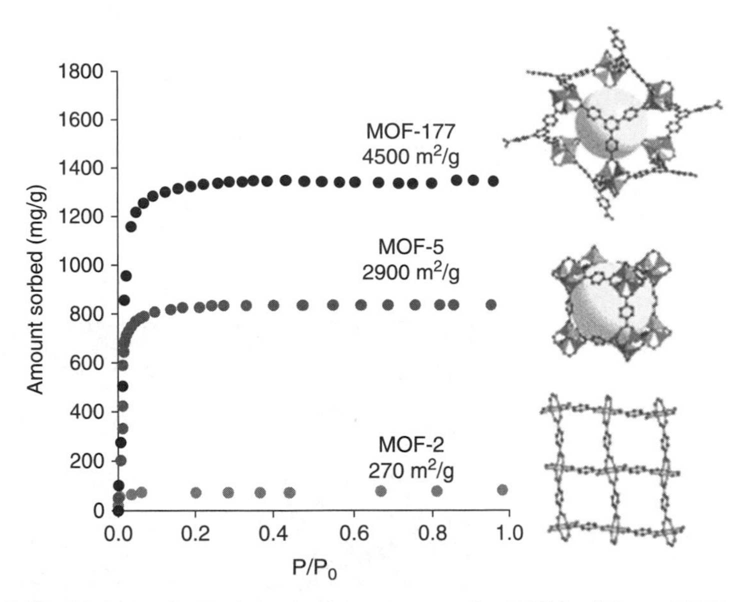

Figure 9.15 Nitrogen physisorption isotherms measured at 77 K for different MOFs. The isotherms display type I behavior expected for materials with uniform micropores. The development of MOF materials has led to an impressive increase in the apparent surface areas as compared to the early measurement of the first isotherm for this family of materials. (Reproduced with permission from J. L. C. Rowsell and D. M. Yaghi, *Microporous Mesoporous Mater*. **2004**, *73*, 3. Copyright 2004 Elsevier.)

flexibility allowing for shrinkage or expansion upon interaction with guest molecules. Frameworks shift upon guest incorporation were probed by x-ray diffraction investigations, and rearrangement of the structure in response to guest uptake was also monitored by gas adsorption. Consequently, the MOF systems may actually couple permanent porosity with flexibility to admit apparently over-sized guests.

Framework flexibility upon guest uptake is nicely demonstrated and rationalized in the studies of Férey and coworkers (132, 133). For example, the material named MIL-53 is a three-dimensional Cr^{3+} (or Al^{3+}) dicarboxylate structure built from bridging terephthalate and doubly bridging hydroxide with one-dimensional pore channels. This MOF shows a very marked and reversible change in its pore dimension in response to the uptake of polar guests such as water or CO_2 (Fig. 9.16). In the case of water, this flexibility seems to be caused by the distortion of the pore resulting from hydrogen bonding between the water guest and the carboxylate and OH components of the framework. MIL-53 undergoes a so-called breathing-type mechanism. This form of dynamical response of the pore windows might be quite general for this type of framework topology, since it is also observed for some isotypic porous terephthalate MOFs based on other metals (AI). On the other hand, flexibility is shown to be dependent on the polarity of the probe molecule. Such an adsorption behavior is not observed in the case of the nonpolar probe CH_4 guest, which opens up the possibility of selective sorption, gas separation (e.g. polar vs nonpolar), or sensing applications.

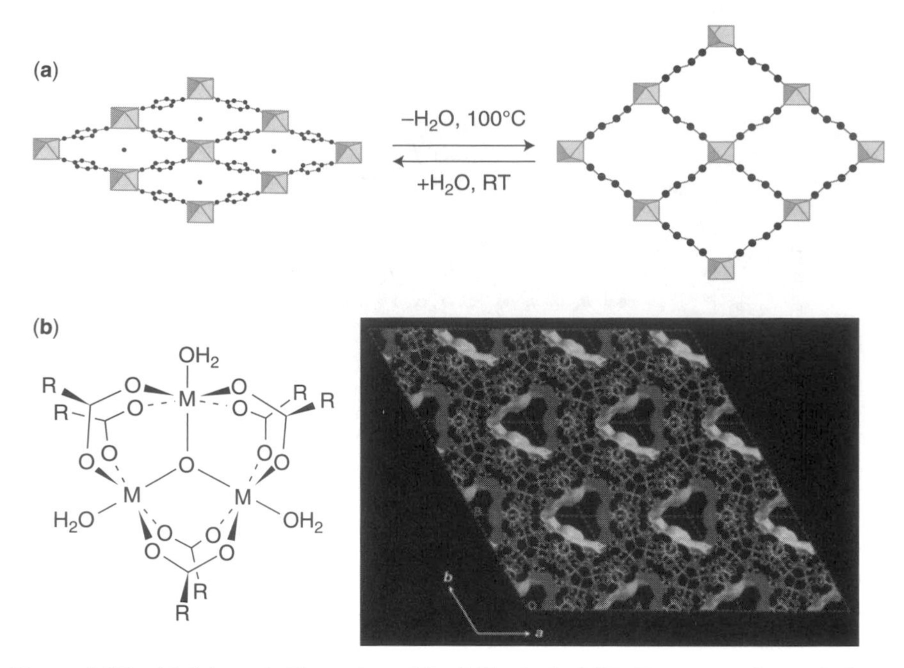

Figure 9.16 (a) Schematic illustration of flexibility in the MIL-53-type metal-organic framework. Shown is the breathing process observed upon reversible hydration and dehydration of MIL-53 (Al, Cr) with change in temperature. (Reproduced with permission from P. L. Llewellyn et al., *Angew. Chem. Int. Ed.* **2006**, *45*, 7751. Copyright 2006 Wiley-VCH Verlag.) (b) Homochiral metal-organic porous material. *Left*: The oxo-bridged trinuclear metal carboxylates (M = divalent or trivalent transition metal ions, O_2CR = organic carboxylate anions). *Right*: The chiral pore system of the open-framework material *D*-POST-1. The view is down the c-axis showing the large chiral channels. (Reproduced with permission from J. S. Seo et al., *Nature* **2000**, *404*, 982. Copyright 2000 Nature Publishing Group.) (See color insert.)

Note that, in general, the possibility of such guest-mediated processes should always be taken into account when these porous MOFs are contacted with fluids.

Chirality. The preparation of chiral molecular sieves for enantioselective sorption and catalysis is an important goal. The assembly of chiral porous networks is, however, not truly possible in the zeolite case. In contrast, the generation of homochiral porous MOFs is possible. To realize this, the use of chiral molecular building blocks appears promising. Kim and coworkers in Korea reported the synthesis of a homochiral metal-organic microporous material which was denoted POST-1. This structure consists of oxo-bridged trinuclear metal carboxylates as the primary building units, and the channels of this framework are chiral (134). More precisely, this layered open framework with large channels, represented schematically in Figure 9.16, is obtained by linking Zn^{2+} with a mixed donor ligand, that is, (4R,5R)-2,2-dimethyl-5-[(4-pyridinyl amino)carbonyl]-1,3-dioxolane-4-carboxylic acid or its enantiomer. In this structure, trigonal prismatic SBUs are generated by linking oxo-centered zinc trimers with six carboxylates groups (Fig. 9.16). Three of the six pyridyl donors connect to neighboring SBUs, while the remaining three hang into trigonal channels, resulting in an aperture of 0.85 nm bounded by available Lewis bases. Accordingly, accessible chiral voids containing exchangeable guests, active for enantioselective separation and catalytic esterification reactions, were first demonstrated. Several related chiral building blocks approaches are now appearing on the basis of binaphtholate and other ligand classes. Interestingly also, a zinc saccharate framework was even made containing both hydrophilic and hydrophobic channels (135).

9.5 ORDERED MESOPOROUS MATERIALS

9.5.1 Ordered Mesoporous Molecular Sieves (M41S Materials)

As introduced in Section 9.3, the characteristic approach for the synthesis of ordered mesoporous materials is the use of liquid crystal-forming templates that enable the specific formation of pores with predetermined size.

The main limitation of microporous molecular sieves (zeolites, carbons, etc.) is their restricted pore size (usually below 1.5 nm), which excludes size-specific processes involving large molecules. Larger pore sizes are necessary for performing catalytic conversion of large molecules, and for applications involving separation of biomolecules and macromolecules. Consequently, numerous attempts were made to extend the hydrothermal synthesis procedures used to prepare zeolites to the mesopore range. However, success was rather limited. In contrast, the introduction of supramolecular micellar aggregates, rather than molecular species, as structure-directing agents in silica synthesis enabled scientists at Mobil Corporation to prepare a new family of mesoporous aluminosilica compounds which they designated as M41S. Among the different materials reported by Mobil in 1992, the one named MCM-41 (*Mobil Composition of Matter* N° 41, the number is added chronologically based on the date of discovery) exhibits a highly ordered hexagonal array of one-dimensional cylindrical pores with a relatively narrow pore size distribution (136, 137). At about the same period, an alternative to the Mobil approach was described by Yanagisawa et al. (138), using kanemite, a layered silicate, serving as the silica source. The resulting material, although quite similar to MCM-41 silica (139), was designated FSM-*n* (*folded sheet mesoporous materials-n*), where *n* stands for the number of carbon atoms in the chain of the surfactant used. This new type of solids is characterized by ordered arrangements of mesopores, but the pore walls are built of amorphous silica. The combination of powder x-ray diffraction, transmission electron microscopy, and gas physisorption analysis allow independent and reliable characterization of ordered mesoporous materials. In particular, the hexagonal arrangement of uniform pores of MCM-41 can be visualized by TEM (Fig. 9.17).

The N_2 sorption isotherm measured for MCM-41 is distinctive, being a type IV isotherm, presenting an unusually sharp capillary condensation step (Fig. 9.17). MCM-41 type mesoporous silicas are highly porous, showing BET surface areas exceeding 1000 $m^2 g^{-1}$ and pore volumes up to 1 $cm^3 g^{-1}$. The pore sizes usually vary between 2 and 10 nm depending on the alkylammonium surfactant chain length, the presence of

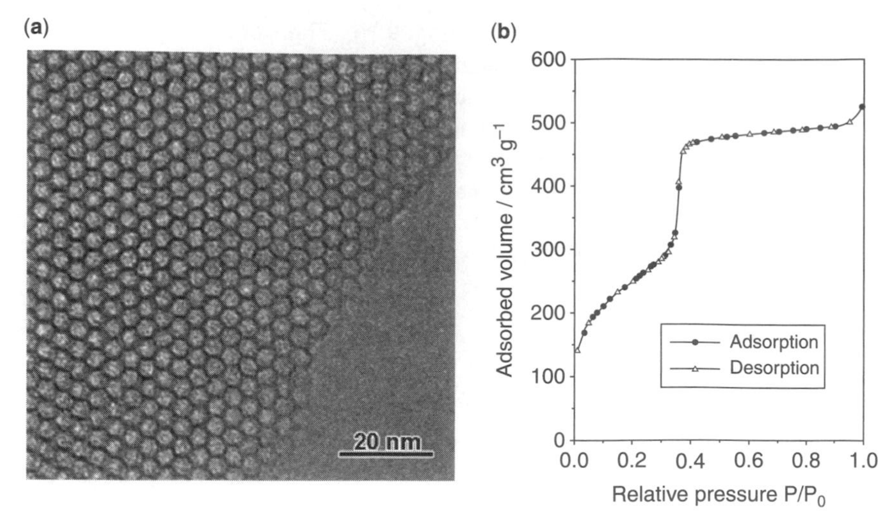

Figure 9.17 (a) Transmission electron microscopy image of ordered mesoporous MCM41 silica. Original image courtesy of Dr. Yasuhiro Sakamoto (Stockholm University, Sweden). (b) Type IV nitrogen physisorption isotherm measured at 77 K for calcined MCM-41. (Reproduced with permission from F. Kleitz, p. 180, in *Handbook of Heterogeneous Catalysis*, 2nd edition, G. Ertl et al., eds. Wiley-VCH, Weinheim, 2008. Copyright 2008 Wiley-VCH Verlag.)

organic additives, or post-treatments as will be discussed later. The pore wall thickness of MCM-41 is estimated to be about 1 nm. Noteworthy is that MCM-41 shows completely reversible type IV isotherms when the pores of the material are narrower than about 4 nm and such isotherms were not known at the time of the initial IUPAC classification. It turned out that hexagonally ordered mesoporous MCM-41 silica and the other related mesostructures are ideal reference materials for experimental and theoretical sorption investigations and crystallographic studies.

9.5.2 Synthesis Pathways for Ordered Mesoporous Materials

The first ordered mesoporous materials (e.g. MCM-41 and FSM-16) were prepared from ionic surfactants, such as quaternary alkylammonium ions, under alkaline conditions. In this case, the formation of the inorganic-organic hybrid mesophase is based on electrostatic interactions between the positively charged surfactant molecules and the negatively charged silicate species in solutions [note that silica species are negatively charged at pH > 2 (8, 22)]. Depending on the synthesis conditions, the silicon source or the type of surfactant, several other mesoporous materials were synthesized according to the principle of the cooperative assembly (Section 9.3.5). Many silica-based mesoporous materials, but also non-silica frameworks have been obtained via these electrostatic assembly pathways and counterion-mediated pathways (e.g. $S^+X^-I^+$). Furthermore, two additional approaches for the synthesis of mesoporous materials on the basis of nonionic organic-inorganic interactions (H-bondings or dipolar) were developed soon after the first reports on ordered mesoporous silica (87, 88). In these latter approaches, neutral surfactants such as primary amines and poly(ethylene oxides) were employed to prepare the materials named HMS (*hexagonal mesoporous silica*) and MSU (*Michigan State University*). The two-dimensional (2-D) hexagonal mesoporous material designated as SBA-15 (for *Santa Barbara* N° 15) was also formed via this pathway. For this SBA-15 silica, the structure direction was achieved under acidic conditions using the triblock copolymer poly(ethylene oxide)-*block*-poly(propylene oxide)-*block*-poly(ethylene oxide)$_x$, $(EO)_{20}$-$(PO)_{70}$-$(EO)_{20}$, with EO = ethylene oxide and PO = propylene oxide.

Besides cooperative pathways, also true liquid crystal templating (TLCT) and the hard template route (Section 9.3.7) have been developed for the synthesis of ordered mesoporous materials. In the case of the TLCT, a preformed surfactant liquid crystalline mesophase is loaded with the precursor for the inorganic materials (140). The nanocasting route, on the other hand, is a clearly distinct method (141). Here, no soft surfactant template is used but, instead, the pore system of an ordered mesoporous solid is used as the hard template serving as a mold for preparing varieties of new mesostructured materials, for example, metals, carbons, or transition metal oxides.

9.5.2.1 Direct Synthesis Methods: Cooperative Self-Assembly versus True Liquid Crystal Templating (TLCT) The original synthesis of mesoporous molecular sieves of the M41S family was performed in aqueous alkaline solution (pH > 8). Beck and coworkers (136, 137) proposed first a liquid-crystal templating mechanism on the basis of the similarities between liquid crystalline surfactant phases and the mesostructured materials. In this case, the structure would be defined by the organization of surfactant molecules into liquid crystal (LC) mesophases which serve as truc templates for the construction of the M41S structures. In the case of MCM-41, the inorganic silicate species would occupy the space between a preformed hexagonal lyotropic crystal phase (Fig. 9.18, pathway 1). However, since the LC structures, which are formed in surfactant solutions, are very sensitive to the characteristics of the solution, it was also proposed that the addition of inorganics could mediate the ordering

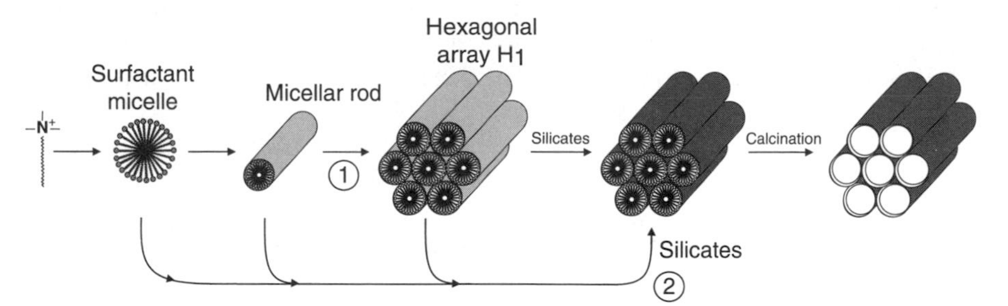

Figure 9.18 Formation of ordered mesoporous MCM-41 silica (liquid crystal templating mechanism (LCT) according to Beck et al. (137). The pathway 1 is liquid crystal-initiated and the pathway 2 is silicate-initiated. (Adapted and reproduced with permission from J. S. Beck et al. *J. Am. Chem. Soc.* **1992**, *114*, 10834. Copyright 1992 American Chemical Society.)

of the surfactants into specific mesophases (Fig. 9.18, pathway 2). In view of that, the self-assembly process of the surfactant template is synchronized with, or followed by, the formation of the inorganic network deposited around the self-assembled substrate. In both pathways, however, the inorganic species interact electrostatically with the charged surfactant head groups and condense into a continuous framework.

Monnier et al. (142) investigated how MCM 41 forms at concentrations where only spherical micelles are present (1 wt%) and established the silicate-initiated mechanism. These authors came to the conclusion that, in fact, the overall mesophase formation of MCM-41 and related materials is most often driven by cooperative assembly of the organic surfactant molecules and the inorganic solution species (that is, pathway 2). Furthermore, they introduced a charge density matching model, based on the cooperative organization of inorganic and organic molecular species into 3-D structures. In this model, three steps are involved in the formation of the silica-surfactant hybrid mesophase: (1) multidentate binding of the silicate polyions to the cationic head groups through electrostatic interactions leading to a surfactant-silica interface (S^+I^-), (2) preferential silicate polymerization in the interface region, and (3) subsequent charge density matching between the surfactant and the silicate. At the initial stage, before the addition of silicon precursors, the surfactant molecules are present in dynamic equilibrium between single molecules and spherical or cylindrical micelles. Upon addition of a silicon source, multicharged silicate species will displace the surfactant counteranions to produce new organic-inorganic ion pairs, which reorganize first into a sort of *silicatropic* mesophase. This is then followed by silica cross-linking upon condensation. In this process, growing silica species cooperatively attach to an increasing number of surfactant molecules and, when the amount of surfactants is large enough, an organized mesostructure precipitates. In conclusion, both surfactant and inorganic soluble species direct the synthesis of mesostructured MCM-41-type materials (see Fig. 9.19).

Alternatively, it is possible, if preferred, to carry out silica sol-gel polymerization specifically into lyotropic liquid crystals. Such a pathway is prevailing when the surfactant is so concentrated that a liquid crystal phase has formed prior to the addition of the precursor for the inorganic framework material. Thus, a liquid crystalline precursor mesophase is used, in which the inorganic species are infiltrated (a method also called true liquid crystal templating, TLCT). This TLCT method enables the preparation of a replica of the organic LC phase. In principle, it should thus be possible to control the final phase by knowing the phase diagram of the surfactant; however, structural features of the liquid crystals may still be subject to subtle changes during the formation of the solid product. The utilization of the TLCT pathway usually leads to mesostructured hexagonal, cubic, or lamellar silica mesostructures as gels or shaped as monoliths or films (140). Moreover, metallic mesoporous materials can also be obtained in this way by chemical or electrochemical reduction of metal salts, dissolved in the hydrophilic region of the LC phase (143).

9.5.2.2 Evaporation-Induced Self-Assembly (EISA) The method called evaporation-induced self-assembly (EISA) was introduced by Jeffrey Brinker and coworkers (144) and describes syntheses producing mesostructured materials starting from dilute

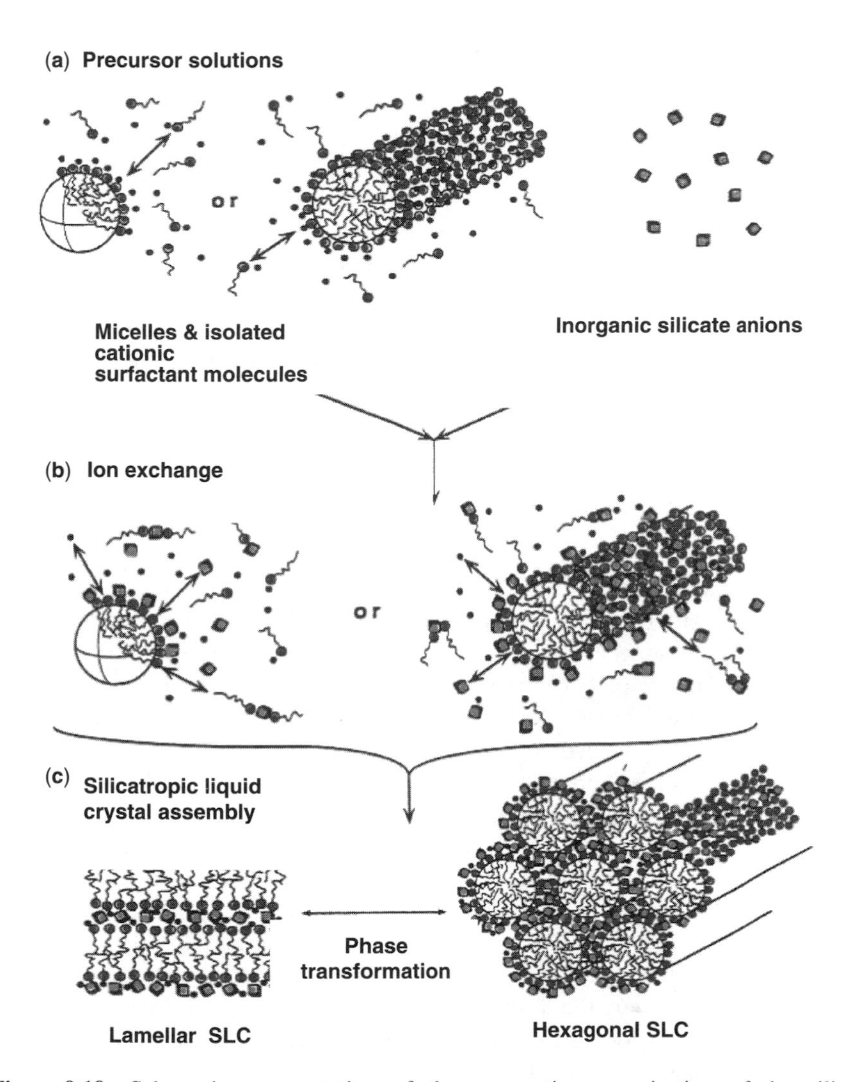

Figure 9.19 Schematic representation of the cooperative organization of the silicate-surfactant mesophase (cooperative self-assembly mechanism). (Reproduced with permission from A. Firouzi et al., *Science* **1995**, *267*, 1138. Copyright 1995 American Association for the Advancement of Science.)

solutions, and followed by solvent evaporation. The EISA method can essentially be regarded as a LC templating-based method [after solvent evaporation, a hybrid LC mesophase is present; (145)]. Starting from surfactant or block copolymer solutions below the critical micelles concentration allows for the preparation of thin films or gels with excellent homogeneity, the high dilution preventing uncontrolled inorganic polymerization. Ordered mesoporous silica thin films and monoliths were obtained via EISA using surfactants or block copolymers. This method is also particularly versatile to prepare non-silica systems in which the condensation has to be thoroughly controlled (Section 9.5.5).

9.5.2.3 Preparation of Ordered Mesoporous Materials by Nanocasting Liquid crystal-like templating methods are considered unsuitable for synthesis of materials with compositions and structures requiring high temperature treatments. Especially, nonsiliceous compositions such as carbon, polymers, nonoxides, or transition metal-based materials are very sensitive to thermal treatment conditions and redox reactions. A possible solution to this problem was brought by a research group in South Korea. Using the approach of hard templating, Ryoo and coworkers succeeded in producing ordered mesoporous carbons with high surface area and unprecedented narrow pore size distribution (141, 146, 147). In contrast to the previous soft templating methods, the hard templating method will make use of pre-prepared ordered mesoporous silica as the rigid nanoscopic mold. In a typical nanocasting experiment using ordered mesoporous silica, suitable precursors are first incorporated (by sorption, ion exchange, covalent grafting, etc.) inside the pores of the mesoporous solid template. The precursor infiltration may be repeated to achieve high loadings which will facilitate the rigidification of the templated framework. After sufficient solidification is achieved within the host pore system, with possible heating treatment applied to form a desired phase, the silica parent matrix is selectively removed and shape-reversed molded mesostructures are released. Evidently, the resulting material composition must be stable either in diluted HF or NaOH solution which will be used to dissolve silica, and the precursor must not react with silica at elevated temperatures. Because a mold with 3-D bridged structure is necessary to maintain a stable replica, only experiments with silica materials having interconnected mesopores leads to ordered mesoporous nanocasts. Thus, ordered mesoporous silicas with 3-D interconnected pore structures (e.g. MCM-48, SBA-15, and KIT-6 silicas) are the most frequently employed hard templates. Conventional MCM-41 silica does not exhibit interconnected channel topology and is therefore not suitable (148; see Section 9.5.3). The nanocasting route has emerged as an especially interesting approach to create frameworks that are too difficult to access using soft liquid crystal-based templates (i.e. reactive oxides, polymers, carbons, nitrides, carbides, sulfides, metals, etc.). This pathway has therefore become an extremely efficient alternative for synthesizing all sorts of ordered nonsiliceous mesoporous materials (92, 93, 141, 146, 147, 149–151).

9.5.3 Ordered Mesoporous Silica

9.5.3.1 Mesostructure Diversity Mesostructured polymorphs of silica prepared under alkaline conditions have attracted the most scientific interest so far. However, ordered mesoporous silicas can also be prepared under acidic and neutral conditions. Several reasons make silica-based materials the most widely studied systems, namely, a large variety of possible mesostructures with various pore network connectivities, narrow pore size distribution, a precise control over hydrolysis-condensation reactions due to lower reactivity of silicates, enhanced thermal stability and a vast variety of

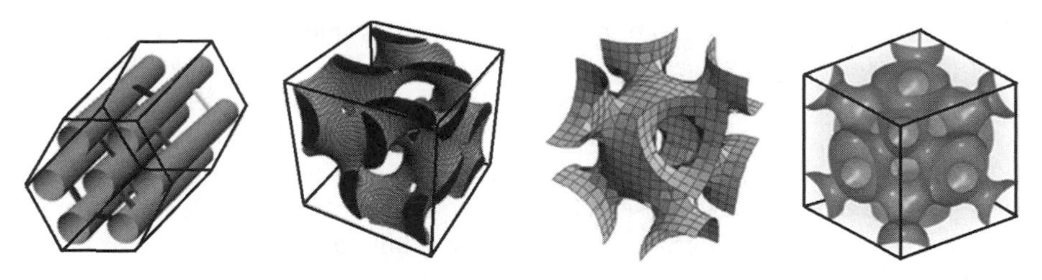

Figure 9.20 Typical pore network topologies found in ordered mesoporous silica materials. From left to right: 2-D hexagonal (*p6mm*) pore structure, bicontinuous 3-D cubic ($Ia\bar{3}d$) pore structure, 3-D interconnected *body-centered* cubic cage-like structure ($Im\bar{3}m$), and 3-D interconnected *face-centered* cubic cage-like structure ($Fm\bar{3}m$).

methods available for functionalization. Some of the typical pore topologies known for ordered mesoporous silica materials are represented in Figure 9.20.

Under alkaline conditions, anionic silicates I^-, and cationic surfactant molecules S^+, cooperatively associate and organize to form hexagonal, lamellar, or cubic structures. Thus, the synthesis in alkaline medium can lead to three well-defined structures: MCM-41, MCM-48, and MCM-50 (137). In this synthesis, the surfactant to silica mole ratio is a key variable in directing the formation of the different M41S materials (57). Using tetraethoxysilane (TEOS) with cetyltrimethylammonium chloride ($C_{16}H_{33}(CH_3)_3NCl$), it was found that increasing the surfactant to silica molar ratio from 0.5 to 2.0 results in hexagonal (<1), cubic $Ia\bar{3}d$ (1 to 1.5) and lamellar (1.2 to 2) composite structures. Moreover, structural phase transformations to lower energy configurations may be induced during hydrothermal treatment of the powder in its mother liquor as changes in the charge density of the silicate occur upon ongoing condensation and subsequent restructuring with changes in surfactant packing. Thus, transition from lamellar to hexagonal phase, from hexagonal to lamellar, and transition from hexagonal to cubic geometries can observed upon condensation (142, 152). In addition, structural evolutions may also be coupled with temperature-dependent changes in hydration, silicate solubility, counterion binding, and migration of organic cosolvent in the system. Another quite efficient route for the synthesis of MCM-41 was proposed by Grün and coworkers (153, 154), based on fast hydrolysis of a molecular silicon source (TEOS) in the presence of ammonia. This method is especially interesting because it provides a convenient route to high-quality products in a short period of time (as short as 1 hour) at room temperature, without sodium traces, and it can further be adapted to prepare colloidal spherical mesoporous particles (Section 9.5.6).

In contrast to MCM-41, the pore network of MCM-48 silica is three dimensional and highly interconnected. The structure of MCM-48 silica belongs to the $Ia\bar{3}d$ space group, which also exists in the binary water/CTAB system (155). This interesting structure is viewed as a bicontinuous system which consists of two mutually intertwined three-dimensional networks of channels (Fig. 9.20; Reference 156). MCM-48 silica can be prepared in high phase purity via hydrothermal synthesis by tuning the silica to CTAB ratio and other synthesis parameters, or at room temperature

(57, 152, 157). The preparation of MCM-48 is also possible by an adjustment of the packing parameter g obtained by the addition of specific organic additives or by employing mixtures of different surfactants (158–160). Assuming a cubic symmetry for the unit cell, the diffraction pattern of MCM-48 is consistent with the body-centered cubic $Ia\,\bar{3}\,d$ space group, in analogy to the respective liquid crystal phase. However, the determination of such a high symmetry only by powder XRD might be uncertain and could need confirmation by additional analyses, for example, TEM and electron diffraction experiments (161).

Huo et al. (85, 86) established that it is possible to synthesize ordered mesoporous silica at low pH. The formation of silica is achievable here by the cooperative assembly of cationic inorganic species with cationic surfactants. The syntheses are carried out under strongly acidic conditions where silicate species are positively charged, according to a halide-mediated cooperative pathway ($S^+X^-I^+$). Mesotructures with 2-D hexagonal $p6mm$, 3-D hexagonal $P6_3/mmc$, cubic $Pm\,\bar{3}\,n$, and lamellar phases are obtained in the presence of HCl (162–164). The 2-D hexagonal material ($p6mm$) prepared according to the acidic route is sometimes named SBA-3 [*Santa Barbara* N° 3; (86, 162, 165)].

Advances in the preparation of the mesoporous silica materials were made by Che and coworkers (166), who proposed specific anionic surfactants in combination with organosilane groups acting as co-structure-directing agents (CSDA). In this system, the mesoporous silica materials are believed to form through electrostatic interactions involving negatively charged carboxylic head groups of anionic surfactants, such as *N*-acyl-alanine, glycine, or N-acyl-glutamic acid, and positively charged amine or ammonium groups of 3-aminopropyltrimethoxysilane (APS) or *N*-trimethoxysilylpropyl-*N*,*N*,*N*-trimethylammonium chloride (TMAPS). These organosilane additives are condensing together with TEOS, building ultimately the silica framework. Different mesostructures are obtained at pH ranging from 5.0 to 9.5 with aging treatment in the interval of 60°C to 100°C. This family of mesoporous materials is designated as AMS-*n* (*anionic surfactant templated mesoporous silicas*). Detailed structural investigations revealed an impressive diversity of materials created as a function of the synthesis parameters (e.g. pH, reagent ratios, temperature, and time) and type of surfactants. Indeed, highly ordered mesoporous silica, with pore sizes ranging between 2.3 and 7.5 nm, with either 3-D hexagonal (AMS-1), 3-D cubic $Pm\,\bar{3}\,n$ (AMS-2), 2-D hexagonal (AMS-3), 3-D cubic $Ia\,\bar{3}\,d$ (AMS-6), or 3-D cubic $Fd\,\bar{3}\,m$ (AMS-8) structures may be synthesized.

Syntheses of silica employing nonionic structure-directing agents are carried out under acidic or neutral conditions and the hybrid inorganic-organic mesophase formation is dictated by weak van der Waals interactions or hydrogen bondings. Tanev and Pinnavaia (87) developed the neutral templating route, which is based on hydrogen-bonding and self-assembly involving neutral amine surfactants and neutral inorganic precursors. Primary amines with alkyl chains from C_8 to C_{18} are used according to the S^0I^0 approach to prepare mesoporous products at pH close to 7. In connection to this, Pinnavaia's group reported in 1995 the use of nonionic poly(ethylene oxide) monoethers (polyoxyethylene-based oligomeric amphiphiles), as well as Pluronic-type polymers, in neutral aqueous media to direct the formation of mesoporous silica (88). The neutral templating pathway (N^0I^0) based on hydrogen-bonding interactions was proposed to operate for this system. Following this pathway, structures denoted MSU-X with worm-like mesopores of uniform diameters ranging from 2 to 6 nm were obtained by varying the size and structure of the surfactant molecules. Using neutral nonionic surfactants instead of the ionic ones offers the following advantages: (1) thicker inorganic walls (1.5 to 4 nm) are usually obtained leading to improved hydrothermal stability, (2) tailoring of the pore diameter is easier, and (3) template removal by solvent extraction is facilitated (167).

A breakthrough in the preparation of ordered mesoporous silica was made by Zhao and coworkers in 1998 (168, 169), who used poly(ethylene oxide)-poly(propylene oxide)-poly(ethylene oxide) triblock copolymers for the synthesis of the large-pore 2-D hexagonal SBA-15 material. The synthesis is simple and based on the use of organic silica sources such as TEOS or tetramethoxysilane (TMOS) in combination with diluted acidic aqueous solution of Pluronic-type triblock copolymer (2 to 7 wt% in water) such as P123 (EO_{20}-PO_{70}-EO_{20}). Here, the hybrid interface formation is sometimes suggested to follow a (N^0H^+) (X^-I^+) model since the block copolymer could be positively charged under the reaction conditions. Silica mesophases related to SBA-15 are, for example, SBA-12 (3-D hexagonal $P6_3/mmc$) and SBA-16 (cubic $Im\bar{3}m$) that are synthesized rather similarly by using different nonionic block copolymers, such as F127 (EO_{106}-PO_{70}-EO_{106}), or nonionic oligomeric surfactants (Brij-type). In general, the use of triblock copolymers expands the accessible range of pore sizes. Mesoporous silicas obtained with triblock copolymers possess uniform large pores with diameters well above 5 nm and thick walls, the latter providing high thermal stability and improved hydrothermal stability compared with mesoporous silicas prepared by using low molecular weight surfactants. Hexagonally ordered SBA-15, particularly, can be synthesized with pore sizes ranging between 4.5 nm and 12 nm and thick walls (3.0 to 6.5 nm in width), depending on the reagent ratios, pH, and aging temperature (170). This material exhibits large apparent surface area around 800 to 1000 $m^2\ g^{-1}$ and pore volume up to 1.5 $cm^3\ g^{-1}$. A TEM image showing the hexagonal structure of SBA-15 is presented in Figure 9.21b.

SBA-15 silica is of growing interest for a wide range of applications (e.g. sorbent, support for catalysts and biomolecules, nanoreactor, and solid template). At first, SBA-15 was thought to be a large-pore equivalent of MCM-41 which has unconnected mesoporous cylindrical channels. However, several studies showed that the pore size distribution of SBA-15 is rather bimodal, whereby the larger, hexagonally ordered structural mesopores are connected by smaller pores (micropores or small mesopores) located inside the silica walls (171, 172). Imaging of platinum replica of the pore structure of SBA-15 allowed direct visualization of the presence of these complementary pores using TEM (171, 173). These pores do not seem to be ordered and most probably originate from the penetration of the PEO blocks of the copolymer inside the silica framework. The presence of this complementary pore system in the walls of SBA-15-type materials is of foremost importance in the development of block copolymer-directed silicas.

In the family of copolymer-templated materials, ordered mesoporous silicas consisting of interconnected large cage-like pores are also of significant interest.

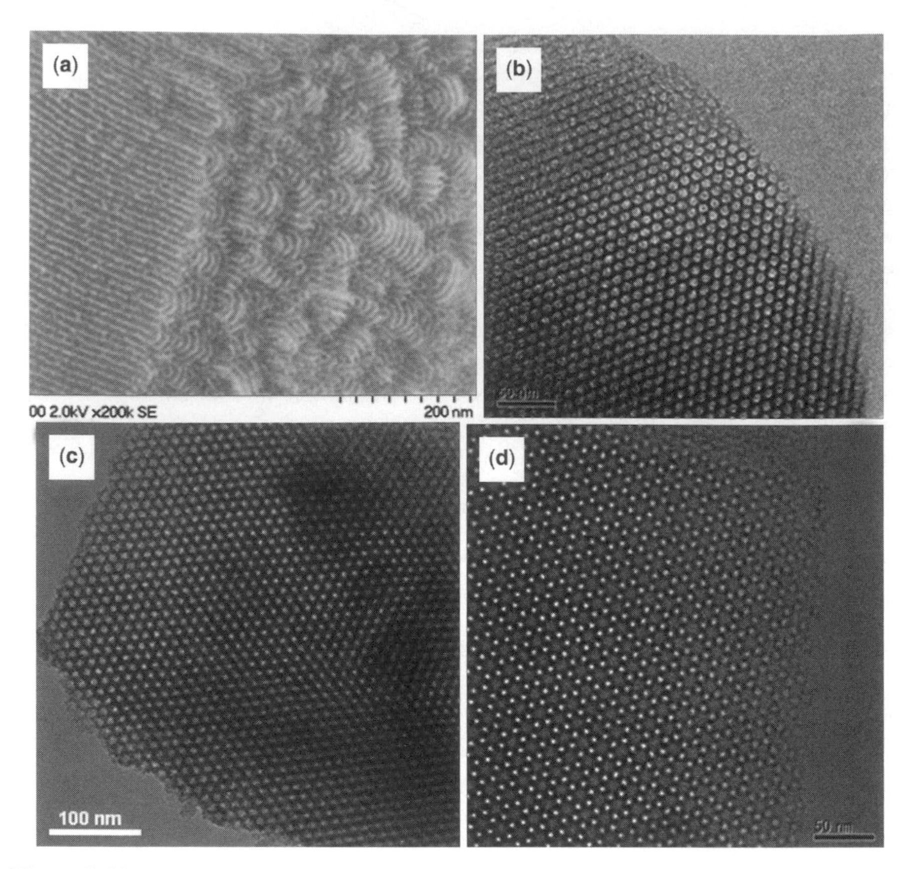

Figure 9.21 (a) High-resolution SEM image revealing the pore network organization of SBA-15 silica. (Reproduced with permission from S. N. Che et al. *Angew. Chem. Int. Ed.* **2003**, *42*, 2182 Copyright 2003 Wiley-VCH Verlag.) (b) TEM image showing the 2-D hexagonal arrangement of the mesopores in SBA-15. (Original image from F. Kleitz, F. Bérubé, and Y. Sakamoto, unpublished results.) (c) TEM image of cage-like SBA-16 silica aged at 100°C, viewed along the [111] direction. (Reproduced with permission from O. C. Gobin et al., *J. Phys. Chem. C* **2007**, *111*, 3053. Copyright 2007 American Chemical Society.) (d) Typical TEM image of the bicontinuous KIT-6 silica structure, viewed along the [111] direction. (Original image from F. Kleitz, S. Bilodeau, and Y. Sakamoto, unpublished results.)

The large-pore silica denoted SBA-16, for instance, consists of spherical cavities of 6 to 11 nm in diameter organized in a body-centered cubic (bcc) array and the cavities are 3-D interconnected through mesoporous openings of 2.0 to 3.5 nm (Fig. 9.21; Reference 174). The aqueous synthesis route to achieve this ordered mesoporous silica exhibiting cubic $Im\bar{3}m$ symmetry is based on the use of F127 (EO_{106}-PO_{70}-EO_{106}), which presents a high hydrophilic to hydrophobic volume ratio (high EO/PO ratio) favorable for the formation of highly curved globular micelles. Another member of the copolymer-based silica mesophases is the large-pore equivalent of MCM-48 known as KIT-6 showing 3-D interconnected pore structure. One of the easiest methods to generate this mesophase was introduced by Kleitz, Choi, and Ryoo (175), who used a blend of Pluronic P123 and *n*-butanol as structure-director along with a fine-tuning of the acid concentration. This KIT-6 material exhibits a structure with cubic $Ia\ \bar{3}\ d$ symmetry and the pore network topology can thus be similarly described as an interpenetrating bicontinuous network of interconnected channels [the gyroid structure shown in Figure 9.20 (176)]. The porosity of KIT-6 is quite similar in nature to that of SBA-15, although subtle differences in the hysteresis behavior are observed (177). The KIT-6 silica shows high pore volume and large accessible pores tailored between 4 and 12 nm (Fig. 9.21d), and supplementary intrawall pores are detected as well. Several other ways to produce a large-pore cubic $Ia\ \bar{3}\ d$ silica have been reported (178–180).

9.5.3.2 Tailoring Structure and Porosity A way to tune the pore size of the surfactant-directed inorganic material is to simply change the length of the surfactant carbon chain. Usually a linear relationship is observed between the pore size and the length of the carbon chain of a molecular template. With C_nTAB ($n = 8$ to 18), the pore size of the as-synthesized MCM-41 materials increases by about 0.45 nm when increasing n by two carbon atoms. It seems that the shortest chain surfactant from which a mesophase can be created is with $n = 8$. On the other hand, long-chain surfactants ($n > 20$) are not easily available and are practically insoluble in water, and the mesophases obtained are sometimes rather poorly ordered (136, 137, 181). Similarly, changes in molecular geometry and chain length of nonionic oligomeric surfactant and block copolymers permit a fine-tuning of the pore size. In these latter cases, the adjustment of pore size can be realized continuously by varying the concentration of the templating agent and by changing the composition of the copolymers or the block size (88, 169, 182, 183). The ratio of hydrophilic to hydrophobic blocks (EO/PO, x/y) can be decisive for the nature of the mesostructure. For instance, block copolymers possessing longer hydrophilic chains, such as F127, lead to materials having highly curved cage-like pores. In addition, the EO/PO ratio of the copolymer has also a marked influence on the pore size and wall thickness of the resulting materials. In general, for a triblock copolymer such as EO_x-PO_y-EO_x, the hexagonal phase is generated if the ratio x/y is within 0.1 and 1.0. Lower ratios typically lead to a lamellar phase, and higher values than 1.0 lead to mesostructures with spherical pores. Simulateneous tailoring the cage dimensions and pore openings of the SBA-16 silica is also possible by employing copolymer blends (P123 mixed with F127) both with a control of the synthesis temperature and time (184).

Another method to tailor pore size of mesoporous solids is to perform restructuring upon hydrothermal treatment. Applying aging treatments at different temperatures and for prolonged periods (from 24 hours up to several days) can efficiently modulate the nature of the mesophase. Such a treatment can either be performed directly in the mother liquor or at a different pH in fresh solutions (typically water or alcohol). For example, Kushalani et al. (185) proved that siliceous MCM-41-type mesophases could be restructured at elevated temperatures in its mother liquor resulting in pore size expansion from 3.7 to 5.9 nm. Also, postsynthesis treatments often improve the thermal stability of samples normally obtained at room temperature (with less shrinkage occurring during calcination) and seem to afford materials with higher structural quality. This improved stability could arise from increased condensation of the silanols within the silicate framework, leading to less silanol groups and thicker walls.

Both synthesis and aging temperature strongly influence the mesostructure formation of ordered mesoporous materials prepared with triblock copolymers, and this by altering micelle hydrophobicity and silica condensation. The solution reactions leading to mesophase formation are usually performed within a range of 35°C to 45°C depending on the block copolymer template. Increasing the synthesis temperature within this range, and beyond, renders the EO groups more hydrophobic, leading to micelles with larger hydrophobic core volume and smaller hydrophilic regions (182). The temperature and duration of the hydrothermal aging that is then applied after this first synthesis step are also critical. The aging temperatures are usually lower than those employed for zeolite synthesis, in the interval of 40°C to 150°C (often 90°C to 100°C). An aging of 12 hours up to several days is normally required to produce silica materials with satisfying quality. Higher aging temperatures in the preparation of SBA-15 always lead to larger pore diameters. Substantially larger pore volumes are obtained and the nature of the intrawall porosity is drastically modified (171, 186). The pore structure of SBA-15 clearly depends on the synthesis conditions and processing (thermal treatment) of the as-synthesized hybrid mesophase. The intrawall porosity, especially, is tuned by modifying the silica-template interactions upon thermal treatment. It is proposed that SBA-15 prepared with aging between 35°C and 60°C exhibits micropores with no apparent connection between mesopores (172, 186). In contrast, at 100°C, SBA-15 shows both the presence of micropores and larger connections between the hexagonally ordered mesopores. At 130°C, the material would exhibit no more micropores, but only much larger interconnections. The mesopore size of SBA-15 is easily tailored from about 5 nm up to 12 nm by increasing the aging temperature from 60°C to 130°C. The mesopore size of the cubic $Ia\bar{3}d$ KIT-6 silica can be varied within a similar range of diameters, as shown by the nitrogen sorption data presented in Figure 9.22 (187). Similarly, the size of the spherical mesopores of cagelike materials can be altered by applying different aging temperatures and times. However, not only the main mesocage is enlarging, but the pore openings of the cages are also becoming wider upon prolonged hydrothermal treatment (Fig. 9.22).

Fine-tuning of the pore size and mesostructure can additionally be performed by adjusting the solution pH upon addition of given amounts of acid or base during the synthesis. This is, for example, well established for MCM-41 and MCM-48 syntheses (188, 189). It is usually explained by the strong influence that the solution pH has on the degree of condensation and polymerization of the inorganic oligomeric species, on the charge density of the polyelectrolyte inorganic species involved, and on the surfactant packing parameter g. Likewise, the addition of salts, such as NaCl or KCl, affects the packing parameter g and the interaction conditions between the surfactant molecules and the silica. The surface charge density of the surfactant micelles can be modified by adsorbed counterions. The surfactant molecules may then self-assemble into a mesophase with a lower surface charge density, and phase transitions could take place upon modification of mesophase surface curvature.

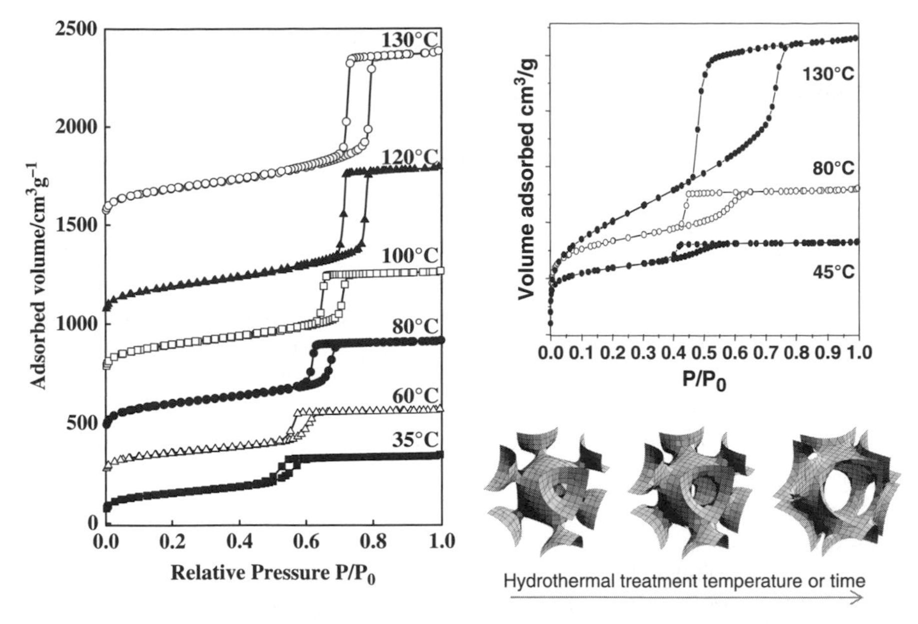

Figure 9.22 (a) Series of N2 adsorption-desorption isotherms measured at 77 K obtained for cubic *Ia* $\bar{3}$ *d* KIT-6 silica aged at the different hydrothermal treatment temperatures as indicated (the isotherms are plotted with offsets). (Reproduced with permission from T.-W. Kim et al., *J. Am. Chem. Soc.* **2005**, *127*, 7601. Copyright 2005 American Chemical Society.) (b) *Top*: N2 sorption isotherms measured at 77 K for cubic SBA-16 silicas obtained after aging at different temperatures. (Reproduced with permission from F. Kleitz et al., *Chem. Mater.* **2006**, *18*, 5070. Copyright 2006 American Chemical Society.) *Bottom*: Corresponding evolution of the cage-like structure of SBA-16 silica upon increasing hydrothermal treatment time and/or temperature. (Adapted and reprinted with permission from T.-W. Kim et al., *J. Phys. Chem. B* **2004**, *108*, 11480. Copyright 2004 American Chemical Society. Original figures courtesy of Dr. Yasuhiro Sakamoto, Stockholm University, Sweden.)

Electrolyte addition is also a way to affect the phase behavior of block copolymer-based mesoporous silicas. Usually, inorganic salts have a strong influence on the values of the CMC and CMT (critical micellar temperature) of the block copolymer micelles, which can be decreased upon salt addition. Salting-out electrolytes (lyotropic ions) such as KCl, NaCl or K_2SO_4, are not adsorbed in the copolymer micelles. These salts dehydrate the hydrophilic portion of the copolymer inducing a pronounced decrease in the preferential interfacial curvature of the micelles. On the other hand, salting-in electrolytes (hydrotropic ions) are adsorbed in the micelles and tend to inhibit their growth and could increase the preferential interfacial curvature (179, 190). Zhao and coworkers were the first to use the salting-out effects caused by the addition of electrolytes in combination with triblock copolymers to produce various well-defined mesostructures (191). It should be kept in mind, however, that not only the aggregation behavior of copolymer micelles is affected by salt additions, but the presence of electrolytes will influence hydrolysis and condensation, and the kinetics of aggregation of the inorganic species.

Finally, the solubilization of hydrophobic additives inside the core of the micellar assembly is largely exploited to increase the pore size of mesoporous silicas. Trimethylbenzene (TMB) has been the most widely used additive, although aliphatic hydrocarbons such as hexane have been used as well (192). For example, it was shown that the pore size of MCM-41 can be varied in a controlled manner between 1.5 nm and 10 nm by addition of TMB (137). Hydrocarbons or hydrophobic aromatics are regarded as swelling agents that are preferentially solubilized in the core of the micelles. Differently, co-surfactant molecules, such as short-chain *n*-alcohols or *n*-amines, are accumulated in the palisade layer of the micellar aggregates and therefore induce more intricate effects, whereby both the mesophase behavior and the *d*-spacing of the mesoscopically ordered material can be affected. In particular, mesopore size may be increased or, alternatively, decreased upon addition of short-chain alcohols or amines (193). Swelling of the SBA-15 mesophase with retention of the 2-D hexagonal mesostructure seems to be limited to a maximum pore size of 15 nm, or slightly above (194). The silica materials exhibiting pores reaching 30 nm that were obtained by addition of TMB to the synthesis mixture were in fact disordered mesocellar foams (MCF). On the other hand, *n*-butanol is also an effective co-surfactant molecule in the triblock copolymer systems. However, to direct a given silica-based mesophase, the adjustment of the HCl concentration to 0.3–0.5 M in the syntheses with butanol is a prerequisite. With this latter method, it became actually possible to synthesize several large-pore cubic silica mesophases ($Ia\bar{3}d$, $Im\bar{3}m$, and $Fm\bar{3}m$) in a wide range of reagent compositions (187, 195).

9.5.4 Functionalization of Mesoporous Materials

Methods for the functionalization of ordered mesoporous materials are numerous, and a considerable number of studies focus on the potential applications of functionalized mesoporous materials, especially in heterogeneous catalysis and separation technologies. Taking into account pore size and shape of the host and the volume accessible, it is possible to introduce varieties of guests and functional groups that may be active inorganic species (metallic cations, clusters, nanoparticles, and oxide layer), organic functional groups (acid, base, ligands, and chiral entities), molecular organometallic catalysts, polymers (fillings, coatings, and dendrimers), biomolecules, etc. Depending on the location of the functionalities, two main modifications are mostly conducted: surface-restricted functionalization and framework functionalization with species placed within the framework walls (54, 100, 196).

9.5.4.1 Surface Properties Since many of the functionalization methods make use of surface silanol groups (SiOH) as anchoring sites for metal species or silane coupling agents, data about surface hydroxyl sites become highly relevant for these modifications based on silylation or organosilane coupling. Silanol density in ordered mesoporous silica (e.g. MCM-41 or SBA-15) is considered to be much lower than that of conventional hydroxylated silica. In general, the silanol density of ordered mesoporous silicas ranges between 1 and 3 SiOH $/nm^2$ (197, 198). In contrast, silanol density of regular hydroxylated silica is about 4 to 6 SiOH $/nm^2$ (199). However, the population of silanols and their distribution on the mesopore surface will strongly depend on thermal treatments, activation under vacuum, and dehydroxylation-rehydroxylation processes. For the example of a standard SBA-15, a silanol density closer to 1 SiOH /nm^2 is observed if the material is heated at 200°C under vacuum (200). Furthermore, at least three different types of silanol groups are distinguished in MCM-41-type silica: single silanol $(SiO)_3SiOH$ (Q^3), hydrogen-bonded $(SiO)_3SiOH$—$OHSi(SiO)_3$, and geminal silanol groups $(SiO)_2Si(OH)_2$ (Q^2). However, only single and geminal silanols seem to be accessible to silylating agents such as trimethylchlorosilane or hexamethydisilazane. Besides, grafting of alkylorganosilanes on the mesopore surface will induce a passivation of silanol groups and thus offer a strategy to improve structural stability of materials, because it renders the walls more hydrophobic (201).

9.5.4.2 Mesoporous Organic-Inorganic Hybrid Materials The association of organic and inorganic units within a single material is attractive because it allows combining the enormous functional diversity of organic chemistry with the advantages of thermally stable and mechanically robust inorganic scaffolds. The inorganic skeleton ensures a stable ordered nanoporous structure. And the organic species integrated to the material allow fine-tuning of both the interfacial and bulk properties of the material (e.g. hydrophobicity, site accessibility, and physical properties). Two major strategies are used for the preparation of mesoporous organic-inorganic hybrid materials: (1) modification of the pore surface of a purely inorganic material by postsynthesis procedures. This is typically done by covalent grafting, but impregnation methods or adsorption are also sometimes employed. (2) One-pot direct incorporation of the organic groups via condensation reactions involving functional organosilane precursors. Figure 9.23a illustrates schematically the outcome of these two different strategies.

Postsynthesis grafting. The abundant methods available for surface modification of ordered mesoporous materials are mostly based on standard techniques that were developed for the surface functionalization of conventional silicas and used in chromatography (100, 202). By the postsynthesis method, organic functions are usually anchored covalently (or tethered) on the oxide walls, leading to hybrid mesostructured materials. Surface silylation is relatively straightforward using trialkylchloro- or trialkylalkoxysilanes. In practice, grafting takes place via a condensation reaction between organoalkoxysilanes of the type $(R'O)_3SiR$, and the free surface silanol groups. In principle, functionalization with a variety of organic groups can be realized this way by varying the nature of the organic residue R. However, the mesoporous hosts must be thoroughly dried before the addition of the organosilane precursors to avoid self-condensation in the presence of water. Post synthesis surface modification of mesoporous silica with basic functional groups is achieved by using condensation of surface silanols with 3-aminopropyltriethoxysilane for instance, resulting in -NH_2 groups placed on the pore surface. Differently, acid functionalities, such as sulfonic acid groups, may also be incorporated by postsynthesis grafting. Anchored functionalities can serve as acid or base catalysts and interaction sites, and also are helpful as ligands for the immobilization of large regio- and enantioselective molecular catalysts (e.g. organometallic complexes and chiral catalysts). However, one should bear in

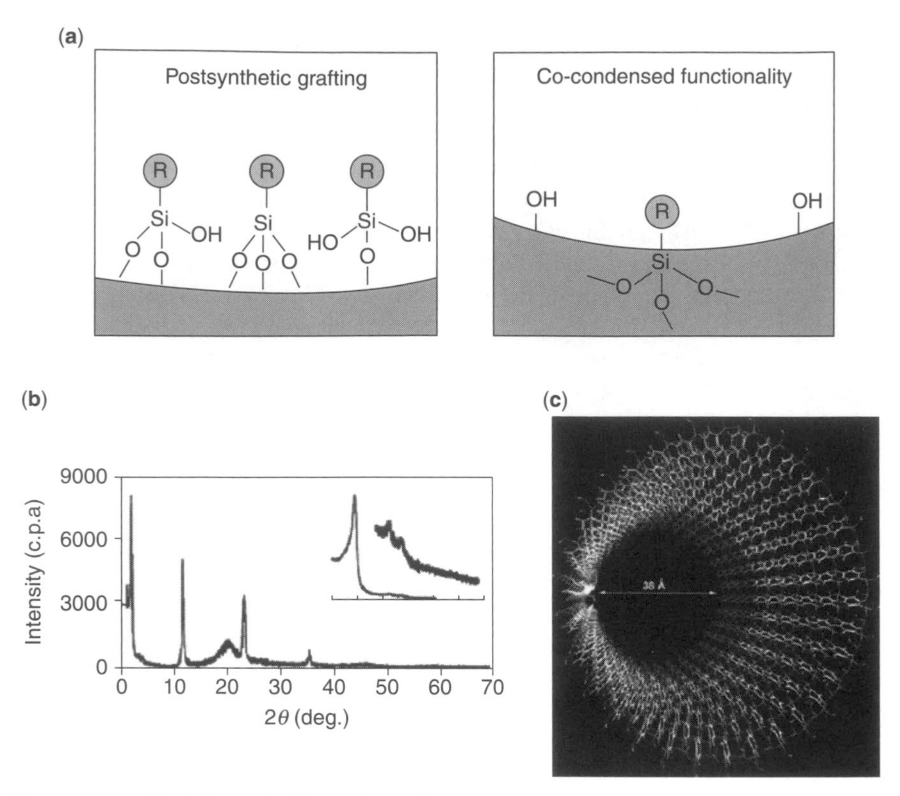

Figure 9.23 (a) Schematic illustration of the outcome of two different methods of pore surface functionalization either via postsynthetic grafting or by one-pot co-condensation. Functionalization may be achieved on regular mesoporous silica exhibiting available Si-OH sites by postgrafting of functional organosilanes (through siloxane coupling on the surface). The other method consists in co-condensation of a tetraalkoxysilane silicon source and a given functional organosilane. (Adapted from Reference 204.) (b) Powder x-ray pattern of a mesoporous benzene-bridged organosilica with crystalline walls. (c) Respective model pore system of the mesoporous benzene-bridged organosilica with crystalline walls. The benzene rings are aligned in a circular manner along the pore, contacted at both sides by silica layers. The benzene and silica layers are alternatively arranged along the pore axis with intervals of 0.76 nm. (Reproduced with permission from S. Inagaki et al., *Nature* **2002**, *416*, 304. Copyright 2002 Nature Publishing Group.)

mind that the concentration and distribution of the organic functionalities is dictated by the density of the surface silanols and their accessibility. Moreover, the grafting yield is known to depend on the precursor reactivity, and it is also limited by diffusion processes and steric effects.

One-pot direct synthesis of mesoporous hybrids. First, ordered mesoporous organosilicas can be obtained by co-condensation of tetraalkoxysilanes (TEOS or TMOS) and terminal trialkoxyorganosilanes of the type $(R'O)_3Si\text{-}R$ (where R′ is either methyl or ethyl and R is a nonhydrolyzable organic group). Here, the functional groups are most often placed dangling on the surface, but also partly inside the framework walls (203–205). Burkett, Sims, and Mann (206) and Macquarrie (207) introduced in 1996 this direct procedure for the formation of templated organic-inorganic hybrids by co-condensation. Macquarrie used 3-aminopropyltriethoxysilane as the organosilica precursor to prepare in one step mesoporous silica modified with pending amine groups. In this one-pot method, the organoalkoxysilane plays two roles since it contributes as building block in the inorganic structure, co-condensing with the tetraalkoxysilane precursor, and it supplies the organic functionality. Several functional groups may be introduced using this pathway (vinyl, phenyl, aminopropyl, imidazole, cyanopropyl, mercaptopropyl, more complex ligands, etc.). The mercaptopropyl groups are especially interesting since these groups can subsequently be oxidized with nitric acid or H_2O_2 to yield sulfonic acid groups. Margolese et al. demonstrated that oxidation of the mercaptopropyl groups can even be achieved *in situ* during the material formation with the addition of H_2O_2 directly in the reaction medium. However, the choice of a suitable organosilane precursor is usually limited by the conditions of synthesis, and furthermore, template removal can only be achieved by solvent extraction or acid treatment.

Another one-pot direct method is the use of bis-silylated organosilicas as single precursors, which allow for the direct incorporation of organic groups as bridging components inside the pore walls. Accordingly, organic functions are incorporated as bridges between silicon atoms within the wall. The use of silsequioxanes such as $[(RO)_3Si\text{-}R'\text{-}Si(OR)_3]$ is widespread in sol-gel chemistry for the preparation of disordered hybrid gels (209). Thus, similarly bridged silsequioxanes may be used through cooperative self-assembly with a structure-directing agent (CTAB or P123) to directly generate organically modified mesoporous silica. The resulting materials are characterized by well-ordered mesopores with narrow pore size distribution, resembling those of the pure silica equivalents. This procedure triggered a fast development of a new class of mesostructured materials, designated periodic mesoporous organosilicas (PMOs). As mentioned above, the organic groups R' are homogeneously incorporated as bridges between silicon atoms inside the mesoporous silica walls, enabling further addition of other functions without pore blocking. The first materials of this PMO class were obtained in 1999 by three independent research groups using 1,2-bis(trimethoxysilyl)ethane (BTME) (210) or 1,2-bis(triethoxysilyl)ethene (211, 212) as the single organosilica precursor combined with alkylammonium surfactants for the hybrid mesophase formation. The preparation of periodic mesoporous organosilicas is now extended to triblock copolymer systems leading to PMOs with much larger pores (213). A great variety of bridging groups are introduced this way, with R' = methylene, ethylene, vinylene, benzene, etc. Most interestingly, simultaneous meso- and molecular-scale periodicity occurs when 1,4-bis(triethoxysilyl)benzene (i.e. R' = phenyl) is used as the organosilica precursor. Inagaki and coworkers (214) demonstrated this feature by the formation of a particularly well-ordered 2-D hexagonal mesostructure showing an additional structural periodicity with a spacing of 0.76 nm along the channel direction (Fig. 9.23b). The crystal-like organization of the walls was confirmed by high resolution TEM images which indicated lattice fringes along the pore axis, the benzene rings being integral parts of the wall structure. Incidentally, the organic groups that are incorporated in the pore walls can themselves be postfunctionalized, for example, post synthesis sulfonation of benzene-bridged PMO.

9.5.4.3 ***Introduction of Heteroelements*** The isomorphous substitution of heteroatoms in mesoporous silica frameworks is especially important with respect to catalytic applications, since substitution of silicon allows fine-tuning of the acidity or creation of redox properties, in a similar way as done for amorphous alumosilicates or zeolites. Again, one can distinguish methods leading to the positioning of inorganic components preferentially on the mesopore surface, mostly as metal oxides, and substitution of silicon atoms by metal ions in the framework (i.e. isomorphous substitution). Generally, pore surface functionalization is accomplished by a postsynthesis grafting of metal centers such as Al, Ti, V, Cr, Fe, Mo, etc., on the surface silanols. On the other hand, modification of the mesoporous framework by substitution of silicon atoms with a tetrahedrally coordinated tri- or tetravalent element (Al, B, Fe, Ga or Ti) is possible via one-pot direct synthesis with adequate mixing of precursors (100, 196b, 215, 216). The outcome of the two different methods is often not identical. While the direct method typically results in a relatively homogeneous incorporation of the heteroelements in the material, postsynthesis treatments lead to higher concentrations of the heteroelement on the mesopore surface. Aluminum incorporation is of interest if, as in zeolites, this would result in the formation of ion-exchange sites and acid sites. Tetrahedrally coordinated aluminum atoms can be substituted for silicon atoms in the mesoporous silica framework, but the degree of substitution depends on the aluminum precursor and method of preparation. Generally, such an incorporation of aluminum in the framework of MCM-41 or other materials is claimed to increase the acid site concentration, ion-exchange capacity, and hydrothermal stability of the material (217). However, in general, the strength of acid sites was found to be similar to that of amorphous aluminosilicates. Nevertheless, they contain larger pores than zeolites and thus are more suitable for catalytic reactions involving bulky substrates (218). In order to create redox properties, incorporation of Ti, for instance, can be achieved via a direct synthesis procedure which involves addition of a titanium source, such as titanium isopropoxide in ethanol, to hydrothermal synthesis gels (219, 220). Postsynthesis addition of Ti is also possible by grafting techniques using $TiCl_4$, titanium alkoxide, or metallocene complexes (221–223). However, metal incorporation does not necessarily lead to the formation of well-defined sites as in zeolites, but rather to a wide variety of different sites with different local environments.

9.5.4.4 ***Introduction of Nanoparticles*** The supported noble metal catalysts that are used extensively in industry typically exhibit metal particle sizes in the nanometer range (4). These materials are very effective as catalysts because of the large surface area and high fraction of atoms present at the surface of the metal particles (i.e. high surface to bulk ratio). However, such small particles are unstable with respect to agglomeration because of their high surface energies. The formation of agglomerated bulk material having much lower surface area is favored by a reduced total free energy especially at elevated temperatures. A stabilization of the nanoparticles against such a sintering process is therefore needed and is typically performed by deposition on suitable support materials. Practically, stabilization of the metal particles (e.g. supported Pt, Ni, and Pd nanoparticles) is most commonly achieved by their deposition onto porous supports such as silica, alumina, or carbon, which also supply high surface area (100, 224–226). Ordered mesoporous materials with their high surface areas are thus perfect candidates for the insertion of nanoparticles with high dispersion. Compared to other supports, ordered mesoporous solids permit the effective stabilization of metal or metal oxide nanoparticles, because nanoparticles are prevented from growing to sizes larger than the pore size unless they move out to the external surface of the support (100). On the other hand, compared to smaller pore systems, such as microporous solids, mesoporous support materials allow for the access of bulky reagents. Due to the fact that the primary particle size of the guest is relatively small compared to the mesopore dimensions in many of these supports, pore blocking is not really severe up to a relatively high loading of nanoparticles.

Several methods are available for depositing the active compounds on mesoporous supports. These are actually applied to introduce all sorts of nanoparticles, for example, metals (Ni, Pt, Pd, Rh, Co, Ru, Au), mixed metals (Ru-Pt, Cu-Ru, Rh-Pt), metal oxides (ZrO_2, Fe_2O_3, In_2O_3, ZnO, TiO_2, etc.) and metal chalcogenides (CdS, ZnS, MoS_2, WS_2). As detailed by Bronstein (226), high resolution TEM imaging techniques and XRD/SAXS are the methods of choice for the characterization of nanoparticles containing mesoporous materials. Supplementary techniques will also be used to assess particle size, particle size distribution, positioning, and interfacial host–guest interactions (chemisorption methods, EXAFS, diffuse reflectance UV-vis spectroscopy, temperature-programmed reduction, etc.).

Classical synthesis methods such as impregnation, adsorption, ion exchange with metal cations, and deposition-precipitation are widespread. Impregnation with metal compounds is usually followed by reduction or thermolysis. However, often the conventional methods yield nanoparticles of various sizes inside the mesopores but also on the external surface of the support. Other strategic options exist in order to favor a more selective and controlled insertion of nanoparticles into mesoporous materials. For instance, it is possible to take advantage of chemical interaction of the metal compounds with functional groups placed on the surface of mesoporous materials. This procedure would involve ion exchange or grafting to the functional groups and a subsequent reduction step for the formation of metal species. Immobilization of polynuclear molecular metal complexes or clusters such as metal carbonyl systems, for example, $Co_2(CO)_8$, $[Pt_3(CO)_6](N(CH_3)_4)_2$, $[H_2Ru_{10}(CO)_{25}]^{2-}$, $[Ru_5PtC(CO)_{15}]$, is also a method put into practice for preparing metal nanoparticles or bimetallic nanoparticles. Differently, direct addition of metal compounds or precursors in the synthetic gel for the formation of the mesoporous material was also studied for some noble metals and oxide systems. Many other methods such as chemical vapor deposition, inclusion of presynthesized nanoparticles, vacuum evaporation impregnation, controlled incipient wetness, and impregnation in the presence of supercritical fluids may also be used to circumvent issues of controlling particle size distribution, dispersion, and localization (100).

9.5.5 Nonsiliceous Mesostructured and Mesoporous Materials

9.5.5.1 ***Oxides*** Compared to silica-based networks, nonsiliceous ordered mesoporous materials have attracted less attention, due to the relative difficulty of applying the same synthesis principles to non-silicate species and their lower stability (227). Nonsiliceous framework compositions are more susceptible to redox reactions, hydrolysis, or phase transformations to the thermodynamically preferred denser crystalline phases. Template removal has been a major issue and calcination often resulted in the collapse of the mesostructure. This was the case for mesostructured surfactant composites of tungsten oxide, molybdenum oxide, and antimony oxide, and mesostructured materials based on vanadia that were obtained at early stages. Because of their poor thermal stability, none of these mesostructures were obtained as template-free mesoporous solids (85, 228, 229).

The first porous transition metal oxide was reported by Antonelli and Ying in 1995 (230). The material was prepared by using an anionic surfactant with phosphate head groups and titanium alkoxy-precursors stabilized by bidentate ligands. After calcination at 350°C, hexagonally ordered porous TiO_2 containing stabilizing phosphate groups in the framework was obtained with a surface area of 200 $m^2\ g^{-1}$ and pore size of 3.2 nm. Subsequently, a variety of other mesoporous materials based on titanium, zirconium, niobium oxide, and tantalum oxide were similarly synthesized (231). For the niobium and tantalum oxides, the redox stability problem was solved using a special ligand-assisted templating pathway, followed by extraction of the template, thus avoiding calcination. This ligand-assisted pathway is based on covalent bonding between the inorganic species and the surfactant head groups (a long chain amine).

An important step forward was brought by Stucky and coworkers (232), who developed a synthesis procedure for a wide range of nonsiliceous oxides (e.g. TiO_2, ZrO_2, Nb_2O_5, and Ta_2O_5) by using the polyalkylene oxide triblock copolymers under nonaqueous conditions. Thermally stable ordered mesoporous oxides were obtained with large pore sizes reaching 14 nm. Interestingly, these mesoporous materials contain some nanocrystalline domains within their thick framework walls, whereas previous materials had amorphous pore walls. While silica walls remain amorphous upon thermal treatment, crystalline phases can arise in transition metal-based frameworks (Fig. 9.24). However, growth of nanocrystals beyond the wall thickness can deteriorate the mesostructure. In the case of materials based on block copolymers, the walls are much thicker, allowing for crystallization without loss of mesostructure. Here, the assembly of these non-silica hybrid mesophase is assumed to proceed through a mechanism which combines the block copolymer self-assembly and alkylene oxide complexation of the metal species.

Obviously, the kinetics of inorganic condensation are very fast for the non-silica systems, and it can represent an issue for mesostructure formation. Nevertheless, there are several known strategies for retarding condensation, such as complexation, acidification, or controlled hydrolysis (69). One way is to selectively favor hydrolysis and limit the condensation by using highly acidic media ($pH < 1$) under EISA conditions (Section 9.5.1) as proposed by Grosso et al. (233) and Crepaldi et al. (234).

This strategy is especially useful for producing thin films and aerosols. For example, ordered mesoporous titania and zirconia thin films are obtained via this pathway. Besides, in the first work of Stucky and Yang mentioned previously, the mesoporous materials were described with only partially crystallized inorganic walls. In contrast, the strategy proposed by Crepaldi and coworkers (235, 236) and Grosso et al. (237) leads to materials with high crystallite fractions by adequately combining structuring approaches with block copolymers and the EISA method, resulting in thermally stable nanocrystalline titania, zirconia, and perovskite thin films.

Alumina is a major support in heterogeneous catalysis and its use is widespread. This is justified by the fact that it is often considered superior to silica because of a higher hydrolytic stability and its different point of zero charge that facilitates loading with different metal species. In addition, alumina is less expensive and available in a wide range of values of surface area and porosity. Consequently, a lot of research focuses on the possibility of controlling structure and porosity of mesoporous alumina and improving thermal stability (227, 238). Mesoporous alumina has been synthesized by a variety of methods using anionic, cationic, and neutral surfactants. The first synthesis of MCM-41-related mesoporous alumina was reported by Bagshaw and Pinnavaia (239), who employed nonionic poly(ethylene oxide) as the structure-directing agent and aluminum tri-*sec*-butoxide as the aluminum source. The specific surface area of the material was around 400 to 500 $m^2\ g^{-1}$. However, the low-angle XRD pattern of the resulting material showed a single broad peak indicative of a rather disordered pore structure. Stucky's approach (Section 9.5.1) also afforded alumina with quite large pore sizes. However, this alumina also exhibited a disordered pore structure. From several other studies, it became clear that in the alumina system the use of surfactant-type molecules for the synthesis does not necessarily follow a real templating route. The possibility that the surfactant does not act as a porogen, but rather stabilizes small alumina particles should be taken into account. Significant progress in the preparation of ordered mesoporous alumina came finally from the group of Sanchez. Again, by combining triblock copolymer structure-directors and the EISA method, they managed to prepare nanocrystalline γ-alumina films exhibiting ordered mesopores in a 3-D cubic structure (240). The mesopores are ellipsoidal, with the shorter pore axis ranging from 3 to 6 nm in size and the larger pore axis ranging from 13 to 24 nm (Fig. 9.24). These ordered mesoporous alumina films show excellent thermal stability up to 900°C.

The nanocasting approach is proven to be highly suitable for the preparation of mesostructured metal oxides of various compositions that are usually impossible or difficult to prepare by direct templating methods because of their poor redox stability. This method is becoming increasingly popular and used widely nowadays for the preparation of functional porous materials aiming at applications in sensing, magnetic materials, energetic materials, catalysis and so forth. As a few examples, ordered mesoporous oxides such as In_2O_3, Co_3O_4, Fe_2O_3, CeO_2, Cr_2O_3, NiO, and WO_3, etc. are all accessible by direct replication of ordered mesoporous silica materials (241–246). Moreover, the framework walls of these templated mesoporous oxides are composed of crystalline nanoparticles in most of the cases reported, the rigid silica scaffold allowing for a sufficient thermal treatment to be performed. Hard

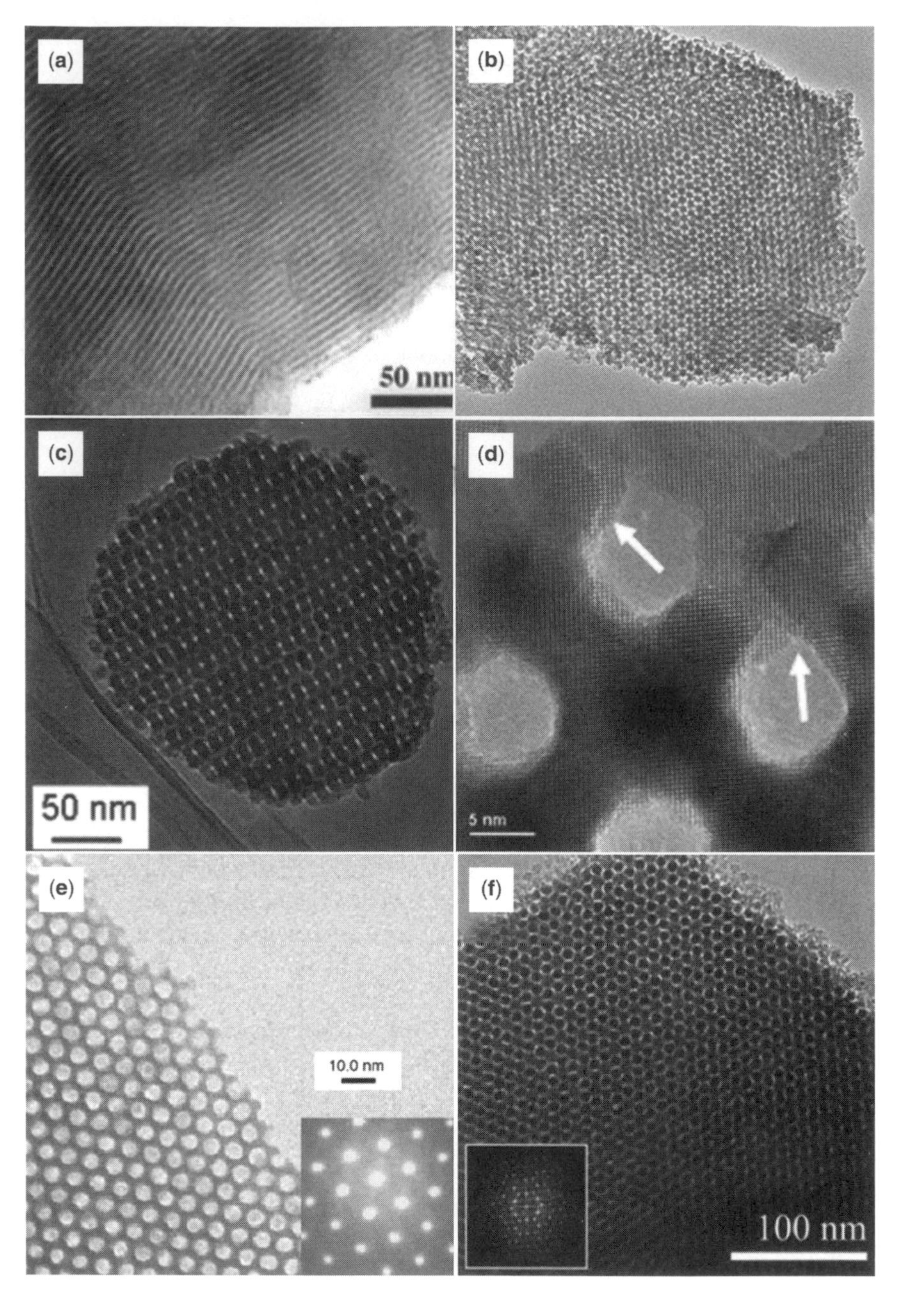

Figure 9.24 (a) TEM image of ordered mesoporous zirconia directed by triblock copolymer micelles. (Reproduced with permission from P. Yang et al., *Chem. Mater.* **1999**, *11*, 2813. Copyright 1999 American Chemical Society.) (b) TEM images of nanocrystalline γ-alumina film showing ordered mesoporosity (3-D face-centered cubic structure). The pores are ellipsoidal, with the shorter pore axis ranging from 3 to 6 nm and the larger pore axis ranging from 13 to 24 nm. (Reproduced with permission from M. Kuemmel et al., *Angew. Chem.* **2005**, *117*, 4665.

templates such as SBA-15, KIT-6, and SBA-16 are most frequently chosen because of their high porosity, large pore walls, and 3-D pore network interconnectivity. In the case of SBA-15, for example, the complementary pores connect sufficiently the hexagonally packed linear mesopores, thus providing the cross-linking required for obtaining a stable replica. Also, we have seen previously that pore size (and pore wall thickness) and degree of mesopores connectivity of the copolymer-templated silicas may be modulated by altering the aging treatment applied during synthesis (time and temperature). Accordingly, the nature of the structure and porosity of the replica can be tailored as a function of the parent silica template. This is well demonstrated in the preparation of carbons (247), but it can also be valid in the case of metal oxides and other compositions. The porosity of a nanocast Co_3O_4, for example, may be tuned according to the network topology of the hard template, for example, SBA-15 and KIT-6 silicas obtained with different aging temperatures, and as a function of the loading in metal precursor species (244). With respect to this, an incomplete or non uniform filling of the template pores can sometimes lead to metal oxide replicas exhibiting an unexpected bimodal mesopore system, especially when using bicontinuous cubic KIT-6 templates (244c). For compositions that are not stable in NaOH or diluted HF solutions, the use of other rigid templates will be necessary, carbon mesostructures for instance, which can be eliminated by a combustion step. This is the case for the amphoteric oxides such as aluminum and zinc oxides, as well as basic oxides like MgO (92, 93).

Another approach, which is related to the hard templating method, can be applied to create crystalline mesoporous transition metal oxides, for example Nb-Ta oxide. The post synthetic filling of the pores of a mesostructured transition metal oxide with carbon (or silica) is a valuable way to increase the thermal stability of the metal oxide framework, thus facilitating high-temperature treatment for inducing crystallization (248). In this approach, the amorphous Nb-Ta oxide is first synthesized by supramolecular soft templating, and subsequently the pore system of this material is filled with a carbon precursor. After solidification and pyrolysis taking place within the metal oxide pores, high-temperature treatment of the composite thus obtained leads to crystallization of the Nb-Ta oxide walls. In this process, the introduced

---

Figure 9.24 (*Continued*) Copyright 2005 Wiley-VCH Verlag.) (c) TEM image of a mesostructured Co_3O_4 particle obtained by the hard templating route (templated from KIT-6 silica). (d) High-resolution TEM image of crystalline Cr_2O_3 nanocast templated from KIT-6 silica (image viewed along the [221] zone axis of the unit cell). The arrows indicate the same crystal orientation at different parts of the mesoporous structure (single-crystal nature). [(c) and (d) reproduced with permission from C. Dickinson et al., *Chem. Mater.* **2006**, *78*, 3088. Copyright 2006 American Chemical Society.] (e) TEM image and electron diffraction pattern of high-temperature crystallized ordered mesoporous Nb-Ta oxide. (Reproduced with permission from T. Katou et al., *Angew. Chem. Int. Ed.* **2003**, *42*, 2382. Copyright 2003 Wiley-VCH Verlag.) (f) TEM image of ordered mesoporous silicon nitride templated from cubic KIT-6 silica. The inset shows the corresponding Fourier diffractogram. (Reproduced with permission from Y. F. Shi et al., *Adv. Func. Mater.* **2006**, *16*, 561. Copyright 2006 Wiley-VCH Verlag.)

carbon skeleton strengthens the mesopore structure sufficiently for the crystallization of the oxide walls at high temperatures. In contrast to standard nanocasting procedure, the parent oxide framework is not removed. Removal of the carbon component, which should rather be regarded as a scaffold instead of a template, leads to an ordered mesoporous niobium-tantalum oxide with crystalline walls having single-crystalline domains and regularly arranged mesopores (Fig. 9.24). It is reasonable to think that this consolidation pathway would be suitable also for other non-silica compositions.

9.5.5.2 Mesotructured Metals and Other Non-Oxide Compositions Templated synthesis of mesoporous metal phosphates, sulfides and selenides, metals, and many other non-oxide compositions is possible (227, 249–251). For example, mesostructured zirconium-based materials showing hexagonal and cubic mesostructures analogous to MCM-41 or MCM-48 are readily produced by using zirconium sulfate as the inorganic precursor and alkylammonium surfactants. Postsynthesis treatment with phosphoric acid is performed to produce zirconium oxo-phosphate materials with improved thermal stability (252, 253). This latter mesostructure is stable up to 500°C owing to the presence of phosphate groups in the structure delaying crystallization, so that a disordered wall structure favorable for mesoporous materials is maintained. In the case of sulfides, Trikalitis et al. (254) suggested a coassembly process involving Ge_4S_{10} units together with Fe_4S_4 clusters in the presence of surfactants, resulting ultimately in a well-ordered hexagonal structure.

Ordered mesoporous noble metals can be prepared by direct supramolecular templating strategies, as reported by Attard and coworkers (255). The syntheses are based on the TLCT mechanism, where the lyotropic liquid crystalline phase is infiltrated with the metal precursor ions, which are then reduced within the liquid crystal. The surfactant, for instance, octaethylene glycol monohexadecyl ether, is usually removed by washing. The porous metals are obtained as powders or as mesostructured metallic thin films. Mesoporous metals, such as platinum or nickel, are hexagonally ordered and have pore sizes around 3 nm and surface area of 20 to 85 $m^2\ g^{-1}$, depending on the metal density. This strategy is also applicable to metal alloys and nonmetallic elements (140, 255). Alternatively, mesostructured metals (e.g. Pt, Au, or Ag) are also prepared by infiltrating mesoporous silica hard templates with metal precursor complexes and subsequent reduction. The resulting nanostructures grow in the confined space of the pores (256, 257). Provided that loadings are high enough, the metal nanocrystals do cross-link properly to form a continuous framework. Finally, hard templating is also a method of choice for the preparation of silicon carbides (Fig. 9.24; Reference 258) and different metal sulfides (259).

9.5.5.3 Ordered Mesoporous Carbons The first ordered mesoporous carbons were synthesized by hard templating methods (141). The use of porous inorganic silica templates allowed the preparation of series of ordered mesoporous carbon materials, designated as CMK-*x* materials (*carbon molecular sieves Korean Advanced Institute of Science and Technology*). The first one, CMK-1, was prepared using MCM-48 as a template and sucrose as the carbon source impregnated in the presence of sulfuric acid, followed by pyrolysis at 900°C under vacuum or nitrogen. After that, the silica framework is dissolved either by an NaOH solution or with HF. CMK-1 actually exhibits a lower symmetry compared to the parent MCM-48 silica, and this is originating from a periodic shift in the framework during the replication process. Differently, the ordered mesoporous carbon, named CMK-3, is synthesized from an SBA-15 rigid template, and this carbon is actually the first faithful

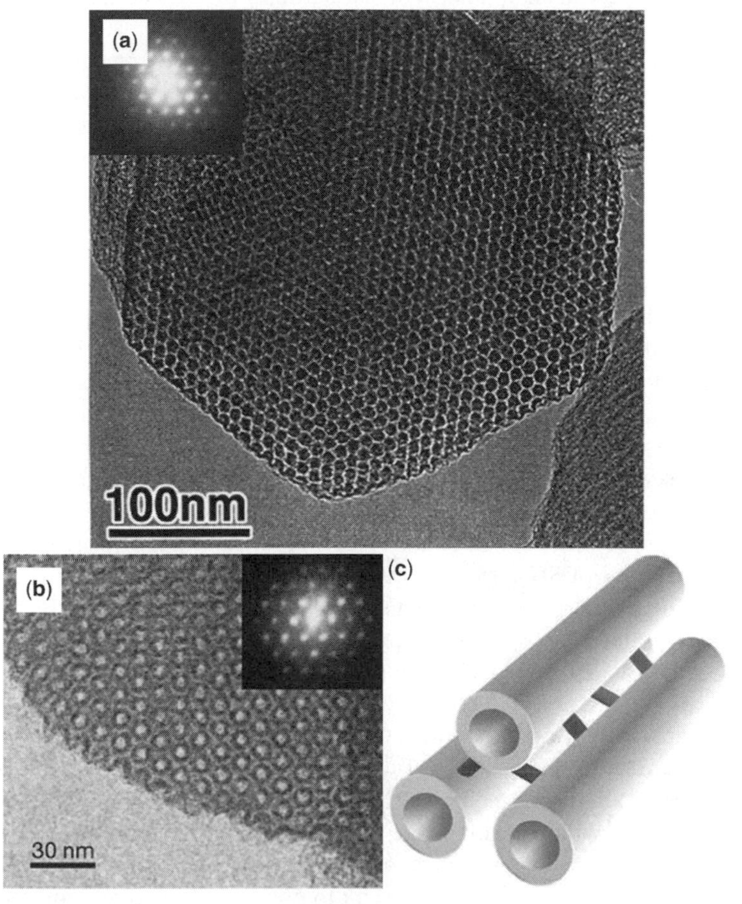

Figure 9.25 (a) A TEM image of ordered mesoporous carbon CMK-3 molecular sieve consisting of a 2-D hexagonal arrangement of carbon rods. The inset shows the corresponding Fourier diffractogram. The structure of the CMK-3 carbon is an exact inverse replica of SBA-15. (Adapted and reproduced with permission from S. Jun et al., *J. Am. Chem. Soc.* **2000**, *122*, 10712. Copyright 2000 American Chemical Society.) (b) TEM image of CMK-5 viewed along the direction of the ordered carbon tubes and the corresponding Fourier diffractogram. (c) Schematic representation of the carbon structure. The structural model shows that the carbon nanopores are rigidly interconnected into an ordered hexagonal array by carbon spacers. (Reproduced with permission from S. H. Joo et al., *Nature*, **2001**, *412*, 169. Copyright 2001 Nature Publishing Group.)

inverse replica of its silica template. More precisely, CMK-3 consists of uniformly sized amorphous carbon rods arranged in a 2-D hexagonal pattern, as shown in Figure 9.25, reflecting the mesostructure of the parent SBA-15. When the interconnecting intrawall pores of SBA-15 are properly filled with the carbon precursor, they allow for structure-supporting links between the carbon rods (146). Mesoporous carbons can be prepared with a variety of pore shape, connectivity, and pore wall thickness, as a function of the pore structure and pore diameter of the parent silica template. In addition, they conserve the shape and morphology of the particles of the chosen silica template. Furthermore, the composition and microstructure of the carbon material is influenced by the nature of the carbon precursor. Monomeric or polymeric precursors supplying reasonable carbon yields can be used as the carbon source, for example, sucrose, furfuryl alcohol, phenol-formaldehyde resin, acenaphthene, polydivinylbenzene, acetylene, acrylonitrile, and pyrrole (147, 150, 260–262), in the presence of an appropriate polymerization catalyst (aluminum sites, oxalic acid, *p*-toluenesulfonic acid, $FeCl_3$, etc.).

Furthermore, the filling degree of the carbon precursor in the pores of ordered mesoporous silica can be controlled, and under appropriate processing conditions, carbon materials with different structure and texture can be generated. Accordingly, the carbon structures formed within the mesopores of silica can be tuned to be either rod-type or tube-type structures, depending on the amount of precursor and the pyrolysis conditions. If the pore system of large-pore silicas such as SBA-15 (or KIT-6) is only coated by the carbon precursor instead of being completely filled, a surface-templated mesoporous carbon, designated CMK-5 (263), is generated instead of the volume-templated CMK-3-type material. Mesoporous CMK-5 carbon is composed of hexagonal arrays of carbon nanopipes that were originally formed within the cylindrical channels of SBA-15. As a consequence, the removal of silica results in a material with bimodal pore system. One pore system is generated in the inner part of the channels that are not completely filled by carbon. The other pore system arises from the spaces previously occupied by the silica walls of the SBA-15. A TEM image of CMK-5 is shown in Figure 9.25, with a schematic representation of the pore structure. CMK-5 has extremely high surface area, reaching $2500\,m^2g^{-1}$, and large pore volume, above 2 cc/g. Tubular CMK-5-type carbon is most easily synthesized by using aluminum-modified SBA-15 or KIT-6 silica and furfuryl alcohol as a carbon precursor, with the carbonization performed by vacuum pyrolysis, which is favorable for the formation of a coated carbon film on the mesopore walls.

Ordered mesoporous carbons are difficult to obtain through direct soft templating strategies. Nevertheless, new methods for the synthesis of mesoporous carbon via cooperative self-assembly of resorcinol or phenol-formaldehyde resin-type polymers in combination with triblock copolymers under sol-gel processing conditions provide an interesting alternative to the nanocasting pathway. Indeed, after a controlled assembly of the ordered mesostructured rigid polymer, the copolymer template that was used to form the system can successfully be removed without degrading the structure. A highly ordered mesoporous carbon material can subsequently be produced by pyrolysis of the resulting mesoporous polymer. Currently, these ordered mesoporous carbons are still obtained mainly as films (264–266).

9.5.6 Morphology

Morphology control is indispensable in many of the advanced applications envisioned for functional mesoporous materials (54, 267). Permselective membranes, microspheres, or monoliths are important for sorption, separation, and chromatography purposes. Porous thin films or fibrous structures are relevant for electronics, optics, low *k*-dielectrics, and sensing applications. Colloidal particles or nanospheres are preferred for biomedical systems to be used in drug delivery or magnetic resonance imaging (MRI) with contrast agents.

The first ordered mesoporous materials that were synthesized were typically finely divided powders consisting of small particles ($<10\ \mu m$) with no well-defined morphology. Later, the TLCT mechanism allowed the preparation of materials with monolithic structures. Since then, a wide variety of shapes, including thin films, spheres, fibers, tubes, and many other complex morphologies have been described for ordered mesoporous materials [see Fig. 9.26 for examples (268–272)]. Porous solids with controlled macroscale morphology can be designed either by processing conditions, such as dip-coating, spin-coating, or emulsion templating, or alternatively, formed spontaneously upon self-organization processes that are mostly based on kinetic regimes in which equilibrium phases may be replaced by high-order organizations determined by local minima. Yang, Coombs, and Ozin (273) have shown that the synthesis of mesoporous silica according the mediated $S^+X^-I^+$ cooperative assembly route under acidic conditions is particularly adapted for the preparation of a large variety of different morphologies. This is attributed to the weaker interactions between surfactant and silica species and less charged framework walls, in comparison to the base-derived composite mesophases, allowing for more flexibility in the mesophase. Ordered thin films of thickness 0.2 to 1.0 μm can be prepared under acidic conditions by heterogeneous nucleation at the air-water and oil-water interfaces as free-standing films (165, 274), and at the solid-liquid interface (275). Also, self-supporting transparent thin films with hexagonal phase can be prepared by slow evaporation of a sol mixture containing TMOS, water, a cationic surfactant (e.g. CTA^+Cl^-), and a mineral acid (276). Dip-coating or spin-coating techniques are methods particularly well suited for the preparation of mesostructured thin films exhibiting various structures of mesophase (277). Following the EISA method, slow and efficient evaporation of the solvent induces the surfactant concentration to cross the CMC, which then triggers mesophase formation by the co-assembly of silicates and ionic surfactants. In addition, postprocessing is possible to improve ordering or even change the nature of the phase. These dip- and spin-coating techniques are also applied using nonionic surfactants and triblock copolymers to prepare oriented mesoporous thin films with 2-D hexagonal, 3-D hexagonal, and cubic structures via similar EISA procedures (233, 234, 278).

Particles with spherical morphology are easily prepared under alkaline conditions. For instance, Huo et al. (279) reported the preparation of hard transparent spheres from an emulsion at room temperature. Grün and coworkers (154, 280) modified the Stöber synthesis (281) of monodisperse spheres in the presence of ethanol and ammonium hydroxide and successfully prepared monodispersed mesoporous MCM-41 and MCM-48 spheres (Fig. 9.26c). Furthermore, *pseudomorphic* transformation of

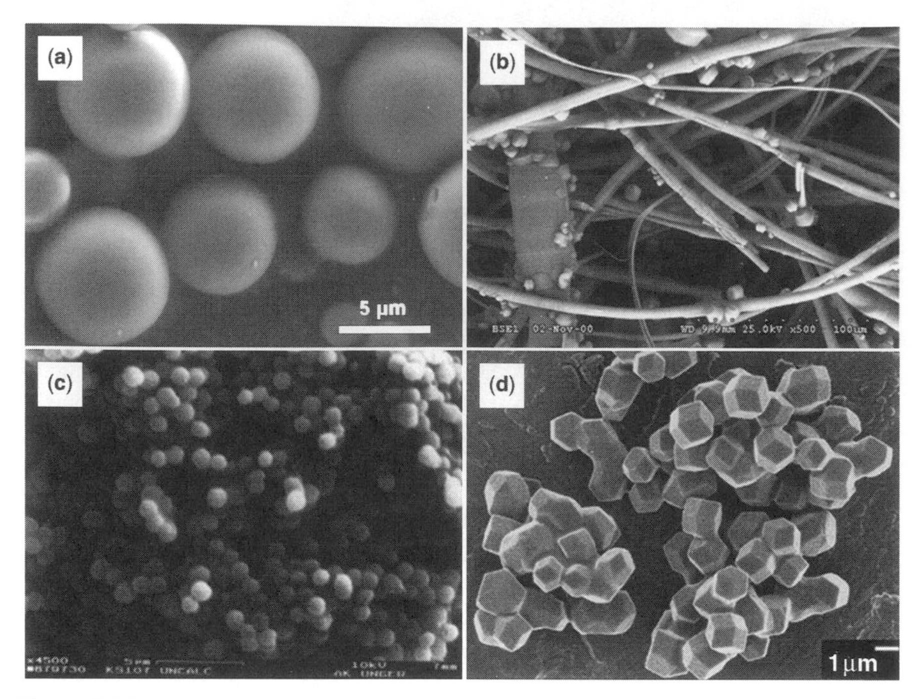

Figure 9.26 (a) SEM image of spherical, benzene-bridged PMO particles with diameter of approximately 5 μm. (Reproduced with permission from F. Hoffman et al., *Angew. Chem. Int. Ed.* **2006**, *45*, 3216. Copyright 2006 Wiley-VCH Verlag.) (b) SEM picture of mesoporous silica fibers synthesized in acidic medium with tetrabutoxysilane as a silicon source and CTAB as structure-directing agent under static conditions. (Reproduced with permission from F. Kleitz et al., *Chem. Mater.* **2001**, *13*, 3587. Copyright 2001 American Chemical Society.) (c) Example of colloidal mesoporous silica spheres obtained by modified Stöber synthesis. (Reproduced with permission from K. Schumacher et al., *Microporous Mesoporous Mater.* **1999**, *27*, 201. Copyright 1999 Elsevier.) (d) SEM image of single crystals of mesoporous silica (cubic Im $\bar{3}m$ mesostructure) prepared with a triblock copolymer template in the presence of K_2SO_4 salt. (Reproduced with permission from C. Z. Yu et al., *J. Am. Chem. Soc.* **2002**, *124*, 4556. Copyright 2002 American Chemical Society.)

commercially available, preshaped spherical silica particles (5 to 800 μm) can be used to produce ordered mesoporous MCM-41 and MCM-48-like particles having spherical morphology. In this latter method, amorphous silica is dissolved progressively and locally under mild alkaline conditions and reprecipitated at the same rate in the presence of the surfactant, without modifying the global spherical morphology (282). The synthesis of hollow spheres is also attracting much attention and it is achievable, for example, by spray-drying techniques, based on very rapid solvent evaporation and retention of preformed shapes (283).

Hollow silica tubes 0.3 to 3 μm in diameter and exhibiting mesopore channels along the tubular axis were described in 1996 (284). Such an organization is produced by careful control of the surfactant to water ratio and the rate of silica condensation at elevated pH. However, more typical sol-gel methods used for preparing silica gel fibers by spinning a high viscosity silica sol or by freezing a solution of silicic acid in water may be applied to produce mesoporous silica fibers, usually under acidic conditions. Bruinsma and coworkers introduced the spinning process to synthesize mesoporous fibers with a cationic alkyltrimethylammonium surfactant (283). In this case, the dry spinning process is based on slow solvent evaporation, driving mesophase formation, and an increased viscosity favored by the addition of polyethylene oxide. The fibers, which are drawn from the mixture into continuous filaments, present pore channels oriented parallel to the fiber axis. Mesoporous silica fibers can also be prepared with larger pores (>5 nm) by drawing the gel strand from a highly viscous amphiphilic block copolymer-silicate solution (285). Nevertheless, spontaneous formation of mesoporous silica fibers at an oil-water interface (e.g. organic silicon source-water interface) is also observed under specific acidic conditions (286, 287). The growth of well-defined ordered silica fibers, which are mesostructured by surfactant templating, takes place preferentially when the oil-water interface is kept static (Fig. 9.26b). The fibers thus obtained are optically transparent in the visible region and are especially appropriate hosts for doping with laser dyes. Interestingly, optical and TEM investigations revealed that these fibers have a peculiar circular inner architecture, consisting of hexagonally organized channels running circularly around the fiber axis.

In general, microporous solids such as zeolites have well-defined crystal morphology that corresponds to their order at the atomic level. However, it is rather unusual to obtain ordered mesoporous materials with well-defined crystal shapes. One of the first examples was the work of Ryoo and coworkers (160). These authors have reported the synthesis of MCM-48 crystals with a truncated rhombic dodecahedral shape, favored through the addition of inorganic salts. It turns out that several other silica materials and even organosilica hybrids can be produced with single crystal morphology, such as for instance cubic crystals with well-defined decaoctahedron shape (174, 288). The possibility of controlling the shape of the particles was also confirmed for triblock copolymer-based silicas. Large-pore cubic mesoporous silica single crystals, equivalent to SBA-16, showing particles with uniform rhombodecahedral shape can be obtained in the presence of inorganic salts, such as K_2SO_4 (Fig. 9.26d; Reference 191).

9.6 CONCLUSION

The huge body of knowledge and data reported in the vast and always growing number of scientific publications, textbooks, and patents dealing with ordered microporous and mesoporous materials could here only be addressed in a rather general and selective overview of some of the important concepts, major developments, and prospects. Readers interested in more comprehensive accounts and specific surveys on the different aspects of these fascinating materials are invited to consult sources listed in Further Readings and References. Nonetheless, it is evident now that the science of nanoporous materials has become a truly interdisciplinary research area, where chemistry, physics, biology, material science, and engineering meet. Although synthesis, characterization and applications of nanoporous materials are becoming mature technologies, the potential for further development and progress is still enormous. This is especially true in the case of the more recently developed metal-organic frameworks and ordered mesoporous materials, for which research will be aiming more and more toward manipulation and targeted application, although progress in synthesis, design, and characterization are still wanted. Even so, despite being massively used in industry, new useful zeolite materials are still being developed in academic or industrial laboratories around the world through innovative strategies for new synthesis, framework design, conception of transport pathways, and processing. Moreover, research in the field of nanoporous materials is attracting growing interest because many of the modern issues concerning environment considerations, energy management, and health may actually be tackled by implementing or integrating specifically designed nanoporous systems.

FURTHER READINGS

J. Cejka, H. van Bekkum, A. Corma, F. Schüth (Eds.), *Introduction to Zeolite Science and Practice*, 3rd revised edition, *Stud. Surf. Sci. Catal.* 168, Elsevier, Amsterdam, 2007.

G. Ertl, H. Knözinger, F. Schüth, J. Weitkamp (Eds.), *Handbook of Heterogeneous Catalysis*, 2nd edition, Wiley-VCH Publishers, Weinheim, 2008.

P. Gómez-Romero, C. Sanchez (Eds.), *Functional Hybrid Materials*, Wiley-VCH, Weinheim, 2004.

S. Lowell, J. Shields, M. A. Thomas, M. Thommes, *Characterization of Porous Solids and Powders: Surface Area, Pore Size and Density*, Springer-Verlag, Berlin, 2004.

G. Q. Lu, X. S. Zhao (Eds.), *Nanoporous Materials: Science and Engineering* (Series on Chemical Engineering), Imperial College Press, London, 2005.

G. A. O. Ozin, A. C. Arsenault, *Nanochemistry: A Chemical Approach to Nanomaterials*, The Royal Society of Chemistry, Cambridge, 2005.

F. Schüth, U. Simon, F. Laeri, M. Wark (Eds.), *Host-Guest-Systems Based on Nanoporous Crystals*, Wiley-VCH, Weinheim, 2003.

F. Schüth, K. S. W. Sing, J. Weitkamp (Eds.), *Handbook of Porous Solids*, Wiley-VCH Publishers, Weinheim, 2002.

Series of Special Issues cover various aspects of microporous molecular sieves, metal-organic frameworks and ordered mesoporous materials. Synthesis principles, templating, formation mechanisms, characterization methods, functionalization strategies, and applications are discussed in excellent and comprehensive review articles and more specific research reports.

- From Organic-Inorganic Nanocomposites to Functional Materials Special Issue, *Chemistry of Materials* **2001**, *13*, 3059 (first page).
- Templated Materials Special Issue, *Chemistry of Materials* **2008**, *20*, 599 (first page).
- Proceedings of the 2nd International Symposium on Mesoporous Molecular Sieves (ISMMS) Special Issue *Microporous and Mesoporous Materials* **2001**, *44–45,* 1 (first page).
- Metal-Organic Open Frameworks Special Issue, *Microporous and Mesoporous Materials* **2004**, *73,* 1 (first page).

REFERENCES

1. K. Klabunde, R. Richards (Eds.), *Nanoscale Materials in Chemistry*, 2nd edition, Wiley-VCH Publishers, Weinheim, Germany, 2009.
2. G. Cao, *Nanostructures and Nanomaterials, Synthesis, Properties and Applications*, Imperial College Press, London, 2004.
3. C. P. Poole, Jr, F. J. Owens, *Introduction to Nanotechnology*, Wiley, New York, 2003.
4. F. Schüth, K. S. W. Sing, J. Weitkamp (Eds.), *Handbook of Porous Solids*, Wiley-VCH Publishers, Weinheim, Germany, 2002.
5. G. Q. Lu, X. S. Zhao (Eds.), *Nanoporous Materials: Science and Engineering*, Series on Chemical Engineering, Imperial College Press, London, 2005.
6. G. Ertl, H. Knözinger, F. Schüth, J. Weitkamp (Eds.), *Handbook of Heterogeneous Catalysis*, 2nd edition, Wiley-VCH Publishers, Weinheim, Germany, 2008.
7. F. Schüth, W. Schmidt, *Adv. Mater.* **2002**, *14*, 629.
8. C. J. Brinker, G. W. Scherer, *Sol-Gel Science*, Academic Press, New York, 1990.
9. C. N. R. Rao, A. Govindaraj, *Nanotubes and Nanowires*, The Royal Society of Chemistry, Cambridge, 2005.
10. G. A. O. Ozin, A. C. Arsenault, *Nanochemistry: A Chemical Approach to Nanomaterials*, The Royal Society of Chemistry, Cambridge, 2005.
11. D. A. Olson, L. Chen, M. A. Hillmyer, *Chem. Mater.* **2008**, *20*, 869.
12. A. Thomas, F. Goettmann, M. Antonietti, *Chem. Mater.* **2008**, *20*, 738.
13. K. S. W. Sing, D. H. Everett, R. H. W. Haul, L. Moscou, R. A. Pierotti, J. Rouquerol, T. Siemieniewska, *Pure Appl. Chem.* **1985**, *57*, 603.
14. H. van Bekkum, E. M. Flanigen, J. C. Jansen (Eds.), *Introduction to Zeolite Science and Practice, Stud. Surf. Sci. Catal.* 58, Elsevier, Amsterdam, 1991.
15. R. M. Barrer, *Hydrothermal Chemistry of Zeolites*, Academic Press, London, 1982.
16. P. A. Jacobs, J. A. Martens (Eds.), *Synthesis of Aluminosilicate Zeolites, Stud. Surf. Sci. Catal.* 33, Elsevier, Amsterdam, 1987.
17. D. W. Breck, *Zeolite Molecular Sieves*, Wiley, New York, 1974.
18. A. Dyer, *An Introduction to Zeolite Molecular Sieves*, Wiley, New York, 1988.
19. W. F. Hölderich, M. Hesse, F. Näumann, *Angew. Chem. Int. Ed.* **1988**, *27*, 226.
20. J. C. Jansen, M. Stöcker, H. G. Karge, J. Weitkamp (Eds.), *Advanced Zeolite Science and Applications, Stud. Surf. Sci. Catal.* 85, Elsevier, Amsterdam, 1994.
21. (a) A. Corma, *Chem. Rev.* **1995**, *95*, 559; (b) A. Corma, *Chem. Rev.* **1997**, *97*, 2373.
22. K. K. Iler, *The Chemistry of Silica*, Wiley, New York, 1979.
23. A. N. Neirmark, K. S. W. Sing, M. Thommes, in *Handbook of Heterogeneous Catalysis*, 2nd edition, G. Ertl, et al. (Eds.), 2008, Chapter 3, pages 721–738.
24. P. Behrens, *Adv. Mater.* **1993**, *5*, 127.
25. T. J. Barton, L. M. Bull, W. G. Klemperer, D. A. Loy, B. McEnaney, M. Misono, P. A. Monson, G. Pez, G. W. Scherer, J. C. Vartuli, O. M. Yaghi, *Chem. Mater.* **1999**, *11*, 2633.
26. J. Rouquerol, D. Avnir, C. W. Fairbridge, D. H. Everett, J. H. Haynes, N. Pernicone, J. D. F. Ramsay, K. S. W. Sing, K. K. Unger, *Pure Appl. Chem.* **1994**, *66*, 1739.
27. M. Thommes, in *Nanoporous Materials: Science and Engineering*, C. G. Lu, X. S. Zhao (Eds.), Imperial College Press, Oxford, 2004, pages 317–364.

28. F. Rouquerol, J. Rouquerol, K. S. W. Sing, *Adsorption by Powders and Porous Solids: Principles, Methodology and Applications*, Academic Press, London, 1998.
29. P. I. Ravikovitch, A. N. Neimark, *Langmuir* **2002**, *18*, 9830.
30. M. Thommes, B. Smarsly, M. Groenewolt, P. I. Ravikovitch, A. N. Neimark, *Langmuir* **2006**, *22*, 757.
31. (a) I. Langmuir, *J. Am. Chem. Soc.* **1913**, *35*, 931; (b) I. Langmuir, *J. Am. Chem. Soc.* **1915**, 37, 1139; (c) I. Langmuir, *J. Am. Chem. Soc.* **1918**, *40*, 1361.
32. S. Brunauer, P. H. Emmett, E. Teller, *J. Am. Chem. Soc.* **1938**, *60*, 309.
33. S. Lowell, J. E. Shields, M. A. Thomas, M. Thommes, *Characterization of Porous Solids and Powders: Surface Area, Pore Size and Density*, 2nd edition, Springer-Verlag, Berlin, 2006.
34. S. J. Gregg, K. S. W. Sing, *Adsorption, Surface Area and Porosity*, Academic Press, London, 1995.
35. O. Franke, G. Schultz-Ekloff, J. Rathouski, J. Starek, A. Zukal, *Chem. Commun.* **1993**, 724.
36. M. Kruk, M. Jaroniec, A. Sayari, *J. Phys. Chem. B* **1997**, *101*, 583.
37. M. Kruk, M. Jaroniec, A. Sayari, *Chem. Mater.* **1999**, *11*, 492
38. P. I. Ravikovitch, S. C. O. Domhnaill, A. V. Neimark, F. Schüth, K. K. Unger, *Langmuir* **1995**, *11*, 4765.
39. A. V. Neimark, P. I. Ravikovitch, *Microporous Mesoporous Mater.* **2001**, *44–45*, 697.
40. E. P. Barrett, L. G. Joyner, P. P. Halenda, *J. Am. Chem. Soc.* **1951**, *73*, 373.
41. P. I. Ravikovitch, A. V. Neimark, *J. Phys. Chem. B* **2001**, *105*, 6817.
42. L. D. Gelb, K. E. Gubbins, R. Radhakrishnan, M. Sliwinska-Bartkowiak, *Rep. Prog. Phys.* **1999**, *62*, 1573.
43. A. V. Neimark, *Langmuir* **1995**, *11*, 4183.
44. B. D. Cullity, S. R. Stock, *Elements of X-Ray Diffraction*, 3rd edition, Prentice Hall, New Jersey, 2001.
45. R. Jenkins, *Introduction to Powder X-Ray Diffractometry*, Wiley, New York, 1996.
46. J. M. Thomas, O. Terasaki, P. L. Gai, W. Zhou, J. Gonzales-Calbet, *Acc. Chem. Res.* **2001**, *34*, 583.
47. A. K. Datye, P. Lenvig Hansen, S. Helveg, in *Handbook of Heterogeneous Catalysis*, 2nd edition, G. Ertl et al. (Eds.), Wiley-VCH Publishers, Weinheim, Germany, 2008, pages 803–833.
48. M. W. Anderson, T. Ohsuna, Y. Sakamoto, Z. Liu, A. Carlsson, O. Terasaki, *Chem. Commun.* **2004**, 907.
49. L. M. Bronstein, *Top. Curr. Chem.* **2003**, *226*, 55.
50. A. H. Janssen, C. M. Yang, Y. Wang, F. Schüth, A. J. Koster, K. P. de Jong, *J. Phys. Chem. B* **2003**, *107*, 10552.
51. S. N. Che, K. Lund, T. Tatsumi, S. Iijima, S. H. Joo, R. Ryoo, O. Terasaki, *Angew. Chem. Int. Ed.* **2003**, *42*, 2182.
52. P. W. Atkins, *Physical Chemistry*, 6th edition, Oxford University Press, Oxford, 1998.
53. F. H. Dickey, *Proc. Natl. Acad. Sci. U.S.A.* **1949**, *35*, 227.
54. G. J. A. A. Soler-Illia, C. Sanchez, B. Lebeau, J. Patarin, *Chem. Rev.* **2002**, *102*, 4093.
55. F. Schüth, *Angew. Chem. Int. Ed.* **2003**, *42*, 3604.

56. K. J. C. Van Bommel, A. Friggeri, S. Shinkai, *Angew. Chem. Int. Ed.* **2003**, *42*, 980.
57. J. C. Vartuli, K. D. Schmitt, C. T. Kresge, W. J. Roth, M. E. Leonowicz, S. B. McCullen, S. D. Hellring, J. S. Beck, J. L. Schlenker, D. H. Olson, E. W. Sheppard, *Chem. Mater.* **1994**, *6*, 2317.
58. N. K. Raman, M. T. Anderson, C. J. Brinker, *Chem. Mater.* **1996**, *8*, 1682.
59. R. L. Wadlinger, G. T. Kerr, E. J. Rosinski (Mobil Oil Corporation), U.S. Patent 3,375,205, 1968 [*Chem. Abstract.* **1968**, *69*, 37667].
60. M. E. Davis, R. F. Lobo, *Chem. Mater.* **1992**, *4*, 756.
61. H. G. Karge, J. Weitkamp (Eds.), *Molecular Sieves, Science and Technology*, Vol. 1, Springer, Berlin. 1998.
62. C. S. Cundy, P. A. Cox, *Chem. Rev.* **2003**, *103*, 663.
63. C. S. Cundy, P. A. Cox, *Microporous Mesoporous Mater.* **2005**, *82*, 1.
64. J.-M. Lehn, *Science* **1985**, *227*, 849.
65. S. Mann, G. A. Ozin, *Nature* **1996**, *382*, 313.
66. G. A. Ozin, *Acc. Chem. Res.* **1997**, *30*, 17.
67. M. T. J. Keene, R. D. M. Gougeon, R. Denoyel, R. H. Harris, J. Rouquerol, P. L. Llewellyn, *J. Mater. Chem.* **1999**, *9*, 2843.
68. (a) F. Kleitz, W. Schmidt, F. Schüth, *Microporous Mesoporous Mater.* **2001**, *44–45*, 92; (b) F. Kleitz, W. Schmidt, F. Schüth, *Microporous Mesoporous Mater.* **2003**, *65*, 1.
69. J. Livage, M. Henry, C. Sanchez, *Prog. Solid State Chem.* **1988**, *18*, 259.
70. D. F. Evans, H. Wennerström (Eds.), *The Colloidal Domain: Where Physics, Chemistry, Biology and Technology Meet*, Wiley-VCH Publishers, Weinheim, Germany, 1984.
71. (a) J. N. Israelachvili, D. J. Mitchell, B. W. Ninham, *J. Chem. Soc. Faraday Trans.* **1976**, *72*, 1525; (b) J. N. Israelachvili, D. J. Mitchell, B. W. Ninham, *Biochim. Biophys. Acta* **1977**, *470*, 185.
72. J. N. Israelachvili, *Intermolecular and Surfaces Forces*, Academic Press, London, 1992.
73. S. T. Hyde, *Pure Appl. Chem.* **1992**, *64*, 1617.
74. (a) G. Clavier, J. L. Pozzo, H. Bouas-Laurent, C. Lière, C. Roux, C. Sanchez, *J. Mater. Chem.* **2000**, *10*, 1725; (b) M. Llusar, L. Pidol, C. Roux, J. L. Pozzo, C. Sanchez, *Chem. Mater.* **2002**, *14*, 5124.
75. S. Kobayashi, K. Hanabusa, N. Hamasaki, M. Kimura, H. Shirai, *Chem. Mater.* **2000**, *12*, 1523.
76. J. H. Jung, Y. Ono, S. Shinkai, *J. Am. Chem. Soc.* **2000**, *122*, 5008.
77. S. Förster, M. Antonietti, *Adv. Mater.* **1998**, *10*, 195.
78. F. S. Bates, G. H. Frederickson, *Phys. Today* **1999**, *52*, 32.
79. G. Wanka, H. Hoffmann, W. Ulbricht, *Macromolecules* **1994**, *27*, 4145.
80. P. Alexandridis, T. A. Hatton, *Colloids Surfaces A: Physicochem. Eng. Aspects* **1995**, *96*, 1.
81. C. Booth, D. Attwood, *Macromol. Rapid Commun.* **2000**, *21*, 501.
82. J. W. Kriesel, T. D. Tilley, *Chem. Mater.* **1999**, *11*, 1190.
83. G. Larsen, E. Lotero, M. Marquez, *J. Phys. Chem. B* **2000**, *104*, 4840.
84. G. J. A. A. Soler-Illia, L. Rozes, M. K. Boggiano, C. Sanchez, C. O. Turrin, A. M. Caminade, J. P. Majoral, *Angew. Chem. Int. Ed.* **2000**, *39*, 4249.

85. Q. Huo, D. I. Margolese, U. Ciesla, P. Feng, T. E. Gier, P. Sieger, R. Leon, P. M. Petroff, F. Schüth, G. D. Stucky, *Nature* **1994**, *368*, 317.
86. Q. Huo, D. I. Margolese, U. Ciesla, D. G. Demuth, P. Feng, T. E. Gier, P. Sieger, A. Firouzi, B. F. Chmelka, F. Schüth, G. D. Stucky, *Chem. Mater.* **1994**, *6*, 1176.
87. P. T. Tanev, T. J. Pinnavaia, *Science* **1995**, *267*, 865.
88. S. A. Bagshaw, E. Prouzet, T. J. Pinnavaia, *Science* **1995**, *269*, 1242.
89. C. Sanchez, G. J. A. A. Soler-Illia, F. Ribot, C. Mayer, V. Cabuil, T. Lalot, *Chem. Mater.* **2001**, *13*, 3061.
90. C. Sanchez, G. J. A. A. Soler-Illia, F. Ribot, D. Grosso, *C.R. Chimie* **2003**, *6*, 1131.
91. J. H. Knox, B. Kaur, G. R. Millward, *J. Chromatogr.* **1986**, *352*, 3.
92. A. H. Lu, F. Schüth, *Adv. Mater.* **2006**, *18*, 1793.
93. M. Tiemann, *Chem. Mater.* **2008**, *20*, 961.
94. (a) A. Stein, *Microporous Mesoporous Mater.* **2001**, *44–45*, 227; (b) F. Marlow, W. T. Dong, *Chem. Phys. Chem.* **2003**, *5*, 549.
95. A. Taguchi, J. H. Smatt, M. Lindén, *Adv. Mater.* **2003**, *15*, 1209.
96. P. M. Arnal, F. Schüth, F. Kleitz, *Chem. Commun.* **2006**, 1203.
97. (a) L. Tosheva, V. Valtchev, *Chem. Mater.* **2005**, *17*, 2494; (b) K. Egeblad, C. H. Christensen, M. Kustova, C. H. Christensen, *Chem. Mater.* **2008**, *20*, 946.
98. M. Schwickardi, T. Johann, W. Schmidt, F. Schüth, *Chem. Mater.* **2002**, *14*, 3913.
99. D. M. Ford, E. E. Simanek, D. F. Shantz, *Nanotechnology* **2005**, *16*, S458.
100. A. Taguchi, F. Schüth, *Microporous Mesoporous Mater.* **2005**, *77*, 1.
101. B. F. G. Johnson, S. A. Raynor, D. S. Shephard, T. Maschmeyer, J. M. Thomas, *Chem. Commun.* **1999**, 1167.
102. D. S. Shephard, W. Z. Zhou, T. Maschmeyer, J. M. Matters, C. L. Roper, J. M. Thomas, *Angew. Chem. Int. Ed.* **1998**, *37*, 2719.
103. (a) J. Sauer, F. Marlow, F. Schüth, *Phys. Chem. Chem. Phys.* **2001**, *3*, 5579; (b) J. Sauer, F. Marlow, B. Spliethoff, F. Schüth, *Chem. Mater.* **2002**, *14*, 217.
104. L. A. Solovyov, O. V. Belousov, R. E. Dinnebier, A. N. Shmakov, S. D. Kirik, *J. Phys. Chem. B* **2005**, *109*, 3233.
105. MFI stands for *Mobil Five* and ZSM stands for *Zeolite Socony Mobil*. Detailed information about the structure-type codes can be found on the homepage of the Structure Commission of the International Zeolite Association (IZA) at http://www.iza-structure.org.
106. F. Schüth, U. Simon, F. Laeri, M. Wark (Eds.), *Host-Guest-Systems Based on Nanoporous Crystals*, Wiley-VCH, Weinheim, Germany, 2003.
107. (a) J. M. Thomas, R. Raja, D. W. Lewis, *Angew. Chem. Int. Ed.* **2005**, *44*, 6456; (b) S. Ray, S. Vasudevan, *Inorg. Chem.* **2003**, *42*, 1711.
108. R. J. Francis, D. O'Hare, *J. Chem. Soc. Dalton Trans.* **1998**, 3133.
109. J. Cejka, H. van Bekkum, A. Corma, F. Schüth (Eds.), *Introduction to Zeolite Science and Practice*, 3rd revised edition, *Stud. Surf. Sci. Catal.* 168, Elsevier, Amsterdam, 2007.
110. H. Gies, B. Marler, *Zeolites* **1992**, *12*, 42.
111. K. J. Chao, J. C. Lin, Y. Wang, G. H. Lee, *Zeolites* **1986**, *6*, 35.
112. (a) S. I. Zones, M. M. Olmstead, D. S. Santilli, *J. Am. Chem. Soc.* **1992**, *114*, 4195; (b) R. F. Lobo, M. Pan, I. Chan, S. I. Zones, R. C. Medrud, P. A. Crozier, M. E. Davis, *J. Phys. Chem.* **1994**, *98*, 12040.

113. C. A. Fyfe, H. Gies, G. T. Kokotailo, B. Marler, D. E. Cox, *J. Phys. Chem.* **1990**, *94*, 3718.

114. (a) A. Corma, M. Diaz-Cabanas, M. Martinez-Triguero, F. Rey, J. Rius, *Nature* **2002**, *418*, 514; (b) A. Corma, F. Rey, J. Rius, M. J. Sabatier, S. Valencia, *Nature*, **2004**, *431*, 287; (c) A. Corma, F. Rey, S. Valencia, J. L. Jorda, J. Rius, *Nature Mater.* **2003**, *2*, 493.

115. F. Schüth, *Annu. Rev. Mater. Res.* **2005**, *35*, 209.

116. M. Choi, H. S. Cho, R. Srivastava, C. Venkatesan, D.-H. Cho, R. Ryoo, *Nature Mater.* **2006**, *5*, 718.

117. (a) O. M. Yaghi, G. Li, H. Li, *Nature* **1995**, *378*, 703; (b) H. Li, M. Eddaoudi, M. O'Keefe, O. M. Yaghi, *Nature* **1999**, *402*, 276; (c) O. M. Yaghi, C. E. Davis, H. Li, *J. Am. Chem. Soc.* **1997**, *119*, 2861.

118. E. A. Tomic, *J. Appl. Polym. Sci.* **1965**, *9*, 3745.

119. S. Kitagwa, R. Kitaura, S. I. Noro, *Angew. Chem. Int. Ed.* **2004**, *43*, 2334.

120. N. L. Rosi, M. Eddaoudi, J. Kim, M. O'Keefe, O. M. Yaghi, *Crystal Eng. Commun.* **2002**, *4*, 401.

121. B. Moulton, M. J. Zaworotki, *Chem. Rev.* **2001**, *101*, 1629.

122. M. J. Rosseinsky, *Microporous Mesoporous Mater.* **2004**, *73*, 15.

123. J. L. C. Rowsell, O. M. Yaghi, *Microporous Mesoporous Mater.* **2004**, *73*, 3.

124. G. Férey, C. Mellot-Draznieks, C. Serre, F. Millange, *Acc. Chem. Res.* **2005**, *38*, 217.

125. O. M. Yaghi, M. O'Keefe, N. W. Ockwig, H. K. Chae, M. Eddaoudi, J. Kim, *Nature* **2003**, *423*, 705.

126. H. K. Chae, D. Y. Siberio-Pérez, J. Kim, Y. Go, M. Eddaoudi, A. J. Matzger, O. M. Yaghi, M. O'Keefe, *Nature* **2004**, *427*, 523.

127. M. Eddaoudi, J. Kim, N. Rosi, D. Vodak, J. Wachter, M. O'Keefe, O. M. Yaghi, *Science* **2002**, *295*, 469.

128. U. Müller, M. M. Schubert, O. M. Yaghi, in *Handbook of Heterogeneous Catalysis*, 2nd edition, G. Ertl, et al. (Eds.), Wiley-VCH Publishers, Weinheim, Germany, 2008, pages 247–262.

129. C. Livage, C. Egger, G. Férey, *Chem. Mater.* **2001**, *13*, 410.

130. O. R. Evans, W. Lin, *Acc. Chem. Res.* **2002**, *35*, 511.

131. G. Férey, C. Mellot-Draznieks, C. Serre, F. Millange, J. Dutour, S. Surblé, I. Margiolaki, *Science* **2005**, *309*, 2040.

132. (a) C. Serre, F. Millange, C. Thouvenot, M. Noguès, G. Marsolier, D. Louer, G. Férey, *J. Am. Chem. Soc.* **2002**, *124*, 13519; (b) K. Barthelet, J. Marrot, D. Riou, G. Férey, *Angew. Chem. Int. Ed.* **2002**, *41*, 281.

133. (a) S. Bourelly, P. L. Llewellyn, C. Serre, F. Millange, T. Loiseau, G. Férey, *J. Am. Chem. Soc.* **2005**, *127*, 13519; (b) P. L. Llewellyn, S. Bourelly, C. Serre, Y. Filinchuk, G. Férey, *Angew. Chem. Int. Ed.* **2006**, *45*, 7751.

134. J. S. Seo, D. Whang, H. Lee, S. I. Jun, J. Oh, Y. J. Jeon, K. Kim, *Nature* **2000**, *404*, 982.

135. (a) Y. Cui, O. R. Evans, H. L. Ngo, P. S. White, W. Lin, *Angew. Chem. Int. Ed.* **2002**, *41*, 1159; (b) R. Robson, B. F. Abrahams, M. Moylan, S. D. Orchard, *Angew. Chem. Int. Ed.* **2003**, *42*, 1848.

136. C. T. Kresge, M. E. Leonowicz, W. J. Roth, J. C. Vartuli, J. S. Beck, *Nature* **1992**, *359*, 710.

137. J. S. Beck, J. C. Vartuli, W. J. Roth, M. E. Leonowicz, C. T. Kresge, K .D. Schmitt, C. T.-W. Chu, D. H. Olson, E. W. Sheppard, S. B. McCullen, J. B. Higgins, J. L. Schlenker, *J. Am. Chem. Soc.* **1992**, *114*, 10834.

138. T. Yanagisawa, T. Shimizu, K. Kuroda, C. Kato, *Bull. Chem. Soc. Jpn.* **1990**, *63*, 988.

139. J. C. Vartuli, C. T. Kresge, M. E. Leonowicz, A. S. Chu, S. B. McCullen, I. D. Johnson, E. W. Sheppard, *Chem. Mater.* **1994**, *6*, 2070.

140. G. S. Attard, J. C. Glyde, C. G. Göltner, *Nature* **1995**, *378*, 366.

141. R. Ryoo, S. H. Joo, S. Jun, *J. Phys. Chem. B* **1999**, *103*, 7743.

142. A. Monnier, F. Schüth, Q. Huo, D. Kumar, D. Margolese, R. S. Maxwell, G. D. Stucky, M. Krishnamurty, P. Petroff, A. Firouzi, M. Janicke, B. F. Chmelka, *Science* **1993**, *261*, 1299.

143. G. S. Attard, P. N. Bartlett, N. R. B Coleman, J. M. Elliot, J. R. Owen, J. H. Wang, *Science* **1997**, *278*, 838.

144. C. J. Brinker, Y. Lu, A. Sellinger, H. Fan, *Adv. Mater.* **1999**, *11*, 579.

145. D. Grosso, F. Cagnol, G. J. A. A. Soler-Illia, E. L. Crepaldi, H. Amenitsch, A. Brunet-Bruneau, A. Bourgeois, C. Sanchez, *Adv. Func. Mater.* **2004**, *14*, 309.

146. S. Jun, S. H. Joo, R. Ryoo, M. Kruk, M. Jaroniec, Z. Liu, T. Ohsuna, O. Terasaki, *J. Am. Chem. Soc.* **2000**, *122*, 10712.

147. R. Ryoo, S. H. Joo, M. Kruk, M. Jaroniec, *Adv. Mater.* **2001**, *13*, 677.

148. Y. Sakamoto, T. Ohsuna, K. Hiraga, O. Terasaki, C. H. Ko, H. J. Shin, R. Ryoo, *Angew. Chem. Int. Ed.* **2000**, *39*, 3107.

149. B. Tian, X. Y. Liu, H. F. Yang, S. H. Xie, C. Z. Yu, B. Tu, D. Zhao, *Adv. Mater.* **2003**, *15*, 1370.

150. (a) A. H. Lu, F. Schüth, *C.R. Chimie* **2005**, *8*, 609; (b) F. Cheng, Z. Tao, J. Liang, J. Chen, *Chem. Mater.* **2008**, *20*, 667.

151. H. F. Feng, D. Y. Zhao, *J. Mater. Chem.* **2005**, *15*, 1217.

152. A. A. Romero, A. D. Alba, W. Zhou, J. Klinowski, *J. Phys. Chem. B* **1997**, *101*, 5294.

153. M. Grün, I. Lauer, K. K. Unger, *Adv. Mater.* **1997**, *9*, 254.

154. M. Grün, K. K. Unger, A. Matsumoto, K. Tsutsumi, *Microporous Mesoporous Mater.* **1999**, *27*, 207.

155. X. Auvray, C. Petipas, R. Anthore, I. Rico, A. J. Lattes, *J. Phys. Chem.* **1989**, *93*, 7458.

156. (a) R. Schmidt, M. Stöcker, D. Akporiaye, E. H. Tørstad, A. Olsen, *Microporous Mater.* **1995**, *5*, 1; (b) V. Alfredsson, M. W. Anderson, *Chem. Mater.* **1996**, *8*, 1141.

157. J. Xu, Z. Luan, H. He, W. Zhou, L. Kevan, *Chem. Mater.* **1998**, *10*, 3690.

158. Q. Huo, D. I. Margolese, G. D. Stucky, *Chem. Mater.* **1996**, *8*, 1147.

159. K. W. Gallis, C. C. Landry, *Chem. Mater.* **1997**, *9*, 2035.

160. J. M. Kim, S. K. Kim, R. Ryoo, *Chem. Commun.* **1998**, 259.

161. V. Alfredsson, M. W. Anderson, *Chem. Mater.* **1996**, *8*, 1141.

162. A. Firouzi, D. Kumar, L. M. Bull, T. Besier, P. Sieger, Q. Huo, S. A. Walker, J. A. Zasadzinski, C. Glinka, J. Nicol, D. Margolese, G. D. Stucky, B. F. Chmelka, *Science* **1995**, *267*, 1138.

163. Q. Huo, D. I. Margolese, G. D. Stucky, *Chem. Mater.* **1996**, *8*, 1147.

164. Q. Huo, R. Leon, P. M. Petroff, G. D. Stucky, *Science* **1995**, *268*, 1324.

165. S. Schacht, Q. Huo, I. G. Voigt-Martin, G. D. Stucky, F. Schüth, *Science* **1996**, *273*, 768.

166. (a) S. Che, A. E. Garcia-Bennett, T. Yokoi, K. Sakamoto, H. Kunieda, O. Terasaki, T. Tatsumi, *Nature Mater.* **2003**, *2*, 801; (b) A. E. Garcia-Bennett, O. Terasaki, S. Che, T. Tatsumi, *Chem. Mater.* **2004**, *16*, 813; (c) A. E. Garcia-Bennett, K. Miyasaka, O. Terasaki, S. Che, *Chem. Mater.* **2004**, *16*, 3587.

167. G. J. A. A. Soler-Illia, E. L. Crepaldi, D. Grosso, C. Sanchez, *Curr. Opin. Colloid Interface Sci.* **2003**, *8*, 109.

168. D. Zhao, J. Feng, Q. Huo, N. Melosh, G. H. Frederickson, B. F. Chmelka, G. D. Stucky, *Science* **1998**, *279*, 548.

169. D. Zhao, Q. Huo, J. Feng, B. F. Chmelka, G. D. Stucky, *J. Am. Chem. Soc.* **1998**, *120*, 6024.

170. M. Choi, W. Heo, F. Kleitz, R. Ryoo, *Chem. Commun.* **2003**, 1340.

171. (a) R. Ryoo, C. H. Ko, M. Kruk, V. Antoschuk, M. Jaroniec, *J. Phys. Chem. B* **2000**, *104*, 11465; (b) M. Kruk, M. Jaroniec, C. H. Ko, R. Ryoo, *Chem. Mater.* **2000**, *12*, 1961.

172. M. Impéror-Clerc, P. Davidson, A. Davidson, *J. Am. Chem. Soc.* **2000**, *122*, 11925.

173. Z. Liu, O. Terasaki, T. Ohsuna, K. Hiraga, H. J. Shin, R. Ryoo, *Chem. Phys. Chem.* **2001**, *4*, 229.

174. Y. Sakamoto, M. Kaneda, O. Terasaki, D. Y. Zhao, J. M. Kim, G. D. Stucky, H. J. Shin, R. Ryoo, *Nature* **2000**, *408*, 449.

175. F. Kleitz, S. H. Choi, R. Ryoo, *Chem. Commun.* **2003**, 2136.

176. Y. Sakamoto, T. W. Kim, R. Ryoo, O. Terasaki, *Angew. Chem. Int. Ed.* **2004**, *42*, 5231.

177. F. Kleitz, F. Bérubé, C. M. Yang, M. Thommes, *Stud. Surf. Sci. Catal.* **2007**, *170*, 1843.

178. X. Liu, B. Tian, C. Yu, F. Gao, S. Xie, B. Tu, R. Che, L.-M. Peng, D. Zhao, *Angew. Chem. Int. Ed.* **2002**, *41*, 3876.

179. Y. Q. Wang, C. M. Yang, B. Zibrowius, B. Spliethoff, M. Lindén, F. Schüth, *Chem. Mater.* **2003**, *15*, 5029.

180. K. Flodström, V. Alfredsson, N. Källrot, *J. Am. Chem. Soc.* **2003**, *125*, 4402.

181. U. Ciesla, F. Schüth, *Microporous Mesoporous Mater.* **1999**, *27*, 131.

182. J. M. Kim, Y. Sakamoto, Y. K. Hwang, Y.-U. Kwon, O. Terasaki, S.-E. Park, G. D. Stucky, *J. Phys. Chem. B* **2002**, *106*, 2552.

183. (a) P. Kipkemboi, A. Fogden, V. Alfredsson, K. Flodström, *Langmuir* **2001**, *17*, 5398; (b) K. Flodström, V. Alfredsson, *Microporous Mesoporous Mater.* **2003**, *59*, 167.

184. T.-W. Kim, R. Ryoo, M. Kruk, K. P. Gierszal, M. Jaroniec, S. Kamiya, O. Terasaki, *J. Phys. Chem. B* **2004**, *108*, 11480.

185. K. Kushalani, A. Kuperman, G. A. Ozin, K. Tanaka, J. Garces, M. M Olken, N. Coombs, *Adv. Mater.* **1995**, *7*, 842.

186. A. Galarneau, H. Cambon, F. DiRenzo, R. Ryoo, M. Choi, F. Fajula, *New J. Chem.* **2003**, *27*, 73.

187. T.-W. Kim, F. Kleitz, B. Paul, R. Ryoo, *J. Am. Chem. Soc.* **2005**, *127*, 7601.

188. R. Ryoo, J. M. Kim, *Chem. Commun.* **1995**, 711.

189. K. J. Edler, J. W. White, *Chem. Mater.* **1997**, *9*, 1226.

190. A. Kabalnov, U. Olsson, H. Wennerström, *J. Phys. Chem.* **1995**, *98*, 6220.

191. C. Z. Yu, B. Tian, J. Fan, G. D. Stucky, D. Y. Zhao, *J. Am. Chem. Soc.* **2002**, *124*, 4556.

192. M. Lindén, P. Ågren, S. Karlsson, P. Bussian, H. Amenitsch, *Langmuir* **2000**, *16*, 5831.

193. F. Kleitz, J. Blanchard, B. Zibrowius, F. Schüth, P. Ågren, M. Lindén, *Langmuir* **2002**, *18*, 4863.

194. J. Sun, H. Zhang, D. Ma, Y. Chen, X. Bao, A. Klein-Hoffmann, N. Pfänder, D. S. Su, *Chem. Commun.* **2005**, 5343.

195. F. Kleitz, T. W. Kim, R. Ryoo, *Langmuir* **2006**, *22*, 440.

196. (a) K. Moller, T. Bein, *Chem. Mater.* **1998**, *10*, 2950; (b) J. Y. Ying, C. P. Mehnert, M. S. Wong, *Angew. Chem. Int. Ed.* **1999**, *38,* 56; (c) A. Stein, B. J. Melde, R. C. Schroden, *Adv. Mater.* **2000**, *12*, 1403.

197. X. S. Zhao, C. Q. Lu, A. K. Whittaker, G. J. Millar, H. Y. Zhu, *J. Phy. Chem. B* **1997**, *101*, 6525.

198. M. Widenmeyer, R. Anwander, *Chem. Mater.* **2002**, *14*, 1827.

199. L. T. Zhuralev, *Langmuir* **1987**, *3*, 316.

200. C. Nozaki, C. G. Lugmaier, A. T. Bell, T. D. Tilley, *J. Am. Chem. Soc.* **2002**, *124*, 13194.

201. (a) K. A. Koyano, T. Tatsumi, Y. Tanaka, S. Nakata, *J. Phys. Chem. B* **1997**, *101*, 9436; (b) J. M. Kisler, M. L. Gee, G. W. Stevens, A. J. O'Connor, *Chem. Mater.* **2003**, *15*, 619.

202. R. Anwander, *Chem. Mater.* **2001**, *13*, 4419.

203. M. H. Lim, A. Stein, *Chem. Mater.* **1999**, *11*, 3285.

204. F. Hoffmann, M. Cornelius, J. Morell, M. Fröba, *Angew. Chem. Int. Ed.* **2006**, *45*, 3216.

205. J. A. Melero, R. van Grieken, G. Morales, *Chem. Rev.* **2006**, *106*, 3790.

206. E. L. Burkett, S. D. Sims, S. Mann, *Chem. Commun.* **1996**, 1367.

207. D. J. Macquarrie, *Chem. Commun.* **1996**, 1961.

208. D. Margolese, J. A. Melero, S. C. Christiansen, B. F. Chmelka, G. D. Stucky, *Chem. Mater.* **2000**, *12*, 2448.

209. P. Gómez-Romero, C. Sanchez (Eds.), *Functional Hybrid Materials*, Wiley-VCH, Weinheim, 2004.

210. S. Inagaki, S. Guan, Y. Fukushima, T. Ohsuna, O. Terasaki, *J. Am. Chem. Soc.* **1999**, *121*, 9611.

211. B. J. Melde, B. T. Holland, C. F. Blanford, A. Stein, *Chem. Mater.* **1999**, *11*, 3302.

212. T. Asefa, M. J. MacLachlan, N. Coombs, G. A. Ozin, *Nature* **1999**, *402*, 867.

213. O. Muth, C. Schellbach, M. Fröba, *Chem. Commun.* **2001**, 2032.

214. S. Inagaki, S. Guan, T. Ohsuna, O. Terasaki, *Nature* **2002**, *416*, 304.

215. R. Mokaya, *Angew. Chem. Int. Ed.* **1999**, *38*, 2930.

216. M. Baltes, K. Cassiers, P. Van Der Voort, B. M. Weckhuysen, R. A. Schoonheydt, E. F. Vansant, *J. Catal.* **2001**, *197*, 160.

217. H. Kosslick, G. Lischke, B. Parlitz, W. Storek, R. Fricke, *Appl. Catal. A: Gen.* **1999**, *184*, 49.

218. (a) A. Corma, M. S. Grande, V. Gonzalez-Alfaro, A. V. Orchilles, *J. Catal.* **1996**, *159*, 375; (b) A. Corma, V. Fornes, M. T. Navarro, J. Perez Pariente, *J. Catal.* **1994**, *148*, 569; (c) A. Corma, A. Martínez, V. Martínez-Soria, J. B. Montón, *J. Catal.* **1995**, *153*, 25.

219. K. A. Koyano, T. Tatsumi, *Chem. Commun.* **1996**, 145.

220. P. T. Tanev, M. Chibwe, T. J. Pinnavaia, *Nature* **1994**, *368*, 321.

221. B. J. Aronson, C. F. Blanford, A. Stein, *Chem. Mater.* **1997**, *9*, 2842.

222. Z. Luan, E. M. Maes, P. A. W. van der Heide, D. Zhao, R. S. Czernszewicz, L. Kevan, **1999**, *11*, 3680.

223. T. Machmeyer, F. Rey, G. Sankar, J. M. Thomas, *Nature* **1995**, *378*, 159.

224. J. M. Thomas, W. J. Thomas, *Principles and Practice of Heterogeneous Catalysis*, VCH, Weinheim, Germany, 1997.

225. I. Chorkendorff, J. W. Niemandsverdriet, *Concepts of Modern Catalysis and Kinetics*, 2nd edition, Wiley-VCH, Weinheim, Germany, 2007.

226. L. M. Bronstein, *Top. Curr. Chem.* **2003**, *226*, 55.

227. F. Schüth, *Chem. Mater.* **2001**, *13*, 3184.

228. T. Abe, A. Taguchi, M. Iwamoto, *Chem. Mater.* **1995**, *7*, 1429.

229. P. Liu, L. Moudrakovski, J. Liu, A. Sayari, *Chem. Mater.* **1997**, *9*, 2513.

230. D. M. Antonelli, J. Y. Ying, *Angew. Chem. Int. Ed.* **1995**, *34*, 2014.

231. (a) D. M. Antonelli, *Microporous Mesoporous Mater.* **1999**, *30*, 315; (b) D. M. Antonelli, J. Y. Ying, *Angew. Chem. Int. Ed.* **1996**, *35*, 426; (c) U. Ciesla, S. Schacht, G. D. Stucky, K. K. Unger, F. Schüth, *Angew. Chem. Int. Ed.* **1996**, *35*, 541; (d) M. S. Wong, J. Y. Ying, *Chem. Mater.* **1998**, *10*, 2067; (e) A. Bhaumik, S. Inagaki, *J. Am. Chem. Soc.* **2001**, *123*, 691.

232. (a) P. Yang, D. Zhao, D. I. Margolese, B. F. Chmelka, G. D. Stucky, *Nature* **1998**, *396*, 152; (b) P. Yang, D. Zhao, D. I. Margolese, B. F. Chmelka, G. D. Stucky, *Chem. Mater.* **1999**, *11*, 2813.

233. D. Grosso, G. J. A. A. Soler-Illia, F. Babonneau, C. Sanchez, P. A. Albouy, A. Brunet-Bruneau, A. R. Balkenende, *Adv. Mater.* **2001**, *13*, 1085.

234. E. L. Crepaldi, G. J. A. A. Soler-Illia, D. Grosso, P. A. Albouy, C. Sanchez, *Chem. Commun.* **2001**, 1582.

235. E. L. Crepaldi, G. J. A .A. Soler-Illia, D. Grosso, C. Sanchez, *Angew. Chem. Int. Ed.* **2002**, *42*, 347.

236. E. L. Crepaldi, G. J. A. A. Soler-Illia, D. Grosso, F. Cagnol, F. Ribot, C. Sanchez, *J. Am. Chem. Soc.* **2003**, *125*, 9770.

237. D. Grosso, C. Boissière, B. Smarsly, T. Brezesinski, N. Pinna, P. A. Albouy, H. Amenitsch, M. Antonietti, C. Sanchez, *Nature Mater.* **2004**, *3*, 787.

238. J. Cejka, *Appl. Catal. A: Gen.* **2003**, *254*, 327.

239. S. Bagshaw, T. J. Pinnavaia, *Angew. Chem. Int. Ed.* **1996**, *35*, 1102.

240. M. Kuemmel, D. Grosso, C. Boissière, B. Smarsly, T. Brezesinski, P. A. Albouy, H. Amenitsch, C. Sanchez, *Angew. Chem.* **2005**, *117*, 4665.

241. (a) H. Yang, Q. Shi, B. Tian, Q. Lu, F. Gao, S. Xie, J. Fan, C. Yu, B. Tu, D. Zhao, *J. Am. Chem. Soc.* **2003**, *125*, 4724; (b) B. Tian, X. Liu, H. Yang, S. Xie, C. Yu, B. Tu, D. Zhao, *Adv. Mater.* **2003**, *15*, 1370; (c) B. Tian, X. Liu, L. A. Solovyov, Z. Liu, H. Yang, Z. Zhang, S. Xie, F. Zhang, B. Tu, C. Yu, O. Terasaki, D. Zhao, *J. Am. Chem. Soc.* **2004**, *126*, 865.

242. S. C. Laha, R. Ryoo, *Chem. Commun.* **2003**, 2138.

243. (a) K. Jiao, B. Zhang, B. Yue, Y. Ren, S. Liu, S. Yan, C. Dickinson, W. Zhou, H. He, *Chem. Commun.* **2005**, 5618; (b) C. Dickinson, W. Zhou, R. P. Hodgkins, Y. Shi, D. Zhao, H. He, *Chem. Mater.* **2006**, *18,* 3088.

244. (a) Y. Wang, C. M. Yang, W. Schmidt, B. Spliethoff, E. Bill, F. Schüth, *Adv. Mater.* **2005**, *17*, 53; (b) E. L. Salabas, A. Rumplecker, F. Kleitz, F. Radu, F. Schüth, *Nano Letters* **2006**, *6*, 2977; (c) A. Rumplecker, F. Kleitz, E. L. Salabas, F. Schüth, *Chem. Mater.* **2007**, *19*, 485.

245. E. Rossinyol, A. Prim, E. Pellicer, J. Arbiol, F. Ramirez, F. Peiro, A. Cornet, J. R. Morante, L. A. Solovyov, B. Z. Tian, B. Tu, D. Zhao, *Adv. Func. Mater.* **2007**, *19*, 657.

246. (a) F. Jiao, K. M. Shaju, P. G. Bruce, *Angew. Chem. Int. Ed.* **2005**, *44*, 6550; (b) F. Jiao, A. Harrison, J. C. Jumas, A. V. Chadwick, W. Kockelmann, P. G. Bruce, *J. Am. Chem. Soc.* **2006**, *128*, 5468.

247. T. W. Kim, L. A. Solovyov, *J. Mater. Chem.* **2006**, *16*, 1445.

248. (a) T. Katou, B. Lee, D. Lu, J. N. Kondo, M. Hara, K. Domen, *Angew. Chem. Int. Ed.* **2003**, *42*, 2382; (b) N. Shirokura, K. Nakajima, A. Nakabayashi, D. Lu, M. Hara, K. Domen, T. Tatsumi, J. N. Kondo, *Chem. Commun.* **2006**, 2188.

249. B. Z. Tian, X. Y. Liu, B. Tu, C. Z. Yu, J. Fan, L. M. Wang, S. H. Xie, G. D. Stucky, D. Y. Zhao, *Nature Mater.* **2003**, *2*, 159.

250. (a) K. K. Rangan, S. J. L. Billinge, V. Petkov, J. Heising, M. G. Kanatzidis, *Chem. Mater.* **1999**, *11*, 2629; (b) K. K. Rangan, P. N. Trikalitis, M. G. Kanatzidis, *J. Am. Chem. Soc.* **2000**, *122*, 10230.

251. P. N. Trikalitis, K. K. Rangan, T. Bakas, M. G. Kanatzidis, *Nature* **2001**, *410*, 6712.

252. U. Ciesla, M. Fröba, G. D. Stucky, F. Schüth, *Chem. Mater.* **1999**, *11*, 227.

253. F. Kleitz, S. J. Thomson, Z. Liu, O. Terasaki, F. Schüth, *Chem. Mater.* **2002**, *14*, 4134.

254. P. N. Trikalitis, T. Bakas, V. Papaefthymiou, M. G. Kanatzidis, *Angew. Chem. Int. Ed.* **2000**, *39*, 4558.

255. (a) G. S. Attard, C. G. Göltner, J. M. Corker, S. Henke, R. H. Templer, *Angew. Chem. Int. Ed.* **1997**, *36*, 1315; (b) P. A. Nelson, J. M. Elliott, G. S. Attard, J. R. Owen, *Chem. Mater.* **2002**, *14*, 524; (c) I. Nandhakumar, J. M. Elliott, G. S. Attard, *Chem. Mater.* **2001**, *13*, 3840.

256. H. J. Shin, R. Ryoo, Z. Liu, O. Terasaki, *J. Am. Chem. Soc.* **2001**, *123*, 1246.

257. Y. J. Han, J. M. Kim, G. D. Stucky, *Chem. Mater.* **2000**, *12*, 2068.

258. (a) P. Krawiec, D. Geiger, S. Kaskel, *Chem. Commun.* **2006**, 2469; (b) Y. F. Shi, Y. Meng, D. H. Chen, S. J. Cheng, P. Chen, T. F. Yang, Y. Wan, D. Zhao, *Adv. Func. Mater.* **2006**, *16*, 561.

259. (a) F. Gao, Q. Lu, D. Zhao, *Adv. Mater.* **2003**, *15*, 739; (b) Y. Shi, Y. Wan, R. Liu, B. Tu, D. Zhao, *J. Am. Chem. Soc.* **2007**, *129*, 9522.

260. J. Lee, S. Yoon, T. Hyeon, S. M. Oh, K. B. Kim, *Chem. Commun.* **1999**, 2177.

261. J. Lee, S. Han, T. Hyeon, *J. Mater. Chem.* **2004**, *14*, 478.

262. A. H. Lu, A. Kiefer, W. Schmidt, F. Schüth, *Chem. Mater.* **2004**, *16*, 100.

263. (a) S. H. Joo, S. J. Choi, I. Oh, J. Kwak, Z. Liu, O. Terasaki, R. Ryoo, *Nature* **2001**, *412*, 169; (b) L. A. Solovyov, T. W. Kim, F. Kleitz, O. Terasaki, R. Ryoo, *Chem. Mater.* **2004**, *16*, 2274.

264. S. Tanaka, N. Nishiyama, Y. Egashira, K. Ueyama, *Chem. Commun.* **2005**, 2125.

265. F. Zhang, Y. Meng, D. Gu, Y. Yan, C. Yu, B. Tu, D. Zhao, *J. Am. Chem. Soc.* **2005**, *127*, 13508.

266. (a) C. D. Liang, K. L. Hong, G. A. Guiochon, J. W. Ways, S. Dai, *Angew. Chem. Int. Ed.* **2004**, *43*, 5785; (b) C. Liang, Z. Li, S. Dai, *Angew. Chem. Int. Ed.* **2008**, *47*, 2.

267. R. C. Hayward, P. Alberius-Henning, B. F. Chmelka, G. D. Stucky, *Microporous Mesoporous Mater.* **2001**, *44–45*, 619.
268. G. A. Ozin, *Chem. Comm.* **2000**, 419.
269. S. H. Tolbert, A. Firouzi, G. D. Stucky, B. F. Chmelka, *Science* **1997**, *278*, 264.
270. P. Yang, T. Deng, D. Zhao, P. Feng, D. Pine, B. F. Chmelka, G. M. Whitesides, G. D. Stucky, *Science* **1998**, *282*, 2244.
271. H. Miyata, K. Kuroda, *Chem. Mater.* **2000**, *12*, 49.
272. H. Fan, S. Reed, T. Bear, R. Schunk, G. P. López, C. J. Brinker, *Microporous Mesoporous Mater.* **2001**, *44–45*, 625.
273. H. Yang, N. Coombs, G.A. Ozin, *Nature* **1997**, *386*, 692.
274. H. Yang, N. Coombs, I. Sokolov, G. A. Ozin, *Nature* **1996**, *381*, 589.
275. H. Yang, A. Kuperman, N. Coombs, S. Mamiche-Afara, G. A. Ozin, *Nature* **1996**, *379*, 703.
276. M. Ogawa, *Chem. Comm.* **1996**, 1149.
277. Y. Lu, R. Ganguli, C. A. Drewien, M. T. Anderson, C. J. Brinker, W. Gong, Y. Guo, H. Soyez, B. Dunn, M. H. Huang, J. I. Zink, *Nature* **1997**, *389*, 364.
278. D. Zhao, P. Yang, N. Melosh, J. Feng, B. F. Chmelka, G. D. Stucky, *Adv. Mater.* **1998**, *10*, 1380.
279. Q. Huo, J. Feng, F. Schüth, G. D. Stucky, *Chem. Mater.* **1997**, *9*, 14.
280. K. Schumacher, M. Grün, K. K. Unger, *Microporous Mesoporous Mater.* **1999**, *27*, 201.
281. W. Stöber, A. Funk, E. Bohn, *J. Colloid Interfaces Sci.* **1968**, *26*, 62.
282. (a) T. Martin, A. Galarneau, F. DiRenzo, F. Fajula, D. Plee, *Angew. Chem. Int. Ed.* **2002**, *41*, 2590; (b) A. Galarneau, J. Iapichella, K. Bonhomme, F. DiRenzo, P. Kooyman, O. Terasaki, F. Fajula, *Adv. Funct. Mater.* **2006**, *16*, 1657.
283. P. J. Bruinsma, A. Y. Kim, J. Liu, S. Baskaran, *Chem. Mater.* **1997**, *9*, 2507.
284. H.-P. Lin, C.-Y. Mou, *Science* **1996**, *273*, 765.
285. P. Yang, D. Zhao, B. F. Chmelka, G. D. Stucky, *Chem. Mater.* **1998**, *10*, 2033.
286. (a) F. Marlow, M. D. McGehee, D. Zhao, B. F. Chmelka, G. D Stucky, *Adv. Mater.* **1999**, *11*, 632; (b) F. Kleitz, F. Marlow, G. D. Stucky, F. Schüth, *Chem. Mater.* **2001**, *13*, 3587.
287. (a) F. Marlow, B. Spliethoff, B. Tesche, D. Zhao, *Adv. Mater.* **2000**, *12*, 961; (b) F. Marlow, A. S. G. Khalil, M. Stempniewicz, *J. Mater. Chem.* **2007**, 17, 2168.
288. S. Guan, S. Inagaki, T. Ohsuna, O. Terasaki, *J. Am. Chem. Soc.* **2000**, *122*, 5660.

PROBLEMS

1. For a material that consists of small particles, the specific surface area increases considerably with decreasing particle size. Similarly, nanoporous materials exhibit high surface areas. Compare the two types of nanostructures (nanoparticles and nanopores) and describe why both lead to high values of specific surface area.

2. In the case of nitrogen physisorption at 77 K, the pore condensation phenomenon takes place in materials exhibiting pores larger than 2 nm (mesopores). At a given temperature, the condensation pressure is a function of pore size. Considering three mesoporous solids having the same pore size but with different pore shapes, such as cylindrical, spherical, and slit-like, indicate whether pore condensation pressure would also be a function of pore shape. Justify your answer.

3. Zeolites are used massively in industrial catalysis, ion-exchange processes, separation technologies, and as desiccants. Summarize briefly the attributes allowing these different applications.

4. The crystallization process occurring during the solvothermal synthesis of MOF materials is sometimes regarded as an esterification-type reaction. Justify such an analogy.

5. In the synthesis of ordered mesoporous materials, some components, termed cosolvent, cosurfactant, and swelling agent, are commonly added to alter the nature of the final product. Specify the possible role of each of these types of reagents. Give a typical example for each.

6. Postsynthetic grafting and cocondensation methods both have advantages for the synthesis of ordered mesoporous organic-inorganic hybrids. However, what are the potential limitations that need to be considered for optimizing the material synthesis following these methods?

7. When an as-made mesostructure has reached sufficient degree of condensation, the organic templating agent is no longer needed and can be removed to open the porous structure. Specify what physical and chemical methods may be implemented to remove the structure-directing species from as-made mesostructured materials.

8. Ordered mesoporous transition metal oxides are usually more difficult to obtain than mesoporous silica. Point out the difficulties that need to be overcome to achieve the synthesis of transition metal oxides with well-developed structural mesoporosity.

9. For each of the following compositions, name what would be a preferred hard template (mesoporous silica or carbon) for preparing mesoporous materials:

   (**a**) manganese oxide, Mn_2O_3;

   (**b**) alumina;

   (**c**) tin oxide, SnO_2;

   (**d**) cadmium sulfide;

   (**e**) tungsten oxide, WO_3;

   (**f**) zinc oxide, ZnO;

   (**g**) nanocristalline zirconium oxide, ZrO_2;

   (**h**) boron nitride, BN;

(**i**) polydivinylbenzene;

(**j**) calcium oxide, CaO;

(**k**) silicon carbide;

(**l**) indium oxide, In_2O_3;

(**m**) cerium oxide, CeO_2.

10. The development of new materials possessing low dielectric constant (k) is crucial for applications in miniaturized integrated circuits (sub-100 nm technology). Porous thin films with low k dielectric could provide solutions to the problems of signal delays caused by interconnect resistance-capacitance, signal crosstalk, and power consumption. Suggest a thin film material that would be suitable for low k dielectric applications.

ANSWERS

1. Materials made of finely divided particles exhibit much higher surface area than bulk materials. Clearly, smaller particles result in an increased dispersion, that is, a higher surface to volume ratio. For pores, the situation is similar, except for the opposite curvature, which is usually negative (concave) for pores, whereas the surface curvature may be seen as mainly positive in the case of nanoparticles. A nanoporous solid is also finely divided but in terms of numbers of cavities or voids. The resulting areas of the exposed surfaces are greatly increased.

2. Pore condensation is evidently a function of pore shape. The adsorption potential will increase with increasing negative curvature. Highly curved cavities will have stronger adsorption potential for a given adsorptive. Thus, for materials having the same pore size but different pore shapes, under otherwise identical conditions, the pore condensation pressure P will thus vary as follows: $P_{spheres} < P_{cylinders} < P_{slit\text{-}like}$.

3. Zeolites contain voids of molecular dimensions which permit molecular sieving. This feature is ideal for selective separation of compounds. Pores of molecular dimension with specific shapes will also enable size- and shape-selective processes in catalytic reactions. In addition, acid-base or redox properties needed for catalytic conversions are introduced in zeolites by isomorphous substitution of Si atoms in the crystalline framework by other elements such as Al, Ti, Ga, etc.

Zeolites usually exhibit negatively charged frameworks in which cations are present as charge balancing elements. These cations may easily be exchanged by other monovalent or divalent cations through rapid ion exchange facilitated by the pore network.

The charge balancing cations are normally accompanied with coordinated water molecules which may be removed by thermal treatment, thereby inducing a change in the cation position. The dehydrated forms of the zeolites will tend to readily adsorb water molecules so that the cations will relocate in the most favored positions.

Zeolites are highly crystalline mineral frameworks that exhibit high thermal and hydrothermal stability and sufficient resistance to attrition. These properties are required for extensive industrial use (especially in catalytic processes).

4. The crystallization of MOF materials can be viewed as an esterification reaction between an inorganic base metal salt and an organic acid. MOF-5 can be taken as a prototypical example. The reaction for the formation of MOF-5 can be expressed stoichiometrically as follows:

$$4[Zn(NO_3)_2 \cdot 4H_2O] + 3[BDC] + 8[OH^-] \rightarrow [Zn_4O(BDC)_3] + 8[NO_3^-] + 23[H_2O]$$

Here, BDC is 1,4-benzene-dicarboxylate (terephthalic acid) and $Zn_4O(BDC)_3$ represents the MOF-5 unit composition. The reaction equilibrium can be shifted toward formation of the MOF product by tuning the concentration profiles of the solvent, the water released, or the nitrate ions produced. Since esterification reactions can be driven in both directions without difficulty, it appears evident that the stability of MOF materials in applications could depend on polar protic environments and pH values (128).

5. A cosolvent is usually present in fairly large amount and miscible with the main solution solvent. A cosolvent will alter micellar behavior and the reactivity of the inorganic components. A cosolvent could lead to increase surfactant solubility and modification of the critical micellar concentration (increased in most cases). The cosolvent is normally not located within the micellar aggregates. However, specific interaction with the polar head groups is possible. Furthermore, the kinetics of hydrolysis and condensation of inorganic precursors (e.g. alkoxysilane) may be affected. Methanol and ethanol are typical examples, acting as cosolvent in aqueous syntheses of mesoporous silica.

A cosurfactant is usually a polar organic molecule that exhibits surface activity (adsorbs at interfaces). The concentration of these organic additives remains low. A cosurfactant comicellizes with the surfactant, being thus located inside the micellar aggregates, mainly at the hydrophobic-hydrophilic interface or the palisade layer. A cosurfactant will induce changes in the micellar aggregate surface curvature (through different hydrophobic-hydrophilic volume ratios) and can stabilize a mesophase with given curvature. The micellar aggregation behavior is mainly affected. Critical micellar concentration can be decreased and mesophase transition to a lower curvature mesophase can occur. Short chain alcohols, such as *n*-butanol, hexanol, or octanol, and short chain amines, such as hexylamine or octylamine, are typical cosurfactants in aqueous micellar solutions.

Most often, swelling agents are apolar organic molecules. These species will be located preferentially inside the hydrophobic core of the micellar aggregates. This results in an increased hydrophobic core volume, leading to an increase in the size of the micelle (swelling). Evidently, since hydrophobic-hydrophilic ratio may also be changed markedly, the addition of swelling agent could also induce mesophase transitions. Common swelling agents added to the synthesis of mesoporous materials are aromatics such as mesitylene or xylene, and hydrocarbons such as hexane.

Note that depending on the species and their location in the micelles, cosurfactant agents can also induce swelling effects. Moreover, depending on the concentration of the organic additives, other effects may be generated, in particular emulsion and phase separation, which may be exploited to produce specific mesostructures (mesocellular foams) and particle morphologies (colloidal spheres and hollow structures).

6. Porous materials obtained via silane grafting may face problems of nonuniform distribution of the functionalities. Pore blocking can occur if the grafted silanes accumulate at the pore entrances. Also, unwanted silane oligomerization can occur, especially in the presence of water.

Cocondensation methods, on the other hand, can sometimes lead to poorly ordered solids, especially at high organic content. Furthermore, the accessibility of the functions located deep in thick walls can be questioned (especially for materials such as SBA-15 with thick silica walls). The functional groups must resist the conditions employed for the synthesis (acid, base, hydrothermal) and for the template removal (chemical extraction). Template removal might sometimes be incomplete in the case of ionic surfactants and silanol alkoxylation by alcohol solvents takes place (*transesterification*). In addition, the bridged silsesquioxane precursors for the synthesis of PMOs are expensive or not always readily available.

7. The most common method used in laboratories to remove the template is calcination. In this method, the as-synthesized materials are alternatively heated in flowing nitrogen, oxygen, or air, burning away the organics. Usually, the heating rates required are slow. It is followed by an extended period of heating at a temperature plateau (4 to 8 h).

An alternative method for surfactant removal is the chemical extraction of the organic template. This can be done either by liquid extraction with organic solvents, acid treatment, or supercritical fluid extraction. Dried as-synthesized samples are usually extracted in acidic solutions, alcohols, or salt solutions. Extraction is obviously necessary in the case of as-made inorganic-organic hybrids (cocondensed or PMO materials) in order to keep intact the organic functionalities. By using solvent extraction, it is possible to extract the structure-directing species without affecting the organic groups present in the silica framework. Various acidic media are used for surfactant extraction, ethanolic HCl solutions being the most commonly employed. It is often suggested that an ion-exchange mechanism occurs during solvent extraction of M41S-type materials. The presence of cationic species in the extraction liquid for charge balance is believed to be essential for the ion exchange. The HMS or SBA type of framework are considered to be relatively neutral, and the resulting framework-surfactant interactions are weak. Such weak electrostatic interactions or hydrogen bonding may be more favorable for surfactant extraction even in the absence of cationic species since counter cations are here not needed. Strongly oxidative treatments using hot concentrated aqueous sulfuric acid solutions or using a H_2O_2/HNO_3 mixture under controlled microwave irradiation can also be used, particularly in the case of copolymer templates. Ozone treatment and supercritical fluid (e.g. CO_2) extraction have been employed occasionally.

8. Compared to silica-based networks, nonsiliceous ordered mesoporous materials have attracted less attention, due to the relative difficulty to apply the principles employed to create mesoporous silica to nonsilica compositions. Other framework compositions are much more sensitive than silica to redox reactions, hydrolysis, or phase transformations. The reactivity of the inorganic precursors is much more difficult to control in the case of transition metal oxides, the reaction kinetics being much faster. Also, crystalline nonsiliceous frameworks are less prone to adapt the curvature of micellar aggregates, whereas the amorphous nature of silica allows for certain flexibility.

The template removal step, needed to achieve porous materials, is one of the most critical points. In contrast to silica, other compositions are usually more sensitive to thermal treatments and calcination can result in breakdown of the mesostructures. Hydrolysis, redox reactions, or phase transformations to the thermodynamically preferred denser crystalline phases account for this lower thermal stability. Many of the transition metal-based mesostructured materials synthesized in the presence of cationic surfactants collapse during thermal treatments. The poor thermal stability observed could be due to the different oxo chemistry of the metals compared to silicon. Several oxidation states of the metal centers may be responsible for oxidation and/or reduction during calcination. In addition, incomplete condensation of the framework is possible.

9. After sufficient solidification is achieved within the host pore system, with possible heating treatment performed to form a desired phase, the solid matrix can selectively be removed and shape-reversed molded structures are obtained. Most importantly for the successful replication from silica, the resulting material composition must be stable either in diluted HF or NaOH solution. Carbon is usually removed by combustion.

(**a**) Mn_2O_3 (silica, dissolved with NaOH);

(**b**) Al_2O_3 (carbon);

(**c**) SnO_2 (silica, preferentially dissolved with HF);

(**d**) CdS (silica, dissolved with NaOH);

(**e**) WO_3 (silica, dissolved with HF);

(**f**) ZnO (carbon);

(**g**) nanocristalline ZrO_2 (silica, dissolved with NaOH);

(**h**) BN (carbon, or silica dissolved with HF);

(**i**) polydivinylbenzene (silica, dissolved with HF or NaOH);

(**j**) calcium oxide, CaO (carbon);

(**k**) silicon carbide (silica, dissolved with HF);

(**l**) In_2O_3 (silica, dissolved with NaOH);

(**m**) cerium oxide, CeO_2 (silica, dissolved with NaOH).

10. A mesoporous hydrophobic organosilica PMO-type thin film should be promising. A large pore volume fraction will be beneficial because of the low dielectric constant of air ($k_{air} \sim 1$). Also, since the dielectric constant scales with electron density, a framework composition with reduced atomic weight and polarizability will be favorable. However, the material should still preserve sufficient mechanical and thermal stability. Therefore, mesoporous organic-inorganic hybrids can bring a compromise between low dielectric constant (lower electron density) and framework stability, although being highly porous. Here, to avoid the presence of humidity in pores (adsorption of moisture), hydrophobicity of the mesoporous film will be a crucial factor for maintaining a reasonably low dielectric constant (10).

10

APPLICATIONS OF MICROPOROUS AND MESOPOROUS MATERIALS

ANIRBAN GHOSH, EDGAR JORDAN, AND DANIEL F. SHANTZ

10.1 Introduction and Overview, 331
10.2 General Background, 333
 10.2.1 Silicate Chemistry, 333
 10.2.2 Zeolite Classification: Composition/Topology, 334
 10.2.3 Shape Selectivity (Zeolites), 335
10.3 Zeolites and Catalysis, 338
 10.3.1 Petrochemical Conversions (Large-Scale Processing), 338
 10.3.2 Methanol to Hydrocarbons (Alternative Fuels), 343
 10.3.3 Dimethylamine Synthesis (Shape Selectivity-Based Catalysts), 345
 10.3.4 Oxidation Chemistry (Heteroatom-Specific Reactions/Fine Chemicals), 346
 10.3.5 OMS-Supported Catalysts, 347
10.4 Ion-Exchange Processes, 348
10.5 Separations, 349
 10.5.1 Trace Gas Removal, 350
 10.5.2 Bulk Gas Separations, 350
 10.5.3 Emerging Areas and Future Directions, 352
References, 353
Problems, 356
Answers, 360

10.1 INTRODUCTION AND OVERVIEW

Solid materials possessing pores, or void spaces, in the size range of 1 to 10 nm have found numerous applications that have immensely benefited society (1). These

Nanoscale Materials in Chemistry, Second Edition. Edited by K. J. Klabunde and R. M. Richards

materials have many properties that render them technologically significant: high surface areas of several hundred square meters per *gram* of solid, strong acid sites that can activate inert substrates such as alkanes, the ability to stabilize small nanometer-sized metal clusters that facilitate chemical reactions, and highly uniform pore sizes that, in some cases, can be exquisitely selective in separating mixtures of molecules based on their size. This chapter focuses on these microporous and mesoporous materials. For the following, microporous will denote materials (per the IUPAC convention) possessing pores 2 nm in size or less, and mesoporous materials will typically possess pores in the 4 to 10 nm size range. The previous chapter by Kleitz is an excellent introduction to these materials. This chapter, after providing some background specifically relevant to the application of these materials, focuses on their use.

The category of microporous materials is dominated by zeolites. Zeolites are technologically important materials related to the natural feldspars and feldsparthoid minerals (2–5). Zeolites are microporous crystalline tectoaluminosilicates comprised of tetrahedral building units (TO_4, T = Al, Si) linked through bridging oxygen atoms. Zeolites possess a diversity of three-dimensional structures because of the numerous ways to link these tetrahedra (6). Zeolites, because they are crystalline, have structures with uniform pore sizes between 3 and 10 Å in diameter and can be made with a wide variety of framework compositions, ranging from a Si/Al = 1 to infinity. Zeolites containing framework aluminum (Al^{3+}) possess anionic frameworks, resulting in the need for charge-compensating extra-framework cations. A wide range of cations, including alkali metal and ammonium cations, as well as transition metals, can be utilized as extra-framework species. It is also possible to substitute other elements for framework aluminum and silicon atoms, including titanium, iron, and gallium. All of these factors give considerable flexibility for tuning the chemical properties of zeolites and explain why zeolites have found widespread use in heterogeneous catalysis, adsorption and separation of gases, and ion-exchange operations. In this chapter, the three-letter codes adopted by the International Zeolite Association (IZA) will be listed in parentheses along with the commercial/trade name of the zeolites (7). Figure 10.1 shows the framework structures of several commercially important zeolites.

The first zeolite mineral, stilbite, was discovered in 1756 when the Swedish mineralogist A.F. Cronstedt observed rock samples releasing steam while heated in a blowpipe flame. Cronstedt coined the name zeolite from the Greek words *zeo*, "to boil" and *lithos*, "stone." Deposits of natural zeolites were first discovered in the late nineteenth and early twentieth centuries. However, natural zeolites have not found widespread industrial use in North America. It was only with the advent of synthetic materials that zeolites were commercially applied to catalysis and separations (6).

Mesoporous materials such as ordered mesoporous silicas (OMS), initially developed by the Mobil labs in the early 1990s have very different properties (8–11). While primarily comprised of silica (SiO_2), the pores are larger (4 to 10 nm) and the materials are amorphous (i.e. noncrystalline). However, the pores of most OMS materials possess long-range ordering. While it is possible in these materials to substitute some silicon atoms in the amorphous matrix as in the case of zeolites, the resulting materials do not typically possess the thermal stability of zeolites, particularly in the presence of water vapor/steam. These materials are currently under intense investigation and their commercial use has been much more limited.

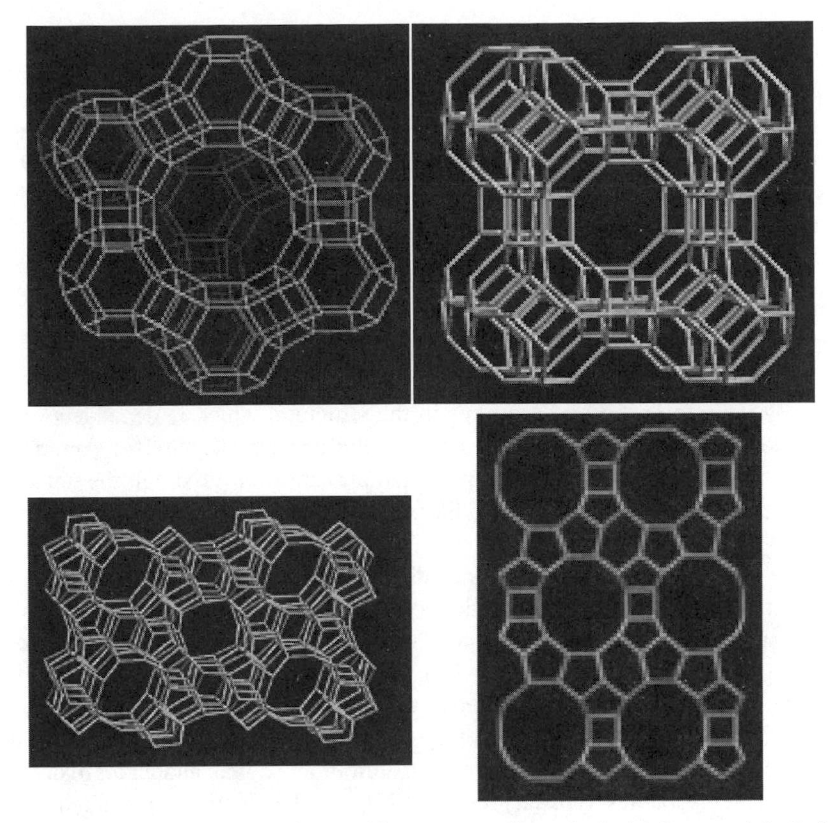

Figure 10.1 Partial structures of several important zeolites, clockwise from top left: faujasite (FAU), linde type A (LTA), zeolite beta (BEA*), and zeolite ZSM-5 (MFI). (See Reference 7.)

The focus of this chapter is porous (alumino)silicates, with an emphasis on zeolitic materials. Porous phosphate materials such as aluminophosphates and metalloaluminophosphates are quite interesting in their own right but beyond the scope of this chapter. There are several excellent overviews of the synthesis, characterization, and application of these materials elsewhere (12, 13). This chapter also will not describe MOFs, or metal-organic framework materials.

10.2 GENERAL BACKGROUND

10.2.1 Silicate Chemistry

Kleitz gave a detailed description of material synthesis in Chapter 9. Nevertheless, a few comments regarding silicate chemistry are in order. Silicon, analogous to carbon, is typically found in zeolites and OMS phases in a tetrahedral coordination environment. While silicon can also be found in an octahedral coordination environment this is not significant to the materials described here. The main elements comprising these materials are silicon, oxygen, and to a much smaller extent aluminum or other heteroatoms. The silicon (Si) and aluminum (Al) atoms are connected to each other through oxygen (O) atoms. The substitution of one Si^{4+} cation by one Al^{3+} cation in the tetrahedron generates a negative charge in the framework. To balance this negative charge, an inorganic or organic cation is required as an extra-framework cation. Thus, the unit cell formula for a zeolite can be represented as $M_{x/n}[(AlO_2)_x(SiO_2)_y] \cdot zH_2O$ where M is the exchangeable cation of valence n used to balance the framework charge. The ratio y/x may have any value ranging from one to infinity.

10.2.2 Zeolite Classification: Composition/Topology

Zeolites can be classified in many ways. Two convenient methods are on the basis of pore size and chemical composition, that is, the Si/Al ratio. The pore diameter is determined by the size of the free apertures in the structure, which is dependent on the number of T atoms (T = Si or Al) that form the aperture. Table 10.1 summarizes some examples of zeolites based on pore size classification. It should be noted that the values typically reported in the literature are determined by crystallographic studies. While these numbers are good guides, it is important to note that the actual pore size depends on many factors, including temperature, framework composition, and the type of extra-framework cations present in the zeolite. These factors can lead to subtle changes in effective pore sizes and subsequently large changes in material properties (adsorption/reactivity).

Table 10.2 summarizes some examples of zeolites based on their classification by chemical composition. Low-silica zeolites ($Si/Al < 5$) are synthesized in basic conditions ($pH > 13$) using a silicon source, an aluminum source, and alkali hydroxides at moderate temperatures, typically less than 120°C. The identity of the alkali species used is a determining factor in which phase is obtained from synthesis, as the relative rates of (alumino)silicate hydrolysis and condensation reactions are dependent on the identity of the alkali cation. It is also believed that hydrated alkali cations effectively direct the assembly of (alumino)silicate precursors into fully connected three-dimensional structures. Sodium and potassium hydroxide have been used most frequently in low-silica zeolite syntheses due to their low cost and high solubility in

TABLE 10.1 Classification of Zeolites Based on Size of Pore Opening Ring

Type	Pore Opening	Nominal Pore Diameter (Å)	Example
Small pore	8-membered ring	4.1 × 4.1, 3.6 × 3.6	Linde Type A (LTA), Zeolite Rho (RHO)
Medium pore	10-membered ring	5.1 × 5.5, 5.3 × 5.4	ZSM-5 (MFI), ZSM-11 (MEL)
Large pore	12-membered ring	7.4 × 7.4, 5.6 × 6.0, 6.6 × 6.7	Zeolite Y (FAU), ZSM-12 (MTW), Beta (BEA*)
Extra-large pore	14-membered ring	8.1 × 8.2, 7.2 × 7.5	UTD-1 (DON), CIT-5 (CFI)

Data from Reference 14.

TABLE 10.2 Classification of Zeolites Based on Si/Al Ratio

Type	Si/Al Ratio	Example
Low silica	1–5	Zeolite A (LTA), Zeolites X, Y (FAU)
High silica	>10	ZSM-5 (MFI), ZSM-11 (MEL), Beta (BEA), Mordenite (MOR)
Siliceous	∞	Silicalite-1 (MFI), Silicalite-2 (MEL)

aqueous solutions. Alkaline earth metals (Ba^{2+}, Sr^{2+}, Ca^{2+}) can also be used. However, due to their poor solubility in aqueous media as compared to alkali metals their use has been limited. Synthesis periods for low-silica materials range from several hours to several days. Several low-silica zeolites, including zeolite A (LTA), faujasite (FAU), zeolite L (LTL), and zeolite P (GIS), have found use industrially.

High-silica zeolites, in contrast, are typically made at lower pH values ($\sim$11), are synthesized at higher temperatures (160°C to 200°C) and in the presence of an organic molecule, typically a quaternary ammonium cation. The role of the organic molecule in determining the zeolite obtained from synthesis has been studied intensely though it is still poorly understood at a molecular level (15–21). While organic molecules used in the synthesis of high-silica zeolites are often referred to as templates, this is not an accurate term, as a molecular imprint of the organic is not reproduced in the zeolite structure obtained. Structure direction is a more appropriate description of the organic molecule's role in synthesis, as often the identity of the organic molecule appears to determine the material obtained from synthesis and in most cases this organocation is occluded in the as-synthesized material. Of the numerous high-silica zeolites reported to date, zeolites Beta (BEA*), ZSM-5 (MFI), and mordenite (MOR) have found the most use industrially.

Whereas zeolites, particularly high-silica zeolites, are made in the presence of simple organic molecules, OMS materials are made in the presence of molecules that aggregate or self-assemble, such as surfactants or block copolymers. Chapter 9 covers the synthesis of OMS in great detail. Most notably for the current discussion, typically OMS materials are made at temperatures comparable to those used to make low-silica zeolites (<100°C) and can be made in either acidic or basic conditions. These materials can be made in anywhere from a matter of hours to a few days.

10.2.3 Shape Selectivity (Zeolites)

Because zeolites have highly uniform pore sizes with dimensions comparable to those of small to medium-sized molecules, they exhibit a phenomenon known as *shape selectivity*. To help put this idea in more concrete terms Table 10.3 lists some Lennard-Jones diameters of various molecules. Since zeolites possess uniform pore sizes small changes in the size of a molecule can mean it is too large to either adsorb into the pores of the material or diffuse through the zeolite pores.

This property also led to zeolites being referred to as "molecular sieves" because of their ability to selectively separate very similar molecules on the basis of their size.

TABLE 10.3 Kinetic Diameters of Various Small Molecules

Molecule	Lennard-Jones Diameter, Å
H_2	2.89
N_2	3.64
CH_4	3.8
C_2H_4	3.9
C_6H_6	5.85

Data from Reference 22.

This has profound implications for the application of zeolites (23, 24). Following the conventions of Weisz and Frilette (25), as well as Csicsery (22), one can delineate three different classes of shape selectivity.

1. *Reactant selectivity*. The reactant molecules are too large to effectively adsorb into the zeolite pores and diffuse.
2. *Product selectivity*. The molecules formed via reaction in the zeolite pores are too large to effectively diffuse out of the zeolite pores, and thus due to subsequent reaction are generally not observed as products.
3. *Transition-state selectivity*. The zeolite structure effectively prohibits certain transition states from forming due to steric or space constraints. Thus, the product that would be formed from that transition state is not observed.

One example of reactant shape selectivity is the dehydration of linear and branched alcohols observed over zeolite A [LTA topology (26)]. Figure 10.2 shows part of the LTA structure. Typically secondary alcohols react more rapidly than primary alcohols. However, secondary alcohols do not react over Ca-LTA due to their inability to adsorb into the zeolite micropores. The same phenomenon is observed with linear and branched alkanes, where linear alkanes are readily adsorbed in the zeolite, whereas

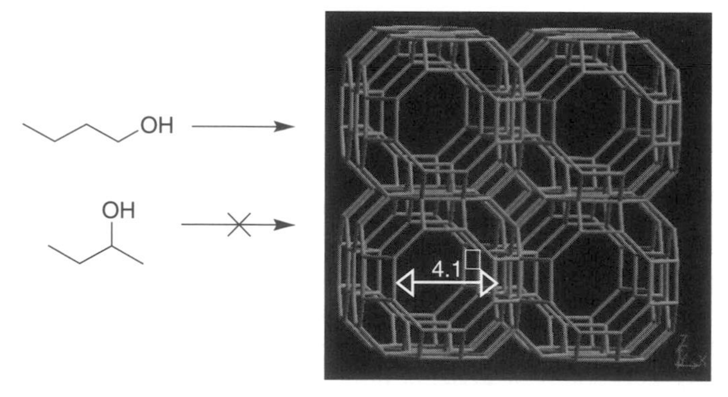

Figure 10.2 Partial LTA structure with schematic indicating the inability of branched alcohols to adsorb (and subsequently react) into the zeolite.

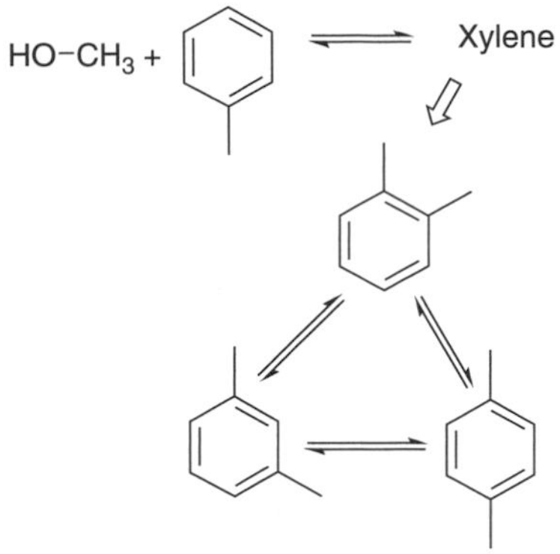

Figure 10.3 Simplified reaction network for the acid-catalyzed alkylation of toluene with methanol, and the isomerization reactions of *o*-, *m*-, and *p*-xylene.

branched alkanes are not. This is the basis for the ISOSIEVE process, an early commercial application of zeolites.

There are numerous examples of product selectivity, many of which involve mono- and disubstituted aromatics formed over ZSM-5 (MFI) catalysts (22–24). One of the early examples was xylene isomerization. Xylene can be formed over MFI catalysts via the acid-catalyzed reaction between methanol and toluene (see Fig. 10.3). According to thermodynamics the equilibrium distribution of *o*-, *m*-, and *p*-isomers is 26 : 51 : 23, which is different from the industrial demand as the *p*-isomer is a feedstock for terephthalic acid, a monomer for PET. However, very high selectivities of *p*-xylene can be obtained over MFI materials primarily because the diffusion of *p*-xylene is faster in the MFI pores compared to the other two isomers.

Transition-state selectivity is fundamentally different from reactant and product selectivity. It is the reaction intermediate, or transition state, that is forbidden from forming due to its size. A well-studied example is the acid-catalyzed transalkylation of dialkylbenzenes. *m*-xylene can undergo a bimolecular reaction to transfer a

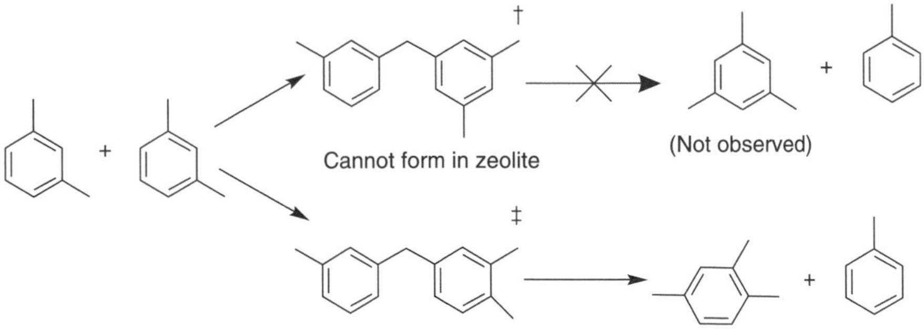

Figure 10.4 Transition-state selectivity illustrated with the transalkylation of *m*-xylene to form toluene and trialkylbenzenes.

methyl group, resulting in the formation of trialkylbenzene and toluene [Fig. 10.4 (27)]. The pore structure of the zeolite (in this case ZSM-5 or mordenite) prohibits certain transition states forming, here resulting in no 1,3,5-substituted products being observed.

10.3 ZEOLITES AND CATALYSIS

Catalysis, or the use of a substance (the catalyst) to accelerate the rate of a chemical reaction, is the most important application of micro- and mesoporous materials, particularly zeolites. The societal impact of zeolites in petroleum refining cannot be overstated. The scope of zeolite catalysis is sufficiently broad that it is impossible to cover here but at the most introductory of levels. The following section will focus on three aspects of zeolite catalysis: (1) impact in the petrochemical industry, (2) nonpetrochemical acid-catalyzed reactions, and (3) reduction/oxidation (redox) reactions.

One point not mentioned above is the thermal stability of zeolites. Most zeolites are thermally stable at elevated temperatures (>200°C to 300°C), with the result that the crystalline structure is not lost. Typically the thermal stability of a zeolite depends strongly on the zeolite Si/Al ratio, with the general trend that increasing Si/Al leads to enhanced stability. There are also exceptions to this; one in particular that will be discussed below is significant for fluid catalytic cracking (FCC) catalysts. The thermal stability of zeolites facilitates catalytic applications as many of the reactions discussed above and below are at elevated temperatures (>200°C).

Finally, it should be noted that the breadth of zeolites catalysts is enormous and in some sense the examples chosen below were arbitrary. That said, each example was chosen to illustrate specific points that are indicated in parentheses next to the section title.

10.3.1 Petrochemical Conversions (Large-Scale Processing)

Oil refineries and petrochemical plants extensively use zeolite catalysts in a number of processes (28, 29). In the most simplistic view an oil refinery's function is to convert its feedstock, crude oil, into a variety of products, perhaps the best known of which is transportation fuels. In reality refineries and petrochemical plants are highly integrated sophisticated operations that produce many products [Fig. 10.5 (28)]. The different types of chemical processing that must occur to achieve this include:

- Fractionation/distillation to separate compounds of different molecular weights.
- Cracking of large molecules into smaller molecules.
- Changing hydrogen content of molecules (dehydrogenation/hydrogenation).

The last two items are described in more depth below. There are numerous excellent texts on various aspects of refineries and thus the focus here will be on a few specific (yet important) chemical transformations that employ zeolites as catalysts (30, 31). Of

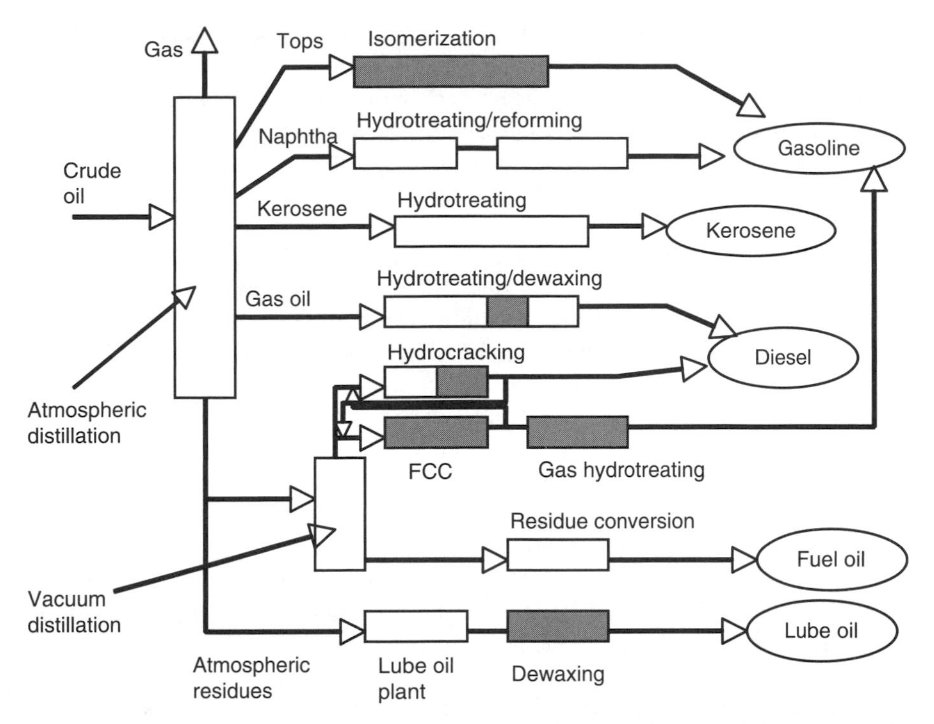

Figure 10.5 Hypothetical conversion refinery. (From J. Weitkamp and L. Puppe, eds., *Catalysis and Zeolites: Fundamentals and Applications*. Springer-Verlag, Berlin, 1999. With permission.)

particular note are References 28 and 32. Reference 28 offers a thorough (but approachable) technical description and Reference 32 a less technical description. The latter would likely be particularly attractive to the undergraduate student without a strong background in reactor design and chemical kinetics.

Zeolites are important refining catalysts for two reasons. The first, and most important, is the presence of strong acid sites. Two types of acid sites are present in zeolites. The first are Brønsted acid sites, shown in Figure 10.6. These are protons that act as charge-compensating cations for framework aluminum. The second are Lewis acid sites, which are less well defined than Brønsted sites and involve extra-framework aluminum species formed by removing framework aluminum (often due to steam). Both of these are extremely active catalytic centers, which can activate relatively inert substrates such as normal alkanes.

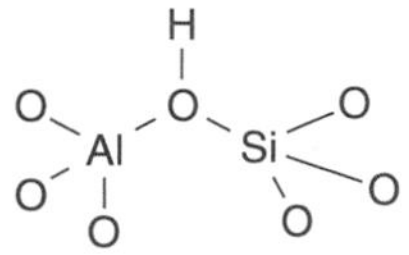

Figure 10.6 A Brønsted acid site in a partial zeolite framework.

10.3.1.1 Fluidized Catalytic Cracking (FCC) Initially the feedstock (crude oil) is separated using atmospheric distillation into a variety of fractions with different boiling points (see Fig. 10.5). The heavy (i.e. high boiling point) residuals are then separated further by vacuum distillation. The "light" components from this separation, typically denoted as vacuum gas oil with a boiling point range of 370°C to 500°C, can be converted into gasoline range (C_5, C_6, naphtha) molecules via catalytic cracking.

10.3.1.1.1 FCC Catalyst The main active component of the FCC catalyst is a rare earth stabilized faujasite zeolite (or RE-Fau, REY) with a fairly high Si/Al value. The zeolite is exchanged or impregnated with a rare earth metal, often lanthanum, to improve its thermal stability. This material is typically 5% to 40% by weight of the actual catalyst, which also includes a matrix material, usually an alumina or alumina-silica material as well as a binder. The importance of the matrix material cannot be overstated as it is often doped with small quantities of specific metals, for instance, to improve catalyst stability and it also imparts mechanical stability on the catalysts as well as determining the catalyst particle size/morphology. The latter is important as it impacts the mass and heat transfer properties of the catalyst. There is significant patent literature on the addition of different zeolites as components of the FCC catalysts. While these innovations are important, it is essential to point out the central role of the rare earth stabilized faujasite material in FCC units.

10.3.1.1.2 Reaction Mechanism There are believed to be two primary reaction pathways for acid-catalyzed cracking over FCC catalysts. The first is through a carbenium ion pathway. In this instance either (1) a Brønsted acid site protonates an olefin or (2) a Lewis acid site extracts a hydrogen from an alkane. In both cases a three-coordinate carbon center ($-CH^+-$) results. This carbenium ion can then undergo one of two propagation reactions: (1) a bimolecular H-transfer (hydride transfer) resulting in a paraffin and another carbenium ion or (2) a unimolecular β-scission step generating an olefin and a carbenium ion. Figure 10.7 shows the mechanism described above. The carbenium ion pathway, while long accepted, cannot completely explain the product distribution observed. Thus, at higher temperatures the possibility of forming reactive intermediates involving penta-coordinated carbon centers, or carbonium ions, has also been postulated. In this case a Brønsted acid site reacts to add a proton to a paraffin, resulting in a five-coordinate ($-CH_3^+-$) carbon center. This intermediate can either lose hydrogen forming a carbenium ion or undergo β-scission resulting in the formation of an olefin and a carbenium ion. While this is a simplified picture given the large number of different molecules present in the FCC feedstock it has been highly successful for describing the FCC unit behavior (e.g. product distribution).

10.3.1.1.3 Reactor Operation The typical FCC unit operates in the range of 480°C to 550°C at pressures slightly above ambient. The feed enters the reactor at 200°C to 300°C and is mixed with hot (750°C) catalyst, resulting in an inlet reactor temperature on the order of 550°C. At these temperatures the cracking reactions outlined above occur rapidly. The vaporization of this feed coupled with the injection of steam

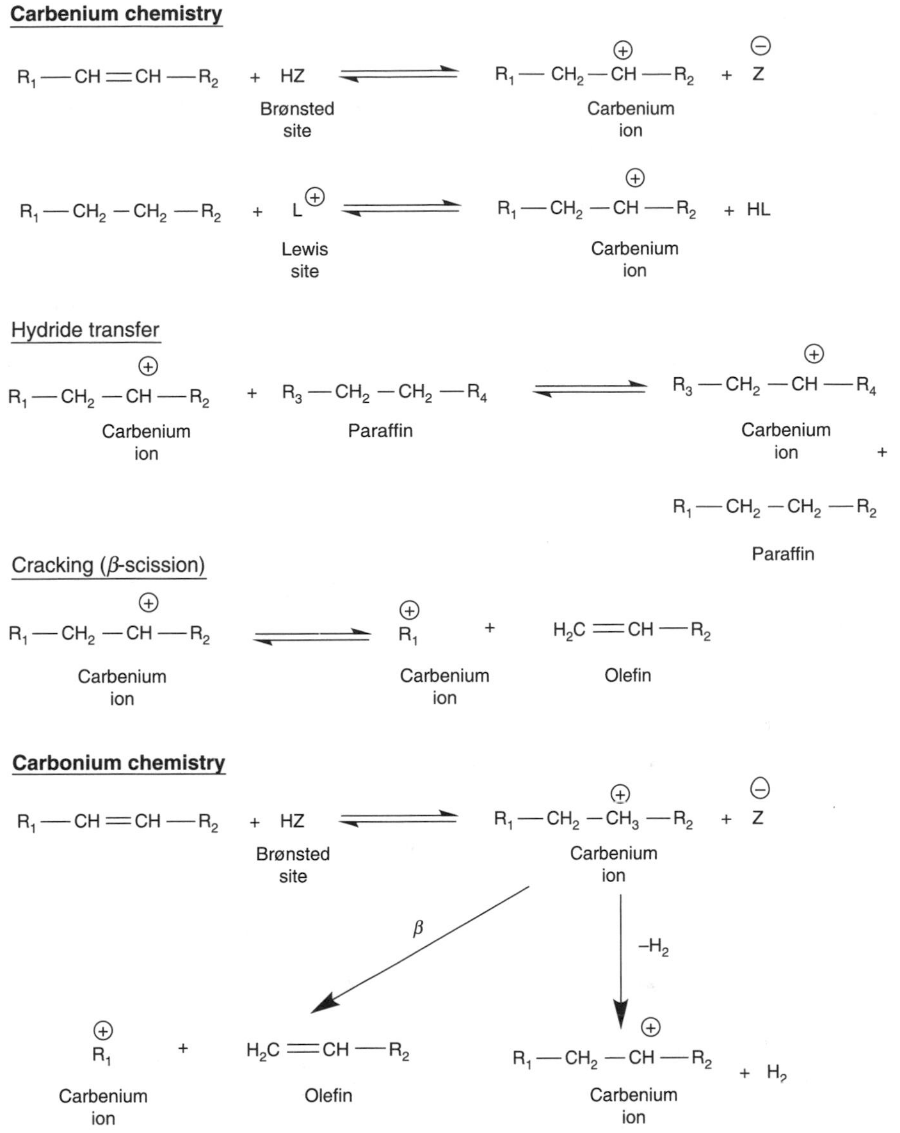

Carbenium chemistry

$$R_1-CH=CH-R_2 + \underset{\text{Brønsted site}}{HZ} \rightleftharpoons \underset{\text{Carbenium ion}}{R_1-CH_2-\overset{\oplus}{C}H-R_2} + \overset{\ominus}{Z}$$

$$R_1-CH_2-CH_2-R_2 + \underset{\text{Lewis site}}{L^{\oplus}} \rightleftharpoons \underset{\text{Carbenium ion}}{R_1-CH_2-\overset{\oplus}{C}H-R_2} + HL$$

Hydride transfer

$$\underset{\text{Carbenium ion}}{R_1-CH_2-\overset{\oplus}{C}H-R_2} + \underset{\text{Paraffin}}{R_3-CH_2-CH_2-R_4} \rightleftharpoons \underset{\text{Carbenium ion}}{R_3-CH_2-\overset{\oplus}{C}H-R_4} + \underset{\text{Paraffin}}{R_1-CH_2-CH_2-R_2}$$

Cracking (β-scission)

$$\underset{\text{Carbenium ion}}{R_1-CH_2-\overset{\oplus}{C}H-R_2} \rightleftharpoons \underset{\text{Carbenium ion}}{\overset{\oplus}{R_1}} + \underset{\text{Olefin}}{H_2C=CH-R_2}$$

Carbonium chemistry

$$R_1-CH=CH-R_2 + \underset{\text{Brønsted site}}{HZ} \rightleftharpoons \underset{\text{Carbenium ion}}{R_1-CH_2-\overset{\oplus}{C}H_3-R_2} + \overset{\ominus}{Z}$$

$$\xrightarrow{\beta} \underset{\text{Carbenium ion}}{\overset{\oplus}{R_1}} + \underset{\text{Olefin}}{H_2C=CH-R_2}$$

$$\xrightarrow{-H_2} \underset{\text{Carbenium ion}}{R_1-CH_2-\overset{\oplus}{C}H-R_2} + H_2$$

Figure 10.7 Carbenium/carbonium ion chemistry (28, 33).

causes the catalyst and feed to effectively become fluidized and rise up the reactor (which is called the riser). The cracking reactions are endothermic in nature, resulting in the temperature decreasing along the length of the reactor. The reactor residence time, or duration the feed is in the reactor, is on the order of one second. At the top of the riser the catalyst and feed/products are separated. During the cracking reactions coke, or polynuclear aromatic compounds, also forms on the catalyst that must be removed. The catalyst, after being separated from the reaction stream, is fed to a regenerator unit where the coke is burned off the catalyst. These reactions are highly *exothermic* and result in a significant increase in the catalyst temperature, typically to around 700°C to 750°C. The hot regenerated catalyst is then fed to the inlet of the riser system. This last point serves to emphasize that heat integration (i.e. the heat generated by catalyst coke removal is effectively used to heat the inlet feed) is essential to the economics of the process. Figure 10.8 shows an example of the reactor configuration, including the actual riser, or reactor, as well as the catalyst regeneration unit.

10.3.1.1.4 Market Demand The market for FCC catalysts is large, with over 350 units worldwide and over 100,000 tons per year of zeolite used. The catalyst composition (e.g. promoters) is highly tailored to the feedstock. Major vendors of FCC catalysts include Engelhard, Albemarle, and Akzo/Nobel. It should be reiterated that the scale of this process commercially is substantial. FCC is the most

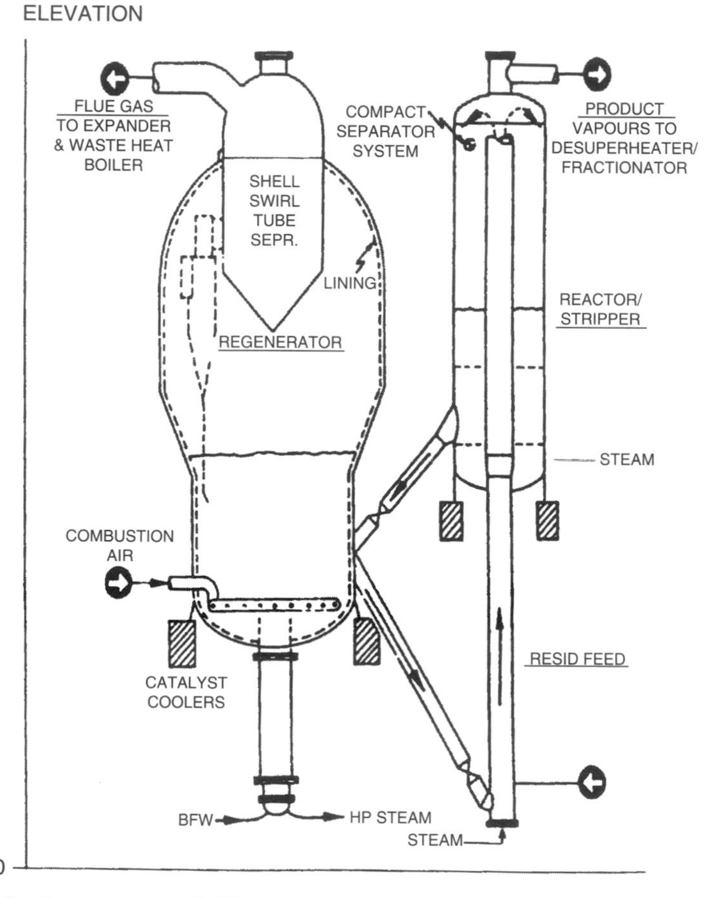

Figure 10.8 Schematic of a Shell FCC unit. (From J. Weitkamp and L. Puppe, eds., *Catalysis and Zeolites: Fundamentals and Applications*. Springer-Verlag, Berlin, 1999. With permission.)

widespread/practiced chemical conversion process in the world, and is the single largest market globally for refining catalysts.

10.3.1.2 Hydrocracking/Hydrotreating (Bifunctional Catalysis) Hydrocracking is another large-scale catalytic process that employs zeolite catalysts. It is included here because it is the quintessential example of bifunctional zeolite catalysis. Hydrocracking catalysts contain a dispersed transition metal so they possess both strong acid sites and the ability to perform hydrogenation/dehydrogenation chemistry. Hydrocracking catalysts can hydrogenate heavy aromatic components in the residual feeds. These aromatics are very difficult to crack, and thus after hydrogenation can be either cracked or used as naphthas in the distillate product pool. Thus, perhaps from an overly simplistic viewpoint, hydrocracking can be viewed as a route to increase the potential FCC feedstock pool. As indicated in Figure 10.5 the exit streams of the hydrocracking and FCC units can be partially recycled to the other unit to reach desired product compositions. The relative rates of hydrogenation/dehydrogenation and cracking over these catalysts are complex yet controllable based on reactor temperature, residence time, metal loading, hydrogen content in the feed, etc. There are several very important differences between FCC and hydrocracking catalysis, which include:

1. The products from hydrocracking tend to have higher hydrogen contents than FCC products.
2. The catalyst deactivation time in hydrocracking is on the order of a few years versus several seconds for FCC catalysts, due primarily to the presence of the transition metal and that the reactor is operated at elevated pressures in the presence of hydrogen.
3. Hydrocracking catalysts must be carefully tailored to handle transition metal poisons such as nitrogen and sulfur.

Hydrocracking, while not practiced at the scale of FCC, is nonetheless an extremely important process. Among other things it facilitates the conversion of aromatics to naphthas that can be used or further processed, allowing the plant another means to manipulate its relative production of benzene/toluene/xylenes (i.e. the BTX pool). More recently hydrocracking of heavier (i.e. higher boiling) residuals to convert them into gasoline range products has become increasingly attractive economically.

10.3.2 Methanol to Hydrocarbons (Alternative Fuels)

Another catalytic process involving zeolites is methanol to hydrocarbons (34). Chang and researchers at Mobil accidentally discovered that methanol could be converted to higher molecular weight products (35, 36). However, the potential of this was quickly realized. Subsequent work over the last 30 years has shown that based on the reactor operating conditions and catalyst that one can either selectivity form olefins or gasoline from methanol. Given the time at which this was discovered, during the energy crisis of the early 1970s, this technology drew significant attention. In fact, given current energy costs (April 2008, \$110/bbl oil) it is likely that this technology will again attract considerable attention as it has been demonstrated at large scale (see below).

The initial patent and literature reports by Chang and coworkers at Mobil were that ZSM-5 zeolite could activate methanol, ultimately leading to aromatics, paraffins, and olefins that fall into the gasoline and distillate ranges (methanol to gasoline, or MTG). This process has been demonstrated at fairly large scales (600,000 ton/year gasoline) in New Zealand. It was also recognized early in the discovery process that olefins are made in reasonable yield; it is in fact olefin oligomerization that leads to the higher molecular weight products mentioned above. Given the significance of new routes to light olefins there has also been extensive investigations of reactor/material design to optimize the selective formation of olefins, also known as methanol to olefins (MTO). Given the scope of this area some major points will be highlighted, emphasizing material properties and a brief discussion of mechanism. There are innumerable reviews on this problem for the interested reader (34–37).

10.3.2.1 Materials The original patent reports in this area employed zeolite ZSM-5 catalysts at elevated temperatures (typically 350°C to 500°C). Given the strong acidity of ZSM-5 it favors the subsequent oligomerization to gasoline products of any olefin intermediates formed. Thus, the optimized methanol to olefin catalyst is a small-pore silicoaluminophosphate, SAPO-34 (CHA) (38). In the context of MTO, SAPO-34 is preferable to ZSM-5 because its acid sites are not as strong, that is, oligomerization is slower, and the smaller pore apertures (formed by 8- not 10-membered rings) favors the selective formation of olefins, particularly C_3 and C_2 olefins (see Fig. 10.9). It is reasonable to say that many zeolites and related materials have been investigated for this chemistry, and that ZSM-5 is generally superior for MTG and that SAPO-34 is superior for MTO.

10.3.2.2 Mechanism Perhaps surprisingly there are still debates about the reaction mechanism of this process, in particular how the first carbon-carbon bond is formed. Figure 10.10 outlines the general mechanism. That the dimethyl ether observed is formed via reaction of methanol with a surface methoxyl group (bottom, Fig. 10.10)

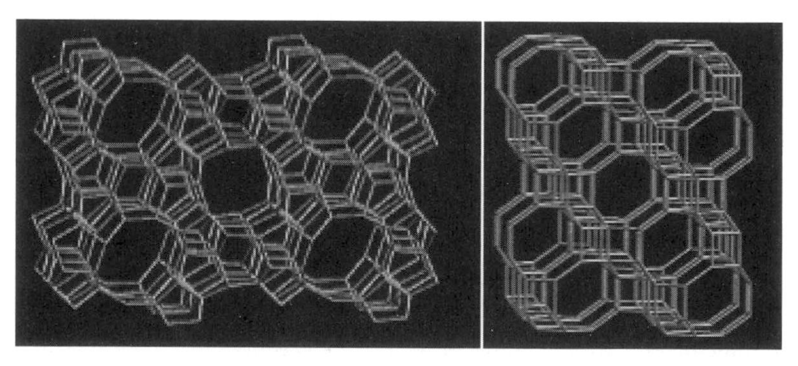

Figure 10.9 Framework structures of ZSM-5 (MFI) (left) and SAPO-34 (CHA) (right).

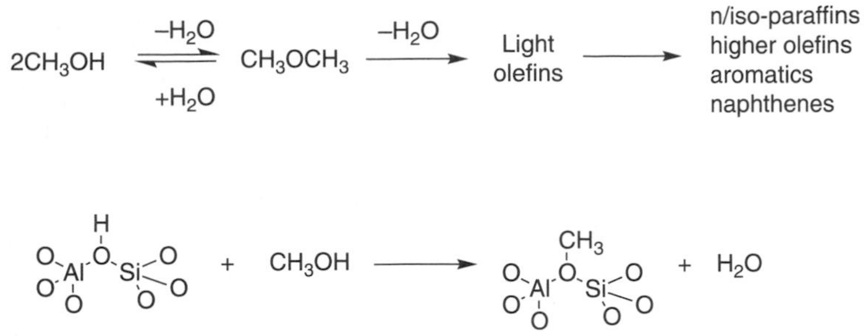

$$2CH_3OH \underset{+H_2O}{\overset{-H_2O}{\rightleftharpoons}} CH_3OCH_3 \xrightarrow{-H_2O} \text{Light olefins} \longrightarrow \text{n/iso-paraffins, higher olefins, aromatics, naphthenes}$$

$$\text{Al–O(H)–Si (zeolite)} + CH_3OH \longrightarrow \text{Al–O(CH}_3\text{)–Si} + H_2O$$

Figure 10.10 Overall reaction scheme in methanol to hydrocarbons (top) and the activation of methanol over an acidic zeolite (bottom). (From M. Stöcker, *Microporous Mesoporous Mater.* **1999**, *29*, 4–38. With permission.)

is well accepted. However, how one goes from the dimethyl ether to the light olefins, that is, how the first carbon-carbon bond is formed, is a matter of considerable discussion (37). There are, in general terms, two classes of mechanisms. The first are so-called "direct" routes or consecutive mechanisms, which invoke a variety of intermediates, including carbenes and/or oxonium ylides. The second are referred to as the "hydrocarbon" pool routes or parallel mechanisms, wherein the methanol reacts with hydrocarbon species within the zeolites. In the hydrocarbon pool route, most of these entrapped species are believed to be small (mono-, disubstituted) aromatics.

10.3.3 Dimethylamine Synthesis (Shape Selectivity-Based Catalysts)

Dimethylamine synthesis (39) was chosen as an example in the current review for two reasons. First, it is a good nonpetrochemical example of zeolites employed as solid acid catalysts. Second, it is also a good example of how zeolites can be used to shift product distributions away from the predicted thermodynamic or equilibrium distribution. Currently dimethylamine is synthesized mainly by reacting methanol and ammonia over a dehydration catalyst, usually alumina. Figure 10.11 gives a simplified reaction scheme. While at thermodynamic equilibrium trimethylamine is the preferred product, industrial demand is highest for dimethylamine, which is used as an intermediate for pesticide and solvent production. Given the high volume (over 1 billion pounds/year) of mono-, di-, and trimethylamine production there are clear driving forces for selective catalysts.

The reaction is well suited to zeolite catalysts. The shape selectivity principles outlined above can be used to shift the selectivity away from the thermodynamically predicted distribution and selectively form the dimethylamine product. More specifically by choosing zeolites with small pore apertures any trimethylamine formed in the pores

$$MeOH + NH_3 \longrightarrow MeNH_2 + H_2O$$
$$MeNH_2 + MeOH \longrightarrow Me_2NH + H_2O$$
$$Me_2NH + MeOH \longrightarrow Me_3NH + H_2O$$

Figure 11.11 Simplified overall reaction scheme for the synthesis of methylamines.

is hindered from diffusing out of the zeolite via product selectivity. Consistent with this faujasite (FAU) zeolites (large pore) led to very high selectivities of trimethylamine. In contrast, reactions over several small-pore zeolites, including zeolite A (LTA), chabazite (CHA), and zeolite rho (RHO) (40) are selective to dimethylamine. The selectivity, it should be emphasized, was highly sensitive to parameters including ion-exchange and post-synthetic treatments. Mordenite (MOR), a material containing pores with 12-membered and 8-membered rings, is also highly active and has been commercialized for this chemistry (41).

10.3.4 Oxidation Chemistry (Heteroatom-Specific Reactions/Fine Chemicals)

The zeolite catalysis examples described above are all instances where the framework elements were either silicon or aluminum. As mentioned in the introduction, it is possible to substitute other elements, such as Ti^{4+}, Fe^{3+}, B^{3+}, and Ga^{3+}, in the framework. Overwhelmingly the inclusion of titanium in the framework has had the largest impact. The synthesis of titanium-containing silicalite-1, or TS-1 (MFI topology) and its commercialization by EniChem in the mid-1980s has spurred enormous interest in investigating framework-substituted zeolites and their potential application (42, 43). The chemistries listed in this section also indicate a transition from petrochemical/commodity chemicals to fine chemicals. There appear to be several unique features that render TS-1 catalytically active that are generally not observed in other Ti-zeolites or with other transition metals:

1. The titanium in the active catalyst has been shown conclusively to be located in the framework in a tetrahedral environment.
2. Syntheses can be optimized to eliminate the formation of extra-framework titania clusters which are deleterious for activity/selectivity.
3. The titanium is stable in the framework under the reaction conditions.
4. The zeolite particle size can be tuned to minimize mass transport resistance into the particle and thus increase activity.

It should be emphasized that in general these points are not valid for other element substitutions in the framework; vanadium is a noted example. Many oxidation reactions employ liquid phase oxidants. This motivates point (3) above. If the metal leaches from the framework it is possible to see effective homogeneous catalysis, and to be misled about where the catalysis is taking place (in solution or in the zeolite). This issue has been discussed extensively in the literature (44).

The original commercial application of TS-1 was the oxidation of phenol in the presence of TS-1 using hydrogen peroxide as the oxidant (45). The main products are a mixture of hydroquinone (*p*-dihydroxybenzene) and catechol (*o*-dihydroxybenzene) (Fig. 10.12). It has been observed that the selectivity and activity of the TS-1 catalyst is tied to the points above. Further, although other titanium-containing zeolites have been investigated [e.g. Ti-Beta (BEA), Ti-ZSM-12 (MTW); Reference 46] none display the activity and selectivity of TS-1. One possible explanation for this is pore size effects; another possibility is that the TS-1 samples, which are quite hydrophobic,

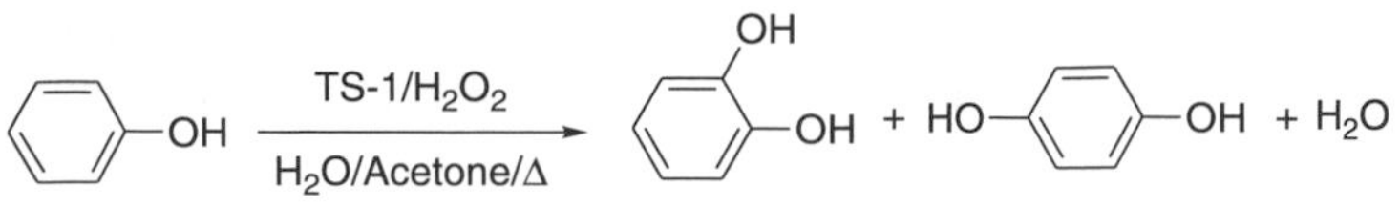

Figure 10.12 Hydroxylation of phenol over TS-1.

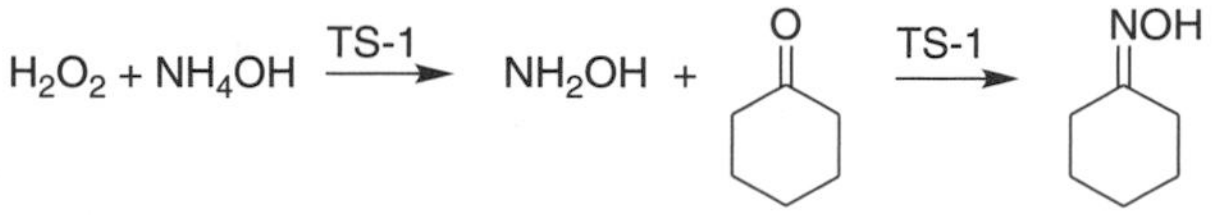

Figure 10.13 Ammoximation of cyclohexanone over TS-1.

effectively partition the water formed during the peroxide decomposition out of the pores increasing both the titanium stability and the catalyst selectivity.

Another reaction commercialized by EniChem is the ammoximation of ketones, particularly the conversion of cyclohexanone to cyclohexanone oxime (47). This latter compound is an intermediate in the manufacturing of caprolactam, the monomer for Nylon 6. This reaction, outlined in Figure 10.13, proceeds with both high conversion and selectivity for the oxime product. Again, TS-1 is uniquely active for this reaction compared to other catalysts, and TS-1 can catalyze this reaction on a variety of substrates. It is believed that in all cases the hydroxylamine is first formed, followed by reaction with the ketone. TS-1 is currently used commercially by EniChem to produce 12,000 ton per year of cyclohexanone oxime.

Finally, titanium silicates have also been extensively investigated for the epoxidation of olefins. The reaction of ethylene over a silver-supported catalyst to ethylene oxide is one of the few large-scale industrial oxidation reactions with molecular oxygen as the oxidant. Numerous studies have shown TS-1 to be effective at selectively forming propylene oxide (PO) from propylene using hydrogen peroxide as the oxidant. This is a more environmentally friendly route to PO than the currently used chlorhydrin route, and it is likely that this process will see commercialization in the near future.

10.3.5 OMS-Supported Catalysts

Ordered mesoporous materials have yet to find significant commercial applications in catalysis. While it was initially hoped that OMS materials containing aluminum would exhibit comparable activity and stability to zeolites that has not come to pass (30). Thus, while it seems likely that OMS phases will not trump zeolites/conventional porous solids used in high temperature catalysis, they do offer intriguing possibilities, particularly in the area of immobilizing homogeneous complexes such as salen complexes (48, 49). The ability to attach a complex (and expensive) homogeneous organic/organometallic catalyst to a surface is quite attractive provided it retains its activity. The ability to easily separate the catalyst from the reaction product would simplify process design, lower catalyst (and separations) costs, and make processes more environmentally friendly. This area is currently seeing tremendous activity in both academic and industrial labs and its future perspectives seem quite positive.

10.4 ION-EXCHANGE PROCESSES

As mentioned in the introduction the presence of extra-framework cations in zeolites impact zeolite properties. Many of the reactions discussed above utilized acid generated when protons are used as charge compensators. More generally zeolites, particularly low-silica zeolites, possess alkali metal cations as extra-framework cations (5, 6, 50). It is possible to (at least partially) exchange one extra-framework cation for another. Consider as an example zeolite LTA containing sodium cations (e.g. $Na_{96} \cdot xH_2O \cdot Si_{96}Al_{96}O_{384}$). It is possible to replace (or exchange) these cations with calcium by repeatedly exposing the zeolite to a calcium salt solution (e.g. $Na_{96\text{-}2y} \cdot Ca_y \cdot xH_2O \cdot Si_{96}Al_{96}O_{384}$, $y \leq 48$). In more general terms this exchange process can be written as

$$A^{+za}(aq) + (n/z_b)B^{+zb}(Al_nO_{2n}Si_mO_{2m}) \rightleftharpoons B^{+zb}(aq) + (n/z_a)A^{+za}(Al_nO_{2n}Si_mO_{2m})$$

where the charge on the cations are z_a and z_b, respectively. An important practical issue to point out is that these exchanges are almost always done in aqueous media, and accurate control of the pH is essential to minimize salt precipitation, zeolite dissolution, etc. Ion-exchange processes can be described using various thermodynamic models that can describe, for instance, the equilibrium uptake of different cations. There are two key points for any application of zeolites in ion exchange: (1) What is the exchange capacity, that is, equilibrium uptake, of the ion? (2) How fast is the ion-exchange process, that is, kinetics of exchange? Given that extra-framework cations are present in zeolites to satisfy charge neutrality point (1) implies that low-silica zeolites are likely to be preferable. The second point is subtler, in that many factors can affect the rate of ion exchange, including where the extra-framework cations are located in the zeolite structure, solution pH, and temperature. Consider the zeolite faujasite (FAU); Figure 10.14 shows the zeolite structure and the typical sites in the structure where extra-framework cations reside. Note that these sites have different accessibility. For instance sites SI and SI′ are very difficult to exchange as the cations are occluded in the so-called sodalite, or β-cages. Also depending on the cation identity the relative populations of the sites will change. Many texts and reviews have discussed both the thermodynamic and kinetic issues related to ion exchange; Barrer's books are of particular interest (2, 51).

In terms of commercial applications by far the largest volume application of zeolites is in detergents (6, 53). Over one million metric tons per year of zeolite 4A (LTA, Na-form) is used in detergent as replacements for phosphate builders. Other low-silica zeolites such as zeolite X (FAU) and zeolite P have found use: however, zeolite 4A dominates the detergent market because it not only has a very high ion-exchange capacity for calcium but the exchange kinetics are quite rapid, essential for successful application. Beyond this, zeolites have also been used as water softening

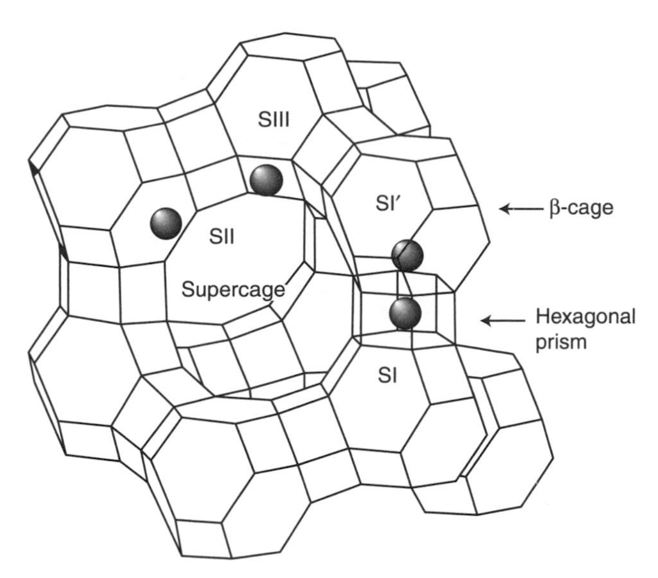

Figure 10.14 Faujasite (FAU) structure showing possible extra-framework cation locations. (From M. Feuerstein and R. F. Lobo, *Chem. Mater.* **1998**, *10*, 2197–2204. With permission.)

agents and in applications in agriculture necessitating the slow release of different compounds, for example, potassium or ammonium. For the latter application clinoptilolite (HEU), a naturally occurring zeolite, has found use.

10.5 SEPARATIONS

Zeolites have also found significant application in the area of separations, particularly gas separations. Consistent with the approach taken by Sircar and Myers (54), it is possible to group the applications into the following categories: (1) removal of dilute impurities from gas streams, (2) bulk gas separations, and (3) gas analysis. The specific examples below are relatively large-scale applications. Zeolites A (LTA), X (FAU), and mordenite (MOR) dominate commercial applications in this area, likely in part due to their relatively low cost. However, particularly in the case of zeolites A and X, it is also due to the presence of a high density of extra-framework cations. Consistent with that, many commercial separations are performed by preferential binding of one component of a mixture to these cations. Thus, relatively few commercial separations employ the sieving property of zeolites to achieve the separation. This will be expounded upon below.

Many commercial separations employing zeolites utilize pressure swing adsorption, temperature swing adsorption, or a combination of the two (55). The basic operating concept is that one component of the mixture preferentially will adsorb into the zeolite leaving an exit gas stream enriched in the other components. The adsorbed component is then removed from the zeolite adsorbent by either heating or reducing pressure once the bed capacity is reached. The key to successful process operation is that one wants to achieve a large difference in the amount adsorbed between the two possible adsorbates within a small pressure or temperature range. Thus, in the case of PSA, at a pressure of say P_1 there would be a large difference in the amount of A and B adsorbed. Then, by slightly reducing the pressure to P_2 one could effectively remove all the A that adsorbed on the zeolite. If the pressure swing needed between adsorption and desorption mode is too large, the associated compression costs render the separation uneconomical. To run PSA/TSA continuously there are usually multiple columns containing the active adsorbent. The columns are cycled such that one column is being regenerated while the other column is being used to perform the separation.

10.5.1 Trace Gas Removal

Zeolite adsorbents have found many applications removing small amounts of impurities or contaminants from gas streams. This is generally achieved by passing the gas stream through a bed of zeolite. The zeolite is chosen so that the trace contaminant or impurity preferentially adsorbs in the zeolite pores. Once the zeolite bed has reached its capacity in terms of impurity uptake, it is regenerated via a change in temperature or pressure. There are many such applications of zeolites, including removing trace amounts of SO_2, NO_x, or other undesirable contaminants. Zeolites, particularly dehydrated forms of zeolite X and A, are exceptionally good drying agents. One of the earliest applications of zeolites was their use to dry gas streams. The strong adsorption of water is due to the high number of alkali and/or alkali earth metals extra-framework cations. These cations bind water strongly (i.e. they have high energies of solvation) and thus are very effective desiccants. In fact, if one holds a few grams of zeolite powder or pellets that have been dehydrated in one's hand they will rapidly heat up as the adsorption of water on these materials is a highly exothermic process.

10.5.2 Bulk Gas Separations

Perhaps the best known of the large-scale zeolite gas separations is the generation of high purity oxygen and nitrogen from air using pressure swing adsorption. This separation is currently performed using lithium-exchanged zeolite X (FAU) materials. The separation is achieved by utilizing the fact that nitrogen preferentially adsorbs on the lithium extra-framework cations versus oxygen due its quadrupole moment. The nitrogen and oxygen adsorption isotherms (amount of gas adsorbed versus gas pressure) on a series of X zeolites are shown in Figure 10.15. While nitrogen adsorbs selectively its heat of adsorption is low enough that the process economics are viable.

The process can ultimately provide both high purity nitrogen (99.0%) and oxygen (90%) from air. Figure 10.16 shows a schematic of the process.

Zeolite adsorbents, in conjunction with activated carbon, are also used in a PSA-type process to produce high purity (>99%) hydrogen. A large-scale reaction used to produce hydrogen is steam reforming of methane, often followed by the water-gas shift reaction (Fig. 10.17).

For many applications (e.g. fuel cells) it is essential to remove the CO, any residual methane or water, and CO_2 produced by the water-gas shift reaction. The typical steam-methane reformer reaction has an exit stream with 70% to 80% H_2. A multiple

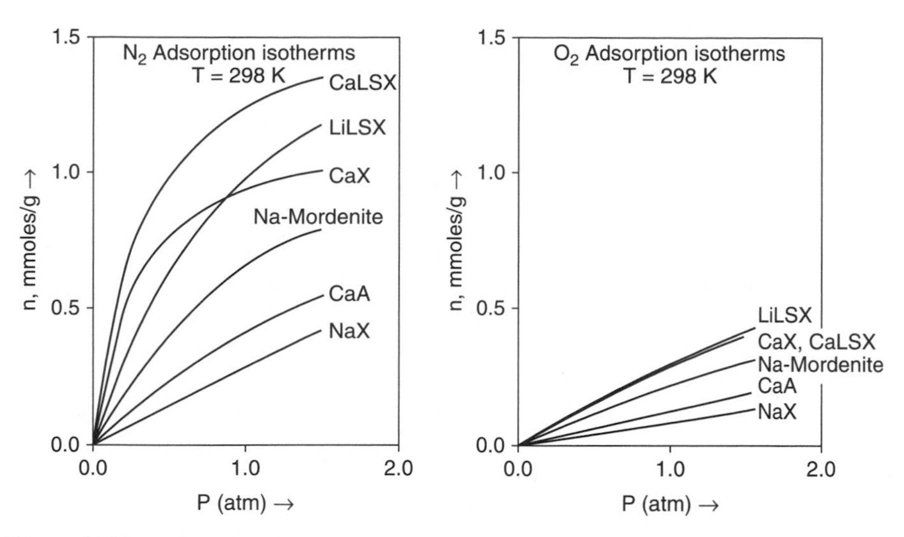

Figure 10.15 Adsorption isotherms of oxygen and nitrogen on various ion-exchanged X zeolites. (From S. Sircar and A. L. Myers, in *Handbook of Zeolite Science and Technology*, S. M. Auerbach et al., eds. Marcel Dekker, New York, 2003. With permission.)

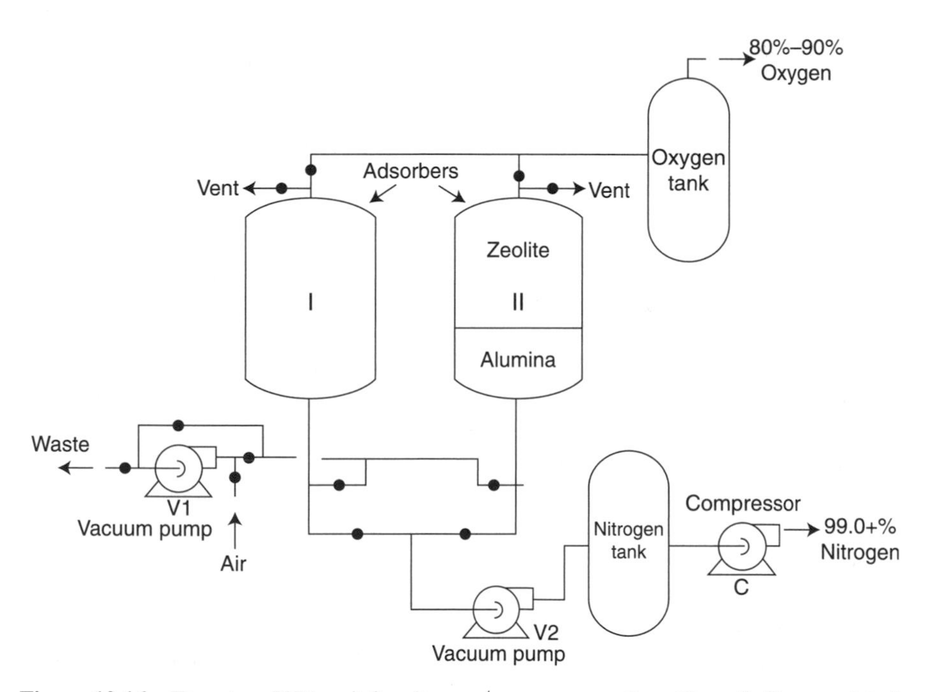

Figure 10.16 Two stage VSA unit for nitrogen/oxygen separation. (From S. Sircar and A. L. Myers, in *Handbook of Zeolite Science and Technology*, S. M. Auerbach et al., eds. Marcel Dekker, New York, 2003. With permission.)

$$CH_4(g) + H_2O(g) \rightleftharpoons CO(g) + 3H_2(g)$$
$$CO(g) + H_2O(g) \rightleftharpoons CO_2(g) + H_2(g)$$

Figure 10.17 Steam reforming of methane and water–gas shift reactions.

bed PSA unit is used to purify this stream to 99+% purity. Zeolite 5A (Ca-LTA) along with an activated carbon is used in the PSA beds.

The separation of olefin and paraffins, particularly the C_2 (ethane/ethene) and C_3 (propane/propene) pairs is an extremely important and demanding separation in the petrochemical industry (56). It is currently performed via cryogenic distillation and is thus energy (and capital) intensive. Thus, the use of zeolites to perform this separation has been studied intensely. While many zeolites have been investigated to selectively adsorb the olefin, PSA-type approaches are not currently used for this separation. More recently, small-pore zeolites have been reported wherein a clear kinetic separation is observed in that the diffusion of propylene is dramatically faster than that of propane (57, 58). This potentially represents a significant breakthrough in the field.

The last example to be discussed was mentioned near the introduction of the chapter, namely the selective separation of normal and iso-paraffins over zeolite 5A (Ca-LTA). While normal paraffins can readily adsorb into the LTA material, the iso-paraffins cannot. This is one of the few examples in gas separations wherein a true sieving effect is utilized. What is observed in practice is that upon exposing the zeolite bed to the feed stream, iso-paraffins are quickly observed at the exit of the bed, as they do not diffuse through the particles but rather around them. At a later time, once the zeolite has been saturated with normal paraffins they are then observed at the bed exit.

10.5.3 Emerging Areas and Future Directions

Hopefully this chapter has conveyed to the reader how porous solids have positively impacted society. It seems appropriate to finish the chapter by identifying some possible areas of growth for these materials, particularly beyond their traditional areas of application. Zeolites, in many senses, are a mature class of materials, yet many challenges remain. By contrast OMS phases are just now starting to become fully appreciated and their potential seems substantial. Three areas where these materials are likely to lead to advanced technologies are discussed briefly here.

10.5.3.1 Catalysis As can be gleaned from the above, zeolites have found significant applications in catalysis, particularly in the area of refining chemistry. It would thus seem logical that zeolites will potentially play a role in the future development of a "biorefinery," in that one can easily envision that at many stages chemical transformations will be necessary (59). This will spur new developments in catalysis, and will likely lead to new applications for zeolites. While most of the processing may be solution based in contrast to the gas-phase reactions above, the oxidation ability of TS-1 described above would seem to indicate that zeolites can be employed successfully as catalysts in polar liquid media. The need to process large biomolecules may also lead to new opportunities for OMS materials. In the shorter term it would seem more probable that suitably functionalized OMS materials will find niche applications in fine chemical productions, via being used as supports for homogeneous catalysts.

10.5.3.2 Membrane-Based Separations and Advanced Devices It is perhaps noteworthy that in the section on separations nearly all the applications discussed were based on selective adsorption and none of the applications involved MFI-type zeolites. Zeolite (or OMS)-based membranes for separations represent, though yet commercially unrealized, the ultimate example of molecular sieving. The ability to make, at large scale, defect-free zeolite thin films that can perform shape-selective separations has long been viewed as one of the holy grails of zeolite science. Researchers are getting ever closer to this goal (60); however, thin films of zeolites are notoriously difficult to make given that silicas, like most ceramics, are very difficult to process continuously in thin-film form due to their brittle nature. The current state of this field is nicely reviewed in a recent publication (61), and shows that high-quality films of certain zeolites can be made at small to modest scale (62). Ongoing work is also exploring nanocomposite membranes. In such membranes small (submicron) zeolite crystals are dispersed in a thin polymeric layer no more than 1 μm thick, which is then supported on a continuous polymeric fibril. The goal is to utilize the thin zeolite boundary layer to perform the separation while at the same time keeping the desirable processing properties of polymers.

The synthesis of OMS films has also advanced tremendously over the last five years (63–65). This work has been less geared towards membranes and more towards using these films as nanocontainers for subsequent growth of metal or metal chalcogenide nanowire arrays. This area, while still in its infancy, is of great interest and will have potentially significant impact as it represents one possible route to mass producing nanowire assemblies that could be used in next generation micro- (or nano)-electronics (66, 67). Related to this, the application of micro- and mesoporous materials in thin film form for use as low k-dielectric materials has attracted considerable interest in recent years (68, 69). The motivation is simple: as the material has a larger void fraction the dielectric constant will decrease. Thus, provided the material has the necessary mechanical properties and processibility, it should be possible to make zeolite- or OMS-based thin films for low k applications.

REFERENCES*

1. Schüth, F.; Sing, K. S. W.; Weitkamp, J., eds. *Handbook of Porous Solids*. Wiley-VCH, Weinheim, 2002.
2. Barrer, R. M. *Hydrothermal Chemistry of Zeolites*. Academic Press, London, 1982.

***Note regarding references** When trying to write an overview of the scope of this chapter one must inevitably be very choosy about references cited. The references thus chosen were used for two primary reasons: (1) they are well referenced and (in the authors' eyes) one of the definitive treatments of the subject, and (2) they will serve as a convenient point for the interested reader to explore a topic in more detail. These two criteria thus shift the references primarily onto textbooks and review articles, with relatively little original literature cited. While this is unfortunate (and unusual), the authors believe it is in the spirit of the overarching goal of the current text.

3. Breck, D. W. *Zeolite Molecular Sieves: Structure, Chemistry, and Use*. Wiley, New York, 1974.
4. Liebau, F. *Structural Chemistry of Silicates*. Springer-Verlag, Berlin, 1985.
5. Auerbach, S. M.; Carrado, K. A.; Dutta, P. K., eds. *Handbook of Zeolite Science and Technology*. Marcel Dekker, New York, 2003.
6. van Bekkum, H.; Flanigen, E. M.; Jansen, J. C., eds. *Introduction to Zeolite Science and Practice*. Elsevier, Amsterdam, 1989; Vol. 59.
7. International Zeolite Association (IZA) website, http://www.iza-online.org.
8. Beck, J. S.; Vartuli, J. C.; Roth, W. J.; Leonowicz, M. E.; Kresge, C. T.; Schmitt, K. D.; Chu, C. T. W.; Olson, D. H.; Sheppard, E. W.; McCullen, S. B.; Higgins, J. B.; Schlenker, J. L. *J. Am. Chem. Soc.* **1992**, *114*, 10834–10843.
9. Kresge, C. T.; Leonowicz, M. E.; Roth, W. J.; Vartuli, J. C.; Beck, J. S. *Nature* **1992**, *359*, 710–712.
10. Zhao, D. Y.; Feng, J. L.; Huo, Q. S.; Melosh, N.; Fredrickson, G. H.; Chmelka, B. F.; Stucky, G. D. *Science* **1998**, *279*, 548–552.
11. Zhao, D. Y.; Huo, Q. S.; Feng, J. L.; Chmelka, B. F.; Stucky, G. D. *J. Am. Chem. Soc.* **1998**, *120*, 6024–6036.
12. Pastore, H. O.; Coluccia, S.; Marchese, L. *Annu. Rev. Mater. Res.* **2005**, *35*, 351–395.
13. Hartmann, M.; Kevan, L. *Chem. Rev.* **1999**, *99*, 635–663.
14. Baerlocher, C.; Meier, W. M.; Olson, D. H. *Atlas of Zeolite Framework Types*, 5th edition. Elsevier, Amsterdam, 2001.
15. Burkett, S. L.; Davis, M. E. *J. Phys. Chem.* **1994**, *98*, 4647–4654.
16. Akporiaye, D. E. *Angew. Chem. Int. Ed.* **1998**, *37*, 2456–2457.
17. Corma, A.; Davis, M. E. *ChemPhysChem* **2004**, *5*, 304–313.
18. Cundy, C. S.; Cox, P. A. *Chem. Rev.* **2003**, *103*, 663–701.
19. Davis, M. E.; Lobo, R. F. *Chem. Mater.* **1992**, *4*, 756–768.
20. Nakagawa, Y.; Lee, G. S.; Harris, T. V.; Yuen, L. T.; Zones, S. I. *Microporous Mesoporous Mater.* **1998**, *22*, 69–85.
21. Wagner, P.; Nakagawa, Y.; Lee, G. S.; Davis, M. E.; Elomari, S.; Medrud, R. C.; Zones, S. I. *J. Am. Chem. Soc.* **2000**, *122*, 263–273.
22. Csicsery, S. M. *Zeolites* **1984**, *4*, 202–213.
23. Chen, N. Y.; Garwood, W. E.; Dwyer, F. G. *Shape Selective Catalysis in Industrial Applications*, 2nd edition. Marcel Dekker, New York, 1996.
24. Chen, N. Y.; Degnan, T. F. J.; Morris Smith, C. *Molecular Transport and Reaction in Zeolites: Design and Application of Shape Selective Catalysis*. Wiley, New York, 1994.
25. Weisz, P. B.; Frilette, V. J. *J. Phys. Chem.* **1960**, *64*, 382–382.
26. Weisz, P. B.; Frilette, V. J.; Maatman, R. W.; Mower, E. B. *J. Catal.* **1962**, *1*, 307–312.
27. Csicsery, S. M. *J. Catal.* **1970**, *19*, 394–397.
28. Weitkamp, J.; Puppe, L., eds. *Catalysis and Zeolites: Fundamentals and Applications*. Springer-Verlag, Berlin, 1999.
29. Wojciechowski, B. H.; Corma, A. *Catalytic Cracking: Catalysts, Chemistry, and Kinetics*. Marcel Dekker, New York, 1986.
30. Corma, A. *Chem. Rev.* **1997**, *97*, 2373–2419.

31. Corma, A.; Garcia, H. *Chem. Rev.* **2003**, *103*, 4307–4365.
32. Magee, J.; Dolbear, G. *Petrochemical Catalysis in Nontechnical Language*. PennWell, Tulsa, 1998.
33. Scherzer, J. *J. Catal. Rev. Sci. Eng.* **1989**, *31*, 215–354.
34. Stöcker, M. *Microporous Mesoporous Mater.* **1999**, *29*, 4–38.
35. Chang, C. D. *Catal. Rev. Sci. Eng.* **1983**, *25*, 1–118.
36. Chang, C. D. *Catal. Rev. Sci. Eng.* **1984**, *26*, 323–345.
37. Haw, J. F.; Song, W.; Marcus, D. M.; Nicholas, J. B. *Acc. Chem. Res.* **2003**, *36*, 317–326.
38. Dahl, I. M.; Kolboe, S. *Catal. Lett.* **1993**, *20*, 329–336.
39. Corbin, D. R.; Schwarz, S.; Sonnichsen, G. C. *Catal. Today* **1997**, *37*, 71–102.
40. Shannon, R. D.; Keane, M. J.; Abrams, L.; Staley, R. H.; Gier, T. E.; Corbin, D. R.; Sonnichsen, G. C. *J. Catal.* **1988**, *113*, 367–382.
41. Mochida, I.; Yasutake, A.; Fujitsu, H.; Takeshita, K. *J. Catal.* **1983**, *82*, 313–321.
42. Clerici, M. G. *Top. Catal.* **2000**, *13*, 373–386.
43. Perego, C.; Carati, A.; Ingallina, P.; Mantegazza, M. A.; Bellussi, G. *Appl. Catal. A: General* **2001**, *221*, 63–72.
44. Sheldon, R. A.; Wallau, M.; Arends, I. W. C. E.; Schuchardt, U. *Acc. Chem. Res.* **1998**, *31*, 485–493.
45. Bellussi, G.; Perego, C., in *Handbook of Heterogeneous Catalysis*, Ertl, G., Knozinger, J., Weitkamp, J., eds. Wiley, Poiters, 1997; Vol. 5.
46. Tuel, A. *Zeolites* **1995**, *15*, 236–242.
47. Petrini, G.; Leofanti, G.; Mantegazza, M. A., in *Green Chemistry: Designing Chemistry for the Environment*, Anastas, P. T., Williamson, T. C., eds. American Chemical Society, New York, 1996; Vol. 626.
48. Li, C. *Catal. Rev.* **2004**, *46*, 419–492.
49. Taguchi, A.; Schuth, F. *Microporous Mesoporous Mater.* **2005**, *77*, 1–45.
50. Townsend, R. P. *Pure Appl. Chem.* **1986**, *58*, 1359–1366.
51. Barrer, R. M. *Zeolites and Clay Minerals as Sorbents and Molecular Sieves*. Academic Press, London, 1978.
52. Feuerstein, M.; Lobo, R. F. *Chem. Mater.* **1998**, *10*, 2197–2204.
53. Yamane, I.; Nakazawa, T. *Pure Appl. Chem.* **1986**, *58*, 1397–1404.
54. Sircar, S.; Myers, A. L., in *Handbook of Zeolite Science and Technology*, Auerbach, S. M., Carrado, K. A., Dutta, P. K., eds. Marcel Dekker, New York, 2003.
55. Ruthven, D. M. *Pressure Swing Adsorption*. VCH, New York, 1993.
56. Eldridge, R. B. *Ind. Eng. Chem. Res.* **1993**, *32*, 2208–2212.
57. Olson, D. H.; Camblor, M. A.; Villaescusa, L. A.; Kuehl, G. H. *Microporous Mesoporous Mater.* **2004**, *67*, 27–33.
58. Olson, D. H.; Yang, X.; Camblor, M. A. *J. Phys. Chem. B* **2004**, *108*, 11044–11048.
59. Huber, G. W.; Iborra, S.; Corma, A. *Chem. Rev.* **2006**, *106*, 4044–4098.
60. Lai, Z. P.; Bonilla, G.; Diaz, I.; Nery, J. G.; Sujaoti, K.; Amat, M. A.; Kokkoli, E.; Terasaki, O.; Thompson, R. W.; Tsapatsis, M.; Vlachos, D. G. *Science* **2003**, *300*, 456–460.
61. Snyder, M. A.; Tsapatsis, M. *Angew. Chem. Int. Ed.* **2007**, *46*, 7560–7573.

62. Carreon, M. A.; Li, S. G.; Falconer, J. L.; Noble, R. D. *Adv. Mater.* **2008**, *20*, 729–732.
63. Doshi, D. A.; Gibaud, A.; Goletto, V.; Lu, M. C.; Gerung, H.; Ocko, B.; Han, S. M.; Brinker, C. J. *J. Am. Chem. Soc.* **2003**, *125*, 11646–11655.
64. Grosso, D.; Babonneau, F.; Albouy, P. A.; Amenitsch, H.; Balkenende, A. R.; Brunet-Bruneau, A. *Chem. Mater.* **2002**, *14*, 931–939.
65. Sanchez, C.; Boissiere, C.; Grosso, D.; Laberty, C.; Nicole, L. *Chem. Mater.* **2008**, *20*, 682–737.
66. Davis, M. E. *Nature* **2002**, *417*, 813–821.
67. Schüth, F.; Schmidt, W. *Adv. Mater.* **2002**, *14*, 629–638.
68. Li, Z. J.; Lew, C. M.; Li, S. A.; Medina, D. I.; Yan, Y. S. *J. Phys. Chem B* **2005**, *109*, 8652–8658.
69. Mitra, A.; Cao, T. G.; Wang, H. T.; Wang, Z. B.; Huang, L. M.; Li, S.; Li, Z. J.; Yan, Y. S. *Ind. Eng. Chem. Res.* **2004**, *43*, 2946–2949.
70. Kärger, J.; Ruthven, D. M. *Diffusion in Zeolites and Other Microporous Solids*. 604 pp. John Wiley, New York, 1992.

PROBLEMS

1. *Interactive/Web Problem.* The International Zeolite Association (IZA) website (http://www.iza-online.org) has several databases, including zeolite structures, synthesis, and the like. You will need to use this website for this question.

(**a**) Pick five different zeolite materials from the IZA website. For these list their name, three-letter structure code, unit cell symmetry and size, and size of the pore apertures. Make sure you pick zeolites with aperatures formed by different numbered (alumino)silicate rings (e.g. six-membered rings, eight-membered rings). Comment on these materials, the variation of the unit cell symmetry and dimensions, and how the pore sizes compare.

(**b**) One complication in studying zeolites is that there are many zeolites that have multiple names used to refer to the same zeolite structure. The ultimate example of this are zeolites with the faujasite topology. For the following zeolites, using the information on the IZA site (particularly the database of zeolite structures), list at least three names for the following zeolites: faujasite (FAU), Linde Type A (LTA), and chabazite (CHA).

2. *Shape Selectivity.* One of the points made in the section on shape selectivity is that particle size effects can alter the product distribution of molecules since the external surface of the crystal can also possess sites to catalyze reactions. As a simple limiting case consider

$$A \longrightarrow B \text{ (in the zeolite micropores)}$$
$$A \longrightarrow C \text{ (on the zeolite external surface)}$$

Assume relative rates are the same *per active site* and the active site density on the surface is the same as in the micropores. Estimate the relative amounts of B

and C made as a function of particle size over the range of particle sizes from 10 to 100 nm in diameter.

3. *Reaction Mechanisms*. Revisit Figure 10.7. An important point about carbenium and carbonium ions is that there are clear stability trends, most notably methyl ions are less stable then primary ions, which are less stable then secondary ions, which are less stable then tertiary ions. Explain why this is the case.

4. *Adsorption*. Consider an equimolar mixture of gases A and B and that each compound obeys Langmuirian adsorption behavior:

$$n_A = \frac{K_A p_A}{1 + K_A p_A + K_B p_B}$$

$$n_B = \frac{K_B p_B}{1 + K_A p_A + K_B p_B}$$

In these expressions n_i is the moles of component i adsorbed, K_i are the equilibrium adsorption constants, and p_i is the partial pressure of component i. Assuming an equimolar gas mixture and that the gases are ideal plot three sets of adsorption isotherms over the *total* pressure range of 0.0001 bar to 1 bar. In case (1) $K_A = 10$ moles/g-bar and $K_B = 2$ moles/g-bar, case (2) $K_A = K_B = 10$ moles/g-bar, and case (3) $K_A = 1$ mole/g-bar and $K_B = 0.2$ mole/g-bar. Discuss the plots of the adsorption isotherms and answer the following two questions: if compression/vacuum costs are unimportant to the economics of the process, which case would be preferable for PSA where it was desired to have B in high purity? If one was constrained to operate a PSA cycle between 1 bar (adsorption) and 0.4 bar (desorption) which case would be preferable and why?

5. *Bifunctional Catalysts*. One reason invoked (there are others) for why metal particles with high dispersion are desirable for catalysis is that the ratio of surface metal atoms to total number of atoms is quite high. Consider the potential cluster size for two zeolites: faujasite and ZSM-12. Faujasite has supercages, which for the purposes of this question can be described as spherical with a diameter of 12 Å, and ZSM-12 has a one-dimensional elliptical pore structure of dimensions 5.6 × 6.0 Å. Assuming the metal atoms of interest have a diameter of 1.0 Å and the cluster has a packing fraction corresponding to an FCC structure (0.74), estimate the number of atoms in a metal cluster in each of the two zeolites mentioned above.

6. *Heat Integration in Reactor Design*. As mentioned in the chapter one of the reasons for the success and efficiency of FCC units is the heat integration. The example cited was that the carbon (coke) burned off the catalyst in the regenerator is used to preheat the catalyst since this is a highly exothermic reaction. If the heat of combustion of carbon can be described as

$$C(s) + O_2(g) \longrightarrow CO_2(g) \quad \Delta H_{RXN} = -393.5\,\text{kJ/mol}$$

(1) Estimate the heat generated when the carbon is removed from 1000 kg of catalyst that has 1 wt% carbon on it. You may assume that the carbon on the catalyst has an effective molecular weight of 12 g/mol.
(2) If the heat capacity of the zeolite can be taken to be approximately 1000 J/kg-°C and the zeolite is at 450°C initially, what is the temperature of the catalyst after the carbon is removed, assuming no heat loss. To simplify the problem analysis you may assume that ΔH and C_p are independent of temperature.

7. *Reaction Mechanisms*. Many reactions over zeolites can be described using the Langmuir–Hinshelwood formalism, which involves describing adsorption, desorption, and reaction over active sites as a series of elementary steps. Consider as an example the hydrogenation of 1-butene:

$$\text{1-butene} + H_2 \longrightarrow \text{butane}$$

over a zeolite metal-impregnated zeolite catalyst. The Langmuir–Hinshelwood formalism involves writing this overall reaction as a series of elementary steps. One step in the mechanism is assumed to be the rate-determining step (RDS) and the rate expression for that step is re-expressed in terms of observable quantities. So for this reaction one possible mechanism would be

$$A(g) + * \overset{k_1,\, k_{-1}}{\longleftrightarrow} A*$$

$$H_2(g) + 2* \overset{k_2,\, k_{-2}}{\longleftrightarrow} 2H*$$

$$A* + H* \overset{k_3,\, k_{-3}}{\longleftrightarrow} AH* + *$$

$$AH* + H* \overset{k_4,\, k_{-4}}{\longleftrightarrow} B* + *$$

$$B* \overset{k_5,\, k_{-5}}{\longleftrightarrow} B(g) + *$$

Reaction 1 is the adsorption of 1-butene (denoted as *A*) on an active site (denoted as $*$), reaction 2 is the dissociative adsorption of hydrogen. Reactions 3 and 4 are the atomic additions of hydrogen to the butene leading to butane (denoted B) and step 5 is the desorption of butane from the surface. The reaction rates for the individual steps can be written as

$$r_1 = k_1 p_A \theta_v - k_{-1}\theta_A$$

$$r_2 = k_2 p_{H2} \theta_v^2 - k_{-2}\theta_H^2$$

$$r_3 = k_3 \theta_A \theta_H - k_{-3}\theta_{AH}\theta_v$$

$$r_4 = k_4 \theta_{AH} \theta_H - k_{-4}\theta_B\theta_v$$

$$r_5 = k_5 \theta_B - k_{-5} p_B \theta_v$$

where θ_v is the fraction of vacant sites, θ_i is the fractional surface coverage of component i, and the ks are rate constants. If one assumes that reaction 4 is the rate-determining step this allows one to set the overall rates of all the other reactions to zero (i.e. rate of forward reaction = rate of reverse reaction). This coupled with the overall site balance that the sum of the θ's must equal one allows one to determine:

$$r_4 = r = \frac{k_4 K_3 K_2 K_1 p_A p_{H2} - k_{-4} K_5 p_B}{\left(1 + K_1 p_A + \sqrt{K_2 p_{H2}} + K_3 K_1 \sqrt{K_2 p_{H2}} p_A + K_5 p_B\right)^2}$$

where the capital Ks are $K_i = k_i / k_{-i}$. Using this worked example as a guide, resolve this problem for two cases: (1) reaction 3 is the rate-determining step, and (2) reaction 5 is the rate-determining step. Comment on the differences in your rate expressions for the two cases.

8. *Diffusion in Zeolites.* Diffusion of aromatic molecules in zeolites is a highly complex process as the size of the molecules is very similar to the size of the zeolite pores. Consider for instance two xylene isomers. Literature data [taken from Kaerger and Ruthven (70)] indicates that the diffusion coefficient of p-xylene in silicalite-1 (MFI) is $4 \times 10^{-14}\,\text{m}^2/\text{s}$, whereas the diffusion coefficient of m-xylene is $4 \times 10^{-16}\,\text{m}^2/\text{s}$. Thus, the difference is two orders of magnitude. Analyzing this experimental data from the absolutely simplest theoretical framework would be to compare this to the diffusion coefficient estimated from a random walk model. While this is oversimplified it will outline some interesting points. From the random walk model it can be shown that the self-diffusion coefficient in a three-dimensional system is given by

$$D = \frac{l^2}{6\tau}$$

where l is the distance between jumps and $1/\tau$ is the jumping frequency. Assuming the jump distance is the same in both cases, 5 Å, find the jumping frequencies for the two isomers based on the diffusion coefficients given. Provide a molecular explanation for why the values are the same or different.

9. *Shape Selectivity.* Analogous to the example in Figure 10.4, one possible route to synthesizing xylenes is the transalkylation of toluene over ZSM-5, that is,

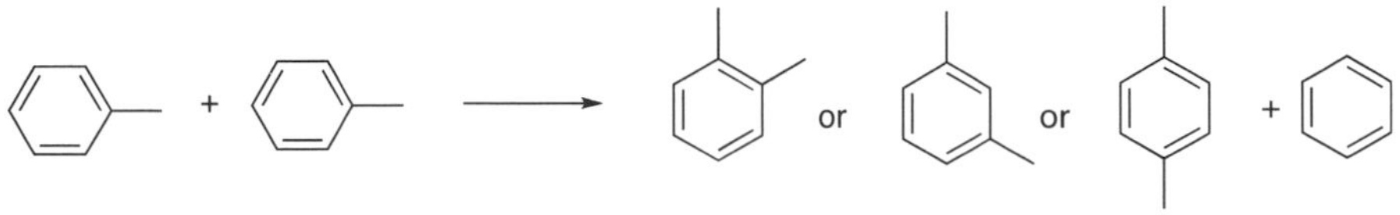

toluene + toluene ⟶ o-xylene or m-xylene or p-xylene + benzene

or or +

For each reaction one molecule of xylene would be generated as well as one molecule of benzene. Draw the different transition states that you believe would lead to the formation of the different xylene isomers. Based on these transition states might you expect to observe transition state selectivity and could you rationalize the selectivity of ZSM-5 for *p*-xylene based on that? What are the pore dimensions of ZSM-5?

10. *Zeolite Chemistry.* It has long been known that zeolites that contain more aluminum in them are much more hydrophilic than zeolites with less or no aluminum in the framework. Based on the discussion in the chapter explain why this is the case. A reasonable answer to this would be one or two paragraphs explaining this in the context of the structural chemistry of zeolites. The issue of hydrophobicity/hydrophilicity is notoriously qualitative. Explain how one might quantify the hydrophilicity of a zeolite? Hint: Think of a series of zeolites of the same framework topology with varying composition (i.e. Si/Al) yet all have the same extra-framework cation type.

ANSWERS

1. (a) There are many frameworks that the student could choose from the website. As an example, consider zeolite ZSM-12

Material Name: ZSM-12
IZA Structure Code: MTW
Space Group: C2/m
Unit cell: $a = 25.55$ Å, $b = 5.25$ Å, $c = 12.11$ Å; $\alpha = 90°$, $\beta = 109.3°$, $\gamma = 90°$

The zeolite has 12-membered ring pores, the pore system is one dimensional, and the pores along [100] are 5.6 × 6.0 Å.

FAU: Zeolite X (low Si/Al ratio), Zeolite Y (higher Si/Al), USY, Linde Type X, LZ-Z10
LTA: Linde type A (typically with a Si/Al = 1), ZK-4 (high silica Si/Al ~ 3 variant), ITQ-29 (pure silica material)
CHA: Chabazite (also a known mineral phase), SAPO-34 (silico-aluminophosphate, DAF-5 (Mg-Aluminophosphate)

2. This problem requires the student to compare the rate at the external versus internal surface. Based on the problem statement the ratio of the products is determined by the ratio of the surface to volume of the particles, that is,

$$\frac{n_B}{n_C} = \frac{r_B}{r_C} = \frac{\text{volume} \times \text{rate}}{\text{surface area} \times \text{rate}} = \frac{(4/3)\pi r^3}{4\pi r^2} = \frac{r}{3}$$

For a 10 nm diameter particle this ratio is $5/3$; for a 100 nm diameter particle it is $50/3$.

3. Why does the following trend hold?

Stability trend

$$\overset{\oplus}{C}H_3 < H_2\overset{\oplus}{C}{-}CH_3 < H_3C{-}H\overset{\oplus}{C}{-}CH_3 < H_3C{-}\overset{\oplus}{C}(CH_3){-}CH_3$$

Carbocations increase in stability from left to right

One generally observes that increasing the degree of substitution on the carbon center increases the stability; more C—C bonds, more stability.

4. If compression cost is not an issue choose case (1) in the accompanying graphs since it has maximal A adsorbed. For a real world system choose case (3) since there is very little B on the adsorbent and the change in A between the two pressures is the largest of the three cases.

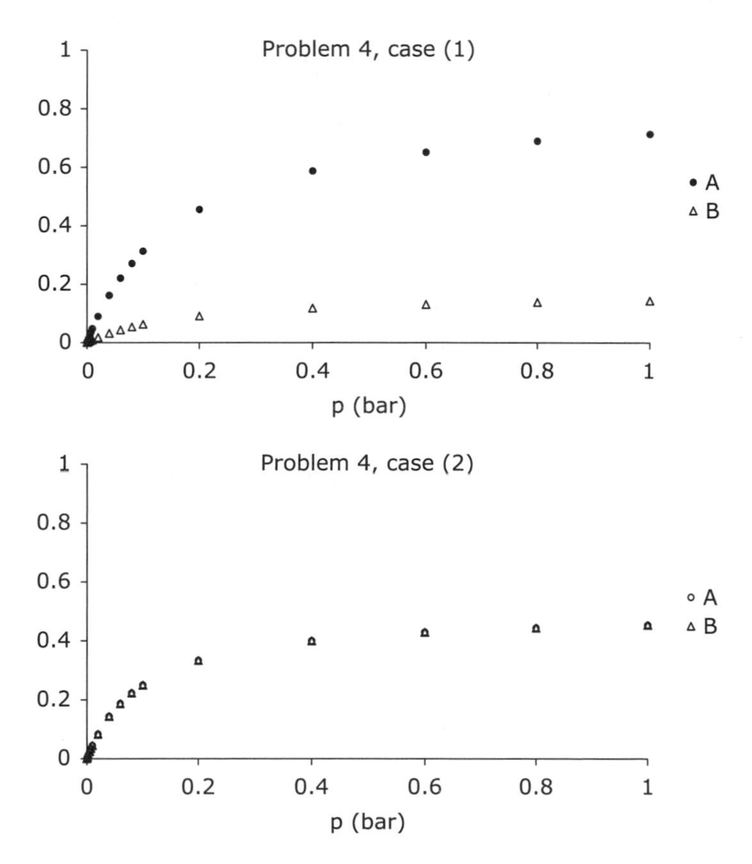

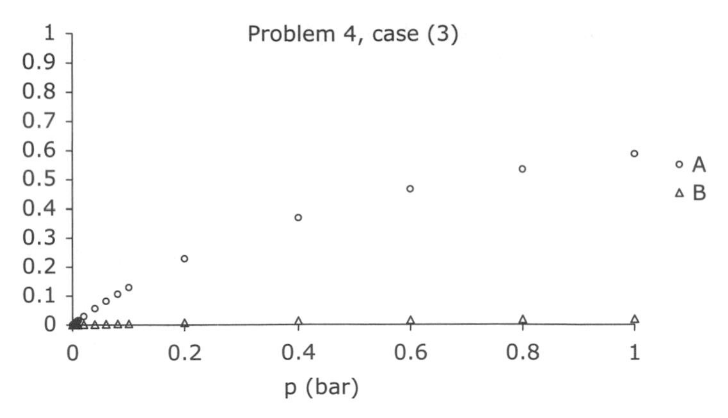

5. Starting point: estimate the void space of the two phases:

For FAU this is $V_{FAU} = (4/3)\pi(12\,\text{Å}/2)^3 = 905\,\text{Å}^3$
For MTW this is $V_{MTW} = (4/3)\pi(5.6\,\text{Å}/2)^3 = 92\,\text{Å}^3$
Actual space for metal atoms is 74% of this based on the problem statement so $V_{FAU,metal} = 670\,\text{Å}^3$; $V_{MTW,metal} = 68\,\text{Å}^3$.

Next piece of information: the volume of one metal atom $V = (4/3)\pi(1/2)^3 = 0.52\,\text{Å}^3$. So the estimated number of metal atoms is

$$n_{FAU} = \frac{V_{FAU,metal}}{V_{1atom}} \cong 1300$$

$$n_{MTW} = \frac{V_{MTW,metal}}{V_{1atom}} \cong 140$$

6. First thing to determine: heat (Q) associated with combustion

$$Q_{TOT} = \Delta H_{RXN} \times (\text{molesCarbon})$$

$$1000\,\text{kg} - \text{cat} \times 0.01\,\text{massC} = 10\,\text{kgC} \times \frac{1\,\text{molC}}{0.012\,\text{kg}} = 833\,\text{molC}$$

$$Q_{TOT} = (-393.5\,\text{kJ/mol})(833\,\text{molC}) = -328{,}000\,\text{kJ}$$

Note that the negative sign on Q indicates that energy is added to the system via the reaction (i.e. T will increase). So what is the temperature rise?

$$Q_{TOT} = C_P m(\Delta T) = C_P m(T_i - T_f)$$
Note: $Q < 0$ so $T_f > T_i$

$$T_f = \frac{-Q}{C_p m} + T_i = \frac{-(-328{,}00\,\text{kJ})}{(1\,\text{kJ/kg} - {}^\circ\text{C})(1000\,\text{kg})} + 450^\circ\text{C}$$

$$T_f = 778^\circ\text{C}$$

So $\Delta T = 230$ C! Note that the mass in the bottom equation is the total mass of catalyst, not mass of coke! In reality the picture is more complex, as there will inevitably be heat loss, and the enthalpy of reaction and heat capacity will be dependent on temperature.

7. Before solving the question posed, work out the example mechanism given in the problem statement:

$$\begin{aligned} A(g) + * &\longleftrightarrow A* \\ H_2(g) + 2* &\longleftrightarrow 2H* \\ H* + A* &\longleftrightarrow HA* + * \\ H* + HA* &\longleftrightarrow B* + * \\ B* &\longleftrightarrow B(g) + * \end{aligned}$$

The rate expression for these are (in order)

$$\begin{aligned} r_1 &= k_1 p_A \theta_v - k_{-1} \theta_A \\ r_2 &= k_2 p_{H2} \theta_v^2 - k_{-2} \theta_H^2 \\ r_3 &= k_3 \theta_A \theta_H - k_{-3} \theta_{AH} \theta_v \\ r_4 &= k_4 \theta_{AH} \theta_H - k_{-4} \theta_B \theta_v \\ r_5 &= k_5 \theta_B - k_{-5} p_B \theta_v \end{aligned}$$

The notation used is stated in the problem. In the problem statement it was assumed that reaction four is the rate-determining step; thus $r_1 = r_2 = r_3 = r_5 = 0$ or in other words all the other reactions are at equilibrium. To solve this one needs to eliminate all the θ_I terms and write them in terms of observable quantities. These turn out to be as follows:

$$\begin{aligned} \theta_A &= K_1 p_A \theta_v \text{ (from rate 1)} \\ \theta_H &= \sqrt{K_2 p_{H2}}\, \theta_v \text{ (from rate 2)} \\ \theta_{HA} &= \frac{K_3 \theta_H \theta_A}{\theta_v} \text{ (from rate 3)} \\ \theta_B &= K_5 p_B \theta_v \text{ (from rate 5)} \end{aligned}$$

and from reaction four one gets

$$r_4 = r = k_4 \theta_{AH} \theta_H - k_{-4} \theta_B \theta_v$$

Using the expressions above for the site occupancies of A, H, HA, and B one obtains

$$r = (k_4 K_1 K_2 K_3 p_A p_{H2} - k_{-4} K_5 p_B)\theta_v^2$$

To finish the problem off one needs to use the site balance to write θ_v in terms of known quantities; upon doing that one obtains:

$$r_4 = r = \frac{k_4 K_3 K_2 K_1 p_A p_{H2} - k_{-4} K_5 p_B}{\left(1 + K_1 p_A + \sqrt{K_2 p_{H2}} + K_3 K_1 \sqrt{K_2 p_{H2}} p_A + K_5 p_B\right)^2}$$

Now, to solve case (1) where the first reaction (step 3) is the RDS:
The same mechanism as above applies; the two main differences are that (1) now $r_4 = 0$, and the rate of the third reaction is given by

$$r_3 = r = k_3 \theta_H \theta_A - k_{-3} \theta_{HA} \theta_v$$

Expressions for the θ_i values are

$$\theta_A = K_1 p_A \theta_v \quad \text{(from rate 1)}$$

$$\theta_H = \sqrt{K_2 p_{H2}}\, \theta_v \quad \text{(from rate 2)}$$

$$\theta_{HA} = \frac{\theta_B}{K_4 \theta_H} \theta_v \quad \text{(from rate 4)}$$

$$\theta_B = K_5 p_B \theta_v \quad \text{(from rate 5)}$$

Using these in the rate expression gives

$$r = k_3 \sqrt{K_2 p_{H2}} K_1 p_A \theta_v^2 - k_{-3} \frac{K_5 p_B}{K_4 \sqrt{K_2 p_{H2}}} \theta_v^2$$

Solving for the value of vacant sites from the site balance below

$$1 = \theta_A + \theta_{HA} + \theta_B + \theta_v$$

gives

$$\theta_v = \frac{1}{1 + K_1 p_A + \sqrt{K_2 p_{H2}} + \dfrac{K_5 p_B}{K_4 \sqrt{K_2 p_{H2}}} + K_5 p_B}$$

which, substituting into the rate, gives

$$r = \frac{k_3\sqrt{K_2 p_{H2}} K_1 p_A - k_{-3}\dfrac{K_5 p_B}{K_4\sqrt{K_2 p_{H2}}}}{\left(1 + K_1 p_A + \sqrt{K_2 p_{H2}} + \dfrac{K_5 p_B}{K_4\sqrt{K_2 p_{H2}}} + K_5 p_B\right)^2}$$

Now for the case where step 5 is the RDS:

The same mechanism as before applies and rates 1 to 4 are equal to zero, so

$$\theta_A = K_1 p_A \theta_v \quad \text{(from rate 1)}$$

$$\theta_H = \sqrt{K_2 p_{H2}}\,\theta_v \quad \text{(from rate 2)}$$

$$\theta_{HA} = \frac{k_3 \theta_H \theta_A}{\theta_v} \quad \text{(from rate 3)}$$

$$\theta_B = \frac{\theta_{HA}\theta_H}{K_3\theta_v} \quad \text{(from rate 4)}$$

so $r_5 = r = k_5\theta_B - k_{-5} p_B \theta_v$, using the same procedure as outlined above, one obtains:

$$r = \frac{k_5 K_2 p_{H2} K_1 p_A - k_{-5} p_B}{(1 + K_1 p_A + K_2 p_{H2} + K_3\sqrt{K_2 p_{H2}} K_1 p_A + K_1 K_2 p_H p_A)}$$

8. Given the equation below and if $l = 5$ Å

$$D = \frac{l^2}{6\tau}$$

$$\frac{1}{\tau} = \frac{6D}{l^2}$$

$$p\text{-xylene: } \frac{1}{\tau} = \frac{6(4 \times 10^{-14}\,\text{m}^2/\text{s})}{(5 \times 10^{-10}\,\text{m})^2} = 960{,}000 \text{ jumps/s}$$

$$m\text{-xylene: } \frac{1}{\tau} = \frac{6(4 \times 10^{-16}\,\text{m}^2/\text{s})}{(5 \times 10^{-10}\,\text{m})^2} = 9600 \text{ jumps/s}$$

Why would you observe this difference? The shape of *m*-xylene likely reduces the number of conformations it can possess (i.e. it makes hopping less likely).

9. Based on the problem statement the three transition states shown below (and the isomer that would be formed) are reasonable.

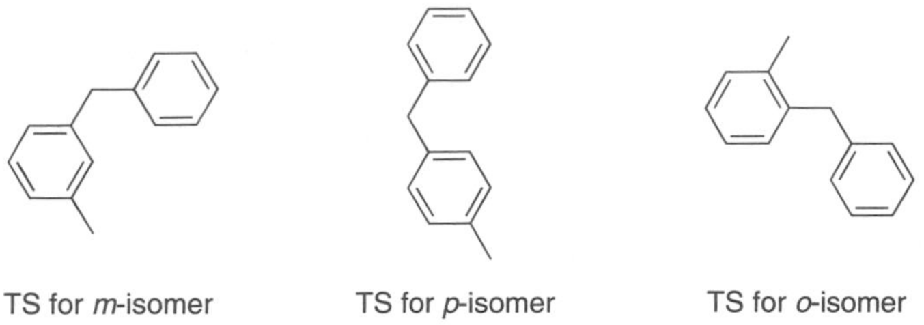

TS for *m*-isomer TS for *p*-isomer TS for *o*-isomer

None of these look to be dramatically unfavored, through one could argue the TS for the *p*-isomer is more linear, and perhaps thus easier to form in the pores of ZSM-5. In reality this reaction is slow relative to the reaction of methanol + toluene to form xylene; toluene transalkylation would not be significant.

10. Concept problem: the explanation is a very simple one in that as one puts more aluminum in the zeolite framework there is a need for more extra-framework cations. These cations bind water very strongly. Thus, as Si/Al decreases, the number of cations increases, and more water should bind.

How to quantify this? One simple idea would be to look at the enthalpy of solvation of cations. For example, consider a Na-containing zeolite of different Si/Al compositions. As the Si/Al ratio decreases, you will have more sodium atoms. Thus

$$\text{Hydrophilicity} = \left(\frac{n_{Al}}{n_{Si} + n_{Al}}\right)\Delta H_{solvation,Na}$$

One could argue that this is overly simplified; however, it does show (correctly) what is observed; increasing cation content increases water uptake and high charge density monovalent cations (e.g. Li^+) bind water more strongly than lower charge density monovalent cations (e.g. K^+).

PART IV

ORGANIZED TWO- AND THREE-DIMENSIONAL NANOCRYSTALS

11

INORGANIC–ORGANIC COMPOSITES

WARREN T. FORD

11.1 Introduction, 369
11.2 Length Scales of Inorganic–Organic Nanomaterials, 371
  11.2.1 Molecular Size, 371
  11.2.2 Nanoscale Materials, 371
  11.2.3 Bulk Materials, 372
  11.2.4 Construction of Materials at Multiple Length Scales, 373
11.3 Molecular Components of Nanoscale Materials, 374
  11.3.1 Organic Materials, 374
  11.3.2 Inorganic Materials, 380
11.4 Organization of the Building Blocks in One, Two, and Three Dimensions, 386
  11.4.1 Components with One Macroscopic Dimension: Wires, Rods, Tubes, and Fibers, 386
  11.4.2 Thin Films: Structures with Two Macroscopic Dimensions, 387
  11.4.3 Nanostructures from Colloidal Particles, 394
11.5 Concluding Remarks and Future Outlook, 397
References, 397
Problems, 400
Answers, 401

11.1 INTRODUCTION

In this chapter nanoscience is defined as the science of materials whose properties depend on size and are those of neither molecules nor bulk materials. This definition is best explained with a few examples.

Nanoscale Materials in Chemistry, Second Edition. Edited by K. J. Klabunde and R. M. Richards

1. The emission spectra of nanoparticles of semiconductors such as CdSe and CdTe, known as quantum dots, depend on size. Quantum confinement of electrons within a particle increases the electronic bandgap with decreasing particle size, so that a series of different sized CdSe particles emits a rainbow of colors.
2. Gold particles are catalysts for oxidation of CO to CO_2 only when nano-sized. The key is the surface of the Au nanocrystal (1).
3. Surface enhanced Raman spectroscopy (SERS) depends on high surface area of nanocrystals of Au and Ag.
4. The mechanical properties of polymer composites having a nano-sized filler whose dimensions are no larger than that of one polymer molecule differ from those of composites with micro-sized filler particles much larger than a polymer molecule because most of the polymer molecules interface with a nano-sized filler but not with a micro-sized filler (2).

Feynman long ago envisioned the field of nanoscale materials. In his 1959 lecture "There's Plenty of Room at the Bottom," he forecast electron beam lithography and writing 24,000,000 volumes (books) with 100 atoms per bit of information into a cube 1/200 inch on a side (3). Klabunde recognized new properties due to nanoscopic size of metal clusters in his research in the late 1970s (4). The tremendous growth of nanotechnology research starting in the 1990s was enabled by the inventions in the 1980s of the scanning tunneling and atomic force microscopes, which complemented scanning and transmission electron microscopy and greatly expanded the numbers and types of materials that could be imaged on a nanometer scale. The terms nanotechnology, nanoscience, and nanoengineering came into wide use after the U.S. National Nanotechnology Initiative (NNI) was announced by President Bill Clinton in January 2000. Since then the United States and other governments have created programs to support research on nanoscale materials, thousands of (mostly small) industries have carried out research on applications, and nanotechnology has become a buzzword. At the time the NNI was launched, most nanotechnology research consisted of construction of particles of materials of <100 nm size. Progress has been rapid. The construction of nanoscale building blocks is now more mature, and much research is devoted to the self-assembly of the building blocks. Not well developed yet is the construction of useful devices from the building blocks and assemblies. This chapter discusses synthesis of the building blocks and their assemblies, for which chemistry is essential, but not device construction. As Ozin remarked, "chemists are at the beginning of the materials food chain" (5).

A nanoscale hybrid inorganic–organic material is defined here as a material having properties that depend on the size of at least one component. The organic part of most hybrid materials is not crystalline but amorphous, usually a surfactant or polymer. Often a surfactant or polymer coating is necessary to prevent nanoscale inorganic materials from coagulating into bulk materials. Polymer composites, in which dispersed inorganic particles or fibers improve the mechanical properties of a plastic or an elastomer, have been commercial since the first carbon black filled rubber tires. Now nanoscopic inorganic particles and fibers in polymers promise a whole new class of composites with improved mechanical, electronic, and even optical properties. The inorganic materials may be dispersed simply by physical methods, but better dispersion and better composite properties are achieved by control of the chemistry at the polymer-filler interface. More specific ways to construct nanoscopic domains in composite materials are to carry out chemical reactions within a preformed solid material by the hydrolysis and condensation of organosilicon compounds into silicon dioxide, or to fill a space vacated by the chemical removal of one component of a multiphase system with a new inorganic material. Examples follow later in this chapter.

11.2 LENGTH SCALES OF INORGANIC–ORGANIC NANOMATERIALS

11.2.1 Molecular Size

Chemistry is devoted to the understanding of macroscopic observations in terms of molecular behavior. Chemists observe a compound with one set of physical properties and by chemical reaction transform it to another compound with a different set of physical properties, such as melting point, boiling point, density, color, odor, and spectra. Chemists explain these macroscopic properties on the basis of molecular structure, regardless of whether they observe the compound in the gas phase, as a dilute solution, or as a solid.

New compounds are synthesized by chemical reactions. A homogeneous or heterogeneous catalyst promotes most reactions. A catalyst increases the rate of a chemical reaction without being consumed. A homogeneous catalyst may be a soluble acid or base, or in the case of inorganic–organic hybrid materials, an organometallic compound. On the other hand, heterogeneous catalysts are solid-state materials. For example active metals such as Pt or Pd are deposited in nano- or microparticulate form onto a solid support such as silica, alumina, or active carbon. Heterogeneous catalysis takes place at the surface of the particles, which have high surface area due to their nano-size. Sometimes the distinction between homogeneous and heterogeneous catalysis is not clear. Pd(0) catalysis of many types of reactions, from hydrogenation to carbon–carbon coupling, may be due to either homogeneous organometallic compounds or colloidal particles (6).

11.2.2 Nanoscale Materials

There is a large range of sizes between molecules (ca. <2 nm for small molecules and tens of nanometers for many polymers) and bulk materials. The properties of solid-state materials larger than molecules and smaller than hundreds of nanometers depend on size. This size dependence often is due to the chemical differences between atoms on the surface and atoms in the core of a solid. A 13-atom, 1-shell cluster has 12 (92%) of its atoms on the surface, whereas a 1415-atom, 7-shell cluster has 35% of the

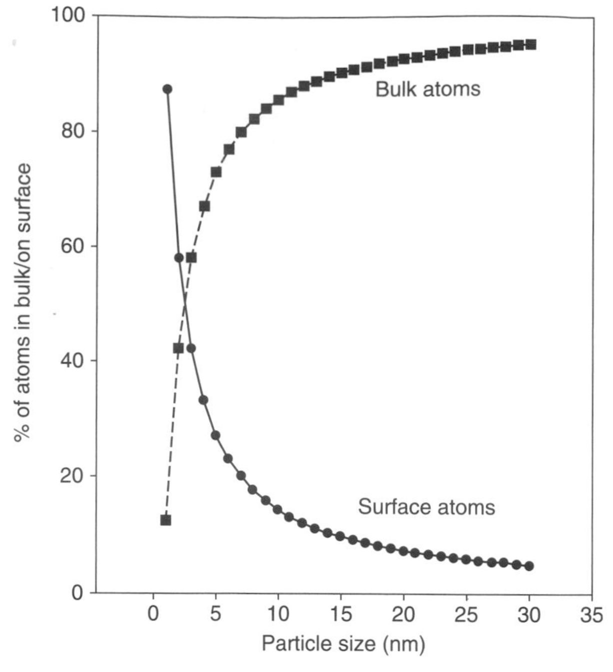

Figure 11.1 Relative numbers of surface and bulk atoms in a metal particle as a function of particle diameter. (Reproduced with permission from K. J. Klabunde et al., *J. Phys. Chem.* **1996**, *100*, 12142.)

atoms on the surface. The surface area per unit volume of a sphere scales inversely with the diameter: a 1000 nm diameter spherical iron crystal has $<1\%$ of its atoms on the surface, whereas a 1 nm sphere has $>90\%$ of its atoms on the surface (Fig. 11.1; Reference 7). In the core of a metal crystal the atoms are bonded to atoms of their own kind. The atoms on the surface have empty orbitals that are electron acceptors or filled nonbonding orbitals that may be electron donors. Because of the high energy of these surface orbitals, freshly created metal surfaces (except gold and platinum) react with atmospheric oxygen to form a metal oxide layer on the surface. Similarly the molecules in the core of a polymer sphere are surrounded by their own kind, whereas those on the surface are exposed to the environment, which could be solvent, a solution, or air. The surface of a polymeric solid has different properties from the bulk because the composition and the packing of the macromolecular chains differ from those in the bulk.

11.2.3 Bulk Materials

When the dimensions of solid-state materials are a micrometer or more, the physical properties are generally those of the bulk material, that is, independent of size. These properties have been the concern of mainly physicists and engineers. Until the late twentieth century, chemical engineers paid most attention to thermodynamics, kinetics, transport phenomena, and unit operations, and left the study of molecular properties and quantum phenomena to chemists and physicists. But now many engineers are conducting research in the chemistry, physics, and biology of materials, and scientists are paying attention to the engineering of materials, especially nanoscale materials. Many of the traditional disciplinary boundaries of research have disappeared.

11.2.4 Construction of Materials at Multiple Length Scales

Many nanoscopic materials are constructed from nanocrystals or molecular assemblies on a repeating basis. Gold nanoparticles 15 to 20 nm in diameter are made by reduction of $AuCl_4^-$ in a citrate solution (8). By high resolution TEM the gold nanoparticles often consist of smaller nanocrystals fused together, as shown in Figure 11.2.

To create a work of art, a sculptor can either chisel away a large block of stone or cement together smaller building blocks. Similarly nanoscale materials can be made by either a top-down or a bottom-up approach. Microlithography to make electronic circuit boards is an example of top-down construction. Construction of nanoscale materials bottom up by assembly of molecules and nanoparticles is just now

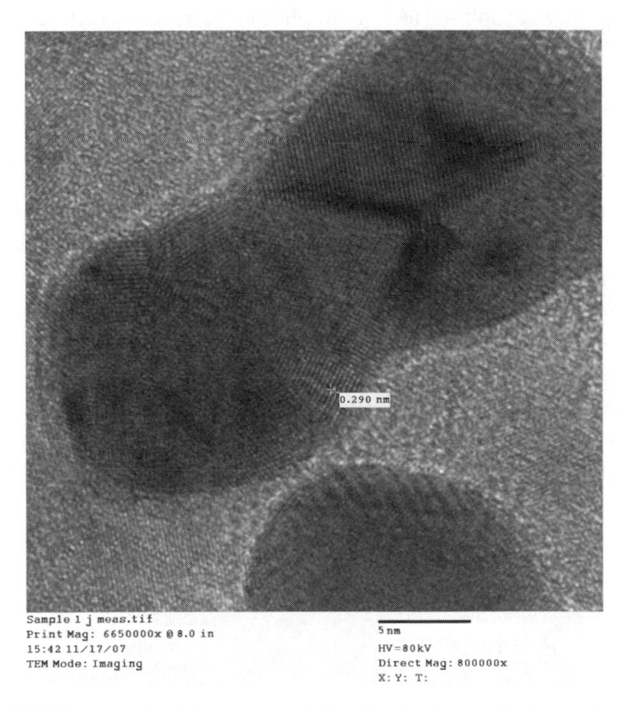

Figure 11.2 TEM image of gold nanocrystal showing lattice resolution and grain boundaries due to the fusion of smaller fused nanocrystals. (Student preparation, Oklahoma State University, November 2007; TEM contributed by Dr. Charlotte Ownby.)

developing, and the engineering of these assemblies into useful macroscopic devices is in its infancy. In principle, but not always in practice, bottom-up construction of materials is less wasteful than top-down construction, and thus is greener chemistry. Due to the orders of magnitude smaller scale of production of nanoscale materials compared with most industrial chemicals, pollution from waste will be less of a problem, but still not negligible. Of greater concern are the unknown health effects of exposure to nanoscale materials due to their size-dependent properties (See Chapters 21 and 22).

Experimental nanotechnology can be divided into "wet" processes and "dry" processes. Casting a dispersion or solution of material onto a substrate and letting the solvent evaporate is a wet method, whereas chemical vapor deposition onto a substrate is a dry method. Wet processes are carried out at moderate temperatures. Dry processes require high temperature to get a material into the vapor phase, and consequently are restricted to thermally stable materials.

11.3 MOLECULAR COMPONENTS OF NANOSCALE MATERIALS

11.3.1 Organic Materials

11.3.1.1 Surfactants Surfactants (soaps and detergents) are compounds or polymers having both hydrophilic and hydrophobic parts (they are *amphiphilic*) that reduce interfacial free energy, such as the surface tension at an air-water interface. Surfactants can be anionic, cationic, or nonionic depending on the structure of the polar end. Typical molecular structures are shown in Figure 11.3. The hydrophobic end dissolves oil or grease, and the hydrophilic end dissolves in water. Surfactants dissolve molecularly in water at low concentrations. As the concentration increases a critical micelle concentration (CMC) is reached at which all further added surfactant forms aggregates called micelles. The surfactant molecules of a micelle are in equilibrium with the molecularly dissolved surfactant. The type of micellar aggregate formed by a surfactant depends on the relative sizes of its polar head group and its nonpolar tail. A single-tailed surfactant such as sodium dodecyl sulfate (SDS) dissolves monomolecularly in water up to its CMC (ca. 8 mM in water at 25°C) and at higher concentrations forms globular micelles. Surfactants having tails longer than C_{12} at concentrations well above the CMC may form cylindrical rather than globular micelles. At still higher concentrations the surfactant may aggregate further into

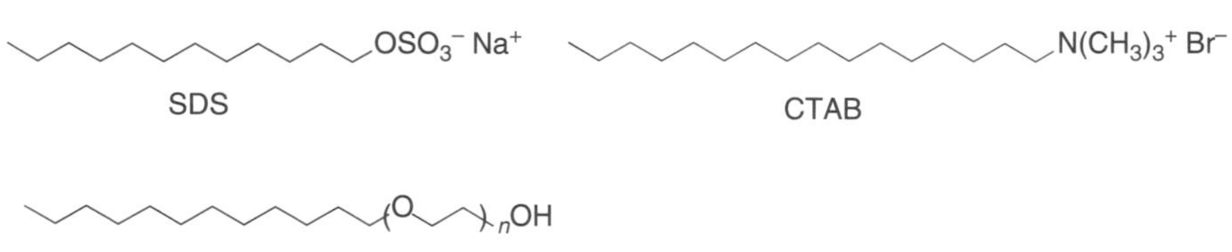

$OSO_3^-\ Na^+$ $N(CH_3)_3^+\ Br^-$

SDS CTAB

$(O\!\!\frown\!\!\frown)_nOH$

Figure 11.3 Structures of typical anionic (sodium dodecyl sulfate, SDS), cationic (cetyltrimethylammonium bromide, CTAB), and nonionic surfactants.

lyotropic liquid crystals. Zeolites are porous silicates formed from precursors of silica in liquid crystalline solutions of surfactants. The size of the surfactant aggregates determines the pore size and structure of the zeolite. The silica source is polymerized to a gel, dried to remove excess water, heated to remove the surfactant template, and sintered to remove the last amounts of the organic material.

The amphiphilic compounds used to stabilize aqueous colloidal dispersions of metals, semiconductors, and organic polymer particles are surfactants, broadly defined. The surfactant replaces the high energy of the particle-water interface with two lower energy interfaces of the surfactant with particles and of the surfactant with water. Amphiphilic polymers also stabilize colloidal dispersions by reducing the interfacial free energies of the system.

Colloidal silica, which normally has a polar surface due to OH groups, can be made dispersable into organic solvents by treatment of its surface with a silane coupling agent, whose structure has at one end a halo or alkoxysilane that reacts with the silica OH groups and at the other end an organic group that mixes with the organic solvent. The surfaces of macroscopic silica particles are modified by the same chemical reactions. For example, octadecyltrimethoxysilane (Fig. 11.4) converts normal phase chromatographic silica to reverse phase silica with C_{18} chains pendant from the silica surface.

Amphiphilic compounds also are used to construct self-assembled monolayers on gold surfaces. The usual agents have a thiol functional group at one end to bind to the gold and a methyl group or some organic functional group at the other end.

11.3.1.2 Polymers The word polymer means a material composed of macromolecules or the macromolecule itself. Most synthetic organic polymers consist of long chains of mostly or all carbon–carbon bonds. To build nanostructured materials, the polymeric building blocks can be linear, highly branched, or amphiphilic. Likewise the far more complex assemblies of living organisms depend on polymers: polysaccharides, proteins, and polynucleotides.

11.3.1.2.1 Linear Polymers Long chains are necessary to confer the mechanical properties of fibers, plastics, and elastomers that make polymers so valuable. Fibers such as cellulose and polyester are semicrystalline materials in which the same chemical structure exists in both rigid microcrystalline and flexible amorphous phases. Plastics may be either semicrystalline, such as poly(ethylene terephthalate) (the same polyester of fibers is also the PET of beverage bottles), or completely amorphous and glassy, such as polystyrene or poly(methyl methacrylate) (PMMA, Plexiglas™ or Lucite™). Elastomers are completely amorphous and flexible and would flow as a viscous rubbery liquid except that the polymer chains are cross-linked to prevent macroscopic flow but allow reversible stretching. As an example, poly(dimethylsiloxane)

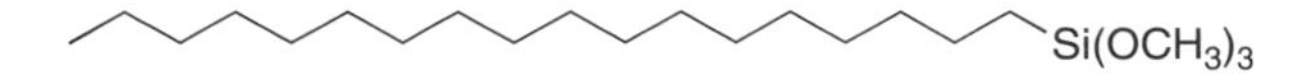

Figure 11.4 Structure of a hydrophobic silane coupling agent.

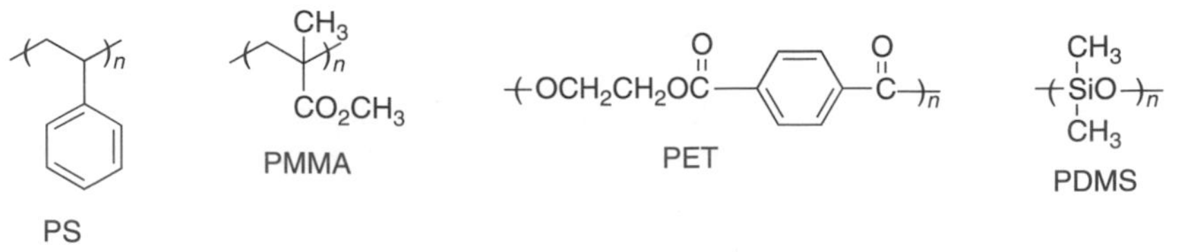

PS

PMMA (CH_3, CO_2CH_3)

PET ($-\!\!\left(OCH_2CH_2OC(=O)-C_6H_4-C(=O)\right)_n-$)

PDMS ($-\!\!\left(Si(CH_3)_2O\right)_n-$)

Figure 11.5 Structures of polystyrene (PS), poly(methyl methacrylate) (PMMA), poly(ethylene terephthalate) (PET), and poly(dimethylsiloxane) (PDMS).

(PDMS) can be anything from an oil to a rubber. PDMS is useful as a liquid (silicone oil) whose viscosity depends on molecular weight, or as an elastomer (silicone rubber) whose mechanical properties depend on the degree of cross-linking of the primary linear polymer chains. The structures of some common linear polymers are shown in Figure 11.5.

Linear synthetic polymers are mixtures of long chain macromolecules having the same repeating structure but different chain lengths, usually with average molecular weights in the range 10^4 to 10^6 Da. In solution or in an amorphous solid state each molecule has a random coil conformation that occupies a volume with dimensions of tens of nanometers. The macromolecules are entangled with one another except in very dilute solutions, as shown in Figure 11.6.

11.3.1.2.2 Dendrimers Since 1985 a new class of polymers called dendrimers has been developed (9–12). Dendrimer molecules are highly branched and lower in molecular weight than most linear polymers. Whereas most polymers are synthesized by chain reactions or by many repetitions of one type of reaction in one pot, dendrimers are synthesized one step at a time, as shown in Figure 11.7. Because branching confers dendrimers with a more compact shape than other polymers, they are valuable for construction of organic or hybrid inorganic–organic nanoparticles.

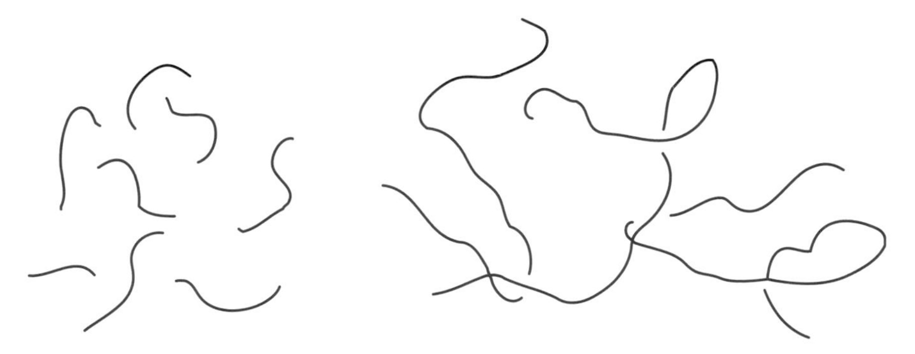

Figure 11.6 Entanglement of chain molecules increases with chain length. (Reproduced with permission from L. H. Sperling, *Introduction to Physical Polymer Science*, 3rd edition, p. 5. Wiley, New York, **2001**.)

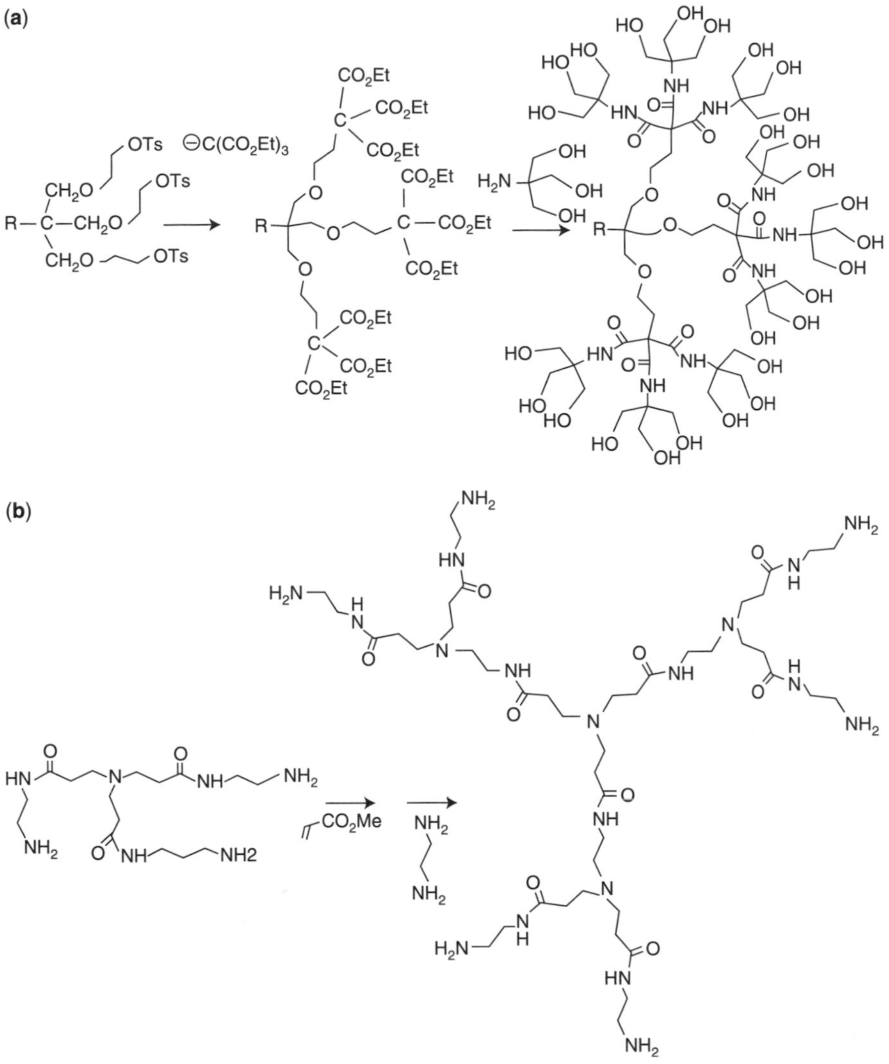

(a)

(b)

Figure 11.7 Syntheses of (a) a Newkome "cascade molecule" and (b) a Tomalia "starburst" dendrimer. Repetition of the synthetic steps produces much larger highly branched polymers. (Reproduced with permission from G. R. Newkome et al., *Dendrimers and Dendrons: Concepts, Syntheses, Applications*, p. 55. Wiley-VCH, Weinheim, Germany, **2001**.)

11.3.1.2.3 Block Copolymers Two or more polymer structures linked end to end by covalent bonds are block copolymers. Due to the low entropy of mixing, seldom are two different polymers miscible. Likewise the components of a block copolymer are immiscible. In the solid state a block copolymer consists of two phases with the nanoscopic domains linked by covalent bonds and the domain sizes controlled by the sizes of the blocks. Depending on the volume fractions of the two phases, the

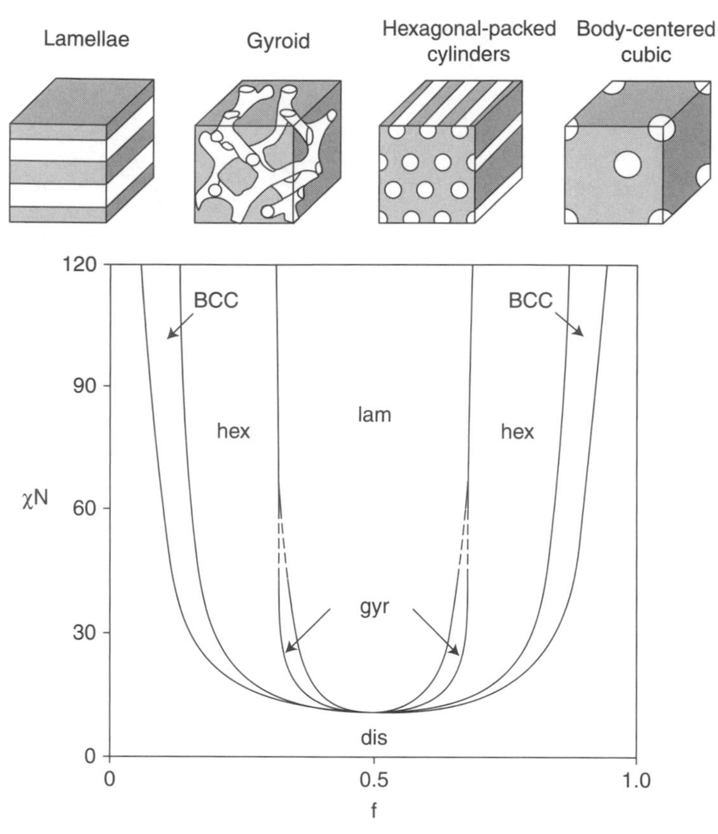

Figure 11.8 Morphologies of two-component block copolymers. In the phase diagram, regions of stability of disordered (dis), lamellar (lam), gyroid (gyr), hexagonal (hex), and body-centered cubic (BCC) phases are indicated. (Reproduced with permission from V. Castelletto and I. W. Hamley, *Curr. Opinion Solid State Mater. Sci.* **2004**, *8*, 426–438.)

observed morphologies of a block copolymer of A and B are spheres of A in a matrix of B, cylinders of A in B, layers of A and B, cylinders of B in A, or spheres of B in A as the fraction of A increases, as shown in Figure 11.8. Gyroid morphologies are possible too, but are not often observed. Still more complex morphologies are possible with triblock copolymers (13). The morphologies of diblock and ABA triblock copolymers are used to create solid nanostructured materials, for example, polymers with cylindrical holes or polymers with embedded cylindrical semiconductors, in which the geometries and the sizes of structural units are controlled by those of the domains of the original block copolymer (14).

11.3.1.3 Polymer Colloids Emulsion polymerization produces monodisperse polymer spheres 50 to 500 nm in diameter by the scheme shown in Figure 11.9. Water-immiscible vinyl monomers such as styrene, acrylic esters, and methacrylic

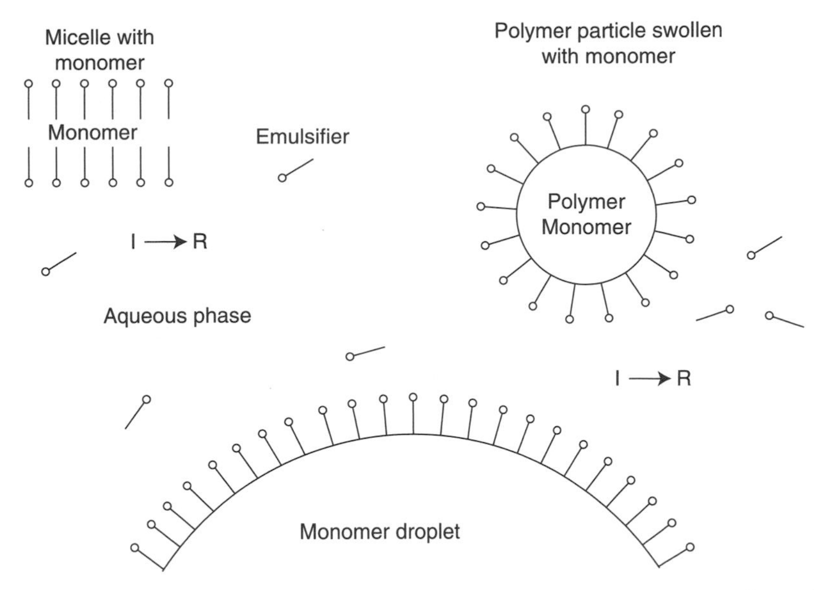

Figure 11.9 Simplified representation of an emulsion polymerization system. (Reproduced with permission from G. Odian, *Principles of Polymerization*, 4th edition, p. 353. Wiley-Interscience, New York, **2004**.)

esters are dispersed into water with either a surfactant such as SDS or with an amphiphilic copolymer produced *in situ* from the nonpolar monomer and a small amount of a water-soluble monomer such as sodium 4-styrenesulfonate. A water-soluble peroxide initiator (I) such as potassium persulfate ($K_2S_2O_8$) decomposes to form radicals (R) and starts polymerization in the aqueous phase. Particles form quickly due to the insolubility of the polymer. The number of particles is fixed after polymerization of a few percent of the monomer. All further polymer growth takes place by diffusion of monomer from monomer droplets to polymer particles and polymerization of the monomer that is swollen into the particles. Seed growth synthesis by addition of more monomer and initiator can enlarge the size of the particles produced in a single batch or continuous flow process. Synthetic rubber and water-based coatings are produced on a large scale by emulsion polymerization. Monodisperse polystyrene spheres are widely used model colloids for electron microscopy calibration standards, medical diagnostics, and fundamental colloid physics.

Amphiphilic block copolymers self-organize in aqueous mixtures to form polymeric micelles having a core of the hydrophobic block and a shell of the hydrophilic block. The cores can absorb additional hydrophobic low molecular weight compounds. The shells can be cross-linked to form capsules that are much more physically robust than the original micelles, as shown in Figure 11.10 (15). The uncross-linked inner phase can be dissolved by a good solvent to produce hollow capsules or serve as a site for trapping metal nanoparticles as catalysts.

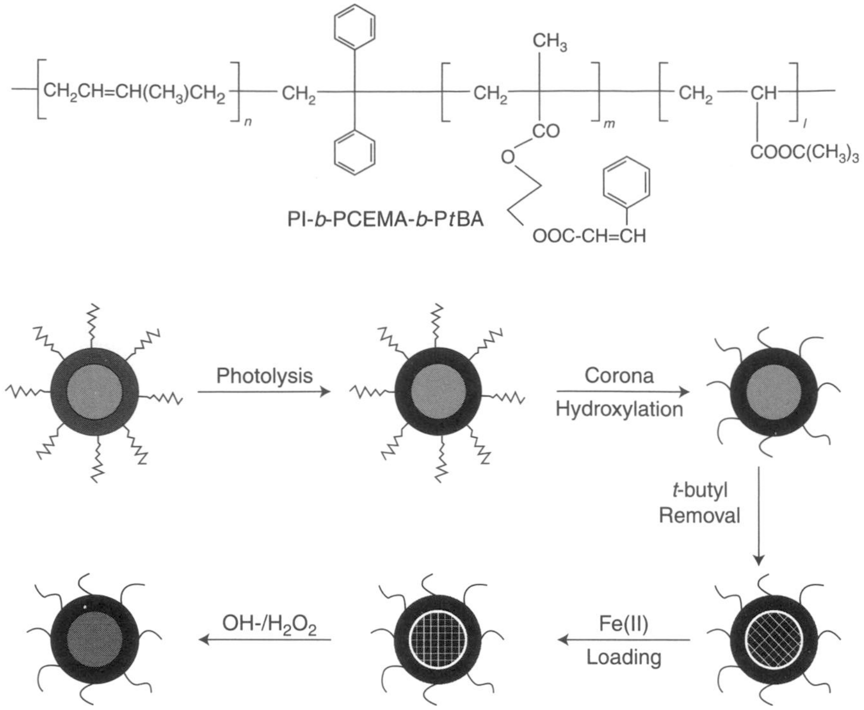

Figure 11.10 Synthesis of a block copolymer capsule containing Fe_2O_3 nanoparticles. Photolysis of the cinnamic ester groups of PCEMA to cyclobutanes cross-links the polymer. Hydroxylation of the double bonds of the PI units makes the particles water-soluble. Hydrolysis of the *tert*-butyl acrylate (PEBA) units gives acrylic acid units which bind Fe(II). Oxidation of Fe(II) in base produces Fe_2O_3. (Reproduced with permission from R. S. Underhill and G. Liu, *Chem. Mater.* **2000**, *12*, 2082.)

11.3.2 Inorganic Materials

A basic fact of nanoscale materials chemistry is that nanoparticles are only metastable. The stable state is bulk material. Surface stabilizers are necessary to prevent the particles in a colloidal dispersion from colliding and sticking together. Stabilizers act by an electrostatic mechanism or a steric mechanism or both (16). Particles with like charged surfaces electrostatically repel one another. By the steric mechanism solvated polymer chains extending from the surfaces of two different particles repel one another because the entropy of the system decreases if the polymer chains occupy the same volume. Colloidal dispersions are stable in water or in organic solvents when the stabilizer coordinates to the particle surface and is well solvated.

11.3.2.1 Semiconductor Nanoparticles The most nearly monodisperse (equal sized), stable, and highest quantum efficiency of nanoparticles of II-VI semiconductors, such as CdSe, are produced in nonpolar organic media at temperatures of 250°C to 300°C with surface coatings of trioctylphosphine oxide [TOPO (17, 18)].

Often a protective inert shell of a large bandgap material (silicon dioxide, zinc oxide, or zinc sulfide) is grown on the surface to retard reactions of the smaller bandgap nanoparticle with the environment. Aqueous dispersions are produced either by replacing the TOPO with a hydrophilic stabilizer, or by carrying out the synthesis under aqueous conditions in the first place using stabilizers such as 2-mercaptoglycolic acid ($HSCH_2CO_2H$) or 2-mercaptoethanol ($HSCH_2CH_2OH$). Colloidal stability in water often requires a large excess of the water-soluble stabilizer in solution, which equilibrates with a much smaller amount of stabilizer on the surface of the particles. In contrast to low molecular weight stabilizers, a large excess of a polymeric stabilizer is not required because the many bonding sites on the polymer add up to much stronger binding of a polymer than of a monodentate ligand. Aqueous dispersions of nanoparticles stabilized initially by 2-mercaptoglycolic acid and 2-mercaptoethanol have poorer optical properties than the TOPO-stabilized nanoparticles due to broader particle size distributions, which result in broader emission bands and surface trapped states that reduce the quantum yield of fluorescence (19). In toluene-water mixtures, 4.6 nm diameter TOPO-coated CdSe particles form a disordered monolayer at a water-toluene interface, but not at a toluene-air interface (20–22).

11.3.2.2 Metal Nanoparticles A sample of colloidal gold made by Faraday in 1857 is still on display at the British Museum in London. Gold nanoparticles stabilized with citrate ions now are made in beginning nanotechnology laboratory courses (see Fig. 11.2). Gold particles stabilized with 4-mercaptobenzoic acid have been prepared in aqueous methanol as single compounds; the structure of $Au_{102}(SC_6H_4CO_2H)_{44}$ has been determined by single-crystal x-ray analysis (23). Gold particles in organic solvents can be stabilized by a variety of coordinating functional groups (24).

In one novel approach to controlling the size of metal nanoparticles, metal ions are coordinated to a dendrimer or other polymeric ligand and then treated with a reducing agent to obtain the metal. A goal of these experiments is to produce monodisperse metal nanocrystals in which the number of metal atoms is controlled by the number of metal ion binding sites in the dendrimer. For example, Cu(II) ions are coordinated by two to four amine and amide groups of a PAMAM dendrimer, and reduced with sodium borohydride to produce Cu nanoparticles [Fig. 11.11a; Reference 25]. The concept of one metal particle per dendrimer works in practice for gold in PAMAM dendrimers having >256 end groups, but smaller dendrimers give gold particles from dendrimer aggregates [Fig. 11.11b; Reference 26]. Pt and Pd metal particles trapped in dendrimers are catalytically active in heterogeneous mixtures (25). For more information about stabilization of metal nanoparticles see chapter 20 in this volume.

11.3.2.3 Carbon Nanotubes Single-walled and multiwalled carbon nanotubes (SWCNT and MWCNT, see Chapter 13) can be grown as a carpet of fibers on a substrate (see Figure 13.12 in Chapter 13) such as a silicon wafer patterned with catalyst. Production of more than a monolayer of SWCNT gives bundles of tubes that are very difficult to separate into individual tubes because the van der Waals forces between parallel tubes in a bundle are stronger than between the tubes and solvent (see

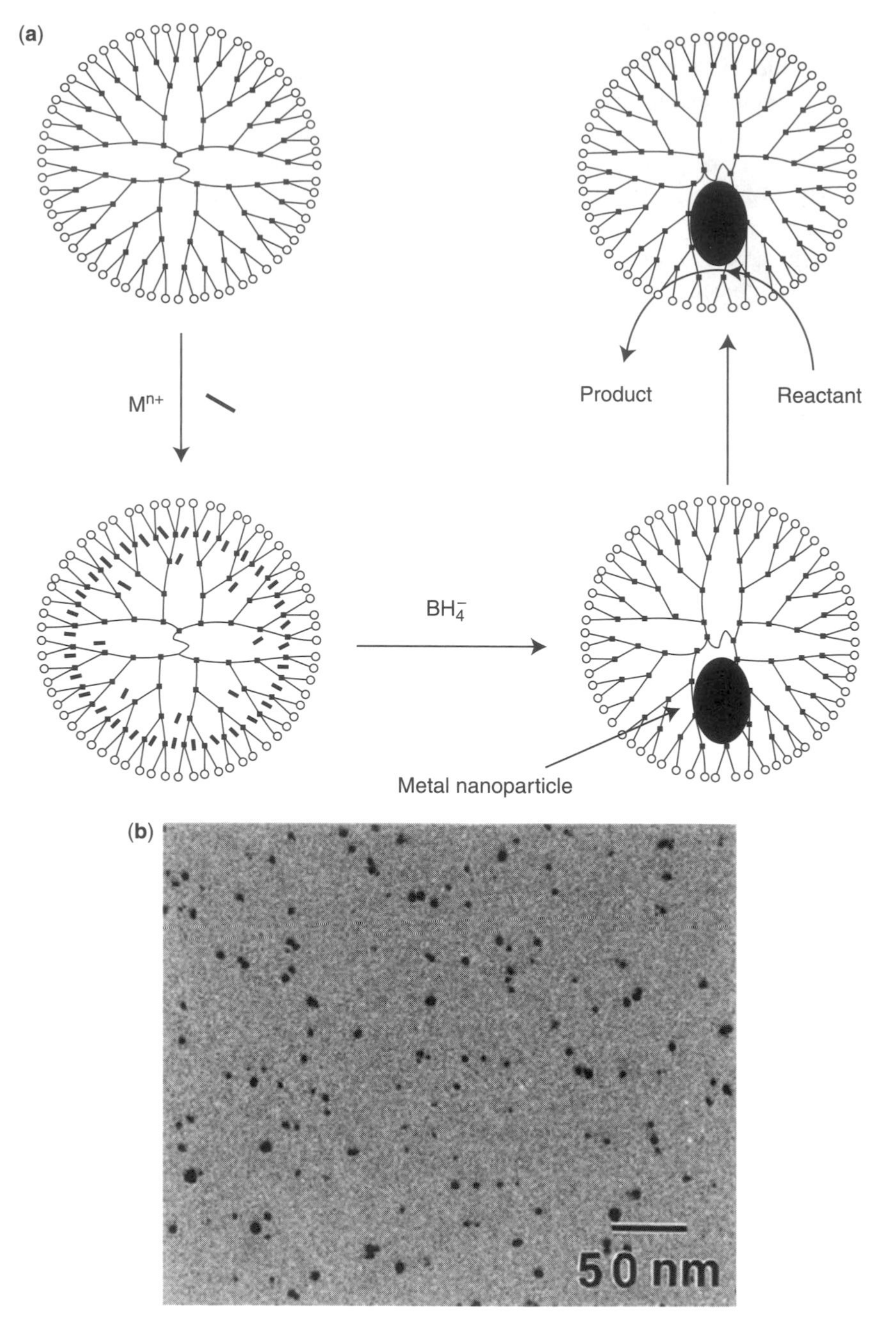

Figure 11.11 (a) Scheme for synthesis of dendrimer-stabilized nanoparticles. (Reproduced with permission from R. M. Crooks et al., *Acc. Chem. Res.* **2001**, *34*, 181.) (b) TEM image of dendrimer-stabilized Au nanoparticles in an 8th generation PAMAM dendrimer. (Reproduced with permission from F. Groehn et al., *Macromolecules* **2000**, *33*, 6042.)

Figure 13.13 in Chapter 13). Consequently most of the literature on individual SWCNT reports properties and images of a minor soluble fraction of the original sample. Great effort has been devoted to the dispersal of SWCNT by use of surfactants and soluble polymers and by first functionalizing the surface of the tubes to improve their compatibility with solvent. As-produced nanotubes are contaminated with varied amounts of catalyst residues, fullerenes, and amorphous carbon, depending on the method of production. Before use the contaminants are removed mostly by treatment with acids to dissolve catalyst residues and by oxidation of carbonaceous impurities, which are more easily oxidized than SWCNT. These purification methods leave carboxylic acid and other oxygen-containing functional groups at the ends and at defects in the sidewalls of the SWCNT. The words "soluble" and "solution" should be used sparingly or not at all when discussing liquid dispersions of SWCNT. Most "stable" dispersions precipitate over periods of days or months. Surfactants such as SDS (27) and sodium cholate (28), and polymers such as poly(sodium 4-styrenesulfonate) (NaPSS) (27, 29) and poly(4-vinyl-*N*-alkylpyridinium halides) (PVPBr) (30), shown in Figures 11.12 and 11.13 have enabled preparation of aqueous dispersions. In some cases the polymers have been covalently grafted to the SWCNT (29, 31). In nonpolar organic solvents such as chloroform, aromatic *m*-phenylenevinylene polymers have enabled the most concentrated dispersions of SWCNT (32, 33). The surfactant and the polymer dispersants adsorb to the SWCNT, overcoming tube-tube van der Waals attractions, and leave solvated chains extending from the tube surface.

11.3.2.4 Nanoparticles Assembled by Phase Separated Block Copolymers Block copolymers serve as templates for the construction of inorganic structures in an organic polymer matrix. Films of block copolymers of 70% styrene and 30%

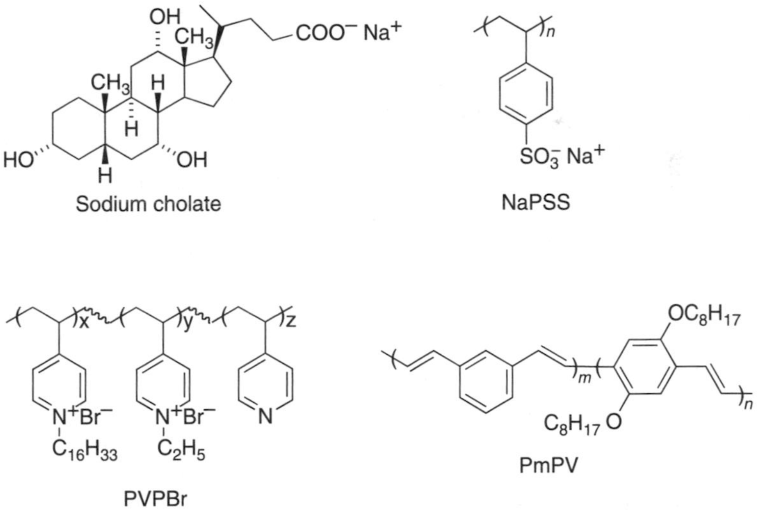

Sodium cholate

NaPSS

PVPBr

PmPV

Figure 11.12 Structures of ionic dispersants of SWCNT in aqueous media and poly(*m*-phenylenevinylene) (PmPV) dispersant for SWCNT in chloroform.

methyl methacrylate by volume consist of PMMA cylinders in a polystyrene matrix. The cylinders can be oriented perpendicular to the substrate either by application of an electric field or by first coating the substrate with a random PS/PMMA copolymer of ~60 vol% styrene. Irradiation degrades the PMMA, making it more soluble, and cross-links the PS, making it insoluble. Dissolution of the degraded PMMA into acetic acid leaves cylindrical holes in the matrix. The holes have been filled with Co metal by electrodeposition (34), Cr and layered Au/Cr nanodots by evaporation (35), or silica by hydrolysis of tetraethyl orthosilicate, as shown in Figure 11.13 (36). Removal of the polystyrene by reactive ion etching and dissolution in toluene leaves wires or posts of the inorganic material with lattice spacings of 24 to 35 nm, depending on the molecular weight of the template block copolymer. The volume fraction of the matrix and the distance between the PMMA cylinders can be increased by addition of low molecular weight homopolystyrene to the block copolymer. Poly(styrene-*b*-4-vinylpyridine), poly(styrene-*b*-ethylene oxide), and other block copolymers can be manipulated similarly. Because CdSe and gold nanoparticles partition to the interface of lamellar poly(styrene-*b*-4-vinylpyridine), thin films of these materials consist of patterns of nanoparticles at the phase boundaries of the block copolymer (37–40).

Another way to template thin films of nano-sized cylinders perpendicular to the surface is to start with a preformed membrane of track-etched polycarbonate or nanoporous alumina. A fluid dispersion of a filler material can be drawn into the pores. Anodized aluminum oxide was the template for construction of lithium ion nanobatteries having many parallel cells filled with the solid state electrolyte PEO-LiOTf (poly(ethylene oxide)-lithium trifluoromethanesulfonate) and the electrodes coated on the top and bottom surfaces of the film (41).

11.3.2.5 Sol-Gel Silica Silica is produced under either acidic or basic conditions at room temperature by the hydrolysis of tetraethyl orthosilicate (TEOS) in water-organic mixtures (42, 43). Under acidic conditions the product is a gel. Slow drying of the gel produces optically clear silica glasses. Both inorganic and organic compounds can be incorporated into the glass by dissolving them into the original TEOS-water-organic mixture. One advantage of sol-gel materials for optical applications is that organic dyes trapped in silica are much more resistant to photobleaching than dyes in organic solutions or organic polymer glasses because the silica matrix is inert. Photobleaching of dyes in organic matrices is due to reaction of the dyes with the matrix.

Hydrolysis of TEOS under basic conditions in ethanol/water/ammonia produces colloidal spheres whose size can be controlled in the range of about 50 to 500 nm (44). The silica spheres have reactive OH groups on the surface and still contain some ethoxy groups. The sizes of the spheres can be increased by seed growth procedures in which more TEOS is added to the dispersion. These silica particles are used as templates in the preparation of photonic band gap materials and porous materials, as shown in Figure 11.14 (45).

Spherical porous TiO_2 particles 1 to 50 μm in diameter have been created from colloidal PMMA particles having PDMS grafted to the PMMA surface. A mixture of

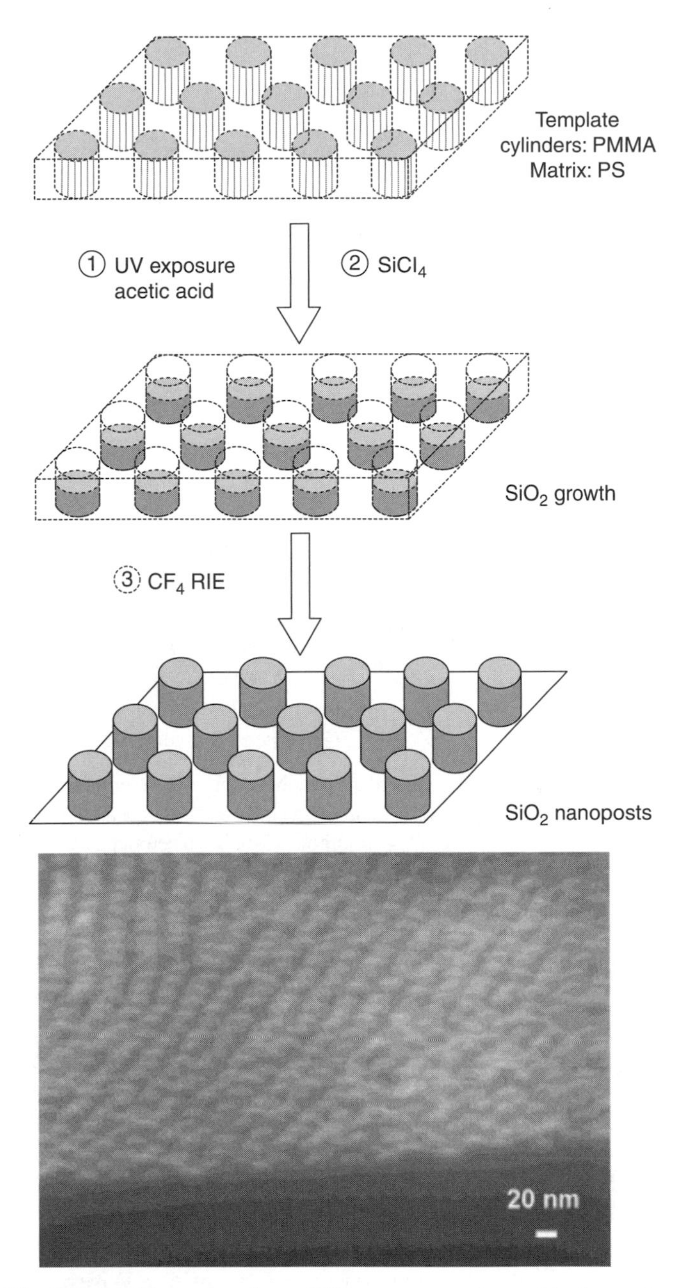

Figure 11.13 Scheme for construction of SiO_2 nanoposts from a block copolymer template and field emission SEM of the nanoposts. (Reproduced with permission from H.-C. Kim et al., *Adv. Mater.* **2001**, *13*, 795.)

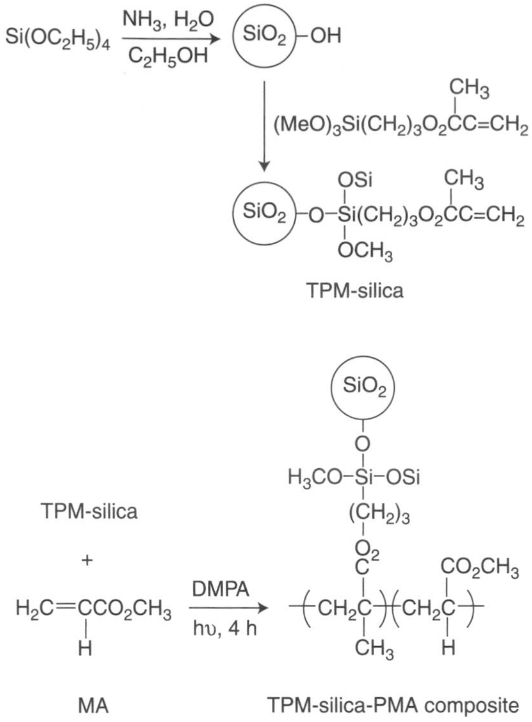

$Si(OC_2H_5)_4$ $\xrightarrow[C_2H_5OH]{NH_3,\ H_2O}$ SiO_2–OH

$(MeO)_3Si(CH_2)_3O_2CC(CH_3){=}CH_2$

SiO_2–O–Si(OSi)(OCH_3)$(CH_2)_3O_2CC(CH_3){=}CH_2$

TPM-silica

TPM-silica + $H_2C{=}C(H)CO_2CH_3$ $\xrightarrow[h\upsilon,\ 4\ h]{DMPA}$ TPM-silica-PMA composite

MA

Figure 11.14 Sol-gel synthesis of colloidal silica spheres, modification of surface with methacrylate groups, and copolymerization with methyl acrylate to form a CCA of silica particles in a poly(methyl acrylate) film. DMPA, 2,2-dimethoxy-2-phenylacetonephenone. (Reproduced with permission from J. M. Jethmalani and W. T. Ford, *Chem. Mater.* **1996**, *8*, 2138.)

hexane and titanium tetrabutoxide was emulsified into formamide with surfactants. Slow removal of the hexane left PMMA particles packed into droplets of the titania precursor, which upon exposure to water and hydrolysis gave particles of titania containing trapped PMMA particles. Pyrolysis of the PMMA at >350°C left porous anatase titania particles that efficiently scattered visible light (46).

11.4 ORGANIZATION OF THE BUILDING BLOCKS IN ONE, TWO, AND THREE DIMENSIONS

11.4.1 Components with One Macroscopic Dimension: Wires, Rods, Tubes, and Fibers

Bottom up construction of electronic circuits requires assembly of molecular wires and switches at the molecular level. Field effect transistors have been constructed with carbon nanotube wires connecting the source and drain electrodes (47). Construction of electronics one element at a time, however, is unlikely to be practical for commercial products. Another model for integrated circuits is a crossed array of 16 nm diameter *p*-doped Si nanowires with 400 parallel nanowires in one direction and 400 nanowires in the orthogonal direction and the layers of nanowires sandwiched around a monolayer of organic molecular switches. Rotaxane molecular switches, such as the structure in Figure 11.15 operate by redox chemistry (48).

Examples of construction from rods, tubes, and fibers are described elsewhere in this and other chapters of this book.

11.4.2 Thin Films: Structures with Two Macroscopic Dimensions

11.4.2.1 Self-Assembled Monolayers The term self-assembled monolayers (SAMs) refers to molecule-thick layers constructed with covalent bonds, such as a carpet of alkyl chains on a gold surface formed by reaction with thiols or disulfides, as shown in Figure 11.16a (49–51). When the alkanethiol or disulfide is monofunctional, the resulting monolayer is a hydrophobically modified gold surface with methyl groups at the air interface. When the thiol or disulfide is ω-substituted with a functional group such as carboxylic acid or alcohol, the film is hydrophilic and has end groups available for further chemical modification. In a tightly packed monolayer the alkyl groups in all-*anti* conformations form a two-dimensional crystal in which the tilt angle of the alkyl chains with respect to the surface depends on the crystalline face of the metal to which the thiol groups bind. The well-defined structures of SAMs enable study of how the end groups affect surface properties such as wetting, adhesion, and tribology.

Similarly silane coupling agents produce monolayers on a silica surface such as a silicon wafer or quartz (52). Careful characterization of the monolayers requires a clean, smooth gold or silicon substrate and exclusion of organic contaminants. Self-assembly has been extended to multilayers by covalent bonding to the ω-functional groups. Multilayers have been prepared from ω-functional trichlorosilanes as shown in Figure 11.16b (52), from zirconium α,ω-alkanebisphosphonates (53), and from donor-acceptor substituted aromatic compounds (54), which produce nonlinear optical thin films having the dipoles of every layer oriented in the same direction to give the entire thin film a large dipole moment. Without covalent bonds to fix the orientation, polar molecules generally form layers with alternating dipoles, and the multilayer structure has little or no net dipole moment.

11.4.2.2 Langmuir–Blodgett Films SAMs resemble Langmuir–Blodgett films that were created first in the 1930s. Benjamin Franklin published in 1774 his experimental observation that a spoonfull of oil on the surface of a pond spreads over a wide area and calms the waves (55). Although Franklin did not report the calculation, his experiment allowed the first estimation of the sizes of molecules from the thickness of the film, the volume of oil, and the area of water covered. (At the time there was no distinction between atoms and molecules, and only a few elements had been identified; Reference 55.)

Langmuir put the spreading phenomenon on a quantitative molecular basis using a film balance, whose essential components are a pan of water with a movable barrier

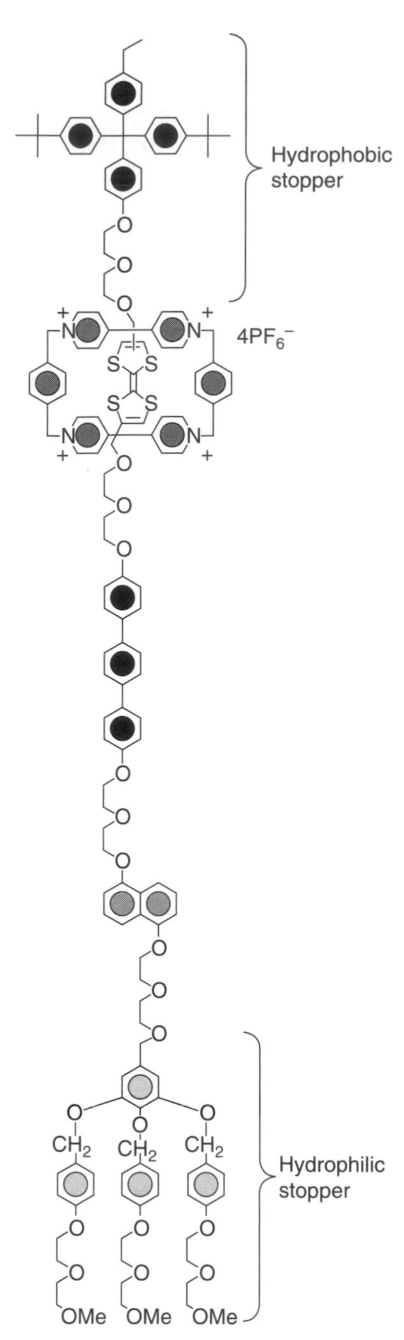

Figure 11.15 Molecular switch constructed from a rotaxane. The tetrapyridinium macrocycle collars the uncharged reduced form of TTF (tetrathiafulvalene) in the low conductance state and slides along the molecular chain to the dioxynaphthalene site when the TTF is oxidized to its cationic high conductance state. (Reproduced with permission from J. E. Green et al., *Nature* **2007**, *445*, 414.)

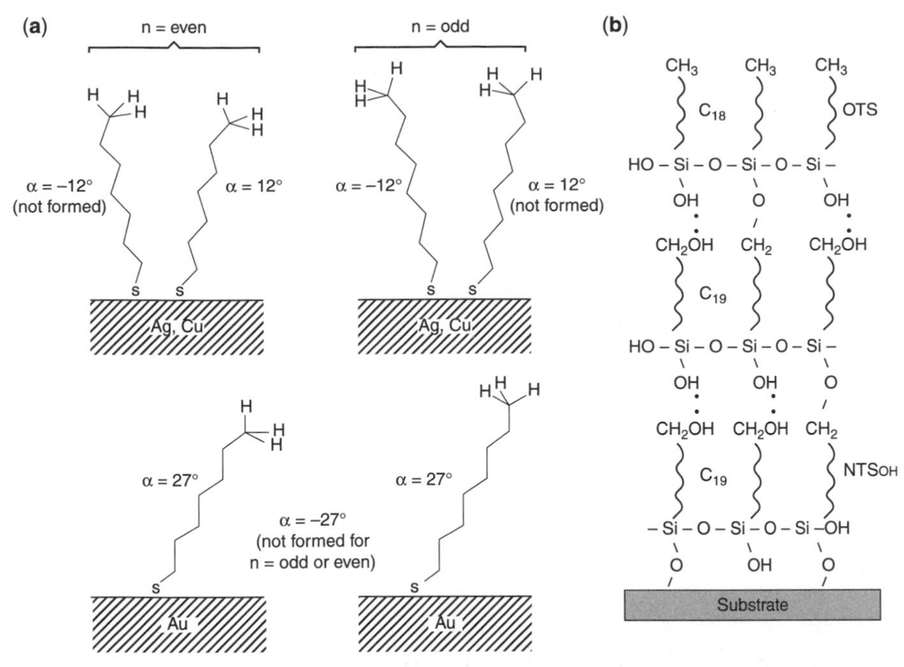

Figure 11.16 (a) Geometry of self-assembled monolayers on gold, silver, and copper. The cant angles of the alkyl chains $-S(CH_2)_nH$ on Au are the same for n = odd and n = even, and on Ag and Cu are of opposite sign for n = odd and n = even. (Reproduced with permission from P. E. Laibinis et al., *J. Am. Chem. Soc.* **1991**, *113*, 7152.) (b) A self-assembled multilayer prepared from a ω-hydroxyalkyltrichlorosilane. (Reproduced with permission from K. Wen et al., *ACS Nano* **2008**, *2*, 579.)

to compress a hydrophobic layer on the surface and a device such as a Wilhelmy plate to measure the surface pressure. When less than a monolayer of a water-insoluble material is deposited on the water and then compressed, there is a rapid increase of surface pressure when the molecules in the hydrophobic layer are packed together. Surfactants orient with their hydrophilic heads on the water surface and their hydrophobic tails projecting into the air to minimize interfacial free energy. Blodgett, Langmuir's colleague, transferred monolayers of fatty acids from the water surface to a solid surface by dipping or drawing a sheet of glass through the air-surfactant-water interface, as shown in Figure 11.17. The hydrophilic end of the surfactant transfers from water to the hydrophilic surface of the glass. Layers of fatty acids are more robust when the water subphase contains divalent metal ions such as Cd or Bi. Bilayers of alternating polarity resembling the phospholipid bilayers of biological membranes often are transferred on the down and up strokes. Repetitive dipping produces Langmuir–Blodgett films tens of bilayers thick (56, 57). Because the area of the substrate coated by transfer often is equal to the decrease in area of the monolayer on water at constant surface pressure, it has been assumed, usually without proof, that the structure of the monolayer on glass is the same as the structure of the

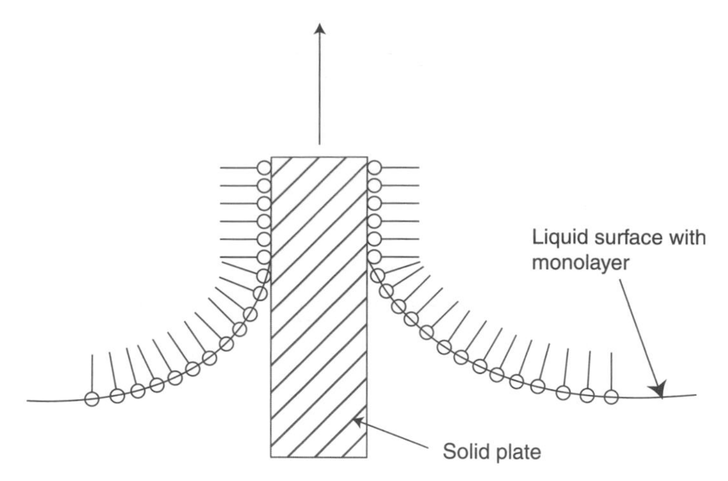

Figure 11.17 Transfer of a monolayer from water to a solid surface by lifting a planar substrate through the liquid-gas interface. (Reproduced with permission from G. L. Gaines, Jr., *Insoluble Monolayers at Liquid-Gas Interfaces*, p. 326. Wiley-Interscience, New York, **1966**.)

monolayer on water. Not only surfactants but also many other hydrophobic compounds and colloidal particles can be coated as monolayers and multilayers on solid substrates by the Langmuir–Blodgett technique. SAMs, however, are more robust mechanically than Langmuir–Blodgett films because of the covalent bonds between the amphiphile and the substrate and between layers of amphiphiles.

11.4.2.3 Layer-by-Layer Self-Assembly Two different polymers usually are immiscible because the entropy of mixing ΔS_m is small, the enthalpy of mixing $\Delta H_m > 0$, and the free energy $\Delta G_m = (\Delta H_m - T\Delta S_m) > 0$. One exception is when electrostatic attraction of oppositely charged polymers makes $\Delta H_m < 0$. Alternating layers of cationic and anionic polyelectrolytes, such as poly(allylammonium chloride) (PAAC) and poly(sodium 4-styrenesulfonate) (PSS), can be deposited on a surface by adsorption from solution in alternating fashion (58). In a typical preparation, a silicon wafer, having a negatively charged surface due to dissociation of H^+ and Na^+ from its oxide surface, is dipped into a dilute solution of PAAC for a few minutes, washed with water, dipped into a dilute solution of PSS for a few minutes, and washed with water to form a polyelectrolyte bilayer, as shown in Figure 11.18. The thicknesses of the layers are usually only 1 to 2 nm, as measured by ellipsometry and by the slope of the linear increase of UV-visible absorbance with the number of layers. The small counterions sodium and chloride are washed out during successive depositions. Each adsorption of a polyelectrolyte leaves a film surface rich in that polyelectrolyte and its counterion. The next layer of oppositely charged polyelectrolyte displaces the excess counterions, forms strong electrostatic bonds, and leaves its own surface excess ready for the next step (59). Repetition of the process tens or hundreds of times produces films thick enough to be picked up and transferred. The films are strong and completely amorphous. Charged colloidal particles also can be incorporated into the films. In fact,

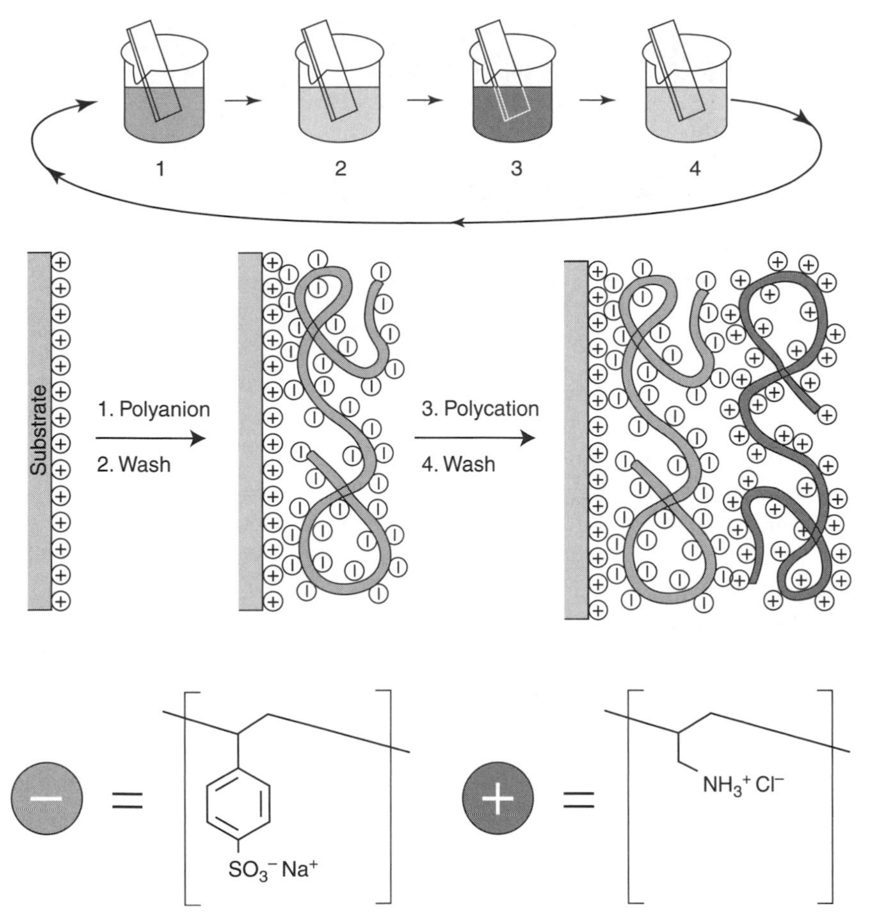

Figure 11.18 Layer-by-layer construction of thin films from polyelectrolytes. (Reproduced with permission from G. Decher, *Science* **1997**, *277*, 1232.)

the principle of layer-by-layer construction of films of colloidal particles of opposite charge dates back to the 1960s (60). A colorful example of alternating polyelectrolytes and colloidal particles is a rainbow film (Fig. 11.19) constructed from CdTe nanoparticles of increasing sizes and therefore increasing emission wavelengths (61). LBL films also can be constructed from strongly hydrogen bonding pairs of polymers such as poly(4-vinylpyridine) and poly(acrylic acid). Electrically conductive LBL films have been interfaced with nerve cells (62). LBL films also have been constructed on the surfaces of colloidal inorganic and polymer particles (63). Dissolution of the core particle leaves an empty capsule of an LBL film.

11.4.2.4 Structures on Patterned Surfaces SAMs can be printed by soft lithography (64). A stamp is made by curing of a PDMS prepolymer on a photoresist having a pattern created by photolithography. The surface of the stamp is inked with an

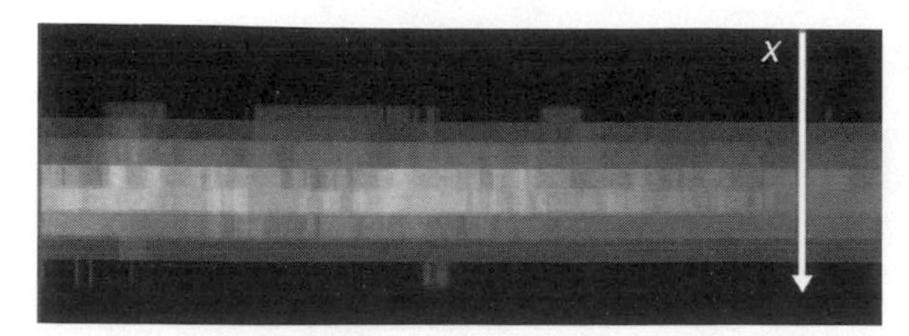

Figure 11.19 Rainbow layer-by-layer film of ten layers each of green, yellow, orange, and red CdTe nanoparticles alternating with polyelectrolytes. (Reproduced with permission from A. A. Mamedov et al., *J. Am. Chem. Soc.* **2001**, *123*, 7738.) (See color insert.)

alkanethiol and then printed onto a gold-coated silicon substrate. Via the ω-functionality of the alkanethiol the surface can be patterned as hydrophilic/hydrophobic, acid/base, or metal-ligand coordinating. One use of a functional SAM is the growth of calcite crystals from the surface of a SAM of a ω-sulfanylalkanecarboxylic acid on gold. The orientation of the calcite crystals depends on the orientation of the COOH groups at the SAM-air interface, which is determined by whether the alkyl chain of the SAM has an odd or even number of carbon atoms (65).

Another approach to growth of inorganic crystals from an organic surface is to incorporate specific nucleating peptides into the filamentous capsid of the M13 bacteriophage coat structure, which provides a template for the synthesis of single crystals of ZnS and CdS and of nanowires of CoPt and FePt (66). Similarly, after determining how hexapeptides of one amino acid, particularly that of histidine (H6), and interdigitated peptides XHXHXHX, in which a second amino acid alternates with histidine, mediate the binding of yeast to CdS, CdSe, ZnS, ZnSe, and Au, peptide sequences were designed specifically for growth of Au and ZnS from peptide-coated surfaces (67). Growth of colloidal crystals of monodisperse silica or colloidal polymer spheres from a surface can be controlled by matching the colloidal crystal structure to a pattern of holes of a solid porous material (68, 69). The pores of anodized alumina serve as templates for growth of cylinders of materials such as poly(styrene-*b*-butadiene), which assume either lamellar or cylindrical microdomain morphologies depending on the volume fraction composition of the polymer (70).

11.4.2.5 Thin Films of Block Copolymers Rapid deposition of a block copolymer on a surface may leave the polymer in a kinetically trapped state. Annealing thermally or with solvent reduces the net energy of the polymer-air, polymer-substrate, and polymer-polymer interfaces. The less polar component of the block copolymer migrates to the air interface, and the component with the lower polymer-substrate interfacial energy migrates to the substrate. The component on the substrate could be the same as at the air interface. The surface of the substrate can be modified prior to deposition of the block copolymer to control which phase deposits. By the method described in Section 11.2.3.4, a PS-*b*-PMMA was used to produce a film of PS having vertical cylindrical pores. Dipping of the polystyrene film into heptane solutions containing CdSe nanoparticles transfers the nanoparticles into the pores to give a hexagonal array of columns of CdSe (71). Other block copolymer formulations have been processed into nanostructured membranes, templates for nanoparticle synthesis, photonic crystals, and high-density information storage media (72). Nanoparticles dispersed into block copolymers segregate to the polymer phase that minimizes the interfacial energy between the polymer and the surface coating on the nanoparticles (37). For example, in poly(styrene-*b*-2-vinylpyridine) TOPO-coated CdSe particles incorporated into the polystyrene domains, while poly(ethylene glycol)-coated CdSe particles incorporated into the P2VP domains. Similarly Au nanoparticles stabilized with thiol-terminated polystyrene partition into the polystyrene phase, and Au nanoparticles stabilized with thiol-terminated P2VP partition into the P2VP (39).

Figure 11.20 illustrates several stages of reconstruction and coating of a thin film of a poly(styrene-*b*-4-vinylpyridine). Annealing in toluene vapor formed vertical cylinders of P4VP in a polystyrene matrix (73). Soaking the film briefly in ethanol, a good solvent for P4VP but a nonsolvent for polystyrene, swelled the cylinders to leave a layer of P4VP on top of the entire film. Annealing of the reconstructed film at 115°C, which is slightly above the glass transition temperature of polystyrene, converted the structure back to the original cylinders of P4VP in a polystyrene matrix. Sputtering of a layer of Au onto each polymer film at glancing angle (5°) decorated the film surfaces, and not the pores, with Au nanoparticles. Heating a film having a <0.5 nm Au surface layer to 115°C drew dots of the Au as well as the P4VP into the nanopores because of the affinity of Au for P4VP and not for PS. Under the same annealing conditions a >0.5 nm Au layer was not drawn into the pores. Annealing the thicker Au-coated film at 180°C created some Au nanoparticles in the pores and a Au film remaining on top, which gave an image of a ring pattern.

Block copolymers in solution form micelles when one block is solvated and the other is not. Spin coating of a solution of poly(styrene-*b*-4-vinylpyridine) and

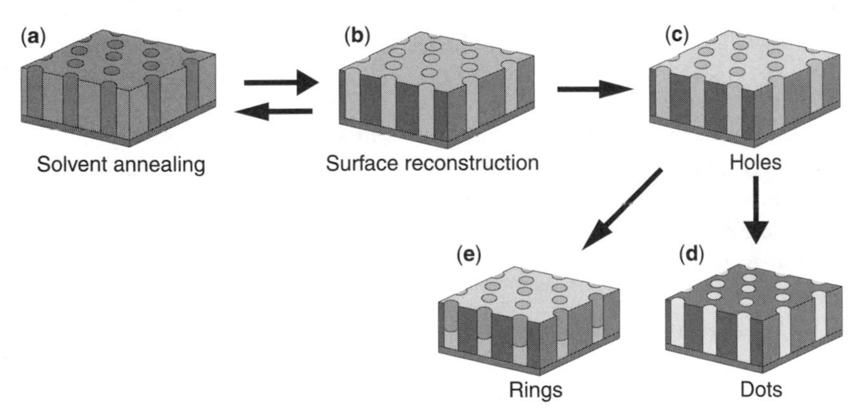

Figure 11.20 Reconstruction of poly(styrene-*b*-4-vinylpyridine) thin films alone and with Au films on the top surface. (a) Solvent annealed film; (b) the film reconstructed from ethanol; (c) nanoporous film from glancing angle deposition of gold; (d) dots of Au made from thin film of gold; (e) rings of Au made from thicker film of gold. (Reproduced with permission from S. Park et al., *ACS Nano* **2008**, *2*, 766.)

gallium(III) nitrate or gallium(III) acetylacetonate in toluene onto silica formed a hexagonally ordered monolayer of micelles with P4VP-Ga(III) cores. Removal of the polymer using oxygen or nitrogen plasma left arrays of Ga_2O_3 and GaN nanoparticles, respectively (74). The metal for an ordered array of nanoparticles also can be introduced as part of the block copolymer itself. A thin film of a block copolymer of poly(styrene-*b*-ferrocenylethylmethylsilane) formed a hexagonal array of vertical cylinders of the ferrocenyl block in a polystyrene matrix. Removal of both polymers by UV and ozone treatment left a hexagonal array of iron particles (75).

11.4.3 Nanostructures from Colloidal Particles

11.4.3.1 Polymer Composites Particles of silica, $CaCO_3$, clay, and carbon black and fibers of glass and carbon have been used for many years to toughen elastomers and plastics (76). Tires are filled with 10 to 100 nm diameter particles of carbon black. Epoxy resins filled with 10 to 20 μm diameter glass fiber are molded into seating, boat hulls, and shower stalls. Composites filled with 5 to 20 μm carbon fiber meet the performance demands for applications as airplane bodies and sporting equipment. In composites containing microparticles or microfibers, the polymer-filler interfacial area is small, and most polymer molecules do not interface with the filler. The interfacial area and the fraction of macromolecules that interface with the filler are much larger with nanoparticles and nanofibers (2). Consequently nano- and micro-scale fillers have much different effects on the mechanical properties of a polymer. Often addition of only 1 to 5 vol% of a nano-scale filler enhances mechanical properties as much as 15% to 40% of a micro-scale filler (76). Dispersion of particles at the nano-scale requires surface modification with chemical groups that are compatible with the polymer. Otherwise the particles tend to cluster and to migrate to cracks in the polymer (77).

Smectite clays consist of 1 nm thick aluminosilicate layers separated by sodium and calcium counterions. As little as 1 to 5 wt% of these layered clays can significantly improve the mechanical properties of nylon, polyolefins, and other polymers (78). Delaminating the clay structure by replacement of the sodium or calcium ions with a polymer-compatible surfactant, a quaternary ammonium ion surfactant, for example, is essential to generate a large polymer-clay interfacial area, as shown in Figure 11.21 (79). Ammonium ion head groups of the dispersant bind to the surface of the clay by Coulombic forces, and the hydrophobic alkyl tails bind to the polymer by van der Waals forces.

11.4.3.2 Colloidal Crystalline Arrays Colloidal spheres of silica and of polymers can be made relatively monodisperse, with standard deviations of 4% of the mean diameter for silica and 1% for polymer latexes. The spheres pack as shown in Figure 11.22a from fluid dispersions into fcc (sometimes hcp or bcc) colloidal crystals (CC) by gravity, by membrane filtration, or by capillary forces at the surface of an evaporating dispersion (80–82). The crystalline order of the materials is strictly at the length scale of the packed colloidal particles: the packing of the atoms and molecules within the silica and polymer particles is totally amorphous. The CCs diffract

Figure 11.21 Molecular dynamics simulation of layers of montmorillonite clay separated by disordered dioctadecylammonium ions. (Reproduced with permission from R. A. Vaia and E. P. Giannelis, *MRS Bull.* **2001**, *26*, 394.)

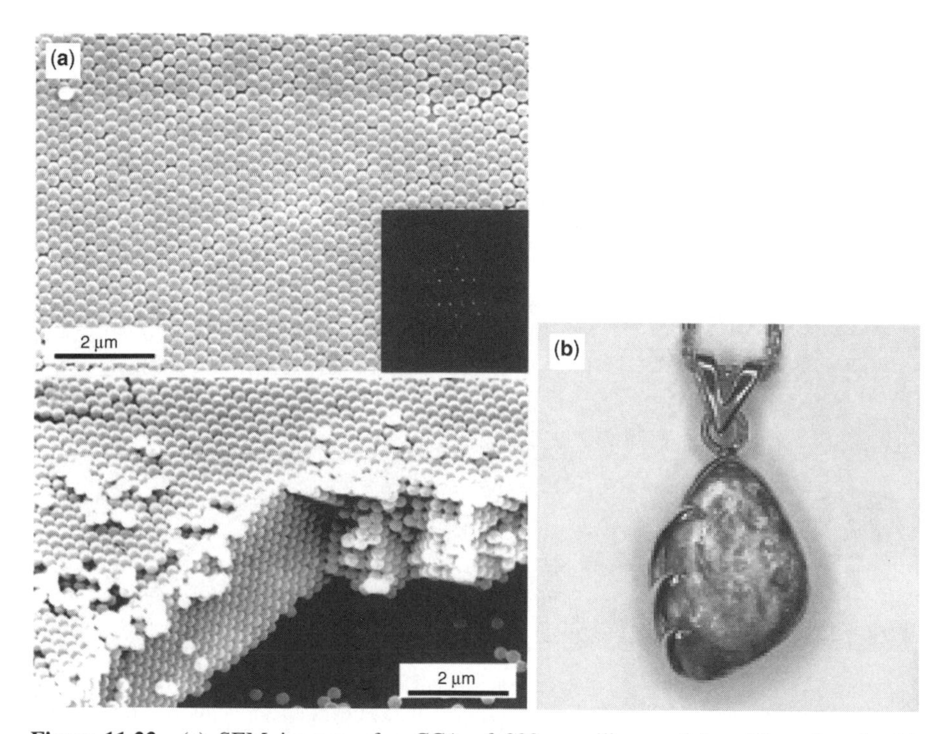

Figure 11.22 (a) SEM images of a CCA of 299 nm silica particles. (Reproduced with permission from P. Jiang et al., *J. Am. Chem. Soc.* **1999**, *121*, 11630.) (b) A natural opal (http://www.costellos.com.au/opals/types.html). (See color insert.)

UV-vis-NIR light according to the Bragg equation ($m\lambda = 2n_{eff}\, d \sin\theta$, where λ and θ are the wavelength and angle of incidence of light, d is the crystal lattice spacing, n_{eff} is the effective refractive index, and m is an integer). Because the particle sizes and the lattice spacings are similar to the wavelengths of visible light, the CCs are opalescent. The color of a natural opal, shown in Figure 11.22b, comes from reflection of light from colloidal crystalline domains of monodisperse silica spheres in an amorphous silica matrix (83, 84). The lattice dimensions can be varied by the size of the particles. Thin films of fcc colloidal crystals usually have the (111) plane parallel to the substrate. Silica CCs can be used to make inverse opals, in which the spaces between the spheres of a dry array are filled with a material of contrasting refractive index. Inverse opals are described more fully below under polymer latexes.

Modification of the surfaces of colloidal silica spheres with silane coupling agents enables transfer of the particles to nonpolar solvents. With 3-methacryloxypropyltrimethoxysilane bonded to the surface, the particles have been transferred from water to the polymerizable monomer, methyl acrylate. Electrostatic repulsion due to a low level of residual charge on the particle surfaces cause the dilute dispersions of particles to form a non-close packed colloidal crystalline array (CCA). Polymerization of the methyl acrylate with 200 nm diameter silica spheres in a CC fixes the positions of the spheres in a plastic film by the reactions shown in Figure 11.14. The diffraction

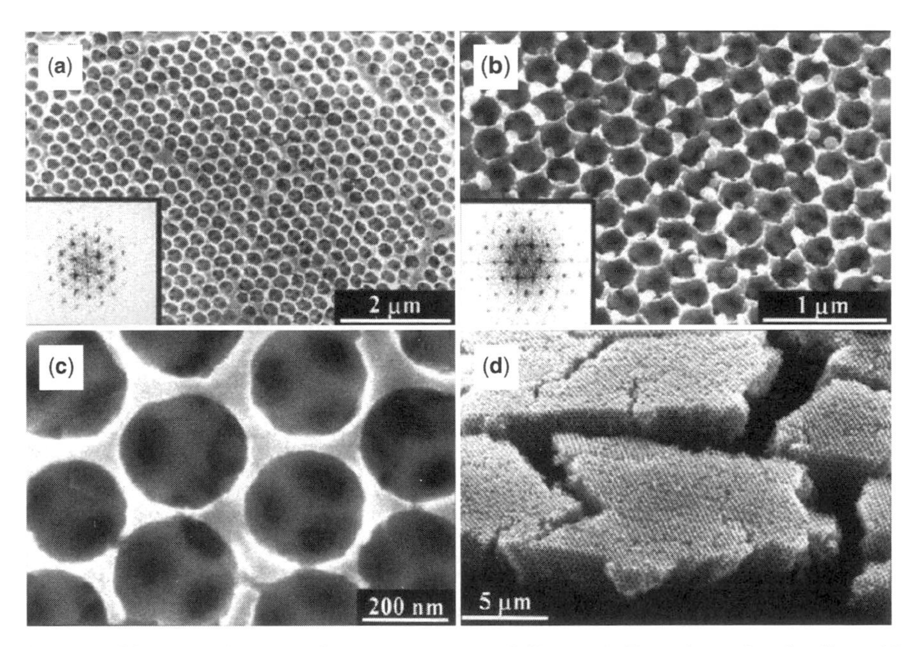

Figure 11.23 TEM images of macroporous metal films. (a) Top view of a Cu film with 325 nm diameter voids. (b) Top view of a Ag film with 353 nm diameter voids. (c) Higher magnification of a Ni film showing the smaller pores interconnecting the air holes. (d) Cu film (a) at lower magnification. (Reproduced with permission from K. M. Kulinowski et al., *Adv. Mater.* **2000**, *12*, 833.)

wavelengths in the visible-NIR spectrum of light normal to the film plane decrease by stretching and increase by swelling of the elastomeric polymer composite film (45, 85).

11.4.3.3 ***Inverse Opals*** Colloidal crystals of polystyrene latexes serve as templates for the synthesis of inverse opals. The interparticle spaces of a dry close packed film of a CC were filled with a colloidal dispersion of much smaller TiO_2 particles, or with a sol-gel precursor of TiO_2. After the TiO_2 solidified, removal of the polystyrene either by solvent extraction or by pyrolysis left a lattice of interconnected spherical holes known as an inverse opal. The inverse opal was a better photonic bandgap material than the parent inorganic–organic hybrid material because of the greater refractive index contrast between TiO_2 and air than between TiO_2 and polystyrene (86–88). Inverse opal materials are formed also by incorporation of metals such as Cu, Ag, and Ni into the voids of a CC, as shown in Figure 11.23 (89).

11.5 CONCLUDING REMARKS AND FUTURE OUTLOOK

Inorganic–organic hybrid materials are ubiquitous on the nano-scale, sometimes by design in order to combine the separate properties of the inorganic and organic components into one material, and sometimes by necessity because an organic surface is the only way an inorganic material can be dispersed in a solvent for processing. Organic materials serve as templates for the synthesis of nanoparticles, such as TOPO-stabilized II-VI semiconductor nanoparticles, for the ordering of inorganic nanoparticles in dendrimers and in block copolymers, and for the synthesis of nanoporous zeolites. The organic template often is removed to obtain a final inorganic product such as a zeolite, or an ordered array of nanoparticles on a surface, and an inverse opal photonic bandgap material. Because the inorganic–organic interface must be stabilized to maintain nanostructures, surfactants and polymers will continue to be essential components of many new hybrid materials.

REFERENCES

1. M. Chen, D. W. Goodman, *Acc. Chem. Res.* **2006**, *39*, 739.
2. R. Krishnamoorti, R. A. Vaia, *J. Polym. Sci., Part B: Polym. Phys.* **2007**, *45*, 3252.
3. R. P. Feynman, in *Lecture at American Physical Society Annual Meeting, Pasadena, CA, December 29,* **1959**. http://www.zyvex.com/nanotech/feynman.html, accessed May 6, 2008.
4. K. J. Klabunde, R. S. Mulukutla, in *Nanoscale Materials in Chemistry*, p. 223, K. J. Klabunde, ed. Wiley-Interscience, New York, **2001**.
5. G. A. Ozin, A. C. Arsenault, in *Nanochemistry*, p. xxix. Royal Society of Chemistry, Cambridge, **2005**.
6. L. N. Lewis, *Chem. Rev.* **1993**, *93*, 2693.
7. K. J. Klabunde, J. Stark, O. Koper, C. Mohs, D. G. Park, S. Decker, Y. Jiang, I. Lagadic, D. Zhang, *J. Phys. Chem.* **1996**, *100*, 12142.
8. J. Turkevich, P. C. Stevenson, J. Hillier, *Discuss. Faraday Soc.* **1951**, *No. 11*, 55.

9. D. A. Tomalia, A. M. Naylor, W. A. Goddard, III, *Angew. Chem. Int. Ed. Engl.* **1990**, *29*, 113.
10. G. R. Newkome, C. N. Moorefield, F. Voegtle, *Dendrimers and Dendrons: Concepts, Syntheses, Applications*, Wiley-VCH, Weinheim, Germany, **2001**.
11. J. M. J. Frechet, *Science* **1994**, *263*, 1710.
12. A. W. Bosman, H. M. Jansen, E. W. Meijer, *Chem. Rev.* **1999**, *99*, 1665.
13. V. Castelletto, I. W. Hamley, *Curr. Opinion Solid State Mater. Sci.* **2004**, *8*, 426.
14. F. S. Bates, G. H. Fredrickson, *Physics Today* **1999**, *52*, 32.
15. R. S. Underhill, G. Liu, *Chem. Mater.* **2000**, *12*, 2082.
16. J. Israelachvili, *Intermolecular and Surface Forces*, 2nd edition. Academic Press, London, **1991**.
17. C. B. Murray, D. J. Norris, M. G. Bawendi, *J. Am. Chem. Soc.* **1993**, *115*, 8706.
18. Y. Yin, A. P. Alivisatos, *Nature* **2005**, *437*, 664.
19. D. J. Norris, A. L. Efros, S. C. Erwin, *Science* **2008**, *319*, 1776.
20. Y. Lin, A. Boeker, H. Skaff, D. Cookson, A. D. Dinsmore, T. Emrick, T. P. Russell, *Langmuir* **2005**, *21*, 191.
21. Y. Lin, H. Skaff, A. Boeker, A. D. Dinsmore, T. Emrick, T. P. Russell, *J. Am. Chem. Soc.* **2003**, *125*, 12690.
22. Z. A. Peng, X. Peng, *J. Am. Chem. Soc.* **2001**, *123*, 1389.
23. P. D. Jadzinsky, G. Calero, C. J. Ackerson, D. A. Bushnell, R. D. Kornberg, *Science* **2007**, *318*, 430.
24. B. L. V. Prasad, S. I. Stoeva, C. M. Sorensen, K. J. Klabunde, *Chem. Mater.* **2003**, *15*, 935.
25. R. M. Crooks, M. Zhao, L. Sun, V. Chechik, L. K. Yeung, *Acc. Chem. Res.* **2001**, *34*, 181.
26. F. Groehn, B. J. Bauer, Y. A. Akpalu, C. L. Jackson, E. J. Amis, *Macromolecules* **2000**, *33*, 6042.
27. M. J. O'Connell, P. Boul, L. M. Ericson, C. Huffman, Y. Wang, E. Haroz, C. Kuper, J. Tour, K. D. Ausman, R. E. Smalley, *Chem. Phys. Lett.* **2001**, *342*, 265.
28. Y. Tan, D. E. Resasco, *J. Phys. Chem. B* **2005**, *109*, 14454.
29. S. Qin, D. Qin, W. T. Ford, J. E. Herrera, D. E. Resasco, S. M. Bachilo, R. B. Weisman, *Macromolecules* **2004**, *37*, 3965.
30. V. A. Sinani, M. K. Gheith, A. A. Yaroslavov, A. A. Rakhnyanskaya, K. Sun, A. A. Mamedov, J. P. Wicksted, N. A. Kotov, *J. Am. Chem. Soc.* **2005**, *127*, 3463.
31. Z. Yao, N. Braidy, G. A. Botton, A. Adronov, *J. Am. Chem. Soc.* **2003**, *125*, 16015.
32. J. Chen, H. Liu, W. A. Weimer, M. D. Halls, D. H. Waldeck, G. C. Walker, *J. Am. Chem. Soc.* **2002**, *124*, 9034.
33. M. N. Tchoul, W. T. Ford, M. L. P. Ha, I. Chavez-Sumarriva, B. P. Grady, G. Lolli, D. E. Resasco, S. Arepalli, *Chem. Mater.* **2008**, *20*, 3120.
34. T. Thurn-Albrecht, J. Schotter, G. A. Kastle, N. Emley, T. Shibauchi, L. Krusin-Elbaum, K. Guarini, C. T. Black, M. T. Tuominen, T. P. Russell, *Science* **2000**, *290*, 2126.
35. K. Shin, K. A. Leach, J. T. Goldbach, D. H. Kim, J. Y. Jho, M. Tuominen, C. J. Hawker, T. P. Russell, *Nano Lett.* **2002**, *2*, 933.
36. H.-C. Kim, X. Jia, C. M. Stafford, D. H. Kim, T. J. McCarthy, M. Tuominen, C. J. Hawker, T. P. Russell, *Adv. Mater.* **2001**, *13*, 795.

37. S. Zou, R. Hong, T. Emrick, G. C. Walker, *Langmuir* **2007**, *23*, 1612.
38. Y. Lin, A. Boeker, J. He, K. Sill, H. Xiang, C. Abetz, X. Li, J. Wang, T. Emrick, S. Long, Q. Wang, A. Balazs, T. P. Russell, *Nature* **2005**, *434*, 55.
39. J. J. Chiu, B. J. Kim, E. J. Kramer, D. J. Pine, *J. Am. Chem. Soc.* **2005**, *127*, 5036.
40. J. J. Chiu, B. J. Kim, G.-R. Yi, J. Bang, E. J. Kramer, D. J. Pine, *Macromolecules* **2007**, *40*, 3361.
41. F. Vullum, D. Teeters, A. Nyten, J. Thomas, *Solid State Ionics* **2006**, *177*, 2833.
42. R. K. Iler, *The Chemistry of Silica*, Wiley, New York, **1979**.
43. C. J. Brinker, G. W. Scherer, *Sol-Gel Science*, Academic Press, San Diego, **1990**.
44. W. Stoeber, A. Fink, E. Bohn, *J. Colloid Interface Sci.* **1968**, *26*, 62.
45. J. M. Jethmalani, W. T. Ford, *Chem. Mater.* **1996**, *8*, 2138.
46. S. M. Klein, V. N. Manoharan, D. J. Pine, F. F. Lange, *Langmuir* **2005**, *21*, 6669.
47. P. Avouris, *MRS Bull.* **2004**, *29*, 403.
48. J. E. Green, J. W. Choi, A. Boukai, Y. Bunimovich, E. Johnston-Halperin, E. DeIonno, Y. Luo, B. A. Sheriff, K. Xu, Y. S. Shin, H.-R. Tseng, J. F. Stoddart, J. R. Heath, *Nature* **2007**, *445*, 414.
49. C. D. Bain, E. B. Troughton, Y. T. Tao, J. Evall, G. M. Whitesides, R. G. Nuzzo, *J. Am. Chem. Soc.* **1989**, *111*, 321.
50. R. G. Nuzzo, L. H. Dubois, D. L. Allara, *J. Am. Chem. Soc.* **1990**, *112*, 558.
51. P. E. Laibinis, G. M. Whitesides, D. L. Allara, Y. T. Tao, A. N. Parikh, R. G. Nuzzo, *J. Am. Chem. Soc.* **1991**, *113*, 7152.
52. K. Wen, R. Maoz, H. Cohen, J. Sagiv, A. Gibaud, A. Desert, B. M. Ocko, *ACS Nano* **2008**, *2*, 579.
53. G. Cao, H. G. Hong, T. E. Mallouk, *Acc. Chem. Res.* **1992**, *25*, 420.
54. S. Yitzchaik, T. J. Marks, *Acc. Chem. Res.* **1996**, *29*, 197.
55. C. Tanford, *Ben Franklin Stilled the Waves*, Duke University Press, Durham, NC, **1989**.
56. G. L. Gaines, Jr., *Insoluble Monolayers at Liquid–Gas Interfaces*, Wiley-Interscience, New York, **1966**.
57. A. Ulman, *An Introduction to Organic Thin Films*, Academic Press, San Diego, **1991**.
58. G. Decher, *Science* **1997**, *277*, 1232.
59. J. B. Schlenoff, S. T. Dubas, *Macromolecules* **2001**, *34*, 592.
60. R. K. Iler, *J. Colloid Interface Sci.* **1966**, *21*, 569.
61. A. A. Mamedov, A. Belov, M. Giersig, N. N. Mamedova, N. A. Kotov, *J. Am. Chem. Soc.* **2001**, *123*, 7738.
62. D. S. Koktysh, X. Liang, B.-G. Yun, I. Pastoriza-Santos, R. L. Matts, M. Giersig, C. Serra-Rodriguez, L. M. Liz-Marzan, N. A. Kotov, *Adv. Funct. Mater.* **2002**, *12*, 255.
63. F. Caruso, R. A. Caruso, H. Moehwald, *Chem. Mater.* **1999**, *11*, 3309.
64. Y. Xia, G. M. Whitesides, *Annu. Rev. Mater. Sci.* **1998**, *28*, 153.
65. Y.-J. Han, J. Aizenberg, *Angew. Chem. Int. Ed.* **2003**, *42*, 3668.
66. C. Mao, D. J. Solis, B. D. Reiss, S. T. Kottmann, R. Y. Sweeney, A. Hayhurst, G. Georgiou, B. Iverson, A. M. Belcher, *Science* **2004**, *303*, 213.
67. B. R. Peelle, E. M. Krauland, K. D. Wittrup, A. M. Belcher, *Langmuir* **2005**, *21*, 6929.
68. A. van Blaaderen, R. Rue, P. Wiltzius, *Nature* **1997**, *385*, 321.

69. A. Van Blaaderen, P. Wiltzius, *Adv. Mater.* **1997**, *9*, 833.
70. H. Xiang, K. Shin, T. Kim, S. I. Moon, T. J. McCarthy, T. P. Russell, *Macromolecules* **2004**, *37*, 5660.
71. M. J. Misner, H. Skaff, T. Emrick, T. P. Russell, *Adv. Mater.* **2003**, *15*, 221.
72. C. Park, J. Yoon, E. L. Thomas, *Polymer* **2003**, *44*, 6725.
73. S. Park, J.-Y. Wang, B. Kim, J. Xu, T. P. Russell, *ACS Nano* **2008**, *2*, 766.
74. S. Bhaviripudi, J. Qi, E. L. Hu, A. M. Belcher, *Nano Lett.* **2007**, *7*, 3512.
75. D. A. Rider, K. A. Cavicchi, L. Vanderark, T. P. Russell, I. Manners, *Macromolecules* **2007**, *40*, 3790.
76. K. I. Winey, R. A. Vaia, *MRS Bull.* **2007**, *32*, 314.
77. A. C. Balazs, T. Emrick, T. P. Russell, *Science* **2006**, *314*, 1107.
78. D. L. Hunter, K. W. Kamena, D. Paul, *MRS Bull.* **2007**, *32*, 323.
79. R. A. Vaia, E. P. Giannelis, *MRS Bull.* **2001**, *26*, 394.
80. Y. Xia, B. Gates, Y. Yin, Y. Lu, *Adv. Mater.* **2000**, *12*, 693.
81. Y. Yin, Y. Lu, B. Gates, Y. Xia, *J. Am. Chem. Soc.* **2001**, *123*, 8718.
82. P. Jiang, K. S. Hwang, D. M. Mittleman, J. F. Bertone, V. L. Colvin, *J. Am. Chem. Soc.* **1999**, *121*, 11630.
83. J. V. Sanders, *Philos. Mag. A* **1980**, *42*, 705.
84. R. K. Iler, *Nature* **1965**, *207*, 472.
85. S. H. Foulger, P. Jiang, Y. Ying, A. C. Lattam, D. W. Smith, Jr., J. Ballato, *Adv. Mater.* **2001**, *13*, 1898.
86. P. Jiang, J. F. Bertone, V. L. Colvin, *Science* **2001**, *291*, 453.
87. V. L. Colvin, *MRS Bulletin* **2001**, *26*, 637.
88. Y. Xia, B. Gates, Z.-Y. Li, *Adv. Mater.* **2001**, *13*, 409.
89. K. M. Kulinowski, P. Jiang, H. Vaswani, V. L. Colvin, *Adv. Mater.* **2000**, *12*, 833.
90. J. Aldana, A. Wang, X. Peng, *J. Am. Chem. Soc.* **2001**, *123*, 8844.

PROBLEMS

1. How could silane coupling agents, $(RO)_3SiCH_2CH_2CH_2X$, be used to modify the surface of particles of TiO_2 for dispersion into nonpolar organic solvents? Explain with a chemical equation.

2. Suggest a procedure to replace the TOPO (trioctylphosphine oxide, $[CH_3(CH_2)_7]_3P{=}O$) stabilizer on CdSe nanoparticles with 3-mercaptopropionic acid, $HSCH_2CH_2CO_2H$, in order to transfer the particles from toluene to water.

3. What is the approximate thickness (to 1 Å) of a self-assembled monolayer that is prepared with $HS(CH_2)_{10}CO_2H$ on a planar gold surface?

4. How does each of the following properties of 5 nm TOPO-coated CdTe nanoparticles compare with that of bulk CdTe? (a) density; (b) melting point; (c) fluorescence.

5. How might the chemical composition of the atoms on the surface of a film of a linear poly(4-vinylpyridine) differ from the bulk composition?

6. Why are the van der Waals attractions between carbon nanotubes much greater than between C_{60} molecules?

7. Where is chemistry located in the food chain of materials research?

8. Suggest a structure of a nonionic surfactant that could be made with a sugar as its polar end.

9. How would the properties of a hypothetical linear poly(amideamine) $[(NHCH_2CH_2NHC(=O)CH_2CH_2)]_n$ differ from those of a dendrimer (Fig. 11.7) of equal molecular weight? (a) glass transition temperature; (b) crystallinity; (c) tensile strength.

10. How might the surface of SiO_2 nanoparticles be modified to trap them in the PEO phase of a poly(styrene-*b*-ethylene oxide)? In the polystyrene phase? Polyethylene oxide has the structure $-(CH_2CH_2O)_n$-.

11. Suggest other common polymers not mentioned in this chapter that might be used to construct LBL thin films.

12. What is the volume in nm^3 of one molecule of a polymer of density 1.0 g cm^{-3} and (a) $M_n = 10{,}000$? (b) $M_n = 1{,}000{,}000$?

13. How many atoms of gold are there in a 10-nm cube (neglecting the stabilizer)? The density of gold is 18.88 g cm^{-3}. The atomic weight is 197.

ANSWERS

1. Metal oxides exposed to air have hydrated surfaces with M-O-H groups. TiO_2 can be treated the same way as SiO_2 to functionalize the surface. In the equation $[TiO_2]$ represents the titania surface. The group X must be nonpolar for dispersion into a nonpolar organic solvent.

$$[TiO_2]\text{-}OH + (RO)_3SiCH_2CH_2CH_2X \longrightarrow ([TiO_2]\text{-}O)_3SiCH_2CH_2CH_2X$$

2. The following abbreviated procedure is a general method in Aldana et al. (90). The TOPO-stabilized nanoparticles in toluene/TOPO were precipitated by addition of methanol and acetone. The supernatant liquid was decanted and the residue was centrifuged to remove more of the solvent. The nanoparticles were added to a solution of 3-mercaptopropionic acid in methanol at pH >10, and the mixture was refluxed. The nanoparticles were precipitated by addition of ethyl acetate and diethyl ether, the supernatant liquid was decanted, and more solvent was removed by centrifugation. The nanoparticles were dried briefly under vacuum and then dissolved in water. All steps were carried out under argon and kept out of light. The basic strategy of any such procedure must be to remove the excess TOPO by precipitating the particles and to redisperse the particles into a basic aqueous solution of the thiol.

3. There are 13 bonds from a gold atom on the surface to an oxygen atom at the end of the undecanoic acid chain. Taking all 13 bond distances in the chain to be 1.54 Å and all bond angles to be 109.5°, each bond projects a distance of 1.54 $\sin(109.5°/2) = 1.26$ Å on the axis of the extended chain. The van der Waals radius of the terminal O atom is about 1 Å, so the extended length of the chain is about $(13 \times 1.26) + 1$ Å $= 17.4$ Å. According to Figure 11.16, the axis of an extended chain makes an angle of 27° from normal to the Au surface. Therefore, the thickness of the monolayer of fully extended chains is 17.4 $\cos(27°)$ Å $= 15.5$ Å. This estimate ignored the O—H bond of the carboxylic acid.

4. (a) The density of the nanoparticles is less than in bulk because the less dense TOPO coating contributes significantly to the volume of the particle. (b) The melting point of the nanoparticles is lower because the surface is less ordered and therefore lower melting than the bulk. (c) Due to quantum confinement the emission band of the nanoparticles is at higher energy and shorter wavelength than in bulk CdTe, and so the fluorescence of the nanoparticle is more to the blue.

5. Nonpolar C-C and C-H components of the polymer have a lower surface energy than the more polar C-N groups. Consequently the aliphatic backbone of the polymer will favor the surface, and the pyridine nitrogen atoms will tend to project away from the surface.

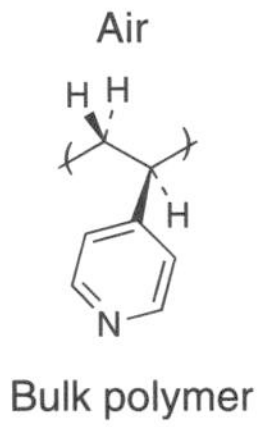

Air

Bulk polymer

6. Parallel carbon nanotubes stick together by van der Waals attraction along the entire length of contact, while the van der Waals attraction between two C_{60} molecules involve only a few carbon atoms on each C_{60}. Fullerenes are soluble in certain solvents. Carbon nanotubes are not known to be soluble in any solvent.

7. Chemistry is the beginning of the food chain of materials research.

8. Any mono- or disaccharide with a long aliphatic substituent is a surfactant; for example,

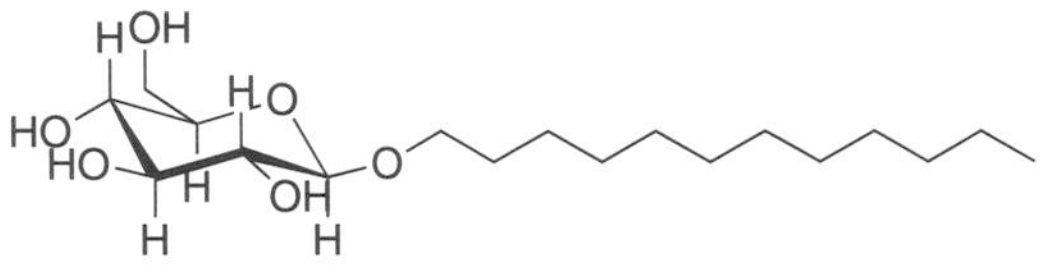

9. Dendrimers are much more disordered than their linear analogues, and consequently (a) have lower glass transition temperature and (b) are completely noncrystalline. Because they lack extendable long chains dendrimers are viscous

liquids at room temperature, whereas linear polyamides are semicrystalline thermoplastics. (c) The dendrimer has no tensile strength.

10. Silane coupling agents can be made with many kinds of terminal groups. Poly(ethylene oxide) (PEO) has the structure $-(CH_2CH_2O)_n$- where the end groups can be both H, both alkyl such as methyl, or one H and one alkyl. To trap SiO_2 particles in the PEO phase of the block copolymer, modify the SiO_2 surface with $(RO)_3SiCH_2CH_2CH_2(OCH_2CH_2O)_nCH_3$. To trap the SiO_2 particles in the polystyrene phase, modify the surface with a less polar silane coupling agent, such as $(CH_3O)_3Si(CH_2)_{17}CH_3$ or $(CH_3O)_3Si(CH_2)_3Cl$.

11. Any ionic polymer with a flexible structure can be used to construct layer-by-layer thin films. Common polyanions are poly(sodium acrylate) and proteins at pH higher than their isoelectric points. Common polycations are protonated polyamines and poly(quaternary ammonium ions) such as poly(4-vinylpyridine hydrochloride) and poly(diallyldimethylammonium chloride).

12. (a) $V = (10^5\ g\ mol^{-1})(1\ cm^3\ g^{-1})(10^{21}\ nm^3\ cm^{-3})/(6.02 \times 10^{23}\ mol^{-1}) = 1.66 \times 10^2\ nm^3$, which is equivalent to a $(5.5\ nm)^3$ cube. (b) $V = 1.66 \times 10^4\ nm^3$, which is equivalent to a $(25.4\ nm)^3$ cube.

13. 577.

12

DNA-MODIFIED NANOPARTICLES: GOLD AND SILVER

ABIGAIL K. R. LYTTON-JEAN AND JAE-SEUNG LEE

12.1 Introduction, 406
12.2 Synthesis and Fundamental Properties of DNA-Modified Gold Nanoparticles, 406
12.2.1 Synthesis of DNA-Modified Gold Nanoparticles, 406
12.2.2 Properties of DNA-Modified Gold Nanoparticles, 407
12.3 Assembly of DNA-Modified Gold Nanoparticles, 410
12.3.1 Nanoparticle-Directed Assembly, 410
12.3.2 Colloidal Crystallization, 410
12.3.3 Assembly Using Asymmetrically Modified Gold Nanoparticles, 411
12.3.4 Template-Based Nanoparticle Assembly, 412
12.4 Diagnostic Applications of DNA-Modified Gold Nanoparticles, 415
12.4.1 DNA Detection Using DNA-Modified Gold Nanoparticles, 416
12.4.2 Protein Detection Using DNA-Modified Gold Nanoparticles, 420
12.4.3 Metal Ions Detection Using DNA-Modified Gold Nanoparticles, 424
12.4.4 Small Organic Molecule Detection Using DNA-Modified Gold Nanoparticles, 427
12.5 Therapeutics: DNA-Modified Gold Nanoparticles as Antisense Agents, 429
12.6 DNA-Modified Silver Nanoparticles, 431
12.7 Conclusion, 433
Further Reading, 434
References, 434
Problems, 438
Answers, 439

Nanoscale Materials in Chemistry, Second Edition. Edited by K. J. Klabunde and R. M. Richards

12.1 INTRODUCTION

The intersection of biomaterials and nanomaterials has witnessed a significant amount of research utilizing both the highly evolved properties of biomaterials and the unique properties of nanomaterials (1–5). One important area of research within the bio-nano-material realm has been the development of DNA-modified gold nanoparticles (DNA-AuNPs) in 1996 by Mirkin et al. (6) and Alivisatos et al. (7). This work demonstrated the first example where oligonucleotides, or short synthetic single strands of DNA, are covalently attached to the surface of AuNPs. This pioneering work initiated a field of research utilizing the properties of these materials and has led to applications in the areas of self-assembly, biodiagnostics, and nanotherapeutics.

The sequence-specific recognition properties of DNA allow DNA-AuNPs to be specifically addressable and controllably assembled. The AuNP core provides an excellent scaffold on which biomolecules or other chemical moieties can be anchored. AuNPs also exhibit intense optical properties that are dependent on the particle size and shape (8, 9). In addition, combining the recognition properties of DNA with the optical properties of AuNPs gives rise to conjugate materials that exhibit new properties not observed by either material individually. Silver nanoparticles (AgNPs) have also been investigated; however, due to the ease of attaching materials to gold surfaces, the majority of the research thus far has been conducted using AuNPs. This chapter will begin by introducing the synthesis and fundamental properties of DNA-AuNPs and continue with a discussion of their applications in nanoparticle assembly, biodiagnostics, and therapeutics. Current research with DNA-AgNPs will also be discussed. The gold and silver nanoparticles discussed in this chapter are faceted particles which appear roughly spherical.

12.2 SYNTHESIS AND FUNDAMENTAL PROPERTIES OF DNA-MODIFIED GOLD NANOPARTICLES

12.2.1 Synthesis of DNA-Modified Gold Nanoparticles

Gold nanoparticles (AuNPs), typically 1 to 100 nm in diameter, are a class of nanomaterials that have become widely studied because they exhibit intense size-dependent optical properties (8, 9), which are typically much greater in magnitude then molecular fluorophores (10). Generally small AuNPs ($\sim$10 nm) exhibit a deep red color when in solution, which shifts to a light purple/pink as the diameter of the AuNPs increases (250 nm). In fact, AuNPs were used in stained glass in the 1700s as a method of creating the intense red and blue colors. These optical properties stem from the surface plasmon of the particles, which is the collective oscillation of the surface electrons as they interact with energy at the resonance wavelength (11). For 13 nm AuNPs, the plasmon band is around 520 nm. As the diameter of the AuNPs increases, the surface plasmon resonance energy red shifts to longer wavelengths and causes the observed color change. Likewise, when AuNPs aggregate the surface plasmon red shifts and a concomitant red to blue color change is observed (10). These optical properties hold for AuNPs in both aqueous and organic solvents (12).

The attachment of DNA to the nanoparticle surface serves two purposes: it (1) provides functionality and recognition properties to the AuNPs and (2) stabilizes AuNPs so they do not aggregate irreversibly in the presences of salt. In the absence of stabilizing ligands such as DNA, AuNPs will aggregate irreversibly in the presence of salt. DNA-AuNPs are synthesized by attaching single-stranded thiol-modified oligonucleotides to the gold surface using sulfur-gold interactions. The attachment is performed by combining thiol-modified DNA with an aqueous solution of AuNPs and then slowly increasing the salt concentration, thereby increasing the shielding between the negatively charged DNA strands and allowing a greater DNA density to be achieved on the AuNP surface. Generally, a single thiol modification at the 3′ or 5′ end of the oligonucleotide is sufficient. However, it has been demonstrated that using a disulfide or a trithiol modification can significantly enhance the stability of the particles (13–15). Typically, DNA is used that was synthesized using standard phosphoramidite DNA synthesis techniques in solid phase. However, these techniques are generally limited to DNA synthesis of less than 100 bases. Efforts have been made to create AuNPs modified with longer DNA sequences. These methods start with standard DNA-AuNPs and then extend the length of the covalently attached DNA using polymerase chain reaction (PCR) (16, 17). The nanoparticle-bound DNA acts as the primer and is extended based on a circular DNA template.

In addition to the DNA length, the number of DNA strands attached to a single AuNP can be carefully tuned as a function of multiple factors, including the salt concentration and AuNP size. For example, a 15 nm AuNP can have anywhere from 20 to 250 strands of DNA covalently attached to the surface (18, 19). In addition, the AuNP size can be varied from a few nanometers to 250 nm in diameter, leading to a variety of nanoparticle sizes and DNA densities (18, 20). To further control the number of DNA strands on each particle, methods have been developed to create AuNPs <10 nm in diameter that have between one and six DNA strands per particle. Particles possessing different numbers of DNA strands can be separated via gel electrophoresis (21–23).

12.2.2 Properties of DNA-Modified Gold Nanoparticles

DNA-AuNPs can be assembled into different structures by designing linker DNA with a complementary sequence. For example, AuNPs modified with either DNA A or DNA B can be assembled in solution by addition of a third DNA sequence A′B′ which is complementary to A and B, respectively. The assembly process is marked by a dramatic red to blue color change and a dampening and red shifting of the surface plasmon (Fig. 12.1). Like DNA duplex formation, the DNA-AuNP assembly process is reversible. By denaturing the interconnecting DNA duplexes within the AuNP assembly, the DNA-AuNPs can be redispersed and a return of the red color is observed. This "melting" process can be induced by an increase in the temperature or a decrease in salt concentration. The melting process can be monitored by measuring the intensity of the surface plasmon as a function of temperature.

The melting transition exhibited by DNA-AuNP assemblies is remarkably different from unmodified DNA duplex melting transitions in two distinct ways: (1) the melting transition of the AuNP assembly is significantly sharper, where the first derivative

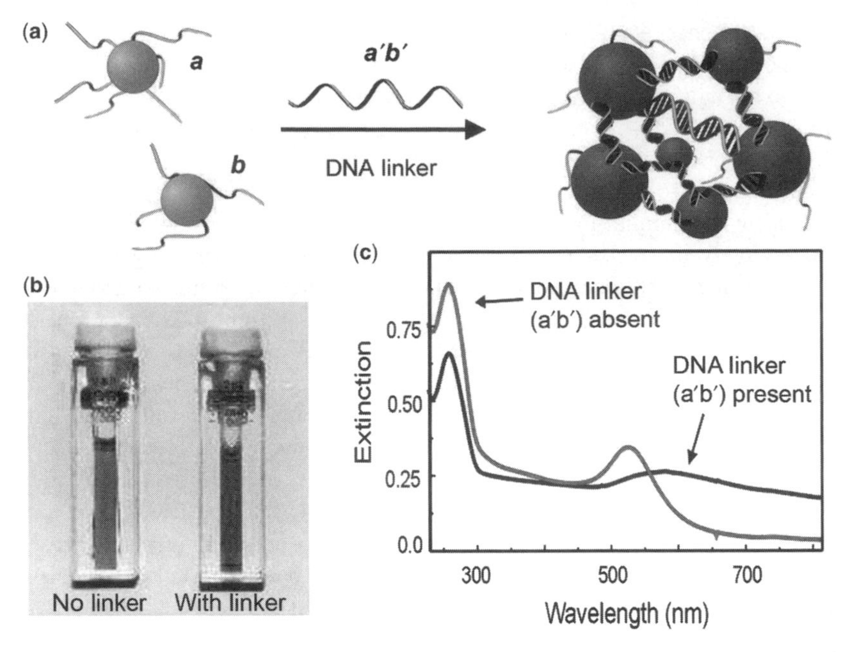

Figure 12.1 (a) DNA-AuNPs assembled using a complementary DNA linker. (b) 13 nm DNA-AuNPs appear red in color without linker DNA and turn blue when linker DNA induces nanoparticle assembly. (Reproduced with permission from C. A. Mirkin et al., *Nature* **1996**, 382, 607–609.) (c) Extinction spectrum of dispersed and assembled DNA-AuNPs. (See color insert.)

displays a full width at half maximum (FWHM) of about 2°C to 4°C compared to a FWHM of about 12°C for DNA duplex, and (2) the melting temperature (T_m) is significantly higher than the T_m of an analogous DNA duplex under identical conditions (Fig. 12.2; Reference 24). Both of these properties are a function of the DNA-AuNP conjugate and are important for the types of applications that will be described later in this chapter. Despite these two notable differences in the melting properties, DNA-AuNPs exhibit other behaviors similar to duplex DNA. For example, increasing the DNA length, DNA concentration, or salt concentration in the system increases the stability of the DNA-AuNP assemblies and the T_m.

The mechanism responsible for the sharp melting transition has been investigated both experimentally and through theoretical analysis (25–32). The sharp melting transition is dependent on multiple factors and can be explained by a combination of two models: a cooperative binding model and a phase transition model. The cooperative binding model is based on a thermodynamic approach that considers the presence of the counterions that associate with DNA when a duplex is formed (25, 30). In this model, the ion cloud associated with each DNA duplex is shared between neighboring duplexes within a DNA-AuNP assembly, thereby adding additional stability to each duplex. As the duplex begins to melt, the ion cloud dissipates, inducing a destabilization of the neighboring duplexes and a cascading melting process. The phase

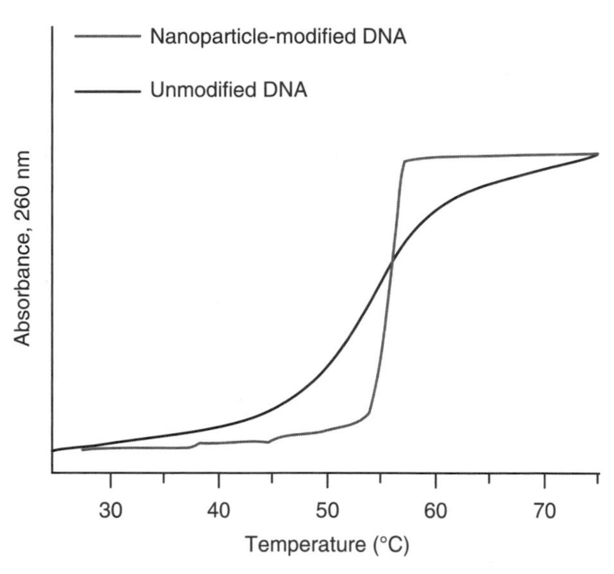

Figure 12.2 Melting transition of unmodified duplex DNA (black) and DNA linked DNA-AuNPs (gray). (Adapted from N. L. Rosi and C. A. Mirkin, *Chem. Rev.* **2005**, 105, 1547–1562.)

transition model, on the other hand, describes the melting transition as dispersion from a gel-like phase to a dilute dehybridized phase where the sharpness stems from the entropy of dissociation of a large number of DNA linkers (28). However, the best explanation of the sharp transition is described by embedding the cooperative model within the overall framework of the phase transition model such that the contributions of both models are combined to determine the overall melting curve (30–32).

In addition to understanding the sharp melting transition, the cooperative interactions between the nanoparticle-bound DNA are also important for describing the increased T_m observed for the DNA-AuNP assemblies. The increased T_m indicates that the DNA-AuNP conjugates possess a greater binding strength than unmodified DNA that is free in solution. The enhanced binding strength is beneficial for some biodiagnostic assays and therapeutic applications utilizing DNA-AuNPs, which will be discussed later in this chapter. Many detection systems are governed under an associative equilibrium. Therefore, DNA-AuNPs that have an increased binding strength between the probe and target promote association to occur at low concentrations and therefore greater detection sensitivities can be achieved. These properties are due to the cooperative binding of the DNA on the nanoparticle surface and are directly dependent on the DNA surface density (33). At the optimal loading density, the DNA-AuNPs exhibit two orders of magnitude enhancement in the binding strength compared to free DNA in solution. This enhancement is a measure of the first binding event between a DNA-AuNP and a complementary DNA sequence in solution. However, because the DNA-AuNP is a polyvalent material, it can undergo several binding events, all of which exhibit different thermodynamics. Indeed, as higher hybridization densities are reached, the binding strength decreases dramatically due to charge repulsion and sterics (34).

12.3 ASSEMBLY OF DNA-MODIFIED GOLD NANOPARTICLES

12.3.1 Nanoparticle-Directed Assembly

The invention and development of DNA-AuNPs has led to the concept of synthetically programmable inorganic materials synthesis. Using single-stranded DNA as molecular tape, scientists are able to control the assembly of materials on the nanoscale that would have otherwise been nearly impossible. The majority of the work carried out thus far has shown one can utilize this programmable assembly concept to effect polymerization with modest control over the placement of, the periodicity in, and the distance between particles within the material. DNA-AuNPs have been assembled in solution using different nanoparticle sizes and arrangements of DNA linkers. Utilizing the sharp melting transition, temperature-programmed assembly can be employed to preferentially assemble DNA-AuNPs that exhibit a higher T_m compared to DNA-AuNPs with a lower T_m (35, 36). By controlling the temperature, particles with longer DNA, for example, can assemble at higher temperatures followed by assembly of particles containing shorter DNA at lower temperature.

Efforts to further control the assembly process have shown that modifying the AuNP surface with additional nonbinding DNA sequences can change the attractive and repulsive forces and influence the overall aggregate size (37). Similarly, partial formation of duplexes on the AuNP surface that do not participate in linking AuNPs together provide rigidity compared to floppy single-stranded DNA and can influence the assembly kinetics (38). Other methods of controlling the assembly process have introduced DNA binding molecules, which can stabilize mismatches in the DNA sequences that would otherwise destabilize the nanoparticle assembly (39, 40).

12.3.2 Colloidal Crystallization

The solution phase AuNP assemblies described thus far have exhibited fractal-like structure, with little internal ordering between the AuNPs (41). In the past, different types of nanoparticles have been assembled into highly ordered crystalline structures, or colloidal crystals, using attractive forces such as electrostatics and van der Waals forces (43). It has long been speculated that the powerful sequence-specific binding properties of DNA should be capable of programming assembly into highly ordered crystalline structures (43). As the synthesis techniques for making monodisperse AuNPs have improved, the ability to assemble highly ordered structures has become possible. To this end, researchers have begun to utilize different DNA sequences to guide the assembly of DNA-AuNPs into macroscopic highly ordered materials, by virtue of programmable base pairing.

The strategy is based on the same organizational packing as atoms in a crystalline lattice, where particles can be designated as A, B, C, etc., based on the DNA sequence. From a thermodynamic standpoint, nanoparticles assembled through DNA linkers will

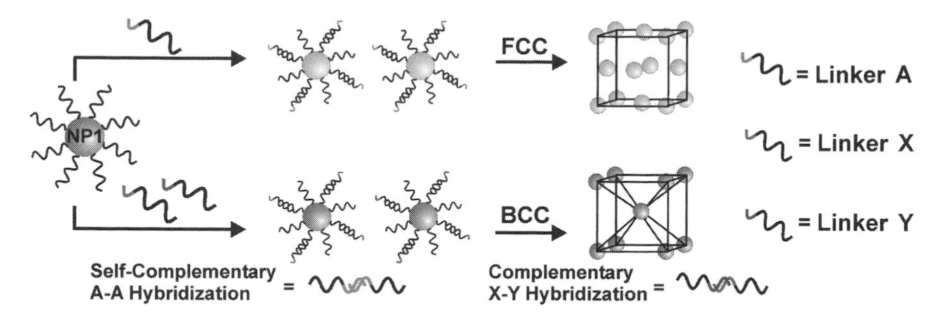

Figure 12.3 The green overhang of linker A is self-complementary; therefore, DNA-AuNPs hybridized to linker A are self-complementary and behave as a single component system. Maximizing the number of particle-particle interactions results in a close-packed structure depicted by an FCC unit cell. The red and blue overhang of linker X and linker Y are complementary and DNA-AuNPs hybridized to linkers X and Y are complementary and behave as a binary system. Maximizing the number of particle-particle interactions results in a non-close-packed structure depicted by a BCC unit cell. (Adapted from S. Y. Park et al., *Nature* **2008**, 451, 553.) (See color insert.)

preferentially assemble into the structure that yields the largest number of DNA hybridization events. If a single component system is employed, where each particle is identical and can bind to every other particle with equal affinity, the most particle-to-particle interactions are achieved through a close-packed crystalline structure. In this type of structure each particle has 12 nearest neighbors and has a face-centered cubic (FCC) unit cell, as depicted in Figure 12.3. However, if a binary system is employed, where each particle is either an A or B particle and only A particles can bind to B particles, the structure with the most particle-to-particle interactions is a non-close-packed structure where each particle has eight nearest neighbors. In this structure the unit cell is body-centered cubic (BCC; Fig. 12.3). If a binary system is forced into a close-packed structure, the number of productive A-B interactions will result in less than eight nearest neighbors and is therefore less energetically favorable.

The first examples demonstrating assembly of DNA-AuNPs into highly ordered crystals have produced both FCC and BCC structures (44, 45). To characterize the crystals in solution, small angle x-ray scattering (SAXS) techniques were employed. The AuNPs act as strong scattering centers and diffract x-rays based on the type of crystalline organization. The Bragg diffraction theory can elucidate the crystal structure of the DNA-AuNP assemblies. DNA is a particularly powerful tool to direct nanoparticle assembly, as it allows a single material, from the standpoint of the inorganic core, to be assembled into multiple structures guided solely by DNA sequence. By adjusting the DNA length, the overall crystal lattice spacing can be tuned without changing the size of the inorganic core particles.

12.3.3 Assembly Using Asymmetrically Modified Gold Nanoparticles

The AuNPs described thus far have been isotropically modified with DNA. To obtain greater control over the AuNP assemblies formed, strategies have been developed to

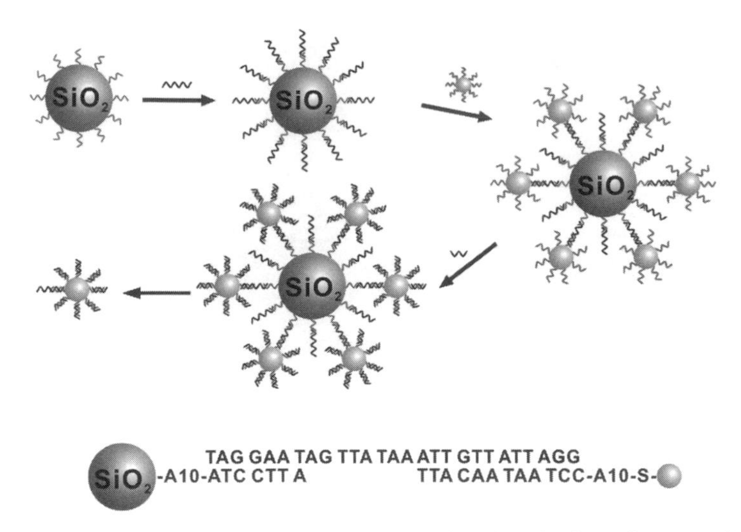

Figure 12.4 Synthetic scheme for the asymmetric functionalization of nanoparticles with DNA. (Adapted from F. Huo et al., *Adv. Mater.* **2006**, 18, 2304–2306.)

build in directionality such that more complicated architectures can be formed. Asymmetrically modified particles have been designed where only one side of the AuNP is modified with a particular sequence (46, 47). These strategies involve the hybridization of DNA-AuNPs to DNA that is attached to a flat surface, thereby blocking one face of the AuNP. The remaining DNA that is not blocked by the flat surface can then hybridize to a particular DNA sequence. Using thermal control, the DNA-AuNPs can then be released from the flat surface, without disrupting the previously annealed sequences, and the "empty" face can be modified with a different DNA sequence. An example of this strategy is described in Figure 12.4. The hybridized sequences can then be attached to the DNA-AuNP permanently using DNA ligase, if desired. The resulting particles have been assembled into different arrangements of satellite-type structures using the built-in directionality of the particle modification and have been verified using transmission electron microscopy (TEM; Fig. 12.5).

12.3.4 Template-Based Nanoparticle Assembly

In addition to nanoparticle-directed assembly, a variety of template materials have been used to guide the assembly of DNA-AuNPs. These materials have included biological materials such as DNA motifs, diatoms, and fungi, as described next. In addition to using DNA as an interconnecting molecule to form three-dimensional structures, DNA has also been used as a template to guide AuNP assemblies into discrete structures. Basic one-dimensional assemblies have been formed by hybridizing DNA-AuNPs to long strands of DNA, either synthetic or synthesized using rolling-circle DNA polymerization (7, 48). More complex structures have been created, such as triangles and rectangles using single-stranded cyclic DNA templates with rigid organic vertices (49). In these examples, DNA-AuNPs are hybridized to the different arms of the cyclic DNA template which anchor the AuNPs at each vertex

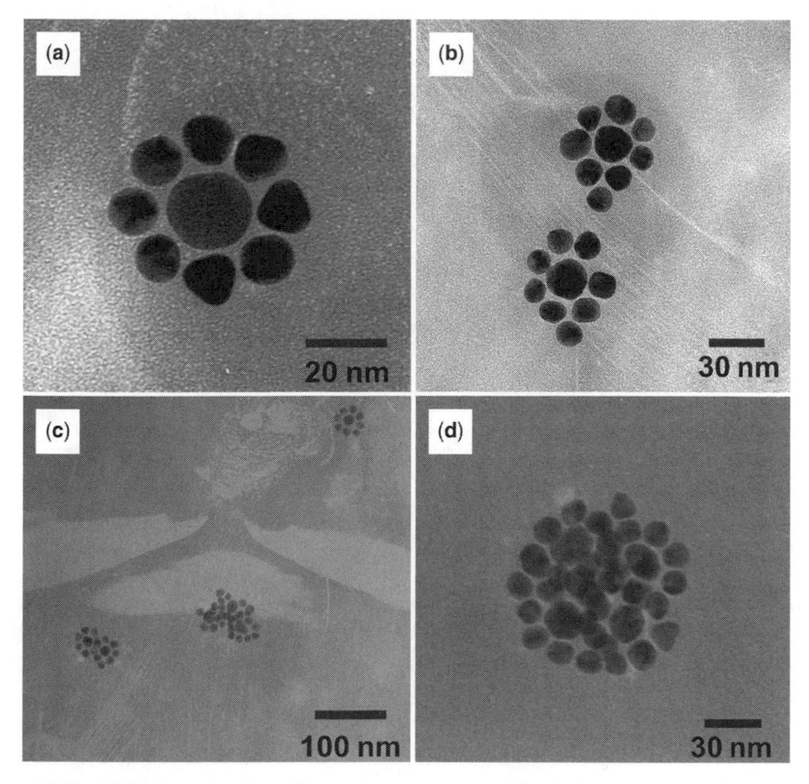

Figure 12.5 TEM images of satellite structures composed of 13 nm asymmetrically functionalized nanoparticles and 20 nm symmetrically functionalized gold nanoparticles. (a) and (b) show the separate satellite structures. (c) and (d) show some clumped satellite structures. (Reproduced with permission from F. Huo et al., *Adv. Mater.* **2006**, 18, 2304–2306.)

point (Fig. 12.6). This assembly process can be modulated to give different shaped structures where the AuNP size and DNA length can be tuned.

Further complexity can be added to these structures by using larger DNA motifs, or DNA tiles. The first example of an artificially constructed assembly composed entirely of DNA was first demonstrated in 1991 (50). Since the initial demonstration, many different shapes and sizes of DNA assemblies have been developed (51). The precisely controlled nature of these assemblies has led to them being used as ideal templates for the assembly of DNA-AuNPs. By designing DNA-AuNPs with sequences complementary to regions in a DNA tile, AuNPs can be assembled over large regions in an ordered and predetermined pattern (Fig. 12.7; Reference 52). Several other structures have been created using these techniques.

In addition to DNA, microorganisms provide a rich source of intricate structures to be used as templates for higher-ordered assembly of nanomaterials. In particular, diatoms have been used to assemble DNA-AuNPs into unique and complicated structures (53). Diatoms are unicellular algae classified by the unique shapes and ornate structural features of their silica walls. To use diatoms as templates, the organic material was digested, leaving only the intricate silica shell. This was followed by

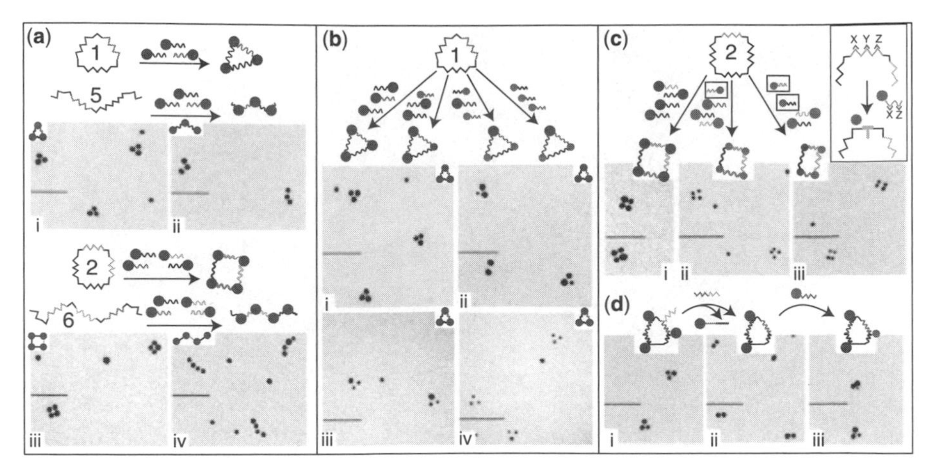

Figure 12.6 (a) **1** and **2** organize Au particles into triangles and squares; **5** and **6** result in open linear assemblies of three and four particles. (b) **1** generates triangles of (i) three large (15 nm, red), (ii) two large/one small (5 nm, purple), (iii) one large/two small, and (iv) three small particles. (c) **2** assembles four Au particles into (i) squares (15 nm particles), (ii) trapezoids, and (iii) rectangles (5 nm). Inset: use of a loop shortens the template's arm. (d) Write/erase function with **1** by (i) writing three Au particles (15 nm) into triangles, (ii) removal of a specific particle using an eraser strand, and (iii) rewriting with a 5 nm particle. Bar is 50 nm. (Reproduced with permission from F. A. Aldaye and H. F. Sleiman, *J. Am. Chem. Soc.* **2007**, 129, 4130–4131.) (See color insert.)

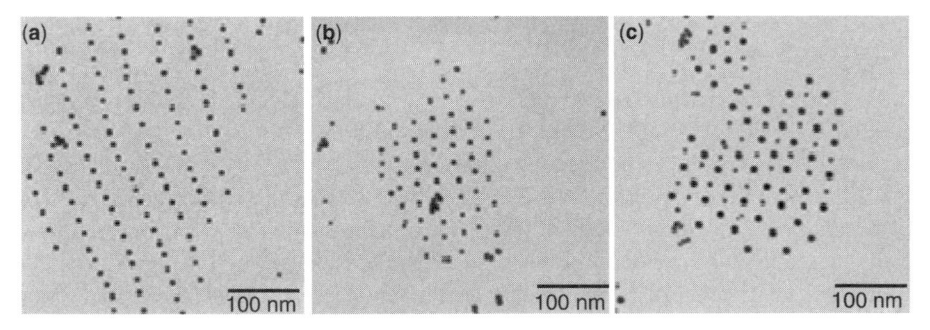

Figure 12.7 Transmission electron micrographs of 2D arrays of organized gold nanoparticles. (a) An array where one tile contains 5 nm particles. It is clear that this arrangement results in one short distance and one long distance. Sometimes a particle is missing. (b) An array where both tiles contain 5 nm particles. The distances between particles are seen to be equal here. (c) An array where one tile contains a 5 nm particle and the other tile contains a 10 nm particle. The alternation of 5 nm particles and 10 nm particles is evident from this image. Note that the spacings are precise in both directions and that the pattern mimics the rhombic pattern of the tile array. (Reproduced with permission from J. Zheng et al., *Nano Lett.* **2006**, 6, 1502–1504.)

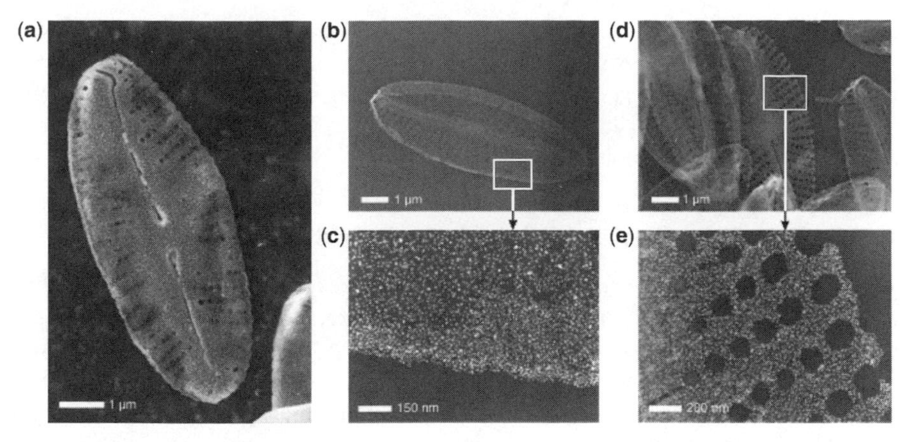

Figure 12.8 Electron microscope images of DNA-functionalized *Navicula* diatoms densely coated with one layer of DNA-modified 13 nm gold particles. SEM (a) and TEM images (b, c) both reveal a dense outer coating of nanoparticles on the *Navicula* surfact. The homogeneous nanoparticle coating of the highly symmetric and nanoscopically detailed *Navicula* template is further illustrated in (d) and (e), which are different magnifications of a single frustule. (Reproduced with permission from N. L. Rosi et al., *Angew. Chem. Int. Ed.* **2004**, 43, 5500–5503.)

modification of the silica surface with single-stranded DNA and the hybridization of complementary DNA-AuNPs. This method resulted in the assembly of DNA-AuNPs into structures portraying the initial diatom template (Fig. 12.8).

Using diatoms as templates requires the destruction of the organic portion of the algae, essentially rendering it an inorganic template. However, fungi have been used as living templates, without such destructive procedures, for the assembly of DNA-AuNPs (54). Upon addition of fungi spores to DNA-AuNPs, the particles begin to adhere to the surface of the fungi, forming a coating around the tubular fungi structure. Multiple layers of particles can be built up by addition of complementary DNA-AuNPs. The presence of the DNA-AuNPs does not inhibit fungi growth. As the fungi continue to grow and reproduce, new tubular extensions form and are available for the addition of new DNA-AuNPs assemblies. Using these types of materials found in nature, DNA-AuNPs can be assembled into complex macro-scale structures.

12.4 DIAGNOSTIC APPLICATIONS OF DNA-MODIFIED GOLD NANOPARTICLES

DNA-AuNPs have been used widely in diagnostics due to their intense optical properties, enhanced binding properties, sharp melting transitions, and high DNA loading (55). Depending on the type of assay, different properties are more or less beneficial and the ability to tailor the DNA-AuNP synthesis for the specific assay can be critical. These materials have been used as probes in biodiagnostic assays for DNA and proteins. They have also been used for the detection of small organic molecules and metal ions. Because of the unique combination of properties exhibited by DNA-AuNPs, many of the assays using these materials have higher sensitivity and selectivity than conventional detection methods. The following sections describe the different types of assays designed using DNA-AuNPs for the detection of DNA, proteins, metal ions, and small organic molecules.

12.4.1 DNA Detection Using DNA-Modified Gold Nanoparticles

The first detection assay using DNA-AuNPs as probes was designed for the colorimetric detection of DNA target sequences (6). Here, the target of the assay is a single-stranded DNA sequence. In a typical assay, two types of probes, each functionalized with DNA sequences complementary to one half of a given target sequence, are mixed with the target strands. The DNA-AuNPs are hybridized to the target sequence, leading to particle aggregation, with a concomitant red to blue color change. The change in optical properties of DNA-AuNPs is indicative of the presence of the target DNA and the nanoparticle hybridization as discussed previously. Unlike unmodified duplex DNA, melting analyses of the hybridized DNA-AuNP aggregates show sharp melting transitions. This sharp melting transition facilitates high selectivity because even slight destabilization is visible as a change in the T_m. This allows one to differentiate even a single base mismatch in the target DNA and is translated to a variety of assay formats. The contribution of the sharp melting transition to the high selectivity will be further described through this section.

The hybridized particle aggregates can be transferred to a reverse-phase silica plate to result in a permanent and easily readable record of the particle hybridization at a given temperature. Further investigation demonstrates that one can selectively detect target DNA sequences with single base imperfections, regardless of the position, using DNA-AuNP probes in the manner of a head-to-tail or tail-to-tail hybridization scheme (56).

In addition to the DNA-AuNP-based homogeneous colorimetric assay discussed above, a number of research groups have used DNA-AuNP probes for surface-based detection schemes, often referred to as chip-based formats. One of the most commonly used chip-based methods, the scanometric DNA array detection method, relies on the light scattering from the AuNPs (57). In this work, a sandwich assay was designed where DNA capture strands were covalently immobilized on the surface of a glass slide in a microarray format. Subsequently, DNA target strands and DNA-AuNP probes were added to the chip surface for hybridization in a stepwise manner. In the scheme (Fig. 12.9), the sequences of the capture strand and the strand on the nanoparticle are each designed to be complementary to one half of the target sequence, respectively. Therefore, the AuNP probes can be immobilized on the surface using the target sequence as a linker. A subsequent silver enhancement step amplifies the light scattering from the AuNPs. The AuNPs are catalytically active and catalyze the reduction of silver ions (Ag^+) to elemental silver (Ag) in the presence of a reducing agent. During the reduction process, the size of the surface-bound nanoparticles increases and they become visible even to the naked eye. The signal readout is recorded with a conventional flatbed scanner or a light scattering measurement

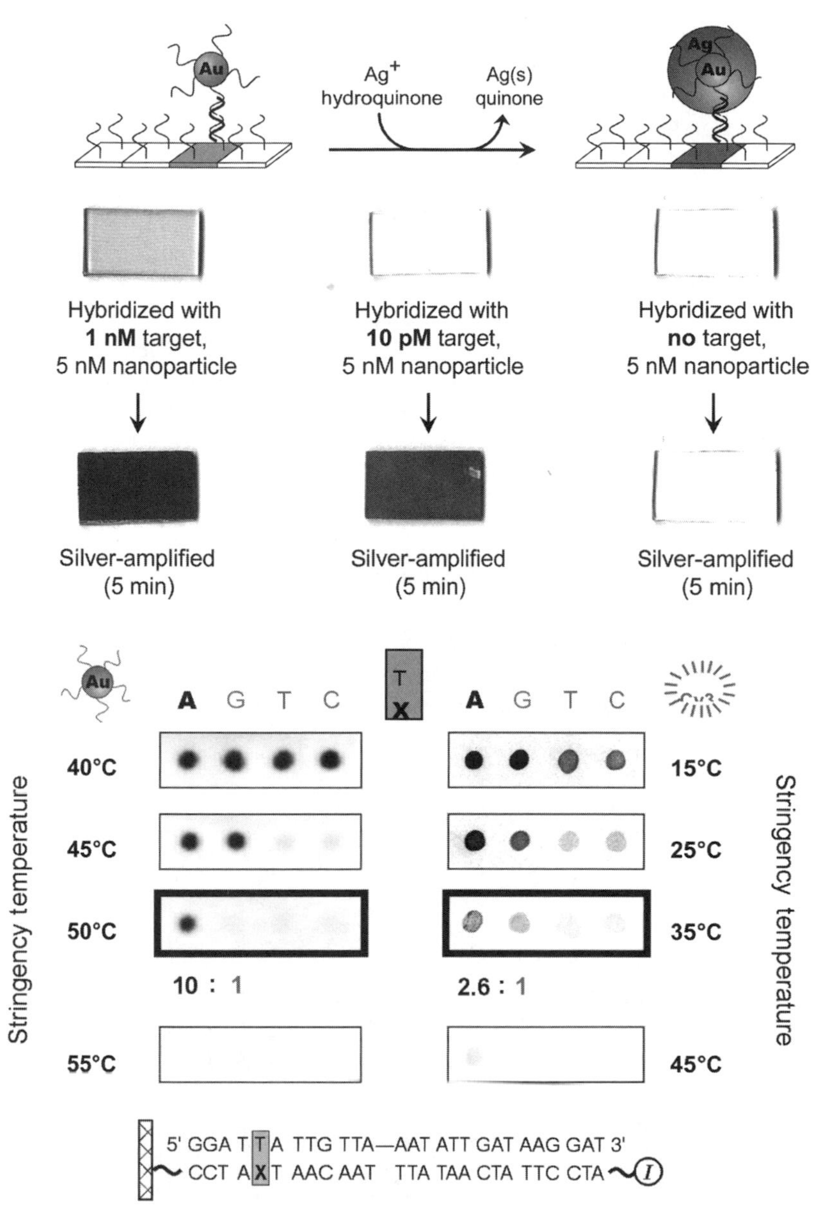

Figure 12.9 Scanometric chip-based DNA detection using silver amplification of the detection signal is illustrated on the left. A direct SNP detection/discrimination comparison of DNA-AuNP probe versus fluorophore-labeled probes is shown on the right. The DNA-AuNP system shows a selective advantage of approximately 4:1. (Reproduced with permission from T. A. Taton et al., *Science* **2000**, 289, 1757–1760. Copyright 2000 American Association for the Advancement of Science.) (See color insert.)

instrument. In some cases the light scattering from the AuNPs is intense enough to eliminate the need for the silver amplification procedure.

Using DNA-AuNP probes compared to standard fluorescence-based probes is two orders of magnitude more sensitive when combined with stringency washes at an elevated temperature. A stringent condition is one under which the majority of the perfectly complementary sequences remain hybridized, while nonspecific and mismatched sequences are washed away. In addition, it was demonstrated that single base imperfections could be detected with increased selectivity (factor of 4) and sensitivity [10,000 times, limit of detection (LOD) = ~50 fM] when compared to the corresponding fluorophore labeling methods (Fig. 12.10).

An alternative to the light scattering signal readout system has been designed using Raman labels coupled with surface-enhanced Raman scattering (SERS) (58). In this detection scheme, DNA-AuNP probes are synthesized using thiol-modified DNA which is also labeled with a Raman-active dye. Because of the unique Raman spectroscopic fingerprint of Raman labels, simultaneously multiplexed detection can be easily facilitated. These materials have been used for both DNA and RNA detection in a chip-based format. Similar to the scanometric detection described above, the signal amplification by the silver enhancement step also amplifies the Raman signal of DNA-AuNPs probes.

For the multiplexing purpose, one can design a large number of DNA-AuNP probes functionalized with Raman dyes as signal reporting groups with unique Raman spectra. The assay typically begins with spotting different capture strands in

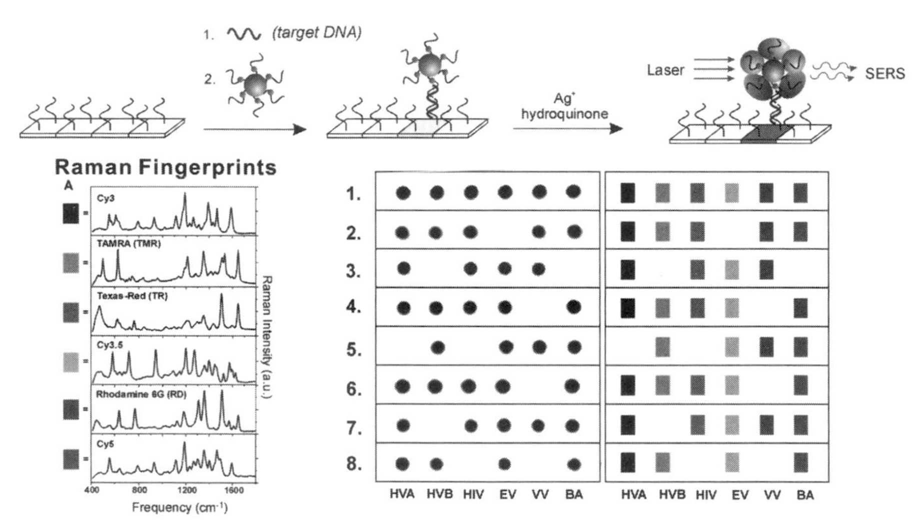

Figure 12.10 Raman-encoded multiplexed detection of six DNA targets using DNA-AuNP, where each Raman spectrum, or color, corresponds to a specific target of interest. Note that after silver staining, all the spots turn black, which makes it impossible to see the difference of the targets. By using SERS, however, the spots can be scanned with a single wavelength excitation laser and observed with dye (and thus, target)-specific Raman spectra. (Reproduced with permission from C. Y. W. Cao et al., *Science* **2002**, 297, 1536–1540. Copyright 2002 American Association for the Advancement of Science.) (See color insert.)

a microarray format to capture one or more DNA target strands in a sample. DNA-AuNPs with unique Raman signatures are hybridized to the appropriate spots after the corresponding target strand has been captured. Silver enhancement, followed by Raman spectroscopic analysis, allows one to quickly analyze which strands are present in the sample in a highly sensitive and selective manner (LOD = 20 fM). In addition, semiquantitative analysis from the correlation of the Raman signal intensity and the relative amount of target present is also possible. As a proof-of-concept, this detection scheme demonstrated simultaneous detection of six different DNA targets as well as two RNA targets with single base mismatches. The detection scheme can be carried out with single-source laser excitation in which photo-bleaching is not a significant problem. Photo-bleaching, however, can lead to a significant loss of signal intensity with other fluorophore labels.

Electrical detection of DNA in a surface-based manner using DNA-AuNP probes was also demonstrated (Fig. 12.11; Reference 59). A device composed of two microelectrodes was designed on a silicon wafer with a thin layer of SiO_2 coating. The capture DNA sequence was immobilized on the exposed SiO_2 surface between the electrodes, and used to sandwich a target sequence with DNA-AuNP probes in a similar manner as the scheme of the scanometric DNA assay described above. After silver development, the enlarged particles fill the gap, generating a measurable electronic signal. Enhanced target selectivity and sensitivity (500 fM) was reported following a salt-stringency wash. This sensitivity is significantly higher compared to conventional molecular fluorescence-based assays.

In addition to the plasmon resonance displayed by AuNPs, gold films also exhibit a surface plasmon resonance (SPR) that fluctuates in response to surface-bound molecules. Using SPR, surface-sensitive analytical techniques have been designed based on the ability of SPR to detect changes in the dielectric constant induced by molecular adsorption onto the metal surface. SPR has been used for the detection of a variety of

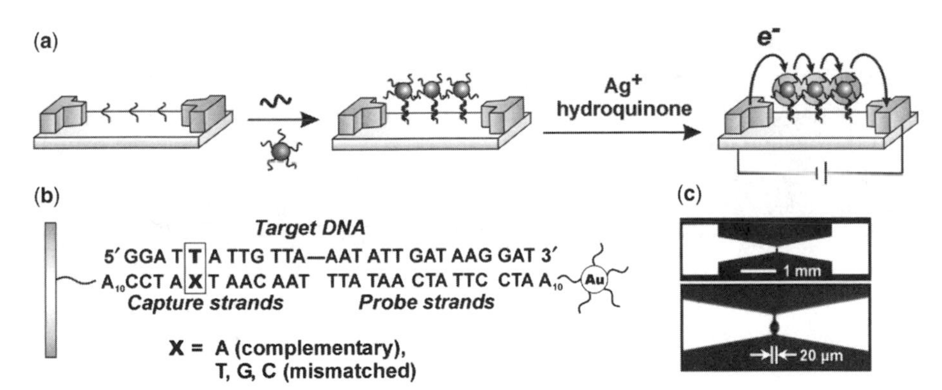

Figure 12.11 Chip-based electrical detection of target DNA strands using DNA-AuNP probes. Immobilized capture strands on the chip are hybridized with the target strands and DNA-AuNP probes followed by silver staining enhancement, which changes the electric conductance for target detection. (Reproduced with permission from S.-J. Park et al., *Science* **2002**, 295, 1503–1506. Copyright 2002 American Association for the Advancement of Science.)

biological molecules for years due to its schematic simplicity. However, this technique generally has poor sensitivity and is not utilized in assays that require high sensitivity. Keating and coworkers (60) designed an assay using DNA-AuNP probes to amplify changes in the SPR of gold films. Similar to the scanometric assay, capture DNA was patterned onto a gold film. A sandwich assay was then performed on the gold surface using a target DNA sequence to immobilize DNA-AuNPs. This led to three orders of magnitude better sensitivity compared to SPR assays without AuNPs (LOD = 10 pM). The amplification is due to three factors: (1) greatly increased surface mass, (2) high dielectric constant of AuNPs, and (3) electromagnetic coupling between AuNPs and the Au film.

Since the initial development of DNA-AuNPs, the technology of nanoparticle-based DNA detection methods has evolved rapidly, improving the selectivity and sensitivity of the assays tremendously. Recently, an ultrasensitive DNA detection method called the biobarcode assay was developed for DNA detection using DNA-AuNPs (61). This assay relies on magnetic microparticles (MMPs) functionalized with a capture sequence and DNA-AuNPs. In this assay, the DNA on the AuNP surface serves dual purposes as both the binding sequence for the target DNA and as barcodes for amplification and signal readout. The MMPs and the DNA-AuNPs capture the target by forming a sandwich structure where the target DNA is the linker (Fig. 12.12). This complex can be separated using the magnetic properties of the MMPs and all unbound DNA-AuNPs and impurities can be washed away. Finally the DNA, or barcodes, on the AuNPs can be released using dithiothreitol (DTT), a common disulfide reducing agent (62). The release of the barcodes relies on the ligand-exchange process between the barcodes and the DTT which is induced by the excess amount of DTT. The released biobarcodes are then detected using the scanometric assay. Because each AuNP can be loaded with hundreds of barcodes, this serves as an amplification process where each initial target DNA is represented by hundreds of barcodes.

The biobarcode assay exhibits very low LOD (7 aM, or 500 zM depending on the conditions and assay schemes studied) which can rival that of polymerase chain reaction (PCR). This high sensitivity is based on three aspects of the assay: (1) Unlike the flat surface-based ELISAs, this pseudo-homogeneous nanoparticle-based assay can push the binding equilibrium towards the hybridized state by increasing the particle probe concentration. (2) Each binding event of the target is amplified with hundreds of biobarcodes. (3) The physical amplification in step (2) is further chemically amplified by the silver enhancement in the scanometric procedure. This assay was further developed to incorporate multiplexed detection of four DNA targets (63) and the detection of genomic DNA extracted directly from bacteria (64). The features of this biobarcode assay can also be utilized for the untrasensitive detection of proteins, which is discussed in the following section.

12.4.2 Protein Detection Using DNA-Modified Gold Nanoparticles

DNA-AuNPs have been used to design powerful detection systems for DNA targets in a variety of assay formats based on their unique chemical and physical properties.

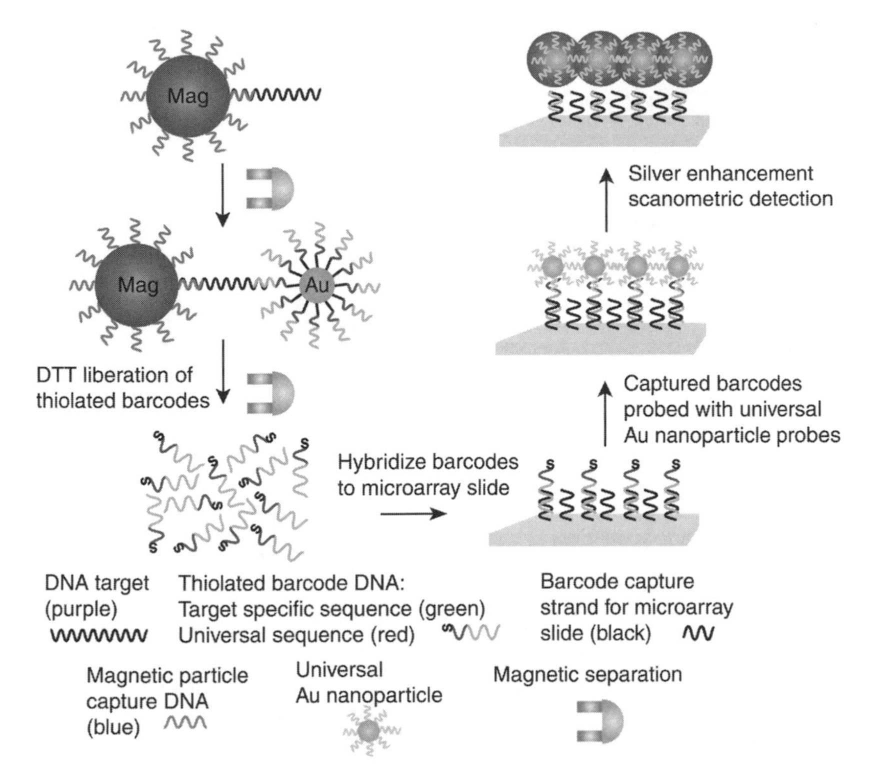

Figure 12.12 Biobarcode assay for DNA detection. Magnetic microparticles (Mag) hybridize to specific targets. AuNPs functionalized with target-specific barcode strands are then added. After magnetic separation, the barcodes are released from the surface of the AuNPs by DTT. Subsequently the barcodes are detected using the scanometric method. (Reproduced with permission from C. S. Thaxton et al., *Anal. Chem.* **2005**, 77, 8174–8178. Copyright 2005 American Chemical Society.) (See color insert.)

These materials, when modified with the appropriate functionalities, can also be utilized for the detection of protein targets. In fact, AuNPs conjugated with proteins were developed decades ago and have been used for diagnostic purposes in histology. When DNA-AuNPs are used for protein detection, however, they have inherent advantages over the conventional protein-conjugated AuNPs. Analogous to DNA targets that can act as a linker in sandwich assays, protein targets can also link DNA-AuNP probes together to facilitate detection. This linkage is based on the biological recognition and binding properties of the protein targets and their recognition moieties.

The first protein assays based on DNA-AuNP probes were designed for the detection of anti-biotin and anti-dinitrophenol antibodies such as immunoglobulin E (IgE) or immunoglobulin G1 (IgG1) (65). In these assays, either biotin or dinitrophenol was used as the recognition element for binding the target. Using a third DNA strand called a barcode, DNA-AuNPs were hybridized to a DNA sequence whose terminal end was modified with the target protein's recognition moieties (Fig. 12.13). In this assay, the

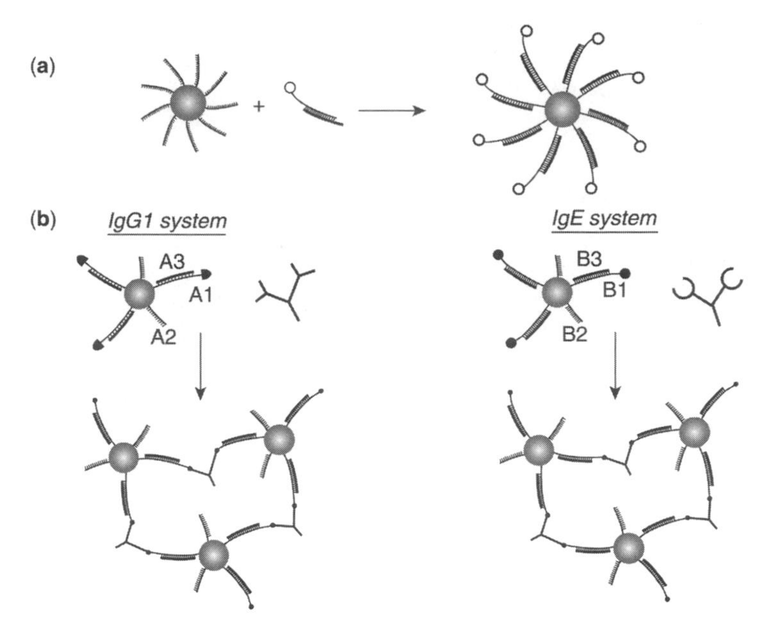

Figure 12.13 A DNA/Au nanoparticle-based protein detection scheme. (a) Preparation of hapten-modified nanoparticle probes. (b) Protein detection using protein binding probes. Notice that there are nine GC pairs in sequence A, and there are only two GC pairs in sequence B. (Reproduced with permission from J.-M. Nam et al., *J. Am. Chem. Soc.* **2002**, 124, 3820–3821. Copyright 2002 American Chemical Society.)

biobarcodes are designed to represent the protein target and act as a surrogate reporting agent. Ideally, an indefinite number of biobarcodes can be produced by designing unique sequences of DNA, which enables one to identify multiple target proteins simultaneously. In the presence of the antibody targets, the DNA-AuNP probes aggregate due to the formation of polymeric network structures. To increase the sensitivity of the assay the barcodes can be isolated and detected using a chip-based DNA detection scheme. The barcodes can be isolated by centrifuging the DNA-AuNP aggregates, collecting the precipitates, and dispersing the barcodes in solution.

While this prototype assay was relatively simple, the LOD was only 20 nM, still far above that of a conventional ELISA method. To overcome the limitations of the prototype protein assay, a new generation of the biobarcode assay for protein detection was developed (66). The fundamental scheme of this assay is similar to the biobarcode assay for DNA detection discussed previously. This assay relies on MMPs functionalized with antibodies specific to the target protein. To enable protein detection and take advantage of the unique properties of DNA-AuNPs, AuNPs are coloaded with both DNA barcodes and a second type of antibody specific for the given target protein. When target protein is combined with both sets of probes, a sandwich structure is formed, similar to the biobarcode assay for DNA detection. The sandwich complex can be separated using the magnetic properties of the MMPs. Following this purification step, the biobarcodes on AuNPs can be released and used for further scanometric identification for quantitative analysis.

This assay exhibits a very low LOD (3 aM) which is six orders of magnitude more sensitive than conventional clinical methods (ELISAs) under the conditions studied for the target protein, prostate specific antigen (PSA). Similar to the biobarcode assay for DNA detection, the high sensitivity is achieved (1) through the ability to push the binding equilibrium towards the bound state using elevated concentrations of probes, (2) by amplifying the protein targets using the DNA barcodes for detection, and (3) by utilizing the catalytic properties of the AuNPs to perform a silver enhancement step in the scanometric assay. Alternatively, an additional amplification of the released biobarcodes by PCR can further improve the sensitivity by one order of magnitude.

The high sensitivity of this assay is significant as it provides the opportunity to evaluate the presence of proteins at very low levels. While nucleic acid detection at attomolar levels has been possible using PCR, this type of assay represents a detection method for proteins with PCR-like sensitivity. Providing the research community with tools for high sensitivity protein detection will benefit discoveries in clinical applications. An initial investigation utilizing this assay addressed the issue of markers for Alzheimer's disease (67). This research investigated the concentration of amyloid-β-derived diffusible ligands (ADDLs), a potential marker for Alzheimer's disease. Although ADDLs had been suspected to be a marker for Alzheimer's disease, the concentration of ADDLs in cerebrospinal fluid was too low to be detected by current techniques such as conventional ELISAs or blotting assays. The biobarcode assay, however, was able to detect clinically relevant levels of ADDLs and to demonstrate the correlation of the ADDL levels with the presence of disease. The format of the biobarcode assay for protein detection was further modified for the multiplexed detection of protein cancer makers (Fig. 12.14; Reference 68) and the detection of p24 protein that is associated with AIDS (69).

In addition to the solution-phase protein detection assays such as biobarcode assays, a surface-based protein detection method was developed by Niemeyer and Ceyhan (70). In this work, a biotin-labeled antibody was conjugated to DNA using streptavidin as a linker molecule. This conjugate was hybridized to complementary DNA-AuNPs and antibody-functionalized AuNPs. These particles are allowed to form a sandwich structure with a target protein and a second antibody immobilized on a flat surface. This was followed by a silver enhancement step catalyzed by the AuNPs which resulted in a LOD of 50 fmol (200 pM). This method is conceptually similar to the corresponding surface-based scanometric detection of DNA. Further work demonstrated a similar protein detection scheme without the silver enhancement (71). Instead, multiple layers of secondary DNA-AuNPs were used to increase signal enhancement [LOD = 0.1 fmol (2 pM)].

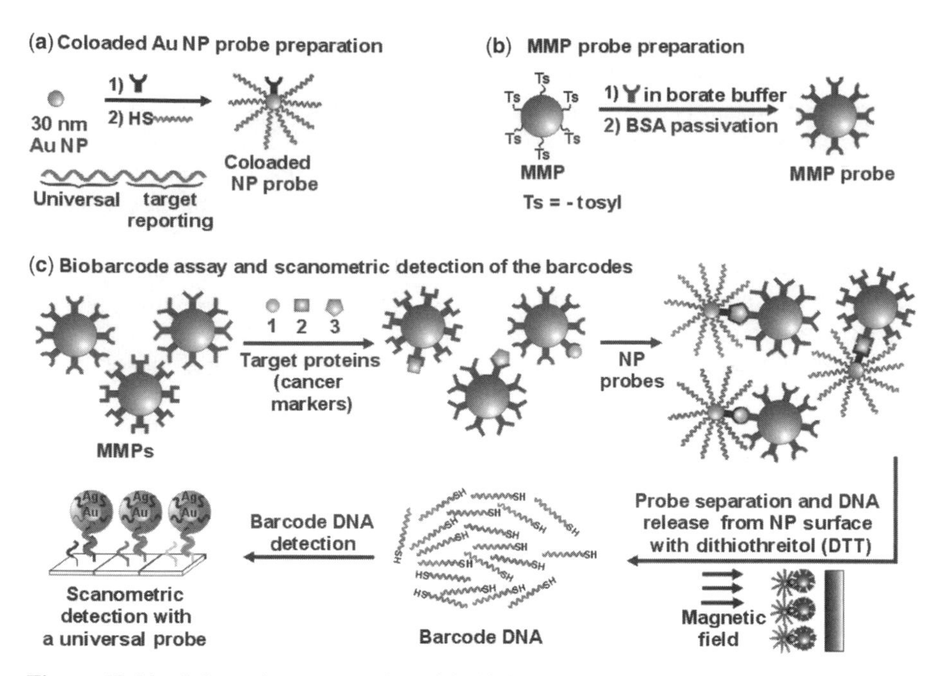

Figure 12.14 Schematic representation of the biobarcode assay for multiplexed protein detection. The general concept for multiplexed DNA detection is similar, with the appropriate recognition elements. (Reproduced with permission from S. I. Stoeva et al., *J. Am. Chem. Soc.* **2006**, 128, 8378–8379. Copyright 2006 American Chemical Society.) (See color insert.)

12.4.3 Metal Ions Detection Using DNA-Modified Gold Nanoparticles

The most commonly used method for detecting metal ions is inductively coupled plasma mass spectrometry (ICP-MS). While comprehensive in both selectivity and sensitivity, ICP-MS often requires complicated instrumentation and sampling, which are known to be the major drawbacks of the technique. To overcome such limitations, several detection methods based on nanotechnology have recently been developed for metal ions. In particular, the unique properties of DNA-AuNPs such as programmable assembly and their optical properties have made them attractive for utilization in optical sensing systems. In this section, several optical sensing systems for metal ions based on DNA-AuNPs are discussed.

In addition to the standard Watson–Crick base pairing it has recently been discovered that thymine can selectively bind to mercuric ion (Hg^{2+}) via the formation of a thymine-Hg^{2+}-thymine (T-Hg^{2+}-T) complex (72). The presence of the Hg^{2+} at the thymine-thymine mismatch leads to a significant increase (~10°C) in the T_m. Mirkin and coworkers (40) incorporated this T-Hg^{2+}-T coordination chemistry in the DNA-AuNP system to design a novel colorimetric assay for detecting Hg^{2+}(Fig. 12.15). In a typical assay, two batches of AuNPs were functionalized with two complementary sequences, respectively, and combined to form aggregates.

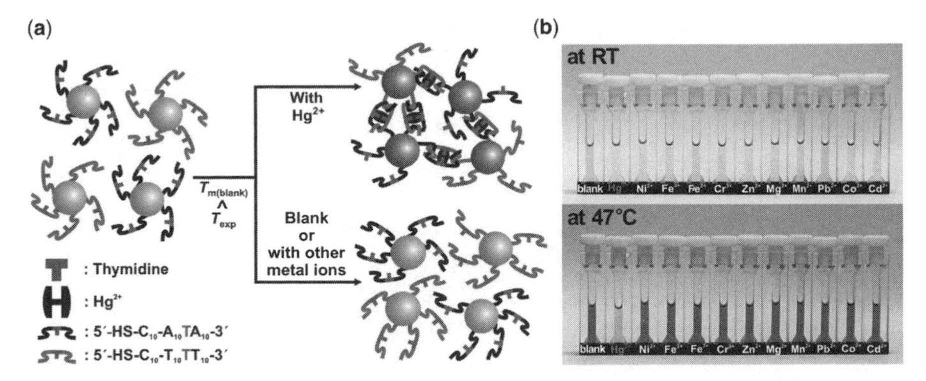

Figure 12.15 (a) Colorimetric detection of mercuric ion (Hg^{2+}) using DNA-AuNPs. (b) Color change of the aggregates in the presence of various representative metal ions (each at 1 μM) on heating from room temperature (RT) to 47°C. (Reproduced with permission from J.-S. Lee et al., *Angew. Chem. Int. Ed.* **2007**, 46, 4093–4096. Copyright 2007 John Wiley & Sons, Inc.)

The aggregation was marked by the standard red to blue color change. These sequences were designed to contain a strategically placed T-T mismatch. In the presence of Hg^{2+}, the T-T mismatch forms a complex with Hg^{2+} (T-Hg^{2+}-T), stabilizing the assembly and increasing the T_m as a function of Hg^{2+} concentration. The same sharp melting transitions are observed in the presence of Hg^{2+}. Therefore, the concentration of Hg^{2+} can be determined quantitatively by the temperature at which the color change takes place. Importantly, the color change can be read out with the naked eye or by a UV-vis spectromcter, which simplifies the assay performance and makes it attractive for on-site applications. This assay also exhibits high sensitivity (LOD = 100 nM) and selectivity. Since the development of this initial assay, it has stimulated the development of further improved Hg^{2+} sensing systems based on the optical properties of AuNPs and the selective T-Hg^{2+}-T coordination chemistry (73, 74).

Another method designed for the detection of metal ions combines DNA-AuNPs with DNAzymes. DNAzymes are a class of functional DNA molecules that are catalytically active. Importantly, the activity of DNAzymes depends on their specific cofactors such as metal ions, which provide high selectivity to the assay system. Liu and Lu (75, 76) utilized the properties of DNAzymes for the detection of lead ions (Pb^{2+}). They first designed a DNA-AuNP system in which two types of DNA-AuNPs are hybridized to a DNA-linking sequence in a head-to-tail manner (Fig. 12.16a, b, and c). The linker sequence is a substrate DNA strand that hybridizes to the DNAzyme and can be cleaved into two pieces, only in the presence of Pb^{2+}. The cleavage of the DNA strand leads to the disassembly of AuNP network and the concomitant blue to red color change. Moreover, this irreversible disassembly process can be controlled by several parameters such as particle size, salt concentration, temperature, and additional complementary DNA sequences. Optimizing these parameters has enhanced the performance of this sensing system and led to detection limits of 100 nM.

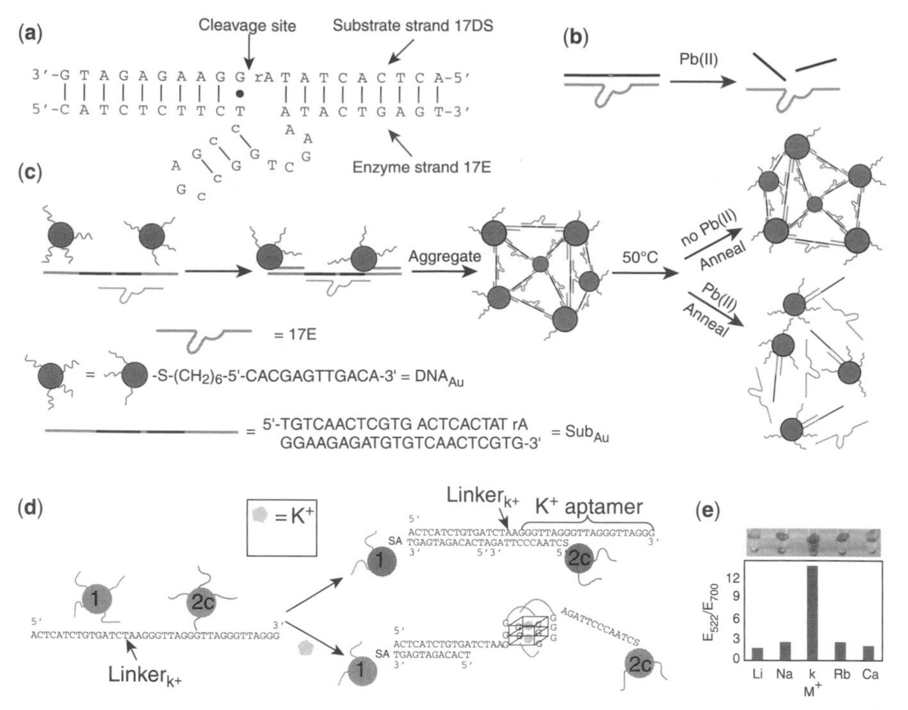

Figure 12.16 (a) Secondary structure of the DNAzyme system that consists of an enzyme strand and a substrate strand. Except for a ribonucleoside adenosine at the cleavage site (rA), all other nucleosides are deoxyribonucleosides. (b) Cleavage of substrate stand by the enzyme strand in the presence of Pb(II). (c) Schematics of DNAzyme-directed assembly of gold nanoparticles and their application as biosensors for metal ions such as Pb(II). In this system, the DNAzyme strand has been extended on both the 3′ and 5′ ends for 12 bases, which are complementary to the 12-mer DNA attached to the 13 nm Au NPs. (d) Materials responsive to K^+. Particles 1 and 2c are assembled by linker$_{K+}$ in the absence of K^+ to form aggregates. In the presence of K^+, the assembly is inhibited and the color remains red. (e) Colorimetric result in the presence of 250 mM of different monovalent metal ions. [Parts (a), (b), and (c) reproduced with permission from J. Liu et al., *J. Am. Chem. Soc.* **2003**, 125, 6642–6643. Copyright 2003 American Chemical Society. Parts (d) and (e) reproduced with permission from J. Liu et al., *Adv. Mater.* **2006**, 18, 1667–1671. Copyright 2006 John Wiley & Sons, Inc.] (See color insert.)

In contrast to the cleavage system, another detection system has used DNAzymes to facilitate DNA ligation for the detection of Cu^{2+} (77). Unlike the cleavage-based DNAzymes, the ligation-based DNAzymes can ligate two DNA sequences into one piece in the presence of Cu^{2+}, which can lead to the permanent assembly of DNA-AuNPs. The signal readout of this method is also based on an easily observed red to blue color change (LOD = $\sim$5 μM).

In addition to Hg^{2+}, Pb^{2+}, and Cu^{2+}, the detection of K^+ was demonstrated by using G-quadruplex (G-quartet) formation in DNA-AuNP systems (78). A G-quadruplex is a unique four-stranded tetrad structure of DNA whose sequence is rich in guanine. The guanine bases are arranged in a square based on Hoogsteen hydrogen bonding, and further stabilized by coordination to monovalent metal ions (typically K^+) located in the center of the tetrads (Fig. 12.16d). To employ G-quadruplex chemistry for detecting K^+, Liu and Lu (78) designed a G-rich sequence as a linker strand, each half of which was complementary to a DNA-AuNPs (Fig. 12.16d). When K^+ was introduced into the DNA-AuNP aggregate system, the linker sequence dehybridized from the particles to preferentially form a G-quadruplex, resulting in the disassembly of the particles and a blue to red color change (Fig. 12.16e). In fact, this scheme involves a similar strategy designed for the detection of adenosine and cocaine using aptamers (see below).

12.4.4 Small Organic Molecule Detection Using DNA-Modified Gold Nanoparticles

Aptamers are single-stranded oligonucleotides (DNA or RNA) that tightly bind a specific target molecule (79, 80). Aptamers are typically selected from millions of randomly generated oligonucleotide sequences via SELEX (systematic evolution of ligands by exponential enrichment; References 79 and 80). Since their discovery in the early 1990s, this emerging class of molecules has been used for a variety of therapeutic and diagnostic applications because they are relatively easy to obtain compared to other molecular recognition moieties such as monoclonal antibodies or synthetic organic or inorganic ligands. Based on these advantages, a competition assay using aptamers and DNA-AuNPs was designed to detect two small organic molecules, adenosine and cocaine, respectively (Fig. 12.17; Reference 81). In the detection system, two types of noncomplementary DNA-AuNPs were combined with a DNA-linking sequence to form aggregates. The linking strand is designed to contain the aptamer sequence that can bind to the target molecule. When the aptamer binds to the target molecule, it dehybridizes from one of the DNA-AuNPs, leading to the dissociation of nanoparticle aggregates and a blue to red color change. This system takes advantage of the competition between the target molecule and the DNA-AuNPs for binding to the aptamer sequence, which converts the binding event of the target into optical signal expression.

Another biologically relevant molecule is cysteine. Cysteine plays physiologically significant roles as the only thiol-containing essential amino acid. Based on the nonspecific interaction of cysteine and unmodified AuNPs, several colorimetric assays have been developed for cysteine detection (82, 83). These assays, however, lack selectivity and have relatively poor sensitivity. To improve the selectivity and sensitivity of cysteine detection, a colorimetric signal readout method has been developed using DNA-AuNPs (84). This work builds on the Hg^{2+} detection method mentioned earlier (see Section 12.4.3) and takes advantage of the interaction between the thiol group in cysteine and Hg^{2+} (40). Hg^{2+} is known to be highly thiophilic, forming complexes with thiol molecules with very high binding affinity and various coordination numbers between one and six (85). As previously described, the formation of T-Hg^{2+}-T complex in DNA-AuNP aggregates increases the melting temperature of the particle aggregates (40). Upon the introduction of cysteine, however, Hg^{2+} is sequestered from

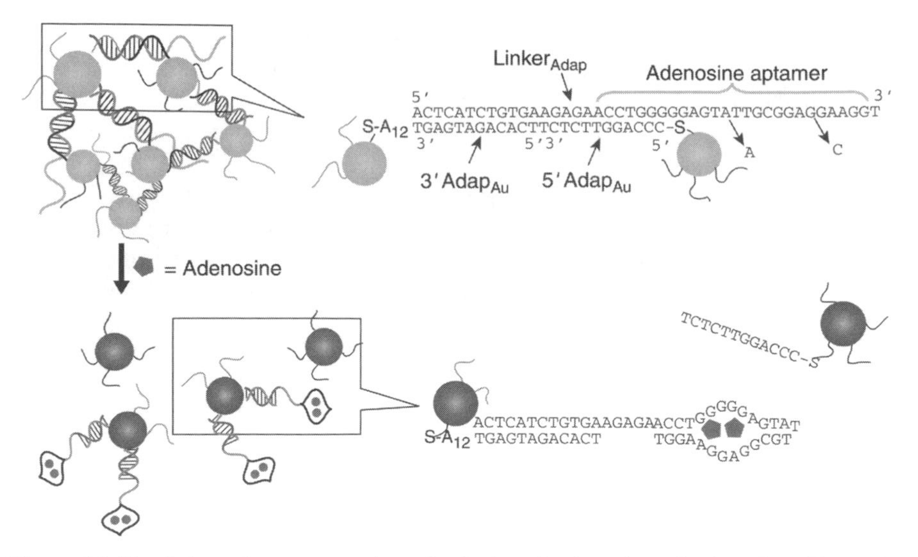

Figure 12.17 Schematic representation of colorimetric detection of adenosine. The DNA sequences are shown on the right side of the figure. The A_{12} in $3'Adap_{Au}$ denotes a 12-mer polyadenine chain. In a control experiment, a mutated linker with the two mutations shown by the two short black arrows was used. The detection of cocaine is performed in a similar manner. The drawing is not to scale. (Reproduced with permission from J. Liu et al., *Angew. Chem. Int. Ed.* **2006**, 45, 90–94. Copyright 2006 John Wiley & Sons, Inc.) (See color insert.)

the T-T mismatch site to preferentially form cysteine-Hg^{2+} complexes. This results in a decrease in the T_m as the T-T mismatch is destabilized and is dependent on the concentration of cysteine present (Fig. 12.18). This assay also takes advantage of the competition of cysteine and the T-T- mismatch for binding Hg^{2+}, generating a signal from the binding event of the target.

Another area of small molecule detection involves the efficient screening of DNA-binding molecules, which can be important for a variety of therapeutic applications. For example, a number of anticancer drugs are known to reversibly bind to duplex DNA, leading to further stabilization of the duplex structure depending on the bioactivity of the drugs. Triplex binders, known to stabilize a triple helix structure, are also of increasing relevance due to their potential application for regulating gene expression. While the development of combinatorial chemistry has successfully built up large libraries of compounds, the conventional methods for screening candidates for drug discovery are limited by the low-throughput capability of current screening technology. To address this bottleneck, DNA-AuNPs were used to develop high-throughput screening methods for detection of DNA-binding molecules (86, 87).

When DNA-binding molecules are combined with the DNA-AuNP aggregates, the DNA-binding molecules stabilize the DNA duplex interconnects between the AuNPs, leading to an increased T_m. Therefore, by monitoring the T_m of DNA-AuNP assemblies, the relative binding affinities of the binding molecules can be determined. This enables the effective screening of the DNA-binding molecules

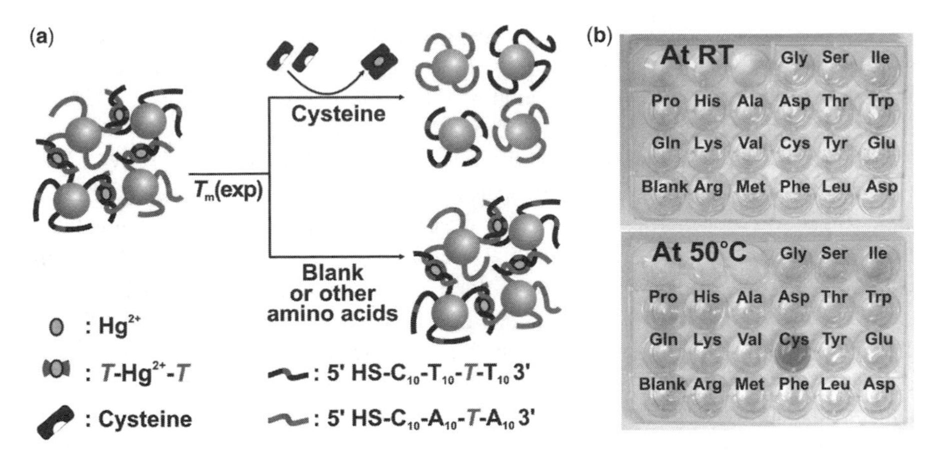

Figure 12.18 (a) Colorimetric detection of cysteine using DNA-AuNPs in a competition assay format. (b) The colorimetric response of the DNA-AuNP/Hg^{2+} complex aggregates in the presence of the various amino acids (each at 1 μM) at RT and 50°C. (Reproduced with permission from J.-S. Lee et al., *Nano Lett.* **2008**, 8, 529–533. Copyright 2008 American Chemical Society.) (See color insert.)

in a high-throughput manner. In addition to this homogeneous screening system, a surface-based heterogeneous assay for screening DNA-binding molecules using the similar chemistry was also demonstrated in a scanometric platform (88).

12.5 THERAPEUTICS: DNA-MODIFIED GOLD NANOPARTICLES AS ANTISENSE AGENTS

Antisense therapy is a method of inducing gene knockdown by inactivating the function of specific mRNA through the introduction of complementary nucleic acids (DNA, RNA, or a chemical analog). The complementary nucleic acids (antisense oligonucleotides) hybridize to target mRNA, blocking the transcription of the corresponding gene. As a powerful method of gene regulation, antisense oligonucleotides have demonstrated strong promise in therapeutic applications to treat a variety of diseases caused by genetic disorders. The history of antisense therapy goes back to the late 1970s, when Zamecnik and Stephenson first discovered the inhibition of Rous sarcoma virus production as a result of the addition of antisense oligonucleotides to the virus-infected tissue (89, 90). Since then, the interest in antisense therapy has increased dramatically, especially in the context of nanotechnology for gene regulation. In fact, many researchers have developed a variety of nanomaterials such as cationic polymers, lipids, peptides, and virus vectors to enhance the delivery and transfection of antisense oligonucleotides to cells (91). However, conventional transfection materials are often limited by a number of disadvantages, including cellular toxicity, low transfection efficiency and low stability of antisense oligonucleotides in the cytoplasm.

Mirkin and coworkers have discovered that DNA-AuNPs can be utilized as an antisense agent (92). This conceptually new strategy takes advantage of the demonstrated stability and the enhanced binding properties of DNA-AuNPs (33). Gold nanoparticles (15 nm in diameter) were functionalized with thiol-modified DNA (antisense oligonucleotide), and introduced to a variety of cell lines (C166, RAW 264.7, HeLa, NIH-3T3, and MDCK). The particles readily entered cells and exhibited gene knockdown properties (up to 20%) observed by fluorescence measurement of EGFP expression (Table 12.1). No toxicity was observed. Positive controls using commercially available transfection reagents such as Lipofectamine 2000 and Cytofectin exhibited less knockdown efficiency and greater toxicity under identical conditions. The high knockdown efficiency is believed to arise from the cooperative binding properties of DNA-AuNPs. It was demonstrated that more densely functionalized DNA-AuNPs provide greater knockdown than DNA-AuNPs with lower DNA density due to the enhanced binding properties. In addition, the densely packed DNA on the AuNP surface increases the *in vitro* resistance to enzymatic degradation compared to naked DNA strands, resulting in an increased lifetime of the antisense DNA.

TABLE 12.1 Performance characteristics of Antisense Nanoparticles

Antisense Experiment[a]	Observed Toxicity	Approximate Binding Constant	% Decrease in EGFP Expression
Antisense particles A (0.024 nmol particles, 1.08 nmol DNA)	No	7.1×10^{20}	11 ± 2
Antisense particles A (0.048 nmol particles, 2.16 nmol DNA)	No	7.1×10^{20}	14 ± 0.4
Antisense particles B (0.024 nmol particles, 2.64 nmol DNA)	No	2.6×10^{22}	14 ± 1
Antisense particles B (0.048 nmol particles, 5.28 nmol DNA)	No	2.6×10^{22}	20 ± 4
Nonsense particles A (0.048 nmol particles)	No	N/A	0 ± 3
Nonsense particles B (0.048 nmol particles)	No	N/A	0 ± 2
Lipofectamine 2000 (0.024 nmol DNA)	No	6.7×10^{20}	6 ± 0.2
Lipofectamine 2000 (2.64 nmol DNA)	Yes	6.7×10^{20}	N/A
Cytofectin (0.024 nmol DNA)	No	6.7×10^{20}	7 ± 0.7
Cytofectin (2.64 nmol DNA)	Yes	6.7×10^{20}	N/A

[a]Antisense particle A and nonsense particle A are functionalized with two cyclic disulfides and antisense particle B and nonsense particle B with an alkyl-thiol anchoring group.

Source: Reproduced with permission from Rosi et al., *Science* **2006**, 312, 1027–1030. Copyright 2006 American Association for the Advancement of Science.

Further investigations reveal a dependence of cellular uptake of DNA-AuNPs on the density of the DNA loading on the AuNP surface (93). Their study began from the following fundamental question: How can the negatively charged DNA-AuNPs enter the cells? In general, conventional transfection agents such as cationic polymers and lipids are positively charged to enhance the interaction with the cellular membrane. The positively charged materials also associate with the negatively charged DNA to facilitate delivery. Therefore, the current explanations of how positively charged materials facilitate transfection do not apply to the negatively charged DNA-AuNPs. To elucidate the cellular uptake mechanism of DNA-AuNPs, the DNA density on the AuNP surface was controlled using oligo (ethylene glycol) as a diluent. Higher numbers of DNA on the AuNP surface correlated with increased cellular uptake. It was proposed that serum proteins in media bind to DNA-AuNPs which leads to a change in the physical properties of the nanoparticle agents. This hypothesis was further supported by the characterization of physical properties of DNA-AuNPs such as the diameter and the surface potential of the particles before and after the exposure to media. Moreover, the quantitative analyses of absorbed proteins per nanoparticle as a function of DNA density indicate their proportional relationship.

The therapeutic application of DNA-AuNPs as antisense agents has led to the development of another class of nucleic acid-nanoparticle conjugates, or locked nucleic acid-gold nanoparticle conjugates (LNA-AuNPs) (94). A locked nucleic acid is a modified RNA where the $2'$ and $4'$ carbons are connected by an extra bridge. The structural confirmation of LNA facilitates the stacking of bases and the organization of the phosphate backbone, resulting in a significant increase in thermal stability of the duplex. In fact, the LNA-AuNPs can be synthesized in a similar manner as DNA-AuNPs based on the thiol-gold interactions. These materials also exhibit high cooperativity, stability, and dense loading of LNA. As expected, the T_m of LNA-AuNPs hybridized with a complementary DNA sequence is increased significantly ($\Delta T_m = \sim 15°C$) compared to the corresponding DNA-AuNPs. This indicates that these materials exhibit even higher binding properties than the analogous DNA-AuNPs. Based on their higher binding properties, downregulation of protein expression in the A549 cell line using antisense LNA-AuNPs was demonstrated, exhibiting higher gene knockdown efficiency (50%) than corresponding antisense DNA-AuNPs (30%) under the conditions studied. This research demonstrates the potential antisense therapeutic applications of nucleic acid-nanoparticle conjugates by synthesizing new materials and controlling their properties.

12.6 DNA-MODIFIED SILVER NANOPARTICLES

The advances in synthesis, characterization, and utilization of DNA-AuNPs have tempted researchers to synthesize DNA-modified silver nanoparticles (DNA-AgNPs). Silver nanoparticles are attractive due to their surface plasmon resonance ($\lambda_{max} = \sim 410$ nm), catalytic activities, high extinction coefficient, and Raman enhancing properties. These properties, when combined with the chemical and physical characteristics derived from the dense DNA loading, are expected to make

DNA-AgNPs a new and potentially useful material in a variety of fields of nanoscience. The extensive research into the chemistry of metal surfaces and thiol coordination is expected to translate to a similar strategy for the synthesis of DNA-AgNPs. However, there are currently only a few examples of DNA-AgNPs reported to date (95, 96). The synthesis of DNA-AgNPs has been limited due to the highly reactive nature of silver compared to gold and the weaker association of thiols to silver. When DNA-AgNPs are synthesized using the standard procedures used for AuNPs, the DNA-AgNPs show reduced yields, reduced salt stability, require a lengthy salt aging procedure of many days, and/or do not exhibit cooperative melting properties (95–97). In certain cases, the DNA sequences are limited to specific bases in order to achieve stability (95, 96). These limitations are believed to be caused by the

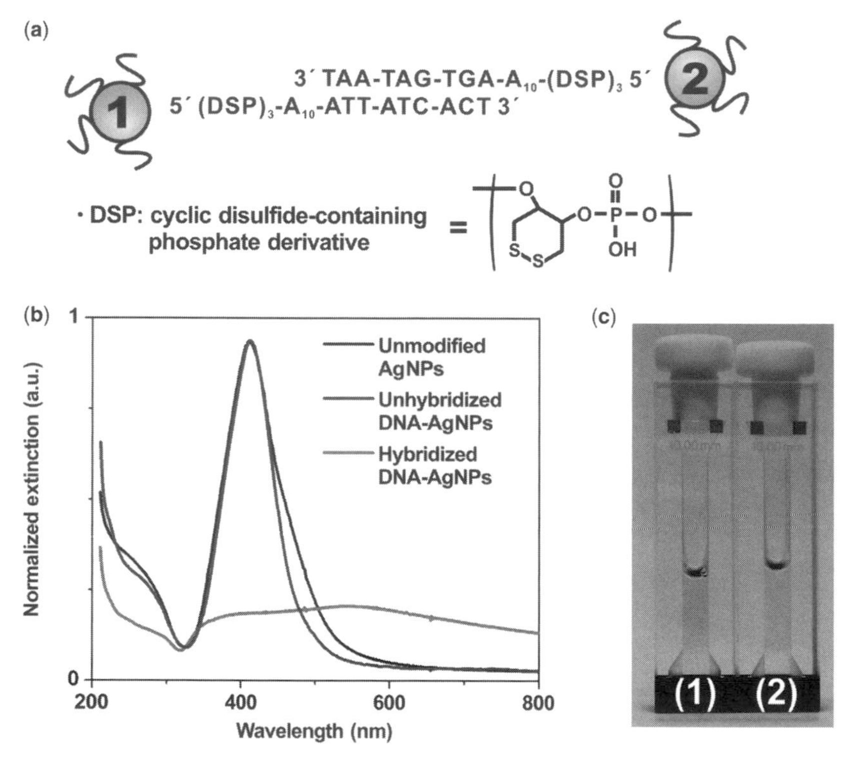

Figure 12.19 (a) Schematic illustration of the hybridization of two complementary DNA-AgNPs. (b) UV-vis spectra of unmodified AgNPs (black line), unhybridized DNA-AgNPs (blue line), and hybridized DNA-AgNPs (red line). Note that the wavelength at which the maximum of the extinction of AgNPs is obtained remains the same after DNA functionalization. After hybridization, however, the band of DNA-AgNPs broadens and red-shifts significantly from 410 nm to 560 nm. (c) Colorimetric change responsible for the assembly process of DNA-AgNPs. The intense yellow color of the unhybridized AgNPs (1) turns to pale red (2) as the particle aggregation proceeds. Heating of (2), however, results in the return of the solution color to yellow (1). (Reproduced with permission J.-S. Lee et al., *Nano Lett.* **2007**, 7, 2112–2115. Copyright 2007 American Chemical Society.) (See color insert.)

chemical degradation of the silver nanoparticles and the susceptibility of the silver surface to oxidation, leading to the weak interaction between the silver surface and the thiol functionality of the DNA strands (98, 99).

This problem has been circumvented by introducing a specially designed, but easily producible, anchoring group to the end of the DNA sequence (100). This anchoring group is composed of three consecutive cyclic disulfide moieties, as a means of strengthening the binding interactions between the anchoring group and the AgNP surface. The design of the anchoring group is inspired by two fundamental concepts: [1] polydentate ligands often form more stable metal-ligand complexes than monodentates, and [2] cyclic disulfide groups bind to metal surfaces stronger than monothiol or acyclic disulfide groups (14, 15). This method allows one to synthesize DNA-AgNPs in a similar manner as DNA-AuNPs. The resulting DNA-AgNPs exhibit very high salt stability, thermal stability, high cooperative melting properties, and reversible assembly and disassembly properties with corresponding optical property changes (Fig. 12.19). These properties of DNA-AgNPs mirror the properties of DNA-AuNPs and indicate that these new materials could be potentially used for a variety of diagnostic and therapeutic applications.

12.7 CONCLUSION

Gold nanoparticles can be modified with synthetic DNA via a terminal thiol-anchoring group. The resulting DNA-AuNP conjugates exhibit programmable DNA recognition properties and the optical properties of the AuNPs. The DNA-AuNPs also display new properties not exhibited by either DNA or gold nanoparticles separately. These properties include greater AuNP stability, enhanced DNA binding properties, sharp DNA melting transitions, and increased resistance to DNA degradation by nucleases.

Based on the properties of these conjugate materials, DNA-AuNPs have been utilized in several different areas, including self-assembly, biodiagnostics, and therapeutics. The sequence-specific recognition properties of DNA-AuNPs have allowed these materials to be assembled into a variety of different architectures. These assemblies can be precisely controlled by considering the design of the DNA-AuNP conjugate and tuning the resulting properties. These assemblies can range from only a few nanoparticles to large assemblies spanning microns in size. These structures have included precisely arranged geometric shapes with DNA-AuNPs at the vertices and micron-sized colloidal crystals where the DNA-AuNPs are organized into defined crystal structures. However, the ability to assemble DNA-AuNPs into highly defined crystal structures has only recently been demonstrated. As research progresses, the control over these processes should continue to develop and we may begin to see these materials become used in platforms that take advantage of the crystalline structures. As methods to synthesize asymmetrically modified particles become more prevalent, this synthetic development should also increase the complexity of the types of structures that can be assembled.

The unique properties exhibited by DNA-AuNPs have allowed them to be used in the design of several different diagnostic systems, including the detection of DNA, proteins, metal ions, and small organic molecules. These diagnostic systems often exhibit increased sensitivity and selectivity compared to conventional detection systems. The enhanced selectivity stems from the sharp melting transition observed for DNA-AuNP assemblies and the programmable recognition properties of DNA. Depending on the type of assay, the increased sensitivity generally arises from the enhanced binding properties compared to unmodified duplex DNA or the high DNA loading on the AuNP surface. In the case of the colorimetric detection, the assays are relatively simple to perform and can be easily read by observing a color change or measuring the change in the absorbance of the sample. These high sensitivity detection methods are particularly exciting for protein detection which is not amenable to PCR. The ability to detect proteins and antibodies at very high sensitivities may open new possibilities of diagnosing protein disease markers that only exist in the body at very low concentrations.

The properties of DNA-AuNPs have also proven useful in nanotherapeutics as antisense materials. These materials can easily transfect a variety of cell types without causing toxicity. Due to the enhanced binding properties of DNA-AuNPs, they act as efficient scavengers of mRNA and induce knockdown of a target gene via an antisense therapeutic mechanism *in vitro*. The future direction and progress of these materials as therapeutics will be highly dependent on how these materials are processed *in vivo* and if these materials can be eliminated from the body.

FURTHER READING

Feldkamp, U. and Niemeyer, C. M., Rational design of DNA nanoarchitectures. *Angew. Chem. Int. Ed.* **2006**, 45, 1856–1876.

He, L.; Musick, M. D.; Nicewarner, S. R.; Salinas, F. G.; Benkovic, S. J.; Natan, M. J.; Keating, C. D., DNA detection using DNA-AuNPs. *J. Am. Chem. Soc.* **2000**, 122, 9071–9077.

Pavlov, V.; Xiao, Y.; Shlyahovsky, B.; Willner, I., Small organic detection using DNA-AuNPs. *J. Am. Chem. Soc.* **2004**, 126, 11768–11769.

Rosi, N. L. and Mirkin, C. A., Nanostructures in biodiagnostics. *Chem. Rev.* **2005**, 105, 1547–1562.

Seferos, D. S.; Giljohann, D. A.; Hill, H. D.; Prigodich, A. E.; Mirkin, C. A., mRNA detection using DNA-AuNPs. *J. Am. Chem. Soc.* **2007**, 129, 15477–15479.

REFERENCES

1. Shenhar, R.; Rotello, V. M., *Acc. Chem. Res.* **2003**, 36 (7), 549–561.
2. Niemeyer, C. M., *Angew. Chem. Int. Ed.* **2001**, 40 (22), 4128–4158.
3. Seeman, N. C.; Belcher, A. M., *Proc. Natl. Acad. Sci. U.S.* **2002**, 99 (90002), 6451–6455.
4. Rao, C. N. R.; Cheetham, A. K., *J. Mater. Chem.* **2001**, 11 (12), 2887–2894.
5. Niemeyer, C. M., *Curr. Opin. Chem. Biol.* **2000**, 4 (6), 609–618.
6. Mirkin, C. A.; Letsinger, R. L.; Mucic, R. C.; Storhoff, J. J., *Nature* **1996**, 382 (6592), 607–609.

7. Alivisatos, A. P.; Johnsson, K. P.; Peng, X. G.; Wilson, T. E.; Loweth, C. J.; Bruchez, M. P.; Schultz, P. G., *Nature* **1996**, 382, 609–611.
8. van de Hulst, H. C., *Light Scattering by Small Particles*. Dover, New York, **1981**.
9. Burda, C.; Chen, X.; Narayanan, R.; El-Sayed, M. A., *Chem. Rev.* **2005**, 105, 1025–1102.
10. Yguerabide, J.; Yguerabide, E. E., *Anal. Biochem.* **1998**, 262, 137–156.
11. Link, S.; El-Sayed, M. A., *J. Phys. Chem. B* **1999**, 103, 8410–8426.
12. Prasad, B. L. V.; Stoeva, S. I.; Sorensen, C. M.; Klabunde, K. J., *Langmuir* **2002**, 18, 7515–7520.
13. Dougan, J. A.; Karlsson, C.; Smith, W. E.; Graham, D., *Nucleic Acids Res.* **2007**, 35 (11), 3668–3675.
14. Li, Z.; Jin, R.; Mirkin, C. A.; Letsinger, R. L., *Nucleic Acids Res.* **2002**, 30, 1558–1562.
15. Letsinger, R. L.; Elghanian, R.; Viswanadham, G.; Mirkin, C. A., *Bioconjugate Chem.* **2000**, 11, 289–291.
16. Zhao, W.; Gao, Y.; Kandadai, S. A.; Brook, M. A.; Li, Y., *Angew. Chem. Int. Ed.* **2006**, 45 (15), 2409–2413.
17. Pena, N. S. R.; Raina, S.; Goodrich, G. P.; Fedoroff, N. V.; Keating, C. D., *J. Am. Chem. Soc.* **2002**, 124 (25), 7314–7323.
18. Hurst, S. J.; Lytton-Jean, A. K. R.; Mirkin, C. A., *Anal. Chem.* **2006**, 78, 8313–8318.
19. Demers, L. M.; Mirkin, C. A.; Mucic, R. C.; Reynolds, R. A.; Letsinger, R. L.; Elghanian, R.; Viswanadham, G., *Anal. Chem.* **2000**, 72 (22), 5535–5541.
20. Lee, J. S.; Seferos, D. S.; Giljohann, D. A.; Mirkin, C. A., *J. Am. Chem. Soc.* **2008**, 130 (16), 5430.
21. Qin, W. J.; Yung, L. Y. L., *Langmuir* **2005**, 21 (24), 11330–11334.
22. Ackerson, C. J.; Sykes, M. T.; Kornberg, R. D., *Proc. Natl. Acad. Sci. U.S.A.* **2005**, 102 (38), 13383–13385.
23. Zanchet, D.; Micheel, C. M.; Parak, W. J.; Gerion, D.; Alivisatos, A. P., *Nano Lett.* **2001**, 1 (1), 32–35.
24. Storhoff, J. J.; Lazarides, A. A.; Mucic, R. C.; Mirkin, C. A.; Letsinger, R. L.; Schatz, G. C., *J. Am. Chem. Soc.* **2000**, 122, 4640–4650.
25. Jin, R. C.; Wu, G. S.; Li, Z.; Mirkin, C. A.; Schatz, G. C., *J. Am. Chem. Soc.* **2003**, 125 (6), 1643–1654.
26. Talanquer, V., *J. Chem. Phys.* **2006**, 125 (19), 194701.
27. Kiang, C.-H., *Physica A* **2003**, 321, 164–169.
28. Lukatsky, D. B.; Frenkel, D., *Phys. Rev. Lett.* **2004**, 92 (6), 068302/1–068302/4.
29. Park, S. Y.; Stroud, D., *Phys. Rev. B* **2003**, 67 (21), 212202/1–212202/4.
30. Long, H.; Kudlay, A.; Schatz, G. C., *J. Phys. Chem. B* **2006**, 110 (6), 2918–2926.
31. Park, S.-Y.; Gibbs, J. M.; Nguyen, S. T.; Schatz, G. C., *J. Phys. Chem. B* **2007**, 111 (30), 8785–8791.
32. Kudlay, A.; Gibbs, J. M.; Schatz, G. C.; Nguyen, S. T.; de la Cruz, M. O., *J. Phys. Chem. B* **2007**, 111 (7), 1610–1619.
33. Lytton-Jean, A. K. R.; Mirkin, C. A., *J. Am. Chem. Soc.* **2005**, 127, 12754.
34. Xu, J.; Craig, S. L., *J. Am. Chem. Soc.* **2005**, 127 (38), 13227–13231.
35. Dillenback, L. M.; Goodrich, G. P.; Keating, C. D., *Nano Lett.* **2006**, 6 (1), 16–23.

36. Lee, J. S.; Stoeva, S. I.; Mirkin, C. A., *J. Am. Chem. Soc.* **2006**, 128 (27), 8899–8903.
37. Maye, M. M.; Nykypanchuk, D.; Lelie, D. V. D.; Gang, O., *Small* **2007**, 3 (10), 1678–1682.
38. Maye, M. M.; Nykypanchuk, D.; VanderLelie, D.; Gang, O., *J. Am. Chem. Soc.* **2006**, 128 (43), 14020–14021.
39. Peng, T.; Dohno, C.; Nakatani, K., *Chembiochem* **2007**, 8 (5), 483–485.
40. Lee, J.-S.; Han, M. S.; Mirkin, C. A., *Angew. Chem. Int. Ed.* **2007**, 46, 4093–4096.
41. Park, S. Y.; Lee, J. S.; Georganopoulou, D. G., *J. Phys. Chem. B* **2006**, 110 (25), 12673–12681.
42. Murray, C. B.; Kagan, C. R.; Bawendi, M. G., *Annu. Rev. Mater. Sci.* **2000**, 30, 545–610.
43. Velev, O. D., *Science* **2006**, 312, 376–377.
44. Park, S. Y.; Lytton-Jean, A. K. R.; Lee, B.; Weigand, S.; Schatz, G. C.; Mirkin, C. A., *Nature* **2008**, 451, 553.
45. Nykypanchuk, D.; Maye, M. M.; van der Lelie, D.; Gang, O., *Nature* **2008**, 451 (7178), 549–552.
46. Huo, F.; Lytton-Jean, A. K. R.; Mirkin, C. A., *Adv. Mater.* **2006**, 18, 2304–2306.
47. Xu, X.-Y.; Rosi, N. L.; Wang, Y.; Huo, F.; Mirkin, C. A., *J. Am. Chem. Soc.* **2006**, 128, 9286–9287.
48. Deng, Z.; Tian, Y.; Lee, S.-H.; Ribbe, A. E.; Mao, C., *Angew. Chem. Int. Ed.* **2005**, 44 (23), 3582–3585.
49. Aldaye, F. A.; Sleiman, H. F., *J. Am. Chem. Soc.* **2007**, 129 (14), 4130–4131.
50. Chen, J.; Seeman, N. C., *Nature* **1991**, 350 (6319), 631–633.
51. Seeman, N. C., *Nature* **2003**, 421 (6921), 427–431.
52. Zheng, J.; Constantinou, P. E.; Micheel, C.; Alivisatos, A. P.; Kiehl, R. A.; Seeman, N. C., *Nano Lett.* **2006**, 6 (7), 1502–1504.
53. Rosi, N. L.; Thaxton, C. S.; Mirkin, C. A., *Angew. Chem. Int. Ed.* **2004**, 43 (41), 5500–5503.
54. Li, Z.; Chung, S. W.; Nam, J. M.; Ginger, D. S.; Mirkin, C. A., *Angew. Chem. Int. Ed.* **2003**, 42 (20), 2306–2309.
55. Rosi, N. L.; Mirkin, C. A., *Chem. Rev.* **2005**, 105 (4), 1547–1562.
56. Elghanian, R.; Storhoff, J. J.; Mucic, R. C.; Letsinger, R. L.; Mirkin, C. A., *Science* **1997**, 277, 1078–1081.
57. Taton, T. A.; Mirkin, C. A.; Letsinger, R. L., *Science* **2000**, 289, 1757–1760.
58. Cao, Y. C.; Jin, R.; Mirkin, C. A., *Science* **2002**, 297, 1536–1540.
59. Park, S.-J.; Taton, T. A.; Mirkin, C. A., *Science* **2002**, 295, 1503–1506.
60. He, L.; Musick, M. D.; Nicewarner, S. R.; Salinas, F. G.; Benkovic, S. J.; Natan, M. J.; Keating, C. D., *J. Am. Chem. Soc.* **2000**, 122, 9071–9077.
61. Nam, J.-M.; Stoeva, S. I.; Mirkin, C. A., *J. Am. Chem. Soc.* **2004**, 126, 5932–5933.
62. Thaxton, C. S.; Hill, H. D.; Georganopoulou, D. G.; Stoeva, S. I.; Mirkin, C. A., *Anal. Chem.* **2005**, 77, 8174–8178.
63. Stoeva, S. I.; Lee, J.-S.; Thaxton, C. S.; Mirkin, C. A., *Angew. Chem. Int. Ed.* **2006**, 45, 3303–3306.
64. Hill, H. D.; Vega, R. A.; Mirkin, C. A., *Anal. Chem.* **2007**, 79, 9218–9223.

65. Nam, J.-M.; Park, S.-J.; Mirkin, C. A., *J. Am. Chem. Soc.* **2002**, 124, 3820–3821.

66. Nam, J.-M.; Thaxton, C. S.; Mirkin, C. A., *Science* **2003**, 301, 1884–1886.

67. Georganopoulou, D. G.; Chang, L.; Nam, J.-M.; Thaxton, C. S.; Mufson, E. J.; Klein, W. L.; Mirkin, C. A., *Proc. Natl. Acad. Sci. U.S.A.* **2005**, 102, 2273–2276.

68. Stoeva, S. I.; Lee, J.-S.; Smith, J. E.; Rosen, S. T.; Mirkin, C. A., *J. Am. Chem. Soc.* **2006**, 128, 8378–8379.

69. Tang, S.; Zhao, J.; Storhoff, J. J.; Norris, P. J.; Little, R. F.; Yarchoan, R.; Stramer, S. L.; Patno, T.; Domanus, M.; Dhar, A.; Mirkin, C. A.; Hewlett, I. K., *JAIDS* **2007**, 46, 231–237.

70. Niemeyer, C. M.; Ceyhan, B., *Angew. Chem. Int. Ed.* **2001**, 40, 3685–3688.

71. Hazarika, P.; Ceyhan, B.; Niemeyer, C. M., *Small* **2005**, 1, 844–848.

72. Miyake, Y.; Togashi, H.; Tashiro, M.; Yamaguchi, H.; Oda, S.; Kudo, M.; Tanaka, Y.; Kondo, Y.; Sawa, R.; Fujimoto, T.; Machinami, T.; Ono, A., *J. Am. Chem. Soc.* **2006**, 128, 2172–2173.

73. Xue, X.; Wang, F.; Liu, X., *J. Am. Chem. Soc.* **2008**, 130, 3244–3245.

74. Willner, I.; Wieckowska, A.; Li, D., *Angew. Chem. Int. Ed.* **2008**, 47, 3927–3931.

75. Liu, J.; Lu, Y., *J. Am. Chem. Soc.* **2003**, 125, 6642–6643.

76. Liu, J.; Lu, Y., *J. Am. Chem. Soc.* **2005**, 127, 12677–12683.

77. Liu, J.; Lu, Y., *Chem. Commun.* **2007**, 4872–4874.

78. Liu, J.; Lu, Y., *Adv. Mater.* **2006**, 18, 1667–1671.

79. Ellington, A. D.; Szostak, J. W., *Nature* **1990**, 346, 818–822.

80. Gold, L.; Tuerk, C., *Science* **1990**, 249, 505–510.

81. Liu, J.; Lu, Y., *Angew. Chem. Int. Ed.* **2006**, 45, 90–94.

82. Zhang, F. X.; Han, L.; Israel, L. B.; Daras, J. G.; Maye, M. M.; Ly, N. K.; Zhong, C.-J., *Analyst* **2002**, 127, 462–465.

83. Zhong, Z.; Patskovskyy, S.; Bouvrette, P.; Luong, J. H. T.; Gedanken, A., *J. Phys. Chem. B* **2004**, 108, 4046–4052.

84. Lee, J.-S.; Ulmann, P. A.; Han, M. S.; Mirkin, C. A., *Nano Lett.* **2008**, 8, 529–533.

85. Jalilehvand, F.; Leung, B. O.; Izadifard, M.; Damian, E., *Inorg. Chem.* **2006**, 45, 66–73.

86. Han, M. S.; Lytton-Jean, A. K. R.; Oh, B.-K.; Heo, J.; Mirkin, C. A., *Angew. Chem. Int. Ed.* **2006**, 45, 1807–1810.

87. Han, M. S.; Lytton-Jean, A. K. R.; Mirkin, C. A., *J. Am. Chem. Soc.* **2006**, 128, 4954–4955.

88. Lytton-Jean, A. K. R.; Han, M. S.; Mirkin, C. A., *Anal. Chem.* **2007**, 79, 6037–6041.

89. Zamecnik, P. C.; Stephenson, M. L., *Proc. Natl. Acad. Sci. U.S.A.* **1978**, 75, 280–284.

90. Stephenson, M. L.; Zamecnik, P. C., *Proc. Natl. Acad. Sci. U.S.A.* **1978**, 75, 285–288.

91. Dias, N.; Stein, C. A., *Mol. Cancer Ther.* **2002**, 1, 347–355.

92. Rosi, N. L.; Giljohann, D. A.; Thaxton, C. S.; Lytton-Jean, A. K. R.; Han, M. S.; Mirkin, C. A., *Science* **2006**, 312, 1027–1030.

93. Giljohann, D. A.; Seferos, D. S.; Patel, P. C.; Millstone, J. E.; Rosi, N. L.; Mirkin, C. A., *Nano Lett.* **2007**, 7, 3818–3821.

94. Seferos, D. S.; Giljohann, D. A.; Rosi, N. L.; Mirkin, C. A., *Chembiochem* **2007**, 8, 1230–1232.

95. Tokareva, I.; Hutter, E., *J. Am. Chem. Soc.* **2004**, 126, 15784–15789.
96. Vidal, B. C.; Deivaraj, T. C.; Yang, J.; Too, H.-P.; Chow, G.-M.; Gan, L. M.; Lee, J. Y., *New J. Chem.* **2005**, 29, 812–816.
97. Thompson, D. G.; Enright, A.; Faulds, K.; Smith, W. E.; Graham, D., *Anal. Chem.* **2008**, 80, 2805–2810.
98. Yin, Y.; Li, Z.-Y.; Zhong, Z.; Gates, B.; Xia, Y.; Venkateswaran, S., *J. Mater. Chem.* **2002**, 12, 522–527.
99. Cao, Y.; Jin, R.; Mirkin, C. A., *J. Am. Chem. Soc.* **2001**, 123, 7961–7962.
100. Lee, J.-S.; Lytton-Jean, A. K. R.; Hurst, S. J.; Mirkin, C. A., *Nano Lett.* **2007**, 7, 2112–2115.

PROBLEMS

1. Explain what SELEX is and how it works.

2. We discussed the colorimetric detection of K^+ using DNA-AuNPs. Describe how the detection scheme for K^+ is similar to those for cocaine and adenosine.

3. Draw the structure of a locked nucleic acid (LNA).

4. Describe the parameters that could be used to achieve stringency conditions to reduce background noise in the scanometric DNA detection scheme.

5. Describe the cooperative model associated with the sharp melting transition of DNA-AuNP aggregates.

6. In biological detection assays where there is an equilibrium between the bound (detected) and unbound (not detected) states, it is generally difficult to achieve both high sensitivity and high selectivity. This is because when more stringent conditions are used to increase selectivity and remove background noise, the 'bound target' is also partially removed causing a decrease in sensitivity. However, assays using DNA-AuNPs such as the scanometric assay are able to achieve high selectivity while maintaining high sensitivity. How is this possible?

7. Explain the two reasons for higher efficiency of DNA-AuNPs as antisensing agents.

8. What are the properties exhibited by DNA-AuNPs?

9. Have AuNPs been synthesized with different shapes?

10. Two dedicated graduate students, Jack and Jill, are modifying 13 nm AuNPs with synthetic DNA. However, Jack forgets to modify the DNA with a terminal thiol-anchoring group. Jill does not realize this and continues to prepare their particles. She adds the DNA to the AuNPs followed by slowly increasing the NaCl concentration up to 1 M NaCl. At what point will Jill notice that something is wrong and how will she notice this?

ANSWERS

1. SELEX (systematic evolution of ligands by exponential enrichment) is a combinatorial technique to generate oligonucleotides that specifically bind to target molecules. The selected sequences are referred to as aptamers. The selection process is conducted by exposing a large library of randomly generated synthetic oligonucleotide sequences to the target molecule. The target molecules can be proteins and small organic molecules. The sequences that do not bind to the target are removed by chromatographic methods and the bound sequences are further amplified by RT-PCR for subsequent rounds of selection. This screening procedure is repeated with increased stringency until the sequence with the highest binding affinity and selectivity is obtained. Finally, the sequence is identified by automated oligonucleotide sequencers or other traditional sequencing methods.

2. In both cases, two types of DNA-AuNPs are connected via a linker sequence which is complementary to the two types of strands but contains an overhang. The overhang and the additional neighboring bases (<10 bases) of the linker sequence is an aptamer sequence for the given targets (K^+, cocaine, or adenosine). In the presence of the target, the aptamer sequence binds to the target and dehybridizes from one of the two particles, leading to a blue to red colorimetric change. In both cases, the target molecule competes with one type of the DNA-Au NPs for binding to the aptamer sequence.

3.

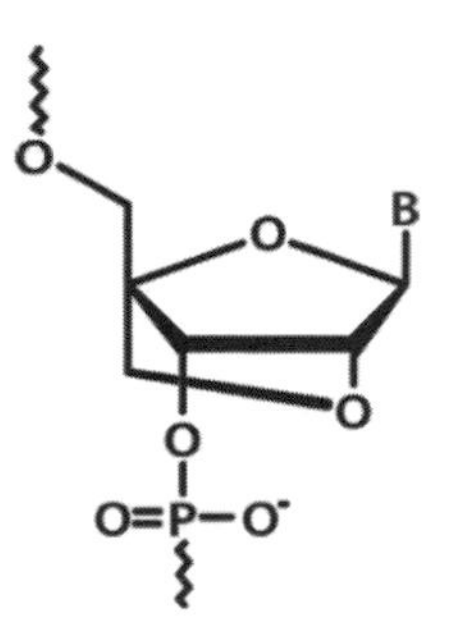

Structure of LNA

4. Stringency conditions include reducing the salt concentration or increasing the temperature at which the assay is run. Both of these options reduce the strength of the duplex and also reduce nonspecific binding.

5. The cooperative binding model is based on a thermodynamic approach that considers the presence of the counterions that associate with DNA when a duplex is formed. In this model, the ion cloud associated with each DNA duplex is shared between neighboring duplexes within a DNA-AuNP assembly, thereby adding additional stability to each duplex. As the duplex begins to melt, the ion cloud dissipates, inducing a destabilization of the neighboring duplexes and a cascading melting process.

6. The ability to maintain sensitivity while also achieving high selectivity is due to the sharp melting transition exhibited by DNA-AuNPs. As the stringency of the system is increased, the bound DNA-AuNPs are not affected until the melting transition is reached.
7. (a) The higher binding efficiency of DNA-AuNPs compared to unmodified oligonucleotides. (b) The higher *in vitro* resistance of DNA to enzymatic degradation due to dense packing of DNA on the gold surface.
8. (a) Optical properties of AuNPs, (b) sequence specific recognition properties of DNA, (c) enhanced binding properties compared to unmodified DNA, (d) increased resistance to DNA degradation, (e) catalyze silver reduction.
9. Several different shapes of AuNPs have been synthesized, including cubes, prisms, and rods.
10. The modification of AuNPs with DNA stabilizes the particles in the presence of salt. If the DNA does not have a terminal thiol group, the DNA will not attach to the surface of the AuNPs and at high NaCl concentrations the AuNPs will aggregate. This will be obvious because the AuNPs will change from red to blue in color.

PART V

NANOTUBES, RIBBONS, AND SHEETS

13

CARBON NANOTUBES AND RELATED STRUCTURES

DANIEL E. RESASCO

13.1 Introduction, 444
13.2 Geometric Structure of SWNT, 445
13.3 Electronic Structure of SWNT, 446
13.4 Characterization of the Geometric and Electronic Structures of SWNT, 447
  13.4.1 Scanning Tunneling Microscopy (STM) and Spectroscopy (STS), 447
  13.4.2 Raman Spectroscopy, 448
  13.4.3 Optical Absorption, 449
  13.4.4 Photoluminescence, 453
13.5 Nanotube Synthesis Methods, 454
  13.5.1 Arc Discharge, 454
  13.5.2 Laser Ablation, 458
  13.5.3 Catalytic Decomposition of Carbon-Containing Compound, 459
13.6 Other Carbon Nanostructures, 462
  13.6.1 Double-Walled Carbon Nanotubes (DWNT), 462
  13.6.2 Multiwalled Carbon Nanotubes (MWNT), 463
  13.6.3 Graphene, 464
  13.6.4 Nanohorns, 465
  13.6.5 Forests of Vertically Aligned Nanotubes, 465
  13.6.6 Horizontally Aligned SWNT, 467
  13.6.7 Yarns of Nanotubes, 468
13.7 Commercial Applications of Carbon Nanostructures, 469
  13.7.1 Conductive Polymer Composites, 470
  13.7.2 High-Strength Composites, 470
  13.7.3 Transparent Electrodes, 471
  13.7.4 Optoelectronic Devices, 473
  13.7.5 Sensors and Actuators, 474
  13.7.6 Nanotube Transistors, 475

Nanoscale Materials in Chemistry, Second Edition. Edited by K. J. Klabunde and R. M. Richards

13.7.7 Field Emission Sources, 475
13.7.8 Interconnects and Vias for Integrated Circuits, 475
13.7.9 Biomedical Applications, 477
13.8 Summary, 478
Acknowledgments, 479
References, 479
Problems, 485
Answers, 487

13.1 INTRODUCTION

The discovery of fullerenes in 1985 (1) sparked the interest of researchers in novel crystalline forms of carbon. As a result, while hollow carbon filaments and nanotube-like structures had been reported many years before (2), the publications of Iijima and Ichihasi (3, 4) and Bethune et al. (5), shortly after the development of fullerenes, generated an unprecedented wave of excitement and intense research in academic as well as industrial labs around the world. Nobel laureate Richard Smalley, co-discoverer of fullerenes, was a strong advocate of nanotube research and his endorsement strengthened the field. The results of such an extensive effort quickly demonstrated the unique properties of these carbon structures, as well as a cornucopia of potential applications in many fields of technology (6).

The unique properties of carbon nanotubes are due to their distinctive structure, which is composed of C—C bonds more closely related to those in graphite than to those in diamond. That is, while diamond has a coordination number of four with sp^3 hybridization, graphite involves three-coordinated carbons, in which three electrons are in sp^2 hybridization and one is delocalized. Fullerenes and nanotubes also have carbon bonds with sp^2 hybridization as graphite, but unlike the graphite structure, which is made up of a flat planar honeycomb, the structures of fullerenes and nanotubes involve a high degree of curvature. A flat graphene sheet is idealized as a single layer of carbon atoms packed in a hexagonal honeycomb structure and it is typically used to describe the structure of sp^2 carbon materials such as graphite and nanotubes. Graphene itself was presumed not to exist in the free form until recently, when it was shown that graphene sheets can be prepared and these sheets can be stable under ambient conditions (7, 8).

The well-known strength of carbon nanotubes is related to the intrinsic strength of the sp^2 carbon–carbon bonds. Young's modulus values near 1000 GPa (9–11) have been predicted or measured on individual nanotubes or nanotube ropes. This is around 50 times higher than the modulus of steel, which, coupled with their light weight, makes carbon nanotubes an attractive filler for high-strength composites. The electronic properties of carbon nanotubes are also astounding. Depending on their geometric structure they can either be metallic or semiconducting since, as discussed below, slight changes in the geometric configuration can result in significant changes in electronic structure. This offers materials scientists the opportunity to design—at the nanoscale—the best material for each specific application.

13.2 GEOMETRIC STRUCTURE OF SWNT

A single-walled carbon nanotube (SWNT) is a seamless cylinder with diameter in the nanometer range that can be visualized as a rolled up sheet of graphene (i.e. two-dimensional graphite plane), as illustrated in Figure 13.1. The structure of any given nanotube is determined by its diameter and the relative orientation of the carbon hexagons with respect to the axis of the tube. Consequently, each SWNT is uniquely identified by two indices, n and m, which are the two integers that determine the chiral vector (also known as wrapping vector) in terms of the primitive vectors of the hexagonal lattice (a_1 and a_2). The chiral vector $C_h = na_1 + ma_2$, connecting two crystallographically equivalent sites on a graphene sheet, determines the circumference of the nanotube's circular cross-section.

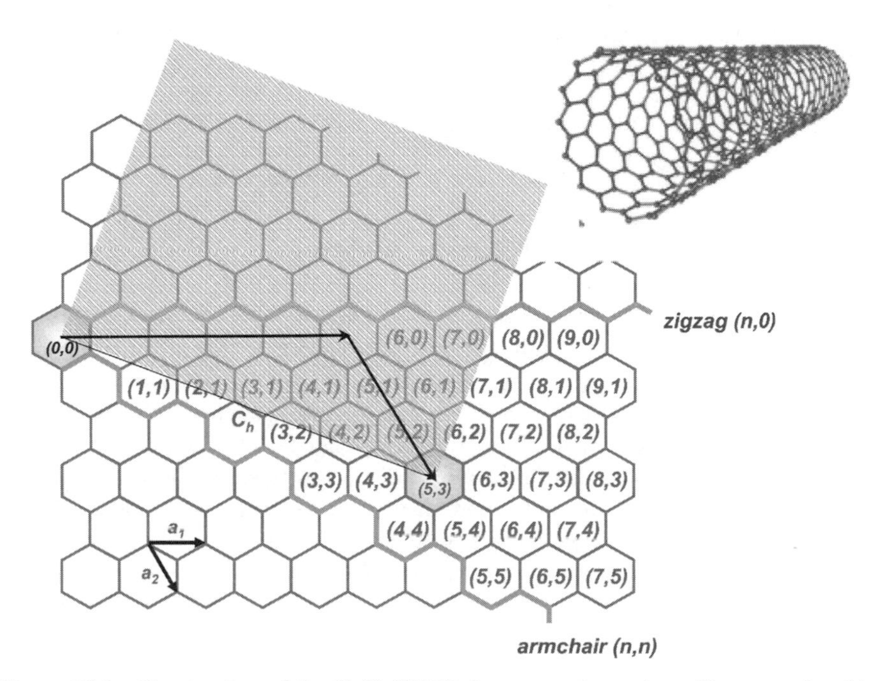

Figure 13.1 Construction of the (5, 3) SWNT from a graphene sheet. First step: the chiral vector C_h is obtained by starting from (0, 0) and moving 5 units in the a_1 direction and then 3 units in the a_2 direction. Second step: the (dashed) sheet determined by the length of the chiral vector C_h is rolled up along the chiral vector, so the origin (0, 0) coincides with the end of C_h, in this example (5, 3). For this nanotube, the chiral angle formed between C_h and the a_1 direction is 21.8 degrees (calculated from Equation 13.2). The inset illustrates the rolled-up structure of a typical SWNT, this is a (14, 0) nanotube with zigzag structure. In this case, the chiral angle is 0.

In the example illustrated in Figure 13.1, the nanotube that would result by rolling up the shaded rectangle would be the (5, 3) since in that case $C_h = 5a_1 + 3a_2$. That is, starting at the (0, 0) site on the plane, the (5, 3) point is reached by moving 5 units in the a_1 direction and 3 units in the a_2 direction. Connecting (0, 0) and (5, 3), the chiral vector C_h is generated, and if a semi-infinite rectangle with base C_h is cut and wrapped, the (5, 3) nanotube is obtained. The angle that is formed between the vectors C_h and a_1 is called the chiral angle. It ranges between 0° and 30° for all the possible structures that vary from zigzag ($m = 0$, angle = 0) to armchair ($n = m$, angle = 30°). All other nanotubes, with chiral angles intermediate between 0° and 30°, are known as chiral nanotubes because their mirror images are not identical, and therefore, they have enantiomeric pairs.

The diameter of a nanotube can be calculated from its (n, m) indices according to the expression:

$$d = a_0 \cdot (m^2 + mn + n^2)^{1/2}/\pi \qquad (13.1)$$

where a_0 is the atomic lattice constant (= 0.246 nm).

Similarly, the chiral angle can be calculated in terms of the (n, m) indices according to the expression:

$$\tan \Theta = (m\sqrt{3})/(2n + m) \qquad (13.2)$$

13.3 ELECTRONIC STRUCTURE OF SWNT

Theoretical calculations of the electronic band structure of single-walled carbon nanotubes have predicted that nanotubes can be metallic or semiconductor in nature depending on their (n, m) indices. The nanotube will be metallic whenever the density of states (DOS) at the Fermi level is nonzero and it will be semiconducting when there is a bandgap at this energy level. In a graphene plane, each C atom is connected to other C atoms via an sp^2 planar bond, which comprises three σ bonds formed from the hybridization of the $2s$, $2p_x$, and $2p_y$ orbitals, while the remaining $2p_z$ orbital, pointing perpendicular to the graphene plane, generates covalent π bonds by overlapping with the $2p_z$ orbitals of the neighbor C atoms. The energy of the system formed by the more stable σ bonds has a lower energy than that of the system formed by the π bonds. As a result, the energy bands for the π system lie in the region of the Fermi level. Therefore, the π system plays a major role in determining the electronic transport properties. In a first approximation, a direct analysis of the π system indicates that either when $n = m$ (i.e. armchair) or when $(n - m)$ is a multiple of 3 the DOS at the Fermi level is nonzero and the nanotubes are metallic. In fact, a closer analysis taking into account the curvature of the tube has shown that mixing of the π system with the σ system for both the bonding and antibonding orbitals can slightly alter the original picture and instead of a nonzero DOS at the Fermi level, these calculations revealed the presence of small gaps. The magnitude of these gaps decreases as the tube diameter increases (12). However, in the case of (n, n) tubes, this gap remains on the sub-band of the nanotube, so they are considered as true metallic.

Therefore, the current view of electronic transport properties with regards to the (n, m) indices can be summarized as follows:

1. Armchair (n, n) tubes are metallic.
2. Bundled armchair tubes exhibit a pseudo-gap at the Fermi level due to tube-tube interactions that affect their rotational symmetry (12).
3. $(n, n + 3i)$ where i is an integer, are small-gap semiconductors (or semimetals) with an energy bandgap proportional to $1/d^2$.
4. The rest (n, m) are semiconducting with energy gaps proportional to $1/d$.

Early on, researchers realized that the one-dimensionality of single-walled nanotubes should result in the appearance of spikes in their DOS, known as van Hove singularities (13). These singularities have been verified experimentally and, in fact, they constitute the theoretical basis for the description of most of the spectroscopic techniques that follow.

13.4 CHARACTERIZATION OF THE GEOMETRIC AND ELECTRONIC STRUCTURES OF SWNT

In addition to the usual microscopy techniques typically employed in nanotechnology research (e.g. TEM, SEM, AFM) several techniques are particularly suitable for characterizing the geometric and electronic structures of SWNT. Here, we briefly describe the type of structural information that can be obtained from scanning tunneling microscopy and spectroscopy (STM/STS), Raman spectroscopy, optical absorption, and photoluminescence.

13.4.1 Scanning Tunneling Microscopy (STM) and Spectroscopy (STS)

The combination of these two techniques offers the possibility of investigating simultaneously the geometric and electronic structures. In essence, STM probes the DOS of a solid by electron tunneling as a conducting tip is placed close to the surface of the solid. When a voltage bias is applied, the resulting tunneling current is a function of the DOS at the Fermi level of the solid. An image is digitally obtained from this current as the probe is scanned over the surface. In the STS mode, by using modulation techniques and a lock-in amplifier dI/dV can be acquired simultaneously with the formation of the image. The detailed STM/STS studies conducted by Odom, Huang, and Lieber (14) have provided a unique experimental confirmation of the DOS predicted by theory. For example, Figure 13.2 illustrates the results of STM/STS measurements conducted in ultrahigh vacuum at 5 K on SWNT samples deposited on a Au(111) single crystal, compared to the theoretical DOS.

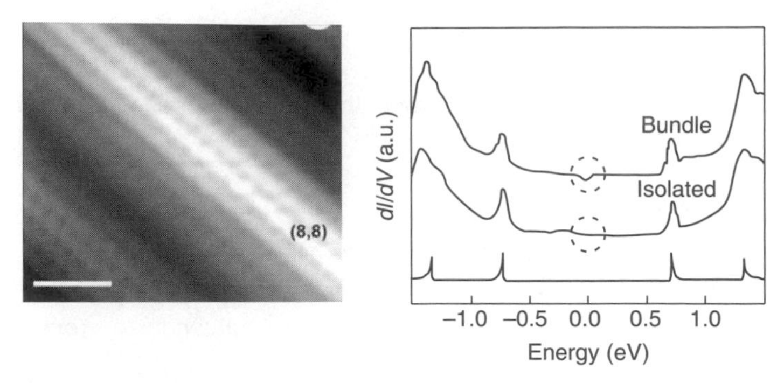

Figure 13.2 (Left) STM image of an armchair (8, 8) SWNT on top of a bundle. Scale bar: 1 nm. (Right) dI/dV recorded on the (8, 8) SWNT in a bundle and as an isolated tube. The theoretical DOS for an isolated (8, 8) tube is included in the figure for comparison. The dashed circles point at the appearance of a pseudogap in the bundled sample, as compared to the isolated tube. (Reproduced with permission from T. W. Odom et al., *J. Phys. Condens. Matter* **2002**, *14*, R145.)

13.4.2 Raman Spectroscopy

Raman spectroscopy has been widely employed to characterize SWNT. Today, Raman is one of the standard techniques used to probe the quality of SWNT. It is very convenient since it requires no specific sample preparation and the analysis is very fast. However, it must be noted that the information resulting from the Raman analysis using only one laser excitation energy does not represent the whole range of nanotubes in the sample. Raman scattering is a weak interaction and typically the observed intensity of light scattered from the bulk solid is low, but if the laser energy is in resonance with an electronic transition, the Raman intensity increases by several orders of magnitude. The sharp transitions exhibited by SWNT have a pronounced effect on the role that resonance plays in Raman spectroscopy for these materials. That is, when the excitation energy matches the energy gap between the spikes of the van Hove singularities, the Raman scattering from the particular tube is greatly enhanced compared to the rest of the tubes, which are not in resonance and, as a result, the Raman spectrum is dominated by resonant contributions.

Consequently, the information obtained for each excitation energy does not reflect the entire nanotubes distribution of the sample, but rather the subset of nanotubes that are in resonance with the laser photons. Therefore, in order to get information that is more representative of the diameter distribution of the SWNTs in the sample, it is necessary to probe the SWNT with several excitation energies.

Raman spectroscopy can be used to probe both structural and electronic features of the nanotubes (15). The Raman spectra of SWNT exhibit several important bands:

1. A tangential mode G band appearing in the 1400 to 1700 cm^{-1} region, which is related to the Raman-allowed phonon mode E_{2g}. Graphite exhibits only one G mode at 1580 cm^{-1}. By contrast, carbon nanotubes have several

contributions to the G band associated with the confinement of wave-vectors along the tube circumference. The higher frequency mode (G+) does not vary with diameter but the lower frequency mode (G−) shifts to lower frequencies as the nanotube diameter becomes smaller.

2. A radial A_{1g} breathing mode (RBM), whose position can be used to calculate the nanotube diameter d_t (16). The energy required for radial deformation increases as the diameter decreases. Therefore, one can expect an inverse relationship between RBM frequency and nanotube diameter. Most of the diameter-frequency relationships developed by researchers have the form

$$\omega\,(\mathrm{cm}^{-1}) = A/d_t\,(\mathrm{nm}) + B\,(\mathrm{cm}^{-1}) \tag{13.3}$$

A number of different values for the A and B parameters have been proposed in the literature, as recently reviewed by Reich, Thomsen, and Maultzsch (17). We typically use $A = 223.5\ \mathrm{cm}^{-1}/\mathrm{nm}$ and $B = 12.5\ \mathrm{cm}^{-1}$ (18), as illustrated in Table 13.1. When the sample contains bundles, accurate estimates can only be obtained if intertube physical interactions are considered, since a significant upshift of the RBM is observed for nanotubes in bundles, with respect to isolated nanotubes (19).

3. The so-called D-band at around $1350\ \mathrm{cm}^{-1}$, which is related to disordered carbon and the presence of carbon nanoparticles and amorphous carbon (20); the D/G intensity ratio has been used as an indication of the degree of disorder in the sample.
4. A G′ band that is a second-order Raman scattering from D-band vibrations.

The D, G, and G′ modes are also found in other forms of carbon, such as graphite; conversely, the RBM mode is uniquely observed in carbon nanotubes.

The analysis of the tangential band offers a method for distinguishing between metallic and semiconducting single-walled carbon nanotubes, since peak broadening and extra bands centered at around $1540\ \mathrm{cm}^{-1}$ are clearly seen when metallic nanotubes are present in the sample. The origin of this extra feature, which can be fitted with a Breit–Wigner–Fano line, is the resonance of the incident or scattered laser photon with the lowest optical transition (E_{11}) between the one-dimensional density of states-singularities in the valence and conduction bands of metallic SWNT (21).

13.4.3 Optical Absorption

As mentioned in previous sections, the one-dimensional structure of SWNT is responsible for singularities in the density of states. Early Raman (22) and STM studies (23) experimentally confirmed the existence of the singularities predicted by theory. The former displayed strong resonance effects that greatly affected the scattering intensity as the excitation energy was varied. The latter directly probed the local density of states, which displayed sharp peaks at specific energies. Shortly after, in good agreement with these concepts, Kataura et al. (24) obtained optical absorption spectra that

TABLE 13.1 Transition Wavelengths and (n, m) Assignments for Semiconducting and Metallic SWNT, Together with Radial Breathing Mode Frequencies for a Selected Number of SWNT Within a Diameter Range of 0.75–1.24 nm and a Chiral-Angle Range of 15–30 Degrees

Semiconducting						Metallic				
(n, m)	d_t (nm)	Chiral Angle	RBM	S_{11} (nm)	S_{22} (nm)	(n, m)	d_t (nm)	Chiral Angle	RBM	M_{11} (nm)
(6, 5)	0.76	27.0	307.6	976	566	(7, 4)	0.76	21.1	302	468
(8, 3)	0.78	15.3	298.3	952	665	(6, 6)	0.81	30	282	458
(7, 5)	0.83	24.5	282.1	1024	645	(8, 5)	0.89	22.4	259	507
(8, 4)	0.84	19.1	278.5	1111	589	(7, 7)	0.95	30	244	504
(7, 6)	0.89	27.5	262.3	1120	648	(10, 4)	0.98	16.1	238	554
(9, 4)	0.92	17.5	256.6	1101	722	(9, 6)	1.02	23.4	228	552
(8, 6)	0.97	25.3	243.9	1173	718	(8, 8)	1.09	30	216	557
(9, 5)	0.98	20.6	241.6	1241	672	(11, 5)	1.11	17.8	211	597
(8, 7)	1.03	27.8	229.1	1265	728	(10, 7)	1.16	24.2	203	600
(10, 5)	1.05	19.1	225.3	1249	788	(9, 9)	1.22	30	194	608
(9, 7)	1.10	25.9	215.1	1322	793	(12, 6)	1.24	19.1	191	643
(10, 6)	1.11	21.8	213.6	1377	754					
(9, 8)	1.17	28.1	203.6	1410	809					
(11, 6)	1.19	20.4	201.0	1397	858					
(10, 8)	1.24	26.3	192.7	1470	869					

displayed the expected optical transitions corresponding to the first and second transitions for semiconducting nanotubes (S_{11} and S_{22}), as well as the first transition for metallic tubes (M_{11}), symmetrically connecting the van Hove singularities above and below the Fermi energy. At the same time, by investigating samples of varying nanotube diameter, Kataura observed that the absorption bands shifted to lower energies as the diameter of the nanotubes increased, in agreement with the theory.

Near-infrared (NIR) and UV-visible spectroscopies are widely available techniques that are ideal for characterizing the optical transitions of SWNTs. Typically, the former covers the range 800 to 1800 nm while the latter covers 200 to 900 nm. Therefore, combining these two techniques one can observe S_{11} and S_{22} for semiconducting nanotubes, as well as M_{11} for metallic nanotubes, for a wide range of diameters, as has routinely been shown in a large number of studies (25–27).

Optical absorption can be used to evaluate the distribution of (n, m) species that a given sample has. For example, Table 13.1 shows the transition wavelengths and (n, m) assignments for semiconducting and metallic SWNT for a selected number

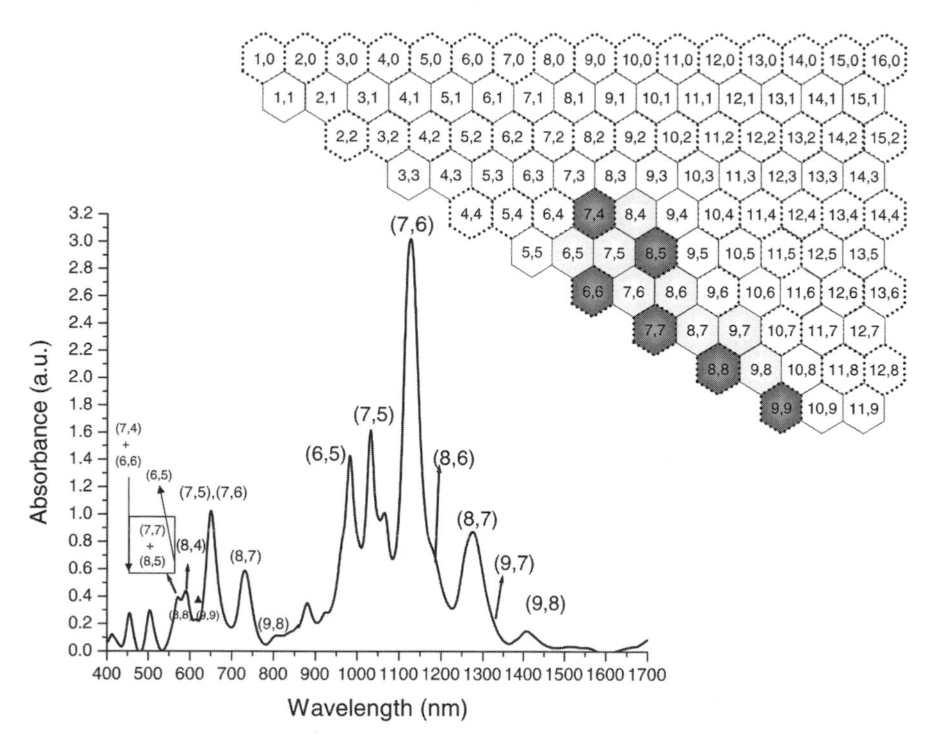

Figure 13.3 (Bottom) Optical absorption spectrum of an SWNT sample produced by CCVD with diameters in the range 0.75 to 1.22 and chiral angles in the range of 19 to 30 degrees. The (n, m) indices corresponding to the different nanotubes present in the sample are noted for each band. In this sample, the (7, 6) semiconducting SWNT is the most abundant species. (Top) The (n, m) nanotubes identified in the sample are indicated in the chirality map. The dark gray hexagons indicate metallic SWNT, the light gray ones indicate semiconducting SWNT.

of SWNT within a diameter range of 0.75 to 1.24 nm and a chiral-angle range of 15 to 30 degrees. At the same time, the table includes the radial breathing mode frequencies corresponding to each nanotube diameter, as calculated with Equation (13.3).

Figure 13.3 illustrates the spectrum of a sample produced by CCVD of CO at 850°C on Co–Mo catalysts at low pressures with diameters in the range 0.75 to 1.22 and chiral angles of 19 to 30 degrees. For comparison, Figure 13.4 shows the spectrum corresponding to the standard CoMoCAT material, obtained by CCVD of CO at 725°C on Co–Mo catalysts at high pressures, which exhibits a much narrower distribution of diameters and chiralities. In this case, the CoMoCAT material exhibits a high concentration of the specific (6,5) type.

Optical absorption can also be used to evaluate the dispersibility and bundle defoliation of SWNT. It has been shown (28, 29) that the ratio of the resonant band area corresponding to the electronic transitions to its nonresonant background, corresponding to π-plasmon absorption from both SWNT and any other carbonaceous species, greatly increases as the purity and dispersibility of the sample increase. At the same time, the width of the resonant band decreases as the SWNT dispersion increases. These two ratios provide a quantitative tool to compare different dispersion parameters (time of sonication, degree of centrifugation, type of surfactant used, etc.) on a given sample, or conversely, to allow for a comparison of different samples suspended in the same way.

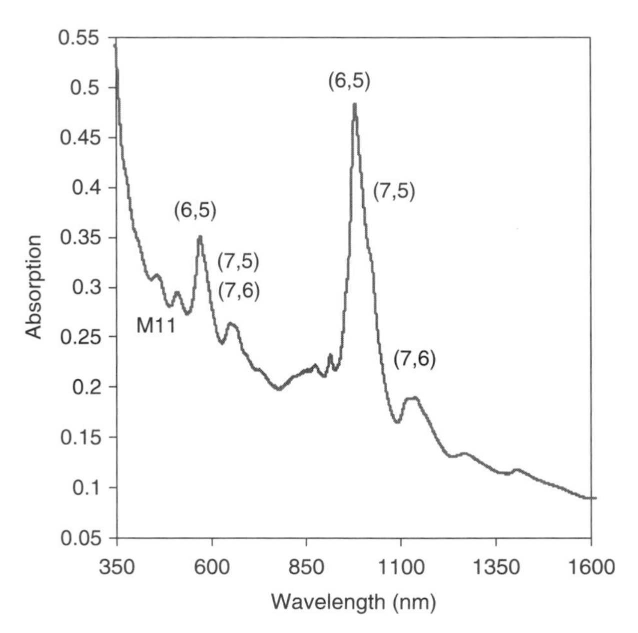

Figure 13.4 Optical absorption spectrum of a typical CoMoCAT material, obtained by CCVD of CO at 725°C on Co–Mo catalysts at high pressures. This material exhibits a high concentration of the specific (6, 5) type. (Reproduced with permission from Y. Tan and D. Resasco, *J. Phys. Chem. B* **2005**, *109*, 14454.)

13.4.4 Photoluminescence

In 2002, O'Connell et al. (30) found that, after excitation with visible light, semiconducting carbon nanotubes suspended in aqueous solution fluoresce at near-infrared (NIR) wavelength. They found that fluorescence occurs when the semiconducting SWNT are debundled as individual nanotubes by ultrasonicating them in the presence of a surfactant. Otherwise, if they are in the form of bundles, the fluorescence is quenched due to the interaction with metallic tubes. Later, Bachilo et al. (31) measured emission intensity as a function of both excitation wavelength (from 300 to 930 nm) and emission wavelength (from 810 to 1550 nm). They explained the observed photoluminescence effects in terms of the one-dimensional nature of the electronic state densities of semiconducting SWNT.

Absorption occurs, as described in the previous section, at a photon energy of S_{22} and this excitation is followed by fluorescence emission at an energy corresponding to S_{11}. A surface contour can be obtained by plotting emission intensity as a function of excitation wavelength and emission wavelength. Since each (n, m) species has a unique pair of S_{11} and S_{22} values, this technique is a powerful tool to obtain a population distribution of (n, m) species in a given sample. For example, Figure 13.5 illustrates the surface contour of fluorescence intensity from a SWNT (HiPCO)

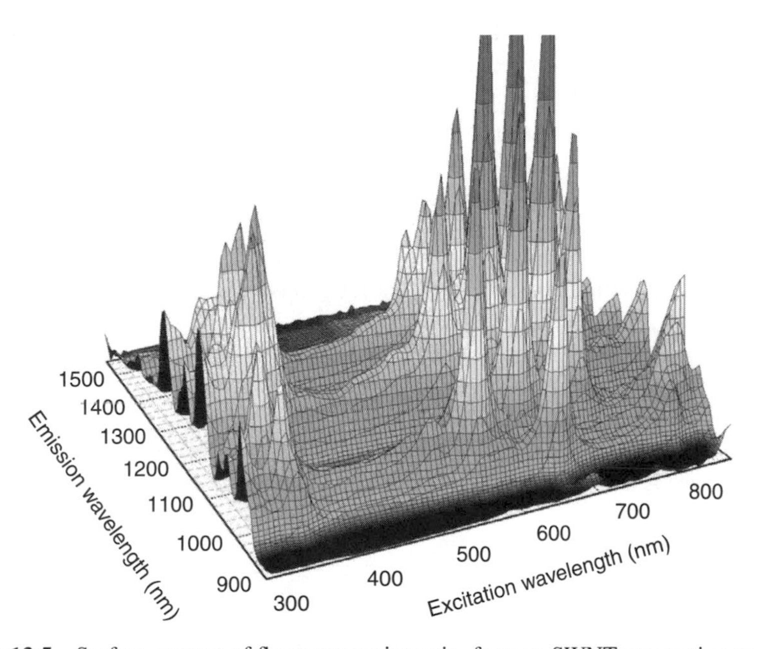

Figure 13.5 Surface contour of fluorescence intensity from an SWNT suspension as a function of excitation and emission wavelengths. Each distinct peak arises from a specific (n, m) species of semiconducting nanotube. (Reproduced with permission from R. B. Weisman, in *Contemporary Concepts of Condensed Matter Science, Vol. 3: Nanotubes*. Elsevier, New York, 2008.)

suspension as a function of excitation and emission wavelengths, in which each peak corresponds to a different (n, m) species (32).

More recently, the application of this technique was extended to the visualization of individual SWNT in different environments (33). Individual SWNT in suspension can be excited with diode laser and its fluorescence detected with a two-dimensional photodiode array, which allows for effective visualization with good spatial resolution and simultaneous spectroscopical identification of the (n, m) type of nanotube being observed. Using this technique, the authors measured nanotube displacement in aqueous surfactant solution and were able to estimate diffusion coefficients. The NIR imaging technique was also applied to investigate the biocompatibility of SWNT in organisms and biological tissues. NIR fluorescence was used to image the SWNT inside living fruit fly larvae (34). It was also observed that the viability and growth of larvae were not reduced by ingestion of SWNT.

13.5 NANOTUBE SYNTHESIS METHODS

Molecular dynamics calculations (35–38) that allow simulating bond breaking and bond forming illustrate how carbon atoms dissolve in the metal cluster and then precipitate on its surface, evolving into various carbon structures and finally forming a cap which eventually grows to a single-wall nanotube, as illustrated in Figure 13.6. The formation of this "embryo" is the important nucleation step that determines whether a nanotube will be formed or the deposition will result in other forms of carbon (amorphous, graphitic, nanofibers, etc.). A common aspect in all the synthesis methods is this nucleation step. Once the cap has been formed, the chemical potential at the nanotube-metal interface is lowered compared to the chemical potential on the rest of the surface (or bulk) of the carbon-bearing particle, as depicted in Figure 13.7. The edge of the growing SWNT then acts as a C sink during the growth process and establishes the required concentration driving force to keep the diffusion-limited process running.

One of the interesting questions that is still open is what governs the termination of nanotube growth. Of course, catalyst deactivation is ultimately responsible for growth termination. However, it is believed that a slowing down in the diffusion process, perhaps due to an increase in the chemical potential at the interface, might trigger the carbon accumulation and final deactivation (39).

Several SWNT synthesis methods have been investigated during the last decade. Arc discharge, laser ablation, and catalytic chemical vapor deposition methods are described below.

13.5.1 Arc Discharge

Arc discharge was the original method used by researchers at NEC (3, 4) and IBM (5) for production of nanotubes. The system differed only slightly from that developed earlier by Kratschmer et al. (40) for the production of fullerenes, which represented an important boost to fullerene research because it opened the possibility of producing in the laboratory significant amounts of fullerenes. A similar system had been

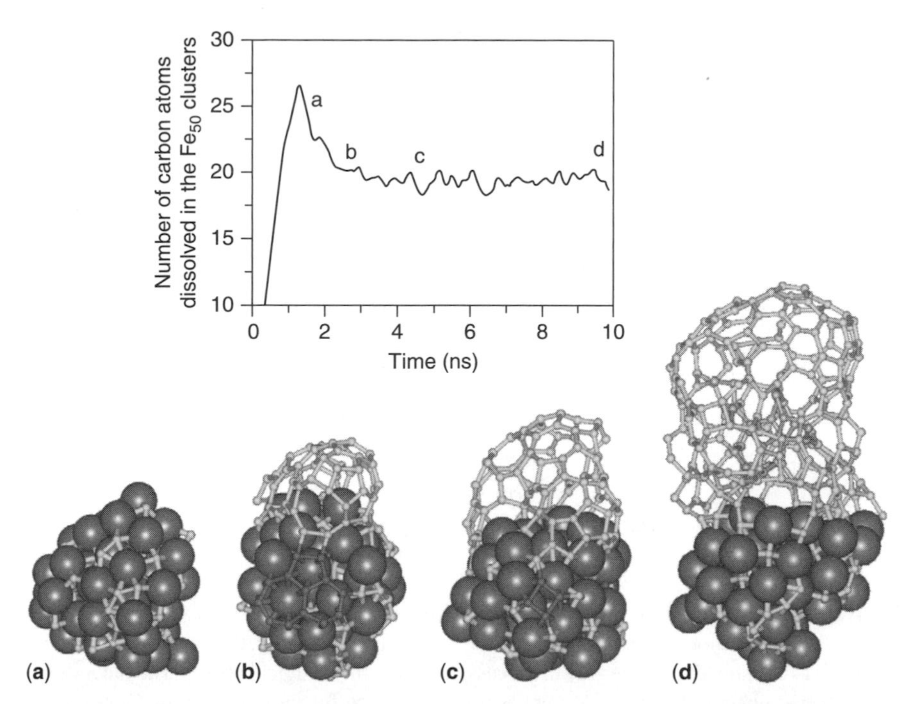

Figure 13.6 (Bottom) Simulated stages during the nucleation and growth of an SWNT on the surface of an iron cluster. The Fe atoms are shown as blue spheres while the C atoms are shown in the ball-and-stick representation mode. The red C atoms form a small graphitic island that later dissolves into the cluster while the grey atoms inside the nanotube are defects. (Top) Number of C atoms dissolved in the iron cluster during SWNT nucleation and growth. The a to d points in the plot refer to the a to d stages shown in the model. (Reproduced with permission from F. Ding et al., *Chem. Phys. Lett.* **2004**, *393*, 309.) (See color insert.)

employed by Bacon more than 30 years earlier for the production of graphite whiskers (41). In the system used by Iijima (3), the graphite electrodes were kept separated by a small gap, rather than being in contact as in Kratschmer's method. It was observed that about half of the carbon evaporated from the anode deposited as nanotubes on the negative electrode while the rest condensed in the form of soot. Onc of the most important observations was that SWNT were only obtained when metal particles were present in the anode.

In a typical arc discharge apparatus (42), the arc is generated between two graphite rods mounted in a stainless steel vacuum chamber equipped with a vacuum line and a gas inlet. As shown in Figure 13.8a, one of the graphite electrodes (cathode) is fixed and it is connected to a negative potential. The other electrode (anode) is moved from outside the chamber through a linear motion feed-through to adjust the gap between the electrodes. A viewport is mounted on the chamber to allow for a direct observation of the arc discharge. In a standard operation, a given background pressure is stabilized within the cell by adjusting the incoming flow of an inert gas, such as helium, and the pumping speed. It has been found that with a continuous flow of

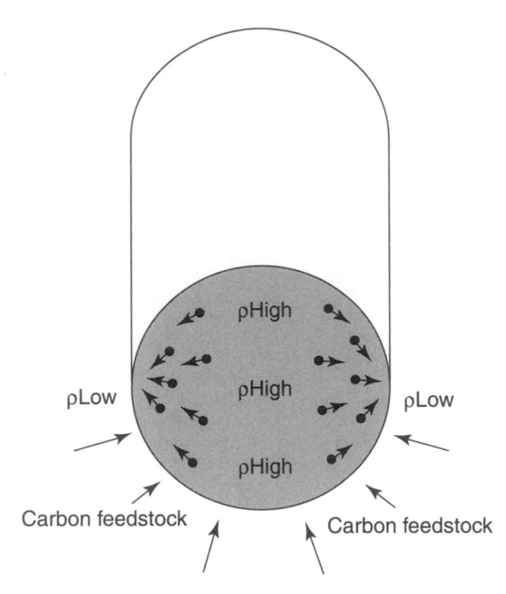

Figure 13.7 Illustration of the formation of a concentration gradient between the gas-metal interface and the nanotube-metal interface. The root of the SWNT (i.e. the nanotube-metal interface) acts as a C sink during the growth process, lowering the chemical potential at that interface. This lowered chemical potential generates a driving force for the diffusion of C atoms from regions of high C concentration to growing SWNT edge, where they incorporate into the growing structure. (Reproduced with permission from F. Ding et al., *Chem. Phys. Lett.* **2004**, *393*, 309.)

He better results are obtained than with a static He pressure. A stabilized voltage of about 20 V is applied while the electrodes are far apart. As the anode is moved in, a gap space is reached (1 to 3 mm) at which arcing occurs. Depending on the He pressure, the rod diameter, and the gap between the electrodes, the electric current can vary between 40 and 250 A.

During the discharge, a plasma is formed generating temperatures of the order of 3700°C. The temperature is particularly high on the anode surface and this electrode is consumed by vaporization. To keep the proper interelectrode gap constant during operation, the position of the anode must be adjusted manually (43) or by an automated system (44). The synthesis process lasts only a few minutes (45). After the synthesis, several types of carbon deposits are obtained: (1) a rubbery soot that condenses on the chamber walls; (2) a web-like structure between the cathode and the chamber walls; (3) a cylindrical deposit on the cathode face; and (4) a small collar around the cathode deposits, which contains the highest concentration of SWNT. Studies have shown that a number of experimental parameters have important effects on SWNT yield and selectivity. For example, increasing the diameters of both electrodes results in yield and selectivity losses, but decreasing the anode diameter while keeping the cathode diameter and current density constant results in an increase in yield (46).

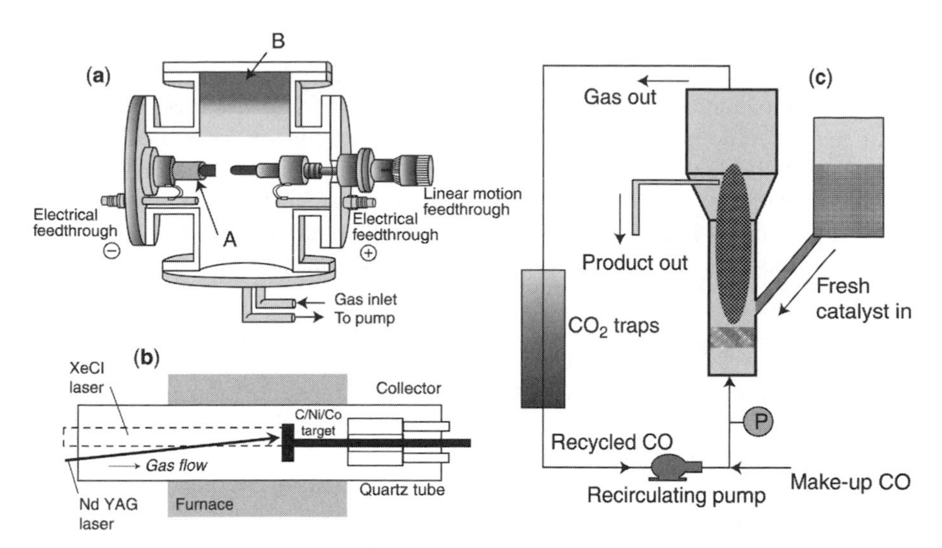

Figure 13.8 Schematic drawings for the different SWNT synthesis methods. (a) Arc discharge (dc). Nanotubes are collected from the cathode surfaces (indicated by A) and from the inner wall of the evaporator (indicated by B). (b) Laser vaporization. The C/Ni/Co target is positioned inside a furnace. The carbon and catalyst vaporization is realized by irradiation with a double laser, in this case, a XeCl and a NdYAG. (c) Fluidized bed CCVD reaction system used in the CoMoCAT process. [Part (a) reproduced with permission from Y. Saito et al., *J. Phys. Chem. B* **2000**, *104*, 2495. Part (b) reproduced with permission from A. A. Puretzky et al., *Appl. Phys. Lett.* **2000**, *76*, 182.]

Another critical parameter that greatly affects the SWNT yield is the pressure of the background gas, usually He.

While some authors indicate that the yield increases with He pressure (46), other authors have found that beyond a certain pressure, further increase in pressure does not result in an increase in yield (42) and it may even result in yield losses (47). Not only the yield of SWNT is affected by the background gas pressure, but also their diameter distribution. In fact, Saito, Tani, and Kasuya (48) observed that the distribution of SWNT diameters shifted systematically to small values as the helium pressure decreased. The most frequently occurring diameters were centered at 1.4 nm for production conducted at 1520 torr, and it shifted to 0.95 nm for production at 50 torr.

Among the various experimental parameters, the choice of catalytic additives is of paramount importance. In the first place, the characteristic feature that is common to all SWNT synthesis methods is the participation of a catalyst. So far, no SWNT has been synthesized without the participation of a catalyst. From the pioneering work of Iijima (3) it was recognized that the use of Ni, Fe, Co, or their mixtures influenced the shape, yield, and perhaps location of the nanotube-containing products. In more recent studies, important differences in SWNT yield and selectivity have been observed when the metals were varied (49). From the very first investigations, researchers realized that the type of catalyst used in the arch discharge had a dramatic effect on the type and quality of nanotube material obtained. For example, Lambert, Ajayan, and Bernier (50) found that while SWNT were produced when Co was used, mostly graphitic nanofibers were obtained with Mn. Most of the additives investigated contain Co or Ni as main ingredients, but in a large number of cases bimetallic or multimetallic mixtures have been used. For example, NiY is a typical composition that has been employed by many authors (51) but other bimetallic pairs such as CoNi (51, 52), FeNi (52), and even noble metals such as PtRh (53) have been used successfully.

13.5.2 Laser Ablation

Single-wall carbon nanotubes have also been produced in high yield by laser vaporization of a graphite rod doped with Co and Ni (54). A typical laser ablation apparatus decribed in Reference 55 is illustrated in Figure 13.8b. Similar to the arc-discharge method, the structure and quality of the SWNT obtained by the laser ablation method strongly depend on the type of catalyst used and the environmental conditions (i.e. surrounding gas and temperature). For example, it has been shown (56) that the structure of the carbonaceous deposits obtained strongly depends on the background argon pressure that is used, and it was found that when the pressure was lower than about 100 torr no SWNT were produced. Apparently, the Ar pressure plays an important role in the heat transfer phenomenon and assists in the metal evaporation. Investigation of the carbon/metal target after the ablation showed that at low pressures, the metal evaporation was inhibited. The texture of the target was also found to have an effect on the quantity of SWNT produced in the laser method. For example, twice as much SWNT were produced when carbon targets containing Ni and Co nitrates were used as when Ni–Co metals or oxides were employed.

This effect was ascribed to the porous structure that results when the Ni or Co nitrates decompose inside the carbon target. At the same time, a better dispersion and smaller particle size of the metals are obtained when using nitrates than when using metals or oxides (57). Likewise, Bandow et al. (58) observed a very interesting effect of the growth temperature on the diameter distribution and chirality of single-wall carbon nanotubes. Working with Fe/Ni and Co/Ni catalysts in the carbon target in Ar gas and controlling the oven temperature they found that as the environment temperature went from 780°C to 1050°C the average diameter was observed to increase from 0.8 to 1.5 nm.

One of the limitations of the laser ablation method is the relatively small quantities that can be produced in each batch. Eklund et al. (59) developed a modification of the pulsed laser vaporization technique that, according to the authors, should result in large-scale production of high-quality SWNT. In this method, ultrafast ablation was achieved by using a high power free-electron laser. The modified setup includes a T-shaped quartz growth chamber placed inside a furnace to keep the chamber at 1000°C. The laser radiation enters from a sidearm that protrudes out the furnace near the center of the hot zone and strikes the carbon target, which is mounted on a rotating/translating rod. A jet of preheated argon deflects the ablation plume away from the incident laser beam, continuously sweeping the target region. The SWNT soot is then collected from a water-cooled copper coldfinger at the end of the quartz tube. A special feature of this method was the use of the free-electron laser operated at a peak laser flux that is about 1000 times greater than the flux used in typical Nd:YAG based systems, but each FEL pulse is only 1/200,000 as long as the typical 10 ns Nd:YAG pulse.

13.5.3 Catalytic Decomposition of Carbon-Containing Compound

The catalytic decomposition of carbon-containing compounds is an extensively investigated method, also known as catalytic chemical vapor deposition (CCVD). One of the advantages of this method is the potential for large-scale production at a lower energy consumption and overall cost than with other methods. The CCVD method is essentially the same as that used for a long time in the synthesis of other filamentous forms of carbon, such as nanofibers or fibrils. The CCVD method involves the catalytic decomposition of hydrocarbons or carbon monoxide on transition metal particles. The major difference with those processes that produce nanofibers is in the structure of the catalyst. To produce SWNT, the size of the metal cluster needs to be very small. Therefore, the success of a CCVD method lies in the design of the catalyst.

The large-scale CCVD process can be conducted on two different types of catalysts, high-surface area powders and "floating" metal clusters produced *in situ* by decomposition of a fluid precursor.

13.5.3.1 High Surface Area Catalysts For large-scale production of nanotubes, the use of a particulate, high surface area catalyst is preferred. In a typical supported catalyst, the active species (e.g. a metal cluster) is stabilized in a high state of dispersion over the surface of a refractory support such as alumina, silica, or magnesia. Such catalyst type is similar to those used in the chemical and petrochemical industry in the production of commodities such as polymers, fuels, and solvents. One of the advantages of using supported catalysts is that the engineering aspects of the possible reactor designs (fluidized bed, fixed bed, transport bed, rotary kiln, etc.) are well known in industry and scaling-up is a mature technology.

Despite uncertainties in the measurements and differences in the synthesis methods employed, it is well recognized that, in an unrestricted state (e.g. during laser ablation), the growth rate of single-walled carbon nanotubes is at least higher than several microns per second (60–63). By contrast, when the growth occurs via catalytic decomposition of carbon-containing molecules on high surface area catalysts, the overall growth process continues in a scale of minutes to hours. It is clear that while the amount of carbon deposits slowly increases with time, this does not necessarily mean that the growth of a given nanotube is that slow (64). That is, the slow rate observed for the overall rate of carbon deposition comprises an induction period followed by a fast nanotube growth rate. Accordingly, new nucleation sites will appear on a high surface area material and each site will give rise to a nanotube that grows relatively fast. The nanotubes that grow later will do it constricted by the presence of those that grew earlier.

A central concept that is common to all CCVD methods that exhibit high selectivity towards SWNT is that nucleation of the nanotube embryo needs to occur before the metal particle sinters. Several approaches have been followed to avoid rapid sintering. The strategy used by our group in the development of the CoMoCAT method (65, 66) was to keep the active species (Co) stabilized in a nonmetallic state by interaction with molybdenum oxide before it is reduced by the carbon-containing compound (CO). When exposed to CO, the CoMo dual oxide is carburized, producing molybdenum carbide and small metallic Co clusters, which remain in a high state of dispersion and result in high selectivity towards SWNT of very small diameter (67, 68). A similar distribution of products at high SWNT selectivity has been obtained at Yale University with Co-based catalysts. Analogous to CoMoCAT, cobalt is stabilized in a nonmetallic state before being exposed to CO. In this case, the stabilizer is a mesoporous silica matrix (MCM-41) (69, 70). In these two methods, the distributions of SWNT diameters and chiralities are very narrow and can be varied in a controlled way by adjusting catalyst formulations and reaction conditions (71).

A similar concept was followed independently by researchers at Toulouse (72), who synthesized dual $CoMgO_x$ oxides, which decomposed under reaction conditions when exposed to methane. While methane is typically less selective than CO for SWNT production, it can also produce a high-quality material with a broader distribution of chiralities and diameters. Methane has also been used as a feed for the production of double-walled nanotubes. A novel type of feed was used by Maruyama et al. (73, 74), who obtained high yield and high selectivity towards SWNT by decomposition of ethanol on Fe-Co catalysts at low pressures and moderate temperatures (i.e. lower than 800°C).

Many other catalyst formulations, reaction conditions, and carbon-containing feeds have been employed using the supported-catalyst CCVD method. A comprehensive review of these efforts has been published (75).

One important aspect in the SWNT production by CCVD method is the reactor design. Large-scale production requires reactor designs amenable to being scaled up. An example of a schematic design is shown in Figure 13.8c for the fluidized bed system employed in the CoMoCAT process.

13.5.3.2 Unsupported "Floating" Metal Cluster Catalysts The term "floating catalyst" implies the incorporation of a catalyst precursor in the form of a vapor into the reaction chamber, where it decomposes and generates the active catalyst under reaction conditions (76). It was employed by Sen, Govindaraj, and Rao (77) in 1997 to prepare carbon nanotubes by decomposition of ferrocene, cobaltocene, and nickelocene under reductive conditions. In this case, the precursor provides both the carbon and the metal to catalyze the synthesis reaction, but later, benzene or hexane was added to the ferrocene precursor to improve the carbon yield (78).

A well-known "floating catalyst" method developed at Rice University is the so-called HiPCO (high-pressure CO) method. In this case, the growth of SWNTs is realized by disproportionation of CO catalyzed by Fe clusters generated *in situ* by decomposition of $Fe(CO)_5$ in continuously flowing CO at high pressure and elevated temperature (79). A mixture of the Fe precursor and CO is injected into the reactor through an insulated injector. The rapid heating of the mixture is important to enhance the formation of nanotubes. Therefore, the CO feed is preheated before entering the mixing/reaction zone and it collides and mixes with the flow emerging from the injector. The temperature, catalyst flow rate, pressure, and CO flow rate were optimized to maximize yield and nanotube quality (80).

13.5.3.3 Commercial Production of SWNT The commercial production of SWNT has began to increase in the last few years. While there are a number of producers in the market, the uncertainty in quality consistency and production capability is still an issue. Several attempts to normalize and quantify the quality of SWNT materials have been attempted and important progress has been made, with involvement of federal institutions such as NIST and NASA. Among the companies that have demonstrated long-term production of SWNT and have quality control in place one could mention Carbolex, Unydim (ex-CNI), SouthWest Nanotechnologies, MER, Thomas Swan, Raymor, NanoLedge, Iljin, etc.

A related issue is that of intellectual property, which is rather complicated, particularly in the area of applications of SWNT. While the number of patents related to the production of single-walled nanotubes is considerably smaller than those related to applications, it is still too large a number to mention all here. Table 13.2 illustrates some of the earlier and more significant U.S. patents for the synthesis of SWNT (excluding MWNT and DWNT). The table includes the type of synthesis method described in each patent.

TABLE 13.2 Some of the Earlier and Most Relevant US Patents on SWNT Synthesis Methods

Method	Patent Number and Year Issued	Inventors
Arc Discharge	5,424,054 (1995)	D. Bethune, R. B. Beyers, C. H. Kiang
Laser Ablation	5,591,312 (1997)	R. Smalley
Arc Discharge	5,747,161 (1998)	S. Iijima
Laser Ablation (G. VIII metals)	6,183,714 (2001)	R. Smalley, D. T. Colbert, T. Guo, A. G. Rinzler, P. Nikolaev, A. Thess
Floating Catalysts (vapor precursor)	6,221,330 (2001)	D. Moy, A. Chishti
Laser Ablation (NiCo)	6,331,690 (2001)	M. Yudasaka, S. Iijima
Supported Catalyst (CoMoCAT)	6,333,016 (2001)	D. Resasco, B. Kitiyanan, J. Harwell, W. Alvarez
Supported Catalyst (CoMoCAT)	6,413,487 (2002)	D. Resasco, B. Kitiyanan, W. Alvarez, L. Balzano
Supported Catalyst	6,692,717 (2004)	R. Smalley, J. H. Hafner, D. T. Colbert, K. Smith
Floating Catalyst (HiPCO)	6,761,870 (2004)	R. Smalley, K. Smith, D. T. Colbert, P. Nikolaev, M. J. Bronikowski, R. K. Bradley, F. Rohmund

13.6 OTHER CARBON NANOSTRUCTURES

In addition to SWNT, several other novel forms of carbon nanostructures have received much attention in recent years and hold great promise to find interesting and important applications; among them, we can mention double- and multiwalled nanotubes, graphene ribbons, and nanohorns. At the same time, researchers have recently developed methods of producing nanotubes in a variety of arrangements. Due to the potentially different properties and applications of these arrangements (vertical forests, parallel arrays, yarns, ribbons, etc.) they are treated in separate sections.

13.6.1 Double-Walled Carbon Nanotubes (DWNT)

Double-walled carbon nanotubes (DWNT) can be described as two coaxial SWNT. In fact, they are closer in morphology, structure, and properties to SWNT than to MWNT. In comparison to SWNT they exhibit some differences that in some cases result in improved properties. For example, in cases in which covalent functionalization is required, the formation of covalent bonds in SWNT breaks the sp^2 configuration and significantly alters the electronic structure, which greatly affects the electronic and optical properties. When DWNT are used, only the outer wall is involved in the chemical bonding, leaving the inner nanotube unaffected. Another interesting difference is that DWNT are stiffer than SWNT, which may be an advantage in some applications.

DWNT were first observed by Smalley and coworkers (81) as a by-product in the synthesis of SWNT by the CCVD method. Later, other researchers were able to produce DWNT with higher selectivity and in larger scale (82, 83). In general, it has been observed that the highest selectivity towards DWNT is obtained from CH_4 decomposition over Fe-based catalysts in contrast with selectivity towards SWNT, which is highest from CO disproportionation.

DWNT have been identified with high-resolution TEM, but sometimes one can find in the literature, particularly when lower resolution microscopes are used, that it is uncertain whether a given nanotube is SWNT or DWNT. Raman provides a good method of confirming the presence of DWNT. It even offers the possibility of obtaining a semiquantitative estimation, as shown by Ren and Cheng (84). Figure 13.9 illustrates the Raman analysis of three different samples. The first sample contains more than 90% DWNTs with outer and inner diameter of 1.85 ± 0.15 and 1.15 ± 0.15 nm, respectively; the second one contains about 30% DWNTs and 70% SWNTs with similar outer diameter of 1.85 nm; the third sample contains only SWNT with average diameter of 1.85 nm.

The RBM region (on the left of Fig. 13.9) exhibits two regions of bands for the sample rich in DWNT, one with frequencies corresponding to the range of 1.7 to 2.0 nm and the other to the range of 1.0 to 1.3 nm, respectively, in good agreement with the diameters observed by HRTEM.

The G′ band (on the right of Fig. 13.9) shows a significant difference when DWNT are present. The band for the samples containing DWNT is composed of several well-resolved contributions while that for SWNTs is a single peak. Similar results are obtained with the analysis of the D band (not shown).

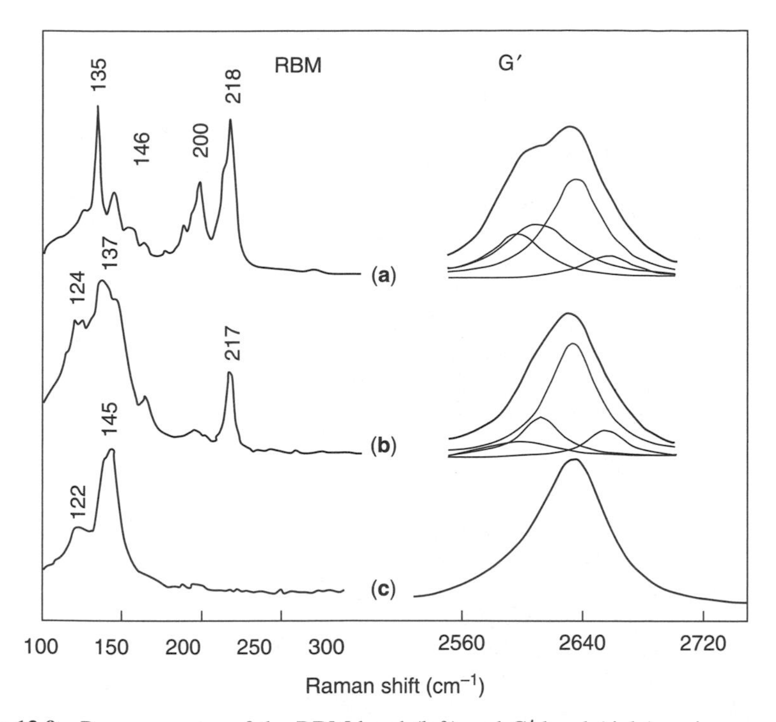

Figure 13.9 Raman spectra of the RBM-band (left) and G′-band (right) regions measured with a laser energy of 1.96 eV of three different samples: (a) sample containing more than 90% DWNTs with outer and inner diameter of 1.85 ± 0.15 and 1.15 ± 0.15 nm, respectively; (b) sample containing about 30% DWNTs and 70% SWNTs with similar outer diameter of 1.85 nm; and (c) SWNTs with average diameter of 1.85 nm. (Reproduced with permission from W. Ren and H. M. Cheng, in *Raman Spectroscopy on Double-Walled Carbon Nanotubes*. Springer, New York, 2008.)

13.6.2 Multiwalled Carbon Nanotubes (MWNT)

As illustrated in Figure 13.10a, a multiwalled carbon nanotube (MWNT) can be visualized as a tube formed by a series of concentric rolled-up graphene layers separated by an interlayer distance close to that between the graphene layers in graphite, that is, approximately 0.3 nm. In fact, in some preparations, like that shown in Figure 13.10b, the observed structure follows the idealized coaxial layer structure.

However, instead of the idealized coaxial layers shown in Figure 13.10a, MWNT prepared by CCVD sometimes exhibit a bamboo-like structure, as that shown in Figure 13.10c (85). This type of MWNT has many more defects than either SWNT or DWNT, and as a result, their intrinsic properties are inferior to those of less defective carbon forms. Also, the variability of properties from tube to tube is much higher than those seen in SWNT, as demonstrated by direct tensile test measurements conducted in a TEM chamber with AFM capability on individual MWNT grown by CCVD (86).

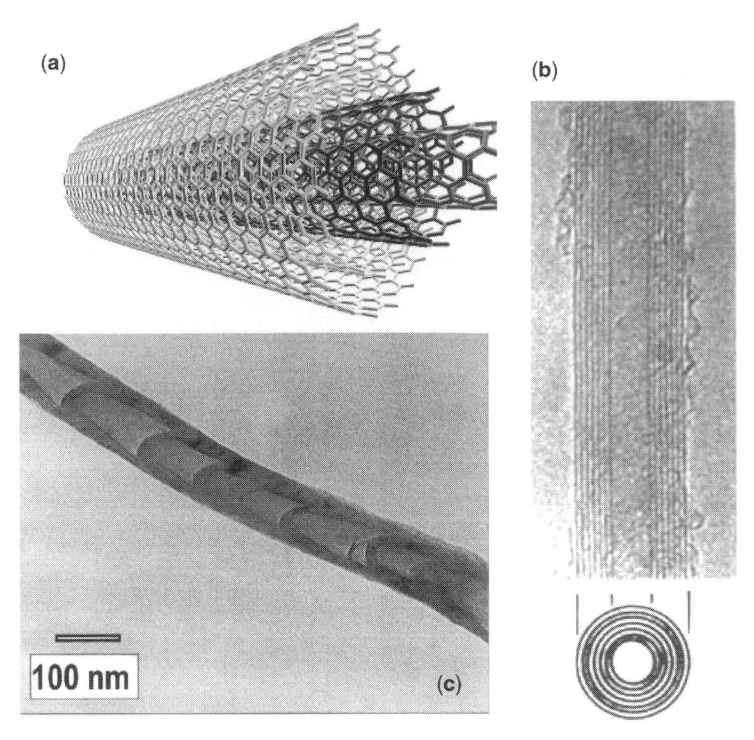

Figure 13.10 (a) Schematic depiction of a multiwalled carbon nanotube (MWNT) as formed by a series of coaxial rolled-up graphene layers; (b) TEM image of multiwall coaxial nanotubes reported by Iijima in 1991 (3). (c) Bamboo-like multiwalled carbon nanotube. (Reproduced with permission from C. J. Lee et al., *Chem. Phys. Lett.* **2000**, *323*, 560.)

MWNT are nowadays produced in large scale by several companies around the world. Bayer (Germany), Hyperion (United States), Nanocyl (Belgium), Iljin (Korea), Arkema (France), and Shenzhen (China) are among the largest producers.

13.6.3 Graphene

Graphene is a single layer of carbon atoms packed into a two-dimensional (2D) honeycomb lattice (see Fig. 13.1). As such, it is the basis for the structure of graphite as well as nanotubes. Until recently (7, 8), graphene was essentially a theoretical structure, presumed not to exist as a free entity. However, free-standing graphene has been produced recently and characterized, which motivated a surge in research and development in many laboratories around the world.

One of the reasons for the soaring interest in graphene is the uniquely high mobility of its charge carriers, which could have an impact on the speed at which an electronic device such as a transistor can be turned on and off. Modern computers that require that transistors switch very rapidly for processing very high frequency signals would certainly benefit from the use of materials of such a high mobility. Also, the extremely small thickness of graphene opens a number of opportunities. Among the potential applications of graphene, researchers have proposed transistors, solid-state gas sensors, biosensors, transparent conductors, photovoltaic cells, electric batteries, and field emitters, all of which clearly parallel those of nanotubes. Graphene can also be patterned into nanoribbons (87), which display interesting properties depending on the orientation and shapes (88), which could lead to applications in devices and interconnections at the nanoscale (89).

So far, graphene has only been produced in small quantities. Therefore, as was the case for nanotubes a few years ago, one of the big challenges for graphene today is the development of production techniques on a scale large enough to attract the interest of industry (90). Several methods have been investigated, mechanical cleaving of graphite, chemical exfoliation, or decomposition of SiC. Under the premise that the most effective method for graphene production should be a chemical approach, Ruoff and colleagues (90, 91) have recently developed a technique by which they initially exfoliate graphite oxide and subsequently chemically reduce it with hydrazine hydrate.

13.6.4 Nanohorns

In 1999, Iijima and coworkers (92) identified cone-shaped single-layered graphitic structures that they called "nanohorns," which were obtained when a pure carbon rod was ablated with a CO_2 laser at atmospheric pressure in Ar. Detailed morphological studies demonstrated that, as illustrated in Figure 13.11 (93), the internal cone angle in these structures was roughly the same for each cone (about 20 degrees). These cones formed dahlia-like aggregates radiating out from the center of a sphere, as seen in the TEM image (94). Later, they developed alternative methods, such as a torch arc in open air that also generated nanohorns in large scale (95).

13.6.5 Forests of Vertically Aligned Nanotubes

The organization of nanotubes is another structural level that needs to be addressed, because organized nanotube arrays may result in interesting properties and applications. Forests of aligned MWNT were obtained on several substrates as early as 1996 (96). Somewhat later, Ren et al. (97), using acetylene as a feed and a plasma-enhanced hot filament were able to obtain hollow nanotubes of various diameters (20 to 400 nm) and lengths up to 50 μm. Recent improvements of this methodology (98) have lowered the synthesis temperature below 400°C, the maximum temperature established by the large-scale integration processes used in the semiconductors industry. This accomplishment has opened a great opportunity for using MWNT as interconnects (see Section 13.7.8).

While vertical alignment of multiwall carbon nanotubes has been mastered by researchers for several years (99, 100), the growth of vertically aligned SWNT (or V-SWNT) is a more challenging task. The special environmental conditions required for the V-SWNT growth have been investigated by several researchers (101–109). Several parameters have been found crucial to obtain V-SWNT. Among them, some researchers have indicated the cleanliness of the vacuum chamber where the

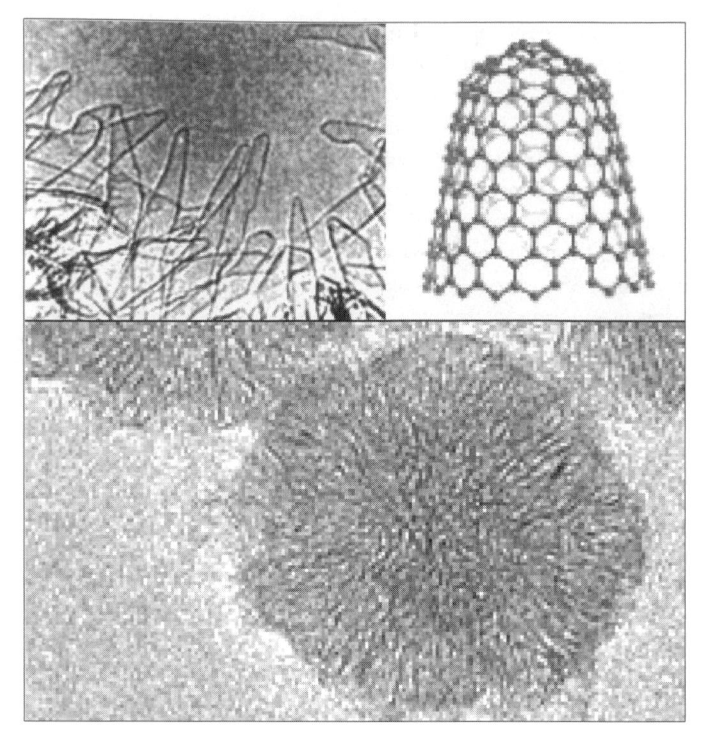

Figure 13.11 Structure of single-walled nanohorns. (Top left) Detailed view of a few nanohorns. (Top right) Schematic depiction of a single-walled nanohorn (note the constant angle of 20 degrees). (Bottom) Dahlia-like aggregate of nanohorns. (Adapted from References 93 and 94.)

growth is conducted (107), others (108) have pointed out the effects of flow rate and addition of water; finally, the density and distribution of the metal particles over the substrate has been found crucial to determine whether the nanotubes grow in a rather oriented fashion ("forest") or a random network parallel to the surface ("grass") (110). SEM and AFM have been employed to investigate the arrangement of catalyst particles on the substrate and how it affects the resulting nanotube morphology, that is, whether the nanotubes grow in vertically oriented fashion ("forest" or V-SWNT) or a random network parallel to the surface ("grass").

It is typically observed that nanotubes grown on flat surfaces are longer than those obtained on a high surface area catalyst. One reason for this difference might be that on a flat surface the nanotube growth is less constricted than in a high surface area catalyst due to the absence of porosity. However, the growth of a nanotube is still constricted by the presence of the other nanotubes, which make them grow vertically oriented (V-SWNT), as shown in Figure 13.12. In fact, it is the formation of a constraining crust at the beginning of the growth that determines the shape and overall morphology of the forest. SEM images of the top of the nanotubes forest reveal the presence of SWNT oriented parallel to the surface. Immediately below this top layer (crust), the

Figure 13.12 SEM image of a forest of vertically aligned SWNT obtained by decomposition of CO at 750°C over CoMo catalyst deposited on a silicon flat substrate. The image illustrates the relative rigidity of the crust on top of the forest, which remains connected while, in this case, a piece has become detached from the rest of the forest. The higher brightness of the crust is due to a higher intensity of secondary electrons emitted by the dense crust. (Reproduced with permission from L. Zhang et al., *Chem. Mater.* **2006**, *18*, 235624.)

orientation is preferentially vertical. Zhang et al. (111) have investigated how this crust develops during the growth process by following the evolution of the forest structure as a function of time by SEM analysis. It was observed that during the first 30 sec, some scattered SWNT were observed on the surface, but a large fraction of the surface remained without nanotubes. More importantly, during the early stages of the growth, there was no indication of a vertically oriented structure. At longer growth times, a thin layer of randomly oriented SWNT started to get woven until a uniform crust was formed. Beyond this point, as the nanotube growth continued from the root, the crust was pushed up, but being rather rigid, the overall growth became concerted and a microscopically uniform growth was imposed. As a result a somewhat vertical alignment of every nanotube occurred guided by the rigid crust. Figure 13.12 illustrates the relative rigidity of the crust, which remains connected while, in this case, a piece has become detached from the rest of the forest.

13.6.6 Horizontally Aligned SWNT

The *in situ* growth of nanotubes on flat substrates is an attractive option for the fabrication of nanotube-based devices such as nanotransistors (112), electron emitters for optoelectronic displays (113), and nanoelectrodes for DNA detection (114). Nanotube alignment is categorized into horizontal (parallel to the surface) and vertical (perpendicular to the surface). Horizontal alignment has been realized by applying electric (115) or magnetic (116) fields or by controlling the direction of the gas flow (117). More recently, horizontal alignment of nanotubes has been accomplished. It is well recognized that despite the high intrinsic conductivity and high carrier mobility of individual nanotubes, when they are arranged on a disordered network on a surface, the observed conductivities and mobilities are significantly lower than those expected. The reason for this discrepancy is the interfacial resistance at the tube-tube contacts in the network which limits the charge transport. As a result, significant efforts have been dedicated to develop methods of synthesis of aligned arrays of long nanotubes. Several research groups have shown success in producing these arrays, admittedly in small scale. For example, Ismach et al. (118) obtained aligned SWNT by producing them by CVD on miscut sapphire crystals and proposed that the aligned growth was templated by the atomic-sized steps. Song et al. (119) observed directional growth on the a-plane and r-plane of a sapphire crystal, but in this case, the order seemed to be unrelated to the arrangement of atomic steps. Similarly, Kocabas et al. (120) found that aligned arrays can be obtained on miscut crystal substrate, but interestingly instead of sapphire they use a low-cost, commercially available, single-crystal quartz substrate. It is possible that the alignment mechanism on quartz may be similar to that on sapphire.

13.6.7 Yarns of Nanotubes

The first attempt at spinning pure nanotube fibers was made by Vigolo et al. in 2000 (121). They assembled SWNT into long ribbons and fibers by first dispersing the nanotubes in surfactant solution and then precipitating them in a flow of a polymer solution. In this way, the flow aligned the tubes to make ribbons. Later, Jiang, Li,

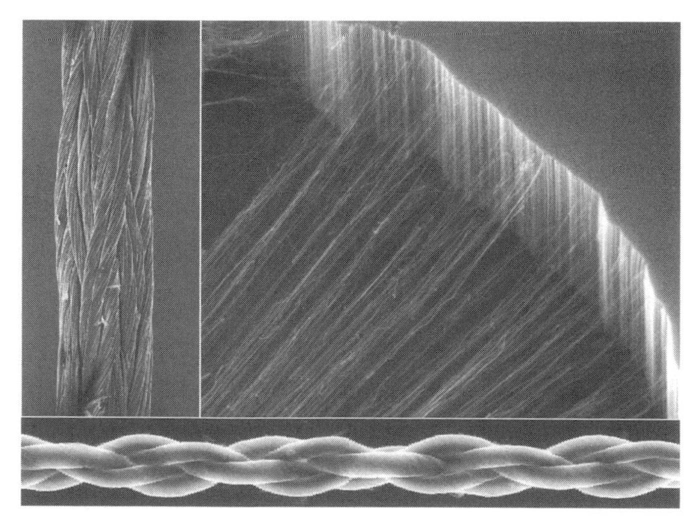

Figure 13.13 SEM images showing the dry spinning of nanotube yarns from a multiwalled carbon nanotube forest. The torque-stabilization applied during spinning results in yarns of superior strength. (Images from References 123 and 124.)

and Fan developed a dry method for spinning continuous nanotube yarns (122). This method was later perfected by Baughman et al. (123, 124) who investigated the effects of introducing different degrees of twist during the spinning process and making torque-stabilized yarns (see Fig. 13.13). By optimizing this method they achieved remarkable improvements in strength and toughness (e.g. yarn strengths greater than 460 MPa). They found that these yarns also retain their strength and flexibility after heating or immersing in liquid nitrogen. Lee, Kinloch, and Windle (125) combined the continuous fiber spinning method with the synthesis of carbon nanotubes and spun fibers and ribbons directly from the chemical vapor deposition (CVD) synthesis reactor. To accomplish this, they continuously withdrew the nanotube product from the reactor with a rotating spindle.

13.7 COMMERCIAL APPLICATIONS OF CARBON NANOSTRUCTURES

Many important applications of nanotubes have been extensively investigated during the last 10 to 15 years. A summary of the range of applications related to the specific properties for which carbon nanostructure present a distinctive advantage is given in Figure 13.14. Some of these applications have become commercial products and others are in the developmental stage. It is expected that, very soon, the cost and availability of nanotubes of consistent quality will become more in line with industrial

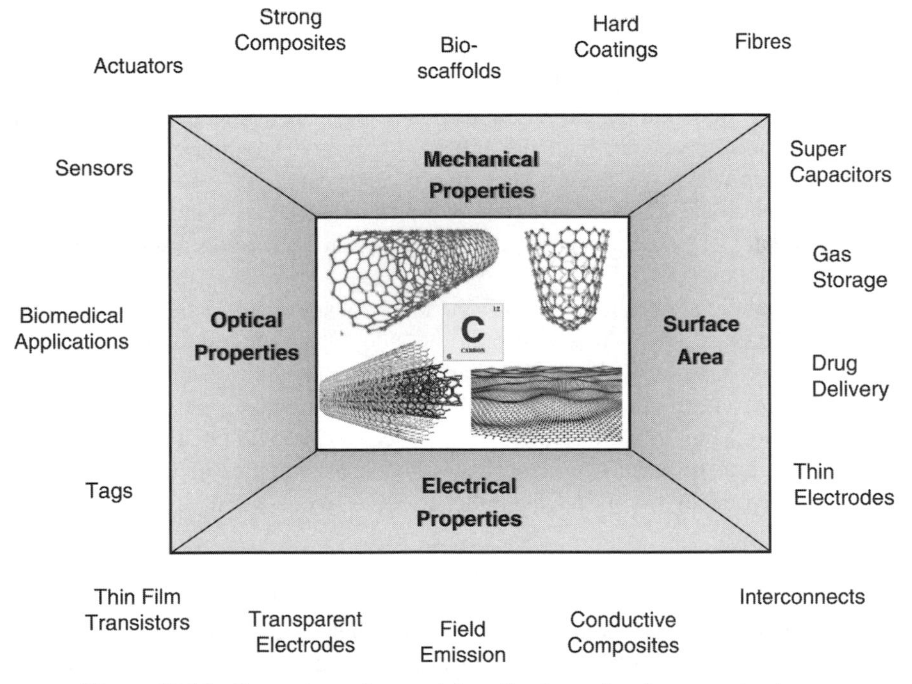

Figure 13.14 Properties and potential applications of carbon nanostructures.

needs, and consequently the pace of development will greatly accelerate. Among the most widely investigated applications, we will mention here conductive and high-strength nanotubes/polymer composites, transparent electrodes, sensors, and nanoelectromechanical devices, field emission displays and radiation sources, semiconductor devices (e.g. transistors), and interconnects.

13.7.1 Conductive Polymer Composites

The covalent bonds present in polymers make them electrical insulators, which is a desirable property for most applications of polymers. However, there are applications in which a minimum level of conductivity is required, such as in those devices in which electrostatic discharge (ESD) can be a problem. Electrostatic charge is typically generated by friction and when a charged body comes into contact with another with a lower electrostatic potential a rapid discharge occurs, with potentially damaging effects (e.g. explosions, fire, and damage to electronic devices). To minimize ESD, electrical resistance cannot exceed the range 10^6 to $10^9\,\Omega$, which is much lower than the typical resistance of polymers, around $10^{14}\,\Omega$. Carbon black is a filler typically used to control ESD. However, the very high aspect ratio of nanotubes results in higher conductivity at much lower loadings, thus having a much lower impact on mechanical and stability properties of polymers. In addition, if they are efficiently dispersed, the nanoscale size of CNT eliminates "hot spots," improves processability, and lowers the specific gravity of the composite, compared to standard carbon black fillers. Therefore, CNT are increasingly being used for ESD protection in electronic circuits, fuel tanks, and fuel lines.

Another important application of nanotube-polymer composites is electromagnetic interference (EMI) shielding of radio frequency radiation to protect circuits from radiation coming from telecommunication equipment. Nanotube-based composites can compete very well with metal-based EMI shielding materials due to their light weight, chemical resistance, and flexibility. The ability of a given filler to impart shielding properties to a composite material depends on several factors, among which high electrical conductivity and high aspect ratio are two of the most important ones in which CNT excel. In fact, very promising EMI shielding results have been obtained with both MWNT (126) and SWNT (127).

A large-volume application of CNT is in the area of automotive painting. In order to minimize the emission of VOCs (volatile organic compounds) during the painting operation, automakers use electrically charged paint that adheres well to the surfaces being painted. Therefore, the polymeric automotive paintable parts such as mirror housings, door handles, bumpers, and fenders need to have conductivity in order for this adherence to be effective. Incorporation of small amounts of CNT makes the part conductive enough, while having minor effects on processability of the composite.

13.7.2 High-Strength Composites

The exceptionally high Young's modulus ($\sim$1.2 TPa), stiffness, and flexibility of carbon nanotubes are widely recognized (128). Therefore, an obvious application that has been widely investigated is structural composites, most commonly with polymer matrices, which would open great opportunities for aircraft and space construction, as well as any function in which a light-weight, high-strength material is needed. However, the resulting structural reinforcement in composites not only depends on the intrinsic strength of the reinforcing fiber, but is also a strong function of dispersion, alignment, and the degree of load transfer from the matrix to the nanotubes. These conditions have not been fully met in many of the numerous attempts made around the world to prepare nanotubes-based composites. As a result, the degree of success has been less than optimal. The difficulty in dispersing of the nanotubes in the polymer matrix is mostly related to their high aspect ratio, which consequently makes the nanotube-nanotube van der Waals attraction a dominant force that keeps nanotubes aggregated in ropes and bundles. Also, processing of nanotubes-based composites by extrusion or shear mixing is complicated by the rapid increase in viscosity observed as the loading of nanotubes in the matrix increases. Finally, the low degree of load transfer between the matrix and the nanotubes is related to the absence of an interfacial bonding between the two components (129).

Haddon et al. (130) have investigated the importance of improving the SWNT-polymer interfacial interaction in a Nylon 6 graft copolymer composite with SWNT of varying degrees of functionalization and thus varying degrees of interfacial interaction. In this study, they produced continuous Nylon composite fibers drawn from the *in situ* polymerized caprolactam in the presence of SWNT having a range of carboxylic acid (SWNT-COOH) and amide (SWNT-$CONH_2$) functional groups. As shown in Figure 13.15, the mechanical performance improved with the concentration of carboxylic acid functional groups in the SWNT, which makes a stronger SWNT-polymer interaction. When the COOH groups were replaced by $CONH_2$, the functionalized SWNT participated in the polymerization reaction and were covalently attached as graft copolymer chains. This method opens the possibility of varying the mechanical properties of the composite, such as ductility and toughness.

An alternative approach has been shown by Mu and Winey (131), who observed that, while the molecular weight of the polymer matrix has no significant effect on the degree of SWNT dispersion or alignment in the composite, it plays an important role in the load transfer at the SWNT-polymer interface. Therefore, even without functional groups attached to the nanotube, when the polymer chain size is large compared to the nanotube diameter the elastic modulus of the composite greatly increases.

13.7.3 Transparent Electrodes

Thin networks of SWNT can be prepared as transparent and highly conducting films which, in addition, have very good mechanical properties. They can be prepared in a reproducible way by a rather simple method of fabrication at room temperature that does not require high vacuum conditions or expensive equipment. Moreover, SWNT thin films have no equal in the tunability of their electrical, optical, and mechanical properties. By varying the type of nanotube used or the film preparation procedures, researchers can vary electrical and optical responses by several orders of

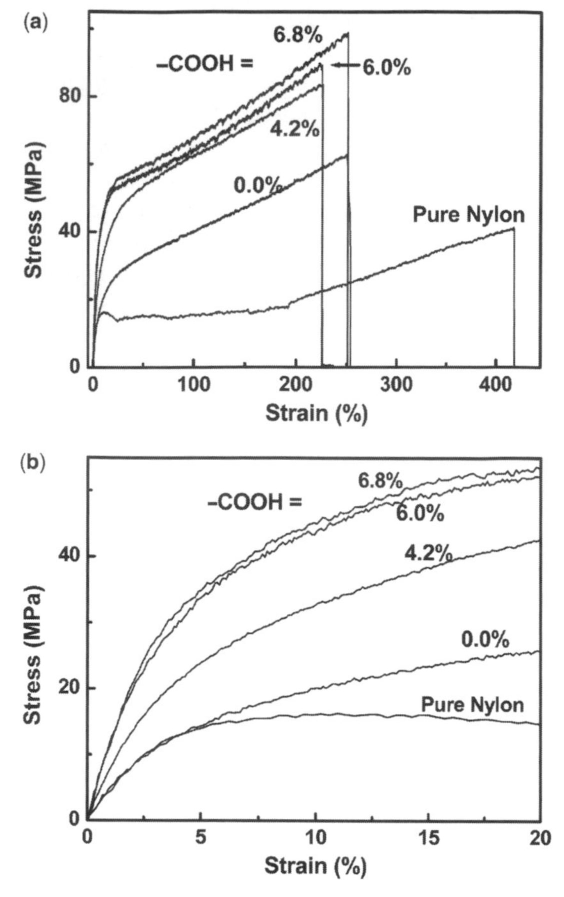

Figure 13.15 Stress-strain curves of composite Nylon/SWNT fibers prepared using SWNTs bearing 0.0%, 4.2%, 6.0%, and 6.8% carboxylic acid groups at the same SWNT loading (0.5 wt%). (a) Strain to fracture; (b) low strain region to compare differences in Young's modulus. (Reproduced with permission from J. Gao et al., *J. Am. Chem. Soc.* **2006**, *128*, 7492.)

magnitude. As a result, extraordinary interest has been generated in SWNT thin films for applications in transparent electrodes and other applications in which flexible, transparent, and conductive coatings together are needed. For example, these conductive films are considered to be used in the LCD (liquid crystal display) industry for flat panel displays and touch screens as well as on emerging technologies such as flexible plastic electronics and smart fabrics and windows (132). Such properties coupled with room temperature deposition from solution ensure that the material will have profound impact on. These optoelectronic properties of SWNT thin films make them a potentially good replacement for indium tin oxide (ITO), used widely in photovoltaics, organic and inorganic light-emitting diodes, displays, touch screens, and smart windows.

13.7.4 Optoelectronic Devices

SWNT and MWNT have been incorporated as essential components in several optoelectronic devices (i.e. devices that produce, detect, or control the transfer of light). Specifically, nanotubes are being considered to be used in solar cells and organic photovoltaics, phototransistors, photomultipliers, photoresistors, injection laser diodes, and light-emitting diodes. For example, in the case of organic solar cells, nanotubes can play an important role. First, if they are used in combination with pi-conjugated polymers, nanotubes can act as electron acceptors and enhance the exciton dissociation. As illustrated in Figure 13.16, by providing a high field at the interface, nanotubes can suppress the electron–hole recombination and efficiently transport the electron towards the corresponding electrode (133).

At the same time, SWNT can be used at the top electrode (that needs to be transparent) as a replacement for the currently used thin films that are also electrical conductors. Typical conductive thin films used today are oxides such as fluorine-doped tin oxide (FTO), Al-doped zinc oxide, and the indium tin oxide (ITO) mentioned above. The disadvantages of these oxide films are that they require expensive deposition procedures at high vacuum, they have poor mechanical properties, and they are not transparent in the infrared. Another interesting difference is that transparent oxide conductors are n-type semiconductors. By contrast, nanotube networks act as p-type semiconductors, which could lead to new designs.

Also, nanotubes can be used as dopants in organic light-emitting diodes (OLEDs). A typical OLED contains an emissive layer and a conductive layer between two electrode terminals. The layers are of electrically conducting organic materials. For example the early OLEDs involved a layer of poly(*p*-phenylene vinylene). When a voltage is applied across the OLED, an electron current is established through the device and the cathode provides electrons to the emissive layer while the anode receives electrons from the conductive layer, which results in the transfer of holes into the conductive layer. As the emissive layer becomes negative and the conductive layer positive, electron–hole recombination in the emissive layer causes an emission

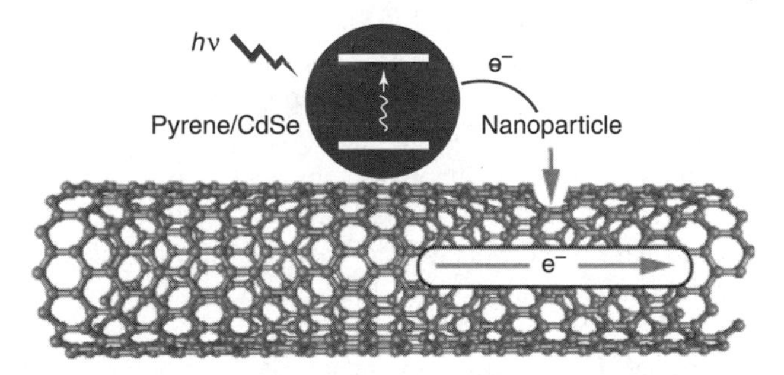

Figure 13.16 Schematic illustration of the light-induced charge excitation (creation of an exciton) in a semiconductor nanoparticle and subsequent charge transfer onto an SWNT, followed by electron transport along the SWNT. (Reproduced with permission with L. Hu et al., *Adv. Mater.* **2008**, *20*, 939.)

of radiation in the visible region. In this process, the holeblocking nature of SWNT added as a dopant enhances charge separation and facilitates charge transport; at the same time, SWNT offer an enhanced carrier mobility (i.e. conductivity normalized by the density of charge carriers), compared to organic semiconductors. For example, the room-temperature hole mobility of poly(3-hexylthiophene) is around 0.1 cm^2/Vs, while the hole mobility in SWNT may be several orders of magnitude higher (134, 135). The net effect of these advantages will be an improvement in the device performance and power conversion efficiency.

13.7.5 Sensors and Actuators

Carbon nanotubes are being considered as gas sensors as they exhibit high surface area ($>1000\ m^2/g$), which is important since gas detection is directly related to the adsorption of gaseous molecules, so small amounts of sensing nanotubes may be needed in the detection process. More importantly, they exhibit significant changes in several properties when put in the presence of specific gases, and those changes are reversed when the gas is removed. For example, it has been observed that their electrical resistance changes in a reversible way when exposed to NO_2 or NH_3 (136). Similarly, the thermoelectric power (137) and the dielectric constant are seen to vary in a systematic way when exposed to different gases (138). For example, Rao et al. (139) found that the resonant frequency of an SWNT-coated disk resonator shifts upon exposure to a few parts per million of NH_3 corresponding to a change in the dielectric constant of the nanotubes. Robinson, Snow, and Perkins (140) have compared capacitance- and conductance-based detection of trace chemical vapors and demonstrated that conductance detection is more susceptible to noise and incomplete reversibility than capacitance detection. The latter is dominated by dielectric effects and better suited for trace-level detection than the former, which has been more commonly used.

SWNT have unique properties that make them attractive as nanoelectromechanical devices. They are hollow and light-weight, but at the same time they are extremely stiff (Young's modulus $\sim$1 TPa). As a result, they are ideal for devices that require fast mechanical action, such as switches and tunable components. For example, a nanorelay (141, 142) has been conceptually designed with conducting nanotubes placed on a Si substrate and connected to a fixed electrode acting as a source. When a gate voltage is applied a charge is induced in the nanotubes, resulting in a capacitive force between the nanotubes and the gate that makes the nanotube bend until it makes contact with the drain electrode, placed on a lower step, closing the circuit (see Fig. 13.17). Similar nanosystems have been conceived to operate as nonvolatile memory devices (143). Experimental demonstrations (144, 145) of electrical actuation and detection of the oscillation modes have been obtained recently on nanotube oscillators. It has been demonstrated that the resonance frequency of these nanotube oscillators can be tuned and that the devices can be used as transducers for very small forces. The nanotubes can be made to detect their own mechanical motion. The nanotubes have no slack and when applying a gate voltage they can be tuned to behave as a vibrating string under tension.

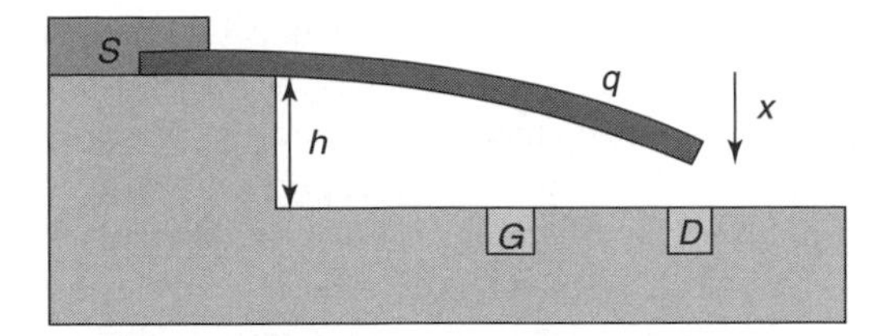

Figure 13.17 Schematic of a nanorelay made of a nanotube deposited on a terrace in a Si substrate (height h). The nanotube is connected to a source (S), while the gate (G) and drain (D) electrodes are placed on a lower terrace. When a charge (q) is applied to the nanotube, it deflects (x) until it touches the drain (D). (Reproduced with permission from J. M. Kinaret et al., *Appl. Phys. Lett.* **2003**, *82*, 1287.)

13.7.6 Nanotube Transistors

The uniquely high mobility displayed by SWNT (146, 147) makes them attractive for applications in nanodevices, such as thin-film transistors (TFT), which could be produced by solution-processed random networks of SWNT. An optimum TFT would be composed entirely of semiconducting nanotubes, since their performance is limited by the presence of metallic tubes. It has been proposed that even below the percolation threshold of metallic tubes, electron hopping or tunneling may occur between neighboring metallic tubes (148). This electron channeling reduces the on/off ratio and therefore, the overall performance of the transistor. Removal of metallic nanotubes from the network has been achieved by electron breakdown (149, 150).

13.7.7 Field Emission Sources

What makes SWNTs good candidates for applications as field emitters is the combination of their natural geometry, chemical stability, and electrical characteristics. They could be used as the electron source in a whole range of devices, including flat panel displays, light elements, and e-beam sources for lithography, either as film emitters or single emitters. In 1999, Choi et al. (151) built the first SWNT-based field emission display (FED) at Samsung. Carbon nanotube-based FEDs are characterized by superior display performances such as fast response time, wide viewing angles, wide operation temperatures, cathode ray tube (CRT)-like colors, ultraslim features, low cost, and low power consumption. FED technology is one of the most promising approaches for direct view displays larger than 60 in. diagonal. Some researchers have investigated the *in situ* growth of vertically aligned nanotubes over a large area of glass substrates at low temperatures. However, while MWNTs can be produced at relatively low temperatures, high synthesis temperatures are often required to produce single-walled carbon nanotubes. Therefore, the common approach is the use of nanotubes produced separately and later deposited on the cathode by techniques such as the screen printing method.

13.7.8 Interconnects and Vias for Integrated Circuits

Interconnects are conductors of power and electrical signals between solid-state devices in integrated circuits. As described by Moore's law the number of transistors that can be placed on an integrated circuit is essentially doubling approximately every two years as the integrated circuits continuously decrease in size. As a result, the number and complexity of interconnects dramatically increases. About 10 years ago, interconnects started to be made out of copper instead of aluminum to lower the resistance and improve the performance, but soon copper will find its limit and new materials will be needed. Carbon nanotubes have been proposed as a replacement for copper due to the high current capacity that they can tolerate (152). Interconnects come in different sizes, according to their function. Global interconnects provide contact between larger blocks, while local interconnects are smaller and typically connect logic units. In addition, the circuits include vias or electrical passages through the die which form a conductive interconnect in a third dimension, as shown schematically in Figure 13.18a. These vias need to conduct very high current densities and bundles of many CNT closely packed could provide the necessary current capacity. Researchers at Fujitisu (153) have developed CNT vias linked to copper interconnects. The CNTs were grown *in situ* on Co catalyst supported on titanium films. The use of tantalum enabled CNTs to be grown on Cu lines and prevented any increase in the sheet

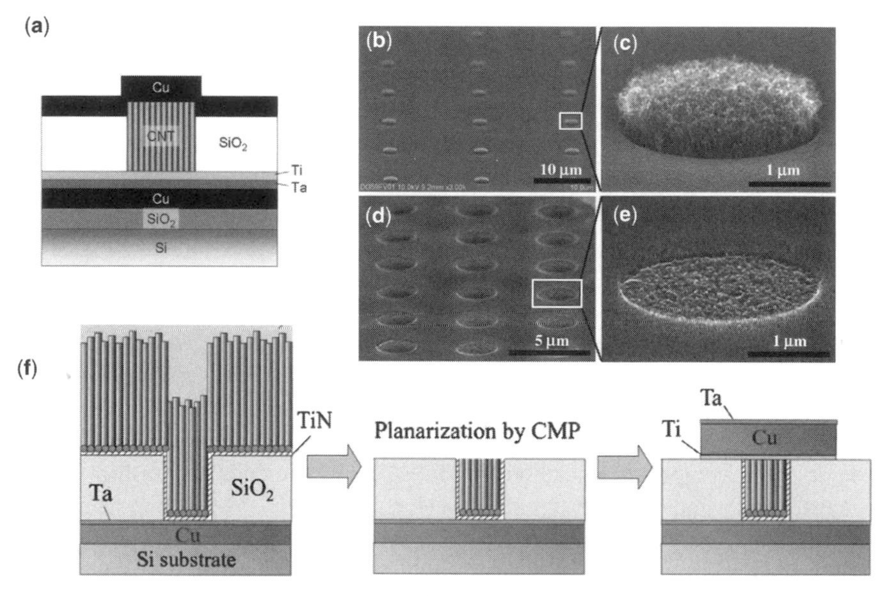

Figure 13.18 (a) Schematics of a through-the-die nanotube via, linking copper interconnects at two different levels (3-D interconnects). (b) to (e) SEM images of the device being constructed. (b), (c) Top surface with the protruding nanotube vias before the chemical-mechanical planarization (CMP) treatment. (d), (e) After the CMP treatment. (f) Schematic of the construction process. The CNTs are grown *in situ* on Co catalyst supported on tantalum films to minimize increase in the sheet resistance of the Cu lines. The CNT are first grown over the whole surface, then CMP is used to remove the excess, leaving only the CNT inside the via conduit. Finally, the new interconnect Cu layer is added, making the new contact with the CNT. (Reproduced with permission from M. Horibe et al., *Jpn. J. Appl. Phys.* **2005**, *44*, 5309.)

resistance of the Cu lines. As described in Figure 13.18f, after the CNT are grown over the whole surface, chemical-mechanical planarization (CMP) is used to remove the excess, leaving only the CNT inside the via conduit. After that, the new interconnect layer is added, making the new contact with the CNT. The SEM images in Figure 13.18b–e illustrate the surface before [(b) and (c)] and after [(d) and (e)] the CMP treatment.

13.7.9 Biomedical Applications

Much interest has grown in recent years in connecting the bio- and nano-fields. The unique physical properties of nanoscale particles used in combination with the intrinsic recognition ability of biomolecular systems open new opportunities for the development of nanobiomedical devices, such as probes, sensors, tracers, tags, etc., with unique characteristics, not only for their size that allows them to access even the intercellular space, but also for their potential in enhanced performance and enhanced biocompatibility.

In the specific case of nanotubes, biofunctionalization for enhanced compatibility has been an area of rapid growth during recent years (154). Several studies have focused on the binding of proteins to SWNT via noncovalent and covalent functionalization schemes. For example, Dai et al. (155) have adsorbed pyrene onto the sidewalls of SWNT and then have made the succinimidyl ester group react with the amine groups in the lysine residues of the proteins to form covalent amide linkages. Similarly, MWNT have been functionalized with 4-hydroxynonenal to induce adsorption of the 4-hydroxynonenal antibody (156).

SWNTs show great potential as biosensors since their electronic structure is very sensitive to changes in the surrounding electrostatic environment. For example, their optical response greatly changes by surface charge transfer or by simple adsorption of molecules, as shown by Barone and Strano (157). They can also be used to detect the dynamics of biological processes in solution, such as the electrooxidation of glucose (158), the reduction of nicotinamide adenine dinucleotide (159), or the hybridization kinetics and thermodynamics of adsorbed DNA (160). Further efforts are underway to maximize selectivity to specific molecules by adding functional groups to the nanotubes that only respond to specific chemical groups.

One of the advantages of SWNT over other biological tags is that nanotubes fluoresce in the NIR and therefore they allow detection at significant penetration since biological tissues are relatively transparent to radiation in the NIR wavelength. Besides, the background due to fluorescence from cells and biological molecules is very low at this wavelength range. Antibody-conjugated SWNTs have been tested as NIR fluorescent labels for probing receptors with high specificity and high sensitivity at the cell surface. In this study, Dai et al. (161) used antibodies that recognize cell surface receptors found on certain breast cancer cells. They detected the NIR photoluminescence from SWNTs and showed that the SWNT-antibody conjugate was indeed selectively attached to the cell surface, where the receptors were (see Fig. 13.19).

The nanohorn dahlias described in Section 13.6.4 appear to be an interesting vehicle for drug delivery, specifically for cancer cell destruction. These carbon

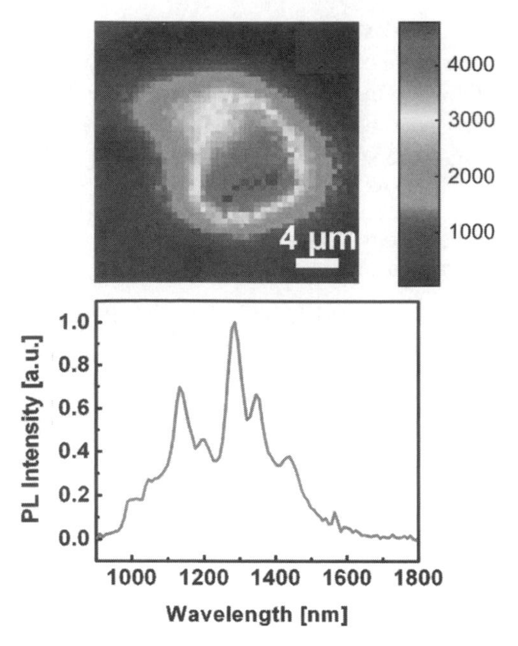

Figure 13.19 (Top) High-magnification NIR fluorescence image of a single Raji cell treated with SWNT-Rituxan conjugate showing NIR fluorescence over the cell. (Bottom) NIR emission spectrum recorded on a SWNT-Rituxan-treated Raji cell. (Reproduced with permission from K. Welsher et al., *Nano Lett.* **2008**, *8*, 586.) (See color insert.)

structures can penetrate the tumor tissue; most importantly, they remain inside this tissue due to the low lymphatic drainage of tumor tissue (162) and have time to slowly release the therapeutic drugs. Their diameter of 2 to 5 nm is particularly suited for both incorporation and slow release, which is important to minimize drug losses before the vehicle has reached the target. Another advantage of nanohorns is the low cytotoxicity exhibited in *in vitro* studies. Moreover, researchers have developed peptide aptamers that selectively recognize proteinous receptors in specific organs and attach them to nanohorns, making the targeted drug delivery field a very promising prospect in medicine (163).

13.8 SUMMARY

Carbon nanostructures display a wide variety of extraordinary properties that greatly depend on their structure. Combined with their small size, these properties can result in devices and emerging applications that have not been possible previously. Advances in controlled synthesis, separation, and characterization of these nanostructures are making possible the reproducible manufacture of carbon nanomaterials with tuned structure for specific applications. Carbon nanostructures may impact diverse fields that range from electronics to materials and biomedical applications.

ACKNOWLEDGMENTS

The Department of Energy is gratefully acknowledged for Grant DE-FG03-02.ER15345 and for funding the Carbon Nanotube Technology Center Program under DE-FG02-06ER64239.

REFERENCES

1. H. W. Kroto, J. R. Heath, S. C. O'Brien, R. F. Curl, R. E. Smalley, *Nature* **1985**, *318*, 162.
2. (a) R. T. K. Baker, M. A. Barber, P. S. Harris, E. S. Feates, R. J. Waite, *J. Catal.* **1972**, *26*, 51; (b) A. Oberun, M. Endo, T. Koyama, *J. Crystal Growth* **1976**, *32*, 335.
3. S. Iijima, *Nature* **1991**, *354*, 56.
4. S. Iijima, T. Ichihashi, *Nature* **1993**, *363*, 603.
5. D. S. Bethune, C. H. Kiang, M. S. de Vries, G. Gorman, R. Savoy, J. Vazquez, R. Beyers, *Nature* **1993**, *363*, 605.
6. R. H. Baughman, A. A. Zakhidov, W. A. de Heer, *Science* **2002**, *297*, 787.
7. K. S. Novoselov, A. K. Geim, S. V. Morozov, D. Jiang, Y. Zhang, S. V. Dubonos, I. V. Grigorieva, A. A. Firsov, *Science* **2004**, *306*, 666.
8. K. S. Novoselov, A. K. Geim, S. V. Morozov, D. Jiang, M. I. Katsnelson, I. V. Grigorieva, S. V. Dubonos, A. A. Firsov, *Nature* **2005**, *438*, 197.
9. E. Hernandez, C. Goze, P. Bernier, A. Rubio, *Phys. Rev. Lett.* **1998**, *80*, 4502.
10. M.-F. Yu, B. S. Files, S. Arepalli, R. S. Ruoff, *Phys. Rev. Lett.* **2000**, *84*, 5552.
11. D. Y. Khang, J. Xiao, C. Kocabas, S. MacLaren, T. Banks, H. Jiang, Y. Y. Huang, J. A. Rogers, *Nano Letters* **2008**, *8*, 124.
12. M. Ouyang, J.-L. Huang, C. L. Cheung, C. M. Lieber, *Science* **2001**, *292*, 702.
13. L. Van Hove, *Phys. Rev.* **1953**, *89*, 1189.
14. T. W. Odom, J.-L. Huang, C. M. Lieber, *J. Phys. Condens. Matter* **2002**, *14*, R145.
15. M. A. Pimenta, A. Marucci, S. A. Empedocles, M. G. Bawendi, E. B. Hanlon, A. M. Rao, P. C. Eklund, R. E. Smalley, G. Dresselhaus, M. S. Dresselhaus, *Phys. Rev. B* **1998**, *58*, R16016.
16. A. M. Rao, E. Richter, S. Bandow, B. Chase, P. C. Eklund, K. A. Williams, S. Fang, K. Subbaswamy, M. Menon, A. Thess, R. E. Smalley, G. Dresselhaus, M. S. Dresselhaus, *Science* **1997**, *275*, 185.
17. S. Reich, C. Thomsen, J. Maultzsch, in *Carbon Nanotubes: Basic Concepts and Physical Properties*. Wiley, New York, 2004.
18. S. M. Bachilo, M. S. Strano, C. Kittrell, R. H. Hauge, R. E. Smalley, R. B. Weisman, *Science* **2002**, *298*, 2361.
19. A. M. Rao, J. Chen, E. Richter, U. Schlecht, P. C. Eklund, R. C. Haddon, U. D. Venkateswaran, Y.-K. Kwon, D. Tomanek, *Phys. Rev. Lett.* **2001**, *86*, 3895.
20. R. Saito, M. Fujita, G. Dresselhaus, M. Dresselhaus, *Appl. Phys. Lett.* **1992**, *60*, 2204.
21. S. D. M. Brown, A. Jorio, P. Corio, M. S. Dresselhaus, G. Dresselhaus, R. Saito, K. Kneipp, *Phys. Rev. B* **2001**, *63*, 155414.

22. A. M. Rao, E. Richter, S. Bandow, B. Chase, P. C. Eklund, K. W. Williams, M. Menon, K. R. Subbaswamy, A. Thess, R. E. Smalley, G. Dresselhaus, M. S. Dresselhaus, *Science* **1997**, *275*, 187.
23. J. Wildoer, L. Venema, A. Rinzler, R. Smalley, C. Dekker, *Nature* **1998**, *39*, 59.
24. H. Kataura, Y. Kumazawa, Y. Maniwa, I. Umezu, S. Suzuki, Y. Ohtsuka, Y. Achiba, *Synth. Met.* **1999**, *103*, 2555.
25. M. E. Itkis, D. Perea, S. Niyogi, J. Love, J. Tang, R. C. Haddon, *J. Phys. Chem.* **2004**, *108*, 12770.
26. M. E. Itkis, D. Perea, S. Niyogi, S. Rickard, M. Hamon, R. C. Haddon, *Nano Lett.* **2003**, *3*, 309.
27. M. E. Itkis, D. Perea, Jung, S. Niyogi, R. C. Haddon, *J. Am. Chem. Soc.* **2005**, *127*, 3439.
28. R. C. Haddon, J. Sippel, A. G. Rinzler, F. Papadimitrakopoulos, *MRS Bull.* **2004**, *29*, 252.
29. Y. Tan, D. Resasco, *J. Phys. Chem. B* **2005**, *109*, 14454.
30. M. J. O'Connell, S. M. Bachilo, C. B. Huffman, V. C. Moore, M. S. Strano, E. H. Haroz, K. L. Rialon, P. J. Boul, W. H. Noon, C. Kittrell, J. Ma, R. H. Hauge, R. B. Weisman, R. E. Smalley, *Science* **2002**, *297*, 593.
31. S. M. Bachilo, M. S. Strano, C. Kittrell, R. H. Hauge, R. E. Smalley, R. B. Weisman, *Science* **2002**, *298*, 2361.
32. R. B. Weisman, "Optical Spectroscopy of Single-Walled Carbon Nanotubes," in *Contemporary Concepts of Condensed Matter Science, Vol. 3: Nanotubes*. Elsevier, New York, 2008.
33. D. A. Tsyboulski, S. M. Bachilo, R. B. Weisman, *Nano Lett.* **2005**, *5*, 975.
34. T. K. Leeuw, R. M. Reith, R. A. Simonette, M. E. Harden, P. Cherukuri, D. A. Tsyboulski, K. M. Beckingham, R. B. Weisman, *Nano Lett.* **2007**, *7*, 2650.
35. F. Ding, A. Rosen, K. Bolton, *Chem. Phys. Lett.* **2004**, *393*, 309.
36. F. Ding, K. Bolton, A. Rosen, *J. Phys. Chem. B* **2004**, *108*, 17369.
37. P. B. Balbuena, J. Zhao, S. P. Huang, Y. X. Wang, N. Sakulchaicharoen, D. E. Resasco, *J. Nanosci. Nanotechnol.* **2006**, *6*, 1247.
38. S. Maruyama, Y. Murakami, Y. Shibuta, Y. Miyauchi, S. Chiashi, *J. Nanosci. Nanotechnol.* **2004**, *4*, 360.
39. A. Monzon, G. Lolli, S. Cosma, M. Sayed-Ali, D. E. Resasco, *J. Nanosci. Nanotechnol.* **2008**, *8*, 1.
40. W. Kratschmer, L. D. Lamb, K. Fostiropoulos, D. R. Huffman, *Nature* **1990**, *347*, 354.
41. R. Bacon, *J. Appl. Phys.* **1960**, *31*, 284.
42. T. W. Ebbesen, P. M. Ajayan, *Nature* **1992**, *358*, 220.
43. Y. Saito, M. Okuda, T. Yoshikawa, *Jpn. J. Appl. Phys.* **1994**, *33*, L186.
44. H. Lange, P. Baranowski, A. Huckzo, P. Byszewski, *Rev. Sci. Instrum.* **1997**, *68*, 3723.
45. Y. Ando, X. Zhao, H. Kataura, Y. Achiba, K. Kaneto, M. Tsuruta, S. Uemura, S. Iijima, *Diamond Relat. Mater.* **2000**, *9*, 847.
46. X. K. Wang, X. W. Lin, V. P. Dravid, J. B. Ketterson, R. P. H. Chang, *Appl. Phys. Lett.* **1993**, *62*, 1881.
47. Y. Ando, S. Iijima, *J. Appl. Phys.* **1993**, *32*, L107.
48. Y. Saito, Y. Tani, A. Kasuya, *J. Phys. Chem. B* **2000**, *104*, 2495.
49. Y. Saito, M. Okuda, T. Koyama, *Surf. Rev. Lett.* **1996**, *31*, 863.

50. J. M. Lambert, P. M. Ajayan, P. Bernier, *Synth. Met.* **1995**, *70*, 1475.
51. C. Journet, W. K. Maser, P. Bernier, A. Loiseau, M. Lamyde la Chapelle, S. Lefrant, P. Deniard, R. Leek, J. E. Fischerk, *Nature* **1997**, *388*, 756.
52. S. Seraphin, D. Zhou, *Appl. Phys. Lett.* **1994**, *64*, 2087.
53. Y. Saito, Y. Tani, N. Miyagawa, K. Mitsushima, A. Kasuya, Y. Nishina, *Chem. Phys. Lett.* **1998**, *294*, 593.
54. A. Thess, R. Lee, P. Nikolaev, H. Dai, P. Petit, J. Robert, C. Xu, Y. H. Lee, S. G. Kim, A. G. Rinzler, D. T. Colbert, G. E. Scuseria, D. Tomanek, J. E. Fischer, R. E. Smalley, *Science* **1996**, *273*, 483.
55. A. A. Puretzky, D. B. Geohegan, X. Fan, S. J. Pennycook, *Appl. Phys. Lett.* **2000**, *76*, 182.
56. M. Yudasaka, T. Komatsu, T. Ichihashi, Y. Achiba, S. Iijima, *J. Phys. Chem. B* **1998**, *102*, 4892.
57. M. Yudasaka, M. Zhang, S. Iijima, *Chem. Phys. Lett.* **2000**, *323*, 549.
58. S. Bandow, S. Asaka, Y. Saito, A. M. Rao, L. Grigorian, E. Richter, P. C. Eklund, *Phys. Rev. Lett.* **1998**, *80*, 3779.
59. P. C. Eklund, B. K. Pradhan, U. J. Kim, Q. Xiong, J. E. Fischer, A. D. Friedman, B. C. Holloway, K. Jordan, M. W. Smith, *Nano Letters* **2002**, *2*, 561.
60. A. A. Puretzky, H. Schittenhelm, X. Fan, M. J. Lance, L. F. Allard, Jr., D. B. Geohegan, *Phys. Rev. B* **2002**, *65*, 245425.
61. A. A. Gorbunov, R. Friedlein, O. Jost, M. S. Golden, J. Fink, W. Pompe, *Appl. Phys. A* **1999**, *69*, S593.
62. S. Arepalli, P. Nikolaev, W. Holmes, B. S. Files, *Appl. Phys. Lett.* **2001**, *78*, 1610.
63. C. D. Scott, S. Arepalli, P. Nikolaev, R. E. Smalley, *Appl. Phys. A* **2001**, *72*, 573.
64. W. E. Alvarez, F. Pompeo, J. E. Herrera, L. Balzano, D. E. Resasco, *Chem. Mater.* **2002**, *14*, 1853.
65. B. Kitiyanan, W. E. Alvarez, J. H. Harwell, D. E. Resasco, *Chem. Phys. Lett.* **2000**, *317*, 497.
66. D. E. Resasco, W. E. Alvarez, F. Pompeo, L. Balzano, J. E. Herrera, B. Kitiyanan, A. Borgna, *J. Nanoparticle Res.* **2002**, *4*, 131.
67. J. E. Herrera, L. Balzano, A. Borgna, W. E. Alvarez, D. E. Resasco, *J. Catal.* **2001**, *204*, 129.
68. W. E. Alvarez, B. Kitiyanan, A. Borgna, D. E. Resasco, *Carbon*, **2001**, *39*, 547.
69. S. Lim, D. Ciuparu, C. Pak, F. Dobek, Y. Chen, D. Harding, L. Pfefferle, G. L. Haller, *J. Phys. Chem. B* **2003**, *107*, 11048.
70. D. Ciuparu, Y. Chen, S. Lim, G. L. Haller, L. Pfefferle, *J. Phys. Chem. B* **2004**, *108*, 503.
71. G. Lolli, L. Zhang, L. Balzano, N. Sakulchaicharoen, Y. Tan, D. E. Resasco, *J. Phys. Chem. B* **2006**, *110*, 2108.
72. E. Flahaut, A. Govindaraj, A. Peigney, Ch. Laurent, A. Rousset, C. N. R. Rao, *Chem. Phys. Lett.* **1999**, *300*, 236.
73. S. Maruyama, R. Kojima, Y. Miyauchi, S. Chiashi, M. Kohno, *Chem. Phys. Lett.* **2002**, *360*, 229.
74. Y. Murakami, Y. Miyauchi, S. Chiashi, S. Maruyama, *Chem. Phys. Lett.* **2003**, *374*, 53.
75. E. Lamouroux, P. Serp, P. Kalck, *Catal. Rev.* **2007**, *49*, 341.
76. H. M. Cheng, F. Li, G. Su, H. Y. Pan, L. L. He, X. Sun, M. S. Dresselhaus, *Appl. Phys. Lett.* **1998**, *278*, 3282.

77. R. Sen, A. Govindaraj, C. N. R. Rao, *Chem. Phys. Lett.* **1997**, *267*, 276.
78. R. Sen, A. Govindaraj, C. N. R. Rao, *Chem. Mater.* **1997**, *9*, 2078.
79. P. Nikolaev, M. J. Bronikowski, R. K. Bradley, F. Rohmund, D. T. Colbert, K. A. Smith, R. E. Smalley, *Chem. Phys. Lett.* **1999**, *313*, 91.
80. M. J. Bronikowski, P. A. Willis, D. T. Colbert, K. A. Smith, R. E. Smalley, *J. Vac. Sci. Technol. A* **2001**, *19*, 1800.
81. J. H. Hafner, M. J. Bronikowski, B. R. Azamian, P. Nikolaev, A. G. Rinzler, D. T. Colbert, K. A. Smith, R. E. Smalley, *Chem. Phys. Lett.* **1998**, *296*, 195.
82. E. Flahaut, R. Bacsa, A. Peigney, Ch. Laurent, *Chem. Commun.* **2003**, 1442.
83. S. C. Lyu, B. C. Liu, S. H. Lee, C. Y. Park, H. K. Kang, C. W. Yang, C. J. Lee, *J. Phys. Chem. B* **2004**, *108*, 2192.
84. W. Ren, H. M. Cheng, in *Raman Spectroscopy on Double-Walled Carbon Nanotubes*. Lecture Notes in Nanoscale Science and Technology. Springer, New York, 2008, 2, 29.
85. C. J. Lee, J. H. Park, J. Park, *Chem. Phys. Lett.* **2000**, *323*, 560.
86. A. H. Barber, R. Andrews, L. S. Schadler, H. D. Wagner, *Appl. Phys. Lett.* **2005**, *87*, 203106.
87. M. Wilson, *Phys. Today* **2006**, *59*, 21.
88. M. Ezawa, *Phys. Rev. B* **2006**, *73*, 045432.
89. B. Obradovic, R. Kotlyar, F. Heinz, P. Matagne, T. Rak-shit, M. D. Giles, M. A. Stettler, D. E. Nikonov, *Appl. Phys. Lett.* **2006**, *88*, 142102.
90. R. Ruoff, *Nature Nanotechnol.* **2008**, *3*, 10.
91. S. Stankovich, D. A. Dikin, R. D. Piner, K. A. Kohlhaas, A. Kleinhammes, Y. Jia, Y. Wu, S. B. T. Nguyen, R. S. Ruoff, *Carbon* **2007**, *45*, 1558.
92. S. Iijima, M. Yudasaka, R. Yamada, S. Bandow, K. Suenaga, F. Kokai, K. Takahashi, *Chem. Phys. Lett.* **1999**, *309*, 165.
93. T. Yamagachi, S. Bow, S. Iijima, *Chem. Phys. Lett.* **2004**, *389*, 181.
94. F. Fernandez-Alonso, F. J. Bermejo, C. Cabrillo, R. O. Loutfy, V. Leon, M. L. Saboungi, *Phys. Rev. Lett.* **2007**, *98*, 215503.
95. H. Takikawa, M. Ikeda, K. Hirahara, Y. Hibi, Y. Tao, P. A. Ruiz, Jr., T. Sakakibara, S. Itoh, S. Iijima, *Physica B* **2002**, *323*, 277.
96. W. Z. Li, S. S. Xie, L. X. Qian, B. H. Chang, B. S. Zou, W. Y. Zhou, R. A. Zhao, G. Wang, *Science* **1996**, *274*, 1701.
97. Z. F. Ren, Z. P. Huang, J. W. Xu, J. H. Wang, P. Bush, M. P. Siegal, P. N. Provencio, *Science* **1998**, *282*, 1105.
98. D. Yokoyama, T. Iwasaki, T. Yoshida, H. Kawarada, S. Sato, T. Hyakushima, M. Nihei, Y. Awano, *Appl. Phys. Lett.* **2007**, *91*, 263101.
99. Y. C. Choi, Y. M. Shin, Y. H. Lee, B. S. Lee, G. S. Park, W. B. Choi, N. S. Lee, J. M. Kim, *Appl. Phys. Lett.* **2000**, *76*, 2367.
100. C. J. Lee, D. W. Kim, T. J. Lee, Y. C. Choi, Y. S. Park, Y. H. Lee, W. B. Choi, N. S. Lee, G. S. Park, J. M. Kim, *Chem. Phys. Lett.* **1999**, *312*, 461.
101. Y. Murakami, E. Einarsson, T. Edamura, S. Maruyama, *Carbon* **2005**, *43*, 2664.
102. G. F. Zhong, T. Iwasaki, K. Honda, Y. Furukawa, I. Ohdomari, H. Kawarada, *Chem. Vapor. Depos.* **2005**, *11*, 127.
103. S. Maruyama, E. Einarsson, Y. Murakami, T. Edamura, *Chem. Phys. Lett.* **2005**, *403*, 320.

104. K. Hata, D. N. Futaba, K. Mizuno, T. Namai, M. Yumura, S. Iijima, *Science* **2004**, *306*, 1362.

105. Y. Murakami, S. Chiashi, Y. Miyauchi, M. Hu, M. Ogura, T. Okubo, S. Maruyama, *Chem. Phys. Lett.* **2004**, *385*, 298.

106. M. Hu, Y. Murakami, M. Ogura, S. Maruyama, T. Okubo, *J. Catal.* **2004**, *225*, 230.

107. Y. Murakami, E. Einarsson, T. Edamura, S. Maruyama, *Carbon* **2005**, *43*, 2664.

108. D. N. Futaba, K. Hata, T. Yamada, K. Mizuno, M. Yumura, S. Iijima, *Phys. Rev. Lett.* **2005**, *95*, 056104.

109. D. N. Futaba, K. Hata, T. Namai, T. Yamada, K. Mizuno, Y. Hayamizu, M. Yumura, S. Iijima, *J. Phys. Chem. B* **2006**, *110*, 8035.

110. L. Zhang, Y. Tan, D. E. Resasco, *Chem. Phys. Lett.* **2006**, *422*, 198.

111. L. Zhang, Z. Li, Y. Tan, G. Lolli, N. Sakulchaicharoen, F. G. Requejo, B. S. Mun, D. E. Resasco, *Chem. Mater.* **2006**, *18*, 235624.

112. W. B. Choi, J. U. Chu, K. S. Jeong, E. J. Bae, J. W. Lee, J. J. Kim, J. O. Lee, *Appl. Phys. Lett.* **2001**, *79*, 3696.

113. L. M. Dai, *Smart Mater. Struct.* **2002**, *11*, 645–651.

114. J. Li, H. T. Ng, A. Cassell, W. Fan, H. Chen, Q. Ye, J. Koehne, J. Han, M. Meyyappan, *Nano Lett.* **2003**, *3*, 597.

115. Y. G. Zhang, A. L. Chang, J. Cao, Q. Wang, W. Kim, Y. M. Li, N. Morris, E. Yenilmez, J. Kong, H. J. Dai, *Appl. Phys. Lett.* **2001**, *79*, 3155.

116. J. E. Fischer, W. Zhou, J. Vavro, M. C. Llaguno, C. Guthy, R. Haggenmueller, M. J. Casavant, D. E. Walters, R. E. Smalley, *J. Appl. Phys.* **2003**, *93*, 2157.

117. S. M. Huang, M. Woodson, R. Smalley, J. Liu, *Nano Lett.* **2004**, *4*, 1025.

118. Ismach, L. Segev, E. Wachtel, E. Joselevich, *Angew. Chem.* **2004**, *116*, 6266.

119. S. Han, X. Liu, C. Zhou, *J. Am. Chem. Soc.* **2005**, *127*, 5294.

120. C. Kocabas, S.-H. Hur, A. Gaur, M. A. Meitl, M. Shim, J. A. Rogers, *Small* **2005**, *1*, 1110.

121. B. Vigolo, A. Pénicaud, C. Coulon, C. Sauder, R. Pailler, C. Journet, P. Bernier, P. Poulin, *Science* **2000**, *290*, 1331.

122. K. Jiang, Q. Li, S. Fan, *Nature* **2002**, *419*, 801.

123. M. Zhang, K. R. Atkinson, R. H. Baughman, *Science* **2004**, *306*, 1358.

124. M. Zhang, S. L. Fang, A. A. Zakhidov, S. B. Lee, A. E. Aliev, C. D. Williams, K. R. Atkinson, R. H. Baughman, *Science* **2005**, *309*, 1215.

125. Y. L. Li, I. A. Kinloch, A. H. Windle, *Science* **2004**, *304*, 76.

126. Y. L. Yang, M. C. Gupta, K. L. Dudley, R. W. Lawrence, *Nano Lett.* **2005**, *5*, 2131.

127. N. Li, Y. Huang, F. Du, X. He, X. Lin, H. Gao, Y. Ma, F. Li, Y. Chen, P. C. Eklund, *Nano Lett.* **2006**, *6*, 1141.

128. M. M. J. Treacy, T. W. Ebbesen, J. M. Gibson, *Nature* **1996**, *381*, 678.

129. P. M. Ajayan, O. Stephan, C. Colliex, D. Trauth, *Science* **1994**, *265*, 1212.

130. J. Gao, B. Zhao, M. E. Itkis, E. Bekyarova, H. Hu, V. Kranak, A. Yu, R. C. Haddon, *J. Am. Chem. Soc.* **2006**, *128*, 7492.

131. M. Mu, K. I. Winey, *J. Phys. Chem. C* **2007**, *111*, 17923.

132. M. Chhowalla, *J. Soc. Inform. Display* **2007**, *15*, 1085.

133. L. Hu, Y.-L. Zhao, K. Ryu, C. Zhou, J. F. Stoddart, G. Grüner, *Adv. Mater.* **2008**, *20*, 939.

134. T. Durkop, S. A. Getty, E. Cobas, M. S. Fuhrer, *Nano Lett.* **2004**, *4*, 35.
135. J. H. Hafner, C. L. Cheung, T. H. Oosterkamp, C. M. Lieber, *J. Phys. Chem. B* **2001**, *105*, 743.
136. J. Kong, N. R. Franklin, C. Zhou, M. G. Chapline, S. Peng, K. Cho, H. Dai, *Science* **2000**, *287*, 622.
137. G. U. Sumanasekera, C. K. W. Adu, S. Fang, P. C. Eklund, *Phys. Rev. Lett.* **2000**, *85*, 1096.
138. O. K. Varghese, P. D. Kichambre, D. Gong, K. G. Ong, E. C. Dickey, C. A. Grimes, *Sensors Actuators B* **2001**, *81*, 32.
139. S. Chopra, A. Pham, J. Gaillard, A. Parker, A. M. Rao, *Appl. Phys. Lett.* **2002**, *80*, 4632.
140. J. A. Robinson, E. S. Snow, F. K. Perkins, *Sensors Actuators A: Physical* **2007**, *135*, 309.
141. J. M. Kinaret, T. Nord, S. Viefers, *Appl. Phys. Lett.* **2003**, *82*, 1287.
142. H. J. Hwang, J. W. Kang, *Physica E* **2005**, *27*, 163.
143. L. M. Jonsson, S. Axelsson, T. Nord, S. Viefers, J. M. Kinaret, *Nanotechnology* **2004**, *15*, 1497.
144. V. Sazonova, Y. Yaish, H. Ustunel, D. Roundy, T. A. Arias, P. L. McEuen, *Nature* **2004**, *431*, 284.
145. B. Witkamp, M. Poot, H. S. J. van der Zant, *Nano Lett.* **2006**, *6*, 2904.
146. A. Javey, J. Guo, Q. Wang, M. Lundstrom, H. J. Dai, *Nature* **2003**, *424*, 654.
147. S. Rosenblatt, Y. Yaish, J. Park, J. Gore, V. Sazonova, P. L. McEuen, *Nano Lett.* **2002**, *2*, 869.
148. H. E. Unalan, G. Fanchini, A. Kanwal, A. Du Pasquier, M. Chhowalla, *Nano Lett.* **2006**, *6*, 677.
149. P. G. Collins, M. S. Arnold, P. Avouris, *Science* **2001**, *292*, 706.
150. T. Fukao, S. Nakamura, H. Kataura, M. Shiraishi, *Jpn. J. Appl. Phys.* **2006**, *45*, 6524.
151. W. B. Choi, D. S. Chung, J. H. Kang, H. Y. Kim, Y. W. Jin, I. T. Han, Y. H. Lee, J. E. Jung, N. S. Lee, G. S. Park, J. M. Kim, *Appl. Phys. Lett.* **1999**, *75*, 3129.
152. J. Robertson, *Materials Today* **2007**, *10*, 36.
153. M. Horibe, M. Nihei, D. Kondo, A. Kawabata, Y. Awano, *Jpn. J. Appl. Phys.* **2005**, *44*, 5309.
154. D. A. Heller, S. Baik, T. E. Eurell, M. S. Strano, *Adv. Mater.* **2005**, *17*, 2793.
155. R. Chen, Y. Zhang, D. Wang, H. Dai, *J. Am. Chem. Soc.* **2001**, *123*, 3838.
156. M. P. Mattson, R. C. Haddon, A. M. Rao, *J. Mol. Neurosci.* **2000**, *14*, 175.
157. P. W. Barone, M. S. Strano, *Angew. Chem. Int. Ed.* **2006**, *45*, 8138.
158. P. P. Joshi, S. A. Merchant, Y. Wang, D. W. Schmidtke, *Anal. Chem.* **2005**, *77*, 3183.
159. F. Valentini, S. Orlanducci, M. Letizia Terranova, G. Palleschi, *Sensors Actuators B* **2007**, *123*, 5.
160. E. S. Jeng, P. W. Barone, J. D. Nelson, M. S. Strano, *Small* **2007**, *3*, 1602.
161. K. Welsher, Z. Liu, D. Daranciang, H. Dai, *Nano Lett.* **2008**, *8*, 586.
162. K. Ajima, M. Yudasaka, T. Murakami, A. Maigne, K. Shiba, S. Iijima, *Mol. Pharmac.* **2005**, *2*, 475.
163. K. Shiba, *J. Drug Targeting* **2006**, *14*, 512.

PROBLEMS

1. Show that the diameter of an (n, m) SWNT is related to the chiral indices by the expression:

$$d = a_0 \cdot (m^2 + mn + n^2)^{1/2}/\pi$$

where a_0 is the atomic lattice constant (=0.246 nm).

2. Show that the chiral angle of an (n, m) SWNT is related to the chiral indices by the expression:

$$\tan \Theta = (n\sqrt{3})/(2m + n)$$

3. The figure shows the unit cell of a (10, 10) SWNT. How long would a (10, 10) SWNT that weighs 1 g be? How does this distance compare with the distance from the Earth to the Sun?

4. The (9, 1) SWNT is a small-diameter semiconducting nanotube

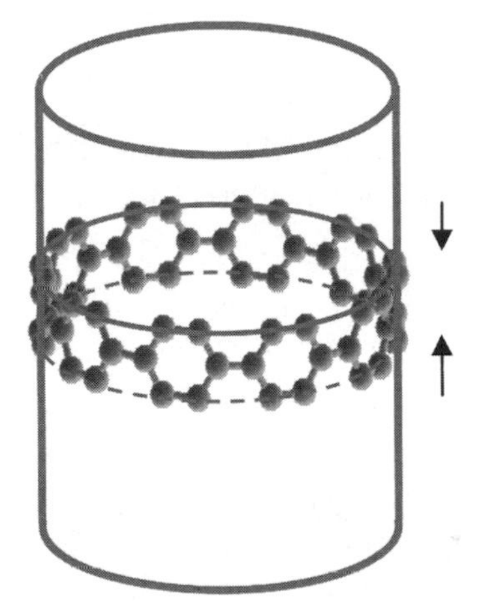

   (a) Calculate its diameter, chiral angle, and radial breathing mode (RBM) vibrational frequency.
   (b) The S11 and S22 bandgap energies for this nanotube are 1.36 and 1.79 eV, respectively. Can the RBM band of this nanotube be observed in a Raman spectrometer that uses a red laser (633 nm) as an excitation source?
   (c) Would the bandgap energies for a (9, 4) nanotube be higher or lower than for (9, 1)?

5. In a hypothetical CCVD process similar to CoMoCAT, SWNT are produced in a fluidized bed reactor by disproportionation of carbon monoxide ($2CO \rightarrow C + CO_2$). Fresh catalyst is added at a continuous rate (CT1) of 10 kg/h. Before purification, the "black sand" product (CT2) contains 30 wt% carbon, while the rest of

the product is composed of the used catalyst. The mass flow rate of pure CO fed to the system (F0) is 35.7 kg/h. To increase the utilization of unreacted CO, a recycle stream (R) is used at a recycle mass ratio of 3 (i.e. R/F0 = 3). Since the above reaction is reversible, the concentration of CO_2 should be kept low. Therefore, a CO_2 membrane separation unit is incorporated in the recycle. The efficiency of the membrane is 80%. That is, of each 1 kg of CO_2 entering the membrane 0.8 kg is purged (P2).

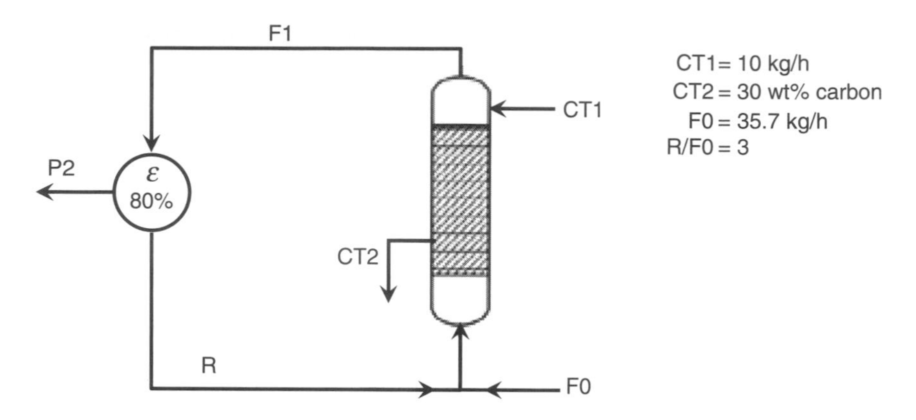

(a) Calculate the molar fraction of CO_2 at the entrance of the reactor.

(b) Calculate the molar fraction of the CO_2 entering the reactor.

6. Use the tensile test curves shown in Figure 13.15.

(a) Calculate the ultimate strength, elongation at fracture, and Young's modulus for pure Nylon and for the SWNT/Nylon composites.

(b) Why is the modulus for the pristine SWNT/Nylon composite the same as that for pure Nylon?

7. In many publications reporting the covalent functionalization of pure SWNT, the D-band in the Raman spectrum is observed to increase significantly upon functionalization. When the functionalized sample is heated in He at 400°C, the band decreases again to a low D/G ratio, similar to that of the pristine sample. Explain these observations.

8. Why is it necessary to use a surfactant in order to see the fluorescence of SWNT? Do you expect that a solid sample of pure SWNT would fluoresce? Why?

9. Give a possible explanation for the calculated maximum in carbon concentration shown in Figure 13.6. Why would the number of C atoms dissolved in a cluster initially increase during growth, reach a maximum, and then decrease to a stable concentration?

10. Nanohorns are being considered as vehicles for therapeutic drug delivery. What physical and chemical modifications should be done on them to optimize their efficiency for this important application?

ANSWERS

1. $a_0 \cdot C_h =$ perimeter of tube $= \pi \cdot d$
From Figure P.1, we see that:

$$C_h^2 = n_j^2 + (n_i + \boldsymbol{m})^2$$

But, in a hexagonal system the angle between n and the y-axis is 30 degrees.

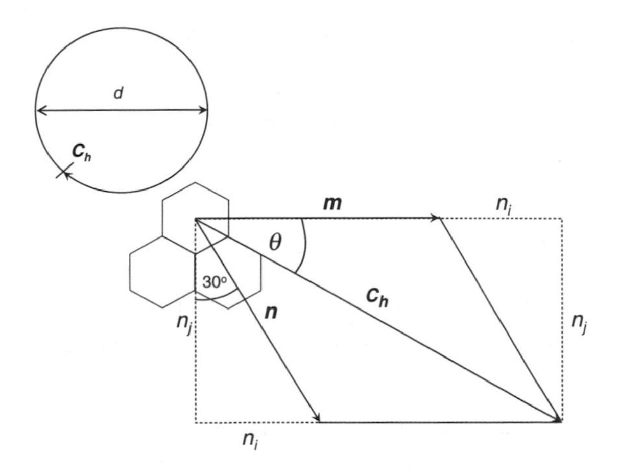

Then,

$$\sin 30 = 0.5 = n_i/\boldsymbol{n} \quad \text{and} \quad \cos 30 = 3^{1/2}/2 = n_j/\boldsymbol{n}$$

Thus,

$$\begin{aligned} C_h^2 &= 3/4\boldsymbol{n}^2 + (0.5\boldsymbol{n})^2 + \boldsymbol{m}^2 + \boldsymbol{mn} \\ &= \boldsymbol{n}^2 + \boldsymbol{m}^2 + \boldsymbol{mn} \end{aligned}$$

Then,

$$d - a_0 \cdot (\boldsymbol{m}^2 + \boldsymbol{mn} + \boldsymbol{n}^2)^{1/2}/\pi$$

2. From Figure P.1 above,

$$\begin{aligned} \tan \Theta &= \frac{n_j}{(n_i + \boldsymbol{m})} \\ &= \frac{(3^{1/2}/2)\boldsymbol{n}}{(\boldsymbol{n}/2 + \boldsymbol{m})} \\ &= \frac{(3^{1/2})\boldsymbol{n}}{(\boldsymbol{n} + 2\boldsymbol{m})} \end{aligned}$$

3. How long is a 1 g (10, 10) nanotube?

   Number of C atoms in the unit cell = 40 C atoms;

   Mass of the unit cell = $40 \times 12/6.0$ E23 g

   Length of the unit cell = 0.246 nm

mass/length	3.2E − 12	g/m
length for 1 gram	3.1E + 11	m
	1.9E + 08	miles
	191.8	million miles
distance of the Earth to the Sun	92955820.0	miles
	93.0	million miles

It is more than twice the distance of the Earth to the Sun!

4. **(a)** The (9, 1) SWNT diameter, chiral angle, and radial breathing mode (RBM) vibrational frequency.

   (b) The S11 and S22 bandgap energies for this nanotube are 1.36 and 1.79 eV, respectively. Can the RBM band of this nanotube be observed in a Raman spectrometer that uses a red laser (633 nm) as an excitation source?

   (c) Would the bandgap energies for a (9, 4) nanotube be higher or lower than for (9, 1)?

5. Doing overall mass balances around the entire system (i.e. F0 + CT1 = P2 + CT2), around each of the two units [(F1 = P2 + R) and (CT1 + F0 + R = F1 + CT2)], as well as species molar balances for CO and CO_2 (using the reaction stoichiometry), we arrive at the following individual mass flow rates in each stream.

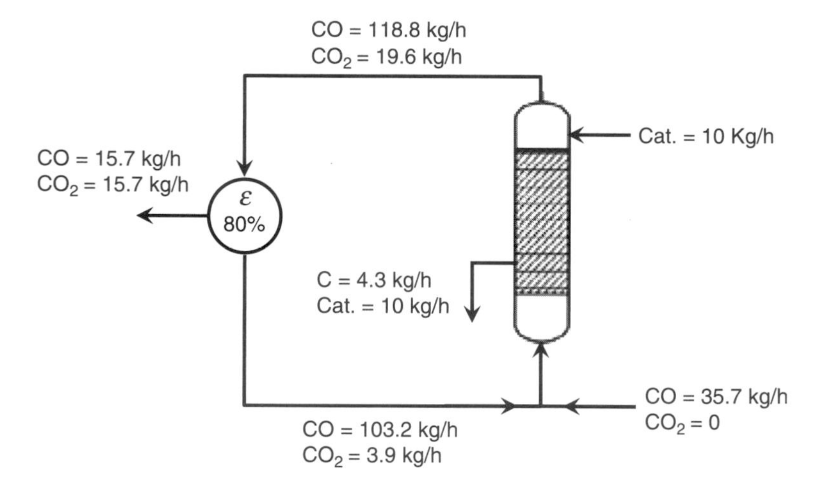

The molar flow rates and molar fractions of CO and CO_2 are summarized in the following table. As shown in the table, the molar fraction of CO_2 at the entrance of the reactor (streams R + F0) is 1.8 mol%.

Stream	CO (kg/h)	CO (mol/h)	CO_2 (kg/h)	CO_2 (mol/h)	y (CO)	y (CO_2)
CT1	0.0	0.0	0.0	0.0	0.000	0.000
CT2	0.0	0.0	0.0	0.0	0.000	0.000
F0	35.7	1275.0	0.0	0.0	1.000	0.000
F1	118.8	4242.9	19.6	445.5	0.905	0.095
R	103.2	3685.7	3.9	88.6	0.977	0.023
P2	15.7	560.7	15.7	356.8	0.611	0.389
F0 + R	138.9	4960.7	3.9	88.6	0.982	0.018

6. To calculate the modulus, one can use the slope of the initial portion of the stress-strain diagram, as shown in the figure.

% COOH in SWNT	Modulus (MPa)
No SWNT	445
0.0	445
4.2	755
6.0	1238
6.8	1238

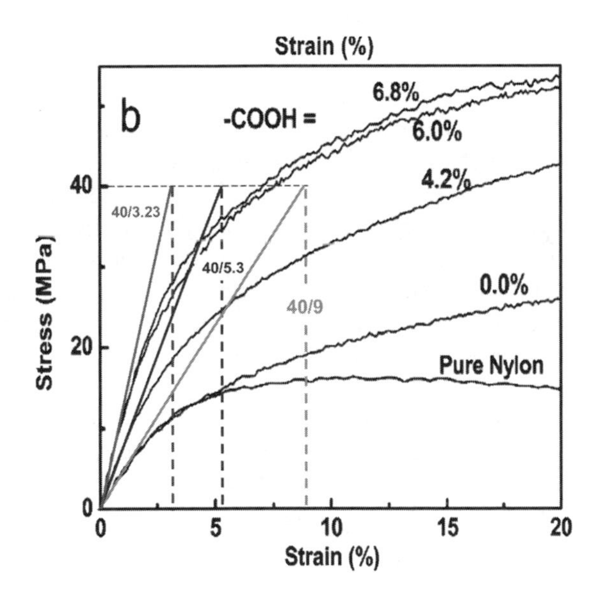

The sample with unfunctionalized SWNT has the same modulus as that without SWNT because in this sample, the SWNT have no interfacial interactions with the matrix. At the other extreme, when the concentration of functional groups exceeds a certain value, there is no much gain in increasing the density of functionalities further.

7. The D-band in Raman spectra is due to the presence of sp^3 carbon on the SWNT wall. Covalent functionalization generates sp^3 bonds between one of the C on the wall and the functional group, which causes the observed increases in D-band. When the functionalized sample is heated in He at 400°C, the sp^3 bond that attaches the functional group to the SWNT is broken. As a result the functional group is detached and the sp^2 C is healed to its original position, causing the observed decrease in D-band. It must be noted that this defunctionalization process is not entirely reversible. It is typically observed that the dispersibility of the SWNT greatly deteriorates after the functionalization/defunctionalization process.

8. Only semiconducting SWNT can fluoresce. Metallic SWNT do not because there is density of states at the Fermi level. In order to see the fluorescence of an SWNT sample they need to be suspended as individual tubes, otherwise if they are in bundles, there will always be a metallic tube in the bundle that will quench the fluorescence.

For the same reason, a solid sample of pure SWNT does not fluoresce because the semiconducting nanotubes will always be in contact with some metallic tubes that quickly deexcite those tubes that have been excited by the incoming radiation. Also, since the excitation energy is in the visible, the penetration of visible light into a solid sample of SWNT is very low, so only those semiconducting tubes on the surface can be excited. By contrast, when SWNT are well dispersed and suspended as individual nanotubes in a solid polymer, this solid can fluoresce.

9. The SWNT growth occurs in several steps, first the C-containing molecule from the gas phase dissociates on the catalyst surface, then C dissolves into the catalyst cluster and distributes between the surface and the bulk. Over the surface, isolated carbon atoms start precipitating and forming small polygons that begin a nucleation process. One can expect that before the nucleation on the surface, C atoms will have a stronger driving force to go into the bulk than staying at the surface since the number of C-metal bonds in the bulk is higher than on the surface. During that time the concentration in the bulk will rapidly grow, as shown in Figure 13.6. However, once the polygons on the surface form a thermodynamically more stable structure (embryo), then the driving force for the C atoms to join the growing nucleus is higher than that pushing it into the bulk of the cluster. As a result, the concentration in the bulk will decrease until it reaches a stable value and will continue at that value until the growth stops.

10. SWNH need to be made more biocompatible and their drug incorporation/release capacity needs to be enhanced. Possible ways of accomplishing this are:

- Create oxygenated groups to enhance biocompatibility.
- Functionalize SWNH via diazonium salt to incorporate amine and carboxylic groups which may interact better with proteins and biological molecules.
- Chemically attack the walls of SWNH to create holes and allow for drug penetration and later delivery.

PART VI

NANOCATALYSTS, SORBENTS, AND ENERGY APPLICATIONS

14

REACTION OF NANOPARTICLES WITH GASES

KEN-ICHI AIKA

14.1 Introduction: Classification of Nanoparticle Reactions, 496
- 14.1.1 Solid with Solid, 496
- 14.1.2 Solid with Liquid, 496
- 14.1.3 Solid with Gas, 497

14.2 Structure of Solid Nanoscale Materials, 497

14.3 Morphology for Understanding of the Mineralization Reaction, 498

14.4 Mineralization of Chlorine Containing Volatile Organic Compounds, 501

14.5 Basic Research on the Mineralization of Organic Chlorides, 501

14.6 CaO-C Composite Materials for Mineralization of Cl-Containing Hydrocarbons (CH_hCl_x) Under Inert Conditions, 502

14.7 Nanopowder CaO Prepared from $Ca(OH)_2$ for Reaction in Air, 505

14.8 Volume Change of Powder During the Reaction, 508

14.9 Reaction System for Mineralization of Cl from Organic Cl Compounds, 511

14.10 Flow and Reaction Rate Analysis for a Column Reactor, 512

14.11 Conclusion, 514

References, 515

Problems, 516

Answers, 516

Nanoscale Materials in Chemistry, Second Edition. Edited by K. J. Klabunde and R. M. Richards

14.1 INTRODUCTION: CLASSIFICATION OF NANOPARTICLE REACTIONS

14.1.1 Solid with Solid

Reactions of nanoscale materials are classified with respect to the surrounding media: solid, liquid, and gas phases. In the solid phase, nanoscale crystals are usually connected with each other to form a powder particle (micron scale) or a pellet (milli scale); see Figure 14.1. Two or more materials (powder or pellet) are mixed and fired to form a new material. The nanosized structure is favored, due to the mixing efficiency and high reaction rate. Alloys (metals), ceramics (oxides), cement (oxides), catalysts (metals and oxide), cosmetics (oxides), plastics (polymers), and many functional materials are produced through solid reaction of nanoscale materials. One recent topic of interest is the production of superconductive mixed oxides, where control of the layered structure during preparation is a key step.

14.1.2 Solid with Liquid

In the liquid phase, nanoscale materials are stabilized as colloids. Colloid particles can react with soluble molecules or two different colloid particles can react with each other. These reactions are usually controlled by the surface charge or the surface functional groups. The reacted colloids are separated, dried, and finally formed into powder or pellets in an attempt to maintain the nanostructure. Nanostructured

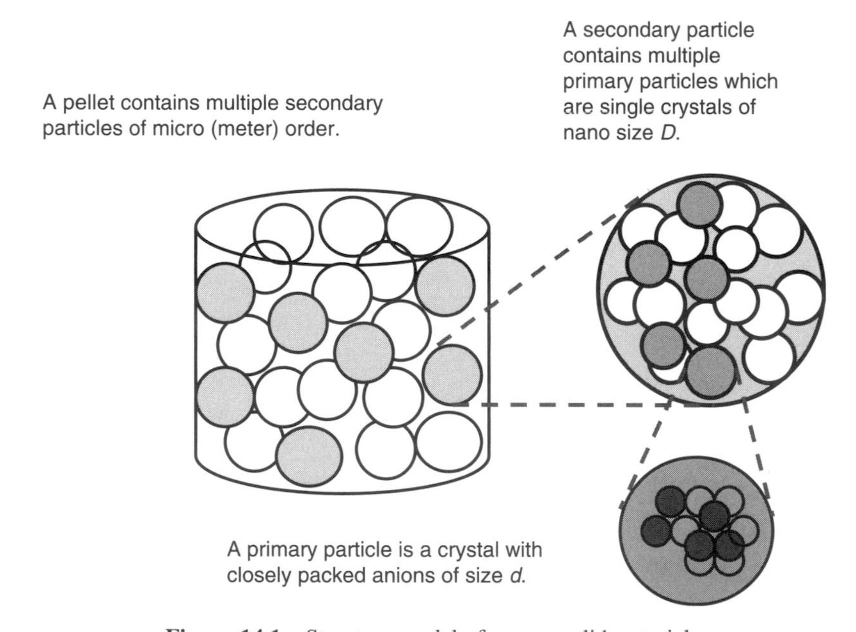

Figure 14.1 Structure model of porous solid materials.

powder or pellets can be covered with functional materials through these wet processes. Pharmaceuticals, foods, fine chemicals, and many other industrial materials are produced using nanoscale technology in liquid phase. Drug-delivery systems using nanoscale materials are a contemporary topic in this field. All of the above areas of application are very broad, and some of them are reviewed in the first edition of this volume (1). This category of reaction is not covered in this chapter.

14.1.3 Solid with Gas

The third category, reactions with gases, is the subject of this chapter. Powder or pellets containing nanoparticles are able to react with gases effectively because of their high surface area. Examples of such reactions are numerous in industry. Many mineral resources are crushed and treated with gases to form industrial products. Zinc and iron sulfides are calcined to form their respective oxides. Crushed magnetite ore is reduced with hydrogen or CO to form steel. Porous carbon materials (coal briquettes) are reacted with air to supply heat for domestic use. Calcium hydroxide powder is supplied to a furnace to absorb hydrogen chloride. In addition to these classic examples, many gas-solid reactions are recently being developed for energy and environmental purposes. Fine iron particles react with water producing hydrogen for use in fuel cell vehicles (2). The resulting iron oxide is then regenerated (reduced) at the hydrogen refilling station. Fine nickel particles (Ni mixed with alumina) are used to produce energy from methane with the fixing of CO_2, but without producing NOx at the reduction column (3). The resulting Ni is transferred and oxidized with air in an oxidation column. The pellet containing nanoscale Ni clusters is cycled between the two columns:

$$CH_4 + NiO \longrightarrow CO_2\ (\text{fixed}) + H_2O\ (\text{condensed}) + Ni + \text{energy} \quad (14.1)$$

$$Ni + O_2 + (N_2) \longrightarrow NiO + (N_2) + \text{energy} \quad (14.2)$$

MgO or CaO reacts with halide-containing molecules [volatile organic compounds (VOC) and Freon; (4, 5)]. These reactions are important for environmental reasons, and such reactions are covered in this chapter. The reported works are mostly based on subjects of academic interest. There are few reviews regarding reactions with respect to nanoscale morphology. In this chapter, reaction engineering and material morphology are stressed, using an example of Cl fixation of organic chlorides into CaO nanoparticles.

14.2 STRUCTURE OF SOLID NANOSCALE MATERIALS

Most high surface area materials are oxides formed by calcining hydroxide that was in nanoscale colloid or sol form in aqueous solution. High surface area oxides (pellets) are usually composed of secondary particles that are composed of nanoscale primary particles, as shown in Figure 14.1. The primary particles are said to maintain their size through the calcination process from hydroxide (colloid or gel) in aqueous solution. The primary particles, usually single crystals, adhere together to form secondary particles with micropores inside. Meso- or macropores are located around the secondary particles. If the secondary particles are in powder form, they are collected and mechanically formed into pellets. If the secondary particles are agglomerated, then crushing and sieving are carried out to form pellets (Fig. 14.1).

The nanoscale materials considered in this chapter are solid materials with primary particles sizes ranging from 1 to 100 nm (10 to 1000 Å), as defined in Chapter 1 of the first edition (1). Materials referred to as powders are simply classified through observation; rough powder (1 mm to 40 μm, optical microscope), fine powder (40 μm to 1 μm, optical microscope), and superfine powder (<1 μm, only visible by electron microscope). The latter, superfine powder or nanoparticles, may correspond to the primary particle. These nanoparticles are usually in the form of smoke, combusted particulates, and dispersed primary particles. Some rough and fine powders may be composed of primary particles if the surface area is high. The microstructure of a pellet or powder is important, because the chemical reaction starts from the surface (favored for small crystals) and gas diffuses through the void channels. The morphology controls the reaction rate, which is one of the main concerns in this chapter.

14.3 MORPHOLOGY FOR UNDERSTANDING OF THE MINERALIZATION REACTION

Powder or pelletized powder is discussed using the model given in Figure 14.1. For convenience, an ideal model is applied, where the size of primary or secondary particles is uniform. First, the crystal structure of Ca and Mg oxides and chlorides is discussed, which is related to the mineralization reaction. CaO and MgO crystals have the salt (NaCl) structure, and $CaCl_2$ and $MgCl_2$ crystals have the distorted rutile and $CdCl_2$ structures, respectively (6). For all of these crystals, the size of the anion is much larger than that of the cation. For a first approximation, we assume that the crystal is composed of only anions, as shown in Figure 14.1 (down right). Any of the four oxide anions (even the salt structure) has a close-packing structure. If we assume the diameter of the anion is d, then the volume allotted for one anion is calculated as $(1/\sqrt{2})\,d^3$. Since the molar volume is known, the imaged diameter d can be calculated, as shown in Table 14.1 where $CaCl_2$ is expressed as $Ca_{0.5}Cl$.

The calculated diameters of oxide anions are 0.34 nm for CaO and 0.296 nm for MgO, while generally accepted sizes from the ionic radii of the oxide ion (0.14 nm) is 0.28 nm (7). Since the cation size of Ca is fairly large (0.20 nm) compared to Mg (0.144 nm), extra space between oxide ions is necessary in the CaO crystal. However, for convenience, the calculated d is used for further discussion. For chlorides, the calculated diameters, 0.394 nm for $Ca_{0.5}Cl$ (calcium chloride) and 0.364 nm for $Mg_{0.5}Cl$ (magnesium chloride), are close to the accepted value of 0.362 nm, twice that of the ionic radii of the chlorine ion (0.181 nm). A better coincidence is observed for chloride, because the chloride ion is much larger than the corresponding cations.

TABLE 14.1 Crystal Data of Related Compounds

One Anion Formula Structure Type	CaO NaCl	MgO NaCl	$Ca_{0.5}Cl$ Rutile	$Mg_{0.5}Cl$ $CdCl_2$	Pt fcc
M (g/mol)	56.1	40.3	55.5	47.6	195
Molar volume (cm^3/mol)	16.64	11	25.8	20.5	9.05
Density (g/cm^3)	3.37	3.65	2.15	2.325	21.5
mp (°C)	2572	2800	772	712	1774
Calc d (anion) (nm)	0.340	0.296	0.393	0.364	0.277
Accepted d (anion) (nm)	0.28	0.28	0.362	0.362	0.277
Cation d (nm)	0.20	0.144	0.20	0.144	—

The reaction of gas molecules (an organic Cl compound) with a nanocrystal oxide first starts at the surface, then ion exchange occurs inside the crystal (O^{2-} to Cl^-). The reactivity depends on the percentage of surface ions exposed, which can be calculated using the model. The actual structure of the crystal is usually cubic, hexagonal, etc. For the condition of the calculation, it is assumed that the spherical nanocrystal oxide has diameter of D, which is packed with spherical oxide ions, as shown in Figure 14.1. The model crystal surface also forms a hexagonal packing structure. The number of surface ions is $\pi D^2/0.5\sqrt{3}d^2$ for a hexagonal structure. The total number of ions in the bulk are $(\pi/6)D^3/(1/\sqrt{2})d^3$ for both the salt structure and closely packed structure. Thus, the surface ion ratios are $2\sqrt{6}d/D$ for any case. The calculated values are shown in Figure 14.2, in which platinum metal is added for comparison. Pt and MgO exhibit almost identical curves, because the Pt atom and oxide

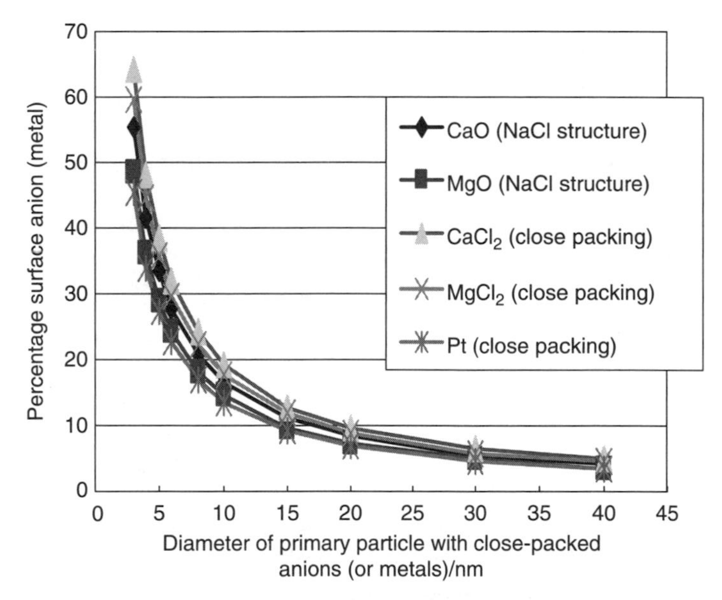

Figure 14.2 Percentage surface anion (or metal atom) as a function of the primary particle diameter D.

ion have almost similar (calculated) d. Thirty percent of the atoms or ions are located on the surface when the size of the nanoparticle is 5 nm. CaO has a high surface % and $Mg_{0.5}Cl$ and $Ca_{0.5}Cl$ have much higher surface % values, because the calculated d is increased in this order (Table 14.1).

Brunauer–Emmett–Teller (BET) surface area is often measured for powder samples. If the above powder model is applied, the relation between the specific surface area (m^2/g) and the crystal size D (m) can be calculated using molar volume (V; m^3/mol) and molecular weight (M; g/mol). The mole number of anions of a particle with diameter D (m) is $(\pi/6)D^3/V$, thus the weight of the particle is $M(\pi/6)D^3/V$. The surface area, πD^2, is divided by the weight to provide a specific surface area of $6V/DM$. The calculated results are shown in Figure 14.3, with platinum added for comparison. It should be noted that the specific surface area of Pt is much lower than those of the oxides or chlorides, due to the high molecular weight of platinum.

The size of Pt nanoparticles supported (dispersed) on oxide or carbon is easily measured using transmission electron microscopy (TEM) or the hydrogen adsorption technique. Oxides or chlorides are not easily observed by TEM. Peak analysis and specific surface area measurements are indirect but important techniques for the estimation of particle size, and the latter is very practical. The surface area and AFM (atomic force microscopy) data of three types of CaO have been reported (8). The specific surface area of commercial CaO, conventionally prepared CaO, and autoclave-prepared CaO were 10, 100, and 120 m^2/g, respectively. AFM observation, which may not cover all the bulk information, indicated crystal sizes of 39, 14, and 7 nm, respectively. The model presented here gives the crystal size (D) as $6V/SM$, where S is the specific surface area. The calculated crystal sizes are 178, 18, and 15 nm, respectively, which fit fairly well with the AFM data considering this technology. Therefore, the crystal sizes (primary particle) of oxide powders can be estimated from the specific surface area.

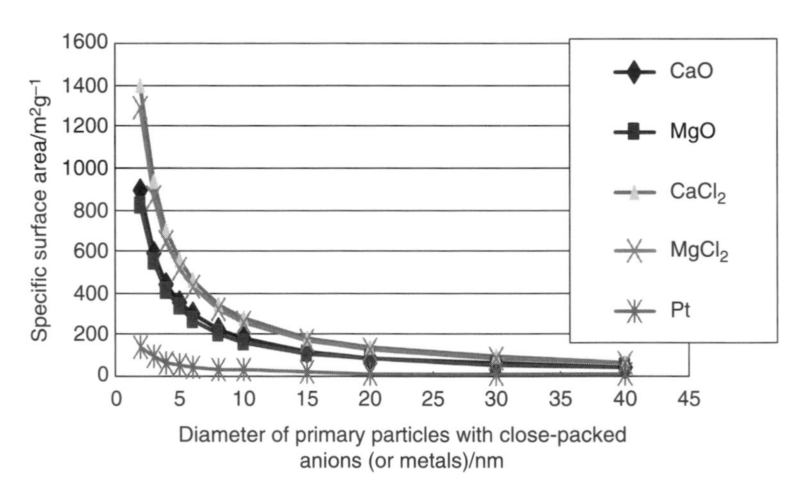

Figure 14.3 Specific surface area of spherical model particles as a function of primary particle diameter D.

14.4 MINERALIZATION OF CHLORINE CONTAINING VOLATILE ORGANIC COMPOUNDS

Dichloromethane (DCM), trichloroethene (TCE), tetrachloroethene (TeCE), chloroform, tetrachloromethane, and 1,2-dichloroethane are nonflammable and powerful solvents for many compounds, and have uses such as cleaning of precision instruments, etc. Most of these solvents are harmful to living creatures, and therefore must be treated carefully and finally decomposed after use. The three major chemicals, DCM, TCE, and TeCE are annually produced at approximately 100 thousand tons/year in Japan, much of which is not recovered and may be disposed of either in air, water, or soil. Ideally, these chemicals should be recovered after use. However, it is not economic when the chemicals are diluted or mixed with other materials and, therefore, the chemicals are often burned. This could produce toxic materials such as dioxin when the burning temperature is low. The reaction during the burning process is hydrolysis, which produces HCl that can cause corrosion of the reactor.

Alkaline earth oxides are known to fix chlorine in organic chlorides, producing common chlorides ($MgCl_2$, $CaCl_2$, etc.), CO_2, water, and carbon. These reactions are referred to as mineralization of hazardous elements, and will be used in the future as an environmentally friendly process in advanced countries, if practical methods are developed. The keys to achieving such processes are the development of active alkaline earth oxide materials that are cheap, and the design of effective reaction systems.

14.5 BASIC RESEARCH ON THE MINERALIZATION OF ORGANIC CHLORIDES

The standard heats of formation (kJ/mol) for oxides and chlorides of calcium, magnesium, and iron are $CaO/CaCl_2$: $-636/-795$, $MgO/MgCl_2$: $-598/-642$, and $FeO_{1.5}/FeCl_3$: $-411/-405$. The stability of iron oxide and chloride are much the same, while magnesium chloride is more stable than the respective oxide, and calcium chloride is the most stable. MgO and especially CaO are expected to be the best reagents for mineralization (9). Mineralization generally replaces oxygen ions with chloride ions and produces stable compounds such as CO_2 and H_2O. Thus, these are energetically favorable reactions and can proceed almost completely, consuming all reactants. An example is given for tetrachloromethane as follows:

$$2CaO + CCl_4 \longrightarrow 2CaCl_2 + CO_2 \quad \Delta H^\circ = -573\,\text{kJ/mol} \tag{14.3}$$

$$2MgO + CCl_4 \longrightarrow 2MgCl_2 + CO_2 \quad \Delta H^\circ = -334\,\text{kJ/mol} \tag{14.4}$$

Although thermodynamically feasible, the reactions do not progress easily, because of the stable C—Cl bond. Thus, mineralization is performed after activation of the C—Cl bond. Photolysis, catalysis, and hydrolysis in a furnace are used as activation processes.

CCl_4 was adsorbed on a MgO film, evacuated at 723°C, and irradiated with laser light (193 nm) at −173°C in a vacuum system (10). The decomposition products (Cl_2, $OCCl_2$, C_2Cl_4, C_2Cl_6, HCl, and CO_2) adsorbed on MgO were then desorbed using temperature-programmed heating (up to 423°C). The mechanism for the initial decomposition reactions and the role of surface MgO or adsorbed water was proposed for the reaction. Mineralization (to $MgCl_2$) was observed, but was not studied in detail.

DCM, 1,2-dichloroethane, 1,2,4-trichlorobenzene, TCE, 1,1,2,2-tetrachloroethane, and tetrachloromethane were decomposed in air below 250°C over metal oxides supported on carbon (11). Oxidation is the main reaction;

$$CH_2Cl_2 + O_2 \longrightarrow CO_2 + 2HCl \tag{14.5}$$

Redox oxides such as CrO_3, TiO_3, $KMnO_4$, and V_2O_5 were active for this reaction. It is interesting to note that the carbon support itself was fairly active. This was explained by the surface acidity (COOH, etc.) and the high surface area (600 to 1200 m^2/g) of the carbon material.

Destructive reaction of Cl-containing compounds with transition metal oxides supported on MgO or CaO has been studied (12–14). The order of activity found for transition metals ($V > Mn > Co > Fe > Ti > \cdots$) is similar to that for typical oxidation reactions. Another interesting finding by the same group was that the transition metals catalyze the exchange reaction between chloride and oxide ions (14). An intermediate metal chloride such as VCl_x may migrate over the MgO nanocrystal to assist the exchange of Cl^- with O^{2-}.

If the CaO has high surface area, then the mineralization reactions of CCl_4, $CHCl_3$, and C_2HCl_3 occur in the absence of transition metals (8, 9, 15). Conventionally prepared CaO (100 m^2/g) reacts with $CHCl_3$ injected with a He carrier to form $CaCl_2$, CO, and H_2O at 400°C (8, 9). Fifty-two percent of CaO was converted to $CaCl_2$ overnight at 400°C (8, 9). CaO was found to be more reactive than MgO for these mineralization reactions (8, 9, 15).

14.6 CaO-C COMPOSITE MATERIALS FOR MINERALIZATION OF Cl-CONTAINING HYDROCARBONS (CH_hCl_x) UNDER INERT CONDITIONS

CaO has been found to react with concentrated Cl-containing hydrocarbons at around 400°C (8, 9). These reactions have been studied under flow conditions with low concentration for practical use (16, 17). A functionalized CaO is reactive with diluted DCE at 450°C (16). The functionalized CaO was prepared by decomposing calcium citrate under N_2 flow at 800°C. The black powder product (160 m^2/g), composed of CaO and amorphous carbon, was pelletized, packed into a flow reactor, and contacted with a N_2 flow containing 2% DCE at 450°C. The reaction products at the outlet were analyzed as a function of time and are shown in Figure 14.4. If the complete reaction

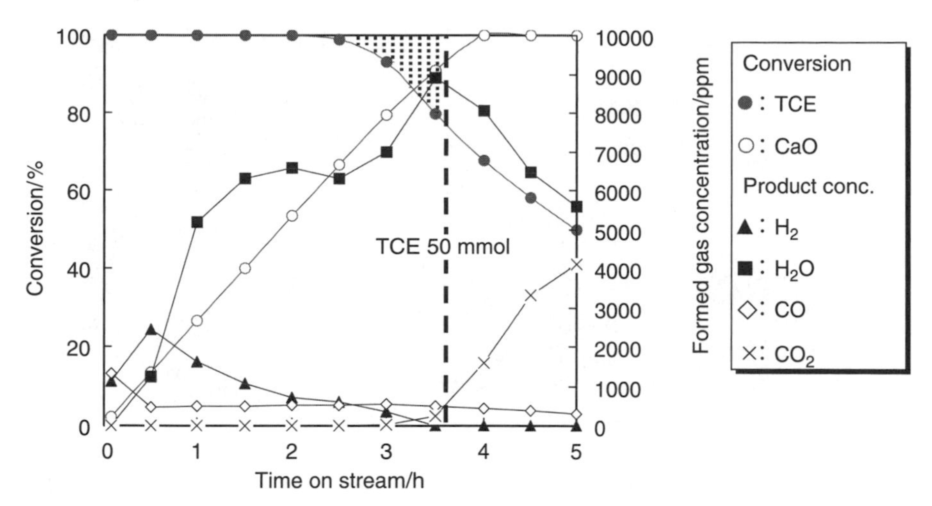

Figure 14.4 Destructive fixation of Cl in TCE (2% in N_2) into C-CaO at 450°C [CaO 100 mmol, TCE is introduced at 50 mmol (stoichiometry) for 3.6 h and 69.4 mmol (excess) for 5 h]. At 3.6 h (TCE/CaO = 1/2), the total TCE conversion was 97% (dotted area corresponds to unreacted 3%). Calculated CaO conversion was 94%. Water was not completely desorbed, but was provably adsorbed on the surface. After the stoichiometric reaction (after 3.6 h), $CaCO_3$ reacts with TCE forming CO_2.

occurs at 450°C, the following equation is expected.

$$C_2HCl_3 + 2CaO \longrightarrow 1.5CaCl_2 + 0.5CaCO_3 + 1.5C + 0.5H_2O \qquad (14.6)$$

CO_2 produced is fixed with CaO forming $CaCO_3$, the decomposition temperature of which is 900°C. Below 700°C, CaO can fix CO_2 almost completely, and CO_2 was not observed before 3 h at 450°C. Up until 3.6 h, 50 mmol of DCE was introduced to 100 mmol of CaO [stoichiometry in Equation (14.6)], and 97% of DCE reacted. CaO conversion was calculated from the amount of products, and was 94% at 3.6 h. After 3.6 h, CaO is almost consumed, and the excess DCE reacts with $CaCO_3$ yielding CO_2 and $CaCl_2$.

$$C_2HCl_3 + 1.5CaCO_3 \longrightarrow 1.5CaCl_2 + 2CO_2 + 1.5C + 0.5H_2O \qquad (14.7)$$

On the other hand, CaO does not fix H_2O at low concentrations of water (<2%) at 450°C, although the decomposition temperature of $Ca(OH)_2$ is as high as 580°C. Water produced is only partly adsorbed, but saturates at the half-way point of the reaction, as shown in Figure 14.4 (see water concentration change). Side reactions such as hydrogen and CO production are mostly due to secondary reactions between the products ($C + H_2O \rightarrow CO + H_2$). Only trace amounts of intermediate Cl compounds were observed during the reaction (16).

Figure 14.5 C-covered CaO (left, 160 m^2g^{-1}, CaO: 0.1 mol) and commercial CaO (right, 7 m^2g^{-1}, CaO: 0.1 mol) samples: C-covered CaO is bulky, fluidal, easily treated, and highly reactive. The reactivity is 10 times more than that of commercial CaO. Carbon (one of the reaction products) is deposited after the reaction on both samples.

The C-covered CaO material, before and after reaction, is shown in the left side of Figure 14.5. A significant increase in volume is observed, due to carbon formation and expansion of the crystal. A pure commercial CaO material (7 m^2/g) was also used for the reaction at 450°C, and was not so active because of the low surface area (right side of Fig. 14.5). The C-covered CaO is also active for most Cl compounds, such as DCM and TeCE. An economic method to produce these materials is expected. These materials may be restricted for use in a reaction system with an inert atmosphere. Alternative materials that could be used in air are also expected to be developed.

A method to fix chloride into CaO under an inert gas (N_2, He) stream has been established. For the purpose of reaction system design (reactor and operation conditions), a convenient way to check the stoichiometry for any kind of organic halides is explored. Suppose CH_hCl_x (any organic halide can be rewritten in this form; x and h are integers) reacts with CaO yielding $CaCl_2$, H_2O, and possibly C, $CaCO_3$, and H_2, depending on the values of x and h. A reaction temperature between 400°C and 600°C is assumed, and CO_2 is fixed as carbonate, while water is evolved from the reactor. The material balance of elements leads to the stoichiometry shown in Table 14.2.

The stoichiometric relationships are illustrated in Figure 14.6. The stoichiometry for the fixation reaction can be obtained from the figure. TCE (trichloroethene,

TABLE 14.2 Stoichiometry for Idealized Mineralization Reaction of CH_hCl_x Molecules with CaO under Inert Atmosphere (at 400°C to 600°C)

	CH_hCl_x	CaO	$CaCl_2$	C	$CaCO_3$	H_2O	H_2
$x > h$	-1	$-0.75x + 0.25h$	0.5x	$1 - 0.25x + 0.25h$	$0.25x - 0.25h$	0.5h	0
$x = h$	-1	$-0.5x$	0.5x	1	0	0.5x	0
$x < h$	-1	$-0.5x$	0.5x	1	0	0.5x	$0.5h - 0.5x$

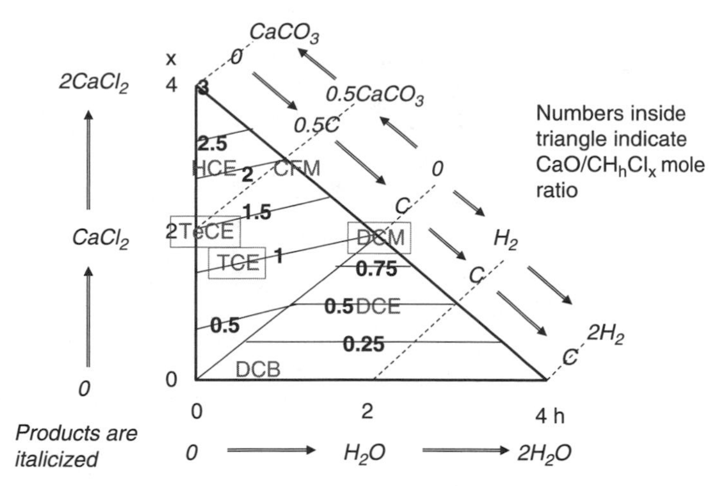

Figure 14.6 Product stoichiometry for the reaction between CH_hCl_x and CaO (CO_2 is fixed and gaseous H_2O produced).

C_2HCl_3 or $CH_{0.5}Cl_{1.5}$) is located on line number 1 inside the triangle, indicating reaction of $CH_{0.5}Cl_{1.5}$ with one mole of CaO. If we start at the position of TCE (Fig. 14.6) going perpendicular to the y-axis, x-axis, and diagonal axis, we reach 0.75 mole of $CaCl_2$, $0.25H_2O$, and 0.75C, and $0.25CaCO_3$. This indicates the following stoichiometry:

$$CH_{0.5}Cl_{1.5} + CaO \longrightarrow 0.75CaCl_2 + 0.25CaCO_3 + 0.75C + 0.25H_2O \quad (14.8)$$

That is practically the same as Equation (14.6).

As expected from Figure 14.6 (diagonal axis), carbon must be produced, except for the case of CCl_4. Actually, carbon is seen as black color on the samples after reaction as is shown in the case of TCE (Fig. 14.5). Carbon is a key material of this reaction under an inert atmosphere, both chemically and physically. Carbon produced might help to activate the C—Cl bond and maintain the nanostructure of CaO. The surface acid site on carbon can activate the C—Cl bond, as previously reported (11). Since it contains no transition metals, it is environmentally benign after use.

14.7 NANOPOWDER CaO PREPARED FROM Ca(OH)$_2$ FOR REACTION IN AIR

Pure CaO is less reactive than transition metal oxides for C—Cl activation. The higher specific surface area provides higher reactivity. DCM (CH_2Cl_2) or TCE (C_2HCl_3) reacts with high surface area CaO (100 or 120 m^2/g) at 400°C under an inert atmosphere (8). High surface area (SA) CaO (30 to 50 m^2/g) conventionally prepared from commercial Ca(OH)$_2$ (50 m^2/g) reacts with diluted DCM or TCE even in air at 450°C to 600°C (17). Reaction of DCM with CaO in air is compared with that in N_2 in Figure 14.7. 200 mmol of CaO was packed in a tube reactor and 2% of DCM

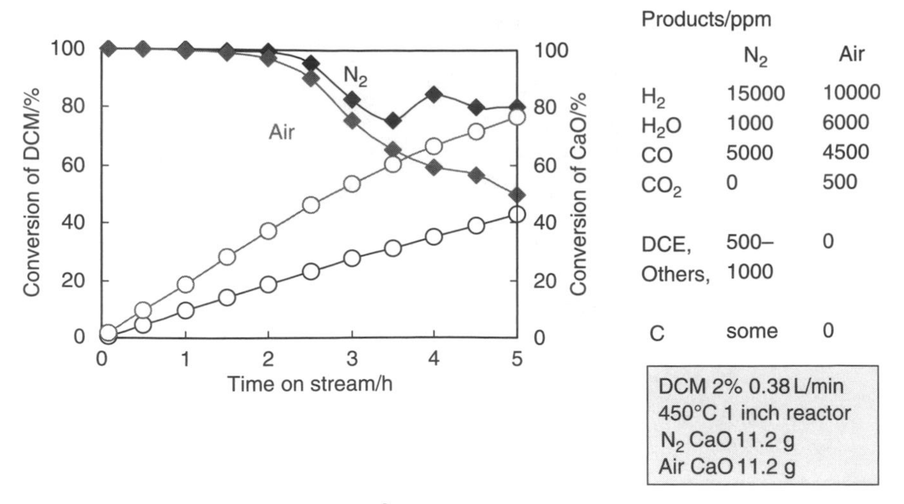

Figure 14.7 DCM reaction with 40 m^2/g CaO [from high SA $Ca(OH)_2$] in N_2 or air at 450°C: CaO 200 mmol; DCM 100 mmol (stoichiometry under air). In 5.4 h, 200 mmol (stoichiometry under N_2), in 10.8 h, DCM conversions are similar. CaO is consumed more in air to form $CaCO_3$. Fewer by-products (DCE, others) are produced in air.

(in air or N_2) was introduced with a flow rate of 0.38 L/min at 450°C. A reaction time of 10.8 h was necessary to supply a stoichiometric amount of DCM (200 mmol) under N_2 atmosphere, while 5.4 h (100 mmol) was sufficient time for reaction in air.

$$CH_2Cl_2 + CaO \longrightarrow CaCl_2 + C + H_2O \tag{14.9}$$

$$CH_2Cl_2 + 2CaO + O_2 \longrightarrow CaCl_2 + CaCO_3 + H_2O \tag{14.10}$$

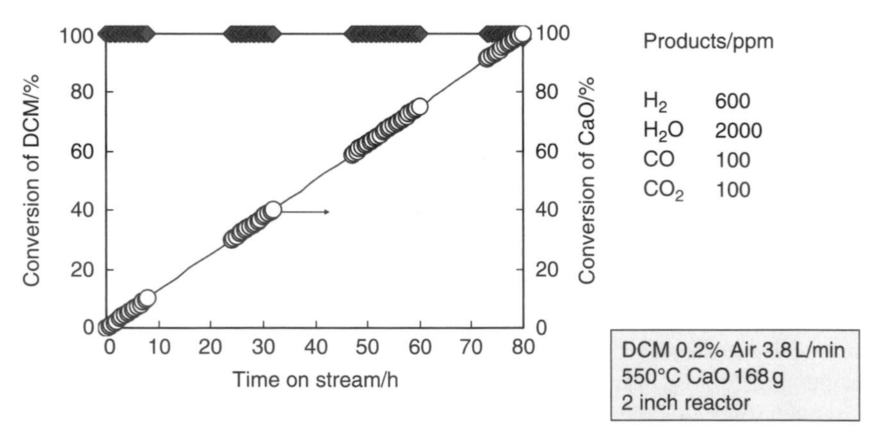

Figure 14.8 Scaled-up run with DCM 0.2% in air at 550°C: CaO 3 mol; DCM 1.5 mol (stoichiometry in air) in 81 h; DCM conversion kept 100% for 80 h at 550°C; CaO reacted highly effectively. H_2 and CO production was controlled.

TABLE 14.3 Stoichiometry for Idealized Mineralization Reaction of CH_hCl_x Molecules with CaO in Air (at 400°C to 600°C)

CH_hCl_x	CaO	O_2	$CaCl_2$	$CaCO_3$	H_2O
−1	−0.5x − 1	−(h − x)/4 − 1	0.5x	1	0.5h

Carbon that might be produced under an inert atmosphere [Equation 14.9)] is combusted to form CO_2 (actually to $CaCO_3$) when in air, where more CaO is consumed [Equation (14.10)]. First, the results of TCE on C-CaO (Fig. 14.4) are compared with those of DCM on CaO (Fig. 14.7), both under N_2 atmosphere at 450°C. The reactivity of CaO is much lower than C-CaO, although the reactants are different. Second, DCM conversion in N_2 and air (Fig. 14.7) are compared. The conversion of DCM appears to be similar, while CaO in air is consumed more than that under N_2. This indicates that the reaction in air is faster. Cl-containing by-products are not observed for the reaction in air. The activity at 450°C is not completely acceptable for industrial use, because oxidation is not complete, so that some amount of CO and hydrogen are produced. The problem is almost solved when the temperature is increased to 550°C. Figure 14.8 shows the scaled-up results at 550°C. Both DCM and CaO conversion reached almost 100% with production of a small amount of by-products (H_2, CO).

For the purpose of designing a reaction process for fixation of any organic chlorides (expressed as CH_hCl_x) in air, a simple method to write the stoichiometry is shown in Table 14.3.

The stoichiometric relationships are illustrated in Figure 14.9. The stoichiometry for the fixation reaction can be obtained from the figure. TCE (trichloroethene, C_2HCl_3 or

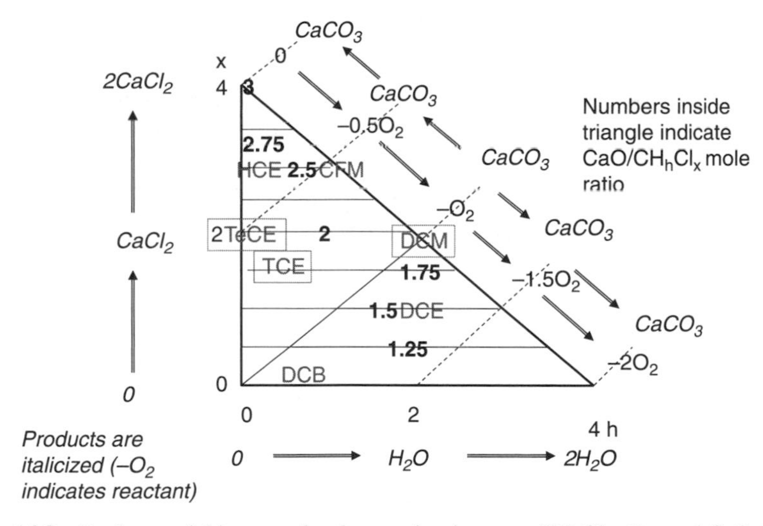

Figure 14.9 Product stoichiometry for the reaction between CH_hCl_x, O_2, and CaO (CO_2 is fixed and gaseous H_2O produced).

$CH_{0.5}Cl_{1.5}$) is located on line number 1.75 inside the triangle, indicating reaction of $CH_{0.5}Cl_{1.5}$ with 1.75 mole of CaO. If we start at the position of TCE (Fig. 14.9) going perpendicular to the y-axis, x-axis, and diagonal axis, we reach 0.75 mole of $CaCl_2$, $0.25H_2O$, and $-0.75O_2$, and $CaCO_3$. This indicates the following stoichiometry:

$$CH_{0.5}Cl_{1.5} + 1.75CaO \longrightarrow 0.75CaCl_2 + CaCO_3 - 0.75O_2 + 0.25H_2O \quad (14.11)$$

That is:

$$C_2HCl_3 + 3.5CaO + 1.5O_2 \longrightarrow 1.5CaCl_2 + 2CaCO_3 + 0.5H_2O \quad (14.12)$$

14.8 VOLUME CHANGE OF POWDER DURING THE REACTION

When nanomaterials are used as catalysts, they do not change in volume. However, if they react with gas molecules, they do change in form and size. This can be a serious problem in the case of chlorine mineralization from Cl hydrocarbons, because oxides and chlorides have much different molar volumes, as shown in Table 14.1. When CaO changes to $CaCl_2$, two Cl^- ions, which are much larger than an O^{2-} ion, replace one O^{2-} ion. The crystal volume then increases more than three times, as illustrated in Figure 14.10.

If the primary particles (spherical balls with diameter D) are closely packed in a secondary particle, the actual crystals occupy 74% of the space in the secondary particle. If the secondary particles are also closely packed, they cannot change in size. The reaction might stop, due to choking of channels through which the gas molecules come

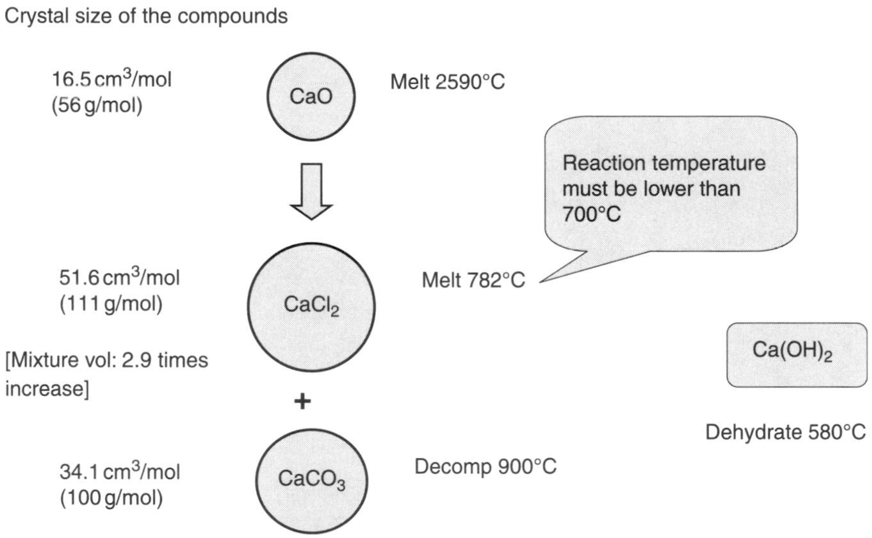

Figure 14.10 Basic data of calcium compounds for the reaction design of Cl-containing VOC: CaO powder with some pore volume is necessary to prevent the problem of adherence due to volume increase.

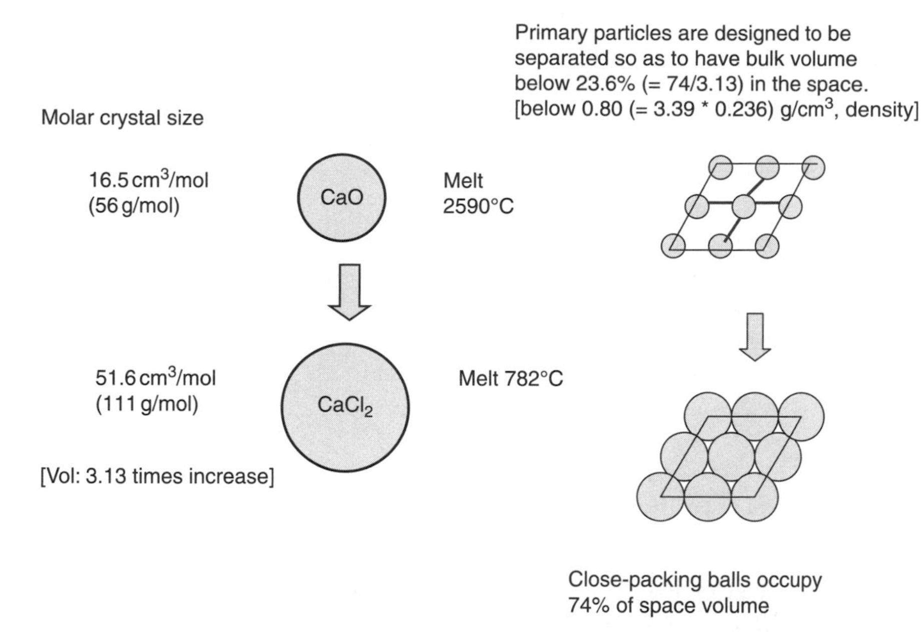

Figure 14.11 Design of CaO nanostructure (powder, pellet) considering the void volume available for expansion.

in. If the sample can be prepared to maintain the space around the primary particles by having a loose connection between primary particles (inside the secondary particle), then the reaction will continue. However, if the sample is prepared to maintain extra space outside the secondary particles, the primary particles adhere together, and the nanostructure is lost. Appropriate powder must have more than 76% void space to allow the (3.13 times) expansion of primary particles without adherence, as illustrated in Figure 14.11. The density of the CaO powder must be below $0.80\ g/cm^3$.

The reaction produces $CaCO_3$ (2.07 times expansion) in addition to $CaCl_2$ (3.13 times), as shown in Figures 14.6 and 14.9, depending on the Cl/C ratio, H/C ratio, and the reaction atmosphere. The maximum densities of CaO powder available for the reaction were calculated and are shown in Figures 14.12 and 14.13.

An example is given for the case of TCE mineralization under inert atmosphere [see Equation (14.6)]; 75% of CaO is converted to $CaCl_2$ and 25% to $CaCO_3$, resulting in 2.85 $(= 3.1 \times 0.75 + 2.1 \times 0.25)$ times volume expansion. The closely packed (primary) particles occupy 74% volume. If the reacted particle swells to this volume, the original volume must be lower than 25.9% $(= 74/2.85)$, that is, the maximum density is 0.88 $(= 3.39 \times 0.259)$. The same value is obtained when we consider x (1.5) and h (0.5) in Figure 14.12.

The reaction of TCE in air is expressed as Equation (14.12). Here, 3/7 of CaO is converted to $CaCl_2$ and 4/7 is to $CaCO_3$, resulting in 2.53 $[= 3.1 \times (3/7) + 2.1 \times (4/7)]$ times volume expansion, that is, a maximum density of 0.99 $(= 3.39 \times 0.74/2.53)$. The same value is obtained when we consider x (1.5) in Figure 14.13.

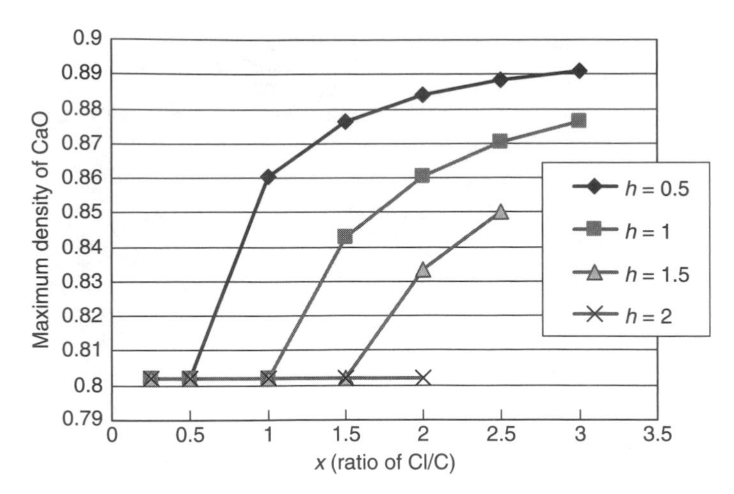

Figure 14.12 Threshold density of CaO for reaction with CH_hCl_x in N_2.

The new key factor here is the high porosity (or low powder density). This factor controls the diffusion of gas molecules for mineralization reactions, and is independent of the specific surface area or primary particle size that may control the rate of reaction. Of course a small-sized primary particle may be a condition to maintain high porosity (low density). The density of material is important both for the laboratory experiment and for the large-scale reactor. The success of complete reaction shown in this text (16, 17) may be due to careful preparation of the samples (for low density). Carbon may have a role maintaining the space between the nanomaterials (16). $Ca(OH)_2$ can be carefully dehydrated to produce CaO with sufficient void space (17). Most earlier research has not given detailed porosity data.

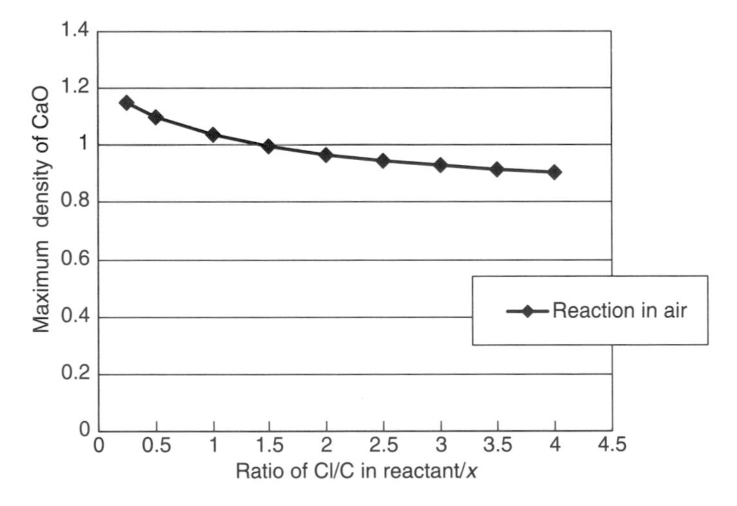

Figure 14.13 Threshold density of CaO for reaction with CH_hCl_x in air.

14.9 REACTION SYSTEM FOR MINERALIZATION OF Cl FROM ORGANIC Cl COMPOUNDS

Typical reaction systems are illustrated in Figure 14.14. A kind of batch reactor with nanopowder spread and pasted on a wall can react with molecules easily (a). A frequently seen system is a fixed bed reactor packed with nanopowder which reacts with a flowing gas (b). The solid material is not removed until the reaction is almost complete, while the reacted gas is removed continuously. This is a kind of semibatch reaction system, which was used to produce the results shown in Figures 14.4, 14.7, and 14.8.

In order to operate the reaction continuously for a long time, both solid and gas must be continuously introduced to the reactor. There are three types of continuous reactor: counter-current (c, d), co-current (e, f), and cross-current (g), depending on the flow direction of solid and gas, as shown in Figure 14.14. A fluidized bed is the most sophisticated method of solid gas reaction, in which the temperature is well controlled, and the solid can effectively contact with the gas. However, caution must be paid to maintain continuous fluidization and not to block the flow system. Continuous flow (h) and batch (i) type reactors are illustrated in Figure 14.14.

A simple case of a batch reactor will be explained briefly. If the powder is mixed well in a container or pasted on a wall (case a), the reaction proceeds continuously throughout the solid particle. If most of the primary particles have similar diameter, the reaction proceeds at the same rate for all the particles. Under such conditions, the reaction rate analysis is rather simple, and two idealized models have been presented; continuous reaction model and unreacted core model. For the former case,

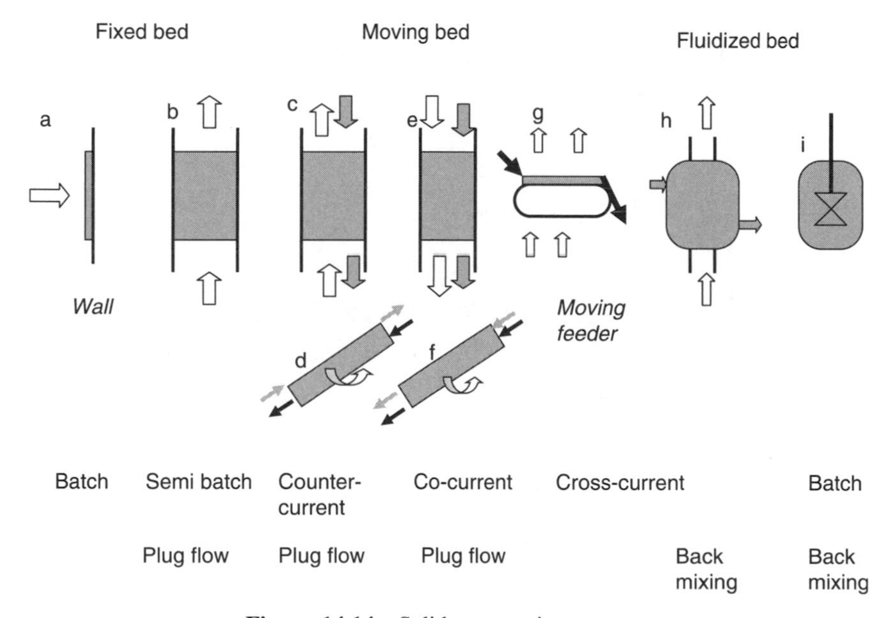

Figure 14.14 Solid gas reaction systems.

the bulk reaction inside the primary particle is fast (e.g. anion exchange between Cl^- and O^{2-} is fast) and concentration of the solid product is progressively increased at the same rate through the particle. For the latter, the gas-solid surface reaction is fast and the reaction front is moving progressively into the core, leaving the solid product outside (e.g. ash remains outside a burning charcoal briquette). A kinetic analysis of the latter case is well described by Levenspiel (18).

A continuous gas flow with a fixed bed of powder (reactor b) is considered to be the most practical for the first stage of development. The principle of this type of reactor is given here by referring to Ruthven's textbook (19), where irreversible adsorption and the fixation reaction are treated in the same way. The material flow analysis is outlined in the next section.

14.10 FLOW AND REACTION RATE ANALYSIS FOR A COLUMN REACTOR

Imagine a column packed with nanocrystals (primary particle, CaO), and a fluid (air) containing a low concentration (c) of reactant (DCM) is introduced with a velocity v (cm^3/s) through the column with porosity (micropores and macropores) of ε (see Fig. 14.15). Due to the gas solid reaction, c decreases and the fixed gas concentration in the solid (q cm^3/cm^3) increases with time at some column length (z cm, ξ; dimensionless length). The maximum q (q_0) corresponds to the stoichiometry of the reaction; 506 (**gas** cm^3 of 1 mol DCM at 298 K divided by **crystal** cm^3 of 2 mol CaO). The concentration of $CaCl_2$ inside the CaO crystal (primary particle) is assumed to be constant through the primary particle (anion exchange between Cl^- and O^{2-} is fast) and ranges between 0 and 506 [see Equation (14.10)].

In the case of Figures 14.4, 14.6, and 14.7, the fluid (N_2 or air) contains only a single reactant with low concentration (e.g. DCM lower than 2%). Therefore, the

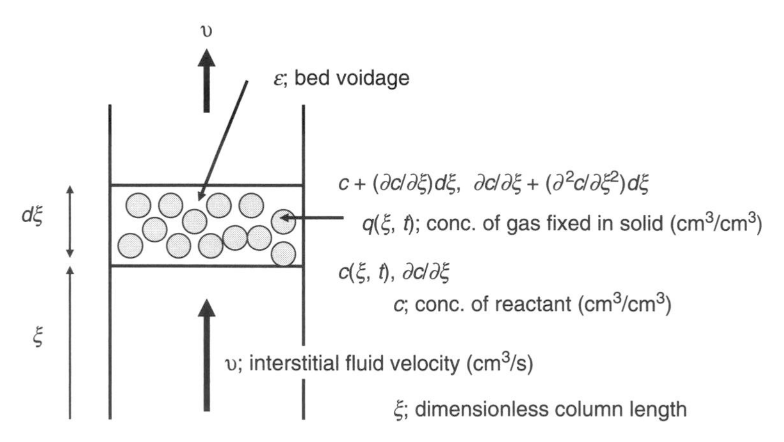

Figure 14.15 Material balance of steady-state flow in a packed bed of nanomaterial for chemical fixation.

differential fluid phase mass balance is written as follows:

$$-D_L(\partial^2 c/\partial z^2) + v(\partial c/\partial z) + \partial c/\partial t + (\partial c/\partial t)(1-\varepsilon)/\varepsilon = 0, \tag{14.13}$$

where D_L is the diffusion constant and t is time (s), as shown in Figure 14.15. For idealized plug flow where the mass transfer resistance is negligible, the first term is deleted.

$$v(\partial c/\partial z) + \partial c/\partial t + (\partial c/\partial t)(1-\varepsilon)/\varepsilon = 0 \tag{14.14}$$

The initial conditions and the boundary conditions for an initially activated material (CaO) subjected to a step change in reactant (DCM) inlet concentration at time zero are;

$$t < 0,\ q(0, z) = c(0, z) = 0 \quad \text{and} \quad t > 0,\ c(t, 0) = c_0 \text{ (e.g. 0.02\% or 2\%)} \tag{14.15}$$

The dimensionless time and bed length are defined as follows:

$$\tau = kc_0(t - z/v) \tag{14.16}$$

$$\xi = kq_0 z/v(1-\varepsilon)/\varepsilon \tag{14.17}$$

The equation has been solved by Bohart and Adams (20), assuming that the rate of fixation is linearly related with the gas concentration, c, and open site concentration for fixation, $(q_0 - q)$.

$$\partial q/\partial t = kc(q_0 - q) \tag{14.18}$$

$$c/c_0 = \exp(\tau)/[\exp(\tau) + \exp(\xi) - 1] \tag{14.19}$$

For the case shown in Figure 14.7, the reaction conditions are as follows; CaO 11.2 g (sample density of 0.5; volume of 22.4 cm^3, porosity $\varepsilon = 0.75$), 1 inch (2.54 cm) reactor length 4.29 cm (z), flow rate of air 380 cm^3/min (6.33 cm^3/s), empty flow velocity 1.21 cm/s for a 1 inch reactor, interstitial velocity 1.62 cm/s (v). Constants are charged into Equations (14.16) and (14.17) under the condition at the exit of the reactor ($z =$ 4.29 cm). Since the value of z/v is much smaller than t (s) operated, the term is neglected. $\tau = 0.02kt$ and $\xi = 447k$ are introduced into Equation (14.19).

The calculated results for DCM concentration [Equation (14.19)] at the exit are shown as a function of rate constant k and time t in Figure 14.16. The simulation results can be compared with the data shown in Figures 14.4, 14.7, and 14.8. Care should be taken, in that the y scales are reversed [conversion % $(1 - c/c_0)$ is used in data figure, but unreacted % (c/c_0) is used for the simulation in Fig. 14.16]. The reaction rate in Figure 14.7 (DCM with CaO at 450°C) seems to be near to $k = 0.01$, whereas those in Figure 14.4 (DCE with C-CaO at 450°C) and 14.8 (DCM with CaO at 550°C) appear to be $k > 0.1$.

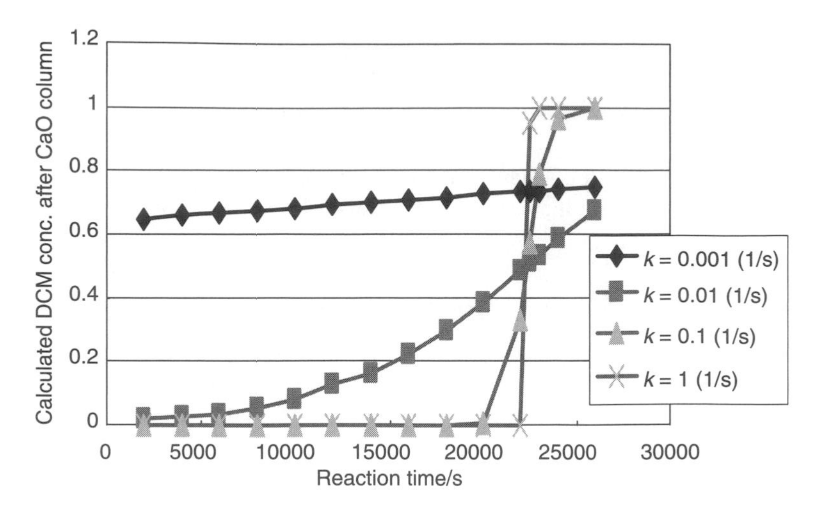

Figure 14.16 Breakthrough curve of 2% DCM over 4.29 cm CaO column in 1.62 cm/s air flow.

The rate is controlled by reaction and not by diffusion, because it depends on the reaction temperature. The reaction rate also depends on the surface area, that is, the size of the nanoparticles. The active material, C-CaO, has higher surface area than CaO. If the activity is low ($k = 0.001$), 60% to 70% of DCM flows out from the exit. If the activity is high enough ($k = 1$), DCM is completely decomposed and fixed until all the CaO is consumed (Fig. 14.16).

14.11 CONCLUSION

Reactions of nanoscale materials with gases are becoming important in the field of energy conversion and environmental technology. As an example, mineralization of Cl from hazardous Cl-containing organic compounds was discussed. A simple concept to characterize oxide and chloride nanoparticles was demonstrated. The cations in the solid can be ignored when discussing the morphology, because of their small size. Spherical crystals composed of only anions is convenient for a model of a primary particle in order to discuss the exposed percentage or surface area as a function of crystal diameter.

CaO nanopowder can fix Cl from Cl-containing hydrocarbons at 450°C to 600°C. The reaction of Cl-containing hydrocarbons with CaO depends on the H/C ratio (h), Cl/C ratio (x), and the reaction atmosphere. Under conditions of an inert atmosphere, C is produced in any case, and hydrogen is produced when h/x is higher than unity ($CaCO_3$ is produced when h/x is lower than unity) in addition to water and $CaCl_2$. Under an air atmosphere, oxygen is consumed and $CaCO_3$ is produced in any case, in addition to water and $CaCl_2$.

When the process design of Cl fixation is discussed, the morphological change must be considered. $CaCl_2$ has a molar volume more than three times that of CaO.

The reactor might become blocked unless the material has a high void ratio. $CaCl_2$ has a melting point of 782°C, thus a temperature more than 100°C lower is suitable for the fixation process. The mineralization reaction is exothermic; therefore, the temperature must be well controlled. In the reactor design for the mineralization reaction, dynamic rate and flow analysis for irreversible adsorption is applicable.

The field of nanoscale material reactions is not well studied; however, the science and technology in the established fields (physical chemistry, inorganic chemistry, and chemical engineering) can be applicable. Developments in this field will most certainly contribute to solving energy and environmental issues in the future.

REFERENCES

1. K. J. Klabunde (ed.), *Nanoscale Materials in Chemistry*. Wiley-Interscience, New York, 2001.
2. S. Takenaka, K. Nomura, N. Hanaizumi, K. Otsuka, *Appl. Catal. A* **2005**, *282*, 333.
3. M. Ishida, H. Jin, *Energy* **1994**, *19*, 415.
4. K. J. Klabunde, R. S. Mulutla, in *Nanoscale Materials in Chemistry*, p. 223. Wiley-Interscience, New York, 2001.
5. (a) T. Tamai, K. Inazu, K. Aika, *Bull. Chem. Soc. Jpn.* **2005**, *78*, 1565; (b) T. Tamai, K. Inazu, K. Aika, *Environ. Sci. Technol.* **2006**, *40*, 823.
6. W. L. Jolly, *The Principles of Inorganic Chemistry*. McGraw-Hill, New York 1976.
7. (a) R. D. Shannon, *Acta Crystalogr.* **1976**, *A32*, 751; (b) J. A. Dean (ed.) *Lange's Handbook of Chemistry*. McGraw-Hill, New York, **1973**.
8. O. Koper, I. Lagadic, K. J. Klabunde, *Chem. Mater.* **1997**, *9*, 838.
9. O. Koper, Y.-X. Li, K. J. Klabunde, *Chem. Mater.* **1993**, *5*, 500.
10. X.-L. Zhou, J. P. Cowin, *J. Phys. Chem.* **1996**, *100*, 1055.
11. S. C. Petrosius, R. S. Drago, V. Young, G. C. Grunewald, *J. Am. Chem. Soc.* **1993**, *115*, 6131.
12. K. J. Klabunde, A. Khalee, D. Park, *High Temp. Mater. Sci.* **1995**, *33*, 99.
13. S. Decker, K. J. Klabunde, *J. Am. Chem. Soc.* **1996**, *118*, 12465.
14. Y. Jiang, S. Decker, C. Mohs, K. J. Klabunde, *J. Catal.* **1998**, *180*, 24, 35.
15. O. Koper, K. J. Klabunde, *Chem. Mater.* **1997**, *9*, 2481.
16. T. Tamai, R. Kojima, T. Furusawa, N. Suzuki, M. Kato, T. Mochida, K. Aika, NEDO (New Energy and Industrial Technology Development Organization, Japan) Report, 06002434-0, 2007.
17. R. Kojima, T. Furusawa, N. Suzuki, M. Kato, T. Mochida, K. Aika, NEDO (New Energy and Industrial Technology Development Organization, Japan) Report, 07000343-0, 2008.
18. O. Levenspiel, *Chemical Reaction Engineering*. Wiley-Interscience, New York, 1962.
19. D. M. Ruthven, *Principles of Adsorption and Adsorption Processes*. Wiley-Interscience, New York, 1984.
20. G. Bohart, E. Adams, *J. Am. Chem. Soc.* **1920**, *42*, 523.

PROBLEMS

1. When we calculate anion diameters of ionic crystals (CaO, MgO, $CaCl_2$, $MgCl_2$), we can simply imagine packed spheres of anions and ignore the cations. What is the justification for this approach? Calculate the diameter of Cl ions in $MgCl_2$ crystals from the molar volume (41 cm^3/mol).
2. Calculate the percentage of ions exposed on the surface of a spherical $MgCl_2$ crystal with a diameter of $D = 10$ (nm).
3. Calculate the specific surface area of a $MgCl_2$ primary particle (spherical crystal) with a diameter of 10 nm using molecular weight and molar volume (Table 14.1).
4. Describe three methods for treating waste Cl-containing volatile organic compounds (VOC) and identify their characteristic features.
5. Transition metals such as Fe and V (added to MgO) are said to promote the mineralization of Cl-containing VOC into MgO. What is the role of the transition metal?
6. Obtain the equation for the reaction of DCM with CaO under N_2 atmosphere using Figure 14.6.
7. Calculate the volume % of spherical balls that occupy a space by the close-packing structure.
8. Identify the characteristics required for a CaO sample that can be used for destructive fixation of DCM in air.
9. Identify advantages and disadvantages by comparing fixed bed [part (b) in Fig. 14.14] and moving bed [part (c) in Fig. 14.14] systems for the reaction between gas and nanoscaled materials for industrial use.
10. Compare reversible adsorption, irreversible adsorption, and mineralization reactions with respect to the process.

ANSWERS

1. Cation sizes are much smaller than those of anions; therefore, the cations can be packed between anions without expanding the structure. Calculation: $MgCl_2$ is replaced with $Mg_{0.5}Cl$ (20.5 cm^3/mol). For close packing of Cl ions with diameter d (nm), the occupied volume of 1 Cl ion is $(1/\sqrt{2})\,d^3 \times (10^{-21})\,cm^3$.

$$(1/\sqrt{2})d^3 \times (10^{-21}) \times (\text{Avogadro number})\,cm^3/mol = 20.5\,cm^3/mol$$

$$d^3 = 0.0483, \quad d = 0.364\,(nm)$$

Comment: The calculated value is a little larger than the literature value (0.362 nm), because the size of the Mg^{2+} ion (0.144 nm) is not so small as to be negligible.

2. The ratio of surface vs. bulk of the close-packed anion sphere model is $2\sqrt{6}d/D$. $100 \times 2\sqrt{6} \times 0.364\,(\text{nm})/10\,(\text{nm}) = 17.8\%$.

Comment: Approximately 20% of ions are exposed on the surface of a 10 nm-sized primary particle.

3. The specific surface area is given as $6V/DM$. $6 \times 20.5\,(\text{cm}^3)/[10 \times 10^{-7}\,(\text{cm}) \times 47.6\,(\text{g})] = 2{,}580{,}000\,\text{cm}^2/\text{g} = 258\,\text{m}^2/\text{g}$.

Comments: The surface area of the powder or pellet is the sum of the surface area of primary particles. The value depends on the density of the material. The value of the specific surface area is quite high (Fig. 14.3), because chlorides are less dense materials. Be careful to compare the specific surface area between different materials.

4. (a) If the waste VOC is collected as a liquid phase and the concentration of Cl-containing material is high, then distillation is possible, and the compound can be reused. (b) Combustion is a practical method; however, it must be carried out at high temperature, so as to not yield dioxin. Since the reaction is hydrolysis, HCl product might corrode the furnace. (c) Direct mineralization of Cl in VOCs into CaO or MgO is an improved process. The key is how to improve the reactivity of CaO or MgO.

5. Transition metals catalyze activation of the C—Cl bond and catalyze ion exchange between O^{2-} and Cl^-.

6. DCM (CH_2Cl_2) is located on the number 1 line inside the triangle, indicating reaction with one mole of CaO. If we start at the position of DCM (Fig. 14.6) going perpendicular to the y-axis, x-axis, and diagonal axis, we reach one mole of $CaCl_2$, H_2O, and C. This indicates the following stoichiometry:

$$CH_2Cl_2 + CaO \rightarrow CaCl_2 + C + H_2O$$

Comments: Any organic halides are expressed as C1 molecules in this method. The illustration appears complicated, but it can cover many reactions at once.

7. Consider the model picture showing closely packed spheres with diameters of d shown at the right bottom in Figure 14.11. The lined rhombus (a layer) contains four spheres that occupy a volume of $(2d)(\sqrt{3}d)(\sqrt{2}d/\sqrt{3})$. The second layer spheres occupy the space at the center of a triangle of spheres. This leads to a space volume of $d^3/\sqrt{2}$ for one sphere. The actual volume of a sphere is $\pi d^3/6$. Thus, $(\sqrt{2}\pi/6 = 0.74)$ of the space is filled with close-packing spheres. 74%.

8. Prepare a sample with high surface area for increased activity, and a sample with low density to maintain a high void ratio and allow expansion of the crystal size to $CaCl_2$. The recommended values can be calculated referring to Figures 14.3 and 14.13.

9. The fixed bed system is simple, but the contents change with time and must eventually be replaced with a new activated material. The nanosized crystal might swell during the reaction, which could cause clogging of the channels in the solid material used for passage of the reactant gas. This can be used for a small reactor system. A moving bed system can be operated under steady-state reaction for a long time if the reaction rate or heat is well controlled. In order to prevent the powder escaping through the gas inlet or outlet (or gas leakage from the powder inlet and outlet), special attention must be paid to the mechanical and physical features of the system. When the amount of gas to be treated is large and continuously evolved, such systems are economical.

10. Although these processes require high surface area materials, adsorption is usually carried out at room temperature, while mineralization is performed at high temperature, because of the slow process. Adsorption does not alter the adsorbate materials, while mineralization results in change of the materials. The mineralization reaction may significantly change the crystal or primary particle morphology, which may cause choking of the gas flow. If the equilibrium of mineralization shifts much to the formation of product, the same rate analysis as that used for irreversible adsorption can be applied.

15

NANOMATERIALS IN ENERGY STORAGE SYSTEMS

WINNY DONG AND BRUCE DUNN

15.1 Introduction, 520
15.1.1 Motivation for Developing United States' Better Batteries and Capacitors, 520
15.1.2 Energy Storage Basics, 520
15.2 General Information: Electrical Energy Storage Devices and Impact of Nanomaterials, 522
15.2.1 Batteries, 522
15.2.2 Capacitors, 523
15.2.3 Gold Standards (State of the Art) for Both Batteries and Capacitors, 523
15.3 Electrochemical Properties of Nanoscale Materials, 524
15.3.1 Aerogels and Structure-Directed Mesoporous and Macroporous Solids, 525
15.3.2 Nanoparticles, 528
15.3.3 Nanotubes, Nanowires, and Nanorolls, 530
15.4 Concluding Remarks: Challenges to Developing Better Batteries and Capacitors, 531
References, 531
Problems, 534
Answers, 534

Nanoscale Materials in Chemistry, Second Edition. Edited by K. J. Klabunde and R. M. Richards

15.1 INTRODUCTION

15.1.1 Motivation for Developing United States' Better Batteries and Capacitors

Concerns for global warming and the dependence on foreign oil have prompted the incorporation of alternative energy sources into the standard energy supply chain. At the end of 2007, the U.S. government passed the Energy Independence and Security Act, which includes a call for the expanded production of renewable fuels (1). In addition, individual states have made specific goals. For example, California has established a target of 33% for state-wide electric power to be supplied by renewable sources by 2020. The portfolios being developed for energy production show that wind and solar account for approximately 60% of the power sources (2). In order to scale-up the production of energy from these nonsteady power sources, electrical energy storage (EES) has become a critical part of the energy management system (3).

The advancement of both batteries and capacitors, two types of EES devices, requires tailored, multifunctional materials whose structure and properties at nanoscale dimensions are thoroughly understood. To underscore this point, the study of electrochemistry at the nanoscale has already broadened the range of materials that can be used in Li-ion secondary batteries. In fact, materials that were previously excluded from consideration under the classical insertion/de-insertion mechanism are now being reevaluated because, as nanoscale materials, they possess beneficial electrochemical properties (4).

15.1.2 Energy Storage Basics

There are two basic types of electrical energy storage: chemical (batteries) and capacitive (electric double layer capacitors, known as EDLCs, and pseudocapacitors). Batteries convert chemical energy into electric current through the electrical potential difference (ΔV) between the two electrodes. The larger the ΔV between the cathode and anode, the larger the voltage that can be provided by the battery. In secondary, or rechargeable, batteries, the chemical reaction that occurs to produce the electrons can be reversed. The chemical reactions differ depending on the cathode and anode. For high energy density batteries, lithium is desirable because it is light weight and has a large potential difference with common cathode materials. In a Li-ion battery, lithium ions are transferred, via the electrolyte, from the anode to the cathode during discharge (Fig. 15.1). The electrons associated with the oxidation and reduction processes occurring at the electrodes flow in the external circuit. This power-generating process is spontaneous because of favorable thermodynamics. To restore the cell to its predischarged state, additional energy is needed to drive the lithium ions back to the anode from the cathode. Thus, both the anode and the cathode need to be able to accept and release lithium ions repeatedly, without significant dimensional change. The cyclability of Li^+ ion batteries depends mainly on the structural stability of the electrodes during insertion and deinsertion of Li^+ and represents a major area of study.

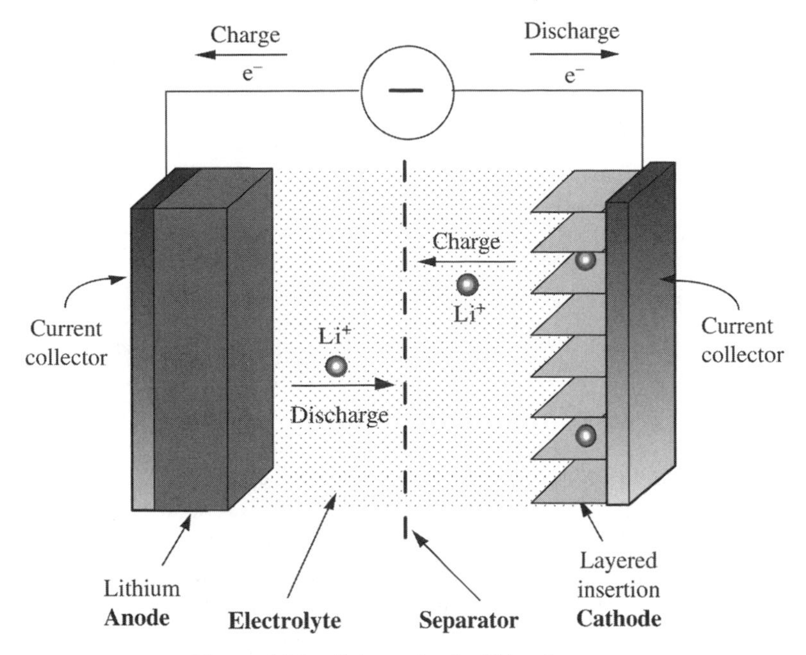

Figure 15.1 Schematic of a Li-ion battery.

EDLCs store energy within the variation of potential at the electrode/electrolyte interface. This variation of potential at a surface (or interface) is known as the electric double layer or, more traditionally, the Helmholtz layer. The thickness of the double layer depends on the size of the ions and the concentration of the electrolyte. For concentrated electrolytes, the thickness is on the order of 10 Å, while the double layer is ~1000 Å for dilute electrolytes (5). In essence, this double layer is a nanoscale model of a traditional capacitor where ions of opposite charges are stored by electrostatic attraction between charged ions and the electrode surface. EDLCs use high surface area materials as the electrode and therefore can store much more charge (higher capacitance) compared to traditional capacitors.

Pseudocapacitors store charge based on reversible (faradaic) charge transfer reactions with ions in the electrolyte. For example, in a metal oxide (such as RuO_2 or IrO_2) electrode, charge storage results from a sequence of redox reactions. Electrochemical capacitors (ECs) based on such pseudocapacitive materials will have both faradaic and nonfaradaic contributions. The optimization of both EDLCs and pseudocapacitors depends on understanding how features at the nanoscale (e.g. pore size distribution, crystallite or particle size) affect ion and electron transport and the fundamental properties of electrochemical interfaces.

Batteries, EDLCs, and pseudocapacitors can often be distinguished from each other by their energy and power characteristics. Energy content can be expressed as specific energy (Wh/kg) or energy density (Wh/L). Power is the amount of energy that can be withdrawn over a certain period of time, measured as specific power (W/kg) or power density (W/L). In general, batteries can be described as providing mid-power and mid-energy while EDLCs are characterized as having lower energy and higher power compared to batteries, and pseudocapacitors offer high power and mid-energy.

15.2 GENERAL INFORMATION: ELECTRICAL ENERGY STORAGE DEVICES AND IMPACT OF NANOMATERIALS

15.2.1 Batteries

There are two principal areas where nanoscale materials offer great potential for dramatically improving secondary batteries. The first is in providing reduced ion and electron transport paths. Nanostructured, high surface area electrode scaffolds can host the fluid electrolyte and provide current collection while also providing dispersed catalysts to promote redox reactions at the electrode (3). In mesoporous architectures, the nanoscale domains of the material are accessible to molecular reactants because of rapid mass transport via diffusion through the continuous volume of mesopores. Additionally, nanoscale materials have led to the design and synthesis of three-dimensional electrodes. The prospect of having active and passive multifunctional components interconnected within a 3-D architecture is envisioned to lead to superior energy storage capacity, high-volume synthesis, and improved safety (6).

The second area for improvement lies in the intercalation/de-intercalation of ions on the nanoscale. The number of electron/redox centers in nanoscale forms of charge storage materials can be significantly superior to their crystalline counterparts for reasons that are still unclear (3). For example, amorphous V_2O_5 prepared with a nanoporous, sol-gel-derived architecture can accept and release four electrons reversibly, compared to two for the crystalline form (7–9). One important factor may be that the high surface area of nanoscale materials amplifies the amount of interface in the material. Since many critical reactions in an electrochemical system occur at the interface, the use of high surface area materials in secondary battery electrodes can take advantage of the different reaction mechanisms occurring at the interface. Another example is the use of phospho-olivines in secondary Li batteries. Traditionally, the Fe redox couple (4+/3+) is metastable, has poor rechargeability, and a low discharge voltage for Li (10). However, the nanophase form of $LiFePO_4$ is now used in commercial, high-rate lithium-ion batteries with excellent stability and safety.

Another area related to the intercalation/de-intercalation of ions is the cycle life of secondary batteries. The volume changes and structural evolution that occur as the electrodes accept and release ions can be quite different for nanoscale materials. For example, it has been shown for rutile TiO_2 that the main kinetic limitation to Li^+ intercalation is related to mechanical strains associated with intercalation. These effects are reduced as particle size decreases, leading to better insertion efficiency (11). Additionally, the ability of nanodomains to stabilize layered $LiMO_2$ and spinel LiM_2O_4 (M = Mn, Ni, Co) electrode structures has been shown (12), while other nanoscale materials exhibit completely new routes of interacting with Li ions. Reversible copper extrusion-insertion electrodes, $Cu_{2.33}V_4O_{11}$, Cu_6Sn_5, and Cu_2Sb, involve a Li-driven displacement reaction that results in the reversible growth of Cu dendrites upon intercalation (13–16).

In contrast to the research described above, advances in nanoscale materials have had less of an impact on the electrolyte and negative electrode of lithium ion batteries. Nonetheless, some interesting results have been reported. In the area of lithium-oxygen batteries, the use of an ultraporous catalyzed electrolyte nanoarchitecture has received renewed attention (17, 18). On the anode side, an alternative to graphite, $Li_4Ti_5O_{12}$, has emerged as a system that does not strain from intercalation/de-intercalation reactions (3). An entirely different approach for negative electrodes is shown in the use of nanoparticles of transition metal oxides (Co, Ni, Cu, Fe), which exhibit high capacities from conversion reactions (19–21). The importance of nanosized particles is particularly evident in conversion reactions with Co_3O_4 anodes (22). Lithium ion insertion in α-Fe_2O_3 anodes showed that nanosized particles suppressed a phase transition leading to much higher Li^+ capacity as well as good reversibility compared to larger particles (23).

15.2.2 Capacitors

Both EDLCs and pseudocapacitors benefit from tailored, high surface area architectures because they each store charge on the surface by electrostatic or faradaic reactions, respectively. There are numerous examples in the literature which show that materials possessing such features as nanodimensional crystallite size and mesoscale porosity exhibit significantly higher specific capacitance as compared to nonporous materials or materials composed of micron-sized powders. The assembly of nanoscale materials is also important. One structure envisioned to be of interest is an array of vertically aligned carbon nanotubes where the spacing between the tubes is matched to the diameters of the solvated electrolyte ions (3).

Understanding the correlation between pore size, ion size, and specific capacitance is crucial to improving electrochemical capacitors. For example, surrounding each electrolyte ion is a layer of solvent molecules that is attracted to the charged ions (solvation layer). It has generally been assumed that the pores of the electrode material must accommodate both the ion and the solvation layer in order for the entire surface area to be accessed by the electrolyte ions. However, it has been shown that capacitance increases anomalously if the pore size is decreased below the size of the solvation shell. The capacitance increase is almost 50% compared with the best performing commercially available activated carbons (24).

15.2.3 Gold Standards (State of the Art) for Both Batteries and Capacitors

This section reviews commercially available systems and the corresponding threshold values that nanomaterials must exceed. These threshold values differ widely depending on the application. In general, commercial batteries require a lifetime of over 300 cycles while retaining above 80% of their capacity (Table 15.1) (25). The energy

TABLE 15.1 Commercially Available Secondary Batteries

Common Name	Nominal Voltage (V)	Energy Density (Wh/L)	Cathode	Anode
Lead acid	2.0	50–100	PbO_2	Pb
Ni-Cd	1.2	80–120	NiOOH	Cd
Ni-metal hydride	1.2	250–350	NiOOH	MH
Alkaline	1.5	150–270	MnO_2	Zn
Li-ion	4.0	250–480	$LiCoO_2$	Li

Source: Data from Reference 25.

density of commercial cells has nearly doubled between 1991 and 2004, from 250 to over 400 Wh/L (26).

The threshold values for other applications are vastly different. For example, MEMS piezoelectric and electrostatic devices require low current (on the order of micro- to nano-amperes) but large voltages (10 to 100 V). There are currently no commercially available batteries on the same dimensional scale as the MEMS piezoelectric and electrostatic devices that meet these power requirements. For MEMS-based sensors, the energy density requirements are more modest, $\sim 1\ J/mm^3$. Li ion-based thin-film batteries have a nominal energy density around $2\ J/mm^3$ (6). However, since these are really 2-D systems, their energy per unit area is only 0.02 to 0.25 J/mm^2 and thus would require a very large surface area in order to meet energy needs (27). One approach to providing power on the micrometer scale is the use of three-dimensional battery architectures. A thorough review and discussion of various 3-D battery designs has been published (6). In biomedical applications, cycle life and safety concerns take precedence. The majority of implantable cardiac defibrillators use $Ag_2V_4O_{11}$ as the cathode for secondary Li batteries. This system has a capacity of over 300 mAh/g and is extremely stable (26).

Most commercial applications for EDLCs and pseudocapacitors are for backup and pulse power sources for electronic devices. More recently, they have been explored as a power source for hybrid vehicles when used in combination with fuel cells or batteries. Their role is in load leveling, start-up, and acceleration (28). The traditional electrodes used in EDLCs are high surface area carbon ($1000\ m^2/g$) which have specific capacitances around 100 F/g for a single electrode (5). Low-voltage devices (2.3 V) with capacitance values of 470, 900, and 1500 F are available commercially (29). By comparison, the specific capacitance of RuO_2 electrode pseudocapacitors is in the range of 720 to 900 F/g (30–33). For the most part, ECs based on this material have yet to reach the market.

15.3 ELECTROCHEMICAL PROPERTIES OF NANOSCALE MATERIALS

One of the limitations in the current generation of lithium ion batteries is the slow diffusion of lithium ions in the solid state ($<10^{-8}\ cm^2\ s^{-1}$), a property that directly determines the charge/discharge rate for the battery. If lithium ion batteries are to meet future requirements for hybrid electric vehicles, their charge/discharge rate must increase by a factor of 10. Nanomaterials offer one approach for reaching this performance standard because the reduced dimensions lower the characteristic time constant for diffusion. A list of general advantages and disadvantages for nanomaterials applied to lithium ion batteries is as follows (34).

Advantages include:

- New reactions are enabled that cannot take place in micron-sized particles.
- Higher rate of lithium ion insertion/de-insertion because of short lithium ion transport distance. The characteristic time constant for diffusion, τ, is given by, $\tau = L^2/D$ where L is the diffusion length and D is the diffusion coefficient. Thus, the time for intercalation decreases substantially as micron-sized powders are replaced by nanometer particles.
- Electron transport within particles is enhanced by the shorter path length; this allows low conductivity materials to be used.
- The large surface area permits more extensive contact area with the electrolyte, leading to higher charge/discharge rates.
- There is better accommodation of the strain from lithium ion insertion/de-insertion along with wider solid solution region of composition.

As for disadvantages, the following have been identified:

- The high electrode/electrolyte surface area may lead to more significant side reactions with the electrolyte, leading to self-discharge and poor cycling.
- The volume of the electrode increases because of poor particle packing.
- Nanoparticles are generally more difficult to synthesize and controlling their dimensions is often challenging.

There are numerous examples in which nanostructured materials have been used in lithium ion batteries. In the sections below we provide various examples according to the form of the material: porous solids, nanoparticles, and nanotubes/nanowires.

15.3.1 Aerogels and Structure-Directed Mesoporous and Macroporous Solids

Porous solids can be categorized by their pore diameter, d, as follows: microporous ($d < 2$ nm), mesoporous (2 nm $< d <$ 50 nm), or macroporous ($d >$ 50 nm). For the most part, research on materials for electrochemical applications has focused on mesoporous and macroporous solids. High surface area (high porosity) solids are in fact two-phase materials (the solid phase and the gas phase, both on the nanoscale). Thus, the synthesis of high surface area materials requires the interruption of single phase formation. This is usually accomplished through the use of interfaces.

Aerogels are typically inorganic solids that use supercritical extraction of a solvent phase to produce extremely low density, mesoporous materials. During the initial

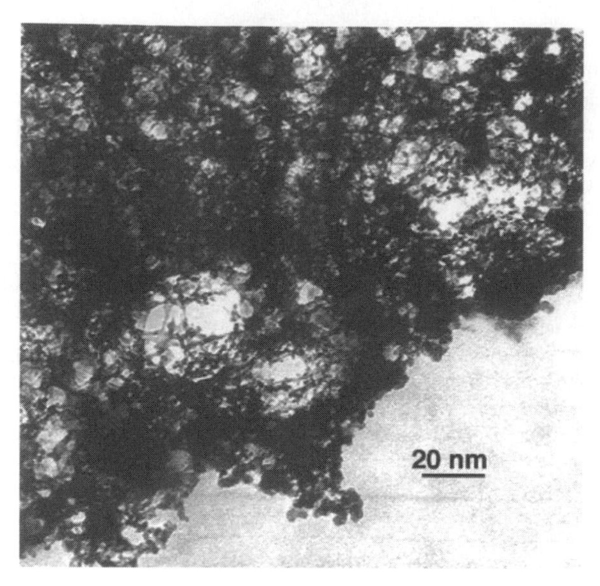

Figure 15.2 TEM image of silica aerogel. (Courtesy of the National Center for Electron Microscopy at Berkeley Lab.)

stage of forming the precursor gel through hydrolysis and condensation reactions, the formation of interpenetrating solid and liquid phases occurs. When the liquid is removed by supercritical methods, the resulting solid network maintains its nanostructure (Fig. 15.2).* Depending on the specific synthesis method, choice of precursors, catalysts, and drying techniques, aerogels can have a variety of surface areas, pore sizes, and pore-size distributions. In general, materials are characterized as aerogels if they are produced through the sol-gel method, have surface areas from 500 to over 2000 m^2/g, and have interconnected mesoporosity. In order to more precisely control the pore size and pore structure, self-assembly and templating techniques with nanocrystals, micelles, etc., can be employed in conjunction with the aerogel process. This type of pore structure control has enabled researchers to synthesize interdigitated, three-dimensional electrochemical cells (6, 35, 36; Fig. 15.3).

One important consideration in electrochemistry is the integrity of the nanostructure when the material is fabricated into a battery electrode. Various processes are involved in electrode preparation, including grinding and pressing, while electrochemical testing requires immersion in an aqueous or nonaqueous electrolyte. Although small-angle neutron scattering (37) and surface area analysis (38) have shown that the nanodimensional porosity and the high surface area are preserved, for the most part, the retention of aerogel morphology must be considered on a case-by-case basis.

A good case study that illustrates how nanostructure affects the electrochemical behavior of a material is in the application of V_2O_5 as a cathode for lithium ion

*If the solvent is removed under ambient conditions, pore collapse occurs producing a xerogel.

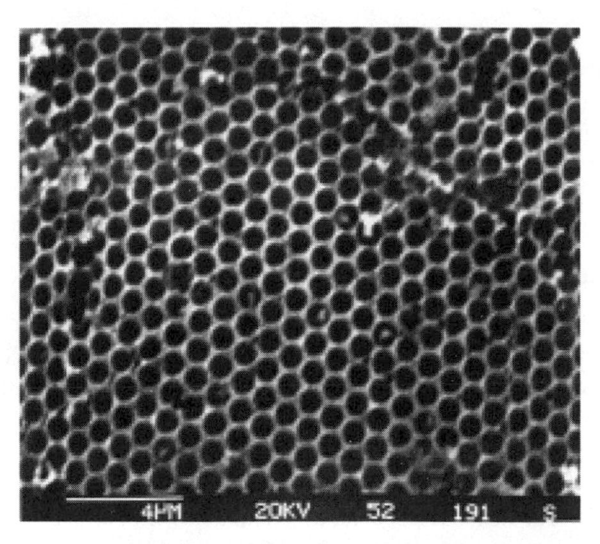

Figure 15.3 Inverse opal of templated vanadium oxide synthesized through the sol-gel method. (Reproduced with permission from J. Sakamoto and B. Dunn, *J. Mater. Chem.* **2002**, *12*, 2859.)

secondary batteries. While crystalline, stoichiometric V_2O_5 can intercalate up to 1 mole of Li^+ per mole of V_2O_5; amorphous mesoporous forms of the material can accept and release up to 6 moles of Li reversibly (7–9). The reduction of V^{5+} to V^{4+} can account for intercalation of up to 2 Li^+ per V_2O_5 (39, 40). However, the mechanism by which up to 6 Li^+ can be incorporated into this oxide structure is still unclear. What is clear, however, is that both the amorphous nature and the nanostructure of the material play important roles. The Li^+ capacity of amorphous V_2O_5 increases with increasing surface area, mainly due to the greater accessibility of the mesoporous network and the reduced diffusion distances for Li ions (8). Additionally, this amorphous, high surface area oxide is able to accept other cations into its structure (up to 4 Mg^{2+}, 3.33 Al^{3+}, and 2.75 Zn^{2+}) (8). Cyclic voltammetry for these polyvalent cations show very similar uptake mechanisms to that of Li^+ (8). This suggests that the additional capacity for Li^+ is highly structure dependent and is based on a surface adsorption (or pseudocapacitance) process, not seen in the crystalline form of V_2O_5. It is proposed that this pseudocapacitance arises from the accessibility of Li^+ (and other polyvalent cations) to surface cation vacancies of the V_2O_5 (41). Because aerogels magnify the surface area by two to three orders of magnitude and the porosity allows for electrolyte access to the surface area, pseudocapacitance can be observed in these nanoscale solids and not in bulk materials.

Another good example of the effects of aerogel nanostructure on the electrochemical behavior of materials is shown in the electroactive behavior of MnO_2. As micrometer-thick films or deposits, MnO_2 has exhibited a specific capacitance of 150 to 250 F/g (42–44). When synthesized as a nanoscale, ultra-thin coating, MnO_2 can deliver a specific capacitance of over 700 F/g (45–47). In a study by Fischer et al. (48), a carbon aerogel (nanofoam) was used as an electrically conductive substrate that can also catalyze the conversion of aqueous permanganate (MnO_4^-) to MnO_2. By coating an electrode with MnO_2, it is possible to increase capacitance without adding volume to the overall design. In a micro- to millimeter scale MnO_2-carbon composite, the area-normalized capacitance is between 0.1 and 0.5 F/cm^2. By comparison, the nanoscale composite had a reported area-normalized capacitance of greater than 1.4 F/cm^2 (48).

15.3.2 Nanoparticles

The positive electrode material $LiFePO_4$ provides one of the best examples of the benefits of using nanoparticles in battery electrode materials. $LiFePO_4$ is of interest as a cathode material in lithium ion batteries because of a number of features, including low cost, environmentally benign chemistry, and electrochemical stability. The material was initially overlooked because of its low electronic conductivity. However, decreasing the particle size to the nanoscale regime increases the electrochemical kinetics so that the material becomes of practical interest. The properties become particularly advantageous when particles are "painted" with a carbon coating, which ensures good electronic transport between particles. The $LiFePO_4$ system serves as an excellent illustration of the importance of decreasing both the ion and electron path lengths so that battery electrode materials do not necessarily require high electronic conductivity or high ion diffusion coefficients.

Another example of how materials at the nanoscale may exhibit entirely different properties is shown in the intercalation behavior of hematite, α-Fe_2O_3. When relatively small amounts of lithium are intercalated (0.005 Li per Fe_2O_3), micron-sized particles of hematite undergo an irreversible phase transformation. However, with 20 nm-sized hematite particles, 0.6 Li per Fe_2O_3 can be reversibly intercalated, as the phase transformation is suppressed. This response underscores an important point for nanoscale materials: compositions that were previously disregarded because of the intercalation response with micron-sized powders should now be reconsidered.

The cathode material $Li_xMn_2O_4$ spinel provides yet another interesting example of how nanodimensional effects lead to new directions in battery electrode materials. In this case, the electrochemical properties are influenced strongly by the nanostructure that develops within a micron-sized particle. Although $Li_xMn_2O_4$ is able to vary in composition from $0 < x < 2$, this composition range has poor electrochemical reversibility because the material passes between cubic ($0 < x < 1$) and tetragonal ($1 < x < 2$) structures. Generally, electrochemical cycling of this material is confined to the cubic phase. However, by starting with the layered form of the compound and converting to the spinel phase, a nanodomain structure is formed within the micron-sized particles. The nanodomains of spinel are able to switch successfully between cubic and tetragonal structures as lithium insertion/de-insertion occurs and still retain excellent capacity on cycling. Researchers have shown that the large change in lattice parameters between the two phases (13%) is accommodated by slippage at the domain wall boundary.

The benefits of using nanoparticles for electrochemical applications have also been noted with capacitor materials. A study of the electrochemical capacitor properties of

TiO_2 nanoparticles ranging in size from 7 nm to 30 nm showed that nanocrystal size affects the amount of capacitance (49). Specifically, a significant pseudocapacitive contribution to charge storage developed for TiO_2 particles of less than 10 nm, leading to higher levels of total stored charge (gravimetrically normalized). Moreover, the smaller particles exhibited much faster charge/discharge kinetics than reactions occurring from intercalation processes. The ability to discharge and charge over 90% of the stored charge within 200 seconds is a very good indication of the excellent power density that occurs with the nanoparticles. These results illustrate the importance of exploiting small particle size to take advantage of capacitive processes. The advantages of small particle size are not only due to the increased surface area for small nanocrystals, but also because of the increasing density of highly redox active edge and corner sites in small nanocrystals.

The electrode materials currently being used in lithium ion batteries are based on lithium intercalation/de-intercalation reactions. Such reactions are inherently tied to crystallographic considerations. The insertion of high concentrations of lithium ions is usually limited to one lithium atom per host atom because lithium concentrations above this level result in phase transformations that may lead to the formation of irreversible phases. Recently, however, another approach has been devised in which the material is not constrained by intercalation and instead accomplishes energy storage through a process based on conversion reactions.

Tarascon and colleagues (20) found that simple binary transition metal oxides with the rock salt structure (e.g. CoO, CuO, NiO) react reversibly with lithium according to the reaction:

$$MO + 2\,Li^+ + 2e^- \longleftrightarrow Li_2O + M^0 \qquad (15.1)$$

The full reduction of the metal oxide enables one to prepare composite materials consisting of nanometer-sized metallic particles dispersed in a Li_2O matrix. The nanodimensional size of the metal particles is instrumental in making the reactions highly reversible. Conversion reactions are not limited to metal oxides, as it is now clear that there are a number of other systems that can be used, including sulfides, nitrides, phosphides, and fluorides. For a binary metal compound (MX_n), with X = F, O, S, N, these reactions proceed as follows:

$$n\mathrm{Li} + \mathrm{MX}_m \longleftrightarrow \mathrm{M} + n\mathrm{LiX}_{m/n} \qquad (15.2)$$

where $m = n$ for X = F, N and $m = n/2$ for X = O, S. Depending on the oxidation state of the 3d metal, conversion reactions can involve multiple electrons per 3d metal, leading to much higher capacities than occurs from the typical single electron intercalation reactions. Storage capacities in excess of 1000 mAh/g have been achieved with several systems through the use of conversion reactions.

One of the limitations with conversion reactions is their relatively poor kinetics. This behavior is shown by the large difference in potential between the voltage on charge versus the voltage on discharge. This polarization may be fundamentally related to the breaking of the M—X bonds. To overcome the kinetics limitations, researchers have used nanoarchitectured electrodes to reduce diffusion distances. The combination of large polarization and fast kinetics is very different behavior than what occurs with intercalation electrodes.

15.3.3 Nanotubes, Nanowires, and Nanorolls

Other nanostructures that have been explored for electrochemical applications include nanotubes, nanowires, and nanorolls. These morphologies have provided important insight into how nanoscale structures affect the electrochemical behavior of materials. Described below are some case studies that illustrate the influence of various parameters at nanoscale dimensions.

Many forms of carbon (nanofoams, nanotubes, etc.) have been investigated in an attempt to further increase capacitance through increased surface area. However, the relationship between capacitance and surface area is not always straightforward, with some extremely high surface area materials actually showing a decrease in capacitance. The most important factor in determining the capacity of a high surface area material is the accessibility of pores (50). Therefore, in addition to considering the surface area, one must also take into account the total pore volume and pore size distribution. Typically, high surface areas are achieved through the presence of a large number of smaller pores. However, this morphology does not necessarily lead to high levels of capacitance because of poor electrolyte accessibility to the electrode or the inability to generate an electrified interface. Multiwalled carbon nanotubes (MWNTs) have considerably less surface area ($\sim$400 m^2/g) compared to carbon aerogels ($\sim$2000 m^2/g). However, once activated, these carbon MWNTs have capacitances around 100 F/g (51). Furthermore, it has been found that the nanostructure of the carbon nanotubes affects their capacitance. For nanotubes with an entangled network and open central canals, higher capacitance was observed ($\sim$70 F/g) compared with much lower capacitance for straight and rigid nanotubes with closed central canals (50).

For many materials, the insertion of lithium ions creates swelling that results in significant mechanical stresses. This generally inhibits increased Li^+ capacity and cyclability even with nanoparticles. For lithium ion-insertion materials, careful design of the nanostructure is needed in order to account for the lattice change. In the case of lithium ion insertion into rutile TiO_2, there is a clear kinetic limitation. A 4.5% expansion of the lattice in the a-b plane is expected and observed for 0.5 moles of Li ion inserted per mole of TiO_2 (52, 53). This mechanical strain prevents further insertion. Even nanoparticles with a size of 50 nm had fairly low specific capacity of $\sim$40 mAh/g. However, when rutile TiO_2 is synthesized in the shape of nanorods (200 nm in length and 10 nm in diameter), with the a-b plane parallel to the cross-section of the nanorod, the Li^+ capacity of the material was improved significantly ($\sim$170 mAh/g). This is due to the main expansion occurring along the smallest dimension of the particle, perpendicular to the particle main axis, limiting the mechanical stresses (11). A similar effect occurs in nanostructured SnO_2 when used as anodes for secondary lithium ion batteries. In typical SnO_2, Li^+ insertion and de-insertion causes structural changes and mechanical stresses that result in loss of capacity with continued cycling. Nanostructured SnO_2 can better accommodate the volume changes and capacities in excess of 700 mAh/g have been reported for as many as 800 cycles (54).

15.4 CONCLUDING REMARKS: CHALLENGES TO DEVELOPING BETTER BATTERIES AND CAPACITORS

This chapter has shown that in the area of electrical energy storage, transforming materials from bulk to the nanoscale can have a significant impact on the properties of electrode materials and the performance of batteries and capacitors. In some instances, the benefits arise from reducing dimensions; in battery electrodes the shorter path length for electrons and ions in combination with the increased electrode/electrolyte contact area leads to faster discharge/charge rates. In other cases, the effects are more subtle, as in nanostructured materials with particular morphologies; control of surface area and pore size distribution leads to considerable improvements in EDLCs. In general, the electrochemical properties of materials are strongly influenced by surfaces and interfaces, features that are amplified in nanodimensional solids. Thus, if nanomaterials are to continue to lead to improvements in electrochemical devices, there is a need to improve our fundamental understanding in a number of related topics, including local structure and reactivity of interfaces, the role of spatial confinement on electrochemical properties, and the changes that occur in thermodynamics and kinetics at nanoscale dimensions. The energy storage field has benefitted from the development of materials over the past decade and with the renewed emphasis on energy storage for a broad range of applications, one can expect continued improvements as new generations of nanodimensional materials emerge.

REFERENCES

1. *Energy for America's Future*. **2008**. Available from http://www.whitehouse.gov/infocus/energy/ (accessed 1/11/2008).
2. Hamrin, J., R. Dracker, J. Martin, R. Wiser, K. Porter, D. Clement, and M. Bolinger, *Achieving a 33% Renewable Energy Target*, C.P.U. Commission, **2005**.
3. Goodenough, J., S. Visco, and M. Whittingham, *Basic Resarch Needs for Electrical Energy Storage*. Department of Energy, Washington, D.C., **2007**.
4. Tarascon, J., S. Grugeon, M. Morcrette, S. Laruelle, P. Rozier, and P. Poizot, New concepts for the search of better electrode materials for rechargeable lithium batteries. *C.R. Chimie* **2005**, *8*, 9.
5. Kotz, R. and M. Carlen, Principles and applications of electrochemical capacitors. *Electrochim Acta* **2000**, *45*, 2483.
6. Long, J., B. Dunn, D. Rolison, and H. White, Three-dimensional battery architectures. *Chem. Rev.* **2004**, *104*, 4463.
7. Passerini, S., D. Le, W. Smyrl, M. Berrettoni, R. Tossici, R. Marassi, and M. Giorgetti, XAS and electrochemical characterization of lithiated high surface area V_2O_5 aerogels. *Solid State Ionics* **1997**, *104*, 195.

8. Rolison, D. and B. Dunn, Electrically conductive oxide aerogels: New materials in electrochemistry. *J. Mater. Chem.* **2001**, *11*, 963.
9. Delmas, C., H. Cognacauradou, J. Cocciantelli, M. Menetrier, and J. Doumerc, The LixV_2O_5 system: An overview of the structure modifications induced by the lithium intercalation. *Solid State Ionics* **1994**, *69*, 257.
10. Padhi, A., K. Nanjundaswamy, and J. Goodenough, Phospho-olivines as positive-electrode materials for rechargeable lithium batteries. *J. Electrochem. Soc.* **1997**, *144*, 1188.
11. Baudrin, E., S. Cassaignon, M. Koesch, J. Jolivet, L. Dupont, and J. Tarascon, Structural evolution during the reaction of Li with nano-sized rutile type TiO_2 at room temperature. *Electrochem. Commun.* **2007**, *9*, 337.
12. Thackeray, M., C. Johnson, J. Vaughey, N. Li, and S. Hackney, Advances in manganese-oxide "composite" electrodes for lithium-ion batteries. *J. Mater. Chem.* **2005**, *15*, 2257.
13. Morcrette, M., P. Rozier, L. Dupont, E. Mugnier, L. Sannier, J. Galy, and J. Tarascon, A reversible copper extrusion-insertion electrode for rechargeable Li batteries. *Nature Mater.* **2003**, *2*, 755.
14. Kepler, K., J. Vaughey, and M. Thackeray, *Electrochem. Solid State Lett.* **1999**, *2*, 307.
15. Thackeray, M., *Nature Mater.* **2002**, *1*, 82.
16. Thomas, J., *Nature Mater.* **2003**, *2*, 201.
17. Abraham, K. and Z. Jiang, A polymer electrolyte-based rechargeable lithium/oxygen battery. *J. Electrochem. Soc.* **1996**, *143*, 1.
18. Ogasawara, T., A. Debart, M. Haolzapfel, P. Novak, and P. Bruce, Rechargeable Li_2O_2 electrode for lithium batteries. *J. Am. Chem. Soc.* **2006**, *128*, 1390.
19. Poizot, P., S. Laruelle, S. Grugeon, L. Dupont, B. Beaudoin, and J. Tarascon, *C.R. Acad. Sci. Paris* **2000**, *3*, 681.
20. Poizot, P., S. Laruelle, S. Grugeon, L. Dupont, and J. Tarascon, Nano-sized transition-metal oxides as negative-electrode materials for lithium-ion batteries. *Nature* **2000**, *407*, 496.
21. Grugeon, S., S. Laruelle, R. Herera-Urbina, L. Dupont, P. Poizot, and J. Tarascon, *J. Electrochem. Soc.* **2001**, *148*, A285.
22. Larcher, D., G. Sudant, J. Leriche, Y. Chabre, and J. Tarascon, The electrochemical reduction of Co_3O_4 in a lithium cell. *J. Electrochem. Soc.* **2002**, *149*, A234.
23. Larcher, D., C. Masquelier, D. Bonnin, Y. Chabre, V. Masson, J. Leriche, and J. Tarascon, Effect of particle size on lithium intercalation into alpha-Fe_2O_3. *J. Electrochem. Soc.* **2003**, *150*, A133.
24. Chmiola, J., G. Yushin, Y. Gogotsi, C. Portet, P. Simon, and P. Taberna, Anomalous increase in carbon capacitance at pore sizes less than 1 nanometer. *Science* **2006**, *313*, 1760.
25. Winter, M. and R. Brodd, What are batteries, fuel cells, and supercapacitors? *Chem. Rev.* **2004**, *104*, 4245.
26. Whittingham, M., Lithium batteries and cathode materials. *Chem. Rev.* **2004**, *104*, 4271.
27. Souquet, J. and M. Duclot, *Solid State Ionics* **2002**, *148*, 375.
28. Reddy, R. and R. Reddy, Analytical solution for the voltage distribution in one-dimensional porous electrode subjected to cyclic voltammetric conditions. *Electroch. Acta* **2007**, *53*, 575.
29. Zubieta, L. and R. Bonert, Characterization of double-layer capacitors for power electronics applications. *IEEE Trans. Industry Applications* **2000**, *36*, 199.

30. Zheng, J., C. Cygan, and T. Jow, Hydrous ruthenium oxide as an electrode material for electrochemical capacitors. *J. Electrochem. Soc.* **1995**, *142*.
31. Miller, J., B. Dunn, T. Tran, and R. Pekala, Deposition of ruthenium nanoparticles on carbon aerogels for high energy density supercapacitor electrodes. *J. Electrochem. Soc.* **1997**, *144*(12), L309.
32. Conway, B., V. Birss, and J. Wojtowicz, The role and utilization of pseudocapacitance for energy storage by supercapacitors. *J. Power Sources* **1997**, *66*, 1.
33. McKeown, D., P. Hagans, L. Carette, A. Russell, K. Swider, and D. Rolison, Structure of hydrous ruthenium oxides: Implications for charge storage. *J. Phys. Chem. B* **1999**, *103*(23), 4825.
34. Bruce, P., B. Scrosati, and J. Tarascon, *Angew. Chem. Int. Ed.* **2008**, *47*, 2930.
35. Ergang, N., M. Fierke, Z. Wang, W. Smyrl, and A. Stein, Fabrication of a fully infiltrated three-dimensional solid-state interpenetrating electrochemical cell. *J. Electrochem. Soc.* **2007**, *154*(12), A1135.
36. Sakamoto, J. and B. Dunn, Hierarchical battery electrodes based on inverted opal structures. *J. Mater. Chem.* **2002**, *12*, 2859.
37. Merzbacher, C., J. Barker, K. Swider, and D. Rolison, *J. Non-Crystalline Solids* **1998**, *224*, 92.
38. Swider, K., P. Hagans, C. Merzbacher, and D. Rolison, *Chem. Mater.* **1997**, *9*, 1248.
39. Passerini, S., W. Smyrl, M. Berrettoni, R. Tossici, M. Rosolen, R. Marassi, and F. Decker, *Solid State Ionics* **1996**, *90*, 5.
40. Giorgetti, M., S. Passerini, W. Smyrl, S. Mukerjee, X. Yang, and J. McBreen, *J. Electrochem. Soc.* **1999**, *146*, 2387.
41. Ruetschi, P. and R. Giovanoli, *J. Electrochem. Soc.* **1988**, *135*, 2663.
42. Lee, H. and J. Goodenough, *J. Solid State Chem.* **1999**, *144*, 220.
43. Long, J., A. Young, and D. Rolison, *J. Electrochem. Soc.* **2003**, *150*, A1161.
44. Nam, K. and K. Kim, *J. Electrochem. Soc.* **2006**, *153*, A81.
45. Pang, S., M. Anderson, and T. Chapman, *J. Electrochem. Soc.* **2000**, *147*, 444.
46. Broughton, J. and M. Brett, *J. Electrochim. Acta* **2004**, *49*, 4439.
47. Toupin, M., T. Brousse, and D. Belanger, *Chem Mater.* **2004**, *16*, 3184.
48. Fischer, A., K. Pettigrew, D. Rolison, R. Stroud, and J. Long, Incorporation of homogeneous, nanoscale MnO_2 within ultraporous carbon structures vis self-limiting electroless deposition: Implications for electrochemical capacitors. *Nano Lett.* **2007**, *7*, 281.
49. Wang, J., J. Polleux, J. Lim, and B. Dunn, *J. Phys. Chem. C* **2007**, *111*, 14925.
50. Frackowiak, E. and F. Beguin, Carbon materials for the electrochemical storage of energy in capacitors. *Carbon* **2001**, *39*, 937.
51. Lafdi, K. and J. Clay, Materials approach to study supercapacitor performance. 22nd Biennial Conf. on Carbon, **1995**, p. 824.
52. Koudriachova, M., N. Harrison, and S. de Leeuw, *Solid State Ionics* **2004**, *175*, 829.
53. Koudriachova, M., S. de Leeuw, and N. Harrison, *Chem. Phys. Lett.* **2003**, *371*, 150.
54. Li, N., C. Martin, and B. Scrosati, Nanomaterial-based Li-ion battery electrodes. *J. Power Sources* **2001**, *97–98*, 240.

PROBLEMS

1. What is the main difference between batteries and capacitors?
2. How do nanostructures play a role in each type of energy storage device?
3. What are some of the important material properties that need to be considered in applying nanoscale materials to electrochemical applications?
4. By decreasing the electrode particle size from microns to nanometers in a battery, how will the intercalation time be affected? By what order of magnitude?
5. If a metal can be reduced from M^{4+} to M^{3+}, is it possible for that metal oxide (MO_2) to incorporate more than one Li^+ ion per metal into the structure? How?
6. Pseudocapacitance and conversion reactions can dramatically increase the energy storage capability of secondary batteries. How are these phenomena dependent on nanoscale structures?
7. Can materials that were thought to be electrochemically irreversible become reversible? Identify two examples where nanoscale domains enable reversible intercalation reactions to occur.
8. A pseudocapacitor stores energy on the surface or subsurface. Why is it that the capacity or specific energy of the electrode does not necessarily increase as a function of the surface area?
9. In comparing V_2O_5 and Fe_2O_3, which material do you expect to have higher levels of capacitive storage from pseudocapacitive processes?
10. From an operational standpoint, how do batteries and capacitors differ?

ANSWERS

1. Batteries store energy based on the electrical potential difference between the two electrodes. This can be converted to electric current through a chemical reaction. Capacitors store energy within the variation of potential at the electrode/electrolyte interface. This is converted to electric current when charges are allowed to return to their equilibrium state.
2. Nanostructures can affect ion and electron transport, accessibility of chemical reaction sites, and the fundamental properties of electrochemical interfaces, among others.
3. Surface area, porosity, pore size distribution, and ability for nanodomains to accommodate dimensional changes in the insertion/de-insertion process.
4. The intercalation time is shorter because of the smaller diffusion distance. Using the equation: $\tau = L^2/D$, by decreasing the diffusion length by 10^3, the intercalation time should decrease by 10^6.

5. Yes, through a pseudocapacitive process such as is the case for V_2O_5, or through conversion reactions where the metal oxide is converted into nanometer-sized metallic particles dispersed in a Li_2O matrix.

6. Pseudocapacitance in normally purely intercalation materials has only been observed in extremely high surface area materials with nanoscale diffusion lengths. In conversion reactions, the nanodimensional size of the metal particles is instrumental in making the reactions highly reversible.

7. $Li_xMn_2O_4$ spinel: nanodomains of spinel are able to successfully switch between cubic and tetragonal structures as lithium insertion/de-insertion occurs.

Rutile TiO_2: when synthesized in the shape of nanorods, with the a-b plane parallel to the cross-section of the nanorod, the previously irreversible structural change associated with lithium intercalation can be avoided. This is due to the main expansion occurring along the smallest dimension of the particle.

8. The accessibility of these surface areas also need to be considered.

9. The V_2O_5 will have higher levels of capacitive storage (gravimetrically normalized) because of its multielectron redox capabilities.

10. Batteries store more energy than electrochemical capacitors, but operate at lower power. This means that electrochemical capacitors can charge much more quickly than batteries.

PART VII

UNIQUE PHYSICAL PROPERTIES OF NANOMATERIALS

16

OPTICAL AND ELECTRONIC PROPERTIES OF METAL AND SEMICONDUCTOR NANOSTRUCTURES

Mausam Kalita, Matthew T. Basel, Katharine Janik, and Stefan H. Bossmann

16.1 Introduction, 540
16.2 Absorption of and Emission from Nanoparticles, 541
  16.2.1 What Is a Surface Plasmon? 541
  16.2.2 The Optical Extinction of Nanoparticles, 542
  16.2.3 The Simple Drude Model Describes Metal Nanoparticles, 545
  16.2.4 Semiconductor Nanoparticles (Quantum Dots), 549
  16.2.5 Discrete Dipole Approximation (DDA), 550
  16.2.6 Luminescence from Noble Metal Nanostructures, 550
  16.2.7 Nonradiative Relaxation Dynamics of the Surface Plasmon Oscillation, 554
  16.2.8 Nanoparticles Rule: From Förster Energy Transfer to the Plasmon Ruler Equation, 558
16.3 Quantum Dots: A Brief Overview, 563
  16.3.1 Electronic and Optical Properties of Quantum Dots, 563
  16.3.2 Type I and Type II Core-Shell Quantum Dots, 565
  16.3.3 The Absorption Cross-Sections of Quantum Dots, 565
  16.3.4 Absorption and Emission Maxima of Quantum Dots, 567
  16.3.5 Luminescence Lifetimes of Q-Dots, 567
16.4 Electrochemistry of Nanoparticles, 569
16.5 Conclusion, 571
Further Reading, 571
References, 572
Problems, 575
Answers, 576

Nanoscale Materials in Chemistry, Second Edition. Edited by K. J. Klabunde and R. M. Richards

16.1 INTRODUCTION

The absorption of a photon leads to the excitation of localized surface plasmons in metals. Localized surface plasmon resonance (LSPR), which occurs upon the excitation of nanometer-sized metallic structures, is the basis for many opto-analytical procedures in the life sciences (e.g. biosensor and lab-on-a-chip sensors) (1). Surface plasmons are the collective oscillations of the free electron gas density in a metal. They often occur at visible frequencies and can be detected by absorption spectroscopy. It is of great mechanistic importance that surface plasmons are surface electromagnetic waves that propagate in parallel along a metal/dielectric interface (2). Since the wave is located on the boundary of the metal and the external medium (or vacuum), these oscillations are very sensitive to any change of the chemical composition of this boundary, such as the adsorption and exchange of ligands and macromolecules [e.g. (bio)polymers] to the metal surface. Because the phenomenon of plasmon resonance is caused by the oscillations of the free electron gas in a metal, it occurs only when metal nanoparticles are excited. Some localized surface plasmons give rise to strong fluorescence, whereas others are only weakly emissive. Surface plasmon (field-enhanced) fluorescence spectroscopy (SPFS) uses the greatly enhanced electromagnetic fields of the surface plasmons' modes for the excitation of organic and inorganic luminophores (3) (Scheme 16.1).

The absorption and emission characteristics of nano-sized semiconductors are based on different principles than that of nano-sized metals. Many semiconductor nanoparticles are strongly emissive and their absorption and emission bands feature narrow bandwidths. Therefore, the term quantum dot (QD) is frequently used (4).

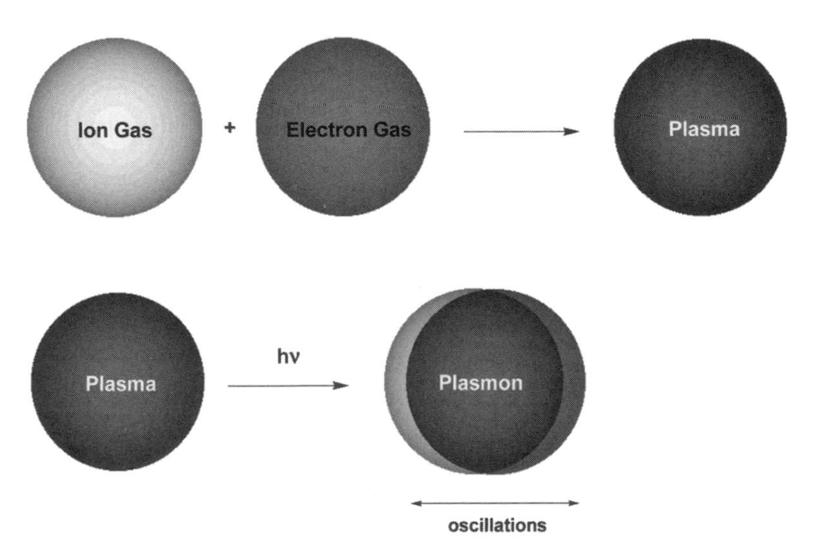

Scheme 16.1 In a metal, the plasma is created by the combination of an ion gas and an electron gas. The plasma is then disturbed by the absorption of a photon. The resulting electrodynamic force is causing an oscillation, which is known as plasmon.

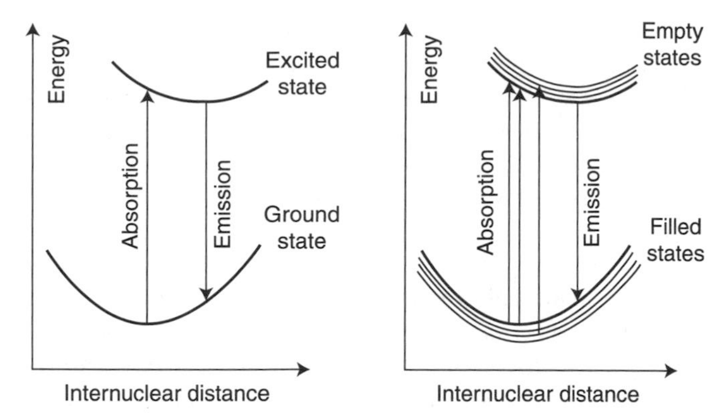

Scheme 16.2 Relationship between absorption and emission energies for molecules (discrete, left) and for semiconductor quantum dots (right). Upon the absorption of a photon, an electron is lifted from the ground state (1) to an electronically excited state (2). The bond order decreases, because excited states are antibinding and the atoms relax to larger internuclear distances (3). From the lowest excited state (only one is shown here), emission of a photon (4) and relaxation to the ground state occurs. In semiconductor quantum dots (e.g. CdSe), more possibilities for the absorption of photons exist, because several/many orbitals can be found. (Reproduced with permission from E. M. Boatman et al., 2005. *J. Chem. Ed.* 82: 1697–1699. Copyright 2005 American Chemical Society.)

A quantum dot is made from a semiconductor nanostructure that confines the motion of conduction band electrons, valence band holes, or excitons (bound pairs of conduction band electrons and valence band holes) in all three spatial directions. A quantum dot contains a small finite number (of the order of 1 to 100) of conduction band electrons, valence band holes, or excitons, that is, a finite number of elementary electric charges (Scheme 16.2). The reason for the confinement is either the presence of an interface between different semiconductor materials (e.g. in core-shell nanocrystal systems) or the existence of the semiconductor surface (e.g. semiconductor nanocrystal). Therefore, one quantum dot or numerous quantum dots of exactly the same size and shape have a discrete quantized energy spectrum. The corresponding wave functions are spatially localized within the quantum dot, but they always extend over many periods of the crystal lattice (5).

16.2 ABSORPTION OF AND EMISSION FROM NANOPARTICLES

16.2.1 What Is a Surface Plasmon?

The fascinating phenomenon of surface plasmon resonance (SPR) occurs when an electromagnetic wave interacts with the conduction electrons of a metal (6). The periodic electric field of the electromagnetic wave causes a collective oscillation of the conductance electrons at a resonant frequency relative to the lattice of positive ions. Light is absorbed or scattered at this resonant frequency. The process of absorption is characterized by the conversion of incident resonant photons into phonons or vibrations of the metal lattice, whereas scattering is the re-emission of resonant photons in all directions. Because of these two processes, the experimentally observable SPR peak of any metal nanostructure features both absorption and scattering components. Gustav Mie (7) was the first scientist to develop a method to calculate the SPR spectra of (noble) metal nanostructures by solving Maxwell's equation for spherical nanoobjects. The Mie theory was the stepwise extension for a variety of objects with simple geometries, such as spheroids and rods. To date, exact solutions to Maxwell's equations have been found only for spheres, concentric spherical shells, spheroids, and infinite cylinders. Therefore, approximation is required to solve the equations for other geometries. The discrete dipole approximation (DDA, see below) is the preferred method of choice, because it can be easily adapted to any geometry (8).

16.2.2 The Optical Extinction of Nanoparticles

The optical extinction $E(\lambda)$ of nanoparticles being smaller than the wavelength of the exciting light source, is given by Equation (16.1).

$$E(\lambda) = S(\lambda) + A(\lambda) \qquad \lambda\text{: wavelength, } S\text{: scattering, } A\text{: absorbance} \tag{16.1}$$

The extinction efficiency factor Q_{ext}, which is the sum of scattering efficiency factor Q_{sca} and the absorption efficiency factor Q_{abs}, is defined as the quotient of C_{ext} and the physical cross-section area πR^2. The scattering and absorption efficiency factors can be calculated according to the general Mie theory, which is explained, in some detail, below. Both can be expressed as infinite series (9).

$$Q_{ext} = \frac{2}{x^2}\sum_{n=1}^{\infty}(2n+1)Re[a_n + b_n] \tag{16.2}$$

$$Q_{sca} = \frac{2}{x^2}\sum_{n=1}^{\infty}(2n+1)[a_n^2 + b_n^2] \tag{16.3}$$

$$Q_{abs} = Q_{ext} - Q_{sca} \tag{16.4}$$

$$a_n = \frac{m\Psi_n(mx)\Psi'_n(x) - \Psi_n(x)\Psi'_n(mx)}{m\Psi_n(mx)\xi_n(x) - m\xi_n(x)\Psi'_n(mx)} \tag{16.5}$$

$$b_n = \frac{\Psi_n(mx)\Psi'_n(x) - m\Psi_n(x)\Psi'_n(mx)}{\Psi_n(mx)\xi_n(x) - m\xi_n(x)\Psi'_n(mx)} \tag{16.6}$$

$$x = \frac{2\pi n_m R}{\lambda} \tag{16.7}$$

Re denotes the real part, m is the ratio of the refractive index of the spherical nanoparticle n to that of the surrounding medium n_m. x is the size parameter. λ is the incident wavelength, R is the diameter of the nanoparticle. Ψ_n and ξ_n are the Riccati–Bessel functions. The prime represents the first differentiation with respect to the argument in parentheses.

$$A(\lambda) = \varepsilon_{abs(\lambda)}cl \tag{16.8}$$

$$E(\lambda) = (\varepsilon_{abs(\lambda)} + \varepsilon_{sca(\lambda)})cl \tag{16.9}$$

$$\varepsilon_{ext} = \frac{N_A C_{ext}}{0.2303} \tag{16.10}$$

Equation (16.8) is known as Lambert–Beer's law. $A(\lambda)$ is the absorbance or optical density of the sample, ε ($M^{-1}\,cm^{-1}$) is the molar absorption (ε_{abs}), scattering (ε_{sca}), or extinction coefficient (ε_{ext}), C (M) is the concentration of the light absorbing and scattering species, and l (cm) is the optical path length (10). Note that the molar absorption and scattering coefficients are directly related to the absorption and scattering cross-section by means of Equation (16.10); (N_A is Avogadro's number).

Metal nanoparticles show remarkably larger absorption cross-sections compared to organic dyes and metal complexes. A typical example are the nanospheres that have been used by El-Sayed et al. (11) for the laser-induced photothermal hyperthermia treatment of cancer cells, which feature an absorption cross-section of $2.93 \times 10^{-15}\,m^2$ ($\varepsilon = 7.66 \times 10^9\ M^{-1}\,cm^{-1}$) at their plasmon resonance maximum of $\lambda = 528$ nm. This is five orders of magnitude larger than that of the commonly used NIR dye indocyanine green ($\varepsilon = 1.08 \times 10^4\,M^{-1}\,cm^{-1}$ at $\lambda = 778$ nm) (12) or the sensitizer ruthenium(II)-*tris*-bipyridine ($1.54 \times 10^4\,M^{-1}\,cm^{-1}$ at 452 nm) (13) and four orders of magnitude larger than rhodamine-6G ($\varepsilon = 1.16 \times 10^5\,M^{-1}\,cm^{-1}$ at $\lambda =$ 530 nm) or malachite green ($\varepsilon = 1.49 \times 10^5\,M^{-1}\,cm^{-1}$ at $\lambda = 617$ nm) (14). Metal nanoparticles possess remarkable light scattering properties as well. Gold nanospheres of 80 nm in diameter have approximately the same Mie-scattering characteristics as polystyrene beads of 300 nm (both feature $C_{sca} = 1.23 \times 10^{14}\,m^2$ at $\lambda = 560$ nm, corresponding to a molar scattering coefficient of $3.22 \times 10^{10}\,M^{-1}cm^{-1}$). It is noteworthy that this strong scattering is five orders of magnitude higher than the light emission (fluorescence) from fluoresceine ($\varepsilon = 9.23 \times 10^4\,M^{-1}\,cm^{-1}$ at $\lambda =$ 521 nm, emission quantum yield $\Phi = 0.98$ at $\lambda = 483$ nm) (14).

Some studies have suggested that multiple-photon absorption of gold nanostructures can be advantageous, for instance in SERS-microscopy (15) and plasmonic heating (photo-induced hyperthermia) (16). It should be noted that noble metal nanostructures (e.g. gold nanorods) can be easily tuned for monophotonic applications in the wavelength range from 600 to 1000 nm. Monophotonic excitation is advantageous, because its absorption cross-section is several orders of magnitude higher!

16.2.2.1 Gold Nanospheres Noble metal nanoparticles that have been synthesized in solution have a strong tendency towards minimizing their free surface energy and, therefore, spherical shapes. Note that further coagulation (Ostwald ripening) is usually prohibited due to the stabilization of the synthesized nanoparticles by strongly surface-bound ligands (17). It is interesting that high resolution TEM often reveals that nanoparticles from solution processes are usually not truly spherical in shape, but quasi-spheres that are multiply twinned particles with facets on their surfaces (18).

In calculating the UV/Vis absorption spectra of nanoparticles, light absorption and scattering have to be understood. Although Rayleigh scattering, otherwise occurring only at very small spherical and dielectric (nonabsorbing) particles/molecules, is observed at metal nanoparticles [named hyper-Rayleigh scattering (19)], Mie scattering is nevertheless the key to understanding the optical properties of (noble) metal nanoparticles. The most accepted form to describe the Mie theory for the interaction of a plane wave with a spherical particle of radius R and the same dielectric constant as the bulk metal is:

$$C_{ext} = \frac{24\pi^2 R^3 \varepsilon_m^{3/2}}{\lambda} \frac{\varepsilon_2}{(\varepsilon_1 + 2\varepsilon_m)^2 + \varepsilon_2^2} \tag{16.11}$$

C_{ext} is the extinction cross-section of the spheres. The extinction efficiency factor Q_{ext}, which is the sum of scattering and absorption, is defined as the quotient of C_{ext} and the physical cross-section area ΠR^2. In all the cases where the spherical nanoparticle is distinctly smaller than the incident wavelength ($2R \ll \lambda$), only dipole oscillations have to be accounted for and higher terms (quadrupole oscillations, etc.) can be neglected.

ε_m is the dielectric constant of the liquid medium (20), $\varepsilon = \varepsilon_1 + \varepsilon_2$ is the complex dielectric constant of the nanoparticle. (The denotation $\varepsilon = \varepsilon_1 + i\varepsilon_2$ is often found in the literature.) In our denotation ε_2 is the imaginary part of the equation. The resonance peak (SPR) occurs at $\varepsilon_1 = -2\varepsilon_m$.

According to Equation (16.11), C_{ext} (and therefore also Q_{ext}) depend on ε_m, the dielectric constant of the surrounding environment. Therefore, the SPR peak positions are dependent on the presence of various ligands or adsorbed (bio)macromolecules on their surfaces. Note that the aggregation of (noble) metal nanoparticles (disregarding whether they are ligand-stabilized or not), leads to more pronounced changes of their absorption spectra due to plasmonic interaction. These shifts are ideal for the design of optical biosensors.

16.2.2.2 Gold/SilverAlloy Formation A simple and usually not quite justified approach to predicting the position of the SPR maximum of an (Au/Ag)-alloy nanoparticle would be to calculate a linear combination of the dielectric functions of gold and silver nanoparticles [$\varepsilon(\alpha) = (1 - \alpha)\varepsilon_{Au} + \alpha\varepsilon_{Ag}$]. Unfortunately, this approach does not lead exactly to the proper prediction of the position of the SPR band in the optical spectrum. Therefore, one should be cautious when working with alloy nanoparticles because a new electronic structure is formed during the formation of the alloy. However, although its exact photophysical properties are difficult to predict, alloy formation is a great strategy to tune the optical properties of metal nanoparticles (21).

16.2.2.3 Rod-Shaped Nanoparticles The important consequence of the change from a sphere to a rod is that the plane of incident light has two distinct orientations with respect to the long axis of the rod. Therefore, nanorods feature two distinct extinction bands, which are due to electron oscillations across and along the long axis. These bands are named transverse and longitudinal modes. The existence of two bands was originally predicted by Richard Gans (22), who modified the Mie equations by introducing the appropriate boundary conditions for rods. In this case, the extinction cross-section can be described as

$$C_{ext} = \frac{2\pi V}{3\lambda}\varepsilon_m^{3/2}\sum_j \frac{\left(\frac{1}{P_j^2}\right)\varepsilon_2}{\left(\varepsilon_1 + \frac{1-P_j}{P_j}\right)^2 + \varepsilon_2^2} \tag{16.12}$$

where V is the volume of the nanorod, P_j [j = A (length), B, C (widths); A > B = C] are the depolarization factors, which are calculated according to

$$P_A = \frac{1-e^2}{e^2}\left[\frac{1}{2e}\ln\left(\frac{1+e}{1-e}\right) - 1\right] \tag{16.13}$$

and

$$P_B = P_C = \frac{1-P_A}{2} \tag{16.14}$$

where e is the rod ellipticity according to $e^2 = 1 - R^{-2}$, and R is the aspect ratio of the nanorod (R = A/B). El-Sayed and coworkers have followed this approach for the calculation of the extinction of various Au-, Ag-, and Au/Ag- nanorods. It is of importance for the medicinal applications of noble metal nanorods that the longitudinal band is continuously shifted from the visible to the near infrared, where human tissue is most transparent, as their aspect ratios (R) increase (23). Contrary to this, the transverse resonance shows a slight blue shift. The aspect ratio is defined as the length of the rod divided by the width of the rod.

16.2.3 The Simple Drude Model Describes Metal Nanoparticles

Many metals exhibit a strong dependence of their UV/Vis/NIR absorption on the behavior of their free electrons up to the so-called bulk plasma frequency (located in the UV). The simple Drude model describes the dielectric response of the metals' electrons (24). Thus, the dielectric function $\varepsilon(\omega)$ can be written as a combination of an interband term $\varepsilon_{IB}(\omega)$, accounting for the properties of the d electrons, and the *Drude term* $\varepsilon_D(\omega)$, which only accounts for the free conduction electrons.

$$\varepsilon(\omega) = \varepsilon_{IB}(\omega) + \varepsilon_D(\omega) \tag{16.15}$$

The Drude term can be expressed as:

$$\varepsilon_D(\omega) = 1 - \frac{\omega_p^2}{\omega^2 + i\gamma\omega} \tag{16.16}$$

where

$$\omega_p^2 = \left[\frac{ne^2}{\varepsilon_0 m_{eff}}\right] \tag{16.17}$$

is the bulk plasmon frequency, n the free electron density in the metal, e the charge of an electron, ε_0 the vacuum permittivity, m_{eff} the electron effective mass. γ is the phenomenological damping constant (it equals the plasmon bandwidth Γ for the case of a perfect free electron gas if $\gamma \ll \omega$). The damping constant γ is related to the lifetimes of all electron scattering processes: electron-electron, electron-phonon, and electron-defect scattering. For a nanoparticle, electron-surface scattering becomes important as well, since the mean free path of the conduction electrons, which is typically in the range of tens of nanometers in noble metals, becomes increasingly limited by the nanoparticles' boundaries. γ therefore becomes dependent on the nanoparticle radius:

$$\gamma(r) = \gamma_0 + \frac{Av_F}{r} \tag{16.18}$$

γ_0 is the bulk damping constant, v_F is the velocity of the electrons at the Fermi energy, A is a parameter that accounts for various scattering processes (e.g. isotropic or diffuse scattering).

According to Equation (16.18), the ratio between the volume and the surface area of a metal nanoparticle is most important for its dielectric properties. For larger gold nanoparticles (>25 nm in diameter), the extinction coefficient explicitly depends on the nanoparticles' sizes (see Fig. 16.1). These nanoparticles experience

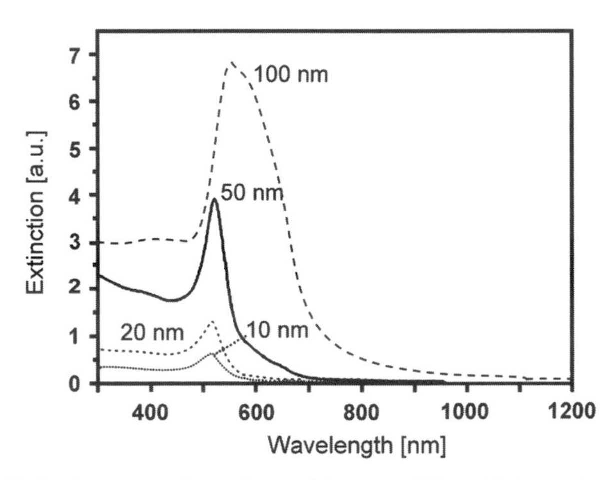

Figure 16.1 Extinction spectra for various gold nanoparticles with increasing radius as calculated according to Equation (16.11). The peak maximum shifts only slightly as the nanoparticles' diameters increase. However, a clearly discernible broadening of the absorption peaks near $\lambda = 520$ nm is observed with increasing radius. (Reproduced with permission from M. Hu et al., 2006 *Chem. Soc. Rev.* 35: 1084–1094. Copyright 2006 The Royal Society of Chemistry.)

inhomogeneous polarization by the electromagnetic field of the incident light. The increased line width is caused by the simultaneous excitation of different multipole modes, which feature peak maxima at different energies. The real and imaginary parts of the dielectric function can be expressed as:

$$\varepsilon_1 = \varepsilon^{\infty} - \frac{\omega_p^2}{\omega^2 + \omega_d^2} \tag{16.19}$$

and

$$\varepsilon_2 = \frac{\omega_p^2 \omega_d}{\omega(\omega^2 + \omega_d^2)} \tag{16.20}$$

where ε^{∞} is the high frequency dielectric constant. The strength of the Drude model, beside its simplicity, is that it explains the plasmon absorption of (noble) metals by using the material properties of the metal and the size of the nanostructures. Note that both Drude and Mie–Gans theories predict that the resonance peak (SPR) occurs at $\varepsilon_1 = -2\varepsilon_m$.

As can be seen in Figure 16.2, the ε_1 is always negative at incident frequencies smaller than ω_p. The physical reason for ε_1 being negative is that the conduction electrons oscillate out of phase with the electric field vector of the light wave. This is also the reason for the strong size-dependent optical properties of metal nanoparticles. In a nanoparticle the dipole, created by the incident electrical field that causes the long wavelength absorption of the bulk metal, is bent around the particle. At the resonance condition $\varepsilon_1 = -2\varepsilon_m$, a single surface-plasmon band is formed around the nanoparticle. The consequences of the nanoparticles' size and chemical composition are demonstrated in Figures 16.3 and 16.4.

16.2.3.1 The Effect of the Solvents' Refractive Index For noble metals (not necessarily metal alloys), which behave in agreement with the Drude model, the

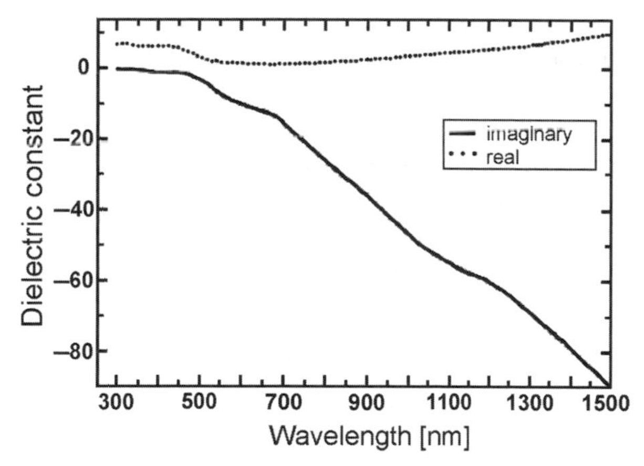

Figure 16.2 ε_1 (real) and ε_2 (imaginary) for gold nanoparticles ($R < 100$ nm). (Data from References 8 and 12.)

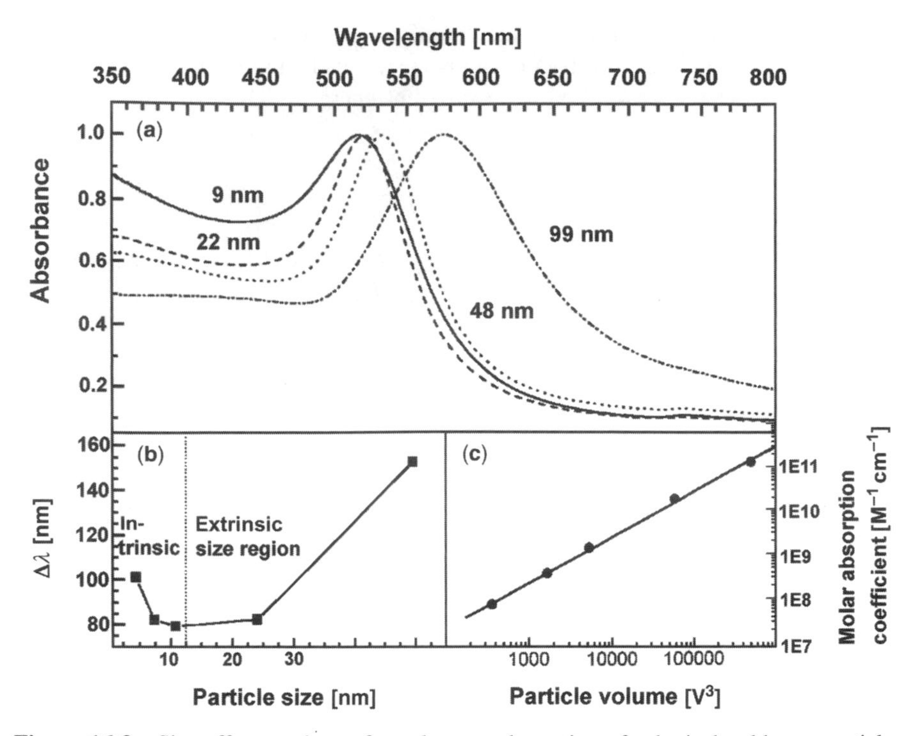

Figure 16.3 Size effect on the surface plasmon absorption of spherical gold nanoparticles. The UV/Vis-absorption spectra of colloidal solutions of gold nanoparticles with diameters varying between 9 and 99 nm show that the absorption maximum red-shifts with increasing particle size in part (a), while the plasmon bandwidth follows the behavior shown in part (b). The bandwidth increases with decreasing nanoparticle radius in the intrinsic size region and also with increasing radius in the extrinsic size region. In (c) the extinction coefficients of the gold nanoparticles at their respective plasmon absorption maxima are plotted vs. their volume on a double-logarithmic scale. The solid line is a linear fit of the data points, illustrating that a linear dependence is observed, as predicted by the Mie theory. (Reproduced with permission from S. Link and M. El-Sayed, 1999. *J. Phys. Chem. B* 103: 8410–8426. Copyright 1999 American Chemical Society.)

position of the SPR peak depends on the refractive index of the surrounding medium. The shifts occur according to Equation (16.21):

$$\lambda^2 = \lambda_p^2(\varepsilon^\infty + 2\varepsilon_m) \tag{16.21}$$

as long as ε_2 is small and, therefore, negligible.

$$\lambda_p^2 = \frac{(2\pi c)^2}{\omega_p^2} \qquad c\text{: speed of light} \tag{16.22}$$

is the metal's bulk plasma wavelength.

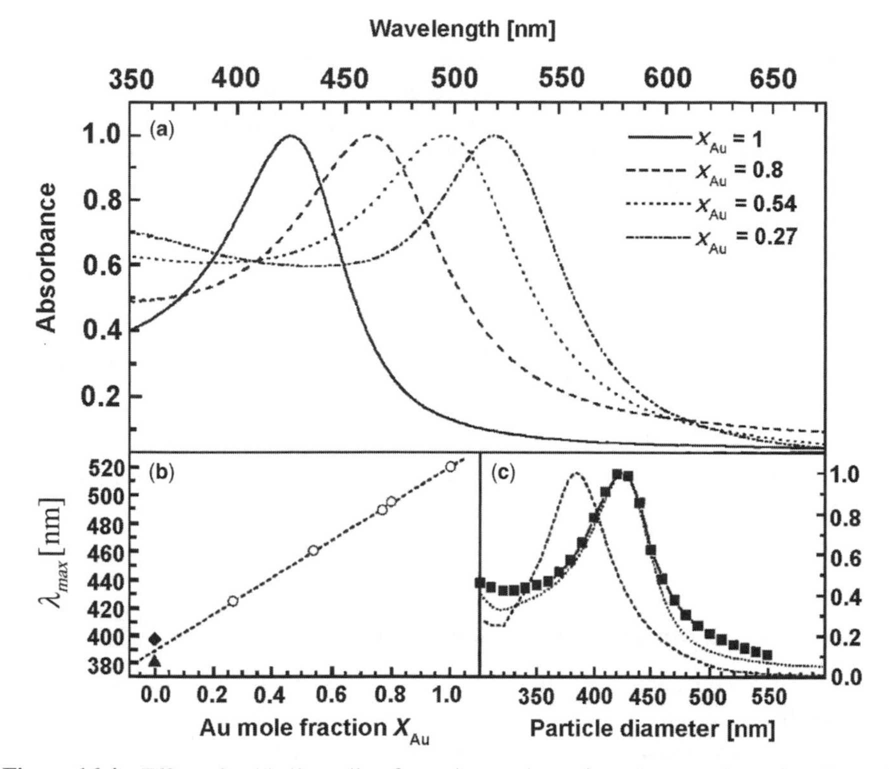

Figure 16.4 Effect of gold-silver alloy formation on the surface plasmon absorption. Part (a) shows the UV/Vis absorption spectra of spherical gold–silver alloy nanoparticles of varying composition. The gold mole fraction X_{Au} varies between 1.0 and 0.27. The dependence of the maximum of the plasmon absorption as a function of the nanoparticle composition and the particle diameter are given in (b) and (c), respectively. (Reproduced with permission from S. Link and M. El-Sayed, 1999 *J. Phys. Chem. B* 103: 8410–8426. Copyright 1999 American Chemical Society.)

16.2.4 Semiconductor Nanoparticles (Quantum Dots)

When *semiconductor* crystallites are studied instead of metal nanoparticles, the free electron concentration is orders of magnitude smaller. Therefore, their SPR bands occur in the infrared region of the electromagnetic spectrum. This is the main reason why the optical spectra of semiconductor crystallites in the UV/Vis region do not change significantly when the size of the particles is decreased below 400 nm. However, when semiconductor crystallites reach the quantum domains ($R \leq 5$ nm), we then call them quantum dots. Their remarkable size dependence of light absorbance and emission results from changes in their electronic band structure, which causes their dielectric function to change.

When a semiconductor particle absorbs a photon, an electron-hole pair is created. The electron quickly loses its excess energy and resides at the lowest energy of the conduction band. In larger particles (diameter >10 nm), the electron-hole recombination can cause the emission of a photon of an energy, which corresponds to the bandgap of the semiconductor. Since real particles are seldom perfect, the imperfections (and impurities) cause a broadening of the emission centered around the bandgap energy. When nanoparticles that are much smaller than 10 nm in diameter are studied, the electron-hole pair becomes confined. This confinement causes a distinct shift ΔE_{qc} (qc: quantum confinement) in the observed emission energy, which increases with decreasing particle size (E_{bg}: bandgap energy).

$$E_{Photon} = \frac{hc}{\lambda_{max}} = E_{bg} + \Delta E_{qc} \tag{16.23}$$

$$\Delta E_{qc} = \frac{h^2}{8R^2}\left(\frac{1}{m_e^*} + \frac{1}{m_h^*}\right) \tag{16.24}$$

R is the nanoparticle radius, m_e^* is the effective mass of the electron, and m_h^* is the effective mass of the hole. The effective masses can be quite different from the mass of the electron because electrons/holes behave differently in semiconductors than they do in discrete molecules.

16.2.5 Discrete Dipole Approximation (DDA)

The numerical DDA approach can be adapted to any arbitrary shape and is, therefore, most useful for all nanostructures that are neither spheres nor rods (8). The nanoparticle, -cluster, -shell, or -sphere is mathematically treated as an array of N cells. Each cell is considered a polarizable point dipole with a moment

$$\vec{P}_i = \alpha_i \vec{E}_{loc}(\vec{r}_i),$$

where $\vec{E}_{loc}$ is the local electromagnetic field of the ith point dipole at the position $\vec{r}_i \times \vec{E}_{loc}$ experiences contributions from the incident light field and all other N dipoles in the nanostructure. Once the position and polarizability of each cell have been specified, the absorption and scattering cross-sections can be calculated. The success of the DDA method is highly dependent on the selection of the right parameters. It is most useful for the interpretation of experimentally obtained data, but only of limited value for the prediction of optical properties. Examples of the successful application of the DDA methods are the absorption spectra of gold nanoshells and gold nanorods. Both are important for biological applications such as plasmon heating for the treatment of various tumors (9).

16.2.6 Luminescence from Noble Metal Nanostructures

The optical absorption properties of noble metal nanoparticles in the visible range of the electromagnetic spectrum are determined by the effect of the boundary condition of the coherent electron oscillations as well as by $d \rightarrow sp$ electronic transitions. Very small gold nanoparticles ($d < 2$ nm), as well as bulk gold do not show a localized surface plasmon absorption band! As discussed earlier (see Fig. 16.5), gold nanorods

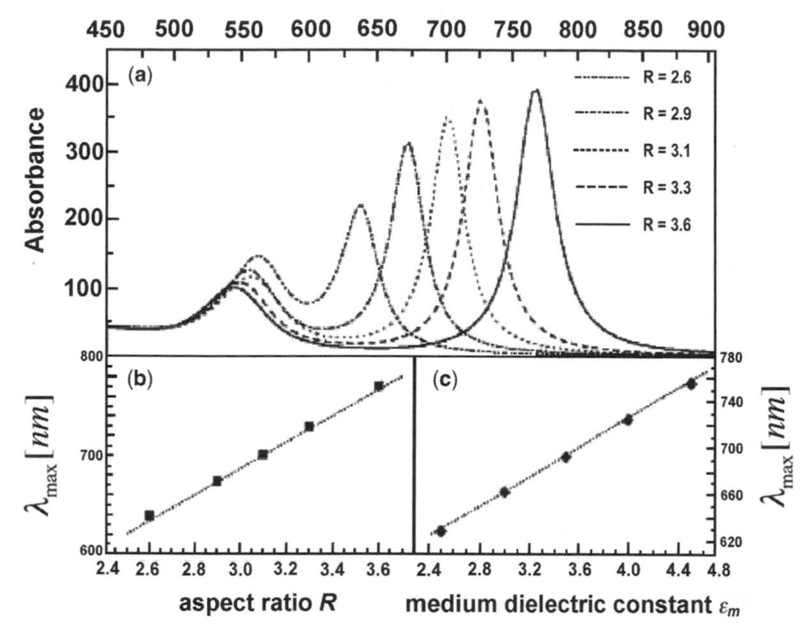

Figure 16.5 Simulation of the surface plasmon absorption for gold nanorods of different aspect ratios. Part (a) shows the calculated absorption spectra of gold nanorods with varying aspect ratios R using the theory developed by Gans. The dependence of the maximum of the longitudinal plasmon absorption as a function of $R = A/B$ and the medium dielectric constant ε_m for a constant value of other parameters are given in (b) and (c), respectively. (Reproduced with permission from S. Link and M. El-Sayed, 1999. *J. Phys. Chem. B* 103: 8410–8426. Copyright 1999 American Chemical Society.)

feature a transverse and longitudinal SPR. Both the absorption maximum and cross-section of the latter increase with increasing aspect ratio (Figs. 16.6 and 16.7). Figure 16.8 summarizes the emission features in the wavelength region of 548 to 590 nm obtained from gold nanorods with a similar average width of 20 nm and various aspect ratios in solution (25). The quantum efficiencies of emission increase with the square of the length of the nanorods! Furthermore, the quantum efficiencies are on the order of 10^{-4} to 10^{-3}, which is more than a million times brighter than the observed emission from planar gold surfaces. The excitation wavelength is 480 nm, which is located in the transverse region of the plasmon absorption band. This spectral region is insensitive to the lengths of the rod. The longitudinal surface plasmon absorption shows a linear red-shift with respect to the length of the nanorod. However, this absorption occurs at much longer wavelengths than the observed emission. Therefore, the observed emission cannot originate from the surface plasmon radiative relaxation. However, the Stokes shift of the luminescence band indicates that there exists a slight overlap between the *sp*-conduction band and the *d*-valence band. Irradiation at 480 nm leads to the excitation of both the surface plasmon coherent

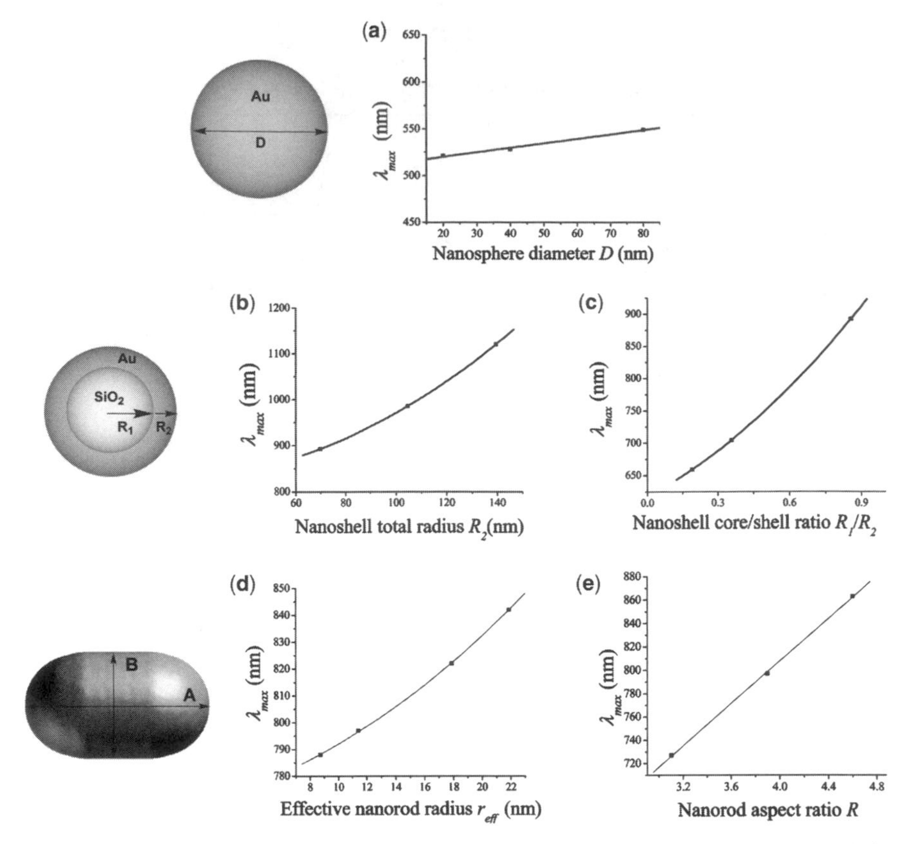

Figure 16.6 Tunability of the plasmon resonance maximum in gold nanostructures. Variation of the surface plasmon extinction maximum λ_{max} with (a) nanosphere diameter D; (b) nanoshell total radius R_2 with fixes $R_1/R_2 = 0.857$; (c) nanoshell core/shell ratio R_1/R_2 at fixed $R_2 = 70$ nm; (d) nanorod effective radius $r_{eff} = (3V/4\Pi)^{1/3}$ at fixed aspect ratio $R = A/B = 3.9$; (e) nanorod aspect ratio R at fixed $r_{eff} = 11.43$ nm. (Reproduced with permission from P. K. Jain et al., 2006. *J. Phys. Chem. B* 110: 7238–7248. Copyright 2006 American Chemical Society.)

electronic motion and the *d*-electrons. The observed emission is then caused by relaxation of both electronic motions, followed by recombination of the *sp*-electrons with holes in the *d*-band. Because of the rod shape, both the exciting and emission fields are greatly enhanced via coupling to the localized plasmon resonances.

The rough surface of the nanorod is assumed to be a random collection of non-interacting hemispheroids of height a, radius b and volume $V = 4/3\Pi ab^2$. The local field correction factor within the spheroid $L(\omega)$ is then expressed by Equation (16.25):

$$L(\omega) = -\frac{L_{LR}}{\varepsilon_1(\omega) + i\varepsilon_2(\omega) - 1 + L_{LR}\left(1 + \dfrac{4\pi^2 iV[1 - \varepsilon_1(\omega) - i\varepsilon_2(\omega)]}{3\lambda^3}\right)} \tag{16.25}$$

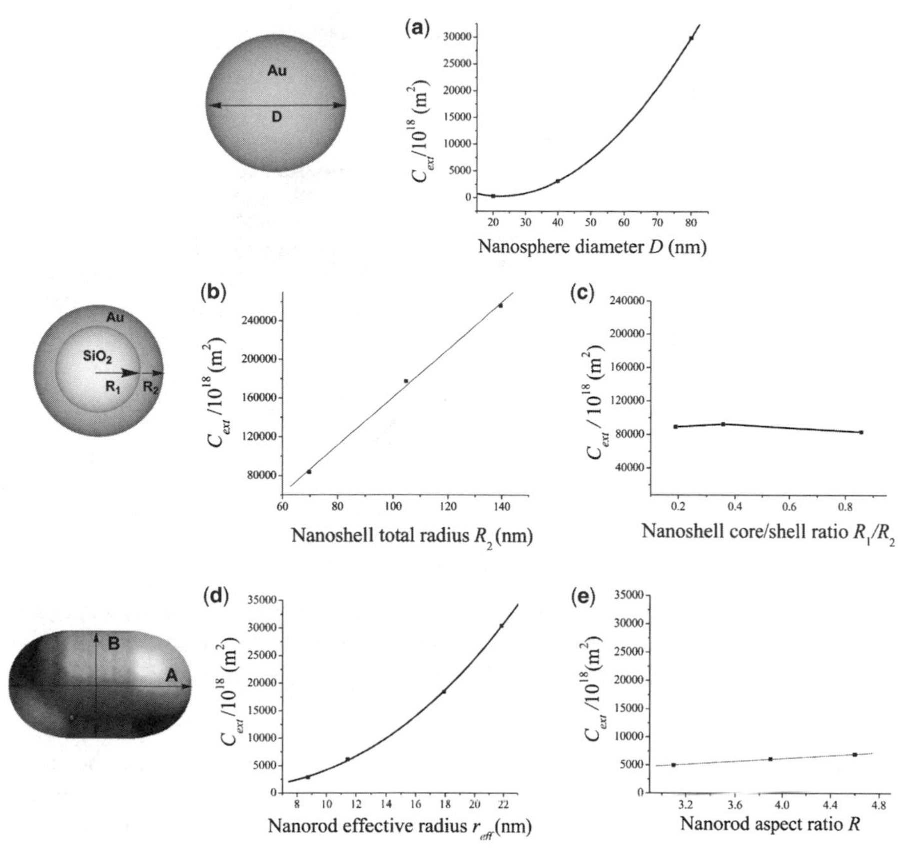

Figure 16.7 Tunability of the extinction cross-section of gold nanostructures. Variation of C_{ext} with (a) nanosphere diameter D; (b) nanoshell total radius R_2 with fixes $R_1/R_2 = 0.857$; (c) nanoshell core/shell ratio R_1/R_2 at fixed $R_2 = 70$ nm; (d) nanorod effective radius $r_{eff} = (3V/4\Pi)^{1/3}$ at fixed aspect ratio $R = A/B = 3.9$; (e) nanorod aspect ratio R at fixed $r_{eff} = 11.43$ nm. (Reproduced with permission from P. K. Jain et al., 2006. *J. Phys. Chem. B* 110: 7238–7248. Copyright 2006 American Chemical Society.)

L_{LR} is the "lightning rod factor," defined as $L_{LR} = 1/A$.

$$A = \frac{1}{1 - \dfrac{\xi Q_1'(\xi)}{Q_1(\xi)}} \tag{16.26}$$

$$\xi = \frac{1}{\sqrt{1 - \left(\dfrac{b}{a}\right)^2}} \tag{16.27}$$

$$Q_1(\xi) = \frac{\xi}{2}\ln\left(\frac{\xi + 1}{\xi - 1}\right) - 1 \tag{16.28}$$

$$Q_1'(\xi) = \frac{dQ_1(\xi)}{d\xi} \tag{16.29}$$

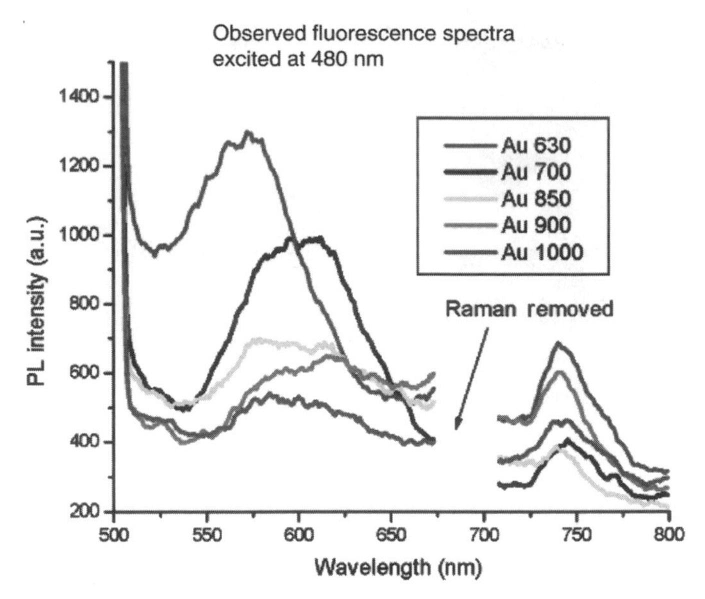

Figure 16.8 Fluorescence spectra of gold nanorods upon $\lambda = 480$ nm excitation: Au630, $R = 2.25$; Au700, $R = 3.25$; Au850, $R = 4.75$; Au900, $R = 5.0$; Au1000, $R = 6.0$. (Reproduced with permission from S. Eustis and M. El-Sayed, 2005. *J. Phys. Chem. B* 109: 16350–16356. Copyright 2005 American Chemical Society.)

Using Equation (16.30), the single-photon luminescence power P_1 can be calculated for the excitation and emission energies $h\omega_{exc}$ and $h\omega_{em}$. E_0 is the incident electromagnetic field and β_1 is a wavelength-dependent proportionality constant, which includes the intrinsic absorption spectrum of gold.

$$P_1 = 16\beta_1 |E_0|^2 V \left| L^2(\omega_{exc}) L^2(\omega_{em}) \right| \tag{16.30}$$

According to this model, which is supported by experimental observations, the luminescence power increases linearly with the square of the length of the nanorod, whereas a simple dependence on the maximal length of the rod is found for the observed emission maxima. $P_1/|E_0|^2\beta_1$ is on the order of 10^7. According to the "lightning rod" model, the upper bound on the luminescence lifetime is approximately 50 fs. This very short lifetime can be regarded as support for the paradigm that the luminescence from gold nanostructures is directly related to the dynamics of the holes created in the d-band immediately after irradiation. It cannot be regarded as a typical radiative lifetime, which would be at least in the nanosecond window. Instead, it is dominated by nonradiative processes.

16.2.7 Nonradiative Relaxation Dynamics of the Surface Plasmon Oscillation

The dephasing of the coherent plasmon oscillation in (noble) metal nanoparticles can be probed by either steady-state absorption spectroscopy (because the bandwidth changes) or by time-resolved pump-probe transmission measurements. The plasmon band absorption is very sensitive with respect to the electron dynamics. Of special interest are the two processes of electron-electron and electron-phonon scattering within the nanoparticle system. It is noteworthy that the pump pulse is able to excite the electrons to temperatures of the electron gas of up to several thousand degrees Kelvin. Because of these high initial temperatures, a non-Fermi distribution is created directly after the absorption of the pump pulse that thermalizes first through electron-electron scattering and then through electron-phonon scattering until the nanoparticle reaches the temperature of its surrounding. The two-temperature model (TTM) is based on the experimental finding that the lattice heat capacity is approximately two orders of magnitude bigger than the electronic heat capacity and is the standard model for explaining the observed kinetics of the nonradiative relaxation dynamics of nanoparticles (26).

Experimentally, the heating of the electron gas leads to spectral broadening of the surface plasmon absorption. Consequently, a transient bleach centered at the plasmon band maximum and two positive absorption wings at the lower and higher energies in the difference spectrum appear. The results obtained with 15 nm spherical gold nanoparticles, 400 nm excitation, and a white light probe pulse are shown in Figure 16.9. The experimental parameters lead to an initial temperature of 4000 K in the system. As

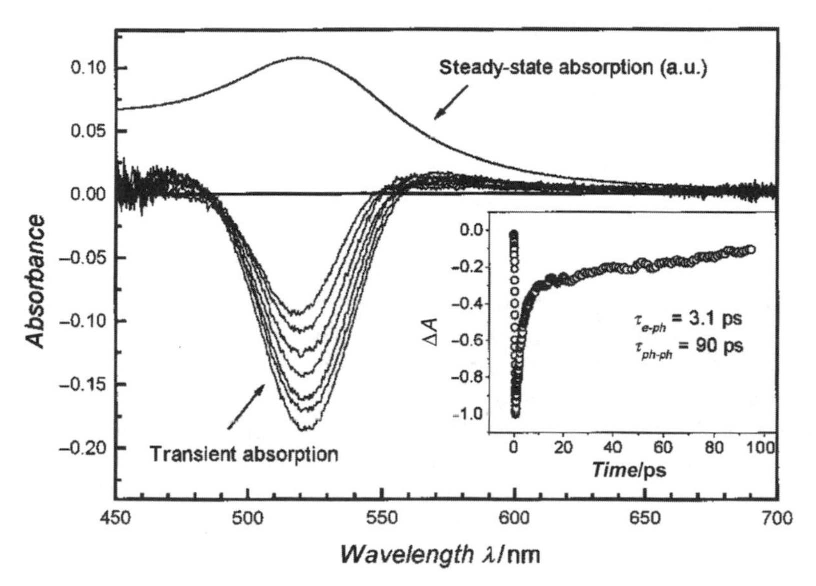

Figure 16.9 Transient absorption spectra of 15 nm spherical gold nanoparticles after excitation at 400 nm with 100 fs laser pulses, recorded as different delay times. Also shown is the steady-state UV/Vis-absorption spectrum of the colloidal gold solution. The inset shows the decay of the transient bleach when the particles are monitored at the bleach maximum at 520 nm. Fitting of the decay curve yields electron-phonon and phonon-phonon relaxation times of 3.1 and 90 ps, respectively. (Reproduced with permission from S. Link and M. El-Sayed, 1999. *J. Phys. Chem. B* 103: 8410–8426. Copyright 1999 American Chemical Society.)

predicted by the two-temperature model, the decay of the transient bleach monitored at 520 nm follows a biexponential fit. The short component of 3.1 ps occurs due to electron-phonon relaxation, whereas the longer component is due to phonon-phonon relaxation with the water molecules.

It is of importance for the understanding of the discussion below to discern between processes of energy relaxation and the dephasing of the coherent plasmon oscillation. (1) The decay of the transient negative absorption signal is assigned to the cooling of the hot electron gas due to inelastic collisions with the phonons of the metal lattice. This process has a picosecond time scale (energy relaxation, T_1). Note that there is no discernible size dependence of T_1 in noble metal nanoparticles smaller than 50 nm. (2) The width of the plasmon band is increased because of the loss of coherence of the free electron oscillation (pure dephasing, T_2). This process comprises both elastic and inelastic electron scattering processes. These processes have a time scale of approximately 10 fs. T_2 is dependent on the nanoparticle size.

16.2.7.1 Pump Power Dependence of the Electron-Phonon Relaxation Dynamics Two coupled differential equations are usually used to describe the time evolution of the electron and lattice temperature T_e and T_l upon laser excitation.

$$\frac{\delta T_e}{\delta t} = -\frac{g}{C_e}(T_e - T_l) + LP(t) \tag{16.31}$$

$$\frac{\delta T_l}{\delta t} = -\frac{g}{C_l}(T_e - T_l) \tag{16.32}$$

C_e and C_l are the electronic and lattice heat capacities, g is the electron-phonon coupling constant, and $LP(t)$ is the intensity profile of the exciting laser pulse dependent on the time t.

It is well established that the electronic heat capacity C_e is a function of the electron temperature T_e. Therefore, the effective rate constant $g/C_e(T_e)$ for the thermal relaxation of the electron gas $\delta C_e/\delta T_e$ decreases with increasing laser pump power. This is the reason why the electron-phonon relaxation time increases with increasing electron temperature/laser intensity.

As can be discerned from Figure 16.10b, a linear fit of the electron-phonon relaxation times $\tau_{e\text{-}ph}$ against the (normalized) pump power can be obtained. The limiting decay time for zero excitation power was determined to be 690 $\pm$ 100 fs when exciting with 400 nm. For an excitation wavelength of 630 nm, the limiting decay time was 830 $\pm$ 100 fs. This corresponds to an electron-phonon coupling constant of 2.5 $\pm$ 0.5 $\times$ 10^{16} W m^{-3} K^{-1}. This value appears to be typical for bulk gold, gold nanoparticles, and Ag/Au alloy nanoparticles.

16.2.7.2 Electron-Electron Thermalization in Noble Metal Nanoparticles In metals the electron-electron scattering rate $1/\tau_{e\text{-}e}$ is inversely proportional to the energy above the Fermi level because this energy difference determines the number

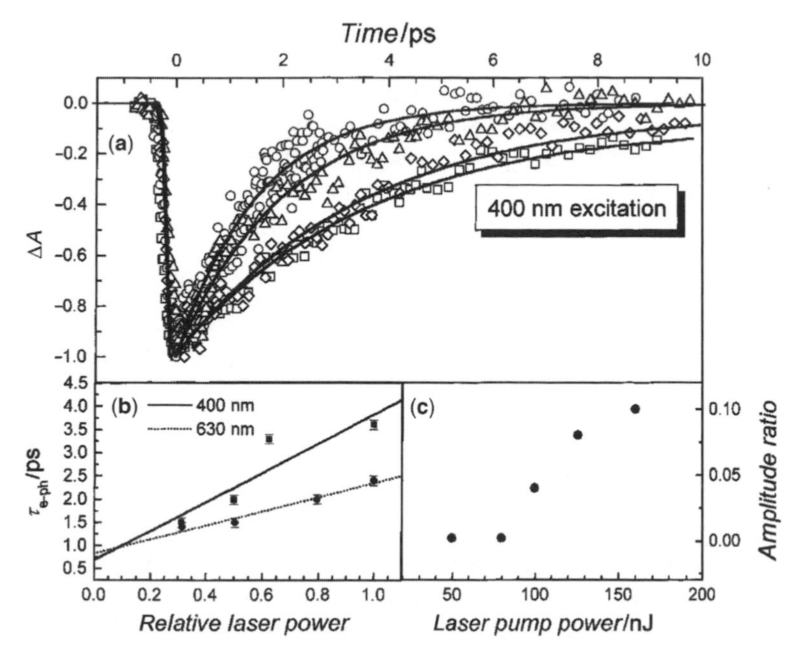

Figure 16.10 Power dependence of the electron-phonon relaxation time. Part (a) shows the results for a 15 nm spherical gold nanoparticle after 400 nm excitation. The bleach recovery is monitored at the bleach maximum of 520 nm. The decay curves were fitted with lifetimes of 1.5, 2.0, 3.3, and 3.6 ps for excitation powers of 50, 80, 100, and 160 nJ, respectively. Part (b) shows a plot of the electron-phonon relaxation times against the relative pump power. Extrapolation to zero laser power yields a decay time of 690 $\pm$ 100 fs for 400 nm excitation. The result for 630 nm excitation is included as well, which features a limiting decay time of 830 $\pm$ 100 fs. This result corresponds to an electron-phonon coupling constant of around $2.5 \pm 0.5 \times 10^{16}\,\mathrm{W\,m^{-3}\,K^{-1}}$. At high laser powers the bleach also shows a long time component as seen by the offset of the decay curves in (a). This corresponds to the phonon-phonon relaxation, which occurs on a time scale of 100 ps. The amplitude ratio of the phonon-phonon to the electron-phonon relaxation time increases with increasing power, as shown in (c). (Reproduced with permission from S. Link and M. El-Sayed, 1999. *J. Phys. Chem. B* 103: 8410–8426. Copyright 1999 American Chemical Society.)

of available unoccupied states, which serve as the final states for the electron-electron scattering processes.

$$\frac{1}{\tau_{e\text{-}e}} \propto (E - E_F)^2 \tag{16.33}$$

E is the energy of the electron gas, E_F is its Fermi energy, $1/\tau_{e\text{-}e}$ is the electron-electron scattering rate.

A ballpark number for the electron-electron scattering time is 10 fs. All electron-electron scattering events occur within a window of several hundred femtoseconds until a Fermi electron distribution is reached. It must be noted that the internal electron thermalization (by electron-electron scattering) and the external electron thermalization (by electron-phonon interactions) occur on comparative time scales. Therefore, it is impossible to clearly separate these two events. The nonthermal electrons already interact with the phonons and, consequently, heat the lattice, while the non-Fermi electron distribution is still decaying by electron-electron scattering.

Sun et al. (27) have proposed the following rate equation:

$$\Delta A(h\nu) = [\Delta A(h\nu)]_{NT} \exp\left(-\frac{t}{\tau'_{th}}\right) + [\Delta A(h\nu)]_{Th} \exp\left(-\frac{t}{\tau'_{e\text{-}ph}}\right)\left\{1 - \exp\left(-\frac{t}{\tau_{th}}\right)\right\} \quad (16.34)$$

$$\frac{1}{\tau'_{th}} = \frac{1}{\tau_{th}} + \frac{1}{\tau_{e\text{-}ph}} \quad (16.35)$$

$\Delta A(h\nu)$ is the change in absorbance as a function of the photon energy $h\nu$. NT and Th represent the nonthermalized and thermalized states. $\tau_{e\text{-}ph}$ is the time constant for the electron-phonon interactions of the thermalized electron distribution and τ_{th} is the internal electron thermalization time. τ'_{th} is the decay time of the nonthermal electron population. Note that τ'_{th} is shorter than τ_{th} because the nonthermalized electron distribution interacts with the lattice during the thermalization process.

$$\frac{\Delta T}{T} = \frac{\delta \ln T}{\delta \varepsilon_1}\Delta\varepsilon_1 + \frac{\delta \ln T}{\delta \varepsilon_2}\Delta\varepsilon_2 \quad (16.36)$$

The transient absorption spectrum of gold nanospheres can be modeled relating the changes in the differential transmittance $\Delta T/T$ to the laser-induced changes in the real and imaginary parts of the dielectric function $\Delta\varepsilon_1$ and $\Delta\varepsilon_2$. T is defined as the ratio of the transmitted to the incident light [$A = -\log(T)$].

Figure 16.11 shows the simulated bleach of the plasmon absorption of spherical gold nanoparticles assuming a temperature change ΔT of either 10 K or 5000 K. It is apparent that the line width of the bleach is much broader for a larger change in the electron distribution.

16.2.8 Nanoparticles Rule: From Förster Energy Transfer to the Plasmon Ruler Equation

The classic method for the determination of distances between (organic) luminophors in the Förster energy transfer, named after the German scientist Theodor Förster (28). This method relies on the presence of a photodonor D^* and an acceptor compound A that is able to quench the fluorescence of the photodonor. The emission spectrum of the photodonor D^* and the absorption spectrum of the acceptor A have to overlap in order to permit radiationless energy transfer (dipole-dipole energy transfer) from D^* to A. The efficiency of the energy transfer depends on the degree of overlap between

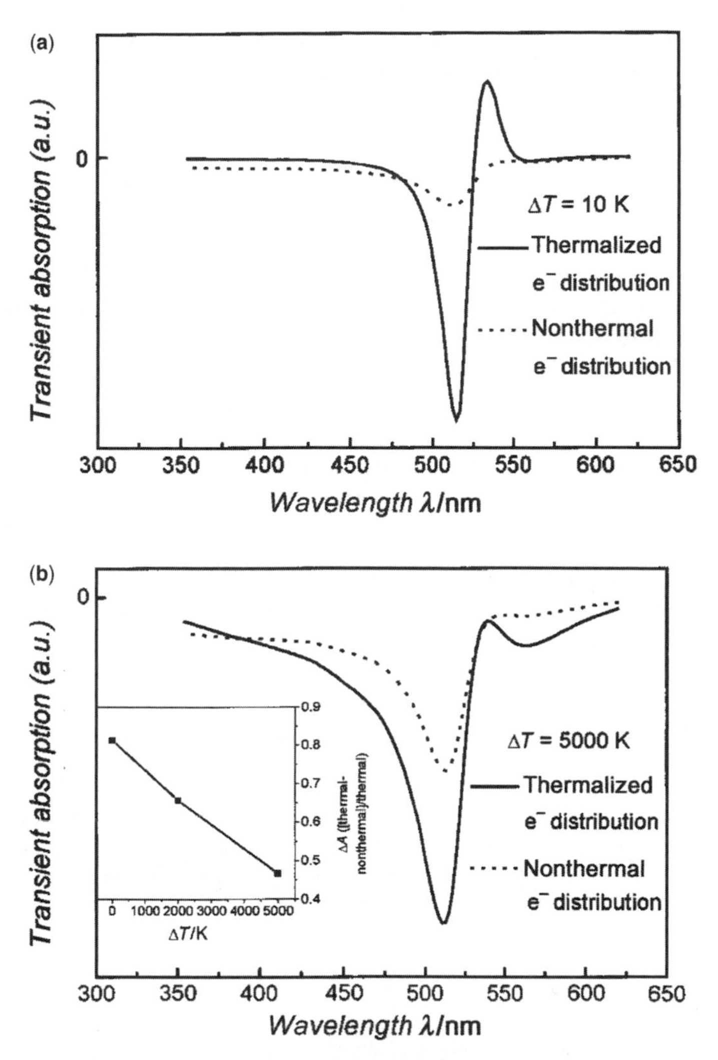

Figure 16.11 Calculated transient absorption spectra of gold nanoparticles for a thermalized (solid line) and nonthermal (dotted line) electron distribution. The temperature change of the electron gas is (a) 10 K and (b) 5000 K. For the non-Fermi electron distribution, excitation with photons of 2 eV energy is assumed. In good agreement with experimental results, the spectral shape of the transient absorption signal arising from a thermalized electron distribution is smaller, whereas the intensity is lower for a nonthermal distribution. Furthermore, the relative difference of the calculated transient absorption intensity originating from a nonthermal electron distribution compared to the thermalized electron gas decreases with increasing electron temperature change ΔT corresponding to higher pump powers [see inset of part (b)]. (Reproduced with permission from S. Link and M. El-Sayed, 1999. *J. Phys. Chem. B* 103: 8410–8426. Copyright 1999 American Chemical Society.)

these two spectra. A suitable acceptor emits a photon at a red-shifted wavelength with respect to the light emission occurring from the photodonor. Förster energy transfer processes can be described by Equation (16.37).

$$\Phi_{ET} = \frac{1}{1 + (r/r_0)^6} \tag{16.37}$$

Φ_{ET} is the quantum efficiency of energy transfer ($\Phi_{ET} < 1$), r is the distance between photodonor and acceptor, r_0 is the Förster radius (distance at which the efficiency of the energy transfer has decreased to $1/e$).

The real disadvantage of the Förster energy transfer is that it is proportional to r^{-6}. Therefore, the decrease of the energy transfer as a function of distance occurs so rapidly that only two cases can be discerned: the distance is either smaller or larger than the Förster radius. However, the measurement of the real distance between the photodonor and acceptor is virtually impossible. A typical Förster radius is 5 ± 1 nm.

16.2.8.1 Surface Energy Transfer (SET) In 2005, Geoffrey F. Strouse and coworkers (29) reported a long-distance energy transfer between a photodonor dye

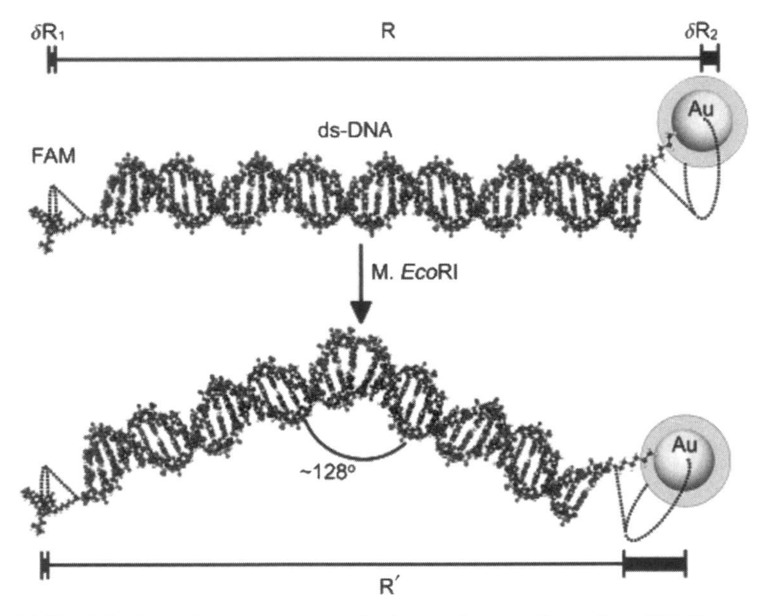

Figure 16.12 Dipole-surface energy transfer from a fluorescein moiety (FAM) appended to ds-DNA of length R with a gold nanoparticle ($d = 1.4$ nm) appended to the other end. The flexible C6-linker causes a cone of uncertainty (δR) for both moieties. Addition of the *Eco*RI DNA methyl transferase (M. *Eco*RI) bends the ds-DNA at its GAATTC site by 128°, producing a new effective distance R'. (Reproduced with permission from C. S. Yun et al., 2005. *J. Am. Chem. Soc.* 127: 3115–3119. Copyright 2005 American Chemical Society.)

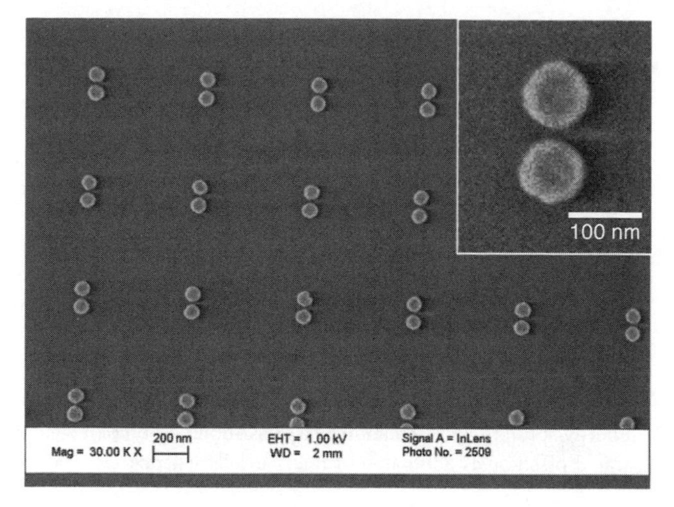

Figure 16.13 Representative SEM image of the array of nanodisc pairs used in the present study having an interparticle edge-to-edge separation gap of 12 nm. The inset shows a magnified image of a single nanodisc pair clearly showing the interparticle gap. Each nanodisk has a diameter of 88 nm and a thickness of 25 nm. (Reproduced from P. K. Jain and M. A. El-Sayed, 2007. *Nano Lett.* 7: 2854–2858. Copyright 2007 American Chemical Society.)

(fluorescein) and a gold nanoparticle of 1.4 nm in diameter. Both were chemically linked to double-stranded DNA oligomers of various lengths and shapes (Fig. 16.12). SET (dipole-surface energy transfer) decreases proportionally to r^{-4}. This is a considerable improvement compared to r^{-6} for dipole-dipole energy transfer.

$$\Phi_{ET} = \frac{1}{1 + (r/r_0)^4} \tag{16.38}$$

16.2.8.2 Plasmon Coupling The changes that occur in the light extinction of two neighboring noble metal nanostructures, compared to their optical properties when being "secluded," is of special interest because they permit the measurements of the internanoparticle distances. The discrete dipole approximation (DDA) method was adapted by Schatz and coworkers (30, 31) to calculate the extinction spectrum of a series of nanostructures featuring different shapes that were subjected to various environments. One of the advantages of DDA calculations is that the optical properties of a single particle pair can be calculated instead of having to deal with the entire array of nanostructures.

Qualitatively speaking, the approach of two nanoparticles can be handled according to simple LCAO (linear combination of atom orbitals) theory as shown in Scheme 16.3 (32). The approach of two nanoparticles of the same energy level (e.g. of the same shape and size) leads to the formation of a lower and a higher energy level. Their combined energies correspond to the sum of the plasmon energies of both nanoparticles.

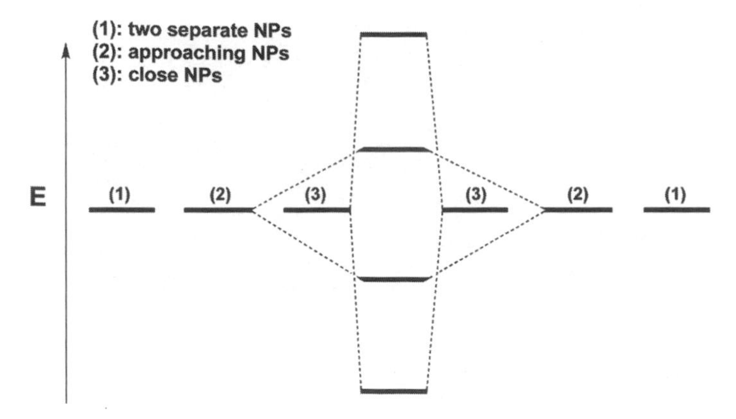

Scheme 16.3 Energy splitting and formation of a quasi-bonding type of state that lowers the energy of the coupled plasmon as a function of interparticle distance.

The lower state is then occupied in analogy to a HOMO (highest occupied molecular orbital) and the upper state remains empty, in analogy to a LUMO (lowest unoccupied molecular orbital). Consequently, quasi-binding occurs between both nanoparticles.

Alivisatos and coworkers (33–35), Jain, Huang, and El-Sayed (36) and Jain and El-Sayed (37, 38) have pioneered the measurement of distances using the plasmon coupling of two nanoparticles. Pairs of gold nanoparticles that are either deposited on surfaces or linked by tethers (e.g. DNA) or kept apart by attached (bio)macromolecules (e.g. DNA) show distinct red-shifts of their plasmon resonance maxima λ_{max}. It is noteworthy that the observed spectral behavior when using parallel and perpendicular light (with respect to the interparticle axis) is very different. Under parallel polarization, the plasmon resonance strongly red-shifts as the interparticle distance is reduced. Conversely, there is a very weak blue-shift for the orthogonal polarization. Apparently, the dipole-dipole interaction is attractive for parallel polarization, which results in the reduction of the plasmon frequency (red-shift of the plasmon band). This red-shift can be observed in the plasmon VIS/NIR absorption bands, as well as in the corresponding emission bands, if the emission quantum efficiency permits.

This universal scaling behavior permits a derivation of a simple empirical equation, which can be used to determine the distance between both nanostructures. Distances in biological systems are of special importance. $\Delta\lambda/\lambda_0$ is the fractional plasmon shift, r the internanoparticle edge-to-edge separation, and D the particle diameter. k_1 and k_2 are empirical factors and in biological systems is equal to approximately 0.2.

$$\frac{\Delta\lambda}{\lambda_0} = k_1 \exp\left(\frac{-\dfrac{r}{D}}{k_2}\right) \tag{16.39}$$

In Figure 16.14a and b, the principal distance-dependent performance of the three types of rulers, Förster energy transfer (FET, dipole-dipole), surface energy transfer

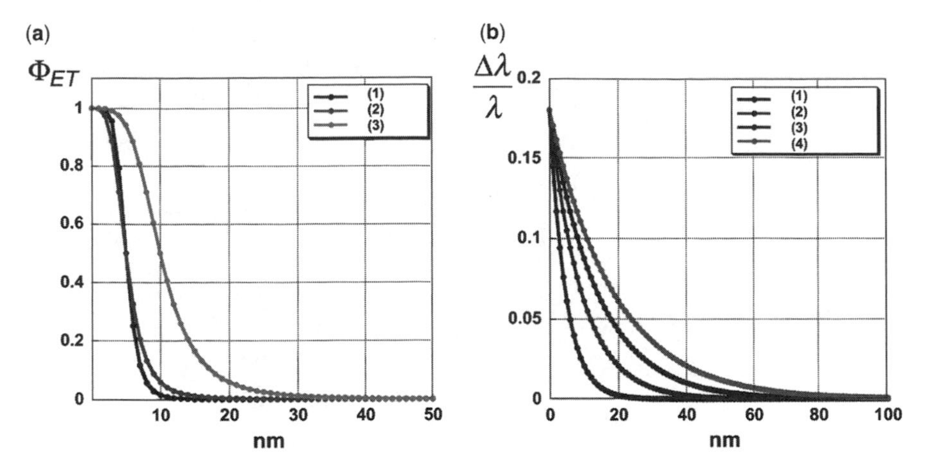

Figure 16.14 Comparison between (a) Förster energy transfer, surface energy transfer and (b) plasmon resonance. a: (1) Förster energy transfer, $r_0 = 5$ nm; (2) surface energy transfer, $r_0 = 5$ nm; (3) $r_0 = 10$ nm. b: Plasmon resonance, (1) D (diameter of both gold nanoparticles) = 20 nm, (2) $D = 40$ nm, (3) $D = 60$ nm, (4) $D = 80$ nm, $k_1 = 0.18$; $k_2 = 0.23$ (according to Reference 36).

(SET, dipole-surface), and plasmon coupling (PC) is compared. The longest achievable distance for FET is approximately 10 nm, whereas for SET up to 30 nm can be achieved if a photodonor/nanoparticle pair featuring a long characteristic energy transfer radius can be found. PC is certainly superior because it permits precise distance measurements up to 60 nm.

16.3 QUANTUM DOTS: A BRIEF OVERVIEW

Quantum dots are semiconductors composed of atoms from groups II–VI or III–V elements of the periodic table, for example, CdSe, CdTe, and InP (39). Their brightness is attributed to the quantization of energy levels due to confinement of an electron in a three-dimensional box. The optical properties of quantum dots can be manipulated by synthesizing a (usually stabilizing) shell. Such Q-dots are known as core-shell quantum dots, for example, CdSe–ZnS, InP–ZnS, and InP–CdSe. In this section, we will discuss the different properties of quantum dots based on their size and composition.

16.3.1 Electronic and Optical Properties of Quantum Dots

In a quantum dot, which is also often called an artificial atom, the excitons are confined in all three spatial dimensions. In a bulk semiconductor, an electron-hole pair is bound within the Bohr exciton radius, which is characteristic for each type of semiconductor. A quantum dot is smaller than the Bohr exciton radius, which causes the appearance of discrete energy levels. The bandgap, ΔE, between the valance and conduction band of the semiconductor is a function of the nanocrystal's size and shape. Q-dots feature slightly lower luminescence quantum yields than traditional organic fluorophors but

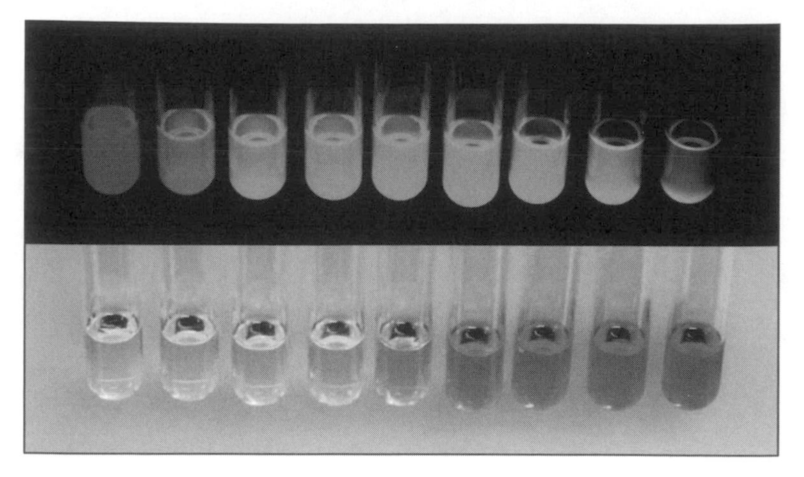

Figure 16.15 Colloidal suspension of CdSe quantum dots of increasing size from left (approximately 1.8 nm) to right (approximately 4.0 nm). Bottom: Samples viewed in ambient light vary in color from green-yellow to orange-red. Top: The same samples viewed under long-wave ultraviolet illumination vary in their color of emission from blue to yellow. (Reproduced with permission from E. M. Boatman et al., 2005. *J. Chem. Ed.* 82: 1697–1699. Copyright 2005 American Chemical Society.) (See color insert.)

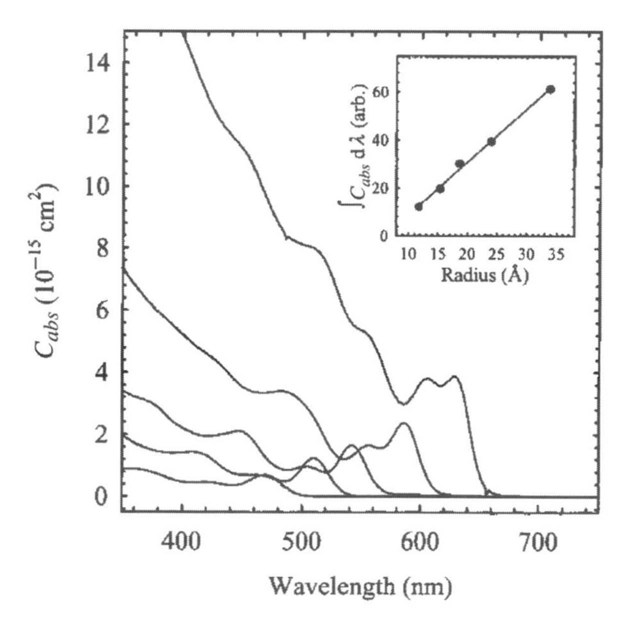

Figure 16.16 Room temperature absorption cross-section per Q-dot for CdSe Q-dots dispersed in hexane. From right to left, spectra are shown for Q-dots with radii of 3.39, 2.40, 1.87, 1.54, and 1.18 nm. In the inset, the integrated oscillator strength per particle for the lowest-energy transition (1Se1S3/2 h) is plotted vs. the particle radius. (Reproduced with permission from C. A. Leatherdale et al., 2002. *J. Phys. Chem. B* 106: 7619–7622. Copyright 2002 American Chemical Society.)

they have much larger absorption cross-sections and very low rates of photobleaching. Molar extinction coefficients of Q-dots are about 10^5 to 10^6 M^{-1} cm^{-1}, which is 10 to 100 times larger than dyes (40) (see Fig. 16.15).

16.3.2 Type I and Type II Core-Shell Quantum Dots

Core-shell quantum dots have higher bandgap shells around their lower bandgap cores, which emit light without any absorption by the shell. This shell passivates surface nonradiative emission from the core, thereby enhancing the photoluminescence quantum yield and preventing natural degradation (41). The shell of the type I QDs has a higher energy conduction band and a lower energy valance band than those of the core, resulting in confinement of both electron and hole in the core, for example CdSe/Zns (c/s) and CdSe/CdS (c/s). On the contrary, the shell of type II QDs has both conduction and valance bands lower (or higher) in energy than those of the core. Thus, the motions of the electron and the hole are restricted to one dimension, for example, CdTe/CdSe (c/s) and CdSe/ZnTe (c/s) (42). Radiative recombination of the exciton at the core-shell interface gives rise to the type II emission. Type II QDs behave as indirect semiconductors near band edges and, therefore, have an absorption tail into the red and near infrared.

Alloyed semiconductor Q-dots (CdSeTe) have been reported; the alloy composition and internal structure, which can be varied, permits tuning the optical properties without changing the size of the particles. These Q-dots can be used to develop near infrared fluorescent probes for *in vivo* biological assays as they can emit up to 850 nm (43).

16.3.3 The Absorption Cross-Sections of Quantum Dots

For optical transitions far from the band edge, where the density of states becomes a quasi-continuum and in the absence of strong resonances, the classic formalism describing the extinction cross-section C_{ext} and the sum of absorption (C_{abs}) and scattering cross-section (C_{sca}), can be used to describe the light absorption of Q-dots (44). For particles possessing radii much smaller than the wavelength in the medium λ/n_o, the scattering processes do not contribute significantly to the optical extinction.

$$C_{abs} = 4\pi k \mathrm{Re}(i\alpha) = \frac{8\pi^2 n_3}{\lambda} \mathrm{Re}(i\alpha) \tag{16.40}$$

where n_3 is the refractive index of the medium, λ the wavelength in vacuum, $k = 2\pi/\lambda$ and α the polarizability of dielectric spheres with radius r.

$$\alpha = \frac{m_1^2 - n_3^2}{m_1^2 + 2n_3^2} r^3 \tag{16.41}$$

$m_1 = n_1 - ik_1$ is the complex refractive index of the semiconductor nanoparticle. An alternative expression for Q-dots is (44):

$$C_{abs} = \frac{\omega}{n_3 c} |f(\omega)|^2 2n_1 k_1 \left(\frac{4}{3} \pi r^3 \right) \tag{16.42}$$

$$f(\omega) = \frac{3n_3^2}{m_1^2 + 2n_3^2} \tag{16.43}$$

($\omega = 2\pi\nu$). C_{abs} is proportional to the product of the "bulk" absorption coefficient ($2n_1k_1$) and a local field factor that is equal to the ratio of the applied field to the electric field inside the sphere.

Figure 16.17 shows the remarkable agreement between theory and experimental findings for CdSe quantum dots. The experimental error of approximately 5 rel. percent results from the inhomogeneous radial size dispersion of the Q-dots. The complex refractive index for bulk CdSe at 350 nm is $m_1 = 2.772 - 0.7726i$. The absorption cross-section $C_{abs} = 5.501 \times 10^5 r^3$ (cm^2) where the particle radius r is in cm.

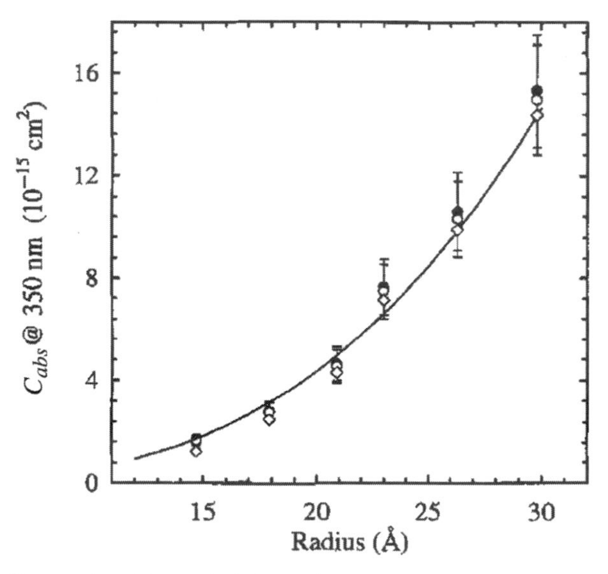

Figure 16.17 Observed and theoretical absorption cross-section as a function of size for CdSe Q-dots dispersed in hexane. (●) indicates C_{abs} based on the integrated absorption spectra; (◇) indicates C_{abs} based on the integrated absorption spectra and corrected for the reaction yield as a function of size; and (○) indicates C_{abs} determined using the absorbances of the growth and fractionated solutions at 350 nm that were corrected for the reaction yield. The solid line is the theoretical curve calculated using Equations (16.39) and (16.40). (Reproduced with permission from C. A. Leatherdale et al., 2002. *J. Phys. Chem. B* 106: 7619–7622. Copyright 2002 American Chemical Society.)

16.3.4 Absorption and Emission Maxima of Quantum Dots

The Stokes shifts between absorption and emission maxima of quantum dots remain relatively constant, as the slope of approximately one in Figure 16.18b indicates (45). Because of additional nearby energy levels, quantum dots can be excited by photons of greater than or only slightly higher energies. Therefore, the effective Stokes shift can be hundreds of nanometers, permitting the excitation of numerous quantum dots possessing various emission maxima with the same excitation wavelength. This allows the simultaneous detection of numerous events.

16.3.5 Luminescence Lifetimes of Q-Dots

The photophysical properties of quantum dots are quite complicated as they are known to exhibit fluorescence intermittency, or blinking, over a wide range of time scales. Furthermore, variability within Q-dot samples leads to nonuniform fluorescence lifetimes. Concerning their excited-state lifetimes, quantum dots differ from traditional organic chromophores in two important ways: (1) extremely long luminescence lifetimes are observed, ranging from tens of nanoseconds at room temperature to microseconds at low temperature; (2) these decays exhibit multiexponential dynamics. Three possible reasons for this behavior could exist: (1) each member of the Q-dot ensemble has its unique single-exponential lifetime; (2) the photoluminescence process is a very complex process so that multiexponential behavior is observed for each individual member of the Q-dot ensemble; (3) the time-averaged photoluminescence decay for each Q-dot is single-exponential, but fluctuated in time.

Experimental evidence suggests that photoluminescence decay fluctuation occurs on time scales up to several seconds. Apparently, the photoluminescence decay rates of quantum dots are correlated to single time-averaged emission intensities.

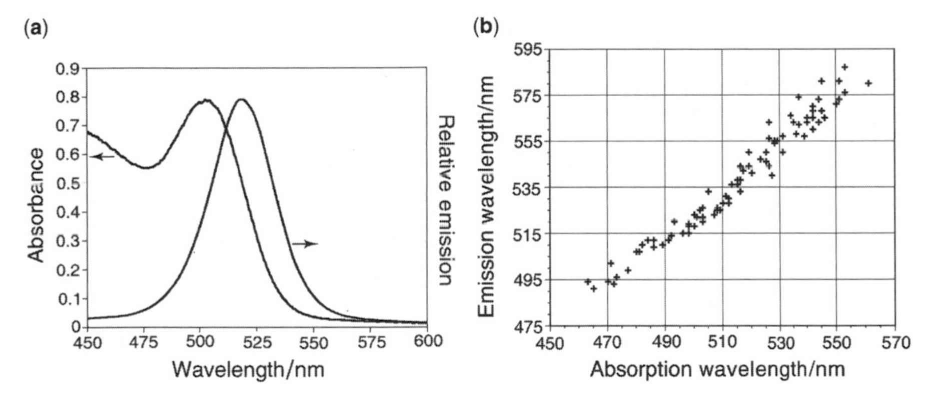

Figure 16.18 (a) For a given size quantum dot there is only a small Stokes shift (difference in absorption and emission maximum) of approximately 30 nm. (b) The absorption maxima of CdSe quantum dots correlate with their emission maxima. (Reproduced with permission from E. M. Boatman et al., 2005. *J. Chem. Ed.* 82: 1697–1699. Copyright 2005 American Chemical Society.)

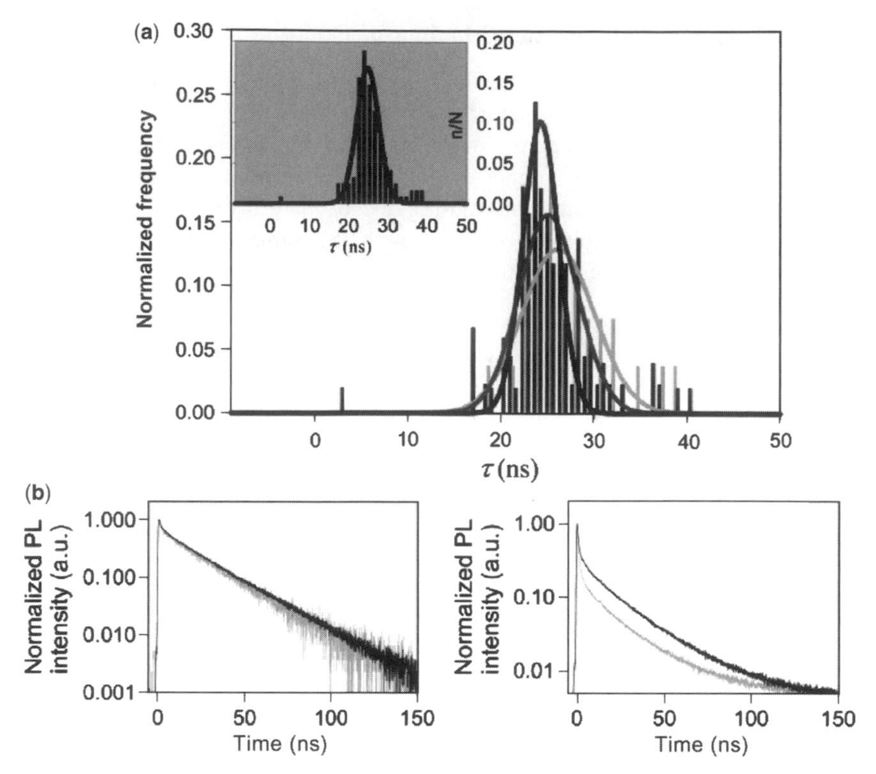

Figure 16.19 (a) Normalized distribution of single-exponential lifetime (τ) values from Q-dots with stretched exponential $\beta > 0.85$ for each of the three samples along with fits to the normal distribution. The mean lifetimes for samples 1 (gray), 2 (dark gray), and 3 (black) were 26.5, 24.6, and 24.6 ns, respectively. The inset of (a) displays the normalized distribution for all of the Q-dots taken together along with a Gaussian fit. The overall mean lifetime is 25.0 ns. (b) Two consecutive photoluminescence (PL) decay measurements for the same Q-dot. On the left, the maximum-intensity PL decays for the two consecutive measurements are shown. On the right are the time-averaged PL decays for the same measurements. (Reproduced with permission from B. R. Fisher et al., 2004. *J. Phys. Chem. B* 108: 143–148. Copyright 2004 American Chemical Society.)

Single-exponential luminescence decays from individual quantum dots can be obtained by discriminating their emission intensity. Therefore, it is likely that single-exponential radiative lifetimes (of CdSe quantum dots at room temperature) can indeed be measured.

Figure 16.19 shows the results from employing a stretched exponential decay model to fit the data (46).

$$y(t) = c + a\exp\left[-\left(\frac{t}{\tau}\right)^{\beta}\right] \tag{16.44}$$

The distribution of the calculated β-values over 180 selected single quantum dot measurements is shown for the maximum-intensity vs. the time-averaged photoluminescence decays. Note that the mean luminescence lifetime for the maximum-intensity photoluminescence decays is of the order of 25 ns, whereas the $1/e$ lifetime for the time-averaged photoluminescence decays is much shorter and less consistent.

16.4 ELECTROCHEMISTRY OF NANOPARTICLES

Small metal clusters possess unique optical and electronic properties that render them particularly useful for nanoscopic data storage devices. This has been attributed to the considerable constraint of the metal's electrons in such a cluster of 10 nm or less in diameter. In this zero-dimensional (0D) situation, metal electrons are beginning to occupy discrete energy levels. Therefore, a nanoparticle, which is small enough, no longer follows the classic laws of (electro) dynamics but rather features quantum-mechanical properties (47). Figure 16.20 shows the relationship between the diameter of a metal quantum dot and the temperature in which the so-called Coulomb-blockade effect occurs. This effect is considered most promising for future memory devices. The voltage V that causes the flow of electrons from the reservoir to the (single) electron storage device decreases by q/C for each transferred electron (48). *If C is sufficiently small, the transfer of a single electron to the storage device impedes further electron transfer (ET) processes.* Based on Coulomb-blockade effects,

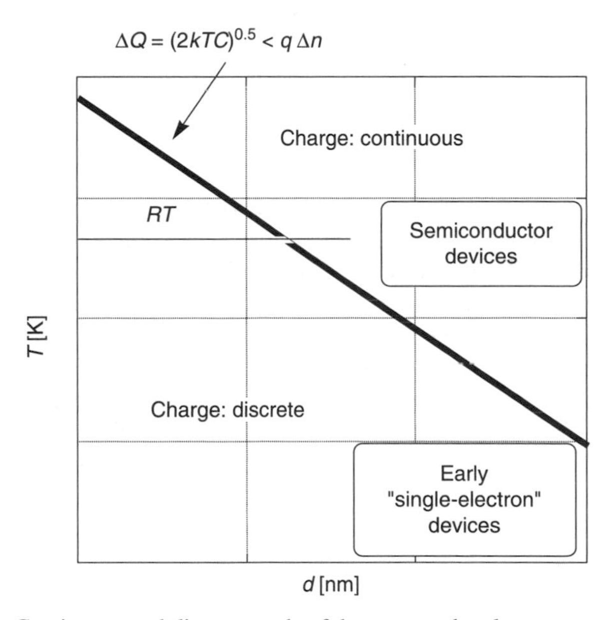

Figure 16.20 Continuous and discrete mode of data storage by electron-transfer. q: absolute charge of the electron; C: capacitance of the storage device; ΔQ: charge fluctuation; Δn: number of exchanged electrons; T: temperature [K], k: Boltzmann constant. (Reproduced with permission from M. Wörner et al., 2007. *Small* 3: 1084–1097. Copyright 2007 Wiley/VCH.)

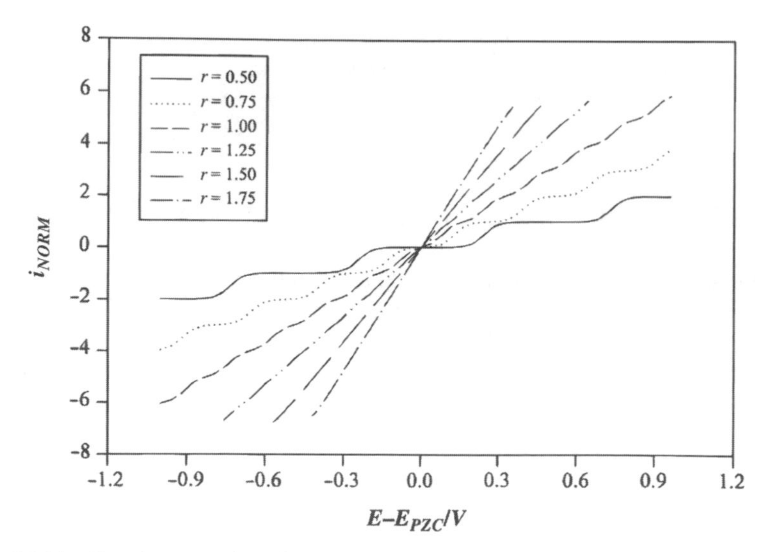

Figure 16.21 Steady-state microelectrode voltammetry simulations for completely monodisperse alkylthiolate-protected ($d = 0.52$ nm) gold nanoparticles with various core sizes (r): $\varepsilon = 3.0$, $T = 298$ K, r values shown in nm. (Reproduced with permission from S. Chen et al., 1998. *J. Phys. Chem. B* 102: 9898–9907. Copyright 1998 American Chemical Society.)

ET-transfer devices with single-electron precision could be designed. Compared to state-of-the-art electronic devices, which still operate in the continuous mode (Fig. 16.20), discrete memory storage would possess several major advantages: current memory devices charge/discharge 10,000 to 100,000 electrons per single digital process. Single-electron devices would reduce this number considerably, even if they would not operate based on the charge/discharge of a single electron. A drastic reduction of the number of electrons per storage event would enable the construction of ultra-low power consuming devices. It is mandatory for the broad technical application of discrete storage devices that they operate at room temperature. This is only possible when the dimension of the nanodevices used for electron storage is 5 nm or less. Finally, their very small physical dimensions permit ultra-large scale operations.

As already pointed out, the collective motion of electrons in a bulk metal obeys Ohm's law $V = RI$, where V is the applied voltage, R is the electrical resistance of the material, and I is the current. As the electronic band structure of the metal changes into discrete energy levels (e.g. as the particle becomes smaller and smaller), Ohm's law cannot describe the electrical properties of the nanoparticle anymore. If one electron is transferred to a nanoparticle, its Coulomb energy increases by $E_c = e^2/2C$, where C is the capacitance of the nanoparticle. If $kT \ll e^2/2C$ (k: Boltzmann constant, T: temperature in K), single electron tunneling processes can be observed. Therefore, the $I–V$ characteristic of a quantum dot (metal or semiconductor) is staircase-like (49, 50). $V_c = \pm e/2C$ is the threshold for the transfer of one electron. This phenomenon is called a Coulomb blockade. An electron tunneling process occurs if the Coulomb energy of the nanoparticle is compensated by an external voltage of $V = \pm ne/2C$. The experimentally observed charging energy increases with decreasing size of the quantum dot. (see Fig. 16.21)

16.5 CONCLUSION

There are two distinctly different mechanisms for metal and semiconductor absorption/emission.

- Nano-sized metals show plasmon resonance, which is very sensitive with respect to surface binding events and, of course, the size, shape, and size distribution of the nanoparticle.
- Semiconductors can be used as quantum dots because their nanosize confines the motion of conduction band electrons, valence band holes, or excitons in all three spatial directions. Quantum dots are often highly emissive, but their absorption and emission is much less sensitive on binding phenomena at their surfaces. The nanoparticles' size and shape is the only effective means to control their optical properties.
- The greatly enhanced electromagnetic field strength of a surface plasmon's mode makes it the ideal "collector" of photon energy for energy transfer processes to organic and inorganic luminophors. Energy transfer processes between metal nanoparticles and to (in)organic luminophors are highly dependent on the size of the nanoparticle and the respective distances.
- The electrochemistry of metal and semiconductor nanoparticles is determined by their size. Quantum effects, such as Coulomb blockades, can be expected for diameters below 5 nm.

FURTHER READING

Huang, Xiaohua; Jain, Prashant K.; El-Sayed, Ivan H.; El-Sayed, Mostafa A. **2007**. Gold nanoparticles: Interesting optical properties and recent applications in cancer diagnostics and therapy. *Nanomedicine* 2(5): 681–693.

Jain, Prashant K.; Huang, Xiaohua; El-Sayed, Ivan H.; El-Sayed, Mostafa A. **2008**. Noble metals on the nanoscale: Optical and photothermal properties and some applications in imaging, sensing, biology, and medicine. *Acc. Chem. Res.* 41(12): 1578–1586.

Jain, Prashant K.; El-Sayed, Mostafa A. **2008**. Surface plasmon coupling and its universal size scaling in metal nanostructures of complex geometry: Elongated particle pairs and nanosphere trimers. *J. Phys. Chem. C* 112(13): 4954–4960.

Liu, Gang L.; Yin, Yadong; Kunchakarra, Siri; Mukherjee, Bipasha; Gerion, Daniele; Jett, Stephen D.; Bear, David G.; Gray, Joe W.; Alivisatos, A. Paul; Lee, Luke P.; Chen, Fanqing Frank. **2006**. A nanoplasmonic molecular ruler for measuring nuclease activity and DNA footprinting. *Nature Nanotechnol.* 1(1): 47–52.

Liu, Wenhao; Howarth, Mark; Greytak, Andrew B.; Zheng, Yi; Nocera, Daniel G.; Ting, Alice Y.; Bawendi, Moungi G. **2008**. Compact biocompatible quantum dots functionalized for cellular imaging. *J. Am. Chem. Soc.* 130(4): 1274–1284.

Malinsky, Michelle Duval; Kelly, K. Lance; Schatz, George C.; Van Duyne, Richard P. **2001**. Nanosphere lithography: Effect of substrate on the localized surface plasmon resonance spectrum of silver nanoparticles. *J. Phys. Chem. B* 105(12): 2343–2350.

Whetten, Robert L.; Price, Ryan C. **2007**. Nano-Golden order. *Science* 318(5849): 407–408.

REFERENCES

1. Odom TW, Nehl CL. **2008**. How gold nanoparticles have stayed in the light: The 3M's principle. *ACS Nano* 2: 612–616.
2. Willets KA, Van Duyne RP. **2007**. Localized surface plasmon resonance spectroscopy and sensing. *Annu. Rev. Phys. Chem.* 58: 267–297.
3. Wiltschi B, Knoll W, Sinner EK. **2006**. Binding assays with artificial tethered membranes using surface plasmon resonance. *Methods* 39: 134–146.
4. Wang X, Ruedas-Rama MJ, Hall EAH. **2007**. The emerging use of quantum dots in analysis. *Anal. Lett.* 40: 1497–1520.
5. Winkler LD, Arceo JF, Hughes WC, DeGraff BA, Augustine BH. **2005**. Quantum dots: An experiment for physical or materials chemistry. *J. Chem. Ed.* 82: 1700–1702.
6. Hu M, Chen J, Li Z-Y, Au L, Hartland GV, Li X, Marquez M, Xia Y. **2006**. Gold nanostructures: Engineering their plasmonic properties for biomedical applications. *Chem. Soc. Rev.* 35: 1084–1094.
7. Mie G. **1908**. Contributions to the optics of turbid media, especially colloidal metal solutions. *Annal. Phys.* 25: 377–445.
8. Kelly KL, Coronado E, Zhao LL, Schatz GC. **2003**. The optical properties of metal nanoparticles: The influence of size, shape, and dielectric environment. *J. Phys. Chem. B* 107: 668–677.
9. Jain PK, Lee KS, El-Sayed IH, El-Sayed MA. **2006**. Calculated absorption and scattering properties of gold nanoparticles of different size, shape, and composition: Applications in biological imaging and biomedicine. *J. Phys. Chem. B* 110: 7238–7248.
10. Protasenko V, Hull KL, Kuno M. **2005**. Demonstration of a low-cost, single-molecule capable, multimode optical microscope. *Chem. Ed.* 10: 269–282.
11. Huang X, Jain PK, El-Sayed IH, El-Sayed MA. **2007**. Gold nanoparticles: Interesting optical properties and recent applications in cancer diagnostics and therapy. *Nanomedicine* 2: 681–693.
12. Fukami Y, Heya M, Awazu K. **2001**. Usefulness of indocyanine green as an infrared marker. *Proc. SPIE—The International Society for Optical Engineering* 4259 (Biomarkers and Biological Spectra Imaging): 157–162.
13. Juris A, Balzani V, Barigelletti F, Campagna S, Belser P, von Zelewsky A. **1988**. Ruthenium(II) polypyridine complexes: Photophysics, photochemistry, electrochemistry, and chemiluminescence. *Coordination Chem. Rev.* 84: 85–277.
14. Murov SL, Carmichael I, Hug GL. (eds.). **1993**. *Handbook of Photochemistry*. Marcel Dekker, New York.

15. Lee S, Kim S, Choo J, Shin SY, Lee YH, Choi HY, Ha S, Kang K, Oh CH. **2007**. Biological imaging of HEK293 cells expressing PLC $\geq$ 1 using surface-enhanced Raman microscopy. *Anal. Chem.* 79: 916–922.
16. Farrer RA, Butterfield FL, Chen VW, Fourkas JT. **2005**. Highly efficient multiphoton-absorption-induced luminescence from gold nanoparticles. *Nano Lett.* 5: 1139–1142.
17. Zhang YX, Zeng HC. **2008**. Surfactant-mediated self-assembly of Au nanoparticles and their related conversion to complex mesoporous structures. *Langmuir* 24: 3740–3746.
18. Link S, Wang ZL, El-Sayed MA. **1999**. Alloy formation of goldsilver nanoparticles and the dependence of the plasmon absorption on their composition. *J. Phys. Chem. B* 103: 3529–3533.
19. Russier-Antoine I, Jonin C, Nappa J, Benichou E, Brevet P-F. **2004**. Wavelength dependence of the hyper Rayleigh-scattering response from gold nanoparticles. *J. Chem. Phys.* 120: 10748–10752.
20. Hamelin J, Mehl JB, Moldover MR. **1998**. The static dielectric constant of liquid water between 274 and 418 K near the saturated vapor pressure. *Int. J. Thermophys.* 19: 1359–1380.
21. Pinchuk A, Kreibig U, Hilger A. **2004**. Optical properties of metallic nanoparticles: Influence of interface effects and interband transitions. *Surface Sci.* 557: 269–280.
22. Gans R. **1912**. The form of ultramicroscopic gold particles. *Annal. der Phys.* 37: 881–900.
23. Bossmann SH. **2009**. Nanoparticles for hyperthermia treatment of cancer, in *Fabrication and Bio-Application of Functionalized Nanomaterials*, Xuemei Wang and Evgeny Katz (eds.). Research Signpost, Trivandrum, Kerala (India).
24. Mulvaney P. **1996**. Surface plasmon spectroscopy of nanosized metal particles. *Langmuir* 12: 788–800.
25. Mohamed MB, Volkov V, Link S, El-Sayed MA. **2000**. The "lightning" gold nanorods: Fluorescence enhancement of over a million compared to the gold metal. *Chem. Phys. Lett.* 317: 517–523.
26. Link S, El-Sayed M. **1999**. Spectral properties and relaxation dynamics of surface plasmon electronic oscillations in gold and silver nanodots and nanorods. *J. Phys. Chem. B* 103: 8410–8426.
27. Sun C-K, Vallee F, Acioli LH, Ippen EP, Fujimoto JG. **1994**. Femtosecond-tunable measurement of electron thermalization in gold. *Phys. Rev. B: Condensed Matter* 50: 15337–15348.
28. Jares-Erijman EA, Jovin TM. **2006**. Imaging molecular interactions in living cells by FRET microscopy. *Curr. Opinion Chem. Biol.* 10: 409–416.
29. Yun CS, Javier A, Jennings T, Fisher M, Hira S, Peterson S, Hopkins B, Reich NO, Strouse GF. **2005**. Nanometal surface energy transfer in optical rulers, breaking the FRET barrier. *J. Am. Chem. Soc.* 127: 3115–3119.
30. Hao E, Schatz GC. **2004**. Electromagnetic fields around silver nanoparticles and dimers. *J. Chem. Phys.* 120: 357–366.
31. Gunnarsson L, Rindzevicius T, Prikulis J, Kasemo B, Kaell M, Zou S, Schatz GC. **2005**. Confined plasmons in nanofabricated single silver particle pairs: Experimental observations of strong interparticle interactions. *J. Phys. Chem. B* 109: 1079–1087.
32. Crozier K, Togan E, Simsek E, Yang T. **2007**. Experimental measurement of the dispersion relations of the surface plasmon modes of metal nanoparticle chains. *Opt. Sur.* 15: 17482–17493.

33. Reinhard B, Siu M, Agarwal H, Alivisatos A, Liphardt J. **2005**. Calibration of dynamic molecular rulers based on plasmon coupling between gold nanoparticles. *Nano Lett.* 5: 2246–2252.
34. Liu G, Yin Y, Kunchakarra S, Mukherjee B, Gerion D, Jett S, Bear D, Gray J, Alivisatos A, Lee L, Chen F. **2006**. A nanoplasmonic molecular ruler for measuring nuclease activity and DNA footprinting. *Nature Nano* 1: 47–52.
35. Reinhard BM, Sheikholeslami S, Mastroianni A, Alivisatos AP, Liphardt J. **2007**. Use of plasmon coupling to reveal the dynamics of DNA bending and cleavage by single EcoRV restriction enzymes. *Proc. Natt. Acad, Sci. U.S.A.* 104: 2667–2672.
36. Jain P, Huang W, El-Sayed M. **2007**. On the universal scaling behavior of the distance decay of plasmon coupling in metal nanoparticle pairs: A plasmon ruler equation. *Nano Lett.* 7: 2080–2088.
37. Jain PK, El-Sayed MA. **2007**. Universal scaling of plasmon coupling in metal nanostructures: Extension from particle pairs to nanoshells. *Nano Lett.* 7: 2854–2858.
38. Jain PK, El-Sayed MA. **2008**. Surface plasmon coupling and its universal size scaling in metal nanostructures of complex geometry: Elongated particle pairs and nanosphere trimers. *J. Phys. Chem. C* 112: 4954–4960.
39. Chan WCW, Maxwell DJ, Gao X, Bailey RE, Han M, Nie S. **2002**. Luminescent quantum dots for multiplexed biological detection and imaging. *Curr. Opinion Biotechnol.* 13: 40–46.
40. Murray CB, Norris DJ, Bawendi MG. **1993**. Synthesis and characterization of nearly monodisperse CdE (E = sulfur, selenium, tellurium) semiconductor nanocrystallites. *J. Am. Chem. Soc.* 115: 8706–15.
41. Dabbousi BO, Rodriguez-Viejo J, Mikulec FV, Heine JR, Mattoussi H, Ober R, Jensen KF, Bawendi MG. **1997**. (CdSe)ZnS core-shell quantum dots: Synthesis and optical and structural characterization of a size series of highly luminescent materials. *J. Phys. Chem. B* 101: 9463–9475.
42. Kim S, Fisher B, Eisler H-J, Bawendi M. **2003**. Type-II quantum dots: CdTe/CdSe(core/shell) and CdSe/ZnTe(core/shell) heterostructures. *J. Am. Chem. Soc.* 125: 11466–11467.
43. Bailey RE, Nie S. **2003**. Alloyed semiconductor quantum dots: tuning the optical properties without changing the particle size. *J. Am. Chem. Soc.* 125: 7100–7106.
44. Leatherdale CA, Woo W-K, Mikulec FV, Bawendi MG. **2002**. On the absorption cross section of CdSe nanocrystal quantum dots. *J. Phys. Chem. B* 106: 7619–7622.
45. Boatman EM, Lisenky GC, Nordell KJ. **2005**. A safer, easier, faster synthesis for CdSe quantum dot nanocrystals. *J. Chem. Ed.* 82: 1697–1699.
46. Fisher BR, Eisler HJ, Stott NE, Bawendi MG. **2004**. Emission intensity dependence and single-exponential behavior in single colloidal quantum dot fluorescence lifetimes. *J. Phys. Chem. B* 108: 143–148.
47. Schmid G, Corain B. **2003**. Nanoparticulated gold: Syntheses, structures, electronics, and reactivities. *Eur. J. Inorg. Chem.* 17: 3081–3098.
48. Wörner M, Lioubashevski O, Niebler S, Gogritchiani E, Egner N, Heinz C, Hoferer J, Cipolloni M, Janik K, Katz E, Braun A, Willner I, Niederweis M, Bossmann S. **2007**. Characterization of nanostructured-surfaces generated by reconstitution of the porin

MspA from *M. smegmatis* into stabilized long-chain-monolayers at gold-electrodes. *Small* 3: 1084–1097.

49. Chen S, Murray RW, Feldberg SW. **1998**. Quantized capacitance charging of monolayer-protected Au clusters. *J. Phys. Chem. B* 102: 9898–9907.
50. Chen S, Ingrma RS, Hostetler MJ, Pietron JJ, Murray RW, Schaaff TG, Khoury JT, Alvarez MM, Whetten RL. **1998**. Gold nanoelectrodes of varied size: transition to molecule-like charging. *Science* 280: 2098–2101.

PROBLEMS

1. Small metal clusters possess unique optical and electronic properties. This has been attributed to the considerable constraint of the metal's electrons in such a cluster of 10 nm or less in diameter. In this zero-dimensional (0D) situation, metal electrons are beginning to occupy discrete energy levels. Therefore, a nanoparticle that is small enough will no longer follow the classic laws of (electro)mechanics but rather features quantum-mechanical properties. Figure 16.20 shows the relationship between the diameter of a metal quantum dot and the temperature in which the so-called Coulomb-blockade effect occurs. This effect is considered most promising for future memory devices. The voltage V, that causes the flow of electrons from the reservoir to the (single) electron storage device decreases by q/C for each transferred electron.

   q: absolute charge of the electron (1.602×10^{-19} C) (C = Coulomb)
   C: capacitancc of the storage device ($C = C\,V^{-1} = m^{-2}\,kg^{-1}\,s^4\,A^2$)
   ΔQ: charge fluctuation
   Δn: number of exchanged electrons
   T: temperature [K]
   k: Boltzmann constant ($k = 1.380 \times 10^{-23}\,J\,K^{-1}$)
   (help: $V = W\,A^{-1} = J\,C^{-1} = m^2 \cdot kg \cdot s^{-3} \cdot A^{-1}$)

   a) What are the units of the fluctuation ΔQ?
   b) Calculate the charge fluctuation ΔQ for $C = 1$ fF ($- 10^{-15}$ F), a typical capacitance for gold nanoparticles possessing a diameter of 10 nm at 10 K!
   c) Calculate the charge fluctuation ΔQ for $C = 1$ fF ($= 10^{-15}$ F) at 273 K!

2. What capacitance is needed to permit the exchange of exactly one electron at 273 K?

3. The absorption coefficients of human (aorta) tissue and of water as a function of the excitation wavelength are shown. The minimum of the tissue absorption can be found at approximately 800 nm. Some studies have suggested that multiple-photon absorption of gold nanospheres at this wavelength might be advantageous for plasmonic heating (photoinduced hyperthermia). What shape

of the gold nanostructures do you suggest instead of spheres? What would be the most notable advantage of the shape you suggest?

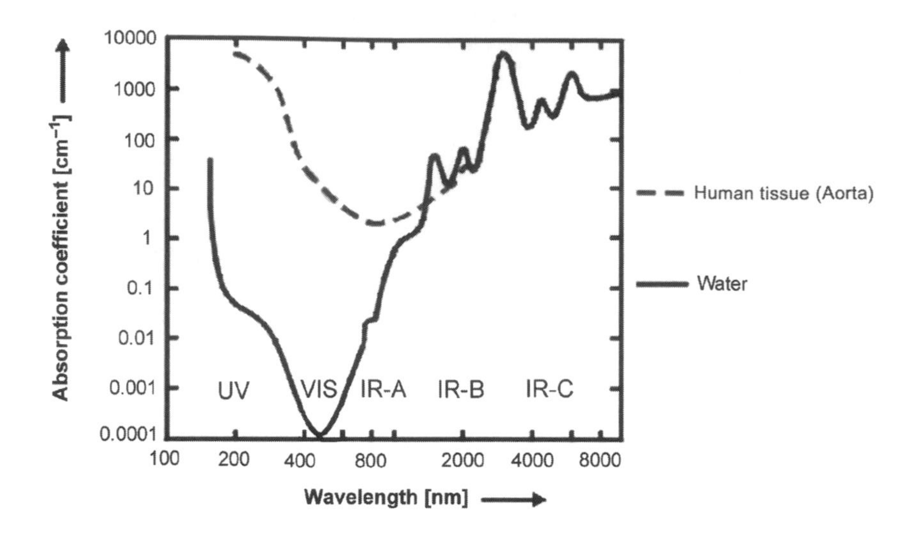

Absorption coefficients of water and a typical human tissue (aorta) as a function of wavelength. The suitable region for optical excitation is between 650 and 1000 nm. This figure is drawn using data from the *Handbook on Industrial Laser Safety*, http://info.tuwien.ac.at/iflt/safety/index.htm.

4. Calculate the absorption *cross-sections* of the following molecules: The commonly used NIR dye indocyanine green ($\varepsilon = 1.08 \times 10^4\ M^{-1}\ cm^{-1}$ at $\lambda = 778$ nm), the sensitizer ruthenium(II)-tris-bipyridine (1.54×10^4 at $M^{-1}\ cm^{-1}$ at 452 nm) and rhodamine-6G ($\varepsilon = 1.16 \times 10^5\ M^{-1}\ cm^{-1}$ at $\lambda = 530$ nm) or malachite green ($\varepsilon = 1.49 \times 10^5\ M^{-1}\ cm^{-1}$ at $\lambda = 617$ nm).

5. Why are quantum dots ideal for simultaneous *in vitro* and *in vivo* fluorescence detection?

6. What are the most important differences between gold nanoparticles and cadmium selenide quantum dots?

7. Why are the luminescence lifetimes of gold nanoparticles/rods in the fs-range?

8. What is the physical basis of the two-temperature model (TTM) describing the cooling of photonically heated gold nanoparticles?

9. Why is *Universal Scaling Behavior* (USB) desirable?

10. Why is it impossible to achieve Universal Scaling Behavior with conventional organic fluorophors?

ANSWERS

1. **a)** Coulomb
b) 5.253×10^{-19} coulomb (this corresponds to the charge of 3.26 electrons).
c) 2.745×10^{-18} coulomb (this corresponds to the charge of 17.25 electrons).

2. $\Delta nq = \sqrt{2kTC}$, $q^2 = 2kTC(\Delta n = 1)$, $C = \dfrac{q^2}{2kT}$, $C_{(\text{required})} = 3.594 \times 10^{-18}\,\text{F}$

3. It should be noted that noble metal nanostructures (e.g. gold nanorods) can be easily tuned for monophotonic applications in the wavelength range from 600 to 1000 nm by adjusting their aspect ratio $R = A/B$. Monophotonic excitation is advantageous, because its absorption cross-section is several orders of magnitude higher than for multiphotonic excitation!

4. $C_{ext} = 0.2303\varepsilon_{ext}/N_A$
Indocyanine green: $2.12 \times 10^{-21}\,\text{m}^2$; ruthenium(II)-tris-bipyridine: $5.89 \times 10^{-21}\,\text{m}^2$; and rhodamine-6G: $4.44 \times 10^{-20}\,\text{m}^2$; malachite green: $5.69 \times 10^{-20}\,\text{m}^2$.

5. The Stokes shifts between absorption and emission maxima of quantum dots remain relatively constant, as the slope of approximately one in Figure 16.18b indicates. Because of additional nearby energy levels, quantum dots can be excited by photons of greater than or only slightly higher energies. Therefore, the effective Stokes shift can be hundreds of nanometers, permitting the excitation of numerous quantum dots possessing various emission maxima with the same excitation wavelength. This allows the simultaneous detection of numerous events!

6. *Surface plasmons* are the collective oscillations of the free electron gas density in a metal, which can be often detected by absorption spectroscopy. *Surface* plasmons are *surface electromagnetic waves* that propagate parallel along a metal/ dielectric interface. Since the wave is partially located on the boundary of the metal and extends partially into the external medium (or vacuum), these oscillations are very sensitive to any change chemical composition occurring at the nanoparticle's surface, such as the adsorption and exchange of ligands and (macro)molecules.

A *semiconductor quantum dot* consists of a nanostructure that confines the motion a finite number of conduction band electrons, valence band holes, or excitons (bound pairs of conduction band electrons and valence band holes) in all three spatial directions. The reason for the observed confinement is either the presence of an interface between different semiconductor materials or the existence of the semiconductor surface (e.g. semiconductor nanocrystal). Therefore, one quantum dot or numerous quantum dots of exactly the same size and shape have a discrete quantized energy spectrum, almost like organic molecules (see Scheme 16.2).

7. The upper bound on the luminescence lifetime of gold nanoparticles and rods is approximately 50 fs. This very short lifetime can be regarded as an experimental proof that the luminescence from gold nanostructures is directly related to the dynamics of the electron holes created in the d-band immediately after irradiation. It does not occur from the surface plasmon! Furthermore, it cannot be regarded as a typical radiative lifetime, which would be at least in the ns window, considering the size of typical nanoparticles/rods.

8. It is noteworthy that the pump pulse is able to excite the electrons to temperatures of the electron gas of up to several thousand degrees Kelvin. Because of these high initial temperatures, a non-Fermi distribution is created directly after the absorption of the pump pulse. The nanoparticle system thermalizes initially through *electron-electron scattering*. After a Fermi distribution is reached, further thermalization through electron-photon scattering is observed until the nanoparticle reaches the temperature of its surrounding. The two-temperature model (TTM) is based on the experimental finding that the lattice heat capacity is approximately 2 orders of magnitude bigger than the electronic heat capacity. The plasmon band absorption is very sensitive with respect to the electron dynamics and permits the direct observation of these processes (Figures 16.9–16.11).

9. Distances in biological systems are of a special importance (e.g. for redox-enzymes, membrane proteins etc.) According to Eq. (39) and Figure 16.14, the fractional plasmon shift ($\Delta\lambda/\lambda_0$) is a function of the inter-nanoparticle edge-to-edge separation (r) and the particle diameter D. This allows the design of plasmon-coupled nanoparticle pairs, which can measure distances up to 60 nm with high precision.

10. Förster energy transfer (FET) between an organic photodonor and acceptor is dependent on r^{-6} (r, inter-fluorophore edge-to-edge separation). Therefore, the decrease of the energy transfer as a function of distance occurs so rapidly that only two cases can be discerned: the distance is either smaller or larger than the Förster radius. The longest achievable distance for FET is approximately 10 nm.

PART VIII

PHOTOCHEMISTRY OF NANOMATERIALS

17

PHOTOCATALYTIC PURIFICATION OF WATER AND AIR OVER NANOPARTICULATE TiO_2

IGOR N. MARTYANOV AND KENNETH J. KLABUNDE

17.1 Introduction, 581
17.2 General Information, 582
17.3 Deep Photocatalytic Oxidation of Organic Compounds: Use of Photocatalysis for Water Purification, 584
17.3.1 Absorption of the Light by Photocatalyst Nanoparticles, 585
17.3.2 Studies of Oxidation of Methyl Viologen in Aqueous Suspension of TiO_2: Influence of pH and Hydrogen Peroxide Concentration, 587
17.3.3 Photocatalytic Oxidation of Methyl Viologen in an Aqueous Suspension of TiO_2: Influence of Light Intensity and Substrate Concentration, 590
17.4 Photocatalysis at TiO_2/Gas Phase Interphase, 595
17.4.1 TiO_2 Films on Quartz, 595
17.4.2 Deep Photooxidation of a Mustard Gas Mimic over Nanoparticulate TiO_2, 596
17.5 Concluding Remarks and Future Outlook, 598
References, 599
Problems, 599
Answers, 600

17.1 INTRODUCTION

Photocatalysis is a quickly evolving area of research. A principal interest in photocatalytic processes lies in the possibility to initiate chemical reactions, which

Nanoscale Materials in Chemistry, Second Edition. Edited by K. J. Klabunde and R. M. Richards

otherwise are difficult or impossible to carry out under ambient temperature and pressure by the methods of conventional chemistry. The driving force of photocatalytic reactions is light energy, which upon absorption promotes an electron from the valence into the conduction band. In addition to the possibility of carrying out reactions under ambient conditions, there are two arguments that favor more extensive use of photocatalytic processes. They include an enormous total energy carried by solar light that reaches the surface of our planet, as well as the clean nature of this energy that would leave no chemical residue after its consumption. Arguably there are two major areas of research in photocatalysis, with the first one aiming at conversion and storage of solar energy in the form of energy of chemical compounds (hydrogen, for example) and the second dedicated to the eradication of environmental pollutants. This chapter is solely focused on the latter subject. It presents an overview of results obtained by the authors over a number of years of studying water and air purification by TiO_2 nanoparticulate photocatalysts. The chapter covers selected general information on the subject, a summary of studies of model photocatalytic reactions in the liquid phase, studies of films of nanoparticulate TiO_2 on quartz in gas phase photocatalytic processes, and results on a practically important process of photocatalytic oxidation of a mustard gas mimic. At the end of the chapter the reader can find problems aimed at emphasizing certain issues crucial for understanding of photocatalytic processes over nanoparticulate TiO_2.

17.2 GENERAL INFORMATION

Photocatalytic oxidation over nanoparticulate TiO_2 is a complex process involving a number of steps. Their simplified sequence is presented in Scheme 17.1. The trigger step is absorption of a photon by a TiO_2 particle. Absorption of the photon with energy greater than the TiO_2 bandgap results in promotion of an electron from the valence band into the conduction band. The vacant state left behind in the valence band behaves like a positively charged particle and is called a hole. The photogenerated

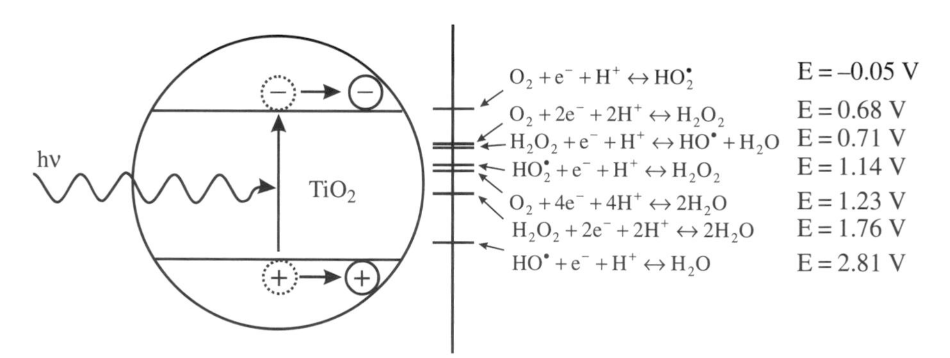

Scheme 17.1 Schematic representation of electron–hole pair formation in TiO_2 nanoparticles and possible redox processes that may occur in air-saturated aqueous medium.

electron and hole move to the solid-liquid interface and get captured by surface sites. Capturing of an electron can be considered a reduction of a surface or subsurface Ti^{4+} into Ti^{3+}, whereas localization of the hole, h, can be considered oxidation of lattice O^{2-} into $O^{\bullet -}$ or oxidation of surface water or hydroxide groups into hydroxyl radicals:

$$\begin{aligned} H_2O + h &\longrightarrow HO^{\bullet} + H^+ \quad \text{or} \\ HO^- + h &\longrightarrow HO^{\bullet} \end{aligned} \tag{17.1}$$

The redox potential of a captured electron is negative enough to reduce oxygen into a superoxide radical, $O_2^{\bullet -}$, which depending on pH of the solution (see Table 17.1) can quickly be protonated forming hydrogen superoxide radical, $HO_2^{\bullet}$. The redox potential of a hole or hydroxyl radical is positive enough to initiate oxidation of organic species:

$$\begin{aligned} HO^{\bullet} + Org &\longrightarrow HO^- + Org^{\bullet +} \quad \text{or} \\ HO^{\bullet} + OrgH &\longrightarrow H_2O + Org^{\bullet} \end{aligned} \tag{17.2}$$

$$Org^{\bullet +} + O_2 + H_2O \longrightarrow Org_{ox} \tag{17.3}$$

where Org_{ox} is a stable oxidation product of Org.

The fate of the superoxide radical, $O_2^{\bullet -}$, or its protonated form hydrogen superoxide, $HO_2^{\bullet}$, is assumed to be further reduced by the second photogenerated electron into hydrogen peroxide:

$$HO_2^{\bullet} + e + H^+ \longrightarrow H_2O_2 \tag{17.4}$$

Hydrogen peroxide is stable enough to be detected as an intermediate, but actively participates in the photocatalytic reaction as an electron acceptor similar to O_2:

$$Ti^{3+} + H_2O_2 \longrightarrow Ti^{4+} + HO^{\bullet} + HO^- \tag{17.5}$$

or as an electron donor competing for holes/hydroxyl radicals with organic moieties:

$$HO^{\bullet} + H_2O_2 \longrightarrow HO_2^{\bullet} + HO^- + H^+ \tag{17.6}$$

TABLE 17.1 pK$_a$ Values of Various Oxygen Intermediates in Photocatalytic Oxidation Processes

Reaction	pK$_a$
$H_2O_2^{\bullet +} \longleftrightarrow HO_2^{\bullet} + H^+$	$pK_a = 1.2$
$HO_2^{\bullet} \longleftrightarrow O_2^{\bullet -} + H^+$	$pK_a = 4.7$
$H_2O_2 \longleftrightarrow HO_2^- + H^+$	$pK_a = 11.58$
$OH^{\bullet} \longleftrightarrow O^{\bullet -} + H^+$	$pK_a = 11.9$
$HO_2^- \longleftrightarrow O_2^{2-} + H^+$	$pK_a > 13.5$

There are a number of events that can be detrimental to the efficiency of the photocatalytic process. They may be referred to as recombination steps that include

1. Recombination of a localized photogenerated electron and a hydroxyl radical (here and later on we will refer to hydroxyl radicals implying localized holes as a possible alternative):

$$Ti^{3+} + HO^{\bullet} \longrightarrow Ti^{4+} + HO^{-} \tag{17.7}$$

2. The oxidation of hydrogen superoxide by a hydroxyl radical:

$$HO_2^{\bullet} + HO^{\bullet} \longrightarrow O_2 + H_2O$$

3. The recombination of hydroxyl radicals into hydrogen peroxide:

$$HO^{\bullet} + HO^{\bullet} \longrightarrow H_2O_2 \tag{17.8}$$

In addition, there is the potential possibility of the reaction of Ti^{3+} with an adsorbed $Org^{\bullet +}$:

$$Ti^{3+} + Org^{\bullet +} \longrightarrow Ti^{4+} + Org \tag{17.9}$$

and the above-mentioned oxidation of hydrogen peroxide by a hydroxyl radical.

The enormous complexity of the photocatalytic process and the presence of many unknowns makes development of a universal mathematical model difficult. Several research groups remain active in this area suggesting new approaches and promoting active discussion in this field. Below we summarize major qualitative results of our studies on photocatalytic water and air purification over nanoparticulate TiO_2.

17.3 DEEP PHOTOCATALYTIC OXIDATION OF ORGANIC COMPOUNDS: USE OF PHOTOCATALYSIS FOR WATER PURIFICATION

Photocatalytic processes over nanoparticulate TiO_2 can be applied for water purification. TiO_2 as a material is particularly attractive in this regard due to its low toxicity, high stability, an ability to produce hydroxyl radicals capable of oxidizing virtually any organic substance to water, carbon dioxide, and corresponding mineral acids.

Typical photocatalytic experiments in aqueous media are usually conducted in a batch reactor illuminated with filtered or unfiltered light of a mercury lamp. The photocatalytic reaction mixture typically consists of a suspension of TiO_2 with a known amount of organic compound added. Bubbling air may be employed to ensure a constant concentration of oxygen in the reaction mixture and to provide continuous agitation. Alternatively, when concentration of an organic compound or its conversion is low, the fall in the concentration of oxygen in the solution can be neglected and agitation can be provided by a rotating Teflon-coated magnetic bar. A typical result of such experiments is a dependence of an initial rate of the organic moiety consumption versus other parameters, such as organic compound concentration, pH of the solution, light intensity, etc. Concentration of organic compounds can be followed by a gas chromatograph equipped with a FID detector. Some large molecules, dyes for example, are not volatile enough and can be tracked with a UV-Vis spectrometer. Caution has to be exercised in this regard as the products of partial oxidation usually adsorb in the same region as an original molecule. Additionally, the products of partial oxidation are often hydroxylated moieties with their UV-Vis absorption spectra pH dependent. Under such circumstances, the change in the absorption may not provide absolute values in the change of concentration of the original organic species but rather information about relative values that should be compared with results obtained at the same pH with an assumption of similar product selectivity. One possible way to circumvent this problem is to assume that concentrations of transient compounds contributing to the absorption are low enough and they do not interfere when conversion of the substrate is reasonably high.

17.3.1 Absorption of the Light by Photocatalyst Nanoparticles

The rate of a photocatalytic process is obviously light dependent. The knowledge of the amount of light absorbed by a photocatalyst is of great importance. Unfortunately, in many important cases this issue cannot be easily resolved. The major problem is scattering. Indeed, for the catalysts used in a form of suspension, a part of the light falling on a suspension of TiO_2 is scattered in different directions, with the rest being transmitted. Standard optical spectrum measurements lead mainly to the capture of transmitted light producing the UV-Vis spectrum of the suspension with contributions from absorption and scattering. But only absorption can drive the photocatalytic reaction! To some extent, the true absorption spectrum of a suspension can be determined with modified UV-Vis spectrum measurements, where scattered light is collected and directed to the spectrophotometer detector together with transmitted light. However, for determination of a quantum yield another approach is often employed. In essence, this method consists of surrounding a light source with a highly concentrated TiO_2 suspension letting no light out. Multiple scatterings do take place, but it is safe to assume that all light will be eventually absorbed by the photocatalyst.

There is a second problem, which is often ignored: the light intensity to which TiO_2 particles in a suspension are exposed is location dependent. It is obvious, for example, that TiO_2 particles in front of the suspension are subjected to higher light intensity than the TiO_2 particles in the back of the solution. Dark areas typically present in the suspension lead to a situation where the intensity of the light varies from the full intensity near the source down to zero light. Even this is not the whole story.

Titanium dioxide P25 from Degussa, which is often considered to be a benchmark in the area of photocatalysis, is a fluffy powder with a surface area of approximately

50 m^2/g. It has a complex morphology consisting of rutile and anatase crystallites and primary particles derived from them, which are on average 30 nm in size. These particles are stuck to each other forming micron-sized aggregates. The aggregates are relatively weakly bound, and if placed in agitated water will float separately. TiO_2 particles in the front of an aggregate should face higher intensity of light than TiO_2 particles in the back. This simple fact that the particles in an aggregate lie in the shade of the particles in front belonging to the same aggregate has rather amazing consequences—the spectrum of a TiO_2 suspension and the amount of light absorbed by it is dependent on the degree of dispersion.

This phenomenon was studied earlier, with the results summarized in Reference 1. As was mentioned above, scattering under normal circumstances contributes a lot to the apparent UV-Vis spectrum of TiO_2. Scattering, however, is a size-dependent phenomenon. It is at its maximum when the sizes of aggregates are comparable to the wavelength of the light, but drops considerably when aggregate sizes fall well below wavelength values. To work in the range at which scattering can be neglected, a colloidal suspension of TiO_2 particles was prepared from TiO_2 Degussa P25. The idea was rather simple and worked well for TiO_2 P25. Aggregates in the suspension vary in their size. Bigger and heavier aggregates precipitate faster than smaller ones. Particles that are small enough will never precipitate and will be in continuous Brownian motion. Consequently, if left undisturbed the TiO_2 suspension should become enriched in small aggregates, while bigger aggregates should settle at the bottom of the flask. Centrifugation is a simple and efficient way to speed up this process through an effective increase in gravitational force. Calculations show that the only aggregates that can stay in the solution after such a procedure should be 30 nm in size and smaller, thus giving a colloidal suspension of TiO_2 consisting of primary particles of Degussa P25.

Stability of the UV-Vis spectrum of the colloidal TiO_2 suspension is pH dependent. The spectrum is quite stable at moderately acidic and basic pH, but unstable in between. The stability of the suspension and consequently its UV-Vis spectrum can be related to the net charge of the particles making up the aggregates. The surfaces of oxides are hydroxylated. Protonation of surface hydroxyls or their deprotonation is pH dependent and results in building a positive or negative charge on TiO_2 particles (2):

$$\begin{array}{llll} \mathrm{Ti-OH_2^+} & \longleftrightarrow & \mathrm{Ti-OH + H^+} & \mathrm{p}K_{a1} = 4.5 \\ \mathrm{Ti-OH} & \longleftrightarrow & \mathrm{Ti-O^-} & \mathrm{p}K_{a2} = 8 \end{array}$$

Positively charged particles in acidic solutions as well as negatively charged TiO_2 particles under basic conditions repel each other, giving a stable suspension. On the other hand when pH of the solution approaches the pH of the point of zero charge [$pH_{zpc} = (pK_{a1} + pK_{a2})/2 = 6.25$], the charge on the TiO_2 particles approaches zero, facilitating their aggregation. The shading phenomenon mentioned above makes it possible to track changes in degree of aggregation by UV-Vis spectroscopy. An example of changes observed on aggregation of TiO_2 particles is shown in Figure 17.1. Addition of a small amount of base to the suspension shifts the pH from 4.7 to 6.1, diminishes the charge of the TiO_2 particles, and dramatically

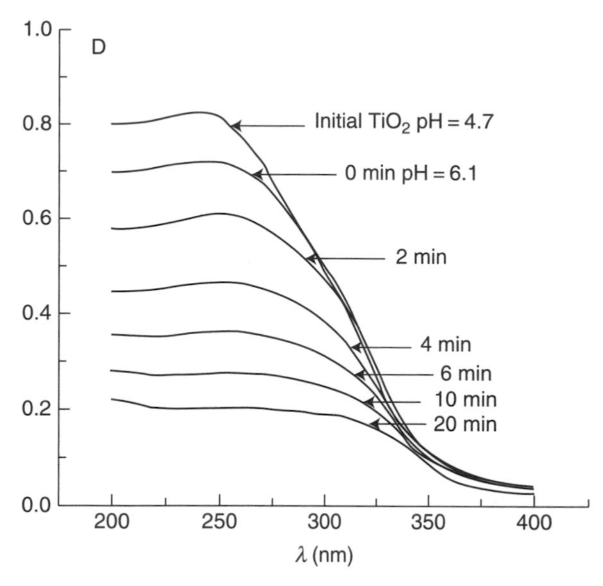

Figure 17.1 Time evolution of optical spectrum of TiO_2 suspension initiated by an addition of a small amount of basic aqueous solution. (Reproduced with permission from I. N. Martyanov et al., *J. Colloid. Interf. Sci.* **2003**, *267*(1), 111. Copyright 2003 Elsevier.)

accelerates the aggregation rate accompanied by optical density drop. Shifting the pH further into the basic region makes TiO_2 particles negatively charged. The repelling forces, however, are not strong enough to initiate spontaneous disaggregation. A little push needed can be provided through the action of ultrasound. The aggregates get dispersed and a starting UV-Vis spectrum shown in Figure 17.1 is restored. A simple mathematical model taking into consideration just absorption by TiO_2 particles is capable of predicting changes in the optical absorption spectra (1). The UV-Vis spectra calculated for aggregates of various sizes are shown in Figure 17.2, which are in good qualitative agreement with the experimental results.

17.3.2 Studies of Oxidation of Methyl Viologen in Aqueous Suspension of TiO_2: Influence of pH and Hydrogen Peroxide Concentration

Arguably, there are three major problems in regard to a practical application of photocatalytic reactions over TiO_2 for water and air purification. The first problem is generally low quantum yields and hence low reaction rates, especially in aqueous media. Second is the formation of intermediate products of partial oxidation with comparable, higher, or unknown toxicity. Third, and this is important particularly for applications aimed at using solar light, is the necessity to use UV light to initiate oxidation over non-doped TiO_2.

For reactions in aqueous media, one of the ways to tackle the first of these problems is to add an acceptor of photogenerated electrons more powerful than oxygen. Among

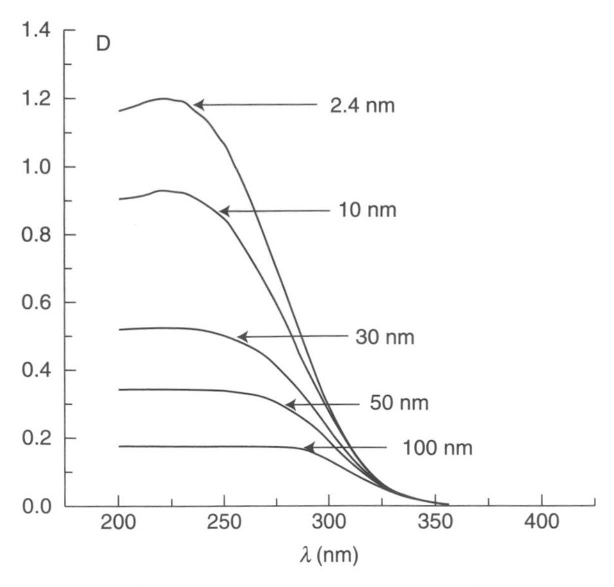

Figure 17.2 Predicted optical spectra of TiO_2 suspension having similar total amount of the material but different aggregate or particle sizes. (Reproduced with permission from I. N. Martyanov et al., *J. Colloid. Interf. Sci.* **2003**, *267*(1), 111. Copyright 2003 Elsevier.)

all the possibilities, the use of hydrogen peroxide looks to be the most attractive one. Indeed, hydrogen peroxide is a relatively benign compound with a long history of medical use. It is a thermodynamically unstable compound. Its excess can be easily decomposed in the dark into oxygen and water over a suitable catalyst.

From a practical point of view, building a detailed model linking observable rates to elemental processes in TiO_2 particles is not necessary. Empiric formulas linking reaction rates to the parameters of the reaction mixture are enough for the determination of optimal conditions under which the reaction rate is at its maximum. An example of such a study, which is devoted to the influence of pH and hydrogen peroxide concentration on the rate of methyl viologen oxidation, is given in Reference 3.

Hydroxyl radicals are highly reactive, short-lived species. Since they have virtually no time to diffuse in the bulk of the solution, organic molecules have to be in the vicinity of the surface of the catalyst to participate in the oxidation reaction. In other words, in a simple situation the reaction rate should increase with an increase in the surface concentration of organic molecules. Here we refer to adsorbed organic species implying, however, that organic molecules may also be located in a double layer surrounding TiO_2.

It should not come as a surprise that the initial rate of oxidation of methyl viologen, MV^{2+}, shown in Figure 17.3 and concentration of surface MV^{2+} shown in Figure 17.4 both measured as a function of pH are quite similar. The surface concentration of MV^{2+} versus pH has a bell-like shape, which is easily explainable. Indeed, as established above, an increase of pH makes TiO_2 particles negatively charged. Negatively

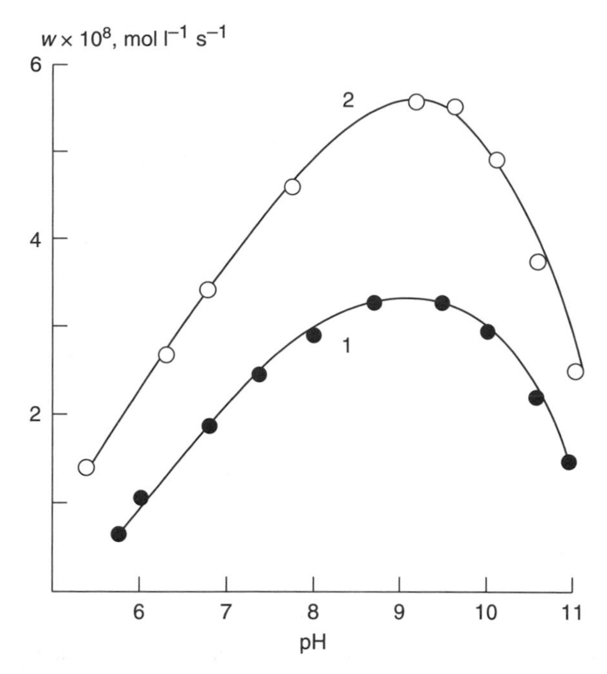

Figure 17.3 Initial oxidation rate of methyl viologen as a function of pH at two concentrations of hydrogen peroxide: (1) $[H_2O_2] = 0$, (2) $[H_2O_2] = 4 \times 10^{-3}$ M. (Reproduced with permission from I. N. Martyanov et al., *Kinetics and Catalysis* **1997**, *38*(1), 70. Copyright 1997 Pleiades Publishing, Inc.)

charged TiO_2 should attract more positively charged MV^{2+}. At the same time, an increase of pH by addition of a solution of NaOH also leads to an increase in ionic strength of the solution. At some concentration of NaOH, the second effect becomes dominant, with more adsorbed MV^{2+} being substituted by Na^+ leading to the reduction in total amount of MV^{2+} adsorbed. At this point an empirical formula for the initial rate of MV^{2+} oxidation can be written as follows:

$$W(\text{pH}, [H_2O_2]) = k_1([H_2O_2]) \cdot \nu(\text{pH}) \qquad (17.10)$$

where W is an initial rate of MV^{2+} oxidation, ν is a concentration of surface MV^{2+} per gram of TiO_2, k_1 is an unknown function of hydrogen peroxide concentration.

It is of interest to link k_1 to the rate of hydrogen peroxide decomposition. As shown in Figure 17.5, for a constant pH, the rate of MV^{2+} oxidation correlates well with the rate of H_2O_2 decomposition. This observation is consistent with the idea of the presence of two channels for MV^{2+} oxidation involving oxygen, and hydrogen peroxide as the second electron acceptor. The total rate of methyl viologen oxidation is the sum of the reaction rates in these two channels:

$$W(\text{pH}, [H_2O_2]) = (k_{O_2} + k_{H_2O_2}[H_2O_2]) \cdot \nu(\text{pH}) \qquad (17.11)$$

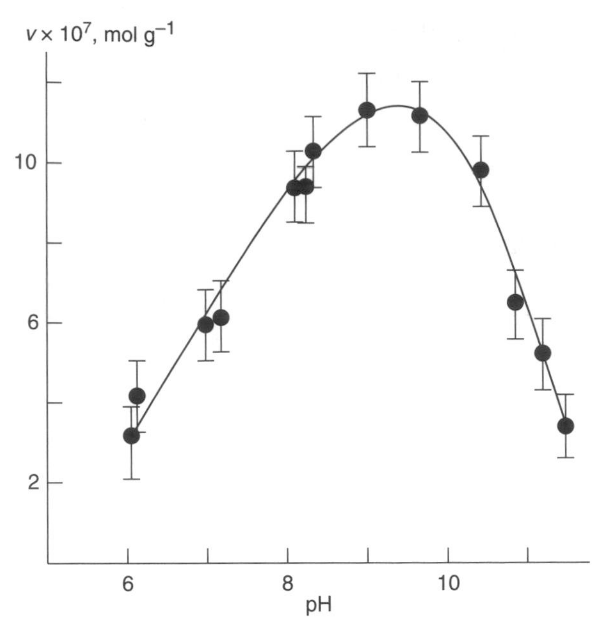

Figure 17.4 Concentration of methyl viologen in the vicinity of TiO_2 surface as a function of pH. (Reproduced with permission from I. N. Martyanov et al., *Kinetics and Catalysis* **1997**, *38*(1), 70. Copyright 1997 Pleiades Publishing, Inc.)

k_{O_2} is a coefficient determining the rate of the reaction in the presence of only oxygen, $k_{H_2O_2}$ is a coefficient determining the rate of the reaction with hydrogen peroxide.

17.3.3 Photocatalytic Oxidation of Methyl Viologen in an Aqueous Suspension of TiO_2: Influence of Light Intensity and Substrate Concentration

In the previous example, the rate of the reaction can be presented as a product of the functions depending on different parameters: concentration of hydrogen peroxide, $[H_2O_2]$, and pH of the solution. The structure of the equation also implies that "the whole picture" can be obtained through studying a response to each parameter when others are held constant followed by multiplication of obtained functions. Unfortunately, this is not always the case for any set of parameters under investigation.

One of the examples of a more complex picture was found when an initial oxidation rate of the MV^{2+} was measured against light intensity and methyl viologen concentration. As we saw earlier, the surface MV^{2+} concentration can be varied by changing pH. Another more straightforward way to do it is to vary the concentration of MV^{2+} in the solution itself. An increase in the solution concentration entails an increase in the surface concentration on TiO_2 particles. At some point, however, it levels off and becomes insensitive to a further solution concentration build up. A typical

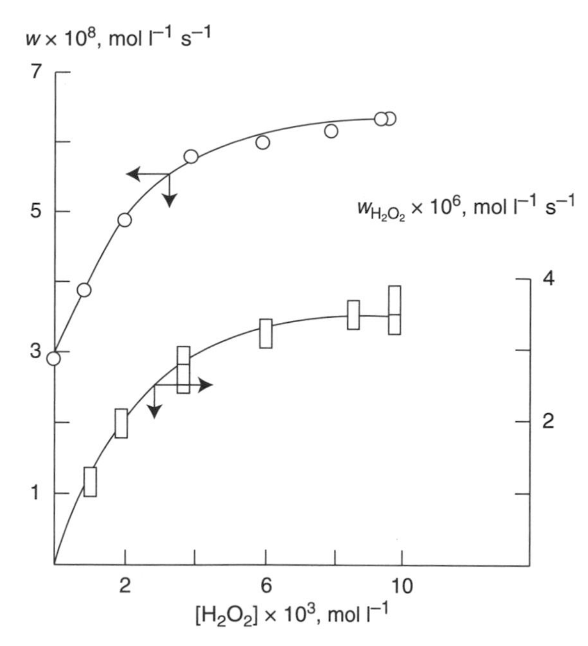

Figure 17.5 Initial rate of methyl viologen oxidation (right scale) and hydrogen peroxide destruction (left scale) as a function of added hydrogen peroxide concentration; pH = 10. (Reproduced with permission from I. N. Martyanov et al., *Kinetics and Catalysis* **1997**, *38*(1), 70. Copyright 1997 Pleiades Publishing, Inc.)

surface versus solution concentration dependence is described by the Langmuir isotherm:

$$\nu = k\frac{K[MV^{2+}]}{1 + K[MV^{2+}]} \tag{17.12}$$

where ν is the concentration of surface MV^{2+} per gram of the TiO_2, k and K are constants. If it is assumed that the initial oxidation rate is just dependent on MV^{2+} surface concentration, the dependencies of initial oxidation rates on MV^{2+} solution concentration measured at two different light intensities would be proportional to each other. The experiment results presented in Figure 17.6 show otherwise. The dependence of an initial oxidation rate versus concentration of MV^{2+} in the solution, $[MV^{2+}]$, measured at light intensity of 0.46×10^{-9} Einstein $cm^{-2}\,s^{-1}$ levels off at significantly lower $[MV^{2+}]$ than the similar dependence measured with more intense light, 4.17×10^{-9} Einstein $cm^{-2}\,s^{-1}$. Accordingly, as presented in Figure 17.7, initial oxidation rates measured as a function of light intensity at two different concentrations of MV^{2+}, 5×10^{-5} M and 0.5×10^{-5} M, are not proportional to each other. The set of the dependences shown in Figures 17.6 and 17.7 was measured at pH = 9. However, the general conclusion holds true also for data obtained at pH = 5.6. What could be a reason for that? In other words, what elemental processes could

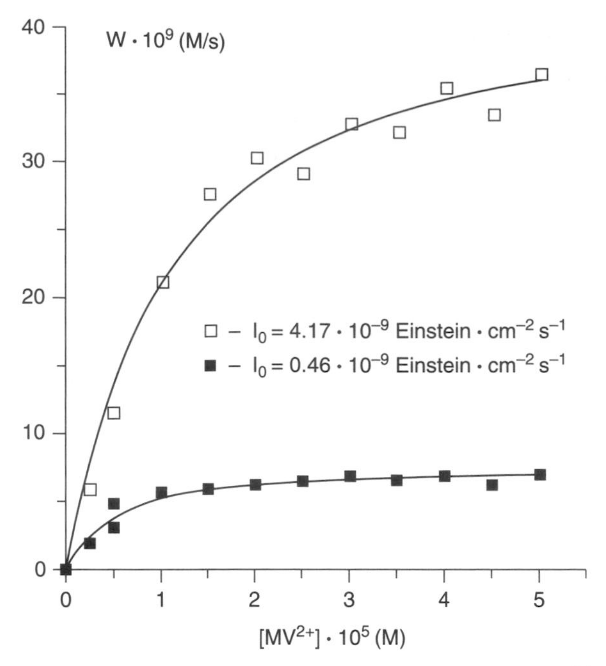

Figure 17.6 Initial rate of methyl viologen oxidation as a function of MV^{2+} concentration at two light intensities. (Reproduced with permission from I. N. Martyanov and E. N. Savinov, *J. Photochem. Photobiol. A: Chem.* **2000**, *134*, 219. Copyright 2000 Elsevier.)

lead to such experimentally observed behavior? The full answer to this question lies in building a mathematical model described elsewhere (4). Here we will focus on the most important qualitative conclusions. For further consideration, it is worth presenting the dependences of initial reaction rate versus light intensity, as photonic efficiency versus intensity of the light. The following is the formula that can be used for this purpose:

$$\varphi = \frac{W \cdot V}{I_0 S_0} \tag{17.13}$$

where φ is photonic efficiency, W the reaction rate, V the volume of the reactor, I_0 the light intensity falling on the reactor, S_0 a cross-section of the light beam. The photonic efficiency is reminiscent of quantum yield. For quantum yield measurements, however, absorbed light has to be used instead. As we saw earlier, scattering creates a great uncertainty leaving quantum yield measurements for specially designed experiments.

Photonic yields as a function of light intensity calculated according to Equation (17.13) are also shown in Figure 17.7. It is important to recognize a general behavior—both photonic yields and quantum yields go up as light intensity diminishes. Since the photonic efficiency by its definition reflects how efficiently photons of

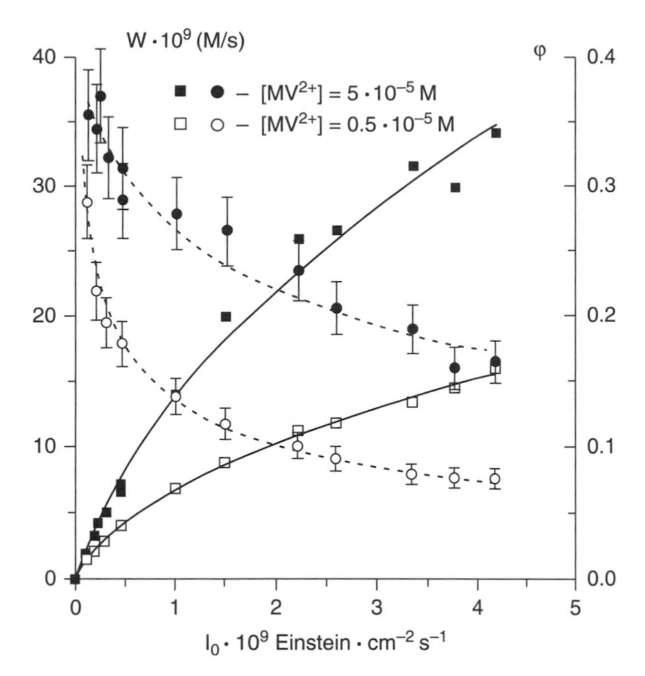

Figure 17.7 Initial oxidation rate of methyl viologen and initial photonic efficiency versus light intensity at two concentrations of MV^{2+}. (Reproduced with permission from I. N. Martyanov and E. N. Savinov, *J. Photochem. Photobiol. A: Chem.* **2000**, *134*, 219. Copyright 2000 Elsevier.)

light are used, it can be concluded that a rise in light intensity results in less effective use of the photons.

A qualitative answer to the increase in photonic efficiency with the light intensity reduction may be inferred from the following consideration. Absorption of a photon with an energy exceeding the bandgap of TiO_2 triggers a sequence of events. This sequence of events ends with desorption of newly formed reaction products from the surface of TiO_2 to the solution. A rise in light intensity increases the number of the sequences of events per second. At some point the second sequence of events begins to start before the first one has been completed, creating interference with each other and causing the photonic yield to drop.

To get a hint about the steps that likely limit the photonic efficiency, it is worthwhile to give a rough estimate of the time scale for individual steps and compare them with an average time between photon absorption. For nanoparticulate catalysts, the time between adsorption of photons can be roughly estimated with the following formula:

$$\tau = 1/[I \cdot d^3 \cdot \alpha(\lambda) \cdot \ln 10]$$

where I is the light intensity, d is a particle size, $\alpha(\lambda)$ is the coefficient of light transmission—a characteristic for a particular material function of wavelength.

Taking $\alpha(\lambda = 334\ \text{nm}) \approx 6 \times 10^4\ \text{cm}^{-1}$ and d to be roughly 30 nm, light intensity 0.46×10^{-9} Einstein $\text{cm}^{-2}\,\text{s}^{-1}$, the time between absorption of photons can be calculated to be roughly $\tau \approx 10^{-3}$ s.

To observe the photonic efficiency drop, the sequence of the events triggered by a photon should take longer than 10^{-3} s. But what step could it be? Is it adsorption, desorption, localization of photogenerated electron/hole, interface charge transfer, or something else? This question is not easy to answer. In fact there is ongoing debate in the scientific literature, with different research groups arguing their point of view.

The limited rate of adsorption of an organic substrate and oxygen can be considered as a rate limiting factor. Indeed, if concentration of the organic substrate is low enough and the hydroxyl radicals frequently generated have nothing to oxidize on the TiO_2 surface, they will degrade in some form of recombination process lowering photonic efficiency. The rate of adsorption as applicable to MV^{2+} under current experimental conditions, however, is easy to estimate with the equation:

$$\tau_{ad} \approx \frac{MV^{2+}_{ad,m}}{2\pi d\chi[MV^{2+}]}, \tag{17.14}$$

where $MV^{2+}_{ad,m}$ is the maximum number of methyl viologen molecules adsorbed on a TiO_2 particle, $\chi \approx 10^{-5}\ \text{cm}^2/\text{s}$ is the diffusion coefficient, $[MV^{2+}]$ is the solution concentration of MV^{2+}. Calculation of $MV^{2+}_{ad,m}$ from previously presented methyl viologen adsorption data leads to the typical adsorption time $\tau_{ad} \approx 10^{-5}$ s. This time is much shorter than $\tau \approx 10^{-3}$ s and therefore the limiting adsorption rate is an unlikely factor that causes a drop in the photonic efficiency.

It is possible to imagine that desorption is the limiting factor. For example, if the products of partial oxidation remained strongly adsorbed on the surface of TiO_2 they would block the surface of a photocatalyst from newly introduced reagents. This, however, is unlikely to be the case for methyl viologen since the products of partial oxidation of MV^{2+} are quite similar to the original compound. One might expect similar adsorption isotherms as for MV^{2+} and the same typical time for achieving adsorption–desorption equilibrium.

What could be other steps that make the sequence of events longer than 10^{-3} s? It cannot be localization of electrons or holes, since these steps take place on the nanosecond time scale. An interfacial charge transfer of the photogenerated electron to oxygen might very well be much longer than 10^{-3} s and rightfully assumed to be the slowest step in the sequence. In our particular case, however, the process should be mediated by methyl viologen:

$$Ti^{3+} + MV^{2+} \longrightarrow Ti^{4+} + MV^{\bullet +}$$

$$MV^{\bullet +} + O_2 \longrightarrow MV^{2+} + O_2^{\bullet -}$$

and proceed within microseconds. Fast removal of the electron from the TiO_2 particle removes the major reaction (recombination) channel for the hydroxyl radical. Now it may live longer and, depending on the concentration of MV^{2+}, likely beyond 10^{-3} s. If the hydroxyl radical can survive that long, it can wait until a newly absorbed photon creates a new electron–hole pair. This can cause recombination with reduced species or with a newly formed hydroxyl radical giving hydrogen peroxide [Equation (17.8)]. The latter assumption was embedded in the scheme, which was found to fit experimental results very well (4). Consistency of the experimental results and the model prediction as well as above-mentioned consideration makes us suggest hydroxyl radical recombination to be the main channel causing complex system behavior.

17.4 PHOTOCATALYSIS AT TiO_2/GAS PHASE INTERPHASE

17.4.1 TiO_2 Films on Quartz

As we saw in the previous sections, fundamental studies of photocatalytic processes in aqueous phases are impeded by uneven light intensity on TiO_2 nanoparticles. This generally holds true for the studies of gas phase processes, when a thick layer of a powdered photocatalyst is employed. One of the possible solutions to this type of problem can be the use of a thin layer of TiO_2 as a model system for gas phase photocatalytic reaction studies. An ideal sample may consist of just one monolayer of TiO_2 particles on flat quartz slides (5). Such samples were prepared by the sol-gel technique and studied from a number of complementary perspectives. In particular, it allowed the collection of a large body of information, including the morphology and sizes of TiO_2 particles, their UV-Vis spectra, crystal structure, as well as their photocatalytic activities in the reaction of acetaldehyde oxidation. Apart from being a great model system, TiO_2 films are quite interesting from a practical point of view. TiO_2 is a very robust material, which can be used as anticorrosive coatings for steel products. TiO_2 films also demonstrate an interesting phenomenon of UV light-induced superhydrophilicity. Exposure to UV light causes a "magical" change in the appearance of foggy glass. Foggy glass becomes clear—a change coming from spreading tiny drops of water into a thin layer. This phenomenon gives rise to a number of related commercial products, such as antifog mirrors, windshields, window glasses, etc.

AFM images of monolayers of TiO_2 on quartz are presented in Figure 17.8. At all calcination temperatures, films prepared by the sol-gel method have a particulate structure. Despite a very high melting point characteristic for the bulk rutile form of TiO_2 (1840°C), the nanoparticles on SiO_2 become mobile and grow at temperatures well below 500°C. The deposition of the TiO_2 particles on SiO_2 alters evolution of their crystal structure with temperature. At low calcination temperatures (70°C and 200°C) nanoparticles on SiO_2, as well as those prepared from the same sol-gel solution, nanosized TiO_2 powders remain amorphous. At 500°C, TiO_2 particles in both samples crystallize in anatase form. Surprisingly, however, TiO_2 on SiO_2 remains in anatase form even after treatment at 800°C (Fig. 17.9). This observation is in sharp contrast with x-ray diffraction data for powdered TiO_2 samples, for which calcination at 800°C results in a compete conversion of anatase into rutile. Stabilization of anatase crystalline structure of TiO_2 by a SiO_2 support is an advantage for photocatalytic

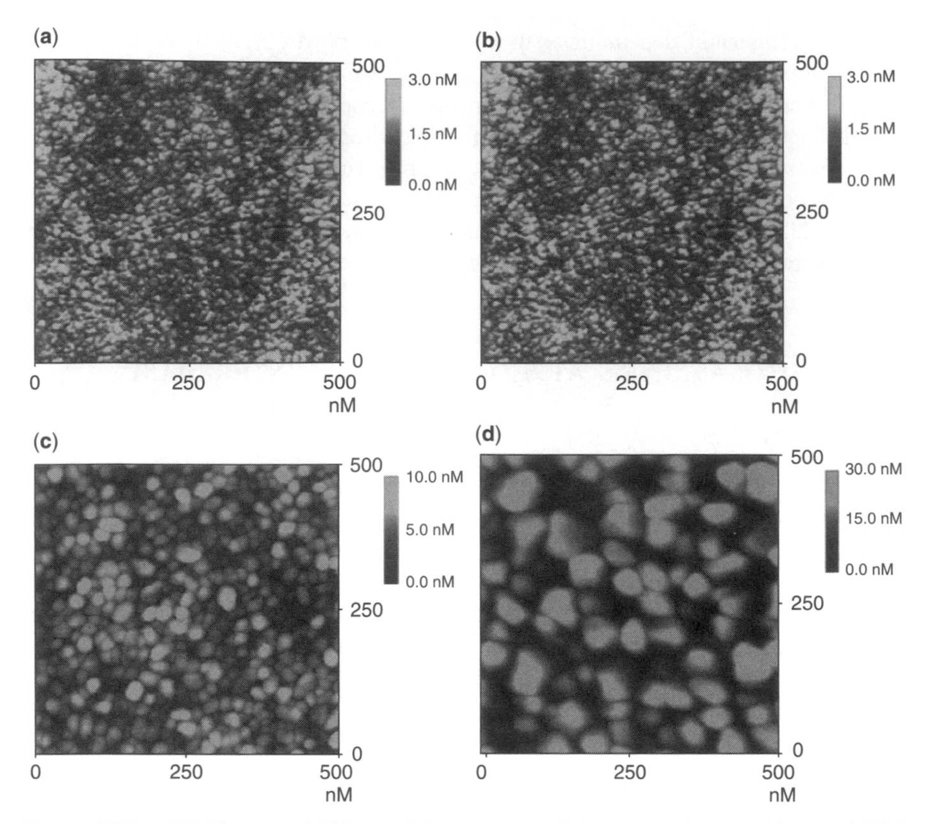

Figure 17.8 AFM images of TiO_2 particles on quartz slides annealed at (a) 70°C, (b) 200°C, (c) 500°C, (d) 800°C. (Reproduced with permission from I. N. Martyanov and K. J. Klabunde, *J. Catal.* **2004**, *225*, 408. Copyright 2004 Elsevier.)

application of TiO_2/SiO_2 samples. Indeed, anatase, a less dense form than rutile TiO_2, is believed to be more photocatalytically active than the thermodynamically more stable rutile form. Increase in calcination temperature though increasing particle size leads to the reduction of the number of internal defects, thus making TiO_2 films progressively more photocatalytically active:

$$TiO_2/SiO_2\ (70^\circ C) < TiO_2/SiO_2\ (200^\circ C) < TiO_2/SiO_2\ (500^\circ C) < TiO_2/SiO_2\ (800^\circ C)$$

17.4.2 Deep Photooxidation of a Mustard Gas Mimic over Nanoparticulate TiO_2

Accumulation of large amounts of chemical warfare agents during the cold war period, aggravated by aged container materials, has created a real threat of severe

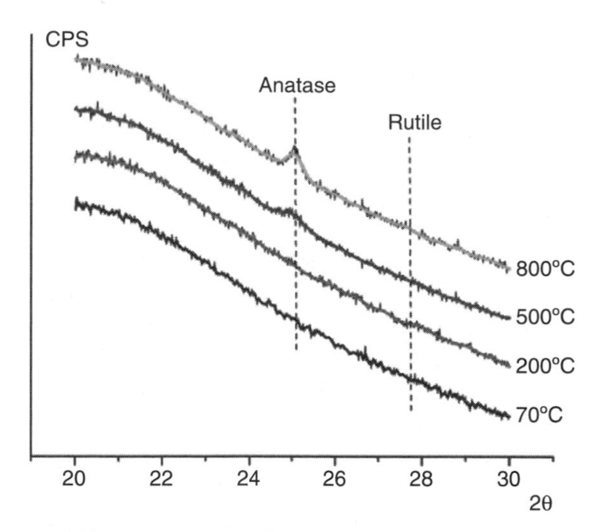

Figure 17.9 X-ray diffraction patterns of TiO_2 films on quartz slides annealed at various temperatures. (Reproduced with permission from I. N. Martyanov and K. J. Klabunde, *J. Catal.* **2004**, *225*, 408. Copyright 2004 Elsevier.)

environmental pollution. The accumulated stockpiles have to be destroyed. Mustard gas, HD, is one of the chemical warfare agents accumulated and stored in large quantities. Sulfur mustard has bis(2-chloroethyl) sulfide ($ClCH_2CH_2SCH_2CH_2Cl$) as a main component. Due to its high toxicity, for laboratory studies bis(2-chloroethyl) sulfide is substituted with 2-choloroethyl sulfide (2-CEES, $ClCH_2CH_2SCH_2CH_3$), which mimics chemical functionalities of the original compound. Encouraging results received with 2-CEES should be later retested with HD, which, however, can be done only in specialized laboratories with all precautions in place.

Potentially there are a number of methods for mustard gas detoxification. Among them are nucleophilic substitution of chlorine moieties with hydroxide group, partial oxidation to the corresponding sulfoxide, or deep oxidation of HD. Photocatalytic oxidation over TiO_2 bears some resemblance to burning but takes place at ambient temperature. Due to the presence of carbon, hydrogen, chlorine, and sulfur, the final products can be expected to be carbon dioxide, water, hydrogen chloride, and sulfuric acid.

As was mentioned above, the formation of toxic intermediates is one of the major problems in application of photocatalytic processes for detoxification purposes. It is certainly true for chemical warfare agents as substrates. Although since chemical warfare agents were designed with the main purpose of creating the most harmful compound possible, formation of even more toxic species is not very likely, the sulfone, which can be formed in a process of HD oxidation, is believed to be as harmful as the original compound.

The process of stepwise conversion of original substance through intermediates into final products can be appreciated through the analysis of evolution of concentrations of volatile compounds in the gas phase in a batch reactor (6). Admission of 2-CEES into the batch reactor containing a certain amount of TiO_2 powder leads to 2-CEES partial adsorption on TiO_2, with the remaining 2-CEES residing in the gas phase. After stabilization of 2-CEES concentration in the gas phase, shining UV light on the TiO_2 results in the partial oxidation of 2-CEES into intermediates, followed by stepwise conversion of the intermediates into final products. Due to the large number of intermediates, an exact path of oxidation of each compound is difficult to follow. It could be reasonably assumed, however, that ethylene (CH_2CH_2) and chloroethylene ($ClCHCH_2$), as well as complementary diethyl sulfide ($CH_3CH_2S_2CH_2CH_3$), 2-chloroethyl sulfide ($ClCH_2CH_2S_2CH_2CH_3$), and bis(2-chloroethyl) disulfide ($ClCH_2CH_2S_2CH_2CH_2Cl$), are primary intermediates, which are stable enough to be detected by GC-MS. The structure of bis(2-chloroethyl) disulfide resembles the structure of mustard gas with the exception that two sulfur atoms of one is present in the molecule. This compound is expected to be of high toxicity. Fortunately, it is the least volatile among the disulfides formed and after its formation, it stays on the surface of TiO_2 until it undergoes further oxidation to more benign products.

Putting aside some ambiguity with nitrogen-containing compounds, the composition of final products in the process of photocatalytic oxidation is determined by elemental composition of the original compound (7). As far as the 2-CEES is concerned, final products such as carbon dioxide, water, hydrogen chloride, and sulfuric acid can be expected. Final products are a matter of considerable concern. Though they are hardly toxic, hydrogen chloride and particularly sulfuric acid tend to accumulate on the surface of TiO_2 in a dissociated form blocking photocatalyst active surface. If untreated, fairly quickly formation of surface sulfates poisons the photocatalyst, which loses its activity towards 2-CEES oxidation. The remedy for such unfortunate behavior could be washing with water, which was found to partially restore the photocatalyst activity. It should be noted here that a nanoparticulate TiO_2 has an obvious and considerable advantage in regard to 2-CEES photooxidation. Being of high surface area, it can detoxify much more 2-CEES than ordinary TiO_2 before regeneration is required.

17.5 CONCLUDING REMARKS AND FUTURE OUTLOOK

Photocatalysis over nanoparticulate materials is a fast developing area of research with direct links to more mature fields, such as material science, photochemistry and photophysics, colloidal and analytical chemistry. Undoubtedly, photocatalysis over nanoparticulate materials has a great potential for an effective use of "free" solar energy to drive selected chemical reactions. Nowadays we can witness emergence of new commercial products on the market: photocatalytic air purifiers, self-cleaning coatings, superhydrophilic glasses, etc. Low quantum yield, poor overlapping of action and solar spectra, formation of toxic products of partial oxidation remain problems. Recently tremendous progress has been made in regard to shifting an action spectrum of TiO_2 into the visible region (8). The solution came in a form of substitution of oxygen in TiO_2 by nonmetals such as nitrogen, carbon, boron, or preparation of defective TiO_2. The possibility of utilization of visible light opens up new exiting perspectives for photocatalytic processes. Such photocatalysts, for example, would be an interesting solution for disinfection of walls in hospitals, where intensity of UV light is negligible, as well as for other indoor and solar light applications. In the long term, the future of photocatalysis will be to a large extent determined by the success of fundamental and applied research conducted at universities and research laboratories. More young scientists are expected to join this field and live up to the task of making new exciting discoveries.

REFERENCES

1. I. N. Martyanov, E. N. Savinov, K. J. Klabunde, *J. Colloid. Interf. Sci.* **2003**, *267*(1), 111.
2. M. R. Hoffman, S. T. Martin, W. Choi, D. W. Bahnemann, *Chem. Rev.* **1995**, *95*, 69.
3. I. N. Martyanov, E. N. Savinov, V. N. Parmon, *Kinetics Catalysis* **1997**, *38*(1), 70.
4. I. N. Martyanov, E. N. Savinov, *J. Photochem. Photobiol. A: Chem.* **2000**, *134*, 219.
5. I. N. Martyanov, K. J. Klabunde, *J. Catal.* **2004**, *225*, 408.
6. I. N. Martyanov, K. J. Klabunde, *Environ. Sci. Technol.* **2003**, *37*(15), 3448.
7. I. K. Konstantinou, T. A. Albanis, *Appl. Catal. B: Eviron.* **2003**, *42*, 319.
8. T. L. Thompson, J. T. Yates, Jr. *Chem. Rev.* **2006**, *106*, 4428.

PROBLEMS

1. Knowing that TiO_2 Degussa P25 has a surface area around 50 m^2/g and primary particle size of 30 nm, derive an approximate formula for estimation of a typical particle size in Hombicat TiO_2 with the surface area of about 300 m^2/g, and a typical particle size of TiO_2 particles in NanoActive TiO_2 from NanoScale Corp. with surface area of approximately 500 m^2/g.

2. Give a rough estimate of the particle size that would never precipitate from an aqueous suspension of TiO_2 left on a shelf.

3. What is an absorption cross-section for a TiO_2 particle 30 nm in size and wavelength of 334 nm? How much is the absorption cross-section in comparison to a geometrical cross-section of the same particle?

4. Give a rough estimate of the time between absorption of photons by TiO_2 particles. Assume that the TiO_2 particle is 30 nm in size, light is monochromatic with a wavelength of 334 nm and light intensity of 0.46×10^{-9} Einstein $cm^{-2}\,s^{-1}$. Give the estimates for 5 nm and 3 nm TiO_2 particles.

5. Predict a pH dependence of an initial oxidation rate of cationic and anionic organic compounds in aqueous suspensions of TiO_2.

6. What final products of photocatalytic oxidation over TiO_2 can you expect for the following substrates: phenol, trichloroethylene, thiophene, sarin?

7. Photocatalytic reactions involve reduction and oxidation redox processes taking place simultaneously on the surface of TiO_2. Suggest two closely related chemical compounds that can be used to selectively follow reduction and oxidation photocatalytic processes.

8. An onset of optical adsorption for semiconductors, large bandgap semiconductors, TiO_2, and isolators is determined by the bandgap of the material. Derive a formula for the conversion of an onset of the absorption expressed in nanometers into the bandgap energy expressed in electron volts. Silicon-based solar cells are widely used for conversion of the solar energy in electricity. On the basis of data available in the literature on the bandgap energy of silicon, predict the optical absorption range of these cells.

9. Suggest additives apart from hydrogen peroxide that can accelerate photocatalytic oxidation of organic compounds in aqueous suspensions of TiO_2.

10. One of the main problems in synthesis of nanoparticles is their sintering on calcination. Suggest a support material for TiO_2 that is most suitable for application in photocatalytic processes.

ANSWERS

1. Assuming a cubic shape of TiO_2 particles, the surface area of the material can be calculated as a ratio of the surface area of an individual particle to its weight:

$$S = \frac{6d^2}{d^3\rho} = \frac{6}{d\rho}$$

where ρ is the density of the material ($3.9\ \mathrm{g/cm^3}$ for anatase TiO_2), d is the size of the cube. Plugging in the surface area for TiO_2 Degussa P25, one can get d to be around 30 nm. For a different type of TiO_2 we have the same formula. Finding a ratio leads to

$$d = d_{Deg}\frac{S_{Deg}}{S}$$

where d_{Deg} and S_{Deg} are the particle size and the surface area of TiO_2 Degussa P25, and S is the surface area of the sample under study. For $300\ \mathrm{m^2/g}$ Hombicat TiO_2 one can expect an average particle size of approximately 5 nm; for $500\ \mathrm{m^2/g}$ NanoActive TiO_2, particle sizes should be around 3 nm. Of course, this is just a rough estimate. A more thorough approach would consist of measuring individual particle sizes with TEM, and finding an average as well as a particle size distribution. This is, however, a time-consuming approach, which is often difficult to follow due to strong particle agglomeration.

2. Translational energy of individual particles in the solution for our purpose can be taken as being close kT, where k is the Boltzmann constant ($k = 1.38 \times 10^{-23}$ J K^{-1}), T is the absolute temperature in Kelvins. Due to finite temperature, the particles are in a continuous motion trying to become evenly distributed in the vessel. The force of gravity, on other hand, tries to pull the particles to the bottom. If a particle moves from the top of the vessel to the bottom, it loses potential energy equal to $ghd^3(\rho - \rho_w)$, where ρ, ρ_w are the densities of the particle and water, d is the particle size, h is the height of the vessel, and g is the normal acceleration due to gravity ($g = 9.8\,\text{m s}^{-2}$). For a particle small enough, the energy of motion will be equal to the potential energy loss due to particle precipitation: $kT = ghd^3(\rho - \rho_w)$, leading to the following estimation of particle size that would never precipitate from the solution:

$$d = \sqrt[3]{\frac{kT}{gh(\rho - \rho_w)}}$$

Taking ρ to be the density typical for anatase, 3.9 g cm^{-3}, h to be 10 cm, T to be 300 K, one can get a particle size of TiO_2 close to about 10 nm.

3. Neglecting scattering, for particles small enough, photoadsorption cross-section, σ, of a photocatalyst particle can be estimated as:

$$\sigma = d^3 \alpha(\lambda) \ln 10$$

where d is the photocatalyst particle size, $\alpha(\lambda)$ is a characteristic for a certain material coefficient of light transmission. Assuming $\alpha(\lambda = 334\,\text{nm}) \approx 6 \times 10^4\,\text{cm}^{-1}$, d is 30 nm, one can find σ to be close to $4 \times 10^{-12}\,\text{cm}^2$. The optical cross-section is lower than the geometrical cross section, $\sigma_g \approx d^2 \approx 9 \times 10^{-12}\,\text{cm}^2$. For 30 nm TiO_2 particles, the above formula can give a rough estimate only. For smaller particles, the results will be much more accurate.

4. The number of photons absorbed by a photocatalyst particle per second is expressed by the formula $\nu = \sigma I$, where σ is photoadsorption cross-section and I is the light intensity. Using σ for a 30 nm TiO_2 particle at $\lambda = 334$ nm estimated in the answer to problem 3 and assuming the light intensity, I, is 0.46×10^{-9} Einstein cm^{-2} s^{-1}, the ν is approximately 10^3 photons/s. Average time between photon absorption is $\tau \approx 1/\nu \approx 10^{-3}$ s. Analogous calculations for 5 nm and 3 nm TiO_2 particles result in an average time between photon absorption to be approximately 0.2 s and 1 s, correspondingly. Supposing for a moment that the rates of all other processes remain the same, the reduction in size of TiO_2 particles should lead to the reduction of recombination of charge originated from different photons and should produce photocatalysts operating with high quantum yield even under high intensities of light.

5. The experimental results (2) support the view that the dependence of the oxidation rate on pH is to a large extent governed by the amounts of organic compounds adsorbed on TiO_2. For cationic compounds, an increase in pH should promote adsorption of the reagent, which in turn is expected to increase its oxidation rate. The opposite is true for anionic compounds, for which a rise in oxidation rate can be expected to occur with pH decrease. It should be noted that variation of ionic strength of the solution and altering the charge of organic molecules due to protonation or dissociation of hydroxyl, amino, or carboxyl groups should be also taken in consideration.

6. Assuming complete mineralization of the organic compounds in photocatalytic oxidation processes, phenol should give only carbon dioxide and water; trichloroethylene should be oxidized into CO_2, H_2O, and HCl; thiophene into CO_2, H_2O, and H_2SO_4 (surface residing sulfate for a gas phase experiment); sarin into CO_2, H_2O, H_3PO_4, and HF. In the latter case, hydrofluoric acid is unlikely to be produced in its pure form, but rather can be expected to react with TiO_2 forming surface fluoride.

7. Carbon tetrachloride, CCl_4, is resistant to oxidation by the photogenerated hole or hydroxyl radical, but susceptible to reduction by the TiO_2 photogenerated electron:

$$CCl_4 + e \longrightarrow {}^{\bullet}CCl_3 + Cl^- \text{ (first step).}$$

Chloroform, on the other hand, can be oxidized by OH radical:

$$CHCl_3 + HO^{\bullet} \longrightarrow {}^{\bullet}CCl_3 + H_2O \text{ (first step).}$$

8. Photon energy, E, is equal to $h\nu$, where h is the Planck constant ($h = 6.6 \times 10^{-34}$ J s), ν is the light frequency. Since $\nu = c/\lambda$, where c is the speed of light (3×10^8 m s^{-1}) and λ is its wavelength, photon energy can be expressed as:

$$E = h\left(\frac{c}{\lambda}\right)$$

Plugging in the value for h and c, and converting energy units into eV, one can get: E (eV) $= 1238/\lambda$ (nm).

The bandgap energy for the anatase form of TiO_2 is close to 3.2 eV which corresponds to the onset of absorption around 390 nm, whereas TiO_2 with crystal structure of rutile (the bandgap is about 3.0 eV) should absorb light with wavelength shorter than 412 nm.

The silicon bandgap is approximately 1.12 eV which implies that silicon-based cells should absorb across all spectra of visible light as well as capture some near IR with wavelength shorter than 1105 nm.

9. Such compounds are expected to be found among oxidizing agents. Commercial Oxone ($2KHSO_5 \cdot KHSO_4 \cdot K_2SO_4$) or sodium persulfate $Na_2S_2O_8$ are two examples.

10. A support material should have low toxicity, high stability in aqueous media, high surface area, and be resistant to oxidation and reduction as well as transparent to UV and Vis light. Carbon-based materials can be of high surface area. Nevertheless, they are not suitable as a support for TiO_2 due to their strong light absorption and concerns with their stability in redox processes. Alumina and silica are typical candidates, with the best results usually obtained with SiO_2.

18

PHOTOFUNCTIONAL ZEOLITES AND MESOPOROUS MATERIALS INCORPORATING SINGLE-SITE HETEROGENEOUS CATALYSTS

Masakazu Anpo, Masaya Matsuoka, and Masato Takeuchi

18.1 Introduction, 606

18.2 The Design of Highly Dispersed Molecular-Sized Titanium Oxide Photocatalysts, 607

- 18.2.1 Highly Dispersed Titanium Oxide Species within Various Oxide Matrices, 607
- 18.2.2 Photocatalytic Reactivity of Titanium Oxide Species Incorporated within Zeolite Frameworks, 609
- 18.2.3 Photocatalytic Reduction of CO_2 by H_2O Using Tetrahedral Titanium Oxide Species Incorporated within Zeolite Frameworks, 611

18.3 Photocatalysis on Mesoporous Materials Containing the V, Mo, or Cr Oxide Single-Site Species, 614

- 18.3.1 Selective Photocatalytic Oxidation of Methane into Methanol on Mesoporous Silica Containing the V Oxide Species (V-MCM-41), 614
- 18.3.2 The Preferential Photocatalytic Oxidation of CO with O_2 in the Presence of Excess Amounts of H_2 on a Mesoporous Silica-Containing Cr Oxide Species (Cr-MCM-41), 617

18.4 Design of the Ag^+/ZSM-5 Catalyst and its Photocatalytic Reactivity for the Decomposition of NO, 621

18.5 Conclusions, 625

References, 625

Nanoscale Materials in Chemistry, Second Edition. Edited by K. J. Klabunde and R. M. Richards

18.1 INTRODUCTION

Environmental pollution and destruction on a global scale, as well as the lack of natural energy resources, have drawn much attention and concern to the vital need for ecologically clean chemical technology, materials, and processes—some of the most urgent challenges facing chemical scientists nowadays. Since the discovery of the photosensitization effect of a TiO_2 electrode on the electrolysis of water by Honda and Fujishima in 1972 (1), pollution-free photocatalysis by TiO_2 semiconductors has received much attention and been widely studied, with the final aim of attaining the efficient conversion of clean solar energy into useful chemical energy (2–9). The effective utilization of clean and unlimited solar energy will lead to promising solutions not only for energy issues resulting from the depletion of natural energy resources but also to address global-scale environmental pollution. With these objectives in mind, the design of photocatalysts that can operate under visible or solar light irradiation is considered essential in establishing a clean and sustainable environment.

Thus far, we have carried out studies on extremely small TiO_2 particles as well as on various titanium oxide-based binary oxides, such as TiO_2/SiO_2, TiO_2/Al_2O_3, and TiO_2/B_2O_3 (10–14). In particular, we have found that nano sized TiO_2 particles of less than 10 nm show significant enhancement in photocatalytic reactivity which can be attributed to the quantum size effect. This phenomenon is due to an electronic modification of the photocatalysts as well as the close existence of the photo-formed electron and hole pairs and their efficient contribution to the photoreaction, resulting in a more enhanced performance over semiconducting TiO_2 powders. These findings have provided us with new insights into the design of more efficient, high performance catalysts such as highly dispersed molecular-sized transition metal oxide photocatalysts. Moreover, the application of an anchoring method enables the preparation of molecular or cluster-sized photocatalysts on various supports, such as SiO_2, Al_2O_3, various zeolites, and mesoporous materials. Highly dispersed titanium, vanadium, chromium, molybdenum oxide single-site species incorporated within the cavities or framework of zeolites are also of great interest due to their unique local structure such as their fourfold coordinated species and efficient photocatalytic properties for the reduction of CO_2 with H_2O, NO decomposition, and the selective photooxidation of alkane or alkene with O_2 when compared with semiconducting photocatalysts (15–20). In addition to these transition metal oxides, transition metal ions such as Cu^+ or Ag^+ ions highly dispersed within zeolite cavities have been reported to show high photocatalytic activity for the direct decomposition of NOx (NO and N_2O) into N_2 and O_2 (19, 20).

In order to prepare such well-defined photocatalytic systems, it is necessary to clarify at the molecular level the chemical features of the photo-formed charge carriers, the reaction intermediate species, as well as the reaction mechanisms using various spectroscopic techniques. These studies, in turn, will necessitate detailed and comprehensive investigations of the photogenerated active sites and the local structures of the photocatalysts (20–27). We have conducted detailed characterizations of potential photocatalysts using various molecular spectroscopies in order to realize two main objectives: (1) A clarification of the dynamics and mechanisms of the photoactive

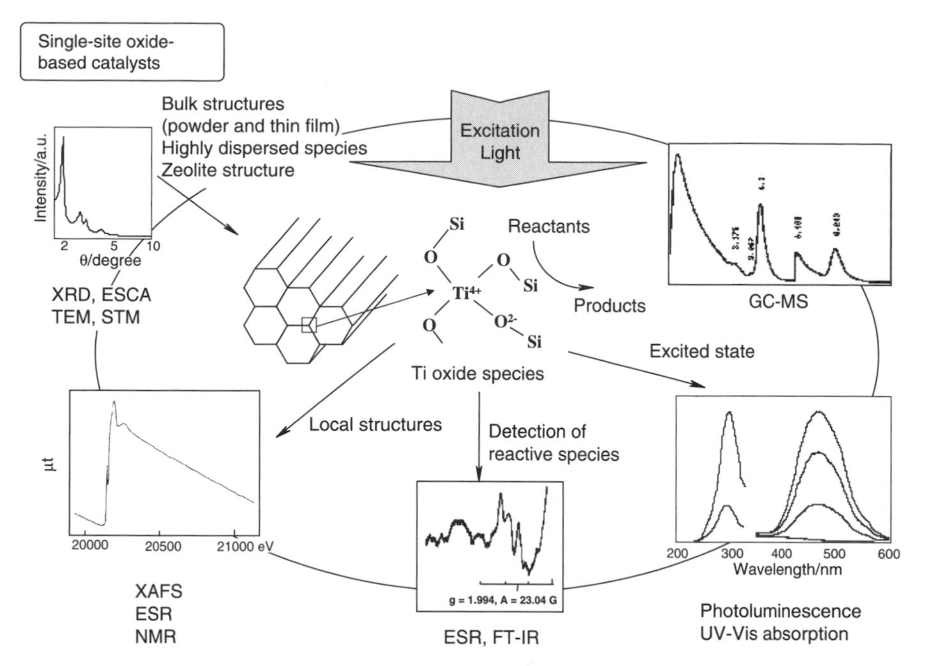

Figure 18.1 Schematic depiction of the relationship between the local structures of the active surface sites and their photocatalytic reactivity at the molecular level.

species and (2) The design of highly efficient, effective, and reactive, photocatalysts. Figure 18.1 shows the outline of our research strategies to achieve these objectives.

In this chapter, the progress being made in the development of highly dispersed transition metal oxides (Ti, V, Cr, Mo) or ions (Ag^+) as single-site heterogeneous photocatalysts within the frameworks or cavities of zeolites and mesoporous materials will be reviewed.

18.2 THE DESIGN OF HIGHLY DISPERSED MOLECULAR-SIZED TITANIUM OXIDE PHOTOCATALYSTS

18.2.1 Highly Dispersed Titanium Oxide Species within Various Oxide Matrices

Highly dispersed titanium oxide species within Al_2O_3 or SiO_2 matrices can be prepared easily by the sol-gel method or a precipitation method (10–14, 28–30). From the X-ray diffraction (XRD) measurements, the particle sizes of the titanium oxide species in the binary oxides become smaller as the Ti content decreases. Figure 18.2 shows the XAFS spectra of Ti/Si binary oxides having different Ti contents. The Ti *K*-edge X-ray absorption near-edge structure (XANES; left) of the binary oxides with lower Ti content of less than 10 wt% exhibited an intense

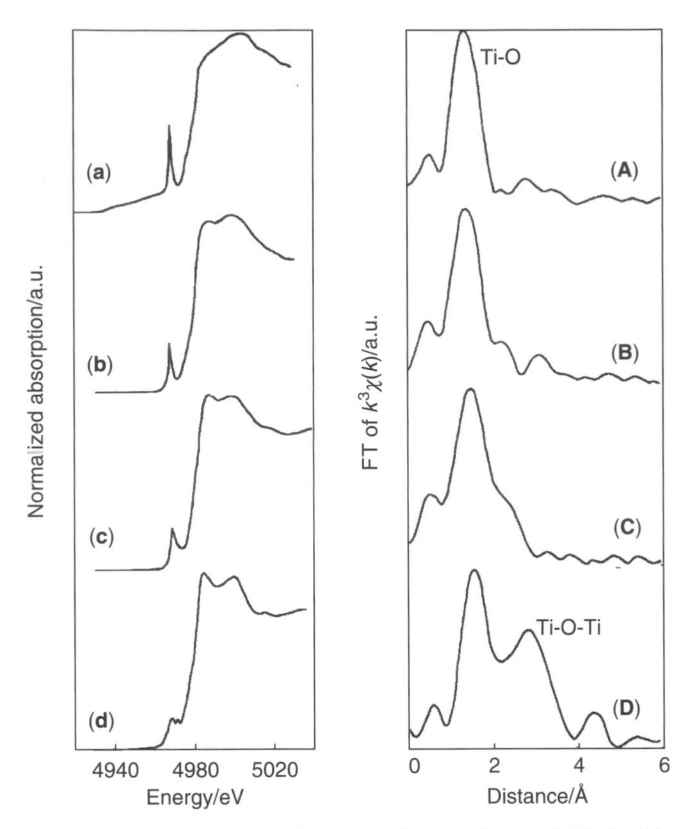

Figure 18.2 Ti *K*-edge XANES (a) to (d) and Fourier transforms of EXAFS (A) to (D) of the Ti/Si binary oxide photocatalysts prepared by the sol-gel method. Ti content (vol%): (a, A) 1; (b, B) 20; (c, C) 50; (d, D) 100.

preedge peak in the XANES region. Curve fitting analysis of EXAFS (right) indicates that these catalysts mainly consist of isolated tetrahedral TiO_4 species having a Ti–O bond distance of approximates 1.83 Å within the SiO_4 matrices. On the other hand, binary oxides having higher Ti content of more than 50 wt% showed three small preedge peaks in the XANES spectrum and a Ti–O–Ti bond at around 3 Å in the Fourier transform of EXAFS, indicating the existence of aggregated TiO_2 nano particles. The highly dispersed titanium oxide species also exhibited unique photoluminescence spectra at around 500 nm upon excitation by UV light irradiation at around 250 nm. As shown in Figure 18.1, this absorption and photoluminescence can be attributed to the charge transfer process and its reverse radiative deactivation from the excited state of the titanium oxide species, respectively. These highly dispersed titanium oxide species show efficient photocatalytic reactivity for various significant reactions, such as the hydrogenolysis reaction of unsaturated hydrocarbons with H_2O, the direct decomposition of NO, and the photoreduction of CO_2 with H_2O.

18.2.2 Photocatalytic Reactivity of Titanium Oxide Species Incorporated within Zeolite Frameworks

The removal of NOx (NO, N_2O, NO_2) from the exhaust of internal combustion engines or industrial boilers is one of the most urgent issues, considering that NOx is a harmful atmospheric pollutant that causes acid rain and photochemical smog. In fact, the direct decomposition of NOx into harmless N_2 and O_2 has been a great challenge for many researchers. To address such concerns, highly dispersed titanium oxides incorporated within the frameworks of zeolites or mesoporous materials can be considered highly promising candidates as effective photocatalysts for the decomposition of NOx (31–36).

As well as the highly dispersed titanium oxide species within Al_2O_3 or SiO_2 matrices, a highly dispersed TiO_4 species incorporated within a zeolite framework also showed unique photocatalytic performance, especially for the direct decomposition of NO. UV light irradiation of these catalysts in the presence of NO at 275 K was found to lead to the effective decomposition of NO to produce N_2 with high selectivity, while the TiO_2 semiconductor photocatalysts were found to decompose NO to produce mainly N_2O. Ti-containing zeolite photocatalysts showed typical UV-vis absorption spectra at around 220 to 250 nm, attributed to the ligand-to-metal charge transfer process of the isolated tetrahedral TiO_4 species. On the other hand, these highly dispersed tetrahedral titanium oxide species showed a unique photoluminescence spectrum at around 450 to 490 nm upon excitation at around 250 nm. Figure 18.3 shows the photoluminescence spectra of the highly dispersed titanium

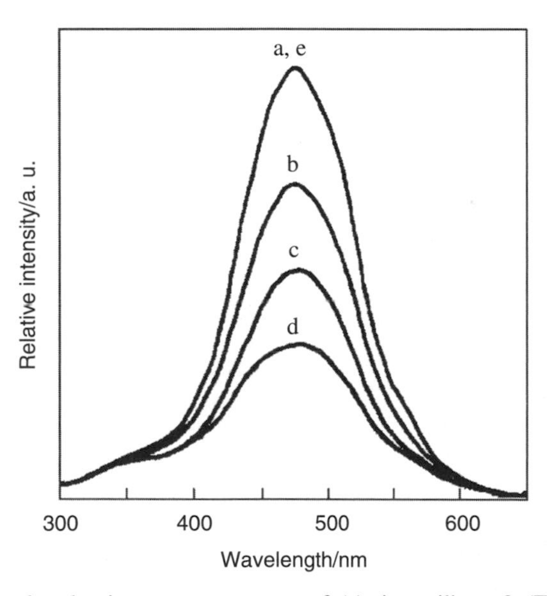

Figure 18.3 The photoluminescence spectrum of (a) titanosilicate-2 (TS-2) catalyst; and (b–d) the effect of the addition of NO measured at 77 K. NO pressure (Torr): (a) 0; (b) 0.05; (c) 0.1; (d) 0.3; (e) evacuation at 295 K after (d).

oxide species incorporated in a zeolite framework (Titanosilicate-2). The addition of NO molecules onto the Ti-containing zeolite catalyst led to a smooth quenching of the photoluminescence as well as a shortening of its lifetime, its extent depending on the amount of NO added. These findings indicate that the tetrahedral titanium oxide species work as active sites accessible to the added NO molecules and, furthermore, the added NO easily interacts with the charge transfer excited state, that is, the $[Ti^{3+}\text{-}O^{-}]^{*}$ electron-hole pairs of the tetrahedral TiO_4 species. Moreover, Ti *K*-edge XAFS measurements revealed that the Ti-containing zeolite photocatalysts include a highly dispersed tetrahedral TiO_4 unit having a Ti–O bond distance of about 1.83 Å within their zeolite framework.

As shown in Scheme 18.1, the reaction mechanism for the photocatalytic direct decomposition of NO over the isolated tetrahedral titanium oxide species can be proposed, that is, two NO molecules are able to adsorb onto these oxide species as weak ligands to form reaction precursors. Under UV irradiation, the charge-transfer excited complexes of the oxides $[Ti^{3+}\text{-}O^{-}]^{*}$ are formed. Within their lifetimes, the electron transfers from the Ti^{3+} site, on which the photo-formed electrons are trapped, into the anti-π^{*}-bonding orbital of the NO molecule, and the electron transfers simultaneously from the π-bonding orbital of another NO molecule into the O^{-} site, where

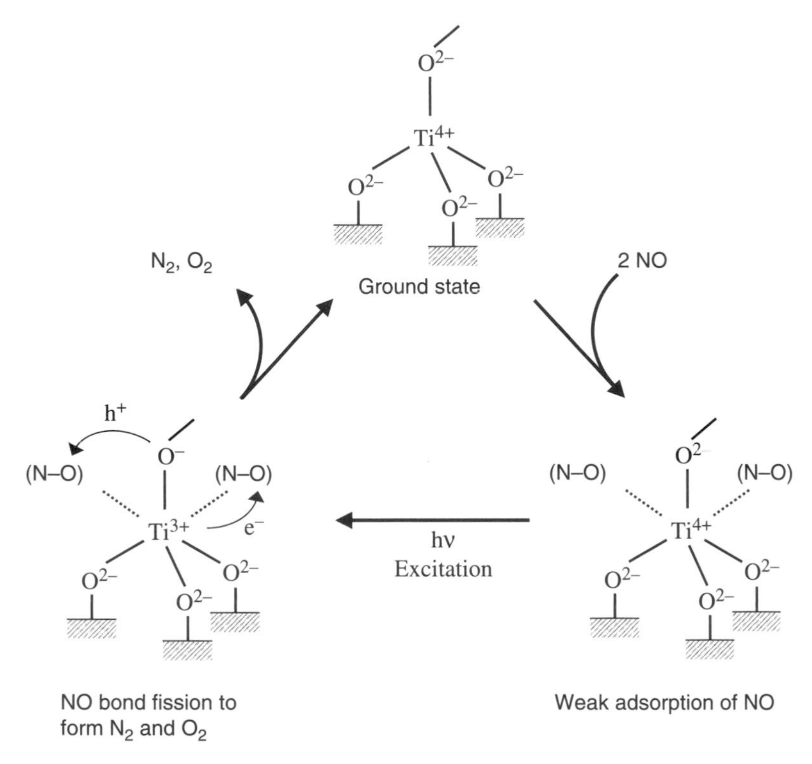

Scheme 18.1 The reaction mechanism of the photocatalytic decomposition of NO into N_2 and O_2 on the tetrahedrally coordinated Ti oxide species under UV light irradiation.

the photo-formed holes are trapped. These simultaneous electron transfers lead to the direct decomposition of two sets of NO into N_2 and O_2 over the tetrahedral titanium oxide photocatalyst under UV light irradiation. These results clearly demonstrate that zeolites and mesoporous materials used as supports enable the anchoring of the tetrahedral TiO_4 species within their frameworks and/or cavities in a highly dispersed state. Such tetrahedral titanium oxide photocatalysts are, thus, promising candidates for unique applications in the reduction of toxic NOx.

18.2.3 Photocatalytic Reduction of CO_2 by H_2O Using Tetrahedral Titanium Oxide Species Incorporated within Zeolite Frameworks

The development of efficient photocatalytic reduction systems of CO_2 with H_2O into chemically valuable compounds such as CH_3OH or CH_4 is a challenging goal in the development of environmentally friendly catalysts. We have found that highly dispersed titanium oxide species within zeolite frameworks, when compared with bulk semiconducting TiO_2 catalysts, exhibit high and unique photocatalytic reactivity for the photoreduction of CO_2 with H_2O (37–45). In this section, special attention has been focused on the preparation of Ti-containing zeolites and their application in the photocatalytic reduction of CO_2.

Titanium oxides incorporated within various zeolites or SiO_2 frameworks contain isolated tetrahedral titanium oxide species having a Ti–O bond distance of about 1.83 Å. These highly dispersed Ti-containing zeolites exhibited a photoluminescence spectrum at around 480 to 490 nm by excitation at around 220 to 260 nm. The addition of H_2O or CO_2 molecules onto the Ti/zeolite catalysts led to the quenching of the photoluminescence as well as shortening of the lifetime of the charge transfer excited state, its extent depending on the amount of gases added. Such efficient quenching of the photoluminescence with H_2O or CO_2 suggests not only that the tetrahedral titanium oxide species locate at positions accessible to these small molecules but also that they interact with these titanium oxides in their excited states. In fact, UV irradiation of the Ti/zeolite catalysts in the presence of CO_2 and H_2O catalyzed the photocatalytic reduction of CO_2 to form CH_3OH and CH_4 as major products, as well as CO, O_2, C_2H_4, and C_2H_6 as minor products.

The photoreduction of CO_2 with H_2O over the Ti-β zeolites, which were synthesized under hydrothermal conditions using the OH^- and F^- anions of structure-directing agents (SDA), have also been investigated. The detailed preparation procedures of these Ti-β zeolites have been reported previously (45). Ti-β(OH) and Ti-β(F) catalysts showed an absorption band at 200 to 260 nm and photoluminescence spectra at around 480 nm, attributed to the highly dispersed tetrahedral titanium oxide species. The photoluminescence yield of Ti-β(OH) is much higher than that of Ti-β(F) and the yields are related to the concentration of the titanium oxide species in the excited state. As shown in Figure 18.4, the addition of CO_2 and H_2O on these Ti-β zeolites led to a quenching of the photoluminescence at 298 K. The lifetime of the charge-transfer excited state was also observed to be shortened by the addition of the CO_2 and H_2O molecules. Such a quenching of the photoluminescence suggests that the added CO_2 and H_2O interact with the titanium oxide species incorporated

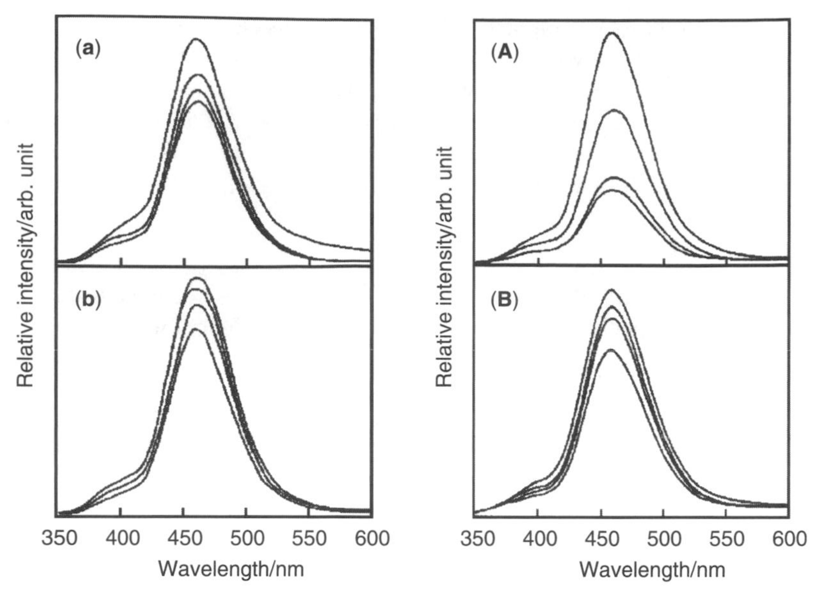

Figure 18.4 Effect of the addition of CO_2 (a) and (b); or H_2O (A) and (B) molecules on the photoluminescence spectra of: (a, A) Ti-β(OH); and (b, B) Ti-β(F) catalysts at 298 K. Excitation: 260 nm. Amount of added (a, b) CO_2 or (A, B) H_2O: (a) 0, 0.05, 0.21, and 1.1 mmol/g; (b) 0, 5.3, 10.5, and 15.8 mmol/g; (A) 0, 0.11, 0.27, and 0.53 mmol/g; (B) 0, 0.27, 0.52, and 1.04 mmol/g, respectively.

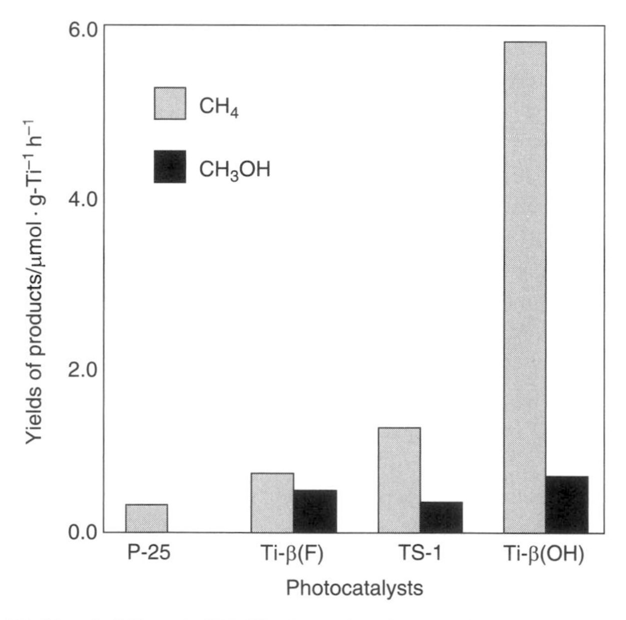

Figure 18.5 Yields of CH_4 and CH_3OH formation in the photocatalytic reduction of CO_2 with H_2O at 323 K on the Ti-β(OH), TS-1, Ti-β(F), and TiO_2 (P-25) catalysts. Intensity of light: 265 μW/cm^2. Reaction time: 6 h.

within the Ti-β zeolite in its excited state. Figure 18.5 shows the yields of the main products in the photocatalytic reduction of CO_2 with H_2O. The Tiβ(OH) catalyst was found to show higher reactivity as compared to other Ti-containing zeolite catalysts. However, the Tiβ(F) catalyst showed the highest selectivity for methanol formation among these Ti-containing zeolites. The higher reactivity of Ti-β(OH) over Ti-β(F) can be explained by the higher concentration of the charge-transfer excited complexes, as observed by photoluminescence measurements. These results clearly indicate that the highly dispersed tetrahedral titanium oxide species exhibit higher selectivity as well as efficiency for methanol formation in the photocatalytic reduction of CO_2 with H_2O as compared with bulk TiO_2 semiconductors.

Figure 18.6 shows the relationship between the coordination number of the titanium oxide species obtained from XAFS measurements and the selectivity for methanol formation in the photocatalytic reduction of CO_2 with H_2O on various Ti-containing zeolites. The clear dependence of the selectivity for methanol formation on the coordination numbers of the titanium oxide species can be observed; that is, the lower the coordination number, the higher the selectivity for methanol formation. TiO_2 semiconducting photocatalysts did not show any reactivity for the formation of CH_3OH from CO_2 and H_2O. From these results, it can be proposed that the highly efficient and selective photocatalytic reduction of CO_2 into CH_3OH by H_2O

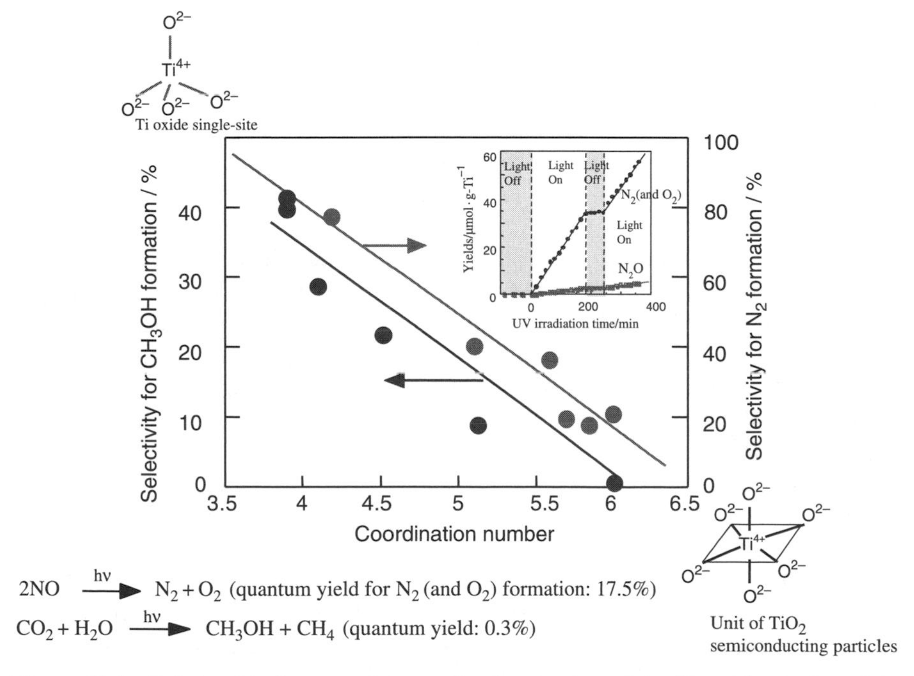

Figure 18.6 Relationship between the coordination number of the Ti oxide species of the Ti/zeolite catalysts and the selectivity for CH_3OH formation in the photocatalytic reduction of CO_2 with H_2O, as well as the selectivity for N_2 (and O_2) formation in the photocatalytic direct decomposition of NO into N_2 and O_2, and the quantum yields for these reactions.

can be achieved using Ti/zeolites involving a highly dispersed tetrahedral titanium oxide species in their frameworks.

18.3 PHOTOCATALYSIS ON MESOPOROUS MATERIALS CONTAINING THE V, Mo, OR Cr OXIDE SINGLE-SITE SPECIES

Mesoporous silica materials containing transition metal oxide species within their frameworks are known to exhibit unique and fascinating properties for applications not only in catalytic but also for various photochemical processes. In fact, V, Mo, or Cr oxide single-site species incorporated within the framework of mesoporous silica such as MCM-41 or FSM-16 exist in a highly dispersed state on a molecular scale and exhibit high photocatalytic activity for such reactions as partial oxidation (19, 20, 27) and the metathesis (19, 20) reaction of hydrocarbons. This section deals with the preparation and characterization of mesoporous silica containing oxide species of V, Mo, or Cr as single-site photocatalysts, as well as their reactivity for various reactions, such as the partial oxidation of methane into methanol by NO or selective oxidation of CO by O_2 in the presence of excess amounts of H_2. Special attention will be focused on the unique reactivity of single-site oxide species in its photoexcited state as well as a detailed elucidation of the reaction mechanism behind the photocatalytic reactions by various in situ spectroscopic techniques.

18.3.1 Selective Photocatalytic Oxidation of Methane into Methanol on Mesoporous Silica Containing the V Oxide Species (V-MCM-41)

Mesoporous silica involving the V oxide single-site species (V-MCM-41) was synthesized under acidic conditions using cetyltrimethylammonium bromide as a template. Tetraethyl orthosilicate (TEOS) and NH_4VO_3 were used as the silicon source and the vanadium ion precursor, respectively. For the basic condition pathway, sodium silicate solution was used as the Si source while NH_4VO_3 was used as the V source. Prior to photocatalytic reactions and spectroscopic measurements, the catalysts were calcined at 773 K and degassed at the same temperature. The results of the XRD patterns and BET surface area show that V-MCM-41 prepared under acidic or basic conditions have a hexagonal lattice with mesopores larger than 20 Å. As shown in Figure 18.7, V-MCM-41 exhibits photoluminescence at around 500 nm under excitation at 300 nm due to the radiative decay process from the charge-transfer excited triplet state of the isolated tetrahedral V^{5+} oxide single-site species (46). The addition of CH_4 or NO molecules onto the catalyst leads to an efficient quenching of the photoluminescence, the extent depending on the amount of added gas, indicating that the CH_4 or NO molecules interact with the excited V oxide species. In fact, UV irradiation of V-MCM-41 in the presence of CH_4 and NO led to an efficient partial oxidation of CH_4 in methanol. Figure 18.8 shows the results of the photocatalytic oxidation of methane on the V-MCM-41 (0.6 wt%) catalyst under UV light irradiation at 295 K. In the presence of O_2, only the complete oxidation of methane into CO_2 and H_2O proceeded; however, the formation of methanol with high selectivity was observed when

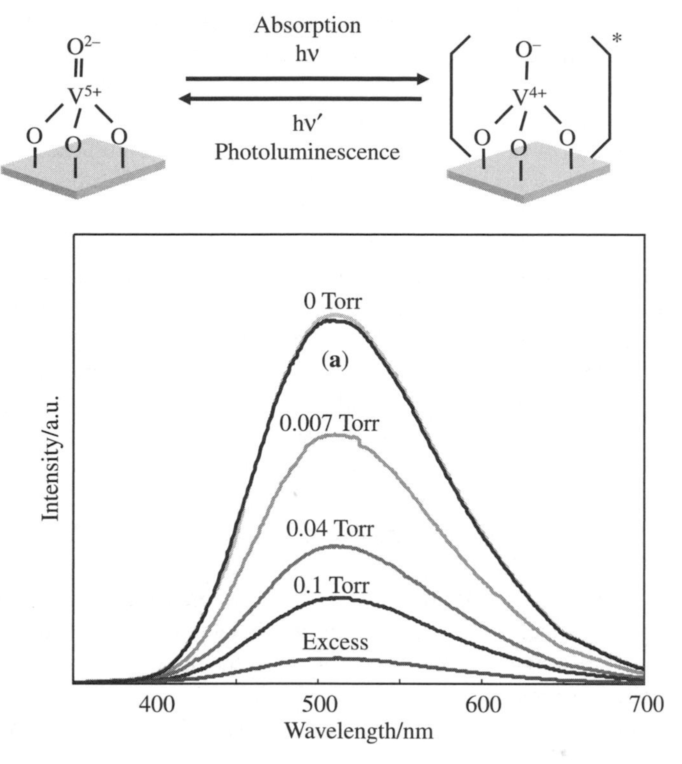

Figure 18.7 Effect of the addition of CH_4 on the photoluminescence spectrum of V-MCM-41 (0.6 wt%) prepared by an acidic pathway. (a) After degassing at room temperature for 30 min.

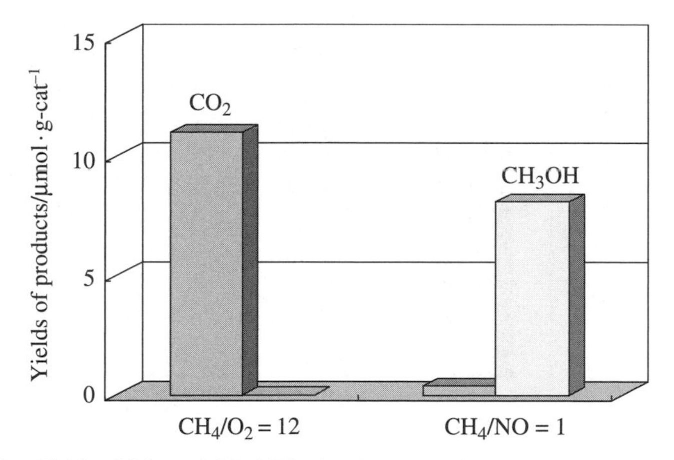

Figure 18.8 Yields of CO_2 and CH_3OH in the photocatalytic partial oxidation of CH_4 with O_2 or NO on V-MCM-41 (0.6 wt%) prepared by an acidic pathway. Amount of reactant: 4 Torr. Reaction time: 3 h.

NO was used as an oxidant. Methane conversion and methanol selectivity reached 6% and 88%, respectively, after UV irradiation for 3 h. As shown in Figure 18.9, the yield of methanol increases with an increase in the V content up to 0.6 wt%, and then decreases with a further increase in the V content. The observed good correspondence between the yield of methanol and the intensity of the photoluminescence indicates that the isolated tetrahedral V^{5+} oxide species act as active sites for the photocatalytic oxidation of methane with NO into methanol.

It was also found that the photocatalytic activity of V-MCM-41 strongly depends on the preparation conditions such as the pH value of the starting solutions and the kind of Si source. A high conversion of methane and high selectivity for methanol formation were observed for V-MCM-41 prepared in acidic solution. On the other hand, for the catalyst prepared in basic solution using sodium silicate as the Si source, almost no methanol formation was observed although methane coupling reaction products and acetaldehyde were observed (46). Moreover, the pre-impregnation of silica with sodium has been reported to strongly diminish the V=O and Mo=O concentrations due to the formation of V oxide monomers with a high degree of tetrahedral symmetry. As a result, the partial oxidation of methane on V and Mo/SiO_2 is strongly inhibited (47, 48). Another reason is that more V oxide species are incorporated into the highly ordered framework of MCM-41 prepared in basic solution (49) and they become inaccessible to reactant molecules such as methane and NO. These results clearly suggest that precise control of the preparation conditions enables the incorporation of highly active single-site V oxide photocatalysts within mesoporous silica materials, thus realizing the selective partial oxidation of CH_4 into methanol using NO as an oxidant.

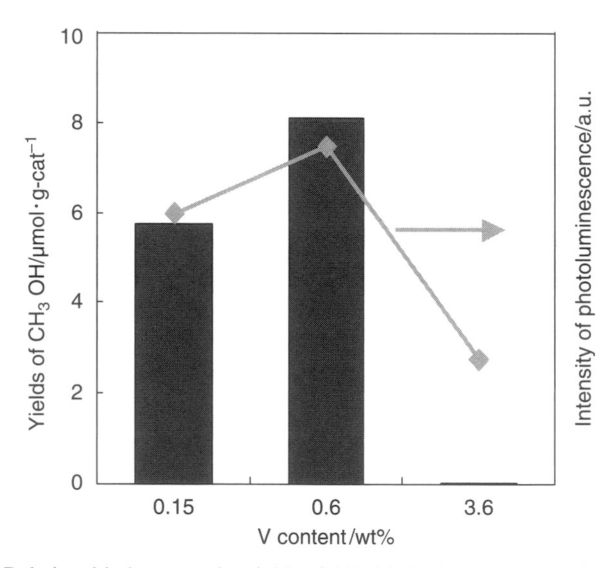

Figure 18.9 Relationship between the yields of CH_3OH in the photocatalytic partial oxidation of CH_4 with NO on V-MCM-41 prepared by an acidic pathway, and the intensity of the photoluminescence spectra.

18.3.2 The Preferential Photocatalytic Oxidation of CO with O_2 in the Presence of Excess Amounts of H_2 on a Mesoporous Silica-Containing Cr Oxide Species (Cr-MCM-41)

Polymer electrolyte fuel cells (PEFC) require a supply of H_2 completely free of CO impurities and the favored method of achieving this is to catalytically oxidize the CO over a supported Pt-rich catalyst, known generally as PROX (preferential oxidation of CO in the presence of H_2). Recent studies have shown that supported Pt-containing, bimetallic nanoparticles such as Pt_5Fe_2 and $PtFe_2$ supported on silica are superior to traditional variants of the Pt/SiO_2 catalysts for effecting the PROX process (50). However, in view of the continuing high cost and ultimate scarcity of Pt, the design of a PROX catalyst for static fuel cell installations that is not only cost efficient but also easily prepared and functions at ambient temperatures is greatly desired. Recently, Cr or Mo oxide species highly dispersed on mesoporous silica (Cr-MCM-41, Mo-MCM-41) have been reported to exhibit photocatalytic activity for the PROX reaction (51–56). This section deals with this PROX reaction on Cr-MCM-41 under visible or solar light irradiation and the mechanism behind the reaction.

The local structure of the Cr oxide species incorporated within an MCM-41 framework structure was investigated by Cr *K*-edge XAFS (XANES and EXAFS) measurements. As shown in Figure 18.10, Cr-MCM-41 exhibits a XANES spectrum with a well-defined preedge peak similar to that of CrO_3, showing that the Cr oxide species exist in a hexavalent (Cr^{6+}) within MCM-41. Furthermore, as shown in Figure 18.10, Fourier transform of EXAFS showed no peak due to the neighboring Cr atoms (Cr-O-Cr) and only exhibited a single peak due to the neighboring oxygen atoms (Cr-O) at approximately 0.8 to 1.5 Å. These results indicate that the Cr^{6+} oxide species exist in a highly dispersed state as a single site. Curve-fitting analysis of the Cr–O peaks revealed that the Cr^{6+} oxide species also exists in a highly distorted tetrahedral coordination with two shorter Cr=O double bonds (bond length: 1.59 Å, coordination number: 2.0) and two longer Cr–O single bonds (bond length: 1.85 Å, coordination number: 2.1) (51, 52). Cr-MCM-41 exhibits three distinct absorption bands below 550 nm due to the ligand to metal charge transfer transition (LMCT: from O^{2-} to Cr^{6+}) of the tetrahedrally coordinated Cr oxide species. Upon excitation at this band, Cr-MCM-41 exhibited a photoluminescence spectrum at around 630 nm at 293 K, as shown in Figure 18.11. The absorption and emission spectra are attributed to the following charge-transfer processes on the Cr=O moieties of the tetrahedral Cr oxide species involving an electron transfer from the O^{2-} to Cr^{6+} ions and a reverse radiative decay from the charge-transfer excited triplet state (51, 52):

$$[Cr^{6+}{=}O^{2-}] \underset{h\nu'}{\overset{h\nu}{\rightleftharpoons}} [Cr^{5+}{-}O^{-}]^{*}$$

The photoluminescence of Cr-MCM-41 was found to be quenched in its intensity by the addition of CO, O_2, and H_2, indicating that the Cr oxide species, in its charge-transfer excited triplet state, easily interacts with CO, O_2, and H_2, as shown in Figure 18.11. In fact, visible light irradiation ($\lambda > 420$ nm) of Cr-MCM-41 in the presence of CO,

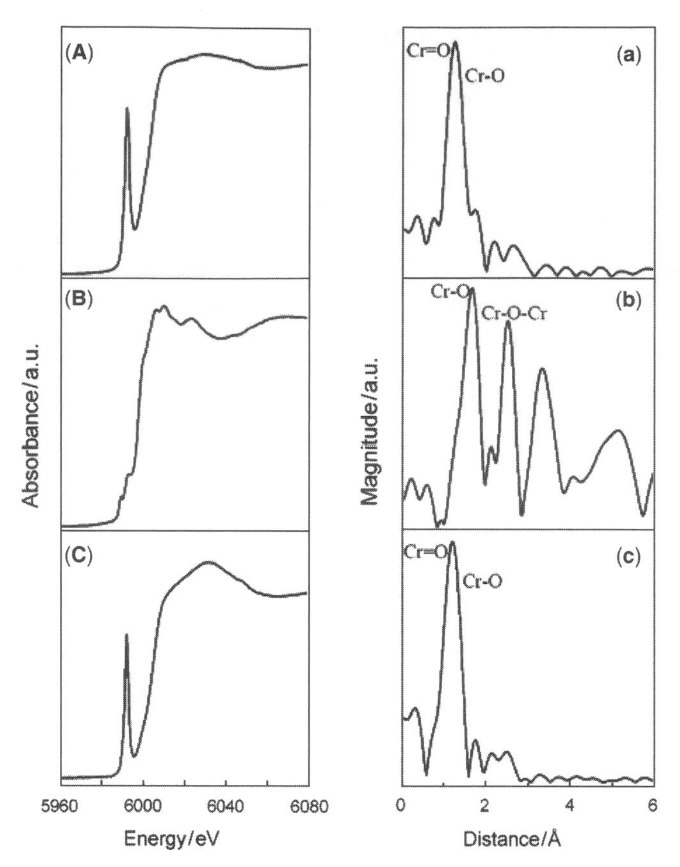

Figure 18.10 (A) to (C) XANES and (a) to (c) Fourier transform of the EXAFS spectra of: (A and a) CrO_3; (B and b) Cr_2O_3; and (C and c) Cr-MCM-41.

O_2, and H_2 led to the efficient oxidation of CO into CO_2, accompanied by the stoichiometric formation and consumption of CO_2 and O_2, respectively, as shown in Figure 18.12. The concentration of the CO gas reached below 8 ppm after light irradiation of 150 min, while the amount of H_2 remained almost constant. CO conversion and selectivity reached $\sim$100% and 97%, respectively, after visible light irradiation of 150 min. The reaction was also found to proceed efficiently even under solar light irradiation (51, 52). After solar light irradiation for 3.5 h (average solar light intensity: 78.5 mW/cm^2), CO conversion and selectivity reached around 100% and 96%, respectively. These results clearly demonstrate that Cr-MCM-41 can be applied for clean and cost-efficient photo-PROX reaction systems using solar light energy at ambient temperatures without the use of Pt or other precious noble metals. Moreover, Mo-MCM-41 incorporating single-site tetrahedral Mo^{6+} oxide species was found to show high activity for the photocatalytic PROX reaction under UV light irradiation (54).

The reactivity of the photoexcited Cr oxide species with reactant gasses (H_2 and CO) was investigated by photoluminescence quenching measurements. The

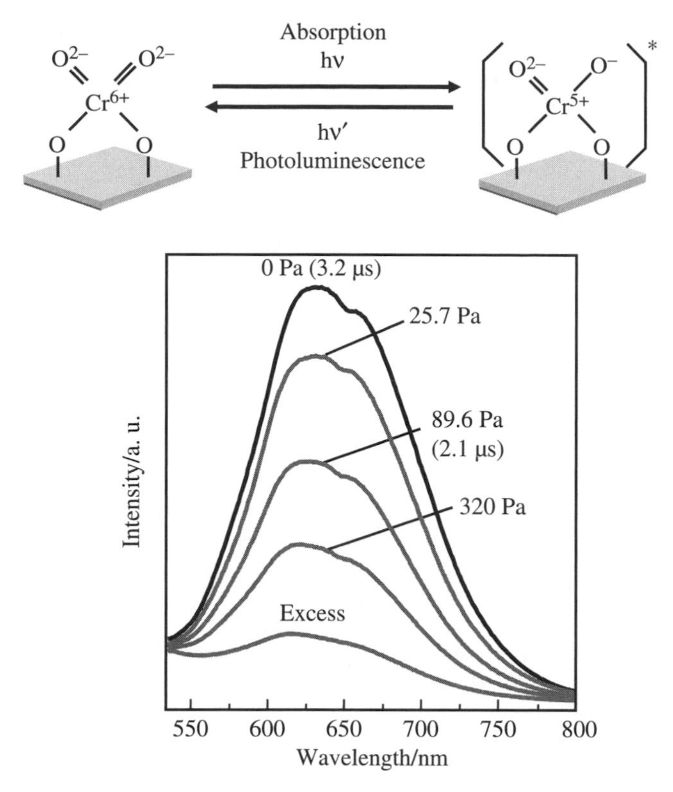

Figure 18.11 Effect of the addition of CO on the photoluminescence spectra of the Cr-MCM-41 catalyst. (Ex = 500 nm, measured at 298 K.)

photophysical processes on Cr-MCM-41 in the presence of the quencher molecules are depicted as follows (51, 52):

$[Cr^{5+}-O^{-}]^*$ → Photoluminescence (k_p)
$[Cr^{5+}-O^{-}]^*$ → Radiationless deactivation (k_d)
$[Cr^{5+}-O^{-}]^*$ → Deactivation by a quencher (k_q)

The value of the absolute quenching rate constants [k_q: (l/mol · sec)] for H_2 and CO were determined to be 8.63×10^5 and 5.91×10^9, respectively. These results show that CO interacts efficiently with the single-site Cr oxide species in its photoexcited state as compared to H_2 (51, 52). The reaction mechanism was also investigated by FT-IR measurements. Visible light irradiation of Cr-MCM-41 in the presence of CO led to the appearance of a typical FT-IR band at 2201 cm^{-1} due to the monocarbonyl Cr^{4+} species [$Cr^{4+}(CO)$], accompanied by the formation of CO_2 (51, 52). The addition of O_2 to these systems under dark conditions led to the complete disappearance of these FT-IR bands. These results clearly suggest that the Cr^{6+} oxide species reacts with CO in its photoexcited state and is reduced into the Cr^{4+} carbonyl species, while these species are easily oxidized by O_2 into the original Cr^{6+} oxide

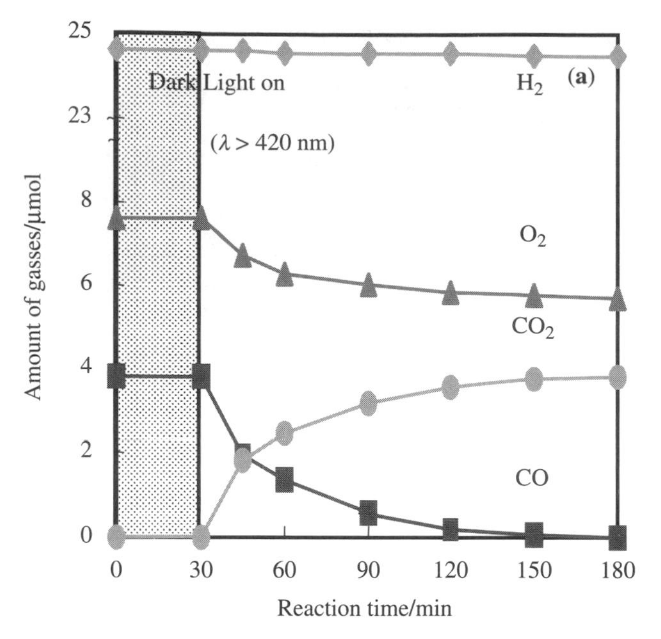

Figure 18.12 Reaction time profiles of the preferential photocatalytic oxidation of CO with O_2 in the presence of H_2 on Cr-MCM-41 under visible light irradiation ($\lambda > 420$ nm) at 293 K.

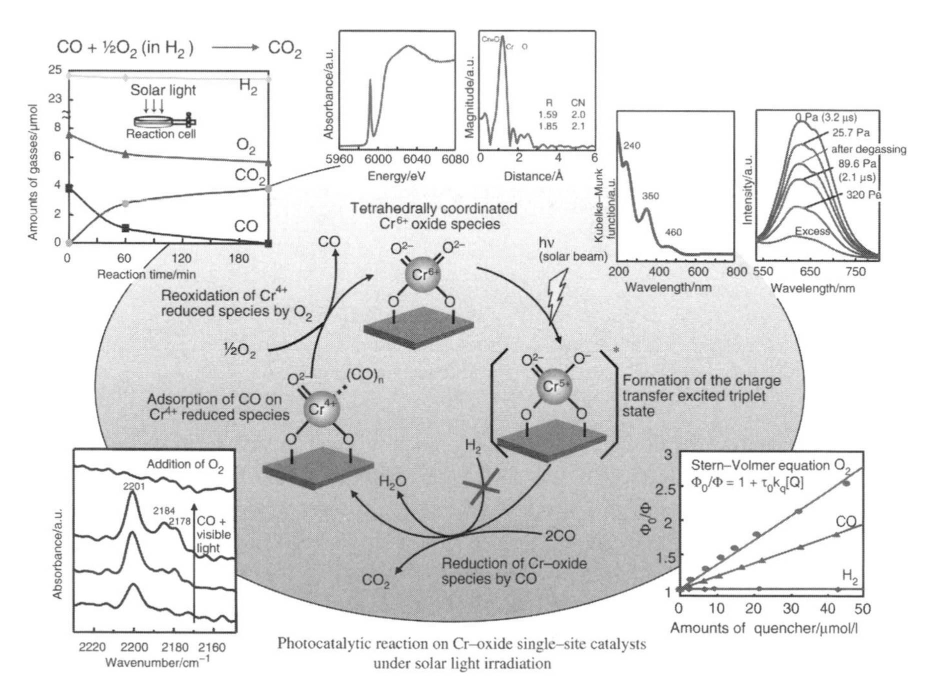

Scheme 18.2 Proposed reaction mechanism in the photocatalytic elimination of CO in excess H_2 (PROX reaction) on Cr-MCM-41 on the basis of various spectroscopic measurements such as UV-vis, EXAFS, photoluminescence, and FT-IR.

species. From these results, the catalytic reaction cycles on Cr-MCM-41 can be proposed, as shown in Scheme 18.2. Initially, the tetrahedral Cr^{6+} oxide species is photoexcited to its charge-transfer excited triplet state and reacts selectively with CO to form CO_2 and a photoreduced Cr^{4+} carbonyl species. The Cr^{4+} oxide species is then efficiently oxidized by O_2 and the original Cr^{6+} oxide species is generated. The high CO selectivity can be attributed to the high and selective reactivity of the photoexcited Cr^{6+} oxide species with CO, as indicated by the high quenching efficiency of CO as compared to H_2. These results clearly show that the unique reactivity of the photoexcited single-site Cr or Mo oxide species enables their practical application to cost-efficient, non-noble metal photo-PROX systems to produce pure H_2 for fuel cells.

18.4 DESIGN OF THE Ag^+/ZSM-5 CATALYST AND ITS PHOTOCATALYTIC REACTIVITY FOR THE DECOMPOSITION OF NO

Catalysis on transition metal ions supported on various oxides or exchanged within zeolites has been studied for applications in efficient photocatalytic reactions (57). It has been reported that Cu^+ ions or Ag^+ ions with a d^{10} electronic configuration and supported on zeolites or silica act as efficient photocatalysts for various reactions such as the decomposition of NOx (NO or N_2O) into N_2 and O_2 under UV irradiation (58–61). In these reactions, the local charge transfer (electron and hole transfer) in the electronic excited state of the Cu^+ or Ag^+ ions with d^9s^1 electronic configuration, that is, thc electron transfer from one *s* electron and hole transfer from one *d* hole, occurs within their lifetimes and a similar charge transfer occurs on semiconducting metal oxides under UV irradiation, playing an important role as the active species in the photocatalysis of metal oxides. This section focuses on characterization studies of the Ag^+ species anchored onto the nanopores of the ZSM-5 zeolite by *in situ* photoluminescence, ESR, XAFS and UV-Vis analyses and their reaction with NO under UV light irradiation.

Ag^+/ZSM-5 was prepared by ion exchange with an aqueous $Ag(NH_3)_2^+$ solution, followed by calcination at 673 K and subsequent evacuation at 473 K. The Ag^0/ZSM-5 catalyst was prepared by heating Ag^+/ZSM-5 at 673 K in the presence of an H_2/H_2O mixture. Figure 18.13 shows the FT-EXAFS spectra of the Ag^+/ZSM-5 catalyst (a′), together with the bulk Ag_2O (b′) and Ag foil (c′) as reference, respectively. The FT-EXAFS spectrum of Ag_2O (b′) and Ag foil (c′) exhibits a peak due to the neighboring Ag atoms at around 3.5 Å (Ag-O-Ag) and 2.5 Å (Ag-Ag), respectively. However, FT-EXAFS spectrum of the Ag^+/ZSM-5 catalyst exhibit only a well-defined peak due to the neighboring oxygen atoms (Ag-O) at around 1.8 Å, suggesting that the silver is anchored within ZSM-5 in an isolated Ag^+ ion as a single site. In fact, the Ag^+/ZSM-5 catalyst exhibits an intense absorption band at around 220 nm which is attributed to the $4d^{10} \rightarrow 4d^95s^1$ electronic transition of the isolated Ag^+ ions (59–61). On the other hand, broad absorption bands due to Ag_n^0 or Ag_m^{n+} clusters were observed for Ag^0/ZSM-5 at wavelengths longer than 250 nm, indicating that reduction and aggregation of the Ag^+ ions occurred.

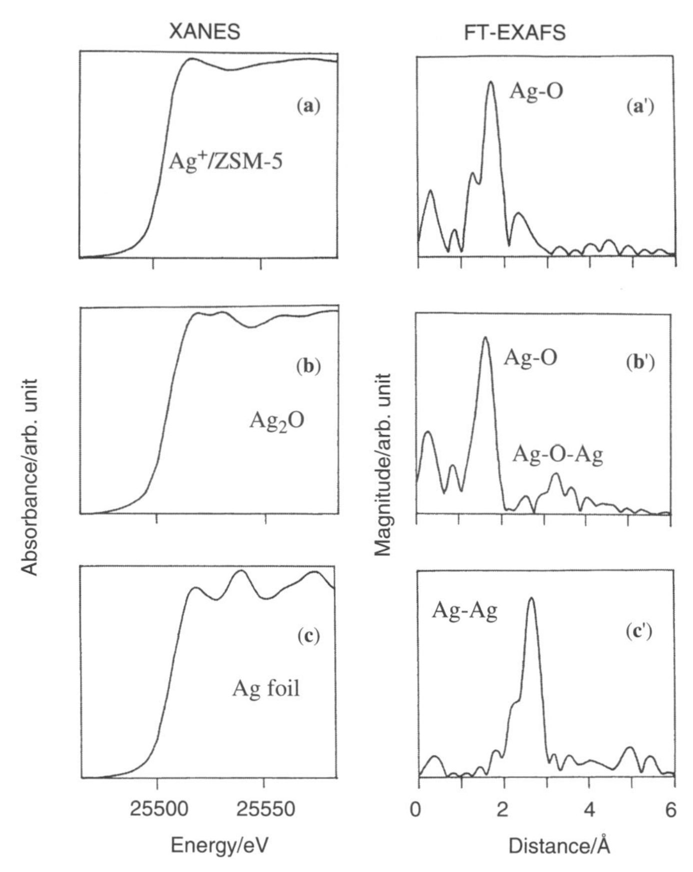

Figure 18.13 XANES (left) and FT-EXAFS (right) spectra of: (a, a′) the Ag^+/ZSM-5 catalyst; (b, b′) Ag_2O powder; and (c, c′) Ag foil.

The addition of NO onto the Ag^+/ZSM-5 catalyst at 77 K led to the appearance of an ESR signal due to the NO adduct species on the Ag^+ ion. The hyperfine splitting of the signal shows the interaction of the electron spin of NO with the nucleus of Ag^+ ($I = 1/2$), indicating that the NO molecules are adsorbed on the Ag^+ ions to form a one-to-one nitrosylic adduct species, that is, $(Ag{-}NO)^+$ (59, 60). UV irradiation of the Ag^+/ZSM-5 catalyst having the $(Ag{-}NO)^+$ adduct species was found to lead to a decrease in the intensity of the ESR signal with the UV irradiation time, while after UV irradiation was discontinued, the signal was found to recover its original intensity. These reversible changes suggest that the $(Ag{-}NO)^+$ adduct species acts as a reaction precursor for the photocatalytic decomposition of NO. Figure 18.14 shows the photoluminescence spectrum (a) and its corresponding excitation spectrum (b) of the Ag^+/ZSM-5 catalyst. A good coincidence of the excitation band position (220 nm) with that of the absorption band due to the isolated Ag^+ ion (220 nm) indicates that these excitation and photoluminescence features can be attributed to the absorption due to the $4d^{10} \rightarrow 4d^9 5s^1$ transition and its reverse radiative deactivation process $4d^9 5s^1 \rightarrow 4d^{10}$, respectively (59, 60). As shown in Figure 18.14, the addition

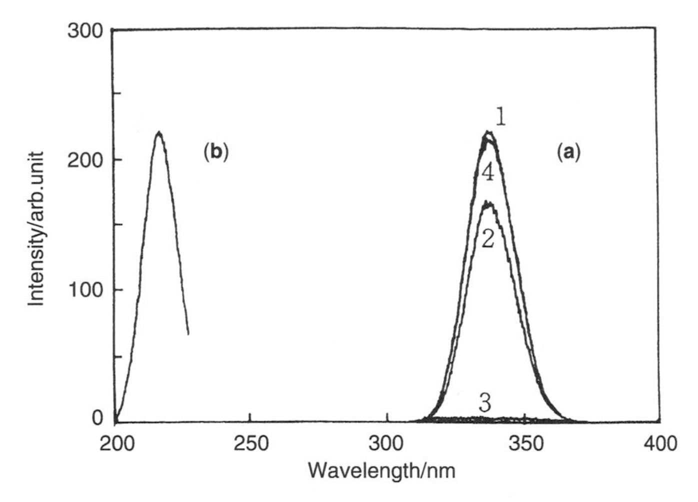

Figure 18.14 The observed ordinary photoluminescence spectrum of: (a − 1) the Ag^{+}/ZSM-5 catalyst; and (b) its excitation spectrum and the effect of the addition of NO on the photoluminescence. The addition of NO was carried out at 298 K. NO pressure (in Torr): 1, 0.0; 2, 0.2; 3, 4.0; 4, after the degassing of NO at 298 K. The excitation spectra were monitored at 340 nm.

of NO leads to an efficient quenching of the photoluminescence and subsequent evacuation of the NO in the system after the complete quenching of photoluminescence leads to a recovery of the photoluminescence to its original intensity level. This means that NO does not reduce the catalysts. The quenching proceeds only through the energy transfer process and not reduction which could be observed with the quenching of the photoluminescence of the V oxide, Cr oxide and Mo oxide single-site catalysts by the addition of CO molecules. The interaction of the NO molecules with the Ag^{+} ions is weak and the added NO easily interacts with the Ag^{+} species in its ground and excited states. The photocatalytic reactivity of Ag^{+}/ZSM-5 for the decomposition of NO was also examined. UV irradiation of Ag^{+}/ZSM-5 catalyst in the presence of NO at 298 K led to the formation of N_2, N_2O, and NO_2, as shown in Figure 18.15. The yields of N_2 and N_2O increased with a good linearity against the irradiation time, suggesting that the reaction is photocatalytic. In fact, after prolonged irradiation, the turnover number (number of N_2 molecules per number of Ag^{+} ions) in the catalyst exceeded 1.0, showing that the reaction proceeds catalytically. Under UV irradiation of Ag^{0}/ZSM-5, the decomposition of NO scarcely proceeded, suggesting that the isolated Ag^{+} ion is responsible for the photocatalytic reaction, while the Ag_n^0 or Ag_m^{n+} clusters are not responsible for the reaction.

From these various findings, as shown in Scheme 18.3, it was concluded that the photoexited electronic state of the highly dispersed Ag^{+} ions ($4d^9 5s^1$) plays a significant role in the photocatalytic decomposition of NO, while an electron transfer from the photoexcited Ag^{+} into the π antibonding molecular orbital of NO initiates the weakening of the N–O bond. At the same time, an electron transfer from the π bonding orbital of another NO to the vacant orbital leads to the weakening of the N–O bond, resulting in the decomposition of NO into N_2, N_2O, and NO_2. The

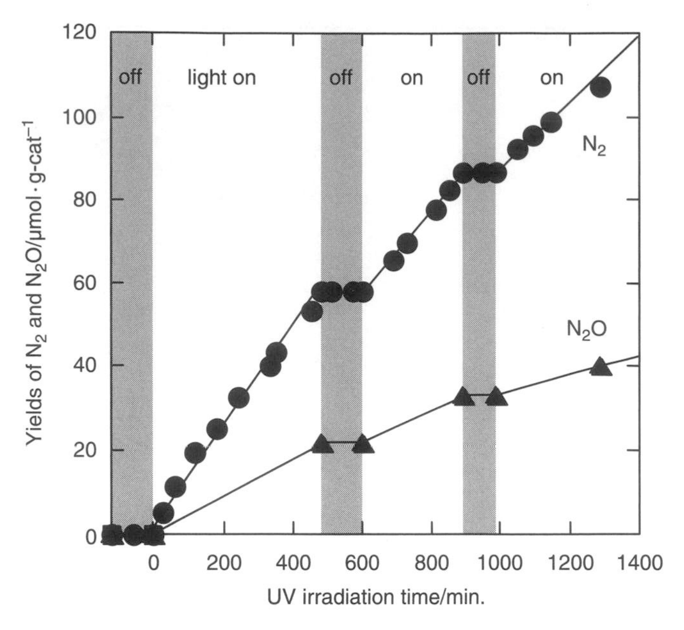

Figure 18.15 The time profiles of the photocatalytic decomposition reaction of NO into N_2 and N_2O on the Ag^+/ZSM-5 catalyst at 298 K.

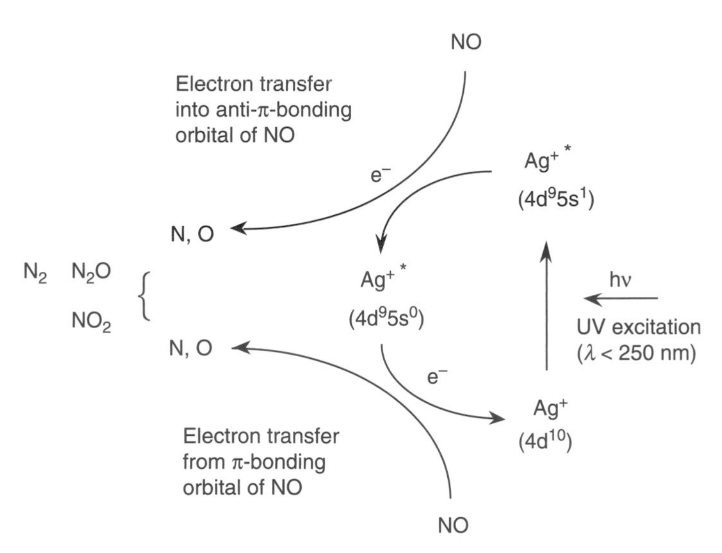

Scheme 18.3 Reaction scheme of the photocatalytic decomposition of NO on the Ag^+/ZSM-5 catalyst at 298 K.

high photocatalytic reactivity of the Ag^+/ZSM-5 catalyst is attributed to the high chemical stability of the Ag^+ ion and the efficient interaction of the excited electronic state of the Ag^+ ion with NO.

18.5 CONCLUSIONS

In this chapter, the local structures of the transition metal oxides, Ti, V, Mo, and Cr oxide single-site species, incorporated within zeolites or mesoporous silica framework structures as well as the local structures of transition metal ions such as Ag^+ exchanged into zeolite cavities were discussed, based on the results of various *in situ* spectroscopic investigations such as ESR, UV-Vis, photoluminescence and XAFS (XANES and EXAFS).

Any of the metal oxide catalysts (Ti, V, Cr, Mo oxide) incorporated within various zeolites or mesoporous materials exists in tetrahedral coordination as a single-site heterogeneous catalyst to act as efficient photocatalysts for the decomposition of NOx (NO, N_2O) into N_2 and O_2, the reduction of CO_2 with H_2O to produce CH_4 and CH_3OH, the preferential oxidation of CO in the presence of H_2 (PROX), and the partial oxidation of various hydrocarbons with O_2 or NO. *In situ* photoluminescence investigations revealed that the efficient and selective interaction of the charge-transfer photoexcited complexes of these oxides, $(Me^{(n-1)+}-O^-)^*$, that is, an extremely localized electron–hole pair state with the reactant molecules plays a significant role in the photocatalytic reactions. On the other hand, highly dispersed Ag^+ ions anchored within zeolite cavities by ion exchange act also as efficient and unique photocatalysts for the decomposition of NO into N_2 and O_2. Various *in situ* spectroscopic investigations have revealed that the unique inner shell type excitation states of Ag^+ ions play a significant role in the photocatalytic decomposition of NO, that is, an electron transfer from the *s* orbital of the photoexcited Ag^+ ions to the π anti bonding orbital of NO was found to initiate the decomposition of NO.

It has thus been elucidated that well-ordered micro- or mesopores of zeolites or mesoporous materials can accommodate transition metal oxides or ions in an isolated state as single-site photocatalysts to realize unique and selective photocatalytic reactions essentially different from those on semiconducting photocatalysts such as TiO_2. It was observed that zeolite or mesoporous frameworks offer one of the most promising molecular reaction fields and approaches in the development of effective new photocatalytic systems that can contribute to the reduction of global air pollution and utilize solar energy as a clean, safe and abundant resource.

REFERENCES

1. K. Honda, A. Fujishima, *Nature* 238 (1972): 37.
2. T. Inoue, A. Fujishima, S. Konishi, K. Honda, *Nature* 277 (1979): 637.
3. M. Anpo, in *Green Chemistry*. Oxford: Oxford University Press (2000), 1, and references therein.

4. M. Anpo, *Pure Appl. Chem.* 72, (2000): 1265, and references therein.
5. M. Anpo, *Stud. Surf. Sci. Catal.* 130 (2000): 157.
6. M. Anpo, M. Takeuchi, H. Yamashita, S. Kishiguchi, A. Davidson, M. Che, in *Semiconductor Photochemistry and Photophysics*. New York: Marcel Dekker (2003), 283, and references therein.
7. M. Anpo, *Bull. Chem. Soc. Jpn.* 77 (2004): 1427, and references therein.
8. M. Anpo, S. Dohshi, M. Kitano, Y. Hu, M. Takeuchi, M. Matsuoka, in John M. Thomas, P. L. Gai, D. R. Clarke, M. Ruhel, J. C. Bravman (Eds.), *Annual Review of Materials Resarch*, Vol. 35 (2005): 1, and references therein.
9. M. Anpo, in G. Ertl, H. Knözinger, J. Weitkamp (Eds.), *Handbook of Heterogeneous Catalysis*. Weinheim: Wiley-VCH (1997), and references therein.
10. M. Anpo, T. Shima, Y. Kubokawa, *Chem. Lett.* (1985): 1799.
11. M. Anpo, T. Shima, S. Kodama, Y. Kubokawa, *J. Phys. Chem.* 91 (1987): 4305.
12. M. Anpo, H. Nakaya, S. Kodama, Y. Kubokawa, K. Domen, T. Onishi, *J. Phys. Chem.* 90 (1988): 1633.
13. M. Anpo, T. Kawamura, S. Kodama, K. Maruya, T. Onishi, *J. Phys. Chem.* 92 (1988): 438.
14. H. Yamashita, S. Kawasaki, Y. Ichihashi, M. Harada, M. Takeuchi, M. Anpo, *J. Phys. Chem. B* 102 (1998): 5870.
15. M. Anpo, H. Yamashita, K. Ikeue, Y. Fujii, S. G. Zhang, Y. Ichihashi, D. R. Park, Y. Suzuki, K. Koyano, T. Tatsumi, *Catal. Today* 44 (1998): 327.
16. H. Yamashita, Y. Fujii, Y. Ichihashi, S. G. Zhang, K. Ikeue, D. R. Park, K. Koyano, T. Tatsumi, M. Anpo, *Catal. Today* 45 (1998): 221.
17. S. G. Zhang, Y. Fujii, H. Yamashita, K. Koyano, T. Tatsumi, M. Anpo, *Chem. Lett.* (1997): 659.
18. M. Anpo, H. Yamashita, Y. Ichihashi, Y. Fujii, M. Honda, *J. Phys. Chem. B* 101 (1997): 2632.
19. M. Matsuoka, M. Anpo, *J. Photochem. Photobiol. C: Photochem. Rev.* 3 (2003): 225.
20. M. Anpo (Ed.), *Photofunctional Zeolites*, NOVA (2000).
21. A. Corma, *Chem. Rev.* 97 (1997): 2373.
22. T. Maschmeyer, F. Rey, G. Sankar, J. M. Thomas, *Nature* 378 (1995): 159.
23. S. Bordiga, S. Coluccia, C. Lamberti, L. Marchese, F. Boscherini, F. Buffa, F. Genoni, G. Leofanti, G. Petrini, G. Vlaic, *J. Phys. Chem.* 98 (1994): 4125.
24. L. Marchese, T. Maschmeyer, E. Gianotti, S. Coluccia, J. M. Thomas, *J. Phys. Chem. B* 101 (1997): 8836.
25. C. Lamberti, S. Bordiga, D. Arduino, A. Zecchina, F. Geobaldo, G. Spano, F. Genoni, G. Petrini, A. Carati, F. Villain, G. Vlaic, *J. Phys. Chem. B* 102 (1998): 6382.
26. S. Dzwigaj, M. Matsuoka, M. Anpo, M. Che, *J. Phys. Chem. B* 104 (2000): 6012.
27. H. Yamashita, K. Yoshizawa, M. Ariyuki, S. Higashimoto, M. Che, M. Anpo, *J. Chem. Soc. Chem. Commun.* (2001): 435.
28. (a) M. Anpo, N. Aikawa, S. Kodama, Y. Kubokawa, *J. Phys. Chem.* 84 (1984): 569; (b) C. Yun, M. Anpo, S. Kodama, Y. Kubokawa, *J. Chem. Soc. Chem. Commun.* (1980): 609.
29. M. Anpo, N. Aikawa, Y. Kubokawa, *J. Phys. Chem.* 88 (1984): 3998.
30. M. Anpo, N. Aikawa, Y. Kubokawa, M. Che, C. Louis, E. Giamello, *J. Phys. Chem.* 89 (1985): 5017.

31. H. Yamashita, M. Anpo, *Surf. Sci. Jpn.* 17 (1996): 30.
32. S. G. Zhang, Y. Ichihashi, H. Yamashita, T. Tatsumi, M. Anpo, *Chem. Lett.* (1996): 895.
33. M. Anpo, S. G. Zhang, S. Higashimoto, M. Matsuoka, H. Yamashita, Y. Ichihashi, Y. Matsumura, Y. Souma, *J. Phys. Chem. B* 103 (1999): 9295.
34. J. Zhang, M. Minagawa, M. Matsuoka, H. Yamashita, M. Anpo, *Catal. Lett.* 66 (2000): 241.
35. J. Zhang, M. Matsuoka, H. Yamashita, M. Anpo, *J. Synchrotron Radiat.* 8 (2001): 637.
36. J. Zhang, M. Minagawa, T. Ayusawa, S. Natarajan, H. Yamashita, M. Matsuoka, M. Anpo, *J. Phys. Chem. B* 104 (2000): 11501.
37. M. Anpo, M. Kondo, S. Coluccia, C. Louis, M. Che, *J. Am. Chem. Soc.* 111 (1989): 8791.
38. S. G. Zhang, M. Ariyuki, H. Mishima, S. Higashimoto, H. Yamashita, M. Anpo, *Micropor. Mesopor. Mater.* 21 (1998): 621.
39. S. G. Zhang, S. Higashimoto, H. Yamashita, M. Anpo, *J. Phys. Chem. B* 102 (1998): 5590.
40. K. Ikeue, H. Yamashita, M. Anpo, *Chem. Lett.* (1999): 1135.
41. K. Ikeue, H. Yamashita, M. Anpo, *J. Phys. Chem. B* 105 (2001): 8350.
42. K. Ikeue, S. Nozaki, M. Ogawa, M. Anpo, *Catal. Lett.* 80 (2002): 111.
43. D. R. Park, J. Zhang, K. Ikeue, H. Yamashita, M. Anpo, *J. Catal.* 185 (1999): 114.
44. M. Ogawa, K. Ikeue, M. Anpo, *Chem. Mater.* 13 (2001): 2900.
45. K. Ikeue, H. Yamashita, M. Anpo, T. Takewaki, *J. Phys. Chem. B* 105 (2001): 8350.
46. Y. Hu, S. Higashimoto, S. Takahashi, Y. Nagai, M. Anpo, *Catal. Lett.* 100 (2005): 35.
47. S. Irusta, A. J. Marchi, E. A. Lombardo, E. E. Miro, *Catal. Lett.* 40 (1996): 9.
48. A. J. Marchi, E. J. Lede, F. G. Requejo, M. Renteria, S. Irusta, E. A. Lombardo, E. E. Miro, *Catal. Lett.* 48 (1997): 47.
49. Y. Hu, N. Wada, M. Matsoka and M. Anpo, *Catal. Lett.* 97 (2004): 49.
50. A. Siani, D. K. Captain, O. S. Alexeev, E. Stafyla, A. B. Hungria, P. A. Midgley, J. M. Thomas, R. D. Adams, M. D. Amiridis, *Langmuir* 22 (2006): 5160.
51. T. Kamegawa, J. Morishima, M. Matsuoka, J. M. Thomas, M. Anpo, *J. Phys. Chem. C* 111 (2007): 1076.
52. M. Anpo, J. M. Thomas, *Chem. Commun.* (2006): 3273.
53. T. Kamegawa, R. Takeuchi, M. Matsuoka, M. Anpo, *Catal. Today* 111 (2006): 248.
54. M. Matsuoka, T. Kamegawa, M. Anpo, *Stud. Surf. Sci. Catal.* 165 (2007): 725.
55. M. Matsuoka, T. Kamegawa, M. Anpo, in G. Ertl, H. Knözinger, F. Schüth, J. Weitkamp (Eds.), *Handbook of Heterogeneous Catalysis*. Weinheim: Wiley-VCH (2008), 1065.
56. M. Anpo and M. Matsuoka, in K. D. M. Harris, P. Edwards (Eds.), *Turning Points in Solid-State, Materials and Surface Science*. RSC Publishing (2008), 496.
57. G. Centi, B. Wichterlova, A. T. Bell (Eds.), *Catalysis by Unique Metal Ion Structures in Solid Matrices*. NATO Science Series, Vol. 13. (2001) and references therein.
58. M. Anpo, M. Matsuoka, Y. Shioya, H. Yamashita, E. Giamello, C. Morterra, M. Che, H. H. Patterson, S. Webber, S. Quellette, M. A. Fox, *J. Phys. Chem.* 98 (1994): 5744.
59. M. Matsuoka, E. Matsuda, K. Tsuji, H. Yamashita, M. Anpo, *J. Mol. Catal. A: Chem.* 107 (1996): 399.
60. M. Matsuoka, M. Anpo, *Curr. Opin. Solid State Mater. Sci.* 7 (2003): 451.
61. W. S. Ju, M. Matsuoka, K. Iino, H. Yamashita, M. Anpo, *J. Phys. Chem. B* 108 (2004): 2128.

19

PHOTOCATALYTIC REMEDIATION

SHALINI RODRIGUES

19.1 Introduction, 629
19.2 Semiconductor Photocatalysis, 631
  19.2.1 Mechanisms of Semiconductor Photocatalysis, 633
  19.2.2 Metal Oxide Semiconductors and TiO_2, 635
  19.2.3 Metal Ion Dopant and Photoreactivity, 636
19.3 Type of Pollutant, 636
  19.3.1 Environmental Application of Photocatalysis, 637
19.4 Concluding Remarks and Future Directions, 642
  19.4.1 Mesoporous Materials, 643
Acknowledgment, 644
References, 645

19.1 INTRODUCTION

Over the years due to chemical disposal in lagoons, underground storage tanks, and dump sites, the underlying aquifers have become contaminated with numerous hazardous chemicals. The urgency in addressing the environmental problems related to remediation of hazardous wastes, contamination of groundwater, and release of toxic air pollutants is an important issue at both the national and international levels. Typical wastes of concern include heavy metals, solvents, degreasing agents, and chemical by-products. Some of these are known to contaminate groundwater

Nanoscale Materials in Chemistry, Second Edition. Edited by K. J. Klabunde and R. M. Richards

and surface water. With clean water resources decreasing gradually it is a major challenge for us in the next few decades to provide clean water.

In response to the growing risk of environmental hazards associated with organic water wastes, several methods have been developed to treat water supplies. Physical techniques such as activated carbon adsorption, ultrafiltration, reverse osmosis, and ion exchange on synthetic resins have been used effectively; however, these methods are inherently flawed. These techniques are nondestructive, as the wastes are simply transferred from the water to another phase, creating secondary wastes. The past couple of decades has witnessed extensive growth in the area of photocatalysis and it is exciting to see that advances are still being made particularly with visible light photocatalysis. Heterogeneous photocatalysis has attracted much attention in the scientific community all over the world due to its high efficiency and generation of harmless by-products. Treatment of polluted air or gas by catalytic treatment is a well-established process. Catalytic combustion or catalytic incineration processes require temperatures of 473 to 1273 K for efficient operations and hence are energy intensive. Further, these traditional methods of treatment are not economically feasible at lower pollutant concentrations. Hence, extensive research is underway to find processes and technologies that operate under ambient conditions of temperature and pressure to treat low pollutant levels.

Photocatalysis has recently emerged as an advanced oxidation process (AOP) (1, 2). Gas phase heterogeneous photocatalysis is an AOP that could be used successfully for improving indoor air quality. The field of photocatalysis was initiated by Fujishima and Honda in 1972, when the photoelectrochemical splitting of water was discovered (3). TiO_2 is the best known semiconductor photocatalyst and performs extraordinarily well using ultraviolet (UV) irradiation for a variety of problems of environmental concern, including water and air purification. It has been widely explored and shown useful for destruction of microorganisms such as bacteria and viruses, odor control, water splitting to produce hydrogen gas, and for clean up of oil spills. Titanium dioxide baked onto the surface of ceramic tiles when exposed to the Sun's ultraviolet rays, decomposes organic compounds, including soot, grime, and oil. The treated surface also becomes hydrophilic, meaning that a thin layer of moisture prevents dirt adhesion. The photocatalytic activity of titania results in thin coatings of the material exhibiting self-cleaning and disinfecting properties under exposure to UV radiation. These properties make the material a candidate for applications such as medical devices, food preparation surfaces, air conditioning filters, and sanitary ware surfaces. There are no limits to the possibilities and applications of titanium oxide photocatalyst as an environmentally harmonious catalyst and/or sustainable green chemical system. A number of major reviews of the academic literature associated with semiconductor catalysis have been reported (4–9). There is also the recently published, excellent online book entitled *Solar Detoxification* (by semiconductor photocatalysis) (10). The fascinating and unique properties of transition metal doped mesoporous materials either in the framework or cavities offer new possibilities not only in catalysis but also in photochemical processes. Mesoporous silica forms the backbone in catalysis, ion exchange, and separation, with progress being made in the area of photocatalysis (11).

19.2 SEMICONDUCTOR PHOTOCATALYSIS

Photocatalysis is defined as the acceleration of a photoreaction in the presence of a catalyst. In catalyzed photolysis, light is absorbed by an adsorbed substrate. In photogenerated catalysis the photocatalytic activity (PCA) depends on the ability of the catalyst to create electron–hole pairs, which generate free radicals (hydroxyl ions; OH^-) able to undergo secondary reactions. Reactions being assisted by light requires that the life time of charge carriers namely holes (h^+) and electrons (e^-) be higher, in order to avoid recombination which otherwise can happen in a few nanoseconds. Metals have a continuum of energy levels which, upon irradiation, can lead to recombination of holes and electrons rapidly prior to charge transfer to any substrate molecule adsorbed. The electronic structure of most semiconductor materials is characterized by the highest occupied band full of electrons called the valence band (VB) and the lowest unoccupied band called the conduction band (CB). These bands are separated by a region that is largely devoid of energy levels, and the difference in energy between the two bands is called the bandgap energy, E_{bg}. Ultrabandgap illumination of such semiconductor materials produces electron–hole pairs, h^+e^-, which can either recombine to liberate heat, Δ, or make their way, via separate pathways, to the surface of the semiconductor material, where they have the possibility of reacting with surface absorbed species. These electrons and holes have a high reductive potential and oxidative potential, respectively, which together cause catalytic reactions on the surface; namely photocatalytic reactions are induced.

Unfortunately, the efficiency-lowering process of recombination, either at the surface, or in the bulk of the semiconductor material, is the usual fate of photogenerated electron–hole pairs. Thus, the efficiencies of most processes involving semiconductor photocatalysis is low; typically $<1\%$. This is particularly true in the case of amorphous semiconductor materials, since electron–hole recombination is promoted by defects. Consequently, in most of the current work involving photocatalysis, the semiconductor is comprised of microcrystalline or nanocrystalline particles and used in the form of a thin film or as a powder dispersion. Photogenerated holes and electrons that are able to make their way to the surface of these semiconductor particles can react either directly or indirectly through slightly less energetic trap surface states, with absorbed species. Thus, if there is an electron donor, D, adsorbed on the surface of the semiconductor particles, then the photogenerated holes can react with it to generate an oxidized product, D^+. Similarly, if there is an electron acceptor present at the surface, that is, A, then the photogenerated conductance band electrons can react with it to generate a reduced product, A^-. The overall reaction can be summarized as follows:

$$A + D \xrightarrow[h\nu \geq E_{bg}]{\text{semiconductor}} A^- + D^+ \tag{19.1}$$

In the purification of water via semiconductor photocatalysis, the electron acceptor, A, is invariably dissolved oxygen, and the electron donor, D, is the pollutant, which is usually organic. Under these circumstances, the overall process is the semiconductor photocatalyzed oxidative mineralization of the organic pollutant by dissolved oxygen, and can be represented by the following equation:

$$\text{Pollutant} + O_2 \xrightarrow[h\nu \geq E_{bg}]{\text{semiconductor}} \text{minerals} \quad (19.2)$$

where minerals are CO_2, H_2O, and, where appropriate, inorganic acids or salts, such as HCl or NaCl.

The initial process involved in heterogeneous photocatalysis is the generation of electron–hole pairs by irradiation of the semiconductor particle. Figure 19.1 is a schematic representation of the electron–hole generation by irradiation of the semiconductor particle with suitable energy (7). As can be seen from the figure the electron and hole migrate to the surface and the semiconductor can donate an electron to reduce an electron acceptor (normally oxygen in aqueous semiconductor suspensions) and the holes can oxidize an electron donor species. To summarize, the valence band holes are powerful oxidants (0.5 to -1.5 V vs. normal hydrogen electrode, NHE) while conduction band electrons are powerful reductants ($+1.0$ to 3.5 V vs. NHE depending on the pH and semiconductor). The electron–hole recombination can happen on the surface or in the volume of the semiconductor particle competes with the charge transfer processes and is thus important in determining the efficiency of the photocatalytic reaction. The efficiency of the photocatalytic reaction is measured in terms of quantum yield. Quantum yield is defined as the number of times that a defined event occurs per photon absorbed by the system (12). Thus, quantum yield is a measure of the efficiency with which absorbed light produces some effect. Often it is very difficult to measure absorbed light in the case of heterogeneous photocatalysis due to scattering and reflection of light by the semiconductor particle.

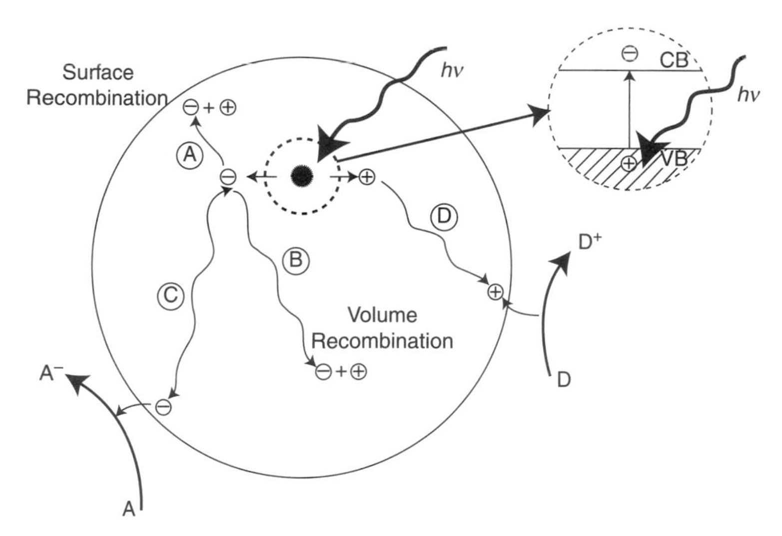

Figure 19.1 Schematic representation of the excitation and deexcitation steps for electrons and holes in a heterogeneous photocatalytic reaction. (Reproduced with permission from A. L. Linsebigler et al., *Chem. Rev.* **1995**, *95*, 735.)

Therefore, it is assumed that all of the light is absorbed and apparent quantum yield is often reported. In the case of homogeneous catalysis it is easier to calculate the number of photons absorbed. It is important to understand that the amount or number of photons absorbed by the semiconductor is important rather than the number of photons that initiate and drive a photocatalytic reaction.

19.2.1 Mechanisms of Semiconductor Photocatalysis

The mechanisms associated with reactions in aqueous and gaseous media are quite different. Hydrogen peroxide, superoxides, hydroxide ions, hydroperoxy radicals, and hydroxyl radicals play an important role in the mineralization of organic compounds. The hydroxyl radical produced in the step shown below is the most active species for reducing most of the organic pollutants. The reducing potential of the conduction band electron is lower than the oxidizing potential of the valence band holes. Most of the reducible molecules cannot compete for the oxidizing power of molecular oxygen in trapping the photogenerated conduction band electrons. The important steps involved with photocatalytic reactions in aqueous medium are as follows:

$$O_2 + e^- \longrightarrow O_2^- \tag{19.3}$$

$$H_2O + h^+ \longrightarrow {}^{\bullet}OH + H^+ \tag{19.4}$$

$${}^{\bullet}OH + {}^{\bullet}OH \longrightarrow H_2O_2 \tag{19.5}$$

$$H_2O_2 + O_2^- \longrightarrow {}^{\bullet}OH + OH^- + O_2 \tag{19.6}$$

Oxygen acts as an electron acceptor in most photocatalytic reactions. Insertion of hydroperoxy and hydroxyl radicals in the C–H bonds leads to the mineralization of a variety of organic pollutants. These O_2^- and OH radicals have very high oxidation potential, including the complete oxidation reactions of various organic compounds such as toxic halocarbons, as shown:

$$C_xH_yCl_z + (x + y - z/4)O_2 \longrightarrow xCO_2 + zH^+ + zCl^- + (y - z/2)H_2O \tag{19.7}$$

Intermediates detected during the photocatalytic degradation of halogenated aromatic compounds have been typically hydroxylated structures and this gives indirect evidence of hydroxyl radicals as the primary reactive oxidant species (13, 14). Such hydroxylated intermediates are also formed when halogenated aromatics are reacted with known precursors of hydroxyl radicals.

Hydrogen peroxide is also formed by an alternate reaction, as follows:

$$O_2 + 2e^- + H^+ \longrightarrow H_2O_2 \tag{19.8}$$

This reaction is very important since electron transfer to oxygen may be the rate-determining step in semiconductor photocatalysis (15–17). Although the formation of hydrogen peroxide leads to depletion of hydroxyl radicals it also causes degradation of inorganic and organic electron donors by acting as a direct electron acceptor (18). The primary processes and associated characteristic time domains for semiconductor

TABLE 19.1 Primary Processes and Associated Characteristic Time Domains in the TiO_2-Sensitized Photomineralization of Organic Pollutants

Primary Processes	Characteristic Times
Charge-carrier generation	
$TiO_2 + h\nu \longrightarrow h^+ + e^-$	fs (very fast)
Charge-carrier trapping	
$h^+ + {>}Ti^{IV}OH \longrightarrow \{{>}Ti^{IV}OH^{\bullet +}\}$	10 ns (fast)
$e^- + {>}Ti^{IV}OH \longleftrightarrow \{{>}Ti^{III}OH\}$	100 ps (shallow trap; dynamic equilibrium)
$e^- + Ti^{IV} \longrightarrow {>}Ti^{III}$	10 ns (deep trap)
Charge-carrier recombination	
$e^- + \{{>}Ti^{IV}OH^{\bullet +}\} \longrightarrow {>}Ti^{IV}OH$	100 ns (slow)
$h^+ + {>}Ti^{III}OH \longrightarrow {>}Ti^{IV}OH$	10 ns (fast)
Interfacial charge transfer	
$\{{>}Ti^{IV}OH^{\bullet +}\}$ + organic pollutant $\longrightarrow {>}Ti^{IV}OH$ + oxidized pollutant	100 ns (slow)
$\{{>}Ti^{III}OH\} + O_2 \longrightarrow {>}Ti^{IV}OH + O_2^{\bullet}$	ms (very slow)

photocatalysis involving TiO_2 are listed in Table 19.1 (6). From this data it appears that the rate-determining step in the overall photocatalytic process is the reduction of dissolved oxygen by trapped photogenerated electrons, that is, by $Ti^{III}OH$. Certainly the supply of oxygen to the particles can control the overall kinetics of the process (10, 19, 20). However, the direct or indirect oxidation step involving the pollutant and the photogenerated holes is also a slow step and, under certain circumstances, can be rate determining. Thus, the kinetics of semiconductor photocatalysis for water purification is complex and often varies from one pollutant to another. Thus, when treated using semiconductor photocatalysis, a complex mixture of pollutants found in most waste streams is likely to produce an equally complex mixture of kinetics. Interestingly, the kinetics of destruction of most simple *single* pollutant systems by semiconductor photocatalysis exhibit very similar features. Thus, the initial kinetics of photomineralization of a general organic pollutant, P, by oxygen, sensitized by titanium dioxide, upon steady-state illumination, usually fit a Langmuir–Hinshelwood kinetic scheme, that is,

$$R_i = -d[P]/dt = k_P K_P[P]/(1 + K_P[P]) \qquad (19.9)$$

where R_i is the initial rate of the substrate removal, $[P]$ is the initial concentration of the pollutant under test and, traditionally, K_P, is taken to be the dark Langmuir adsorption constant of species P on the surface of titanium dioxide. A rough but useful guide to efficiency in semiconductor photocatalytic systems is photonic efficiency, δ, which is defined as follows:

$$\delta = \text{Overall rate of photocatalytic process/Intensity of incident light} \qquad (19.10)$$

In semiconductor photocatalysis, the value of δ has often been found to approach unity, at low incident light levels of ultrabandgap light. However, at the typical light levels generated by black light bulbs or germicidal lamps (approximately 6 mW cm^{-2} for an 8 W lamp), the value of δ is usually about 0.01 for most semiconductor photocatalysis systems. This low efficiency is due to a number of factors, including reflection and scattering losses and significant electron–hole recombination at the light intensities typically found in most photoreactors (i.e. 0.1 to 10 mW cm^{-2}).

19.2.2 Metal Oxide Semiconductors and TiO_2

Ideally, a semiconductor photocatalyst for the purification of water and air should be chemically and biologically inert, photocatalytically active, easy to produce and use, and activated by sunlight. Not surprisingly, no semiconductor perfectly fits this demanding list of requirements, although the semiconductor titanium dioxide, TiO_2, comes close, failing in its inability to absorb visible light. In fact titanium dioxide has a large bandgap energy, $E_{bg} \cong 3.2$ to 3.0 eV, and so is only able to absorb UV light (typically <380 nm) which represents a small fraction, about 6%, of the solar spectrum. However, the other very positive features of titanium dioxide as a semiconductor photocatalyst, such as high photoactivity, chemical and photochemical robustness, and low cost, far outweigh its deficient spectral profile overlap with the solar spectrum. As a result, titanium dioxide has become *the* semiconducting material for research and for use in commercial photocatalytic reactors in the field of semiconductor photocatalysis for water purification. Although titanium dioxide exists in three crystalline forms, namely anatase, rutile, and brookite, the common form used in semiconductor photocatalysis is anatase, as this is the most photocatalytically active and easiest to produce. Apart from TiO_2 ($Eg = 3.2$ eV) several other oxides and sulfide semiconductor materials, such as ZnO ($Eg = 3.2$ eV), WO_3 ($Eg = 2.8$ eV) $SrTiO_3$ ($Eg = 3.2$ eV), α-Fe_2O_3, ($Eg = 3.1$ eV), ZnS (3.6 eV), CdS, and CdSe, have bandgap energies sufficient for catalyzing a wide range of chemical reactions of environmental interest.

The most important criteria for any semiconductor material to be a photocatalyst for the degradation of organic compound are that it should (1) be stable over prolonged periods of irradiation and (2) the reducing potential of $H_2O/^{\bullet}OH$ lie between the conduction band edge and valence band edge of the semiconductor. Stability issues arise in the case of metal sulfide semiconductors which readily undergo photoanodic corrosion while iron oxide polymorphs, in spite of their high bandgap energies, are not suitable because they can readily undergo photocathodic corrosion. ZnO is unstable, thus leading to inactivation of the catalyst over time. Degussa P-25 TiO_2 is the standard for comparing photoreactivity of materials used as photocatalyst. It is a non-porous material comprising a 70% anatase to 30% rutile mixture with surface area of 50 m^2/g. The high photoactivity of Degussa P25 compared to pure anatase or rutile is attributed to high absorption affinity of organic compounds, high surface area, and lower recombination rates due to lattice electron trapping sites in both anatase and rutile phases. In the recombination of charge carriers in Degussa P25 TiO_2 it is seen that the photogenerated holes are trapped on particle surfaces and the photogenerated electrons are trapped at the lattice sites.

19.2.3 Metal Ion Dopant and Photoreactivity

Modification by selective surface treatments such as surface chelation, platinization, and selective doping of the crystalline matrix can enhance the interfacial charge transfer reactions of TiO_2. Doping with various metal ions changes the electronic properties of titanium oxide catalyst. For example, doping of TiO_2 by Fe^{3+} increases the quantum efficiency for photoreduction of N_2 and methyl viologen, also to inhibit the electron hole pair recombination. The enhanced photoactivity is due to the rutile phase of TiO_2, which is more photoactive than the anatase phase. This is surprising since the conduction band of anatase lies at a more negative potential compared to rutile. It therefore suggests that the reducing ability should have only a marginal increase in the yield of ammonia for semiconductor oxides that have a similar flat band potential at a particular pH. Chromium doping in TiO_2 has produced enhanced photoactivity for water cleavage and N_2 reduction in the gas-solid interface. There are also negative effects of doping as seen in the case of Mo and V doping in TiO_2. It should be recognized, however, that the results of experiments conducted in the field are often quite different from those in a laboratory. Most importantly, it is clear that a TiO_2 photocatalyst will not work very well, if at all, if the water under test contains any strongly UV absorbing pollutants, such as dye stuffs and humic acid. Nor will most photocatalysts work if metal ions, such as ferric and ferrous ions, are present and able to precipitate out as insoluble and inert oxides and hydroxides onto the surface of the semiconductor material to form a passivating layer. Although the latter may be removed by using an acid wash, it appears a clear limitation of the technology. As a consequence, semiconductor photocatalysis, as a method of water treatment, is most likely limited in application to water that has been prefiltered to remove any solid material that is likely to deposit on the surface of the semiconductor and any strong UV-absorbing species that may be present, and, preferably deionized, so that it is fairly free of metal ions that may form insoluble oxides or hydroxides on the surface of the semiconductor.

19.3 TYPE OF POLLUTANT

Research into semiconductor photocatalysis has focused largely on the removal of organic pollutants from water, since it appears that a wide range of such substrates are prone to complete mineralization by dissolved oxygen, photosensitized by titanium dioxide. Table 19.2 provides a list of a wide number of classes of compounds, and examples of each, that have been shown to be completely mineralized by semiconductor photocatalysis using TiO_2 (8, 9). Notable compounds in Table 19.2 include haloalkanes, such as chloroform, and tetrachloromethane, since they are carcinogens, and often found in small but detectable amounts in drinking water purified by chlorination. Indeed, it appears likely that semiconductor photocatalysis may find initial commercial application as an "afterchlorination" step in the production of drinking water. Other compounds listed in Table 19.2 that are highly relevant to the treatment of wastewater include surfactants, hormones, herbicides, pesticides, and dyes, all of which can be considered as relatively common pollutants from farms, households, and industry. Semiconductor photocatalysis is ideal for treating refractory, hazardous,

TABLE 19.2 Some Examples of TiO_2-Sensitized Photomineralization of Organic Substrates

Class	Examples
Alkanes	Methane, *iso*-butane, pentane, heptane, cyclohexane, paraffin
Haloalkanes	Mono-, di-, tri-, and tetrachloromethane, tribromoethane, 1,1,1-trifluoro-2,2,2-trichloroethane
Aliphatic alcohols	Methanol, ethanol, *iso*-propyl alcohol, glucose, sucrose
Aliphatic carboxylic acids	Formic, ethanoic, dimethylethanoic, propanoic, oxalic acids
Alkenes	Propene, cyclohexene
Haloalkenes	Perchloroethene, 1,2-dichloroethene, 1,1,2-trichloroethene
Aromatics	Benzene, naphthalene
Haloaromatics	Chlorobenzene, 1,2-dichlorobenzene, bromobenzene
Nitrohaloaromatics	3,4-dichloronitrobenzene, dichloronitrobenzene
Phenols	Phenol, hydroquinone, catechol, 4-methyl catechol, resorcinol, *o*-, *m*-, *p*-cresol
Halophenols	2-,3-,4-chlorophenol, pentachlorophenol, 4-fluorophenol, 3,4-difluorophenol
Aromatic carboxylic acids	Benzoic, 4-aminobenzoic, phthalic, salicyclic, *m*- and *p*-hydroxybenzoic, chlorohydroxybenzoic acids
Polymers	Polyethylene, PVC
Surfactants	SDS, polyethylene glycol, sodium dodecyl benzene sulfonate, trimethyl phosphate, tetrabutyl ammonium phosphate
Herbicides	Methyl viologen, atrazine, simazine, prometron, propetryne, bentazon
Pesticides	DDT, parathion, lindane
Dyes	Methylene blue, rhodamine B, methyl orange, fluorescein, reactive black 5

toxic, and nonbiodegradable pollutants. In addition to the removal of organic pollutants, semiconductor photocatalysis has been used to sensitize the photoconversion of toxic inorganic substrates to harmless or less toxic ones: a number of examples of such photoreactions, sensitized by titanium dioxide, are listed in Table 19.3 (8). Thus, semiconductor photocatalysis can be used to oxidize nitrite, sulfite, and cyanide anions to form relatively harmless products, such as NO_3^-, SO_4^{2-}, and CO_2. Titanium dioxide is also able to photosensitize the decomposition of bromate to bromide and oxygen, even when the level of bromate ions is as low as 50 ppb. Bromate is a carcinogen and is found in ppb levels in drinking water that has been purified by chlorinating water sources that have a reasonably high background level of bromide ions. The removal of bromate ions at the ppb level represents a major problem for the potable water production industry but semiconductor photocatalysis appears to offer a simple and effective solution (21).

19.3.1 Environmental Application of Photocatalysis

19.3.1.1 Photodegradation of Trichloroethylene (TCE) The field of photocatalysis has its major application in environmental remediation. The increasing

TABLE 19.3 TiO_2-Sensitized Photosystems for the Removal of Toxic Inorganics

Overall reaction $A + D \longrightarrow A^- + D^+$
$5O_2 + 6NH_3 \longrightarrow 2N_2 + N_2O + 9H_2O$
$O_2 + 2NO_2^- \longrightarrow 2NO_3^-$
$O_2 + 2SO_3^{2-} \longrightarrow 2SO_4^{2-}$
$2O_2 + H_2O + S_2O_3^{2-} \longrightarrow 2SO_4^{2-} + 2H^+$
$O_2 + 2CN^- \longrightarrow 2OCN^-$
$5O_2 + 4H^+ + 4CN^- \longrightarrow 2H_2O + 4CO_2 + 2N_2$
$2BrO_3^- \longrightarrow 2Br^- + 3O_2$

contamination of land, water, and air has raised serious environmental problems (22–29). To reduce the impact of environmental pollution many efforts have been made in green chemistry and purification technologies. The photocatalytic degradation of organic pollutants using TiO_2 has been demonstrated to be successful in various remediation systems of polluted water and air (22, 24–26). It is well known that in the presence of atmospheric oxygen and at room temperature, chlorinated molecules undergo progressive oxidation to complete mineralization, resulting in the formation of CO_2, H_2O, and HCl. For destruction of indoor air pollutants (office buildings, aircrafts, factories, automobiles) photocatalytic treatment leads most often to complete oxidation into CO_2 and H_2O. Photocatalytic remediation using TiO_2 and transition metal ion supported mesoporous materials is discussed in this section.

Trichloroethylene (TCE) is the most common and abundant pollutant in groundwater in the United States. It is primarily used as a solvent to remove grease from metal parts, as a solvent for extraction of waxes, oil, fats, tar, and in several consumer products such as paints, carpet cleaning fluid, etc. It is estimated that between 9% and 34% of drinking water supply sources are contaminated with TCE. Several epidemiological studies link TCE exposure to health problems related to congenital heart disease, spontaneous abortion, cancer, etc.

The heterogeneous semiconductor-based photocatalytic degradation of trichloroethylene using TiO_2 was first reported by Pruden and Ollis (30). The complete mineralization results in formation of CO_2 and HCl. Gas phase photocatalytic oxidation of TCE is much faster compared to the aqueous phase. Various mechanisms are currently proposed for TCE photooxidation on TiO_2 surfaces. In aqueous solution hydroxyl radicals initiate the reaction and dichloroacetaldehyde (DCAAD) is formed as an intermediate (30). The disappearance of TCE and the intermediate dichloroacetaldehyde is represented by a Langmuir expression. Molecular oxygen is required for complete mineralization of chlorinated hydrocarbons to CO_2 and HCl (31). The corresponding stoichiometric equation is given by Equation (19.11):

$$CCl_2CClH + H_2O + 3/2O_2 \longrightarrow 2CO_2 + 3HCl \qquad (19.11)$$

In the absence of molecular oxygen the reaction is self-poisoning. TCE could be oxidized to CO_2 and HCl directly in a photolytic way. However, ultraviolet light is necessary for its degradation (27). In addition to the $OH^{\bullet}$ initiated oxidative reaction channels, a parallel reductive pathway involving conduction band electrons of TiO_2 also plays an important role (32). In studies of the TiO_2 photocatalyzed gaseous TCE oxidation a mechanism involving $OH^{\bullet}$ and a monochloroacetate as an intermediate has been proposed (33). In contrast there is evidence that TCE is oxidized in a chain reaction initiated by chlorine atoms (34). When using Cr-AlMCM-41 as the photocatalyst, the primary intermediates produced during photooxidation using UV irradiation are dichloroacetylchloride (DCAC, $CHCl_2COCl$), phosgene, and CO_2 (35). Upon extensive UV irradiation most of the DCAC gets converted to CO_2 and phosgene. DCAC formed as an intermediate gets converted into CO_2 and HCl. Prolonged illumination using UV light leads to the disappearance of all the detected reaction intermediates, to yield CO_2 and water as the final reaction products. The surface hydroxyl radicals can cause the reaction to proceed initially. On photoexcitation the holes are utilized for oxidation of TCE while the e^- is trapped by the O_2. Since the reaction is carried out in a closed reactor, once the O_2 concentration is depleted the electron concentration builds up resulting in the reduction of Cr^{6+} to Cr^{3+}. The diffuse reflectance spectra results indicate that the chromium ions are highly dispersed and exist in an isolated state on the MCM-41 support. These highly dispersed chromium ions can be excited under UV light irradiation to form the corresponding charge-transfer excited state involving an electron transfer from O^{2-} to Cr^{6+}, as shown by Equation (19.12):

$$Cr^{6}\text{–}O^{2-} \xrightarrow{h\nu} [Cr^{5+}\text{–}O^{-}]^{*} \tag{19.12}$$

These charge-transfer excited states have high reactivities due to the presence of electron–hole pairs localized next to each other, compared to electron–hole pairs produced in traditional semiconductors such as TiO_2, ZnO, or CdS (36). Thus, the localized electron–hole pairs under UV and/or visible light irradiation initiate unique photocatalytic activities that cannot usually be found on dispersed metal oxide semiconductors. Surface hydroxyl groups are ubiquitous on oxide surfaces and especially on MCM-41 type of materials. The density of OH groups on these surfaces has been estimated to be as high as 3 OH per nm^2 at room temperature (37). Thus, the surface hydroxyl groups can react with the hole produced by excitation to form hydroxyl radicals that are highly reactive:

$$OH^{-} + h^{+} \longrightarrow {}^{\bullet}OH \tag{19.13}$$

The ${}^{\bullet}OH$ radical generated then can react with TCE to generate $Cl^{\bullet}$ radicals via Equations (19.14) and (19.15):

$$Cl_2C{=}CHCl + {}^{\bullet}OH \longrightarrow Cl_2C^{\bullet}\text{–}CHClOH \tag{19.14}$$

$$Cl_2C^{\bullet} - CHClOH \longrightarrow Cl_2C{=}CHOH + Cl^{\bullet} \tag{19.15}$$

The $Cl^\bullet$ radicals are highly reactive species too and can react with TCE through a chain mechanism in propagation steps [Equations (19.16) to (19.19)] as proposed initially by Nimlos (18):

$$Cl_2C{=}CHCl + Cl^\bullet \longrightarrow Cl_2HC{-}CClCl^\bullet \tag{19.16}$$

$$Cl_2HC{-}CClCl^\bullet + O_2 \longrightarrow Cl_2HC{-}CCl_2OO^\bullet \tag{19.17}$$

$$2Cl_2HC{-}CCl_2OO^\bullet \longrightarrow 2Cl_2HC{-}CCl_2O^\bullet + O_2 \tag{19.18}$$

$$2Cl_2HC{-}CCl_2O^\bullet \longrightarrow Cl_2HC{-}COCl + Cl^\bullet \tag{19.19}$$

The DCAC formed can further decompose, forming phosgene, carbon dioxide, and hydrochloric acid.

$$Cl_2HC{-}COCl + O_2 \longrightarrow COCl_2 + CO_2 + HCl \tag{19.20}$$

The presence of chlorine radical chain reactions in a photocatalytic reaction may significantly increase reaction rates and photocatalytic efficiencies. These enhancements would appear to have the potential to overcome shortcomings typically associated with photocatalytic oxidation of aromatic contaminants if a chlorine radical chain reaction could be initiated in conjunction with an aromatic photocatalytic reaction. *In situ* solid-state nuclear magnetic resonance spectroscopy is advantageous since an examination of the reaction occurring in both the gas phase and on the catalyst surface is possible. Accordingly, three new intermediates, oxalyl chloride (ClCOCOCl), trichloroacetic acid (Cl_3CCO_2H), and trichloroacetyl chloride (Cl_3CCOCl, TCAC) have been identified. Molecular oxygen is the primary initiating species and mechanisms supporting the involvement of surface bound H_2O or hydroxyl groups are not important. Thus, various mechanisms are currently proposed for TCE photooxidation on TiO_2 surfaces. The role of water vapor in the photocatalytic degradation of trichloroethylene (TCE) on anatase titanium dioxide films have been studied. It is found that TCE conversion is not affected by relative humidities up to 20% but as the gas mixture approached saturation with respect to water vapor, conversion efficiency of TCE decreased. Moderate concentration of water vapor (12% relative humidity) resulted in greater conversions due to increase in the rate of hydrogen extraction reactions. At high concentrations of water vapor, TCE conversions decreased because of a deteriorating rate effect of water vapor on chlorine extraction reactions.

19.3.1.2 Photocatalytic Degradation of Chlorophenols Chlorophenols represent an important class of pollutants that are toxic to a variety of organisms (38–48). They are used extensively as wood preservatives, herbicides, and fungicides and hence are found in soil and aquatic environments. The most common exposure of chlorinated organics to human beings is through drinking water treated with chlorine. The chlorophenols that are most likely to be present in drinking water are 2,4-dichlorophenol and 2,4,6-trichlorophenol. These are present due to chlorination of phenol, as degradation of phenoxy herbicides, or as by-products of the reaction of phenolic acid and hypochlorite. The C–Cl bond is so strong that chlorinated compounds are persistent in the environment. They have been classified as toxic by the U.S. Environmental Protection Agency and the European Commission. The reaction pathway for photodegradation of 4-chlorophenol (4-CP) has three concurrent pathways (8). Two of these pathways lead to the formation of stable intermediates, including hydroquinone and 4-chlorocatechol (4-CC). The third pathway involves the formation of an unstable intermediate that undergoes ring cleavage and subsequent decarboxylation and dechlorination.

A detailed kinetic and mechanistic study on two different commercial titania photocatalysts have been examined, that is, Degussa P25 and Sachtleben Hombikat UV 100 (49). The oxidation products of 4-chorophenol (4-CP) are 4-chlorocatechol (4-CC), benzoquinone, and hydroquinone. The formation of 4-CC is explained by the addition of hydroxyl radical to the ortho position of the hydroxyl group of 4-CP followed by the abstraction of a hydrogen atom to recover the aromatic ring. Dechlorination of 4-CP results in hydroquinone and benzaquinone. These compounds are further oxidized to secondary intermediates, hydroxyhydroquinone (HHQ) and hydroxybenzoquinone (HBQ). HHQ and HBQ are the final aromatic intermediates before ring cleavage to the final mineralization product. The kinetic data are examined in terms of Langmuir–Hinshelwood parameters. The rate of photochemical degradation is expressed in terms of either the oxidant or the reductant (organics) by the following equation:

$$-d[\mathrm{Red}]/dt = -d[\mathrm{Ox}]/dt = k_d \theta_{\mathrm{Red}} \theta_{\mathrm{Ox}} \qquad (19.21)$$

where [Red] is reductant and [Ox] is oxidant and k_d is the photodegradation rate constant. θ_{Red} represents the fraction of the electron-donating reductant (organics) and θ_{Ox} is the fraction of the electron-accepting oxidant (oxygen), sorbed to the surface. The equilibrium constant K_{ads} for the reactants is determined from a Langmuir sorption isotherm.

19.3.1.3 Photodegradation of Chloroform

Chloroform is also known as trichloromethane and methyl trichloride and has myriad uses as a reagent and a solvent. It is also considered an environmental hazard. As might be expected for an anesthetic, inhaling chloroform vapors depresses the central nervous system. It is immediately dangerous to health and life at approximately 500 ppm according to the U.S. National Institute for Occupational Safety and Health. Breathing about 900 ppm for a short time can cause dizziness, fatigue, and headache. Chronic chloroform exposure may cause damage to the liver (where chloroform is metabolized to phosgene) and to the kidneys. During prolonged storage hazardous amounts of phosgene can accumulate in the presence of oxygen and ultraviolet light.

The mechanism for oxidation of chloroform after generation of an electron–hole pair due to excitation under UV light is proposed as follows (50):

$$>\mathrm{TiOH}^{+} + \mathrm{HCCl_3} \longrightarrow \mathrm{TiOH_2^{+}} + >{}^{\bullet}\mathrm{CCl_3} \qquad (19.22)$$

$$>\mathrm{Ti^{III}OH^{-}} + >\mathrm{TiOH_2^{+}}{:}\mathrm{O_2} \longrightarrow >\mathrm{Ti^{IV}OH} + >\mathrm{TiOH_2^{+}} - {}^{\bullet}\mathrm{O_2^{-}} \qquad (19.23)$$

$$>{}^{\bullet}\mathrm{CCl_3} + \mathrm{O_2} \longrightarrow {}^{\bullet}\mathrm{O_2CCl_3} + > \qquad (19.24)$$

$$2^{\bullet}O_2CCl_3 \longrightarrow 2^{\bullet}OCCl_3 + O_2 \tag{19.25}$$

$$^{\bullet}OCCl_3 + HO_2^{\bullet} \longrightarrow Cl_3COH + O_2 \tag{19.26}$$

$$Cl_3COH \longrightarrow Cl_2CO + H^+ + Cl^- \tag{19.27}$$

$$Cl_2CO + H_2O \longrightarrow CO_2 + H^+ + Cl^- \tag{19.28}$$

Similar mechanisms can be operative for a wide range of oxidizable chlorinated hydrocarbons with abstractable hydrogen atoms. In the case of chlorinated hydrocarbons with no abstractable hydrogen atoms or with tetravalent carbon in C(IV) state, reactions can be initiated by direct hole or electron transfer as in the case of trichloroacetic acid or carbon tetrachloride, respectively:

$$>Ti-h_{vb} + {}^{-}OH + >CCl_3CO_2^- \longrightarrow >TiOH + {}^{\bullet}CCl_3 + CO_2 \tag{19.29}$$

$$>Ti-e_{cb}^- -OH + >CCl_4 \longrightarrow >TiOH + {}^{\bullet}CCl_3 + Cl^- \tag{19.30}$$

The $CCl_3^{\bullet}$ radical then continues to react via Equations (19.24) to (19.28).

19.4 CONCLUDING REMARKS AND FUTURE DIRECTIONS

Photocatalysis has become a promising method for several important applications ranging from splitting of water to purification of contaminated water, degradation of toxic pollutants (in air and water), and destruction of pathogenic microorganisms. There are no limits to the possibilities and applications of titanium oxide photocatalyst as an environmentally harmonious catalyst and sustainable green chemical system. The past couple of decades have witnessed extensive growth in photocatalysis. It is exciting to see that advances are still being made particularly with visible light photocatalysis and development of new catalysts that meet the demand for cleaner environment in the twenty-first century. Semiconductor photocatalysis is still quite young with regard to commercialization and this may appear slightly surprising given the intense level of academic research that has been conducted over the last decade. Although semiconductor photocatalysis appears to be rejected at present by most major water purification and disinfection companies as an economically viable alternative to current methods, such as chlorination and ozonolysis, it *has* been taken up by companies and government agencies that have small, specific water purification needs, such as those found in the electronics industry (high purity water), the military, and petrochemical industries. Commercialization in Japan has been initiated on a large scale and recently in China, the United Kingdom, and the United States products have started appearing that rely on the principle of photocatalysis. No semiconductor perfectly fits this demanding list of requirements, although the semiconductor titanium dioxide, TiO_2, comes close, failing in its inability to absorb visible light. In fact titanium is only able to absorb UV light (typically <380 nm), which represents a small fraction, about 6%, of the solar spectrum. However, the other, very positive features of titanium dioxide as a semiconductor photocatalyst, such as high photoactivity, chemical and photochemical robustness, and low cost, far outweigh its deficient spectral profile overlap with the solar spectrum. The hope is that sustained research will lead to the broad scale application of photocatalysis in the coming years.

Mesoporous silica forms the backbone in catalysis, ion exchange and separation, with progress being made in the structural, compositional, and morphological aspects. The next section will briefly discuss this interesting new class of materials known as mesoporous materials doped with transition metal ions and possible application in the area of photocatalytic remediation.

19.4.1 Mesoporous Materials

Mesoporous materials have created a lot of interest in photocatalysis because of their unique properties such as high surface areas and thermal stability. The synthesis of hexagonally packed mesoporous silicate and aluminosilicate materials with uniform pore sizes evenly distributed throughout the material was first shown by researchers at Mobil in 1992 (51, 52). They have been synthesized as silicates and aluminosilicates in basic media using a cationic alkyltrimethylammonium surfactant system, in acidic media (53, 54), using neutral amines (55), with substitution of various heteroatoms in the silicate structure. The cubic structure includes MCM-48 (surface area $\sim$1400 m^2/g), which was first synthesized in the siliceous form in basic media. However, MCM-41 with its unidirectional pore arrangement restricts the transport of products and reactants, whereas MCM-48 possesses a continuous three-dimensional pore system that provides favorable mass transfer kinetics and seems to be a more attractive candidate for catalytic applications than MCM-41. Owing to the lack of active sites in the matrices of pure silica, its use as a catalyst is limited. Active sites can be generated via chemical modification, that is, by introduction of a heteroatom into the silica matrix. Mesoporous materials with encapsulated transition metal cations either in the framework or cavities (such as Cr^{6+}, Co^{2+}, Ag^{+}, Mn^{2+}, V^{5+}, and Ti^{4+}) possess very high surface areas comparable to the parent mesoporous materials and their role as photocatalysts has been established. Numerous review articles and books have been published in this area; readers may refer to them for additional information.

The redox properties of mesoporous materials have been seriously considered only after an isomorphous replacement of silicon and aluminum by transition metal ions was demonstrated. For catalytic purposes transition metal ions are particularly attractive. For a given metal ion M^{n+} to be incorporated into the framework of the molecular sieve the ionic radii is most important as well as the tendency of the ion to assume tetrahedral coordination in its oxide. Incorporation of the metal ion leads to partial substitution of the silicon in the framework of the mesoporous material. The transition metal ions are considered to be finely dispersed at the atomic level at very low M/Si ratio. However, at higher M/Si ratio the metal ions can get into the extra framework position and might even form isolated metal oxide that can coat the walls of the pore channels. This is of great significance in the dispersion of transition metal ions which, when excited by visible light irradiation, form the corresponding ligand to metal charge-transfer excited state involving electron transfer from O_2^- to M^{n+}, as shown by Equation (19.31)

$$[M^{n+}O_2^-] \longrightarrow M^{(n-1)}-O^-]^* \qquad (19.31)$$

These charge-transfer excited states have high reactivities and are localized quite close to each other. It should be noted that the mechanism of these highly dispersed transition metal ions in mesoporous materials is quite different from the conventional bulk semiconductor photocatalysis. The highly dispersed metal ions show exceptional photocatalytic activity (36). An example of how these materials find application as indoor air purifiers using visible and UV illumination is acetaldehyde, a common toxic indoor air pollutant. It is regulated by the Environmental Protection Agency (EPA) and classified in its Toxic Release Inventory (TRI) as a suspected carcinogen. It is emitted from incomplete combustion from fireplaces, and consumer products such as adhesives, nail polish, etc. Acetaldehyde can be photocatalytically oxidized either under UV or visible light irradiation. Mixed metal oxides are known to be employed for decomposition of acetaldehyde under visible and UV irradiation. Cr, Co, Mn, V, Ni, and Fe incorporated titania–silica aerogels have been effective. Microporous titanium and transition metal incorporated ETS have proven useful candidates both under visible and UV irradiation for decomposition of acetaldehyde. A common approach to enhance the photocatalytic activity of TiO_2 is to use supported TiO_2 catalysts. The Ti–Cr–MCM-48 photocatalyst prepared in a single step exhibits far superior photocatalytic activity (56) compared to the TiO_2–Cr–MCM-48 prepared by the postimpregnation method. The high activity of the Ti–Cr–MCM-48 photocatalyst is attributed to the synergistic interaction between the Cr ions dispersed in the silica framework and the nanocrystalline nature of titania crystallites anchored onto the pore walls. The role of Cr ions is to impart visible light functionality. The highly dispersed Cr ions could be excited by visible light to form charge transfer excited states which involves transfer of an electron from oxygen anion to the chromium cation, as shown in the equation:

$$Cr^{6}-O^{2-} \xrightarrow{hv} [Cr^{5+}-O^{-}]^{*} \tag{19.32}$$

The charge-transfer states have unique reactivities because of the close proximity of the electron hole pairs that are localized next to each other in contrast to bulk semiconductors. Many transition metal ions enable visible light absorption by providing defective states in the bandgap. However, there is the possibility that the interband states can serve as recombination centers and therefore decrease the overall photocatalytic efficiency. Visible light activity by sub-bandgap illumination can be seen because of the electronic transition from the valence band to the defect sites or from defect sites to the conduction band.

ACKNOWLEDGMENT

I am thankful to my research advisor Prof. K. J. Klabunde at Kansas State University, Manhattan, KS for initiating research in photocatalysis and also for giving me the opportunity to be a part of the team involved in creating this book.

REFERENCES

1. H. Yamashita, M. Takeuchi, M. Anpo, *Encyclopedia of Nanoscience and Nanotechnology*, **2004**, *10*, 639.
2. A. Corma, H. Garcia, *Chem. Commun.* **2004**, 1443.
3. A. Fujishima, K. Honda, *Nature*, **1972**, *238*, 37.
4. D. F. Ollis, C. Turchi, *Environ. Prog.* **1990**, *9*, 229.
5. D. F. Ollis, E. Pelizzetti, N. Serpone, *Environ. Sci. Technol.* **1991**, *25*, 1523.
6. M. R. Hoffmann, S. T. Martin, W. Choi, D. W. Bahnemann, *Chem. Rev.* **1995**, *95*, 69.
7. A. L. Linsebigler, G. Lu, J. T. Yates, *Chem. Rev.* **1995**, *95*, 735.
8. A. Mills, S. Le Hunte, *J. Photochem. Photobiol. A: Chem.* **1997**, *108*, 1.
9. D. S. Bhatkhande, V. G. Pangarkar, A. A. C. M. Beenackers, *J. Chem. Technol. Biotechnol.* **2001**, *77*, 102.
10. J. Galvez, S. M. Rodriguez, Online publication: http://www.unesco.org/science/wsp/publications/solar.htm, 2002.
11. H. G. Karge, J. Weikamp (Eds.), *Molecular Sieves Science and Technology*, Vol. 1, 1998.
12. N. Serpone, A. Salinaro, *Pure Appl. Chem.* **1999**, *71*, 303.
13. C. S. Turchi, D. F. Ollis, *J. Catal.* **1989**, *119*, 480.
14. C. S. Turchi, D. F. Ollis, *J. Catal.* **1990**, *122*, 178.
15. H. Gerischer, A. Heller, *J. Phys. Chem.* **1991**, *95*, 5261.
16. J. Schwitzgebel, J. G. Ekerdt, H. Gerischer, A. Heller, *J. Phys. Chem.* **1995**, *99*, 5633.
17. C. M. Wang, A. Heller, H. Gerischer, *J. Am. Chem. Soc.* **1992**, *114*, 5230.
18. E. Pelizzetti, C. Minero, *Electrochim. Acta* **1993**, *38*, 47.
19. N. S. Foster, C. A. Koval, J. G. Sczechowski, R. D. Noble, *J. Electroanal. Chem.* **1996**, *406*, 213.
20. S. Upadhia, D. F. Ollis, *J. Phys. Chem.* **1997**, *101*, 2625.
21. A. Mills, A. Belghazi, D. Rodman, *Water Res.* **1996**, *30*, 1973.
22. M. A. Fox, M. T. Dulay, *Chem. Rev.* **1993**, *93*, 341.
23. D. F. Ollis, H. Al-Ekabi (Eds.), *Photocatalytic Purification and Treatment of Water and Air*, Elsevier, Amsterdam, 1993.
24. N. Serpone, E. Pelizzetti (Eds.), *Photocatalysis: Fundamentals and Applications*, Wiley-Interscience, New York, 1989.
25. O. Legrini, E. Oliveros, A. M. Braun, *Chem. Rev.* **1993**, *93*, 671.
26. J. Peral, X. Domenech, D .F. Ollis, *J. Chem. Technol. Biotechnol.* **1997**, *70*, 117.
27. Annual Report on Carcinogenesis Bioassay of Chloroform, National Cancer Institute, Bethesda, MD, March 1, 1976.
28. M. W. Stein, F. B. Snsome, *Degradation of Chemical Carcinogens*, Van Nostrand Rienhold, New York, 1968, 116.
29. J. J. Rook, *Water Treat. Exam.* **1974**, *23*, 234.
30. A. L. Pruden, D. F. Ollis, *J. Catal.* **1983**, *82*, 404.
31. C.-Y. Hsiao, C.-Y. Lee, D. F. Ollis. *J. Catal.* **1982**, *82*, 418.
32. W. H. Glaze, J. F. Kenneke, J. L. Ferry, *Environ. Sci. Technol.* **1993**, *27*, 177.

33. M. A. Anderson, S. Yamazaki-Nishida, S. Cervera-March, *in Photocatalytic Purification and Treatment of Water and Air*, D. F. Ollis, H. Al-Ekabi (Eds.), Elsevier, Amsterdam, 1993, 405.
34. M. R. Nimlos, W. A. Jacoby, D. M. Blake, T. A. Milne, *Environ. Sci. Technol.* **1993**, *27*, 732.
35. S. Rodrigues, K. T. Ranjit, S. Uma, I. N. Martyanov, K. J. Klabunde, *J. Catal.*, **2005**, *230*, 158.
36. M. Matsuoka, M. Anpo, *J. Photochem. Photobiol. C: Photochem. Rev.* **2003**, *3*, 225.
37. X. S. Zhao, G. Q. Lu, A. J. Whittaker, G. J. Millar, H. Y. Zhu, *J. Phys. Chem. B* **1997**, *101*, 6525.
38. S. Tunesi, M. Anderson, *J. Phys. Chem.* **1991**, *95*, 3399.
39. J. C. D'Oliveira, G. Al-Sayyed, P. Pichat, *Environ. Sci. Technol.* **1990**, *24*, 990.
40. J. C. D'Oliveira, C. Minero, E. Pelizzetti, P. Pichat, *J. Photochem. Photobiol A: Chem.* **1992**, *96*, 2226.
41. T. Pandiyan, O. M. Rivas, J. O. Martínez, G. B. Amezcua, M. A. Martínez-Carrillo, *J. Photochem. Photobiol. A: Chem.* **2002**, *146*, 149.
42. B. Serrano, H. de Lasa, *Chem. Eng. Sci.* **1999**, *54*, 3063.
43. U. Stafford, K. A. Gray, P. V. Kamat, *Chem. Phys. Lett.* **1993**, *205*, 55.
44. S. M. Rodríguez, J. B. Gálvez, M. I. M. Rubio, P. F. Ibáñez, W. Gernjak, I. O. Alberola, *Chemosphere* **2005**, *58*, 391.
45. M. P. Titus, V. G. Molina, M. A. Baños, J. Jiménez, S. Esplugas, *Appl. Catal. B: Environ.* **2004**, *47*, 219.
46. D. Robert, S. Malato, *Sci. Total Environ.* **2002**, *291*, 85.
47. G. Pecchi, P. Reyes, P. Sanhueza, J. Villaseñor, *Chemosphere* **2001**, *43*, 141.
48. A. G. Agrios, K. A. Gray, E. Weitz, *Langmuir* **2004**, *20*, 5911.
49. J. Theurich, M. Lindner, D. W. Bahnemann, *Langmuir* **1996**, *12*, 6368.
50. C. Kormann, D. W. Bahnemann, M. R. Hoffmann, *Environ. Sci. Technol.* **1994**, *28*, 786.
51. C. T. Kresge, M. E. Leonowicz, W. J. Roth, J. C. Vartuli, J. S. Beck, *Nature*, **1992**, *359*, 710.
52. J. S. Beck, J. C. Vartuli, W. J. Roth, M. E. Leonowicz, C. T. Krege, K. D. Schmitt, C. T.-W. Chu, D. H. Olson, E. W. Sheppard, S. B. McCullen, J. B. Higgins, J. L. Schlenker, *J. Am. Chem. Soc.* **1992**, *114*, 10834.
53. A. Monnier, F. Schuth, Q. Huo, D. Kumar, D. Margolese, R. S. Maxwell, G. D. Stucky, M. Krishnamurthy, P. Petroff, A. Firouzi, M. Janicke, B. F. Chmelka, *Science*, **1993**, *261*, 1299.
54. Q. Huo, D. I. Margolese, U. Ciesla, P. Feng, T. E. Gier, P. Sieger, R. Leon, P. Petroff, F. Schuth, G. D. Stucky, *Nature*, **1994**, *368*, 317.
55. P. T. Tanev, T. J. Pinnavaia, *Science* **1995**, *267*, 865.
56. S. Rodrigues, K. T. Ranjit, S. Uma, I. N. Martyanov, K. J. Klabunde, *Adv. Mater.* **2005**, *17*, 2467.

PART IX

BIOLOGICAL AND ENVIRONMENTAL ASPECTS OF NANOMATERIALS

20

NANOMATERIALS FOR ENVIRONMENTAL REMEDIATION

Angela Iseli, HaiDoo Kwen, and Shyamala Rajagopalan

20.1 Introduction, 650
20.2 Nanoscience in Environmental Remediation, 651
20.3 Groundwater Remediation, 652
    20.3.1 Removal of HOCs Using Monometallic and Bimetallic Nanoparticles, 652
    20.3.2 Removal of Inorganic Metals, 657
20.4 Drinking Water Purification, 659
    20.4.1 Removal of Bacteria, 659
    20.4.2 Removal of Organic Compounds, 660
    20.4.3 Removal of Arsenic, 662
20.5 Soil Remediation, 662
    20.5.1 General Soil Composition, 662
    20.5.2 Common Soil Contaminants, 663
    20.5.3 Nanotechnology in Soil Remediation, 663
20.6 Air Purification, 668
    20.6.1 Removal of VOCs, 668
    20.6.2 Removal of NOx and SOx, 669
20.7 Conclusions and Future Outlook, 669
Further Reading, 670
References, 670
Problems, 674
Answers, 676

Nanoscale Materials in Chemistry, Second Edition. Edited by K. J. Klabunde and R. M. Richards

20.1 INTRODUCTION

A nanoparticle is broadly defined as "a sub-classification of ultrafine particle with lengths in two or three dimensions greater than 0.001 μm (1 nm) and smaller than about 0.1 μm (100 nm), and which may or may not exhibit a size-related intensive property" (1). The seemingly arbitrary size cutoff of 100 nm or less is based on the observation that novel properties that differentiate nanoparticles from the bulk material typically develop at a critical length scale of under 100 nm. This effect is due primarily to both the large surface area of the nanoparticle relative to its volume and the fact that at the nanometer scale, quantum effects occur (2). The area of research into nanoparticle and nanomaterial properties is known as nanoscience, and it has grown from the desire to understand and manipulate the unique properties of the nanoparticles. Nanotechnology is the design and production of structures, devices, and systems by controlling shape and size at the nanometer scale. To date, an enormous variety of nanomaterials has been manufactured for an equally enormous number of medical, industrial, consumer, construction, and military uses. Nanoparticles are already present in some consumer goods, such as nanoparticle zinc oxide in sunscreens and cosmetics and the nanomaterial stain-resistant coatings on select items of clothing (3). Nanosurfaces and nanostructures are being researched for use in orthopedic implants; nanofibers strengthen many industrial composite materials; and the U.S. government is particularly interested in energy storage devices composed of nanomaterials (4).

Nanotechnology also promises to be a powerful tool in the field of environmental remediation. Currently, planet Earth is home to approximately 6.6 billion human beings, plus an inestimable number of plants, animals, insects, and microbes (5). All of these living creatures share the same air, soil, and water. Recent research has demonstrated that anthropogenic environmental changes (changes caused by human activity) can have an extremely harmful effect on all living organisms. The United States Environmental Protection Agency (U.S. EPA) has issued bulletins outlining how global warming and changing weather patterns are assisting in the spread of malaria and the West Nile virus by creating conditions favorable to the mosquito vectors of those diseases (6). Pollution caused by human activities such as mining, manufacturing, farming, and even simply dwelling has a profound effect on the environment. The most common methods for controlling pollution resulting from mining or manufacturing are removal, sequestration, or confinement: large amounts of contaminated soils or water are kept contained behind dams, earthen berms, or other sizable structures, or are removed entirely (7). These confinement areas can be expensive and labor intensive to maintain, particularly since contaminants must be kept from leaching into the groundwater or causing runoff into surface water. Frequently, these requirements necessitate a lined containment area of considerable size to prevent leaching, overflow, and subsequent runoff (8). The contaminants generally do not break down by themselves at any appreciable rate, even with long-term environmental exposure, and the resulting breakdown products or adducts may be even more toxic than the original contaminant (9). An effective means of environmental remediation would allow the contaminated soils and waters to be reclaimed and reused.

20.2 NANOSCIENCE IN ENVIRONMENTAL REMEDIATION

Pollution from human activities is the largest single threat to the environment. Improper waste disposal of environmentally persistent chemicals such as polychlorinated biphenyls (PCBs) and heavy metals causes soil and water pollution that lasts for years and is extremely difficult to clean up. As of September 2003, there were a total of 1244 Superfund sites on the U.S. EPA's National Priorities List (10). This is an average of nearly 25 sites per state where the soil, air, and/or water are so polluted as to pose an imminent health threat to anyone living in the vicinity. The amount of pollution worldwide is unknown, but it is clearly a greater threat to residents of developing countries, where environmental standards tend to be lax and enforcement is sometimes nonexistent. Further, because of the natural patterns of air and water movement (prevailing winds, ocean currents, and water cycling), pollution does not tend to stay confined to its area of origin but is frequently dispersed far from its original source. Research by both the Commonwealth Scientific and Industrial Research Organization Marine and Atmospheric Research in Australia (CMAR) and the National Aeronautics and Space Administration (NASA) in the United States has confirmed a link between air pollution and severe drought (11, 12). Severe droughts, in turn, cause hardship and harm to animals and plants in the affected areas, and lead to epidemics of disease in the weakened human population.

Because of the tendency of pollution to spread and the difficulty involved in cleaning contaminated air, soils, and water, pollution should ideally be removed at its source prior to its release into the environment. The ability to remove toxic contaminants from the surface water sources to a safe level rapidly, efficiently, and within reasonable costs is important. Adsorbent-based technology is generally used for water and air purification (13, 14). The adsorbents are usually high surface area materials such as carbons, metals, metal oxides, or zeolites. In particular, a number of nanomaterials are available for use in air and water filtration, including nanoscale zero-valent iron ($Fe^{0}_{(NP)}$) for removal of As(III) from groundwater (15, 16) and the trapping and decomposition of nitrous oxide on carbon nanotubes (17). A variety of nanoparticles are in various stages of research and development, each possessing unique functionalities that are potentially applicable and environmentally acceptable for the remediation of drinking water. Examples of various nanoparticles and nanomaterials that have been used in water remediation include zeolites (18), carbon nanotubes (19), self-assembled monolayers on mesoporous supports (20), biopolymers (21), single-enzyme nanoparticles (22), zero-valent iron nanoparticles (23), bimetallic iron nanoparticles (24), and nanoscale semiconductor photocatalysts (25).

$Fe^{0}_{(NP)}$ has also been shown to remove As(III) from contaminated soils (15, 16), and assist in the remediation of aquifers contaminated with polycyclic aromatic hydrocarbons (26). Other nanomaterials have been developed to act as pollutant sensors, such as the trinitrotoluene (TNT) sensor, which can detect TNT at concentrations as low as 20 ng/mL (27).

In summary, nanotechnologies can be used to detect and monitor, as well as remediate or enhance the remediation of environmental contamination. However, this chapter focuses specifically on remediation based primarily on high surface area metal oxides.

20.3 GROUNDWATER REMEDIATION

Nanostructured materials offer several advantages over their bulk counterparts that can be potentially exploited for groundwater remediation. For example, the higher surface area to volume ratio of nanostructures can lead to an enhanced reactivity with environmental contaminants, resulting in their destruction by reactive adsorption (2, 28). Nanostructured materials, such as Si(IV), III-V semiconductors (GaN, GaP, GaAs, InN, InP, InAs, etc.), and II-VI semiconductors (ZnS, ZnSe, ZnTe, CdS, CdSe, CdTe, etc.), can also display quantum size effects, wherein the bandgap (the separation between the valence band maximum and the conduction band minimum) energy increases with a decrease in particle size (29). The increased bandgap means that light excitation produces a stronger reductant and/or oxidant than the corresponding bulk semiconductor. This is especially important for photo-driven environmental remediation reactions. In addition to increased surface to volume ratio and quantum size effects, the properties of nanostructures can be potentially tuned toward specific environmental remediation applications through surface modification.

Remediation specifically refers to efforts focused on converting hazardous materials accidentally released into the environment into benign end products. Groundwater remediation schemes generally involve pumping the polluted groundwater to the surface and passing it through a treatment system. This helps with the degradation of the pollutant, as in the case of advanced oxidation systems, or transfers them to another medium, as in the case of air stripping and granular activated carbon. Methods that degrade the contaminants are preferable to those that simply transfer them to another medium and it is expected that nanoparticles will play a decisive role in this aspect of remediation. As described in the following sections, most of the nanotechnology-based approaches are currently targeted at removing halogenated organic compounds (HOCs), with a few in the area of heavy metal removal (30).

20.3.1 Removal of HOCs Using Monometallic and Bimetallic Nanoparticles

Disposal of industrial chemicals is a problem all over the world. Traditionally, toxic chemicals such as HOCs were disposed of by burying containers in the ground or simply dumping liquids onto the ground. However, these methods of disposal are highly unsatisfactory. Buried containers sometimes degrade, leaking their contents into the environment, and liquids dumped on the ground tend to seep into the soil, where they eventually find their way into water systems, thereby contaminating the environment and domestic water supplies.

Destruction of HOCs by zero-valent metals, particularly iron, represents an emerging technology in environmental remediation. It has been shown that granular iron can degrade many chlorinated compounds, including chlorinated aliphatics, chlorinated aromatics, and polychlorinated biphenyls (PCBs), as well as nitroaromatic compounds. Early studies by Gillham and O'Hannesin examined the utility of zero-valent iron (Fe^0) in the degradation of 14 chlorinated aliphatic hydrocarbons (31). Significant rates (rapid degradation with t_{50} values 5 to 15 orders of magnitude lower than half-lives reported for abiotic degradation processes) of degradation were observed for all compounds tested, with the exception of dichloromethane. The observed results were attributed to reductive dechlorination, with iron serving as the source of electrons as shown in Equation (20.1).

$$Fe^0 + X - Cl + H_2O \longrightarrow Fe^{2+} + OH^- + X - H + Cl^- \qquad (20.1)$$

Further studies by Orth and Gillham on the dechlorination of trichloroethene (TCE) using Fe^0 confirmed the presence of degradation products (32). The major degradation product was ethene, followed by ethane, with appreciably smaller amounts of other C1-C4 hydrocarbons. In addition, three dichloroethene (DCE) isomers and vinyl chloride (VC) accounted for about 3.0% to 3.5% of the initial TCE concentration. Based on the low concentrations of chlorinated degradation products in the solution phase, it is proposed that most of the TCE remains sorbed to the iron surface until complete dechlorination is attained.

Even though Fe^0 is effective for treatment of a variety of organic and inorganic contaminants, specifically when employed in permeable reactive barriers (PRBs) for remediation of contaminated groundwater, improvements to this technology are needed to overcome a variety of limitations, including minimization of the formation of toxic compounds such as DCE and VC from tetrachloroethene (PCE) and TCE, enhanced degradation of otherwise recalcitrant compounds such as PCBs, and ready deployment where PRBs are not feasible, such as in the deep subsurface (33).

In regard to this, nanoscale iron particles were expected to offer true advantages over granular iron because they provide a greater surface area and hence higher reactivity. Additionally, higher mobility due to their smaller size may make them highly suitable for *in situ* application by injection. In fact, $Fe^0_{(NP)}$ is reported to be more reactive than commercial iron powders towards halogenated hydrocarbons and PCBs (34). Unlike granular zero-valent iron, $Fe^0_{(NP)}$ is not commercially available. Generally, it is synthesized by reduction of an aqueous solution of ferric iron (Fe^{3+}) using sodium borohydride. This produces $Fe^0_{(NP)}$ with 90% of the particles having a primary particle size in the range of 1 to 100 nm. BET specific surface area of the particles is ~37 times larger than the commercially available Fe^0 (33.5 m^2/g versus 0.9 m^2/g). Samples of $Fe^0_{(NP)}$ have also been coated with a layer of Pd and have been shown to be very effective as well in dechlorination reactions. Degradation of TCE by various metal particles is presented in Figure 20.1.

As seen, freshly synthesized $Fe^0_{(NP)}$ as well as nano Pd/Fe bimetallic samples were more reactive than the corresponding commercial samples. For example, TCE was degraded in 1.7 h when reacted with Fe nanoparticles, while under the same reaction conditions it took only 0.25 h with Pd/Fe nanoparticles. Interestingly, when bulk iron was used to degrade TCE, other toxic products such as DCE and VC were produced. None of these toxins were detected with either Fe or Pd/Fe nanoparticles. Similar reactivity differences were noticed with PCB.

In addition to Pd/Fe other higher surface area nanoscale bimetallic particles such as Pt/Fe, Ni/Fe, and Pd/Zn have been synthesized and tested (35). Rapid and complete dechlorination was observed with several chlorinated organic solvents and chlorinated

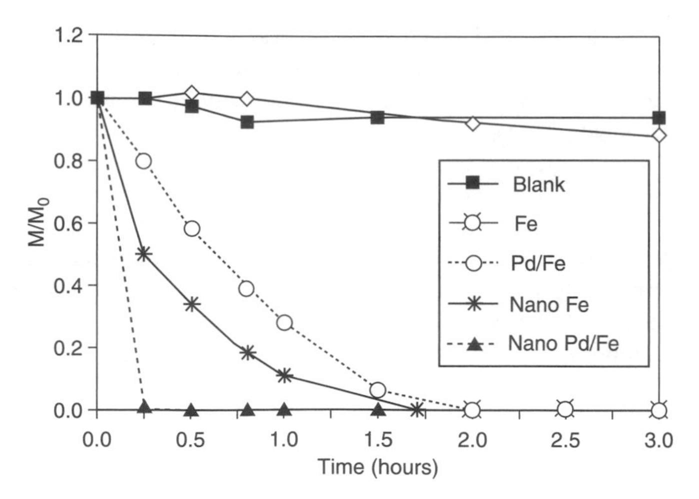

Figure 20.1 Reactions of TCE with commercial Fe powders (Fe), Pd-modified commercial Fe powders (Pd/Fe), nanoscale Fe particles (nano Fe), and nanoscale Pd/Fe particles (nano Pd/Fe). Initial TCE concentration was 20 mg/L. Metal to solution ratio was 2 g/100 mL. Copyright © 2008 American Chemical Society.

aromatic compounds. Similar studies with Fe/Ag nanoparticles reveal them to be highly effective for dechorination of hexachlorobenzene (HCB) to terta-, tri-, and dichorobenzenes (TeCB, TCB, and DCB, respectively) within 24 h. The rate of dechlorination was positively correlated to the silver loading of the bimetallic particles (36). Although the exact mechanism of the enhanced reactivity of the nanoscale bimetallic particles is not known, a reasonable explanation appears to be that with nano bimetallics, one metal (Fe or Zn) serves mostly as electron donor reacting with the contaminants, while the other metal (Pt, Ni, or Pd) acts as the catalyst further promoting degradation (35). More importantly, production of chlorinated by-products is considerably reduced due to the presence of a catalyst. Based on their studies on the catalytic reduction of chlorobenzenes with Pd/Fe nanoparticles, Zhu and Lim (37) concluded that aged particles exhibited a significant decrease in dechlorination activity. The loss in reactivity was attributed to either Pd leaching from the surface of iron or Pd encapsulation by the iron oxide films developed over the aging period.

During their investigation on the dechlorination of carbon tetrachloride (CT) and chloroform (CF), Lien and Zhang (38) found that in chlorinated organic water solution or mixtures, $Fe^0_{(NP)}$ particles were predominantly suspended in the liquid phase while Fe^0 (microscale) usually settled at the bottom of the bottle. Therefore, initial diffusion and subsequent transport of the chlorinated organics to the suspended $Fe^0_{(NP)}$ may have resulted in the faster degradation achievable with $Fe^0_{(NP)}$ compared to Fe^0.

An alternative cost-effective method for the preparation of $Fe^0_{(NP)}$ involves reduction of $FeCl_2$ using technical grade sodium dithionite (39). $Fe^0_{(NP)}$ produced by this method is significantly cheaper, easier, and safer to produce than the particles produced by borohydride reduction.

Li, Vipulanandan, and Mohanty (40) explored the degradation of TCE using two types (microemulsion and solution methods based) of $Fe^0_{(NP)}$ particles. The size of particles produced by the solution method ranged from 20 to 70 nm, with most of the particles in the range of 50 nm. Interestingly, microemulsion methodology provided particles that were uniform in size, with the average size less than 10 nm. The time needed for complete reduction of TCE was different for different initial TCE concentrations, $Fe^0_{(NP)}$ loading levels, and particle sizes. For example, reduction time proved to be lower with lower initial TCE concentrations, smaller particle sizes, and higher iron loading.

In a recent study, Liu et al. (41) used two types of $Fe^0_{(NP)}$ particles in exploring the TCE degradation: particles synthesized by aqueous phase reduction of ferrous or ferric iron by sodium borohydride (Fe^{BH}; average primary particle size = 30 to 40 nm) and commercially available reactive nanoscale iron particles (RNIP), synthesized by gas phase reduction of iron oxides in hydrogen (Fe^{H2}; average primary particle size = 40 to 60 nm). Both particles had a Fe(0) core structure with similar specific surface areas (30 m^2/g). The shell structure for Fe^{BH} was found to be oxide or oxyhydroxide of some sort while Fe^{H2} contained Fe_3O_4 as the only oxide in the shell. Employed dechlorination reaction conditions included high TCE concentration (limited iron) and low TCE concentration (excess iron). The measured surface area normalized pseudo first-order rate constant for Fe^{BH} (1.4×10^{-2} L h^{-1} m^{-2}) was approximately fourfold higher than for Fe^{H2} (3.1×10^{-3} L h^{-1} m^{-2}). The final products of TCE degradation by Fe^{BH} were similar resulting in ethane and C3-C6 products. On the other hand, the primary products from Fe^{H2} were acetylene and ethene under limited iron conditions while ethene and ethane were observed under excess iron conditions. Their study also concluded that faster reaction could be achieved through lower TCE concentrations and smaller particle sizes. In a study directed towards understanding the structure-reactivity relationships of micro and various $Fe^0_{(np)}$ particles, Nurmi et al. (42) have examined Fe^{BH} and Fe^{H2}samples using benzoquinone (BQ) and CT as the contaminants. Based on the observed data, it was concluded that both types of $Fe^0_{(NP)}$ react more rapidly than micro Fe^0 based on mass-normalized rate constants, but surface area-normalized rate constants do not show a significant nanosize effect. The apparent lack of an "intrinsic" nanosize effect is attributed to the complexity in defining or measuring the applicable specific surface area for highly reactive materials whose properties alter with time and environmental conditions.

Lowry and Johnson (43) explored the efficiency of dechlorination of dissolved PCBs by Fe^0 (particle size = <150,000 nm) and $Fe^0_{(NP)}$ (particle size = 30 to 50 nm) particles in water/methanol solutions. With commercial Fe^0, no PCB dechlorination was observed even after 180 days, suggesting that this form of iron is not reactive. On the other hand, $Fe^0_{(NP)}$ resulted in effective dechlorination of PCBs within 45 days. Wang and Zhang (34) also studied the dechlorination of PCBs in an ethanol-water mixture under ambient conditions using $Fe^0_{(NP)}$ and Pd coated $Fe^0_{(NP)}$. Over a 17 h experiment, partial degradation (~25%) of PCB to biphenyl was observed with $Fe^0_{(NP)}$ while Pd/Fe resulted in complete dechlorination.

One of the issues with $Fe^0_{(NP)}$ particles prepared by traditional methods is their tendency to either agglomerate rapidly or react quickly with the surrounding media, resulting in considerable loss in reactivity. Since agglomerated Fe^0 particles are often in the range of micron scale, they are not readily transportable or deliverable in other media and thus can't be used for *in situ* applications. In order to solve this problem, low-cost, food-grade polysaccharides such as starch, cellulose, and their derivatives have been used as "green" stabilizers to yield nanoparticles suitable for the *in situ* uses (44). Starch-stabilized Fe/Pd nanoparticles were distinct, well dispersed, and displayed greater dechlorination capability than those prepared without a stabilizer. The presence of ~0.2 wt% starch prevented agglomeration of the Fe/Pd nanoparticles and helped maintain their high surface area and associated reactivity. For example, at an initial TCE concentration of 52 mg/L and Fe dose of 0.1 g/L, the starch-stabilized Fe/Pd nanoparticles removed ~98% TCE within 1 h, while the pristine Fe/Pd samples degraded ~78% within 2 h. Similarly, carboxymethyl cellulose (CMC) stabilized Fe/Pd nanoparticles eliminated TCE 17 times faster than their nonstabilized counterparts.

Despite the much-studied reactivity of $Fe^0_{(NP)}$ their transport mechanisms and long-term stability appear to be relatively untested. In 2003, Zhang reported the outcome of a field-scale pilot test using $Fe^0_{(NP)}$ for *in situ* groundwater treatment (45). The results observed were very similar to the laboratory results, with concentrations of PCE, TCE, and DCE reduced by $Fe^0_{(NP)}$ to levels near or below groundwater quality standards within six weeks. Similarly, the applicability of the bimetallic nanoparticles was demonstrated in a field study in which Pd/Fe nanoparticles were injected into groundwater contaminated by chlorinated hydrocarbons. Near complete dechlorination of TCE to benign hydrocarbons was reported by using ~1.7 kg sample of Pd/Fe nanoparticles (test area dimensions: 4.5 m × 3.0 m; thickness 6.0 m) during a four week monitoring period (46).

One study evaluated the performance of an emulsified zero-valent iron ($Fe^0_{(ENP)}$) technology to specifically improve the cleanup time achievable with the use of conventional Fe^0 (47). $Fe^0_{(ENP)}$ is composed of food-grade surfactant, biodegradable vegetable oil, and water, which form emulsion particles that contain the Fe^0 (micro or nano) particles in water surrounded by an oil-liquid membrane. Groundwater samples were collected before and after treatment and analyzed for volatile organic compounds (VOCs) to evaluate the changes in VOC mass, concentration, and mass flux. Significant reductions in TCE groundwater concentrations (57% to 100%) were observed at all depths targeted with $Fe^0_{(ENP)}$. The decrease in concentrations of TCE in groundwater is believed to be due to abiotic degradation associated with the Fe^0 as well as biodegradation enhanced by the presence of the oil and surfactant in the $Fe^0_{(ENP)}$ emulsion.

Quite recently, the effect of aging on the structure and reactivity of Fe^{H2} in aqueous solution was studied (48). As mentioned earlier, aging appears to be one of the most critical and potentially limiting factors in the use of $Fe^0_{(NP)}$ in the remediation of groundwater contaminants. Results based on structural characterization and reactivity studies show that Fe^{H2} becomes more reactive between 0 and ~2 days exposure to water and then slowly loses reactivity over the next several days. This decline in reactivity is attributed to the rapid destruction of the original Fe(III) oxide film on Fe^{H2}

during immersion in water and the subsequent passivation of the surface by a mixed valence Fe(II)-Fe(III) oxide shell.

As seen from the above discussion, mono- and bimetallic nanoparticles are promising materials for *in situ* remediation of halogenated hydrocarbons. These materials offer ease of use, effectiveness against a large number of contaminants, and deployment methods that are potentially easier and less disruptive than the technology currently in use. While the production of these nanoparticles and nanoparticle-containing structures is currently not cheap, improved production methods can make them more economical. However, more detailed studies are required to establish the structure-activity relationships of various nano formulations as well as the mechanism of dehalogenation of chlorinated hydrocarbons on the surface of mono- and bimetallic nanoparticles. Information emerging from such studies will aid in the rational design of novel metal-based nanoparticles for remediation of specific environmental pollutants.

20.3.2 Removal of Inorganic Metals

Inorganic contamination is a significant hazard to the environment and more specifically to water supplies. Metallic ions such as chromium (Cr), lead (Pb), and arsenic (As) are significant threats to the environment as well as human health. These metallic ions are introduced into the environment through various natural processes (natural erosion, volcanic emissions, biochemical reactions, etc.) and human activities (industrial processes such as steel mill runoff, leather tanning and wood preservation industries, mining, coal burning, household plumbing, auto exhaust, etc.). In September 2003, there were 1244 Superfund sites listed on the NPL; of those, 77% needed to be remediated for metals. Arsenic, chromium, and lead are the most prevalent metals found at Superfund sites, with zinc, nickel, and cadmium also common. EPA estimates that at 456 sites where remediation is planned, $\sim$66 million cubic yards of soil, sludge, and sediment are contaminated with metals (10).

20.3.2.1 Removal of Chromium and Lead A few studies have examined the use of Fe^0 for the reduction of Cr(VI) in aqueous solutions and very little work has been performed on the reduction of aqueous Pb(II) (49). The removal mechanism for Cr(VI) by Fe^0 is through the reduction of Cr(VI) to Cr(III), coupled with oxidation of Fe^0 to Fe(II) and Fe(III), and the subsequent precipitation of a sparingly soluble Fe(III)-Cr(III)(oxy)hydroxide phase.

One important parameter in the rate at which remediation of the undesirable contaminants occurs is the surface area of the Fe^0 particles. Ponder and Coworkers (50) have specifically addressed this question by using $Fe^0_{(NP)}$ particles, 10 to 30 nm in diameter, in the presence of a support material such as polymeric resin, silica gel, or sand. The use of a support prevents agglomeration of the iron and therefore presents a higher specific surface area of iron to the aqueous stream. These supported iron particles, referred to as Ferragel, were tested for their ability to separate and immobilize Cr(VI) and Pb(II) ions from aqueous solutions. Rates of remediation of Cr(VI) and

Pb(II) were up to 30 times higher for Ferragels than for iron filings or iron powder on a Fe molar basis.

In a related approach, starch- or cellulose-stabilized $Fe^0_{(NP)}$ were shown to be highly effective in the reductive removal of Cr(VI) from water. While 1 g of nonstabilized $Fe^0_{(NP)}$ was able to reduce 84.4 to 109 mg Cr(VI) in the groundwater, under the same conditions stabilized $Fe^0_{(NP)}$ removed 252 mg Cr(VI), attesting to their enhanced reactivity (51). Recently, nanoscale iron/carbon composites (C-Fe^0) have been shown to exhibit both rapid remediation of Cr(VI) and good transport properties (52).

In an effort to increase the reactivity of Fe^0 towards chromium removal, acid washed zero-valent iron [AW-Fe(0)] was evaluated under groundwater geochemistry conditions. It was found that AW-Fe(0) could remove Cr(VI) from synthetic groundwater in the absence of bicarbonate, magnesium, and/or calcium ions. The presence of bicarbonate alone had the mildest impact on adsorption while co-presence with calcium had a pronounced negative impact. In comparison with unwashed Fe(0), the AW-Fe(0) displayed a poorer Cr(VI) removal capability [~57% to 77% drop in capacity (53)].

20.3.2.2 Removal of Arsenic Arsenic (As), a metal that is a common constituent of the Earth's crust, is also a well-known carcinogen. Depending on redox and pH conditons, it is usually present in water in different oxidation states and acid-base species. In groundwater arsenic exists predominantly as inorganic arsenite, As(III) (H_3AsO_3, $H_2AsO_3^{1-}$, $HAsO_3^{2-}$), and arsenate, As(V) (H_3AsO_4, $H_2AsO_4^{1-}$, $HAsO_4^{2-}$). The natural occurrence of As in groundwater in of huge concern due to its toxicity and the potential for chronic exposure. Application of laboratory-synthesized $Fe^0_{(np)}$ (by sodium borohydride reduction) for the remediation of As(III)

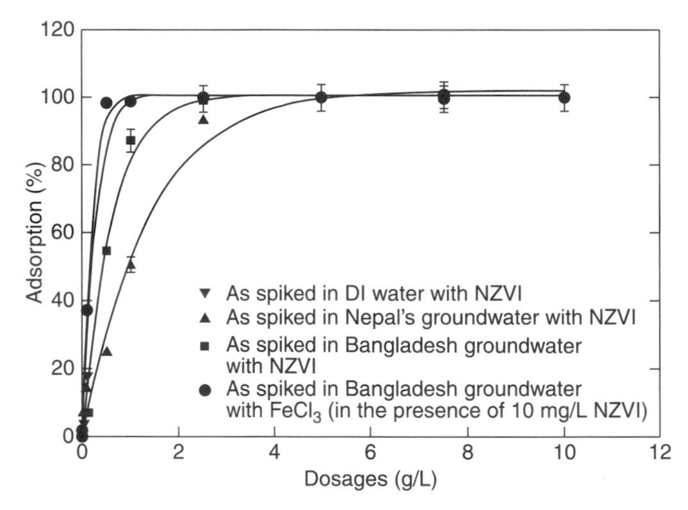

Figure 20.2 As(III) removal from groundwater using NZVI and $FeCl_3$ from groundwater of Nepal and Bangladesh. As(III): 1 mg/L in 0.01 M NaCl; NZVI: 1 g/L, pH 7, 25°C. Copyright © 2008 American Chemical Society.

in groundwater has been investigated in some detail (15). By using a variety of characterization techniques this group of researchers has characterized the particle size, surface morphology, and corrosion layers formed on pristine $Fe^0_{(NP)}$ and As(III)-treated $Fe^0_{(NP)}$. Batch adsorption studies on the removal of As(III) were carried out on groundwater obtained from Bangladesh and Nepal. As shown in Figure 20.2, As(III) was removed with 100% efficiency with 1.0, 2.5, and 4.5 g/L $Fe^0_{(NP)}$ for As spiked in DI water, in the Bangladesh groundwater, and in the Nepal groundwater, respectively.

It was also found that As removal by $FeCl_3$ required only a very low concentration of $Fe^0_{(NP)}$ (10 mg/L). The need for greater amounts of $Fe^0_{(NP)}$ for complete removal of As(III) from field-collected groundwater is attributed to the presence of anions such as HCO_3^{1-} and SO_4^{2-} and possibly trace amounts of silica and $H_2PO_4^{1-}$. The actual surface site on $Fe^0_{(NP)}$ that is reactive towards As(III) appears to be either an iron(II), mixed iron(II)/(III), or iron(III) oxide, hydroxide, or oxyhydroxide corrosion product. Similar studies with As(V)-contaminated water samples clearly established the potential of $Fe^0_{(NP)}$ for adsorptive removal of this contaminant as well (54).

20.4 DRINKING WATER PURIFICATION

There is a need for potable water in all areas of the world. In developed countries, this water is purified and typically supplied from large-scale municipal water systems that rely on surface water sources such as reservoirs, lakes, and rivers. However, in developing and underdeveloped nations many people still draw their drinking water from private water sources such as a village well, natural springs, and aquifers that contain bacteria, viruses, toxic metal ions, inorganic and organic solutes, and radionuclides that are harmful to humans, in many cases rendering the water unfit for consumption. As a result, a considerable number of people die each year as the direct result of drinking contaminated water. In the United States, the EPA regulates the quality of the nation's drinking water by enforcing safe drinking water standards and also protects the nation's drinking water by regulating the release of pollutants into the environment (55).

20.4.1 Removal of Bacteria

The primary sources of bacteria in water are human and animal wastes, seepage or discharge from septic tanks, sewage treatment facilities, and natural soil and plant bacteria. Pathogenic organisms can cause intestinal infections, dysentery, hepatitis, typhoid fever, cholera, and other illnesses. The EPA requires that all public water suppliers regularly test for coliform bacteria and deliver water that meets the EPA standards. The EPA Maximum Contaminant Level (MCL) for coliform bacteria in drinking water is zero total coliforms per 100 mL of water (56).

It has been reported that single or mixed metal oxide nanoparticles, such as zinc oxide, copper oxide, aluminum oxide, or titanium oxide, incorporated into a filtration medium containing a binder matrix, can destroy bacteria (57). The metal oxide nanocrystals are included in amounts ranging from approximately 0.1% up to about 10% by weight, based on the entire filtration medium. In a series of studies, it has been shown that MgO nanoparticles are very effective biocides against Gram-positive and Gram-negative bacteria (*Bacillus megaterium* and *Escherichia coli*) and bacterial spores (*Bacillus subtillus*) (58).

In another study *Escherichia coli* cells treated by Ag nanoparticles were found damaged, showing formation of "pits" in the cell wall of the bacteria (59). Jain and Pradeep have studied the efficacy of silver nanoparticles as a drinking water filter where there is bacterial contamination of the surface water (60). Silver nanoparticles were utilized to make stable, silver-coated filters from common polyurethane (PU) foams. The performance of the material as an antibacterial water filter was checked and no bacterium was detected in the output water when the input water had a bacterial load of 1×10^5 colony-forming units (CFU) per milliliter. The antibacterial action was also checked inline for a flow rate of 0.5 L/min and no bacterium detected, which suggests that domestic use of this technology is possible.

Pal, Joardar, and Song (61) found that the modification of activated carbon (AC) by aluminum hydroxychloride (AHC contains 42% to 44% Al_2O_3 and substantial amounts of tridecamer Al_{13}) at pH 6.0 to 6.5 and coating diatomaceous earth (DE) with zinc hydroxide at pH 6.5 to 7.0 changed the zeta potentials of the filter medium from negative to positive. Since most bacteria are negatively charged due to the abundance of the anionic groups present within the cell wall a positively charged filter medium should enhance bacteria removal. AHC-modified AC samples were further treated with silver halide and two antibacterial compounds to prevent microbial growth on filter media. *In situ* precipitation of silver bromide on AC resulted in the formation of nanosized AgBr (25.1 to 27.9 nm) crystals. Thirty grams of either AHC-treated AC or nano-AgBr supported AC could provide >6 log *E. coli* removal over ~1000 L of input water with a bacterial load of 10^7 CFU/mL.

All of these approaches offer advantages over current treatment technologies, which rely heavily on the addition of disinfectants and sanitizers, plus various flocculants to settle out large particulate contaminants. However, more research and development is still needed to clearly establish the advantages of nanotechnology for water purification.

20.4.2 Removal of Organic Compounds

Various organic compounds are common contaminants in both wastewater and drinking water and must be removed before the water can be discharged or consumed. Organic compounds can enter drinking water supplies through leaking underground tanks, agricultural and urban runoff, or improperly disposed waste. Because these contaminants are soluble in water at low concentrations, standard coagulation and sedimentation techniques are not effective for their removal. The presence of natural organic matter (NOM) in the source water is also a cause of concern because the reaction of NOM with disinfectants, such as chlorine, results in the formation of disinfection by-products (DBPs). These include trihalomethanes (THMs) and haloacetic acids (HAAs). Because of their toxicity, the THMs and HAAs are regulated by the U.S. EPA. Activated carbon or charcoal has been used for many years to remove VOCs from water. However, the process of removal is by adsorption of organic material onto active sites on or in the pores of the carbon, resulting in eventual saturation of the carbon with the adsorbed species. Hence, reliable nanotechnology is needed to destroy the organics during removal, to prolong the life of the filtration media.

Moeser et al. studied a class of water-based magnetic fluids that are specifically tailored to extract soluble organic compounds from water (62). These magnetic fluids consist of a suspension of ~7.5 nm magnetite (Fe_3O_4) nanoparticles coated with a 9 nm bifunctional polymer layer comprising of an outer hydrophilic poly(ethylene oxide) (PEO) region for colloidal stability and an inner hydrophobic poly(propylene oxide) (PPO) region for solubilization of organic compounds. The particles exhibit a high capacity for organic solutes with partition coefficients between the polymer coating and water on the order of 103 to 105, which is consistent with values reported for solubilization of these organics in PEO-PPO-PEO block copolymer micelles.

Choi et al. investigated the effect of photocatalytic coagulation on the fouling of nanofiltration (NF) membranes using copper-doped TiO_2 nanoparticles (63). Results indicated that the use of photocatalytic coagulation in conjunction with NF membrane processes can potentially provide benefits in terms of both NOM removal and fouling reduction. As shown in Figure 20.3, under bandgap excitation, semiconductor nanoparticles such as TiO_2 undergo charge separation and initiate oxidation of the organics at the interface.

ZnO is another important large bandgap semiconductor with band energies very similar to those of TiO_2. By using 4-chlorocatechol (4-CC) as a model compound, the dual role of ZnO semiconductor film as a sensor and a photocatalyst was demonstrated (64).

In a recent study, a commercial membrane sample was coated 20 or 40 times with 4 to 6 nm iron oxide nanoparticles (65). With this membrane, the concentration of dissolved organic carbon was reduced by >85% and the concentrations of simulated distribution system total THMs and HAAs decreased by up to 90% and 85%, respectively. When the coated membranes were used, the concentrations of aldehydes, ketones, and ketoacids in the permeate were reduced by >50% as compared to that obtained with the uncoated membranes. Hydroxy or other radicals produced at the

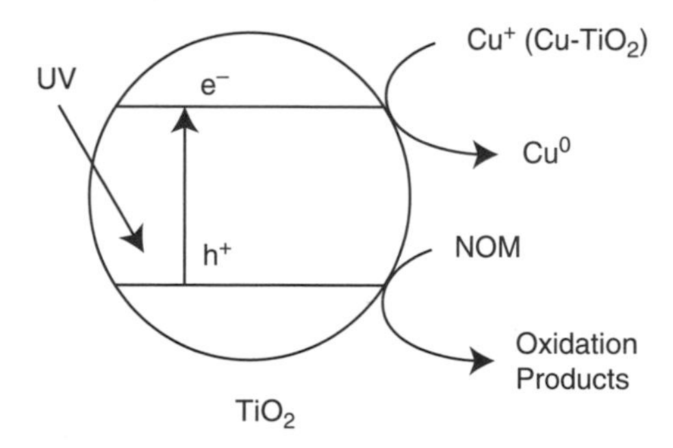

Figure 20.3 Representation of charge separation within a photocatalytic coagulant ($Cu\text{-}TiO_2$) following bandgap excitation.

iron oxide-coated membrane surface are believed to have enhanced the degradation of the NOM, thereby reducing the concentration of DBPs.

20.4.3 Removal of Arsenic

As mentioned in Section 20.3.2.2, arsenic is a toxic element that occurs naturally in soils, rocks, and groundwater. It enters drinking water supplies from natural deposits in the Earth or from agricultural and industrial practices. Long-term drinking water exposure can cause serious health problems such as skin, lung, bladder, and kidney cancer. The U.S. EPA has set the arsenic standard for drinking water at 10 parts per billion (10 $\mu g/L$), which has been effective since January 2006 (66).

Many methods have been used for removing As from drinking water sources, including anion exchange, reverse osmosis, lime softening, microbial transformation, chemical precipitation, and adsorption (67). As(III) is readily oxidized by MnO_2 followed by adsorption of the As(V) onto the MnO_2 surface (68). Similarly, high surface area activated alumina prepared by thermal dehydration of aluminum hydroxide effectively removed As from drinking water at pH 5.5 (69). Removal of As by combined ion exchange procedure has also been investigated. It was found that As removal on an adsorbent prepared by precipitation of $Fe(OH)_3$ onto Al_2O_3 showed a breakthrough capacity of 0.10 mg of As/g adsorbent at 0.05 mg As/L (70). The ability of nanocrystalline TiO_2 to oxidize As(III) photocatalytically has been evaluated. Nanocrystalline TiO_2 synthesized in the laboratory of Pena, et al. (71) was effective for As(V) removal at pH <8 and maximum As(III) removal occurred at pH ~ 7.

Gu, Fang, and Deng (72) developed a granular activated carbon-based, iron-containing adsorbent (As-GAC) to remove arsenic effectively from drinking water. Granular activated carbon was used primarily as a supporting medium for ferric iron that was impregnated by ferrous chloride ($FeCl_2$) treatment, followed by chemical oxidation with NaClO. Structural analyses by scanning electron microscope (SEM) and energy dispersive spectroscopy (EDS) indicated that the distribution of iron in the adsorbents was mainly on the edge of As-GAC in the low iron content ($\sim 1\%$ Fe) sample but extended to the center at the higher iron content ($\sim 6\%$ Fe). When the iron content was $>7\%$, an iron ring formed at the edge of the GAC particles. As-GAC could remove arsenic most efficiently when the iron content was approximately 6%; further increase of iron decreased arsenic adsorption.

20.5 SOIL REMEDIATION

20.5.1 General Soil Composition

A soil is essentially an agglomeration of mineral and organic matter, with pore spaces containing air, water, and nutrient solutions (73). Specific soils develop under the influence of five different factors: parent material, climate, living organisms, topography, and time (74). The mineral portion of soils frequently includes clays such as montmorillonite and kaolinite, and crystalline minerals such as quartz, whereas the organic portion of soils includes both nonhumic and humic substances (75). There are thousands of different types of soils in the world, each with their own composition and varying amounts of mineral, organic, and aqueous components. These varying compositions determine the chemistry of the soil and other characteristics such as fertility and physical properties.

20.5.2 Common Soil Contaminants

There are a large number of common soil contaminants, some of which are naturally occurring and some of which are the result of human activity. These include heavy metals such as arsenic and mercury, halogenated compounds such as PCBs, anthropogenic compounds such as estrones, and radioactive materials such as thorium (76). The elemental contaminants, such as the radioactive materials and the heavy metals, may originate either from human activity such as mining and manufacturing, or may occur naturally as the result of weathering or leaching of minerals rich in those materials; for example, the minerals orpiment, lorandite, and smithite are rich in arsenic (77). Historically many metal-containing minerals have been valued by humans, and some, such as cinnabar, have been mined since ancient times. Of course, the activity of mining exposes even more of the naturally occurring mineral deposit to weathering and leaching, thus accelerating soil contamination. For example, the area of the Almadén del Azoque, Ciudad Real, Spain, where cinnabar has been mined since ancient times, is still heavily contaminated by mercury (78–80).

The nonelemental contaminants originate almost exclusively from human activity. Estrones, steroid hormones that are endocrine disrupting chemicals (EDCs), find their way into the environment as metabolites from birth control pills and fertility medications. They have a strong affinity for humic acid, which leads them to bind to dissolved organic matter and be deposited when the matter drops out of solution (81). PCBs were widely used as coolants and lubricants in capacitors, transformers, and other electrical equipment because of their nonflammable and insulating properties. While PCBs are no longer manufactured in the United States, they are still present in old transformers and capacitors, and are pollutants in soils and sludge in more than 400 U.S. sites (82).

20.5.3 Nanotechnology in Soil Remediation

Conventional soil remediation techniques include bioremediation, incineration, removal, chemical treatments, and sequestration and burial. No one conventional method works perfectly on every contaminant and some, such as sequestration and burial, require a great deal of monitoring. In the early 1990s, Gillham and O'Hannesin used Fe^0 as a reductant to degrade halogenated aliphatics (31). Since that time, there has been considerable interest in $Fe^0_{(NP)}$ as an aid to environmental remediation (45). Other useful iron-based nanoparticles include iron phosphate nanoparticles (83, 84), magnetite nanoparticles (85), bimetallic iron-containing nanoparticles (86), and ferritin nanoparticles (87). Palladium nanoparticles (88, 89), titanium dioxide nanoparticles (90), and multiwalled carbon nanotubes (91) have also been reported to have specific contaminant remediation properties. The iron and iron-based nanoparticles may be used as free nanoparticles, as membrane-bound or incorporated nanoparticles, and as stabilized nanoparticles. Free iron and iron-based nanoparticles have been demonstrated to be highly effective at removing heavy metals, chlorinated organic solvents, and PCBs; however, they tend to aggregate rapidly and demonstrate poor mobility in porous media (15, 92, 93). Consequently, there is considerable interest in membrane-bound or incorporated iron and iron-based nanoparticles for use in PRBs (94). These barriers would be deployed as filters in the path of contaminant plumes in groundwater and contaminated sediments. Another approach to overcome the agglomerative nature of Fe^0 and $Fe^0_{(NP)}$ is nanoparticle stabilization. Iron and iron-based nanoparticles may be stabilized by different means, including preparation in a CMC matrix (84, 93), modification with surfactant (16), and in colloidal solution with activated carbon or colloidal poly(acrylic acid) (15, 95, 96). These stabilized nanoparticles remain unagglomerated in solution for extended periods of time, which increases their mobility in contaminated soils, and are as reactive as free nanoparticles (96).

In addition to the laboratory produced iron nanoparticles there are also reports of naturally occurring iron-based nanoparticles composed of schwertmannite, ferridydrite, and goethite (97, 98). These particles may be produced by a number of bacterial species and families, including *Klebsiella oxytoca* (98). Samples taken from an inactive sulfide mine in Finland demonstrated that these naturally occurring schwertmannite and ferridhydrite nanoparticles were active in scavenging arsenic from the contaminated waters at the mine site (97). Other biogenic iron-based nanoparticles have been produced by members of the Geobacteriaceiae family and by *Aneromyxobacter*, *Paenibacillus*, and *Brevibacillus* species in acidic sediments contaminated with uranium and/or technetium (99, 100).

Different bacterial species, including *Shewanella oneidensis*, *Desulfovibrio desulfuricans*, and *Bacillus sphaericus*, may produce or be used to produce Pd(0) nanoparticles (88, 101, 102). The Pd(0) nanoparticles obtained by these processes have been shown to reductively dehalogenate PCBs, and may prove to be a useful tool for removing large PCB contaminations (101).

Pd(0) nanoparticles are used as both free nanoparticles and as supported nanoparticles, where the supporting substrate may be various foams. Free Pd(0) nanoparticles produced by the bacterium *Shewanella oneidensis* were shown to reductively dechlorinate PCBs in solution, and supported Pd-Fe nanoparticles have been shown to assist in reductively dechlorinating 1,2,4-trichlorobenzene (45, 101). For the most part, Pd(0) nanoparticles appear to resist the agglomeration and oxidation problems that plague Fe(0) nanoparticles (86, 101).

20.5.3.1 Removal of Heavy Metals A number of heavy metals may be removed from contaminated soils and soil slurries by the use of $Fe^0_{(NP)}$. In most cases, the $Fe^0_{(NP)}$ reduces the toxic metals to less toxic, less soluble forms, or adsorbs the metals to form stable, relatively insoluble mineral products.

Both free $Fe^0_{(NP)}$ and surface-modified $Fe^0_{(NP)}$ have been shown to be effective at removing As(III) (16, 45). Removal of As(III) by free $Fe^0_{(NP)}$ resulted in the rapid formation of the minerals magnetite and lepidocrocite, which arose from corrosion products of the $Fe^0_{(NP)}$ and adsorbed As(III). At pH values ranging from 4 to 10, between 88.6% and 99.9% of the aqueous As(III) was removed from solution. The presence of NO_3^-, SO_4^-, or HCO_3^- negatively affected the ability of the $Fe^0_{(NP)}$ to adsorb As(III), as did the presence of H_4SiO_4 and $H_2PO_4^-$. These results demonstrate that the efficiency of As(III) removal by free $Fe^0_{(NP)}$ is influenced by the presence of other ionic species, and may be lessened in actual usage by the presence of these species.

A second metallic contaminant of interest is Cr. Remediation of Cr-contaminated soils is centered around the reduction of Cr(VI), a more toxic and soluble form, to Cr(III), which is considerably less soluble and which readily precipitates out of solution. A number of different nanoparticles have been shown to possess this capability, including ferritin protein cages containing ferrihydrite (Fe(O)OH), palladium (Pd) nanoparticles, and carboxymethyl cellulose (CMC)-stabilized $Fe^0_{(NP)}$ (87, 89, 93). In both the ferritin cages and the CMC-$Fe^0_{(NP)}$, the active species was a form of Fe, either Fe(0) in the case of the CMC-$Fe^0_{(NP)}$, or Fe(II) in the case of the ferritin cages. The reaction known to reduce Cr(VI) utilizing Fe(II) is shown in Equation (20.2).

$$Cr^{6+} + 3Fe^{2+} \longrightarrow Cr^{3+} + 3Fe^{3+} \tag{20.2}$$

This reaction is believed to be the mechanism underlying the reduction of Cr(VI) to Cr(III) by the ferritin cages. This reduction only takes place in the presence of light, suggesting a role for photoinduced electron transfer from the Fe(O)OH catalyst to the Cr(VI) (87). In contrast, the CMC-$Fe^0_{(NP)}$ reaction with Cr(VI) did not require the presence of light. In this reaction, the underlying mechanism is believed to be as shown in Equation (20.3).

$$Fe(0) + CrO_4^{2-} + 4H_2O \longrightarrow Cr(OH)_3 + Fe(OH)_3 + 2OH^- \tag{20.3}$$

The CMC-$Fe^0_{(NP)}$ also engaged in a side reaction with water, as shown in Equation (20.4).

$$Fe(0) + 2H_2O \longrightarrow Fe^{2+} + H_2 + 2OH^- \tag{20.4}$$

This side reaction could potentially consume some of the Fe(0) reactivity, and so dissolved oxygen and high oxidation/reduction potential should be minimized to maximize the reduction of the Cr(VI). Experimentation with CMC-$Fe^0_{(NP)}$ and contaminated sandy loam soil demonstrated that all of the leached Cr(VI) was converted to Cr(III), and that the addition of CMC-$Fe^0_{(NP)}$ did not result in the increased mobilization of Cr from the soil (93). Cr leachability testing of treated and untreated soils, conducted using both EPA Method 1311 and California HML Method 910, showed that even brief treatment with CMC-$Fe^0_{(NP)}$ reduced the soil's leachability by 90% and 76%, respectively. Further, the leached Cr was all in the form of the less toxic Cr(III).

Pd nanoparticles loaded with sulfur (PdNPs/S) at a ratio of 1.6:10 displayed a reduction efficiency of 99.8% at 130°C when challenged with 400 μM Cr(VI) for 1 hour (89). Testing with real soil samples demonstrated 97.2% Cr(VI) reduction by the PdNPs/S when challenged with 0.17 mg/g Cr(VI) for 1 hour at 130°C. Similarly, a 1 hour challenge at 130°C with 0.50 mg/g of Cr(VI) led to a 92.0% removal. The soil matrix did not influence the rate of Cr(VI) removal by the PdNPs/S until ~3.36 mg/g of Cr(VI) was reached. It must be noted that the reduction of Cr(VI) to Cr(III) by the PdNPs/S was slower and less efficient at 100°C than at 130°C, and that both of these temperatures are greatly elevated compared to the room temperature reactions characteristic of the $Fe^0_{(NP)}$ reductions; however, removal efficiency for this process is still greater than conventional removal efficiencies, which are ~66% for much lower Cr(VI) concentrations.

Soils contaminated with Pb may also be remediated by iron-based nanoparticles. Iron phosphate (vivianite) nanoparticles stabilized with CMC have been reported to reduce the toxicity-characteristic leaching procedure (TCLP) and physiologically based extraction test (PBET) bioaccessibility in calcareous, neutral, and acidic soils (84). While phosphate addition has been known since at least 1993 to immobilize Pb(II) in soils, phosphate addition can cause its own problems in that it easily leaches into surface and groundwaters, where it causes problems related to excessive nutrient input. CMC-vivianite nanoparticles release 50% less phosphate into the environment than more traditional phosphate soil amendments, partly because of the insolubility of vivianite. Unlike the PdNPs/S mixture, the Pb-sequestering reactions take place at ambient temperatures. At acidic pH values, the reaction sequence is as shown in Equations (20.5a) and (20.5b).

$$Fe_3(PO_4)_2 \cdot 8H_2O + 4H^+ \longrightarrow 3Fe^{2+} + 2H_2PO_4^- + 8H_2O \quad (20.5a)$$

$$5Pb^{2+} + 3H_2PO_4^- + X^- \longrightarrow Pb_5(PO_4)_3X + 6H^+ \quad (20.5b)$$

(X is F, Cl, Br, or OH)

At pH >7.2, the reaction sequence is as shown in Equations (20.6a) and (20.6b).

$$Fe_3(PO_4)_2 \cdot 8H_2O + 2H^+ \longrightarrow 3Fe^{2+} + 2HPO_4^{2-} + 8H_2O \quad (20.6a)$$

$$5Pb^{2+} + 3HPO_4^{2-} + X^- \longrightarrow Pb_5(PO_4)_3X + 3H^+ \quad (20.6b)$$

The solubility product values for the $Pb_5(PO_4)_3X$ may be as low as $10^{-84.4}$, effectively removing the Pb. Pyromorphite ($Pb_5(PO_4)_3X$) is highly stable, and Pb fixed there is only very poorly leachable or bioavailable (84). In addition to remediating Pb, the CMC-vivianite nanoparticles can also be used to immobilize Cu(II) *in situ* (83). Treatment of Cu-contaminated acidic, neutral, or calcareous soils with the CMC-vivianite nanoparticles resulted in 58%, 50%, and 78% reductions in the TCLP extraction results, demonstrating the value of this technology.

Finally, Fe-based nanoparticles may be utilized to immobilize Hg contamination and prevent it from being acted upon by bacteria and becoming methylmercury. CMC-stabilized FeS nanoparticles have been shown to react with Hg to produce HgS, which is a highly insoluble mineral that offers little or no Hg bioavailability (103). The addition of CMC-FeS nanoparticles to low Hg content and high Hg content sediments resulted in ~46% and 67% reduction in leachable Hg respectively.

20.5.3.2 Removal of Organic Contaminants Remediation of PCB-contaminated soils by iron-based nanoparticles has been reported (82). At ambient temperatures, $Fe^0_{(NP)}$ mixed with dry samples of contaminated soil or acidic previously filtered soil resulted in the destruction of only 38% and 26% of the PCBs contained in the respective samples, whereas 90% destruction has been observed for PCBs in water. In these dry samples, 98% PCB destruction was achieved in the presence of air at 300°C. The lower destruction at ambient temperatures was attributed to the lower diffusion rate of PCBs from the soils to the $Fe^0_{(NP)}$ catalyst. These results suggest that $Fe^0_{(NP)}$ remediation of PCB-contaminated soils may be maximized by making a heated aqueous slurry of the contaminated soil, which would be difficult to achieve *in situ*. Different results were achieved with the catalysis of PCBs by Pd nanoparticles produced on *Shewanella oneidensis* by bioreduction. PCB 21 (2,3,4-chlorobiphenyl) solubilized in M9 microbiological medium at a concentration of 1 mg/L was reduced to undetectable levels within 1 h when 500 mg/L of palladized *S. oneidensis* (bioPd) was added to the mixture and incubated at 28°C (101). Further testing with PCB-contaminated sediments and 50 mg of bioPd/L demonstrated that PCB concentrations could be reduced from 2 mg/kg of soil to ~0.5 mg/kg soil in 48 h, demonstrating the potential of this technology.

PAHs are frequently found contaminating soils in industrial areas, where they are spilled or leaked from oil and gas storage tanks. In laboratory testing utilizing pyrene-contaminated clay loam soil, $Fe^0_{(NP)}$ particles were demonstrated to remove 62% of the pyrene within 60 minutes, using a $Fe^0_{(NP)}$ concentration of 0.15 g/g of soil (26). Acid treatment of the $Fe^0_{(NP)}$ immediately prior to usage (to remove Fe oxides formed by reactions with moist air) was necessary to achieve this result.

20.5.3.3 Removal of Radioisotopes Some data also exists for the removal of the radioactive element thorium (Th) by nanomaterials. The experiment utilizing TiO_2 nanoparticles to remediate Th(IV) contamination was conducted specifically with contaminated soils. Experiments conducted in the presence or absence of soil humic acid and fulvic acid demonstrated that both the fulvic and humic acids increased the sorption of Th(IV) to TiO_2 nanoparticles at acidic pH (90). Bare TiO_2 nanoparticles by themselves were able to form surface complexes with 94% of the available Th(IV); this percentage increased to 97%–98% in the presence of fulvic or humic acids and remained stable with increasing pH. This technique could be utilized to remediate soils contaminated with Th(IV), and in fact would likely be more effective in soils than in water.

20.6 AIR PURIFICATION

Many VOCs, nitrogen oxides (NOx), and sulfur oxides (SOx) in air contribute to smog and high ozone levels, which harm human health (104). The sources of these pollutants are various human activities, including combustion engines, large combustion and industry sources, and consumer products (105). Human activity also has increased the concentration of various greenhouse gases that trap heat in the atmosphere.

Nanotechnology is being used to reduce air pollution in more than one way: catalysts, which are currently in use and constantly being improved upon (106, 107), and nanostructured membranes, which are under development (108–110). Catalysts can be used to enable a chemical reaction at lower temperatures or make the reaction more effective and it appears that nanotechnology can improve the performance and cost of catalysts used to transform toxic vapors into harmless gases. As mentioned earlier catalysts made from nanoparticles have a greater surface area to interact with the reacting chemicals than catalysts made from larger particles.

20.6.1 Removal of VOCs

Traditionally, adsorbents such as activated carbon or ozone-promoted oxidation have been used for air purification system. Klabunde and coworkers (111) have prepared nanoparticles of metal oxides (MgO and CaO) and core-shell binary oxides (Fe_2O_3/MgO or V_2O_3/MgO) and tested them with typical air pollutants such as acetaldehyde, propionaldehyde, perfluoropropene, and a number of other polar organic compounds. Nanocrystalline MgO (specific surface area = 350 to 400 m^2/g) showed faster kinetics and a much higher capacity for acetaldehyde adsorption than commercial activated carbon adsorbents. It was also observed that MgO or CaO having a monolayer coating of Fe_2O_3 showed enhanced reactivity for the destruction of chlorocarbons, organophosphates, and acid gases (112, 113).

Sinha et al. investigated the efficacy of high surface area (>300 m^2/g) mesoporous manganese oxide (γ-MnO_2) in the removal of VOCs at ambient temperature (114). It was found that γ-MnO_2 modified with gold nanoparticles efficiently eliminated a range of VOCs such as acetaldehyde, toluene, and *n*-hexane. The removal capacity was almost doubled after Au deposition (2.8 wt%, particle size 3 to 6 nm) on the MnO_2. Acetaldehyde removal was rapid and complete within 1 h at room temperature. The pristine γ-MnO_2 showed approximately 60% toluene removal at 25°C but less than 2% hexane removal. However, after Au deposition, the efficacy increased considerably. The γ-MnO_2/Au materials removed about 95% toluene and 30% *n*-hexane at 25°C. Increasing the reaction temperature to 85°C resulted in 99% toluene and 77% *n*-hexane removal.

Wei, Radhakrishnan, and Vanderspurt (104) have developed a coating material based on nanosized manganese oxide/titanium dioxide that oxidizes VOCs and decomposes ozone. Hydroxyl radicals produced on the sorbent surface oxidize the VOCs to water and carbon dioxide. It is postulated that manganese oxide lowers the energy barrier required for ozone decomposition to molecular oxygen.

Similarly, mesoporous high surface area (350 to 370 m^2/g) ceria-titania mixed metal oxide showed greater efficacy for removal of toluene at room temperature compared to the corresponding nonporous mixed metal oxide or the mesoporous titania or ceria (107). Interestingly, the toluene removal ability almost doubled after Pt impregnation into the base mesoporous ceria-titania.

Suzuki and Sinha have prepared novel bimodal mesoporous crystalline ceria nanoparticles and evaluated their performance in VOC removal (106). The mesoporous ceria showed 92% acetaldehyde removal with 33% CO_2 conversion at ambient temperature after 24 h. This acetaldehyde removal performance is nearly twice as high as that for conventional VOC removal using materials such as activated carbon or mesoporous silica.

NanoScale Corporation has performed air filtration testing of several metal oxide nanoparticles against select toxic industrial chemicals (TICs). Initial results show that these formulations outperform the more commonly used commercial activated carbons (115). Additional in-depth studies are currently continuing to unambiguously establish the utility of nanoparticles in air filtration. It is noteworthy that the metal oxides produced by NanoScale Corporation were found to be no more toxic than their non-nano commercial counterparts.

20.6.2 Removal of NOx and SOx

Titanium dioxide is one of the most commonly used photocatalysts since it has a high oxidative ability and extremely good chemical stability. TiO_2 can generate hydroxyl radicals under UV radiation and the hydroxyl radicals generated can then oxidize pollutants on the catalyst surface to form less harmful products. For example, oxidation of NOx forms NO_3^- on the catalyst surface (116). Akbari and Berdahl reported on the potential for removing NOx from polluted air using anatase TiO_2 nanoparticles (117). In another study, it was found that mesoporous γ-MnO_2 displayed a good performance in the removal of NOx (72 mg/g) and SOx (700 mg/g) (114). Klabunde et al. reported that nanocrystalline CaO reacted with SO_2 at relatively lower temperature and generated a mixture of calcium sulfite, calcium sulfate, and calcium sulfide (113). It was also found that the presence of a small amount of Fe_2O_3 on the surface of CaO enhanced the ability of the CaO to act as a destructive adsorbent for SO_2.

20.7 CONCLUSIONS AND FUTURE OUTLOOK

As seen from the above discussion, nanomaterials offer substantial potential and possibilities for remediating contaminated water, soils, and air. Among the advantages offered by nanomaterials are ease of deployment, multifunctional remediation ability against multiple contaminants, and "greener" reaction by-products. While additional research and development are necessary to make nanomaterial production more economical, the ease of nanomaterial deployment and the widespread need for environmental remediation will eventually develop a large commercial market for these materials. These remediation technologies offer particularly large benefits for developing nations, since having no requirement for an outside power source make them ideal for use in remote areas by relatively untrained persons. Further, most of these nanomaterials lend themselves well to *in situ* remediation, as many of the pertinent chemical reactions take place at room temperature without the need for additional resources. In conclusion, we anticipate that the use of nanomaterials in environmental remediation will continue to be actively explored. This in turn will result in large-scale use and commercial sale of nanomaterials. Specifically, given the unique reactivity of nanomaterials, we expect their use for remediation purposes will become commonplace in the very near future.

FURTHER READING

Nigel Cameron and M. Ellen Mitchell (Eds.), *Nanoscale: Issues and Perspectives for the Nano Century*. Wiley, Hoboken, NJ, 2007.

Ronald A. Hites (Ed.), *Elements of Environmental Chemistry*. Wiley, Hoboken, NJ, 2007.

M. C. Roco, R. S. Williams and P. Alivisatos (Eds.), *Nanotechnology Research Directions*. Kluwer Academic Publishers, Norwell, MA, 2001.

REFERENCES

1. ASTM E 2456-06, Terminology for Nanotechnology, ASTM International 2006.
2. K. T. Ranjit, G. Medine, P. Jeevanandam, I. N. Martyanov, K. J. Klabunde, Nanoparticles in environmental remediation, in *Environmental Catalysis*, V. H. Grassian (Ed.), 391. CRC Press, Boca Raton, FL, 2005.
3. (a) S. Singh, H. S. Nalwa, *J. Nanosci. Nanotechnol.* **2007**, *7*, 3048; (b) http://www.nanotechproject.org/inventories/consumer/.
4. Q. Sun, Q. Wang, P. Jena, Y. Kawazoe, *J. Am. Chem. Soc.* **2005**, *127*, 14582.
5. http://www.census.gov/ipc/www/popclockworld.html.
6. U.S. EPA, *Climate Change and Public Health*. EPA 236-F-97-005. U.S. EPA, Washington, D.C., 1997.
7. U.S. EPA. *In Situ Treatment Technologies for Contaminated Soil*. EPA 542-F-06-013. U.S. EPA, Washington, D.C., 2006.
8. U.S. EPA. NATO/CCMS Pilot Study. *Evaluation of Demonstrated and Emerging Technologies for the Treatment of Contaminated Land and Groundwater (Phase III)*. EPA 542-R-02-001. U.S. EPA, Washington, D.C., 2002.
9. U.S. EPA. *Abandoned Mine Site Characterization and Cleanup Handbook*. EPA 910-B-00-001. U.S. EPA, Washington, D.C., 2000.
10. U.S. EPA. *Cleaning Up the Nation's Waste Sites: Markets and Technology Trends*. EPA 542-R-04-015. U.S. EPA, Washington, D.C., 2004.
11. CSIRO (Common Wealth Scientific and Industrial Research Organisation). *Climate Change and the Risk to Water Supply*. CSIRO press release 06/191. CSIRO, Clayton, S. Victoria, Australia, 2006.

12. NASA. *NASA Ties El Nino-Induced Drought to Air Pollution from Fires*. NASA press release 03-128. NASA, Washington, D.C., 2003.
13. U.S. EPA. *Guidance Manual for Compliance with the Filtration and Disinfection Requirements for Public Water Systems Using Surface Water Sources*. EPA 68-01-6989. U.S. EPA, Washington, D.C., 2001.
14. U.S. EPA. *Air Pollution Control Orientation Course*. EPA SI-422. U.S. EPA, Washington, D.C., 2007.
15. S. R. Kanel, B. Manning, L. Charlet, H. Choi, *Environ. Sci. Technol.* **2005**, *39*, 1291.
16. S. R. Kanel, D. Nepal, B. Manning, H. Choi, *J. Nanopart. Res.* **2007**, *9*, 725.
17. S. Namuangruk, P. Khongpracha, Y. Tantirungrotechi, J. Limtrakul, *J. Mol. Graph. Model.* **2007**, *26*, 179.
18. N. Moreno, X. Querol, C. Zyora, *Environ. Sci. Technol.* **2001**, *35*, 3526.
19. Y. H. Li, J. Ding, Z. K. Luan, Z. C. Di, Y. F. Zhu, C. L. Xu, D. H. Wu, B. Q. Wei, *Carbon* **2003**, *41*, 2787.
20. W. Uantasee, Y. Lin, G. E. Fryxell, B. J. Busche, J. C. Brinbaum, *Seapr. Sci. Technol.* **2003**, *38*, 3809.
21. B. W. Stanton, J. J. Harris, M. D. Miller, M. L. Bruening *Langmuir* **2003**, *19*, 7038.
22. http://www.cluin.org/download/studentpapers/K_Watlington_Nanotech.pdf.
23. S. R. Kanel, B. Manning, L. Charlet, H. Choi, *Environ. Sci. Technol.* **2005**, *39*, 1201.
24. B. Schrick, J. L. Blough, A. D. Jones, T. E. Mallouk, *Chem. Mater.* **2002**, *14*, 5140.
25. R. Asahi, T. Morikawa, T. Ohwaki, K. Aoki, Y. Taga, *Science* **2001**, *293*, 269.
26. M. Chang, H. Shu, *Air Waste Manage. Assoc.* **2007**, *57*, 221.
27. S. Hrapovic, E. Majid, Y. Liu, K. Male, J. H. T. Luong, *Anal. Chem.* **2007**, *78*, 5504.
28. (a) S. O. Obare, G. J. Meyer, *J. Env. Sci. Heal. A* **2005**, *39*, 2549; (b) A. Vaseashta, M. Vaclavikova, S. Vaseashta, G. Gallios, P. Roy, O. Pummakarnchana, *Sci. Technol. Adv. Mat.* **2007**, *8*, 47; (c) J. Z. Zhang, *Acc. Chem. Res.* **1997**, *30*, 423; (d) K. T. Ranjit, K. J. Klabunde, Nanotechnology: Fundamental principles and applications, in *Riegel's Handbook of Industrial Chemistry and Biotechnology*, 11th edition, James A. Kent (Ed.), 328. Springer, New York, **2007**.
29. (a) A. P. Alivisatos, *Science* **1996**, *217*, 933; (b) A. P. Alivisatos, *J. Phys. Chem.* **1996**, *100*, 13226; (c) C. C. Yang, S. Li, *J. Phys. Chem.* **2008**, *112*, 2851.
30. (a) J. S. Narr, T. Viraraghavan, Y.-C. Jin, *Fresenius Environ. Bull.* **2007**, *16*, 320; (b) P. G. Tratnyek, R. L. Johnson, *Nanotoday* **2006**, *1*, 44; (c) L. Li, M. Fan, R. C. Brown, J. Van Leeuwen, J. Wang, W. Wang, Y. Song, P. Zhang, *Crit. Rev. Environ. Sci. Technol.* **2006**, *36*, 405.
31. R. W. Gillham, S. F. O'Hannesin, *Ground Water* **1994**, *32*, 958.
32. W. S. Orth, R. W. Gillham, *Environ. Sci. Technol.* **1996**, *30*, 66.
33. P. G. Tratnyek, M. M. Scherer, T. L. Johanson, L. J. Matheson, *Chemical Degradation Methods for Wastes and Pollutants: Environmental and Industrial Applications*. Marcel Dekker, New York, 2003.
34. C.-B. Wang, W. Zhang, *Environ. Sci. Technol.* **1997**, *31*, 2154.
35. (a) W. Zhang, C.-B. Wang, H.-L. Lien, *Catal. Today* **1998**, *40*, 387; (b) H.-L. Lien, W. Zhang, *Colloids Surf. A* **2001**, *191*, 97; (c) B. Shrick, J. L. Blough, A. D. Jones, T. E. Mallouk, *Chem. Mater.* **2002**, *14*, 5140.

36. Y. Xu, W. Zhang, *Ind. Eng. Chem. Res.* **2000**, *39*, 2238.
37. B.-W. Zhu, T.-T. Lim, *Environ. Sci. Technol.* **2007**, *41*, 7523.
38. H. I. Lien, W. Zhang, *J. Environ. Eng.* **1999**, *125*, 1042.
39. A. Feitz, J. Guan, D. Waite, *U.S. Patent Appl. Pub.* # U.S. 2006/0083924 A1, **April 20, 2006**.
40. F. Li, C. Vipulanandan, K. K. Mohanty, *Colloids Surf. A: Physicochem. Eng.* **2003**, *223*, 103.
41. Y. Q. Liu, S. A. Majetich, R. D. Tilton, D. S. Sholl, G. V. Lowry, *Environ. Sci. Technol.* **2005**, *39*, 1338.
42. J. T. Nurmi, P. G. Tratnyek, V. Sarathy, D. R. Baer, J. E. Amonette, K. Pecher, C. Wang, J. C. Linehan, D. W. Matson, R. Lee Penn, M. D. Driessen, *Environ. Sci. Technol.* **2005**, *39*, 1221.
43. G. V. Lowry, K. M. Johnson, *Environ. Sci. Technol.* **2004**, *38*, 5208.
44. (a) F. He, D. Zhao, *Environ. Sci. Technol.* **2005**, *39*, 3314; (b) F. He, D. Zhao, J. Liu, C. B. Roberts, *Ind. Eng. Chem. Res.* **2007**, *46*, 29.
45. W. Zhang, *J. Nanoparticle Res.* **2003**, *5*, 323.
46. D. W. Elliott, W. Zhang, *Environ. Sci. Technol.* **2001**, *35*, 4922.
47. J. Quinn, C. Geiger, C. Clausen, K. Brooks, C. Coon, S. O'Hara, T. Krug, D. Major, W.-S. Yoon, A. Gavaskar, T. Holdsworth, *Environ. Sci. Technol.* **2005**, *39*, 1309.
48. V. Sarathy, P. G. Tratnyek, J. T. Nurmi, D. R. Baer, J. E. Amonette, C. L. Chun, R. L. Penn, E. J. Reardon, *J. Phys. Chem. C* **2008**, *112*, 2286.
49. (a) R. M. Powell, R. W. Puls, D. K. Hightower, D. A. Sabatini, *Environ. Sci. Technol.* **1995**, *29*, 1913; (b) D. W. Blowes, C. J. Ptacek, J. L. Jambor, *Environ. Sci. Technol.* **1997**, *31*, 3348; (c) E. H. Smith, *Water Res.* **1996**, *30*, 2424.
50. (a) S. M. Ponder, J. G. Darab, T. E. Mallouk, *Environ. Sci. Technol.* **2000**, *34*, 2564; (b) S. M. Ponder, T. E. Mallouk, *U.S. Patent 6,232,663*, **June 5, 2001**; (c) S. M. Ponder, T. E. Mallouk, *U.S. Patent 6,689,485*, **January 17, 2002**.
51. D. Zhao, Y. Xu, *U.S. Patent Appl. Pub # US2007/0256985 A1*, **November 8, 2007**.
52. L. B. Hoch, E. J. Mack, B. W. Hydutsky, J. M. Hershman, J. M. Skluzacek, T. E. Mallouk, *Environ. Sci. Technol.* **2008**, *42*, 2600.
53. K. C. K. Lai, I. M. C. Lo, *Environ. Sci. Technol.* **2008**, *42*, 1238.
54. S. R. Kanel, J.-M. Greneche, H. Choi, *Environ. Sci. Technol.* **2006**, *40*, 2045.
55. "EPA drinking water standards" at http://www.water-research.net/standards.htm.
56. "Sources of bacteria" at http://www.water-research.net/bacteria.htm.
57. (a) E. Levy, *U.S. Patent 7,264,726*, **September 4, 2007**; (b) E. Levy, *U.S. Patent 7,288,498*, **October 30, 2007**.
58. (a) P. K. Stoimenov, R. L. Klinger, G. L. Marchin, K. J. Klabunde, *Langmuir* **2002**, *18*, 6679; (b) O. Koper, J. Klabunde, G. Marchin, K. J. Klabunde, P. Stoimenov, L. Bohra, *Curr. Microbiol.* **2002**, *44*, 49.
59. I. Sondi, B. Salopek-Sondi, *J. Coll. Interf. Sci.* **2004**, *275*, 177.
60. P. Jain, T. Pradeep, *Biotechnol. Bioeng.* **2005**, *90*, 59.
61. S. Pal, J. Joardar, J. M. Song, *Environ. Sci. Technol.* **2006**, *40*, 6091.
62. G. C. Moeser, K. A. Roach, W. H. Green, P. E. Laibinis, T. A. Hatton, *Ind. Eng. Chem. Res.* **2002**, *41*, 4739.

63. I. H. Choi, S. M. Lee, I. C. Kim, B. R. Min, K. H. Lee, *Ind. Eng. Chem. Res.* **2007**, *46*, 2280.
64. P. V. Kamat, R. Huehn, R. Nicolaescu, *J. Phys. Chem. B* **2002**, *106*, 788.
65. B. S. Karnik, S. H. Davies, M. J. Baumann, S. J. Masten, *Environ. Sci. Technol.* **2005**, *39*, 7656.
66. D. Mohan, C. U. Pittman, Jr., *J. Hazard. Mater.* **2007**, *142*, 1.
67. J. F. Ferguson, J. Garvis, *Water Res.* **1972**, *6*, 1259.
68. (a) B. A. Manning, S. E. Fendorf, B. Bostick, D. L. Suarez, *Environ. Sci. Technol.* **2002**, *36*, 976; (b) B. A. Manning, M. Hunt, C. Amrhein, J. A. Yarmoff, *Environ. Sci. Technol.* **2002**, *36*, 5455.
69. S. W. Hathaway, F. Rubel, Jr., *J. Am. Water Works Assoc.* **1987**, *79*, 61.
70. H. Hodi, K. Polyak, J. Hlavay, *Environ. Int.* **1995**, *21*, 325.
71. M. E. Pena, G. P. Korfiatis, M. Patel, L. Lippincott, X. Meng, *Water Res.* **2005**, *11*, 2327.
72. Z. Gu, J. Fang, B. Deng, *Environ. Sci. Technol.* **2005**, *39*, 3833.
73. "Soil Composition" at http://student.britannica.com/comptons/article-208229/soil.
74. "Soils" at http://esa21.kennesaw.edu/activities/soil/soilcomposition.pdf.
75. "Soil Components" at http://www.landfood.ubc.ca/soil200/components/mineral.html.
76. "Soil Contaminants" at http://www.epa.gov/ebtpages/pollsoilcontaminants.html.
77. B. Minceva-Sukarova, G. Jovanovski, P. Makreski, B. Soptrajanov, W. Griffith, R. Willis, I. Grzetic, *J. Mol. Struct.* **2003**, *651*, 181.
78. J. J. B. Nevado, L. F. G. Bermejo, R. C. R. Martin-Dolmeadios, *Environ. Pollut.* **2003**, *122*, 261.
79. R. Ferrara, B. E. Maserti, M. Andersson, *Atmos. Environ.* **1998**, *32*, 3897.
80. J. E. Gray, *U.S. Geological Survey Circular 1248*, **2003**.
81. X. Jin, J. Hu, S. L. Ong, *Water Res.* **2007**, *41*, 3077.
82. P. Varanasi, A. Fullana, S. Sidhu, *Chemosphere* **2007**, *66*, 1031.
83. R. Liu, D. Zhao, *Chemosphere* **2007**, *68*, 1867.
84. R. Liu, D. Zhao, *Water Res.* **2007**, *41*, 2491.
85. M. Hull, *EPA Research Project Database*, **2005**, contract # EPD05027.
86. B. Zhu, T. Lim, J. Feng, *Chemosphere* **2006**, *65*, 1137.
87. I. Kim, H. Hosein, D. R. Strongin, T. Douglas, *Chem. Mater.* **2002**, *14*, 4874.
88. K. Pollmann, M. Merroun, J. Raff, C. Hennig, S. Selenska-Pobell, *Lett. Appl. Microbiol.* **2006**, *43*, 39.
89. I. O. K'Owino, M. A. Omole, O. A. Sadik, *J. Environ. Monit.* **2007**, *9*, 657.
90. X. Tan, X. Wang, C. Chen, A. Sun, *Appl. Rad.* **2006**, *65*, 375.
91. X. Wang, C. Chen, W. Hu, A. Ding, D. Xu, X. Zhou, *Environ. Sci. Technol.* **2005**, *39*, 2856.
92. S. R. Kanel, H. Choi, *Water Sci. Technol.* **2007**, *55*, 157.
93. Y. Xu, D. Zhao, *Water Res.* **2007**, *41*, 2101.
94. E. E. Nuxoll, T. Shimotori, W. A. Arnold, E. L. Cussler, Iron nanoparticles in reactive environmental barriers, *paper presented at AIChE Annual Meeting*, 2003.
95. K. Mackenzie, A. Schierz, A. Gerogi, F. Kopinke, *Hemiholtz Centre for Environmental Research* **2007**.

96. B. Schrick, B. W. Hydutsky, J. L. Blough, T. E. Mallouk, *Chem. Mater.* **2004**, *16*, 2187.
97. L. Carlson, J. M. Bigham, U. Schwertmann, A. Kyek, F. Wagner, *Environ. Sci. Technol.* **2002**, *36*, 1712.
98. J. M. Senko, T. A. Dewers, L. R. Krumholz, *App. Environ. Microbiol.* **2005**, *71*, 7172.
99. L. Petrie, N. N. North, S. L. Dollhopf, D. L. Balkwill, J. E. Kostka, *Appl. Environ. Microbiol.* **2003**, *69*, 7467.
100. M. M. Michalsen, A. D. Peacock, A. M. Spain, A. N. Smithgal, D. C. White, Y. Sanchez-Rosario, L. R. Krumholz, J. D. Istok, *Appl. Environ. Microbiol.* **2007**, *73*, 5885.
101. W. De Windt, P. Aelterman, W. Verstraete, *Environ. Microbiol.* **2005**, *7*, 314.
102. K. Pollman, J. Raff, M. Merroun, K. Fahmy, S. Selenska-Pobell, *Biotechnol. Adv.* **2006**, *1*, 58.
103. D. Zhao, Z. Xiong, M. Barnett, R. Liu, W. F. Harper, F. He, *U.S. Patent Appl. Pub # 2007/0203388 A1*, **August 30, 2007**.
104. D. Wei, R. Radhakrishnan, T. H. Vanderspurt, *U.S. Patent Appl. Pub # 2004/0258581*, **December 23, 2004**.
105. T. Koch, H. Bohr, *International Patent Application # WO/2006/005348*, **January 19, 2006**.
106. K. Suzuki, A. K. Sinha, *J. Mater. Chem.* **2007**, *17*, 2547.
107. A. K. Sinha, K. Suzuki, *J. Phys. Chem. B* **2005**, *109*, 1708.
108. Y. O. Park, S. K. Jeong, W. J. Shim, S. H. Lee, http://lib.kier.re.kr/balpy/clean5/5b.pdf.
109. B. Gauthier-Manuela, T. Pichonat, *J. Nanotechnol. Online* **2005**, *1*, 1.
110. S. Zorne, A. David, *International Patent Application # WO/2007/117274*, **October 18, 2007**.
111. A. Khaleel, E. Lucas, S. Pates, O. B. Koper, K. J. Klabunde, *NanoStruct. Mater.* **1999**, *11*, 459.
112. S. Decker, K. J. Klabunde, *J. Am. Chem. Soc.* **1996**, *118*, 12465.
113. K. J. Klabunde, J. Stark, O. Koper, C. Mohs, D. G. Park, S. Decker, Y. Jiang, I. Lagadic, D. Zhang, *J. Phys. Chem.* **1996**, *100*, 12142.
114. A. K. Sinha, K. Suzuki, M. Takahara, H. Azuma, T. Nonaka, K. Fukumoto, *Angew. Chem. Int. Ed.* **2007**, *46*, 2891.
115. Nanoparticles for Air Filtration of TICs (Final Report for Contract R43ES12279), NanoScale Corporation, USA, 2003.
116. Deactivation and Regeneration of Environmentally Exposed Titanium Dioxide (TiO_2) Based Products (Testing Report of the Environmental Protection Department, HKSAR), at http://www.epd.gov.hk/epd/english/resources_pub/publications/pub_reports_epr.html.
117. H. Akbari, P. Berdahl, PIER Final Project Report, January 2008 at www.energy.ca.gov/2007publications/CEC-500-2007-112/CEC-500-2007-112.PDF.

PROBLEMS

1. Nitrate is a naturally occurring compound that is formed in the soil when nitrogen and oxygen combine. Small amounts of nitrate are normal, but excess amounts can pollute supplies of ground and surface water. For most people, consuming small amounts of nitrate is not harmful. However, it can cause health problems for infants, especially those six months of age and younger, by interfering with their blood's ability to transport oxygen. This causes an oxygen deficiency, which results in a dangerous condition called methemoglobinemia, or "blue baby syndrome." The Environmental Protection Agency (EPA) has set the Maximum Contaminant Level (MCL) of nitrate as nitrogen (NO_3-N) at 10 mg/L (or 10 ppm) for the safety of drinking water. A few reports discuss the utility of Fe^0 for the removal of nitrate from water. Based on the dechlorination chemistry displayed by Fe^0, write equations to show plausible pathways for Fe^0 based nitrate remediation from water.

2. $Fe^0_{(NP)}$ is a highly effective reagent for treatment of hazardous and toxic chemicals. As a powerful reductant, it can decompose a wide range of pollutants by adsorption and chemical reduction. It is generally accepted that $Fe^0_{(NP)}$ has a core-shell structure with a central iron core surrounded by iron oxide/hydroxide shell. However, a detailed structural characterization is essential to understand structure-reactivity relationships. What are some techniques that can be used to measure the thickness of the $Fe^0_{(NP)}$ oxide shell?

3. Selenium-contaminated water is highly harmful to human health. According to U.S. EPA regulations, long-term selenium exposure above maximum contaminant level (0.01 mg/L) has the potential to cause fingernail and hair loss, as well as damage to kidneys, liver, and the circulatory and nervous systems. In an x-ray absorption near-edge structure (XANES) study of the reaction of selenate (SeO_4^{2-}) with iron filings (Fe^0), the product profile showed the following selenium (Se) distribution: 74% elemental, 17% Se(IV), and 9% Se(VI). Based on the redox chemistry of Fe^0, write balanced equations to illustrate possible reactions for the reduction and deposition of selenium on iron surfaces.

4. Agglomeration is one of the most important and potentially limiting factors in the use of $Fe^0_{(NP)}$ to reduce groundwater contaminants. As discussed in Section 20.3.1, stabilizers such as starch, cellulose, etc., are used to minimize the deterioration in the performance of nanoparticles. Discuss plausible pathways by which these stabilizers help preserve the reactivity of $Fe^0_{(NP)}$.

5. Two methods of remediating soils contaminated by PCBs were given in this chapter. Which method appears to be more feasible for *in situ* remediation? Why might the party performing the remediation choose the other, more labor-intensive method?

6. Arsenic contamination in drinking water is a global problem. It is usually present in water in two different oxidation states depending on redox and pH conditions.

   (a) State the two forms of commonly encountered arsenic species in contaminated groundwater and their key properties.

   (b) Discuss one adsorption method for simultaneous removal of As(III) and As(V) from contaminated water.

7. In the late 1800s and early 1900s the extreme southeastern corner of Kansas was the site of several coal, lead, and zinc mining operations and smelters. The mining was conducted by the room-and-pillar method, and took place primarily in rock layers that were also aquifers. Over 2.9 million tons of zinc and 650,000 tons of lead were produced from these Kansas mines until the last one closed in 1970. Today much of the area is a Superfund site, slated for environmental cleanup because of the extensive soil and water pollution from the mining. What material mentioned in this chapter could be utilized to remediate Pb-contaminated soil, and what would the reaction sequence be in the alkaline Kansas soils?

8. Former industrial areas can be extremely difficult to reclaim for residential or public use because of long-lasting contamination of the soil. These contaminated soils can subsequently become sources of air and water pollution. One particularly recalcitrant group of contaminants are the PAHs, which frequently become soil contaminants after spills or leaks from oil and gas storage tanks. What nanomaterial could be used for remediation of spilled PAH, and how much removal could be expected after 1 h if 0.15 g material/g soil were used?

9. You have just built your dream home on some land you purchased in the Wet Mountains area of Colorado. However, while browsing a local history museum you discover that your property is sited very near a former Th mine. You consult with a county geologist, who tells you that Th is relatively insoluble in water; but soil contamination from Th(IV) is not unusual, particularly if there are mine tailings present. The county geologist goes on to mention that there are two companies in the area that perform soil remediation and testing, and she hands you their brochures. One states that they use a three-step approach where they first enrich the soil with humic and fulvic acids, then apply TiO_2 nanoparticles to complex the Th(IV), and finally remove the complexes. The second company adds TiO_2 nanoparticles to the soil to complex and stabilize the Th(IV) but does not remove the complexes. Which company's method would remove more Th(IV) and best take care of your potential problem?

10. TiO_2 is one of the most commonly used photocatalysts. In Hong Kong, a study titled "Ambient Air Treatment by Titanium Dioxide Based Photocatalyst" was conducted in 2002 (116). The purpose of the study was to determine the durability of TiO_2 catalyst. Extensively used TiO_2-coated paving blocks were tested for their use in the photocatalytic oxidation of NOx.

(a) State the properties of TiO_2 that are critical for its photocatalytic activity.

(b) Write equations to show NO_x removal by photoactive TiO_2.

ANSWERS

1. Fe^0 may reduce nitrate to nitrogen (Equation 1), nitrite (Equation 2), or ammonia (Equation 3), depending on the reaction conditions.

$$10Fe^0 + 6NO_3^- + 3H_2O \longrightarrow 5Fe_2O_3 + 6OH^- + 3N_2 \qquad (1)$$

$$Fe^0 + NO_3^- + 2H^+ \longrightarrow Fe^{2+} + H_2O + NO_2^- \quad (2)$$

$$NO_3^- + 6H_2O + 8e^- \longrightarrow NH_3 + 9OH^- \quad (3)$$

For references related to this topic see: D. P. Siantar, C. G. Schreier, C.-S. Chou, M. Reinhard, *Water Res.* **1996**, *30*, 2315; S. Choe, Y.-Y. Chang, K.-Y. Hwang, J. Khim, *Chemosphere* **2000**, *41*, 1307; M. J. Alowitz, M. M. Scherer, *Environ. Sci. Technol.* **2002**, *36*, 299.

2. (1) **High resolution transmission electron microscopy (TEM) imaging**: The transmission electron microscope (TEM) is a scientific instrument that uses electrons instead of light to analyze objects at a high magnification and at very fine resolutions. HRTEM is an imaging mode of the TEM that allows resolutions in the nanometer range. HRTEM analysis may be used to provide distinct images of the core-shell structure and the dimension and variation of the shell thickness in the area examined. (2) **High resolution x-ray photoelectron spectroscopy (HR-XPS)**: HR-XPS is a powerful analytical technique that depends on the measurement of the energies of photoelectrons that are emitted from atoms when they are irradiated by soft x-ray photons (1 to 2 keV). The photoelectrons travel only a relatively short distance (~3 nm) in solids before impacting the lattice and suffering an inelastic collision. This makes the technique inherently surface sensitive. At low energy resolution it provides *qualitative* and *quantitative* information on the elements present. At high energy resolution it gives information on the chemical state and bonding of those elements. The shell thickness can be determined through comparison of the relative integrated intensities of metallic iron core and oxidized iron shell with a geometric correction applied to account for the curved layer. (3) **Measurement of $Fe^0_{(NP)}$ content**: By performing a suitable chemical reaction such as evolution of hydrogen with water or reduction of Cu(II), the mass fraction of Fe^0 in the nanoparticles can be calculated. Then, by using a median diameter (obtained by HRTEM measurements) and by using the known bulk densities of the core iron (0) and shell FeOOH, the thickness of the oxide shell can be estimated.

For more details on the utility of the three approaches in determining the shell thickness of $Fe^0_{(NP)}$ see J. E. Martin, A. A. Herzing, W. Yan, X. Li, B. E. Koel, C. J. Kicly, W. Zhang, *Langmuir* **2008**, *24*, 4329.

3. In the redox reaction (Equations 1 and 2), selenate can be the electron acceptor while Fe^0 acts as an electron donor. The selenate removal mechanism will be reductive precipitation followed by sorption on the corroded iron surface.

$$Fe^0 + SeO_4^{2-} + H_2O \longrightarrow Fe^{2+} + SeO_3^{2-} + 2OH^- \quad (1)$$

$$3Fe^0 + SeO_4^{2-} + 4H_2O \longrightarrow 3Fe^{2+} + Se^0_{(s)} + 8OH^- \quad (2)$$

4. Agglomeration of $Fe^0_{(NP)}$ takes place mostly through direct interparticle interactions such as van der Waals forces and magnetic interactions. Agglomeration reduces the specific surface area and the interfacial free energy, thereby diminishing particle reactivity. A stabilizer can reduce agglomeration of nanoparticles through (1) electrostatic repulsion and (2) steric hindrance. Adsorption of charged stabilizer molecules to the metal core results in an enhanced electrical double layer which in turn causes increased Coulombic repulsion between the stabilized particles. Also, sterically bulky polymer-based stabilizers impede particle attractions. Besides weakening the physical interactions, coating nanoparticles with stabilizers can also passivate the highly reactive surface from reacting with the surrounding media such as dissolved oxygen and water. Even though the surface deactivation effect might also inhibit reaction with target pollutants, the net reactivity gain toward a target compound can be significant because of the enormous particle size reduction and the gain in surface area from the particle stabilization.

5. Utilizing Pd nanoparticles formed by bioreduction is more feasible for *in situ* remediation of PCB-contaminated soils because the reactions take place at room temperature. However, the use of $Fe^0_{(NP)}$ might be preferred despite the need for elevated temperatures, because of the multiple reactions of $Fe^0_{(NP)}$ with other soil contaminants, including As, Cr, and PAH.

6. **(a)** Arsenite [As(III)] and arsenate(V) are the two commonly encountered forms of arsenic in contaminated water. As(III) species is more toxic and mobile than As(V). Thus, the oxidation of As(III) species is required in common arsenic removal technologies to reduce toxicity and enhance immobilization.

(b) Oxidation in combination with adsorption-based methodologies can be used for removing both As species from water. For example, As(III) can be readily oxidized to As(V) followed by adsorption of As(V) using MnO_2.

7. CMC-vivianite nanoparticles could remediate the Pb-contaminated soils. As the Kansas soils are alkaline, the soil pH would be >7.2 and consequently the reaction sequence would be:

$$Fe_3(PO_4)_2 \cdot 8H_2O + 2H^+ \longrightarrow 3Fe^{2+} + 2HPO_4^{2-} + 8H_2O \quad \text{(a)}$$

$$5Pb^{2+} + 3HPO_4^{2-} + X^- \longrightarrow Pb_5(PO_4)_3X + 3H^+ \quad \text{(b)}$$

where X is F, Cl, Br, or OH.

8. $Fe^0_{(NP)}$ could be utilized to remove the PAH, and 62% removal could be expected after 1 h of treatment at the concentration utilized.

9. The first company's method, where the soil is enriched with humic and fulvic acids, then treated with TiO_2 nanoparticles, and the TiO_2-Th(IV) complexes removed, would be preferable. The Th(IV) removal could be as high as 98%, and since Th(IV) is a radioactive element, removal of the TiO_2-Th(IV) complexes is highly desirable.

10. **(a)** TiO_2 has a high oxidative ability and extremely good chemical stability. TiO_2 capacity in oxidizing pollutants arises from the generation of high oxidizing power hydroxyl radicals under UV radiation (wavelength <400 nm). The hydroxyl radical generated can oxidize pollutants on the catalyst surface and generate less harmful products.

$$
\begin{aligned}
TiO_2 + \text{UV radiation} &\longrightarrow TiO_2\,(e^-) \\
TiO_2\,(e^-) + O_2 &\longrightarrow O_2^- \\
O_2^- + H_2O &\longrightarrow HO_2^\bullet + OH^- \\
HO_2^\bullet + H_2O &\longrightarrow 2OH^\bullet \\
OH^\bullet + \text{pollutants} &\longrightarrow CO_2 + H_2O
\end{aligned}
$$

(b) Oxidation of NOx leads to the formation of NO_3^- on the surface of the catalyst by way of NO_2:

$$
\begin{aligned}
NO_x + 1/2O_2 &\longrightarrow NO_2 \\
2NO_2 + 1/2O_2 + H_2O &\longrightarrow 2HNO_3
\end{aligned}
$$

21

NANOSCIENCE AND NANOTECHNOLOGY: ENVIRONMENTAL AND HEALTH IMPACTS

SHERRIE ELZEY, RUSSELL G. LARSEN, COURTNEY HOWE, AND VICKI H. GRASSIAN

21.1 Introduction: Environmental Health and Safety Concerns of Nanomaterials, 682
21.2 Overview of Environmental and Health Impacts of Nanomaterials, 684
  21.2.1 Nanomaterials and the Environment, 684
  21.2.2 Life Cycle Assessment of Nanomaterials, 686
  21.2.3 Interdisciplinary Approaches toward Understanding the Environmental, Health and Safety Impacts of Nanomaterials, 690
21.3 Fate, Transport, and Health Concerns of Nanomaterials, 695
  21.3.1 Carbon-Based Nanomaterials in Aquatic Environments, 696
  21.3.2 Properties of Commercial Oxide Nanoparticles that Affect Their Removal in Water, 699
  21.3.3 Transport and Retention of Nanomaterials in Soils and Porous Media, 700
  21.3.4 Environmental Interactions of CdSe Quantum Dots with Biofilms, 705
  21.3.5 Potential Toxicity and Health Hazards of Nanomaterials, 709
21.4 Concluding Remarks and Future Outlook, 712
Acknowledgments, 713
Further Reading, 713
References, 713
Problems, 718
Answers, 722

Nanoscale Materials in Chemistry, Second Edition. Edited by K. J. Klabunde and R. M. Richards

21.1 INTRODUCTION: ENVIRONMENTAL HEALTH AND SAFETY CONCERNS OF NANOMATERIALS

The ability to manipulate and fabricate matter on the nanometer length scale is far better than ever before, with new capabilities to prepare and synthesize highly uniform nanoscale materials, and the advent of new instrumentation to interrogate nanoscale materials. On the nanometer length scale, properties of matter can differ substantially and can exhibit size-dependent behavior. For example, on the nanoscale electronic, optical, and magnetic properties can be size dependent. In addition, because of the high surface to volume ratio, surface properties, surface free energies, and surface coatings are also important and can change the properties of nanomaterials. With these new properties, a host of new materials has now become available for energy and environmental applications, and for improving human health (e.g. for disease detection, drug delivery, etc.) as discussed in detail in earlier chapters.

Given that on the nanoscale, the properties of matter depend on size, it can be asked, how well do we understand the environmental health and safety risks of nanomaterials? For example, can material safety data sheets developed for bulk graphite be used for nano-based carbon materials such as carbon nanotubes or buckyballs? What necessary measures are needed to ensure that the implications of nanomaterials are well understood as these materials are developed and applied in a wide range of applications? Furthermore, what can be done to reduce the uncertainties in our understanding of the environmental health and safety implications of nanotechnology?

In the United States, the Environmental Protection Agency (EPA) provides money for cleanup at Superfund sites and supports Superfund centers to provide a scientific basis for understanding the toxicological impacts of these sites. A much better approach for future technologies is to fund centers for the study of environmental implications as technologies are being developed. This new paradigm of developing technology and simultaneously developing an understanding of the potential risks is beginning to occur, as is evident by the recent partnership of the EPA with the National Science Foundation to fund a national center for the environmental implications of nanomaterials. In addition to a national center, there are several current funding initiatives that support the efforts of individual researchers and small research teams through several federal funding agencies in the United States, Canada, Europe, Australia, and other nations. In the United States, federal agencies taking part in these initiatives include the National Institute of Environmental Health Science and the National Institute of Occupational Safety and Health. As noted above, this represents a new paradigm for balancing technology development with environmental health and safety concerns.

An important goal of ongoing research activities is to provide a strong scientific basis of understanding of the environmental health and safety of nanomaterials. A high level of scientific understanding is essential so that sound policies, if needed, can be developed and implemented with certainty while nanoscience and nanotechnology continue to grow. The avoidance of environmental health and safety problems, such as those that occurred with the use of chlorofluorocarbons (CFCs), polychlorinated biphenyls (PCBs), asbestos, and others is imperative. In addition, from an economic standpoint even the avoidance of perceived risks is essential if nanotechnologies are to be accepted by the public. So from many perspectives—occupational health: will this be the next asbestos?; environmental health: will this be the next PCB or CFC?; manufacturing/marketing: will this be the next genetically modified foods?—it is critical that issues related to the environmental and health impacts of nanomaterials be evaluated and understood.

Manufactured nanomaterials are currently found in cosmetics, lotions, coatings, and used in environmental remediation applications. As these materials develop further and become even more widespread, there are many questions as to the consequences that nanomaterials may have on the environment and human health (1). In fact it is clear from some of the recent literature that the full impact, or even partial impact, of manufactured nanomaterials on human health and the environment has yet to be fully understood (1–12).

Nanomaterials are of varying chemical complexity (bulk and surface), size, shape, and phase (see Fig. 21.1; Reference 13). Therefore, there exist large challenges in understanding the environmental health and safety of nanomaterials and truly interdisciplinary efforts are needed. This chapter reflects the interdisciplinary nature of the research on the environmental and health impacts of nanoscience and nanotechnology. The research discussed in this chapter represents a compilation of some of the most recent studies and the current state of the science of the environmental and health impacts of nanoscience and nanotechnology.

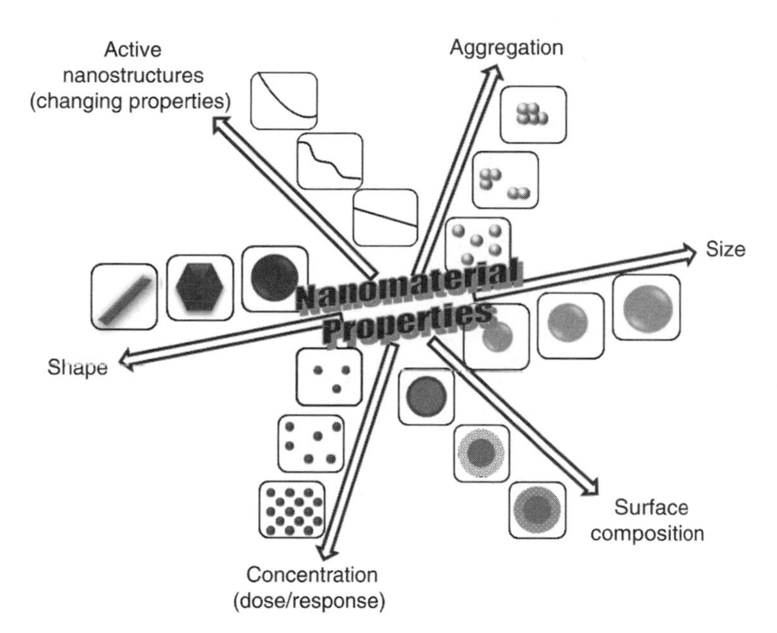

Figure 21.1 Nanomaterial properties vary depending on a number of important characteristics, such as shape, concentration (dose/response), surface composition, and aggregation for passive and active (changing) nanostructures. (Modified and adapted from A. P. Tinke et al., *American Pharmaceutical Review*, **2006**, 9(6), September/October, 1.)

21.2 OVERVIEW OF ENVIRONMENTAL AND HEALTH IMPACTS OF NANOMATERIALS

As the production of engineered nanomaterials continues to increase, an understanding of the impact these materials will have on the environment and human health remains a pressing concern. To fully understand the potential risks and rewards of nanomaterials, a thorough characterization of the size, state, and particle and surface composition is necessary. Studies of the environmental fate and transport of nanomaterials, as well as human health and safety, need to be correlated to nanomaterial characterization to understand how nanomaterials may behave in the environment and the cause of any toxicity or hazards attributed to nanomaterials. Nanomaterial properties must be studied in a variety of conditions to consider how they might interact in the environment and biological systems. There are current challenges in identifying or developing proper tools to evaluate the risks of nanomaterials, and the proper dose metric for measuring nanoparticle concentration. For instance, materials of larger particle size are measured on a mass basis, but nanomaterials typically have low mass concentration but a large particle number concentration (14–17). Another challenge is evaluating the effectiveness of current personal protective equipment and safety devices, such as respirators, filters, and ventilation systems, to see if they are sufficient for protection from nanomaterial exposure.

Thus, there is a need for an interdisciplinary approach to tackle the complex problem of understanding the environmental and health impacts of nanomaterials. Only when all areas of nanomaterial interaction with the environment and human health are understood, and proper equipment is recognized for evaluating and protecting against nanomaterials, can there be an accurate risk assessment and recommendations for appropriate guidelines, safety standards, and protocols regarding nanomaterials. This section focuses on nanomaterial exposure and transformation in the environment, life cycle assessment of nanotechnology, and an interdisciplinary approach that highlights current methods of characterizing nanomaterial properties and strategies for toxicological studies.

21.2.1 Nanomaterials and the Environment

It can be expected that nanomaterials will become present in the environment, including air, water, and soil, at some point during their life cycle of production, distribution, usage, or disposal. For example, nanoparticles, the primary building blocks of many nanomaterials, may become suspended in air during production, distribution, and/or use. Therefore, manufactured nanoparticles can become a component of indoor and outdoor environments and thus the air we breathe. Several years ago an abundance of SiO_2 nanoparticles around 10 nm in size were detected with a single particle mass spectrometry in Houston, Texas (18). Although the exact source of these SiO_2 nanoparticles was unclear, it appeared to be from manufactured or industrial processes. As manufacturing of nanomaterials becomes more commonplace, we can expect that these manufactured materials will get into the environment during manufacturing, distribution, or use. Since nanoparticles and their agglomerates are likely to be in the respirable size range, it is important to investigate the potential health effects of these particles suspended in air as aerosols (12, 19, 20). Commercial engineered nanoparticles join a class of particles known as ultrafine particles whose size is below 100 nm. Ultrafine particles are known to have greater adverse health effects than larger particles (11, 20, 21) because of their extremely high surface areas and the ability to deposit in the alveoli (12, 20, 21). Given that manufactured nanoparticles are a specific subset of ultrafine particles, it is reasonable to surmise that they may have similar deleterious health effects if inhaled.

Nanoparticles often form stable agglomerates or aggregates, resulting in particles that are much larger than the primary particle size. An agglomerate is a grouping of particles held together by weak forces that may be broken into smaller particles easily, while an aggregate is a group of particles held together by stronger interactions, such as bonding, that are not easily broken apart (22). Since nanoparticles possess unique properties due to the size of their primary particles, it cannot be assumed they will behave in the environment the same way as materials of larger particle size, even when they form agglomerates or aggregates. Therefore, an assessment of the physical and chemical properties of nanoparticles in each state (aerosols, dissolved ions, or liquid suspensions) must be made in order to consider their exposure, fate, transport, and transformation in the environment. For example in aqueous environments, nanoparticles may undergo greater aggregation, or in some cases deaggregation, dissolution, precipitation, or sedimentation. These processes will depend on the type of nanomaterial and its properties, as well as conditions such as aqueous phase composition (e.g. ionic strength, pH, and the presence of organic matter) and temperature. Studies have shown nanoparticle fullerene crystals will dissolve or precipitate based on the salinity of their aqueous environment (23, 24).

Water solubility is an important factor affecting the fate and transport of nanomaterials in the environment and their bioavailability in the human body. Many nanomaterials are coated to suppress or enhance their solubility and bioavailability in specific surroundings. However, these coatings may also affect the way the particles interact with the natural environment upon exposure to air or water. For example, quantum dots are coated to enhance activity in biological systems (7), but this could also lead to greater activity for chemical interactions or dispersion in the environment. As shown in Figure 21.2, these coatings not only change the surface composition of the quantum dots, but may also have a large impact on the particle size, resulting in a three- to fourfold increase compared to the bare particle size (25). This size change could affect the properties of the nanoparticles and influence their interactions in the environment and biological systems.

As nanomaterials are transported through the environment or the human body, they may undergo transformations that could influence their properties. Oxidation-reduction (redox) reactions are known to cause environmentally and biologically relevant transformations for certain nanomaterials. Titanium dioxide (TiO_2), for example, can oxidize organic compounds in the environment and inactivate microorganisms (26, 27). This oxidative property of TiO_2 has led to applications such as skin care and water treatment, where antimicrobial properties are desirable, but it may be detrimental to microorganisms that are beneficial to the natural environment. Solar

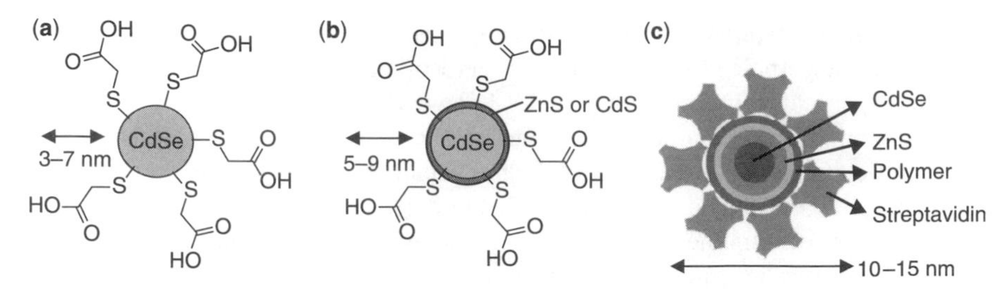

Figure 21.2 Chemical composition of the most commonly used quantum dots in biological applications. (a) CdSe quantum dots functionalized with mercaptoacetic acid, whose —SH bonds directly to the semiconductor, leaving the carboxylate group free to interact with aqueous solution. (b) CdSe quantum dots with a 1 to 2 nm thick layer of ZnS or CdS, functionalized with mercaptoacetic acid. (c) CdSe/ZnS quantum dots coated with polymers and the protein streptavidin. The overall nanocrystal size is a function of the surface coating and/or functionalization. (Repoduced with permission from V. H. Grassian, *Nanoscience and Nanotechnology: Environmental and Health Impacts*. Copyright © 2008 John Wiley & Sons Inc.)

radiation can additionally be responsible for nanomaterial transformations in the environment. Solar radiation can cause photochemical reactions or photoactivated reactive species. It can also contribute to the degradation of nanoparticles or their surface coatings. The bioactive coatings of quantum dots can be photosensitive, degrading to expose the metal core to the surrounding environment (7, 28).

Nanoparticle exposure to biological systems can occur through skin contact, ingestion, and inhalation. Table 21.1 (29) highlights exposure routes for some common products containing nanomaterials. Many skincare products contain TiO_2 or ZnO, and quantum dots and metal nanoparticles are being developed for drug delivery applications, ensuring dermal and ingestion exposure routes. Inhalation will also be of concern as nanoparticles become suspended in air, particularly in indoor environments in occupational settings. Once nanoparticles are absorbed in a biological system they may be able to translocate to other parts of the body, as illustrated in Figure 21.3 (11). The distribution of nanoparticles will depend on their properties, such as size, shape, and chemical composition. Studies have shown that TiO_2 inhaled into the lungs can be distributed to other organs (30), and some nanoparticles are even able to cross the blood-brain barrier (31, 32).

The transport and transformation of nanomaterials as they interact, react, and break down in the natural environment and biological systems will determine their fate and toxicity. Although there is currently a lack of data regarding environmental and human health effects of nanomaterials, recent studies have shed light on the path future directions should take to enhance understanding of the impacts of nanotechnology.

21.2.2 Life Cycle Assessment of Nanomaterials

Environmental and human exposure to nanomaterials may occur during manufacturing, distribution, use, disposal, or decomposition. In other words, exposure will occur

TABLE 21.1 Examples of Potential Sources of General Population and/or Consumer Exposure for Several Product Types

Product Type	Release and/or Exposure Source	Exposed Population	Potential Exposure Route
Sunscreen containing nanoscale materials	Product application by consumer to skin	Consumer	Dermal
	Release by consumer to water supply (e.g. washing with soap)	General population	Ingestion
	Disposal of container (with residual sunscreen) after use (to landfill or incineration)	General population	Inhalation or ingestion
Metal catalysts in gasoline for reducing vehicle exhaust[a]	Release from vehicle exhaust to air (then deposition to surface water)	General population	Inhalation or ingestion
Paints and coatings	Weathering, disposal	Consumers, general population	Dermal, inhalation or ingestion
Clothing	Wear, washing, disposal	Consumers, general population	Dermal, inhalation, ingestion from surface or groundwater
Electronics	Release at end of life or recycling stage	Consumers, general population	Dermal, ingestion from surface or groundwater
Sporting goods	Release at end of life or recycling stage	Consumers, general population	Dermal, inhalation, ingestion from surface or groundwater

[a]Metal catalysts are not currently being used in gasoline in the United States. Cerium oxide nanoparticles are being marketed in Europe as on- and off-road diesel fuel additives.

Source: U.S. Environmental Protection Agency, *Nanotechnology White Paper*. U.S. EPA, Washington, D.C., 2007.

sometime during the life cycle of the nanomaterial. The interconnected life cycle processes that lead to nanomaterial exposure are shown in Figure 21.4 (29). New technologies may offer environmental benefits during a certain stage of their life cycle, while presenting drawbacks during another stage. Failure to identify the risks associated with nanotechnology in the later stages of the life cycle may lead to unexpected surprises with harmful consequences. Therefore, in order to achieve a complete understanding of the impact a specific nanomaterial will have on the environment a life cycle assessment (LCA) must be completed.

LCA is a quantification of the environmental impact of a product or process throughout its entire life cycle. The standard method for LCA is ISO 14000, consisting

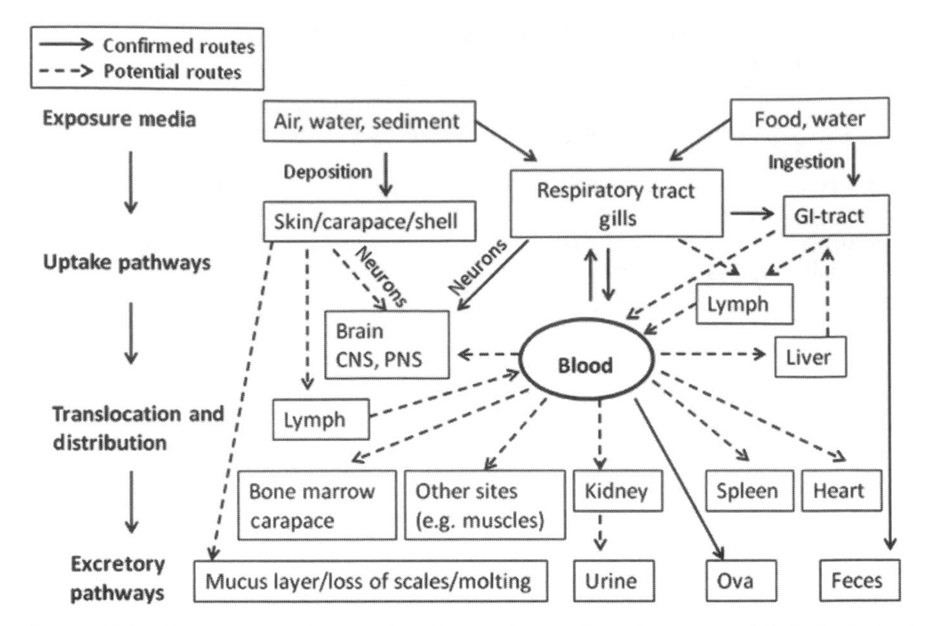

Figure 21.3 Exposure, uptake, translocation, and excretion of nanomaterials in biological systems. (Reproduced with permission from G. M. Oberdorster et al., *Environmental Health Perspectives* **2005**, *113*, 823.)

of four steps: goal definition and scope, inventory analysis, impact assessment, and improvement analysis (33). The goal definition and scope defines the cradle-to-grave analysis in which the LCA practitioner states the boundaries and processes to be considered in the study. Examples of this phase could be a comparison of a nanoproduct and an alternative, an evaluation of various synthesis methods for manufacturing a nanomaterial, or considering a nanomaterial for a new application. The appropriate functional units for comparing different products, as well as input and output boundaries, would be defined in this stage of the LCA.

Inventory analysis consists of collecting the input and output data for each process included in the scope of the LCA. Such data is often collected from public or commercial databases. Input data could include material resources, energy requirements, labor, and equipment. Output data could include products, by-products, and emissions into the environment. In many cases, extensive input and output data are not available, especially in the case of an emerging technology such as nanotechnology.

Impact assessment classifies the emissions of a product based on its impact in various categories, known as midpoint indicators. Impact categories commonly used are global warming potential, ozone layer depletion potential, acidification potential, human toxicity potential, etc. The impact of a certain chemical for a category is often quantified based on a metric of relativity. For example, the global warming potential of carbon dioxide is used as a reference, and the impact of other chemicals in this category are reported relative to carbon dioxide.

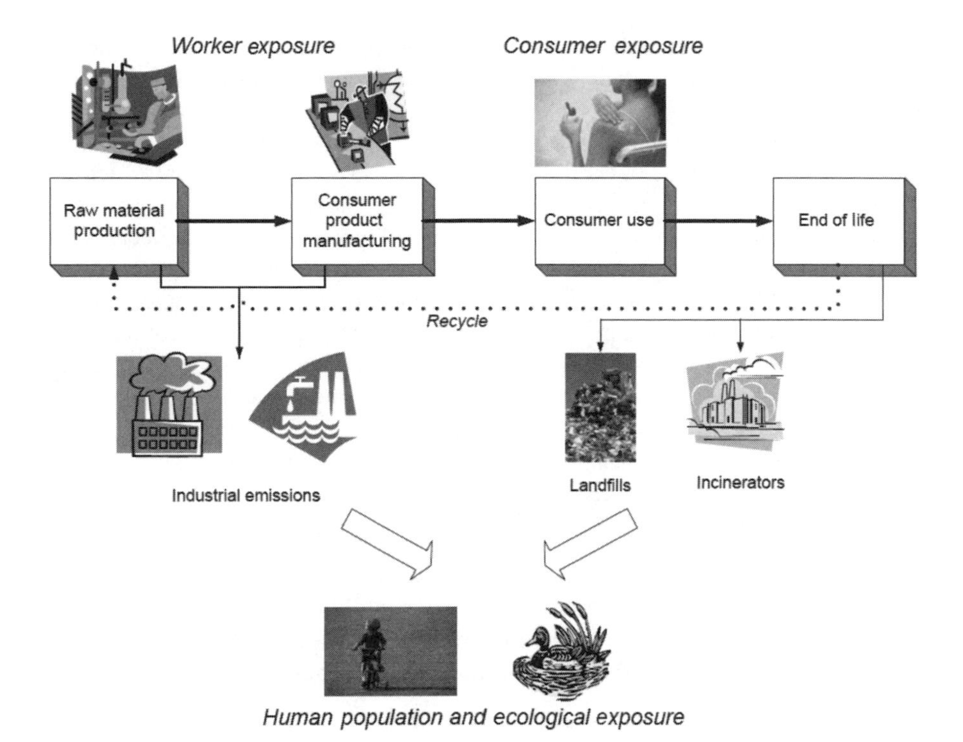

Figure 21.4 Life cycle risk assessment for human and environmental exposure to nanomaterials. (From U.S. Environmental Protection Agency, *Nanotechnology White Paper*. U.S. EPA, Washington, D.C., 2007.)

Midpoint indicators can be grouped into larger categories known as endpoint indicators, which include humans, natural resources, plants, and animals. Endpoint indicators are further grouped into three damage indicators: human health, ecosystem quality, and resources (34). Damage to human health is measured in terms of disability adjusted life years (DALYs), based on years of life lost (YLL) and years lived disabled (YLD) due to exposure to a specific emission. Damage to ecosystem quality is measured in terms of potentially disappeared fraction (PDF), expressed as the percentage of species threatened or that disappear in a certain area and time. Damage to resources is measured in terms of megajoules (MJ) surplus, based on the idea that extracting resources will result in increased energy for future extractions. Thus megajoules surplus per kilogram of material extracted is the energy increase per kilogram to extract material in the future.

Interpretation is the final phase of LCA. In this phase, results from the previous phases are combined so that conclusions and recommendations can be made. To ensure the validity of results used for interpretation, statistical analysis is often performed to quantify uncertainty. Despite a standardized method for LCA, many challenges remain for LCA in environmental evaluation of nanotechnology.

One challenge for LCA of nanomaterials is that most of the current data is proprietary, making it difficult to obtain inputs and outputs for manufacturing processes. The data that is available is often based on laboratory-scale processing and may be difficult to translate into large-scale manufacturing processing, or only include common products and processes. Since nanotechnology is still in its infancy, data on toxicity and ecosystem impacts is lacking. Attempts to predict the impacts of nanotechnology are difficult because this technology is rapidly developing, expanding, and changing.

The challenges of nanotechnology LCA make a complete cradle-to-grave analysis nearly impossible for many new nanomanufacturing processes. Therefore, a predictive approach for LCA may prove more useful during the early development of nanotechnology. While output data is lacking for nanotechnology, input data is more readily available. Input data, combined with nanoparticle properties (size, shape, and composition), can be used as an indicator to predict the impact a nanomaterial may be expected to have on human health and the environment. Thus, a predictive approach for LCA of nanotechnology can provide a proactive means of understanding the toxicity and environmental impact of nanomaterials while the output data necessary for LCA remains pending.

21.2.3 Interdisciplinary Approaches toward Understanding the Environmental, Health and Safety Impacts of Nanomaterials

Due to the broad scope of understanding the environmental, health and safety impacts of nanomaterials, interdisciplinary strategies are necessary. The approach must integrate complete characterization of nanomaterials, toxicity investigations, and environmental impact studies to draw conclusions and make recommendations for safe nanotechnology practices. The need for an interdisciplinary approach has been recognized in several workshops and reports, including publications in *Environmental Health Perspectives* (35), *Toxicological Sciences* (36–42) and an EPA Nanotechnology White Paper (29). These reports and others agree that the physical and chemical properties of nanomaterials must be correlated to the causes of toxicity and compatibility in biological systems and harmful or beneficial effects on ecosystems. Strategies must be employed that look at classes of compounds that probe the impact of physicochemical properties on toxicity for example. One such strategy is shown in Figure 21.5, which demonstrates how nanomaterial type (classification) and nanomaterial properties, including functionalization and shape, must be probed to better understand biological response. Thus, an important component of these studies is the complete characterization of the properties of nanomaterials as an important first step toward understanding the environmental, health, and safety impacts of nanomaterials.

Complete characterization includes determination of both the bulk and surface properties of nanomaterials, since both can influence impacts on the environment and biological systems. Bulk characterization consists of studying size, shape, phase, electronic structure, and crystallinity, while surface characterization looks at surface area, atomic structure, surface composition, and functionality. Specific examples of bulk and surface characterization methods are described below.

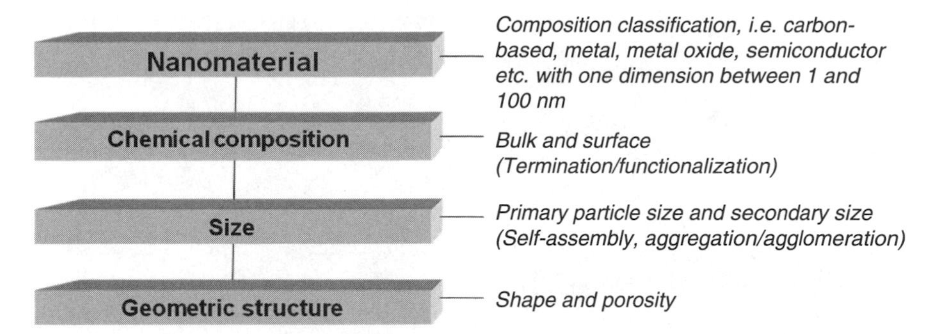

Figure 21.5 Strategies for understanding nanomaterial environmental health and safety, toxicity and biological response include nanomaterial classification, that is, compositional classification (metal, metal oxide, polymer, semiconductor, carbon-based, etc.) for a material that has one dimension between 1 and 100 nm; chemical composition in terms of bulk and surface; size considerations, primary and secondary (aggregate) sizes and geometric structure which includes shape and porosity.

X-ray diffraction (XRD) is a bulk characterization method used to determine the phase and crystallinity of nanomaterials. The size of a particle can affect the thermodynamic stability of its phase because the surface contribution to total free energy, G, becomes increasingly significant as particle size decreases:

$$G = G_{surface} + G_{bulk} \tag{21.1}$$

due to the increasing percentage of total atoms that are surface atoms. The crystallite size of a nanomaterial can also be measured using XRD according to Scherrer's equation:

$$t = \frac{\kappa\lambda}{\beta \cos \Theta} \tag{21.2}$$

where t is crystal size, κ is the shape factor, λ is the wavelength, Θ is the Bragg angle, and β is the full width at half maximum of the diffraction peaks. The crystalline size is inversely proportional to the width of the peaks, and line broadening occurs as crystallite size decreases.

Transmission electron microscopy (TEM) and scanning electron microscopy (SEM) techniques are commonly used to gain information about the size, shape, and tendency toward particle aggregation of a nanomaterial. The TEM images shown in Figure 21.6 highlight different nanomaterials with a variety of sizes, shapes, and particle interactions, including high aspect ratio nanomaterials, spherical particles, and agglomerated particles. The image of the iron spherical particles shows an oxide layer surrounding the iron particle. Microscopy techniques cannot always detect organic surface coatings on nanoparticles due to a lack of contrast in the images. Techniques for measuring particle size are being coupled with microscopy

Figure 21.6 TEM images of different nanomaterials with a variety of sizes, shapes, and particle interactions. The upper images show high aspect ratio nanomaterials, including aluminum oxide whiskers (a) and iron oxide rods and tubes (b). The lower images show spherical particles, including iron (c) and titanium oxide highly agglomerated nanoparticles (d).

techniques to maximize information on particle size due to surface modification and coatings.

One potential method for measuring the size of aerosol nanoparticles is a scanning mobility particle sizer (SMPS), consisting of a differential mobility analyzer (DMA) and a condensation particle counter (CPC). Aerosol particles enter the DMA where they are charged using a radioactive source and their size is classified based on the electrical mobility, Z_p, of the particles in the applied electrical field:

$$Z_p = \frac{neC_c}{3\pi\eta D_p} \tag{21.3}$$

where n is the number of charges on the particle, e is the elementary charge, C_c is Cunningham slip correction, η is the dynamic viscosity of air, and D_p is the diameter of the particle. Particle concentration is then measured by the CPC. The coupled particle sizing of the DMA and particle counting of the CPC allow the SMPS to provide a particle size distribution for a nanomaterial sample. The SMPS method has been used successfully to characterize the thickness of DNA surface coatings on manufactured gold nanoparticles (45).

For larger particles or nanoparticle aggregates, SMPS measurements can be coupled with an aerodynamic particle sizer (APS). For spherical particles, it is easy to relate the measured diameters from the SMPS and APS because no corrections need to be made for shape and volume, but for irregularly shaped particles the APS reports an aerodynamic diameter, D_a, by comparing the settling velocity to a spherical particle with a density of 1 g cm^{-3} to compute the particle size. A volume equivalent diameter, D_{ve}, which is defined as the volume of a sphere with the same volume as a particle with an irregular shape, is used to relate the aerodynamic diameter from the APS with mobility diameter, D_m, from the SMPS (46):

$$D_{ve} = D_a \sqrt{x \frac{\rho_o}{\rho_p} \frac{C_s(D_a)}{C_s(D_{ve})}} \tag{21.4}$$

$$D_{ve} = D_m \frac{C_s(D_{ve})}{x C_s(D_m)} \tag{21.5}$$

$$D_m = D_a x^{3/2} \sqrt{\frac{\rho_o}{\rho_p}} \frac{C_s(D_m)\sqrt{C_s(D_a)}}{C_s(D_{ve})^{3/2}} \tag{21.6}$$

where x is the dynamic shape factor, ρ_o is the reference density (1 g/cm^3), ρ_p is the density of the particle, and $C_s(D_m)$, $C_s(D_a)$, and $C_s(D_{ve})$ are the Cunningham slip factors for the mobility, aerodynamic, and volume equivalent diameters, respectively. For spherical particles the dynamic shape factor, x, is equal to one, and the volume equivalent diameter (D_{ve}) is equal to the measured mobility diameter (D_m).

A method for measuring the size of aggregates in aqueous environments is dynamic light scattering (DLS). This technique uses scattered light to measure diffusion rates (Brownian motion) of particles in stable suspensions to determine a size based on the Stokes–Einstein equation:

$$D_H = \frac{kT}{f} = \frac{kT}{3\pi\eta D} \tag{21.7}$$

where D_H is the hydrodynamic diameter, k is the Boltzmann constant, T is absolute temperature, f is the particle frictional coefficient, which consists of solvent viscosity, η, and the diffusion constant, D. This equation assumes particles are hard spheres, and the size of nonspherical hydrated or solvated particles is related to a hard sphere that would behave the same as the measured particles. DLS allows the hydrodynamic diameter of nanoparticle aggregates to be measured.

Surface characterization techniques measure surface area and composition and study surface reactivity of nanomaterials. The Braunner–Emmet–Teller (BET) method is used to measure surface area by adsorbing nitrogen on the surface of a powdered sample. Surface composition can be determined using X-ray photoelectron spectroscopy (XPS) to obtain information on the elements present near the surface, oxidation states of the elements, and surface functional groups on the surface of nanoparticles. Attenuated total reflectance Fourier transform infrared (ATR-FTIR) spectroscopy is a technique used to investigate surface reactivity of nanoparticles (43, 44). This technique allows *in situ* studies of solution phase reactions on the surface of the thin films containing nanoparticles. Surface functionality is an important component of the biological response of nanomaterials.

Strategies for investigating the biological response of nanomaterials consist of an initial screening, where *in vitro* and short-term *in vivo* exposure are studied for inhalation and instillation exposure methods. The biological response to exposure can be monitored using biological markers. Bronchoalveolar lavage (BAL) fluid is commonly extracted to recover markers such as macrophages, neutrophils, lymphocytes, and total protein counts. The presence of an increased number of cells is a possible sign of inflammation, and activity of a cytoplasmic enzyme, LDH, is an indicator of cytotoxicity. LDH is present in all cells of major organs and is rapidly released when the plasma membrane is damaged.

Markers for cytotoxicity not measured from the BAL include cytokines. Cytokines are proteins that indicate specific types of cellular responses. Monitoring various cytokine levels in exposure studies can provide information about cells that are targeted because each cytokine possesses a specific function. There are more than 18 cytokines that have been identified, and cytokines commonly examined in toxicity studies can be grouped into pro-inflammatory, anti-inflammatory, and cell proliferation and differentiation stimuli subsets (47). Measuring cytokine levels along with BAL fluid markers allows a better understanding of the types of cells being activated and the types of injuries induced by the introduction of the nanomaterial.

Deposition of nanoparticles in the lungs may lead to translocation into the brain and other organs [see Section 21.3.5; (48, 49)]. This potential for serious health consequences of nanomaterial exposure creates a need for effective, fast, and affordable toxicity screening methods. An *in vitro* approach to screen nanomaterials results in simpler, faster, and less inexpensive studies for screening compared to *in vivo* testing. However, correlation between *in vitro* and *in vivo* testing has not been successful. Challenges for comparing *in vitro* and *in vivo* testing include the particle dose, marker selection, interactions of cells with particles in different biological media, time course effects (acute versus chronic), and end points for evaluation (50).

Manufactured nanomaterials must be evaluated for potential threats to the environment and human health to ensure safe implementation of nanotechnology. Due to the size-dependent properties of nanomaterials, the standard approaches for toxicology studies need to be assessed for nanomaterial-based products. Correlations between physicochemical properties of nanomaterials with environmental impacts and biological responses can only be accomplished through collaboration of investigators in a variety of scientific disciplines. This necessitates an interdisciplinary approach involving a research team to understand the environmental, health, and safety impacts of nanomaterials.

21.3 FATE, TRANSPORT, AND HEALTH CONCERNS OF NANOMATERIALS

Due to the certainty that nanomaterials will be exposed to the environment, it is imperative to understand the subsequent fate, transport and transformation that will determine their impact on the environment. A model of the movement of nanoparticles through the air, soil, and water is shown in Figure 21.7 (11). The size-dependent properties of nanomaterials can influence how they will be transported and transformed in the environment; therefore, the behavior of larger-sized materials cannot be used to predict nanomaterial behavior. The type of nanomaterial exposed and the medium it enters are important factors for determining the fate and toxicity of the nanomaterial. These factors will also be important when considering measures for exposure control, waste treatment, or removal of nanomaterials from the environment or biological systems.

This section will consider in greater detail specific examples of particular types of nanomaterials interacting with different media in the environment. The fate and transport of carbon-based nanomaterials, including carbon nanotubes and fullerenes, in aqueous environments and the properties of commercial oxide nanoparticles that affect their removal in water will be discussed. Nanomaterial exposure to soils and porous media, focusing on transport and retention, as well as environmental interactions of cadmium selenide (CdSe) quantum dots with biofilms will be presented. These specific examples provide an idea of the types of environmental interactions that must be considered, and illustrate that environmental impacts of nanomaterials cannot be generalized, but rather, are dependent on properties of the material in question and the environment to which it is exposed or transported.

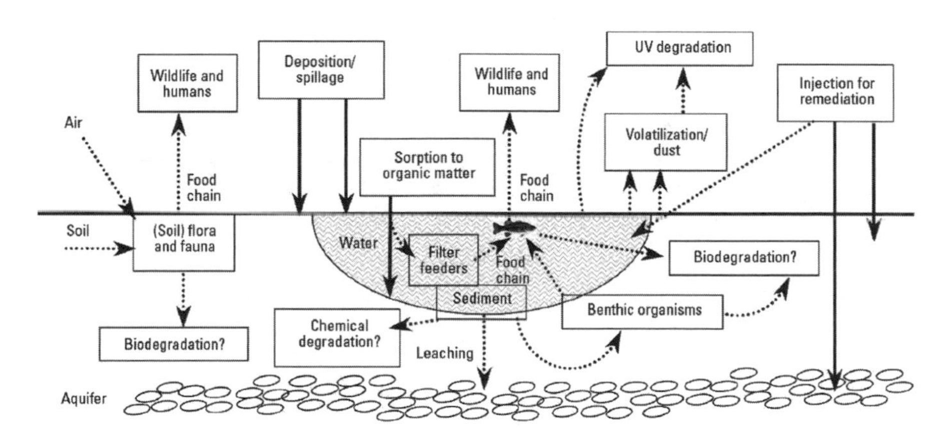

Figure 21.7 Model of nanoparticle movement through the environment. As nanoparticles are transported through the air, soil, and aquifer, they may be transformed physically and chemically. Their transport and transformation can affect the biological and ecological systems they encounter. (Reproduced with permission from G. M. Oberdorster et al., *Environmental Health Perspectives*, **2005**, *113*, 823.)

21.3.1 Carbon-Based Nanomaterials in Aquatic Environments

Commercial applications of carbon nanotubes (CNTs) are rapidly expanding, but their fate and transport in the environment is not well understood (6, 29, 51). Carbon nanotubes are sheets of graphite rolled into hollow tubes with lengths ranging from 1 to 100 μm and diameters from 1 to 25 nm (52). Single-walled carbon nanotubes (SWCNTs) consist of one graphite tube, while multi walled carbon nanotubes (MWCNTs) may consist of two or more concentric graphite tubes. Many of the potential applications of CNTs will require stability in aqueous suspensions, for example, gene therapy agents (53), drug delivery agents (54), and sorbents for wastewater treatment (55, 56). The hydrophobicity of the CNT surface inhibits dispersion in aqueous conditions, and the surface of CNTs is often functionalized with hydrophilic oxides to promote dispersion for aqueous suspensions (57–59). Figure 21.8 (25) lists processes and common oxidants used for oxidizing the surface of CNTs. Oxidation of the CNT surface can also occur in the natural environment through photolysis in surface water.

Surface functionality and oxygen concentration can influence chemical properties of carbon-based materials in aquatic environments, including surface potential, pH, surface reactivity, and sorption properties (60). Since most of the atoms in CNTs are surface atoms the effects of surface chemistry can be expected to have a large impact on the fate and transport of CNTs in aqueous environments. Surface chemistry is thought to be the primary factor regulating biological toxicity and compatibility of nanomaterials (61), and many potential applications of CNTs are in the biomedical field.

The extent of surface oxidation, type, and relative ratio of surface oxide functional groups on CNTs, and solution chemistry of the aqueous environment will dictate the sorption properties of CNTs and whether the CNT agglomerates form stable suspensions or settle out of solution. The tendency of CNTs to settle in an aquatic suspension as a function of the extent of oxidation is shown in Figure 21.9 (25). The CNTs were vortexed in a salt solution and allowed to settle for two hours. After two hours, the

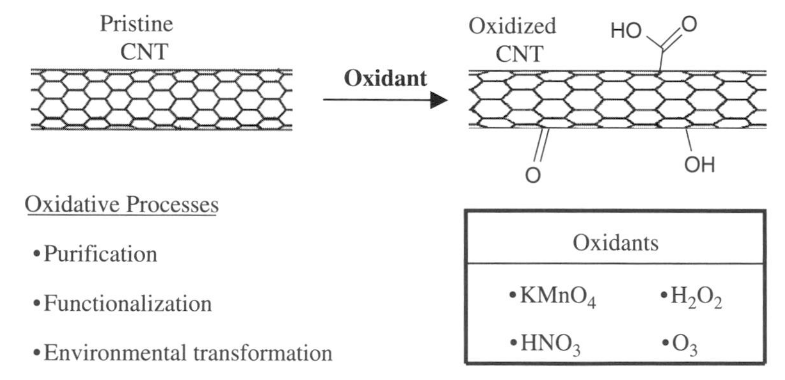

Figure 21.8 Methods for oxidizing carbon nanotubes. Oxidation results in surface functional hydroxyl, carbonyl, and/or carboxylic acid groups. (Reproduced with permission from V. H. Grassian, *Nanoscience and Nanotechnology: Environmental and Health Impacts*. Copyright © 2008 John Wiley & Sons, Inc.)

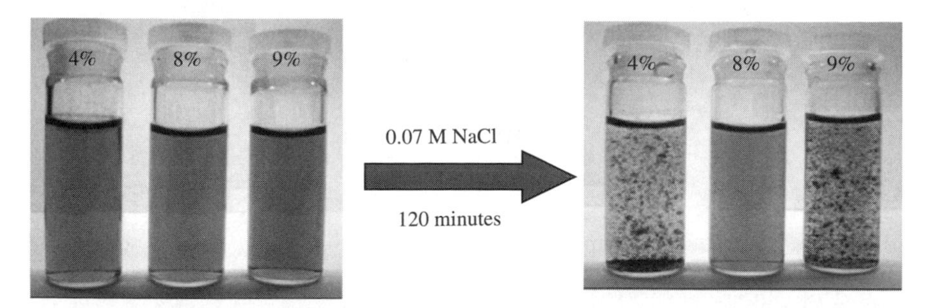

Figure 21.9 Influence of oxygen concentration on the aquatic stability of oxidized MWCNTs. Each vial contains an equal concentration of MWCNTs in 0.07 M NaCl, with the oxygen concentration shown on the cap of the vial. The samples were vortexed and allowed to settle for 120 minutes. (Reproduced with permission from V. H. Grassian, *Nanoscience and Nanotechnology: Environmental and Health Impacts*. Copyright © 2008 John Wiley & Sons, Inc.)

CNTs with the lowest oxygen concentration showed the greatest settling, the highest oxygen concentration also showed significant settling, and the CNTs with an oxygen concentration in the middle still appeared to be suspended. The high ratio of surface area to volume for CNTs will allow even low concentrations of these materials to participate in significant interactions with aquatic environments.

The effects of oxidation on the physiochemical properties of CNTs can be characterized using a number of techniques. TEM can provide information about the structural integrity post oxidation, BET measurements quantify the amount of surface area oxidized, infrared spectroscopy is used for functional group identification, and XPS can quantify the concentrations of various surface functional groups. These techniques have shown surface oxide groups, including hydroxyl, carbonyl, and carboxylic acid groups, result from oxidation of CNTs, and the relative percentages of each oxide group present can depend on the extent of surface oxygen (62). The atomic percent of surface oxide groups as a function of total oxygen concentration for pristine and oxidized CNTs is shown in Figure 21.10 (25).

The surface oxide groups and oxygen concentration can influence the sorption properties of CNTs, with unique effects for organic and inorganic contaminants found in aquatic environments. For organic contaminant adsorption, an increase in the surface oxygen concentration of the CNTs results in a decrease of contaminant adsorption. In contrast, inorganic metal cation adsorption increases with increasing surface oxygen concentration of CNTs (63). While the extent of surface oxidation and surface oxide functional groups exert a great influence on the stability and sorption properties of CNTs, there is no significant affect on the physical or structural characteristics of the CNTS due to oxidation.

Fullerenes are another carbon-based nanomaterial that may be exposed to aquatic environments. There has been increased commercial interest in C_{60} (buckyballs) for a variety of applications, including fuel cells and drug delivery agents. There is a lack of knowledge relating to environmental and health effects of C_{60}, but concern is rising due to fullerene's ability to bind and cleave DNA and inhibit enzyme activity (64, 65).

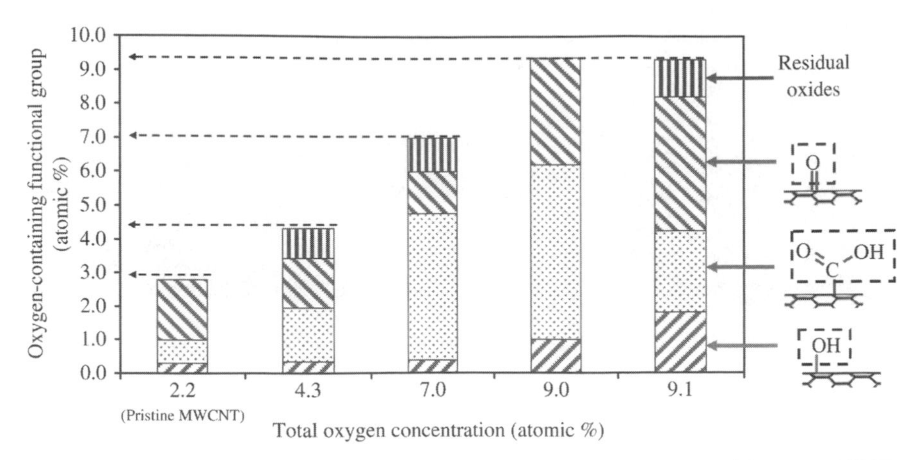

Figure 21.10 Distribution of surface oxide groups for pristine and oxidized MWCNTs. Atomic percent determined by chemical derivatization for hydroxyl, carbonyl, and carboxylic acid groups. (Reproduced with permission from V. H. Grassian, *Nanoscience and Nanotechnology: Environmental and Health Impacts*. Copyright © 2008 John Wiley & Sons, Inc.)

The structure of C_{60} is a cage of 60 carbon atoms consisting of 12 pentagons each isolated by 5 hexagon rings. Physical, structural, electrical, and optical properties of C_{60} are listed in Table 21.2 (66). Although C_{60} is highly hydrophobic, its solubility can be altered by modifying the surface chemistry with hydrophilic functional groups (67), adding surfactants or polymers to the aqueous solution to block the C_{60} surface from water (68, 69), or the formation of stable C_{60} aggregate suspensions. Stable C_{60} aggregates in water can be achieved with mixing at high sheer rates for

TABLE 21.2 Selected Solid-State Properties of C_{60}

Property	Value[a]
Density	1.65 g/cm^3
Graphite density[b]	2.3 g/cm^3
Diamond density[b]	3.5 g/cm^3
Crystal structure (>255 K)	Face centered cubic (FCC)
Crystal structure (<255 K)	Simple cubic (SC)
Nearest neighbor distance	10.04 Å (FCC)
Cage diameter	7.1 Å
Lattice constant	14.198 Å (FCC)
Electrical conductivity	Nonconductor (neutral state)
Index of refraction (RI)	2.2 at 630 nm λ

[a]All values are at standard temperature and pressure unless otherwise noted.
[b]Allotrope comparison.
Source: Reproduced with permission from Donald Huffman, *Physics Today* **1991**, *44*, 26. Copyright 1991 American Institute of Physics.

extensive mixing times (70) to produce particles in the 5 to 500 nm size range (71). Even at low concentrations, C_{60} aggregates have been shown to cause a biological response (72); therefore, it is imperative to understand the physiochemical properties of C_{60} in aquatic environments.

The photoexcitation of C_{60} can result in reactive oxygen species (ROS) responsible for cytotoxicity (72). While ROS are efficiently produced by photoexcitation of C_{60} in organic solvents, production is not efficient for C_{60} dispersed in aqueous environments and becomes undetectable as C_{60} aggregates. The degree of C_{60} aggregation in aquatic environments is dependent on the solution chemistry (surfactants, polymers, etc.) and the dispersion method (73). Therefore, in aquatic environments, C_{60} is unlikely to cause cytotoxicity through ROS production, but further investigation is needed to understand the potential of C_{60} to produce ROS in specific biological environments.

21.3.2 Properties of Commercial Oxide Nanoparticles that Affect Their Removal in Water

Commercial nanomaterials may enter aquatic environments through direct discharge, wastewater, runoff, atmospheric deposition, or other exposure routes. Commercial oxide nanoparticles (titanium dioxide, zinc oxide, iron oxide) are used in paints and sunscreens for their pigmentation and antibacterial properties. Oxide nanoparticles purchased commercially as powders or liquid suspensions contain aggregates of the primary nanoparticles, formed during the manufacturing process. While the primary particles may be ~5 to 25 nm in diameter, aggregates greater than 100 nm tend to dominate aqueous suspension even after extensive sonication (74, 75). Oxide nanoparticles can be manufactured with surface functional groups to allow dispersion of the primary particles, thus, oxide nanoparticles may exist in aquatic environments as primary particles or aggregates of larger sizes.

Water treatment plants (WTPs) are used to remove bacteria, viruses, and particles from water through chemical and physical processes, including coagulation, flocculation, sedimentation, and filtration processes. The properties of the nanoparticles in an aquatic environment will affect their behavior during each of these removal processes. In particular, the size of the particles and particle aggregates and surface charge play a large role in oxide nanoparticle removal in water.

DLS instruments can be used to measure the size of particle aggregates in water, as described previously. DLS can also measure the movement of nanoparticles in an electric field to determine the zeta potential, which provides information about the surface charge on a particle. The zeta potential will affect the distribution of the nanoparticles in solution and influence surface reactive properties. The zeta potential can be calculated using the Henry equation:

$$U_E = \frac{2\varepsilon z f(ka)}{3\eta} \tag{21.8}$$

where U_E is electrophoretic mobility, z is the zeta potential, ε is the dielectric constant, η is the viscosity of the suspension and $f(ka)$ is Henry's function, which is approximated as either 1.0 or 1.5 depending on particle size and ionic strength. The surface potential of oxide nanoparticles may be positive, neutral, or negatively charged, but may vary with conditions of their aquatic environment such as pH and natural organic matter (NOM) content.

The first process involved in particulate removal from water is coagulation. In this step, chemical salt coagulants (most commonly aluminum sulfate) are added to the water to destabilize suspended particles, causing them to aggregate or precipitate and form larger particles. The stability of particles in water can be described by Dejaguin–Landau–Verwey–Overbeek (DLVO) theory, where the total force between particles is the sum of the van der Waals attraction (Φ_{vdW}) and the electrical double layer repulsion (Φ_{EDL}),

$$\Phi_{\mathrm{Total}} = \Phi_{\mathrm{vdW}} + \Phi_{\mathrm{EDL}} \tag{21.9}$$

which describes the variation of electric potential near the surface of a particle. If the net interaction (Φ_{Total}) between two particles is repulsive the nanoparticles will be stable in water, and if the interaction is attractive the particles will aggregate and destabilize. In DLVO theory, the particle size and surface charge are the most important factors that determine the stability of particles in water.

Many experimental studies have shown DLVO theory is not sufficient to describe particle stability in environmental systems, and additional non-DLVO interactions (hydrophobic/hydrophilic and steric) have been applied to an extended DLVO theory. The unique and size-dependent properties of nanoparticles may require additional modification of DLVO theory to model their interactions in aqueous environments.

The destabilization of nanoparticles during the coagulation process allows easier removal during the subsequent processes. The second processing step consists of flocculation and sedimentation. Flocculation occurs when the particles that were destabilized during coagulation aggregate and form large flocs. The water is often mixed to promote the aggregation of destabilized particles. For particles in the nanometer size range, flocculation occurs due to the Brownian motion of the particles and is controlled by their diffusion rate in water. The larger aggregates formed during flocculation can be removed by sedimentation of the nanomaterial in water, or a subsequent filtration process. Filtration is accomplished using membrane filters or granular filters with sand media. As filtration occurs, particle removal improves because the previously captured particles help to retain additional particles. Filtration is a highly successful method of removing nanoparticles from water, with removal efficiencies greater than 99.99%. To date, complete removal of 100% of the nanoparticles in water has never been achieved, and drinking water may expose humans to low concentrations of the nanoparticles that remain.

21.3.3 Transport and Retention of Nanomaterials in Soils and Porous Media

In addition to air and water, nanomaterials may also be exposed to the environment by entering the subsurface. The region of the environment beneath the Earth's surface consists of a solid phase composed of minerals and organic matter, a liquid aqueous phase, and a gas phase within the pores of soil. The diagram in Figure 21.11 (25) shows the distinct regions of the subsurface, water distribution, and flow, and illustrates the solid, liquid, and gas phases present in unsaturated and saturated soil. The pores of soil may account for up to 50% of the soil volume and are usually greater than 1 μm in diameter. Figure 21.12 (25) shows the particle diameter range of nanomaterials is on the same order or magnitudes of order smaller than the pore diameter range of natural soil components (25). Thus, the porosity of soil can be expected to accommodate nanoparticle transport in the subsurface environment. Exposure routes for nanomaterials to reach the land surface and enter the subsurface include atmospheric deposition, spills, leakage from underground storage or pipeline equipment, and landfills.

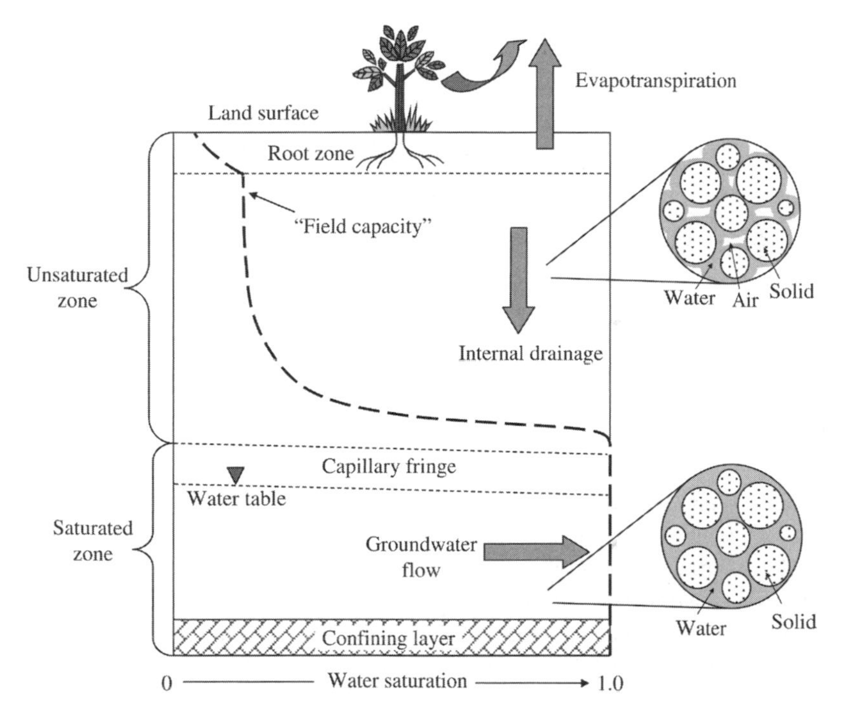

Figure 21.11 Schematic diagram of the subsurface. *Root zone*: region immediately below the land surface occupied by plant roots, typically on the order of 0.5 to 1.0 m in depth. *Unsaturated zone*: region extending from the land surface to the top of the capillary fringe, void space usually contains a gas phase and is therefore "unsaturated" with respect to water. *Saturated zone*: region extending from the top of the capillary fringe to a lower confining layer; void space is completely filled or "saturated" with water. *Capillary fringe*: region immediately above the water table; void space is completely saturated with water, but the water is under suction or negative pressure. *Confining layer*: a geologic unit that does not transmit significant quantities of water (low permeability). The inset shows a three phase system consisting of soil grains, water, and air. (Reproduced with permission from V. H. Grassian, *Nanoscience and Nanotechnology: Environmental and Health Impacts*. Copyright © 2008 John Wiley & Sons, Inc.)

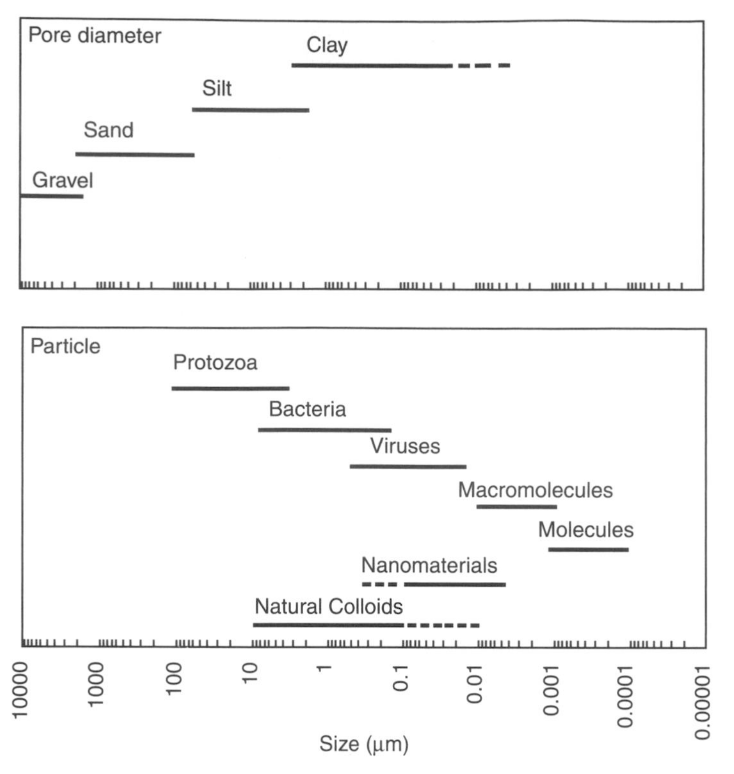

Figure 21.12 Comparison of the diameter of representative particles and the effective pore diameter of natural soil materials. (Reproduced with permission from V. H. Grassian, *Nanoscience and Nanotechnology: Environmental and Health Impacts*. Copyright © 2008 John Wiley & Sons, Inc.)

Once nanoparticles enter the subsurface, their transport will be highly dependent on the flow of water through the porous soil media. Nanomaterials may be retained in the soil as they migrate, or they may continue to migrate and enter drinking water sources or streams and lakes. Thus, transport of nanoparticles in the subsurface could present a potential route for human exposure to nanoparticles. The transport and retention of nanoparticles in porous media must be investigated to understand their fate as it relates to the environmental, health, and safety effects of nanomaterials. A study comparing the transport of nanomaterials of different sizes through porous glass bead media found that smaller nanoparticles were transported through the pores in higher concentrations (Fig. 21.13; Reference 76).

The extent of water saturation in the subsurface is a critical factor affecting the transport and retention of nanomaterials in soil. Completely saturated soil has a water saturation = 1, and unsaturated soil contains some fraction of gas phase with water saturation <1. As nanomaterials migrate in the subsurface they may encounter varying saturation conditions. Transport and retention of nanomaterials in saturated soil will be discussed first.

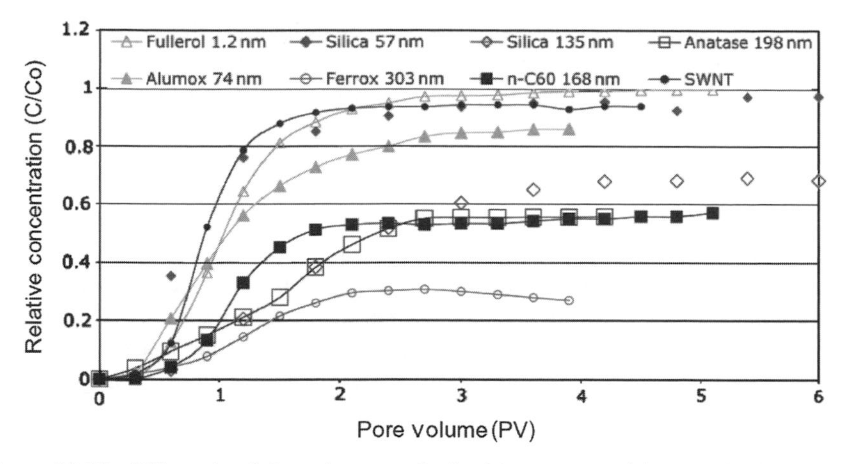

Figure 21.13 Effluent breakthrough curves obtained for nanomaterial suspensions (10 mg/L) introduced at a flow rate (Q) of 12 mL/min in columns (2.5 cm diameter × 9.25 cm length) packed with spherical glass beads (diameter = 0.355 mm). (Reproduced with permission from H. F. Lecoanet et al., *Environ. Sci. Technol.* **2004**, *38*, 5167.)

Mechanisms of transport and retention in saturated porous media include solid-water interface attachment and pore straining. The movement and interactions of particles in these mechanisms are illustrated in Figure 21.14 (25). Solid-water interface attachment is the dominant mechanism of filtration for nanomaterials in saturated media. In this process, nanoparticles or aggregates collide with solid media and adhere to the surface. Nanoparticle movement, such as Brownian motion or

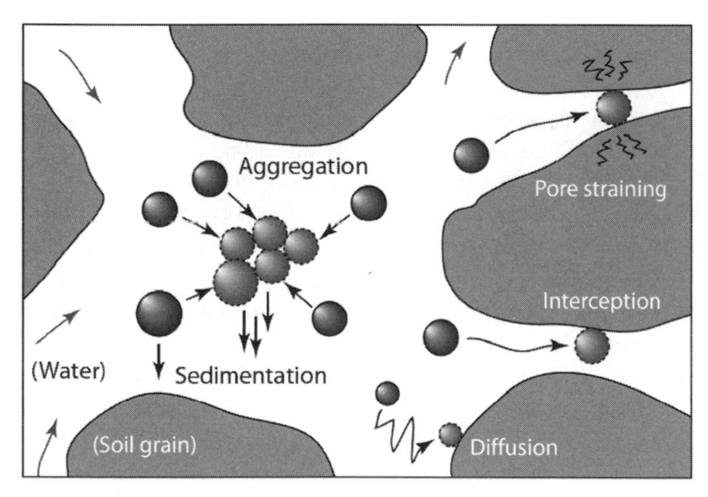

Figure 21.14 Major saturated transport mechanisms. (Reproduced with permission from V. H. Grassian, *Nanoscience and Nanotechnology: Environmental and Health Impacts.* Copyright © 2008 John Wiley & Sons, Inc.)

sedimentation, and particle-solid interactions, such as hydrophobicity or electrostatic forces, affect the solid-water interface attachment (77, 78). When particle movement and particle-solid interactions do not result in attachment, pore straining can lead to particle filtration in saturated porous media. Pore straining is a process that traps particles that are too large to pass through pores, and it depends on particle size distributions and pore size distributions (79, 80). Properties of nanoparticles or saturated porous media that influence either mechanism of transport will affect the migration and retention of the nanoparticles.

Although the pores of soil media are on the micron scale, there is considerable retention of nanoparticles with diameters greater than 100 nm in saturated conditions (74). Nanoparticle aggregation, therefore, has a large effect on the transport and retention of nanomaterials in soil. Surface charge and ionic strength are two other important properties that influence nanomaterial transport and retention in soil. DLVO theory can be applied to predict nanoparticle stability and surface interactions in aqueous saturated conditions. The presence of ions and salts in the soil media can change the surface charge of the nanoparticles and aggregates from positive to negative (or negative to positive) and alter the ionic strength of the suspension, affecting the transport properties of the nanoparticles (81).

The ionic strength of a nanoparticle suspension can influence the retention of the nanoparticles in saturated porous media. Current investigations of ionic strength and nanoparticle transport in porous media use DLVO theory. Experiments that increase ionic strength of salt solutions containing C_{60} nanoparticle aggregates result in greater retention by porous media due to a greater number of collisions between the particles and media (82). Theoretical calculations find a higher ionic strength results in a decreased repulsive force and increased attractive force between particles and porous media for C_{60} suspensions in salt, consistent with the greater retention of particles in porous media. DLVO theory is a useful tool for modeling the transport and retention of nanomaterials in water-saturated porous media, but research is needed to understand processes in the natural environment where minerals, organic matter, and air may be present.

Porous soil media containing air are unsaturated, and the air-water interface can affect the transport and retention of nanomaterials. In addition to the transport mechanisms of solid-water interface attachment and pore straining for saturated media, air-water interface attachment and film straining are important transport mechanisms in unsaturated media. As the water content of unsaturated porous media decreases the retention of particles increases due to attachment to the air-water interface. Thus, total retention of particles in unsaturated porous media is greater than retention in saturated porous media (83). Film straining refers to the straining of particles by the thin film of water that surrounds solid grains in unsaturated porous media. The water film can be as thin as 10 nm, and is thought to be the dominant mechanism responsible for immobilizing particles in unsaturated porous media (84, 85). However, film straining is reversible while air-water interface attachment is irreversible. As with saturated porous media, mechanisms of transport in unsaturated porous media can be affected by the surface charge of nanoparticles and the ionic strength of the suspension.

The water distribution in unsaturated porous media is controlled by capillary forces, and pore sizes will vary as water flows and drains through the pores. The pore size can be determined from the capillary pressure using the Young–Laplace equation:

$$P_c = \frac{2\gamma \cos \theta}{r} \tag{21.10}$$

where P_c is the capillary pressure, γ is the surface tension of the pore solution, θ is the contact angle between the air-water interface and the solid surface, and r is the pore radius of the largest saturated pore at a given P_c. From Equation (21.10) it can be seen that higher capillary pressure corresponds to a smaller saturated pore radius. The capillary pressure-saturation relationship is a hydraulic property of unsaturated porous media that controls flow and transport.

Transport in saturated and unsaturated porous media in the subsurface is governed by many of the same mechanisms and affected by many of the same properties of nanomaterials and suspension chemistry. The extent of saturation will affect the retention of nanoparticles as they migrate through a porous medium, with a greater total retention expected for unsaturated porous media. Theoretic models and mathematical relationships are helpful for predicting transport of nanomaterials in porous media, but experimental data in natural conditions is needed for life cycle assessments and environmental standards.

21.3.4 Environmental Interactions of CdSe Quantum Dots with Biofilms

Quantum dots (QDs) are semiconductor nanocrystals with unique and size-dependent electronic and fluorescent properties. The properties of QDs have sparked interest in many commercial applications, including computing, photovoltaic cells, and biological labeling. CdSe QDs are the most common type of QDs considered for use in biological applications. The CdSe core is often surrounded by another semiconductor shell, such as ZnS, to enhance emission properties. The shell of QDs can be capped with functionalized thiol molecules to make them water soluble and may be additionally coated with polymer or protein coatings to make them biologically compatible.

Commercially available QDs are not expected to pose an inhalation hazard since they are synthesized and stored in liquids. The main concerns are for human ingestion and water and soil contamination of QDs. Therefore, the environmental and biological interactions of QDs must be investigated to ensure the safety of medical applications throughout the life cycle of QDs. Varying the size and composition of QDs allows control of the fluorescent properties for targeted binding to biological molecules (86, 87). CdSe has not shown cytotoxicity over the short time scale for biological labeling and imaging, which takes place on the time scale of hours (7). However, CdSe could interact with the natural environment over much longer time scales, potentially resulting in accumulation of toxic levels of the heavy metals Cd, Se, and Zn. In addition to the risks posed by the metal components of QDs are size-specific properties that distinguish them from bulk semiconductors, namely the ability to cross cell membranes and act as strong oxidizing and reducing agents.

In an abiotic environment QD interactions may be affected by factors such as light, pH, and chemical composition of the environment or QD surface modifications. Functional groups commonly attached to the surface of QDs (thiol molecules, citrate) can be sensitive to variations in pH and become ineffective as capping agents. Photochemical instability of QDs can lead to oxidation of surface caps and subsequent oxidation of the metal shell and core. Toxic levels of metal cations (Cd^{2+}, Zn^{2+}) may accumulate in solution over the course of several hours due to oxidation (88). QDs can also participate in redox reactions with agents in their surrounding environment. The oxidation and reduction potential of QDs is greater than that of bulk semiconductors due to the size confinement of electrons in QD nanocrystals. Redox reactions may also contribute to nanoparticle breakdown, releasing toxic metal material from the QD core.

The environmental interactions of CdSe QDs in abiotic environments are expected to be enhanced in biological environments due to the presence of biofilms (89). Biofilms are bacterial cells attached to surfaces where the cells multiply and produce extracellular polymeric substances (EPS) (90). Thus, a biofilm is a combination of cells, polysaccharides, nucleic acids, proteins, and inorganic material that may form structured channels (91, 92). The biophysiochemical interactions between biofilms and CdSe QDs can include binding and uptake, decomposition, and expulsion. These processes may have toxic repercussions for biological organisms.

CdSe QDs functionalized with biomolecules can bind to biofilms and become internalized in the microorganisms. A conceptual illustration of the interactions between QDs and biofilms, including binding, internalization, and expulsion, is shown in Figure 21.15 (25). This process is being studied for applications in fluorescent labeling of targeted cells with QDs. Once the QDs are internalized, peak shifts in fluorescence spectra have been observed, indicating breakdown of the nanocrystals. These peak shifts are similar to what is seen when salts of cadmium or selenium are present rather than the nanocrystals. The breakdown of CdSe nanocrystals is dependent on the type of biofilm and enzymes present and may be direct or indirect breakdown. Direct breakdown refers to degradation processes occurring within the cell, and indirect breakdown refers to biological molecules that remove QD surface ions outside the cell or on the cell membrane (93). Expulsion of Cd^{2+} after hours of QD exposure has been observed for *Escherichia coli* cells and other bacteria, indicating that certain cells may be resistant to Cd^{2+} ions (94).

Cells exposed to metal ions of CdSe or ZnS QDs can expel the Cd metal and form Cd nanocrystals or create ZnS nanocrystals from Zn metal (95, 96). Thus, in addition to uptake and breakdown of QDs by biofilms, reformation of nanocrystals from the metal ions can also occur via microbial processing. This formation of semiconductor nanoparticles from ions using microorganisms may have commercial applications, but the same processes may also have detrimental impacts on biofilms in the human body and the natural environment.

Toxicity of CdSe QDs has not been observed on the time scale of biomedical applications, as the results shown in Figure 21.16 confirm (25). In a study comparing the growth of *Psuedomonas aeruginosa* over 50 hours with CdSe QDs present, Cd metal present, or no metals present, the presence of QDs had no significant effect.

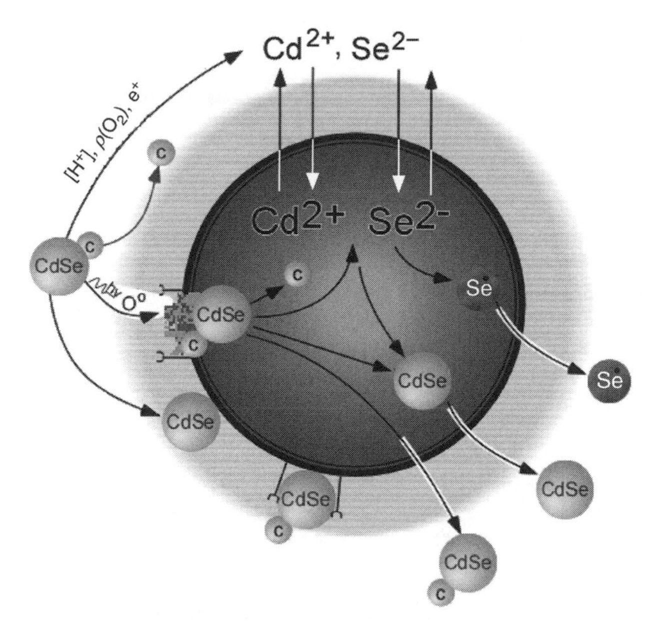

Figure 21.15 Conceptual diagram of CdSe quantum dot (QD, light gray spheres with small sphere conjugated attached) interactions with a bacterial cell. Bioconjugated CdSe QDs are subject to decomposition processes that can lead to heavy metal and metalloid release; released metals and metalloids may be intracellularized, causing toxicity and possibly reassembling inside cells, where they are retained or expelled. (Reproduced with permission from V. H. Grassian, *Nanoscience and Nanotechnology: Environmental and Health Impacts*. Copyright © 2008 John Wiley & Sons, Inc.)

However, dissolved Cd^{2+} ions have been known to bind to DNA and adsorb on membranes of bacterial cells (92). Cellular response (cell viability, gene expression) to QD exposure is highly dependent on QD surface functionalization and capping, particle size, and the nature of the biofilm (97–99). Preliminary research concludes that the biophysiochemical interactions between biofilms and QDs must be understood before biological testing for medical applications can be accurately interpreted (100).

In addition to potential human health toxicity affects, QD exposure may also have environmental implications. QD breakdown in the natural environment can lead to metal accumulation over longer time scales, resulting in toxic levels of Cd and Zn. Metal exposure can disrupt carbon and nitrogen interactions in soil biofilms (101). Sorption of metals to biofilms in soil can result in transport of metals, possibly into water systems. Metals have a higher affinity for EPS associated with biofilms than for geological surfaces (sand, soil), resulting in enhanced metal transport (102, 103). However, if metals adsorb to immobilized biofilms, sequestration of the metals may occur (92). The processes of CdSe activation, migration, often mediated by cellular interactions, and sequestration are shown in Figure 21.17 (25).

Efforts to understand QD interactions with biofilms have highlighted the need for further research to understand biophysiochemical interactions and factors affecting

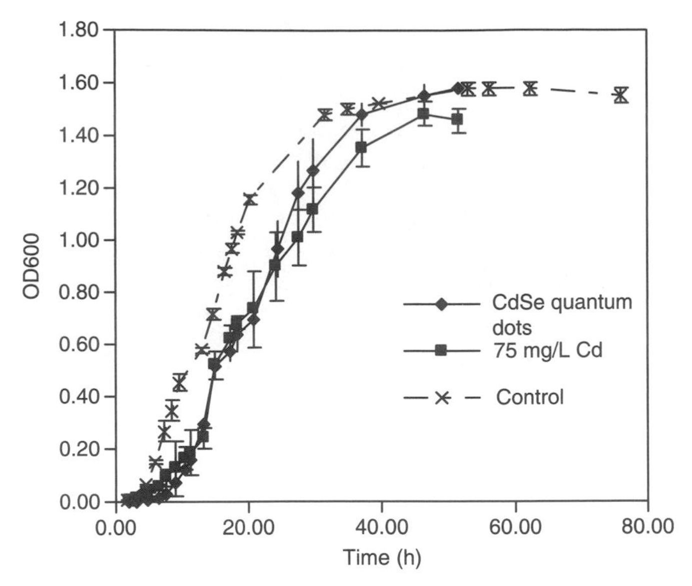

Figure 21.16 Growth curve (optical density at 600 nm = OD600 vs. time) for *Pseudomonas aeruginosa* cultivated with no metals (control), 75 mg/L Cd(II), or bare citrate-stabilized CdSe QDs (75 mg/L as Cd). (Reproduced with permission from V. H. Grassian, *Nanoscience and Nanotechnology: Environmental and Health Impacts*. Copyright © 2008 John Wiley & Sons, Inc.)

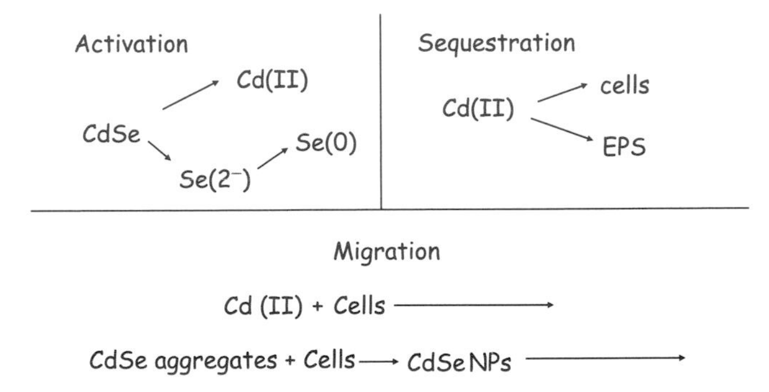

Figure 21.17 Three potential environmental implications of research to date regarding bacterial interactions with CdSe QDs. Activation (upper left) refers to the increase in toxicity of a substance, in this case resulting from Cd(II) ion release during bacterial decomposition of CdSe QDs. Sequestration (upper right) occurs when released Cd(II) binds with cells, or to EPS associated with bacterial biofilms. Sequestration could limit migration. Migration (lower panel) could also be enhanced if Cd(II) ions or CdSe QDs are transported with cells when the metals or particles would otherwise stick to stationary inorganic surfaces in the environment. Lastly migration of primary particles could be enhanced if aggregated CdSe QDs distribute onto cells as primary particles wherein cellular, and thus QD, transport could occur. (Reproduced with permission from V. H. Grassian, *Nanoscience and Nanotechnology: Environmental and Health Impacts*. Copyright © 2008 John Wiley & Sons, Inc.)

toxicity in various media. Size effects of QD nanoparticles on cellular response need to be investigated in more detail. Aggregation tendencies of QD will play a large role in size effects, and could determine whether QD nanoparticles can become internalized in cells. Another property to consider is the stability of QD nanoparticles both outside and inside cells, and also the processes that affect stability and lead to breakdown. Furthermore, the mechanisms and products of QD breakdown need to be identified. In addition to investigating how biofilms affect the stability, breakdown, and transport of QDs, it is important to understand how QD exposure affects biofilm formation.

There are obviously many questions yet be answered concerning the interactions between QDs and biofilms. The complex nature of this research and extensive investigations needed to understand these interactions require an interdisciplinary approach. This approach will need to connect unique QD properties with observed cellular responses. A thorough recognition of the correlation between QD properties and human and environmental toxicity will allow the benefits of QD applications to be realized and potential risks minimized.

21.3.5 Potential Toxicity and Health Hazards of Nanomaterials

A further concern regarding engineered nanomaterials is that the rate of production of nanomaterial applications is not balanced with adequate toxicology testing. The wide range of nanomaterial variability complicates the assessment of potential health risks and health implications. A multitude of considerations need to be taken into account to accurately assess the potential nanoparticle toxicology. Furthermore, in order to be effective with policy decisions and human health standards additional factors need to be considered. These include identifying the most appropriate exposure/dose metric (e.g. particle mass, surface area, particle concentration, etc.). Recently it has been suggested that a comparative analysis be made to assess the current status of the known/unknown information pertaining to particle characteristics, exposure/dose metrics, and route of exposure among different nanomaterials.

First, it is important that a low cost, high throughput assay be developed to evaluate the toxicity of nanomaterials so that toxicological testing can keep up with the nanoparticle advances. Although animal or human exposures will always remain as the authoritative method, *in vitro* testing is affordable and efficient. Furthermore, *in vitro* cell cultures are easily manipulated in ways ideal for testing mechanistic biochemical hypotheses, or the genetic role of specific biomolecules mediating a response. Recent work has suggested that *in vitro* testing be used for preliminary screening strategies, which would also reduce animal use. This approach complies with the ethical and toxicological demands and minimizes the use of animal tests.

Many different cell types are currently being used for *in vitro* studies relating to nanoparticle toxicology. Some cells are obtained as donor cells which behave similar to normal cells when in culture. Other cells are derived from immortalized lines which have infinite growth capabilities in culture. Techniques have also been applied to construct cell cultures that mimic a specific location within the body (e.g. dermal cell, epithelial lung cells).

It is also important to recognize the portal of entry of nanoparticles when considering human toxicity. Nanoparticles can enter the human body through several different ways. The two most recognized pathways are through the airway or the skin. Since inhalation is the pathway of greatest concern and one of the most studied pathways, the discussion here is focused on inhalation. As nanoparticles have been shown to translocate to other organs (104, 105), not only does inhalation impact the pulmonary system, it can also impact the central nervous system.

Various techniques of particle delivery are used to deduce pulmonary and systemic effects from the wide parameters of potential toxicological influences. The specific techniques that are currently employed in these studies include intratracheal instillation, intratracheal aspiration, and intratracheal inhalation. Of these different delivery techniques, intratracheal installation is a useful technique to assess the potential health effects of different particles efficiently and cost effectively. Intratracheal instillation is characterized by saline suspended particles administered directly into the trachea of the animal being tested. Intratracheal installation provides a relatively easy way to compare the toxicology between different materials. However intratracheal installation is not able to assess particle deposition. Intratracheal aspiration involves droplet administration of suspended particle matter in the form of a puff of air. Intratracheal inhalation is the most relevant for toxicity and risk assessment. Intratracheal inhalation involves nanoparticulate aerosol formation at constant concentrations during the exposure.

In all of the methods noted above, it is typical to then measure a number of inflammatory markers and compare these to control animals, that is, animals not exposed to nanoparticles. Lungs are lavaged and the recovered bronchoalveolar lavage (BAL) fluid is used for enumeration of total and differential cell counts. Furthermore, lavage supernatants are analyzed for total protein, lactate dehydrogenase (LDH) activity, and cytokine levels. Cellularity, total protein, lactate dehydrogenase activity, and cytokines, as well as lung histopathology, are evaluated to assess responses.

Typically, the presence of foreign particles within the lung activates the surveying alveolar macrophage cells. Inactivated alveolar macrophages patrol alveolar sites for the presence of foreign debris that passed through upper respiratory barriers. When activated, the cells elicit phagocytosis and destroy foreign matter, and thus can prevent entry into the systemic circulation. Studies have demonstrated that the ability of macrophages to sequester foreign particles is influenced by particle size. For example, when 500 nm diameter particles were compared with 20 and 80 nm diameter particles, the macrophages were 55% more effective at retaining the 500 nm particles. This barrier mechanism is most effective when particle diameter exceeds 100 nm (106).

Although many studies on nanoparticle toxicology have focused primarily on pulmonary implications, recent work has begun to identify an association between nanoparticle exposure and neurological damage. Translocation of nanoparticles into the central nervous system (CNS) is closely associated with the mechanisms of inhalation. The CNS is strictly regulated in order to facilitate operational control. Any insult or damage to the CNS can result in impairment. Therefore, it is important to recognize the potential neurological threats of nanoparticle exposure. Several studies have demonstrated <100 nm diameter nanoparticles have a greater chance of translocating to the systemic circulation and traveling to the CNS (11, 33, 106–108). In order for particles within the systemic system to enter the brain, they must pass through the blood-brain barrier (Fig. 21.18; Reference 109).

In addition to the systemic circulation, there are two other ways nanoparticles can translocate to the CNS. A study examining the translocation of manganese dioxide to the CNS discovered a significant nanoparticle accumulation in the olfactory bulb compared to other areas of the brain (striatum, cerebellum, and frontal cortex). The concentration difference between the olfactory bulb and other regions of the brain suggests that translocation occurred along the axon of the olfactory nerve into the CNS (Fig. 21.19; Reference 110). The other pathway over which the nanoparticles migrate is around the nerve across the ethmoid bone into the cerebrospinal fluid (108). These two pathways bypass the lung and systemic translocation, entering the CNS directly through the nasal cavity. As can be seen in Figure 21.19, the brain and nasal cavity are in close spatial proximity.

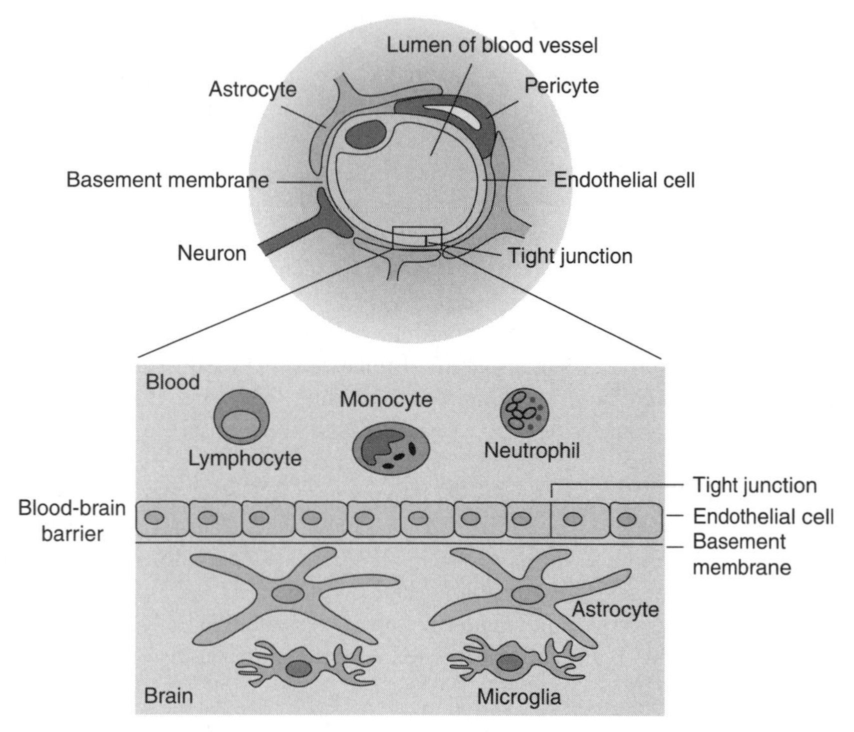

Figure 21.18 Diagram of the blood-brain barrier (BBB). The endothelial cells and tight junctions that form the BBB significantly restrict transport of virtually all molecules from the blood to the brain. (Reproduced with permission from K. Francis et al., *Expert Reviews in Molecular Medicine*, **2003**, *5*(15), 1. Copyright © 2003 Cambridge University Press.)

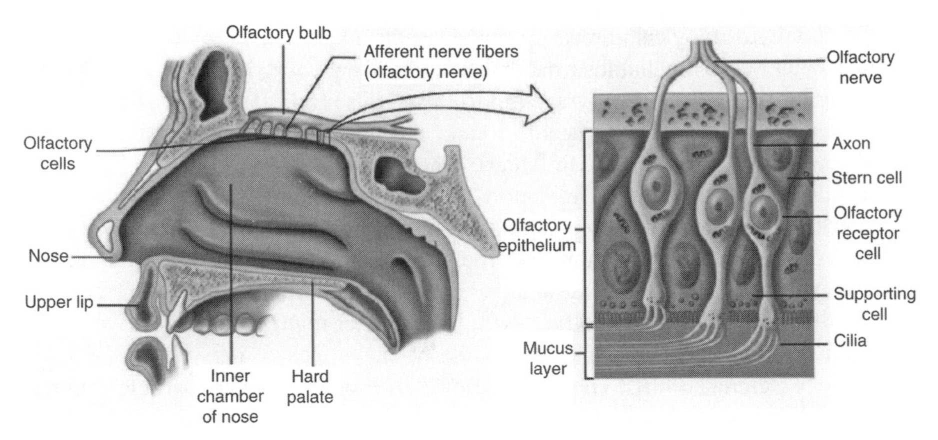

Figure 21.19 Close proximity of the olfactory cells to the olfactory bulb of the central nervous system (CNS). Inhaled nanoparticles, especially below 10 nm in diameter, deposit efficiently on the olfactory cells by diffusion. Subsequent uptake and translocation along axons of the olfactory nerve have been demonstrated in nonhuman primates. (Reproduced with permission from E. P. Widmaier et al., *Vander, Sherman & Luciano's Human Physiology: The Mechanisms of Body Functions*, 9th edition, 2004. Copyright © 2004 McGraw-Hill.)

Injury or insult to the brain has the potential to activate a cascade of events that ultimately lead to inflammation. The first cells to respond to insult or injury to the brain are microglia cells (111). Microglia cells function as macrophages that monitor the parameters of the CNS. Microglia are activated directly by pathogens and/or foreign material, or they can be activated by pro-inflammatory signals in a secondary response (112). Microglia have essential roles in development and in maintaining homeostasis in the CNS. However, overactivation of microglia can have negative implications associated with neurodegenerative disorders (113).

Evidence indicates exposure to nanoparticles can induce an inflammatory response in the CNS. For example, when a sample of mice were exposed to airborne particle matter, increased levels of pro-inflammatory cytokines (TNF-α & IL-1α), transcription factor, and nuclear factor-kappa beta (NF-$\kappa\beta$) were observed (114). TNF-α serves a neuroprotective function (115), but given certain pathogens TNF-α can be neurotoxic (116–120). IL-α activates cyclooxygenase (COX)-2, phospholipase A_2, and inducible nitric oxide synthase (iNOS) activity, which are all associated with inflammation and immune response (121). IL-α is also partially responsible for increasing the permeability of the blood-brain barrier (122, 123). Thus, there is great interest in better understanding how nanoparticles enter the body and translocate as this will impact all organs and thus the toxicity of nanomaterials.

21.4 CONCLUDING REMARKS AND FUTURE OUTLOOK

The development of nanotechnology-based consumer products is predicted to grow substantially in the next 10 years and beyond. Along with this growth, it is clear that there will be many issues and questions that need to be addressed related to the potential impact this technology will have on the environment, living organisms, and human health. We hope that this chapter inspires some readers to rise to the challenges that are faced so that the environmental and health impacts of nanoscience and nanotechnology can be understood, and therefore properly controlled as new commercial uses and applications emerge.

ACKNOWLEDGMENTS

Although the research described in this chapter has been funded in part by the Environmental Protection Agency through grant number EPA RD-83171701-0 to VHG, it has not been subjected to the EPA's required peer and policy review and therefore does not necessarily reflect the views of the EPA and no official endorsement should be inferred. This material is based on work partially supported by the National Science Foundation under Grant CHE0639096 and the National Institutes of Health under Grant R01OH009448. Any opinions, findings, and conclusions or recommendations expressed in this material are those of the authors and do not necessarily reflect the views of the National Science Foundation or the National Institutes of Health. Some of the research presented here was supported by the Department of Defense (DoD) through the National Defense Science & Engineering Graduate Fellowship (NDSEG) Program to Sherrie Elzey.

FURTHER READING

Vicki H. Grassian (Ed.). *Nanoscience and Nanotechnology: Environmental and Health Impacts*. Wiley, New York, 2008.

Challa S. S. R. Kumar (Ed.). *Nanomaterials. Toxicity, Health and Environmental Issues*. Nanotechnologies for the Life Sciences, Vol. 5. Wiley-VCH, New York.

Nancy A. Monteiro-Riviere, C. Lang Tran (Eds.). *Nanotoxicology: Characterization, Dosing and Health Effects*. Informa Health Care, New York, 2007.

Jo Anne Shatkin (Ed.). *Nanotechnology: Health and Environmental Risks*, Perspectives in Nanotechnology, Vol. 2. CRC Press, Boca Raton, FL, 2008.

Mark Wiesner, Jean-Yves Bottero (Eds.). *Environmental Nanotechnology: Applications and Impacts of Nanomaterials*. McGraw-Hill, New York, 2007.

Yuliang Zhao, Hari Singh Nalwa (Eds.). *Nanotoxicology: Interactions of Nanomaterials with Biological Systems*. American Scientific Publishers, 2006.

REFERENCES

1. A. D. Maynard, R. J. Aitken, T. Butz, V. Colvin, K. Donaldson, G. Oberdorster, M. A. Philbert, J. Ryan, A. Seaton, V. Stone, S. S. Tinkle, L. Tran, N. J. Walker, D. B. Warheit, *Nature* **2006**, *444*, 267.
2. P. J. A. Borm, *Inhalation Toxicology* **2002**, *14*, 311.

3. V. L. Colvin, *Nature Biotechnology* **2003**, *21*, 1166.
4. R. Dagani, *Chemical & Engineering News* **2003**, *81*, 30.
5. Y. Gogotsi, *Materials Research Innovations* **2003**, *7*, 192.
6. K. A. D. Guzman, M. R. Taylor, J. F. Banfield, *Environmental Science & Technology* **2006**, *40*, 1401.
7. R. Hardman, *Environmental Health Perspectives* **2006**, *114*, 165.
8. K. Kleiner, J. Hogan, *New Scientist* **2003**, *177*, 14.
9. T. M. Masciangioli, W. X. Zhang, *Abstracts of Papers of the American Chemical Society* **2003**, *225*, U952.
10. A. Nel, T. Xia, L. Madler, N. Li, *Science* **2006**, *311*, 622.
11. G. Oberdorster, E. Oberdorster, J. Oberdorster, *Environmental Health Perspectives* **2005**, *113*, 823.
12. G. M. Oberdorster, Andrew, K. Donaldson, V. Castranova, J. Fitzpatrick, K. Ausman, J. Carter, B. Karn, W. Kreyling, D. Lai, S. Olin, N. Monteiro-Riviere, D. Warheit, H. Yang, *Particle and Fibre Toxicology* **2005**, *2*, 8.
13. A. P. Tinke, R. Govoreanu, K. Vanhoutte, *American Pharmaceutical Review* **2006**, *9*(6), September/October, 1.
14. K. W. Powers, M. Palazuelos, B. J. Moudgil, S. M. Roberts, *Nanotoxicology* **2007**, *1*, 42.
15. J. G. Teeguarden, P. M. Hinderliter, G. Orr, B. D. Thrall, J. G. Pounds, *Toxicological Sciences* **2007**, *95*, 300.
16. O. R. Moss, V. A. Wong, *Inhalation Toxicology* **2006**, *18*, 711.
17. T. M. Sager, D. W. Porter, V. A. Robinson, W. G. Lindsley, D. E. Schwegler-Berry, V. Castranova, *Nanotoxicology* **2007**, *1*, 118.
18. D. J. Phares, K. P. Rhoads, M. V. Johnston, A. S. Wexler, *Journal of Geophysical Research: Atmospheres* **2003**, *108*. DOI: 10.1029/2001JD000660.
19. J. J. Bang, L. E. Murr, *Journal of the Minerals Metals & Materials Society* **2002**, *54*, 28.
20. R. Wilson, J. Spengler, Emissions, dispersion and concentration of particles. In *Particles in Our Air: Concentrations and Health Effects*, Wilson, R., Spengler, J. (Eds.), 41. Harvard University Press, Boston, MA, **1996**. ISBN 0-674-24077-4.
21. C. C. Daigle, D. C. Chalupa, F. R. Gibb, P. E. Morrow, G. Oberdorster, M. J. Utell, M. W. Frampton, *Inhalation Toxicology* **2003**, *15*, 539.
22. ASTM International Committee E56 on Nanotechnology. ASTM E2456-06 Standard Terminology for Nanotechnology is available at www.astm.org.
23. J. P. Simonin, *Journal of Physical Chemistry B* **2001**, *105*, 5262.
24. S. Deguchi, R. G. Alargova, K. Tsujii, *Langmuir* **2001**, *17*, 6013.
25. V. H. Grassian, *Nanoscience and Nanotechnology: Environmental and Health Impacts.* Wiley, New York, **2008**, p. 199.
26. T. E. Agustina, H. M. Ang, V. K. Vareek, *Journal of Photochemistry and Photobiology C: Photochemistry Reviews* **2005**, *6*, 264.
27. K. Sunada, Y. Kikuchi, K. Hashimoto, A. Fujishima, *Environmental Science & Technology* **1998**, *32*, 726.
28. A. M. Derfus, W. C. Chan, S. N. Bhatia, *Nano Letters* **2004**, *4*, 11.
29. US Environmental Protection Agency, *Nanotechnology White Paper*. U.S. EPA, Washington, D.C., **2007**.

30. R. P. F. Schins, R. Duffin, D. Hohr, A. M. Knaapen, T. M. Shi, C. Weishaupt, V. Stone, K. Donaldson, P. J. A. Borm, *Chemical Research in Toxicology* **2002**, *15*, 1166.
31. M. Osier, G. Oberdorster, *Fundamental and Applied Toxicology* **1997**, *40*, 220.
32. G. S. Z. Oberdorster, V. Atudorei, A. Elder, R. Gelein, W. Kreyling, C. Cox, *Inhalation Toxicology* **2004**, *16*, 437.
33. G. Rebitzer, T. Ekvall, R. Frischknecht, D. Hunkeler, G. Norris, T. Rydberg, W. P. Schmidt, S. Suh, B. P. Weidema, D. W. Pennington, *Environment International* **2004**, *30*, 701.
34. M. Goedkoop, R. Spriensma, The Eco-indicator 99: a damage oriented method for life cycle impact assessment. Available at http://www.pre.nl/eco-indicator 99/ei99-reports.htm.
35. J. Balbus, A. D. Maynard, V. L. Colvin, V. Castranova, G. P. Daston, R. A. Denison, K. L. Dreher, P. L. Goering, A. M. Goldberg, K. M. Kulinowski, N. A. Monteiro-Riviere, G. Oberdorster, G. S. Omenn, K. E. Pinkerton, K. S. Ramos, K. M. Rest, J. B. Sass, E. K. Silbergeld, B. A. Wong, *Environmental Health Perspectives* **2007**, *115*, 1654.
36. K. W. Powers, S. C. Brown, V. B. Krishna, S. C. Wasdo, B. M. Moudgil, S. M. Roberts, *Toxicological Sciences* **2006**, *90*, 296.
37. K. Thomas, P. Aguar, H. Kawasaki, J. Morris, J. Nakanishi, N. Savage, *Toxicological Sciences* **2006**, *92*, 23.
38. T. Thomas, K. Thomas, N. Sadrieh, N. Savage, P. Adair, R. Bronaugh, *Toxicological Sciences* **2006**, *91*, 14.
39. J. S. Tsuji, A. D. Maynard, P. C. Howard, J. T. James, C. W. Lam, D. B. Warheit, A. B. Santamaria, *Toxicological Sciences* **2006**, *89*, 42.
40. D. M. Balshaw, M. Philbert, W. A. Suk, *Toxicological Sciences* **2005**, *88*, 298.
41. M. P. Holsapple, W. H. Farland, T. D. Landry, N. A. Monteiro-Riviere, J. M. Carter, N. J. Walker, K. V. Thomas, *Toxicological Sciences* **2005**, *88*, 12.
42. K. Thomas, P. Sayre, *Toxicological Sciences* **2005**, *87*, 316.
43. L. F. Pease III, D. H. Tsai, R. A. Zangmeister, M. R. Zachariah, M. J. Tarlov, *Journal of Physical Chemistry C* **2007**, *46*, 17155.
44. P. K. Hudson, E. R. Gibson, M. A. Young, P. D. Kleiber, V. H. Grassian, *Aerosol Science and Technology* **2007**, *41*, 701.
45. S. J. Hug, B. Sulzberger, *Langmuir* **1994**, *10*, 3587.
46. T. H. Yoon, S. B. Johnson, C. B. Musgrave, G. E. Brown, *Geochimica et Cosmochimica Acta* **2004**, *68*, 4505.
47. J. M. Decker, *Immunology*, University of Arizona Press, Tucson, AZ, **2007**. Available at http://microvet.arizona.edu/Courses/MIC419/Tutorials/cytokines.html.
48. Z. Chen, H. A. Meng, G. M. Xing, C. Y. Chen, Y. L. Zhao, G. A. Jia, T. C. Wang, H. Yuan, C. Ye, F. Zhao, Z. F. Chai, C. F. Zhu, X. H. Fang, B. C. Ma, L. J. Wan, *Toxicology Letters* **2006**, *163*, 109.
49. L. E. Yu, L. L. Yung, C. Ong, Y. Tan, K. S. Balasubramaniam, D. Hartono, G. Shui, M. R. Wenk, W. Ong, *Nanotoxicology* **2007**, *1*, 235.
50. C. M. Sayes, R. Wahi, P. A. Kurian, Y. P. Liu, J. L. West, K. D. Ausman, D. B. Warheit, V. L. Colvin, *Toxicological Sciences* **2006**, *92*, 174.
51. J. Muller, F. Huaux, D. Lison, *Carbon* **2006**, *44*, 1048.
52. S. Iijima, *Nature* **1991**, *354*, 56.

53. G. Pastorin, K. Kostarelos, M. Prato, A. Bianco, *Journal of Biomedical Nanotechnology* **2005**, *1*, 133.
54. A. Bianco, K. Kostarelos, M. Prato, *Current Opinions in Chemical Biology* **2005**, *9*, 674.
55. X. Peng, Y. Li, Z. Luan, Z. Di, H. Wang, B. Tian, Z. Jia, *Chemical Physics Letters* **2003**, *376*, 154.
56. Y. Li, S. Wang, Z. Luan, J. Ding, C. Xu, D. Wu, *Carbon* **2003**, *41*, 1057.
57. Y. Peng, H. Liu, *Industrial & Engineering Chemistry Research* **2006**, *45*, 6483.
58. I. Rosca, F. Watari, M. Uo, T. Akasaka, *Carbon* **2005**, *43*, 3124.
59. D. Yang, J. Rochette, E. Sacher, *Journal of Physical Chemistry B* **2005**, *109*, 7788.
60. A. Franchi, C. R. O'Melia, *Environmental Science & Technology* **2003**, *37*, 1122.
61. C. M. Sayes, F. Liang, J. L. Hudson, J. Menedez, W. Guo, J. M. Beach, V. C. Moore, C. D. Doyle, J. L. West, W. E. Billups, K. D. Ausman, V. L. Colvin, *Toxicology Letters* **2006**, *161*, 135.
62. L. A. Langley, D. Villanueva, D. H. Fairbrother, *Chemical Materials* **2006**, *18*, 169.
63. H. Cho, B. A. Smith, J. Wnuck, H. Fairbrother, W. P. Ball, *Environmental Science & Technology* **2008**, *42*(8), 2899.
64. X. C. Zhao, A. Striolo, P. T. Cummings, *Biophysical Journal* **2005**, *89*, 3856.
65. Y. Yamakoshi, S. Sueyoshi, K. Fukuhara, N. Miyata, *Journal of the American Chemical Society* **1998**, *12363*.
66. Donald Huffman, *Physics Today* **199l**, *44*, 26.
67. L. Y. Chiang, J. B. Bhonsle, L. Wang, S. F. Shu, T. M. Chang, J. R. Hwu, *Tetrahedron* **1996**, *52*, 4963.
68. C. N. Murthy, S. J. Choi, K. E. Geckeler, *Journal of Nanoscience and Nanotechnology* **2002**, *2*, 29.
69. C. Ungurenasu, A. Airinei, *Journal of Medicinal Chemistry* **2000**, *43*, 3186.
70. X. K. Cheng, A. T. Kan, M. B. Tomson, *Journal of Chemical and Engineering Data* **2004**, *49*, 675.
71. W. A. Scrivens, J. M. Tour, K. E. Creek, *Journal of the American Chemical Society* **1994**, *116*, 4517.
72. C. M. Sayes, J. D. Fortner, W. Guo, D. Lyon, A. M. Boyd, K. D. Ausman, Y. J. Tao, B. Sitharaman, L. J. Wilson, J. B. Hughes, J. L. West, V. L. Colvin, *Nanoletters* **2004**, *4*, 1881.
73. R. V. Bensasson, E. Bienvenue, M. Dellinger, S. Leach, P. Seta, *Journal of Physical Chemistry* **1994**, *98*, 3492.
74. H. F. Lecoanet, J. Y. Bottero, M. R. Wiesner, *Environmental Science & Technology* **2004**, *38*, 5164.
75. D. Heymann, *Fullerene Science and Technology* **1996**, *4*, 509.
76. H. F. Lecoanet, J. Y. Bottero, M.R. Wiesner, *Environmental Science & Technology* **2004**, *38*, 5167.
77. C. R. O'Melia, *Environmental Science & Technology* **1980**, *14*, 1052.
78. M. Elimelech, C. R. O'Melia, *Environmental Science & Technology* **1990**, *24*, 1528.
79. S. A. Bradford, J. Simunek, M. Bettahar, M. van Genuchten, S. R. Yates, *Environmental Science & Technology* **2003**, *37*, 2242.
80. S. A. Bradford, J. Simunek, S. L. Walker, *Water Resources Research* **2006**, *42*, W12S12.

81. J. Brant, H. Leocoanet, M. Hotze, M. Wiesner, *Environmental Science & Technology* **2005**, *39*, 6351.
82. B. Espinasse, E. M. Hotze, R. Weisner, *Environmental Science & Technology* **2007**, *41*, 7369.
83. J. Wan, J. L. Wilson, *Water Resources Research* **1994**, *30*, 857.
84. J. J. Lenhart, J. E. Saiers, *Environmental Science & Technology* **2002**, *36*, 769.
85. G. Gargiulo, S. A. Bradford, J. Simunek, P. Ustonal, H. Vereecken, E. Klumpp, *Environmental Science & Technology* **2007**, *41*, 1265.
86. W. C. W. Chan, D. J. Maxwell, X. H. Gao, R. E. Bailey, M. Y. Han, S. M. Nie, *Current Opinions in Biotechnology* **2002**, *3*, 40.
87. L. H. Qu, X. G. Peng, *Journal of the American Chemical Society* **2002**, *124*, 2049.
88. S. J. Clarke, C. A. Hollmann, Z. J. Zhang, D. Suffern, S. E. Bradforth, N. M. Dimitrijevic, W. G. Minarik, J. L. Nadeau, *Nature Materials* **2006**, *5*, 409.
89. P. Borm, F. C. Klaessig, T. D. Landry, B. Moudgil, J. Pauluhn, K. Thomas, R. Trottier, S. Wood, *Toxicological Sciences* **2006**, *90*, 23.
90. J. W. Costerton, K. J. Cheng, G. G. Geesey, T. I. Ladd, J. C. Nickel, M. Dasgupta, T. J. Marrie, *Review of Microbiology* **1987**, *41*, 435.
91. C. B. Whitchurch, T. Tolker-Nielsen, P. C. Ragas, J. S. Mattick, *Science* **2002**, *295*, 1487.
92. J. H. Priester, S. G. Olson, S. M. Webb, M. P. Neu, L. E. Hersman, P. A. Holden, *Applied and Environmental Microbiology* **2006**, *72*, 1988.
93. G. E. Brown, V. E. Henrich, W. H. Casey, D. L. Clark, C. Eggleston, A. Felmy, D. W. Goodman, M. Gratzel, G. Maciel, M. I. McCarthy, K. H. Nealson, D. A. Sverjensky, M. F. Toney, J. M. Zachara, *Chemical Reviews* **1999**, *99*, 77.
94. D. H. Nies, *Plasmid* **1992**, *27*, 17.
95. W. Bae, R. Abdullah, R. K. Mehra, *Chemosphere* **1998**, *37*, 363.
96. W. Bae, R. Abdullah, D. Henderson, R. K. Mehra, *Biochemical and Biophysical Research Communications* **1997**, *237*, 16.
97. T. T. Zhang, J. L. Stilwell, D. Gerion, L. H. Ding, O. Elboudwarej, P. A. Cooke, J. W. Gray, A. P. Alivisatos, F. F. Chen, *Nano Letters* **2006**, *6*, 800.
98. J. Lovric, H. S. Bazzi, Y. Cuie, G. R. A. Fortin, F. M. Winnik, D. Maysinger, *Journal of Molecular Medicine* **2005**, *83*, 377.
99. J. S. Chang, K. L. B. Chang, D. F. Hwang, Z. L. Kong, *Environmental Science & Technology* **2007**, *41*, 2064.
100. A. Thill, O. Zeyons, O. Spalla, F. Chauvat, J. Rose, M. Auffan, A. M. Flank, *Environmental Science & Technology* **2006**, *40*, 6151.
101. P. H. Kao, C. C. Huang, Z. Y. Hseu, *Chemosphere* **2006**, *64*, 63.
102. R. M. Miller, *Environmental Health Perspectives* **1995**, *103*, 59.
103. A. Vecchio, C. Finoli, D. Di Simine, V. Andreoni, *Fresenius Journal of Analytical Chemistry* **1998**, *361*, 338.
104. K. S. Saladin, *Human Anatomy*, 1st Ed., McGraw-Hill, **2004**. ISBN: 0070390800.
105. W. G. Kreyling, M. Semmler, F. Erbe, P. Mayer, S. Takenaka, H. Schulz, G. Oberdörster, A. Ziesenis, *Journal of Toxicology and Environmental Health A* **2002**, *65*, 1513.
106. A. Shimada, N. Kawamura, M. Okajima, T. Kaewamatawong, H. Inoue, T. Morita, *Toxicologic Pathology* **2006**, *34*, 949.
107. G. Oberdörster, Z. Sharp, V. Atudorei, A. Elder, R. Gelein, A. Lunts, W. Kreyling, C. Cox, *Journal of Toxicology and Environmental Health A* **2002**, *65*, 1531.

108. A. Elder, R. Gelein, V. Silva, T. Feikert, L. Opanashuk, J. Carter, R. Potter, A. Maynard, Y. Ito, J. Finkelstein, G. Oberdörster, *Environmental Health Perspectives* **2006**, *114*, 1172.
109. K. Francis, J. van Beek, C. Canova, J. W. Neal, P. Gasque, *Expert Reviews in Molecular Medicine* **2003**, *5*(15), 1.
110. E. P. Widmaier, H. Raff, K. T. Strang, *Vander, Sherman & Luciano's Human Physiology: The Mechanisms of Body Functions*, 9th edition. McGraw-Hill, New York, **2004**.
111. M. J. Hatcher, P. D. Jones, W. G. Miller, D. K. Pennell, Neurotoxicity of manufactured nanoparticles. In *Nanoscience and Nanotechnology: Environmental and Health Impacts* **2008**, Chap. 16.
112. C. E. Hamill, A. Goldshmidt, O. Nicole, R. J. McKeon, D. J. Brat, S. F. Traynelis, Special lecture. Glial reactivity after damage: Implications for scar formation and neuronal recovery. *Clinical Neurosurgy* **2005**, *52*, 29.
113. B. Liu, H. M. Gao, J. S. Hong, *Environmental Health Perspectives* **2003**, *111*, 1065.
114. A. Campbell, M. Oldham, A. Becaria, S. C. Bondy, D. Meacher, C. Sioutas, C. Misra, L. B. Mendez, M. Kleinman, *Neurotoxicology* **2005**, *26*, 133.
115. W. Pan, J. E. Zadina, R. E. Harlan, J. T. Weber, W. A. Banks, A. J. Kastin, *Neuroscience & Biobehavioral Reviews* **1997**, *21*, 603.
116. S. W. Perry, S. Dewhurst, M. J. Bellizzi, H. A. Gelbard, *Journal of Neurovirology* **2002**, *8*, 611.
117. Y. S. Kim, T. H. Joh, *Experimental and Molecular Medicine* **2006**, *38*, 333.
118. L. Minghetti, M. A. Ajmone-Cat, M. A. De Berardinis, R. De Simone, *Brain Research Reviews* **2005**, *48*, 251.
119. P. B. Rosenberg, *International Review of Psychiatry* **2005**, *17*, 503.
120. M. N. Woodroofe, *Neurology* **1995**, *45*, S6.
121. R. M. Gibson, N. J. Rothwell, R. A. Le Feuvre, *Veterinary Journal* **2004**, *168*, 230.
122. C. A. Dinarello, *FASEB Journal* **1994**, *8*, 1314.
123. C. A. Dinarello, *Clinical and Experimental Rheumatology* **2002**, *20*, S1.
124. A. C. Mitropoulos, *Journal of Colloid and Interface Science* **2008**, *317*, 643.
125. Y. Sakatani, D. Grosso, L. Nicole, C. Boissière, G. Soler-Illia, C. Sanchez, *Journal of Materials Chemistry* **2006**, *16*, 77.
126. Z. Chen, H. Meng, G. Xing, C. Chen, Y. Zhao, G. Jia, T. Wang, H. Yuan, C. Ye, F. Zhao, Z. Chai, C. Zhu, X. Fang, B. Ma, L. Wan, *Toxicology Letters* **2006**, *163*, 109.
127. R. C. Murdock, L. Braydich-Stolle, A. M. Schrand, J. J. Schlager, S. M. Hussain, *Toxicological Sciences* **2008**, *101*, 239.

PROBLEMS

1. A variety of techniques are necessary to characterize nanomaterials. Often these techniques are referred to using acronyms. For each acronym listed below, provide the full name and a brief description of the technique: APS, ATR-FTIR, BET, DLS, SEM, SMPS, TEM, XPS, and XRD.

2. Due to their size-dependent optical properties, quantum dots (QDs) have found many applications in biological labeling applications. In these applications, often the core nanometer-sized semiconductor is capped by a molecular coating. This coating not only serves to protect the core, but can also be used to impart specific solubility and binding properties to the QDs. Suppose two CdSe quantum dots that have been synthesized with a trioctylphosphine oxide (TOPO) capping layer are characterized by TEM and DLS. For sample 1, the apparent diameter in TEM is 2.2 nm and in DLS the diameter is 4.7 nm, and for sample 2, the apparent diameter in TEM is 3.7 nm and in DLS is 6.1 nm. Give an explanation for the apparent disparity in the sizes provided by the two methods.

3. Polychlorinated biphenyls (PCBs) were used in this chapter as an example of a technological development that resulted in unforeseen adverse consequences to the health of the environment. Briefly summarize: why PCBs were developed, what types of environmental problems were discovered, and what was done to mitigate the problems once they were identified. (You may need to use Internet or other sources to gather this information.) Suggest at least one characteristic of nanomaterials that should be defined in order to prevent analogous adverse consequences.

4. There is a growing realization that particle size can have an impact on the biological response to inhaled particulate matter. For regulatory and other reasons, the size fractions of particulate matter are often separated into three categories: PM_{10}, $PM_{2.5}$, and UFP. The inhalable coarse fraction, PM_{10}, includes particulate matter with diameters in the range of 2.5 to 10 μm. The fine fraction, PM 2.5, includes particulate matter within the range of 0.1 to 2.5 μm, and the ultrafine fraction, UFP, includes particles of less than 0.1 μm. Particulate matter is among the six air quality criteria standards set by the EPA, and in 2006 the EPA tightened the 24 hour $PM_{2.5}$ standard from 65 $\mu g/m^3$ to 35 $\mu g/m^3$ (and retained the annual standard of 15 $\mu g/m^3$) based on the growing evidence of health problems associated with particles in this size range. At the same time, the EPA decided to retain the existing 24 hour PM_{10} standard of 150 $\mu g/m^3$, but dropped the annual PM_{10} standard due to a lack of evidence linking long-term exposure to coarse particle pollution to health problems. Although the regulations are given in mass concentrations, the health consequences may be more closely linked to the particle number concentrations or particle surface area within a given range. For a mass concentration of 35 $\mu g/m^3$ calculate the corresponding particle number concentrations (particles/m^3) for three samples composed of uniform spheres of 0.1 μm, 2.5 μm, and 10 μm diameter, respectively, each composed of a material with a density of 1.0 g/cm^3.

5. As the size of a particle decreases, the contribution of the surface to the total free energy becomes increasingly significant. This trend is evident in the Kelvin equation, a classic thermodynamic equation relating the vapor pressure of a droplet to its physical size. This equation is often applied in the determination of the pore size distribution through adsorption porosimetry. For this application, the equation can be written:

$$\ln \frac{P}{P_o} = -\frac{2\gamma M}{\rho r R T}$$

where P is the vapor pressure of the droplet, P_o is the vapor pressure of the bulk liquid, γ is the surface tension, M is the molar mass of the liquid, r is the radius of the droplet, ρ is the density, R is the gas constant, and T is the absolute temperature. This result has implications for the filling of small pores via capillary condensations. In a study using x-ray scattering to measure the pore size of Vycor porous glass, Mitropoulos (124) concluded that a simple model based on the Kelvin equation was useful for liquids with a radius of curvature as small as 4 nm. To investigate the impact of size on vapor pressure, calculate the relative pressure, P/P_o, of dibromomethane (CH_2Br_2) droplets of 4 μm, 40 nm, and 4 nm. In each case assume a temperature of 25.0°C, a density of 2.477 g/cm^3, and a surface tension of 35.33 dyne/cm = 35.33×10^{-3} N/m.

6. Thin films of TiO_2 are of interest as photocatalyzed depolluting layers. In such applications particle size and porosity are important material characteristics to be measured. In one such study (125) the crystallite sizes of TiO_2 samples with different thermal treatments were measured. A sample treated at 600°C yielded a $2\theta = 1.2°$, a full width at half maximum of the (101) diffraction peak at $2\theta = 25°$, and a sample treated at 700°C yielded a $2\theta = 2.4°$ for the same (101) peak. For spherical particles ($\kappa = 1$ and $\lambda = 0.154$ nm) use Scherrer's equation to calculate the particle size at each treatment temperature. In the same study, the pore diameter of the 600°C and 700°C samples were found to be 5.5 nm and 8.5 nm, respectively. Are these pore sizes reasonable given the particle size results?

7. DLVO theory has often been applied to gain insight into the stability of colloidal systems. For colloidal montmorillonite with a particle radius of $R = 200 \times 10^{-9}$ m in an aqueous NaCl solution, the electrical double layer repulsion can be approximated as

$$\Phi_{\mathrm{EDL}}(\text{in J}) = -2.55 \times 10^{-18} \ln[1 - \exp(-3.28 \times 10^9 \sqrt{I}\, d)]$$

where I is the ionic strength (in M) and d is the separation distance between particle surfaces in meters. For $R \gg d$, the van der Waals attraction can be expressed as

$$\Phi_{\mathrm{vdW}}(\text{in J}) = -\frac{2.3 \times 10^{-20} R}{12\, d}$$

For an ionic strength of 0.25 M, create a plot of the net interaction energy as a function of the separation distance over the range 0.01 nm to 10 nm.

A general rule is that a barrier greater than 10 k_bT is sufficient to prevent colloidal aggregation. Use your plot to estimate the barrier energy that separates the primary minimum from the secondary minimum. Is the barrier greater than 10 k_bT at $T = 295$ K with $k_b = 1.38 \times 10^{-23}$ J/K?

8. Direct ingestion is one potential entry pathway for nanoparticles into the body. In one study designed to assess the role of size in the toxicity of copper, Chen et al. (126) orally administered copper to mice, as microparticles (17 μm), as nanoparticles (24 nm), and as cupric ions. One conclusion of this study was that the nano and ionic forms of copper were significantly more toxic than the micro form, with median lethal doses (LD_{50}) determined to be >5000, 414, and 110 mg/kg body mass for micro, nano, and ionic forms, respectively.

(**a**) In this study, surface area was claimed to be a key physical characteristic of the nanocopper that was responsible for the elevated toxicity of the nanomaterials. Assuming spherical particles and equal metal densities (8.92 g/cm^3), what is the ratio of the surface area of 414 mg of nanoparticles (diameter = 24 nm) to 5000 mg of microparticles (diameter = 17 μm)?

(**b**) You will need to read the original paper before answering the following questions: In the proposed mechanism of toxicity the nanocopper was thought to be dissolved by gastric juice in the stomach. Why should this process depend on the particle size? How does this process help explain the enhanced toxicity of the nanoparticles? What other evidence do the authors present to support the proposed mechanism?

9. For nanometer-sized particles, both size and shape may have an impact on toxicity. For many materials TEM can be used to determine the size, shape, and distribution of sizes of primary nanoparticles. However, the high vacuum environment of the TEM is drastically different than most biologically relevant environments. For this reason, studies targeting potential size-specific nanotoxicity often need to characterize materials *in vitro* in media that approximate those *in vivo*. In one such study, Murdock and coworkers use a combination of TEM and DLS to assess the relevance of TEM measured size when nanoparticles are dispersed in aqueous fluids (127). Some data from this work are summarized in the table below.

Various Nanoparticles Dispersed in Deionized Water (25 μg/mL)

Particle[a]	TEM Diameter (nm)	DLS Diameter (nm)	Zeta Potential (mV)
Al_2O_3	30	210	43.0
Al_2O_3	40	237	36.2
HC-Ag	25	208	−20.6
Ag	80	1810	−20.4
PS-Ag	25	128	−25.9
PS-Ag	80	250	−20.3
Cu	40	1120	−4.75
SiO_2	35	29	−23.1
SiO_2	51	53	−30.1
TiO_2 (anatase)	39	486	−4.95
TiO_2 (39% anatase, 61% rutile)	39	796	17.7
TiO_2 (amorphous)	40	1300	−12.1

[a]HC, hydrocarbon; PS, polysaccharides.

(a) From this data, what can you conclude generally about the use of TEM as the sole measure of particle size?

(b) For these samples, does zeta potential appear to correlate with the tendency of the primary particles to aggregate?

(c) From these data, does coating the silver particles with hydrocarbons (HC) and polysaccharides (PS) appear to significantly affect the tendency to aggregate compared to bare Ag particles?

10. In soil samples, pores are sometimes classified by how easily water can be removed. Transmission pores (>50 μm) are readily drainable; storage pores (0.2 to 50 μm) maintain water available for plant root systems, and residual pores (<0.2 μm) have the water tightly bound and therefore unavailable. Soils with high sand components would typify the first classification and soils with high clay components would typify the latter, with other soil types being intermediate in value and distributions of pore sizes. Assume that soil samples are hydrophilic with a contact angle $\theta = 10^{\circ}$, and determine the capillary pressure for water ($\gamma = 72.0 \times 10^{-3}$ N/m) capillaries with radii of 50 μm, 5 μm, and 50 nm.

ANSWERS

1. APS (aerodynamic particle sizer)—measures the aerodynamic particle diameter, D_a, by monitoring the settling velocity of namomaterials. Results are referenced to a spherical particle with a density of 1 g/cm^3.

ATR-FTIR spectroscopy (attenuated total reflectance Fourier transform infrared spectroscopy)—IR spectroscopy uses the absorption of infrared radiation to probe the vibrational frequency of molecular motions. Attenuated total reflectance method uses a crystal of high refractive index to channel the infrared light (using total internal reflectance) into the crystal and causes only a thin layer of a sample in contact with the exterior of the crystal to be sensitively detected.

BET (Braunner–Emmet–Teller)—measures the surface area of powdered solids using the physical absorption of a gas, usually molecular nitrogen. The analysis uses the Braunner, Emmet, and Teller model of isothermal adsorption.

DLS (dynamic light scattering)—in dynamic light scattering laser light is scattered by the nanoparticles. Due to the Brownian motion of the particles, a time-dependent fluctuation is imparted to the scattered light intensity. Analysis of the signal intensity yields information about the diffusional motion of the particles, which is in turn related to the hydrodynamic size via the Stoke–Einstein equation.

SEM (scanning electron microscopy)—an electron beam is focused onto a small region of a sample contained within a vacuum. The beam is scanned across the sample and the backscattered electrons (those undergoing elastic collisions) and the secondary electrons (those electrons released through inelastic collisions) are collected on a positively charged detector to obtain an image on a scintillation screen or electronic detector.

SMPS (scanning mobility particle sizer)—combines a DMA (differential mobility analyzer) and a CPC (condensation particle counter) to measure the mobility diameter, D_m, of aerosol nanoparticles. In the DMS, the particles are first caused to pick up a charge. The charged particles of the aerosol are then carried by a gas into an electric field where the mobility is measured; mobility depends on particle size. Because DMS is nondestructive to the sample, the particles, which have been sorted by size, emerge within the gas stream of the DMS and can then be counted by the CPC.

TEM (transmission electron microscopy)—an electron beam is passed through a small region of a sample contained within a vacuum. The transmitted electrons are magnified and focused to obtain an image on a scintillation screen or electronic detector.

XPS (x-ray photoelectron spectrometer)—the method uses x-ray irradiation to cause core electrons to be ejected from elements near the surface of a sample. Analysis of the kinetic energies of these ejected electrons, also known as photoelectrons, yields information about the elemental composition and chemical state of the surface.

XRD (x-ray diffraction)—a reciprocal-space method of structural analysis that measures the structure of a lattice based on the diffraction pattern generated when x-ray radiation passes through a regular crystal lattice.

2. In TEM only the inorganic semiconductor provided enough contrast to be resolved, whereas in the case of DLS the full (hydrodynamic) diameter was measured. The difference between the two measurements can therefore be taken as an estimate of the TOPO layer depth $(4.7 - 2.2)/2 = 1.25$ nm and $(6.1 - 3.7)/2 = 1.2$ nm. The TOPO layer appears to contribute about 1.2 nm.

3. PCBs were developed as dielectric fluids (first manufactured by Monsanto in 1929). The fluids display low reactivity, high electrical resistance, and are stable when exposed to heat and pressure. Based on these highly desirable chemical and physical characteristics many applications such as transformer fluids, fire retardants, hydraulic fluids, plasticizers, adhesives, and many others were developed.

In the life cycle of the products that were based on PCBs, varying amounts of PCBs were released into the environment in the manufacture, use, and disposal period of the product life cycle. In the environment, as in their use, PCBs proved to be very stable, and this became the problem. PCBs did not break down in the environment, and therefore spread and entered the food chain. In the food chain, PCBs were found to build up in the fatty tissue of animals that consumed PCB-contaminated food. This was particularly true of animals at the top of the food chain due to bioaccumulation in the tissue at each level of the food chain and biomagnification due to the build up at each step of the food chain. PCB in sufficient concentration are toxic to wildlife. As a result of the adverse environmental impact, PCB production has been halted and several large-scale contamination sites have been remediated. PCB contamination remains a challenge, such as for the fish in the Great Lakes area.

One issue that has recently been widely publicized is that detectable levels of pharmaceuticals have often been found in drinking water supplies across the United States. If nanopharmaceuticals become widespread, issues like bioaccumulation and biomagnification will need to be assessed for these nanomaterials. For example, if nanopharmaceuticals are designed to target specific organs, it may be that this specificity provides a mechanism for bioaccumulation. In looking at this issue one key question to address therefore is "what is the lifetime of a nanopharmaceutical in the environment?"

4. The total volume of material is give as the mass times the density to give 35×10^{-6} cm^3 or 3.5×10^{-11} m^3.

Each 0.1 μm sphere corresponds to a volume of

$$V = \frac{4\pi}{3}\left(\frac{1 \times 10^{-7}\,\text{m}}{2}\right)^3 = 5.2 \times 10^{-22}\,\text{m}^3$$

therefore the number of particles is given by

$$\frac{3.5 \times 10^{-11}\,\text{m}^3}{5.2 \times 10^{-22}\,\text{m}^3} = 6.7 \times 10^{10}\ \text{particles per m}^3\ \text{of air.}$$

Similarly, there are 4.2×10^6 particles/m^3 in the case of the 2.5 μm particles and 6.7×10^4 particles/m^3 in the case of the 10 μm particles.

5. Variable given as: $g = 3.533 \times 10^{-2}$ N/m, M = 0.17383 kg/mol, $\rho = 2477$ kg/m^3, and $T = 298.15$.

$$\begin{aligned}\frac{P_{4\,\mu\text{m}}}{P_o} &= \exp\left(\frac{-2\gamma M}{\rho r RT}\right)\\ &= \exp\left(\frac{-2(3.533 \times 10^{-2}\,\text{N/m})\ (0.17383\,\text{kg/mol})}{(2477\,\text{kg/m}^3)(4 \times 10^{-6}\,\text{m})(8.3145\ \text{J/mol K})(298.15\,\text{K})}\right)\\ &= 0.9995\end{aligned}$$

$$\begin{aligned}\frac{P_{40\,\text{nm}}}{P_o} &= \exp\left(\frac{-2\gamma M}{\rho r RT}\right)\\ &= \exp\left(\frac{-2(3.533 \times 10^{-2}\,\text{N/m})\ (0.17383\,\text{kg/mol})}{(2477\,\text{kg/m}^3)(40 \times 10^{-9}\,\text{m})(8.3145\,\text{J/mol K})(298.15\,\text{K})}\right)\\ &= 0.9512\end{aligned}$$

$$\begin{aligned}\frac{P_{4\,\text{nm}}}{P_o} &= \exp\left(\frac{-2\gamma M}{\rho r RT}\right)\\ &= \exp\left(\frac{-2(3.533 \times 10^{-2}\,\text{N/m})\ (0.17383\,\text{kg/mol})}{(2477\,\text{kg/m}^3)(4 \times 10^{-9}\,\text{m})(8.3145\,\text{J/mol K})(298.15\,\text{K})}\right)\\ &= 0.6065\end{aligned}$$

6. Given: $\kappa = 1$, $\lambda = 0.154$ nm, and $\cos(25°/2) = 0.976$, the $600\,°$C sample result with $2\theta = 2.4\,°$ and β (in rad) =

$$\left(\frac{0.6° \times 2\pi}{360°}\right) = 0.0052 \text{ rad}$$

obtains $t = \dfrac{0.154 \text{ nm}}{0.0052 \times 0.976} = 7.5$ nm, and at $700°$C, $t = 15$ nm.

The values or pore sizes are reasonable in that the larger particles yield larger pores, and that the near match in size is consistent with the pores being formed due to the interstitial regions between reasonably well-packed particles.

7.

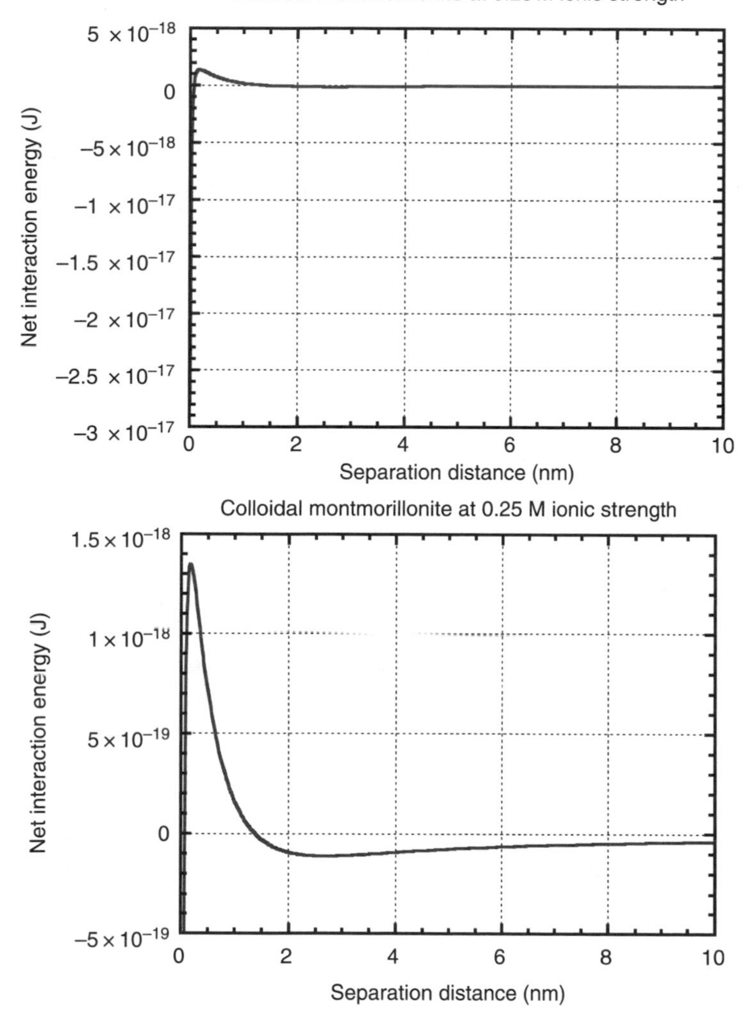

The barrier energy appears to be approximately 1.35×10^{-18} J and $>300\,kT$, therefore the colloid would not be expected to aggregate.

8. **(a)** For nanoparticles, the surface area of each particle can be found as $4\pi r^2 = 4\pi(12 \times 10^{-9}\ \text{m})^2 = 1.8 \times 10^{-15}\ \text{m}^2/\text{particle}$.
The total volume is

$$\frac{414 \times 10^{-3}\ \text{g}}{8.92 \times 10^{6}\ \text{g/m}^3} = 4.6 \times 10^{-8}\ \text{m}^3$$

and the volume per particle is $4/3\ \pi r^3 = 7.24 \times 10^{-24}\ \text{m}^3$ which gives the total number of particles as

$$\frac{4.6 \times 10^{-8}\ \text{m}^3}{7.24 \times 10^{-24}\ \text{m}^3} = 6.4 \times 10^{15}\ \text{particles.}$$

Therefore, the total surface area is $(1.8 \times 10^{-15}\ \text{m}^2\ \text{particle})(6.4 \times 10^{15}\ \text{particles}) = 12\ \text{m}^2$. An analogous calculation for 5000 mg of microparticles obtains a total surface area of $2.0 \times 10^{-1}\ \text{m}^2$. Therefore, the ratio of the surface areas is about 60.

(b) The reaction of copper metal with hydrogen ions causes the dissolution of the metal to form copper ions. Since dissolution takes place at the liquid-solid interface, a greater surface area is expected to enhance the rate of reaction and dissolution. Once the copper is transformed to cupric ions, toxicity similar to the ionic copper should be expected and is found (LD_{50} of 413 vs. 110 mg/kg). Further support for a mechanism involving the reaction of copper with the gastric juices include: (1) the observation of metabolic alkalosis due to the production of bicarbonate as a by-product of the copper reaction with the stomach acid; (2) the observation of a sex dependence of the toxicity, with nanocopper exerting a stronger toxicity on male mice. Since it is known that male mice have a higher capacity to produce stomach acid, this observation supports the mechanistic claim.

9. **(a)** TEM is insufficient as a sole measure since the aggregation of particles may be very different in vacuum compared to in solution.

(b) No, the aggregation is not simply related to zeta potential.

(c) Although the data is quite limited, it does appear that both coatings reduce the tendency of the primary particles to aggregate. For bare silver, the ratio of the DLS diameter is 23 times the TEM diameter, in comparison to the coated systems that show ratios of 8, 5, and 3, for the HC-Ag, PS-Ag(25), and PS-Ag(80), respectively.

10. Applying the Young–Laplace equation gives:
For a 50 μm capillary:

$$P_c = \frac{2\gamma \cos\theta}{r} = \frac{2(7.2 \times 10^{-3}\,\text{N/m})\,0.985}{50 \times 10^{-6}\,\text{m}} = 283\,\text{Pa (or 0.0028 atm).}$$

For a 5 μm capillary:

$$P_c = \frac{2\gamma \cos\theta}{r} = \frac{2(7.2 \times 10^{-3}\,\text{N/m})\,0.985}{5 \times 10^{-6}\,\text{m}} = 2830\,\text{Pa (or 0.028 atm).}$$

For a 50 nm capillary:

$$P_c = \frac{2\gamma \cos\theta}{r} = \frac{2(7.2 \times 10^{-3}\,\text{N/m})0.985}{50 \times 10^{-9}\,\text{m}} = 283000\,\text{Pa (or 2.8 atm).}$$

22

TOXICITY OF INHALED NANOMATERIALS

John A. Pickrell, L. E. Erickson, K. Dhakal, and
Kenneth J. Klabunde

22.1 Introduction, 730
22.2 Particle Size, 731
 22.2.1 Physical Characterization, 732
22.3 Characterization of Biological (Health) Effects, 741
 22.3.1 Urban Dust, 741
 22.3.2 Manure Dust, 744
 22.3.3 Crustal, Desert, and Mountain Dust, 744
 22.3.4 Metal Oxide Fumes or Dust, 745
 22.3.5 Summary, 750
22.4 The Effect of Particle Structure, 751
 22.4.1 Physical Characterization, 751
 22.4.2 Characterization of Biological and Health Effects, 751
22.5 The Effect of Particle Shape, 753
 22.5.1 Physical Characterization, 753
 22.5.2 Characterization of Biological and Health Effects, 753
22.6 The Potential for Adverse Effects on The Health of Individuals, Animals, The Environment, and Major Ecosystems, 754
 22.6.1 Case 1: Human Studies, 755
 22.6.2 Case 2: Animal Studies, 755
Acknowledgments, 758
Glossary, 758
References, 761
Problems, 766
Answers, 766

Nanoscale Materials in Chemistry, Second Edition. Edited by K. J. Klabunde and R. M. Richards

22.1 INTRODUCTION

Inhalation of noxious gases or particles in sufficient quantities can cause different abnormal pulmonary responses at different doses. From high dose to low dose particles cause acute edema, fibrosis (healing by secondary intent), emphysema (tissue loss and airway enlargement), metaplasia (barely controlled growth), and neoplasia (tumors or uncontrolled growth). Alternatively, such inhalation can act in concert to worsen a variety of lung insults, including infectious diseases. This chapter targets noninfectious lung diseases resulting from inhalation of particles in the fine particle size range—particulate mass median aerodynamic diameters (MMAD) <2.5 microns (2500 nanometers, nm).

Particle dimensions—mass, size, relative surface area, surface activity, total number, shape, and composition—can have profound effects on pulmonary responses after their inhalation in animals or man. Aerosol scientists are most interested in describing physical characteristics and motion, engineers in measuring how particles were delivered, for example, particle number, surface area, and mass. Allied health scientists study health effects of animals or persons exposed to the particles. Teams composed of aerosol scientists, engineers, and health scientists often combine to examine the broad implications of a type of particle, for example, nanoparticles, how man or animals are exposed, and the responses that develop from this sustained exposure. Since direct measurements of aerosol mass, size, surface area, and their distributions are difficult, each group develops measurements they consider most important. Aerosol scientists develop ways to directly measure particle numbers. Health scientists measure response to exposures of different particles both *in vitro* (lung epithelial cells and alveolar macrophages) and *in vivo* (rat or mouse exposures). Engineers develop ways to measure aerosol mass, particle surface area and aggregate surface area and total particle measurements (total number and size distributions) that include a combination of these that are often valuable in estimating risks to persons, animals, or the environment and correlating to health effects following particle inhalation.

Interpretation allows each group to add value to the actual measurements by predicting the degree to which each aerosol measurement or group of aerosol measurements reflects positively or negatively on the environment and the tendency of those individuals in this environment to develop health effects. For example, it is not recommended that we either live or work in environments with greater than a certain mass of a specific toxic chemical because it would have an increased probability of being associated with systemic or pulmonary disease. Such limits are called occupational limits—recommended exposure limits (REL) are limits of chemical mass that it is recommended we do not exceed. Permissible exposure limits (PEL) are limits of chemical mass required for us to be below, for our occupation; these are enforced by the Environmental Protection Agency (EPA) and persons responsible for the violation are fined, on a case by case basis. Threshold limit values (TLV) are similar in magnitude to PELs, and relate to maximum allowable occupational levels of the chemicals in question. It is not necessary to measure the effects in each instance. Moreover, if the effect were measured, it would be too late for that group of people, who are already exposed. Engineers, specifically chemical engineers, measure the elevated chemical concentration with standardized instruments, because the chemical composition of the particles affects the physiologic response. Aerosol scientists describe the particles, their size, shape, and reactivity and health scientists describe health effects that result or might result from such exposure. Most research is anthropomorphic (human centered), so human systemic lung disease is of the greatest importance. Health of our domestic animals and plants is also of considerable importance to maintain the food supply. The quality of our wildlife and wilderness areas is also important. Aerosol scientists, engineers, and health scientists working together provide a cohesive picture not possible when we work in isolation.

Aerosol size is an important determinant of where in the respiratory tract the greatest mass of particles will deposit. Size can influence how rapidly particles are cleared. Once particles have deposited in the deep lung and entered the epithelial lining fluid (ELF), hydrodynamic size determines the level of interaction of the phagocytes and the epithelial cells that are associated with ELF. If particles are sufficiently small and insoluble, phagocytes will interact less than they do with larger particles. Such particles will persist in the lung interstitial space, not be cleared, and have potential to injure the lung.

Particle surface area will increase the potential for interaction with lung epithelial cells. Interaction and activation of epithelial cells can lead to the release of cytokines, which cause inflammation of lung cells. If the cytokine release is excessive, it can have the potential to cause lung cell injury. If persistent and progressive, that injury will be one from which the lung will recover with difficulty, or not at all. Responses in cultured lung epithelial cells are expected to closely correspond to responses in whole live animals or persons exposed to aerosols.

This chapter relates aerosol and hydrodynamic particle size to *in vivo* deposition, phagocytosis, and release of cytokines. We are interested in comparing these properties in ultrafine particles (nanoparticles; <100 nm) with those in the accumulation fraction (100 to 600 nm), the intermodal fraction (600 to 2500 nm) of fine particles and the coarse fraction (2500 to 10,000 nm). We expect the most interesting comparison to be between the minimally aggregated ultrafine particles (nanoparticles) and moderately aggregated accumulation fraction of the fine particles.

22.2 PARTICLE SIZE

Particles that can be inhaled, those less than 10 μm (10,000 nm; PM10) can be separated by the high volume cascade impactor (HVCI) into four fractions. When the HVCI was used to collect organic urban aerosols presumably from transportation, combustion, and the Earth's crust, the breathable particulate matter (PM) was divided into PM 10 to 2.5 μm (coarse aerosols, which are mechanically produced), and the PM 2.5 to 1 μm (intermodal) fraction, which is expected to have particles that contain properties of both coarse (larger) and fine (smaller) aerosols. In addition it separates the PM 1 to 0.2 μm (1000 to 200 nm; accumulation) fraction (just larger than nano- or ultrafine particles with properties similar to those particles) and PM 0.2 (particles <200 nm diameter in air) fraction. The cutoff size (200 nm) was chosen for convenience.

HVCI separations were thought to have moderate aggregation. PM 0.2 μm (ultrafine or nanoparticles) would be expected to have at least moderate aggregation, relative to the larger fractions (1–4). Ultrafine particles include those thought to occur naturally. Nanoparticles denote those that were manufactured; because of high surface area and stickiness, nanoparticles may aggregate or agglomerate to a greater degree. PM 2.5 to 0.2 μm (2500 to 200 nm) were said to be fine particles compared to PM 0.2 μm (200 nm), ultrafine (natural) or nanoparticles (manufactured). The upper end of fine particle sizes was thought to more nearly resemble both the physical size and properties of coarse aerosols. The lower end of this distribution more nearly resembled the physical size and properties of the smaller ultrafine aerosols, or nanomaterial aerosols.

Ultrafine particles were a complex fraction, containing mostly the short-lived nanoparticles that condensed from gas or liquid (25 to 100 nm). The upper end of this fraction contained lower particle numbers in the accumulation fraction (100 to 600 nm). The ultrafilterable particle properties were heavily dependent on local sources (1–3). This chapter compares the properties of fine (intermodal-accumulation) and ultrafine (nano) particles.

22.2.1 Physical Characterization

22.2.1.1 Urban Dust

22.2.1.1.1 Proportional Change with Size: Surface to Mass Surface Area, and Activity With decreasing particle diameter, particle surface to mass ratios (S/M) progressively increase; nanoparticles have a greater portion of their mass available for reaction at any point in time and generally are more reactive than accumulation particles, intermodal particles, or coarse particles. In contrast, coarse PM 10 to 2.5 would be expected to have had the lowest S/M, the least fraction of particle mass available for reaction and be less reactive. Thus, nanoparticles (<200 nm in diameter) would be expected to be most reactive with each other (aggregated or agglomerated) and with substances in the atmosphere and in the cell. The next most reactive would be expected to be the next larger accumulation particles (200 to 1000 nm in diameter).

Nanosized particles emphasize different properties than do macrosized particles (4a). Important properties are those of Brownian motion and stickiness. At the nanoscale, Brownian motion is influenced more by viscosity than by displacement of the medium in which it is placed, often fluid or air (4a). The viscosity effect depends more on absolute temperature than the medium (4a). Stickiness can be manipulated by polymer coating (gum Arabic for India Ink; polyethylene glycol for specific manmade nanoparticles). Both provide randomly flexible surface chains that help to control stickiness. With respect to mechanical properties at the nanoscale, strength (bonding) is more important than stiffness (brittleness) (4a). Finally, color may change at the nanoscale—cadmium selenide quantum microdot fluorescences can change from blue through yellow to red (4a). The size of the microdot constrains spatial uncertainty, but loosens momentum uncertainty (4a).

22.2.1.1.2 Quality of measurement When the Harvard high volume cascade impactor was compared to the virtual impactor, fine (PM 2.5 to 0.2; 2500 to 200 nm) material and coarse (PM 10 to 2.5; 10,000 to 2500 nm) material masses in the respective fractions were within 10% of agreement with values from other methods of collection and particle sizing. Measurements were simpler to make with the Harvard impactor. However, chemical constituents appeared to be sufficiently variable within fractions that activity should be measured on the material used to expose whole animals, or cells (5).

22.2.1.1.3 Particle Mass from Transportation Vehicles Released into the Atmosphere Smaller particle masses are composed mostly of exhaust from fuel combustion. To estimate the impact of the fuel cycle on total particle mass and fine particle mass released into the atmosphere, 10 school buses were sampled. The largest particles from the buses were found to be less than 6000 nm (6 μm) in aerodynamic diameter. PM 2.5 (fine) particles with median diameters 200 to 500 nm, and whose largest particles were less than 6000 nm were used to measure particle mass released into the atmosphere (6). Restarting the buses frequently reduced masses of fine atmospheric particles if the bus time of idling was <2 to 4 minutes (6). Released particle masses suggested a more optimal practice. Combustion gases (carbon monoxide and oxides of nitrogen) were also measured.

22.2.1.2 Road Dust

22.2.1.2.1 Metal Surface Composition To determine which metal bound to surfaces of road dust, soil dust was collected that contained larger particles (400 to 3000 nm diameter in air) and that compared closely to amorphous silica dust (6, 7). This dust was thought to be composed of both particles ground from the Earth's crust and particles generated by transportation vehicles that drive on the roads from which road dust was collected. Dust was collected near a road in four fractions. <56 nm, <100 nm, both ultrafine (nano) particle aerosols, a fraction <2500 nm (a fine aerosol), and one <10,000 nm (a coarse aerosol). Different elements were enriched on the surface of different-sized particles (6, 7).

The highest quartile (25%) fractions of elements richest in the two ultrafine fractions (<56 nm and <100 nm) relative to the fine and coarse fractions were Sb > Cd > Si > V = Ag. With the exception of silicon, these chemicals had at least moderate occupational toxicity as dusts. Silicon particles enriched on the surface suggested the aerosols' crustal origin. These data suggested that these elements could have been enriched on the surface of ultrafine road dust (6–8).

A possible source of Sb, Cd, and V is the exhaust of the vehicles driving on the road. That silicon was among these elements was somewhat surprising, but suggested that some of the ultrafine dust was from the Earth's crust. This finding suggests that road dust may be ground by continuous heavy vehicular traffic to contain ultrafine silicon dust particles on which Sb, Cd, and V condense on the particle surface. The smallest diameter fraction of the particles was likely to have been resuspended.

Although ultrafine particles have the possibility of long range, or even global transport, most of the particle masses settled within 300 m of the road (9).

The bottom quartile of elements least rich in the ultrafine fractions relative to the fine and coarse fractions were Fe > Sr = Cu = Mg > Mn. These elements had minimal to moderate toxicity as dusts and were likely to be a part of the crust that was ground by traffic. They had 6% to 11% of bulk mass concentration, suggesting that concentrations of the ultrafine fractions would have been lower, if it had been found in the bulk fraction of the larger fine and/or coarse road dust. The authors suggested that a large portion of the ultrafine road dust was likely condensate from motor vehicle exhaust (6, 7).

22.2.1.3 Metal Oxide Dust Zinc oxide metal fumes generated at 550°C aggregated into chains of ultrafine primary particles, whose secondary particles had a mass median aerodynamic diameter (MMAD) of 0.3 μm (300 nm) (10). Fumes would contain both fine accumulation particles (100 to 600 nm) and ultrafine particles (3, 11).

Fine zinc oxide had a median diameter of 111 nm and airborne diameters of less than 1000 nm (7). These particles would contain both fine accumulation particles (100 to 600 nm) and the smaller ultrafine particles (3, 11). Nanosized zinc oxide had measured median diameters of 90 nm and an airborne diameter of 50 to 90 nm (7). Similar samples of particles contained mostly ultrafine particles (3, 11). All samples of these particles were relatively pure zinc oxide but had similar metal compositions to the ultrafine road dust and would be expected to have similar hydrodynamic behaviors. However, zinc, is a moderately toxic element, relative to elements in ordinary road dust.

22.2.1.4 Hydrodynamic Sizes of Dusts Relative to Airborne Sizes What happens when ultrafine particles encounter the 100% humidity of progressively smaller lung airways—bronchi, bronchioles, and alveoli—during gas exchange? In aqueous solutions, fine and nanosized zinc oxide particles aggregate modestly. Particle distributions in both distilled water and lung tissue culture media capable of supporting growth of lung cells appeared nearly monomodal. Thus, particles had a single size distribution whose average size was less than 500 nm (7).

22.2.1.4.1 Hydrodynamic Sizes Relative to Airborne Sizes of Metal Oxide Particles Particles in dry form were found to be smaller than 1000 nm for fine zinc oxide dust and 50 to 70 nm for ultrafine zinc oxide dust (7). Particles were sonicated for 30 minutes. When suspended in distilled water, ~20% of the ultrafine (nano) zinc oxide particles were suspended, with an average hydrodynamic diameter of ~170 to 250 nm. When suspended in tissue culture media 45% had an average hydrodynamic diameter of ~300 to 400 nm. Tissue culture media suspended or extracted two to three times more particle mass than did dilute buffer or deionized distilled water, because of hydrodynamic action of media minerals. Hydrodynamic diameters were only modestly larger in tissue culture media than in distilled water (7).

Increased hydrodynamic diameters suggest that greater masses of these particles had grown into accumulation-sized particles in the presence of Ham's F-12 tissue culture media, and that fluids that resembled epithelial lining fluid (ELF) might have facilitated this process, relative to distilled water. Sonication-agitation for longer times did not disaggregate the fine or nano-sized zinc oxide particles. In both distilled water and tissue culture media the size distribution of particles contained significant numbers of both ultrafine particles and accumulation-fraction particles (7), suggesting that the minerals and nutrients of tissue culture fluid did not influence hydrodynamic size.

Modest changes in diameter (d/d_0) reflect larger changes in surface area and even larger changes in mass or volume. Nanoparticles or fine zinc oxide particles in Ham's F-12 k media had average hydrodynamic diameters ~60% larger; surface area $(d/d_0)^2$ and volume or mass $(d/d_0)^3$ of particles were approximately twofold and fourfold, respectively, that of the same particles in distilled water sonicated for the same time, and under the same conditions (7). While these particles are slightly larger, most of the particles were smaller than 500 to 1000 nm in diameter, small enough to be ignored by phagocytes (7). Thus, media composition favored progressive aggregation into the accumulation of fractions with larger sizes over the 30 minute sonication period (7, 12). We expect that when these small particles are deposited in the deep lung and enter the epithelial lining fluid (ELF), modeled by tissue culture fluid, they have a high potential either to be dissolved or to persist as a particle, with injury potential for the lung epithelial cells. There were no recognizable differences in the hydrodynamic diameters of nano-sized and fine zinc oxide particles (7). Thus, in biological fluids both zinc oxide nanoparticles and fine particles appear to reach the same sizes (<0.5 to 1 μm, 500 to 1000 nm), still small enough to be in the accumulation fraction and of sizes that we would expect to escape phagocytosis.

22.2.1.5 Titanium Dioxide Particles Airborne titanium dioxide particles ranged in size from 200–400 nm to 4000–17,000 nm (13, 14). Geometric mean diameters were 1290 to 1440 nm and geometric standard deviations (sigma g; σg) were an average of 1.7 (range 1.5 to 2.6). These are much larger than the P25 Degussa titanium dioxide unit particle in dry powder of 21 nm (13, 14), reflecting particle size formation in the atmosphere. These data imply that considerable particle aggregation and agglomeration of the titanium dioxide particles occurs during airborne generation or suspension, as they grow 100 to 1000 times in size in the atmosphere. Only a small fraction (<5% to 10% of the mass) would be expected to escape phagocytosis.

An additional study (12), contrasts the behavior of titanium dioxide sonicated for 10 minutes in distilled water. The particles had a hydrodynamic diameter of ~20 nm. If one combines this with the BET diameter determined in dry powder at the time of density measurements, aggregation would have been expected to be minimal, less than four particles per unit diameter (12). Further studies of these fine dusts (6), ranging up to 2000 nm, were used to assess cellular response.

Inhalation exposure has been explored for TiO_2 and other ultrafine particles (14a). When an inhalation chamber was used before animal exposure and polytetrafluoroethylene (PTFE; Teflon) ultrafine particles were used, 15 to 26 nm ultrafine particles

(σg 1.4 to 1.7) quickly grew to >100 nm (accumulation) diameter particles for PTFE, TiO_2, carbon, iron, iron oxide, vanadium, and vanadium oxide (14a). Approximately 50 mg/m^3 were used for inhalation exposure (discussed in part 3).

These findings can be compared to those for titanium dioxide particles whose diameters increase in biological fluids both over time and with increased concentrations, presumably by aggregation. Hydrodynamic diameters were measured for P25 Degussa titanium dioxide nanoparticles (unit diameter 21 nm) in Hanks Balanced Salt Solution (HBSS) or Dulbecco's Modified Eagles Medium (DMEM) with approximately the same minerals and enriched with amino acids (15). These initial hydrodynamic diameters, measured at 10 minutes, increased with increasing TiO_2 concentrations from 500 nm (0.5 μm) at a concentration of 5 ppm to 1350 nm (1.4 μm) at a concentration of 120 ppm titanium dioxide. The final average diameter was not only concentration dependent, but somewhat time dependent. For example, steady-state hydrodynamic diameters taken at 18 hours equilibration grew to hydrodynamic sizes 1.6-fold larger (volume is $1.6 \times 1.6 \times 1.6$, ~4.1-fold larger). These sizes ranged from 830 nm (0.8 μm) at concentrations of 5 ppm to 2370 nm (2.4 μm) at concentrations of 120 ppm. Thus, a concentration of 24-fold higher increased the hydrodynamic diameter 2.8-fold and the aggregate volume or mass 22-fold. The concentration factor (2.8-fold) was greater than the time factor to equilibrium (1.6-fold). All particles were measured within the fine particle range. Most particles grew to become the size of particles in the accumulation fraction or intermodal fraction (16, 17). The relative increase in size over 18 hours of incubation and settling was similar (1.6-fold) regardless of the concentration of TiO_2 (15). The reason for this increase in diameter and mass by aggregation for titanium dioxide particles, in contrast to the relative absence of change in these quantities with particles of other metal oxides, is unclear.

In another paper (18) diameter and mass were reported to be progressively increased and extended at even higher concentrations (5 mg/mL; 5000 ppm or 5 μg/cm^2 of *in vitro* tracheal cell culture surface area). Hydrodynamic agglomerate size is greater (~3000 nm diameter) with 21 nm unit sized titanium dioxide dust particles than with 120 nm titanium dioxide anatase dust particles (590 nm hydrodynamic size), suggesting an active surface (18). The tendency of titanium dioxide particles to aggregate in tissue culture media was greater than their tendency to disaggregate. Moreover, aggregation was not a matter of sonication intensity or duration. Sonication for 30 minutes in an attempt to disaggregate particles only succeeded in increasing the average hydrodynamic diameters 10% to 20%, rather than reducing their size (18).

As expected, the larger hydrodynamic sizes (aggregates-agglomerates) of titanium dioxide settled more rapidly than did smaller hydrodynamic sizes (15). If we examine only the dilute concentrations, during the first 2 hours >90% of the 5 to 30 ppm TiO_2 remained in suspension. However, by 18 hours most of these particles had settled so that only 15% to 25% of the original TiO_2 at 5 and 10 ppm suspensions remained in suspension. The 5 and 10 ppm suspensions were most relevant to a relatively heavy aerosol exposure (15). The very high concentrations of titanium dioxide were greater than that expected for a field exposure to lung cells *in vivo* or *in vitro*.

Sonication increased particle aggregation, rather than disaggregating them, suggesting that titanium dioxide may have a strong tendency to aggregate. This caused Churg and coworkers to speculatively raise the question of direct particle effects (18). This proved prophetic with regard to later work with carbonaceous particles.

If those particles had been of moderate solubility, like the calcium phosphate, iron oxide, or zinc oxide particles, they would have been suspended sufficiently long to be dissolved. Alternatively, if they had been nearly insoluble, as were the titanium oxide, cerium oxide, and zirconium oxide particles, we would expect them to persist. Thus, the particles could either dissolve directly or persist as insoluble particles. Previous work with titanium dioxide has shown it to be virtually insoluble *in vitro* and to persist *in vivo*.

22.2.1.5.1 Size of Single (Unit) Particles Relative to Aggregated Particle Clumps How many particles clumped together to give aggregates of these sizes? Particles that had been ultrasonically agitated in aqueous media were mostly smaller than 200 nm in diameter, with average diameters of 19 to 50 nm (12). Average diameters were 34 to 50 nm. Particle aggregates were calculated to contain less than eight particles, based on size calculated as powder and hydrodynamic size measured in solution aggregates. Less than 3% to 20% of the particles were 200 nm. Comparable data was obtained with insoluble cerium oxide, titanium dioxide, or zirconium oxide. Particles had slightly smaller average diameters (19 to 20 nm). Less than 10% to 11% of the particles exceeded 200 nm and clearly separated into the accumulation fraction (12). Accumulation fraction particles were minimal in both mass and number for particles in solution. Specific surface areas were similar (average surface area 100 m^2/g; range 57 to 188 m^2/g) (12). At this early time, in distilled water, particles were ultrafine with minimal aggregation. Particles do aggregate modestly in aqueous buffers, similar to epithelial lining fluid interactions.

22.2.1.5.2 Summary Persistence of particles is one of four major properties of hazardous (organic) particles; the others are bioaccumulation, long-range transport, and adverse effects. Particles that have small enough hydrodynamic diameters to avoid phagocytosis may accumulate in the interstitial space. Ultrafine airborne particles are capable of long-range transport. We will comment on potential adverse effects later in the chapter.

Persistence is a function of solubility, surface area, and mass. Ultrafine or nanoparticles have high surface area, reflecting the potential for a high fraction of chemical ions on the surface.

1. High surface area particles have high surface activity which has a tendency to focus in the lipid-aqueous interface and act as a surfactant by surrounding water particles and maintaining lipid water interface (19).
2. Highly active surface area causes particles to clump with each other, increasing particle size and settling velocity from the atmosphere, and binding with lipid cell membranes. This effect enhances its potential to remove microbes from the air, and to bind with lung epithelial cells in the lung interstitial space.

3. Clumping is so rapid and haphazard that after nanoparticle clumping, there is little reduction in total surface area and considerable internal space remains. However, more of the surface area is internal. The geometry (internal volume) of the internal surface may limit access for binding based on size of the molecule.
4. Phagocytes and most epithelial cells internalize particles $\geq$500 nm more efficiently than smaller ones (16, 17). A549 lung epithelial cells do not take up most of the ultrafine TiO_2 (20). However, ultrafine particles that were taken up remained as aggregates 24 hours after exposure, even at lower doses (20). In rare instances, two or three particles were seen within a single membrane-bound vesicle, indicating that disaggregation is possible, but rare in 24 hours (26).
5. The processes by which ultrafine particles disaggregate are poorly understood, and have not been studied to any great extent (21). Gold nanoparticles linked to carboxylated dextran chains have more freedom of movement and less aggregation in a hydrated environment (21). Licorice root saponin caused type 1 or type 2A protein phosphatases mediated connexin (CX) 43 dephosphorylation, that was coincident with the disassembly of gap-junction cellular plaques of communicating epithelial cells (22). Bone oxidized by hypochlorite can take two forms, one where crystalline plate aggregates, 25 by 45 nm, are intimately associated with collagen fibrils; and a second denser form where crystals are mostly compressed between fibrils (23). Hypochlorite oxidized bone with crystals intimately associated with collagen can be disaggretated by sonication into crystalline aggregates. Denser bone with crystals compressed between collagen fibrils persisted as "fused" aggregates (24). Disaggregation is less difficult when aggregates are less fused and more internal surface area remains. Nanoparticle interaction with a surfactant may facilitate this process (19, 19a).

22.2.1.6 Magnesium Oxide In some cases, metal oxides are converted to a more soluble chemical and dissolve more rapidly. For example, magnesium oxide is relatively insoluble in water, but is converted to a more soluble carbonate form by exposure to body fluids whose buffer is bicarbonate (25). This would be expected to enhance the dissolution of magnesium oxide nanoparticles (25). However, titanium dioxide's most soluble form is the acid chloride. If the particles were porous, the porosity (access to the internal surface area) could enhance solubility still further for MgO. Empirical evidence suggests that the aggregated-agglomerated metal oxide nanoparticles have minimal reduction of measured surface area, or solubility properties that parallel such surface areas.

We have reported that dissolution of magnesium oxide nanoparticles was more rapid in a bicarbonate buffer than in distilled water (26); rate of solubility was proportionate to bicarbonate quantity in HBSS and DMEM. The concentrations of DMEM and HBSS were approximately those anticipated in lung ELF at the end of expiration and the end of inspiration, respectively. *In vitro* solubility studies revealed a uniform pH increase (pH change ~1 to 2 units; data not shown). This increase is consistent with the formation of magnesium hydroxide. However, the rate of solubility and the saturated solution concentration were consistent with magnesium carbonate. It was likely that magnesium carbonate trihydrate (nesquehonite) was formed, because the trihydrate form is stable at body temperature and a P_{CO_2} ~40 torr (mmHg). Its pH, P_{CO_2} stabilities were consistent with conditions supporting human life. Moreover, it is soluble in these *in vitro* incubations and the *in vivo* conditions in the human lung (18). It is a fortunate coincidence to have surface activity in the atmosphere that cleans it and conversion to a more soluble form in the epithelial lining fluid which renders active MgO nanoparticles soluble, reducing persistence and health effects consequences.

Nanoactive MgO Plus particles when suspended at 50 to 250 ppm in HBSS or DMEM and gently stirred formed a 10 to 50 cm halo that macroscopically resembled a tenuous and porous lattice. The lattices persisted for 2 to 4 day solubility studies conducted at 22°C or 37°C. We interpreted this process as the formation of aggregates marginally smaller than the size needed to settle out of solution with gentle stirring. The porous lattices focused particles that deposited in deep lung and moved to the ELF near their point of dissolution (25, 26).

22.2.1.6.1 Solubility of Manure Dust Ground manure mat dust from cattle feedlot dust was assayed for aqueous solubility at 50 and 150 mg/L (ppm) in HBSS and DMEM and measured as transmitted light (luminols) (24a, 24b). In the presence of gentle stirring, this suspension formed a halo similar to that for the magnesium oxide nanoparticles. This halo persisted for the 72 hour solubility study and constitutes ~25% to 50% of the luminol reduction (24a, 24b). Those magnesium oxide and manure dust particles would appear to have three properties. First, they had less probability of settling out of ELF and irritating the lung epithelial cells. Considering the convolution of the ELF space and the continuous inhalation and exhalation motion of the lung, this difference may be minimal. Second, their porosity implied higher surface area and a greater probability of dissolving than would be the case with less porous particles. Thus, they would be expected to have less persistence than if they were less porous. Finally, their tenuousness and porosity implied a greater tendency to disaggregate and reform than would less porous and tenuous particle structures (24a, 24b).

22.2.1.6.2 Particles in Extrapulmonary Tissues Manganese oxide nanoparticles were 39 ± 12 nm, the size of a nucleation fraction. In PC12 cells they ranged from 24 to 57 nm in size. This size was still in the ultrafine (nano) particle range. However, some cell agglomerates ranged from 1 to over 10 μm in diameter (27, 28). Single particles were irregular in shape and could be readily measured; as a group, particles were below 1 μm in diameter and small enough to be ignored by brain phagocytes (27). In tissue, the particles showed considerable ability to aggregate that was not present in the *in vitro* solutions.

Multiple-walled carbon nanotubes (MWCNT) (0.3 to 5.3 mg/m^3; 0.7 to 1 μm; 700 to 1000 nm diameter; geometric standard deviation [GSD] 2 for the two lowest levels; and 1.8 μm; 1800 nm diameter; GSD 2.5 for the high level) were used to expose rats and mice for 7 or 14 days (6 hours/day) (28a). At 1 mg/m^3, the number mode (most frequent value) was ~0.4 μm (400 nm) analyzed by differential mobility–time of flight analysis. These levels produced no significant lung inflammation or tissue damage but caused a systemic immune function alteration (reduced response to sheep erythrocytes and nonspecific natural killer cell activity at 1 mg/m^3). There were no changes in gene expression, but interleukin 10 and NADPH oxido-reductase 1 mRNA levels were increased in the spleen (28a).

To investigate potential harmful effects of inhaling nanoparticles, 49,000 ± 2000 and 93,000 ± 5000 coated ferrite particles/cm^3 (49 to 51 nm diameter), the investigators scanned them with confocal laser microscopy and compared them to sham controls. Weighted spin-echo MR images showed that these particles could penetrate the blood-brain barrier. Their work has shown that the blood-brain barrier can be penetrated with these particles without producing detectable toxicity. One possibility for penetration would be along the fibers of the olfactory bulb. These particles also passed the blood-testes barrier. The authors recommended further study of the potential violation of the immune precluded status (28b). Application of nanotechnologies should not produce adverse effects on human health and the environment (28b, 28c). Subacute exposure of C57B1/6 mice to 2–5 nm TiO_2 nanoparticles in a whole body chamber caused a moderate but significant inflammatory response in the lung within the first two weeks of exposure, beyond which the inflammation resolved without permanent damage (reviewed in Reference 28d).

Almost no models of pulmonary deposition take dispersion of particles into account. With compartmental modeling it is easy to represent modeling in an anatomically accurate airway network. When the airway network is represented accurate anatomically, pulmonary deposition, especially deposition of fine mode particles, is calculated to be increased (29).

Shaw et al. (28e) have approached *in vitro* perturbational profiling of the biologic activity of nanomaterials in a generalizable and systematic fashion. The assessment was multidimensional, using multiple cell types and multiple assays that reflected different aspects of cellular physiology. Heirarchial clustering of these data identified nanomaterials with similar patterns of biologic activity across a broad sampling of cellular contexts, as opposed to extrapolating from any single *in vitro* assay. The approach yielded robust and detailed structure-activity relationships. For example, it would be much less sensitive to peculiarities of an individual cell line.

A subset of nanoparticles were tested in mice (28e). Nanoparticles with similar activity profiles *in vitro* exerted similar effects on monocyte numbers *in vivo*. Alterations of leukocyte subsets, including an increased monocyte fraction, can be a sign of proinflammatory or other toxic exposure. Second, monocytes are phagocytic and take up certain nanoparticles more than other cell types. Finally, nanomaterials have been shown to cause pleiotropic effects on immune cells that are very sensitive to the materials' composition and surface. Because this analysis compares biologic activity across nanoparticles, we can still draw conclusions as to similarities and differences among nanoparticles based on this *in vitro* data. The authors' emphasis on comparing across nanomaterials becomes progressively more useful as the number of well-characterized nanomaterials increases (28e).

22.3 CHARACTERIZATION OF BIOLOGICAL (HEALTH) EFFECTS

22.3.1 Urban Dust

22.3.1.1 Potential Exposure to Airborne Particles Atmospheric particles of different sizes have been shown to act predominantly as inflammatory or cytotoxic-apoptotic particles (1, 2). Testing urban samples collected using HVCI showed coarse particles (PM 10 to 2.5) to be more capable of causing cells to release inflammatory cytokines such as tumor necrosis factor alpha (TNFα), and interleukins 6 and 8 (IL6, IL8) than were fine (PM 2.5 to 0.2) particles. The smaller fine particles (PM 2.5 to 0.2) have less inflammatory potential (1, 2, 30). Perhaps the reason is because of less mass, or a greater possibility of altering the active surface of smaller particles. For example, if surface area is increased it will increase the potential to produce inflammation. For this reason fine particles have quite variable inflammatory potential. The degree to which specific chemical composition plays a role is not understood sufficiently at this time.

Ultrafine particles or nanoparticles were more cytotoxic by apoptosis. This is because the process of apoptosis (cell death) reduces the tendency of cells to produce cytokines while still alive (1, 2).

Barcelona and Athens fine particles (PM 2.5 to 0.2) collected during the summer had the highest levels of inflammatory potential of all those sampled. Conversely, wintertime Prague particles released during coal and biomass burning were smaller ultrafine or nanoparticles. These particles had the highest potential to release substances leading to cytotoxicity or apoptosis. It would appear that high S/M, surface area, and probably surface activity relate most closely to cytotoxicity and apoptosis (1, 2).

Long-range transport (LRT) of particles from wildfires greatly increase in small fine particles (PM 1 to 0.2 particles) or ultrafine particles. LRT is the second of four criteria of hazardous (organic) particles. Other properties of particle hazard are persistence (previously discussed), bioaccumulation (particles sufficiently small to avoid phagocytosis), and adverse biological (biomedical) effects. PM 1 to 0.2 particles are the smallest portion of fine particles, which we call the accumulation mode (1, 2). They are the closest to ultrafine (nano) particles in size (structure) and function.

Overall airborne ultrafine particles are present in larger numbers and with large surface area per cubic meter of air. They have significant potential to cause inflammation and apoptosis. If artificially aspirated into the deep lung (31–33), they cause intense pulmonary inflammation and damage. However, surface changes during aging tend to reduce damaging effects. Less effect was noted per particle on cytotoxicity, related to PM 0.2, ultrafine particles with high S/M, surface area and surface activity (1, 2). Surface modifications during particle aging are not uncommon. Alternatively, agglomeration and/or aggregation with active surfaces may cause particles to grow to larger size; for example, an accumulation particle may agglomerate into a larger particle before deposition into lung.

Increased overall potential to produce inflammation, toxicity, and apoptosis per cubic meter of air can be magnified if danger is imminent and time is short.

First, surface aging effects are less. Second, the average human with a breath volume of 0.5 L and 15 breaths per minute will require 3.4 hours to breathe 1 m^3 of air. Stressed or exercising humans, such as those working near a forest fire, may breathe deeper and more frequently (hyperventilation). Individuals who are hyperventilating will require far less time to breathe 1 m^3 of air. If one breathes deeper, more particles will be deposited. When this factor is multiplied by the increased potential to produce inflammation and toxicity-apoptosis, lung disease may be increased disproportionately by 5- to 10-fold more potential for injury in a stressful situation such as a forest fire. To the extent that toxins are soluble, toxic chemicals on surfaces can desorb and be carried away by diffusion and other general processes, rendering the remaining particles less toxic. If the particles are not soluble, they will persist and remain as potential sources of toxicity.

22.3.1.2 Effect of Particle Size on Potential for Lung Response

22.3.1.2.1 Lung Deposition Particle size is an important metric for entry into the lung. Particles larger than 10 μm (10,000 nm) are not considered to be inhalable. These particles have minimal potential to cause health effects in the lung. Particles larger than 3.5 to 4.0 μm (3500 to 4000 nm) will be trapped in the upper airways by impaction or settling (31, 32). These largest particles will contain ~5% of the surface area. Because flow is still primarily directional, only the smallest particles (<0.03 to 0.05 μm; 30 to 50 nm) will be trapped by diffusion in the upper airway. These particles will contain ~10% to 15% of the total surface area. Neither the surface area of the very large particles nor the surface area of the very small particles will be of significance in a potential health response. Both are trapped in the upper airway, the former by impaction or settling, and the latter by diffusion among the intricate twists and turns of the upper airway (31–33).

Particles between 1000 and 2500 nm—the intermediate fraction—will settle and also have some component of random diffusion where they contact a bronchiole wall (25, 26). Particles smaller than 1000 nm will also settle, but as they become progressively smaller, <500 nm, they will be able to slip in between gas molecules and become driven (randomly) by them (32). They will contact the airway wall increasingly by diffusion only. In many distributions, more than 80% of the total surface area is contained in the smaller fine particles (less than 1000 nm) and the ultrafine or nanoparticles (less than 100 to 200 nm). Approximately half of this surface area is small enough to be ignored by phagocytes and will go to the pulmonary interstitial space, with potential for injury or adverse effect. In this context, adverse effect is taken to mean that from which the cell, tissue, or animal cannot completely or efficiently recover. If the particles are persistent they will remain in the interstitial lung space with potential for adverse effect (31, 32).

22.3.1.2.2 Lung Surface Area Surface area, organics, and metals are important metrics for potential for lung response of inflammation, toxicity, and apoptosis (16, 17), and the surface area of particles between 50 nm and 1000 nm is expected to be the most influential in causing lung injury.

22.3.1.2.3 Hydrodynamic Size, Chemical Composition, and Particle Coating Several nanoparticles were compared for nanocrystal size (at manufacture), hydrodynamic size (size in water, or physiological aqueous buffer) or airborne size (33a). In general, nanocrystal size at manufacture was smaller than hydrodynamic size and that in turn was smaller than airborne size of aggregated-agglomerated particles (33a). A high percentage of particles in physiological aqueous buffer was small enough to be ignored by phagocytes. Four of 13 compounds were coated to see the effect of coating on toxicity to cells. Silver particles coated with polysaccharide or hydrocarbon were toxic to cells in buffer, but not toxic to the same cells in the presence of serum. These data indicate that coating can affect toxicity to cells (33a). Titanium dioxide, copper, and carbon black (mildly toxic in bulk) were more toxic to cells than aluminum and magnesium (minimally toxic in bulk). Thus, even when chemicals are present as nanoparticles, they appear more toxic to cells if they were toxic in bulk. Silicon dioxide had minimal toxicity, whether crystalline or amorphous (33a). We would anticipate that amorphous silica would have minimal toxicity, but crystalline silica would be more toxic. However, this was not observed in this study.

22.3.1.2.4 Lung Hazard Evaluation Ultrafine or nanoparticles can have varying persistence, but organic particles are usually especially persistent (34). Other factors that increase hazard are bioconcentration facility, long-range transfer ability, and toxicity. Organic particles have a high efficiency to concentrate in predators because they are fat soluble and tend to concentrate in the fat as they are consumed by progressively larger predators. However, this is not the only mechanism of bioconcentration. For example, small particles less than 500 nm, are largely ignored by phagocytes and remain in the lung interstitial space. Interstitial particle concentrations would be expected to increase with persisting nanoparticle exposures. Thus, if they do not dissolve, nanoparticles may concentrate in the pulmonary interstitial space (31–34). Alternatively, particles that are larger would be expected to be cleared with progressively increasing efficiency by lung phagocytes, reducing the potential for their adverse effect. Thus, small, durable particles with the highest surface area will gather in the pulmonary interstitial space with continued exposure. Nanoparticles and the smallest fine particles have the greatest potential for long-range transport. Inherently toxic chemicals are more toxic to cells *in vitro* than are nontoxic chemicals in nanoparticle form (27a). They also have significant potential for adverse effect (31–34).

Reviewing hazard criteria of POP (34), because of their size nanoparticles have the greatest potential for long-range transport (criterion number 3), and relative surface area for adverse effect (criterion number 4). If they are persistent (criterion number 1), and most carbonaceous nanoparticles are, they extend the time of their exposure and the cumulative potential for adverse effect (criterion number 4). All nanoparticles have potential to accumulate in lung interstitial space (criterion number 2) with continued exposure. However, only organic particles whose constituents are fat soluble will bioconcentrate (criterion number 2) progressively, as the compounds are consumed up the food chain by increasingly large predators.

Organic coating of nanoparticles because of lipid solubility can facilitate nanoparticle entry into cells. Thus, we would anticipate that particles of diesel exhaust coated with surface organics would be potentially more toxic than just carbonaceous particles. Inhaling diesel exhaust in humans with heart disease has been shown to reduce tissue plasminogen activator (tPA). Similar changes have been seen in humans with fully stabilized coronary heart disease. Likewise, metals or metal oxides (criterion number 3), could produce adverse effects by remaining in the lung interstitial space or by entering lung cells and combining with organics, so that they can enter cells more easily than metals or metal oxides alone (16, 17, 31–34).

22.3.2 Manure Dust

Ground manure mat from cattle feedlot dust was used to expose calves in individual chambers and was studied for potential to produce cytokines and endotoxin activity (24a, 24b). This dust was extracted in physiological saline and diluted to 1% (1:99 saline), 5%, 10%, 25%, and 50% (1:1 saline) strength and measured for potential to produce TNF alpha, and interleukins (IL6 and IL8) in BEAS 2B lung cells. The results showed that 50% solutions were cytotoxic and 10% and 25% solutions had the highest inflammatory potential, as indicated by production of IL6 at 6 to 24 hours of culture. These data suggest that lung cells need to be exposed to considerably more dust, to release more IL6 and increase potential for lung inflammation and angiogenesis, and heal by fibroplasia. In addition, 1% or 5% concentration of soluble manure mat extract was required to increase the amount of IL8 released by BEAS 2B LEC. IL8 can attract neutrophils (polymorphonuclear leukocytes; PMN) by chemotaxis, amplify (extend), and maintain pulmonary inflammation after the dust was inhaled (24a, 24b). These data correspond to size analysis that shows the manure mat contains a high mass percentage of coarse particles. Size analysis indicates that lower mass percentages of the particles were sufficiently small to be enriched in endotoxin. Endotoxin was enriched in PM4 aerosols collected from swine confinement buildings, indicating ~2% to 10% of that aerosol to be PM4 (24a, 24b). With the manure mat dust, about half of the potential of the dust to release cytokines in lung cells was blocked if endotoxin was inhibited. When endotoxin was inhibited in the soluble extract of manure mat dust, peak IL8 releases were seen only at 12 to 24 hours, but the amount was roughly equivalent to the amount released when endotoxin was not inhibited, suggesting that extracted endotoxin alone was not responsible for the release of IL8 (24a, 24b). Heat treatment at 120°C inhibited it still further, indicating that endotoxin and some other component were destroyed by heat (24a, 24b).

22.3.3 Crustal, Desert, and Mountain Dust

When soluble extracts of dust from the Earth's crust, desert, wind-generated, foothills, and high mountain (Uinta) dust were cultured with BEAS 2B lung epithelial cells, more IL6 and IL8 were released than when endotoxin alone was cultured (6). These data suggested that soluble material extracted from this dust caused cytokine release from more than endotoxin alone, compared with soluble extracts from manure mat dust. This supports the hypothesis that some ambient dusts from geological sources can cause cell death and cytokine release in a lung cell line that is widely used as an *in vitro* model to study mechanisms of environmental respiratory injury (6).

22.3.4 Metal Oxide Fumes or Dust

Both fine and ultrafine TiO_2 particles can upregulate mediators of fibroplasia and fibrosis (19). However, only ultrafine titanium dioxide particles have been shown to enhance production of procollagen. Although the cause is unknown, this enhanced production is associated with increased tissue peptide hydroxyproline, suggesting the production and deposition of increased collagen, apparently without detection of any structural change. Differential upregulation of platelet derived growth factor (PDGF) receptors does not explain the dust-induced fibrosis in this model. Chronic exposure to PM10 can lead to the development of chronic airway obstruction. Issues of dose are complex. For example, high levels of particles are required to enhance a chronic respiratory response such as the production of procollagen in the respiratory tract. But very little uptake occurs into tracheal explants (18). It is possible that increased procollagen expression might be a result of ultrafine particles penetrating the epithelium, reaching the interstitial space, and interacting directly with fibroblasts without the requirement of an intermediate agent (18).

When mice were exposed to potentially toxic zinc oxide metal fumes (300 nm MMAD aggregates), mouse lung metallothionein and tolerance to increases in bronchoalveolar lavage fluid (BALF) protein were induced; increased protein was used to indicate edema (31, 32). No tolerance was induced to influxes of polymorphonuclear leukocytes, signifying pulmonary inflammation (31, 32).

Fine and nanosized zinc oxide particles caused *in vivo* releases of LDH, indicating cell killing; cell numbers, indicating inflammation and increased polymorphonuclear leukocytes (PMN), indicating acute inflammation in rat BALF; this response reduced quickly for LDH and BALF cell numbers; BALF PMN persisted to 1 week after initial exposure (7). These particle responses resembled those of particles that did not persist (carbonyl iron and amorphous silica). However, crystalline silica while having similar intensities of response, persisted to 3 months after initial exposure, suggesting that the inflammation continued (7). Responses of factory workers to continued zinc fumes more closely resembled that of crystalline silica chronic fibrosis (16, 17), or healing by secondary intent (fibroplasia and fibrosis) (31, 32, 35–37).

In vitro responses only correlated qualitatively, not quantitatively with the *in vivo* responses (7). Specifically, cells responded indicating cell killing; this response quickly returned to normal, except for the higher doses of crystalline silica (7). Responses of co-cultures of epithelial cells and macrophages more closely resembled each other *in vivo*, in that there was evidence of toxicity of each at higher concentrations. However, the correlation was only qualitative (7).

22.3.4.1 Extrapulmonary Tissues Nanoparticles (unit size 39 ± 12 nm) bound to PC 12 cells led to depletion of dopamine (28). The nanoparticles aggregated to much larger sizes in these cells. Nanoparticles can travel to brain tissue through the nasal bulb and were shown to produce oxidative changes in fish brains (16, 17). They were able to deplete serotonergic compounds from the PC 12 cells. Manganese is the mineral associated with changes in the nigro-pallidal area in

Parkinson's disease. The extent to which this *in vitro* response is relevant to Parkinson's disease remains undefined (28).

Intratracheal instillation of combustion particulate matter from an eastern power plant was sufficiently high to induce rapid increased lung lobe wet weight (~20%; Reference 32). Vanadium, nickel and lead from the particulate matter rapidly translocated to extrapulmonary organs. Clearance of vanadium, nickel, and lead appeared to have a half-life of 1 to 2 days from lungs, plasma, and heart. Half-life from liver appeared significantly longer for all three metals. Although the mechanism is incompletely understood, the toxicity of particulate matter (PM) associated metals may relate to their total bioavailable mass (38). A potential limitation of this study is that the aerosol characterization was only reported with respect to the metal composition (38).

22.3.4.2 Effect of Particle Surface: Reaction with Both Epithelial Growth Factor and Integrins Cells interact with either extracellular matrix or carbonaceous nanoparticles (Table 22.1; Fig. 22.1). The surface area of carbon nanoparticle cluster events leads to cellular phosphorylation and cell division of healthy or damaged cells following particle interaction with lung epithelial cell (LEC) integrins and epithelial growth factor receptors (EGFR) (33). Phosphorylation activates either the phosphoinisitol (PI) 3 kinase (3K) (healthy LEC) or the NFκB cascade (damaged LEC) (40–50). This involves the extracellular kinase cascade (37) and protein kinase C epsilon (ε) (36). PKCε led to the release of IL6 and IL8 (40–42), cytokines responsible for maintaining potential for inflammation, oxidant load, and healing by secondary intent (IL6). IL8 maintains potential for amplifying inflammation, and chemoattraction of neutrophils (40–42). PKC alpha (α) appears to be associated with LEC survival movement (42), and with either healthy or damaged cells. Interleukin 10 (IL10) opposes inflammation or induces tolerance (44).

IL8 increases, and is followed progressively by an increase in IL10 to its capacity to increase (38). The balance between IL8 and IL10 is important to lung integrity and healing. Measuring IL8 and IL10 enables one to evaluate the balance between the two cytokines and the degree to which we expect inflammatory or allergic injury. Healthy cells (LEC) generate sufficient IL10 so that the major histocompatibility complex (MHC-1, thymocyte [Th] 1-acute inflammation) can shift to MHC Th2, allergic inflammation (38–41), and avoid the proteolysis and tissue loss of acute lung tissue injury. If the limit of IL10 to anti-proteolysis is exceeded and the MHC-1 (inflammation) to MHC-2 (allergen) shift cannot be made, proteolysis and tissue loss become progressively more important.

These reactions led to an increased matrix metalloproteinase (MMP) and a subsequent increase in tissue inhibitor of metalloproteinase 1 (TIMP 1), with some tissue remodeling (50–61). Speculatively, TIMP-1 may not be sufficient to balance the effects of the MMP-1, leading to tissue injury, apoptosis, remodeling, and loss of lung epithelium. Connexin connects epithelial cells and expands the epithelial cell response. Connection to connexin 43 can allow the apoptosis and tissue loss to be expanded among communicating epithelial cells (51). Alternatively, if TIMP-1 is sufficient to balance injury, such reactions will be minimized, favoring healing.

TABLE 22.1 A Model for Nanoparticle Interaction with Lung Epithelial Cells

Carbon nanoparticle surface area (**SA**) focuses phosphorylation after particle interaction with lung epithelial cell (LEC) integrins and epithelial growth factor receptors (EGFR) (33)

Triggers LEC **Phosphorylation** (33), and Releases the apoptic (protein) kinase (**Akt**) cascade (33) and the interleukin **IL6** (34)

- IL6 is a prooxidant driven by NFκB (reviewed in Reference 28d); Toll-like receptor 2 (TLR2) is a key mediator in the inflammation (24c) that increases reactive oxygen substances (ROS) and inhibits microbe clearance (34). IL6 is angiogenic and profibrotic.
- **Akt** includes phosphatidyl inositol 3 kinase (**PI3K**), protein kinase C alpha (33), (**PKCα**) (35), and E-cadherin, and B catenin to activate membrane proteins downstream (33, 36).

If LEC are healthy, maintenance and healing will be by LEC hyperplasia.

The response will stimulate extracellularly regulated kinase ERK 1/2 (37), and release **PKCε**, **IL8**, BCl_2 ***cell survival factor*** (40–55)

- TGFB1- **IL10** DC-TCR (34–45)
- Shift major histology compatibility complex MHC-1 (Th1) response to MHC-2 (Th2) (47) and
- Increase matrix metalloproteinase (MMP-1) relative to tissue inhibitor metalloproteinase (TIMP-1) (54, 55).

If LEC are damaged, they will attempt to heal by LEC apoptosis and fibroplasia (39, 42)

Damage models include: Bleomycin, LEC scratch, mechanical stretch, BAP and nanoparticles; asbestos is proportional to vitamin C yellowing (42–47). This will stimulate Jun nuclear kinase (JNK1/2)

- **PKCδ**-NFκB; BAX ***cell death factor***; caspase-3
- Can't stimulate enough **PKC BII**; TGF-B- **IL10**-DC-TCR to shift MHC 1Th 1 response to MCH 2-Th 2 (39) OR TIMP-1 (44–61)
- GAP connexin (CX) 43 expands LEC apoptosis (55) (emphysema) (56, 57) OR
- From added extensive particle surface area (45–52), p53 is activated and the changes mimic **radiation carcinogenesis** (47).

Adult male mice (30 g; 2 months old) given intratracheal instillation (ITI) of high doses (0.1 or 0.5 mg) of TiO_2 (50 m^2/g; 19 to 21 nm; 0.005 m^2 and 0.025 m^2, respectively, rutile) developed pneumonitis, type II cell hyperplasia, apoptosis, and an upregulation of cytokines (51a). Cytokines that were upregulated included placenta growth factor PIGF (glycosyl PI glycolipid anchor), CXCL1 (IL8 receptor A), CXCL5 (IL8 receptor B), CCL3 (macrophage inhibitory protein MIP-1) by PCR; they may cause inflammation, neutrophil infiltration, lung tissue destruction, and pulmonary emphysema (56). Cultured THP-1 cell-derived macrophages treated with this nano-TiO_2 had increases of the same cytokines *in vitro* (56), suggesting at least a partial source of the excess cytokine release and lung damage. Comparison of 22 nm and 280 nm ferric oxide particles given by ITI to rats at 0.8 mg/kg body weight led to alveolar macrophage overloading with particles (57), and inflammatory reactions, increased numbers of immune cells, protein effusion, pulmonary capillary vessel hyperemia, alveolar lipoproteinosis, and lung emphysema with no signs of fibrosis (57), in nano-TiO_2,

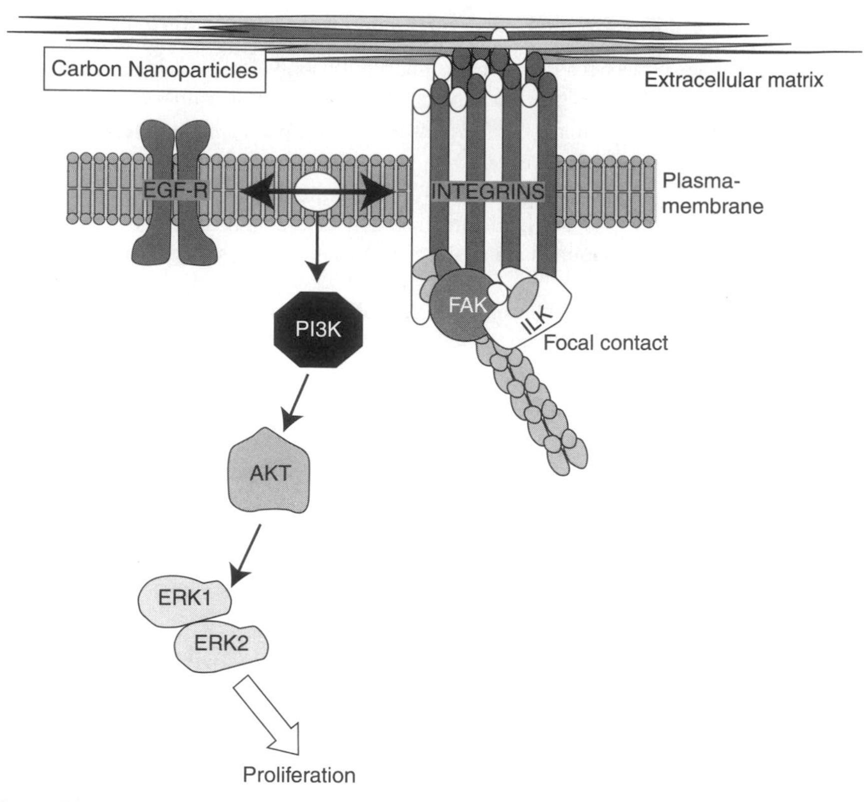

Figure 22.1 Signaling cascade induced by NPCB (carbon black nanoparticles) in rat lung epithelium (RLE)-6TN cells. Both membrane receptors, EGF-R and β_1-integrin, probably cooperatively induce proliferative signaling via extracellular ERK1/2. Data from this study demonstrate that PI3K and Akt are main mediators of this signaling between membrane receptors and MAPKs. Furthermore, no evidence for an involvement of focal contact proteins FAK and ILK could be obtained. (Modified after Reference 4 and used with permission of K. Unfried et al., *Am. J. Physiol. Lung Cell. Mol. Physiol.* **2008**, 294: L358–L367.)

but not fine TiO_2 particles, reflecting greater permeability for these particles and greater particle surface area. Carbonaceous particles interact to depress reactive oxygen substances (ROS) in young rats, but increase them in old rats. Thus, under some conditions, exposure can increase the oxidative stress, pulmonary inflammation, and damage (57a).

Female rats, mice, and hamsters were exposed to ultrafine TiO_2 at 0.5 to 10 mg/m^3 for 6 hours/day, 5 days/week for 13 weeks and allowed to recover for 4 to 52 weeks (14). Degussa particles (P25 TiO_2 particles; 25 nm nominal diameter) were used. Median aerodynamic diameter was 1.44 μm (1440 nm) suggesting that considerable aggregation-agglomeration of the 25 nm particles had taken place. Geometric standard deviation for rat chambers was 2.60, indicating that this aggregation-agglomeration was somewhat variable (~95% size range 200 nm to 10,000 nm; accumulation, intermediate, and coarse particles). Approximately 10% to 15% of the particle mass was 500 nm with a higher probability of becoming available to the interstitial space and being retained long term (16, 17). A larger portion of the particle surface area would be expected to be less than 500 nm in diameter, possibly as much as 60% or more based on comparison with similar but not identical distributions (26, 57b).

Several investigators have discussed that the surface area may be the more appropriate metric (16, 17, 47, 54, 69, 70). A high fraction of the surface area of particles below diameter of 500 nm suggests that a large surface area would likely be available in the pulmonary interstitial space to cause irritation or damage. Since TiO_2 is insoluble, the particles would be expected to persist in that location for some time. Consistent indicators of acute lung injury were seen at the highest exposure levels (LDH, cell death; increased macrophages and neutrophils). At the highest dose level, rats (and presumably lung epithelial cells) were most severely injured, compared to mice and hamsters. They developed epithelial metaplastic and fibroproliferative changes in their lung. Both rats and mice had impaired clearance of particles from lung. These changes duplicate an earlier study; the fibroproliferative changes increased in severity with increasing time after exposure, suggesting persistence of the particles. Clearance was not affected at lower doses (14).

Exposure to ultrafine TiO_2 particles at 130 $\mu g/mL$ was used to assay cytotoxicity (tetrazolium MMT) and population growth, as well as for comet assay (58a). There was an increase in the *in vitro* mutation rate as indicated by the comet assay and hypoxanthine guanosine phosphoribosyl transferase (HPRT) assays of 2.5- to 5-fold relative to controls, suggesting that mutations had occurred from the nanoparticle exposure.

If repair is inadequate, p53 may be involved with tumor induction that mimics the molecular changes of radiation-induced carcinogenesis (41). Investigations into this exciting area are only just beginning, but this work did involve a known carcinogen benz [*A*] pyrene and nanomaterials (41). Other models of tissue injury included bleomycin lung injury, scratch of cells at LEC confluency, repeated mechanical stretch of LEC *in vitro*, and BAP in the presence of nanoparticles; asbestos are proportional to vitamin C yellowing (42–47).

Rats exposed to low solubility, low toxicity particles (LSLTP) have been reported to develop chronic degenerative disease, for example fibrosis and lung cancer (48). The particles are regulated on a mass basis, but this may not be the best metric to describe the response (16, 17, 47, 69, 70). Using an epithelial-like type II cell (A549) they measured proinflammatory markers (IL8 mRNA and protein; glutathione depletion) and found TiO_2 ultrafine cobalt and nickel, as well as carbon black nanoparticles produced much stronger responses than the same mass of conventional fine particles of TiO_2 and carbon black. Glutathione depletion confirmed that oxidative stress was involved. The threshold suggested by these data for surface area of particles relative to exposed cells was 1 to 10 cm^2/cm^2 of lung surface area (54, 55), when nanoparticles are plentiful enough to more than cover lung epithelial cells. Both sets of data suggested that surface area was a more appropriate metric (54), agreeing with the data of others (16, 17, 57). In addition, the mechanism of induction of tumors in rats appears to be a secondary genotoxic mechanism associated with persistent inflammation. Finally, the inflammatory response showed a nonzero threshold (50). The 95% lowest confidence limit (0.7 to 6 mg/m^3) expressed a magnitude of uncertainty similar to the *in vitro* estimates (54, 56). Although lung cancer risk is consistently elevated among workers with occupations where diesel engines have been used, quantification of that elevated risk has not been possible, to date (57).

Risk analysis of nanoparticles for health damage contains many of the same elements as macrosized compounds: description of the material in its specific application, profiling lifestyle (properties, hazards, and exposure), and evaluating risks. Risk management is assessed at this point. In a recent Environmental Defense–DuPont nano partnership, they advocated that deciding, documenting and acting, and especially reviewing and adapting would add value to the risk estimation and management (57d).

Lung epithelial cells (LEC) are damaged; for NFκB driven reactions, PKCδ replaces the PKCε of healthy LEC; PKC βII appears to be required for the replacement. Speculatively, either the amount of PKC βII or the ratio of PKCδ/PKCβII may be important to the release of IL8 and IL10 and the capacity for IL10 release that limits healing to secondary intent (fibroplasia, fibrosis). Future research will determine if this is of importance (58–62).

22.3.5 Summary

Healthy LEC are stimulated to maintain or heal by hyperplasia (39–55). Phosphorylation is stimulated, activating PKCα and PKCε, releasing the cytokines IL6 and IL8, supporting this reaction. The MHC-1 Th1 reaction is shifted to an allergic inflammation MHC-2 Th2. As IL8 increases, IL10 anti-inflammatory reaction increases to balance IL8 release. LEC (lung cells) are maintained or healed by cell division or hyperplasia. Alternatively, damaged or injured LEC become apoptotic and healing is by secondary intent, fibroplasia or fibrosis, depending on severity. Phosphorylation is activated for cell division. In this instance, it is driven by PKCδ and PKCβII. The capacity to increase IL10 can be exceeded, depending on damage or injury (39–61).

Reviewing, it would appear that degree of damage to LEC exposed to nanoparticles determines which path to healing the lung will take (33–55). This is of special importance to animals or humans exposed to ultrafine (nano) particles in the air from transportation vehicles or cigarettes. The optimum response, in healthy LEC, maintaining or healing by hyperplasia is the most common. Alternatively, damaged LEC can undergo apoptosis, necessitating healing by fibroplasia. Failing that, connexin can extend apoptosis of LEC (45) and chronic exposures potentiate it; emphysema can develop. If healing is a little more successful, fibroplasia can lead to fibrosis. Under just the right conditions of injury, mutations can occur. LEC can involve the p53 response and neoplastic change is induced (47), promoted (achieved independence), progressed (become invasive) and a lung tumor develops, as well as metastasizes (31, 32).

In addition in rare instances with sensitive proteins, nanoparticles enhance the rate of protein fibrillation by decreasing lag time for nucleation (4). The nature of the surface can be important as well—lipid bilayers, collagen fibers, polysaccharides, and other liquid surfaces can have a specific and significant effect, promoting amyloid formation or protein self-assembly. Such interactions are important in describing the extrapulmonary, specifically nervous system, responses (4). Since particles have been shown to enter the central nervous system through the olfactory bulb, this is an important potential mechanism of injury to animals and man receiving high or persistent exposures. Chronic degenerative diseases of the central nervous system, *Alzheimer-like diseases*, are important in today's world and increasing in incidence (62).

Repeatedly, we come back to the strongest driver of good pulmonary health, reduction of particle exposure. Respiratory protective masks can reduce lung exposure to their wearers (63, 64). In swine confinement buildings N-95 masks removed >50% of the respirable aerosol (PM4); RPF were ~3–5 fold reductions in potential danger (57, 58). This translated into an improved forced expiratory volume (FEV_1). For example shift change/8 hours work—unexposed improved FEV_1 1.6 ± 0.5%; workers not wearing a mask FEV_1 worsened (−1.0%) over the same shift and on an invervention day when workers wore a mask improved to 0.32 ± 0.62%. A similar trend was noted for the concentration of methacholine needed to produce a 20% fall in FEV_1 (baseline 328 ± 34 mg/L; wearing a mask 242 ± 38 mg/L [*74% unexposed*]; not wearing a mask 130 ± 37 mg/L [40% of unexposed]) (58). Lower concentrations of methacholine causing a fall in FEV_1 indicate more sensitive airways in swine workers than in unexposed humans or humans with an N-95 respirator. Decreased FEV_1 is consistent with more sensitive airways in swine workers not wearing a mask. Respiratory protection of three- to fivefold is consistent with the improved pulmonary function (FEV_1 and methacholine challenge) in workers wearing a mask (57, 58). Minimization of particle exposure by these N-95 respiratory protective masks is preferable to healing by primary intent (hyperplasia). This approach is important to recognizing and assuring continued good lung health (57–59).

22.4 THE EFFECT OF PARTICLE STRUCTURE

22.4.1 Physical Characterization

Phosphorylation of phosphokinase A in lung epithelial cells by calcium response elements (CREB) is important to CREB translation and the development of inflammation after lung exposure to crocidolite asbestos (54, 55). Concentrations used in these studies have formerly been shown to cause *in vitro* apoptosis and compensatory epithelial cell proliferation. It is difficult to determine how *in vitro* concentrations relate to airborne asbestos concentrations. However, generation of reactive oxidant substances (ROS) during the phagocytic burst and frustrated phagocytosis of the long fibers may be responsible for these changes (54, 55). The extent to which each of these processes is involved in persisting injury is not known.

22.4.2 Characterization of Biological and Health Effects

Nanoparticles of titanium dioxide 3 to 10 nm in diameter were found to correlate toxicity, as indicated by cytokines in lung epithelial cells, most closely to the fraction of anatase (as opposed to rutile fractions) in the mixture and the ability of ultraviolet

(UV) light to produce relative luminosity in these compounds (7, 8, 23). Assays for cellular viability and release of lactate dehydrogenase (LDH) indicate cell killing and mitochondrial activity and half-life of congo red, ability to reduce an intravital dye supported these measurements (7, 8, 29). As an example, anatase samples were 100 times faster at degrading organic dyes and two orders of magnitude more acutely cytotoxic than rutile samples of comparable surface areas (7, 8, 29). The quantity of RS (reactive surface) increased when titania was suspended in the culture mixture and increased still further when the titania particles were irradiated with UV light for 20 minutes. This report highlights the important role that *ex vivo* measures of reactive species production can play in developing screens for cytotoxicity and relating the cytotoxicity to inflammation (7, 8, 29). These data suggested that oxidative damage was involved in cytotoxicity of these cells. Unlike aggregated fullerenes, titania particles are capable of disrupting the mitochondrial activity of cells (29). Perhaps titanium hydroxide and/or anatase surfaces in the presence of appropriate donors may oxidize and damage cells (7, 8, 29).

Toxicity of structure was evaluated by comparing alpha quartz toxicity in rats to the toxicity of two ultrafine and one fine TiO_2 particles, intratracheally instilled into rats (38). Ultrafine anatase TiO_2 was less toxic than alpha quartz, but significantly more toxic than ultrafine TiO_2 or fine TiO_2, as indicated by histopathology and markers of inflammation (release of cellular lactate dehydrogenase and alkaline phosphatase) or cell proliferation (cell labeling with bromodeoxyuridine) in lung. Responses were evaluated 24 hours to 3 months postinstillation. Ultrafine anatase TiO_2 was uncharacteristically more toxic than were different mixtures of either fine or ultrafine TiO_2, suggesting that crystalline structure of TiO_2 may play a role in the pathogenesis of particulate lung inflammation (38).

Aqueous extracts (500 μg/mL; high doses) from ambient aerosols in the central Utah valley caused the production of IL6 and IL8 and their respective mRNAs in BEAS 2B lung epithelial cells. The steel mill was shut down for about 13 months, providing an opportunity for this effect to be assessed. During that time aerosol mass decreased to half of that while the mill was operating, and the production of IL6 and IL8 was not significantly above control. During year three both were elevated; concentrations were intermediate during year one. Deferoxamine, chelating iron, and dimethylthiourea scavenging free radicals both reduced the elevation of these cytokines to control levels (66). Thus, aerosols from steel mills contain metals important to the lung response.

IL6 knockout mice and labeling with bromodeoxyuridine were used to establish the importance of IL6 to pulmonary inflammation from environmental oxidant and particulate pollution and its progressive response. Increased IL6 was associated with reduced Clara Cell secretory protein that protects the lung (67). IL6 antibodies may modulate environmental airborne toxicity of particulates (67). The higher potency of ultrafine particles is not verified by the data that these authors reviewed (68). Particles associated with high concentrations of biological components (endotoxins, beta-glucans, and mold spores) are relatively scarce outdoors. PM 10 to 2.5 had higher endotoxin causing the increase in cytokines and proinflammatory reaction. Blockers suggested the cytokine importance. Moreover components from Gram-positive bacteria induced a higher IL8 (cytokine) release. Particle analysis revealed high surface metal concentrations of zinc, copper, vanadium, iron, and nickel, favor cytokine release (62). Crystalline silica has more cytotoxicity than the less persistent amorphous silica, suggesting that persistence, especially that due to a crystalline structure, enhanced toxicity (62). These data demonstrate the importance of particle associated metals, and raise the question of the degree to which release of metals in just the right internal location may enhance toxicity.

22.5 THE EFFECT OF PARTICLE SHAPE

22.5.1 Physical Characterization

Fibers sufficiently thin can deposit in deep lung by aligning themselves parallel to the airflow in airways (69, 70). Fibers sufficiently long will not be cleared from the lung easily. Fibers sufficiently durable, usually those with native crystalline structure, will remain where they locate, usually in the lung or mesothelial space, as a source of chronic injury. Alternatively, noncrystalline structures will dissolve slowly and long nonpersistent fibers will be cleared rapidly. Reviewing these concepts, fibers that are thin enough to deposit, long enough to be retained, and durable enough to resist dissolution and persist as a source of potential injury will maximize the possibility of fiber injury (69, 70).

22.5.2 Characterization of Biological and Health Effects

Nanoparticles in lungs at critical dose levels can initiate the oxidative stress-responsive transcription factors, such as nuclear factor κB (NFκB) and activator protein-1, and then translate the oxidative stress to release proinflammatory proteins such as IL6 and IL8 (64). This may lead to chronic degenerative pulmonary and cardiovascular disease. It is possible but not as yet proven that surface factors may trigger oxidative stress (64).

Carbon nanotubes (CNT) are fiber-like (ratio length of diameter $>3:1$; often $>20-100:1$), but thin (diameter <20 nm). As their actual length exceeds 5000 nm and approaches 20,000 nm we expect them to become more biopersistent and a more efficient trigger of inflammation. Current fiber paradigms state that composition is important if it causes biopersistence. CNT have more surface area per gram than most other fibers. Surface area is a nanoparticle issue. Because of their great surface area, nanotubes have a tendency to clump in *ropes*. Singly dispersed or small aggregates of CNT can move to the distal lung, blood, extrapulmonary locations, and the nervous system (56, 63–66). Inhalation of pathogenic fibers associated with pulmonary fibrosis (asbestosis) can lead to lung cancer or mesothelioma (63, 64).

Very small particles may evade detection and phagocytosis (63, 64). CNT can gain access to lung tissue, lung cells, blood, and the nervous system. Long fibers are doubly dangerous; first they are more likely to be retained, and second, activation of the transcription factor for inflammation (NFκB) is several times higher for longer fibers (63, 64). Their great surface area provides them the ability to activate inflammation at levels that exceed that of crystalline silica or crodidolite asbestos; they form granulomas with elevated efficiency. Two distinct responses were noted. With foci of single-wall carbon nanotubes (SWCNT), foci of granulomatous inflammation and discrete granulomas with hypertrophic epithelial cells developed (33). In areas that did not have these aggregates, diffuse interstitial fibrosis was present (33). Subsequent discussions revealed that both trends increased with increasing time after instillation (64a). However, the mice did not have progressive decrements of physiological respiratory function or greatly increased mortality (64a). The reason for this disparity was not clear.

In the first of two studies by the same laboratory, inhalation exposures to SWCNT were compared to pharyngeal aspirates of these SWCNT (64b, 64c). Inhalation produced a more intense granulomatous response and modestly more alveolar thickening (64b). Because a 4000 nm cut filter was operated in the line, the average mass size was 4200 nm and the average number size was 240 nm. In a study where superior dispersion of SWCNT was performed (DSWCNT) and pharyngeal aspirate was given, there was virtually no granulomatous inflammation, but greater alveolar thickening (64c). In this study, carbon nanotubes in the lung had presumably interacted with surfactant and dispersed, so that the injury and fibrosis were more diffuse. If carbon nanotubes were dispersed, the reduced intensity slowed or stopped lung injury from reaching a critical point. If pulmonary architecture was not changed, it could result in healing the existing injury (33, 36, 64), physiologically, anatomically, and biochemically, in the absence of significant additional exposure. The key to this dose pattern of DSWCNT is that it caused less granulomatous injury, and presumably less injury (64c). The degree to which carbon nanotubes are just too small to sustain lung injury, upon dispersal, or the lung actually heals itself (36), remains to be determined.

22.6 THE POTENTIAL FOR ADVERSE EFFECTS ON THE HEALTH OF INDIVIDUALS, ANIMALS, THE ENVIRONMENT, AND MAJOR ECOSYSTEMS

Nanoparticles represent the newest possible solution to a variety of technical problems; previous attempts at solutions included radioactive materials, heavy metals, and insecticides. Unlike natural materials, solutions to which a system has adapted, manmade materials may considerably perturb the systems with which they interact.

Radiation differs from the others, because it is sequestered to specific locations in the environment, walled off, and tightly controlled, for the most part, especially weapons grade radioactive material. Radioactive materials represented material with known deleterious chronic health effects that underwent much study, although they had numerous benefits to medicine and for power generation (36, 37).

Alternatively, heavy metals, insecticides, and nanoparticles are clearly released into the environment, represented technical solutions, and had the power to do good for the environment. For our purposes, it is best to focus on nanoparticles because they represent technical solutions to many medical and environmental problems. Moreover, nanomaterials have the potential for progressively expanding use. This chapter is evaluating their potential for adverse health effects.

Of course, the question "Are nanoparticles toxic," is like asking the question "Are chemicals toxic?" But, the specific answer is dependent on the specific chemical and the specific dosage or exposure. The same reasoning is true for nanomaterials. Indeed, some nanomaterials would, on first consideration *not* be thought of as toxic. An example is magnesium oxide. This material is not toxic in its bulk form, and is slightly water soluble. We anticipate that it would dissipate as magnesium and hydroxyl or carbonate ions. Furthermore, milk of magnesia is used as therapy in human and animal medicine. Although not especially soluble in distilled water (0.7 mg/100 mL), it appears rapidly soluble in tissue culture media, roughly proportional to its bicarbonate concentration (24b, 25, 26). These data suggested conversion to another magnesium compound, a carbonate; magnesium carbonate trihydrate is stable at room temperature and 40 torr (mmHg) P_{CO_2} and soluble enough to explain the rapid dissolution of magnesium oxide nanoparticles in physiological fluids.

The airborne diameter of MgO nanoparticles was 15,000 nm, similar to an urban aerosol (57b). At that size, less than 5% of the mass and less than 35% of the surface area would enter the deep lung, evade phagocytosis, and be available to the interstitial space. Moreover, that which was available to the lung's interstitial space would not persist (24b, 25), irritate, or injure the lung tissue. Likewise if MgO nanoparticles were spilled in the environment, conventional sodium bicarbonate would facilitate their rapid dissolution into aqueous magnesium salts. In the human body, excess magnesium is used for many biochemical reactions, and is cleared within a short time (57c).

In both of the following studies (Case 1 and Case 2), MgO and companion metal oxide nanoparticles were given at high doses for short times and the subjects watched for clinical signs, weight loss, and other changes that could be observed *in vivo*. Animals were sacrificed at 14 days after exposure as part of an acute toxicity study and their lungs observed for visible pathologic lesions (70a, 70b).

22.6.1 Case 1: Human Studies

Exposure to air polluted with particles less than 2.5 μm (2500 nm) in size was associated with epidemiologically adverse cardiopulmonary health consequences in humans. In this study, six healthy volunteers were exposed to fine and ultrafine but nontoxic magnesium oxide (MgO) particles (70a). Exposure levels were: mean ± standard deviation for exposure was $\sim$4,138 ± 2163 minutes × mg/m^3. Ninety-eight percent of the particles were fine (<2.5 μm; 2500 nm) and 28% were ultrafine (<100 nm). No differences were noted from controls, as indicated by bronchoalveolar lavage (BAL) inflammatory cell concentrations, interleukin concentrations (IL1, IL6, IL8), tumor necrosis factor, pulmonary function, or peripheral blood neutrophil concentrations (70a). Thus, volunteers who inhaled nontoxic MgO at high levels over considerable time periods showed no indications of inflammation in the lung as indicated by BAL washings.

22.6.2 Case 2: Animal Studies

By way of comparison, in animal studies, rats were exposed *in vivo* to levels (1.5 to 2 g/m^3) of nanoactive magnesium oxide, titanium dioxide, or FAST ACT (combined

MgO and TiO_2) that were as high as technically feasible for ~4 hours as part of an initial acute toxicity study, observed for clinical signs daily, and sacrificed at 14 days as part of an acute toxicity study for these metal oxide nanoparticles. These rats had no reductions of body weight between 3 and 14 days after initiation of exposure, and no rats died. The rats had no clinical signs. As part of the protocol, when none died, they were sacrificed at 14 days after the initiation of exposure. They had no macroscopically visible pathologic changes on examination at necropsy at 14 days. Since no rats died, no histopathology was needed or performed. Based on this evidence, manufactured magnesium, titanium, or combined magnesium and titanium nanoparticles had no *in vivo* effects, or macroscopically visible pathology at very high doses for short exposure times with a 2-week period to allow lesions to develop (70b).

Summarizing, with magnesium oxide nanomaterials we have something with great enough surface area and reactivity to be beneficial in environmental remediation (adsorption/destruction of toxic gases and chemicals). Moreover, even if these nanomaterials aggregate-agglomerate they maintain most of the active surface area for cleaning the atmosphere. Thus, we have an economically efficient particle. Only 5% of the mass or 35% of the surface area of the nanomaterials can be inhaled into the deep lung, avoid phagocytosis, and be available to the lung interstitial space. MgO in the presence of excess bicarbonate converts to nesquehonite (magnesium carbonate trihydrate) and rapidly dissolves (18). Thus, our risk benefit analysis of MgO nanomaterials showed that it is an economically useful material with little if any health cost to individuals, the environment, or ecosystems.

On the other hand, recent *in vitro* and *in vivo* work suggests that inhaled carbon nanotubes may escape the lung's defenses, go to the draining lymph nodes and spleen and cause an inappropriately increased IL10 (28a). The effects were somewhat less than exposure to normal wood smoke. They noted, among other findings, that such effects had been related to atopy (28a). Nanoparticles are able to escape the lung and go to heart, liver (71), draining lymph nodes, and spleen (28a). If the lung is bypassed, the predominant response would have been anti-inflammatory, Th2, IL10 in the spleen, which the authors judged as inappropriately high (28a). Cultured BEAS 2B lung epithelial cells reduced prooxidant IL6 and IL8, which the authors judged to be a nonspecific surface response of undefined meaning, possibly specific to cell. If both of these followed nanoparticle exposure to a carbon-based aerosol, we would expect to note a spleen-driven atopy-like increase in IL10 similar to that observed by Mitchell et al. (28a). The extent to which bypass of lung defenses by carbon nanotubes was dependent on particle size [1 to 2 mass median diameter, 300 to 400 nm number median diameter for Mitchell et al. (28a) inhalation; and 4.2 mass median diameter, 240 number median diameter for Shvedova et al. inhalation (64b)] or shape remained an important question.

In an additional study, small carbon nanoparticles (printex 90 and SRM 1680, St. Louis, MO) elicit an IL1 release, which when added to cocultures of cardiomyocytes and cardiofibroblasts at high doses elicit increased secretion of IL6, a prooxidant cytokine (70c), but not to either cell cultured separately, or in the absence of conditioned media. Several caveats should be pointed out. The particles were carbon based and presumably persistent. They were nanoparticles. The atopy-like response to inhaled carbon nanotubes was in the absence of significant retention or reaction in lung.

Speculatively, the cardiac response seems important, and cardiac deaths have been noted *in vivo*, following very high concentrations of carbon black in London smog episodes and in related events. However, this response does require a significant response, from a carbon-based aerosol in the presence of considerable air pollution. The response required co-culture of cardiomyocytes and cardiofibroblasts with conditioned media from what seemed to be an acute lung injury as indicated by IL1 and the ability to block it with anti-IL1. The *in vitro* dose levels appeared to have been high to both lung and heart. In our opinion the requirements for all this to happen make it possible for cardiac toxicity to occur, with high doses of a carbon-based aerosol to lung and high doses of carbon-based nanoparticles escaping to lung. The frequency of occurrence would be expected to correspond closely to the state of health of the cardiopulmonary system and the cardiopulmonary dose.

Titanium does not form a highly soluble chemical compound. However, lung responses to titanium exposures alone have been minimal. Acute, high level exposures to TiO_2 nanoparticles alone, or in combination with MgO for 4 hours with a 14 day observation period, followed by sacrifice and observation of any macroscopic lesions, suggested that no *in vivo* (clinical signs, body weight) or macroscopic pathologic changes had occurred (70b). Although the same physical size and distribution would have roughly the same surface area available to the interstitial space, particles would persist in the interstitial space much longer. They would have the possibility of progressive degenerative lung disease if sufficient dose were deposited and a lung response elicited. TiO_2 nanoparticles of the same size distribution would be expected to be dissolved only slightly as hydroxide and oxychloride (73).

A recent study has shown this to be the case, as indicated by dissolution of more than background, but less than the amount of MgO when incubated at 37°C, with low agitation to simulate the lung's movement, using lung stimulant fluids (tissue culture fluids—Hanks Balanced salt solution, a minimal medium and Dulbecco's Modified Eagle's medium—a modestly enriched medium, both capable of culturing BEAS 2B lung epithelial cells); dissolution was minimal ($<5\%$), independent of time and concentration (74).

Zinc carbonate (0.6 to 5 ppb seawater; 5 to 10 ppb river water) has moderate solubility as a carbonate (21 mg/100 mL) (73). Zinc compounds are not particularly hazardous; above certain limits they may be toxic and irritating. The maximum level allowed in sludge from wastewater is 3 g/kg. The predicted no effect concentration (PNEC) for ecotoxicological effects was 50 μg/L (5 μg/100 mL) dissolved zinc (73). Current values are not a very serious risk. High zinc levels protect against cadmium intoxication and reduce lead absorption. Pockets of historical contamination exist (73).

Thus, it is possible to choose nontoxic, nonpersistent nanomaterials with little projected health effects that serve a technical solution with benefit to society. Such nanomaterials will likely prove beneficial in the overall balance, prove themselves sustainable, and have an acceptable real cost for society. From a different point of view, response is proportional to dose. Nanomaterials less widely available to the environment (bound to lanolin in skin creams, resin coating to permanent press pants, coating walls as self-cleaning paints) will provide less opportunity for exposure, retention, and injury and an increased positive and sustainable balance for society.

Because nanomaterials have large surface areas, the toxicity of chemicals that adsorb to the surfaces of nanoparticles may be important as well. A product that is designed to adsorb organic compounds that might be spilled may have much greater toxicity after it is applied to clean up a spill. While the greater toxicity may be obvious in this case, there may be other applications such as those in indoor air quality where the greater toxicity of the used material may not be immediately obvious.

ACKNOWLEDGMENTS

We acknowledge research supported with the M2 Technologies through contracts with The Marine Corps Systems Command, U.S. Marine Corps, Department of Defense, U.S. Marine Corps, Quantico, VA; and the support of NanoScale Corporation, Manhattan, Kansas.

GLOSSARY

Allied health scientists Assist doctors of medicine or veterinary medicine in surgical or medical interventions.

Alzheimer's disease Common, degenerative, terminal cause of dementia causing loss of short-term memory.

Apoptosis Programmed cell death to minimize tissue damage.

Apoptic kinase Akt or protein kinase B is important to mammalian cellular signaling in cell responses such as inflammation.

Autoimmunity Immunity against self-tissues.

BALF (bronchoalveolar lavage fluid) Lung washings when animal anesthetized and alive (pre-mortem; before death) and post mortem (after death).

BEAS 2B slightly transformed lung epithelial cells.

CCL1 Interleukin 8 receptor A.

CCL3 Macrophage inhibitory protein (MIP) that inhibits the motility of macrophages and favors functions at the next level.

CCL5 Inteleukin 8 receptor B.

Cytokine Secreted signaling proteins and glycoproteins that, like hormones and neurotransmitters, are used extensively in cellular communication.

Cytotoxic Toxic to cells (in this case LEC).

Dopamine A hormone, neurotransmitter that has several receptors.

Epithelial growth factor receptor EGFr is an epithelial cell receptor for signaling that contains tyrosine kinase.

Epithelial lining fluid ELF is the fluid that lines the lung epithelial cells.

ERK 1/2 Extracellularly regulated kinase.

Endotoxin Cell wall of *Escherichia coli* contains this inflammatory promoting substandard reaction.

Fibroproliferative change Change involving proliferation of fibroblasts to heal lung epithelium by fibrosis.

Forced expiratory volume The maximum amount of air that can be expelled in 1 minute.

Gap junction Small intracellular molecule that contacts lung cell cytoplasm.

Inflammation Cellular reaction that allows cells to defend theirselves.

Integrin Cell surface receptors that define cell shape and mobility and regulate the cell cycle.

Interleukin (IL) A group of cytokines (secreted signaling molecules) expressed by white blood cells or their equivalent in inflammation and allergy.

IL6 Interleukin 6 is made by thymocyte (Th) 2 and B lymphocytes, macrophages, epithelium, and endothelium, leading to acute allergenic inflammation.

IL8 Interleukin 8 is made by lymphocytes, macrophages, epithelial and endothelial cells and is important in acute inflammation and chemotaxis (attraction) of neutrophils.

IL10 Interleukin 10 is made by thymocyte 2 lymphocytes, macrophages, epithelial and endothelial cells, leading to inhibition of thymocyte 1 cytokine production [interferon gamma, IFNγ; tumor necrosis factor alpha, TNFβ; IL2, thymocyte 1 (Th1) producing cells] or tolerance to acute inflammation.

Lactate dehydrogenase LDH is an enzyme that metabolizes lactate.

LEC Epithelial cells in the lungs that line the respiratory airways.

Lung alveoli Blind microscopic sacs in lung parenchyma for exchange of the gas in the atmosphere to the plasma of the blood capillaries.

Lung intercellular space The space between the endothelial and epithelial cells. Structure is provided by fibroblasts that elaborate connective tissue.

Metaplasia A change in form, often to a less differentiated form of the same cell type or less frequently to another cell form (fibrous metaplasia). Metaplasia is not synonymous with dysplasia or carcinogenesis (neoplasia).

Neoplasia Uncontrolled abnormal disorganized growth within a tissue. This growth is called a neoplasm or tumor.

Neoplasm (tumor) initiation Believed to follow a multistep, stepwise process with abnormal growth in a disorganized manner. In some cases the cell may undergo mutation.

Neoplasm (tumor) promotion The neoplasm is made to grow at the expense of other tissues.

Neoplasm (tumor) progression Gradually loses tissue identity, becomes a tumor, but remains at the site of initiation.

Neoplasm (tumor) metastasis Leaves the site of initiation, usually spreading through the lymph nodes. Specific lymph nodes can act as an *early warning, sentinel*. However, by that time it is late enough that it has entered the blood stream, but earlier than the tumor cell's attachment to receptors at distant tissue sites. It is possible to create a surgical or medical intervention when spread is still minimized. Once attached, tumor cells continue neoplastic growth in the new location.

Nigro-pallidal area A dopaminergic projection distinct from the nigrostriatial pathway that may have involvement in Parkinson's disease.

Parkinson's disease Degenerative disorder of the central nervous system that impairs the sufferer's motor skills.

PI Phosphatidyl inositol, a phospholipid in membranes.

PI3K Phosphatidyl inositol 3 kinase, an enzyme that phosphorylates phosphatidyl inositol.

PIGF PI glycan anchor biosynthesis, class F is a protein involved in glycosylphosphatidylinositol.

Particle diameter Diameter of the particle.

Hydrodynamic Diameter Diameter in aqueous media (buffer or tissue culture fluid).

Aerodynamic diameter Diameter in air.

Diameter mode The most frequent diameter in the distribution.

Diameter median That diameter where 50% of the particles measured are above and 50% below the diameter.

Mass median diameter (MMD) That diameter where 50% of the particle masses are above and 50% below the diameter.

Volume median diameter (VMD) That diameter where 50% of the particle volume is above and 50% below the diameter.

Surface median diameter (SMD) That diameter where 50% of the surface areas are above and 50% below the diameter.

Count median diameter (CMD) That diameter where 50% of the particle count (number) is both above and below the diameter.

PM Particulate matter, dust.

PMN Polymorphonuclear leukocytes, neutrophils that clean up the cell debris from acute inflammation.

RS Reactive surface (area).

Sigma g (σg) The geometric standard deviation. σg is the arithmetic antilog of the standard deviation of a normal population of logs of aerosol diameters, surface areas, volumes, or masses. Each of these different measures of aerosol size is log-normally distributed. σg is dimensionless. Using antilogs to transform them to arithmetic numbers makes the distribution easier to understand.

To obtain a 68% confidence interval of the log-normally distributed measure (log median diameter $\pm$ log standard deviation; log median diameter $\pm \sigma g$), one calculates the median ± 1 standard deviation. Adding a log means multiplying a number. Subtracting a log means dividing a number. Thus the confidence interval is the median divided by the σg to the median multiplied by σg. For log values -1.69 ± 0.24 this would be $0.02/1.74$ to 0.02×1.74. This would be 0.011 to 0.035 μm (11 to 35 nm; median 20 nm). Thus, 11 to 35 nm would have a 50% chance of containing 68% of the particle sizes.

In a second example, a 95% confidence interval is the log median measure ± 2 standard log deviations. Arithmetically, 2 log standard deviations equal the σg^2. Thus, the 95% confidence interval (CI; log median $\pm 2\ \sigma g$) is the median divided by σg^2 to the median multiplied by σg^2 (log diameter -1.69 ± 0.24; arithmetic 0.02, σg 1.74; 95% CI = median 0.02 μm divided by σg^2 3.01 to median multiplied by the same number; $0.02/3.01$ to 0.02×3.01; 0.007 to 0.06 μm; 7 to 60 nm in size would have a 50% chance of containing 95% of the particles).

Surfactant (lung) Surface active lipoprotein formed by type II alveolar cells with both a hydrophobic (touching cells) and hydrophilic (touching aqueous compounds) region.

Tissue inhibitor of metalloproteinase Matrix (interstitial connective tissue) metalloproteinase (degrades proteins with metal cofactor).

Tracheal explants Finely minced trachea cultured *in vitro* from which biochemicals including procollagen can be released.

TNF Tumor necrosis factor.

TNF alpha (TNFα) Tumor necrosis factor alpha is the most frequently referred to of the TNF and when referred to without additional designation, means TNFα. TNFα is a factor that triggers acute inflammation, fever, and wasting. It is also called the wasting factor.

TNF beta (TNFβ) Also referred to as lymphotoxin.

REFERENCES

1. Jalava, PI; Salonen, RO; Hälinen, AI; Penttinen, P; Pennanen, AS; Sillanpää, M; Sandell, E; Hillamo, R; Hirvonen, MR. *Toxicol. Appl. Pharmacol.* **2006**, 215: 341–353.
2. Jalava, PI; Salonen, RO; Pennanen, AS; Sillanpää, M; Hälinen, AI; Happo, MS; Hillamo, R; Brunekreef, B; Katsouyanni, K; Sunyer, J; Hirvonen, MR. *Inhal. Toxicol.* **2007**, 19: 213–225.
3. Lin, C-C; Chen, S-J; Huang, K-L; Hwang, W-I; Chang-Chien, G-P; Lin, W-Y. *Environ. Sci. Technol.* **2005**, 39: 8113–8122.
4. Linse, S; Cabaleiro-Lago, C; Wei-Feng, X; Lynch, I; Lindman, S; Thulin, E; Radford, SA; Dawson, KA. *Proc. Natl. Acad. Sci. U.S.A.* **2007**, 104 (21): 8691–8696.

4a. Jones, RAL. In Soft Machines: Nanotechnology and Life; **2004**; Oxford; Toronto; CAN 218 p; pp 54–87.

5. Pennanen, AS; Sillanpää, M; Hillamo, R; Quass, U; John, AC; Branis, M; Hůnová, I; Meliefste, K; Janssen, NA; Koskentalo, T; Castaño-Vinyals, G; Bouso, L; Chalbot, MC; Kavouras, IG; Salonen, RO. *Sci. Total Environ.* **2007**, 374: 297–310.

6. Veranth, JM; Reilly, CA; Veranth, MM; Moss, TA; Langelier, CR; Lanza, DL; Yost, GL. *Toxicol. Sci.* **2004**, 82: 88–96.

7. Sayes, CM; Reed, KL; Warheit, DB. *Toxicol. Sci.* **2007**, 97: 163–180.

8. Sayes, CM; Wahi, R; Kurian, PA; Liu, Y; West, JL; Ausman, KD; Warheit, DB; Colvin, VL. *Toxicol. Sci.* **2006**, 92: 174–185.

9. Prospero, JM. *Proc. Natl. Acad. Sci. U.S.A.* **1999**, 96: 3396–3403.

10. Wesselkamper, SC; Chen, LC; Gordon, T. *Toxicol. Sci.* **2001**, 60: 144–151.

11. Hussein, T; Hameri, K; Aalto, P; Asmi, A; Kakko, L; Kumala, M. *Scand. J. Work Environ. Health* **2004**, 29 (Suppl 2): 54–62.

12. Brunner, TJ; Wick, P; Manser, P; Spohn, P; Grass, RN; Bruinink, A; Stark, WJ. *Environ. Sci. Technol.* **2006**, 40: 4374–4381.

13. Bermudez, E; Mangum, JB; Ashgarian, B; Wong, BA; Reverdy, EE; Janszen, DB; Hext, PM; Warheit, DB; Everitt, JI. *Toxicol. Sci.* **2002**, 70: 86–97.

14. Bermudez, E; Mangum, JB; Wong, BA; Ashgarian, B; Hext, PM; Warheit, DB; Everitt, JI. Pulmonary responses of rats, mice and hamsters to subchronic inhalation of ultrafine titanium dioxide particles. *Toxicol. Sci.* **2003**, 70: 86–97.

14a. Oberdorster, G; Finkelstein, JN; Johnston, C; Gelein, R; Cox, C; Baggs, R; Elder, AC. *Resp. Rep. Health Eff. Inst.* **2000**, 96: 5–74.

15. Long, TC; Navid, S; Tilton, RD; Lowry, GV; Veronesi, B. *Environ. Sci. Technol.* **2006**, 40 (14): 4346–4352.

16. Oberdorster, G; Oberdorster, E; Oberdorster, J. *Environ. Health Perspect.* **2005**, 113: 823–839.

17. Oberdorster, G; Oberdorster, E; Oberdorster, J. *Environ. Health Perspect.* **2007**, 115 (6): A290.

18. Churg, A; Gilks, B; Dai, J. *Am. J. Physiol.* **1999**, 277: L975–L982.

19. Gindy, ME; Panagiotopoulos, AZ; Prud'homme, RK. *Langmuir* **2008**, 24: 83–90.

19a. Nassar, NN; Husein, MM. *Langmuir* **2007**, 23: 13093–13103.

20. Stearns, RC; Paulauskis, FD; Godleski, JJ. *Am. J. Respir. Cell. Mol. Biol.* **2001**, 24: 108–115.

21. Lee, S; Perez-Luna, VH. *Langmuir* **2007**, 23: 5097–5099.

22. Kruth, HS. *Curr. Opin. Lipidol.* **2002**, 13: 483–488.

23. Guan, X; Wilson, S; Schlender, KK; Ruch, RJ. *Mol. Carcinogen.* **1996**, 16: 157–164.

24. Weiner, S; Price, PA. *Calcif. Tissue Int.* **1986**, 39: 365–375.

24a. Dhakal, M. Cattle Feedlot Dust: Solubility in Lung Stimulant and Stimulation of Cytokine Release from Lung Epithelial Cells. MS Thesis, **2008** (in final preparation).

24b. Dhakal, M; Pickrell, JA; Castro, SD; Klabunde, KJ; Erickson, LE. Comparative Solubility of Nanoparticles and Bulk Oxides in Lung Simulant Fluids. *Presented at the Society of Toxicology Annual Meeting*, **2007**; Charlotte, NC, March 27 (Abstract).

24c. Bailey, KL; Poole, JA; Mathisen, TA; Von Essen, SG; Romberger, DJ. *Am. J. Physiol. Lung Cell. Mol. Physiol.* **2008**, 294: L1049–1054.

25. Langmuir, D. *J. Geol.* **1965**, 73: 730–754.

26. Pickrell, JA. In *Veterinary Toxicology, Basic and Clinical Principles*, Gupta, RC, ed., 305–311. Academic Press, New York, **2007**.

27. Pickrell, JA. In *Veterinary Toxicology, Basic and Clinical Principles*, Gupta, RC, ed., 177–192. Academic Press, New York, **2007**.

28. Hussain, SM; Javorina, AK; Schrand, AM; Duhart, HM; Ali, SF; Schlager, JJ. *Toxicol. Sci.* **2006**, 92: 456–463.

28a. Mitchell, LA; Gao, J; Vander Wal, R; Gigliotti, A; Burchiel, SW; McDonald, JD. *Toxicol. Sci.* **2008**, 100: 203–214.

28b. Kim, JS; Yoon, TJ; Yu, KN; Kim, BG; Park, JP; Kim, HW; Lee, KH; Park, SB; Lee, J-K; Cho, MH. *Toxicol. Sci.* **2006**, 89: 338–347.

28c. Kwon, J-T; Hwang, S-K; Jin, H; Kim D-S; Tehrani, AM; Yoon, H-J; Choi, M; Yoon, T-J; Han, D-Y; Kang, Y-W; Yoon, B-I; Lee, J-K; Cho, M-H. *J. Occup. Health* **2008**, 50: 1–6.

28d. Li, N; Xia, T; Nel, AE. *Free Radic. Biol. Med.* **2008**, 44: 1689–1699.

28e. Shaw, SY; Westly, EC; Pitet, MJ; Subramanian, A; Schreiber, SL; Weissleder, R. *Proc. Natl. Acad. Sci. U.S.A.* **2008**, 105: 7387–7392.

29. Sayes, CM; Gobin, AM; Ausman, KD; Mendez, J; West, JL; Colvin, VL. *Biomaterials* **2005**, 26: 7487–7595.

30. Happo, MS; Salonen, RO; Hälinen, AI; Jalava, PI; Pennanen, AS; Kosma, VM; Sillanpää, M; Hillamo, R; Brunekreef, B; Katsouyanni, K; Sunyer, J; Hirvonen, MR. *Inhal. Toxicol.* **2007**, 19 (3): 227–246.

31. Witschi, H; Last, JO. In *Casarret and Doull's Toxicology: The Basic Science of Poisons*, 6th edition, Klaassen, CD, ed., 515–534. McGraw-Hill, New York, **2001**.

32. Witschi, H; Pinkerton, KP; van Winkle, LS; Last, JA. Toxic responses of the respiratory system. In *Casarret and Doull's Toxicology: The Basic Science of Poisons*, 7th edition, Klaassen, CD, ed., 609–630. McGraw-Hill, New York, **2008**.

33. Shvedova, AA; Kisin, ER; Mercer, R; Murray, AR; Johnson, VJ; Potapovitch, AI; Tyurina, YY; Gorelick, OO; Arepalli, S; Schwegler-Berry, D; Hubbs, AF; Antonini, J; Evans, DE; Ku, B-K; Ramsay, D; Maynard, A; Kagan, VE; Castronova, V; Baron, P. *Am. J. Physiol. Lung Cell. Mol. Physiol.* **2005**, 289: L698–L708.

33a. Murdock, RC; Brayditch-Stolle, L; Schrand, AM; Schlager, JJ; Hussain, SM. *Toxicol. Sci.* **2008**, 101: 239–253.

34. Gebbink, WA; Sonne, C; Dietz, R; Kirkegaard, M; Born, EW; Muir, DC; Letcher, RJ. *Environ. Sci. Technol.* **2008**, 42: 752–759.

34a. Tornqvist, H. Press release September, 2007 University of Edinburg, Edinburg, Scotland; Dissertation June 2008; Umea University; Umea, Sweden; **2008**; Part Fibre Toxicol 5: 1833 (4019; PSEB).

35. Pickrell, JA; Mauderly, JL. In *Lung Connective Tissue: Location, Metabolism and Response to Injury*, Pickress, JA, ed., 131–157. CRC Press, Boca Raton, FL, **1981**.

36. Pickrell, JA; Diel, JH; Slauson, DO; Halliwell, WH; Mauderly, JL. *Exp. Mol. Pathol.* **1983**, 38 (1): 22–32.

37. Pickrell, JA; Abdel-Mageed, AB. In *Pulmonary Fibrosis*, Phan, SH; Thrall, RS, eds., 363–381. *Lung Biology in Health and Disease*, Vol. 80. Marcel Dekker, New York, **1995**.

38. Wallenborn, JG; McGee, JK; Schladweiler, MC; Ledbetter, AD; Kodavanti, UP. *Toxicol. Sci.* **2007**, 98: 231–239.

39. Unfried, K; Sydlik, U; Weissenberg, A; Abel, J. *Am. J. Physiol. Lung Cell. Mol. Physiol.* **2008**, 294: L358–L367.

40. Wu, Q; Martin, RJ; LaFasto, S; Efaw, BJ; Rino, JG; Harbeck, RJ; Chu, HW. *Am. J. Respir. Crit. Care Med.* **2008**, 177: 720–729.

41. Sun, Y; Wu, F; Sun, F; Huang, P. *J. Immunol.* **2008**, 180: 4173–4181.

42. Wyatt, TA; Slager, RE; DeVasure, J; Auvermann, BW; Mulhern, ML; Von Essen, S; Mathiesen, T; Floreant, AA; Romberger, DJ. *Am. J. Physiol. Lung Cell. Mol. Physiol.* **2007**, 293: L1163–L1170.

43. Sydlik, U; Bierhals, K; Soufi, M; Abel, J; Schins, RP; Unfried, K. *Am. J. Physiol. Lung Cell. Mol. Physiol.* **2006**, 291: L725–L733.

44. Lee, H-S; Wang, Y; Maciejewski, BS; Esho, K; Sharma, S; Sanchez-Esteban, J. *Am. J. Physiol. Lung Cell. Mol. Physiol.* **2007**, 294: L225–L232.

45. Scheffold, A; Murphy, KM; Hofer, T. *Nature Immunol.* **2007**, 8: 1285–1287.

46. Lau, AWT; Biester, S; Cornall, RJ; Forrester, JV. *J. Immunol.* **2008**, 180: 3889–3899.

47. Mroz, RM; Schins, RPF; Jimenez, LA; Drost, EM; Holownia, A; MacNee, W; Donaldson, K. *Eur. Respir. J.* **2008**, 31: 241–251.

48. Monteiller, C; Tran, L; Macnee, W; Faux, S; Jones, A; Miller, B; Donaldson, K. *Occup. Environ. Med.* **2007**, 64: 607–615.

49. Moss, OR; Wong, VA. *Inhal. Toxicol.* **2006**, 18: 711–716.

50. Dankovic, D; Kuempel, E; Wheeler, M. *Inhal. Toxicol.* **2007**, 205–212.

51. Wichmann, HE. *Inhal. Toxicol.* **2007**, 19 (suppl 1): 241–244.

52. Barlow, CA; Barrett, TF; Shukula, A; Mossman, B; Lounsbury, KM. *Am. J. Physiol. Lung Cell. Mol. Physiol.* **2007**, 292: L1361–L1369.

53. Barlow, CA; Kitiphongspattana, K; Siddiqui, N; Roe, MW; Mossman, BT; Lounsbury, KM. *Apoptosis* **2008** (e-publ ahead of print).

54. Amara, N; Bachoual, R; Golda, S; Buichard, C; Lanone, S; Aubier, M; Ogier-Denis, E; Bocakowski, J. *Am. J. Physiol. Lung Cell. Mol. Physiol.* **2007**, 293: L170–L181.

55. Koval, M. *Am. J. Physiol. Lung Cell. Mol. Physiol.* **2002**, 283: L875–L893.

56. Chen, H-W; Su, S-H; Chien, C-T; Lin, W-H, Yu, S-L; Chou, C-C; Chen, JJW; Yang, P-C. *FASEB J* **2006**, E1732–E1741.

57. Zhu, MT; Feng, WY; Wang, TC; Gu, YQ; Wang, M; Wang, Y; Ouyang, H; Zhao, YL; Chai, ZF. *Toxicology* **2008** (e-pub ahead of print).

57a. Elder, AC; Gelein, R; Finkelstein, JN; Cox, C; Oberdorster, G. *Inhal. Toxicol.* **2000**, 12 (suppl 4): 227–246.

57b. Pun, BK; Wu, SY; Seigneur, C; Seinfeld, JH; Griffin, RJ; Pandis, SN. *Environ. Sci. Technol.* **2003**, 37: 3647–3661; http://cloudbase.phy.umist.ac.uk/people/dorsey/Aero.htm introduction to aerosol; Urban Aerosol.

57c. Batemen, S. In *Fluid, Electrolyte, and Acid-Base Disorders*, DiBartola, SP, ed., 210–226. Saunders-Elsevier, St. Louis, MO, **2006**.

57d. Environmental Defense–Dupont Nano Partnership. *NANO Risk Framework* Dupont, Wilmington, DE, **2007**.

58. Zhao, Y; He, D; Stern, R; Usatyuk, PV; Spannhake, EW; Salgia, R; Natarajan, V. *Cell. Signaling* **2007**, 19: 2329–2338.

58a. Wang, JJ; Sanderson, BJ; Wang, H. *Mutn. Res.* **2007**, 628: 99–106.

59. Page, K; Li, J; Zhou, L; Iasvoyskaia, S; Corbit, KC; Soh, J-W; Weinstein, IB; Brasier, AR; Lin, A; Hershenson, MB. *J. Immunol.* **2003**, 170: 5681–5689.

60. Park, J-W; Kim, HP; Lee, S-J; Wang, X; Ifedigbo, E; Watkins, SC; Ohba, M; Ryter, SF; Vyas, YM; Choi, AMK. **2008**, 180: 4668–4678.

61. Contreras, X; Bennasser, Y; Bahroaoui, E. *Microbes Infection* **2004**, 6: 1182–1190.

62. Oberdorster, G; Sharp, Z; Elder, AP; Gelein, R; Kreyling, W; Cox, C. *Inhal. Toxicol.* **2004**, 16: 437–445.

63. Pickrell, JA; Heber, A; Murphy, JP; Henry, SC; May, M; Nolan, D; Gearhart, S; Cederberg, B; Oehme, FW; Schoneweis, D. *Vet. Hum. Toxicol.* **1995**, 37: 430–435.

64. Dosman, JA; Senthisilvan, A; Kirychuk, SP; Lemay, S; Barber, EM; Wilson, P; Cormier, Y; Hurst, TS. *Occup. Environ. Lung Disease* **2000**, 118: 852–860.

64a. Castronova, V. Research Communication on mechanisms of action of inhaled fibers, particles, and nanoparticles in lung and cardiovascular disease. Co EPA Conference Center, Research Triangle Park, October 26–28, **2005**.

64b. Shvedova, AA; Kisin, E; Murray, AR; Johnson, VJ; Gorelik, O; Arepalli, S; Hubbs, AF; Mercer, RR; Keohavong, P; Sussman, N; Jin, J; Yin, J; Stone, S; Chen, BT; Deye, G; Maynard, A; Castronova, V; Baron, P; Kagan, VE. *Am. J. Physiol. Lung Cell. Mol. Physiol.* **2008**, 295: 552–565.

64c. Mercer, RR; Scabilloni, J; Wang, L; Kisin, E; Murray, AR; Schwegler-Berry, D; Shvedova, AA; Castranova, V. *Am. J. Physiol. Lung Cell. Mol. Physiol.* **2008**, 294: 87–97.

65. Pickrell, JA; Heber, A; Murphy, JP; Henry, SC; May, M; Nolan, D; Oehme, FW; Gillespie, JR; Schoneweis, D. *Vet. Hum. Toxicol.* **1993**, 35: 421–428.

66. Frampton, MW; Ghio, AJ; Samet, JM; Carson, JL; Carter, JD; Devlin, RB. *Am. J. Physiol. Lung Cell. Mol. Physiol.* **1999**, 277: 960–967.

67. Yu, M; Zheng, X; Witschi, H; Pinkerton, K. *Toxicol. Sci.* **2002**, 68: 488–497.

68. Schwarz, PE; Overik, J; Lag, M; Refnes, M; Nafstad, P; Hetland, RB; Dybing, E. *Hum. Exp. Toxicol.* **2006**; 25: 559–579.

69. Donaldson, K; Aitken, R; Tran, L; Stone, V; Duffin, R; Forrest, G; Alexander, A. *Toxicol. Sci.* **2006**, 92: 5–22.

70. Donaldson, K. *Yonsei Med. J.* **2007**, 48 (4): 561–572. Review.

70a. Kuschner, WG; Wong, W; D'Alessandro, A; Quinlan, P; Blanc, PD. *Environ. Health Perspect.* **1997**, 105: 1234–1237.

70b. Koper, OB; Bergmann, J; Klabunde, KJ; Wilson, CM; Pickrell, JA. Research Communication, Acute Inhalation of Nanoactive® Metal Oxides in rats. Kansas State University and NanoScale Corporation, Manhattan, KS, (September 18, 2008) (Male and female Sprague-Dawley rats were studied according to the USACHPPM Toxicology Directorate, TOS SOP No. 29.02. Acute Inhalation Study 37, based on the guidelines from the Environmental Protection Agency Office of Prevention, Pesticides and Toxic Substances Health Effects Test Guidelines OP PTS 870.1300, Acute Inhalation Toxicity.)

71. Nemmar, A; Vanbilloen, H; Hoylaerts, MF; Hoet, PH; Verbruggen, A; Nemery, B. *Am. J. Respir. Crit. Care Med.* **2001**, 164: 1665–1668.

72. Nemmar, A; Hoet, PH; Vanquickenborne, B; Dinsdale, D; Thomeer, M; Hoylaerts, MF; Vanbiblloen, H; Mortelmans, L; Nemery, B. *Circulation* **2002**, 105: 411–414.

73. Raiswell, RW; et al. Environmental Chemistry: The Earth-Air-Water Factory, Edward Arnold, London, 1980. http://www.lenntech.com/elements-and-water/zinc-and-water.htm LynnTech: the chemical elements and water.
74. Pickrell, JA; Dhakal, K; van der Merwe, D; Koper, OB; Winter, M; Erickson, L; Klabunde, KJ. Research Communication, Kansas State University and NanoScale Corporation, Manhattan, KS, October 8, 2008.

PROBLEMS

1. Physical size is a parameter that helps predict particle deposition in the deep lung. Differentiate the accumulation fraction of fine particles from ultrafine particles; describe where in the lung they are most likely to deposit.
2. Using the language of risk, define REL and PEL and describe their use in analyzing the risks of inhaling particulates.
3. Define hydrodynamic particle size and particle size in air. How are the differences in definition important to predicting risk to lung?
4. Nanoparticles can aggregate and subsequently agglomerate in both air and aqueous media. Can such particles disaggregate? If they disaggregate, what is the mechanism of their disaggregation?
5. What relevance does the state of health of lung epithelium have for its healing from carbonaceous nanoparticle exposure and injury?
6. Does nanomaterial exposure always result in tissue response?
7. In normal lung epithelial cells, what is the relationship of macrophages or fibroblasts to epithelial cells?
8. What mass or surface area fraction does the targeted pulmonary interstitial space see for long-term exposure? Which is the best predictor of lung toxicity?
9. Which is most important for toxicity, the high surface area or the inherent toxicity of nanoparticles?
10. Speculatively, how might nanoparticle disaggregation affect lung healing?

ANSWERS

1. Ultrafine particles are particles with real diameters less than or equal to 100 nm (0.1 μm) in diameter. These particles are randomly buffeted by gas molecules and deposit by diffusion in the alveoli. Accumulation fractions are the first sizes beyond ultrafine particles (100 to 500 nm), are deposited mostly by diffusion, and have only minimal settling velocity. Both particles with real physical size are small enough to go into the deep lung to deposit in the alveoli.

2. Recommended exposure level (REL) and permissible exposure levels (PEL) are levels below which it is uncommon to see health effects. REL are only recommended, while PEL can be strictly enforced.

3. Hydrodynamic particle size is particle size in aqueous fluids such as water, buffers, serum, or epithelial lining fluid (a thin layer of buffer and soluble protein which bathes the lung alveolar epithelial cells in fluid). Typically, hydrodynamic particle sizes are less than 1 μm in diameter. Particles that are sufficiently small will be ignored by phagocytes and go to the interstitial lung space. They will have potential for chronic lung injury for the length of time they persist. Their risk to lung is proportional to the time of their persistence.

   Particle size in air is the diameter typical of the aggregate-agglomerate volume. Because nanoparticles have a large and active surface area, airborne diameters can typically be quite large. Large aggregates-agglomerates are typically excluded from the deep lung (respiratory parenchyma) unless they disaggregate. If airborne particles retain their airborne diameter, they constitute minimal risk to lung unless they can penetrate into the deep lung.

4. Phagocytes and most epithelial cells internalize particles $\geq$500 nm more efficiently than they do those less than 500 nm. When cultured lung epithelial cells are given ultrafine TiO_2, they do not take up most of the ultrafine TiO_2. However, ultrafine particles that were taken up remained as aggregates through 24 hours after exposure, even at lower doses. In rare instances, two or three particles were seen within a single membrane-bound vesicle, indicating that disaggregation of particle clumps was possible, but rare in 24 hours. One would expect that disaggregation would continue with increasing time.

   The processes by which ultrafine particles disaggregate are poorly understood. Gold nanoparticles linked to carboxylated dextran chains have more freedom of movement and less aggregation when they are in a hydrated environment (27). Licorice root saponin caused mediated connexin 43 dephosphorylation and the disassembly of gap-junction cellular plaques of communicating epithelial cells.

   Hypochlorite-oxidized bone with crystals intimately associated with collagen can be disaggretated by sonication into crystalline aggregates. Bone with crystals compressed between collagen fibrils did not disaggregate, but remained as "fused" aggregates. These data demonstrate that disaggregation is less difficult when aggregates are less fused and more internal surface area remains. Action of Nanoparticles as a surfactant may facilitate this process.

5. Healthy lung epithelial cells (LEC) are stimulated to maintain or heal by hyperplasia (39–55). Phosphorylation is stimulated, activating protein kinase C (PKC) α and PKCε, releasing the cytokines interleukin (IL) 6 and IL8, supporting this reaction. The major histocompatibility complex (MHC) -1 Th1 inflammatory reaction is shifted to an allergic inflammation MHC-2 Th2. As IL8 increases, IL10 anti-inflammatory reaction increases to balance IL8 release. LEC (lung cells) are maintained or healed by cell division or hyperplasia. Alternatively, damaged or injured LEC become apoptotic (undergo programmed cell death) and healing is

by secondary intent, fibroblast proliferation, or fibrosis, depending on severity. Phosphorylation is activated for cell division. In this instance, it is driven by PKCδ and PKCβII. The capacity to increase IL10 can be exceeded, depending on damage or injury. When IL10 can no longer balance progressively increasing IL8, damage and risk to lung cells increase rapidly (39–55).

6. Response is to particles via a cooperative interaction with epithelial growth factor receptor and integrins. This response will maintain or heal, depending on extent of lung epithelial cell injury if the cell is healthy. In maintaining, no lung injury is sustained. In healing only minor injury is sustained that is healed by proliferation of lung epithelial cells. There are no known consequences of epithelial hyperplasia to respiratory function or gas exchange.

   If the cell is not healthy, it will undergo programmed cell death, which may become widespread so that clinical signs of emphysema are seen. If the cell is not healthy and is initiated, a tumor may slowly or rapidly develop.

7. Normal lung epithelial cells can heal by hyperplasia without assistance from macrophages or fibroblasts. Macrophages may provide a healing protease (degrades protein) which will facilitate hyperplasia in a more friendly environment. Fibroblasts and their potential to heal by secondary intent are not needed, because healing is by epithelial hyperplasia. Fibroblasts can provide auxiliary growth factors to epithelial cells which help their well being.

8. Those particles below 500 nm in real diameter will be phagocytized with minimal efficiency and tend to translocate to the pulmonary interstitial space. We will examine the mass fraction and the surface area fraction of the aerosol translocated to the pulmonary interstitial space. We expect that the fraction of the total surface area will be the most important metric to examine (16, 17, 47, 54, 69, 70). Let us calculate the fraction of the mass and the surface area in particles <500 nm diameter.

   The median aerodynamic diameters for rat TiO_2 aerosol exposures (14) were 1.44 μm (1440 nm) suggesting that considerable aggregation-agglomeration of the 25 nm particles had taken place. Geometric standard deviations for rat chambers was 2.60, indicating that this aggregation-agglomeration was somewhat variable. The 68% confidence interval size range (median ±1 standard deviation) was ~550 to 3700 nm containing accumulation, intermediate, and coarse particles. The 95% confidence interval size range (median +2 standard deviations) was ~200 to 10,000 nm, containing accumulation, intermediate, and coarse particles). The lower end of the 68% confidence interval (median –1 standard deviation), 550 nm, would be 17% of particle mass. The lower end of the 95% confidence interval, 200 nm, would be 2.5% of the particle mass. 500 nm particles lie below the lower end of the 68% confidence interval (550 nm), above the lower end of the 95% confidence interval (200 nm), but nearer 550 nm. Thus, 10% to 15% of the particle mass was <500 nm.

   However, surface area is a better metric to predict toxicity to the lung interstitial space (16, 17). Bermudez et al. (14) did not publish a distribution of surface area, because this is not routinely measured in inhalation exposures. However, we can approximate it within limits of uncertainty we put on the fraction of mass, if we compare information on distributions of surface area as an urban aerosol available on the Internet (mass median diameter ~8000 to 9000 nm; larger than our aerosol; surface area median diameter 150 to 300 nm; Reference 57b). In this instance <500 nm would contain $\geq 65\%$ of the surface area available to lung interstitial space.

A second comparison can be made if one compares to a still larger aerosol, with a mass median diameter of 20,000nm and a surface area median diameter of 200 to 400 nm (described in Reference 26). In this case, <500 nm would also contain $\geq 65\%$ of the surface area available to the lung interstitial space. Since the aerosol in this exposure is only 1440 nm, an even higher percentage of the surface area should be less than 500 nm. Thus, most of the surface area would likely locate in the pulmonary interstitial space. Since TiO_2 is insoluble, the particles would be expected to persist in that location for some time, and serve as a possible source of irritation, inflammation, and injury.

9. Toxicity is a function of the inherent toxicity of the chemical (copper salts are more irritating than magnesium salts), the surface activity (nanoparticles are worse than conventionally sized particles), and mass-focus-persistence (smaller active crystals tend to have less persistence and less focus than larger particles. In the presence of some healing, small dispersed active nanocrystals may induce minimal injury in any one spot). The answer to this question depends on how the nanocrystals interact with lung epithelial cells and future results will more adequately define the response.

10. It is possible that nanotubes in the lung are dispersed, so that injury and fibrosis are not as focal. Nanotubes are aggregated by a variety of forces, some of which resemble static electricity. If carbon nanotubes were dispersed, the reduced intensity would slow or stop lung injury from reaching a critical point, the point at which that injury could progressively change pulmonary architecture. If pulmonary architecture was not changed, it might result in healing the existing injury, in the absence of additional exposure (33, 36).

INDEX

Adverse effects on health, 754
  animal studies, 755
  human studies, 755
Aerodynamic diameter, 730, 734
Aerogels, 209
  applications
    batteries, 225
    catalysis, 223–225
    cerenkov radiation detection, 218–219
    magnetic separation, 225–226
    space dust separation, 218–219
    thermal insulation, 219–220
  compositions
    Al_2O_3, 223
    Au-TiO_2, 224–225
    carbon, 227–228
    metal sulfides/selenides, 229–233
    $NiFe_2O_4$, 225–226
    RuO_2-SiO_2, 221
    SiO_2, 216–220
    TiO_2, 224
    V_2O_5, 225
  formation, 212
  organic–inorganic, 221
  properties
    elastic, 221–222
    electronic conductivity, 221, 225, 227–229
    heavy ion sorption, 232–233
    hydrophobicity, 217, 220
    optical, 210, 216, 219–220, 231
    porosity/surface area, 210–212, 217, 219, 221, 226–230, 232
    sensing, 222
    thermal insulation, 219–220
  synthesis
    rapid supercritical extraction (RSCE), 214, 217, 223
    supercritical drying, 214–223
    surfacc modification, 214–217
  types
    biological composites, 221–222
    carbon, 227–228
    chalcogenide, 229–233
    mixed metal oxide, 220–221, 223
    organic–inorganic hybrids, 221–222
    oxide, 216–226

Nanoscale Materials in Chemistry, Second Edition. Edited by K. J. Klabunde and R. M. Richards

Aggregated particles, 736
Airborne particles, 741
Air purification, 668, 669
Alloys, 157
Alumina, 272, 303
Amphiphile, 262
Amyloid-β-Derived Diffusible Ligand (ADDL), 423
Anisotropic structure, 83
Antisense nanoparticles, 430
Antisense therapy, 429
Aptamer(s), 427
Asymmetric DNA modification, 412
Asymmetrically modified gold nanoparticles, 411
  antibiofilm activity, 98
  antibiotic, 80
  antimicrobial, 95
  bacterial biofilm, 80
  bacterial colonization, 96

Batteries, 520
BET, 251, 280
Biobarcode, 423
  assay, 420–424
Biofilm interaction with quantum dots, 705
Biofilms, 80
Biological health effects, 741
  hydrodynamic size, 734
  particle shape, 753
  particle structure, 751
  solubility, 738, 756
Biometric synthesis, 174
Biosensors, 477
Barrett, Joyner, Halenda (BJH Model), 251
Block copolymer, 264, 285, 377, 392
Body-centered cubic (BCC) unit cell, 411
Burst nucleation, 129–130

Cages, 262, 294
Calcium oxide nanopowder, 499, 505
Capacitors, 520
Capping, liquid, 44
Carbenium ion, 341
Carbon, 267, 306
Carbon aerogels, 227
Carbon dioxide reduction, 611
Carbonium ion, 341
Carbon molecular sieves/Korean, 306
Carbon nanotubes, 159, 161–163, 197, 381, 443
Catalytic growth method, 170
CdSe nanocrystals, 142
Charge trapping, 24
  centers, 25
Chemical bath deposition, 181, 185–186
Chemical effects, 20, 24
Chemical potential, 138
Chemical vapor deposition, 161, 181, 197
Chromium oxide photocatalysts, 614
Cluster, 16
Cluster compounds, 41
Colloidal, 47
Colloidal crystalline arrays, 394
Colloidal crystallization, 410
Colorimetric DNA detection, 416
Cooperative binding model, 408–409, 430, 438
Commercial application of carbon nanotubes, 469
  conductive polymers, 470
  high-strength composites, 470
  nanotube transistors, 475
  optoelectronic devices, 473
  sensors and activators, 474
  transparent electrodes, 471
Computational methods, 19
Cooperative self-assembly, 265, 285
Core-shell, 184
  nanorods, 181, 183–186, 202, 204
Co-surfactant, 296
Critical micelle concentration (CMC), 262, 295
Cysteine, 427

Delocalized electrons, 4
Dendrimers, 376
Destructive adsorption, 501
Destructive reaction, 502
Diatom, 413
Diffusion controlled growth, 133, 140
Digestive ripening, 40, 132
Dimethyl amine, 345
Dipolar forces, 53
Dissolution, 134
Dithiothreitol, 420

DNA, 406, 562, 563
barcodes, 423
detection, 416
melting, 408
modified gold nanoparticles, 405–407
assembly, 410–412
colloidal crystallization, 410
diagnostic applications, 415
metal ion detection, 424
silver nanoparticles, 431
tile, 413
DNAzyme, 425
Double walled carbon nanotubes, 462
Drinking water remediation, 659
arsenic, 662
bacteria, 659
organic compounds, 660
Drude model, 547
Dust, 733, 744

EISA, 286, 309
Electrical DNA detection, 419
Electrochemical method, 165–166
Electrochemistry, 571
Electrodeposition, 198
Electron diffraction, 255
Electron effects, 20, 24
Electron gas, 557
Electron–hole pair, 582
Electron microscopy, 255
Electrophoresis, 160
Electrospinning, 188–189
Embedded atom method, 27
Empirical potentials, 19, 23
Emulsion polymerization, 378
Energy storage, 519
Enhanced binding strength, 409
Environmental remediation, 582, 606, 637, 649
Epithelial growth factor, 746
Escherichia, 80
Evaporation-induced self assembly, 286, 309
Excluded volume, 50

Face-centered cubic (FCC) unit cell, 411
Fermi level, 27
Fibers, 310
Fluidized catalytic cracking (FCC), 340, 342
Förster energy transfer, 560
FSM-16, 284
Fullerenes, 444
Full width at half maximum (FWHM), 408
Functionalization, 187, 203–205, 296
Fungus/fungi, 415
Furfuryl alcohol, 308

Galvanic exchange, 183, 185
Gas physisorption, 249
Geometrical effects, 20, 24
Glossary, 758
Gold, 26, 29–31, 43, 57, 373
Gold cluster molecules, 43
Gold nanoparticles, 26, 406–434
Gold nanospheres, 545
Gold/silver alloy, 546
Graphene, 444, 464
Green chemical synthesis, 173
G-Quadruplex, 426
Groundwater remediation, 652

Hard templating, 258, 267, 288, 305, 306
Health impacts, 681, 729, 690
Heat-up method, 145
Heteroelement, 300
Heterogeneous catalysis, 29
Heterogeneous nucleation, 129
Hierarchal nanostructures, 189
High volume cascade impactor, 733
Homogeneous nucleation, 129
Host–guest interactions, 248, 301
Hot injection method, 142
Hydrid, 297
Hybrid interface, 264, 291
Hydrocracking, 343
Hydrodynamic particle size, 734
Hydrothermal, 166
Hydrothermal synthesis, 166, 173, 179, 262
Hydroxyl radial, 583
Hysteresis, 250–251, 293

Inductively coupled plasma mass spectroscopy (ICP-MS), 424
Inhalation of noxious gases, 730

Inhaled nanomaterials, 729
  crustal, desert, mountain dust, 744
  manure dust, 744
  metal oxide dust, 745
  urban dust, 741
Inorganic–organic, 369
Inorganic polymerization, 260
Instrumentation, 8
Insulators, 4
Interleukins, 747–749
Interparticle interactions, 46
Ion exchange, 348
Ionic liquid, 76
  anion, 77
  application, 77–78
  capping, 78
  cation, 77
  ionic-conductive, 76
  polarizability, 76
  properties, 77
  structure directing template, 79
  template, 79
IRMOF, 277
Iron fluoride, 94
Iron-oleate complex, 146
Iron oxide nanocrystals, 128, 145
Iron phosphate, 666
Isotherms, 251, 280, 295

Kinetics of growth, 127–128, 140, 145
KIT-6, 292–293, 305

LaMer model, 149–150
Langmuir–Blodgett films, 387, 390
Layer-by-layer assembly, 163, 178
Lennard–Jones potential, 19
Life cycle/nanomaterials, 686
Ligand–ligand interactions, 53
Light absorption, 585
Liquid crystals, 163, 183
Lithium batteries, 520, 524
  CoO, CuO, NiO, 529
  life PO_4, 528
Localized electrons, 5
Locked nucleic acid (LNA), 431
London forces, 50
Luminescence, 552, 569
Lung effects, 742
Lung epithelial cells, 747
Lung surface area, 742

Macroporous, 247
Madelung potential, 25
Magnesium fluoride, 91, 95, 98
Magnesium oxide, 21, 498, 499
Magnesium oxide particles, 738
Manure dust, 739, 744, 745
MAS-NMR, 256
MCM-41, 254, 283, 296
MCM-48, 254, 289, 311
MCM materials, 211, 252, 283, 286
Melting temperature (T_m), 408–410, 416, 424–425, 428
Melting transition, 407–410, 415, 416, 425, 433
Mesocellular foam (MCF), 296
Mesophase, 284, 287, 290
Mesoporous, 243, 283, 288, 296, 302, 306, 331
Mesoporous materials, 246, 283
Mesoporous materials in photocatalysis, 643
Mesoporous nanorod, 204
Mesostructure, 284, 306
Metal carbides, 113, 116, 120, 158
Metal fluorides, 74
Metal ion dopants/photocatalysts, 636
Metal nitrides, 111, 116, 120, 158
Metal organic frameworks (MOF), 5, 274, 279
Metal oxides, 157, 162, 167, 172, 190
Metal oxide semiconductors, 635
Metals, 4, 381
Metal selenides, 157, 167
Metastable, 380
Metal sulfides, 157, 163, 167
Methanol, 343
Methyl viologen oxidation, 587, 589–590, 592–593
MgO nanoparticles, 21
Micelles, 165, 205, 262
Microporous, 243, 269, 331
Microporous materials, 246, 269, 280
Microwave, 75
  dielectric heating, 76
  electromagnetic, 75–76
  electric field, 75
Microwave synthesis, 177

Mineralization, 498
  carbon tetrachloride, 501
  dichloromethane, 501, 506
  organic halides, 501, 504
  tetrachloroethene, 501
  trichloroethene, 501
Molecular components, 374
Molecular sieves, 211, 283
Molecular switch, 388
Molybdenum oxide photocatalysts, 614
Monodispersity, 128
Monolayers, 387
Monomer, 134
Monte Carlo, 19
Morphology, 309
Multiwalled carbon nanotubes, 463
Mustard gas mimic, 596

Nanobars, 83
Nanocasting, 258, 267, 288, 303, 308
Nanocomposites, 157, 188
Nanocrystals, 16, 266
Nanohorns, 465
Nanomaterials/environment, 684
Nanomaterials in energy, 519
  advantages, 525
    disadvantages, 525
  aerogels, 525
  batteries, 522
  capacitors, 523
  electrochemical properties, 524
  macroporous solids, 525
  mesoporous solids, 525
  nanoparticles, 528
  nanorolls, 530
  nanotubes, 530
  nanowires, 530
Nanoneedles, 84
Nano-ovals, 90
Nanoparticle, 16, 300
Nanoparticle interactions, 50
Nanoparticle molecules, 38
Nanopores, 244
Nanoporous, 245, 248
Nanorods, 155, 157, 173, 187, 188
Nanosheets, 173
Nanostructured porous solids, 257
  consolidation processes, 260
  hydrothermal, 262
  inorganic polymerization, 260
  self-assembled, 262
Nanotechnology products, 687
  catalysts, 687
  clothing, 687
  electronics, 687
  paints/coatings, 687
  sporting goods, 687
  sun screen, 687
Nanotube synthesis methods, 454
  arc discharge, 454
  catalytic, 457, 459–460
  laser ablation, 458
Nanotubes, 173
National Nanotechnology Initiative, 3
Nitric oxide, 610, 624
Nitrogen oxides, 669
Non-hydrolytic sol-gel process, 172
Nonsiliceous oxides, 302, 304
Nuclear magnetic resonance (NMR), 256
Nucleation, 49, 129, 134, 150
Nucleation rate equation, 134

Opals, 395, 397
Optical extinction, 544
Optical properties, 31
Organic molecule detection, 427
Ostwald ripening, 20, 140, 141

Particles as molecules, 37
Particle size, 731–732, 742
Patterned surfaces, 391
Periodic mesoporous organosilicas, 299
Perovskites, 269
Phase transition model, 408, 409
Photocatalytic oxidation, 581
Photocatalytic remediation, 629, 637
Photochemical synthesis, 168–170
Photofunctional, 605
  heterogeneous catalysts, 621
  mesoporous materials, 614, 617, 643
  oxide matrices, 607
  titanium oxide, 607
  zeolites, 609, 611
Photo-mineralization of pollutants, 637
Photonic efficiency, 594
Photooxidation, 596
Physisorption, 249, 280, 283
Plasmon coupling, 564

Plasmons, 31, 543
Polymer colloids, 378
Polymer composites, 394
Polymers, 375
Pore networks, 289, 293, 305
Pore size, 249
Pore volume, 248
Porogen, 244, 248
Porosity, 244, 248
Porous materials, 211
Porous solids, 243–244, 247, 249, 258
Post-synthetic grafting, 298
Potential for adverse effects, 754–758
Powder X-ray diffraction, 253
Precipitation, 134
Precursor method, 177
Prostate specific antigen (PSA), 423
Protein detection, 420
Pseudocapacitors, 521
Pseudomonas aeruginosa, 80

Quantum dots, 551, 565
 absorption cross section, 567
 core-shell, 567
 electronic properties, 565
 emission, 569
 luminescence, 569
 optical properties, 565
Quantum mechanics/molecular mechanics, 25

Radioisotopes, 667
Reaction-controlled growth, 140
Reaction rate analysis, 512
Reactions, 495
 solid-solid, 496
 solid-liquid, 496
 solid-gas, 497
Recommended exposure limits, 730
Redox processes, 582
Rolling circle DNA polymerization, 412

Safety concerns, 681, 690, 729
Sandwich assay, 416, 419–421, 423
SBA-15, 291, 294, 296, 305
SBA-16, 291, 293, 305
Scanning electron microscopy (SEM), 255
Scanometric DNA detection, 416–421, 423, 429
Secondary building unit (SBU), 275
Seed-mediated synthesis, 129, 158, 163, 179, 181, 183, 185, 196
Self-assembly, 178–179, 181, 187, 201, 262, 264
Self-catalysis, 171
Semiconductor cores, 40
Semiconductor nanoparticles, 551
Semiconductor photocatalysis, 631, 633
Semiconductors, 4, 380
Shape selectivity, 335
Shell model potential, 22
Silanols, 261, 296
Silica, 260, 288, 310
Silica aerogels, 216
Silica composites, 220
Silver, 44
Silver enhancement, 416, 418–421, 423
Silver nanoparticle, 406, 431–433
Silver/ZSM-5 catalyst, 621
Single-walled nanotubes (SWNT), 444, 445
 characterization of SWNT, 447
 electronic structure, 446
 geometric structure, 445
Size distribution, 143, 147
Size focusing, 132
Small angle X-ray scattering (SAXS), 411
Soil remediation, 662–663
Solar detoxification, 630
Sol-gel, 160, 165, 172, 181, 188, 200, 248, 260
Sol-gel reactions, 212
 cluster addition (chalcogenide gels), 232–233
 Co-doping/"nanogluing", 215–216, 221–222
 epoxide method (oxide gels), 220–221
 general, 212–213
 nanoparticle condensation (chalcogenide gels), 230–232
 Ostwald Ripening, 217
 resorcinal-formaldehyde method (organic gels), 227–228
 thiolysis (sulfide gels), 229–230
Sol-gel synthesis, 172, 384
Sol-gel template, 160
Sol-gel/vanadium oxide, 527

Solid state NMR, 256
Solubility of nanoparticles, 49
Solvent forces, 51
Solvothermal synthesis, 166–167, 184, 186, 199–200, 262, 279
Sonochemical synthesis, 178, 185, 188
Specific surface area, 248, 280
Spheres, 311
Spinels, 269
Statistics, 44
Stoichiometric cluster compounds, 39–40
Structure-directing agent, 158, 199, 202, 258, 262, 274, 290
Sulfur oxides, 669
Superlattice, 183–185
Superlattice melting, 59
Superlattices, 55, 183, 394
Superoxide radical, 583
Supersaturation, 130, 135
Supramolecular aggregates, 262
Surface-enhanced raman scattering, 418
Surface plasmon, 543, 556
Surface plasmon resonance, 419
Surface roughness, 248
Surfactant, 142, 262, 286
Systematic evolution of ligands by exponential enrichment (SELEX), 427
Swelling agent, 296

Template-based method, 159, 188
Templating, 258, 268, 274, 285, 305
Therapeutics with DNA modified gold nanoparticles, 429
Thermal decomposition, 142, 146
Thin films, 310
Thymine-Hg^{2+}-thymine (T-Hg^{2+}) coordination chemistry, 424, 431, 434
Titanium dioxide particles, 735–736
Titanium oxide, 587, 590
  aggregates, 588
  effect of pH, 587
  light intensity, 590, 593
  optical spectra, 587–588
  photonic efficiency, 594
Titanium oxide/highly dispersed, 607
  alumina, 607
  carbon dioxide reduction, 607
  nitric oxide oxidation, 610
  silica, 607
Titanium oxide on quartz, 595
Tortuosity, 248
Toxicity of inhaled nanomaterials, 729
  magnesium oxide, 738
  titanium oxide, 735
Toxicity of nanomaterials, 709, 729
Transmission electron microscopy (TEM), 255, 283
Transport in the environment/nanomaterials, 695
  aquatic/carbon based, 696
  aquatic/metal oxides, 699
  porous media, 700
  soils, 700
Trapped electrons, 24
True liquid crystal templating, 285, 306

Uniformly sized nanocrystals, 128

Vanadium oxide photocatalysts, 614
Van Der Waals, 50
Vertically aligned nanotubes, 465
Volatile organic compounds, 668
Volume change during reaction, 508

Water purification, 584
Wulff construction, 18

Xerogels, 211, 244, 269, 272
X-ray diffraction, 253

Yarns of nanotubes, 468

Zeolite Y, 271
Zeolites, 248, 258, 269, 272, 332, 334, 338, 344, 349, 352
Zeotypes, 248, 258, 269
Zero-valent iron, 651, 653, 655, 656–659, 663–665, 667
ZSM-5, 270